ıces in Engineering Structures, Mechanics & Construction

SOLID MECHANICS AND ITS APPLICATIONS
Volume 140

Aims and Scope of the Series

The fundamental questions arising in mechanics are: *Why?, How?,* and *How much?* The aim of this series is to provide lucid accounts written by authoritative research giving vision and insight in answering these questions on the subject of mechanics as relates to solids.

The scope of the series covers the entire spectrum of solid mechanics. Thus it includ the foundation of mechanics; variational formulations; computational mechanics statics, kinematics and dynamics of rigid and elastic bodies: vibrations of solids and structures; dynamical systems and chaos; the theories of elasticity, plasticity and viscoelasticity; composite materials; rods, beams, shells and membranes; structural control and stability; soils, rocks and geomechanics; fracture; tribology; experimental mechanics; biomechanics and machine design.

The median level of presentation is the first year graduate student. Some texts are monographs defining the current state of the field; others are accessible to final year undergraduates; but essentially the emphasis is on readability and clarity.

For a list of related mechanics titles, see final pages.

Advances in Engineering Structures, Mechanics & Construction

Proceedings of an International Conference on Advances in Engineering Structures, Mechanics & Construction, held in Waterloo, Ontario, Canada, May 14–17, 2006

Edited by

M. PANDEY

University of Waterloo, Ontario, Canada

WEI-CHAU XIE

University of Waterloo, Ontario, Canada

and

LEI XU

University of Waterloo, Ontario, Canada

Springer

A C.I.P. Catalogue record for this book is available from the Library of Congress.

ISBN-10 1-4020-4890-4 (HB)
ISBN-13 978-1-4020-4890-6 (HB)
ISBN-10 1-4020-4891-2 (e-book)
ISBN-13 978-1-4020-4891-3 (e-book)

Published by Springer,
P.O. Box 17, 3300 AA Dordrecht, The Netherlands.

www.springer.com

Printed on acid-free paper

Printed in the Netherlands.

This volume of research papers is dedicated by the authors to

Professor Donald Edward Grierson

on the occasion of his retirement from the University of Waterloo

TABLE OF CONTENTS

Structural Optimization

Engineering Reliability

Construction Management

PREFACE

This volume is the proceedings of the *International Conference on Advances in Engineering Structures, Mechanics & Construction*, convened at the University of Waterloo on May 14–17, 2006.

The conference was held to celebrate forty years of related research achievement at the University of Waterloo. The Solid Mechanics Division (SMD) of the university was founded in 1966 as a research centre of excellence. During the next three decades, SMD hosted many scientific visitors from around the world and held more than a dozen international conferences, lecture series and symposia on a diverse range of topics in mechanics and structural engineering. Upon the retirement of many of its founding members in the 1990s, SMD was renamed the Structures, Mechanics & Construction Division (SMCD) to reflect the changing research interests of its newer members.

The conference also celebrated the academic career of Professor Donald E. Grierson as a distinguished researcher and educator. Professor Grierson has been a faculty member of the Department of Civil Engineering and SMD/SMCD of the University of Waterloo since 1968. He was an early contributor to the Solid Mechanics Division in the 1960s, and was pivotal in the reformation of the Structures, Mechanics & Construction Division in the mid of 1990s. After a productive career of thirty-eight years in research and teaching, Professor Grierson retired from the university at the end of 2005.

This book contains seventy-three papers in the areas of structural engineering, applied mechanics, and construction management. Most of the papers present original research results, while some papers present reviews of specific research areas. The contributors are from various parts of the world, yet most of them have a close tie with Waterloo. Some of them are former or current members of SMD/SMCD or their collaborators; some were graduate students, postdoctoral fellows, or visitors of the Division; some delivered courses to our graduate students; and others presented their research in the famous SMD/SMCD seminar series. The papers cover the entire research spectrum of the Structures, Mechanics & Construction Division: concrete structures, steel structures, cold-form steel structures, engineering mechanics, structural optimization, applied mathematics, engineering reliability, and construction management. The papers reflect the achievements and influence of SMD/SMCD over the past four decades.

We three editors of this volume are privileged to be members of SMD/SMCD since our graduate years in the 1980s. We are fortunate to be associated with Professor Grierson as former students, colleagues, and friends, and are delighted to be able to celebrate the 40th anniversary of SMD/SMCD and to honour Professor Grierson with old and new friends.

We acknowledge the Department of Civil Engineering, the Faculty of Engineering, and the University of Waterloo for their support of SMD/SMCD since its inception in 1966. We also appreciate their financial support for this conference and proceedings.

Mahesh Pandey, Wei-Chau Xie & Lei Xu
May 2006

Professor Donald Edward Grierson

DEDICATION TO PROFESSOR DONALD EDWARD GRIERSON

This volume is dedicated to Professor Donald E. Grierson to celebrate his long career as a distinguished engineer, researcher, and educator on the occasion of his retirement. Professor Grierson has been a faculty member in the Department of Civil Engineering at the University of Waterloo, Canada, since receiving his Ph.D. degree there in 1968.

Professor Grierson has held visiting appointments at various engineering institutions in North America and abroad, including the University of California at Berkeley in the USA in 1970/71, Imperial College in England in 1974, University of Liege in Belgium in 1975, Politecnico di Milano in Italy in 1975, University of California at Los Angeles (UCLA) in the USA in 1979/80, National Defense Academy of Japan in 1986, Heriot-Watt University in Scotland in 1991, the University of Brescia in Italy in 1992, the Technion-Israel Institute of Technology in 2001, and the Federal Institute of Technology, Lausanne (EPFL) in Switzerland in 2004.

In his thirty-eight years of academic career, Professor Grierson has carried out research and made significant contributions in many areas of structural engineering; among these are structural plasticity, structural optimization, evolutionary computing, failure-load analysis, fail-safe design, performance-based seismic design, design under abnormal loading, and computational conceptual design. Some of his notable contributions include deformation analysis of elasticplastic frames, optimal design of structural steel frameworks, optimal sizing, geometrical and topological design, and progressive-failure analysis of buildings subjected to abnormal loading. His erudite knowledge of structural optimization and profound insight into structural plasticity, engineering computing, and structural design have been amply demonstrated through his approximately 200 research articles in internationally renowned scientific and engineering journals, and conference proceedings. His research work and that of his students is the basis for structural steel design software used in North American engineering institutions and consulting offices since the 1980s.

Professor Grierson received the Canadian Society for Civil Engineering *E. Whitman Wright Award* in 1995 for excellence in computer-aided design. He was the co-recipient of the American Society of Civil Engineering *State-of-the-Art Award* in 1998 and 2004 for his contributions in structural optimization and optimal structural design. He was honoured by the Structural Engineering Institute of the American Society of Civil Engineering with the *Distinguished Service Award* in 2003. He has been a member of a number of professional societies, editorial boards of international scientific journals, international scientific committees and organizing committees for international meetings.

Not only is Professor Grierson a leading researcher in applied mechanics and structural engineering, he is also an eminent educator. Professor Grierson has given invited lectures at many universities and institutions, and has been a keynote speaker in a goodly number of international conferences. Professor Grierson has devoted his academic career to the promotion of the discipline of computer-aided structural analysis and design through teaching and training of highly qualified researchers. He has supervised nearly forty graduate students, many undergraduate research assistants, postdoctoral fellows, and visiting scholars. Professor Grierson's influence on students was enormous. Innumerable students and researchers have benefited through his lectures, presentations, discussions, and writings. Professor Grierson was an early contributor to the *Solid Mechanics Division*, and was instrumental in the reformation of the *Structures, Mechanics & Construction Division* of the

University of Waterloo. He has demonstrated exceptional leadership in mentoring young faculty members. His passion for teaching and dissemination of knowledge was honored by the Sandford Fleming Foundation *Teaching Excellence Award* in 2002 in recognition of an exemplary record of outstanding teaching, concern for students and a commitment to the development and enrichment of engineering education.

Professor Grierson's achievements over many years and in many ways demonstrate that his qualities as a scholar, educator, and contributor are of the highest level. His academic career as a researcher and teacher is exemplary for his colleagues and engineers of younger generations in terms of teaching and learning, research and scholarship, leadership and service to society.

Plenary Presentations

FORTY YEARS OF ENGINEERING STRUCTURES, MECHANICS AND CONSTRUCTION RESEARCH AT THE UNIVERSITY OF WATERLOO

Mircea Z. Cohn and Donald E. Grierson
Civil Engineering Department, University of Waterloo, Canada

"In order that they may live and bring forth life, generations must continue to meet, and the teaching assume the form of a human link, awakening and activating our bond with our Father. The spark that leaps from him who teaches to him who learns rekindles a spark of that fire which lifted the mountain of revelation to the very heart of heaven..." (Buber, 1956).

Initial planning for the formation of the Solid Mechanics Division (SMD) at the University of Waterloo began in 1966, and it was implemented into a formal entity following the award of a National Research Council of Canada Development Grant in 1968. Upon the departure and retirement of many of its initial members over the next thirty years, SMD was reconstituted in the mid-1990s as the Structures Mechanics & Construction Division (SMCD) to reflect the changing research interests of its more recent and new members. The collective research work of both SMD and SMCD has resulted in many publications and valuable contributions to the progress of engineering structures, mechanics and construction. This paper presents a 40-year historical perspective through to 2006 of this activity at the University of Waterloo.

1. Introduction

In the early 1960s, the University of Waterloo was only a few years old and its Faculty of Engineering was just about to produce its first graduates. D.T. Wright, the founding dean and subsequently the University's third president, brought to the Faculty a number of promising scholars from around the world, and thus opened the prospects of vigorous progress in its innovative cooperative programs in engineering.

The Civil and Mechanical Engineering Departments acquired at the time a group of young faculty members eager to realize their academic potential and widen their research horizons. Under the leadership of A.N. Sherbourne, the next Dean of Engineering, new ways for enhancing existing capabilities of the two departments were explored. Their parallel or intersecting areas of activity in Mathematics, Theoretical and Experimental Mechanics, Soil Mechanics, Structural and Machine Analysis and Design, Material Science and Technology, etc., suggested that considerable benefits could be expected from closer cooperation and coordination of activities for groups with similar interests in the two departments.

Dean Sherbourne advanced the concept of bringing the above disciplines under a broad umbrella organism, that came to be known as the SOLID MECHANICS DIVISION (SMD), and initiated efforts to define its goals, structure and mode of operation, as well as means of formalizing its creation.

This paper describes the beginning, growth and maturing of SMD, as well as that of its successor, the STRUCTURES MECHANICS and CONSTRUCTION DIVISION (SMCD). The presentation gives a brief historical overview of these entities, offers a permanent record of results of their activities, examines the causes of related successes and failures, and draws some possible lessons from the first four decades of this academic experience.

M. Pandey et al. (eds), Advances in Engineering Structures, Mechanics & Construction, 3–29.

2. Solid Mechanics Division (SMD)

2.1. Beginnings

Prompted by Dean Sherbourne, a group of UW faculty decided to formalize the voluntary cooperation of members in the Civil and Mechanical Engineering Departments into a research unit, the Solid Mechanics Division (SMD). Its goal was to establish a centre of excellence in solid mechanics at the University of Waterloo (UW) by enriching the existing research and teaching programs. While the main emphasis was placed on the human factor (i.e., attracting outstanding visiting professors, research fellows and graduate students), serious consideration was also given to modernizing laboratory facilities and expanding the working environment and supporting staff.

Implementation of these general objectives was detailed in an application to the National Research Council (NRC) for a major negotiated grant to UW, aiming at the creation and development of SMD. In April 1968, NRC announced the three-year award of Grant D-10 in the amount of $600,000. The funds were to be expended in installments of $100,000, $200,000 and $300,000 in each year of the award term. It was the understanding of the grant applicants and of the Dean of Engineering that, whereas Grant D-10 was to help establish SMD as a centre of excellence, by the very nature of the grant UW was committed to ensuring its continuing existence and development through appropriate support and funding at the end of the award term.

Thus, SMD was brought to life in the spring of 1968 by a decision of the University Board of Governors and the generous NRC award of Development Grant D-10. One of the first tasks of the SMD steering group was to establish an administrative structure that could implement the stated objectives under the terms of Grant D-10. After considering to offer the chair to a distinguished foreign academic, the group ultimately decided to nominate, and the Faculty of Engineering appointed, M. Z. Cohn to fill that position for three years.

The founding SMD membership included Professors S. T. Ariaratnam, D. J. Burns, R. Green, N. C. Lind, J. T. Pindera, H. B. Poorooshasb, A. N. Sherbourne, T. H. Topper, and the founding SMD Chair, M. Z. Cohn. Other members that joined SMD soon thereafter and over the next few years were Professors E. F. P. Burnett, R. Dubey, G. M. L. Gladwell, D. E. Grierson, K. Huseyin, H. H. E. Leipholz, G. M. McNeice, T. P. Prasad, J. Roorda, R. M. Schuster and J. C. Thompson. It is of interest to note that the initial and early SMD membership was comprised of individuals from several different departments of the faculty of engineering, including Civil, Mechanical and Systems Engineering.

The SMD founding group and chairman agreed on a few basic operational principles:

- Minimal administration and formalism;
- Expectation of performance from key members with authority and responsibility;
- Clarity of objectives for efficient implementation.

The SMD operation was to be ensured by the following elements:

- Executive Committee, responsible for planning, funding and deciding on all activities;
- Publication Committee, in charge of policy matters for production of SMD publications;
- Area Representatives, responsible for initiating and overseeing activities in each area of concern (i.e., structures, computer analysis and design, geotechnical, mechanics, experimental mechanics, etc);
- Task Officers, responsible for three SMD major tasks: Publications, Seminars and Library.

Main operational features were the complete decision transparency of its working committees and total initiative and authority of each task officer. Monthly SMD meetings - not exceeding one hour - were planned and rigorously held in order to keep members informed in a timely manner on all matters of general concern.

D. E. Grierson was appointed as Publications Officer, a capacity in which he served throughout the life of SMD. Over the years he was assisted by a dedicated production staff, A special contribution was brought by D. Bartholomew of UW Graphic Services who designed the SM Studies jackets and all publication formats. His response to our exigencies produced the SM Logo shown in Figure 1, that soon become our well-recognized and proud emblem on all SMD printed output.

Figure 1. SM Logo

Bi-monthly SMD Seminars, initiated and regularly held on Mondays at 3:30-5:00 p.m., became a permanent entry in our calendars. SMD staff ensured the conservation and growth of library collections that included our own publications series, as well as solid mechanics volumes and journals donated by members.

2.2 Growth

SMD vision and ambitious program became apparent with the number and variety of its printed and verbal expressions. The main evidence of success in its activities was the body of SM publications that began to be produced within a short time, including the SM Studies Series, SM Papers/Report Series, SM Special Publications and the SM Archives Journal.

2.2.1. SM Studies Series

This, probably the most important and widely known of SMD products, is a collection of publications intended to record major contributions by the SMD at UW in Applied Mathematics, Solid Mechanics, Structural Analysis and Synthesis, Experimental Mechanics, and Philosophy of Design. The SM Studies Series makes available works of academic merit which, because of their content or volume, may not be published by scientific journals or as independent books. These include proceedings of seminars and symposia, monographs, research studies and significant doctoral dissertations. (The jacket cover for the first SM Study is shown in Figure 2).

The SM Studies Series was conceived in order to confer a character of permanence to valuable work initiated and completed by SM members and their associates, work that might otherwise be lost or forgotten. The series was published under the direction of an Editorial Committee consisting of S.T. Ariaratnam, M.Z. Cohn, H.E. Leipholz, N.C. Lind, J.T. Pindera and A.N. Sherbourne, with the cooperation of a group of external Consulting Editors of international reputation.

Members of the Editorial Committee led their colleagues by example, by making repeated contributions to the Study Series and other SMD publications in their areas of expertise as follows:

S.T. Ariaratnam: Dynamics, Stochastic Processes, Continuum Mechanics
M. Z Cohn: Structural Plasticity and Optimization, Concrete Structures
N.C. Lind: Reliability, Codified Design and Risk Analysis
H.E. Leipholz: Stability and Mechanics of Continua
J. T. Pindera: Experimental Mechanics
A. N. Sherbourne: Steel Structures, Shell Analysis and Design

Other SMD members at the time also made contributions to the various SMD publications in their areas of expertise as follows:

E. F. P. Burnett: Building Sciences
D. J. Burns: Engineering Mechanics
R. Dubey: Mechanics of Materials
G. M. L. Gladwell: Applied Mathematics
R. Green: Bridge Engineering
D. E. Grierson: Structural Optimization
K. Huseyin: Stability
G. M. McNeice: Finite Element Analysis
H. B. Poorooshasb: Soil Mechanics
T. P. Prasad: Applied Mathematics
J. Roorda: Structural Stability
R. M. Schuster: Cold-Formed Steel
J. C. Thompson: Experimental Mechanics
T. H. Topper: Materials Engineering

Figure 2. SM Study No. 1

Among the numerous Consulting Editors and friends of SMD and its publications, we remember the names of E. F. Masur (University of Illinois at Chicago), N. J. Hoff (Stanford University), C. P. Siess (University of Illinois), Ch. Massonnet (University of Liège), and M.R. Horne (Manchester University).

As the SM Studies Series became widely recognized, foreign scholars found it a congenial medium for disseminating results of their work and began organizing and publishing conference proceedings using SMD and UW Press facilities (e.g. see Appendix 1: Studies 12, 15, 16 and 17).

The reputation of SMD and its SM Studies Series may be attributed to its:

- Broad range and depth of areas covered;
- Quality of contributors and their contributions;
- Balance between theory and practice of engineering, classical and modern views, and fundamental and applied research;
- Innovative and outstanding doctoral dissertations (e.g., see Appendix 1: Studies 2 and 4).

It is worth mentioning that the graphic excellence of the SM Studies Series has been a major factor in its broad recognition, due in no small measure to the striking jacket design sporting the SM Logo in different colours for each volume. As listed in Appendix 1, eighteen volumes were published in this series between 1969 and 1988. (The jacket covers for two representative SM Studies are shown in Figure 3).

Figure 3. SM Studies 9 and 14

2.2.2. SM Papers, Reports and Notes Series

A second series of publications was the *white-cover* SM Reports/Papers Series, intended to ensure rapid dissemination, reaction and comment on papers prior to formal publication in refereed technical and scientific journals. As listed in Appendix 2, this series ran for 199 issues from 1969 to 1985. The series was initially called SM Reports , but its name was changed to SM Papers beginning with issue 101 in order to better reflect the nature of its contents. (The front cover of the first issue in the series is shown in Figure 4).

Figure 4. SM Report/Paper No. 1

The *blue-cover* SM Technical Notes/Reports Series was devoted to materials that deserved preserving for specialized study, but which, because of the type of information or volume, would not be normally published in standard periodicals. As listed in Appendix 3, this series ran for 31 issues between 1971 and 1980, and includes texts presented at various congresses, documents with extensive numerical or experimental data, and computer programs. The series was initially called SM Technical Notes, but its name was changed to SM Reports beginning with issue 15 in order to better reflect the nature of its contents. (Part of the front cover of the first issue in the series is shown in Figure 5)

Figure 5. SM Technical Note/Report No. 1

Figure 6. Contents of First and Last Issues of SM Archives Journal

2.2.3. Solid Mechanics Archives

In 1976, SMD initiated and directed the publication of SM Archives, a quarterly International Journal, dedicated to original contributions to solid mechanics. It was printed with the cooperation of Noordhoff International Publishing, The Netherlands, and Oxford University Press, U.K. The Journal was under the direction of internal and external editorial boards, and appeared in 14 volumes until 1989. (The Contents for the first and last journal issues are shown in Figure 6).

2.2.4. SM Special Publications

Over the years a number of volumes were occasionally published outside the range of the regular SM series. These somewhat randomly-produced texts are of two distinct types:

- Lectures Sets given at UW by SMD members or visiting guests;
- Textbooks, Manuals, and Conference Proceedings sponsored by external agencies.

Some of these texts were organized and edited by SMD members, but produced by commercial publishers. A heterogeneous collection, the SM Special Publications cover a wide spectrum of interests within recognized SMD areas of expertise, while addressing an audience reflective of the international sponsoring organizations. A partial listing of the Special Publications is given in Appendices 4 and 5. (The partial front cover of a representative SM Special Publication is shown in Figure 7).

Figure 7. Representative SM Lecture Set

2.3. Maturity

Processes of learning and sharing technical knowledge constituted the permanent focus of SMD activities, not only through the printed medium, but also by direct personal interaction of Division associates and visitors. Such interaction found its expression in various seminars, workshops, conferences and symposia, as well as lectures by invited guests on topics within the stated SMD mission.

A systematic listing of all SMD live-events is rather difficult to achieve because of the variety of venues (Waterloo or elsewhere), sponsorships (SMD, other, combined), publishers (UW or other), and orientation (research or teaching). A sample of the major events with SMD involvement is given in Appendix 6. The publication activities for these events have been surveyed in the preceding section. We now turn our attention to communication events of either research or teaching orientation.

2.3.1. Research-oriented Events

From the late 1960s to the early 1980s a number of topics in solid mechanics captured the interest of investigators around the world. At the time SMD was a centre of exciting developments in structural plasticity and optimization, reliability and codification, and stability and experimental mechanics. Interaction of SMD faculty, associates, students and friends allowed the planning and organizing at Waterloo of some notable international meetings on these topics. Images of the delegate participation at three such events are shown in Figures 8, 9 and 10.

Figure 8. NATO ASI- *Engineering Plasticity by Mathematical Programming*, Waterloo 1977

Figure 9. International Symposium-*Nonlinear Design of Concrete Structures,* Waterloo 1979

Figure 10. Int'l Symposium-*Nonlinearity and Continuity in Prestressed Concrete*, Waterloo 1983

The first SMD meeting was the *Colloquium on Limit Design for Structural Concrete,* September 6-8, 1967. The purpose of the meeting was to clarify the theory and practical design application of three methods of Limit Design under debate by the joint ASCE-ACI Committee 428. The premises and feasibility of the three (European, American and Canadian) approaches were each discussed for one full day, with the basic principles presented by their authors during the morning sessions. Three representative structure examples, fully worked out by the Waterloo group of faculty and graduate students, were presented in the afternoon sessions. The latter results are summarized in the first issue of the SM Paper/Report Series (Cohn et al, 1969). The 1967 meeting was the beginning of an intense activity of live-events, whose contents and output is reflected by the Special Publications reviewed in the preceding section (see Appendix 6).

2.3.2. Teaching-oriented Events

The extent of advanced teaching by SMD and visiting faculty members may be assessed from the related printed documents listed in Appendix 4. While these lecture sets were mainly addressed to graduate audiences, SMD was equally interested in enhancing the overall quality of undergraduate teaching of solid mechanics at UW, consistent with its basic mandate. In this regard, during the 1969 summer semester SMD organized a weekly seminar cycle designed to stimulate undergraduate teaching excellence so that it would match our high graduate programs standards. Some of the problems and dilemmas identified but unsolved were mainly consequences of the distinctive features of UW's engineering education, particularly the cooperative work-study program and the related academic calendar. Also of concern were the handling of the general knowledge explosion, new directions of study, and their accommodation within the curricula of standard four-year study programs. A fresh outlook on the design of undergraduate teaching seemed unavoidable if the following topics were to be properly addressed:

- Role of solid mechanics in CE and ME curricula;
- Nature of SM teaching output (engineers or engineering scientists);

- Optimal teaching techniques (lecturing, problem solving, laboratory work, projects, TV sessions, etc.);
- Optimal mix of various areas of study (mathematics, physics, computer techniques, theoretical and applied mechanics, etc.)
- Objective teaching evaluation ("delivery" or "retention" models).

Specific questions pertaining to objectives, major topics, minimal ingredients, significance and optimal instructional media for each individual course were explored. Some of the major flaws identified in solid mechanics courses were: coverage of basic statics, insufficient laboratory work, poor understanding of physical problems and role of modelling, inefficient tutorials and insufficient engineering project work. Furthermore, it was strongly felt that lack of coordination of CE and ME offerings was manifest in unnecessary repetitions, omissions or redundancy, along with unsuitable textbooks and inadequate lecture-tutorial coordination.

A comprehensive discussion of the above problems took place in the framework of 12 weekly seminars during the 1969 summer term. Each seminar was devoted to one course presentation and full debate by faculty members and students with a view to:

- Establishing quality standards for the teaching process;
- Developing lecture notes, laboratory manuals, problem sets and other teaching aids tailored to the specific needs of UW students and their cooperative programs;
- Developing and completing within a few years an integrated set of *SM Textbooks* that encapsulated UW needs and standards, to be known as the SM Text Series.

The 1969 summer term SM Teaching Seminars were well prepared, attended and debated and generated productive discussion by the participants. Regrettably, the perseverance and long-term motivation required for bringing this major project to fruition were lost with time and the concept of a complete, modern SM Text Series, unified in format and educational perspective, failed to materialize. However, a lonely sample of the potential of this ambitious project saw the light of day in 1972 (Cohn 1972), and remains a prototype of what might have been the outcome of staying on course with our initial vision. (The partial jacket cover for the sole publication in the SM Text Series is shown in Figure 11).

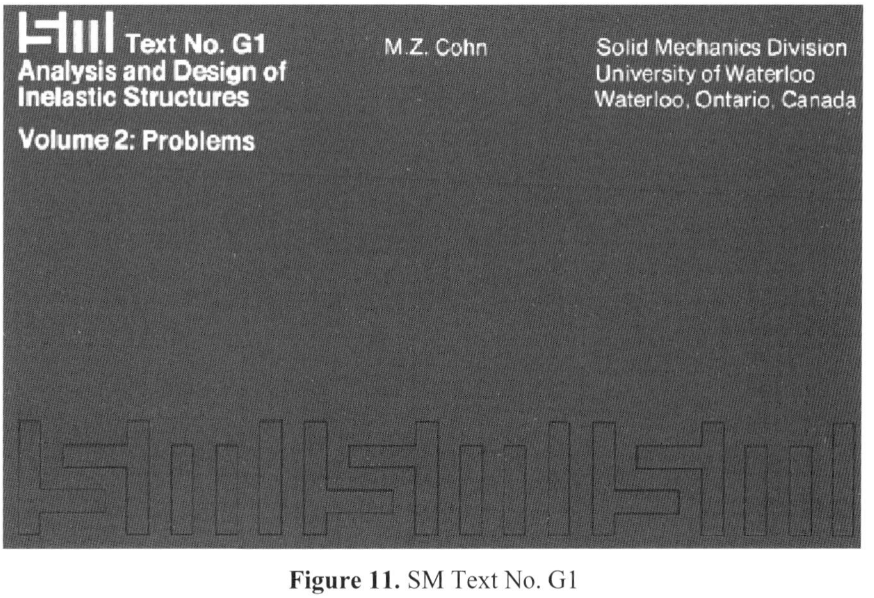

Figure 11. SM Text No. G1

2.3.3. Visiting Scholars

The maturing of SMD is convincingly illustrated by its wide recognition by scholars from all continents and their interest in visiting UW for cooperative projects with our faculty and students. As it would be difficult to list all our distinguished guests and to identify the centres of learning with which SMD has established and developed most productive contacts over the years, we mention below only some of the most prominent guests.

Among many others, SMD was honoured to host such eminent personalities as W. Prager (University of San Diego), A. L. L. Baker (Imperial College, London), F. Levi (University of Torino), C. Menn (ETH, Zurich), J. Muller (Paris, France), G. Maier (Politecnico, Milano), F. Moses (Case Western University), Z. Mroz (Technical Research Institute, Warsaw), N. Khachaturian (University of Illinois at Urbana), and L. A. Schmit (UCLA).

2.3.4. Late Years

Three years after its promising beginnings and depletion of the funds awarded through NRC Grant D-10, the SMD autonomy started to be eroded; the promised continuing financial support by UW administration, implied by the negotiated grant, did not materialize; CE and ME departments favoured the parochial management of their traditional territories; SMD involvement in the planning and delivery of undergraduate programs was grossly misunderstood.

A 1972 UW Senate decision put an end to the formal existence of SMD. Its ongoing activities were to be carried out through departmental channels, with a small stipend granted from the CE department budget. SMD chairmanship was reassigned to H. E. Leipholz, who served from 1972 to 1979, when he assumed a senior position in the university administration. With the era of major initiatives and innovations being over, Leipholz must be credited for having maintained some of the activities and original SMD ways.

As a sign of the emerging hard times, SMD members volunteered to make individual financial contributions in order to sustain on a limited scale the planned projects and events. With SMD turning from a living organism to a surviving name and a bright memory, it became the task of G. M. L. Gladwell in 1979 to serve as the chair of the once proud and independent research group.

3. Critical Remarks

3.1. Successes

Statistical data regarding SMD activities can only give a limited view of its successes. Yet, it is noteworthy that in just the first three years of existence its printed output amounted to six SM Studies, one hundred SM Papers and fourteen SM Reports. During the same period it organized four major symposia and lecture series, and established a bi-weekly seminar series as a permanent forum for scholarly presentation and discussion. These accomplishments were multiplied within the next two decades of rather fragile survival, as shown by Appendices 1 to 6.

Beyond statistical facts, major successes must be credited to A.N. Sherbourne's initiative of integrating individual strengths into a cross-departmental focused entity, and to SMD's founding members for shaping it so that the division could represent more than the sum of its individual components. Indeed, in a short time span, the collective formal unit that embodied the members' confidence, dedication and pride produced a noticeable rise in the level and dissemination of research results.

The principal SMD achievement consisted of living up to an uncompromising commitment to excellence, marked by a constant promotion of new young talent, individual growth, group development and international cooperation. The SMD reputation can be measured by the calibre of people associated with the wide range and high quality of its activities.

These were distinctive not only in substance, but also in a style expressed by the graphic standards and formats of its publications, as well as by the sense of occasion, attention to detail and meticulous organization of conferences. Not to be forgotten are the musical themes used as inspirational motifs, calls to order for working sessions or relaxing interludes.

What were the means that made possible the above successes? The essential factor was the rich reservoir of individual talent. But existing potential might not have materialized without the financial support of NRC Grant D-10.This award confirmed the viability of the SMD visionary program, energized its members and stimulated their creativity. Although relatively modest even by 1970s standards, the award funds were sufficient for initiating a large number of projects in a short time. However, we believe that, even with proper funding, SMD program could not have been implemented had it not been animated by the concepts of personal autonomy and responsibility, along with collective involvement and transparent leadership.

The establishment of the SMD Centre in 1972 in the E-4 (Pollock) Building had a very positive impact on the efficiency of its operation. Conceived, designed and furnished with active SMD participation, the Centre provided adequate space for SM faculty and graduate students, secretariat offices, library, lounge and a fully equipped seminar room. These facilities ensured precious proximity and improved the interaction among faculty, students and staff, as well as with laboratories and computer rooms.

3.2. *Failures*

Successes and failures can be assessed by comparison to either adopted or desirable objectives. By the first criterion SMD has to be characterized as an overall success, because it delivered on 100% of all contracted obligations.

By the second criterion, the projected self-imposed development and production of the SM Text Series, following the summer 1969 teaching seminars (see *2.3.2*), ended as a serious failure. Completion of this project might have represented a unique contribution to the teaching of solid mechanics at the University of Waterloo. Ironically, while this initiative gave a constructive response to the often heard criticism that faculty members were more interested in research than teaching, it probably generated a political malaise responsible for the negative attitudes of Civil and Mechanical Engineering Departments toward SMD and its future. SMD efforts were aimed at reviewing the curricula, rationalizing inter-departmental programs and improving their implementation, *without* involvement in, or change to existing administrative structures. Sadly, departments misread our intentions, fearing an encroachment of their jurisdiction, power, and decision-making authority.

Beside the above, the critical SMD failure was an existential one. Although decisions on its fate were beyond the will and power of our group, a more outspoken response could have been given when SMD formal status was questioned and finally revoked. The UW reneging on the initial commitment to sustain its hard won centre of excellence brought disbelief, consternation, and eventual resignation to SMD membership.

It is fair to note that the political winds changed considerably during the short period of time since SMD inception: whereas the dominant orientation in the late 60's was toward fundamental research and outward-looking, the early 1970s saw a switch to applied, nationally profitable research and development. UW followed the trend of chasing funding opportunities by downsizing long-term pursuits in favour of reaching tangible material benefits in the shortest possible time. It was in this climate that extensive debates took place on research policies related to basic vs. mission-oriented development, federal vs. provincial jurisdictions, the role of research in university vs. government labs, or industry. UW's position may be inferred from the words of its Academic Vice-President at the time (Petch 1973):

> There are three areas in which debate is still hot, with little indication that agreement will soon be reached: funding university research, organization of government research and stimulation of innovation in industry. (p.40)
>
> Basic research has been one of the chief targets of those critical of the government's policy, or lack of policy, on scientific matters. In moments of passion, some critics have suggested that Canada should divert all financial support to applied research and development and phase-out basic research. (p.36).
>
> The demands for more mission-oriented basic research, applied research and development, arise because of a wide-spread feeling that Canada must utilize science and technology more in striving to achieve national goals, coupled with a new understanding that benefits do not necessarily flow automatically from free basic research, but require a conscious effort at application. (p.38).

3.3. Lessons

SMD was a memorable episode in the UW Faculty of Engineering: great talent and work pre-existed its creation and subsisted its demise. But its early years energized and brought out the best of existing potential and raised it to a level of high academic distinction and achievement. Not only did SMD live up to its commitments, but it amply exceeded the expectations of its founding members and caused our university to be recognized as a real force in the world of solid mechanics.

While we started this survey with the intention of factually recording the SM activities over approximately the last forty years, we may be excused if the intended writing objectivity has suffered occasionally: some of the old passion for SMD ideals could not be entirely eliminated while undertaking this memory trip.

Perhaps we may also be indulged in a little literary flight of fancy by viewing the SMD story as a real academic saga with not a few rather dramatic ingredients: indeed, on a close look, the evolving process of creation, development, stagnation and renewal are quite obvious. However, the less obvious conflicts of authority between an old departmental structure and a superimposed (perceived) rival new entity, jealousy and personality clashes, tradition versus innovation, loyalty and deception, growth, decay and redemption, all major themes of the classic drama are, or may be assumed, as part of this narrative. And so might have been pain, dissatisfaction, disappointment, and resistance to change, which are seldom absent at transitions from old to new.

Yet, in spite of all the ups and downs, a dominant leit-motif is discernible throughout the entire SMD project: fostering learning and inspiring human interaction between the old and the young, regardless of contingent circumstances. Whether funded adequately or not at all, the commitment to excellence was, and apparently will remain, the leading motivation of the UW group of scholars active in the fields of structures, mechanics and construction.

4. Structures, Mechanics and Construction Division (SMCD)

4.1. An Ongoing Commitment

G. M. L. Gladwell served as chair of SMD from 1979 to 1992. A thirteen year period over which, despite the noted difficulties, the membership continued to act as a loosely-knit research group; collaborating on research projects, coordinating seminars, pooling resources to support visitors, and participating in the hiring of new faculty. The group has maintained an ongoing commitment to research and its related activities through to the present day. D. E. Grierson served as chair from 1992 to 2002, while G. W. Brodland continues as chair to 2006 and beyond.

Upon the departure and retirement of nearly all of its initial and early members over almost 30 years since its inception, SMD was renamed in the mid-1990s as the Structures, Mechanics and Construction Division (SMCD) to reflect the changing research interests of its more recent and new members. Professors Brodland, W.-C. Xie and M. A. Polak joined SMD/SMCD during the late 1980s and early 1990's; M. Pandey, K. Soudki, T. Hegazy and L. Xu joined in the mid-1990s; J. Straube, J. West and S. Potapenko joined in the late 1990s and early 2000s; while S. Narasimhan and S. Walbridge became new members as of May 1, 2006.

4.2. SMCD Member Research Interests

The present-day membership of SMCD consists of individuals from the Department of Civil Engineering. A brief sketch of each member's research interests is given in the following.

G. Wayne Brodland: advanced computational models for tissue and cell mechanics; unique instrumentation and software for data collection and analysis; and development of a unified theory of embryo mechanics.

Wei-Chau Xie: structural dynamics and random vibration, reliability and safety analysis of engineering systems, dynamic stability of structures, nonlinear dynamics and stochastic mechanics, localization in randomly disordered engineering structures, vibration design of residential floor systems using cold-formed steel joists, and seismic analysis and design of engineering structures.

Marianna Polak: punching-shear strengthening techniques for RC slab-column connections, development of constitutive models and nonlinear finite element formulations, non-destructive evaluation of structural members, high-density polyethylene pipes installed by trenchless technology, and interdisciplinary research on characterization of polymeric materials

Mahesh Pandey: risk-based decision support systems for infrastructure management, time-dependent reliability analysis models, risk-based dynamic health monitoring systems, socio-economic cost-benefit analysis for sustainable development, and probabilistic load and strength models.

Khaled Soudki: rehabilitation of corroded reinforced concrete beams with FRP, shear strengthening slab-column connections using FRP, durability of structural members strengthened by FRP sheets, experimental and analytical evaluation of the bond strength of reinforcement in FRP-wrapped concrete beams, and fatigue behaviour of corroded concrete beams strengthened with FRP sheets.

Tarek Hegazy: management of infrastructure networks, decision support for project scheduling and control, time and cost and resource optimization, artificial intelligence applications in construction, and internet-based collaboration in design and construction.

Lei Xu: performance-based building seismic analysis and design, advanced analysis and design for structures under abnormal loading, lightweight steel framed residential building design, analysis and design of cold-formed steel structures, stability analysis, design and optimization of steel framed structures, and floor vibration analysis and design.

John Straube: building enclosure design, moisture physics, and whole-building performance.

Jeffrey West: behaviour of precast structures, durability of precast and post-tensioned concrete structures, use of stainless steel reinforcement, development of high performance self-consolidating concrete, and use of carbon fibre reinforced polymer materials for strengthening of existing structures.

Stanislav Potapenko: modeling the mechanical behaviour of advanced materials, singular integral equations, potential methods in elasticity, mechanics of rubber-like solids, mechanics of thin-walled structures, composite materials, and fluid mechanics

Sriram Narasimhan: risk and reliability analysis of floating structures, control of seismic responses in base-isolated buildings, development of base isolation systems for buildings, algorithms for control and damage detection, and on-line structural health monitoring methods.

Scott Walbridge: small- and large-scale testing of structural elements, stability of structural systems, fatigue analysis of welded structures, fatigue assessment of bridges, probabilistic methods to model deterioration processes, and probabilistic design criteria for structures.

5. Conclusion

Trying to draw some lessons from life or academic experience may prove a rather subjective undertaking as it depends not only on objective factors, but, to a large extent, on such personally-chosen criteria as *survival* and *progress.* In collective endeavors these requirements are dominant at both individual and group levels. We assume for simplification that if the requirements are satisfied for the group, they are implicitly satisfied for the individual. Also assumed is that in an academic group individual members have the necessary intellectual strength and motivation for pursuit of higher learning.

The above observations follow from the analysis of SMD experience, and their general validity should be viewed with caution. However, it seems fair to believe that they are quite apt when considering the future of SMCD.

Group *survival* is influenced by external factors (time, prevailing policies, university orientation, funding sources, etc.), but is critically determined by subjective factors; in addition to its make-up and prerequisite academic strength, a group's existence largely hinges on:

- Strong motivation (conviction that the group adds up to more than its summed parts)
- Clear, realistic objectives
- Firm, properly focused leadership
- Internal cohesion (positive interaction at the operational level)
- Group resilience (will to survive) and adaptability
- Steadfastness in pursuit of objectives

Group *progress* requires unanimous understanding and consistent work on programs and cannot be achieved without:

- Translation of general vision into specific goals

- Choice of correct priorities
- Efficient planning and monitoring of results
- Preservation and promotion of talent (by ensuring true bond and member collegiality)
- Reduction of administrative tasks to minimum necessary

In brief, our main lesson is that group survival and progress are paramount considerations arising out of SMD history. While survival is to a large extent dependent on external factors, progress is mostly an internal product, the essence of which appears to be *making a difference*. Given their essential ingredients, both survival and progress may be achieved with a high probability if membership bond thrives under a far-reaching leadership guided by ambitious, yet realistic goals.

One cannot but hope that, in the spirit of this paper's motto (Buber, 1956), the torch passed from SMD to SMCD will continue to rekindle the sparks of discovery for generations to come.

Acknowledgements

This paper is dedicated to the memory of H.E. Leipholz, J.T. Pindera, T.P. Prasad, J. Roorda and A.N. Sherbourne, our colleagues and friends no longer among us. We gratefully acknowledge the invaluable contributions of all SMD and SMCD members, staff and UW support services.

APPENDIX 1: SM Studies Series

SM Study No. 1
An Introduction to Structural Optimization
Seven Special Lectures at the University of Waterloo
September 23 to December 16, 1968
By: N. Khachaturian, L.A. Schmit, Jr., R.L. Fox, F. Moses, N.C Lind, C.A. Cornell and W. Prager
Editor: M.Z. Cohn

SM Study No. 2
Theory of Structural Design Decisions
C.J. Turkstra
Ph.D. Thesis, 1962

SM Study No. 3
Structural Reliability and Codified Design
Eight Special Lectures presented at the University of Waterloo
October 6 to December 15, 1969
By: Fred Moses, Jack R. Benjamin, G.G. Meyerhof, C.J. Turkstra,
Masanobu Shinozuka, C. Allin Cornell, F. Legerer and H.C. Shah
Editor: N.C. Lind
SM Study No. 4
Stress-Strain Relationships for Concrete and
the Analysis of Structural Concrete Sections
M. Sargin
Ph.D. Thesis, 1971

SM Study No. 5
Computer Aided Engineering
Proceedings of the Symposium held at the University of Waterloo, May 11-13, 1971
Editor: G.M.L. Gladwell

SM Study No. 6
Stability
Fourteen Special Lectures, University of Waterloo
October 1970 to September 1971
By: H. Ziegler, K. Huseyin, A.H. Chilver, M. Sewell, F.K.G. Odqvist, H.H.E. Leipholz,
R. Reissig, S. Nemat-Nasser, V.V. Bolotin, E. Adams, R.H. Plaut
Editor: H.H.E. Leipholz

SM Study No. 7
Applications of Solid Mechanics
Proceedings of the Symposium held at the University of Waterloo, June 26 - 27, 1972
Editors: R.G. Charlwood, University of Waterloo, D.S. Weaver, McMaster University, and B. Tabarrok, University of Toronto

SM Study No. 8
Inelasticity and Non-Linearity in Structural Concrete
International Symposium,
University of Waterloo, January – June 1972
Editor: M.Z. Cohn

SM Study No. 9
Experimental Mechanics In Research and Development
Proceedings of the International Symposium on Experimental Mechanics held at the University of Waterloo, June 12-16, 1972
Editors: J.T. Pindera, H.H.E. Leipholz, D.E. Grierson, University of Waterloo, and F.P.J. Rimrott, CSME – Canadian Society of Mechanical Engineers

SM Study No. 10
Stochastic Problems in Mechanics
Proceedings of the Symposium of Stochastic Problems in Mechanics
University of Waterloo, September 24-26, 1973
Editors: S.T. Ariaratnam, H.H.E. Leipholz

SM Study No. 11
Mechanics in Engineering
Proceedings, First ASCE-EMD Specialty Conference on Mechanics in Engineering
University of Waterloo, May 26-28, 1976
Editors: R.N. Dubey, H.H.E. Leipholz

SM Study No. 12
Continuum Models of Discrete Systems
Proceedings, Second International Symposium on Continuum Models of Discrete Systems
Mont Gabriel, Quebec, June 26 – July 2, 1977
Editor: J.W. Provan, McGill University
SM Study Editor: H.H.E. Leipholz

SM Study No. 13
Advances in Analysis of Geotechnical Instabilities
International Symposium
University of Waterloo, September 1976 – October 1977
Editor: J.C. Thompson

SM Study No. 14
Nonlinear Design of Concrete Structures
CSCE-ASCE-ACI-CEB International Symposium
University of Waterloo, August, 7-9, 1979
Editor: M.Z. Cohn

SM Study No. 15
Continuum Models of Discrete Systems (CMDS3)
Proceedings, Third International Symposium on Continuum Models of Discrete Systems
Freudenstadt, Germany, June 24-30, 1979
Editors: E. Kröner, K.H. Anthony, University of Stuttgart, Germany
SM Study Editor: H.H.E. Leipholz

SM Study No. 16
Fuel-Air Explosions
Proceedings, International Conference on Fuel-Air Explosions
McGill University, Montreal, Canada, November 4-6, 1981
Editors: J.H.S. Lee, C.M. Guirao, McGill University
SM Study Editor: D.E. Grierson
University of Waterloo Press, 1982

SM Study No. 17
New Problems in Mechanics of Continua
Proceedings, Third Swedish-Polish Symposium on Mechanics,
Jablonna, Poland, May 10-14, 1982

Editors: O. Brulin, R.K.T Hsieh, Royal Institute of Technology, Sweden
SM Study Editor: D.E. Grierson
University of Waterloo Press, 1983

SM Study No. 18
Studies on Prestressed Concrete
M.Z. Cohn
University of Waterloo, 1988

APPENDIX 2: SM Reports/Papers Series

1. "Application of Limit Design to reinforced Concrete Structures", by M.Z. Cohn, E.F. Burnett, D.E. Grierson, O. Dutt, R. Francis, H.C. Parameswar, S. Talwar (April 1969).
2. "Frame Buckling: An Illustration of the Perturbation Technique", by J. Roorda and A.H. Chilver (May 1969).
3. "A General Formulation or the Stress-Strain Properties of Concrete", by M. Sargin and V.K. Handa (May1969).
4. "Evaluation of the Fatigue Strength of Integrated Tanks for LNG Ships", by D.J. Burns, R.G. Jackson and J.G. Kalbfleisch (May 1969).
5. "A Fundamental Study of Consolidation and the Estimation of Settlement", by B. LeLievre and H.B. Poorooshasb (June 1969).
6. "Concave Programming in Structural Optimization" by G.I.N. Rozvany (June 1969).
7. "Comparison of Finite Element and Unique Limit Analysis Solutions for Certain Reinforced Concrete Slabs", by G.N. McNiece (July 1969).
8. "The Morphology and Growth of Creep Cavities in Alpha-Iron", by A.L. Wingrove and D.M.R. Taplin (July 1969).
9. "Variational Method for Non-Conservative Problems", by R.N. Dubey (July 1969).
10. "Computer Design of Multi-Component Hydrostatic Extrusion Vessels Containing Autofrettaged Components", by R.J. Pick and D.J. Burns (July 1969).
11. "A Theorem on the Limit Analysis of Plates and Shells", by G.I.N. Rozvany (August 1969).
12. "Limit Design for Reinforced Concrete Structures: An Annotated Bibliography", by M.Z. Cohn (August 1969).
13. "On the Studies of Phenomena Related to the Formation of Residual Stress States in Glass Plates", by J.T. Pindera and N.K. Sinha (August 1969).
14. "Inter-granular Creep-Cavitation of Alpha-Brass" by I.J. Spark and D.M.R. Taplin (August 1969).
15. "Further Results on the Equilibrium Method of Limit Design", by M.Z. Cohn and D.E. Grierson (August 1969).
16. "An Elastic-Plastic Finite Element Analysis for Plates with Edge Beams", by G.M. McNeice (August 1969).
17. "Response of Photoelastic Systems", by J.S. Pindera (August 1969).
18. "A Uniqueness Theorem for Non-Linear Plans Frameworks", by O. Dutt and N.C. Lind (September 1969).
19. "Fatigue and Cyclic Thermal Softening of Thermoplastics", by I. Constable, J.G. Williams and D.J. Burns (September 1969).
20. "The Influence of Polycrystal Grain Size upon Creep Ductility", by G.J. Cooks and D.M.R. Taplin (September 1969).
21. "A Stress-Strain Function for the Fatigue of Metals", by K.N. Smith, P. Watson and T.H. Topper (October 1969).
22. "An Automatic Approach to the Analysis of Plastic Frames Under Fixed and Variable Loading", by M.Z. Cohn and D.E. Grierson (October 1969)
23. "Probabilistic Evaluation of Safety Factors", by M.K. Ravindra, A.C. Heaney and N.C. Lind (October 1969)
24. "Studies of Intergranular Creep Cavitation", by G.L. Dunlop, G.J. Cocks, A.L. Wingrove and D.M.R. Taplin (October 1969)
25. "Simple Nonparametric Failure Probability Estimates", by N.C. Lind (October 1969)
26. "Application of Liapunov's Direct Method to the Stability Problem of Rod's Subject to Follower Forces", by H.H.E. Leipholz (December 1969)
27. "Instability of Symmetric Structural Systems with Independent Loading Parameters", by K. Huseyin (December 1969)
28. "Influence of Inertial parameters on the Stability of Non-Conservative Systems", by K. Vepa and J. Roorda (January 1970).
29. "Superplastic Copper Alloys", by D.M.R. Taplin, G.L. Dunlop, S. Sagat and R.H. Johnson (January 1970).
30. "A Lower Bound Approach to the Optimal Design of Concrete Frames and Slabs", by G.I.N. Rozvany and M.Z. Cohn (January 1970).
31. "On the Energy Criterion Applied to Continuous Systems and Based on Liapunov's Method" by H. Leipholz (January 1970)
32. "Variational Finite Element Calculation of the Acoustic Response of a Rectangular Panel", by G.M.L. Gladwell (February 1970)
33. "The Effect of Induced Axial Resistance Due to Geometry Changes in Simple Elastic-Plastic Systems", by N.K. Srivastava and A.N. Sherbourne (February 1970)

34. "On the Estimation of the Stability Boundary of Symmetric Structural Systems", by K. Huseyin (February 1970).
35. "The Application of Liapunov's Direct Method to Stability Problems of Continuous, Nonconservative Systems", by H.H.E. Leipholz (February 1970).
36. "Consistent Partial Safety Factors", by N.C. Lind (February 1970).
37. "The Mechanical Behaviour of a Nickel-free Austenitic Stainless Steel", by V.V.P. Kutumbarao, D.M.R. Taplin and P. Rama Rao (March 1970).
38. "Application of a Generalized Principle of Hamilton to Non-Conservative problems", by H.H.E. Leipholz (March 1970).
39. "Studies of Physical and Mathematical Models of Some Flanged Connections", by Jerzy T. Pindera and Yulun Sze (March 1970).
40. "Elastic Plastic Bending of Circular Plates", by N.K. Srivastava and A.N. Sherbourne (March 1970).
41. "On the Stability of continuous Systems in the First Approximation", by H.H.E. Leipholz (March 1970).
42. "Influence of the Bolt Tensile Stiffness and the Bolt Pretensioning on the Behaviour of the Fact-to-Face Flanged Connections", by Jerzy T. Pindera and Yulun Sze (March 1970).
43. "The Loading-Frequency Relationship in Multiple Eigenvalue Problems", by K. Huseyin and J. Roorda (June 1970)
44. "Thermal Instability in Differentially Heated Inclined Fluid Layers – A Generalization of the Buoyancy Driven Stationary Instability", by T.E. Unny (June 1970)
45. "The Influence of Cyclic Strain Amplitude on the Fatigue Behaviour and Sub-Structure of Pure Iron", by H.A. Raouf, P.P. Benham and A. Plumtree (June 1970).
46. "Prediction of Strain Rate for Constant Strain Rate Drained Triaxial Tests", by A Thurairajah (June 1970).
47. "Experiments on Plastic Limit Behaviour of Shell-Nozzle Junctures Subjected to Nonsymmetric Loading by N.C. Lind and S. Palusamy (June 1970).
48. "The Characteristics and Application of the Clough-Felippa Element in Slab Analysis", by J.C. Jofriet, G.M. McNeice and R. Green (June 1970).
49. "Determination of Dominant Radiation in Polar-scope by Using Babinet Compensator", by J.T. Pindera and N.K. Sinha (June 1970)
50. "Finite Element Analysis of Reinforced Concrete Slabs with Progressive Cracking", by J.C. Jofriet and G.M. McNeice (June 1970).
51. "Limit Analysis of Non Symmetrically Loaded Spherical Shells", by S. Palusamy and N.C. Lind (July 1970)
52. "Elastic-Plastic Bending of Restrained Pin-Ended Circular Plates", by A.N. Sherbourne and N.K. Srivastava (July 1970)
53. "On the Studies of Residual Stresses in Glass Plates, by J.T. Pindera and N.K. Sinha (July 1970)
54. "A Simplified Model for Moderately Thick Plates in Bending", by A.N. Sherbourne and N.K. Srivastava (July 1970)
55. "Nonlinear Optimisation Programmes – Write-up and Performance", by S. Palusamy (July 1970)
56. "Plastic Limit Analysis Using Linear Programming", by D.E. Grierson and G.M.L. Gladwell (July 1970)
57. "The Response of a System with Double-Bilinear-Hysterics to Random Excitation", by B. Sahay and S. Oen (July 1970)
58. "The Role of the Engineer and Technologist in Mechanical Design", by D.C. Ferguson (August 1970)
59. "Development of a 100,000 lb/in^2, 100 cu. In/min Fluid Support Intensifier", by U. Mohaupt and D.J. Burns (September 1970)
60. "Behaviour of Sand During Undrained Triaxial Tests", by A. Thurairajah and B. LeLievre (September, 1970)
61. "On the Stability of the Critical Equilibrium States", by K. Huseyin (November 1970)
62. "Sustained and Cyclic Loading of Concrete Columns", (1970)
63. "On the Sufficiency of the Energy Criterion for the stability of Certain Non-Conservative Systems of the Follower-Load Type", by H. Leipholz (Oct. 1970)
64. "On a Common Theory for the Stability of Self-Adjoint and Non-Self-Adjoint Mechanical Systems" by H. Leipholz (December 1970)
65. "Stability Criteria for Plüger's Rod with Internal and External Damping via Liapunov's Second Method", by H. Leipholz and R. Kapoor (December 1970)
66. Investigation of the Stability of Continuous Non-conservative Systems with Damping by means of Liapunov's Method", by H. Leipholz and R. Kapoor (December 1970)
67. The Plates method and its Application in the Shell Theory", by L. Fischer (December 1970)
68. "An Approximate Method for Stability of Nonconservative Systems", by R.N. Dubey, H. Leipholz, R. Vepa and T.C. Chen (January 1971)
69. "Divergence Instability of Multiple Parameter Circulatory Systems", by K, Huseyin and H. Leipholz (January 1971)
70. "On the Stability of One-Dimensional Continuous Systems with Polygenic Forces", by H. Leipholz and K. Huseyin (January 1971)
71. "Finite Element Analysis of Threaded End Closures of Thick-Walled Vessels", by R.J. Pick and D.J. Burns (January 1971)

72. Optimization of a Structural Code", by M.K. Ravindra and N.C. Lind (January 1971)
73. Towards Practical Application of Structural Reliability Theory", by N.C. Lind and A.G. Davenport (February 1971)
74. "Singular Critical Points in the General Theory of Elastic Stability", by K. Huseyin (March 1971)
75. "On the Uniqueness of Plastic Optimal Design", by M.Z. Cohn, S.R. Parimi and S.K. Ghosh (March 1971)
76. "The Reinforced Concrete Beam-Column Connection", by E.F.P. Burnett and R. Jojoo (May 1971)
77. "The Influence of the Column Load on the Reinforced Concrete Beam-Column Connection", by E.F.P. Burnett and R.J. Trenberth (May 1971
78. "A Unified Approach to the Theory of Plastic Structures", by M.Z. Cohn, S.K. Ghosh and S.R. Parimi (May 1971)
79. "On a Maxwell-Type Solution of Three Dimensional Elasticity", by H. Leipholz (March 1971
80. Random Buckling Behaviour in Axially Loaded Cylindrical Shells with Axisymmetric Imperfections", by J. Roorda and J.S. Hansen (May 1971)
81. "Strength Characteristics of Reinforced Concrete Columns Under Sustained Loading", by J. Hellesland and R. Green (May 1971)
82. "On a Hybrid Type of Galerkin's Equation", by H. Leipholz (May 1971)
83. "The Tensile Properties of a Superplastic Aluminium Bronze", by G.L. Dunlop and D.M.R. Taplin (June 1971)
84. "Multi-Criteria Probabilistic Structural Design", by M.Z. Cohn and S.R. Parimi (June 1971)
85. "Textures and Anisotropic Flow of a Superplastic Aluminium Bronze", by G.L. Dunlop and D.M.R. Taplin (June 1971)
86. "Pattern Loading on Reinforced Concrete Flat Plates", by J.C Jofriet and G.M. McNeice (July 1971.
87. "Equivalent Axisymmetric Imperfections in Axially Compressed Cylindrical Shells", by J. Roorda (June 1971)
88. "Elastic-Plastic Deformation Analysis Using Linear Programming", by D.E. Grierson (July 1971)
89. "The Design of Structural Design Norms", by N.C. Lind (August 1971)
90. "On the Dynamic Stability of Fluid-Conveying Pipes", by D.S. Weaver and T.E. Unny (August 1971)
91. "The Elastic Stability of Two-Parameter Nonconservative Systems", by K. Huseyin and R.H. Plaut (August 1971)
92. "Superplasticity and Cavitation in Alpha-Beta Brass", by S. Sagat, P. Blenkinsop and D.M.R. Taplin (September 1971)
93. "A Metallographic Study of Superplasticity in a Micrograin Aluminium Bronze", by G.L. Dunlop and D.M.R. Taplin (September 1971)
94. "Potential Flow of a Vortex Row with Statistically Varying Spacings", by F.J. Legerer and U. Lemmin (September 1971)
95. "Influence of the Width of Loading Plate on the Rotation Capacity of Reinforced Concrete members", by C.S. Chandrasekhar and H.A. Falkner (October 1971)
96. "A Rotary Coupling for Pressures up to 100,000 psi", by F.M. Smith and U.H. Mohaupt (October 1971)
97. "The Multiple-Parameter Perturbation Technique for the Analysis of Non-Linear Systems", by K. Huseyin (November 1971)
98. "Optimal Design of Plastic Structures for Fixed and Shake-Down Loadings", by M.Z. Cohn and S.R. Parimi (November 1971)
99. "Equilibrium (Serviceability) Methods of Limit Design for Concrete Frames", by M.Z. Cohn (November 1971)
100. "The Flexural Ductility of Reinforced Concrete Sections", M.Z. Cohn and S.K. Ghosh (November 1971)
101. "Some Re-Interpretations of Structural Analysis", by G.M.L. Gladwell (January 1972)
102. "Stability of Hauger and Pfluger Systems with Kelvin Model and External Damping", by H.H.E. Leipholz and Ravi Kapoor (January 1972)
103. "A Note on Effect of Stiffeners in Light Gage Members", by N.C. Lind (January 1972)
104. "Qualitative Stability Theory of Completely Supported Roads Subjected to Follower Forces", by H.H.E. Leipholz and R.N. Dubey (January 1972)
105. On the Calculation of Flutter Loads by Means of a Variational Technique", by H.H. E. Leipholz (February 1972)
106. "Finite Element Analysis of the Axisymmetric Vibrations of Cylinders", by G.M.L. Gladwell and U.C. Tahbildar (February 1972)
107. "On the Analysis of Elgenvalue Curves of Roads Subjected to Conservative and Non-Conservative Loads", by H.H.E. Leipholz (February 1972)
108. "Mechanical Behaviour of Polycrystals at Elevated Temperatures", by D.M.R. Taplin, P. Rama Rao and V.V.P.K. Rao (March 1972)
109. "Multi-Parameter Optimal Design of Plates and Shells", by Z. Mroz (March 1972)
110. "Influence of Kelvin Model Damping in Flutter-Divergence Transitional Polygenic Systems", by H.H.E. Leipholz and Ravi Kapoor (March 1972)
111. "Structural Analysis for Idealized Nonlinear Material Behaviour", by Graham M.L. Gladwell and Donald E. Grierson (April 1972)
112. "On Extremum Properties of the Generalized Rayleigh Quotient Associated with Flutter Instability", by K. Huseyin and R. Plaut (September 1972)

113. "an Invariant Second Moment Reliability Format", by N.C. Lind (November 1972)
114. "On the Solution of Plane Stress Problems Using Complex Potentials and Photoelastic Data", by H.A. Atta and J.C. Thompson (December 1972)
115. "Bending by Beginners", by E.F.P. Burnett (February 1973)
116. "Some Aspects of Office Design Live Load", by W.W.C. Siu and N.C. Lind (February 1973)
117. "Second-Order Collapse Load Analysis: a LP Approach", by S. Baset, D.E. Grierson and N.C. Lind (March 1973)
118. "Analysis of Plastic Frames Allowing for the Effect of Axial forces", by M.Z. Cohn and T. Rafay (March 1973)
119. "An Exact and Invariant First-Order Reliability Format", by A.M. Hasofer and N.C. Lind (May 1973)
120. "Adaptive Control of Design Codes", by G. Schorn and N.C. Lind (May 1973)
121. "On the Method of Trefftz in the Light of a Generalized Variational Principle", by H. Leipholz (June 1973)
122. "Effect of Grain Size on the Tensile Properties of a Cr-Mn-N Austentific Strainless Steel in the Temperature Range 25 to 1000 degrees C", by V.V. Ktumbarao, P. Rama Rao and D.M. R. Taplin (October 1973)
123. "Floor Beams in Orthotropic Bridges", by K.N. Smith and J. Mastofrancesco (October 1973)
124. "Design of Reinforced Concrete Structures for Strength and Deformability", by Mir M. Ali and Donald E. Grierson (January 1974)
125. "On a Modified Rayleigh"s Quotient for a Certain Nonconservative Elgenvalve Problem", by H.H. E. Leipholz (January 1974)
126. "Shear Strength of Reinforced Concrete Beams with Short Shear Spans", by S.M. Fereig and K.N. Smith (February 1974)
127. "Inelastic Behaviour of Prestressed Concrete Structural Members", by M. Balabanian, S.K. Ghosh and M.Z. Cohn (May 1974)
128. "A Formulation of Probabilistic Design", by N.C. Lind (April 1974)
129. "On Conservative Elastic Systems of the First and Second Kind", by H.H.E. Leipholz (June 1974)
130. "The Post-Yield Strength of R.C. Flexural Sections", by E.F.P. Burnett and S.K. Ghosh (June 1974)
131. "Optimal Solutions in Probabilistic Structural Design", Part 1: Theory; Part 11: Applications, by S.R. Parimi and M.Z. Cohn (June 1974)
132. "Some Remarks on Liapunov Stability of Elastic Dynamical Systems", by H.H.E. Leipholz (July 1974)
133. "A Linear Programming Approach to the Problem of Moments", by W.W.C. Siu, S.S. Sengupta and N.C. Lind (August 1974)
134. "On Destabilizing Parameters, the Degree of Stability of Mechanical Systems, and General Stability Criteria", by H.H.E. Leipholz (February 1975)
135. "Fatigue Studies of O.F.H.C. Copper and a Copper Dispersion Alloy at 300-750 degrees C", by G.J. Cocks and D.M.R. Taplin (May 1975)
136. "On Extension of Steklov's Inequality and a Lower Bound Theorem for Pflüger's Rod" by H.H.E. Leipholz (December 1975)
137. "Mathematical Programming Methods for Deformation Analysis At Plastic Collapse", by G. Maier, D.E Grierson, M.J. Best (March 1976)
138. "Four Papers on Catalogue Optimization", by N.C. Lind et al (August 1976)
139. "Plastic Analysis Under Combined Stresses", by Donald E. Grierson, M.ASCE, and Saleh B. Abdel-Baset (August 1976)
140. "Morphology and Fracture of Bone", by K.R. Piekarski (September 1976)
141. "On the Application of the Energy Method to the Stability Problem of Nonconservative Autonomous and Nonautonmous Systems", by H.H.E. Leipholz (September, 1976)
142. "A General Computer Program For Plastic Analysis" by D.E. Grierson and S.B. Abdel-Baset (July 1977)
143. "Local Buckling of Thin Walled Channels" by K.R. Venkataramaiah and J. Roorda (November 1977)
144. "Fracture of a Petrochemical Furnace Tube Steel at High Temperature" by G.L. Dunlop, R.J. Twigg and D.N.R. Taplin (November 1977)
145. "On a Variational Principle For Beck's Rod", by H.H.E. Leipholz (January 1978)
146. "On an Extension of Hamilton's Variational Principle to Nonconservative Systems Which are Conservative in a Higher Sense", by H.H.E. Leipholz (January 1978)
147. "Fracture Research in Canada" by D.J. Burns and D.M.R. Taplin (May 1978)
148. "Effective Width of Stiffened Steel Plate Components" by J. Roorda and K.R. Venkataramaiah (May 1978)
149. Optimization of Finite element Studies of Stress Concentration Regions Using Asymptotic Analysis" by D.D. Curtis and J.C. Thompson (September 1978)
150. "Mathematical Programming and Nonlinear Finite Element Analysis" by D.E. Grierson, A. Franchi, O. DeDonato, and L. Corradi (October 1978)
151. "Effective Width of Stiffened Cold-Formed Steel Plates" by J. Roorda and K.R. Venkataramaiah (October 1978)
152. "Pressurised Ring Stiffened Shells: A Model Study" by John Roorda (November 1978)
153. "Non-Stationary Narrow-Band Response and First Passage Probability", by Steen Krenk, Risö National Laboratory, Roskilde, Denmark (December 1978)

154. "On the Elastic Constants of Plane Orthotropic Elasticity" by Steen Krenk, Risö National Laboratory, Roskilde, Denmark (December 1978)
155. "Computer Analysis of Elastic-Plastic Structures" by M.Z. Cohn and A. Franchi (January 1979)
156. "Bifurcation in Elastic-Plastic Plates" by R.N. Dubey (April 1979)
157. "An Elastic-Plastic Analysis Computer System" by A. Franchi, D.E. Grierson, M.Z. Cohn (August 1979)
158. "Plastic Design Under Combined Stresses", by Donald E. Grierson and Ahmed A. Aly (September 1979)
159. "On the Use of the Hamiltonian for an Evaluation of the Stability of Elastic Systems Subjected to Follower Forces" by H.H.E. Leipholz (July 1980)
160. "On Topological Aspects of Liapunov Stability of Continuous Systems" by H.H.E. Leipholz" (July 1980)
161. "Plastic Design Under Multiple Loads" by Donald E. Grierson, M.ASCE and Ahmed A. Aly (August 1980)
162. "On a Generalization of the Lower Bound Theorem for Elastic Rods and Plates Subjected to Compressive Follower Forces" by H.H.E. Leipholz (September 1980)
163. "Bayesian Approach to Prototype Testing" by P. Hauge Madsenand Niels C. Lind (January 1981)
164. "Probabilistic Models for Prototype Testing", by Mircea Grigoriu M.ASCE and Niels C. Lind (May 1981)
165. "Synthesis Under Service and Ultimate Performance Constraints" by D.E. Grierson and L.A. Schmit Jr. (July 1981)
166. "Active Vibration Control of a Simple Beam" by D. Camotim and J. Roorda (August 1981)
167. "The High Temperature Fatigue and Tensile Fracture of Cu-Cr Based Alloys", by N.Y. Tang and D.M.R. Taplin
168. "Nonlinear Flexural Response of Partially Prestressed Concrete Sections" by M.Z Cohn and M. Bartlett (October 1981)
169. "The Analysis and Prediction of Lateral Instability in Highly Stressed, Near-Surface Rock Strata", by J. Roorda, J.C. Thompson and O.L. White (October 1981)
170. "A Review of the Role of Stress State, Environment and Section Size on Creep Deformation and Fracture", by M.C. Pandey, B.F. Dyson and D.M.R. Taplin (October 1981)
171. "A New Method of Fast Probability Integration", by X. Chen and N.C. Lind (June 1982)
172. "Synthesis of Frameworks Under Multilevel Performance Constraints" by Donald E. Grierson and Thomas C.W. Chiu (July 1982)
173. "Modelling of Uncertainty in Discrete Dynamical Systems" by N.C. Lind (August 1982)
174. "Dynamics of a Mobile Tank Partially Filled with Liquid: Equations of Motion and Their Linearization", by Aleksander Kornecki (September 1982)
175. "Inelastic Behavior of Continuous Prestressed Concrete Beams" by M.Z. Cohn, F. ASCE and Y. Frostig (November 1982)
176. "The Power Method in Finite Element Analysis of Plastic Bifurcation Plate Problems", by H. El-Ghazaly and A.N. Sherbourne (1983)
177. "Structural Design Using Intelligent Structural Analysis" by Don Grierson (January 1983)
178. "Inelastic Interactive Distortional Buckling of W-Shape Steel Beams", by H.A. El-Ghazaly A.N. Sherbourne and R.N. Dubey (1983)
179. "Optimal Synthesis of Steel Frameworks Using Standard Sections" by D.E. Grierson and W.H. Lee (October 1983)
180. "Degenaly, Hopf Fifurcation", by S.T. Ariaratnam and N. Sri Namachchivaya (September 1983)
181. "Optimization of Structural Concrete Beams", by M.Z. Cohn and A.J. MacRae (September 1983)
182. "The Inverse Problem for the Vibrating Beam", by G.M.L. Gladwell (September 1983)
183. "External Disturbances in Transient One-Dimensional Diffusion", by E. Rosenblueth (September 1983)
184. IEM Paper No. 1, "Theory of Elastic and Photoelastic Isodynes", by Jerzy-Tadeusz Pindera and Bogdan Roman Krasnowski (October 1983)
185. IEM Paper No. 2, "Analysis of Models of Stress in the Regions of Crack, Using Isodyne Photoelasticity", by J.T. Pindera, B.R. Krasnowski and M.J. Pindera (October 1983)
186. "Partial Prestressing: From Theory to Practice, Introductory Report", by M.Z. Cohn (October 1983)
187. "On the Inverse Problem of the Vibrating String or Road", by G.M.L. Gladwell and J.A. Gbadeyan (November 1983)
188. "IEM Paper No. 3 "Basic Theories and Experimental Methods of Gradient Photoelasticity", by J.T. Pindera, F.W. Hecker and B.R. Krasnowski (January 1984)
189. "Computer-Automated Synthesis of Building Frameworks", by Donald E. Grierson and Gordon E. Cameron (February 1984)
190. IEM Paper No. 4, "The Interplay and Use of Experimental and Mathematical Models in Research", by Peter G. Glockner (August 1984)
191. "IEM Paper No. 5, "The Impact of Research on Socio-Economic Development", by Karl Hans Laermann (August 1984)
192. "Discussion of Four Theories on the Penetration of Metallic Targets" by K. Nixdorff (September 1984)
193. "On the Effect of Residual Stresses in the Plastic Buckling of Columns – a Model Study" by Dinar Camotim and John Roorda (October 1984)
194. "Bifurcation of Elastic-Plastic Solids with Residual Stresses" by D. Camotim and J. Roorda (December 1984)

195. "Buckling Behaviour of Ring-Stiffened Cylinders: Experimental Study", by S.S. Seleim, J. Roorda (January 1985)
196. "The Stability of Cylindrical Air-Supported Structures", by R. Maaskant and J. Roorda (March 1985)
197. "Theoretical and Experimental Results on the Post-Buckling of Ring-Stiffened Cylinders" by S.S Seleim and J. Roorda (May 1985)
198. "On the Modified S-N Curve for Metal Fatigue Prediction and its Experimental Verification", by H.H.E. Leipholz (May 1985)
199. "Analysis of Flexible Frames by Energy Search", H.A. El-Ghazaly and G.R. Monforton (November 1985)

APPENDIX 3: SM Technical Notes/Reports Series

1. "Yield Surfaces", by Antoni Sawczuk (March 1971)
2. "Optimal Limit Design for Concrete Structures", by M.Z. Cohn (April 1971)
3. "An experience in Equilibrium and Stability", by J. Roorda (May 1971)
4. "Cyclic Deformation and Fatigue Behaviour of Axial and Flexural Members – a Method of Simulation and Correlation", by H.R. Jhansale and T.H. Topper (June 1971)
5. "Utilization of Cold Work in Light Gage Steel", by N.C. Lind and D.K. Shroff (June 1971)
6. "A Review of the Effective Width Formula", by N.C. Lind, M.K. Ravindra and J. Power (June 1971)
7. "Optimum Edge Stiffeners for Thin-Walled Structural Elements", by K.R. Venkataramaiah (July 1971)
8. "How Safe are the Eccentrically Loaded Thin-Walled Steel Struts", by K.R. Venkataramaiah (August 1971)
9. "Ultimate Strength of Eccentrically Loaded Thin-Walled Open Channel Columns", by K.R. Venkataramaiah (August 1971)
10. "Fatigue and Creep Failure of Alpha-Iron at 450-700 degrees C", by H.J. Westwood and D.M.R. Taplin (September 1971)
11. "Creep and Fatigue Fracture Behaviour of Inconel Alloy X-750", by P.K. Venkiteswaran, D.C. Ferguson and D.M.R. Taplin (October 19710
12. "The Influence of Grain Size and Dispersion on the Creep Ductility of Copper Bass Alloys", by R.G. Fleck and D.M.R. Taplin (October 1971)
13. "Curvature of Cracked Reinforced Concrete Sections", by S.K. Ghosh and C.S. Chandrasekhar (November 1971)
14. "Effects of Temperature and Deformation Rate on Cyclic Strength and Fracture of Low-Carbon Steel", by H. Abdel-Raouf, A. Plumtree and T.H. Topper (November 1971)
15. "Compression in thin-Walled Members", by K.R. Venkataramaiah and N.C. Lind (January 1972)
16. "The Analysis and Synthesis of Resonators", by G.M.L. Gladwell (February 1972)
17. "Some Examples of Code Calibration", by N.C. Lind, M.K. Ravindra and W. Siu (February 1972)
18. "The Algebraic Elgenvalue Problem – A Set of Fortran Subroutines", by G.M.L. Gladwell and U.C. Tahbildar (March 1972)
19. "Safety Level Decisions", by N.C. Lind and E. Basier (February 1972)
20. "On Simplified Theories of Cyclic Plasticity", by Z. Mroz and N.C Lind (June 1972)
21. "Post-Buckling Behaviour of Thin Compression Elements with Edge Stiffeners", by K.R. Venkataramaiah (July 1972)
22. "Structural Reliability and the Invariance Problem", by Ove Ditlevsen (March 1973)
23. "A computer Programme for the Non-Linear Analysis of Reinforced Concrete Sections", by M. Sargin, S.K. Ghosh and M.Z. Cohn (March 1973)
24. "Deformational Analysis of Reinforced Concrete Structure-Problems and Prospects", by C.S. Chandrasekhar and S.K. Ghosh (March 1973)
25. "The Algebraic Elgenvalue Problem: A Commentary on Some Computer Programs", by G.M.L. Gladwell (May 1973)
26. "The Computer Programme DAPS for the Design and Analysis of Plastic Structures", by S.R Parimi, S.K. Ghosh and M.Z. Cohn (June 1973)
27. "Optimization of Economic Activities Under Public Regulation", by Niels C. Lind (April 1974)
28. "Stochastic Models for Fatigue Failure of Materials", by S.K. Srinivasan (June 74)
29. "Some Reinterpretations of Structural Analysis" by G.M.L. Gladwell (December 1975)
30. "On Stochastic Load Combination", by Henrik O. Madsen (October 1978)
31. "Khacian's New Linear Programming Algorithm by Richard J. Caron (August 1980)

APPENDIX 4: SM Lecture/Seminar Series

Six Lectures on Stability of Elastic Systems
H.H.E. Leipholz
1971

Theory of Codified Structural Design
N.C. Lind
A set of invited lectures, Polish Academy of Sciences, Warsaw, Poland
June 5-15, 1972

Mathematical Models of Inelastic Material Behaviour
Z. Mroz
Institute of Fundamental Technical Research,Warsaw, Poland
1973

Stochastic Point Processes
S.K. Srinivasan
Indian Institute of Technology,Madras, India
1973

Design in Cold Formed Steel
R.M. Schuster (Editor)
1974

Newmark's Numerical Method
N.C. Lind
1975

Six Lectures on Variational Principles in Structural Engineering
H.H.E. Leipholz
1978

Buckling of Elastic Structures
J. Roorda
1980

A Brief Course on the Theory of Buckling
L.M. Kachanov
University of Delaware
H.H.E. Leipholz (Editor)
1981

Buckling of Thin-Walled Rods
L.M. Kachanov
Boston University, Boston, Ma.
H.H.E. Leipholz (Editor)
1983

APPENDIX 5: SM Texts, Manuals, Proceedings, and Other Publications

SM Text No. G1
Analysis and Design of Inelastic Structures
Volume 2: Problems, 1972
M. Z. Cohn

Cold Formed Steel Design Manual
R.M. Schuster and N.C. Lind
1975

Fracture 1977
Proceedings, 4th International Conference on Fracture
Waterloo, Ont., June 1977
D. M. R. Taplin (Editor)

Engineering Plasticity by Mathematical Programming
Proceedings, NATO Advanced Study Institute
Waterloo, Ont., August 2-12, 1977
M.Z. Cohn, G. Maier and D.E. Grierson (Editors)
Pergamon Press, New York, Toronto, 1979

B. M. Fraeijs de Veubeke Memorial Volume: Selected Papers
SM Archives Special Publication

M. Geradin (Editor)
1980

STRUPL 1: User's Manual
M.Z. Cohn, F. Erbatur and A. Franchi
1982

Micromechanisms of Plasticity and Fracture
M.H. Lewis, Warwick University, England and
D.M.R. Taplin, Trinity College, Dublin, Ireland (Editors)
University of Waterloo Press, Waterloo, Ont.,1983

Nonlinearity and Continuity in Prestressed Concrete
Preliminary Publication (4 Volumes), International Symposium
ACI, ASCE, CEB, CPCI, CSCE, FIP, IABSE, PCI, SMD
Waterloo, Ont., July 4-6, 1983
M.Z. Cohn (Editor)

Contact Mechanics and Wear of Rail and Wheel Systems
Proceedings, First International Symposium
University of British Columbia, Vancouver, B.C., July 6-9, 1982
J. Kalousek, R.V. Dukkipati and G.M.L. Gladwell (Editors)
University of Waterloo Press, 1983

Partial Prestressing: From Theory to Practice
Proceedings, NATO Advanced Research Workshop (2 Volumes)
Paris, France, June 18-22,1984
M.Z.Cohn (Editor)
Martinus Nijhoff Publishers, Dordrecht, The Netherlands,1986

Contact Mechanics and Wear of Rail and Wheel Systems
Proceedings, Second International Symposium
University of Rhode Island, Kingston, R.I., July 8-11.1986
G.M.L. Gladwell, H. Ghonem and J. Kalousek (Editors)
University of Waterloo Press, 1987

*APPENDIX 6: SM Workshops, Conferences, and Symposia **

Colloquium on Limit Design for Structural Concrete
ACI, ASCE, SMD Colloquium (see Appendix 2, No.1)
September 6-8, 1967

Computer-Aided Engineering
SMD Symposium (see Appendix 1, No.5)
May 11-13, 1971

Applications of Solid Mechanics
SMD Symposium (see Appendix 1, No 7)
June 26, 27, 1972

Experimental Mechanics in Research and Development
International Symposium (see Appendix 1, No.9)
June 12-14, 1972

Stochastic Problems in Mechanics
International Symposium (see Appendix 1, No.10)
September 24-26, 1973

Mechanics in Engineering
First ASCE, EMD Specialty Conference (see Appendix 1, No.11)
May 26-28, 1976

Fracture 1977
Fourth International Conference (see Appendix 5, No.3)
June 1977

Continuum Models of Discrete Systems (CMDS 2)
Second International Symposium (see Appendix 1, No.12)
June 28-July 2, 1977

Engineering Plasticity by Mathematical Programming
NATO Advanced Study Institute (see Appendix 5, No.4)
August 2-12, 1977

Nonlinear Design of Concrete Structures
ASCE, ACI, CEB, CSCE International Symposium (see Appendix 1, No.14)
August 7-9, 1979

Continuum Models of Discrete Systems (CMDS 3)*
Third International Symposium, Freudenstadt, Germany (see Appendix 1,No.15)
June 24-30, 1979

Fuel-Air Explosions *
International Conference, Montreal, Que. (see Appendix 1, No.16)
November 4-6, 1981

New Problems in Mechanics of Continua *
Third Polish-Swedish Symposium, Jablonna, Poland (see Appendix 1, No.17)
May 10-14, 1982

Contact Mechanics and Wear of Rail and Wheel Systems (I) *
International Symposium, Vancouver, B.C. (see Appendix 5, No.9)
July 6-9, 1982

Nonlinearity and Continuity in Prestressed Concrete
ACI, ASCE, CEB, CPCI, CSCE, FIP, IABSE, PCI, SMD
International Symposium (see Appendix 5, No.8)
July 4-6, 1983

Partial Prestressing: From Theory to Practice *
NATO Advanced Research Workshop, Paris, France (see Appendix 5, No.10)
June 18-22, 1984

Contact Mechanics and Wear of Rail and Wheel Systems (II) *
International Symposium, Kingston, R.I. (see Appendix 5, No.11)
July 8-11, 1986

References

Buber M. (1956) "Teaching and Deed", in *The Writings of Martin Buber*, W. Herberg (Ed.), Meridian Books, NY, p. 324.

Cohn M.Z. (1972) *Analysis and Design of Inelastic Structures*, SM Text G1, Vol. 2, 521 pp.

Cohn M.Z., Burnett E.F., Grierson D.E., Dutt O., Francis R., Parameswar H.C. and Talwar S. (1969) "Application of Limit Design to Reinforced Concrete Structures", SM Report (Paper) No. 1, 56 pp.

Petch H.E. (1973) "Policy Issues for Canadian Science and Technology", in *Experimental Mechanics in Research and Development*, J.T. Pindera et al. (Eds.), SM Study No. 9, pp. 33–42.

* NOTE: Listed conferences were held at Waterloo, or had editorial/publishing SMD involvement when star-marked *.

ISOSPECTRAL VIBRATING SYSTEMS

G.M.L. Gladwell
Department of Civil Engineering, University of Waterloo,
Waterloo, Ontario, Canada N2L 3G1
E-mail: ggladwell@uwaterloo.ca

Abstract

Two vibrating systems are said to be *isospectral* if they have the same natural frequencies. This paper reviews some recent results on isospectral conservative (i.e., undamped) discrete vibrating systems. The paper centres around two ways of creating isospectral systems: by QR factorisation with a shift, and by using the concept of isospectral flow. Both these procedures are illustrated by using FEM models.

Keywords: vibration, isospectral, QR factorisation, isospectral flow

1. Introduction

An undamped vibrating system has certain frequencies, called *natural* frequencies, at which it can vibrate freely, without the application of forces. An actual physical system has theoretically an infinity of such frequencies. A model of such a system may be either *continuous* or *discrete*, having respectively an infinity or a finite number of natural frequencies; in this paper we consider only discrete systems. Two systems with the same set of natural frequencies, *spectrum*, are said to be *isospectral*. In general, the spectrum of a system *mirrors* the system, but does not *specify* it completely: there can be many systems, an *isospectral family*, with the same spectrum. There are two broad classifications of problems relating to a system and its spectrum. In *inverse* problems, one attempts to construct a system with a given *spectrum*; in *isospectral* problems, one attempts to find another system, maybe a family of systems, having the same spectrum as a *given* system. In some ways, isospectral problems are easier than inverse problems: at least one is sure that there exists at least one system, the given system, with the specified spectrum; this is not always the case with inverse problems. For an in-depth discussion of inverse and isospectral problems, see Gladwell (2004).

Modelling of a physical system is usually done by means of some *finite element method* (FEM): the system is treated as a set of *elements*, connected in some way. Each element is specified by a set of generalised displacements, an *element stiffness matrix* $\mathbf{K}_e$, and an *element mass (inertia) matrix* $\mathbf{M}_e$. In the process of assembling the elements, one constructs overall or *global* stiffness and mass matrices, $\mathbf{K}$ and $\mathbf{M}$, and assembles the element displacements into a displacement vector $\mathbf{u}$. The natural frequencies of the system appear as (the square roots, $\omega = \lambda^{\frac{1}{2}}$, of) the eigenvalues $(\lambda_i)_1^n$ of the generalised eigenvalue problem

$$(\mathbf{K} - \lambda\mathbf{M})\mathbf{u} = \mathbf{0}. \tag{1}$$

A system is thus defined by a pair of matrices $(\mathbf{K}, \mathbf{M})$. We say that two systems $(\mathbf{K}, \mathbf{M})$ and $(\mathbf{K}', \mathbf{M}')$ are isospectral if $(\mathbf{K}' - \lambda\mathbf{M}')\mathbf{u}' = \mathbf{0}$ has the same spectrum of eigenvalues as (1).

In practice, the matrices $\mathbf{K}, \mathbf{M}$ have specific forms. Let M_n denote the set of square matrices of order n, and S_n denote the subset of symmetric matrices. If the system is conservative, then

M. Pandey et al. (eds), Advances in Engineering Structures, Mechanics & Construction, 31–38.

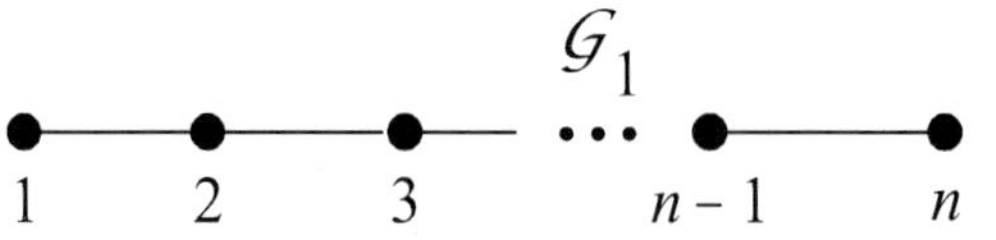

Fig. 1. The matrix J lies on the path $\mathcal{G}_1$.

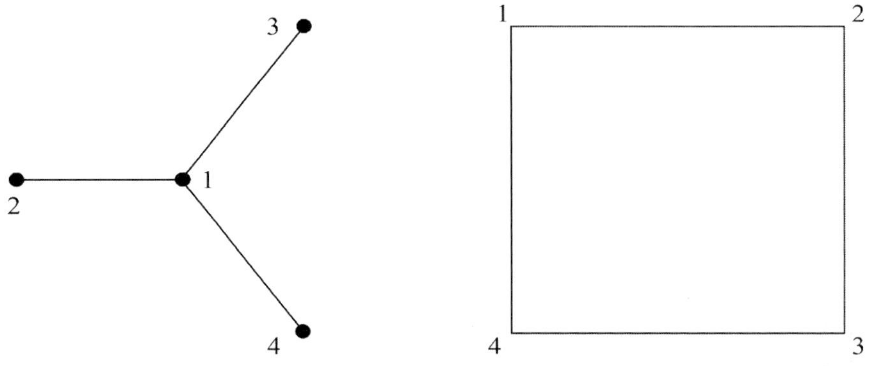

Fig. 2. A star, $\mathcal{G}_2$, on 4 vertices, and a ring, $\mathcal{G}_3$, on 4 vertices.

$\mathbf{K}, \mathbf{M} \in S_n$, $\mathbf{K}$ is *positive semi-definite* (PSD), i.e., $\mathbf{u}^T\mathbf{K}\mathbf{u} \geq 0$; $\mathbf{M}$ is positive definite (PD), i.e., $\mathbf{u}^T\mathbf{M}\mathbf{u} > 0$, for all $\mathbf{u} \neq \mathbf{0}$.

The matrices $\mathbf{K}$, $\mathbf{M}$, and in particular their *structure*, i.e., the pattern of zero and non-zero entries, will depend on the choices of finite elements, and on how these elements are connected. It is convenient to use concepts from graph theory: a (simple, undirected) graph $\mathcal{G}$ is a set of vertices P_i in a *vertex set* $\mathcal{V}$, connected by edges $(i, j)(= (j, i))$ in an *edge set* $\mathcal{E}$.

A matrix $\mathbf{A} \in S_n$ is said to lie *on* $\mathcal{G}$ if $a_{ij} = 0$ whenever $(i, j) \notin \mathcal{E}$. For example, a symmetric tridiagonal matrix, sometimes called a Jacobi matrix $\mathbf{J}$, and written

$$\mathbf{J} = \begin{bmatrix} a_1 & b_1 & & & \\ b_1 & a_2 & b_2 & & \\ & \ddots & \ddots & \ddots & \\ & & \ddots & \ddots & b_{n-1} \\ & & & b_{n-1} & a_n \end{bmatrix} \tag{2}$$

lies on a graph $\mathcal{G}_1$ that is a *path* with $\mathcal{V} = \{1, 2, \ldots, n\}$ and $\mathcal{E} = \{(1, 2), (2, 3), \ldots, (n-1, n)\}$, as shown in Figure 1.

Figure 2 shows two simple graphs: a star, $\mathcal{G}_2$, and a ring, $\mathcal{G}_3$. The matrices $\mathbf{A}_2$ and $\mathbf{A}_3$ lie on $\mathcal{G}_2, \mathcal{G}_3$ respectively.

$$\mathbf{A}_2 = \begin{bmatrix} a_1 & b_2 & b_3 & b_4 \\ b_2 & a_2 & 0 & 0 \\ b_3 & 0 & a_3 & 0 \\ b_4 & 0 & 0 & a_4 \end{bmatrix}, \quad \mathbf{A}_3 = \begin{bmatrix} a_1 & b_1 & 0 & b_4 \\ b_1 & a_2 & b_2 & 0 \\ 0 & b_2 & a_3 & b_3 \\ b_4 & 0 & b_3 & a_4 \end{bmatrix}. \tag{3}$$

The matrix $\mathbf{A}_2$ is an example of a *bordered* matrix, $\mathbf{A}_3$ is called a *periodic Jacobi matrix*. A Jacobi matrix is a particular case of a *band* matrix – it is a symmetric matrix with bandwidth 1. An important

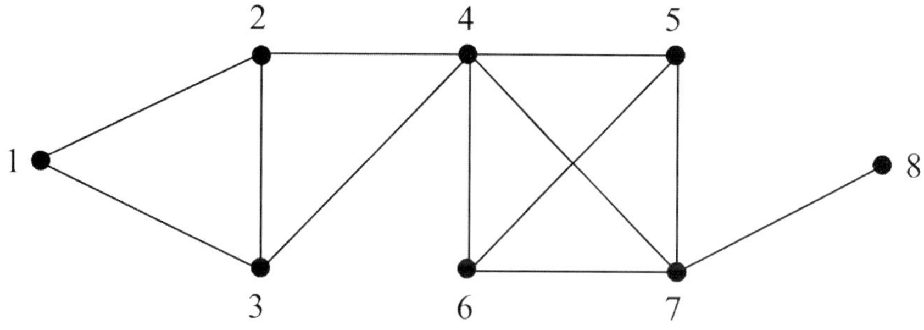

Fig. 3. The matrix $\mathbf{A}_4$ lies on $\mathcal{G}_4$.

subset of band matrices is the set of *staircase* matrices, an example of which is shown in (4); $\mathbf{A}_4$ lies of the graph $\mathcal{G}_4$.

$$\mathbf{A}_4 = \begin{bmatrix} x & x & x & & & & & \\ x & x & x & x & & & & \\ x & x & x & x & & & & \\ & x & x & x & x & x & x & \\ & & & x & x & x & x & \\ & & & x & x & x & x & \\ & & & x & x & x & x & x \\ & & & & & & x & x \end{bmatrix}. \tag{4}$$

The general isospectral problem is this: Given a system $(\mathbf{K}, \mathbf{M})$ with $\mathbf{K}, \mathbf{M} \in S_n$, and $\mathbf{K}, \mathbf{M}$, lying on a graph $\mathcal{G}$, find another (or all) isospectral system(s) $(\mathbf{K}', \mathbf{M}')$ with $\mathbf{K}', \mathbf{M}'$ lying on the same graph $\mathcal{G}$.

This problem is very difficult, and is still open. To simplify it somewhat, we suppose that the systems have lumped mass, so that $\mathbf{M}, \mathbf{M}'$ are *diagonal*. In that case, if $\mathbf{K}$ is PD (PSD) then the λ_i will be positive (non-negative) and $\mathbf{K}'$ will be PD (PSD).

In practice, the problem is even more difficult because, instead of being just PD (PSD), $\mathbf{K}$ will have to satisfy other, usually *positivity* constraints, that state that the system is physically realisable. We need some more concepts from matrix theory.

Suppose $\mathbf{A} \in M_n$. Let $\alpha = \{i_1, i_2, \ldots, i_k\}$ be a sequence of k numbers taken from $\{1, 2, \ldots, n\}$. The *submatrix* of $\mathbf{A}$ with rows taken from $\alpha = \{i_1, i_2, \ldots, i_k\}$, and columns taken from $\beta = \{j_1, j_2, \ldots, j_k\}$ is denoted by $A(\alpha|\beta)$. The determinnant

$$\det(A(\alpha|\beta)) = A(\alpha; \beta)$$

is called a *minor* of A. A minor $A(\alpha; \alpha)$ is called a *principal* minor of $\mathbf{A}$.

Suppose $\mathbf{A} \in M_n$:

- $\mathbf{A}$ is *totally positive*, TP, if all its minors are *positive*,
- $\mathbf{A}$ is *totally non-negative*, TN, if all its minors are non-negative; $\mathbf{A}$ is NTN if it is *non-singular* and TN,
- $\mathbf{A}$ is *oscillatory*, O, if A is TN and a power of $\mathbf{A}$, $\mathbf{A}^P$, is TP.

It may be shown that $\mathbf{A}$ is O iff it is NTN, *and* its immediately off-diagonal entries $a_{i,i+1}$ and $a_{i+1,i}$, $i = 1, 2, \ldots, n-1$ are *positive*, see Gladwell (1998). Note that TP is much stronger than PD: $\mathbf{A} \in S_n$ is PD iff its *principal* minors are positive.

Note that the definition of TP, TN and O matrices applies to *any* matrix in M_n, not just to symmetric matrices, those in S_n. Such matrices have many important properties.

Define $\mathbf{Z} = diag(+1, -1, \ldots, (-)^{n-1})$; the operation $\mathbf{A} \to \mathbf{ZAZ} = \tilde{\mathbf{A}}$ changes the signs of the entries of $\mathbf{A}$ in a chequered pattern.

We list three properties:

- if $\mathbf{A}$, $\mathbf{B}$ are O, so is $\mathbf{AB}$,
- $\mathbf{A}^{-1}$ is O iff $\tilde{\mathbf{A}}$ is O; we say $\mathbf{A}$ is *sign-oscillatory*, SO,
- if $\mathbf{A}$ is O then it *has n* eigenvalues, and they are *positive* and *distinct*.

The last property is particularly important: recall that if $\mathbf{A} \in S_n$, all we can say is that it *has n* real eigenvalues; they may not be distinct. If we know only that $\mathbf{A} \in M_n$, then we do not know *a priori*, how many eigenvalues it has.

2. QR Factorisation

Recall that a matrix $\mathbf{Q} \in M_n$ is *orthogonal* if $\mathbf{QQ}^T = \mathbf{I} = \mathbf{Q}^T\mathbf{Q}$. If $\mathbf{A} \in M_n$ is non-singular, then it may be factorised in the form $\mathbf{A} = \mathbf{QR}$, where $\mathbf{Q}$ is orthogonal, and $\mathbf{R}$ is upper triangular with *positive* diagonal terms. This factorisation is equivalent to the Gram-Schmidt process in which the columns of $\mathbf{A}$ are expressed as linear combinations of orthonormal vectors $\mathbf{q}_i$, the columns of $\mathbf{Q}$.

QR factorisation gives us a procedure for getting another matrix, $\mathbf{A}^*$, isospectral to $\mathbf{A}$: if $\mathbf{A} = \mathbf{QR}$, then $\mathbf{A}^* = \mathbf{RQ}$ may be written $\mathbf{A}^* = \mathbf{Q}^T(\mathbf{QR})\mathbf{Q} = \mathbf{Q}^T\mathbf{AQ}$: $\mathbf{A}^*$ is orthogonally equivalent to $\mathbf{A}$. If $\mathbf{A}$ is symmetric, so that $\mathbf{A}^T = \mathbf{A}$, then $(\mathbf{A}^*)^T = (\mathbf{Q}^T\mathbf{AQ})^T = \mathbf{Q}^T\mathbf{A}^T\mathbf{Q} = \mathbf{Q}^T\mathbf{AQ} = \mathbf{A}^*$: $\mathbf{A}^*$ is symmetric; $\mathbf{A}$ and $\mathbf{A}^*$ are isospectral.

We can apply $QR \to RQ$ transformation with a *shift*. Suppose μ is *not* an eigenvalue of $\mathbf{A} \in S_n$, write

$$\mathbf{A} - \mu\mathbf{I} = \mathbf{QR} \tag{5}$$

and construct $\mathbf{A}^*$ from

$$\mathbf{A}^* - \mu\mathbf{I} = \mathbf{RQ}. \tag{6}$$

Again, $\mathbf{A}^* = \mu\mathbf{I} + \mathbf{RQ} = \mathbf{Q}^T(\mu\mathbf{I} + \mathbf{QR})\mathbf{Q} = \mathbf{Q}^T\mathbf{AQ}$, so that $\mathbf{A}$ and $\mathbf{A}^*$ are isospectral; we write $\mathbf{A}^* = \mathcal{G}_\mu(\mathbf{A})$.

The simplest application of this result is to isospectral in-line spring-mass systems. Now $\mathbf{K}$ is a Jacobi matrix with negative off-diagonal and $\mathbf{M}$ is diagonal: $\mathbf{M} = diag(m_1, m_2, \ldots, m_n)$. We write $\mathbf{M} = \mathbf{P}^2$, where $\mathbf{P} = diag(p_1, p_2, \ldots, p_n)$; Equation (1) may be written

$$\mathbf{P}^{-1}(\mathbf{K} - \lambda\mathbf{P}^2)\mathbf{P}^{-1}\mathbf{Pu} = \mathbf{0},$$

or

$$(\mathbf{A} - \lambda\mathbf{I})\mathbf{x} = \mathbf{0}, \quad \mathbf{A} = \mathbf{P}^{-1}\mathbf{KP}^{-1}, \quad \mathbf{x} = \mathbf{Pu}.$$

The matrix A is a Jacobi matrix with negative off-diagonal. We form $\mathbf{A}^*$ from (5), (6) and then factorise $\mathbf{A}^*$ in the form $\mathbf{A}^* = \mathbf{P}^{*-1}\mathbf{K}^*\mathbf{P}^{*-1}$ to obtain a new in-line spring-mass system. The details of the analysis may be found in Gladwell (1995). The extension to a system in which $\mathbf{K}$, $\mathbf{M}$ are both Jacobi matrices, with negative, positive off-diagonals respectively, may be found in Gladwell (1999).

This application depends on the fact that the operation $\mathcal{G}_\mu$ defined by (5), (6) changes a Jacobi matrix $\mathbf{A}$ with negative (positive) off-diagonal into a Jacobi matrix $\mathbf{A}^*$ with negative (positive) off-diagonal. It may be verified that $\mathbf{A}^*$ is PD (PSD) iff $\mathbf{A}$ is PD (PSD). This is a special case of a general result, proved in Gladwell (1998): Suppose $\mathbf{A} \in S_n$, and μ is *not* an eigenvalue of $\mathbf{A}$ (both these conditions are necessary), then $\mathbf{A}$, $\mathbf{A}^*$ have the same staircase pattern, and $\mathbf{A}^*$ is TP, NTN, O or SO, iff $\mathbf{A}$ is TP, NTN, O or SO, respectively.

This general result may be used to find an isospectral family of FEM models of a thin straight rod in longitudinal vibration. Now $\mathbf{K}, \mathbf{M}$ are Jacobi matrices with negative, positive, off-diagonals respectively. We start from Equation (1), factorise $\mathbf{M} = \mathbf{B}\mathbf{B}^T$ where $\mathbf{B}$ is a bi-diagonal upper triangle matrix, and reduce (1) to standard form

$$(\mathbf{B}^{-1}\mathbf{K}\mathbf{B}^{-T} - \lambda\mathbf{I})\mathbf{B}^T\mathbf{u} = \mathbf{0}.$$

It may be shown that $\mathbf{A} = \mathbf{B}^{-1}\mathbf{K}\mathbf{B}^{-T}$ is SO. We find $\mathbf{A}^* = \mathcal{G}_\mu(A)$, and then factorise $\mathbf{A}^* = \mathbf{B}^{*-1}\mathbf{K}^*\mathbf{B}^{*-T}$ to form a new FEM model, $\mathbf{K}^*$, $\mathbf{M}^*$ with $\mathbf{M}^* = \mathbf{B}^*\mathbf{B}^{*T}$; $\mathbf{K}^*$, $\mathbf{M}^*$, like $\mathbf{K}, \mathbf{M}$ are Jacobi matrices with negative, positive, off-diagonals respectively, corresponding to a FEM model of a new rod, as described in Gladwell (1997).

The operation $\mathcal{G}_\mu$ is essentially tied to staircase matrices: $\mathbf{A}^*$ is a staircase iff $\mathbf{A}$ is a staircase. To obtain a wider class of isospectral matrices, we must consider the concept of *isospectral flow*.

3. Isospectral Flow

If $\mathbf{A}$ is symmetric, i.e., $\mathbf{A} \in S_n$, it has n eigenvalues $(\lambda_i)_1^n$ and n corresponding orthonormal eigevectors $\mathbf{q}_i$ that span R^n. The matrix $\mathbf{A}$ may be written

$$\mathbf{A} = \mathbf{Q}\wedge\mathbf{Q}^T,$$

where

$$\wedge = diag(\lambda_1, \lambda_2, \ldots, \lambda_n), \quad \mathbf{Q} = [\mathbf{q}_1, \mathbf{q}_2, \ldots, \mathbf{q}_n], \quad \mathbf{Q}\mathbf{Q}^T = \mathbf{Q}^T\mathbf{Q} = \mathbf{I}.$$

The family of matrices with all possible orthogonal matrices $\mathbf{Q}$ forms an isospectral family, the complete family with the given spectrum $(\lambda_i)_1^n$.

Instead of seeking the complete family, we look for a family in which $\mathbf{Q}$ depends on a single parameter t.

$$\mathbf{A}(t) = \mathbf{Q}(t)\wedge\mathbf{Q}^T(t).$$

Differentiating w.r.t. t we find

$$\dot{\mathbf{A}} = \mathbf{Q}\wedge\dot{\mathbf{Q}}^T + \dot{\mathbf{Q}}\wedge\mathbf{Q}^T.$$

Since $\mathbf{Q}$ is orthogonal, we may write

$$\dot{\mathbf{A}} = (\mathbf{Q}\wedge\mathbf{Q}^T)(\mathbf{Q}\dot{\mathbf{Q}}^T) + (\dot{\mathbf{Q}}\mathbf{Q}^T)(\mathbf{Q}\wedge\mathbf{Q}^T).$$

Put $\mathbf{Q}\dot{\mathbf{Q}}^T = \mathbf{S}$ then, since $\mathbf{Q}\wedge\mathbf{Q}^T = \mathbf{A}$, we have

$$\dot{\mathbf{A}} = \mathbf{A}\mathbf{S} + \mathbf{S}^T\mathbf{A}.$$

Now, $\mathbf{Q}$ is orthogonal, so that $\mathbf{Q}\mathbf{Q}^T = \mathbf{I}$, and hence

$$\mathbf{Q}\dot{\mathbf{Q}}^T + \dot{\mathbf{Q}}\mathbf{Q}^T = \mathbf{0} : \mathbf{S} + \mathbf{S}^T = \mathbf{0}.$$

This means that

$$\dot{\mathbf{A}} = \mathbf{AS} - \mathbf{SA} \tag{7}$$

and that $\mathbf{S}$ is a skew-symmetric matrix. We note that (7) is a so-called *autonomous* differential equation: the parameter t does not appear explicitly; it appears implicitly because $\mathbf{S}$ depends on $\mathbf{A}$, i.e., $\mathbf{S} = \mathbf{S}(\mathbf{A})$, and $\mathbf{A}$ depends on t.

Most importantly, we may argue conversely: if $\mathbf{S}$ is skew symmetric and $\mathbf{A}$ varies according to (7) then $\mathbf{A}(t)$ keeps the same eigenvalues, those for $\mathbf{A}(0)$. We may choose $\mathbf{S}$ in many ways; different choices will lead to different isospectral families.

The autonomous differential equation (7), called the *Toda flow* equation, was investigated first for tridiagonal matrices, with the choice

$$\mathbf{A}(t) = \begin{bmatrix} a_1 & b_1 & & & \\ b_1 & a_2 & b_2 & & \\ & \ddots & \ddots & \ddots & \\ & & \ddots & \ddots & b_{n-1} \\ & & & b_{n-1} & a_n \end{bmatrix}, \quad \mathbf{S}(t) = \begin{bmatrix} 0 & b_1 & & & \\ -b_1 & 0 & b_2 & & \\ & \ddots & \ddots & \ddots & \\ & & \ddots & \ddots & b_{n-1} \\ & & & -b_{n-1} & 0 \end{bmatrix}, \tag{8}$$

as in Symes (1982). Watkins (1984) gives a survey of the general theory. See also Chapter 7 of Gladwell (2004). In this case, it may easily be shown that $\mathbf{AS} - \mathbf{SA}$ is tridiagonal, so that if $\mathbf{A}(0)$ is tridiagonal, then $\mathbf{A}(t)$ will be tridiagonal. This is a special case of he result that if $\mathbf{A}(0)$ is a staircase matrix and if

$$\mathbf{S}(t) = \mathbf{A}^{+}(t) - \mathbf{A}^{+T}(t), \tag{9}$$

where $\mathbf{A}^{+}(t)$ denotes the upper triangle of $\mathbf{A}(t)$, then Equation (7) constrains $\mathbf{A}(t)$ to remain a staircase matrix, with the same staircase dimensions as $\mathbf{A}(0)$. In general, even if $\mathbf{A}(0)$ is a staircase with holes, these holes will eventually be filled in. Gladwell (2002) showed that the Toda flow (7), with S given by (9) maintains the properties TP, NTN, O and SO.

There are two important engineering structures for which the stiffness matrix is a staircase matrix: the rod in logitudinal vibration, already mentioned; an Euler–Bernoulli beam in flexure, for which the stiffness matrix is pentadiagonal, see Gladwell (2002b).

To obtain an isospectral flow for more general cases, we must pose the question: How may we construct an isospectral flow that constrains $\mathbf{A}$ to lie on a given graph $\mathcal{G}$?

Consider a very simple graph, the star on n vertices, as shown in Figure 4; a matrix on $\mathcal{G}$ has the form $\mathbf{A}_5$ in Equation (10); the only non-zero entries are those on the *borders*, i.e., the first row and column, and the diagonal. The matrix S must be skew-symmetric, so that we need consider only its upper triangle. We choose $s_{ij} = a_{ij}$ for entries in the first row, and then find the remaining entries below the first row and above the diagonal by making $\dot{a}_{ij} = 0$ for those entries; there are $m = (n-1)(n-2)/2$ algebraic equations for the m unknown s_{ij}.

$$\mathbf{A}_5 = \begin{bmatrix} a_1 & b_2 & b_3 & \dots & b_n \\ b_1 & a_2 & & & \\ b_3 & & a_3 & & \\ \vdots & & & \ddots & \\ b_n & & & & a_n \end{bmatrix}, \quad \mathbf{S}_5 = \begin{bmatrix} 0 & b_2 & b_3 & \dots & b_n \\ & 0 & s_{23} & \dots & s_{2n} \\ & & 0 & & \\ & skew & & \ddots & s_{n-1,n} \\ & & & & 0 \end{bmatrix} \tag{10}$$

The equations for the s_{ij} are separable, that for s_{ij} is

$$(a_i - a_j)s_{ij} + 2b_i b_j = 0 \quad i = 2, \dots, n, \quad j = i+1, \dots, n. \tag{11}$$

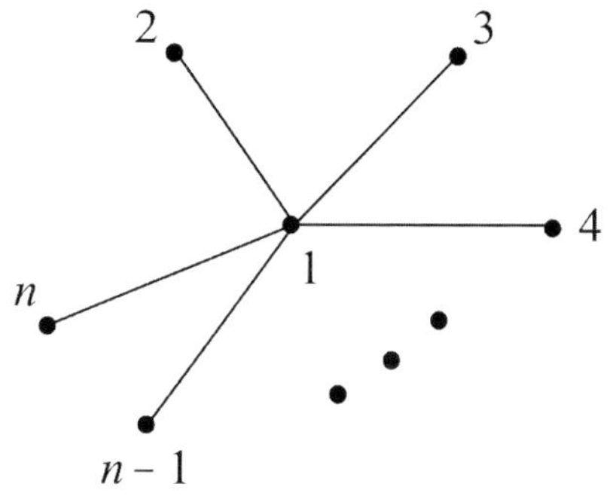

Fig. 4. A star on n vertices.

These m algebraic equations are combined with the equations

$$\dot{a}_1 = -2\sum_{i=2}^{n} b_i^2, \quad \dot{a}_j = 2b_j^2, \quad j = 2, \ldots, n \tag{12}$$

$$\dot{b}_j = (a_1 - a_j)b_j + \sum_{i=2}^{j-1} s_{ij}b_i - \sum_{i=j+1}^{n} s_{ji}b_i, \quad j = 2, \ldots, n. \tag{13}$$

On substituting for s_{ij} from (11) into (13), we find

$$\dot{b}_j = b_j\left\{a_1 - a_j + 2\sum_{k=2}^{n}{}' \frac{b_k^2}{a_j - a_k}\right\}, \quad j = 2, \ldots, n, \tag{14}$$

where $'$ denotes $k \neq j$. It may be shown that if the $a_i(0)$, $i = 2, \ldots, n$ are *distinct*, and the $b_i(0)$, $i = 2, \ldots, n$ are *non-zero*, then the $a_i(t)$, $i = 2, \ldots, n$ will be distinct, and the $b_i(t)$ will be non-zero, so that the denominators in (14) will remain non-zero.

This procedure may be generalised: in the upper triangle of S, take $s_{ij} = a_{ij}$ when $(i, j) \in \mathcal{E}$; find s_{ij} for the remaining entries $(i, j) \notin \mathcal{E}$ by demanding that $\dot{a}_{ij} = 0$ when $(i, j) \notin \mathcal{E}$; this gives a set of p algebraic equations for the p entries, s_{ij} which are combined with the remaining equations for $\dot{a}_{ij}$, $(i, j) \in \mathcal{E}$, and $\dot{a}_{ii}$, $i = 1, 2, \ldots, n$. The algebraic equations for the s_{ij} will have coefficients that are linear combinations of the a_{ij}, as in Equation (11) for the star. In general, unlike for the star, we cannot assume that the p equations for the p entries s_{ij} will always admit a solution; we will have to fall back on continuity arguments – if they admit a solution when $t = 0$, they will admit a solution for some small interval of t around $t = 0$.

The application of this procedure to a typical FEM model, a triangular model of a membrane, is the topic of a forthcoming paper.

References

Gladwell, G.M.L. (1995), On isospectral spring-mass systems, *Inverse Problems*, **11**, 591–602.

Gladwell, G.M.L. (1997), Inverse vibration problems for finite element models, *Inverse Problems*, **13**, 311–322.

Gladwell, G.M.L. (1998), Total positivity and the QR algorithm, *Linear Algebra Appl.*, **271**, 257–272.

Gladwell, G.M.L. (1999), Inverse finite element vibration problems, *J. Sound Vibration*, **211**, 309–342.

Gladwell, G.M.L. (2002a), Total positivity and Toda flow, *Linear Algebra Appl.*, **350**, 279–284.

Gladwell, G.M.L. (2002b), Isospectral vibrating beams, *Proc. Roy. Soc. London* **A**, **458**, 2691–2703.

Gladwell, G.M.L. (2004), *Inverse Problems in Vibration*, Kluwer Academic Publishers, Dordrecht.

Symes, W.W. (1982), The QR algorithm and scattering for the finite non-periodic Toda lattice, *Physica d.*, **4**, 275–280.

Watkins, D.S. (1984), Isospectral flows, *SIAM Review*, **26**, 379–391.

COLD-FORMED STEEL RESEARCH AT THE UNIVERSITY OF WATERLOO

R.M. Schuster
Department of Civil Engineering, University of Waterloo, Canada

Abstract

Active cold-formed steel research started in the early 70s at the University of Waterloo with Profs. N.C. Lind, A.N. Sherbourne, J. Roorda, and R.M. Schuster, and their respective graduate students. More recently, Prof. L. Xu and his graduate students have also contributed in this area of research. Numerous publications have been documented in National and International sources over the past 35 years. Presented in this paper are the many noteworthy contributions that have been made as a result of the cold-formed steel research activities at the University of Waterloo. What is of particular interest is the number of design recommendations that have been adopted by National and International cold-formed steel design Standards and Specifications. With the recently formed Canadian Cold-Formed Steel Research Group, cold-formed steel research is very much alive at the University of Waterloo.

Introduction

The use of cold-formed steel as a structural building material dates back to the 1850s in the USA and the UK. Initially, the acceptance of cold-formed steel in the building construction industry faced difficulties because no published Specification for the design of cold-formed steel structures existed at that time. For such a Specification to be published, research had to be carried out. In 1939, the American Iron and Steel Institute (AISI) sponsored a research project at Cornell University under the direction of Professor George Winter with the objective of developing the necessary technical information that would lead to the development of the first cold-formed steel design Specification, which was published by AISI in 1946. This design Specification has been revised subsequently by AISI a number of times to reflect the technical developments resulting from the continuing research in the field of cold-formed steel. The Canadian Standards Association (CSA), following a number of subsequent editions, published the first cold-formed steel design Standard, S136, in Canada in 1963.

In comparison to hot-rolled structural steel members, cold-formed steel is uniquely different in that structural panel/deck sections and individual profile members are made of rather thin-coated steel plate material. This allows for the production of a great variety of geometric shapes, as can be seen in Figure 1. Most of the cold-formed steel shapes are produced economically by using roll-forming operations. Cold-formed steel structural members have one of the highest strength to mass ratios of any structural building material. The use of cold-formed steel in the construction industry in North America is steadily growing and gaining momentum, including cladding, roofing, composite decking, lightweight steel framing, corrugated steel pipe, storage racks, transmission towers and numerous other applications. Due to the thinness of the steel plate material, cold-formed steel products are subject to local buckling, resulting in more detailed and complicated calculations for the structural designer. Ongoing research in this field is extremely important in order to update and improve the governing cold-formed steel structural design Specifications and Standards.

The 1970s

Cold-formed steel research started in the early 70s at the University of Waterloo with Prof. Lind as one of the leading Canadian researchers.

Lind and Schroff (1971, 1975) dealt with the utilization of cold work of forming in cold-formed steel. The mechanical properties of cold-formed sections can be substantially different from those of flat sheet steel before the cold forming operation. When a piece of flat sheet steel is bent about a radius, the yield strength and tensile strength will increase as a result of this forming operation, but at the same time decreasing the ductility. Lind and Schroff (1971, 1975) used a linear strain-hardening

M. Pandey et al. (eds), Advances in Engineering Structures, Mechanics & Construction, 39–52.

Fig. 1 Various Cold-Formed Steel Shapes (Yu 2000)

model and concluded that the increase in yield strength depends only on the inside bend radius ratio, r/t, and the hardening margin ($F_u - F_y$). Furthermore, to take the cold work strengthening into account, it is only necessary to replace the virgin yield strength by the virgin ultimate strength over a length of 5t in each 90° corner. The researchers concluded that the r/t ratio is not a significant parameter, which was established by examination of test data. Based on the work by Lind and Schroff (1971, 1975), the following simplified equation resulted:

$$F'_y = F_y + \frac{5D}{W^*}(F_u - F_y) \qquad (1)$$

where F'_y is the calculated average tensile yield strength of the full cold-formed section of tension or compression members, or the full flange of flexural members. F_y and F_u are the tensile yield and tensile strengths of the virgin steel, respectively. D is the number of 90° corners – if other angles are used, D is the sum of the bend angles divided by 90°. W* is the ratio of the length of the centerline of the full flange of flexural members, or of the entire section of tension or compression members, to the steel thickness, t. Equation 1 was adopted by CSA Standard S136-74 – Cold Formed Steel Structural Members.

Another important work that was undertaken by Lind et al. (1971) and Lind et al. (1976) was investigating the effective width formula used at that time. Their study showed that, based on experimental evidence, the effective width formula for stiffened compression elements can be expressed independent of the flat width ratio, w/t, as follows.

$$b/t = 1.64\sqrt{\frac{E}{f_{max}}} \qquad (2)$$

One can conclude that this expression is simpler and offers greater computational advantages, often eliminating the usual time-consuming iteration process with the existing effective width formula that is a function of w/t.

Venkataramaiah (July 1971) carried out an extensive experimental investigation on different edge stiffeners that can be used with cold-formed sections instead of the typical simple lip stiffener. The objective of his work was to obtain optimum shape and optimum size edge stiffeners for thin-walled compression elements. Another extensive experimental study was carried out by Venkataramaiah (August 1971), Lind and Venkataramaiah (1972) into the behaviour of straight-lipped and L-lipped simply-supported channel sections, eccentrically loaded, thin-walled short columns. It was concluded that the ultimate strength of eccentrically loaded thin-walled short channel columns can be expressed by a simple but general formula involving critical stress.

Schuster (1972) started his research at the University Waterloo by investigating composite steel-deck reinforced concrete slabs, in short, "composite slabs". In order for a steel deck to achieve the required composite action between the deck and concrete, the deck must be capable of resisting horizontal shear and prevent vertical separation between the concrete and deck. The most common of composite deck products in the marketplace today are decks that utilize a fixed pattern of embossments/indentations rolled into the deck, thus providing a mechanical interlocking device between concrete and deck so that the required composite action can be developed. These types of floor systems are typically used in high rise buildings because of their numerous inherent advantages, one of the major advantages being, the steel deck serves as a form for the concrete during the construction stage and it remains permanently in place as the positive reinforcement. Based on tests, "shear-bond" is the most common mode of failure with composite slab systems. Schuster (1972) developed the following shear-bond expression:

$$\frac{V_{uc}}{bd} = K_5 \frac{\sqrt{f_c'}d}{L'} + K_6 \rho \tag{3}$$

where V_{uc} is the ultimate calculated transverse shear; b is the unit slab width; d is the effective slab depth; f_c' is the concrete compressive strength; L' is the shear span; ρ is the percent of steel; K_5 and K_6 are shear-bond coefficients that have to be obtained from a linear regression analysis of test data.

Lind (1973) investigated the buckling of multiple closely spaced intermediate longitudinal stiffened plate elements. When the stiffeners are spaced so closely that the compression sub-elements have no tendency to buckle individually, overall plate buckling of the entire element can occur. It was assumed in the 1963 edition of the CSA S136 Standard that the entire element was comprised of a rectangular plate having an equivalent steel thickness of t_s. While this was a simple design method, Lind (1973) developed a more comprehensive expression by treating the entire stiffened plate element as an orthotropic plate, resulting in the following equation:

$$t_s = t\left[\frac{w_s}{2\rho} + \left(\frac{3I_s}{\rho t^3}\right)^{1/2}\right]^{1/3} \tag{4}$$

where t is the plate thickness; w_s is the overall plate element width; I_s is the moment the inertia of the full area of the multiple-stiffened element, including the intermediate stiffeners, about its own centroidal axis; p is the perimeter length of the multiple-stiffened element (between edge stiffeners). Equation 4 is valid up to the initial buckling stress, however, it was demonstrated by Sherbourne et al. (1971), Sherbourne et al. (1972) that some post-buckling strength is available. Based on this, Eq. 4 is meant to also apply in the post-buckling range as an approximation, and it was adopted in the 1974 Edition of CSA S136.

In 1974, the Solid Mechanics Division published a special publication, entitled "Design in Cold Formed Steel", edited by Schuster. Contained in this document are a number of noteworthy papers dealing with the topic of cold-formed steel. Lind (Strength, Deformation and Design of Cold Formed Steel Structures), Lind (Design Procedures for Flexural Members), Lind (Recent Developments in

Cold Formed Steel Design Requirements), Lind (Additional Design Examples); Roorda (On the Buckling Behaviour and Design of Thin-Walled Beams and Columns); Schuster (Composite Steel-Deck Reinforced Concrete Floor Systems), Schuster (Current Design Criteria of Steel-Deck Reinforced Concrete Slabs), Schuster (Proposed Ultimate Strength Design Criteria for Steel-Deck Reinforced Concrete Slabs). Presented in this publication were important papers on cold-formed steel to date, providing structural engineers with the latest design information in cold-formed steel.

Schuster, in collaboration with Lind (1975) published the book, entitled, "Cold Formed Steel Design Manual", which was the first document of its kind in Canada, containing the 1974 Edition of CSA Standard S136 and its respective Commentary. The purpose of this book was to assist engineers, designers, manufacturers and educators in the design of cold-formed steel structures. Including in the book were numerous design examples, helpful design aids, and many useful tables.

Lind et al. (1975) investigated the economic benefits of the connection safety factor used in the CSA S136-74 Standard and the AISI – 1968 Specification. Using an analysis based on economic principles of equal marginal returns, they concluded that the current safety factors in CSA S136-74 and AISI-1968 are not far from the economic optimum.

Knab and Lind (1975) developed a rational, reliability based design criteria for determining allowable stresses for temporary (2–5 years) cold-formed steel buildings. The method involves comparing differences in reliability levels between permanent and temporary building design criteria. It was shown that allowable stresses can be established that take into account the temporary nature of the buildings and, at the same time, maintain comparable permanent building reliability levels.

Parimi and Lind (1976) the objective of this paper was to explain the new limit states design option in the CSA S136-74 Standard for cold-formed steel design. It was concluded that in the case of cold-formed steel design, the limit states cases could be reconstructed rather easily from the working stress design format. As well, it was found that the stress format was also quite suitable for the limit states design format of cold-formed steel, which allows for ease of calibrating the respective resistance factors. Probabilistic and deterministic approaches were used in the resistance factor calibrations, resulting in identical results.

Schuster (1976) presented a comprehensive overview of the state of the art of composite steel-deck concrete floor systems in the US and Canada, outlining the important inherent attributes of such floor systems in the construction industry. Also presented in the paper is an overview of the latest research findings to-date.

Lind (1978) carried out a study relating to the buckling strength of steel plate assemblies. The analysis is reduced to an eigenvalue problem of an ordinary differential equation that is solved by a Vianello-Stodola procedure, using the tabular numerical approach presented by Newmark for buckling analysis of columns. Two example problems are presented, illustrating the simplistic approach.

The 1980s

Schuster and Ling (1980) developed a new shear-bond expression for composite slabs on the basis of the mechanical interlocking capacity of concrete and steel deck within the shear span. Equation 5 is similar to *Eq. 3*, however, the strength of concrete and the percent of steel terms are no longer in the expression. Based on the available data, it was found that these two parameters are not important terms. The percent of steel parameter is not required since separate tests have to be carried out for each steel deck thickness anyway. See *Eq. 3* for the description of terms.

$$\frac{V_{uc}}{bd} = K_5 \frac{1}{L'} + K_6 \qquad (5)$$

Equation 5 was adopted in 1984 by the Canadian Sheet Steel Building Institute (CSSBI) for the calculation of the shear-bond strength of composite slabs. A number of updated editions have since followed [Schuster and Trestain (2002)]. Schuster (1980) presented a paper on composite steel deck concrete floor slabs, highlighting the detailed development of *Eq. 5*.

Venkataramaiah et al (1980) did a study on the experimental determination and the statistical evaluation of the elastic modulus of elasticity of cold-formed sheet steel. In addition to some column compression data found in the literature, tensile coupon tests were carried out as part of this study. After a statistical study, it was found that the mean value of the elastic modulus of cold-formed steel is 30 071 ksi (207 332 MPa), which is greater than the value of 29 500 ksi (203 395 MPa) that is used by the AISI Specification and the CSA S136 Standard.

Wing and Schuster (1981, 1982, 1986) started to carry out research in the field of web crippling of cold-formed steel members, followed by numerous other studies over the years. This work dealt with web crippling and the interaction of bending and web crippling of multi-web cold formed-steel deck sections. One of the major reasons for this study was to correct some of the inconsistencies that existed in the two North American cold-formed steel design documents at that time. Since web crippling of cold-formed steel sections is an extremely complex analytical problem, experimental testing had to be carried out. Web crippling without the influence of bending is typically divided into the following four categories: 1) End one-flange loading (EOF), 2) Interior one-flange loading (IOF), 3) End two-flange loading (ETF), 4) Interior two-flange loading (ITF). Only with the IOF loading case was some moment present, however, the specimens were short enough so that the moment influence was kept to an absolute minimum. In the case of the interaction of bending and web crippling, the specimens were of larger length to account for different degrees of bending influence. The work by Wing on web crippling of multi-web cold formed-steel deck sections was adopted in the 1984 edition of CSA S136.

Schurter et al. (1982) carried out a study on three different cold-formed steel C-section type studs, i.e., one section had solid webs and the other two had lip-reinforced trapezoidal holes in the webs. The tests consisted of 1) Stub column tests and 2) Full scale wall assembly panel test that were subjected to compression load and combined compression and lateral load. The test results were compared to their respective calculated values, resulting in good correlations.

Seleim and Schuster (1982, 1985) developed a new shear-bond expression for composite slabs. The reason for this work was to reduce the number of composite slab tests required to establish the shear-bond capacity of any given composite floor system. Typically, four different steel deck thicknesses are produced by any given composite slab manufacturer. Using *Eq. 5* and having to test four composite slab specimens for each steel deck thickness [Schuster, Trestain (2002)], 16 composite slabs would have to be tested to establish the shear-bond capacity of any given composite floor system. This can be quite costly. The work by Saleim produced the following shear-bond expression, which has been adopted by the CSSBI [Schuster, Trestain (2003)]:

$$\frac{V_{uc}}{bd} = K_1 \frac{t}{L'} + K_2 \frac{1}{L'} + K_3 t + K_4 \tag{6}$$

where (t) is the steel deck thickness and K_1 to K_4 are shear-bond coefficients that have to be obtained from a multiple linear regression analysis of the test data. In this case, only eight slab specimens have to be carried out to establish the shear-bond capacity.

Schuster and Suleiman (1986) carried out a study on composite slabs subjected to repeated point loading. Composite slabs are sometimes subjected to repeated point loading, such as resulting from forklift trucks. In this study, experimental testing was carried out on single span and double span composite slabs, using only one particular company's composite slab product. To establish the ultimate capacity of the composite slabs, first static load tests were carried, followed by a number of respective repeated load tests subjected to cyclic loading. It was found that shear-bond was always the mode of failure, and the composite slabs were able to carry 75% of the static load up to 1.25 million cycles without failure.

McCuaig and Schuster (1988) did a similar study as Suleiman (1986), except the composite steel deck was from a different manufacturer. Based on these test results, the conclusions were that both simple and double span specimens were able to sustain repeated point loads of 55% of static ultimate for at least 1.25 million cycles and the mode of failure in all cases was by metal fatigue in the steel deck.

Schurter and Schuster (1986) were involved in investigating the shear-bond capacity of composite slabs relating to the surface conditions/coatings of the steel decks. Twenty pull-out tests involving four different metallic coatings and four different surface conditions were carried out. As well, six full-scale slab tests with two different metallic coatings (AZ150 Galvalume and Z275 galvanized) were performed. It was concluded that the shear-bond capacity of the AZ150 Galvalume steel slab was on average 32 % greater than the Z275 galvanized steel slab.

Schuster et al. (1986), Fox et al. (1986) and Schuster (1987, 1991) presented papers on the Canadian Standard for the Design of Cold-Formed Steel Structural Members, highlighting the new Limit States Design approach used in Canada.

The 1990s

McCuaig and Schuster (1990) performed resistance factor calibrations for the shear-bond failure mode of composite slabs. 196 test data, representing nine different product types, were used in the reliability calibrations based on a dead load factor of 1.25 and a live load factor of 1.5, as specified in the National Building Code of Canada. It was concluded that a resistance factor of 0.7 can be used for shear-bond of composite slabs, resulting in an average calculated safety index of 3.71, which corresponds well with the target safety index of 3.5.

Schuster (1991, 1992) presented a paper on lightweight steel framing in the 90s, focusing on the use of cold-formed steel in residential construction. In 1992, Schuster presented a paper, entitled "The 1989 Edition of the Canadian Cold Formed Steel Design Standard".

Dinovitzer et al. (1992) made some important observations pertaining to the CAN/CSA-S136-M89 cold-formed steel design Standard. In the case of partially stiffened compression elements, discontinuities in effective width estimates for sections with similar flanges and stiffeners were observed, resulting from a sudden change in the behavioural states. Recommendations are presented to rectify this discontinuity and at the same time simplifying the analysis procedure. These recommendations are now contained in the North American Design Specification. In addition, a simplification was presented for deflection determination of multi-web deck sections. A new plate buckling coefficient was developed to be used in the basic effective width expression.

Papazian et al. (1994) carried out an experimental investigation on multiple intermediate longitudinally stiffened deck profiles. The primary objective of this work was to substantiate the design approach contained in the CSA S136 Standard for closely spaced intermediate longitudinal stiffeners. It was the belief that this approach was quite conservative in that it only addressed the elastic plate buckling capacity. All tests were subjected to a uniform bending load using a vacuum chamber. It was concluded that indeed the ultimate capacities were conservative.

Schuster et al. (1995) undertook an investigation of perforated cold-formed steel C-Sections in shear. The reason for this work was because no specific design provisions were contained in the design Standards for perforated web elements subjected to shear, even though such sections are frequently found in practice. Experimental testing was carried out to substantiate the analytical formulation. The work resulted in a design approach that is now contained in the North American Design Specification.

Rogers and Schuster published four different papers in 1996, all resulting from the research by Rogers. 1) Effective Width of a Single Edge Stiffener Subjected to a Stress Gradient; 2) Interaction of Flange/Edge-Stiffened Cold Formed Steel C-Sections; 3) Cold Formed Steel Flat Width Ratio Limits, d/t and d_i/w; 4) Test Results and Comparison of Cold Formed Steel Edge Stiffened C and Z-sections. Every one of these papers resulted in meaningful design information.

Hancock et al. (1996) published a paper entitled, "Comparison of the Distortional Buckling Method for Flexural Members with Tests" and in 1997, Rogers and Schuster followed with a paper on the flange/web distortional buckling of cold-formed steel sections in bending.

Prabakaran and Schuster (1998) developed a new web crippling expression for cold-formed steel members that can be used for different geometric shapes and the four loading cases, and it is non-dimensional. The following expression for the nominal web crippling strength was adopted by the 94 edition of CSA S136 and the North American Specification for the Design of Cold-Formed Steel Structural Members:

$$P_n = Ct^2F_y \sin\theta \left(1 - C_R\sqrt{\frac{R}{t}}\right)\left(1 + C_N\sqrt{\frac{N}{t}}\right)\left(1 - C_h\sqrt{\frac{h}{t}}\right) \tag{7}$$

where (t) is the web steel thickness; F_y is the design yield point of the steel; C, C_R, C_N and C_h are coefficients that have been established based on experimental data found in the literature. These coefficients are given in the North American Specification for the different geometric section types and load cases. This work by Prabakaran was actually completed in 1993.

Gerges and Schuster (1998) carried out an experimental investigation on the web crippling of single web cold-formed steel members subjected to end one-flange loading (EOF). The primary focus of this study was on single-web C-sections with large inside bend radius to thickness ratios. There was insufficient data in the literature to properly establish the web crippling coefficients. This study produced the additional data and the appropriate web crippling coefficients were established.

Acharya and Schuster (1998, 1998) undertook a study of hat-type sections with multiple intermediate longitudinal stiffeners subjected to uniform bending load. Similar to Papazian et al. (1994), additional tests were carried out in a vacuum chamber to complete the data pool for the establishment of a new plate buckling coefficient to be used with the basic effective width expression.

Fox and Schuster (1998) investigated the use of bearing stiffeners with C-section floor joists. Tests were carried out involving three different types of stiffeners, i.e., stud section, track section and a fully effective bridging channel section. The governing design Standards required that a stiffener must be fully effective, which is a restriction that is rather difficult to meet in practice. Based on this investigation, it was concluded that the governing design provisions for bearing stiffeners do not accurately predict the capacity of the stiffeners currently used in lightweight steel framing.

Schuster (1998, 1998) presented a paper on the advantages of having one North American Specification for the Design of Cold Formed Steel Structures. In the paper entitled, "Cold-Formed Steel - the Construction Material of the Future", Schuster presented

The 2000s

Sloof and Schuster (2000) investigated the yield strength increase of cold-formed sections due to cold work of forming. Extensive testing was performed to substantiate the simplified design approach developed by Lind and Schroff in the early 70s at the University of Waterloo, which is *Eq. 1*. This additional test data provided the needed information to conclude that *Eq. 1* is indeed the best predictor expression when compared to the more complex approach used by AISI.

Beshara and Schuster (2000) performed additional web crippling tests to obtain data of C- and Z-sections subjected to interior two-flange loading and end two-flange loading. In addition, new web crippling coefficients were generated using the data from this study and all of the available published data. Calibrations for resistance factors and factors of safety were also carried out (Beshara and Schuster, 2002). The results of this work have been adopted by the North American Specification for the Design of Cold-Formed Steel Structural Members.

Craig and Schuster (2000) carried out calibrations of the cold-formed steel shear equations in an effort to correct a discontinuity in the nominal shear equation that was being considered by the North American Specification for the Design of Cold-Formed Steel Structural Members. Again, the results were adopted by the Specification.

Fox and Schuster (2000) investigated the lateral strength of wind load bearing wall stud-to-track connections. Typically in this case, stud sections are framed into a channel track section, resulting in the possibility of web crippling at the end of the stud connection due to the action of the wind load. The governing cold-formed steel Design Standards/Specifications in North America do not cover this case specifically. The web crippling provisions are based on the member being supported by a rigid bearing plate. Based on this, testing was carried out to develop a design approach for this specific end condition load case. The recommended procedure recognizes two different observed failure modes, i.e., 1) web crippling of the stud and 2) punch-through of the track. The basic web crippling expression was used (*Eq. 7*) with different web crippling coefficients and a new expression was developed for the punch-through case. Again, the results were adopted by the North American Specification for the Design of Cold-Formed Steel Structural Members.

Fox and Schuster (2000, 2003) continued their work of 1998 to establish a simplified design approach for bearing stiffeners in cold-formed C-Sections. It was found that in all of the tests with stud or track stiffeners the failure mode was local buckling of the stiffener in compression. There are a number of variables affecting the strength of the assembly, but in general, the study found that a simple expression for the nominal bearing resistance of the stiffener types tested can be used.

Xu et al. (2000) carried out a study on the optimum design of cold-formed steel residential roof trusses. A computer-based optimal design approach for residential roof trusses was developed, using cold-formed steel C-sections. The truss design was based on the CSA S136-94 and the truss design guide published by the American Iron and Steel Institute and the Canadian Sheet Steel Building Institute. A generic algorithm was adopted to obtain the minimum cost design with due consideration to truss topology and member size simultaneously.

Xu et al. (2000, 2005) investigated the dynamic/vibration behaviour of floors with cold-formed steel joists. Both static and dynamic tests were conducted on cold-formed C-section floor joists with different span lengths based on different design criteria. The static tests were done to obtain the stiffness and the degree of load sharing between the joists, and the purpose of the dynamic tests was to establish the frequencies of the floor systems. To identify the critical parameters that contribute to the control of floor vibration, tests were also carried out on floors without attached ceiling materials, with different bridging and blocking patterns, and with different support conditions. Test results are presented in comparison with the analytical results obtained from different design methods.

Xu et al. (2001) carried out compressive tests of cold-formed corrugated steel curved panels. In the absence of a standard test protocol, presented in this paper are two types of compressive tests, i.e., 1) corner and flange-section tests, and 2) full-panel tests. The purpose of these tests was to investigate the influences of panel curvatures and transverse corrugations on the buckling behaviour of the cold-formed corrugated steel curved panels. The full section tests yielded consistent results, with the deviation of the individual test ultimate load less than 6% and 8% of the average ultimate load with regard to CSA S136-94 and AISI-96, respectively. For corner and flange section tests, the deviation of the ultimate load was generally less than 7%, except one group. Based on this, both test approaches and associated experimental set-ups could be regarded as reliable.

Xu and Cui (2002) undertook a study relating to the connection flexibility of cold-formed steel C-shape connections" Current design practice on cold-formed steel trusses assumes that web-to-chord connections of trusses are ideally pin-connected. The most recent revision of the *AISI Standard for Cold-Formed Steel Framing - Truss Design* permits the connection flexibility to be taken into account in the analysis and design of such trusses. However, no specific guidelines on how to incorporate this connection flexibility into the design process are provided, which is primarily due to the lack of information on the moment-rotation behaviour of such connections. This paper presents results from a series of tests on web-to-chord connections of cold-formed steel trusses. A mathematical model was developed to represent the behaviour of the connections. The objective of the investigation was to assess connection flexibility via the moment-rotation behaviour for the purpose of design of cold-formed steel trusses.

Tangorra et al. (2002) performed calibrations for resistance factors and factors of safety on cold-formed steel welded connections, using all of the published data. The North American Specification again adopted these results for the Design of Cold-Formed Steel Structural Members.

Wallace and Schuster (2002) carried out tests on bolted cold-formed steel tension member connections in bearing (with and without washers). Additional data was needed to complete the pool of data required to properly establish the best design method for bolted tension members failing in bearing. Comparisons were made with the two design methods used in North America and the resulting recommended the North American Specification adopted design approach for the Design of Cold-Formed Steel Structural Members. Calibrations for resistance factor and factor of safety were also carried out as part this study (Wallace et al., 2002).

Wallace et al. (2002) investigated the bending and web crippling interaction of cold-formed steel members. With the recent adoption of the new web crippling design approach *(Eq. 7)*, the web crippling and bending interaction expressions contained in the North American Specification for the Design of Cold-Formed Steel Structural Members need to be re-evaluated. In addition, changes in the bending strength calculation (effective web method) have been introduced, hence, possibly affecting the interaction evaluation. The Specification contains interaction expressions for single web geometry, I-section geometry, and two nested Z-shapes. Based on the recommendations of this study, the North American Specification for the Design of Cold-Formed Steel Structural Members adopted the design approach.

Wallace and Schuster (2004) carried out tests on web crippling of cold-formed steel multi-web deck sections subjected to end one-flange loading. The resulting data was virtually the only data in this category. New web crippling coefficients have been developed along with the respective calibrated resistance factors and factors of safety. And once again, the North American Specification has adopted the results of this work for the Design of Cold-Formed Steel Structural Members.

Conclusions

It has been clearly demonstrated in this paper that in the past 35 years a great deal of research in the field of cold-formed steel has been produced at the University of Waterloo. What is even more remarkable is, that so many of the resulting information has been adopted by North American and International cold-formed steel design Standards/Specifications. I have truly enjoyed having played a small part at this remarkable University over the past 35 years and I look forward to being around and involved in future research in cold-formed steel.

References

Lind, N.C., and D.K. Schroff: "Utilization of Cold Work in Light Gage Steel," *Proceedings of the 1st Specialty Conference on Cold-Formed Steel Structures,* University of Missouri-Rolla, Aug. 1971.

Lind, N.C., M.K. Ravindra, and J. Power: "A Review of the Effective Width Formula," *Proceedings of the 1st Specialty Conference on Cold-Formed Steel Structures,* University of Missouri-Rolla, Aug. 1971.

Venkataramaiah, K.R.: "Optimum Edge Stiffeners for Thin-Walled Structural Elements," technical note 7, *Solid Mechanics Division*, University of Waterloo, Ont., Canada, July 1971.

Venkataramaiah, K.R.: "Ultimate Strength of Eccentrically Loaded Thin-Walled Open Channel Columns," technical note 9, *Solid Mechanics Division*, University of Waterloo, Ont., Canada, August 1971.

Sherbourne, A.N., C. Marsh, and C.Y. Liaw: "Stiffened Plates in Uniaxial Compression," *International Association of Bridge and Structural Engineering,* Vol. 31, Issue I, 1971

Sherbourne, A.N., and R.M. Korol: "Postbuckling of Axially Compressed Plates," *Journal of the Structural Division, ASCE Proceedings,* vol. 98, Oct. 1972.

Lind, N.C., and K.R. Venkataramaiah: "Compression in Thin-Walled Members," *Proceedings, Canadian Structural Engineering Conference,* Montreal, 1972.

Schuster, R.M.: "Composite Steel-Deck Reinforced Concrete Systems Failing in Shear-Bond", *9th Congress - International Association for Bridge and Structural Engineering,* Amsterdam, Holland, May, 1972.

Lind, N.C.: "Strength, Deformation and Design of Cold Formed Steel Structures," *Design in Cold Formed Steel* – Edited by R.M. Schuster, Solid Mechanics Division, University of Waterloo, Waterloo, Ont., Canada, 1974.

Lind, N.C.: "Design Procedures for Flexural Members," *Design in Cold Formed Steel* – Edited by R.M. Schuster, Solid Mechanics Division, University of Waterloo, Waterloo, Ont., Canada, 1974.

Lind, N.C.: "Recent Developments in Cold Formed Steel Design Requirements," *Design in Cold Formed Steel* – Edited by R.M. Schuster, Solid Mechanics Division, University of Waterloo, Waterloo, Ont., Canada, 1974.

Lind, N.C.: "Additional Design Examples," *Design in Cold Formed Steel* – Edited by R.M. Schuster, Solid Mechanics Division, University of Waterloo, Waterloo, Ont., Canada, 1974.

Roorda, J.: "On the Buckling Behaviour and Design of Thin-Walled Beams and Columns," *Design in Cold Formed Steel* – Edited by R.M. Schuster, Solid Mechanics Division, University of Waterloo, Waterloo, Ont., Canada, 1974.

Schuster, R.M.: "Composite Steel-Deck Reinforced Concrete Floor Systems," *Design in Cold Formed Steel* – Edited by R.M. Schuster, Solid Mechanics Division, University of Waterloo, Waterloo, Ont., Canada, 1974.

Schuster, R.M.: "Current Design Criteria of Steel-Deck Reinforced Concrete Slabs," *Design in Cold Formed Steel* – Edited by R.M. Schuster, Solid Mechanics Division, University of Waterloo, Waterloo, Ont., Canada, 1974.

Schuster, R.M.: "Proposed Ultimate Strength Design Criteria for Steel-Deck Reinforced Concrete Slabs," *Design in Cold Formed Steel* – Edited by R.M. Schuster, Solid Mechanics Division, University of Waterloo, Waterloo, Ont., Canada, 1974.

Lind, N.C., and D.K. Schroff: "Utilization of Cold-Work in Cold Formed Steel," *Journal of the Structural Division, ASCE Proceedings,* vol. 101, Jan. 1975.

Lind, N.C., L.I. Knab, and W.B. Hall: "Economic Study of the Connection Safety Factor," *Proceedings of the 3rd International Specialty Conference on Cold-Formed Steel Structures,* University of Missouri-Rolla, Nov. 1975.

Knab, L.I., and N.C. Lind: "Reliability Based Design Criteria for Temporary Cold-Formed Steel Buildings," *Proceedings of the 3rd International Specialty Conference on Cold-Formed Steel Structures,* University of Missouri-Rolla, Nov. 1975.

Schuster, R.M.: "Cold Formed Steel Design Manual," *Solid Mechanics Division,* University of Waterloo, Waterloo, Ont., Canada, 1975.

Parimi, S.R., and N.C. Lind: "Limit States Basis for Cold Formed Steel Design," *Journal of the Structural Division, ASCE Proceedings,* vol. 102, March 1976.

Schuster, R.M.: "Composite Steel-Deck Concrete Floor Systems," *Journal of the Structural Division, ASCE Proceedings,* vol. 102, May 1976.

Lind, N.C., M.K. Ravindra, and G. Schorn: "Empirical Effective Width Formula," *Journal of the Structural Division, ASCE Proceedings,* vol. 102, Sept. 1976.

Lind, N.C.: "Numerical Buckling Analysis of Plate Assemblies," *Journal of the Structural Division, ASCE Proceedings,* vol. 104, Feb. 1978.

Schuster, R.M.: "Composite Steel Deck Concrete Floor Slabs," *Proceedings of the Cold-Formed Steel Structures Conference,* University of Windsor, Ont., Canada, Apr. 1980.

Schuster, R.M., and W.C. Ling: "Mechanical Interlocking Capacity of Composite Slabs," *Proceedings of the 5th International Specialty Conference on Cold-Formed Steel Structures,* University of Missouri-Rolla, Nov. 1980.

Venkataramaiah, K.R., J. Roorda, and K.R. Srinivasaiah: "Elastic Modulus of Cold-Formed Sheet Steel," *Proceedings of the 5th International Specialty Conference on Cold-Formed Steel Structures,* University of Missouri-Rolla, Nov. 1980.

Wing, B.A., and R.M. Schuster: "Web Crippling and the Interaction of Bending and Web Crippling of Unreinforced Multi-Web Cold Formed Steel Sections," University of Waterloo, Waterloo, Ont., Canada, 1981.

Wing, B.A., and R.M. Schuster: "Web Crippling of Decks Subjected to Two Flange Loading," *Proceedings of the 6th International Specialty Conference on Cold-Formed Steel Structures,* University of Missouri-Rolla, Nov. 1982.

Schurter, P.G., R.M. Schuster, and A.S. Zakrzewski: "Combined Compression and Lateral Loads on Load-Bearing Steel Stud Walls," *Proceedings of the 6th International Specialty Conference on Cold-Formed Steel Structures,* University of Missouri-Rolla, Nov. 1982.

Seleim, S.S., and R.M. Schuster: "Shear-Bond Capacity of Composite Slabs," *Proceedings of the 6th International Specialty Conference on Cold-Formed Steel Structures,* University of Missouri-Rolla, Nov. 1982.

Queen, D.J., and N.C. Lind: "Interactive Buckling of Thin-Walled Compression Members," *Proceedings of the SSRC Annual Technical Session*, 1985.

Seleim, S.S., and R.M. Schuster: "Shear-Bond Resistance of Composite Deck-Slabs," *Canadian Journal of Civil Engineering*, vol. 12, Number 2, June 1985.

Wing, B.A., and R.M. Schuster: "Web Crippling of Multi-Web Deck Sections Subjected to Interior One Flange Loading," *Proceedings of the 8th International Specialty Conference on Cold-Formed Steel Structures,* University of Missouri-Rolla, Nov. 1986.

Schuster, R.M., S.R. Fox, and D.L. Tarlton: "An LRFD Standard for Cold Formed Steel Design," *Cold-Formed Steel, Workshop Proceedings of the 3rd International Conference on Tall Buildings,* Jan. 1986.

Fox, S.R., R.M. Schuster, and D.L. Tarlton: "The Canadian LRFD Standard for Cold-Formed Steel Design," *Proceedings of the 8th International Specialty Conference on Cold-Formed Steel Structures,* University of Missouri-Rolla, Nov. 1986.

Schurter, P.G., and R.M. Schuster: "Aluminum-Zinc Alloy Coated Steel for Composite Slabs," *Proceedings of the 8th International Specialty Conference on Cold-Formed Steel Structures,* University of Missouri-Rolla, Nov. 1986.

Schuster, R.M., and R.E. Suleiman: "Composite Slabs Subjected to Repeated Point Loading," *Proceedings of the 8th International Specialty Conference on Cold-Formed Steel Structures,* University of Missouri-Rolla, Nov. 1986.

Schuster, R.M.: "Excerpts From the New Canadian Cold Formed Steel Design Standard," *Proceedings of the International Conference on Steel and Aluminum Structures,* Cardiff, Great Britain, 1987.

McCuaig, L.A., and R.M. Schuster: "Repeated Point Loading on Composite Slabs," *Proceedings of the 9th International Specialty Conference on Cold-Formed Steel Structures,* University of Missouri-Rolla, Nov. 1988.

Schuster, R.M. et al., "Standard For Composite Steel Deck," Canadian Sheet Steel Building Institute, CSSBI 12M -96, Cambridge, Ontario, October 1996.

McCuaig, L.A., and R.M. Schuster: "Safety Indexes for Shear-Bond Failure of Composite Slabs," *Proceedings of the 10th International Specialty Conference on Cold-Formed Steel Structures,* University of Missouri-Rolla, Oct. 1990.

Schuster, R.M.: "The New Canadian Cold Formed Steel Design Standard," *Trends in Steel Structures for Mining and Building,* Johannesburg, South Africa, 1991.

Schuster, R.M.: "Lightweight Steel Framing in the 90s," *Trends in Steel Structures for Mining and Building,* Johannesburg, South Africa, 1991.

Dinovitzer, A.S., M. Sohrabpour, and R.M. Schuster: "Observations and Comments Pertaining to CAN/CSA-S136-M89," *Proceedings of the 11th International Specialty Conference on Cold-Formed Steel Structures,* University of Missouri-Rolla, Oct. 1992.

Schuster R.M.: "The 1989 Edition of the Canadian Cold Formed Steel Design Standard," *Proceedings of the 11th International Specialty Conference on Cold-Formed Steel Structures,* University of Missouri-Rolla, Oct. 1992.

Papazian, R.P., R.M. Schuster, and M. Sommerstein: "Multiple Stiffened Deck Profiles," *Proceedings of the 12th International Specialty Conference on Cold-Formed Steel Structures,* University of Missouri-Rolla, Oct. 1994.

Schuster, R.M., C.A. Rogers, and A. Celli: "Research into Cold Formed Steel Perforated C-Sections in Shear," *Progress Report No. 1,* CSSBI/IRAP Project, University of Waterloo, Waterloo, Ont., 1995.

Rogers, C.A., and R.M. Schuster: "Effective Width of a Single Edge Stiffener Subjected to a Stress Gradient," *Proceedings of the 13th International Specialty Conference on Cold-Formed Steel Structures,* University of Missouri-Rolla, Oct. 1996.

Rogers, C.A., and R.M. Schuster: "Interaction of Flange/Edge-Stiffened Cold Formed Steel C-Sections," *Proceedings of the 13th International Specialty Conference on Cold-Formed Steel Structures,* University of Missouri-Rolla, Oct. 1996.

Rogers, C.A., and R.M. Schuster: "Cold Formed Steel Flat Width Ratio Limits, d/t and d_i/w," *Proceedings of the 13th International Specialty Conference on Cold-Formed Steel Structures,* University of Missouri-Rolla, Oct. 1996.

Hancock, G.J., C.A. Rogers, and R.M. Schuster: "Comparison of the Distortional Buckling Method for Flexural Members with Tests," *Proceedings of the 13th International Specialty Conference on Cold-Formed Steel Structures,* University of Missouri-Rolla, Oct. 1996.

Rogers, C.A. and R.M. Schuster: "Test Results and Comparison of Cold Formed Steel Edge Stiffened C and Z-sections," *1st CSCE Structural Specialty Conference*, Edmonton, May 29 to June 1, 1996.

Rogers, C.A., and R.M. Schuster: "Flange/Web Distortional Buckling of Cold Formed Steel Sections in Bending," *Thin-Walled Structures Journal,* vol. 27 Number 1, Elsevier, January 1997.

Prabakaran, A., and R.M. Schuster: "Web Crippling Behavior of Cold Formed Steel Members," *Proceedings of the 14th International Specialty Conference on Cold-Formed Steel Structures,* University of Missouri-Rolla, Oct. 1998.

Gerges, R.R., and R.M. Schuster: "Web Crippling of Single Web Cold Formed Steel Members Subjected to End-One-Flange Loading," *Proceedings of the 14th International Specialty Conference on Cold-Formed Steel Structures,* University of Missouri-Rolla, Oct. 1998.

Acharya, V.V., and R.M. Schuster: "Bending Tests of Hat Sections with Multiple Longitudinal Stiffeners," *Proceedings of the 14th International Specialty Conference on Cold-Formed Steel Structures,* University of Missouri-Rolla, Oct. 1998.

Acharya, V.V., and R.M. Schuster: "Analysis of Hat Sections with Multiple Intermediate Longitudinal Stiffeners," *Proceedings of the 14th International Specialty Conference on Cold-Formed Steel Structures,* University of Missouri-Rolla, Oct. 1998.

Fox, S.R., and R.M. Schuster: "Testing of Cold Formed Steel Floor Joists with Bearing Stiffeners," *Proceedings of the 14th International Specialty Conference on Cold-Formed Steel Structures,* University of Missouri-Rolla, Oct. 1998.

Schuster, R.M., "One North American Specification for the Design of Cold Formed Steel Structures," *XVI National Steel Conference*, Buenos Aires, Argentina, Sept. 1998.

Schuster, R.M., "Cold Formed Steel – the Construction Material of the Future," *XVI National Steel Conference*, Buenos Aires, Argentina, Sept. 1998.

Sloof, P.A., and R.M. Schuster: "Yield Strength Increase of Cold Formed Sections Due to Cold Work of Forming," *Proceedings of the 15th International Specialty Conference on Cold-Formed Steel Structures,* University of Missouri-Rolla, Oct. 2000.

Beshara, B., and R.M. Schuster: "Web Crippling of Cold Formed Steel Members," *Proceedings of the 15th International Specialty Conference on Cold-Formed Steel Structures,* University of Missouri-Rolla, Oct. 2000.

Fox, S.R., and R.M. Schuster: "Strength of Bearing Stiffeners in Cold Formed C-Sections," *Proceedings of the 15th International Specialty Conference on Cold-Formed Steel Structures,* University of Missouri-Rolla, Oct. 2000.

Craig, B., and R.M. Schuster: "Calibration of Cold Formed Steel Shear Equations," *Proceedings of the 15th International Specialty Conference on Cold-Formed Steel Structures,* University of Missouri-Rolla, Oct. 2000.

Fox, S.R., and R.M. Schuster: "Lateral Strength of Wind Load Bearing Wall Stud-to-Track Connection," *Proceedings of the 15th International Specialty Conference on Cold-Formed Steel Structures,* University of Missouri-Rolla, Oct. 2000.

Xu, L., H. Min, and R.M. Schuster: "Optimum Design of Cold Formed Steel Residential Roof Trusses," *Proceedings of the 15th International Specialty Conference on Cold-Formed Steel Structures,* University of Missouri-Rolla, Oct. 2000.

Xu, L., Z. Ling, W.-C. Xie, and R.M. Schuster: "Dynamic Behaviour of Floors with Cold-Formed Steel Joists," *Proceedings of the 15th International Specialty Conference on Cold-Formed Steel Structures,* University of Missouri-Rolla, Oct. 2000.

Yu, W.W.: *Cold-Formed Steel Design*, 3rd Edition, John Wiley & Sons, Inc. New York, NY., 2000.

Xu, L., and J. Cui: "Connection Flexibility of C-Shape Cold Formed Steel Connections," *Proceedings of the 6th Pacific Steel Structures Conference,* Beijing, China, Oct. 2002.

Xu, L., Y. Gong, and P. Guo: "Compressive Tests of Cold Formed Steel Curved Panels," *Journal of Constructional Steel Research*, 57, 2001.

Schuster, R.M., and T. Trestain: "Criteria for the Testing of Composite Slabs," Canadian Sheet Steel Building Institute, CSSBI S2-2002, Cambridge, Ontario, March 2002.

Beshara, B., and R.M. Schuster: "Web Crippling of Cold Formed Steel Members," *Proceedings of the 16th International Specialty Conference on Cold-Formed Steel Structures,* University of Missouri-Rolla, Oct. 2002.

Tangorra, F.M., R.M. Schuster, and R.A. LaBoube: "Calibrations of Cold-Formed Steel Welded Connections," *Proceedings of the 16th International Specialty Conference on Cold-Formed Steel Structures,* University of Missouri-Rolla, Oct. 2002.

Wallace, J.A., and R.M. Schuster: "Testing of Bolted Cold Formed Steel Connections in Bearing (With and Without Washers)," *Proceedings of the 16th International Specialty Conference on Cold-Formed Steel Structures,* University of Missouri-Rolla, Oct. 2002.

Wallace, J.A., R.M. Schuster, and R.A. LaBoube: "Calibrations of Bolted Cold Formed Steel Connections in Bearing (With and Without Washers)," *Proceedings of the 16th International Specialty Conference on Cold-Formed Steel Structures,* University of Missouri-Rolla, Oct. 2002.

Wallace, J.A., R.M. Schuster, and R.A. LaBoube: "Bending and Web Crippling Interaction of Cold-Formed Steel Members," *Proceedings of the 16th International Specialty Conference on Cold-Formed Steel Structures,* University of Missouri-Rolla, Oct. 2002.

Fox, S.R., and R.M. Schuster: "Bearing Stiffeners in Cold-Formed Steel C-Sections," *International Conference on Advances in Structures*, Sidney, Australia, 2003.

Schuster, R.M., and T. Trestain: "Criteria for the Design of Composite Slabs," Canadian Sheet Steel Building Institute, CSSBI S3-2002, Cambridge, Ontario, September 2003.

Wallace, J.A., and R.M. Schuster: "Web Crippling of Cold Formed Steel Multi-Web Deck Sections Subjected to End One-Flange Loading," *Proceedings of the 17th International Specialty Conference on Cold-Formed Steel Structures,* University of Missouri-Rolla, Nov. 2004.

Xu, L.: "Vibration Evaluation of Lightweight Floors Supported by Cold Formed Steel Joists," *Proceedings of the 4th International Conference on Advances in Steel Structures,* Shanghai, June, 2005.

Concrete Structures

SULPHIDE, SULPHATE AND SULPHURIC ACID CORROSION OF CONCRETE IN LABORATORY TESTS

V. Assaad Abdelmseeh, J.C. Jofriet, S.C. Negi and G. Hayward

School of Engineering, University of Guelph, Guelph, ON, Canada N1G 2W1
E-mail: jjofriet@uoguelph.ca

Abstract

Portland cement (PC) concrete is generally a highly durable structural material. Nevertheless, certain chemical actions and aggressive environments in a livestock building can cause deterioration and total collapses of structures have occurred long before they have reached their design life.

The sulphide and sulphate resistance of three replicates of eight different reinforced concrete mixes were investigated in a laboratory study in which one half of the 48 specimens were half submerged in a sodium sulphate solution (20,000 ppm SO_4^{2-}) and also exposed to hydrogen sulphide gas (1,000 ppm H_2S). The other half of the 48 specimens was subjected to hydrogen sulphide gas only. The mixes included PC concrete with W/CM ratios of 0.4 and 0.5 and six mixes with cement replacements of slag, fly ash or silica fume, all with water/cementitious material (W/CM) ratio of 0.4.

After 23 cycles of testing over about 36 months, the electrochemical potential results and visual inspection of the reinforcing bars indicate that the PC concrete with 0.5 W/CM ratio was the least resistant against steel corrosion. Corrosion of the concrete was more critical than that of the steel. All treatments containing silica fume performed much better than PC40. Treatments that contained fly ash performed worse than plain PC concrete with the same W/CM ratio. Sulphate resistant cement concrete was more resistant than Type 10 Portland cement concrete, in both sets of tests. In general the samples that were exposed to hydrogen sulphide and sulphate corroded at a higher rate then those exposed to the H_2S gas only.

In subsequent tests 6 of the 8 mixes were exposed to 7% sulphuric acid for about one year. Preliminary results indicate the greatest mass loss for the concrete with 0.5 W/CM ratio and very similar amount of loss for the 5 mixes with W/CM ratio of 0.4. The least amount of mass loss was experienced by the mix with sulphate resistant cement.

Keywords: concrete corrosion, livestock building, manure, hydrogen sulphide, sulphate, sulphuric acid

Introduction

Reinforced concrete for the use in agricultural facilities often exposed to aggressive environment conditions. Temporary storage of liquid manure underneath barn floors produce corrosive agents generated from aerobic and anaerobic fermentation, causing premature corrosion of reinforcement steel and degradation of the concrete. The rehabilitation of reinforced concrete structures due to corroding steel reinforcing bar is quite expensive compared with the use of good quality concrete at the time of construction.

For Canadian climatic conditions, storage of manure for a period of six months, or even longer, is generally required so that manure spreading can be avoided during winter. Anaerobic fermentation produces several gases of which hydrogen sulphide is the most corrosive leading to rapid deterioration of concrete floors in barns (Frénay and Zilverberg, 1993). Sulphate-reducing bacteria, known to thrive in animal confinement buildings, generate sulphuric acid as the end product of their metabolism. The sulphuric acid strength has been measured in the laboratory as equivalent to 7% (by volume) H_2SO_4 (Hewayde 2005).

High humidity, the concentration of various gases above the liquid manure and the continuous wetting of concrete floor slats are all contributing factors (Svennerstedt et al., 1999). As a

M. Pandey et al. (eds), Advances in Engineering Structures, Mechanics & Construction, 55–65.

consequence of the concrete degradation in some instances slatted floor have deteriorated to the point of requiring replacement in less than five years. In some swine barns in Ontario a 50% loss of expected service life was reported, when regular concrete mixes were used. It is estimated that an average annual cost of depreciation on all structures is about $250 million and about $100 to 150 million is spent on repairs.

In order to improve the durability of concrete recent investigations have been made in an attempt to reduce the rate of deterioration by changing the concrete composition (De Belie et al. 1997; Idriss 2000; Jiang 2002; Berge and Verhardsson, 2002). This study involves concrete made with various combinations of Portland cement, slag, fly ash and silica fume, subjected to hydrogen sulphide gas and sulphate solution.

The experimental part of the project will be described in this paper. The results of this research will help finding the most cost-effective solution in Ontario for reducing concrete corrosion to a minimum and enhance the service life of reinforced concrete in livestock buildings there.

Materials and Methods

The concrete cover quality is the most effective protection and first defense against concrete corrosion. Thus, the study looked at the comparable durability of of eight different concrete mixes that are practical and cost effective alternatives in the construction of livestock buildings and manure storage structures. One of the mixes was Portland cement type 10 with water/cementing material (W/CM) ratio of 0.5, which is considered a commonly used mix in farm building structures. Common wisdom is to lower the W/CM ratio to increase durability. So the same cement type was used with a W/CM ratio of 0.4. However, lowering the W/CM ratio invariably means increasing the cement content of the mix thereby increasing the C_3A content of the hardened cement paste and thus decreasing the sulphate resistance.

On order to reduce the Portland cement content six other mixes with various supplementary cementing materials, like slag, fly ash and silica fume, were included. The use of supplementary cementing materials to replace a portion of the Portland cement in the mix contributes to the properties of the hardened concrete through hydraulic or pozzolanic activity. It reduces the concrete permeability, decreases the cement content, and the W/CM ratio. It also, in general, improves the resistance of the concrete to sulphate attack by lowering the C_3A content of the hardened cement paste.

Concrete used for the construction of liquid manure handling and storage structures is subjected to sulphides, sulphates and sulphuric acid. In some locations the concrete is submerged continuously, in other places it is submerged some of the time and some locations are never submerged. Field observations indicate that the latter two situations lead to more severe corrosion than the totally submerged condition.

To reach the objectives of the research a combination of laboratory research and numerical simulation was selected. The laboratory research consisted of accelerated corrosion testing of concrete specimens by subjecting them to hydrogen sulphide gas and a sulphate solution in high concentrations. One half of the specimens was partially immersed in sodium sulphate (20,000 ppm SO_4^{2-}) and also subjected to hydrogen sulphide gas (1,000 ppm H_2S). This condition was chosen to simulate units of construction that are partially submerged in liquid manure. The second set was subjected to hydrogen sulphide gas only, a condition that may occur for concrete permanently above the manure. Each set consisted of the eight different mixes. The sulphuric acid corrosion was measured on those six mixes not containing slag by applying 2 cm^3 of 7% sulphuric acid three times per week to disc-shaped specimens.

The corrosion of the concrete and of the embedded reinforcing steel was measured separately. For the purpose of the embedded steel corrosion study, all treatments were tested with the half-cell potential between the concrete surface and the reinforcing steel to define the corrosion state. The reinforcing steel was visually inspected at three stages of the sulphide/sulphate experiment.

For the purpose of the study on concrete corrosion due to sulphate and/or sulphide, the compressive strength of all mixes was measured at 28 days. The sulphate solutions, in which the mixes were partially immersed, were analyzed and their pH's measured to follow the leaching of the alkalis from the concrete paste and the effect of H_2S gas on the solution acidity and SO_4^{2-} concentration. The actual corrosion damage to the concrete was measured using the volume loss of the concrete at the end of the test period. The corrosion of concrete is either caused by a direct chemical reaction of the corrosive agents with the concrete at its surface, or by reactions below the surface after diffusion of the corrosive gases or liquids into the concrete. To try to measure the extent of the latter the total sulphur profile was determined across the concrete cover thickness. Finally, the mineralogy of the concrete was studied using powder x-Ray test to confirm the nature of the chemical reactions that took place and to confirm that these were similar to those that happened in a field situation where a pig barn collapsed occurred due to corrosion of the concrete structure.

The concrete corrosion from exposure to sulphuric acid was measured by weight loss. That aspect of the study is in progress at the time of writing.

Liquid swine manure has sulphate concentrations in the order of 1500 to 2000 ppm. Table 1 lists the concrete requirements in Canadian concrete standards for four degrees of exposure to sulphate (Kosmatka et al. 2002).

Table 1. Requirements for concrete subjected to sulphate attack

Degree of exposure	Water-soluble sulphate in soil (%)	Sulphate in groundwater (mg/L)	Max. water-cementing materials ratio	Portland cement type to be used
Very severe	Over 2.0	Over 10,000	0.40	Type 50[1] plus a pozzolanic admixture
Severe	0.20 to 2.0	1500 to 10,000	0.45	Type 50
Moderate	0.10 to 0.20	150 to 1500	0.50	Type 20[2]
Negligible	Below 0.10	Below 150	No restriction	No restriction

[1] CSA A5 (eq. to ASTM C1157 type HS) [2] CSA A5 (eq. to ASTM C1157 type MS)

The replacement of 35% of type 10 cement by slag is considered sufficient to provide Type 50 equivalence. The relative improvement in sulphate resistance through the use of fly ash is greater for low cement content mixes and in high sulphate exposure. Class F fly ash is effective in improving the sulphate resistance of concrete when used at suitable replacement rates. The higher the calcium content of a Class C fly ash, the less likely it is that it will provide benefits with regard to sulphate resistance (A Publication of Lafarge Canada Inc.). Unfortunately, in Ontario the available fly ash is type C.

Another effective mix is the use of silica fume. Due to reduced permeability, silica fume cement provides excellent sulphate resistance to concrete. Recent research shows that ternary blends containing slag cement or fly ash, along with silica fume and Portland cement, can be effective for sulphate resistance concrete (A Publication of Lafarge Canada Inc.).

It is intended that the objectives of the research be attained by the combination of laboratory research and numerical simulation. The corrosion testing in the laboratory are accelerated tests using hydrogen sulphide gas, and a sodium sulphate solution as the only corrosive agents. Acceleration is achieved by using much higher concentration than those experienced in the field, 1,000 ppm H_2S and 20,000 ppm SO_4^{2-}.

Eight mixes, which could reasonably be used in the construction of liquid manure tanks, floors and slats, were examined for corrosion resistance. Concrete cylinders (100 mm diameter by 100 mm high) made of Portland cement, limestone, sand and water, each with a 10 mm diameter by 90 mm long reinforcing steel bar embedded in the center. Six replicates for each mix were made. In all mixes a superplasticizer (CATEXOL 1000 SP-MN) is used (625 ml/100 kg of cementitious material) to reduce the water requirements in concrete and attain the necessary workability without the use of excess water. Also an air-entraining admixture (CATEXOL A.E.260) was used in all mixes (50 ml/100 kg of cementitious material) to increase concrete durability, improve workability and reduce bleeding. A further five larger replicates of each mix, 100 mm diameter and 200 mm height, were cast without a steel bar for compressive strength determination at 28 days (three replicates) and for the sulphuric acid corrosion tests. The mix proportions and materials used for all eight mixes are provided in Table 2. The coarse aggregate was crushed limestone.

Table 2. Mix proportions for all eight mixes

	PC50	PC40	SR	SC	SFC	FAC	SSFC	FASF
	PC with W/CM ratio 0.5	PC with W/CM ratio 0.4	Sulphate resisting cement	Slag cement	Silica fume cement	Fly ash cement	Silica fume & slag cement	Silica fume & fly ash cement
Cement type	10	10	50	10	10	10	10	10
Cement (kg/m^3 of concrete)	340	425	425	276	391	319	293	293
Water (kg/m^3 of concrete)	170	170	170	170	170	170	170	170
W/CM ratio	0.5	0.4	0.4	0.4	0.4	0.4	0.4	0.4
Additive (% of cementitious material content)	---	---	---	35% slag	8% silica fume	25% fly ash	25% slag 6% silica fume	25% fly ash 6% silica fume

The top of all specimens, except those for the compressive strength tests, and the exposed ends of the steel bars were coated with an epoxy coating (TRU-GLAZE 4508 Chemical Resistant Epoxy Coating 4508-1000A and 4508-9999B with a ratio of 1:1), in order to prevent the diffusion of the corrosive ions through that surface.

For the sulphide/sulphate corrosion study the 48 specimens from 6 replicates were divided into two sets, each set had three replicates. The first set (i.e. 24 specimens) was tested partially (50%) immersed in sodium sulphate (20,000 ppm SO_4^{-2}) and at the same time subjected to hydrogen sulphide gas (1,000 ppm H_2S) above the surface of the sodium sulphate solution. Each treatment was submerged in a separate container, and placed in the upper level of a two-storey test chamber. The second set (i.e. the remaining 24 specimens) was subjected only to the hydrogen sulphide gas, nitrogen and air (0.1% H_2S, 9.9% N_2, and 90% air) in the lower level of the same chamber.

To keep the gas in the sealed Plexiglas test chamber at the required concentration a control circuit consisting of H_2S sensor, solenoid valves, flow meter, control program and a gas cylinder (1% H_2S, 99% N_2) was used. The hydrogen sulphide gas cylinder lasted about 3 weeks, keeping the concentration of the gas inside the chamber at 1000 ppm. After this the chamber was left closed for one more week in order to lower the H_2S concentration back to 0 ppm. Then all specimens were taken out of the chamber for the half-cell potential measurement, and thus they were subjected to air for another week. This test cycle was repeated approximately every five weeks.

The specimens for the sulphuric acid tests were 25 mm thick discs cut from the 100 mm diameter by 200 mm high unreinforced cylinders described earlier. Six mixes that did not include fly ash were included in these tests. The reason for excluding the fly ash containing mixes was that the type C fly ash that was used is not a viable solution for corrosion resistance in an acidic environment.

Standard 28-day compressive strength test were carried out for the 100 mm diameter by 200 mm high cylindrical specimens. The splitting tensile strength test (CSA A23.2-13C 1994) was carried out after 11, 15 and 23 cycles of exposure to the corrosive environment. This allowed the removal of the reinforcing bar for inspection of corrosion and provided a rough indication of the specimen's strength about 26, 32 and 41 months after the specimens were cast and after the exposure to a corrosive environment for most of that period.

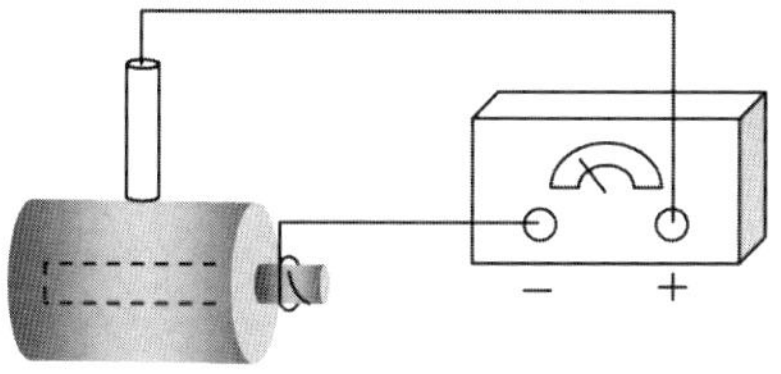

Figure 1 Half-cell potential measurement

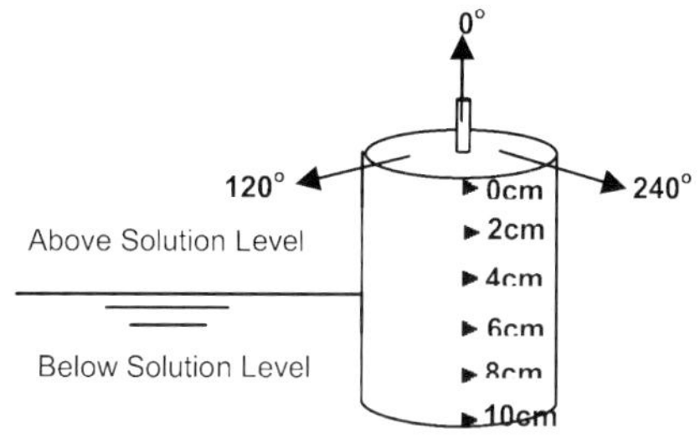

Figure 2 Positions where the potential measurements were taken

The half-cell potential, E_{corr}, was used to define the corrosion state of reinforcement bars. A copper-copper sulphate electrode (CSE) and a high impedance voltmeter were used to read E_{corr}. The half-cell potential measurement connection is shown in Fig. 1. The first half-cell potential measurements were taken after three cycles of exposure to the corrosive environment. Measurements were made every test cycle of about 5 weeks thereafter until the end of the experiment. Six potential readings were taken every 20 mm (at the air end, 20, 40, 60, 80, and 100 mm) along three meridians at 0°, 120°, and 240°, for a total of 18 readings per specimen (Fig. 2). The 18 readings were averaged. Corrosion potential measurements provide an indication of the oxidizing power of the environment in which a specimen is exposed.

At the end of the test period (after the 23rd test cycle), specimens were split open and concrete cover was removed to inspect the steel in the specimens. A visual assessment of the corrosion of reinforcing steel was performed. The corrosion was rated as three corrosion levels. Level 1 represents very slight or no visible rust on the surface. Level 2 shows a little rust at the surface. Level 3 indicates relatively heavy corrosion.

After the exposure of concrete specimens to the sulphate solution and/or sulphide gas, for the study period, the last set of specimens were air dried and brushed to remove all loose material. The volume loss of each specimen was determined using the water displacement method.

X-Ray powder diffraction patterns provided information on the phase, chemical and crystal structure. For the XRD test, 50 mg samples of the concrete paste were obtained from the outer surface of each mix and were well-ground to a uniform particle size to below 10 μm. The test was carried out for three concrete mixes: PC40, SR, and SFC. The samples were taken from the last set of replicates for both corrosive environment exposures, a total of six samples. For the ones that were submerged the samples were taken from above the solution level. Also, samples were taken from a barn that collapsed in Innerkip in 2001, reportedly as a result of corrosion of the supporting piers. One sample was taken from the surface of a beam that supported a solid slab over the manure pit. Another sample was obtained from the outer surface of one of the piers that caused the failure.

The sulphuric acid corrosion tets were carried out on 25 mm thick by 100 mm diameter discs positioned on the flat. The present tests use discs that have the top surface of the cylinders from which they were cut on top, the cut surface on the bottom. Three times per week 2 cm^3 of 7% (by volume) H_2SO_4 is dripped slowly on the surface of the discs. After each 10 applications (about 3 weeks) the discs are washed, scrubbed to remove loose material, dried in an oven for one week, and weighed. The weight losses are used as a comparative measure of corrosion.

Results and Discussion

The compressive 28-day strengths of the specimens are shown in Table 3. The PC50 mix with W/CM ratio of 0.5 had much lower strength than the other seven mixes with W/CM ratio of 0.4. Of those the silica fume concrete, SFC, had the highest strength of 49.1 Mpa.

Table 3. 28-day compressive strength for the eight mixes

Mix	**PC50**	**PC40**	**SR**	**SC**	**SFC**	**FAC**	**SSFC**	**FASF**
W/CM ratio	0.5	0.4	0.4	0.4	0.4	0.4	0.4	0.4
Compressive Strength (MPa)	34.0	48.6	47.4	42.3	49.1	47.0	43.5	46.0

The splitting strengths were variable. The specimens exposed to H_2S only gained some strength, about 1 kPa on average, between cycles 11 and 15, but lost about the same amount of strength between cycles 15 and 23. The specimens exposed to both H_2S and sulphate gained similar amounts of tensile strength in the first part of the testing period but retained these strength until the end of testing, despite the loss of concrete volume to be discussed later.

The electrochemical potentials for the different mixes have been taken 19 times since the start of the experiment. They are plotted in Fig. 3 for the 8 mixes exposed to hydrogen sulphide and half submerged in sulphate solution. Fig. 4 presents the results for the 8 mixes exposed to hydrogen sulphide only. In examining the curves it should be noted that a potentials of –350 mV may be considered a 90% probability of active steel corrosion. All values are the mean of all available replicates.

In Fig. 3 the potential measurements of PC50 were consistently the highest. It crossed the –350 mV boundary at cycle 17. The values measured for the sulphate resistant concrete mix, SR were almost consistently the lowest, indicating excellent protection against steel corrosion.

They did not cross the –350 mV till near the end of testing. All other mixes performed in between; mix FASF crossed the –350 mV value at cycle 18, SC between cycles 18 and 19, mix SFC at about cycle 19 and SSFC at cycle 20.

The results in Fig. 4 indicate that the specimens exposed to H_2S only experienced far less steel corrosion than did those exposed to both sulphates and H_2S. Potentials were roughly one half. Indeed, none of the former reached the –350 mV value. But as before, the PC50 mix exhibited the highest values almost consistently. Mix SSFC shows the lowest values up to cycle 18 after which mix SR gave the lowest value. In all mixes except PC50 there appeared to be a distinct increase in the rate at cycle 16.

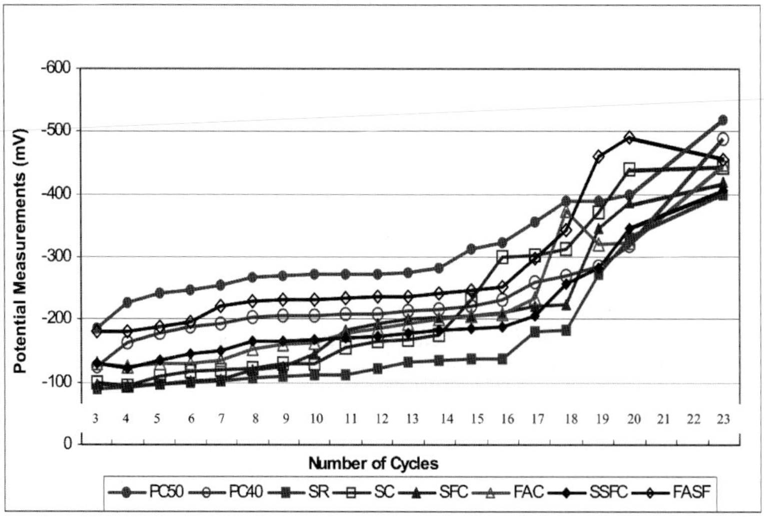

Fig. 3 Electrochemical potential of mixes exposed to sulphate and H_2S

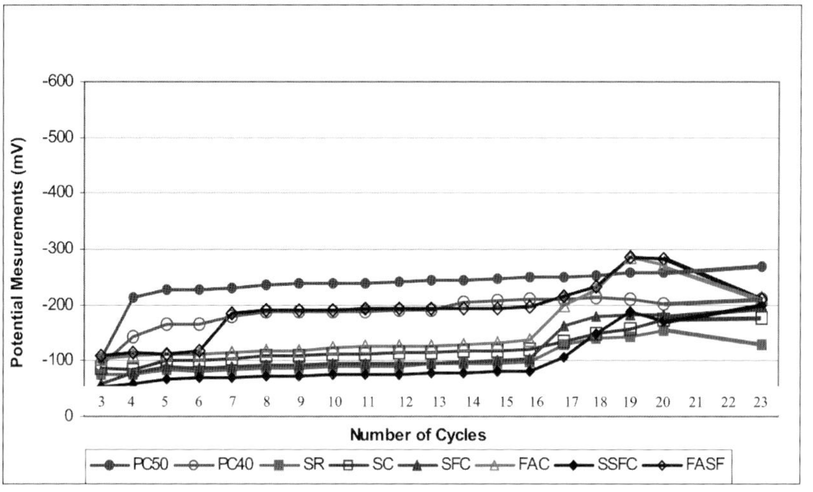

Fig. 4 Electrochemical potential of mixes exposed to H_2S only

The degree of corrosion of the reinforcing steel was evaluated visually using numerical rating system of 1 to 3, 1 indicating no evidence of visible rust on the surface, and 3 indicating heavy corrosion. Table 4 presents the visual rating of steel surface corrosion of the 16 steel bars for the last set of specimens after the 23rd test cycle. The corrosion of the steel reinforcement was quite uniform in the specimens exposed to H2S only, regardless of the mix. All had medium amount of corrosion except mix FASF which had severe corrosion. There was considerable variability in the mixes exposed to H2S and sulphate. Mix SR had visibly the least amount of steel corrosion whereas in mixes PC50, PC40 and FAC the steel was heavily corroded. The visual corrosion ratings are in reasonable agreement with the electrochemical potential results.

Table 4. Visual rating of steel corrosion

Mix	H_2S and sulphate	H_2S only
PC50	3	2
PC40	3	2
SR	1	2
SC	2	2.5
SFC	2	2
FAC	3	2.5
SSFC	1.5	2
FASF	2	3

The volume loss experienced in the concrete because of corrosion is illustrated in Fig. 5. All concrete specimens exposed to sulphate solution and hydrogen sulphide gas showed higher volume loss than those exposed to hydrogen sulphide gas only. The volume loss was the greatest in mix PC40 exposed to both H_2S and sulphate. The least amount of concrete lost to corrosion was in mix SR; SFC was second lowest.

Photographs of all specimens that experienced the full 23-cycle 3-year exposure to sulphate and/or sulphide are shown in Figs. 6a and 6b. Again, it is obvious from the photographs that all concrete specimens exposed to sulphate solution and hydrogen sulphide gas showed higher volume loss than those exposed to hydrogen sulphide gas only. And it is also obvious that most of the

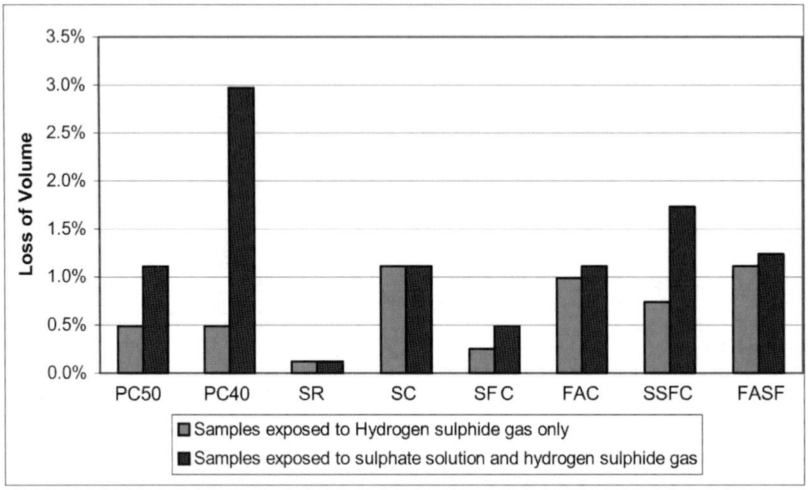

Fig. 5. Loss of volume for all mixes after 23 cycles of testing

concrete loss is located above the sulphate solution level. As was discussed earlier, the evaporation of the water from the concrete surface that contains high amount of alkali sulphates causes crystallization of salts, which in turn generates a disruptive pressure (Skalny et al. 2002). This is in good agreement with field observations of the piers partially submerged in liquid manure in the barn that collapsed in 2001 as a result of corrosion, after only 12 years of service.

X-ray diffraction results were obtained from surface scrapings from mixes PC40, SR and SFC exposed to H_2S only and those exposed to both H_2S and sulphate after 23 cycles of exposure. The scrapings were taken above the level of the sulphate solution in the case of the latter three specimens. Scrapings from the surfaces of these six specimens contained gypsum, but no ettringite. Samples taken from the piers and beams over the manure pit of the barn that collapsed in 2001 as a result of corrosion, after only 12 years of service, showed very similar diffraction patterns; again gypsum was the predominant corrosion product. As gypsum is the expected corrosion product at sulphate concentrations over 8,000 ppm it is clear that very high concentrations of sulphate (over 8,000 ppm) do occur in parts of manure structures.

After 110 applications of H_2SO_4 over a period of approximately one year the PC50 experienced the greatest loss in mass of 5.2% of the original dry mass. The 5 specimens with a W/CM ratio of 0.4 had losses ranging from 4.0 to 4.7%. The least amount of mass loss was experienced by the sulphate resistant cement, SR. These preliminary results are not conclusive. What is evident from visual observation of the specimens that the limestone aggregate helps to reduce concrete deterioration from cement paste loss only and breakdown of the concrete structure. After a year of testing no pieces of coarse aggregate have become dislodged.

Conclusions and Recommendations

After 23 cycles of exposure to sulphate and/or sulphide over about 36 months, the electrochemical potential results and visual inspection of the reinforcing bars indicate that the PC concrete with 0.5 W/CM ratio (PC50) is the least resistant to steel corrosion, as might be expected. All treatments containing silica fume are performing much better than PC40. Treatment FAC, and FASF stand out as performing worse than plain PC concrete with the same W/CM ratio confirming the fact that Class C fly ash, most commonly available in Ontario, does not provide improvement in sulphate resistance. SR was more resistant than Type 10 Portland cement, in both sets of tests. In general the samples that were exposed to hydrogen sulphide and sulphate corroded at a higher rate then those exposed to the H_2S gas only.

Fig. 6a. Pictures of specimens after the 23rd cycle showing the concrete loss.

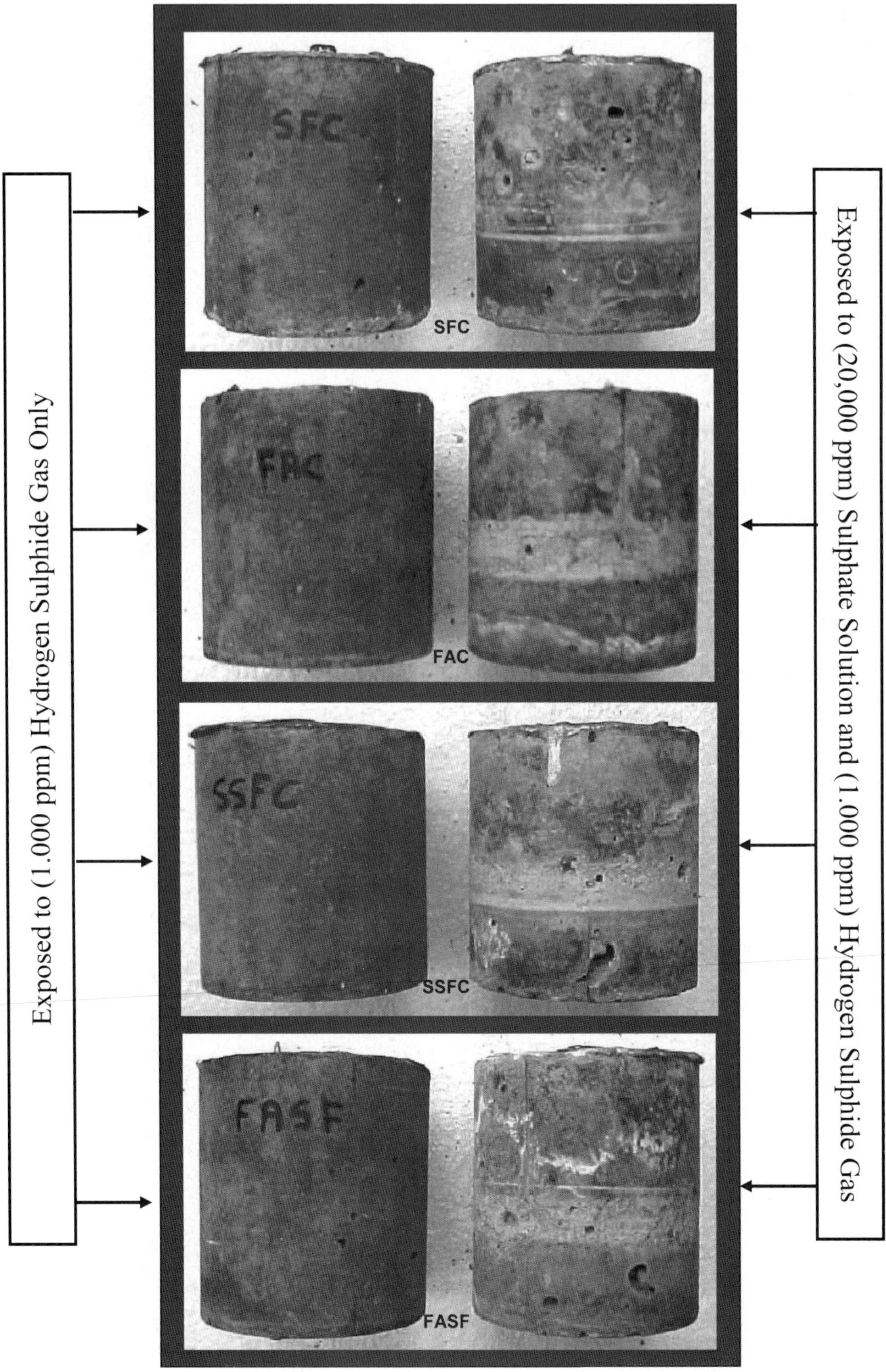

Fig. 6b. Pictures of specimens after the 23rd cycle showing the concrete loss.

The results of this study show that, in the sulphide/sulphate environment of manure storage and handling structures made of reinforced concrete, concrete corrosion is the critical factor, steel corrosion is less of a problem. The results indicate that the quality of concrete is the key to the best protection for the steel against sulphate and sulphite attack. It is relatively simple to produce such concrete by making an impermeable, well-cured concrete using a reasonably low W/CM ratio. However, the results also indicate that high Portland cement content creates a problem. Therefore, partial replacement of the Portland cement or the use of sulphate resistant cement will benefit the durability of the concrete of manure structures against the sulphates present in manure.

In summary, it is recommended that reinforced concrete in areas in a livestock building exposed to corrosive materials from manure should be specified to have:

1. high quality, with a maximum W/CM ratio of 0.45, 0.40 if feasible;
2. a maximum Portland cement content of 350 kg/m^3 in order to limit C_3A content;
3. 8% of the cementitious material to be silica fume, or 35% to be slag;
4. adequate (50 mm) concrete cover over the reinforcement

Acknowledgements

The authors acknowledge the financial support from the Natural Sciences and Engineering Research Council of Canada and the Ontario Region of the Cement Association of Canada for the research work that is reported in this paper. In addition, Lafarge Canada Inc. for all their technical assistance and the materials they provided for this project.

References

Canadian Standards Association. 1994. Concrete materials and methods of concrete construction, CAN3-A23.1-94 and CAN3-A23.2-94, Canadian Standards Association, Toronto, ON.

De Belie, N., De Coster, V., Van Nieuwenburg, D. 1997. Use of fly ash or silica fume to increase the resistance of concrete to feed acids, Magazine of Concrete Research, 49:181

De Belie, N., Monteny, J., and Vincke, E. 2002. The effect of chemical and biogenic sulphuric acid on the different types of concrete for sewer pipes, Proc. IVth Int. Symposium for Concrete for a Sustainable Agriculture - Agro-, Aqua and Community Applications, Ghent, Belgium.

Berge, E. and Vernhardsson, C. 2002. Durability of manure stores. Proc. IVth Int. Symposium for Concrete for a Sustainable Agriculture - Agro-, Aqua and Community Applications, Ghent, Belgium.

Frénay J. W., Zilverberg H. 1993. Durability of concrete in agricultural environment, IMAG-DLO Report 93-17, IMAG –DLO, Wageningen, Netherland.

Hewayde, Esam H. 2005. Investigation on degradation of concrete sewer pipes by sulfuric acid attack. PhD thesis, Dept. of Civil Engineering, Univ. of Western Ontario, London, ON.

Idriss, A.F. 2000. Corrosive effect of hydrogen sulphide on cement mortar. Ph.D thesis, School of Engineering, University of Guelph, Guelph, ON, Canada.

Jiang, S. 2002. Use of blast furnace slag cement for the durability of precast concrete in agricultural application. Proc. IVth Int. Symposium for Concrete for a Sustainable Agriculture - Agro-, Aqua and Community Applications, Ghent, Belgium.

Kosmatka, S.H., Kirkhoff, B., Panarese, W.C., MacLeod, N.F. and McGrath, R. 2002. Design and control of Concrete Mixtures, 7th Ed. Canadian Portland Cement Association, Ottawa.

Lafarge Canada Inc. 2001. Publication No 16.

Skalny, J., Marchand, J., and Odler, I. 2002. Sulphate attack on Concrete, Spon Press.

Svennerstedt, B., de Belie, N., Braam, C.R., Lenehan, J.J., Richardson, M., Snock, B. 1999. Durability of Building Materials and Components in Agricultural Environment, Report 119, Swedish University of Agricultural Sciences Department of Agricultural Biosystems and Technology; Sweden.

Tuthil, L. H. 1978. Resistance to chemical attack, ASTM STP-169-A, pp. 369-387.

SUMMARY OF DEVELOPMENT AND USE OF CSA 2004 SHEAR DESIGN PROVISIONS

Evan C. Bentz
Civil Engineering, University of Toronto, Canada
E-mail: bentz@ecf.utoronto.ca

Introduction

In the design of concrete structures, the engineer is tasked with selecting structural geometry, member sizes and reinforcement details to ensure that the applied loads to which the structure is subjected can be safely carried to the foundation. For design against axial load and moment there is essentially global unanimity concerning the relationships that should be used to achieve this design. Specifically, the well-known rules of engineering beam theory or "plane sections remain plane" as originally explained by Robert Hooke are used; see Figure 1 (Hooke, 1678). This simple, general and accurate theory allows the engineer to design a member with confidence even when confronted with unusual geometry or new materials.

In contrast to the agreement that exists for flexural design, the mechanisms of shear resistance and provisions based on these mechanisms have not yet reached international consensus. In some jurisdictions purely empirical methods are used while others use more theoretically derived methods of various complexity and understandability. While the use of different methods in different places is not of direct concern, that these different methods produce transverse reinforcement requirements that vary by as much as a factor of six suggests that engineers' understanding of shear is far behind their understanding of flexure.

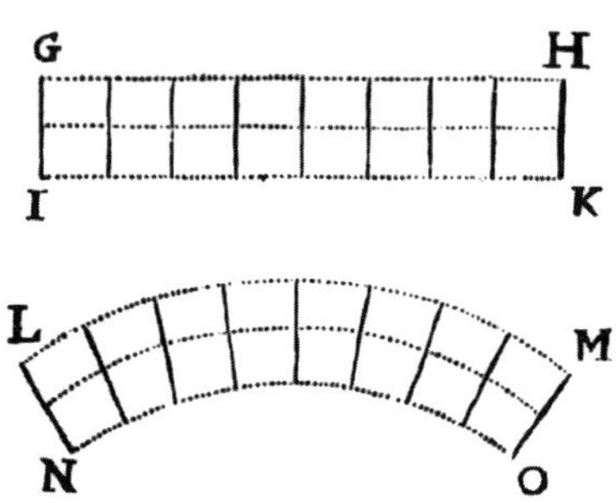

Fig. 1. Robert Hooke's figure demonstrating that plane sections remain plane published in 1678.

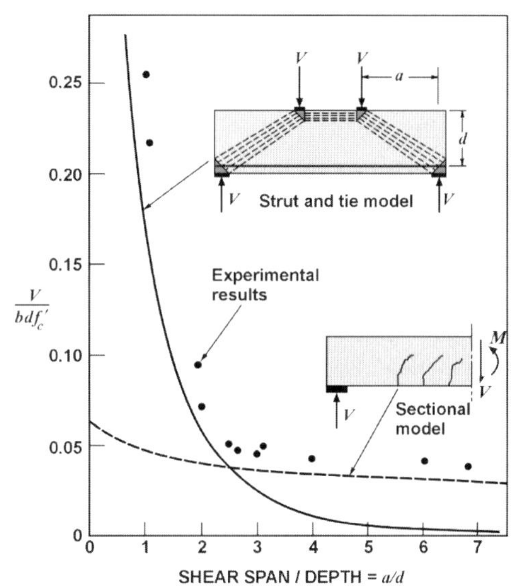

Fig. 2. Different modes of shear failure modelled in Canadian code.

M. Pandey et al. (eds), Advances in Engineering Structures, Mechanics & Construction, 67–80.

As shown in Figure 2, shear forces in beams are modelled in the Canadian code (CSA, 2004) with either a strut-and-tie model, suitable for short shear spans or a sectional model suitable for longer shear spans. As the strut-and-tie provisions of the 2004 concrete design code are largely unchanged from the 1994 standard, they will not be described further in this document. The sectional provisions of the 2004 concrete design code, on the other hand, have changed significantly and this paper will summarize the development and use of these provisions. In the author's opinion, the new shear provisions of the 2004 standard represent a breakthrough in the modelling of shear behaviour and should help foster international discussion towards the generation of a consensus on shear behaviour.

Historical Considerations

Up to the 1980s, the Canadian shear design provisions directly paralleled the US provisions of the American Concrete Institute (ACI) Building Code Requirements (ACI Committee 318, 2005). These provisions, which are still in use in the United States today, were empirically based on test results from a large number of shear experiments, generally of small size and with a large amount of longitudinal reinforcement.

Staring in 1984, the Canadian provisions (CSA, 1984) began to diverge from the ACI rules. Specifically, the Canadian provisions became divided into a "simplified method" and a "general method". The simplified method was essentially the same as the ACI provisions but the general method was a new set of design rules based on the Compression Field Theory (CFT) (Collins, 1978; Collins & Mitchell, 1980). This general method required the explicit evaluation and satisfaction of constitutive and compatibility equations. While the method was significantly more complex than the ACI provisions, it did apply to general loading conditions and would usually produce more economical designs. Unlike the ACI provisions, but like the then equally new CSA strut-and-tie provisions, the engineer needed to explicitly select the angle of diagonal compression in the web of a member. Because of this, different engineers would produce different designs for the same applied loading. While some engineers appreciated the freedom that this allowed, others were uncomfortable in that it became more difficult to check each others work; more than one solution to the same problem could be equally correct. An additional limit to the method was that it could not determine the strength of members without transverse reinforcement such as slabs.

In the 1994 CSA shear design provisions (CSA, 1994), a number of changes were made. For the simplified method provisions, a size effect was added to the shear strength of members without transverse reinforcement. This resulted from the experimental observation that members of a larger overall depth tended to fail at lower shear stresses, which was relevant for thick slabs for example. The 1994 general method of shear design was based on the Modified Compression Field Theory (MCFT) (Vecchio and Collins, 1986) and now provided a single design solution for a given set of applied loads. The use of the MCFT instead of the CFT also extended the general method to the design of members without transverse reinforcement, including the size effect. To use the new general method, however, the engineer needed to consult tables of values of β and θ, the two main variables for the shear strength of members. Partly due to the difficulty of using table-lookups with spreadsheets, the general method was still significantly more difficult to use than the simplified method, particularly for the analysis of existing designs. Despite this, the method has allowed many impressive structures to be built that exceed the limits of the ACI code. As just one example, Figure 3 shows a 176 metre tall tower in Toronto designed by Yolles Partnership which has high demands on the lateral load resisting system due to the narrow nature of the tower (a height to width ratio of 11:1). Were this tower designed according to the ACI shear provisions, it likely could not have been built as economically since significantly larger coupling beams would have been required by that standard. The necessary increase in floor-to-floor height would have lowered the sellable number of condominium units for the same height of building. Similar shear provisions to the CSA general method were also adopted by the Ontario and Canadian Highway Bridge design codes (OHBDC, 1993; CHBDC, 2000), and the US AASHTO LRFD Bridge Design Specifications (AASHTO, 2004).

Despite the advantages of the CSA 1994 general method, there remained difficulties with its practical use. As noted above, the difficulty of using tables of values with spreadsheets made the method difficult to automate for day-to-day design work. In addition, it was difficult for engineers to develop confidence in the values given by the tables of β and θ as it was not clear where these values came from or why they varied the way that they did.

For the generation of the 2004 shear design provisions, a conscious decision was made to make the provisions as general as the 1994 general method, but as easy to apply as the 1994 simplified method. The intention was to make design with the general method non-iterative and to replace the tables of behaviour with simple equations. As an intentional side effect of doing this, it became possible for the simplified method to be a special case of the general method rather than being based on the ACI provisions. The 2004 provisions remain based on the MCFT but are now simple enough to explain and use that they represent a significant advance in research on shear.

Fig. 3. CSA 1994 general method allowed this tower to be more economical.

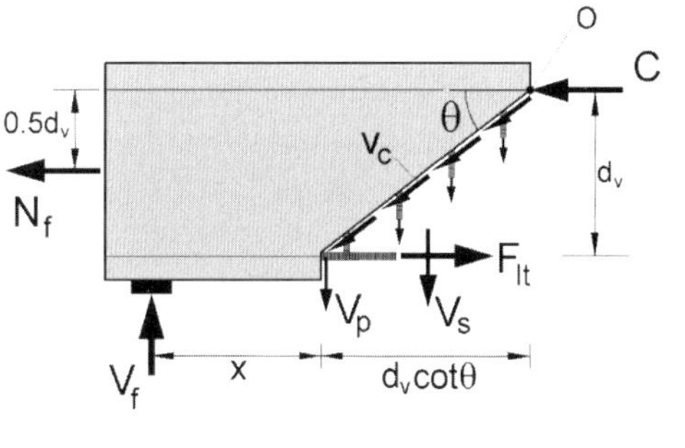

Fig. 4. Basic shear resisting mechanism assumed in 2004 code.

The Strength of Members Subjected to Shear

The 2004 CSA shear provisions are based on the shear resisting mechanism shown in Figure 4. This free body diagram shows the end portion of a beam, cuts through the longitudinal reinforcement and stirrups, follows the diagonal shear crack and then cuts the top flexural compression region. Flexural moments (M_f) and axial tension (N_f) are resisted by the force couple between the flexural compressive force C and the tension in the reinforcement F_{lt}, here calculated at the crack. Shear forces (V_f) are resisted by three mechanisms. Shear stresses (v_c) on the crack surface itself are resisted by aggregate interlock, which is the primary method of strength resistance observed for members without stirrups (Fenwick, 1968; Taylor, 1970; Kani, 1979). Stirrups that cross the diagonal crack provide a steel contribution, V_s, and any vertical component to prestressing forces, V_p, will also resist shear forces. Even from such a simple diagram as that in Figure 4, useful deductions may still be made. In this case, note that the shear on the crack, v_c, has a vertical component that resists shear, but also has a horizontal component that must be balanced by additional strain in the longitudinal reinforcement. This is one of the causes of shear-moment interaction in that shear force cause stress in the flexural reinforcement.

If the contribution due to V_p is ignored, the following equations are the basic shear strength terms used in the CSA code for both the general and simplified methods:

$$\begin{aligned} V_r &= V_c + V_s \leq 0.25\phi_c f_c' b_w d_v, \\ V_r &= \phi_c \lambda \beta \sqrt{f_c'} \cdot b_w d_v + \phi_s \frac{A_v}{s} f_y d_v \cot\theta, \end{aligned} \tag{1}$$

where V_r is the factored shear strength, $b_w d_v$ is the shear area taken as the web width multiplied by the shear depth shown in Figure 4 and taken simply as $0.9d$. The term λ relates to the use of lightweight concrete. The equation is written in metric notation (MPa, mm, Newtons), with $\phi_c = 0.65$ and $\phi_s = 0.85$ as the material reduction factors for concrete and steel. The ability of the member to resist aggregate interlock stresses is represented by the variable β, and the angle of the crack, θ, indicates how many stirrup legs will cross the crack, and is necessary to determine V_s.

Development of shear provisions can therefore be boiled down to finding a method to quantify the parameters β and θ. The 1994 provisions used tables to estimate these values, but the 2004 code uses simple equations to do this. As these two variables are conceptually independent, they will be treated in separate sections.

Modelling Aggregate Interlock to Determine β

Aggregate interlock is assumed to carry all shear forces for members without stirrups, and a potentially significant proportion of the total shear force for members with stirrups. The ability of a crack to resist these stresses is predicted by the MCFT to decrease with decreasing concrete strength; decreasing maximum specified coarse aggregate size, representing crack roughness; and increasing absolute crack width. The suggestion that wider cracks are less able to resist sliding shear stresses is hopefully intuitive. To estimate a crack width, w, the following relationship can be used:

$$w = \varepsilon \cdot s, \tag{2}$$

where s is the crack spacing and ε is the average strain perpendicular to the crack. As wider cracks will be associated with lower aggregate interlock strength, anything that increases the value of w can be expected to result in decreased shear capacity. Thus if the crack spacing increases due to, say, the construction of a larger member, it can be expected that the shear strength will decrease. This is called the size effect and is an important part of the behavior of members without stirrups. Secondly, if the average strain in the concrete increases due to, say, applied tension then the shear strength is also predicted to drop. This is called the strain effect in shear and is less well known than the size effect, though it is of comparable importance.

Fig. 5. Crack pattern in member without stirrups. s_z is spacing at mid-depth.

Overall, then, a size effect and a strain effect are predicted to be important aspects of the concrete component of shear strength of members. Experiments show that these are indeed the two most important aspects influencing shear stress at failure and should be included in any state of the art shear provisions.

The full derivation of the proposed method is given elsewhere (Bentz, 2006; Bentz et al., 2006), but the important concepts are explained here. Determining shear strength will depend on the terms s and ε in Equation (2). The value of the crack spacing will depend largely on the size of the member. The crack spacing in the longitudinal direction, s_z, is taken as $s_z = jd = 0.9d$ if no stirrups are provided. If the member is constructed with an aggregate size, a_g, different from 20 mm, the aggregate interlock capacity will be affected, and this is accounted for by using an effective crack spacing, s_{ze}, given by:

$$s_{ze} = \frac{35d_v}{15 + a_g} \geq 0.85d_v. \tag{3}$$

For high strength concrete, the aggregate fractures and does not contribute to crack roughness. To account for this, take $a_g = 0$ for $f'_c > 70$ MPa. To avoid a discontinuity, linearly interpolate a_g from the specified value at $f'_c = 60$ MPa down to zero at $f'_c = 70$ MPa. For members with stirrups, the stirrups will control the crack spacing and the term s_{ze} may be simply taken as 300 mm.

Figure 5 shows the test of a 300 mm wide strip taken from a 1500 mm (5 foot) thick slab. It can be seen that the spacing of the cracks at the mid-depth of the member is much greater than the spacing of the cracks at the flexural tension face. The parameter s_z refers to the longitudinal spacing of the cracks at mid-depth of the member where the shear stress is generally critical.

The value of ε is slightly more complex to determine compared to the effective crack spacing as it depends on the currently applied load level, amount of prestress, material properties of the flexural reinforcement, etc. Consider that a given amount of applied load will be associated with a given strain in the longitudinal reinforcement based on a free body diagram such as that in Figure 4. This value of strain will be present in the reinforcement, whereas Equation (2) requires the strain at 90° to the diagonal crack. As such, the equations of the MCFT are employed to derive a relationship between the width of a diagonal crack, for a given crack spacing, at shear failure given that the longitudinal strain is a known quantity. This involves the simultaneous solution of 15 nonlinear equations and is described elsewhere (Bentz, 2006).

Figure 6 shows the results of the calculation of diagonal crack width for various longitudinal strains by the MCFT. As the longitudinal strain increases, the critical crack width also increases. The analysis results are nonlinear, but a simplified equation is also shown that conservatively approximates the nonlinear behaviour. This equation is intentionally selected to provide a good match to the MCFT for mid-depth strains that are expected for a member reinforced with 400 MPa flexural

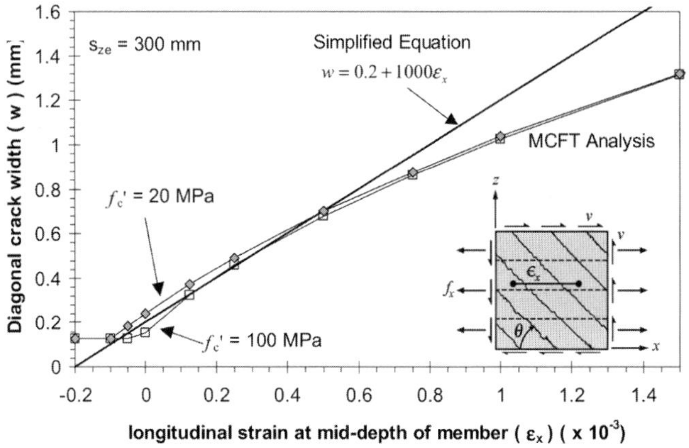

Fig. 6. Diagonal crack widths at shear failure for members without stirrups.

reinforcement. If a member is to be subjected to larger longitudinal strains, say with FRP reinforcement, these provisions should be conservative as the crack width will be overestimated by the simplified equation.

When the simplified equation in the figure is substituted into the MCFT equation for aggregate interlock, and a size effect terms is also added (Bentz, 2006), the following equation is obtained for the value of β which defines the concrete contribution based on the MCFT:

$$\beta = \frac{0.40}{(1 + 1500\varepsilon_x)} \cdot \frac{1300}{(1000 + s_{ze})}. \tag{4}$$

The first term in this equation accounts for the strain effect whereby members with smaller longitudinal strains are stronger in shear. The second term accounts for the size effect which cancels out for members with stirrups (s_{ze} = 300 mm). The estimation of the longitudinal strain in the member, ε_x, will be discussed below.

Modelling of Diagonal Crushing of Concrete to Determine θ

Members with at least a minimum quantity of well-anchored stirrups are predicted not to fail by sliding on the crack, but by yielding of the stirrups and eventual crushing of the concrete in the web. In the 1984 shear provisions which allowed the engineer to select the value of θ, equations were provided to ensure that the concrete did not crush before reaching the design shear strength, which provided a lower limit on θ, and additional rules were provided to ensure that the stirrups would yield at design shear failure, which provided an upper limit on θ. At low applied shear forces, the range over which the value of θ could be selected was large, but as the shear stress increased, the range became more restrictive. For the 2004 shear provisions, it was decided to maintain the maximum shear limits that were present in the 1994 standard, so this high shear loading would control the selection of θ for all applied loading levels.

Figure 7 shows the limits on allowable angle of principal compression, θ, based on the MCFT for members heavily loaded in shear and for different strains in the member at mid-depth (ε_x). As can be seen, the range of allowable angles to select from at this high shear loading is rather narrow. Members designed based on angles in the upper shaded region would be expected to fail in shear before yielding of the transverse steel making the use of Equation (1) unconservative. Members designed based on angles from the lower shaded region would also be unconservative as here the

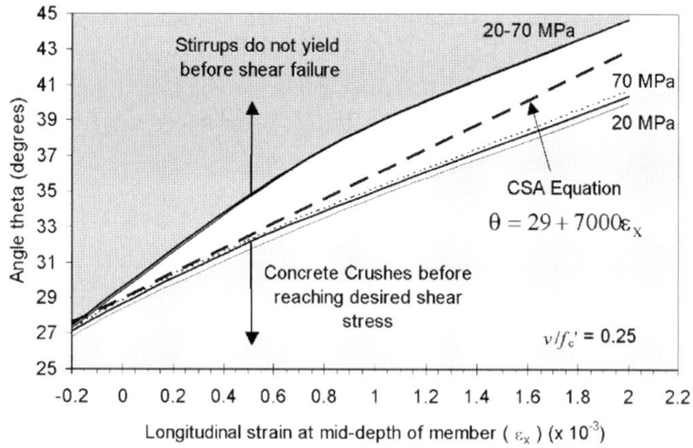

Fig. 7. Selection of equation for θ.

member is predicted to fail by crushing of the concrete in diagonal compression before achieving the design shear strength. Only within the unshaded region would a member as heavily loaded as this be predicted to be able to resist the applied shear force. Shown in Figure 7 is a simple equation that lies within the allowable range as:

$$\theta = 29^\circ + 7000\varepsilon_x. \tag{5}$$

Longitudinal Strain in Member

Embedded in Equations (4) and (5) is the strain term ε_x which represents the average longitudinal strain in the member at the mid-depth. Figure 5 shows that the crack spacing, s_z, is calculated in the longitudinal direction of the member at mid-depth as well as it is appropriate to calculate behaviour based on coexisting strains and crack spacings. For reinforced concrete members subjected only to moment and shear, the equation to determine ε_x can be generated from a simple free body diagram such as that shown in Figure 4 and can be taken as:

$$\varepsilon_x = \frac{M_f/d_v + V_f}{2E_s A_s}. \tag{6}$$

In the use of this equation, the values of moment and shear are always taken as positive. The numerator of the equation estimates the force that must be resisted in the flexural reinforcement of a member including the shear-moment interaction mentioned above. The denominator converts this force to a strain in the tension reinforcement by dividing by the quantity and stiffness of flexural reinforcement. As a final simplification, the strain at mid-depth of the member, ε_x, is simply taken as one half of the value in the flexural tension reinforcement. For prestressed concrete or members subjected to axial loads, additional terms are added to the numerator of Equation (6) which are not shown here.

Checking for Flexure-Shear Failures

As a final check that a given structural member is safe, it is necessary to ensure that the flexural reinforcement does not yield under the combined effect of applied loading including the interaction of shear and moment. The member will have been designed against flexural forces, but the addition of the shear term in the numerator of Equation (6) means that it may still be dominated by yield of the longitudinal reinforcement. To ensure that this does not occur, the following equation is derived from Figure 4 by taking moments about point O and solving for the necessary force in the reinforcement:

$$F_{lt} = \frac{M_f}{0.9d} + 0.5N_f + (V_f - 0.5V_s - V_p)\cot\theta. \tag{7}$$

The reinforcement that is provided at the location where this equation is being checked must be sufficiently developed to resist the force calculated by Equation (7). This force need not be taken as greater than the reinforcement force calculated under maximum moment alone at the maximum moment location of the same member. The reason for this limit is that near applied loads, where the moment is maximum, the shear is carried by a "compression fan", similar to a strut-and-tie model rather than by the sectional behaviour assumed in Figure 4.

If Equation (7) is not satisfied, then the applied loads (M_f, N_f, V_f), can be proportionately reduced until it is satisfied, and these new values are the final estimate of the strength of the member. A member which is controlled by this equation is predicted to fail in a flexure-shear mode rather than simply in flexure or shear alone.

Solving for the Strength of Members by the CSA General Method

Overall, then, solving for the shear strength of a member by the 2004 CSA shear equations require the solution of Equation (1) using the value of β from Equation (4) and θ from Equation (5). Equation (7) must also be checked at the end of the process to ensure that a flexure-shear case does not control. These calculations are all performed at the critical shear location, which is generally taken as d_v back from the edge of the column, pedestal or wall that applies the load. For uniformly loaded members, multiple sections along the span should be checked to determine where the critical location is. To use either Equation (4) or Equation (5), however, the value of ε_x must be determined. A number of different cases exist for performing the estimate of ε_x which differ depending on if they represent a design or analysis case:

Design by General Method

For the design of members to resist a known combination of M_f, and V_f, it is appropriate to substitute these values directly into Equation (6) and explicitly determine the value of ε_x. This value is then substituted into Equation (5) to determine θ. If it is not clear whether stirrups are required, the shear strength may then be determined without stirrups first by substitution of ε_x into Equation (4) with the term $s_{ze} = d_v$ to see whether the calculated value of $V_r = V_c$ is greater than or equal to the factored load V_f. If the calculated V_c is insufficient, then the engineer may conservatively use this with Equation (1) and solve for the quantity of stirrups (A_v/s) which is necessary so that $V_r \geq V_f$.

For more accurate results when stirrups are required, the engineer should recalculate V_c as the presence of at least minimum stirrups mitigates the size effect. Equation (4) should therefore be reevaluated with $s_{ze} = 300$ mm to determine the value of V_c for members with stirrups. Then Equation (1) is again used to determine the required quantity of stirrups (A_v/s) so that $V_r = V_f$.

Equation (7) will generally only control near the ends of members where it is necessary to check the anchorage of flexural reinforcement. If the reinforcement detailing is sufficient this check will usually not control here or elsewhere in the member.

Overall, the process above is non-iterative and easy to perform either by hand or with a spreadsheet. The steps required are identical to those used with the 1994 general method, but now there is no iteration required for design and the use of equations for β and θ makes the design much easier to perform, particularly with a spreadsheet.

Design by Simplified Method

As was noted above, the general method and the simplified method in the 2004 CSA code are in fact now the same thing. The key to the simplified method is the appreciation that the value of the longitudinal strain at the mid-depth of the beam, ε_x, is in fact a "constrained" parameter for shear failures. Consider that for 400 MPa steel the flexural reinforcement will yield and cause a flexural failure rather than a shear failure when the strain in this steel equals $f_y/E_s = 400/200{,}000 = 0.002$. With the strain at mid-depth taken as one half of the strain in the bottom chord of the member for simplicity, the ε_x strain associated with flexural yield will equal 0.001, or 1×10^{-3} for this reinforcement. Thus for 400 MPa steel, if the strain ε_x exceeds 1×10^{-3}, the member will not fail in shear, but will fail in flexure and none of the equations in this paper will govern.

The simplified method is derived by assuming that the strain term, ε_x, is equal to 0.85×10^{-3}, or slightly less than the value associated with the yield of 400 MPa reinforcement. Substituting this value into Equation (5) produces a constant angle $\theta = 35°$. Substituting ε_x into Equation (4) will produce a constant value of β for members with stirrups, and a size effect equation for members without stirrups. Thus the shear strength of members with stirrups can be determined by the simplified method as:

$$V_r = 0.18 \cdot \phi_c \lambda \sqrt{f_c'} \cdot b_w d_v + 1.43 \phi_s \frac{A_v}{s} f_y d_v, \tag{8}$$

which is no more complicated than the traditional ACI shear strength provisions. The major difference from the ACI provisions is that the V_c term is slightly lower and the beneficial effect of adding stirrups is estimated to be higher in Canada. As the Canadian provisions allow the design of members without stirrups up to the full value of V_c whereas the ACI code requires stirrups for members loaded in excess of $V_c/2$, these differences seem acceptable.

For members without stirrups, the same assumptions about the value of ε_x are made and the final version of Equation (1) with all simplifications reduces to:

$$V_r = \frac{230}{1000 + d_v} \cdot \phi_c \lambda \sqrt{f_c'} \cdot b_w d_v. \tag{9}$$

This equation indicates that when stirrups are not provided that a size effect remains which must be taken into consideration at design time. At the same time, similar to the 1994 code, this equation is not iterative as all the terms in this equation will be known once the flexural design has been completed.

A significant benefit of basing the simplified method on the general method is that it becomes clear what to do as the limits of the simplified method are exceeded. As an example, Equations (8) and (9) are only appropriate when used with 400 MPa reinforcement. For designs based on, for example, 500 MPa reinforcement, the code specifies that Equations (8) and (9) may not be used. In this situation, however, it is clear how to generate new simplified equations for use with this higher strength steel. For this case, the strain at mid-depth associated with flexural yield will be $f_y/2E_s = 1.25 \times 10^{-3}$. Taking the same factor of 85% of this yield strain produces a value of $\varepsilon_x = 1.06 \times 10^{-3}$ for use with 500 MPa reinforcement. This too can be substituted into the previous equations to produce safe equations for use with this steel. The value of θ, for example, would be taken as 36.4°, and the numerator in Equation (9) would be replaced by the value of 200 rather than 230. Design by this newly derived simplified method could then proceed. Note that the use of this technique implies that members with high strength reinforcement will be weaker in shear than members with normal strength reinforcement. This is not necessarily the case, however. What is predicted is that when higher strength reinforcement is used, that the shear strength will be lower at flexural failure and so this lower, safe value must be used for all designs if a simplified type equation is desired.

Analysis by the General Method

The concepts of design and analysis can be thought of as opposite circumstances. In the design case the geometry of a member is selected to make it strong enough which means the engineer solves for many of the terms on the right side of Equation (1). For analysis, the engineer solves for the one term on the left side of Equation (1) instead. The general method equations have been intentionally arranged so that they produce design results, as above, with no iteration. That is, the equations have been selected so that they make design easy, but analysis slightly more difficult. This was a philosophical decision as the design code is intended for the construction of new structures.

To solve for the strength of a member by the CSA general method, two options are available. Firstly, for members without stirrups Equations (6), (4), and (1) may be solved simultaneously to produce a closed form solution for the shear strength of a member without stirrups. This derivation is reasonably easy and requires the solution of a quadratic equation.

Alternatively, an iterative process, suitable for spreadsheet calculations can be easily implemented which works for all types of applied loading and all types of members. The process can be explained as follows. The spreadsheet will allow the determination of the shear strength of one beam/slab/column per row of the spreadsheet. The left side of the spreadsheet would include the various material, geometric, and loading properties, again, one beam per row. On the right side of the spreadsheet would be the iterative determination of the shear strength.

Considering the small sample spreadsheet below, the iteration begins with an initial estimate of the strain at the mid-depth of the member, say $\varepsilon_{x-1} = 1 \times 10^{-3}$. For convenience, it is helpful to determine the values of strains in the units of parts per thousand to make mistakes easier to visually identify. Directly to the right of this value, Equations (1), (4), and (5) would be evaluated within a single spreadsheet cell to determine the shear strength for that assumed value of ε_x, listed as V_{r-1} below. This is listed as Equation type B and as the value ε_x is known, it is a direct substitution of values from the left side of the spreadsheet. To the right of this value, Equation (6) would then be evaluated to determine the new estimate of ε_x from the given applied loading, shown as ε_{x-2}. For the sake of numerical stability, it is appropriate to take the new estimate of ε_x as the average of the previous estimate and the newly calculated value, or $\varepsilon_{x-2} = (\varepsilon_{x-1} + \text{Equation (6)})/2$. To avoid any manual iteration actions by the user, the cells which contains the estimates of the shear strength and ε_x terms (Equations B, C) can be copied perhaps 10 times to the right to provide 10 iterations which is usually more than sufficient to converge. The final spreadsheet would look something like this:

b_w	d	rho	etc.	ε_{x-1}	V_{r-1}	ε_{x-2}	V_{r-2}	ε_{x-3}	V_{r-3}
300	925	0.83	etc.	1	200	1.2	183	1.24	177
Equation type				A	B	C	B	C	B

where Equation A is the constant initial guess of 1.0 part per thousand; Equation B is the direct substitution into Equations (1), (4), and (5); and Equation C is Equation (6) in this paper. The final value of the shear strength, perhaps V_{r-10}, can be copied back near the left side of the spreadsheet for easy access. Using this technique, the spreadsheet is iterative but is fully "live" in that no macros or special routines need to be called. When a value is changed on the constitutive properties on the left of the spreadsheet, the iterative equations will automatically update the predicted shear strength. That the spreadsheet requires no macros is curiously powerful as well in that nothing is hidden from the engineer with this spreadsheet: no "black-box" calculations or interpolation routes are present. From experience, engineers quickly become much more confident with the new general method partly due to this transparency.

The method above is specified in such a way that the equations (A, B, C, etc.) can be copied down over multiple rows of the spreadsheet to determine the strength of multiple members in a building, multiple experimental results, or to determine sensitivity plots by systematically varying the various beam properties on the left side of the spreadsheet.

Discussion is provided below concerning the predictions of the general method compared to experimental test results.

Analysis by the Simplified Method

An alternative to estimating beam strengths with the general method is to use the simplified method. This alternative is not recommended, however, as it will ignore the important consequences of the strain effect. Simple strength predictions can be made with the simplified method and these will generally be safe, but as the strain effect is ignored and only the size effect is considered, poor results should be expected, particularly in comparison to statistical databases.

Experimental Support for the CSA Shear Provisions

Extensive comparisons to experimental results have been made for the CSA shear design provisions and only a very small subset of these will be shown here. For members without transverse reinforcement, two main variables are predicted to control behaviour: the size effect and the strain effect. To demonstrate the quality of the predictions on a graph with only a single independent variable, Equation (4) will be used with β as the experimentally measured strength and both sides of the equation divided by either the strain effect term or the size effect term. Thus a 2D plot can be made showing the influence of one variable at a time. When results are normalized by the strain effect, the strain used is that predicted by the CSA provisions rather than that calculated from the experimental results.

Figure 8 shows 411 members of various depths and aggregate sizes without stirrups to demonstrate the size effect and the quality of Equation (4) at predicting this effect. The vertical axis shows the value of β in psi units which are 12 times that given by Equation (4). As seen, the CSA equations do an excellent job and generally provide a conservative estimate of shear strength across the range of depths. Recall that these provisions were not based on curve fits to experimental beam test results, but were based on the MCFT which itself resulted from more fundamental shear and material tests. The line on the plot is therefore a true prediction and not a "post-diction" or curve fit result. In addition, the experiments upon which the MCFT is based had values of the s_{ze} parameter that were generally in the range of 50 to 100 mm. The plot therefore shows a significant extrapolation from the data originally used to calibrate the MCFT. That the prediction is as good as the figure shows confirms both that the MCFT is a good model for the behaviour of cracked reinforced concrete and that the assumptions used in generating Equation (4) are appropriate.

Figure 9 shows the same data but normalized by the size term to highlight the strain effect. Again, to calculate this, the experimentally observed value of β was divided by the size effect term so that the predicted behaviour would only vary by one independent variable and a 2D plot could be generated. As with Figure 8, the overall behaviour of the strain effect is seen to be well modelled by Equation (4) for members without stirrups. As expected from the derivation in Figure 6, as the strain is increased beyond a value of about 1.0×10^{-3}, the method begins to become more conservative as the crack width at failure is overestimated. The strain effect explains why FRP reinforced members have lower shear strengths than members reinforced with steel. Such members generally have a

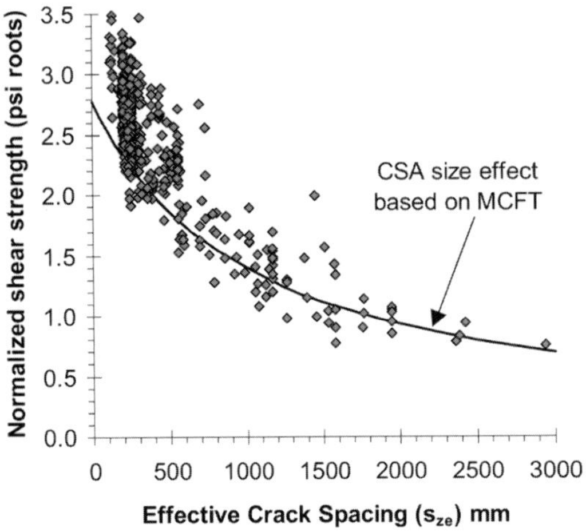

Fig. 8. The size effect in shear.

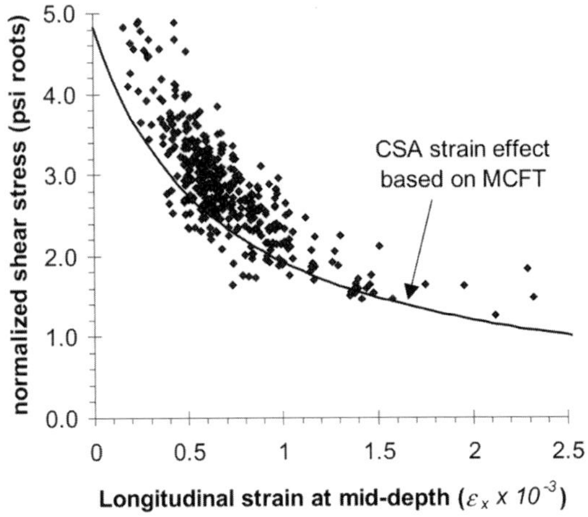

Fig. 9. The strain effect in shear.

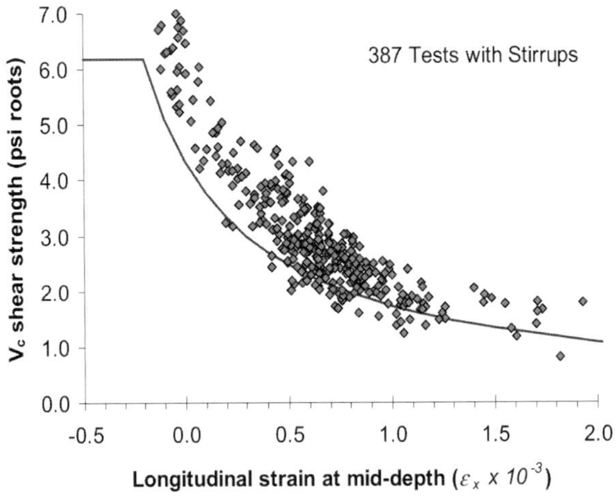

Fig. 10. The strain effect for members with stirrups .

lower stiffness of flexural reinforcement and are therefore subjected to higher strains at shear failure. Due to the strain effect, these members will therefore be weaker in shear. Members reinforced with cast in place FRP are modelled conservatively by the general method shear equations of the 2004 A23.3 concrete code.

As a final comparison, Figure 10 shows a plot of the strain effect for members with transverse reinforcement. This plot includes members of various depths, with and without axial loads and with or without prestressing. The effect of axial compression or prestress is to lower the value of the strain term ε_x and in this figure it can be seen that shear strength is modelled well even for negative values of ε_x. It may not be intuitively obvious that the strain effect should work just as well for members with stirrups, which can undergo redistribution of shear stresses as it does for the members in Figure 9. As seen in Figure 10, however, the effect applies equally well to all members when a realistic estimate of V_s is made as with Equation (5). This figure also shows why the simplified method (which assumes $\varepsilon_x = 0.85 \times 10^{-3}$) will produce poor statistical fits to databases: the systematic effect of strain will be ignored when using the simplified method to predict member strengths.

Conclusions

This paper has provided a brief description of the derivation of the CSA shear design provisions of the 2004 CSA Design of Concrete Structures code. The code continues to employ a simplified and a general method, but now the general method is based on simple equations derived from the MCFT rather than tables derived from the MCFT. Because of this, the simplified method is now a clear derivation from the general method. The use of equations means that the general method is now much easier to apply, particularly with spreadsheets. Design with the general method is now a non-iterative process. Analysis with the general method requires iteration for one variable, and a process for implementing this into a spreadsheet with no programming or macros is provided. The new equations are shown to work well for experimental predictions.

In the generation of the 2004 general method, it was expected that the simplifications necessary to make practical equations would come at the cost of a significant reduction in accuracy or generality. It has been a pleasant surprise to find that the resulting simplified MCFT equations, while much easier to apply, have essentially the same ability to accurately predict shear strength for a very wide range of different parameters. It is believed that the new equations presented in this paper represent a breakthrough in the understanding and modelling of shear. It is hoped that these equations can encourage the necessary discussion to move towards international consensus on shear behaviour just as it currently exists for flexural behaviour.

References

AASHTO, 2004, LRFD Bridge Design Specifications and Commentary, Third Edn., American Association of State Highway Transportation Officials, Washington, D.C.

Bentz, E.C., 2006, Development of the 2004 CSA A23.3 Shear Provisions for Reinforced Concrete, Canadian Journal of Civil Engineering, in press, February 2006.

Bentz, E.C., Vecchio, F.J., and Collins, M.P., 2006, The Simplified MCFT for Calculating the Shear Strength of Reinforced Concrete Elements, ACI Structural Journal, in press, February 2006.

ACI Committee 318, 2005, Building Code Requirements for Reinforced Concrete (ACI 318-05) and Commentary ACI 318 R-05. American Concrete Institute, Detroit.

Canadian Standards Association, 1984, CAN CSA A23.3-M84 Design of Concrete Structures for Buildings. CSA, Rexdale, Ontario.

Canadian Standards Association, 1994, CAN CSA A23.3-94 Design of Concrete Structures. CSA, Rexdale, Ontario.

Canadian Standards Association, 2000, S6-00: Canadian Highway Bridge Design Code (CHBDC). CSA, Rexdale, Ontario, Canada.

Canadian Standards Association, 2004, CAN CSA A23.3-04 Design of Concrete Structures. CSA, Rexdale, Ontario.

Collins, M.P., 1978, Towards a Rational Theory for RC Members in Shear, Journal of the Structural Division, ASCE, Vol. 104, pp. 649–666.

Collins, M.P. and Mitchell, D., 1980, Shear and Torsion Design of Prestressed and Non-Prestressed Concrete Beams, PCI Journal, Vol. 25, No 5, pp. 32–100.

Fenwick, R.C. and Paulay, T., 1968, Mechanisms of Shear Resistance of Concrete Beams, Journal of the Structural Division, ASCE, Vol. 94, No. ST10, pp. 2235–2250.

Hooke, R., 1678, Lectures de potentia restitutiva, or of spring explaining the power of springing bodies. Printed for John Martyn printer to the Royal Society, Bell in St. Paul's church-yard, 24 pp.

Kani, M.W., Huggins, M.W., and Wittkopp, R.R., 1979, Kani on Shear in Reinforced Concrete, University of Toronto Press, Toronto.

Ministry of Transportation, Ontario, 1993, Ontario Highway Bridge Design Code (OHBDC). MTO, Downsview, Ontario, Canada

Taylor, H.P.J., 1970, Investigation of Forces Carried Across Cracks in Reinforced Concrete Beams in Shear by Interlock of Aggregate, TRA 42.447, Cement and Concrete Association, London, 22 pp.

Vecchio, F.J. and Collins, M.P., 1986, The Modified Compression Field Theory for Reinforced Concrete Elements Subjected to Shear, ACI Journal, Vol. 83, No. 2, pp. 219–231.

DAMAGE ASSESSMENT OF REINFORCED CONCRETE BRIDGE DECKS USING TAM NETWORK

Hitoshi Furuta[1], **Hiroshi Hattori**[2] **and Dan M. Frangopol**[3]
[1]*Department of Informatics, Kansai University, Takatsuki 569-1095, Japan*
E-mail: furuta@res.kutc.kansai-u.ac.jp
[2]*Graduate School of Informatics, Kansai University, Takatsuki 569-1095, Japan*
[3]*Department of Civil, Environmental, and Architectural Engineering, University of Colorado, Boulder, CO 80309-0428, USA*

Abstract

In order to establish a rational management program for bridge structures, it is necessary to evaluate the structural damage of existing bridges in a quantitative manner. However, it is difficult to avoid the subjectivity of inspectors when visual data are used for the evaluation of damage or deterioration. In this paper, an attempt is made to develop an optimal bridge maintenance system by using a health monitoring technique. The damage of Reinforced Concrete (RC) bridge decks is evaluated with the aid of digital photos and pattern recognition. So far, neural networks have been applied to judge the damage state of RC bridge decks. However, there are still some problems that learning data are not enough and recognition accuracy is not satisfactory. In order to solve these problems, TAM network is applied here, which is an optical system. Though the numerical examples using actual data, it is shown that the recognition rate is increased.

Keywords: damage assessment, RC slab, pattern recognition, TAM network

Introduction

In order to establish a rational management program for bridge structures, it is necessary to collect enough data about the material and structural characteristics and to evaluate the structural damage of existing bridges in a quantitative manner. However, it is often seen to lose the drawings or design specifications. Moreover, it is difficult to avoid the subjectivity of inspectors when visual data are used for the evaluation of damage or deterioration. In this paper, an attempt is made to develop a new system that can evaluate the damage condition of existing structures by using the visual information given by digital photos (Furuta et al., 2004a). The proposed system is based upon such new technologies as image processing, photogrammetry, pattern recognition, and artificial intelligence (Furuta et al., 2004b). The damage of Reinforced Concrete (RC) bridge decks is evaluated with the aid of digital photos and pattern recognition. Using the proposed system, it is possible to automatically evaluate the damage degree of RC bridge decks and therefore avoid the subjectivity of inspectors. Several numerical examples are presented to demonstrate the applicability of the proposed system.

Damage Evaluation of RC Deck by Pattern Recognition

In this study, the damage of Reinforced Concrete (RC) bridge decks is evaluated with the aid of digital photos and pattern recognition (Gonzales and Woods, 2002; Yagi, 2000). In general, the procedure for extracting the characteristics of cracks showing up on concrete decks through digital images and classification based on the damage levels are used in the typical pattern recognition system. Figure 1 shows the procedure of such a pattern recognition system. First, the input data to

M. Pandey et al. (eds), Advances in Engineering Structures, Mechanics & Construction, 81–86

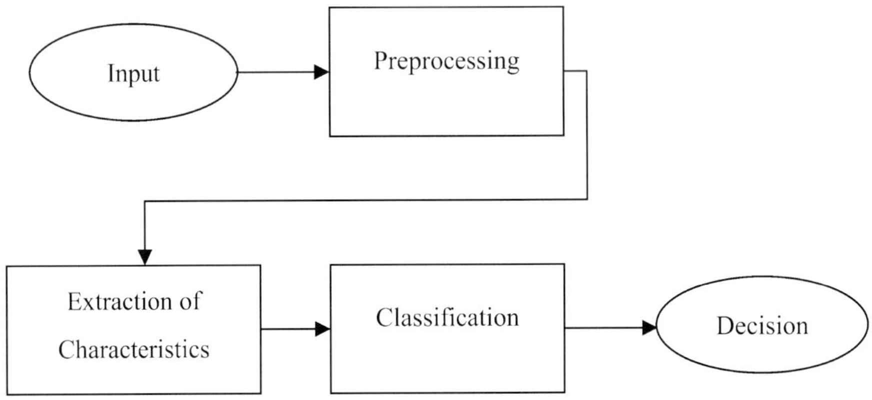

Fig. 1. The procedure of pattern recognition system.

this system consists of the digital images of the concrete decks taken by a digital camera. Next, the linear pattern of cracks is extracted from the digital images of the concrete decks through image processing techniques. Moreover, the characteristics of cracks such as the projection histograms are extracted. Finally, the digital images of cracks are classified into different damage levels based on the extracted characteristics through the TAM (Topographic Attentive Mapping) network (Seul, 2001).

To obtain the test material, digital images of concrete decks taken by a digital camera are used. If input data that can be acquired in low resolution and by using common digital camera is used, the costs for the assessment of integrity can be reduced and input data can be acquired easily. The total number of digital images is 47 and each image is scanned with the resolution of 360 pixels per inch in both directions. In this resolution, each image is normalized to the 768×480 pixel rectangle and converted to greyscale image. The digital images used in this study are obtained by marking the cracks with white chalk. The damage levels for all digital images are classified into three categories by an expert. Some examples for each damage level are shown in Figure 2.

Cracks in a digital image of concrete deck are detected in accordance with the following procedure: First, the digital image undergoes geometric transformation to extract a rectangular part containing a crack zone. The binarization is a method for transforming greyscale image pixels into either black or white pixels by selecting a threshold. Because the crack zone existed in only a small part of digital image, and also the brightness is not uniform through the crack zone due to the uneven lighting, the extracted rectangular part is divided into smaller blocks. The method proposed by Ohtsu (Yagi, 2000) is applied to the block unit to determine the threshold for binary-coding processing. Then, each block is divided into sub-blocks and the binary-coding processing is applied to each sub-block. These binary images are reduced some noise such as spots and holes after the binary-coding processing. The aim of thinning processing is to reduce the crack zone pixels to lines one pixel width. The crack pattern can be easily recognized by such a thinning processing. Finally, the smoothing processing such as the reduction of insufficient points and the addition of missing line is implemented. After all of these processing, the crack pattern is obtained and used for extracting the characteristics of digital images. The procedure is shown in Figure 3.

In this study, characteristics are extracted based on four criteria; continuity, concentration, directionality (unidirectional or bi-directional), and types (hexagonal or linear) of cracks. The crack pattern of thin lines can be considered a set of directional linear elements and hence characteristics extraction by the projection histogram (Seul, 2001; Sakai, 2002; Duda et al., 2001) would be effective. Because the characteristics of projection histogram of a crack pattern provide information

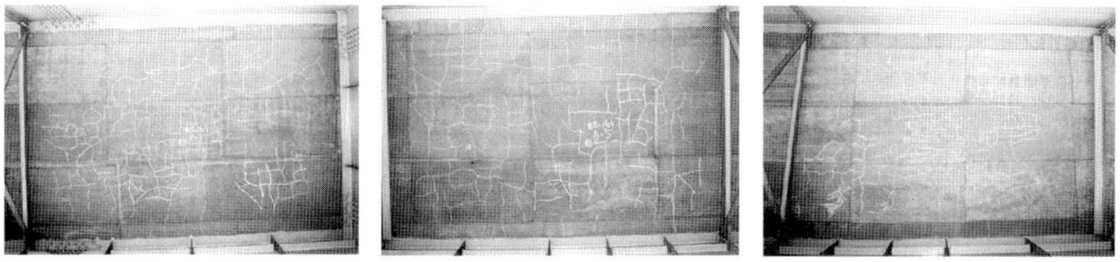

(a) Images of damage level I

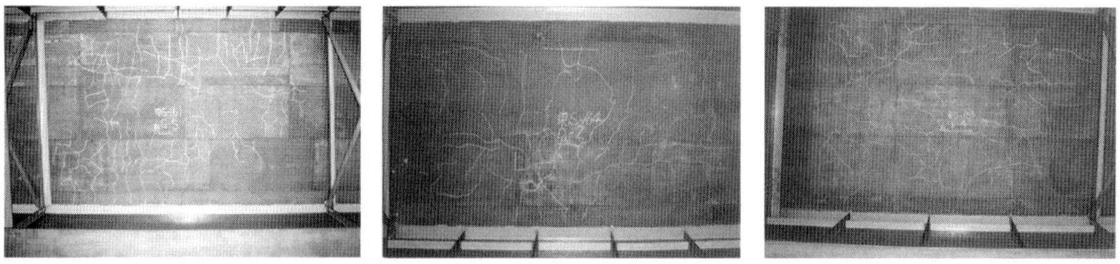

(b) Images of damage level II

(c) Images of damage level III

Fig. 2. Examples of images associated with different damage levels in concrete decks.

on the positions and the quantities of cracks, they can be used as the quantitative characteristics representing the continuity and the concentration of cracks, for the classification of crack patterns. The histograms projected on two directions are computed for extracting a crack pattern; one is the horizontal direction and the other is the vertical direction. The projection histograms are data structures used to count the number of crack pixels when the image is projected on the vertical and horizontal axes.

The characteristic values in each dimension are the number of crack pixels in row for the horizontal histogram, in column for the vertical histogram, and are the quantum numbers in accordance with the dimensionality of characteristics vectors. Figure 4 shows an example of horizontal and vertical projection histograms extracted from a crack pattern.

TAM Network

TAM network (Hayashi et al., 2003) is a one of the neural network and it is modeled from the primary visual area to prestriate cortex. TAM network has four layers; input layer, undimensional basis layer, category layer and output layer. Figure 5 shows the structure of TAM network.

TAM network is available at image of pattern recognition problem because TAM network is a modeled vision system.

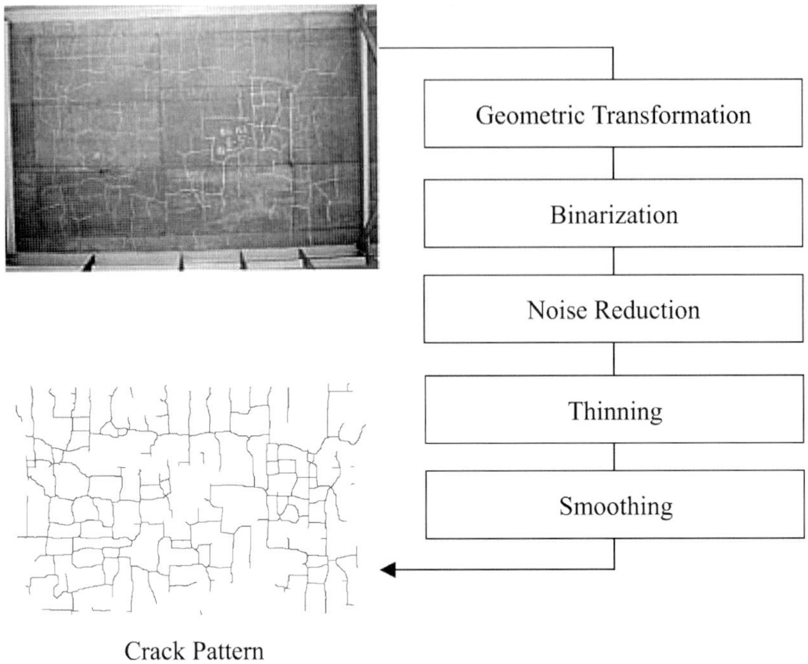

Fig. 3. Procedure of image pre-processing.

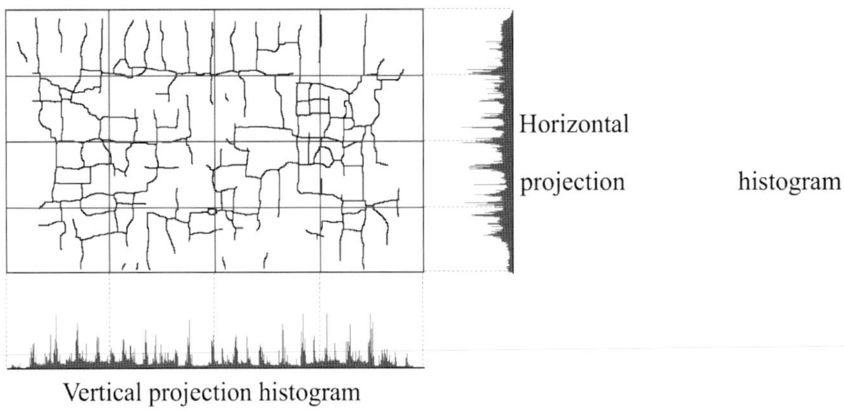

Fig. 4. Example of horizontal/vertical projection histograms extracted from a crack pattern.

Experimental Results

The classification of the digital images of cracks is implemented by using the TAM network. Twenty images of cracks are used for learning of TAM network and the remaining 27 images are used for evaluating the classification results. This implementation is repeated many times by changing the learning data every times. In this learning stage, three damage levels judged by an expert for each image are used as the teaching signal. Also, the teaching data include the same number of each damage levels data. In this numerical example, it is evaluated by the recognition rate of non-learning data, and by comparing with the neural network, the effectiveness is examined. The used learning

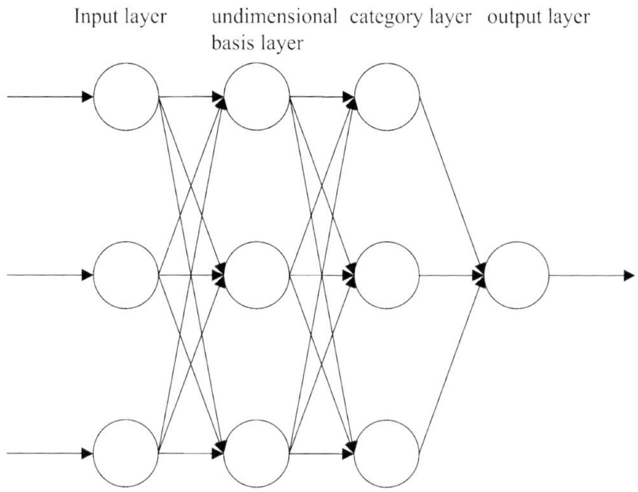

Fig. 5. Structure of TAM network.

Table 1. Learning parameters.

Learning factor	0.3
Learning time	100,000

parameters of TAM network and neural network are shown Table 1. In this research, the digital image is divided to 256 blocks, and detected the element of directionally.

The learning parameter is the same as the neural network to compare the performance. Table 2 shows the classification results with the distribution of directionality. This result is average of all trials.

Only about 70% recognition accuracy is obtained by using neural network. Especially, the recognition rate of B rank is very low (38.5%). It is caused that same digital image of B rank are close to the A rank or C rank. So, recognition rate of B rank is low. On the other hand, by using TAM network, over 90% recognition accuracy is obtained and the recognition rata of all rank are over 90%. Mainly, the recognition rate of B rank is improvement. From this result, a TAM network can recognize the complex problem that a neural network cannot recognize. It is considered that the proposed system can recognize similar digital image by using TAM network that is modeled ocular system.

Table 2. Recognition accuracy.

Method	Recognition accuracy (%)			
	A (10 entries)	B (13 entries)	C (24 entries)	TOTAL
Neural network	60.0	38.5	87.5	68.1
TAM network	90.0	92.3	95.8	93.6

Conclusions

In this paper, an attempt was made to develop a new system that evaluates the damage condition of existing structures by using the visual information given by digital photos. The proposed system is based upon such new technologies as image processing, pattern recognition, and artificial intelligence. A new measuring system was developed by using two digital cameras. Moreover, the system for extracting the characteristics of cracks showing up on concrete slabs through digital images was developed and classification based on damage levels was attempted by using these results. First, the linear pattern of cracks is extracted from the digital images of the concrete slabs through image processing techniques. Next, the characteristics such as the projection histograms that are often applied in the field of optical character recognition, and the feature points in the border expression are extracted. Finally, the digital images of cracks are classified into different damage levels based on the extracted characteristics through TAM (Topographic Attentive Mapping) system.

References

Duda, R.O., Hart, P.E. and Stork, D.G., 2001, *Pattern Recognition*, A Wiley-Interscience Publication.

Furuta, H., Hirokane, M., Dogaki, M. and Frangopol, D.M., 2004a, Optimal bridge maintenance using health monitoring, in *Proceedings of the Second European Workshop on Structural Health Monitoring*, Munich, Germany, July.

Furuta, H., Hirokane, M., Tanaka, S., Muraki, H. and Frangopol, D.M., 2004b, Bridge management system using digital photos, in *Proceedings of the Second International Workshop on Structural Health Monitoring of Innovative Civil Engineering Structures*, Winnipeg, Canada, September.

Gonzales, R. and Woods, R., 2002, *Digital Image Processing*, Prentice Hall.

Hayashi, I. and Maeda H., 2003, A formulation of fuzzy TAM network with gabor type receptive fields, in *Proceedings of International Symposium on Advanced Intelligent Systems*.

Sakai, K., 2002, *Introduction to Digital Image Processing*, CQ Publication [in Japanese].

Seul, M., O'Gorman, L. and Sammon, M., 2001, *Practical Algorithms for Image Analysis*, Cambridge University Press.

Yagi, N., 2000, *Digital Image Processing*, Ohmsha, pp. 55–60 [in Japanese].

AN APPROXIMATE DAMAGE MODEL FOR CONCRETE UNDER FINITE DEFORMATION

S. Khajehpour[1], G.D. Morandin and R.G. Sauvé

Computational Mechanics Development
Reactor Engineering Services
Atomic Energy of Canada Ltd.
2251 Speakman Dr.,
Mississauga, On, Canada L5K 1B2

Abstract

As the computing power of computers is constantly increasing, more accurate finite element analysis and detailed modelling of structures are sought. The critical issue of concern at hand is the characterization of complex material constitutive behaviour using numerical techniques. Finite element analysis of reinforced concrete structures under severe and reversible loadings requires a proper representation of concrete material behaviour. Abnormal loads such as impact, blast and seismic are generally reversible and cause structures to vibrate. To arrive at a reasonable approximation of damage in reinforced concrete structures under abnormal loading, the cracking of the concrete and its direction must be addressed. The inclusion of a mechanism that accounts for crack closure should be considered to include the compression strength of the cracked concrete if the load direction is reversed and the crack is closed. Thus, development of an improved material model for concrete and its implementation in a non-linear finite element code that is well suited to this class of problem is undertaken. In the work described in this paper, the methodology used in the development of this new material model for concrete is discussed. A sample case is analysed and the results of these FE analyses are discussed. The new concrete material model predicts the location and the direction of the cracks accurately and also allows for the inclusion of the compression strength of the cracked material in directions parallel to crack plane. In addition the closure of the crack and re-activation of the compression strength of the concrete orthogonal to the crack plane when the crack is closed is achieved.

1. Introduction

The work described in this paper covers the methodology used to evaluate the structural integrity of a reinforced concrete structure (shown in Figure 1) following a postulated handling accident scenario. To achieve this objective, a material model that approximately represents true behaviour of concrete material is developed. The primary structure considered in this work will be used for the purpose of facilitating the loading of a special container with radiated waste material. It is postulated that during the various handling operations, a loading accident occurs when the container is suspended over the structure at a certain vertical position above the water surface. The objective of this work is to determine, using full-scale explicit transient analysis with modelling of reinforced concrete, if the mentioned reinforced structure is capable of withstanding the pressure pulse generated by the accidental drop of the container onto the surface of the water contained in the loading bay. The loading considered in this work arises from dropping the container from the maximum handling height of 0.354 m onto the surface of the water assuming that the concrete structure is filled with water. This evaluation was achieved by employing the concrete material model described in this paper as part of the state-of-the-art three-dimensional non-linear continuum computer code H3DMAP (Sauvé, R.G. et al. 2004) for the numerical simulation of the fluid-structure interaction response of the container drop-generated shock wave. Pertinent modelling details include the hydrodynamic and acoustic

[1] khajehpours@aecl.ca, Corresponding author

M. Pandey et al. (eds), Advances in Engineering Structures, Mechanics & Construction, 87–98.

effects of the fluid with surface waves, discrete attachments, finite deformation material constitutive laws (i.e., large displacement and large strain), concrete reinforcing bar and large motion sliding/contact surfaces between the water and containing structures. Description of the proposed concrete material model, the method of analysis, key assumptions, and detailed results are presented in this paper.

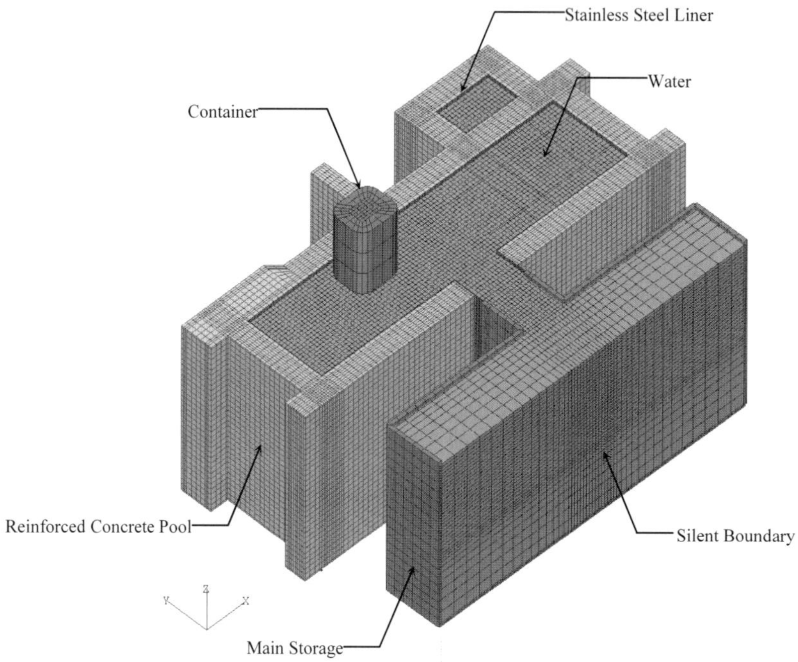

Figure 1: Overall View of Reinforced Concrete Structure (FE Model)

2. Methodology for Inclusion of Cracks in Concrete Material Model

The major methodologies used in finite element modelling of fracture in material are namely: a) the discrete and b) the smeared approaches (Gerstle W, et al.). In a smeared model, cracks are represented by changing the constitutive properties of the finite elements rather than changing the topography of the finite element meshing (Rashid) whereas the discrete model treats a crack as a geometric entity (Ngo et al.). The material model proposed in this paper is based on the smeared methodology. To achieve an accurate representation of cracked concrete material, it is imperative that the cracked plane is established. The methodology for the proposed material model is employed to find the crack plane in an element based on the principal stresses at the instant that cracking in the element occurs. In dynamic problems, where structures experience large displacements, the relativity of the crack plane to the element axis must be preserved as the element is displaced in three-dimensional space. The steps necessary to account for cracking of an element and its displacement in three-dimensional space are briefly described in this section followed by a brief discussion on how the compression failure of concrete is considered.

In Figure 2, the sequences of steps taken to establish the fracture of each element is shown. The first step is to verify if the element of concern has been cracked in previous time steps. Depending on the crack status of the element the following steps are taken,

1. If the element has never cracked before:

a. The maximum principal stress is found to establish if element is under tensile normal stress,

b. If the maximum principal stress is greater than tensile strength of material then the Jacobi iterative method is used (Golub G. et al.) to find the rotation tensor that relates the global to principal axes. This tensor transforms the element stress tensor to the principal stress tensor (cracked plane stress tensor); the same tensor can be used to relate the cracked plane to the element axis,

c. The tensile principal stress that exceeds the tensile strength of material is set to zero (cracking has occurred); the stress perpendicular to crack plane is released and reapplied to the structure as residual loads,

d. The tensile strength of the material in the element is set to zero,

e. The Cracked and Ever-Cracked flags are activated,

f. The principal stress (tensile crack component removed) tensor is rotated back to the element axis using the rotation tensor from step "b",

2. If the element has cracked at any point during static or dynamic loading stages:

a. The rotation tensor from step "b" is updated by using an average element spin (rotational velocity of the element) tensor in each time step,

b. The updated rotation tensor is then used to rotate the stresses in the element axis to the cracked plane,

c. If stress perpendicular to the crack plane is positive (tensile) then the value of the stress in that direction is set to zero,

d. The crack plane stress (tensile removed) tensor is rotated back to the element axis using the rotation tensor from step "a",

e. The Cracked flag is activated.

When the status of the element regarding the crack is established, a compressive failure using a failure model as shown in Figure 3 is checked. Failure of the element is based on the element hydrostatic pressure and deviatoric stresses. For concrete, it is proposed to use a hydrodynamic pressure-dependent material model in conjunction with a failure/damage model. Since the reinforcing steel is explicitly modelled, the need for assumptions regarding the use of a mixture rule is avoided. In this proposed material model, a pressure-dependent flow rule is defined with the attendant parabolic form of the yield function for compression as:

$$\varphi = J_2 - [a_0 + a_1 p] \tag{1}$$

where p is the hydrostatic stress (pressure), J_2 is the second invariant of the deviatoric stress tensor, a_0 and a_1 are the constants that are defined based on the concrete material. At yield $\varphi=0$ and $J_2=\sigma^2_y/3$ where σ_y is the yield point corresponding to hydrostatic stress p in the concrete. The basic specification data for the concrete is obtained in references (Bangash, M.) and (Winter, G. et al.).

If J_2 is less than $[a_0+a_1p]$, then no compression failure in the element has occurred. In cases, when J_2 is slightly larger than $[a_0+a_1p]$, to account for a relatively small ductility in concrete, some ductility might be defined as a percentage of the yield (shaded area in Figure 3). For these cases the deviatoric stresses are scaled down using equation (2).

$$\beta = \frac{\sigma_y}{\sqrt{3J_2}} \tag{2}$$

If J_2 is significantly larger than $[a_0+a_1p]$, the element is considered failed in compression. Theoretically, as hydrostatic pressure increases, yield stress of the material increase. However, two cap models are considered to limit the extent at which the yield strength can be increased (Figure 3).

These cap models ensure that the material properties will degrade after it has experienced large hydrostatic pressures. This is necessary due to the fact that voids in concrete material collapse and micro cracks are formed in the material.

For reversible loadings such as blast or impact, it is important to limit the strength of the element after it has passed the softening hydrostatic pressure (location I on Figure 3) to the minimum value it experienced during the previous loading phases. As an example if an element experiences hydrostatic pressures up to the value of location II on Figure 3 and the elliptical cap model is chosen, then the yield strength of that element is limited to σ_{yl}. This mechanism will ensure that elements that have previously experienced softening in their material strength will not carry loads beyond their set limits.

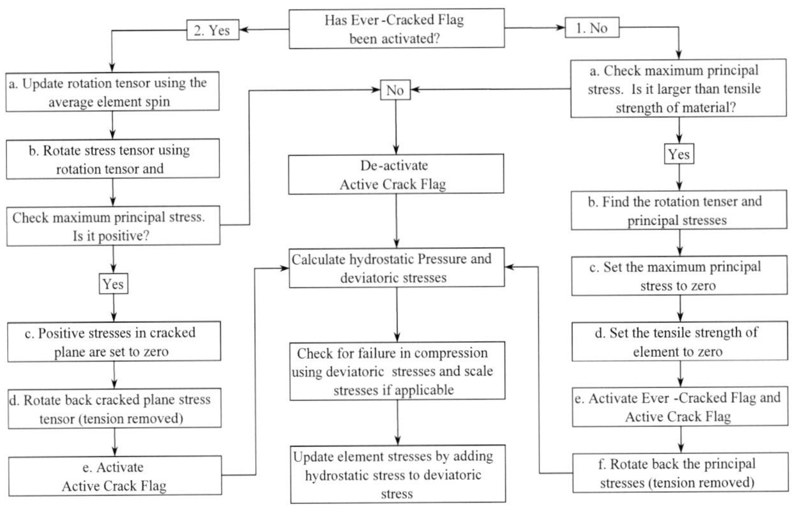

Figure 2: Methodology Flowchart

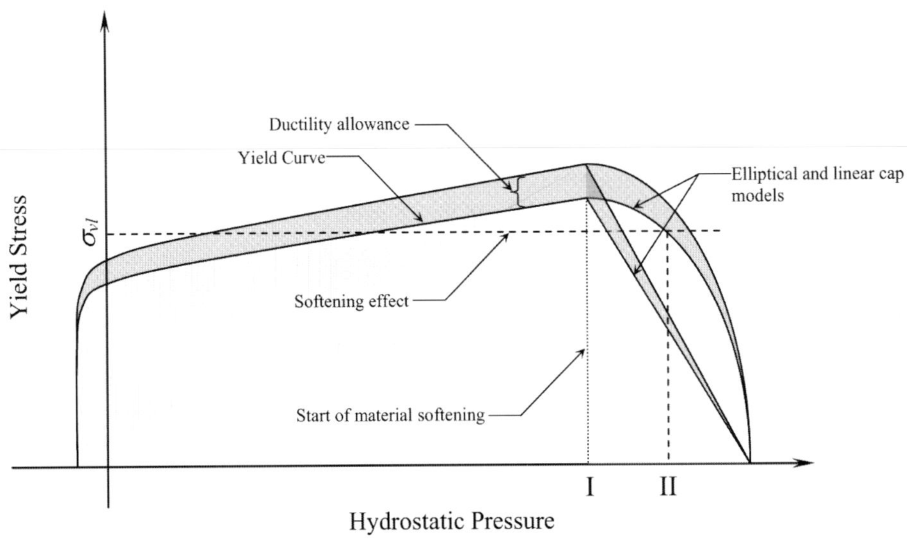

Figure 3: Yield Curve

An equation of state is employed for the proposed concrete material that relates the volumetric strain to the hydrostatic pressure in each element. This equation of state is used as part of the proposed material model in a form of a table lookup. The equation of state for the proposed concrete material is schematically shown in Figure 4.

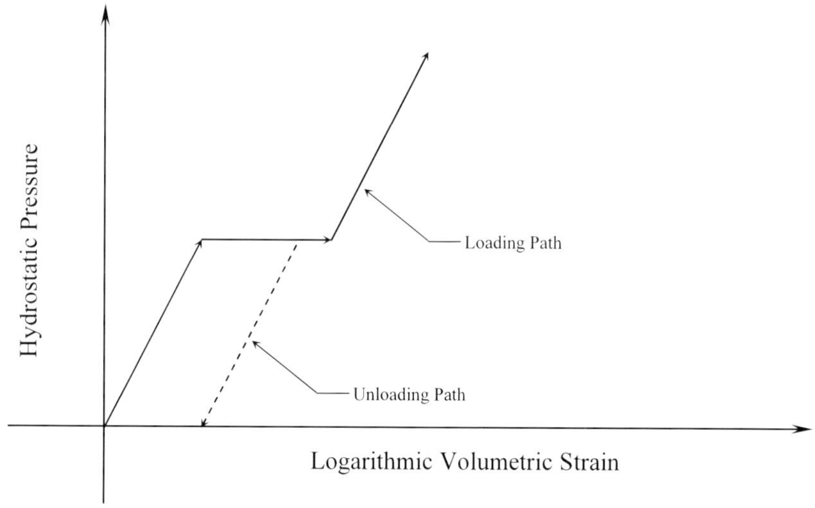

Figure 4: Equation of State

3. Example

The methodology described in Section 2 is employed for an assessment of a postulated accident scenario. The work described in this paper covers the methodology used to evaluate the integrity of the structure shown in Figure 1 following a postulated handling accident scenario. As described in Section 1, it is postulated that during the various operations, a loading accident occurs when the container is suspended at a certain vertical position above the water surface. The objective of this work is to determine, using full-scale analysis, if the reinforced concrete pool and neighbouring structures are capable of withstanding the pressure pulse generated by the accidental drop of the container onto the surface of the water contained in the pool.

3.1 Finite Element Model Description

In the analysis of the structure, a three-dimensional non-linear transient fluid-structure shock wave analysis is carried out to assess the structural integrity of the reinforced structure following a postulated handling accident. In Figure 1, the schematic representation of the structure of intent and part of the surrounding structures are shown. Given the localized nature of the drop scenario considered, only part of the water in the adjacent structure is included in the assessment as shown in Figure 1. Since the main storage is not directly connected to the pool, it is modelled using rigid shell elements. To ensure that the main storage contribution to the lateral stability and rigidity of the adjacent pool is minimized, the channel that connects the main storage to the pool is modelled in a manner that will minimize the transfer of load to the pool. The discretised components of the full model are shown Figure 1. The embedded reinforcing bars are explicitly modelled considering their respective diameters. The reinforcing bar topology is shown in Figure 5.

Given the complexity of the model, a variety of element types are used for each of the structural components and are listed in Table 1. In addition to the non-linear structural components, the water transmitting the shock wave within the containing structure is modelled using a hydrodynamic finite element formulation. The fluid-structure interaction is accommodated using a large deformation contact/sliding surface algorithm applied to all fluid-structure boundaries. At the boundary interfaces of the main storage structure and fluid, silent (i.e., non-reflective) viscous elements are utilized. This prevents artificial wave reflections due to modelling constraints by permitting incident waves to travel through the boundary to the portion not included in the model.

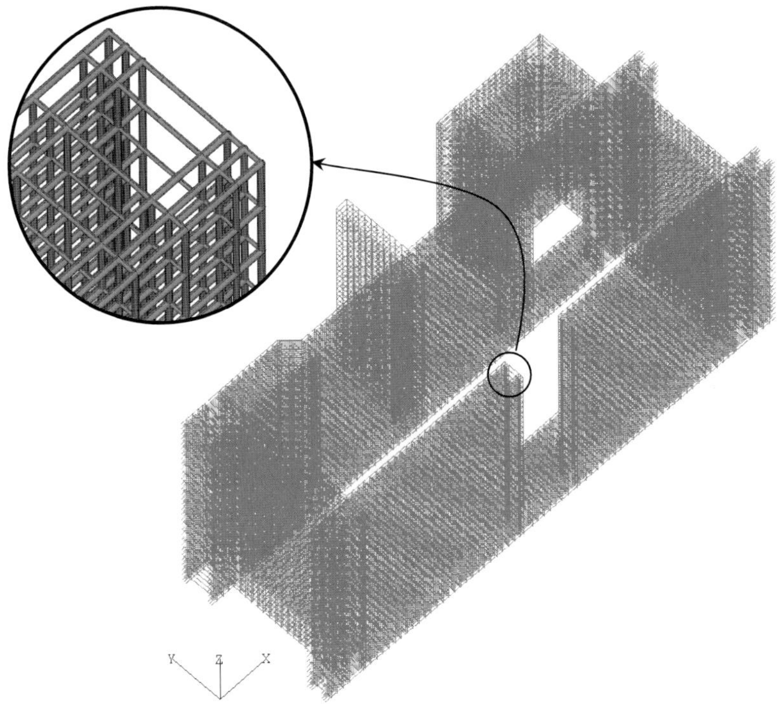

Figure 5: Steel Reinforcement in Wet Cask Handling Bay

Table 1 Details of Simulation Model Topology

Model Details		
Component	**Element Type**	**Number of Elements**
Concrete	3D Continuum	86,408
Rebar	3D Beam	132,422
S.S. Steel Liner	3D Shell	7,138
Inspection Platform	3D Shell	4,648
Inspection Platform	3D Beam	5,712
Water	3D Continuum (Hydrodynamic)	133,456
Container	3D Continuum (Rigid)	252
Bottom of the main storage and pool	3D Shell	2,572
Contact Pairs	Contact	6
Total Number of Elements		372,614
Total Number of Nodal Points		254,576

3.2 Material Model

Details of the material input data, used to model each of the components outlined in Table 1, are given in the following:

3.3 Concrete

The coefficients are obtained to give yield stress versus pressure consistent with values given for concrete with $f_C^{'} = 25\ MPa$ in the literature (Bangash, M.) and (Winter, G. et al.). In the event of failure, the concrete material is permitted to fail around the reinforcing bar. The specific data for the concrete used in the analysis is listed in the following.

Compressive strength $f_c^{'}$	= 25 MPa
Weight per unit volume	= 2400 kg/m^3
Modulus of Elasticity	= 23,400 MPa
Shear modulus	= 10,174 MPa
Poisson's ratio	= 0.15
Modules of rupture (tensile strength) f_r	= 2.5 MPa
Cracking Failure Criterion	= principal stress > $f_r/3$
Pressure-hardening yield function:	$\varphi = \sigma_y^2 - [226.5 + 29.2p]$

3.4 Reinforcing Steel

The reinforcing bars are explicitly modelled (Figure 5). The diameters of the bars are included as per the specification.

Weight per unit volume	= 7800 kg/m^3
Modulus of Elasticity	= 200,000 MPa
Poisson's ratio	= 0.3
Initial yield strength	= 400 MPa
Hardening Modulus	= 0.0 (elastic-perfectly plastic).

3.5 304L Stainless Steel Liner (5 mm) and Inspection Platforms

For this material, a bi-linear elastic-plastic isotropic hardening model is applied. This material model is applicable to finite deformation and calculated on a co-rotational material frame in the beam and shell elements. This provides for objectivity in the updated configuration. Shell thinning is included and failure of the liner plate structure is defined when the maximum plastic strain is exceeded at all the fibres through the thickness of the shell. Three integration points through the thickness are used to sample the non-linear response and obtain the thinning. For HSS beam members, sixteen integration points around the cross section of the member is used to capture the non-linear response of the beam members. Weight per unit volume = 7800 kg/m^3

Modulus of Elasticity	= 200,000 MPa.
Poisson's ratio	= 0.3
Initial yield strength	= 170 MPa
Hardening Modulus	= 2000 MPa

3.6 Water

The water is modelled using a hydrodynamic formulation with an equation of state (Sauvé R.G. et al. 2004). Shock discontinuities at the shock front are handled using an adaptive bulk viscosity treatment that correctly spreads the shock over a number of elements as a propagated steady wave and eliminates the numerical noise associated with this type of analysis.

Mass per unit volume	= 1000 kg/m^3
Bulk Modulus	= 2071 MPa.

3.7 Loading Conditions

The loading on the model occurs from 1) hydrostatic/dynamic pressure of water and 2) the interaction of the container with the water at the water-container interface. The nodes defining the rigid container model impact the water surface with an impact velocity V_0, determined from rigid body mechanics:

$$V_0 = \sqrt{2gh} \qquad \textbf{(3)}$$

Where h is the drop height and g is the gravitational constant.

The relative location of the container to the pool walls and loading platforms are given in Figure 6. The topological information of the drop is given in the following.

Weight of container	= 80 metric ton
Drop Orientation in Pool	= bottom down flat end
Distance of container from pool short wall	= 3.52 m
Distance of container from pool long Wall	= 1.05 m
Drop height above water surface, h	= 0.354 m
Impact velocity from (3)	= 2.64 m/s

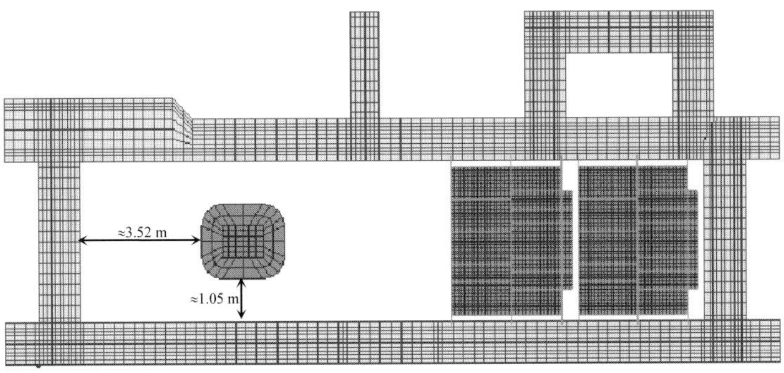

Figure 6: Plan view of Reinforced Concrete Structure

4. Assumptions

The key assumptions made in the simulation are as follows:

i) Given the localized nature of the drop scenario considered, only part of the water in the main storage in the proximity of the pool is included in the assessment. Since the water mass in the main storage is relatively large, a silent (non-reflective) boundary as shown in Figure 1 ensures that erroneous reflective waves are not included in the simulation.

ii) Since the main storage is not directly connected to the pool the main storage is modelled using rigid shell elements.

iii) To ensure that the main storage's contribution to the lateral stability and rigidity of the adjacent pool concrete wall is minimized, the channel that connects the main storage to the pool is modelled in a manner that will minimize the transfer of load to the pool. This is warranted, as the pool is not connected to the main storage.

iv) Failure in the concrete occurs when the principal stress exceeds a fraction (1/3) of the maximum tensile stress for the concrete material. This assumption ensures that the final results are conservative and no extra credit is given to concrete tensile strength.

v) Failure is assumed to occur in the stainless steel components and reinforcing rebar when the material continues to flow plastically beyond a specified effective plastic failure strain.

vi) No damping is assumed. The material's hysteresis arising from the non-linear response and failure provides the damping.

vii) It is assumed that the stainless steel liner is flush against the pool's concrete wall. The stainless steel liner is modelled explicitly to include its effect on the transfer of the shock wave from the water to the reinforced concrete wall. However, it is prevented from contributing to the pool's rigidity and lateral stability.

viii) Inspection platforms are modelled with beam and shell elements that interact with water elements through common nodes. This ensures that the deformation of the platforms due to shock wave and water movement is conservatively captured.

5. Method of analysis and Results

The coupled fluid-structure model for the intended application must account for non-linear geometric and material behaviour. This includes elastic as well as finite deformation elastic-plastic constitutive laws for both three dimensional continuum and thin shells. The explicit solution module of the general-purpose, three-dimensional, non-linear in house finite element code H3DMAP (Sauvé R.G. et al. 2004) is used to simulate the shock wave propagation and resulting structural response. This module is based on the explicit hydrodynamic finite element formulation and includes the conservation of mass, momentum and energy equations. The element technology accounts for finite deformation (large displacement and strain) and includes shell thinning along with finite elastic-plastic materials and hydrodynamic material formulations. The efficiency and robust nature of the associated element technology and explicit algorithms make it particularly well suited to this class of stress wave propagation problem. The shell element is based on a co-rotational formulation that accounts for non-linear variations in through-thickness strains and large deformation (Sauvé R.G. et al. 1995). It is compatible with an under-integrated, eight-noded, three dimensional continuum element that includes a unified stabilization algorithm. A three-dimensional sliding/contact algorithm (Sauvé R.G. et al. 2005) provides the capability for modelling sliding interfaces with friction including interpolation of applicable quantities such as fluid forces and structural motions along a prescribed fluid-solid interface. A variety of failure models are used in conjunction with an erosion model that provides adaptive element deletion when element/material failure is detected.

The simulation is run to 800 millisecond (ms) beyond the initial impact at which time all pressure waves and dynamic effects resulting from the initial drop of the container onto the water surface are essentially dissipated. At approximately 10 to 15 ms after the impact, the initial shock wave is dissipated and the container is displacing the water in the pool as it descends. Initially a shock wave, imparted to the water, occurs under the container. A comparison of the pressure in the water, at a location in the centre of the container/water interface at the instant following the initial impact, with the peak pressure obtained from an idealization of the shock wave (i.e., water hammer) provides a useful, albeit approximate, check on the results of the simulation. Just following the initial drop, and before dissipation of the initial shockwave inspection of the peak pressure at the interface is predicted to be 3.95 MPa (shown in Figure 7). This compares to the peak value of 3.81 MPa obtained from the idealized acoustic wave equation.

In order to effectively track any damage to the containing structure during the course of the container drop generated pressure pulse, irrecoverable state variables such as cracking and effective plastic strain are stored at specified time intervals throughout the transient. In Figure 8 the crack locations in the concrete are identified before impact (under sustained hydrostatic load). As observed, localized cracking in the surface elements occurs in the concrete at the junction of the wall/wall and wall/foundation for the unsupported wall. These areas, being corners (wall to wall and wall to foundation), are subject to higher bending moments and hence higher tensile stress. In Figure 9 the locations where concrete has been cracked at least at one point in time during the simulation are shown. No crack on the outside surface of the pool concrete wall due to the pressure pulse generated by the drop is observed. Also from Figure 9, it is concluded that the cracks shown in Figure 8 marginally grow due to the impact-induced shock wave. It should be noted that since no plastic strain in the reinforcing rebar is observed as a result of the shock wave, the state of cracking in concrete after the shock wave would be the same as that shown in Figure 8 (i.e., tensile cracks due to the shock wave will close).

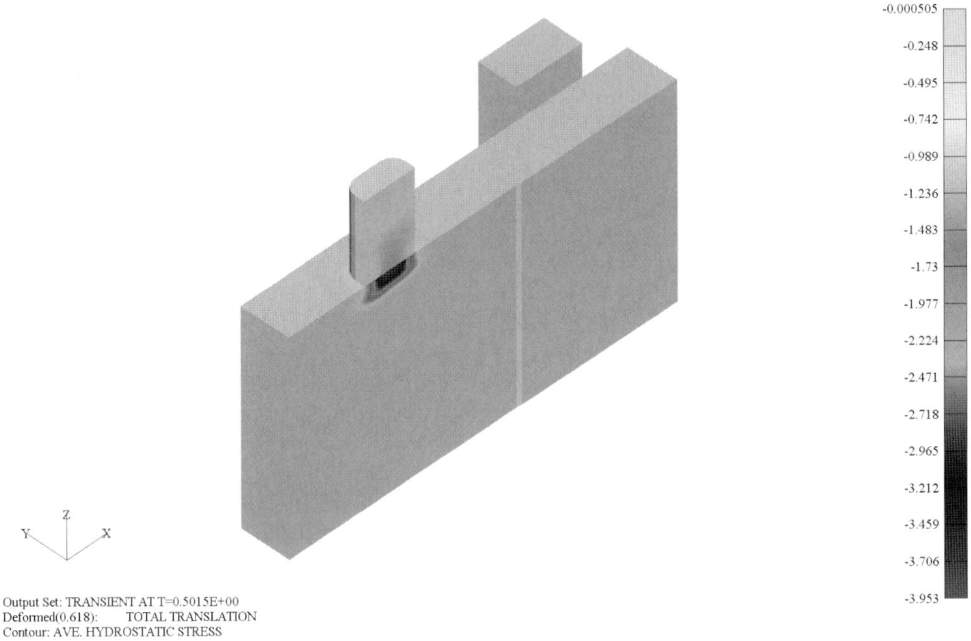

Figure 7: Hydrostatics Pressure in Water at Impact

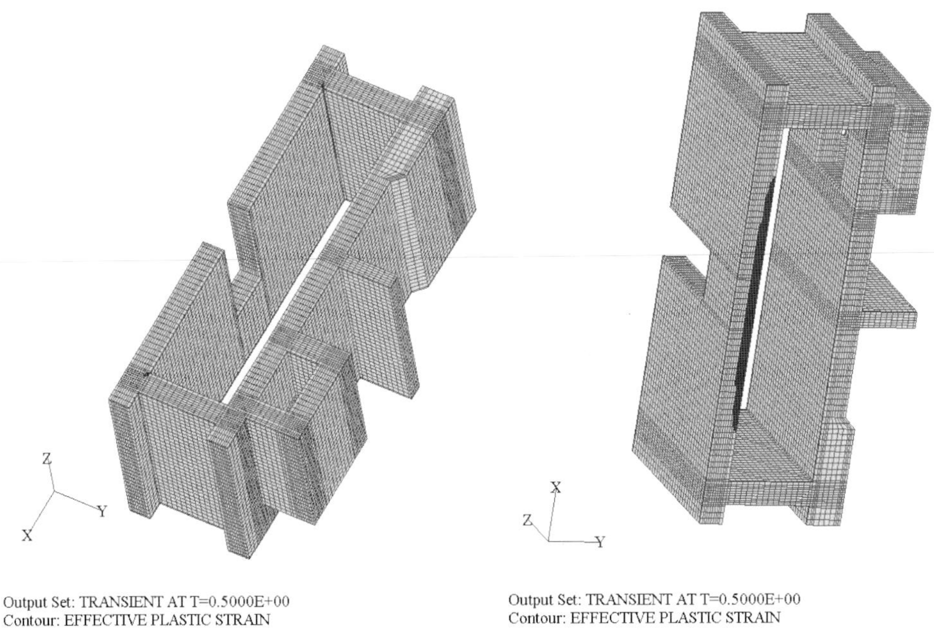

Figure 8: Accumulated Cracking in Concrete Before Impact

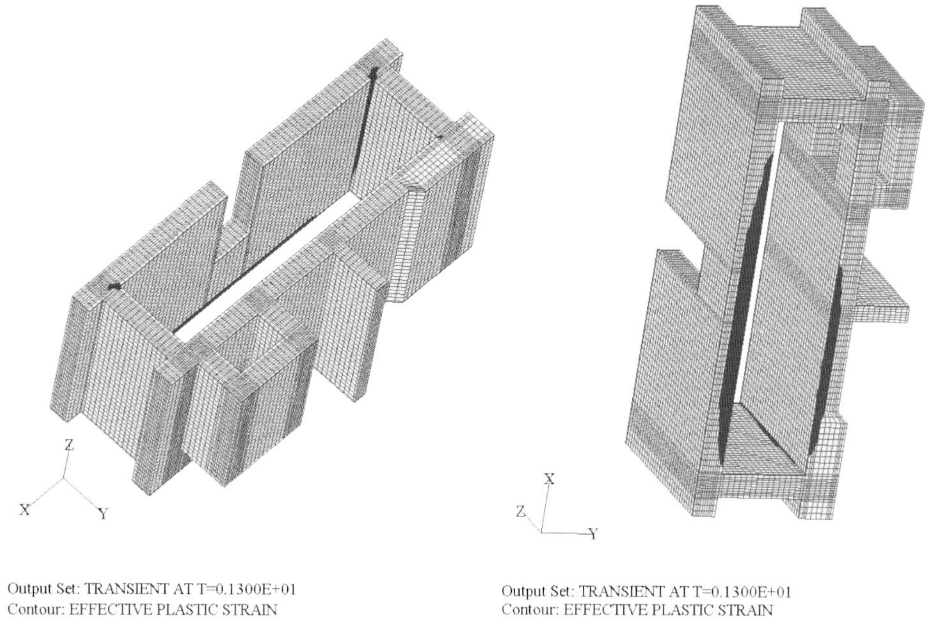

Figure 9: Accumulated Cracking in Concrete at End of Simulation

6. Conclusions

A simulation of a postulated container handling accident into a pool has been performed. The results indicate that for this accident scenario, no significant cracking of the concrete structure occurs due to the fluid shock wave propagation. While the stainless steel liner experiences virtually no damage and maintains its function, some localized cracking in the concrete is observed in the walls of the pool at the wall/wall and wall/foundation junctions. However, these cracks are a result of the sustained loading (hydrostatic water pressure) with some general and reversible increased crack widths and depths due to shock wave. There is no indication of plasticity in the rebar used in the reinforced concrete walls or in the inspection platforms. The material model described in this paper in conjunction with explicit finite element engines can be used to track the fracture of concrete material and will help analysts in prediction of damage level to reinforced concrete structures subjected to abnormal loadings.

References

Bangash, M.Y.H., 1989, *Concrete and Concrete Structures*, Elsevier Applied Science.

Gerstle W., et al., "Finite Element Analysis of Fracture in Concrete Structures: State-of-the-Art", ACI 446.3R-97.

Golub Gene H. and Van Loan Charles F., 1989, *Matrix Computations*, Second Edition, Johns Hopkins University Press.

Ngo, D. and Scordelis, A.C., 1967, "Finite Element Analysis of Reinforced Concrete Beams", *ACI Journal*, Proceedings Vol. 64, No. 3, March.

Rashid, Y.R., 1968, "Ultimate Strength Analysis of Prestressed Concrete Pressure Vessels", *Nuclear Engineering and Design*, Vol. 7.

Sauvé, R.G. and Metzger, D., 1995, "Efficient Unified Hourglass Stabilization for One Point Quadrature Three Dimensional Shell Elements'', In: Cory JF, Gordon JL, editors. *Current Topics in Computational Mechanics*, ASME PVP-305; pp. 3–10, July.

Sauvé, R.G. and Morandin, G., 2004, Computer Program Documentation – User Manual, Programmer Manual, H3DMAP Version 7: A Three Dimensional Finite Element Computer Code for Linear and Nonlinear Continuum Mechanics, AECL Report No. CW-114515-225-001 R0, May.

Sauvé, R.G. and Morandin, G.D, 2005, "Simulation of Contact in Finite Deformation Problems – Algorithm and Modelling Issues", *International Journal of Mechanics and Materials in Design*, Vol. 1, pp. 287–316.

Westbroke, H., 1990, "Hydrostatic Pressure Testing of Samples from the Concrete Integrated Container", Ontario Hydro Research Department, December.

Winter, G. and Nilson, A., 1979, *Design of Concrete Structures*, McGraw-Hill.

DYNAMIC RESPONSE ANALYSIS FOR A LARGE-SCALE RC GIRDER UNDER A FALLING-WEIGHT IMPACT LOADING

N. Kishi[1], T. Ohno[2], H. Konno[3], and A.Q. Bhatti[1]
[1]*Department of Civil Engineering and Architecture, Muroran Institute of Technology, Muroran 050-8585, Japan; E-mail: kishi@news3.ce.muroran-it.ac.jp*
[2]*Department of Civil Engineering, National Defense Academy, Yokosuka 239-8686, Japan*
[3]*Structural Division, Civil Engineering Research Inst. of Hokkaido, Hiragishi, Toyohira-Ward, Sapporo 062-8602, Japan*

Abstract

In order to establish a rational impact resistant design procedure for prototype reinforced concrete (RC) structures, not only experimental study but also numerical analysis study should be conducted. However, numerical analysis method on impact response analysis for those structures has not been established yet. Here, in order to establish a rational numerical analysis method for prototype RC structures under impact loading, a falling-weight impact test was conducted for prototype RC girder with 8 m clear span. Referring to the experimental response waves, numerical accuracy was investigated varying major parameters. From this study, following results were obtained as: (1) fine mesh should be used near supporting gigues; (2) Drucker–Prager yield criterion should be applied which gives better results than von Mises one; and (3) appropriate system damping constant should be set to $h = 0.015$.

Keywords: prototype RC girder, falling-weight impact test, impact response analysis, LS-DYNA, Drucker–Prager yield criterion

1. Introduction

In Japan, design method concerning infrastructures tends to be shifted from allowable stress design method to performance based design one. Under such a situation, however, impact resistant design for RC structures has been still performed based on the allowable stress design concept. To accomplish the shift of design method for RC structures under impact resistant design, experimental and numerical researches for small-scale members have been conducted (JSCE, 2004; Kishi et al., 2001; Kishi et al., 2002). On the other hand, impact research sub-committee of JSCE in Japan was managed round-robin pre/post analysis, which was conducted to confirm the numerical accuracy and characteristics of the method applied by each institution. From those researches, following conclusions were obtained: (1) from the pre-analysis results, it is too difficult to precisely predict numerically both impact force and reaction force waves obtained from the experimental results; and (2) on the contrary, displacement wave can be better simulated comparing with above two response waves.

In general, a basic concept for designing RC members should be constituted based on the research findings for small-scale members and the applicability should be confirmed conducting large-scale experiments. In the cases of RC structures under shock and impact, however, it is not easy to conduct the experiments because of expenses for preparing of specimens and experimental set-up fittings. On the other hand, numerical simulations for large-scale impact test for RC members may be easily performed by using personal computers because of an advanced development of computer technology. Then, a rational impact resistant design procedure for RC members may be better established by

M. Pandey et al. (eds), Advances in Engineering Structures, Mechanics & Construction, 99–109.

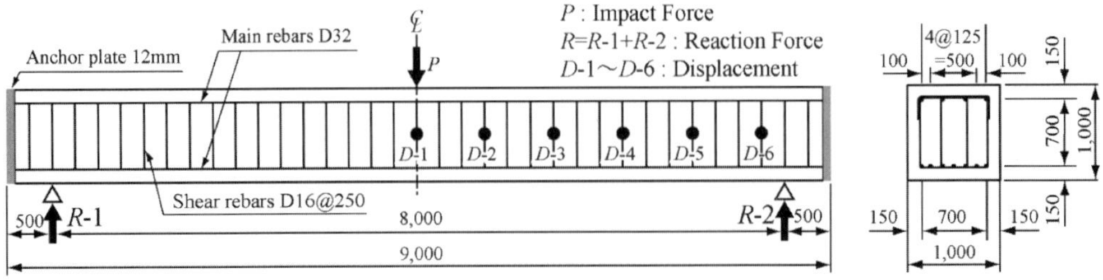

Fig. 1. Dimensions of RC girder and measuring items.

supporting of numerical simulations. To accomplish this, a rational numerical simulation method for large-scale impact test must be established. Authors have discussed numerical simulation method using three-dimensional elasto-plastic finite element technique for small-scale impact test of RC beams (JSCE, 2004; Kishi et al., 1999). However, numerical investigations for large-scale impact tests have not been conducted yet.

From this point of view, in this paper, in order to establish a more precise numerical analysis method for analyzing elasto-plastic impact behavior of prototype RC structures, numerical accuracy of the results obtained using a three-dimensional elasto-plastic finite element method was investigated for each input parameter referring to the experimental results obtained from a falling-weight impact test for large-scale RC girder which was similar to that of designing roof of RC rock-sheds constructed over the highway to ensure people and vehicles safety. In this study, a general-purpose program LS-DYNA code, which is developed based on finite element method is used for those investigations (Hallquist, 2000).

2. Experimental Overview

2.1. *Dimensions and Static Design Values of RC Girder*

In this study, a RC girder, which is for designing roof of real RC rock-sheds, was taken for falling-weight impact test of prototype RC structures. The girder is of rectangular cross section and the dimensions are of 1 m × 1 m and clear span is 8 m long, which is similar to the width of real RC rock-sheds. Figure 1 shows dimensions of the RC girder, distribution of rebar, and measuring points for each response wave. In this figure, it is confirmed that 7#D32 rebars are arranged as main rebar assuming 0.65% of main rebar ratio corresponding to designing of real RC rock-sheds and 4#D32 rebars are arranged as the upper axial rebar to be about a half of main rebar ratio. Thickness of concrete cover is assumed to be 150 mm as well as real rock-sheds. D16 stirrups are arranged with intervals of 250 mm, which is less than a half of an effective height of the girder. In this study, arranging interlayer stirrups and upgrading in shear load-carrying capacity, the RC girder was designed to be collapsed with flexural failure mode. Axial rebars were welded to 12 mm steel-plates at the ends to save the anchoring length of the rebars.

The displacements of the girder were measured at mid-span ($D-1$) and the other five points ($D-2$ to $D-6$) with the intervals of 750 mm from the mid-span. Impact force P was estimated using deceleration of the heavy weight, which is measured using accelerometer set at its top surface. Reaction force R ($= R_1 + R_2$) was also measured using load-cells installed in the supporting gigues. The detailed static design parameters of the RC girder are listed in Table 1. Static flexural and shear load-carrying capacities P_{usc} and V_{usc} were calculated based on Japanese concrete standards (JSCE, 1996). From this table, it is confirmed that the RC girder designed here will collapse with flexural

Table 1. Static design parameters of RC girder.

Shear rebar ratio	Static shear depth ratio	Static shear capacity	Static bending capacity	Shear-bending capacity ratio
ρ_t	a/d	V_{usc} (kN)	P_{usc} (kN)	α
0.0065	4.71	1651	892	1.85

Table 2. Material properties of concrete.

Age (days)	Compressive strength f_c' (MPa)	Young's modulus E_c (GPa)	Poisson's ratio ν_c
36	30.4	27.6	0.186

failure mode under static loading because shear-bending capacity ratio α is larger than unity. The static material properties of concrete and rebars during experiment are listed in Tables 2 and 3, respectively.

2.2. *Experimental Method*

In the experiment, a 2,000 kg heavy weight was lifted up to the prescribed height of 20 m by using the track crane, and then dropped freely to the mid-span of girder with a desorption device. A heavy weight is made from steel outer shell with 1 m in the diameter, 97 cm in height, and spherical bottom with 80 cm in radius as shown in Figure 2(a) and its mass is adjusted filling concrete and steel balls. Figure 2(b) shows supporting gigue including load-cells and Figure 2(c) shows gigue for preventing RC girder from jumping up. RC girder was set on the supporting gigues, which are made so as to freely rotate but not to move toward each other.

The ends of RC girder is fixed in the upward direction using steel rods and beams to prevent from jumping up at the time of impacted by a heavy weight as shown in Figure 2(c). In this experiment, impact force wave (P), reaction force wave (R), and displacement waves (D) at six points along the girder were measured. Impact force wave was estimated using a deceleration of heavy weight, which is measured using accelerometers set at the top-surface of weight.

The accelerometer is of strain gauge type and its capacity and frequency range for measuring are 1,000 times gravity and DC through 7 kHz, respectively. Each load-cell for measuring reaction force are of 1,500 kN capacity and more than 1 kHz measuring frequency. For measuring displacements, laser-type variable displacement transducers (LVDTs) were used which are of 200 mm maximum stroke and 915 Hz measuring frequency. Analog signals from those sensors were amplified and converted to digital ones.

Table 3. Material properties of rebar.

Rebar type	Grade	Yield stress σ_y (MPa)	Young's modulus E_s (GPa)	Poisson's ratio ν_s
D16	SD345	390	206	0.3
D32	SD345	375		

Fig. 2. Pictorial views of heavy weight and supporting gigues.

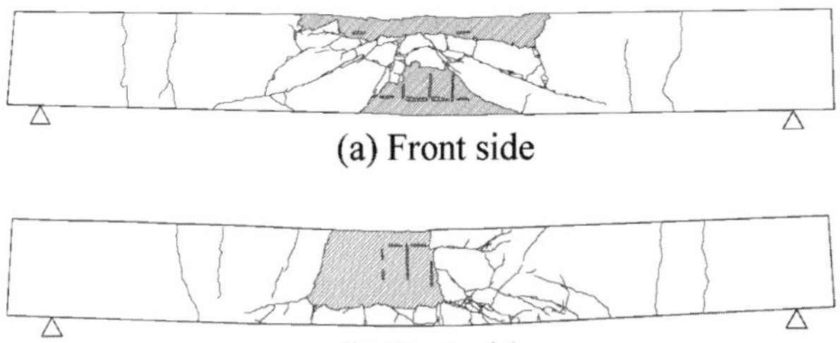

Fig. 3. Crack patterns of RC beam sides after experiment.

The digital data were continuously recorded with 0.1 ms time intervals by using digital data recorders. After that, impact force wave was numerically filtered by means of rectangular moving average method having 0.5 ms time. After experiment, pictures for views of crack patterns occurred around impacted area and on side-surface of RC girder, and a view of peeling and spalling of concrete cover were took. Figure 3 shows crack distributions occurred in the both side-surfaces of the girder.

3. Analytical Overview

3.1. *Finite Element Model*

One quarter of RC girder was three-dimensionally modeled for numerical analysis with respect to the two symmetrical axis. Figure 4(a) shows the mesh geometry of the girder, which is finally used for numerical analysis with an appropriate design accuracy investigated here. A geometrical configuration of heavy weight was modeled as the real one. Supporting gigues including load-cells and gigue for protecting the girder from jumping up were also precisely modeled corresponding to the real ones. In this model, axial rebar and stirrup were modeled using beam element having equivalent axial stiffness, cross sectional area and mass with those of real ones. The other were modeled using eight-node and/or six-node solid elements. The mesh geometries for axial rebar and stirrup are shown in Figure 4(b).

Total number of nodal points and elements for the whole structure shown in Figure 4(a) are 13,963 and 12,360, respectively. Number of integration points for solid and beam elements are one and four, respectively. In order to take into account of contact interface between adjoining concrete and a head of heavy weight elements and between adjoining concrete and supporting gigue elements, contact surface elements for those are defined, in which contact force can be estimated by applying penalty methods for those elements but friction between two contact elements were neglected.

A head of heavy weight was set so as to contact the impacting point of the upper surface of RC girder and predetermined impact velocity was applied to all nodal points of the weight model. The

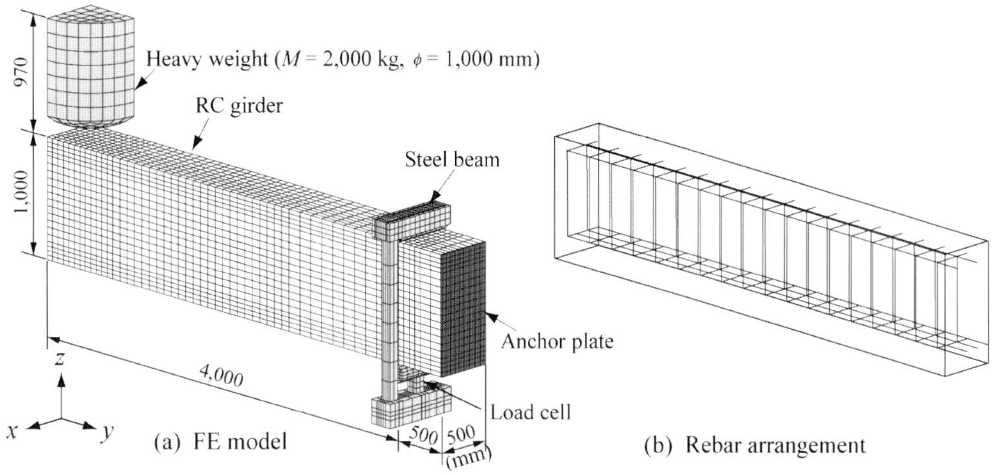

Fig. 4. FE numerical analysis model.

elasto-plastic impact response behavior was analyzed during 400 ms from the beginning of impact to the RC girder reaching steady state. The time increment of numerical analysis has been determined calculating in the LS-DYNA code based on a Courant numerical stability condition and was about 0.8 μs for all numerical analysis conducted here.

3.2. *Modeling of Materials*

The stress and strain characteristic of concrete used for the finite element analysis are shown in Figure 5. For the compression region, assuming that concrete is yielded at 1,500μ strain, perfect elasto-plastic bilinear model was used. In this study, finally, yielding of concrete has been judged based on the Drucker–Prager's yield criterion. For the tension region, linear model was applied, but it is assumed that the stress cannot be transferred when a tensile pressure acted in the element reaches the breaking point. Here, the pressure is evaluated as an average of three normal stresses acted in each element and the tensile strength of concrete is assumed to be 1/10th of compressive strength similarly to the case of the numerical analysis for small-scale RC beams conducted by authors. Stress-strain relationship for main rebar and stirrup was defined using a bilinear isotropic hardening model. Plastic hardening coefficient H' was assumed to be 1% of Young's modulus E_s.

Yield of rebar and stirrup was calculated following von Mises yield criterion. Heavy weight, supporting gigues and anchor plates for axial rebars set at the both ends of RC girder were assumed to be elastic body because of no plastic deformation for those being found.

4. Analytical and Experimental Modeling Accuracy

In this study, numerical accuracies due to element size in the region near supporting gigue in the longitudinal direction, yield criterion for concrete material, system damping constant were investigated from the viewpoint of wave configuration of response. Numerical accuracies should be investigated for not only each parameter but also interaction between each two parameters considered here. However, those investigations need too much computation time. Then, here, the accuracy was investigated for each parameter considered here based on the finally determined input data by conducting preliminary analysis.

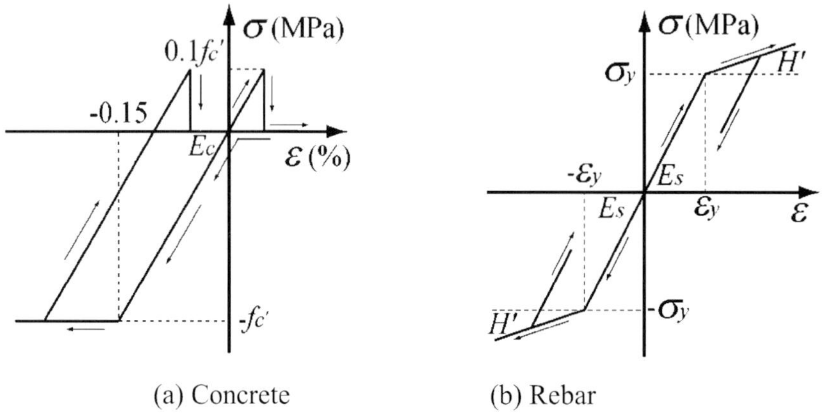

Fig. 5. Stress-strain relation of constitutive law model for materials.

Table 4. Analytical cases for investigating numerical accuracy by element size near support point.

Division near support point	Reinforced concrete modeling	Concrete yield criterion	Damping constant
4 Division			
6 Division	Beam element	Drucker–Prager	1.50%
8 Division			

4.1. *Element Size in Longitudinal Direction near Supporting Point*

Element size in the longitudinal direction near supporting point is basically set as one-half of an interval of stirrups and is 125 mm long. However, since an interaction between RC girder and the steel beam set on the upper surface of the girder for protecting jump up may influence to the numerical analysis on impact response behavior, numerical accuracy due to element size in the longitudinal direction was investigated, in which the element size in the region of 500 mm long back and forth of supporting point was considered. Here, three kinds of element size were taken as listed in Table 4, which are corresponding to those defined in the longitudinal direction of the RC girder, and element size for each four and eight division is the same for the elements near supporting part and the mid-span part, respectively, as shown in Figure 6.

Figure 7 shows the comparisons of analytical results for each response wave in case varying the mesh size of concrete elements near supporting point with the experimental results. From Figures 7(a) and 7(b), it is understood that the impact force wave at the beginning of impact can not be influenced by the element size near supporting point, but second dominant wave after 30 ms passed from the beginning of impact tends to be high amplitude with decreasing down of element size. From Figures 7(c) and 7(d), it is confirmed especially from Figure 7(d) that reaction force wave at the one supporting point tends to be high amplitude for high frequency component waves corresponding to element size decreasing and maximum amplitude in case of eight division was almost similar to that of the experimental results. From the comparison for displacement waves at the points $D - 1/2$ shown in Figures 7(e) and 7(f), it is observed that the wave during impact load surcharging behaves almost similar in spite of the element size near supporting point. Then, it is

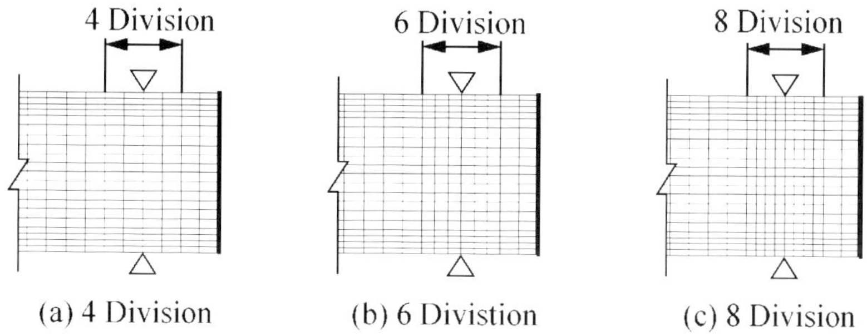

Fig. 6. FE division near 500 mm supporting point in longitudinal direction.

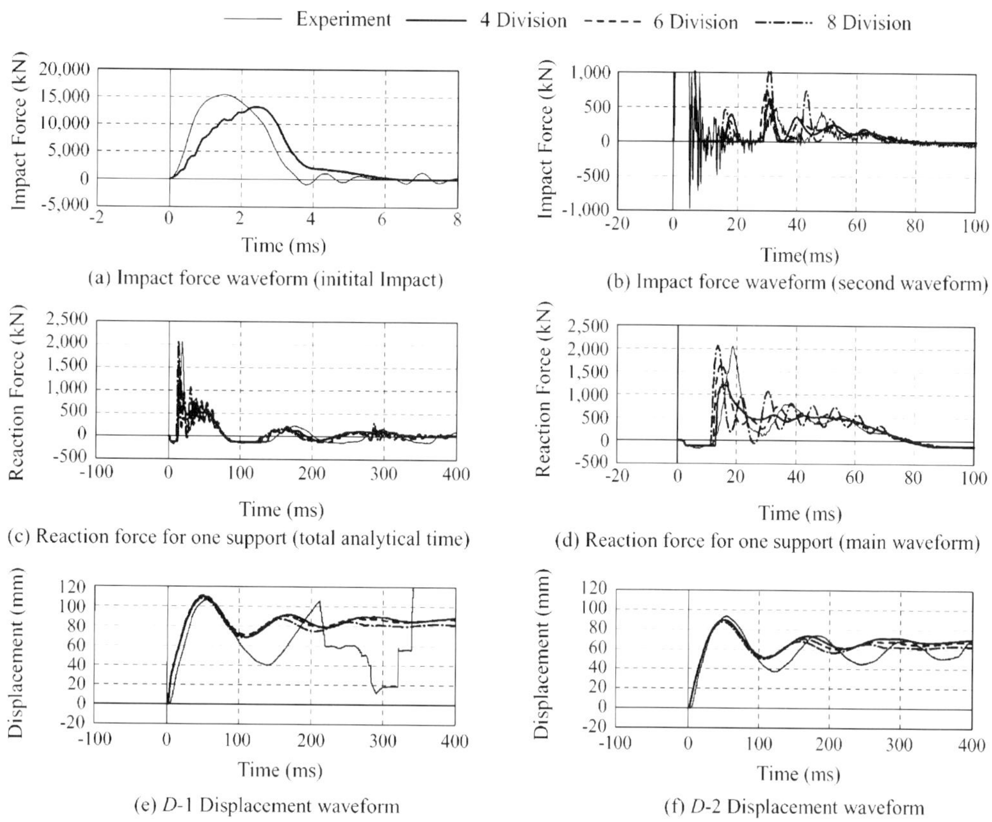

Fig. 7. Comparison of each response waveforms in case varying element size near support point.

confirmed from this investigation that the element size near supporting point will influence to the response of reaction force and maximum amplitude can be corresponded to the experimental results by dividing the 500 mm region near supporting point into eight elements in the longitudinal direction, in which the longitudinal length of those elements is the same to a quarter of an interval of stirrups.

Table 5. Analytical case for investigating numerical accuracy by yield criterion for concrete.

Concrete yield criterion	Division near support point	Reinforced concrete modeling	Damping constant
von Mises Drucker–Prager	8 Division	Beam element	1.50%

4.2. *Yield Criterion for Concrete Material*

In case analyzing concrete structure, since the confined effects of concrete material can be clearly appeared, it is generally pointed out that Drucker–Prager's yield criterion should be used. However, it has been seen by authors that elasto-plastic impact response behavior for small-scale RC beams under falling-weight impact loading can be better simulated using von Mises yield criterion (Kishi et al., 2005). In this study, an applicability of both yield criterion on impact response analysis for prototype RC girder was investigated with the parameters listed in Table 5 by comparing with the experimental results. The Drucker–Prager type yield criterion can be written as shown in Equation (1):

$$f(I_1, J_2) = \alpha I_1 + \sqrt{J_2} - k = 0, \tag{1}$$

where I_1 is the first invariant of stress tensor, J_2 is the second invariant of deviatoric stress tensor, α is a material constant, and k is a yield strength under pure shear. Coefficients α and k were determined referring to the book by Chen (1982) as $\alpha = 0.472$ and $k = 3.19$ MPa assuming that the tensile strength of concrete is 1/10th of compressive strength which is $f_t = 3.04$ MPa.

Figure 8 shows the comparisons of the impact response waves of the girder obtained using both yield criterion with the experimental results. From Figures 8(a) and 8(b), it is observed that time increment of the first dominant wave obtained using Drucker–Prager yield criterion is greater than that obtained using von Mises one. It is supposed that yield strength of concrete near impacted area is estimated so as to be upgraded due to three-dimensional confined effects in case using Drucker–Prager's yield criterion. However, even though Drucker–Prager's yield criterion is used, time increment of the wave cannot be increased up to that of the experimental results and the maximum amplitude is also a little smaller than experimental one.

From the comparisons of second dominant wave shown in Figure 8(b), it is observed that: (1) in case using von Mises yield criterion, impact force wave has not been excited during 65 ms after first dominant wave being excited, and then a half sine wave with high amplitude was excited. This wave configuration is greatly different from that of experimental results; and (2) on the other hand, in case using Drucker–Prager's yield criterion, four half-sine waves with a few ms duration time were excited with the time intervals of 10 to 15 ms and then the wave has been decreased to zero level. The wave configuration is some different from that of experimental results but is more similar to the experimental one than that obtained using von Mises yield criterion.

From Figure 8(c), it is seen that reaction force waves for one supporting point obtained using both yield criterion are almost the same to each other. The numerically estimated period for free vibration excited during a heavy weight being rebounded is shorter than that obtained from the experimental results in spite of yield criterion of concrete material. From Figure 8(d) of enlarged wave configuration in the beginning of impact, configurations of the first dominant wave obtained using both yield criterion are almost the same to each other. The maximum amplitude for those wave configurations is similar to that of experimental results but time increment of the wave at the

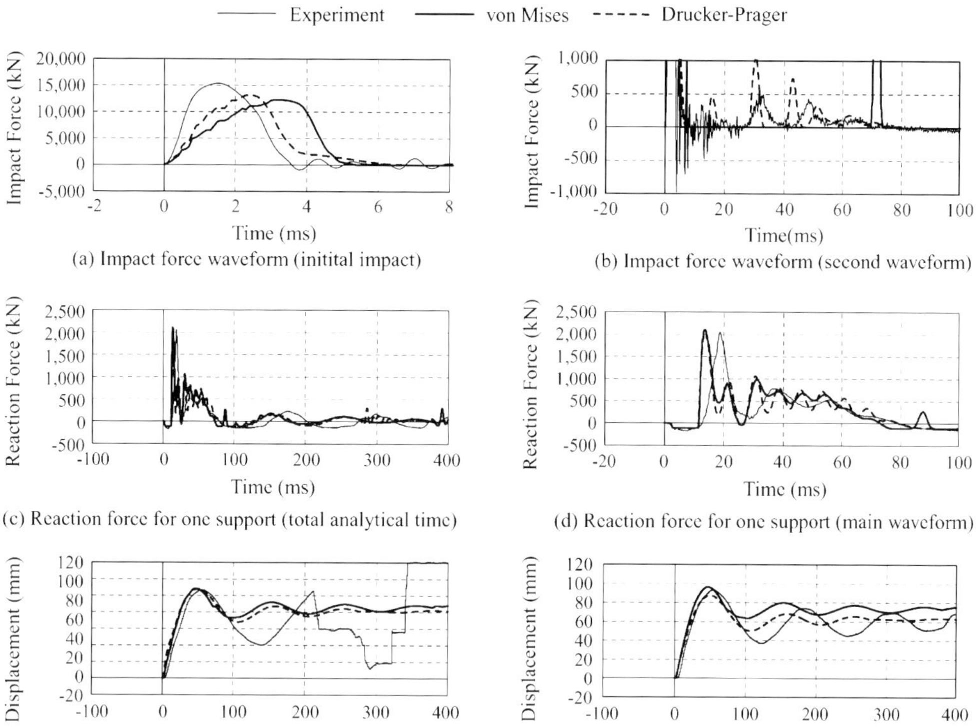

(a) Impact force waveform (initital impact)

(b) Impact force waveform (second waveform)

(c) Reaction force for one support (total analytical time)

(d) Reaction force for one support (main waveform)

(e) *D*-1 Displacement waveform

(f) *D*-2 Displacement waveform

Fig. 8. Comparison of each response waveforms for changing yield criterion of concrete.

beginning of impact is larger than that of the experimental results and is contradictory to the case of impact force wave.

From Figures 8(e) and 8(f) for displacement waves at the points $D - 1/2$, it is confirmed that numerical response wave during the impact load surcharging to the RC girder is similar to that of the experimental results irrespective of yield criterion of concrete material considered here. From Figure 8(f), it is observed that numerically estimated period for free vibration during a heavy weight being rebounded is shorter than that from the experimental results as well as reaction force wave. Residual displacement obtained using Drucker–Prager yield criterion is almost the same to the experimental result but that obtained using von Mises yield criterion is estimated larger than the experimental result. This implies that the local stress concentration and confined effects of concrete near impacted area of the girder can be better evaluated by using Drucker–Prager's yield criterion. Then, in case of spherical head of a heavy weight impacting for the prototype RC girder, impact response behavior of the girder may be better analyzed using Drucker–Prager yield criterion for concrete material.

4.3. *Damping Constant h*

In the impact response analysis for prototype RC girders, system damping effects of the structure should be considered as well as the hysteretic damping effects of concrete and rebar material. Here, an appropriate viscous damping constant, h (hereinafter, damping constant) was determined with referring to the first natural vibration frequency. In this study, four cases were considered: $h =$

Table 6. In case of varying damping constant for concrete.

Damping constant	Division near support point	Reinforced concrete modeling	Concrete yield criterion
0.50% 1.00% 1.50% 2.00%	8 Division	Beam element	Drucker–Prager

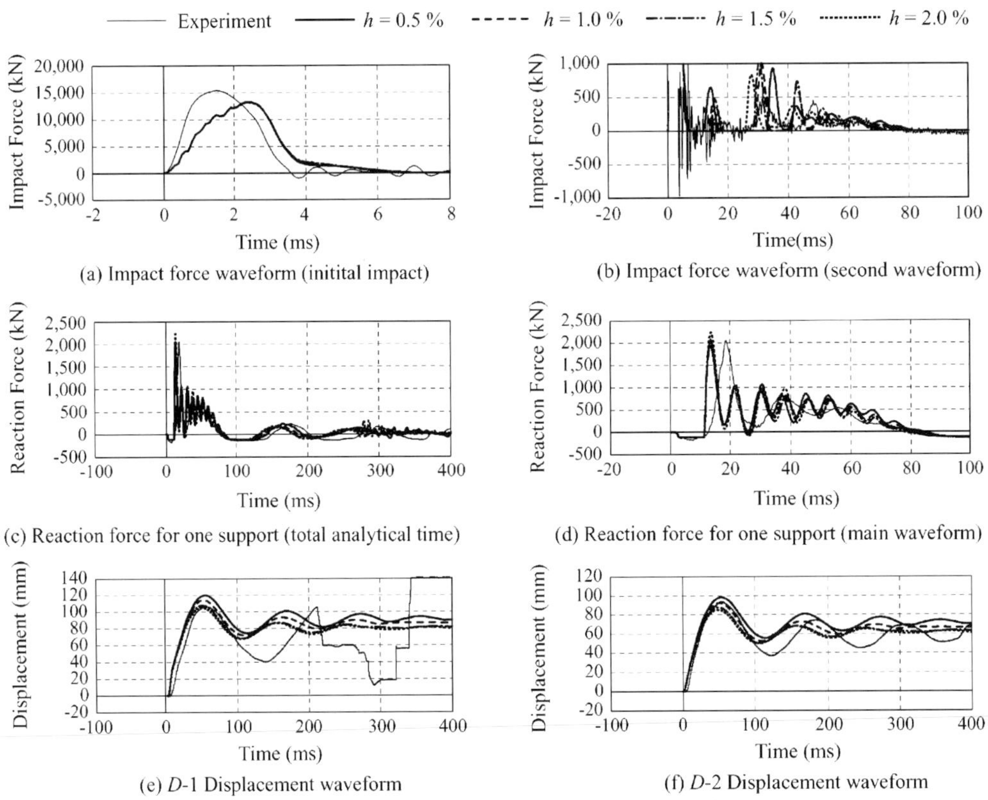

Fig. 9. Comparison of each response waveforms in case varying damping constant h.

0.005, 0.01, 0.015, and 0.02 with the design parameters as listed in Table 6. Figure 9 shows the comparison of impact response waves obtained varying damping constant h with those from the experimental results. From Figures 9(a) through 9(d), it is seen that: (1) maximum amplitudes of impact force and reaction force at one supporting point are hardly influenced by damping constant h; and (2) phase of the second dominant wave of impact force and reaction force wave during impact load surcharging tends to be large with increasing of damping constant h. From Figures 9(e) and 9(f) for the comparison of displacement waves at points $D - 1/2$, it is confirmed that maximum displacement and residual displacement tend to be large with decreasing of damping constant h and an appropriate damping constant may be $h = 0.015$.

5. Conclusions

To establish a simple elasto-plastic impact analysis method of flexure-failure-type RC girder, a falling-weight impact test and three-dimensional FE analyses were conducted for full-scale RC girder. The results obtained from this study are as follows:

(1) Changing yield criterion for concrete material from von Mises model to Drucker–Prager model, and applying system damping constant of $h = 0.015$, the maximum amplitudes and the configuration of primary wave for three waves: weight impact force; reaction force; and mid-span displacement, can be better simulated;
(2) To better simulate the characteristics of mid-span displacement wave, the region near supporting point should be modeled with fine mesh which is the same to a quarter of an interval of stirrups;
(3) Stress-strain relation of concrete assumed for analysis of small-scale RC beams can be applicable for the analysis of large-scale RC girders; and
(4) However, the initial gradient of weight impact force wave and frequency of secondary displacement wave cannot be properly estimated whatever any parameter is changed.

References

Chen, W.F., 1982, *Plasticity in Reinforced Concrete*, McGraw-Hill.
Hallquist, J.O., 2000, *LS-DYNA User's Manual*, Livermore Software Technology Corporation.
Japan Society of Civil Engineers, 1996, *Japan Concrete Standard*, JSCE [in Japanese].
Japan Society of Civil Engineers, 2004, *Practical Methods for Impact Tests and Analysis*, Structural Engineering Series 15, JSCE [in Japanese].
Kishi, N., Mikami, H., Matsuoka, K.G. and Ando, T., 1999, Elasto plastic impact behavior analysis of flexure failure type RC beam, *Journal of Structural Mechanics and Earthquake Engineering, JSCE*, 619(47), 215–233 [in Japanese].
Kishi, N., Nakano, O., Matsuoka, K.G. and Ando, T., 2001, Experimental study on ultimate strength of flexural-failure-type RC beams under impact loading, in *Transactions of the 16th International Conference on Structural Mechanics in Reactor Technology (SMIRT)*, Paper #1525.
Kishi, N., Mikami, H., Matsuoka, K.G. and Ando, T., 2002, Impact behavior of shear-failure-type RC beams without shear rebar, *International Journal of Impact Engineering*, 27, 955–968.
Kishi, N., Mikami. H, Namba. K. and Bhatti A.Q., 2005, Impact response analysis of light weight shear-failure type RC beams with stirrups, *Journal of Applied Mechanics JSCE*, 8, 471–480 [in Japanese].

DAMAGE STATES FOR REINFORCED CMU MASONRY SHEAR WALLS

Jia Li and Terence A. Weigel
Department of Civil Engineering, University of Louisville, Louisville, KY 40292, USA

Abstract

In order to assess the damage loss of reinforced masonry walls under earthquake loading, it is helpful to have a set of commonly accepted damage states. HAZUS gives detailed description for the qualitative damage states and assigns threshold drift ratios for achievement of each damage state. The HAZUS damage states are assigned based on expert opinion and judgment, and performance and experience data. Unfortunately, application of the HAZUS damage states is limited by the fact that they do not differentiate failure mode: flexure, shear and possible mixed flexure/shear. Furthermore, drift ratios defined in HAZUS have not been fully verified by experiment or experience. As a step toward addressing these deficiencies, this paper examines experimental results from three experimental programs and assesses the accuracy of the HAZUS methodology. Drift ratios at achievement of defined damage states are compared with HAZUS definitions. Results show that the HAZUS methodology tends to overestimate the drift ratio achieved by a wall at a given level of damage. In this paper, only experimental results for concrete masonry unit (CMU) walls are considered.

Introduction

Shear walls are the primary lateral load-resisting elements in reinforced masonry structures. When excited seismically, they are usually subjected to simultaneous in-plane loads, out-of-plane loads, axial loads and overturning moments. Design of shear walls for in-plane loads, as well as estimation of damage and associated economic loss upon occurrence of a seismic event, are important issues to all stakeholders: architects, engineers, building officials, insurance companies and owners. A key requirement in establishing a basis for accurate design provisions and assessment of the economic impact of seismic events is having an accurate set of quantitative damage descriptions for critical wall conditions. Also important is a corresponding set of qualitative descriptors that can be related to the quantitative damage descriptions.

A damage state is defined as a specified level of damage under earthquake, corresponding to some critical condition of the wall. Damage states have been defined differently by different agencies / researchers. For example ATC-40 (1996) defines the performance levels (consistent with damage) of Operational, Immediate Occupancy, and Life Safety and Structural Stability. HAZUS (FEMA, 1999) defines four qualitative damage states: slight, moderate, extensive, collapse, as shown in Table 1. HAZUS is software that will estimate potential regional earthquake losses. Loss estimates are used to plan and assist in reduction of risks from earthquakes, and to prepare for pre-earthquake emergency response and post-earthquake recovery. In addition to qualitative damage state, HAZUS also provides quantitative assessment of damage based on drift ratio and a "code level", as is shown in Table 2. Other quantitative damage indicators used by researchers (Hwang, 2001; Park, 1985; Mander, 1999) include displacement, ductility, energy, restoration time and money. In this paper, the HAZUS damage states are used. Much of this research is done for concrete and may not be directly applicable to the performance of masonry walls.

An issue not addressed by researchers and the HAZUS methodology, at least not directly, is the influence of the mode of response of the wall on the relationship between qualitative and quantitative damage measures. A reinforced masonry wall may respond to a seismic event in a shear critical or

M. Pandey et al. (eds), Advances in Engineering Structures, Mechanics & Construction, 111–120.

Table 1. Qualitative definition of damage states in HAZUS.

Damage States	Building Type (RM1L/RM2L*)
Slight	Diagonal hairline cracks on wall surfaces; large cracks around door and window openings in walls with large proportion of openings; minor separation of walls from the floor and roof diaphragms.
Moderate	Most wall surfaces exhibit diagonal cracks; some of the shear walls have exceeded their yield capacities indicated by larger diagonal cracks. Some walls may have visibly pulled away from the roof.
Extensive	Most shear walls with large openings have exceeded their yield capacities and some of the walls have exceeded their ultimate capacities indicated by large, through-the-wall diagonal cracks and visibly buckled wall reinforcement. Partial collapse of the roof may result from failure of wall to diaphragm connections.
Complete	Structure has collapsed or in imminent danger of collapse due to failure of the wall anchorages or the wall panels. Approximately 13(low-rise) of the total area of the building is expected to be collapsed.

*RM1L/RM2L is used to denote low-rise reinforced masonry bearing walls, generally ranging from 1–3 stories, with a total height less than 20 feet.

Table 2. Quantitative definition of damage states in HAZUS.

Seismic Design Level	Building Type (Low-Rise)	Drift Ratio at Damage State Threshold			
		Slight	Moderate	Extensive	Complete
High-Code*	RM1L/RM2L	0.004	0.008	0.024	0.070
Moderate-Code	RM1L/RM2L	0.004	0.007	0.019	0.053
Low-Code	RM1L/RM2L	0.004	0.006	0.016	0.044
Pre-Code	RM1L/RM2L	0.003	0.005	0.013	0.035

*High-Code, Moderate-Code, Low-Code correspond to the "quality" of the design code to which the building was designed. Pre-Code is used to indicate that the building was not designed for seismic loading.

flexurally critical mode. Some walls may exhibit a mixed flexural/shear critical mode. There are other possible response modes (for example, base sliding), but only flexure and shear are considered in this paper.

Flexurally critical walls exhibit yielding of vertical reinforced and crushing of the compression toe as they experience cycles of large excursions of reversed cyclic deformation. Shear critical walls exhibit diagonal tensile cracking when they experience the same type of loading. Examination of Tables 1 and 2 shows that the HAZUS methodology does not differentiate damage states based on behavior mode.

The principal objective of this research is to develop realistic qualitative and quantitative damage states and relate these states to behavior modes. In this paper four damage states are used as found in HAZUS: Slight, Moderate, Extensive, Collapse. The assessment of damage states and their relationship to behavior mode is based on experimental work for in-plane loading done by Shing (1988, 1990), Ibrahim (1999) and Eikanas (2003).

Damage States for Reinforced Masonry Walls

Test Results

Shing tested 16 reinforced concrete masonry walls, each with an aspect ratio of 1. Walls were reinforced both horizontally and vertically, and the amount of reinforcement was varied in order

Table 3. Test results – Shing*.

No	Mode	Str (psi)	ρ_v (%)	ρ_h (%)	Y (k)	Y_d (in)	M (k)	M_d (in)	U (k)	U_d (in)	D (k)	D_d (in)
1(P)	Flexure	200	0.38	0.24	60	0.15	82	0.56	87	0.82	68	1.22
1(N)					N/A	N/A	76	0.35	78	0.59	58	0.85
2(P)	Flexure	270	0.38	0.24	66	0.15	82	0.35	83	0.50	60	0.90
2(N)					N/A	N/A	84	0.50	98	0.54	82	0.90
3(P)	Shear	270	0.74	0.14	N/A	N/A	80	0.17	100	0.60	93	0.92
3(N)					N/A	N/A	N/A	N/A	105	0.70	80	1.10
4(P)	Shear	0	0.74	0.14	65	N/A	55	0.17	72	0.34	N/A	N/A
4(N)					N/A	N/A	51	0.12	87	0.48	N/A	N/A
5(P)	Shear	100	0.74	0.14	82	0.35	60	0.16	89	0.40	70	0.46
5(N)					N/A	N/A	N/A	N/A	84	0.40	65	0.80
7(P)	Shear	100	0.74	0.14	83	0.25	65	0.15	97	0.45	N/A	N/A
7(N)					N/A	N/A	60	0.10	97	0.61	72	0.73
9(P)	Shear	270	0.38	0.14	76	0.10	92	0.30	96	0.30	80	0.35
9(N)					N/A	N/A	N/A	N/A	96	0.30	N/A	N/A
10(P)	Mixed	100	0.38	0.14	48	0.10	60	0.15	69	0.80	60	0.85
10(N)					N/A	N/A	58	0.18	67	0.85	64	1.15
12(P)	Flexure	100	0.38	0.24	46	0.10	69	0.38	71	0.62	70	0.92
12(N)					N/A	N/A	70	0.57	71	0.87	66	1.10
13(P)	Shear	270	0.54	0.24	90	0.17	109	0.30	109	0.40	98	0.70
13(N)					N/A	N/A	115	0.50	115	0.64	N/A	N/A
14(P)	Shear	270	0.54	0.14	85	0.12	98	0.25	98	0.40	95	0.58
14(N)					N/A	N/A	105	0.32	112	0.40	80	0.50
15(P)	Mixed	100	0.54	0.24	60	0.19	67	0.25	82	0.55	66	1.30
15(N)					N/A	N/A	80	0.30	88	0.55	84	1.20
16(P)	Shear	270	0.74	0.24	100	0.20	87	0.14	120	0.60	105	0.70
16(N)					N/A	N/A	85	0.15	120	0.55	N/A	N/A

*See the Appendix for explanation of table notation.

to produce either shear dominated or flexurally dominated response. A summary of Shing's test results are shown in Table 3. Of Shing's walls, specimens 1, 2, 10, 12 and 15 exhibited primarily flexural response (10 and 15 actually exhibited mixed response) and specimens 3, 4, 5, 7, 9, 13, 14 and 16 exhibited shear response. Shing's specimens 6, 8, and 11 exhibited significant base sliding and results for those walls are not included in this paper.

Eikanas tested seven reinforced concrete masonry walls. Aspect ratios varied from 1.0 to 2.63. Walls were reinforced both horizontally and vertically and the amount of reinforcement was varied from wall to wall. All of Eikanas's walls exhibited primarily flexurally critical behavior. A summary of Eikanas's test results are shown in Table 4.

Ibrahim tested five reinforced concrete masonry walls. The walls were all 55 in tall and aspect ratios varied from 0.467 to 1.00. Walls were reinforced both horizontally and vertically. All of Ibrahim's walls exhibited shear critical behavior. A summary of Ibrahim's test results are shown in Table 5.

Table 4. Test results – Eikanas.

No	AR	Str (psi)	ρ_v (%)	ρ_h (%)	Y (k)	Y_d (in)	S (k)	S_d (in)	U (k)	U_d (in)	D (k)	D_d (in)
1(P)	1.29	27	0.29	0.18	27.2	0.10	36.0	0.16	49.7	0.73	45.1	1.12
1(N)					28.1	0.11	40.4	0.31	47.6	0.90	39.0	1.15
2(P)	1.87	27	0.29	0.18	9.9	0.19	8.5	0.13	29.7	1.58	29.4	1.99
2(N)					25.1	0.37	31.6	0.58	35.1	0.99	28.0	1.65
3(P)	2.63	27	0.31	0.18	N/A	N/A	N/A	N/A	N/A	N/A	N/A	N/A
3(N)					4.5	0.48	2.9	0.19	18.9	3.19	18.7	5.00
4(P)	1.29	27	0.51	0.18	35.8	0.16	47.9	0.28	64.8	0.57	49.8	0.69
4(N)					41.2	0.34	50.1	0.50	53.6	0.70	41.3	0.90
5(P)	1.87	27	0.51	0.18	21.5	0.28	30.9	0.38	46.2	0.99	36.8	1.38
5(N)					26.1	0.28	34.3	0.53	45.3	1.19	33.8	1.78
6(P)	2.63	27	0.52	0.18	13.7	0.23	20.7	0.48	25.3	1.12	20.8	2.24
6(N)					15.2	0.25	23.3	0.75	25.6	1.13	18.3	1.81
7(P)	1.00	27	0.29	0.18	45.7	0.08	60.5	0.20	75.3	0.50	57.6	0.70
7(N)					38.1	0.11	49.7	0.20	64.3	0.50	50.6	0.80

Table 5. Test results – Ibrahim.

No	AR	Str (psi)	ρ_v (%)	ρ_h (%)	Y (k)	Y_d (in)	M (k)	M_d (in)	U (k)	U_d (in)	D (k)	D_d (in)
1(P)	1.000	100	0.4	0.2	34.4	0.13	38.6	0.23	44.9	0.31	40.4	0.59
1(N)					34.4	0.15	37.3	0.17	47.0	0.59	30.8	0.79
2(P)	0.636	100	0.4	0.2	56.4	0.16	56.8	0.20	83.1	0.59	N/A	N/A
2(N)					61.6	0.13	64.3	0.16	91.7	0.39	76.4	0.59
3(P)	0.467	100	0.4	0.2	106.5	0.16	108.3	0.18	124.3	0.39	N/A	N/A
3(N)					118.2	0.16	116.4	0.16	118.9	0.26	101.1	0.39
4(P)	0.636	100	0.6	0.2	81.3	0.15	85.2	0.18	93.3	0.22	73.0	0.63
4(N)					79.5	0.16	74.4	0.20	91.9	0.31	67.4	0.59
5(P)	0.636	250	0.4	0.2	75.0	0.10	75.7	0.12	78.0	0.26	62.9	0.59
5(N)					80.0	0.12	92.6	0.16	105.2	0.31	73.0	0.59

Qualitative Damage States

Because HAZUS fails to differentiate between flexurally critical and shear critical walls, it was necessary to create such a distinction when evaluating test results. Table 6 shows the classification selected.

For flexurally dominated behavior, initial yielding of flexural steel is associated with "Slight" damage. For this damage state repairs can be made relatively easily by grout injection and/or cosmetic repairs. The value of masonry strain used to define the end of the "Moderate" damage region (0.0025) is based on what the MSJC Code (2005) calls the "maximum usable strain".

Table 6. Qualitative definition of damage states based on experimental results.

Damage States	Flexurally Dominated Behavior	Shear Dominated Behavior
Slight	First yield of vertical steel achieved	First yield of vertical steel achieved
Moderate	Masonry compressive strain 0.0025 achieved	Major diagonal cracking
Extensive	Toe crushing or ultimate load achieved	Ultimate load achieved
Complete	20% load degradation	20% load degradation

For shear dominated behavior, major diagonal cracks extend at an angle of approximately 45°, from the top of the wall toward its base. When large amounts of flexural reinforcement are provided, the vertical steel may not yield, and the first indication of damage will be diagonal cracking, and the wall enters the "Moderate" damage state after vertical steel yielding. Shing's specimens 5, 7, and 16 exhibited this type of behavior.

"Complete" damage is defined as 20% strength degradation. Although collapse may not occur at load degradation of 20% (or larger), repair costs will typically be such that the structure is classified as a total economic loss. Therefore the load degradation of 20% or larger is be considered "Complete" damage.

Quantitative Damage States

In order to compare directly with HAZUS methodology, drift ratio is used as the quantitative damage indicator. Tables 7 through 9 show measured drift ratio (DR) from experimental results.

Parameters that influence the behavior mode of a reinforced masonry wall include aspect ratio, amount of flexural steel, amount of shear steel and masonry strength. HAZUS methodology presented in this paper focuses on low-rise masonry bearing walls. Most of the walls tested in the three experimental programs had axial loads consistent with low-rise construction. One of Ibrahim's walls had an axial load of 250 psi and some of Shing's walls (see Table 3) had axial loads of 200 or 270 psi. These levels of axial load would be consistent with medium-rise or high-rise construction. Despite this inconsistency between axial load used in some of the experimental tests and the low-rise HAZUS methodology used in this paper, axial load is not considered as a parameter in this study. Future work may incorporate effects of axial load in damage assessment.

For the experimental database examined, walls with aspect ratio less than 1 exhibited shear critical response, regardless of the amount of reinforcement present. The specimens falling in this category are 2, 3, 4 and 5 from Ibrahim's test program.

When the aspect ratio was greater than or equal to 1, behavior can be either flexurally critical, shear critical or mixed mode. Ibrahim's Wall 1 had an aspect ratio of 1, and exhibited shear critical behavior. All of Shing's walls had an aspect ratio of 1. His Walls 1, 2, 10, 12 and 15 exhibited primarily flexural response (10 and 15 actually exhibited mixed response) and specimens 3, 4, 5, 7, 9, 13, 14 and 16 showed shear critical behavior. The specific behavior mode depended on the amount of flexural and shear reinforcing present in the wall. All of Eikanas's walls exhibited flexural response.

Drift ratio versus the corresponding damage states are plotted in Figure 1. The flexurally critical specimens ($1.0 < AR < 2.6$, $AR = 1.0$) and shear critical specimens ($AR = 1.0$, $AR < 1.0$) are plotted separately. In these figures, numerical designations 1, 2, 3, 4 correspond to damage states Slight, Moderate, Extensive and Collapse, respectively.

In Figure 1, the aspect ratio (AR) is used as a discriminator. The aspect ratio is defined as the ratio of wall height divided by wall length, without regard to wall support conditions. Some

Table 7. Drift ratio – Shing.

No	H (in)	Y_d (in)	DR	Avg	M_d (in)	DR	Avg	U_d (in)	DR	Avg	D_d (in)	DR	Avg
1(P)	72	0.15	0.21	0.21	0.56	0.78	0.64	0.82	1.14	0.98	1.22	1.69	1.44
1(N)		N/A	N/A		0.35	0.49		0.59	0.82		0.85	1.18	
2(P)	72	0.15	0.21	0.21	0.35	0.49	0.59	0.50	0.69	0.72	0.90	1.25	1.25
2(N)		N/A	N/A		0.50	0.69		0.54	0.75		0.90	1.25	
3(P)	72	N/A	N/A	N/A	0.17	0.24	0.24	0.60	0.83	0.90	0.92	1.28	1.41
3(N)		N/A	N/A		N/A	N/A		0.70	0.97		1.10	1.53	
4(P)	72	N/A	N/A	N/A	0.17	0.24	0.21	0.34	0.47	0.57	N/A	N/A	N/A
4(N)		N/A	N/A		0.12	0.17		0.48	0.67		N/A	N/A	
5(P)	72	0.35	0.49	0.49	0.16	0.22	0.22	0.40	0.56	0.56	0.46	0.64	0.88
5(N)		N/A	N/A		N/A	N/A		0.40	0.56		0.80	1.11	
7(P)	72	0.25	0.35	0.35	0.15	0.21	0.18	0.45	0.63	0.74	N/A	N/A	1.01
7(N)		N/A	N/A		0.10	0.14		0.61	0.85		0.73	1.01	
9(P)	72	0.10	0.14	0.14	0.30	0.42	0.42	0.30	0.42	0.42	0.35	0.49	0.49
9(N)		N/A	N/A		N/A	N/A		0.30	N/A		N/A	N/A	
10(P)	72	0.10	0.14	0.14	0.15	0.21	0.23	0.80	1.11	1.15	0.85	1.18	1.39
10(N)		N/A	N/A		0.18	0.25		0.85	1.18		1.15	1.60	
12(P)	72	0.10	0.14	0.14	0.38	0.53	0.66	0.62	0.86	1.03	0.92	1.28	1.41
12(N)		N/A	N/A		0.57	0.79		0.87	1.20		1.10	1.53	
13(P)	72	0.17	0.24	0.24	0.30	0.42	0.56	0.40	0.56	0.73	0.70	0.97	0.97
13(N)		N/A	N/A		0.50	0.69		0.64	0.89		N/A	N/A	
14(P)	72	0.12	0.17	0.17	0.25	0.35	0.40	0.40	0.56	0.56	0.58	0.81	0.75
14(N)		N/A	N/A		0.32	0.44		0.40	0.56		0.50	0.69	
15(P)	72	0.19	0.26	0.26	0.25	0.35	0.39	0.55	0.76	0.76	1.30	1.81	1.74
15(N)		N/A	N/A		0.30	0.42		0.55	0.76		1.20	1.67	
16(P)	72	0.20	0.28	0.28	0.14	0.19	0.20	0.60	0.83	0.80	0.70	0.97	0.97
16(N)		N/A	N/A		0.15	0.21		0.55	0.76		N/A	N/A	

researchers define an "effective" aspect ratio, which, for example, would be calculated as half the wall height divided by its length, when fixed-fixed support conditions exist. However, in order to keep the quantitative classification simple, and because the current experimental database is limited in size, only the aspect ratio is used in Figure 1.

Based on experimental results and results shown in Figure 1, quantitative damage states are classified as shown in Table 10.

Results Compared with HAZUS

From Table 10 we observe that comparable levels of damage will be produced at much lower drift ratios when the wall behavior is dominated by shear, as opposed to flexure. This implies that a basing damage estimates on a single drift ratio for all low-rise masonry walls, as does HAZUS, is not likely to be particularly accurate.

Table 8. Drift ratio – Eikanas.

No	H (in)	Y_d (in)	DR	Avg	S_d (in)	DR	Avg	U_d (in)	DR	Avg	D_d (in)	DR	Avg
1(P)	52	0.10	0.19	0.20	0.16	0.31	0.46	0.73	1.40	1.57	1.12	2.15	2.18
1(N)		0.11	0.21		0.31	0.60		0.90	1.73		1.15	2.21	
2(P)	84	0.19	0.23	0.34	0.13	0.15	0.42	1.58	1.88	1.53	1.99	2.36	2.16
2(N)		0.37	0.44		0.58	0.69		0.99	1.18		1.65	1.96	
3(P)	84	N/A	N/A	0.57	N/A	N/A	0.22	N/A	N/A	3.80	N/A	N/A	5.95
3(N)		0.48	0.57		0.19	0.22		3.19	3.80		5.00	5.95	
4(P)	52	0.16	0.31	0.48	0.28	0.54	0.75	0.57	1.10	1.24	0.69	1.33	1.53
4(N)		0.34	0.65		0.50	0.96		0.70	1.35		0.90	1.73	
5(P)	84	0.28	0.33	0.33	0.38	0.45	0.54	0.99	1.18	1.30	1.38	1.64	1.88
5(N)		0.28	0.33		0.53	0.63		1.19	1.42		1.78	2.12	
6(P)	84	0.23	0.27	0.29	0.48	0.57	0.73	1.12	1.33	1.34	2.24	2.67	2.41
6(N)		0.25	0.30		0.75	0.89		1.13	1.35		1.81	2.15	
7(P)	52	0.08	0.15	0.18	0.20	0.38	0.38	0.50	0.96	0.96	0.70	1.35	1.45
7(N)		0.11	0.21		0.20	0.38		0.50	0.96		0.80	1.54	

Table 9. Drift ratio – Ibrahim.

No	H (in)	Y_d (in)	DR	Avg	M_d (in)	DR	Avg	U_d (in)	DR	Avg	D_d (in)	DR	Avg
1(P)	55	0.13	0.24	0.26	0.23	0.41	0.36	0.31	0.57	0.82	0.59	1.07	1.25
1(N)		0.15	0.27		0.17	0.30		0.59	1.07		0.79	1.43	
2(P)	55	0.16	0.29	0.27	0.20	0.36	0.33	0.59	1.07	0.89	N/A	N/A	1.07
2(N)		0.13	0.24		0.16	0.29		0.39	0.71		0.59	1.07	
3(P)	55	0.16	0.29	0.29	0.18	0.32	0.31	0.39	0.71	0.59	N/A	N/A	0.71
3(N)		0.16	0.29		0.16	0.29		0.26	0.46		0.39	0.71	
4(P)	55	0.15	0.27	0.28	0.18	0.32	0.34	0.22	0.39	0.48	0.63	1.14	1.10
4(N)		0.16	0.29		0.20	0.36		0.31	0.57		0.59	1.07	
5(P)	55	0.10	0.18	0.20	0.12	0.21	0.25	0.26	0.46	0.52	0.59	1.07	1.07
5(N)		0.12	0.21		0.16	0.29		0.31	0.57		0.59	1.07	

Table 10. Quantitative damage states.

Damage States	Flexurally Dominated Behavior		Shear Dominated Behavior	
	Drift Ratio (AR = 1.0)	Drift Ratio (1.0 < AR < 2.6)	Drift Ratio (AR = 1.0)	Drift Ratio (AR < 1.0)
Slight	<0.25	0.20–0.50	<0.25	0.20–0.30
Moderate	0.25–0.70	0.40–0.80	0.25–0.55	0.25–0.35
Extensive	0.70–1.20	1.20–1.60	0.55–0.80	0.40–1.00
Complete	1.10–1.75	1.50–2.40	0.80–1.50	1.00–1.25

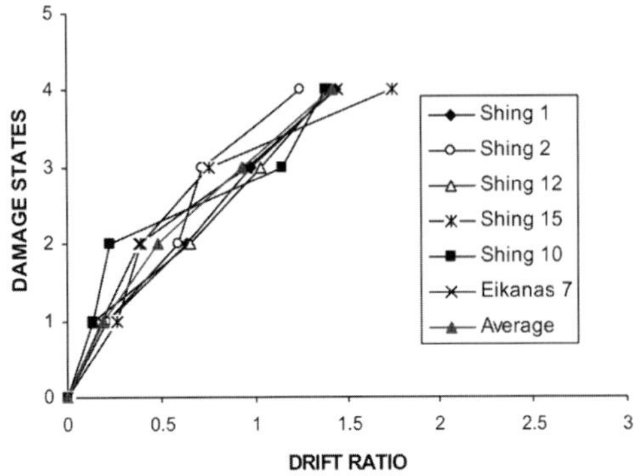

(a) Flexurally critical behavior (AR = 1.0)

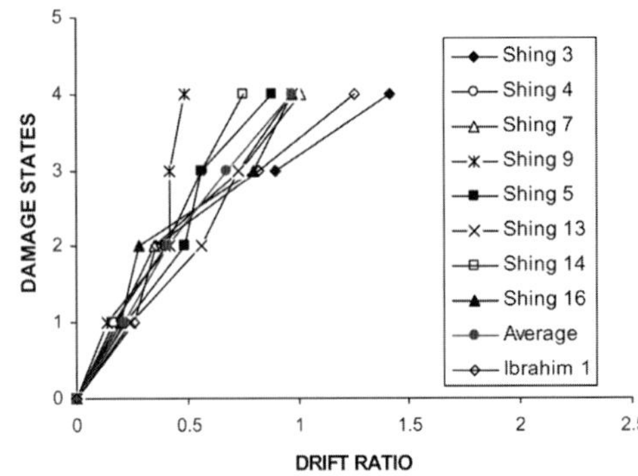

(b) Shear critical behavior (AR = 1.0)

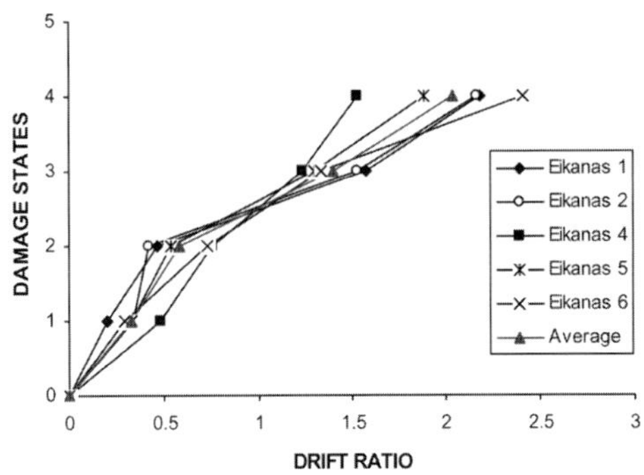

(c) Flexurally critical behavior (1.0 < AR < 2.6)

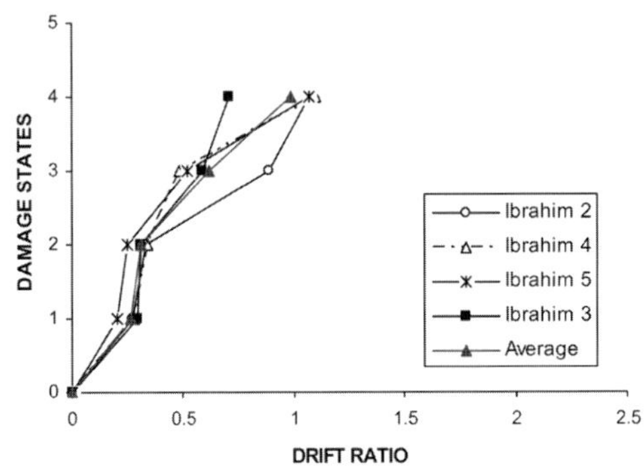

(d) Shear critical behavior (AR < 1.0)

Fig. 1. Drift ratio versus damage states.

Table 11. Drift ratio comparison with HAZUS.

Damage States	HAZUS				Flexure		Shear	
	High-Code	Moderate-Code	Low-Code	Pre-Code	AR = 1.0	1.0 < AR < 2.6	AR = 1.0	AR < 1.0
Slight	0.40	0.40	0.40	0.30	<0.25	0.20–0.50	<0.25	0.20–0.30
Moderate	0.80	0.70	0.60	0.50	0.25–0.70	0.40–0.80	0.25–0.55	0.25–0.35
Extensive	2.40	1.90	1.60	1.30	0.70–1.20	1.20–1.60	0.55–0.80	0.40–1.00
Complete	7.00	5.30	4.40	3.50	1.10–1.75	1.50–2.40	0.80–1.50	1.00–1.25

Table 11 compares HAZUS drift ratios with those selected in this paper, as discriminated by qualitative damage state. Note that HAZUS predicts much larger drift ratios than do experimental results. The difference is quite high in the Extensive and Complete damage states. Current HAZUS methodology compares best with high aspect ratio, flexurally-critical walls.

Conclusions

In this study, quantitative and qualitative damage states for low-rise reinforced masonry walls are defined based on experimental results. Drift ratio is used as the quantitative damage indicator, and separate classifications are provided depending on whether the behavior of the wall is dominated by flexure or by shear. Comparison of the drift ratios selected in this paper indicates large differences when compared to the HAZUS methodology, which does not differentiate based on wall behavior mode. Current HAZUS provisions best correlate with high aspect ratio, flexurally critical walls.

While it is certain that masonry wall damage is significantly influenced by behavior mode, limitations of the experimental database restrict the confidence that can be place in any classification system. Although the authors do believe classification system presented in this paper is superior to the current HAZUS methodology, more testing is required before any classification system can be made reliable.

Appendix

No. = specimen number
(P) = value in positive test direction
(N) = value in negative test direction
AR = aspect ratio
H = wall height, in
DR = drift ratio, %
Avg = average drift ratios in positive and negative directions
ρ_v = wall vertical steel, %
ρ_h = wall horizontal steel, %
Y = lateral load at first yield of vertical steel, k
Y_d = wall displacement at Y, in
M = lateral load at major diagonal crack, k
M_d = wall displacement at M, in
S = lateral load when masonry achieved compression strain of 0.0025, k
S_d = wall displacement at S, in
U = maximum lateral load attained, k
U_d = wall displacement at U, in
D = lateral load at 20% degradation, k
D_d = wall displacement at D, in
Str = axial stress, psi
N/A = value is not available

References

ATC-40, 1996,, *Seismic Evaluation and Retrofit of Concrete Buildings, Volume 1*, Applied Technology Council, Redwood City, CA.

Eikanas, I.K., 2003, Behavior of Concrete Masonry Shear Walls with Varying Aspect Ratio and Flexural Reinforcement, M.S. Thesis, Washington State University.

Federal Emergency Management Agency (FEMA), 1999, *FEMA's Tool for Estimating Potential Losses from Natural Disasters*, HAZUS Technical Manuals.

Hwang, H., Jingbo, L. and Yi-Huei, C., 2001, Seismic Fragility Analysis of Highway Bridges, MAEC RR-4 Projects.

Ibrahim, K.S. and Suter, G.T., 1999, Ductility of Concrete Masonry Shear Walls Subjected to Cyclic Loading, in *Proceedings of the Eighth North American Masonry Conference*, The Masonry Society, University of Texas, Austin, TX.

Mander, J.B., 1999, Fragility Curve Development for Assessing the Seismic Vulnerability of Highway Bridges, University at Buffalo, State University of New York.

Masonry Standards Joint Committee, 2005, Building Code Requirements for Masonry Structures, ACI 530-05/ASCE 5-05/TMS 402/05.

Park, Y.-J. and Ang, A.H.S., 1985, Mechanistic Seismic Damage Model for Reinforced Concrete, *ASCE Journal of Structural Engineering*, 111(4), 722–739.

Shing, P.B., Noland, J.L., Klamerus, E. and Spaeh, H., 1988, Inelastic Behavior of Concrete Masonry Shear Walls, *ASCE Journal of Structural Engineering*, 115(9), 2204–2225.

Shing, P.B., Schuller, M. and Hoskere, V.S., 1990, In-Plane Resistance of Masonry Shear Walls, *ASCE Journal of Structural Engineering*, 116(3), 619–639.

SEISMIC BEHAVIOUR OF PRECAST COLUMN-TO-FOUNDATION GROUTED SLEEVE CONNECTIONS

Paolo Riva

Structural Engineering, Department of Design and Technology,
University of Bergamo, Italy
E-Mail: paolo.riva@unibg.it

Abstract

The results of a set of experimental tests concerning the cyclic behaviour of prefabricated column-to-foundation connections is presented. The tests allow to compare the response of cast-in-place connections against pocket foundation and grouted sleeve solutions. The results demonstrate that grouted sleeves ensure a ductility similar to the one of cast in situ column-foundation connections and of pocket foundations, although a slightly smaller dissipation capacity is observed. It is found that in grouted sleeves connections the damage is localized at the column base, in the thin grout layer existing between the prefabricated column and the foundation. As a result, very little damage may be observed in the column, allowing an easier post-seismic column repair.

Introduction

Warehouses and commercial malls in Italy are generally built using precast reinforced concrete elements. The typical structural layout consists of cantilever columns, connected by simply supported precast and prestressed beams, supporting prestressed concrete roof elements. The foundations are usually made of isolated precast cup-footings, in which the columns are inserted and grouted in-situ. Such a structural layout is extremely cost effective, and sensibly reduces the construction time. However, its effectiveness is seriously hampered when it is intended for construction in seismic areas, particularly if Capacity Design (CD) based codes are adopted in the design process.

According to the European Code (EC8, 2003) or to the new Italian Code (OPCM 3274, 2003), the described layout is defined as an "inverted pendulum" system, which has the characteristic of being statically determinate, with potential plastic hinge regions located only at the base of the columns.

The design of the column footing is carried out by: i) assuming that a plastic hinge might develop at the column base section for the design earthquake event; ii) preventing the formation of any inelastic mechanism in the footing. The latter goal is reached by designing the foundation to resist to the design axial force and to the maximum possible resisting bending moment of the column base section, computed by considering an appropriate overstrength factor.

Adopting a CD approach, the foundation base becomes easily very large, also for medium-sized columns. This issue might be hardly relevant for cast in-situ structures, but could seriously limit the cost effectiveness of pre-cast concrete structures such as those previously described. In fact, due to the dimensions of the foundations, the footings have to be cast-in-situ, making the use of isolated cup-foundations less attractive.

Hence, mat foundations become often more convenient than isolated footings. In this case, the column-to-foundation connection is usually done: i) by using steel base plates; ii) by adopting column pockets, grouted in-situ; iii) by means of grouted sleeves.

Both from the point of view of prefabrication and of seismic response, steel base plates appear to be the least attractive of the three connection types, as they require small tolerances for the on-site placing, and adequate ductility might not be easily ensured in the column base section.

M. Pandey et al. (eds), Advances in Engineering Structures, Mechanics & Construction, 121–128.

Pocket foundations are the most used, at least in Italy, as they ensure ease of placement, and adequate ductility of the column base section after grouting. As a matter of fact, the behaviour of the connection is very much similar to that of cast-in-situ structures. On the other hand, either a thick mat foundation or a collar are required to accommodate the column, with consequent increase of costs related to the construction of the pocket.

Grouted sleeves appear to be the cheapest, hence most convenient, type of connection. However, the seismic response of such column-to-foundation connections is not well documented. In fact, while a large amount of experimental tests have been carried out on columns subjected to cyclic loading (e.g. CEB, 1996), no experimental results concerning the cyclic response of grouted sleeve column-to-foundation connections are available in the literature, at least to the Author's knowledge.

In order to investigate the cyclic behaviour of grouted sleeve connections and to compare the behaviour of such connections with cast-in-situ and grouted pocket column-to-foundation connections, an experimental campaign was set-up at the University of Brescia. The response of five columns (section 400x400mm, height 3200mm), subjected to a cyclic top horizontal displacement history, was investigated, considering different connection details. A constant axial force equal to 600kN was applied to all columns.

Experimental Tests

The experimental tests concerned five specimens with different column-to-foundation connections and approximately the same maximum bending moment capacity. All the specimen tested had a 400x400mm column cross section and a clear height from the foundation to the top equal to 3200mm. The geometry of the tested specimens and the mechanical characteristics of concrete and reinforcing steel are shown in Fig. 1.

Specimens CS and PF are representative of a typical cast-in-situ column-to-foundation connection and grouted pocket foundation, respectively. Specimens GS4 and GS4B are both characterized by having four grouted sleeves. The difference between the two specimens consist in the anchorage length of the ϕ26 bars in the foundation: the former has straight anchored bars, whereas the latter has 90° hooks at the bar ends. Specimen GS8 has 8 grouted sleeves with ϕ22 bars.

The experimental setup adopted is shown in Fig. 2. For all the tests, the axial force, equal to 600kN, was first applied by means of two hydraulic jacks. A cyclic horizontal displacement was then applied at the top of the column by means of a 1000 kN electromechanical screw jack having a 500 mm maximum stroke.

The applied displacement history is shown in Fig. 3. It is observed that the maximum imposed displacement is equal to 200 mm, corresponding to a 6.0% drift, much larger than the maximum drift commonly accepted for prefabricated columns under a design seismic event, equal to 2.5%.

Experimental Results

The experimental load-displacement curves for the tested specimens and the pictures of the critical section at 1% and 2.5% drift, corresponding, respectively, to the drift under the 50% in 50yr and 10% in 50yr probability earthquakes, are illustrated in Fig. 4. Based on the results, the following observations may be made:

- all the columns had almost the same maximum force capacity, equal to 75kN, as expected;
- the cast in situ column (CS) collapsed during the second cycle at 5% drift (160mm), due to the tensile failure of one of the reinforcing bars. At the time of collapse, the residual column strength was 30% smaller than the maximum column strength. Considerable pinching appears in the cycles after the cycles at 2.5%. Up to the cycles at 2.5% drift, representative of the maximum drift under the design earthquake, the column behaviour was stable, and little damage could be observed in the column, as demonstrated by the pictures at 1% and 2.5% drift;

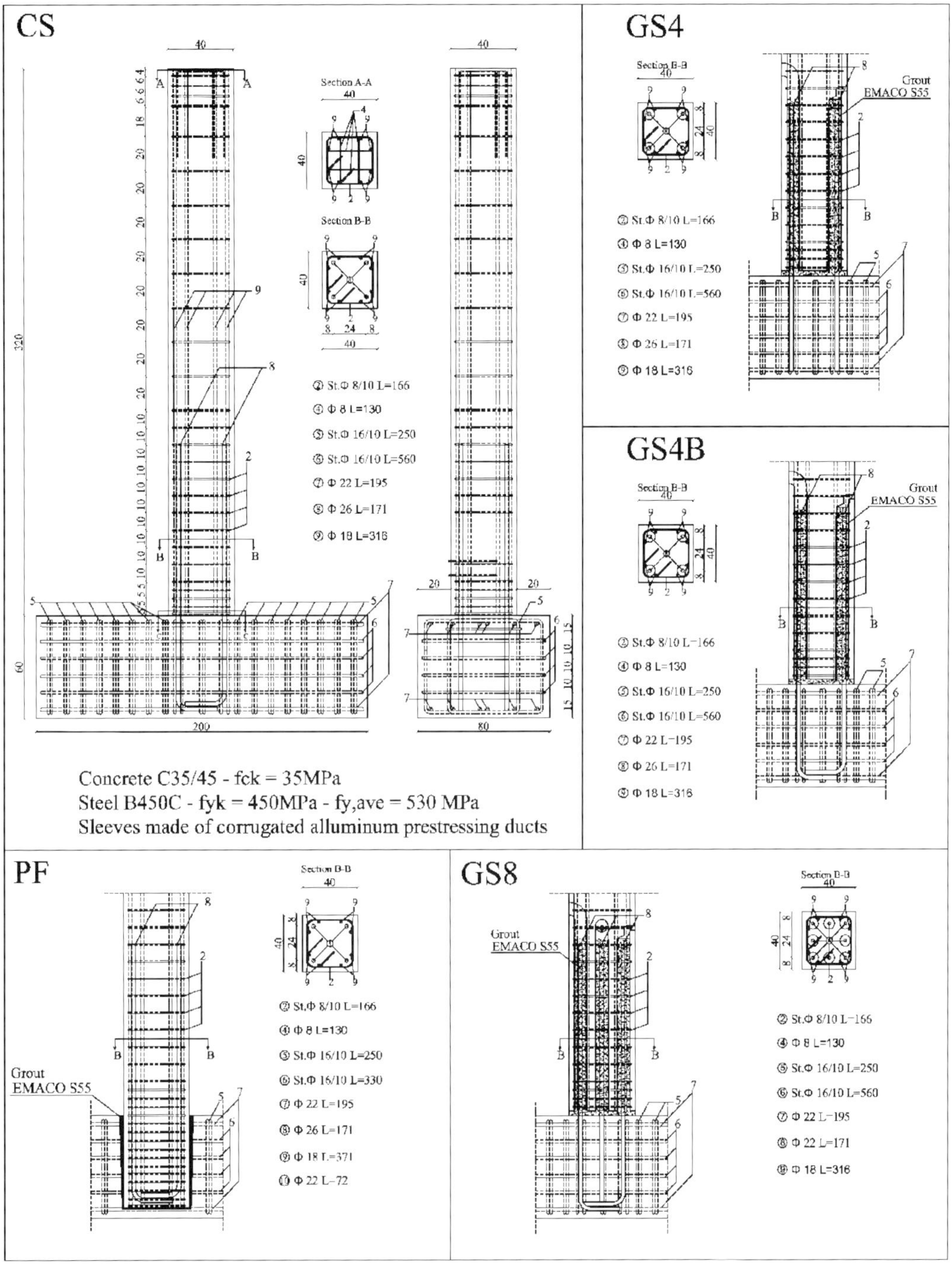

Fig. 1 – Tested specimens (Dimensions expressed in cm): (a) CS, cast in situ column; (b) GS4, column with four grouted sleeves; (c) GS4B, column with four grouted sleeves and 90° hooks in the anchored bars; (d) PF, pocket foundation; (e) GS8, column with eight grouted sleeves.

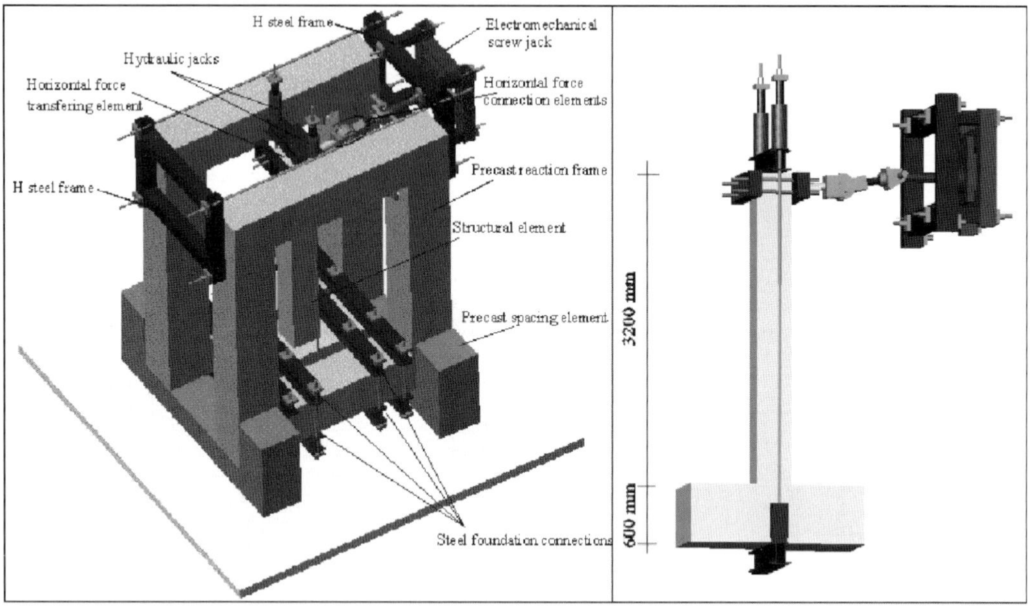

Fig. 2 – Experimental setup: (a) reaction frame; (b) loading system.

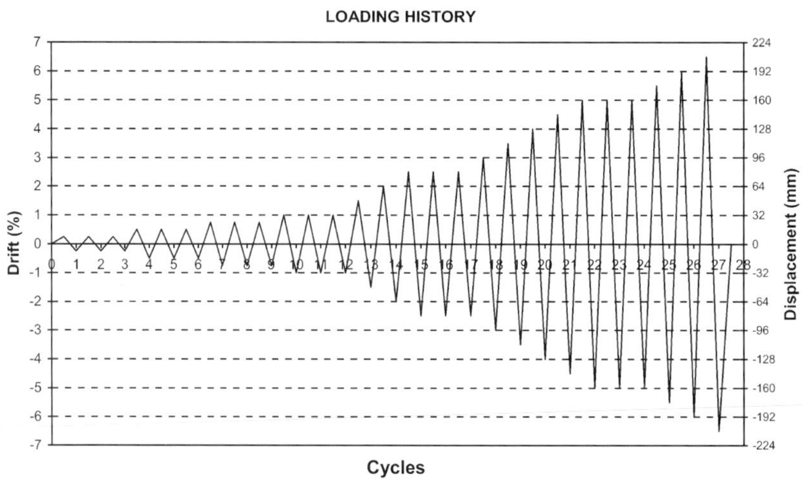

Fig. 3 – Loading history.

- the grouted pocket foundation (PF) solution showed the smallest strength degradation during the cycles. The column strength at the 5% drift cycles is equal to 82% the maximum column strength. Collapse was reached during the first cycle at 5.5% (168mm), due to buckling of the longitudinal bars. Specimen PF showed the most stable behaviour up to collapse among all of the specimen tested;
- all of the grouted sleeves (GS) column-to-foundation connections showed a considerable strength degradation during the cycles. In all cases, the strength of the column at the 5% drift cycle was approximately equal to 2/3 the maximum strength. The maximum strength at the 2.5% drift was approximately equal to 90% the maximum strength. The observed strength degradation is due to the progressive damage of the 20 mm grout layer existing between the precast column base and the foundation. This grout layer eventually crushed and was expelled

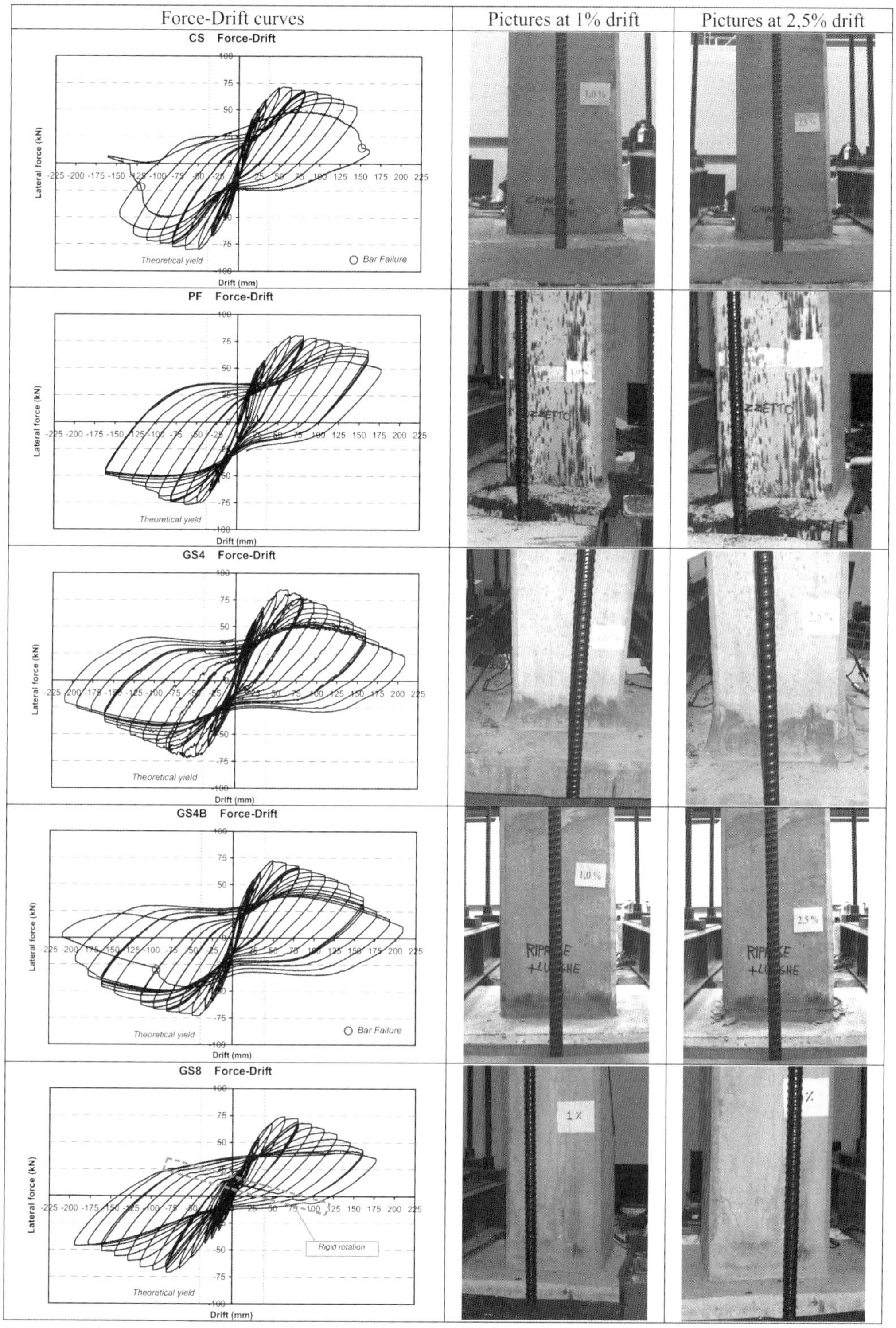

Fig. 4 – Experimental results.

Fig. 5 – Pictures at 5% drift.

from the base, leaving the role to resist compressive forces due to cyclic bending to the grout confined by the aluminium sleeves and to the vertical rebars only;

- A slightly more pronounced pinching is observed in the cycles of the GS specimens. This is due to the aforementioned progressive damage of the base grout layer, leading to a larger strain localization at the column base;
- for all of the GS specimens, the test could be carried out up to the 6.5% drift cycle. The failure of one rebar during the cycle at 6.5% drift was observed only in specimen GS4B. The higher displacement capacity of grouted sleeves compared to cast-in-situ and pocket foundation specimens is due to the heavy confinement induced by the aluminium sleeves on the grout;
- it is observed that, due to an imperfect clamping of specimen GS8 to the reaction frame, a rigid rotation occurred after each loading reversal during the test of this specimen. The observed rigid rotation did not, however, affect the overall response of specimen GS8.

Fig. 5 shows the damage at the base column section corresponding to the 5% drift cycle (maximum displacement equal to 160mm). The following observations may be made;

- specimen CS shows a large localized crack at the base, some minor cracks along the column, and some spalling. The high strain localization observed in this specimen justifies the premature bar failure observed;
- specimen PF is affected by several large cracks spreading along a length approximately equal to the column base dimension. This behaviour is typical of reinforced concrete columns;
- all of the grouted sleeve connections show concrete spalling at the corners, next to the aluminium sleeves, and a considerable crushing of the grout layer at the column base. On the other end, grouted sleeve specimens showed no other noticeable sign of damage;
- although the strain localization at the base of GS specimens should in principle lead to an anticipated collapse of the columns, the existence of heavily confined grout columns within the sleeves effectively prevented an early failure of the connections. Furthermore, the sleeve, and the confined grout within, prevented buckling of the vertical rebars, anchored in the foundation;
- the observed damage patterns allow to conclude that, being the damage limited to the grout layer existing at the base, the remaining part of the column being mostly undamaged, grouted sleeve connections are more easily repairable after a seismic event.

Finally, Fig. 6 illustrates the comparison of the energy dissipated during each cycle by all of the specimens tested.

It is observed that no significant differences exist between specimens CS, PF, GS4, and GS4B up to a 2% drift (64mm). Starting from the 2.5% cycles, specimen PF shows a larger energy dissipation, due to a smaller strength degradation and to a smaller strain localization next to the column base section (Figs. 4 and 5). Negligible differences exist between the remaining specimens up to the collapse of specimen CS, occurring during the first cycle at 5% drift.

Specimen GS8 systematically showed a considerably smaller energy dissipation than the remaining specimens. This is due to the top drift resulting from the sum of two terms, the first related

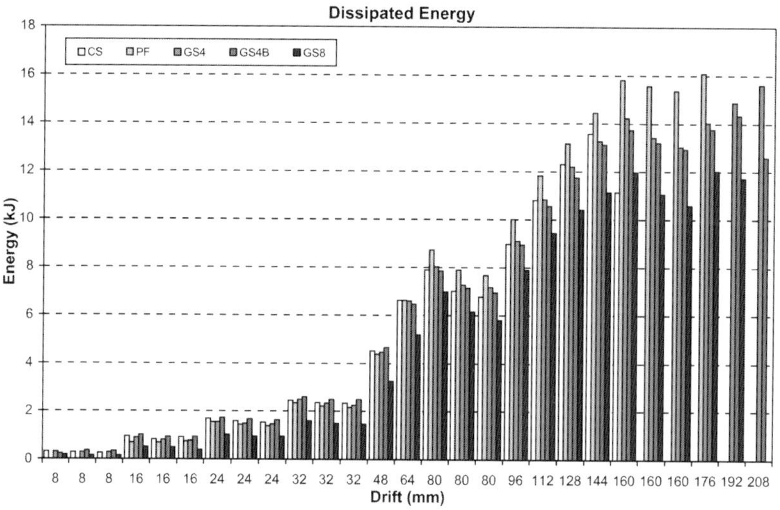

Fig. 6 – Dissipated energy.

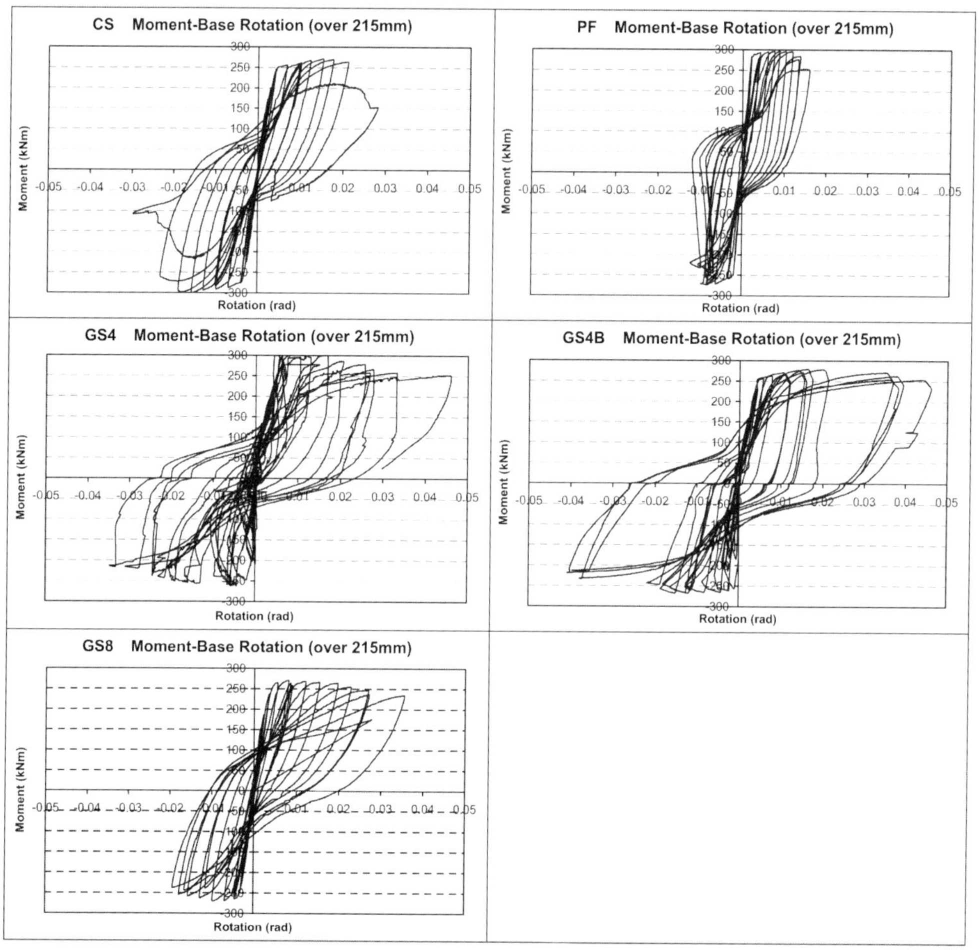

Fig. 7 – Moment-base rotation diagrams (rotation taken over 215mm gauge length).

to the column deformation, and the second due to the aforementioned rigid rotation at the base, the latter implying no energy dissipation.

Fig. 7 shows the moment-rotation diagrams at the column base sections, where the rotation is measured over a 215mm gauge length. The results presented allow the following observations:

- the base rotation is consistently much higher for the grouted sleeve specimens than for the cast-in-situ and pocket foundation specimens. This effect once more demonstrates that a much larger strain localization occurs in the former than in the latter;
- the maximum rotation in the GS specimens is only 20% smaller then the value it would exhibit considering a rigid rotation of the column around its base. This result demonstrates that the behaviour of the column outside the base section is mostly linear elastic, and that very little cracking and damage occurs outside the base section;
- deformations in the CS specimen show a greater localization in the base section with respect to those in the PF specimen. This effect lead to the early rebar failure observed.

Conclusions

The experimental results presented allow to conclude that grouted sleeves ensure a ductility similar to the one of cast in situ column-foundation connections and of pocket foundations, although a slightly smaller dissipation capacity is observed.

The high ductility of the grouted sleeve solutions is related to the high confining effect of the corrugated aluminium sleeves on the grout columns contained within. Furthermore, the presence of a highly confined grout prevents longitudinal reinforcing buckling.

It was shown that, in grouted sleeves connections, the damage is localized at the column base, in the 20 mm grout layer existing between the prefabricated column and the foundation. As a result, very little damage may be observed in the column outside of the base section.

The damage of the base grout layer results in a higher strength degradation for the grouted sleeve connections with respect to more traditional cast-in-situ and pocket foundations solutions.

Due to the damage localization observed, and to the consequent small damage existing along the column, an easier post-seismic column repair has to be expected for the grouted sleeve column-foundation connections, with respect to cast-in-situ or pocket foundation solutions.

Acknowledgements

The experimental tests were carried out within a research program on precast column-foundation connections financed by Moretti SpA, Erbusco (BS), Italy.

The cooperation of ing. Cristian Ratti, ing. Andrea Zini, and ing. Andrea Belleri, in setting up the reaction frame, carrying out the experimental tests, and post-processing all the experimental test results, is gratefully acknowledged.

References

EC8 (2003), "Eurocode 8: Design of structures for earthquake resistance - Part 1: General rules, seismic actions and rules for buildings," PrEn 1998-1, European Committee for Standardization, December 2003.

OPCM 3274 (2003), "First elements concerning general criteria for the seismic classification of the Italian territory, and seismic code provisions (in Italian)," march, 2003.

CEB (1996), "Frame Members in bending with or without axial force," CEB Bullettin N. 231, May 1996.

AIR-VOID STABILITY IN FRESH SELF-CONSOLIDATING CONCRETES INCORPORATING RICE HUSK ASH

Md. Safiuddin, G.R. FitzGerald, J.S. West* and K.A. Soudki
Department of Civil Engineering, University of Waterloo,
200 University Avenue West, Waterloo, Ontario, Canada N2L 3G1

Abstract

This paper presents the results of experimental study on air-void stability in fresh self-consolidating concretes. Two series of self-consolidating concrete were undertaken for conducting laboratory tests. Each series of concrete included three different fresh mixtures. The air-void stability in fresh concretes was investigated with respect to post-mixing and agitation. The air content of fresh concretes was determined at various test stages and adjusted considering aggregate correction factors. The flowing ability of the fresh concretes was also examined with regard to slump and slump flow. The entire testing period involved four stages extended to 60 and 90 minutes for series 1 and 2, respectively. Test results reveal that the slump and slump flow of the concrete mixtures were consistent in all test stages, and the loss of air content was minimal. The maximum loss of air content over the period of 60 and 90 minutes was less than 1.0%. Rice husk ash did not affect the air-void stability in fresh concretes. However, it increased the demand for high-range water reducer and air-entraining admixture. The overall test results indicate that the air-void stability in all fresh self-consolidating concretes was satisfactory.

Keywords: Air content, Air-void stability, Post-mixing, Rice husk ash, Self-consolidating concrete.

Introduction

Self-consolidating concrete (SCC) is a flowing concrete that spreads through congested reinforcement, fills every corner of the formwork and achieves consolidation under its own weight (Khayat 1999). In order to provide better durability performance and extended service life in freezing and thawing environments, SCC must contain an appropriate air-void system. For this, an adequate amount of entrained air-voids with proper specific surface and spacing factor should be retained in SCC. Usually, limits on volume of air-voids or air content are specified although the role of spacing factor is significant. This is because air content can be determined more easily and quickly than spacing factor. Canadian Standards Association (CSA) has specified various ranges of air content between 3 and 9% depending on maximum aggregate size and exposure conditions (CSA A23.1, 2004). These ranges of air content are recommended to create an adequate air-void system required for freeze-thaw and scaling resistance. However, when the specifications do not enforce any given air content but rather a spacing factor, an air content of 6% is usually suggested (Lessard et al. 1995). This air content generally maintains a spacing factor less than 230 μm if an effective air-entraining admixture is used in concrete mixture.

Air-voids are widely used for improving the freeze-thaw durability of concrete. The entrained air-voids improve the frost resistance of concrete in freeze-thaw environments, and thereby increase the service life of concrete structures (Cohen et al. 1992, Hayakawa et al. 1994, Siebel 1989). Although the mechanism of concrete deterioration due to freezing and thawing is still a subject of debate, the need for entrained air-voids has become indisputable. In general, the network of entrained air-voids offsets the dilating pressure posed by freezing water, and thus improves the performance of concrete

* Corresponding author: e-mail. jswest@uwaterloo.ca

M. Pandey et al. (eds), Advances in Engineering Structures, Mechanics & Construction, 129–138.

in freezing and thawing environments (Neville 1996, Chatterji 2003). The dosage of air-entraining admixture is the most important parameter that controls the air-void system in concretes. A sufficient dosage of air-entraining admixture should be added to the fresh mixture to create the required volume of air-voids in hardened SCC. However, determining the correct dosage is not very straightforward. There are numerous factors such as mixture proportions, aggregate grading, cement composition, type of high-range water reducer, type and composition of supplementary cementing materials, quality of mixing water, mixing or placing methods, and temperature etc., that might affect air-entrainment, and therefore achieving the required air content in hardened concrete becomes much more difficult (ACI Committee 201, 2001; Du and Folliard 2005, Pigeon 1994). Nevertheless, the proper air-void system must be maintained in concrete to ensure a good resistance to freezing and thawing. Proper air-void system means that the entrained air-voids remain stable until the concrete is set, and becomes permanent in the hardened paste. This is particularly important for SCC, as the presence of high-range water reducer tends to destabilize the entrained air-bubbles during transport and placement of concrete (Saucier et al. 1990, Khayat and Assaad 2002). Many research reports indicate that high-range water reducers can cause some loss of air content during intermittent agitation (Jana et al. 2005, Johnston 1994). This is perhaps attributed to the production of greater large-size bubbles, which could easily disappear with time. Also, the large dosage of high-range water reducer could induce excessive fluidity and segregation, and thus may cause some loss of entrained air. Consequently, the air-void system in hardened concrete could be affected, and the freeze-thaw durability would be reduced.

Instable air-voids affect the air present in concrete, and thus air-void stability also influences other concrete properties such as strength and porosity. The air content of concrete is reduced due to the loss of instable air-voids. As a result, the total porosity of concrete is decreased and the concrete gains greater strength. In general, the average gain of compressive strength can be 3 to 5% for each percentage loss of air content (Neville 1996). However, the increase in compressive strength does not produce any significant benefit for air-entrained concrete since the desired level of compressive strength is already considered in mixture design. Instead, the loss of air content results in a significant reduction in freeze-thaw durability that cannot be overcome in any way. Therefore, the air-void stability in fresh SCC is significantly more important for freeze-thaw durability than strength.

Comprehensive studies have been conducted on air-void stability of low to medium slump concretes in presence of high-range water reducer (Baalbaki and Aïtcin 1994, Pigeon et al. 1989, Saucier et al. 1990). These studies report the effects of high-range water reducer, supplementary cementing materials, and cement-admixture compatibility on air-void stability in normal and high strength concretes with some contradictory results. In addition, Baekmark et al. (1994) investigated the air-void stability of medium-slump concretes with respect to post-mixing, re-dosing of high-range water reducer and pumping operation. They observed that the air-void stability in concrete is greatly affected by the type of air-entraining admixture. While conducting the air content test in all of the aforementioned studies, the concretes were consolidated by external means, which can affect the air-void system. External means of consolidation, including vibration, was found to produce a detrimental effect on quality of air-void system in air-entrained concrete (Stark 1986). Some air-voids can be lost during consolidation with rodding or vibration. Therefore, the air content obtained from an air-meter test may not represent the actual air content of the parent concrete mixture. This discrepancy is eliminated in SCC since it does not require any external means of consolidation. Still the air-void stability problem could occur in SCC due to other factors such as excessive fluidity and segregation, and destabilization caused by HRWR. Yet very few studies have been carried out to investigate the air-void stability in SCC.

Bouzoubaâ and Lachemi (2001), and Lachemi et al. (2003) developed air-entrained SCC and examined various fresh properties including air content. However, they did not study the air-void stability in SCC. Recently, Khayat (2000), and Khayat and Assaad (2002) investigated the air-void stability of SCC. They showed that the air-void stability could be ensured in SCC by a proper mixture composition including a suitable combination of chemical admixtures. An increase in the amount of fine material at a low water-binder ratio or the use of viscosity-modifying admixture can secure the appropriate air-void system during transport, placement and setting of SCC. However, none of the above studies investigated the effect of rice husk ash (RHA). The present study produced two series of SCC incorporating RHA and examined the air-void stability in fresh mixtures with respect to effects of post mixing and agitation. The effects were observed through subsequent determination of air content at different test stages.

Experimental Investigation

Experimental investigation was carried out through selection and testing of materials, determination of concrete mixture proportions, and preparation and testing of various fresh concretes.

Materials

Ordinary (ASTM Type I or CSA Type 10) portland cement, crushed granite stone, pit sand, non-crystalline amorphous rice husk ash (RHA), tap water, poly-carboxylic acid-based high-range water reducer (HRWR), and a synthetic air-entraining admixture (AEA) were used to produce various SCC. Crushed granite stone and pit sand were selected as coarse and fine aggregates, respectively, whereas rice husk ash was used as a supplementary cementing material. Prior to use in preparing concrete mixtures, the component materials were tested for a number of physical properties. These properties were useful to judge the suitability of the constituent materials. In addition, most of these properties were directly used in mixture proportioning of different concretes. The physical properties of the concrete materials are given in Table 1.

Table 1: Physical Properties of Constituent Materials

Material	Properties
Crushed granite stone	Maximum size: 19 mm Total evaporable moisture content: 0.1% Oven dry basis bulk density: 1670 kg/m^3 Void content: 37% Saturated surface-dry basis relative density: 2.71 Absorption: 1.5% Fineness modulus: 6.78
Pit sand	Maximum size: 4.75 mm Total evaporable moisture content: 0.1% Oven dry basis bulk density: 1860 kg/m^3 Void content: 28% Saturated surface-dry basis relative density: 2.62 Absorption: 1.0% Fineness modulus: 2.74
Ordinary portland cement	Relative density: 3.16 Blaine fineness: 412 m^2/kg
Rice husk ash (RHA)	Relative density: 2.07
Tap water	Density (24^{0}C): 997.28 kg/m^3
High-range water reducer (HRWR)	Relative density: 1.069 Solid content: 41%
Air-entraining admixture (AEA)	Relative density: 1.01 Solid content: 12.8%

Concrete Mixture Proportions

Two series of concretes, as shown in Table 2, were undertaken for testing of air-void stability. The fresh concrete mixtures were designed based on a weight-basis water-binder (W/B) ratio of 0.35, an air-dry and weight-basis sand-aggregate ratio of 0.50, and nominal air contents of 4 and 8%. For both air contents, the content of rice husk ash was varied with 0, 15 and 20% of binder by weight. The basic mixture proportions for the two series of concretes, as shown in Table 3, were determined based on the absolute volume of the constituent materials. The dosages of HRWR and AEA are not included in basic mixture proportions. HRWR and AEA were used in concrete as additives, and their dosages have been presented in Table 4. The amounts of HRWR shown in Table 4 are the saturation dosages for various concretes. These dosages were obtained by testing the flow of various binder pastes, which are not detailed here. Also, the proportions of coarse and fine aggregates given in Table 3 are based on their absolute volume in saturated surface-dry condition. But, the aggregates were batched in air-dry condition and hence they absorbed some mixing water. In addition, the liquid HRWR contributed some water during mixing. Therefore, the proportions of coarse and fine aggregates, and water presented in Table 3 were adjusted before concrete batching.

Table 2: Different Types of Self-consolidating Concrete for Testing of Air-void Stability

Concrete Series	Concrete Designation	W/B	Rice Husk Ash (% B)	Design Air Content (%)	Time of Testing (minute)
S1	C35/0/4	0.35	0	4	15, 30, 45, 60
	C35/15/4		15		
	C35/20/4		20		
S2	C35/0/8	0.35	0	8	15, 30, 60, 90
	C35/15/8		15		
	C35/20/8		20		

Table 3: Basic Mixture Proportions of Different Self-consolidating Concretes (W/B = 0.35)

Concrete Mixture	Coarse Aggregate kg/m^3	Fine Aggregate kg/m^3	Cement kg/m^3	Rice Husk Ash kg/m^3	Water kg/m^3
C35/0/4	902.7	898.3	422.3	0	147.8
C35/15/4	888.6	884.2	359.0	63.3	147.8
C35/20/4	883.9	879.5	337.8	84.5	147.8
C35/0/8	849.4	845.2	422.3	0	147.8
C35/15/8	835.3	831.2	359.0	63.3	147.8
C35/20/8	830.6	826.5	337.8	84.5	147.8

Preparation and Testing of Concretes

The fresh concretes were prepared using a pan-type mixer. Constituent materials were batched and mixed to produce the concretes. The total initial mixing time was 10 minutes. At first, fine and coarse aggregates were mixed for 1 minute with first quarter of the mixing water. Mixing of aggregate blend was continued for another 1 minute with the addition of AEA dispensed in second quarter of the mixing water. After 2-minute mixing, the mixer was stopped to add the binding material (cement alone or cement with RHA). Immediately, the mixer was restarted, third quarter of mixing water was added, and the mixing was continued for 2 additional minutes. Then the mixer was stopped again, the pan was covered with wet burlap and the aggregate-binder mixture was allowed for 3-minute rest. Thereafter, the mixer was restarted and the concrete materials were mixed for 3 more minutes with the gradual addition of HRWR dispensed in fourth quarter of mixing water. The saturation dosage of HRWR was used to produce the required flowing ability. For all concretes, the saturation dosage was split into four parts such as 70, 10, 10 and 10%, and added sequentially to maintain similar slump and slump flow over the whole testing period. The initial dosage of HRWR was equivalent to 70% of the saturation dosage. The remaining three parts of the saturation dosage were added to the concrete mixture at subsequent mixing stages, as shown in Table 2.

Table 4: Dosages of HRWR and AEA for Various Self-consolidating Concretes

Concrete Series	Concrete Designation	Saturation Dosage of HRWR		Dosage of AEA	
		(%B)	l/m^3	(%B)	ml/m^3
S1	C35/0/4	1.0	3.95	0.015	64.6
	C35/15/4	2.0	7.90	0.031	129.3
	C35/20/4	2.5	9.88	0.041	172.4
S2	C35/0/8	1.0	3.95	0.026	107.7
	C35/15/8	2.0	7.90	0.093	387.8
	C35/20/8	2.5	9.88	0.124	517.1

The air-void stability in fresh concretes was investigated with respect to post mixing and agitation. The entire testing task was comprised of four stages, as shown in Table 2. In both series of concrete, the primary mixtures were tested after 15 minutes from the start of mixing to determine the air content. Then the concrete mixtures were covered with wet burlap and allowed for rest followed by further mixing and testing. The rest period varied from 10 to 25 minutes depending on testing time shown in Table 2. The post-mixing was conducted for 2 minutes for each test stage. Similar flowing ability was maintained in fresh concretes during the entire test period. For this, an additional HRWR by 10% of the saturation dosage was added during post-mixing. The subsequent measurements for air content were taken after 30, 45 and 60 minutes for series 1. In case of series 2, these measurements were carried out after 30, 60 and 90 minutes. The air content of fresh concretes was determined in accordance with ASTM C231 (2004) with an exception that concrete sample was placed in the measuring bowl without any consolidation. The readings obtained from the air meter were corrected based on aggregate correction factors. The flowing ability of the fresh concretes was also examined during each test stage. For this, the slump and slump flow were determined using the Abram's slump cone specified in ASTM C143/C143M (2004). The freshly mixed concrete was placed in the slump cone in one layer and without any consolidation. Then the slump cone was raised vertically and the concrete was allowed to subside. The fall in height of initial concrete sample and the average diameter of the deformed concrete were measured and recorded as the slump and slump flow, respectively.

Test Results and Discussion

Various self-consolidating concretes were produced with desired flowing ability and air content. HRWR produced the self-consolidation effect, whereas AEA provided the desired air contents. The initial and successive dosages of HRWR worked very well to fulfil the performance criteria for flowing ability with respect to slump and slump flow. Also, consistent slump and slump flow were maintained during all test stages since the air-void stability could be affected by the flowing ability of the concretes. The ability of AEA to reduce the surface tension is generally decreased at lower flowing ability (Khayat and Assaad 2002). Thus, bigger and less stable air-voids could be produced. Also, more air-voids can be entrapped at lower flowing ability resulting in higher total air content. These drawbacks were eliminated in the present study, as the flowing ability of the concretes was almost the same at all test stages.

The measured slump and slump flow have been presented in Table 5 and Table 6, respectively. It can be seen from Table 5 that the slump was always consistent for each concrete. On the whole, the slump has varied from 270 to 280 mm, with an average of 275 mm and with a standard deviation of only about 2%. Moreover, Table 6 shows that the slump flow varied in the range of 670 to 720 mm. For each concrete, the slump flow was also consistent throughout the testing period. The average slump flow for all concretes was about 697 mm with a standard deviation of 11%. The slump of SCC usually varies from 250 to 280 mm (Ferraris et al. 2000). Again, the slump flow of SCC generally ranges between 600 and 800 mm (Khayat 2000, Xie et al. 2002). These criteria were maintained in the present study by split use of the saturation dosages of HRWR. For this, higher saturation dosages of HRWR were required in the presence of 15 and 20% RHA, as can be seen from Table 4. This is due to an increase in flocculation forces resulting from greater amount of fine particles and higher specific surface area. Therefore, the demand for HRWR was increased to improve the dispersion of the binding materials.

Table 5: Variation of Slump with Time for Different Self-consolidating Concretes

Concrete Series	**Concrete Designation**	**Slump (mm)**			
		T = 15 min.	T = 30 min.	T = 45 min.	T = 60 min.
S1	C35/0/4	275	275	275	270
	C35/15/4	280	275	275	275
	C35/20/4	275	275	275	275
		T = 15 min.	T = 30 min.	T = 60 min.	T = 90 min.
S2	C35/0/8	275	275	275	275
	C35/15/8	280	275	275	275
	C35/20/8	275	275	275	270

Table 6: Variation of Slump Flow with Time for Different Self-consolidating Concretes

Concrete Series	**Concrete Designation**	**Slump Flow (mm)**			
		T = 15 min.	T = 30 min.	T = 45 min.	T = 60 min.
S1	C35/0/4	700	700	700	680
	C35/15/4	720	705	700	700
	C35/20/4	690	695	700	695
		T = 15 min.	T = 30 min.	T = 60 min.	T = 90 min.
S2	C35/0/8	670	680	700	695
	C35/15/8	720	700	710	695
	C35/20/8	695	690	695	690

The results for air content of various SCC have been presented in Figures 1 and 2. It can be seen from Figure 1 that the air content varied from 3.5 to 4.3% for concretes under series 1. Conversely, the air content ranged from 7.5 to 8.6% for concretes under series 2, as can be seen from Figure 2. Hence, the actual air contents deviated from the design air contents (4 and 8%) within the range of ±0.6%. This is below the acceptable tolerance for air content measurement of ±1.5% (ACI Committee 201, 2001). In a few cases, the air content observed during second stage of testing at 30 minutes from concrete batching was slightly higher than the initial air content. This is possibly due to enhanced dispersion of the entrained air under more mixing action. The overall test results indicate that the air-void stability in all fresh self-consolidating concretes was good. The maximum loss of air content over the period of 60 and 90 minutes was less than 1.0% for all concretes. This is below the generally occurring air loss of 1 to 2% due to transportation of concrete (Kosmatka et al. 2002). Also, there was no significant difference in loss of air content between series 1 and 2 due to post-mixing and agitation. It suggests that the concrete placement can be delayed up to 60 to 90 minutes from the time of batching, while maintaining the desirable air content. This time length is adequate for the transport of concrete from ready-mixed plant to the construction site.

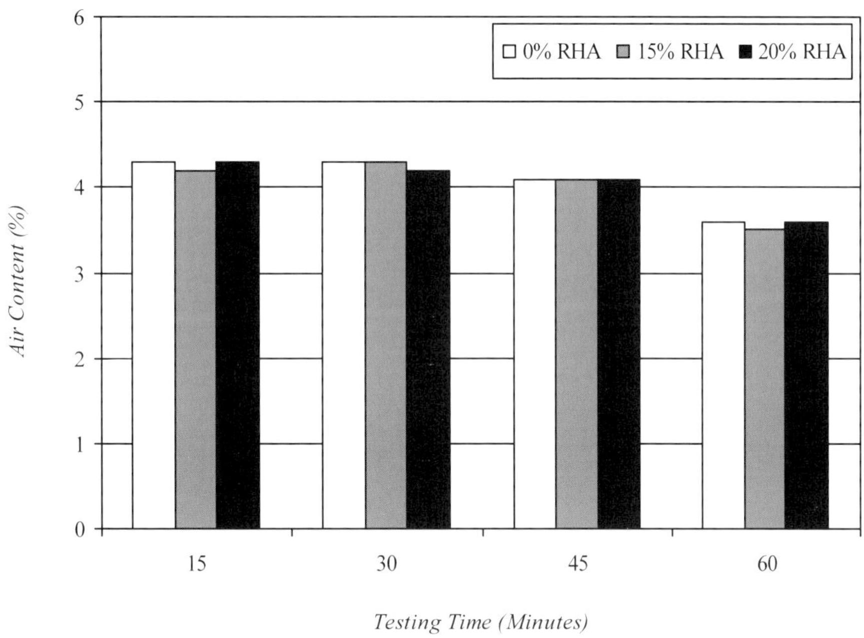

Figure 1: Variation of Air Content with Time for Self-consolidating Concretes under Series 1

The air-void stability observed in various fresh SCC is attributed to the relatively high binder content, low water-binder ratio and high sand-aggregate ratio. These properties are not only conducive to the flowing ability of concretes, but also enhance air-void stability (Khayat and Assaad 2002). In the present study, the concretes were prepared using a relatively low coarse aggregate content with a higher amount of binder. In addition, the saturation dosages of HRWR dispersed the binding materials and maintained good fluidity in the concrete mixtures. As a result, the inter-particle friction and collision of aggregates were reduced during mixing, handling and placement, which might produce more stable air-voids in fresh concrete. Moreover, the split addition of HRWR did not affect the air content of fresh concrete. Schemmel et al. (1994) observed similar results in case of high-performance concrete. However, it might have some effects on air-void system in hardened concrete. Higher dosages of HRWR can drive some of the entrained air-voids to coalesce resulting in increased spacing factor for a given air content (Pigeon et al. 1989, Siebel 1989). Nevertheless, this issue is beyond the scope of the present study.

The air-void stability in fresh concretes was not affected by RHA. This is evident from Figures 1 and 2. However, the presence of RHA increased the demand for AEA for a given range of air content. It can be seen from Table 4 that greater AEA dosages were required for concretes with 15 and 20% RHA in both series. This is mostly due to increased paste viscosity and reduced attachment of entrained air-voids. The paste viscosity was increased due to fine particle size and extremely high specific surface area of RHA. The increased paste viscosity increases the internal pressure in air-voids that causes collapse of some entrained air-voids (Khayat and Assaad 2002). Consequently, a reduction in fresh air content occurs and more AEA is required to compensate for this loss. Some AEA molecules could also be adsorbed or absorbed in porous and honeycombing microstructure of RHA. Thus, a lower amount of AEA will be available for air-void formation and stabilisation. Furthermore, HRWR can impede the attachment of entrained air-voids onto binding materials by reducing the attachment sites (Khayat and Assaad 2002). Some HRWR, particularly poly-carboxylic acid-based HRWR, also produce a steric repulsion with long grafted side chains (Mindess et al. 2003) that may further reduce the attachment of air-voids on cement and RHA particles. Hence, some air-voids could become less stable and combine to seep out of the concrete mixture, resulting in a higher demand for AEA. Both attachment site reduction and steric repulsion effects of HRWR could be more pronounced in case of 15 and 20% RHA, as greater dosages were used to obtain the desired flowing ability of concretes.

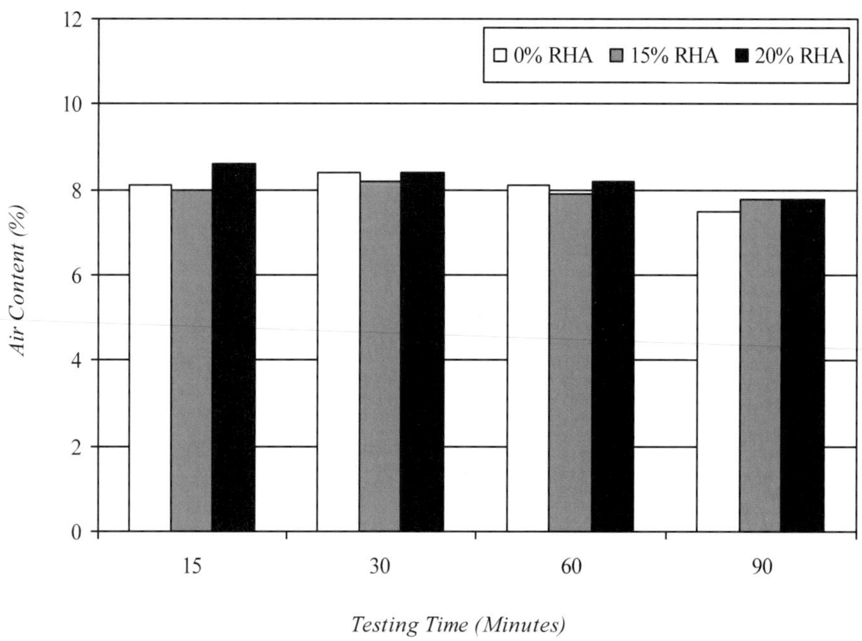

Figure 2: Variation of Air Content with Time for Self-consolidating Concretes under Series 2

Conclusions

1. Air-void stability in various fresh self-consolidating concretes was not affected by post-mixing and agitation, as the air content at various test stages did not differ significantly.
2. Air-void stability in various fresh self-consolidating concretes was good over the period of 60 to 90 minutes, as the maximum loss of air content remained below 1%.
3. Rice husk ash increased the demand for air-entraining admixture for a given air content but it did not affect the overall air-void stability in fresh self-consolidating concretes.
4. The flowing ability of various self-consolidating concretes had no effect on air-void stability, as the slump and slump flow of the concretes were kept consistent in all test stages.
5. The split addition of the saturation dosage of HRWR maintained consistent flowing ability at all test stages but it did not affect the air-void stability in fresh self-consolidating concretes.

Acknowledgements

The authors are thankful to Mr. Cameron Monroe, Manager of Technical Services of Degussa Master Builders, Ltd. for supplying chemical admixtures and to Mr. Michael Rich for obtaining the rice husk ash required for the experimental investigation. The authors are also grateful to Lafarge North America Inc. for the supply of the cement needed for the entire project.

References

ACI Committee 201, "Guide to Durable Concrete", *ACI 201.2R-01*, American Concrete Institute, Farmington Hills, Michigan, USA, 2004, 41pp.

ASTM C143/C143M, "Standard Test Method for Slump of Hydraulic-Cement Concrete", *Annual Book of ASTM Standards*, Vol.04.02, American Society for Testing and Materials, Philadelphia, USA, 2004, pp.95-98.

ASTM C231, "Standard Test Method for Air Content of Freshly Mixed Concrete by the Pressure Method", *Annual Book of ASTM Standards*, Vol.04.02, American Society for Testing and Materials, Philadelphia, USA, 2004, pp.146-154.

Baalbaki, M., and Aïtcin, P.-C., "Cement/Superplasticizer/Air-entraining Agent Compatibility", *Proceedings of the Fourth CANMET/ACI International Conference on Superplasticizers and Other Chemical Admixtures in concrete*, ACI SP-148, V.M. Malhotra, ed., American Concrete Institute, Farmington Hills, Michigan, USA, 1994, pp.47-62.

Baekmark, K., Hansen, H., and Reichert, J., "Stabilized Air Void System in Superplasticized Concrete", *Concrete Technology – Past, Present, and Future: Proceedings of V. Mohan Malhotra Symposium*, ACI SP-144, P.K. Mehta, ed., American Concrete Institute, Farmington Hills, Michigan, USA, 1994, pp.177-190.

Bouzoubaâ, N., and Lachemi, M., "Self-compacting Concrete Incorporating High Volumes of Class F Fly Ash: Preliminary Results", *Cement and Concrete Research*, Vol. 31, No. 3, 2001, pp.413-420.

Chatterji, S., "Freezing of Air-entrained Cement-based Materials and Specific Actions of Air-entraining Agents", *Cement & Concrete Composites*, Vol.25, No.7, 2003, pp.759-765.

Cohen, M.D., Zhou, Y., and Dolch, W., "Non-air-entrained High-strength Concrete – Is it Frost Resistant?", *ACI Materials Journal*, Vol.89, No.2, 1992, pp.406-415.

CSA A23.1, *Concrete Materials and Methods of Concrete Construction*, Canadian Standards Association, Mississauga, Ontario, Canada, 2004.

Du, L., and Folliard, K.J., "Mechanisms of Air Entrainment in Concrete", *Cement and Concrete Research*, Vol.35, No.8, 2005, pp.1463-1471.

Ferraris, C.F., Brower, L., Ozyildirim, C., and Daczko, J., "Workability of Self-compacting Concrete", *Proceedings of the International Symposium on High Performance Concrete*, PCI/FHWA/FIB, Orlando, Florida, USA, 2000, pp.398-407.

Hayakawa, M., Matsuoka, Y., and Shindoh, T., "Development and Application of Super-workable Concrete", *Special Concretes: Workability and Mixing, Proceedings of the International RILEM Workshop*, P.J.M. Bartos, ed., E & FN Spon, London, UK, 1994, pp.183-190.

Jana, D., Erlin, B., and Pistilli, M.F., "A Closer Look at Entrained Air in Concrete", *Concrete International*, Vol.27, No.7, 2005, pp.31-34.

Johnston, C.D., "Deicer Salt Scaling Resistance and Chloride Permeability", *Concrete International*, Vol.16, No.8, 1994, pp.48-55.

Khayat, K.H., "Workability, Testing and Performance of Self-consolidating Concrete", *ACI Materials Journal*, Vol.96, No.3, 1999, pp.346-353.

Khayat, K.H., "Optimization and Performance of Air-entrained Self-consolidating Concrete", *ACI Materials Journal*, Vol.97, No.5, 2000, pp.526-535.

Khayat, K.H., and Assaad, J., "Air-void Stability of Self-consolidating Concrete", *ACI Materials Journal*, Vol.99, No.4, 2002, pp.408-416.

Kosmatka, S.H., Kerkhoff, B., Panarese, W.C., MacLeod, N.F., and McGrath, R.J. *Design and Control of Concrete Mixtures*, Seventh Canadian Edition, Cement Association of Canada, Ottawa, Ontario, Canada, 2002.

Lachemi, M., Hossain, K.M.A., Lambros, V., and Bouzoubaâ, N., "Development of Cost-effective Self-consolidating Concrete Incorporating Fly Ash, Slag Cement, or Viscosity-modifying Admixtures", *ACI Materials Journal*, Vol.100, No.5, 2003, pp.419-425.

Lessard, M., Baalbaki, M., and Aïtcin, P.-C., "Mix Design of Air-entrained High-performance Concrete", *Concrete under Severe Conditions: Environments and Loading*, Vol. II, K. Sakai, N. Banthia, and O.E. Gjørv, eds., E & FN Spon, London, UK, 1995, pp.1025-1034.

Mindess, S., Young, J.F., and Darwin, D., *Concrete*, Second Edition, Prentice Hall, Pearson Education Inc., New Jersey, USA, 2003.

Neville, A.M., *Properties of Concrete*, Fourth and Final Edition, John Wiley & Sons, Inc., New York, USA, 1996.

Pigeon, M., "Frost Resistance, A Critical Look", *Concrete Technology – Past, Present, and Future: Proceedings of V. Mohan Malhotra Symposium*, ACI SP-144, P.K. Mehta, ed., American Concrete Institute, Farmington Hills, Michigan, USA, 1994, pp.141-158.

Pigeon, M., Plante, P., and Plante, M., "Air-void Stability, Part I: Influence of Silica Fume and Other Parameters", *ACI Materials Journal*, Vol.86, No.5, 1989, pp.482-490.

Saucier, F., Pigeon, M., and Plante, P., "Air-void Stability, Part III: Field Tests of Superplasticized Concretes", *ACI Materials Journal*, Vol.87, No.1, 1990, pp.3-11.

Schemmel, J.J., Arora, V., and Williams, J., "Split Addition of a HRWRA and Its Effect on High-Performance Concrete", *Proceedings of the Fourth CANMET/ACI International Conference on Superplasticizers and Other Chemical Admixtures in Concrete*, ACI SP-148, V.M. Malhotra, ed., American Concrete Institute, Farmington Hills, Michigan, USA, 1994, pp.301-316.

Siebel, E., "Air-void Characteristics and Freezing and Thawing Resistance of Superplasticized Air-entrained Concrete with High Workability", *Proceedings of Third International Conference on Superplasticizers and Other Chemical Admixtures in Concrete*, ACI SP-119, V.M. Malhotra, ed., American Concrete Institute, Michigan, USA, 1989, pp.297-319.

Stark, D.C., "Effect of Vibration on the Air-void System and Freeze-thaw Durability of Concrete", *Research and Development Bulletin*, RD092.01T, Portland Cement Association, Illinois, USA, 1986, 10 pp.

Xie, Y., Liu, B., Yin, J., and Zhou, S., "Optimum Mix Parameters of High-strength Self-compacting Concrete with Ultrapulverized Fly Ash", *Cement and Concrete Research*, Vol.32, No.3, 2002, pp.477-480.

BEHAVIOR OF CONCRETE BRIDGE DECKS REINFORCED WITH MMFX STEEL

Hatem Seliem, Gregory Lucier, Sami Rizkalla and Paul Zia
Civil Engineering Department. North Carolina State University, USA
E-mail: sami_rizkalla@ncsu.edu

Introduction

Corrosion of steel reinforcement is considered to be one of the leading causes of deterioration of concrete bridges. This fact has led to the development of numerous technologies such as corrosion-resistant steel that attempt to mitigate this expensive problem. The recent development of high-strength, highly corrosion-resistant steel, commercially known as Micro-composite Multi-Structural Formable (MMFX) steel, is a promising technology. MMFX steel offers its high corrosion resistance without the use of the coating technologies. This characteristic was achieved by proprietary alteration of the steel composition and microstructure. In addition, the control of MMFX steel's morphology of its microstructure has resulted in its higher strength. Use of MMFX steel could lead to potential savings through using less reinforcement ratios due to its higher strength characteristics and longer service life of structures because of its high corrosion resistance. Recently, many state transportation departments have begun to use MMFX steel as a direct replacement for conventional Grade 60 steel in concrete bridge decks. However, despite these field applications, there is insufficient information about the behavior of such concrete bridge decks utilizing MMFX steel as main reinforcement.

This paper evaluates the use of MMFX steel as main flexural reinforcement in concrete bridge decks in light of test results. Assessment of the effect of the arching action on the strength of bridge decks due to the use of this new steel is also presented.

Experimental Program

Specimen Details

The experimental program consisted of testing three full-scale bridge decks with a span-to-depth ratio of 12.5 to evaluate the structural performance of MMFX steel as main flexural reinforcement in comparison to the use of conventional Grade 60 steel. The bridge decks were identical in all aspects except for the type and amount of steel used in each. The bridge decks consisted of two spans and two cantilevers, supported in composite action by three pre-cast post-tensioned concrete girders having cross-sectional dimensions of 24 × 10 in. (610 × 254 mm). The overall nominal dimensions of the bridge decks were 21' – 10" × 13' – 2" × 8 $^5/_8$" (6655 × 4013 × 220 mm). The supporting girders were post-tensioned using deformed prestressing bars of 1 in. (25 mm) diameter with ultimate strength of 150 ksi (1034 MPa). Each girder was prestressed by four bars resulting in a total prestressing force of 360 kips (1601 KN) per girder.

The first and third bridge decks were reinforced with MMFX steel, while the second bridge deck was reinforced with conventional Grade 60 steel for comparison purposes. The test matrix is given in Table 1 and the reinforcement details for the three bridge decks are shown in Figure 1. It should be noted that the reinforcement ratio (ρ) is calculated using the total slab thickness. The first and second bridge decks were constructed with the same reinforcement ratio using MMFX and conventional Grade 60 steel, respectively. However, the third bridge deck was reinforced with MMFX steel using only two-third of the reinforcement ratio used for the first deck in an attempt to utilize the higher

M. Pandey et al. (eds), Advances in Engineering Structures, Mechanics & Construction, 139–152.

Table 1. Bridge decks test matrix.

Bridge Deck	Steel Type	Bottom Reinforcement		Top Reinforcement	
		Transverse	Longitudinal	Transverse	Longitudinal
First	MMFX	#5 @ 6.75" (#16 @ 170) ρ = 0.54%	#5 @ 10" (#16 @ 250) ρ = 0.36%	#5 @ 6.75" (#16 @ 170) ρ = 0.54%	#4 @ 14" (#13 @ 360) ρ = 0.17%
Second	Grade 60	#5 @ 6.75" (#16 @ 170) ρ = 0.54%	# 5 @ 10" (#16 @ 250) ρ = 0.36%	#5 @ 6.75" (#16 @ 170) ρ = 0.54%	#4 @ 14" (#13 @ 360) ρ = 0.17%
Third	MMFX	#5 @ 10" (#16 @ 250) ρ = 0.36%	#5 @ 10" (#16 @ 250) ρ = 0.36%	#5 @ 10" (#16 @ 250) ρ = 0.36%	#4 @ 14" (#13 @ 360) ρ = 0.17%

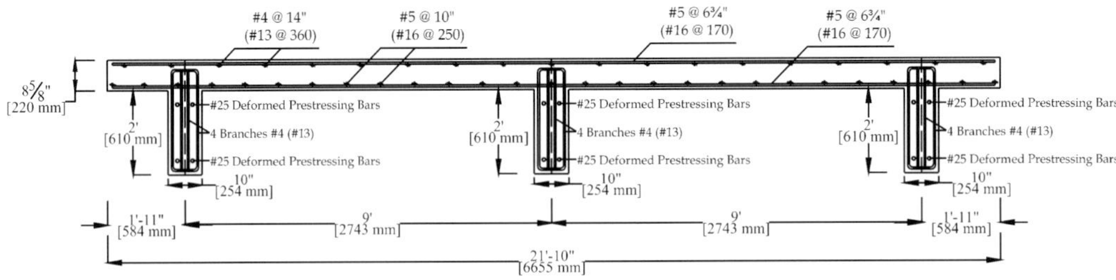

Reinforcement Details of the First and Second Bridge Decks

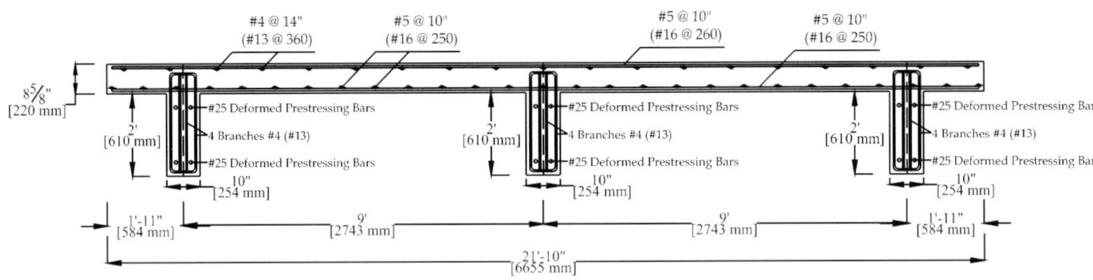

Reinforcement Details of the Third Bridge Deck

Fig. 1. Reinforcement details for the three bridge decks.

tensile strength of MMFX steel. It should be noted that the first bridge deck was designed to simulate the same reinforcement ratio of an actual bridge that was built in Johnston County, North Carolina in 2004. The three bridge decks had the same span and thickness of the bridge, and supported by girders designed to have the same torsional stiffness as the actual bridge.

Material Properties

The average concrete compressive strength at the day of testing for the three bridge decks was 7000, 4500, and 5278 psi (48.2, 31, and 36.4 MPa), respectively. The concrete compressive strengths were determined using 4 × 8 in. (102 × 204 mm) concrete cylinders cast for each bridge deck and cured under the same conditions as the deck. Concrete was provided by a local ready-mix supplier.

Tension coupons of MMFX and Grade 60 steel were tested according to ASTM-A370 specifications. The measured stress-strain characteristics of the MMFX and grade 60 steel are shown in Figure 2. The MMFX reinforcing bars exhibit a linear stress-strain relationship up to 100 ksi (689 MPa) followed by a non-linear behavior with ultimate strength of 173 ksi (1193 MPa). According

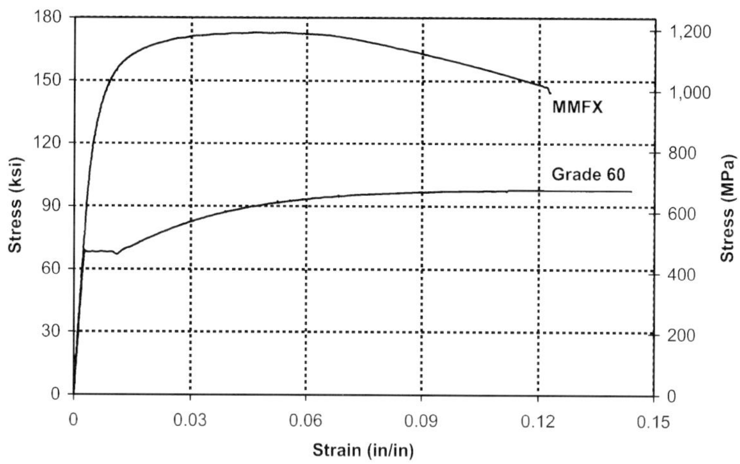

Fig. 2. Stress-strain characteristics of Grade 60 and MMFX steel.

to the ASTM-A370 offset method (0.2% offset) the yield strength of MMFX steel was found to be 120 ksi (827 MPa). The initial modulus of elasticity was determined to be 29,000 ksi (200 GPa), followed by a non-linear behavior and reduction in the modulus of elasticity at stress higher than 100 ksi (689 MPa). The yield strength of the Grade 60 steel was determined to be 68 ksi (469 MPa).

Test Setup and Instrumentation

Two 440 kips (1957 KN) MTS hydraulic actuators were used to apply a concentrated load to each span simultaneously to simulate the effect of a truck wheel load. Two 10 × 20 in. (254 × 508 mm) steel plates were used to transfer the load from the actuator to comply with the AASHTO LRFD Bridge Design Specifications for tire contact area. A 1/2 in. (13 mm) thick neoprene pad was placed under each loading plate to prevent possible local crushing of the concrete. The supporting girders were supported by concrete blocks to transfer the applied load to the strong floor resulting in a clear span of 96 in. (2438 mm). The clear span of supporting girders was determined based on the equivalency of the torsional stiffness of the supporting girders to that of the steel girders used in the actual bridge. Figure 3 shows an isometric view of the test setup and the first bridge deck prior to testing.

A total of 72 channels were used for instrumentation of each bridge deck. A 440 kips (1957 KN) load cell was mounted to each actuator to measure the applied load. Twenty-four string potentiometers (string pots) were used to measure the bridge deck deflection profiles along the longitudinal and transverse directions. In addition, six linear potentiometers were used to measure the girders deflections and rotations. Twenty PI gages were used to measure the concrete strain at various locations. The measured strains were used to determine the strain profiles of the sections at the measured locations. Twenty electrical resistance strain gages of 120 ohm and 6 mm gage length were attached to selected reinforcing bars to determine the strains in these bars. Data were electronically recorded by an Optim Megadac data acquisition system. Figure 4 shows the locations of the PI gages used and establishes the notation adopted hereafter.

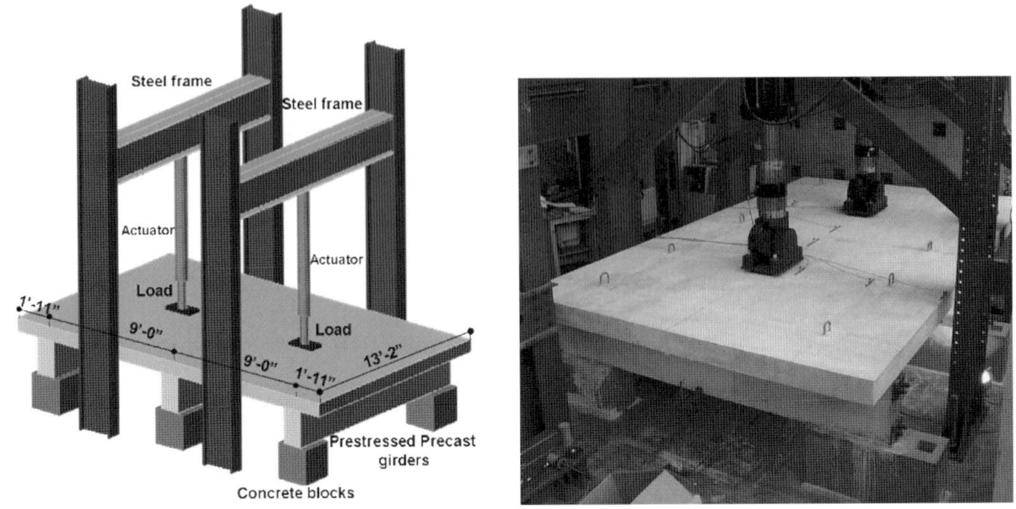

(a) Test setup (b) First bridge deck prior to testing

Fig. 3. Test setup and the first bridge deck prior to testing.

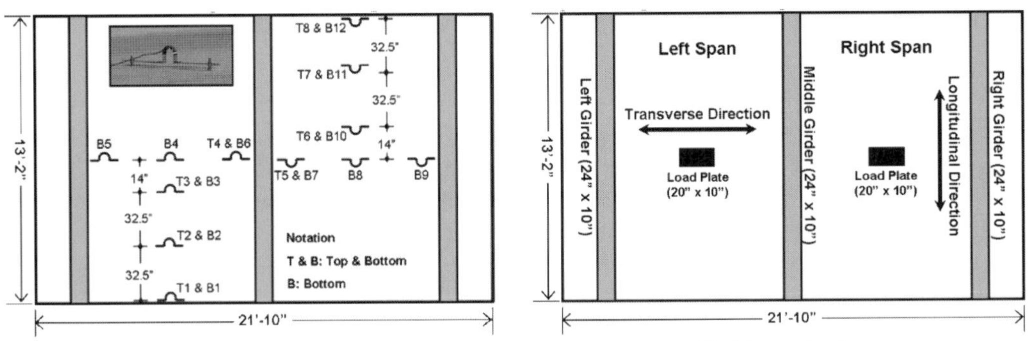

(a) Location of PI gages (b) Schematic plan

Fig. 4. PI gages locations and notations for the three bridge decks.

Experimental Results

Load-Deflection Behavior

The load-deflection envelopes up to failure for three bridge decks are given in Figure 5. It should be noted that deflection plotted in Figure 5 is measured at the center of the respective deck span directly under the applied load. It is readily apparent from Figure 5 that the first bridge deck reinforced with MMFX steel using the same reinforcement ratio as used for the actual bridge exhibited smaller deflection in comparison to the other two bridge decks. Due to the use of higher reinforcement ratio in the first bridge deck, stiffness was higher than the other two decks; this could also be due to the higher compressive strength of the concrete used for the first deck. Despite the lower reinforcement ratio used for the third bridge deck (33% less than the first two decks), it was capable of sustaining the same load as the second bridge deck of the Grade 60 steel. This behavior is attributed to the utilization of the higher tensile strength of MMFX steel. The slight increase of the deflection measured

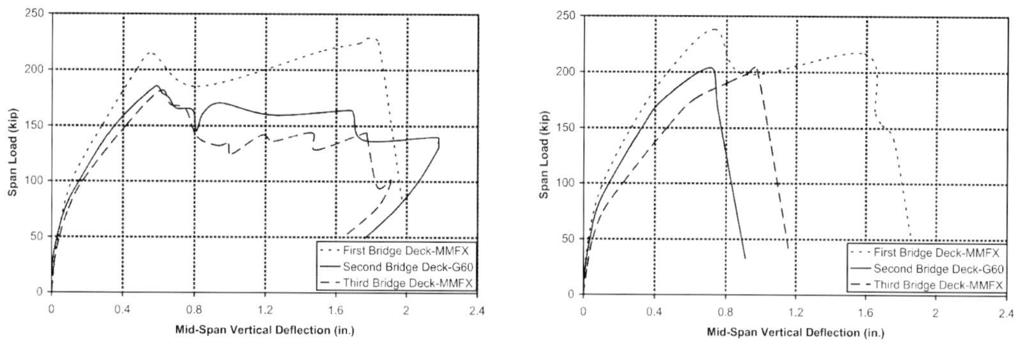

(a) Left Span (b) Right Span

Fig. 5. Load-Deflection envelope of the three bridge decks.

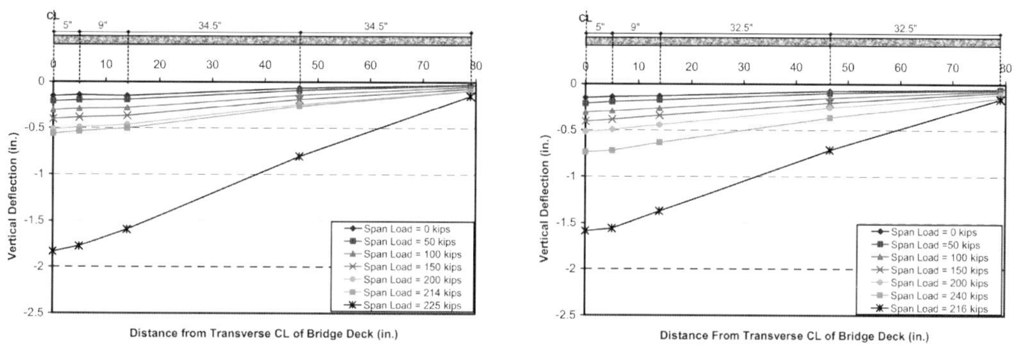

(a) Left span (b) Right span

Fig. 6. Longitudinal deflection profile for the first bridge deck.

for the third bridge deck in comparison to the second deck is possibly due to the slight reduction of the modulus of elasticity of MMFX steel at high stress levels. According to the AASHTO LRFD Bridge Design Specifications (1998), the design tandem consists of a pair of 25 kips (111 KN) axles. Therefore, at a load level of 25 kips (111 KN), which is less than the cracking load; the deflection at service load was almost identical for the three bridge decks.

Deflection Profile

The deflection profiles along the longitudinal direction of the three bridge decks are given in Figures 6, 7, and 8, respectively. It should be noted that the deflection profiles are plotted for the final loading cycle only. The deflections shown for each deck represent the residual deflection from previous loading. The deflection profiles for the three bridge decks indicate that the deflection at the edge of the bridge decks was very small. This implies that selection of the length of the model is effective for carrying the total load, and therefore, representative to the actual bridge deck.

The deflection profiles along the transverse direction of the three bridge decks are given in Figures 9, 10, and 11, respectively. It should again be noted that the deflection profiles are plotted for the last loading cycle only, therefore residual deflections are shown at the beginning of the loading cycle (zero load). The deflection profiles indicate that the maximum deflection occurred at the mid-span under the applied load. Also, it is clear that the spans failed in punching shear (right span) exhibited less deflection than the spans failed due to flexural as will be discussed in the following sections.

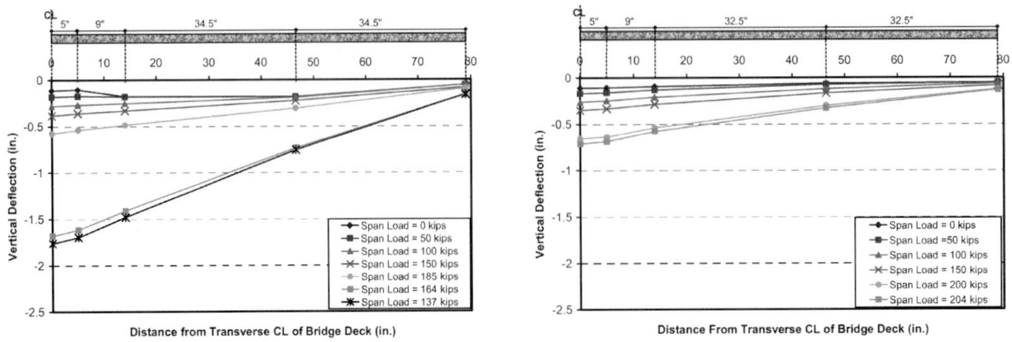

(a) Left span (b) Right span

Fig. 7. Longitudinal deflection profile for the second bridge deck.

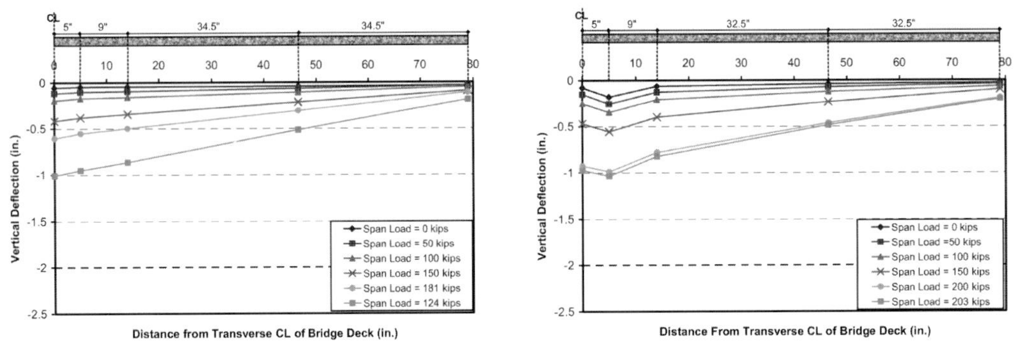

(a) Left span (b) Right span

Fig. 8. Longitudinal deflection profile for the third bridge deck.

Mode of Failure

In general, the behavior was two-way flexural mode followed by development of an arching action supported by membrane forces developed in the bottom layer of the reinforcement. At first peak load of the first bridge deck a sudden drop in the load occurred due to the formation of flexural-shear

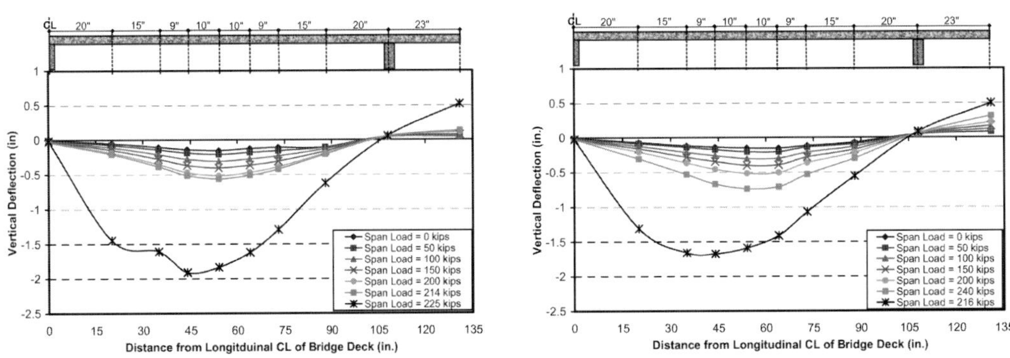

(a) Left span (b) Right span

Fig. 9. Transverse deflection profile for the first bridge deck.

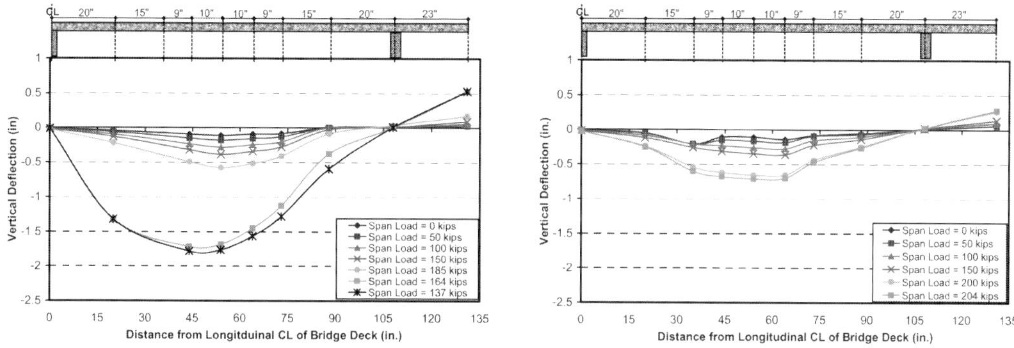

(a) Left span (b) Right span

Fig. 10. Transverse deflection profile for the second bridge deck.

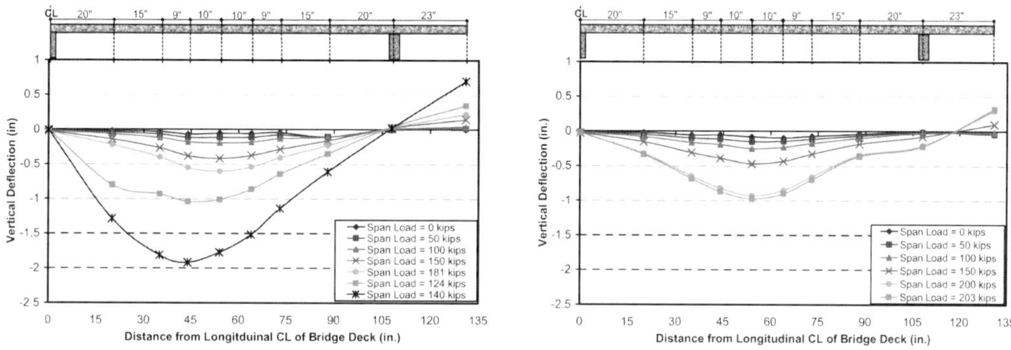

(a) Left span (b) Right span

Fig. 11. Transverse deflection profile for the second bridge deck.

cracks along the top surface of the bridge deck on both sides of the middle girder. Further loading led to the widening of those cracks associated with slight increase in the load resistance until punching failure occurred. Punching failure of both spans occurred simultaneously at a load level of 229 kips (1019 N) and 216 kips (961 N) for the left and right spans, respectively. Figure 12 shows the first bridge deck at the conclusion of the test, where the punching areas under the two loads and the shear cone at the bottom of the left span can be seen clearly.

The behavior of the second bridge deck, reinforced with grade 60 steel using the same reinforcement ratio was similar to the first deck. At the peak load of the left span, a sudden drop in the load occurred due to the formation of a flexural-shear crack on the top surface of the bridge deck to the left of the middle girder only (left span only). This drop in the load made the left span incapable to carry higher load equivalent to the punching shear capacity of the deck. The test was terminated due to excessive deflections in the left span. The smooth decrease of the load carrying capacity of the left span reveals that flexural-shear failure was the mode of failure of the left span. The maximum measured load for the left span was 185 kips (823 KN) and a deflection of 2.2 in. (56 mm) prior to termination. Failure of the right span was due to punching shear at a load level of 204 kips (907 KN).

Similar to the second bridge deck the right span of the third deck failed by punching shear prior to the failure of the left span. A flexural-shear crack formed in the left span causing a sudden drop in the load which made the left span incapable to carry more load equivalent to its punching shear resistance. Flexural-shear failure was the mode of failure of the left span as revealed by the smooth

(a) Punching shear of the two spans

(b) Bottom surface of the punching area for left span

Fig. 12. Failure of the first bridge deck.

(a) Second bridge deck

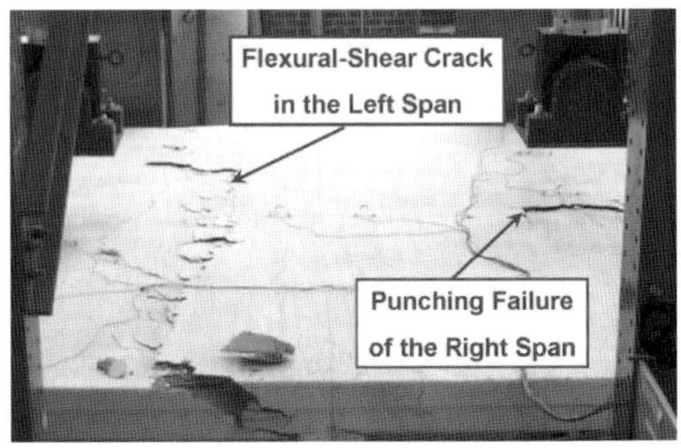

(b) Third bridge deck

Fig. 13. Second and third bridge decks at failure.

decrease in the load carrying capacity of the load, whereas the right span failed in punching shear at a load level of 203 kips. The test terminated due to excessive deflections in the left span and the maximum recorded load for the left span was 181 kips (805 KN). Figure 13 shows the second and third bridge decks at failure, where the punching area under the actuator in the right span and the flexural-shear crack formed in the left span are clearly visible.

Crack Pattern

No cracks were observed up to a load level of 50 kips (222 KN) for any of the three bridge decks. However visible top cracks started to occur at a load level of roughly 60 kips (444 KN) for each deck. Figure 14 shows the top cracks at a load level of 100 kips. Negative flexural cracks formed before the positive cracks due to the higher values of negative moments in comparison to the positive moments.

Positive moment flexural cracks at load levels of 100 and 150 kips (444 and 666 KN) for the first bridge deck are shown in Figure 15, where the crack pattern confirms the two way distribution of the load. Further loading led to spreading and widening of the flexural cracks until the formation of the flexural-shear crack at the top surface of the deck close to the middle girder. The formation of the flexural-shear crack led to a sudden drop in the load as previously discussed. However, the flexural-shear crack formed symmetrically on both sides of the middle girder of the first bridge

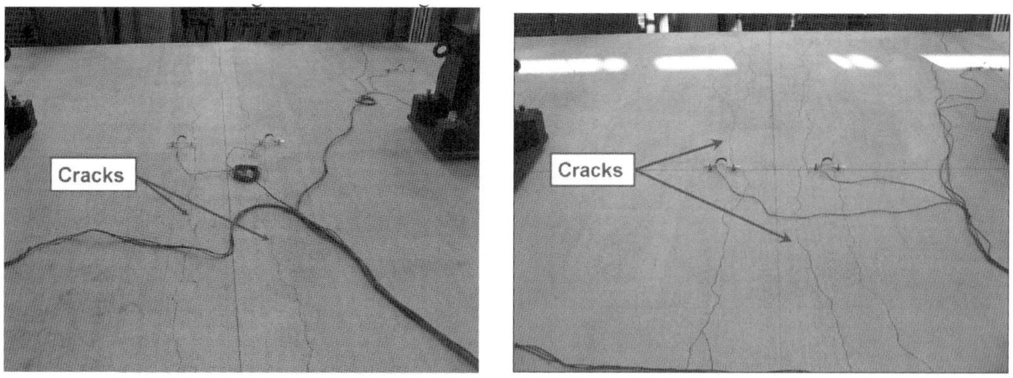

(a) Second bridge deck (b) Third bridge deck

Fig. 14. Negative flexural cracks at load level of 100 kips for the second and third bridge decks.

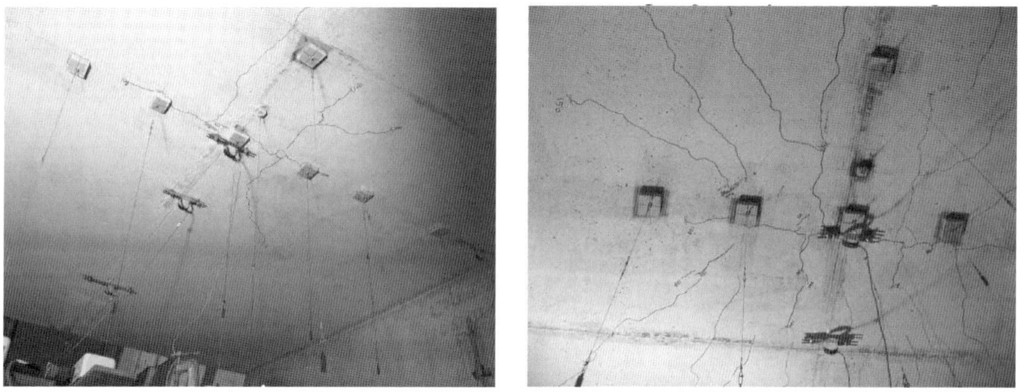

(a) At load level of 100 kips (b) At load loevel of 150 kips

Fig. 15. Positive flexural cracks for the first bridge deck.

deck, therefore allowed increase of the load to cause punching shear of both spans. For the second and third bridge decks, the flexural-shear crack occurred on the left side of the middle girder only which allowed the load to increase in the right span causing punching shear at the right span only as shown in Figure 13.

Strain Profiles and Girders Rotation

Based on the deformations measured by the PI gages, strain profiles were determined using the measured strain at the extreme top and bottom fibers of each bridge deck. It should be noted that all the strain profiles are plotted for the final loading cycle only, and therefore residual strains are shown at zero load. The strain profile obtained from the two PI gages located in the right span at 14 in. (356 mm) from the centerline of the deck (T6 and B10 in Figure 4) is depicted in Figure 16. The strain profile indicates that the top surface of the concrete at the vicinity of the punching area exceeded the limiting compressive strain value. The limiting compressive strain at a distance y from the edge of the loading plate reported by Kinnunen and Nylander (1960) was 0.0019, Marzouk and Hussein (1991) was 0.00215, and Mufti and Newhook (1998) was 0.002, where y is the distance from the bottom surface of the slab to the root of the shear crack at failure. The strain profile obtained from

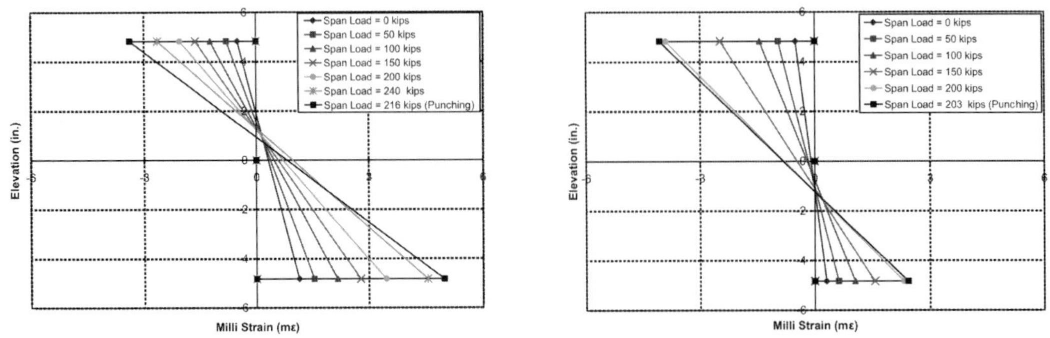

Fig. 16. Strain profile from T6 and B10 PI gages for the first and third bridge decks.

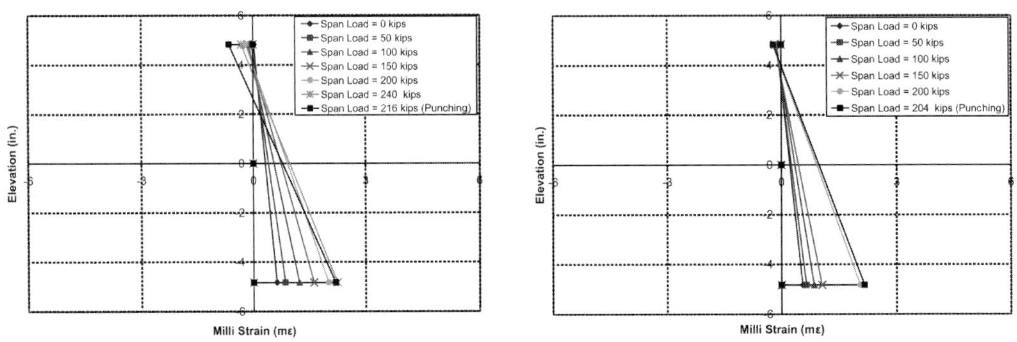

(a) First bridge deck (b) Second bridge deck

Fig. 17. Strain profile from T8 and B12 PI gages for the three bridge decks.

the PI gages at the edge of the right span (T8 and B12 in Figure 4) is shown in Figure 17. The strain profile shows that the strain values were very small, which is another indication that the length of the bridge deck is effective and representative to the behavior of typical bridges.

Rotation of the three supporting girders was monitored throughout the test of the three bridge decks as shown in Figure 18. For all three decks the two outside girders exhibited larger rotations in comparison to the middle girder due to the unbalanced moment effect

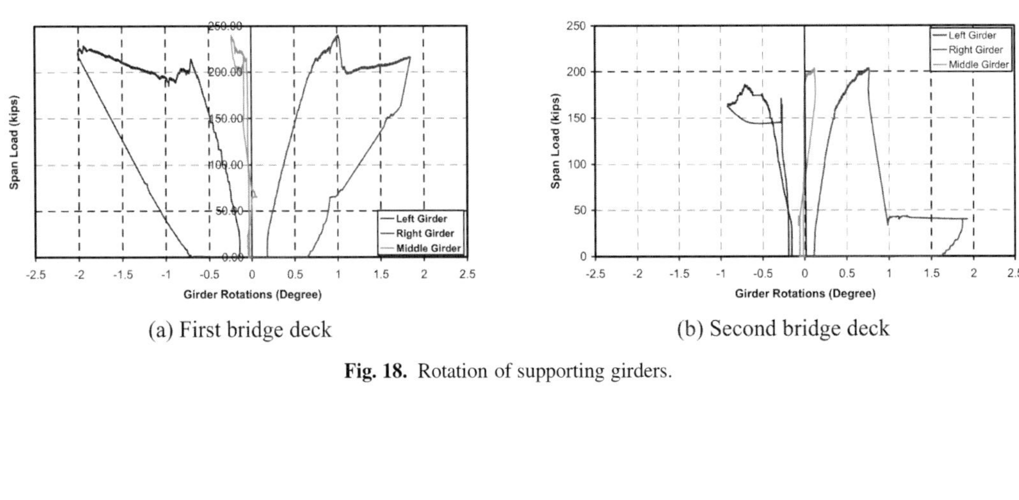

(a) First bridge deck (b) Second bridge deck

Fig. 18. Rotation of supporting girders.

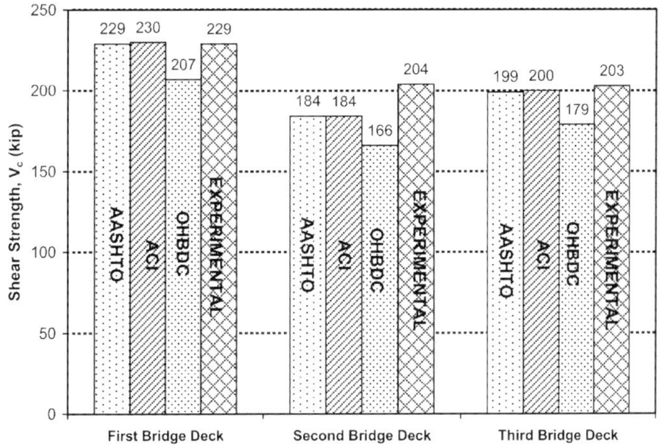

Fig. 19. Predicted and experimental shear strength of bridge decks.

Predicted Strength

The predicted shear strengths for the three bridge decks according to the different design codes are given in Figure 19 as well as the experimental values. The design codes included are: American Association of State Highway and Transportation Officials (AASHTO), American Concrete Institute (ACI 318-02), and Ontario Highway Bridge Design Code (OBBDC, 1991). The equations used are as follows:

$$\text{AASHTO:} \quad V_c = \min\left[0.063 + \frac{0.126}{\beta_c}; 0.126\right]\sqrt{f_c'}\, b_0 d; \text{ units: kips \& in.} \tag{1}$$

$$\text{ACI:} \quad V_c = \min\left[2 + \frac{4}{\beta_c}; 4; \frac{\alpha_s d}{b_0} + 2\right]\sqrt{f_c'}\, b_0 d; \text{ units: lbs \& in.} \tag{2}$$

$$\text{OHBDC:} \quad V_c = [0.6 f_r + 0.25 f_{pc}] b_0 d + 0.9 V_p; \text{ units: N \& mm,} \tag{3}$$

where V_c = punching shear capacity of bridge deck; β_c = ratio of long side to short side of loading plate; f_c' = concrete compressive strength; b_0 = perimeter of critical section at a distance of $d/2$ from loading plate; d = effective section depth; α_s = constant; f_r = concrete tensile strength; f_{pc} = compressive stress in concrete due to prestressing; and V_p = component of effective prestressing force in direction of applied shear.

It is clearly seen from Figure 19 that the predicted values according to the AASHTO and ACI design codes predict very well the measured values for the bridge decks using MMFX and Grade 60 steel.

Analytical Modeling

General

The three bridge decks were modeled using the finite element analysis program ?ANACAP? (Anatech Concrete Analysis Program) Version 3.0, (James, 2004). The concrete material model is based on smeared cracking methodology developed by Y. R. Rashid, 1960. Within the concrete constitutive

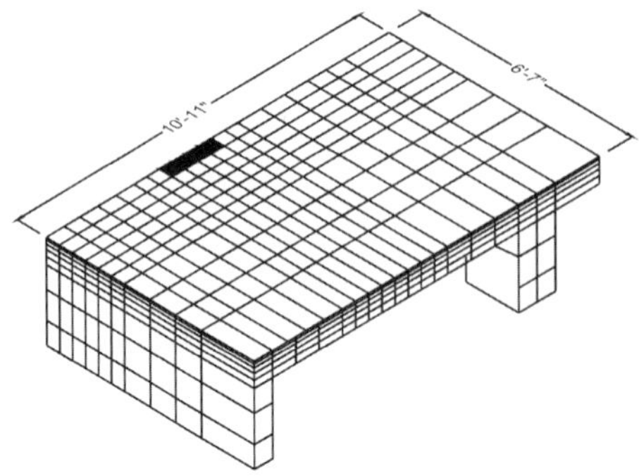

Fig. 20. Mesh used for modeling the three bridge decks.

model, cracking and all other forms of material non-linearity are treated at the finite element integration points. Cracks are assumed to form perpendicular to the principal tensile strain direction in which the criterion is exceeded and they are allowed to from at each material point. When cracking occurs, the normal stress across the crack is reduced to zero and distribution of cracks around the crack is recalculated. Cracks may close or re-open under load reversals. Concrete modeling also included residual tension stiffness for the gradual transfer of load to the reinforcement during crack formation. In addition, the program accounts for the reduction in shear stiffness due to cracking and further decay as the crack opens. The reinforcement is modeled as individual sub-elements within the concrete elements. The stiffness of the bar sub-element is superimposed on the concrete element stiffness in which the bar resides. The anchorage loss is modeled as an effective stiffness degradation of the bar as a function of the concrete strain normal to the bar.

A 3-D analysis was conducted for the three bridge decks using 20-node hexahedral continuum elements. Only one quarter of the deck was modeled due to its symmetry about both axes. The depth of the deck was divided into five layers within its thickness with a total number of elements of 1040, as shown in Figure 20.

Analytical Results

The predicted and experimental load-deflection envelopes for the three bridge decks are compared as shown in Figure 21. It can be seen that the predicted load-deflection behaviors of the three bridge decks compared very well with the measured values. The initial and post-cracking stiffnesses were very accurately predicted by the analytical model. In addition, the ultimate load was very reasonably predicted considering the fact the two spans of the second and third bridge decks failed in two different modes. However, the predicted ultimate deflection was slightly less than the experimental values; this is due to the nature of the smeared cracking methodology adopted by the program. For validation purposes the contours of the principal strain at failure and the portion of the first bridge deck that failed due to punching failure are shown in Figure 22. The strain contours depict the punching shear cone developed which matches very well the experimental failure shear cone.

Conclusions

1. The ultimate load carrying capacity of the three bridge decks tested in this investigation was eight to ten times the service load specified by AASHTO Design Specifications (1998).
2. Punching shear failure was the primary mode of failure for the three bridge decks tested.

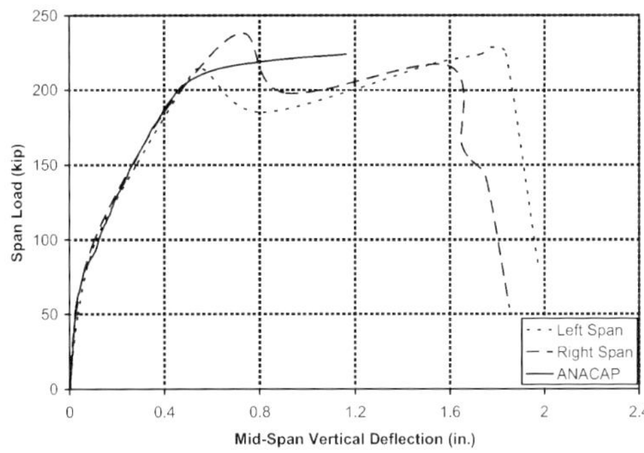

(a) First bridge deck

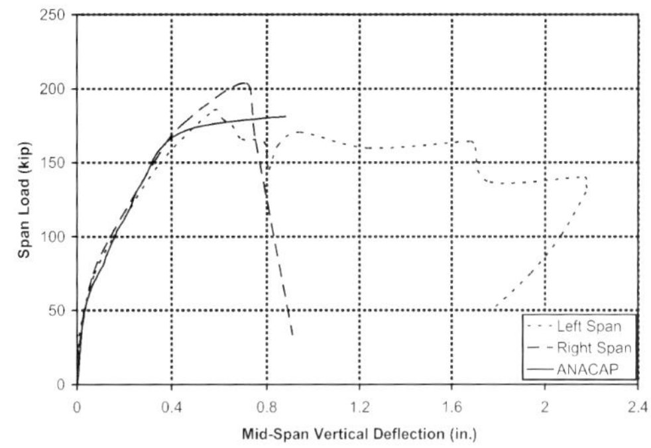

(b) Second bridge deck

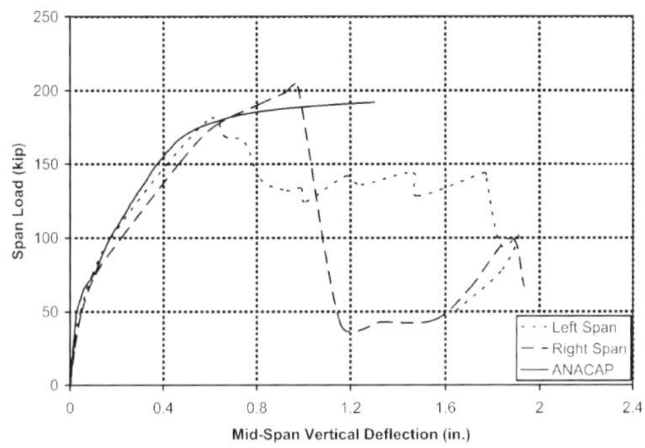

(c) Third bridge deck

Fig. 21. Analytical and experimental load-deflection envelopes for the three bridge decks.

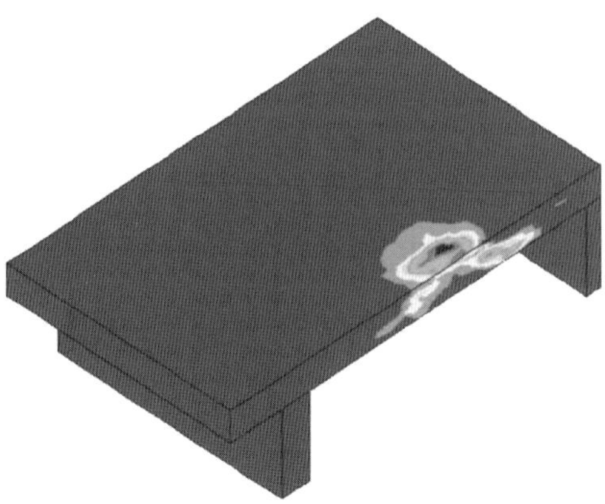

(a) ANACAP

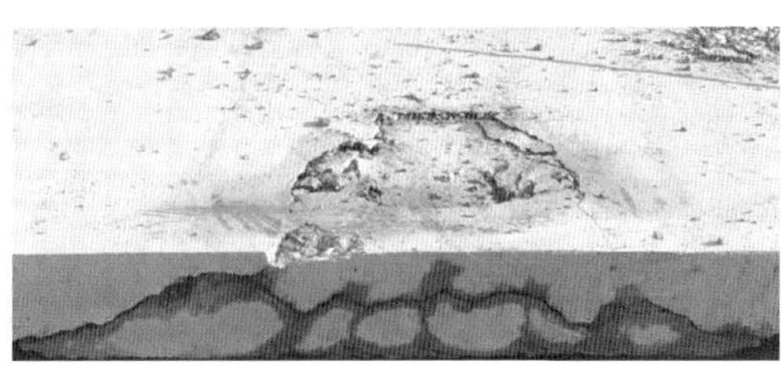

(b) Experimental

Fig. 22. Principal strains contours at failure for the first bridge deck.

3. Punching failure resulted in sudden decrease of the load carrying capacity, while flexural failure resulted in gradual decrease of the load carrying capacity.
4. Bridge decks reinforced with MMFX steel exhibited same deflection at service load but developed more load carrying capacity when compared to decks reinforced with Grade 60 steel with the same reinforcement ratio.
5. Bridge decks reinforced with 33% less MMFX steel developed the same ultimate load carrying capacity and deflection at service load as those reinforced with Grade 60 steel. This is attributed to the higher strength of the MMFX steel compared to Grade 60 steel.

References

American Association of State Highway and Transportation Officials, 1998, ? "AASHTO LRFD Bridge Design Specifications", Washington, D.C.

American Concrete Institute (ACI), 2002, "Building Code Requirements for Structural Concrete", ACI 318-02, Farmington Hills, Michigan.

James, R.G., 2004, "ANACAP Concrete Analysis Program Theory Manual", Version 3.0, Anatech Corporation, San Diego, CA, 2004.

Kinnunen, S. and Nylander, H., 1960, "Punching of Concrete Slabs without Shear Reinforcement", *Transactions of the Royal Institute of Stockholm*, Sweden, No. 158.

Marzouk, H. and Hussein, A., 1991, "Punching Shear Analysis of Reinforced High-Strength concrete Slabs", *Canadian Journal of Civil Engineering*, 18, 954–963.

Mufti A.A. and Newhook, J.P., 1998, "Punching Shear Strength of Restrained Concrete Bridge Deck Slabs", *ACI Structural Journal*, 95(4), July–August, 375–381.

OHBDC, 1991, "Ontario Highway Bridge Design Code", Ministry of Transportation of Ontario, Downsview, Ontario

Rashid, Y.R., 1960, "Ultimate Strength Analysis of Prestressed Concrete Pressure Vessels", *Nucl. Eng. & Design*, pp. 334–344.

EVALUATION OF SHEAR DESIGN METHODS FOR LARGE, LIGHTLY-REINFORCED CONCRETE BEAMS

E.G. Sherwood, E.C. Bentz and M.P. Collins
Department of Civil Engineering, University of Toronto, 35 St. George St., Toronto, Ontario, Canada M5S 1A4
E-mail: sherwoo@ecf.utoronto.ca

Abstract

This paper describes one phase of an extensive experimental program that has recently been completed at the University of Toronto. In this phase, eighteen lightly-reinforced shear-critical reinforced concrete beams were loaded to failure. The abilities of the ACI-318 shear design method and a simplified design method based on the Modified Compression Field Theory to predict the failure loads are compared. It is found that the ACI design method is dangerously unconservative when applied to large beams and one-way slabs constructed without stirrups, while the simplified MCFT design method is both safe and accurate. Studies of the mechanism of shear transfer indicate that approximately one quarter of the shear in a reinforced concrete beam constructed without stirrups is transferred in the compression zone, with the rest carried primarily by aggregate interlock. The development of theoretically-sound shear design methods must therefore be based on the fact that aggregate interlock plays a critical role in the shear behaviour of reinforced concrete structures.

Introduction

It has long been a goal of code writers to improve the quality of reinforced concrete design procedures for shear. Unlike flexural failures, shear failures in reinforced concrete are brittle and sudden, and occur with little or no warning. Furthermore, they are less predictable than flexural failures, due to considerably more complex failure mechanisms. While flexural design provisions are based on the rational assumption that plane sections remain plane, the search for equally rational design provisions for shear continues.

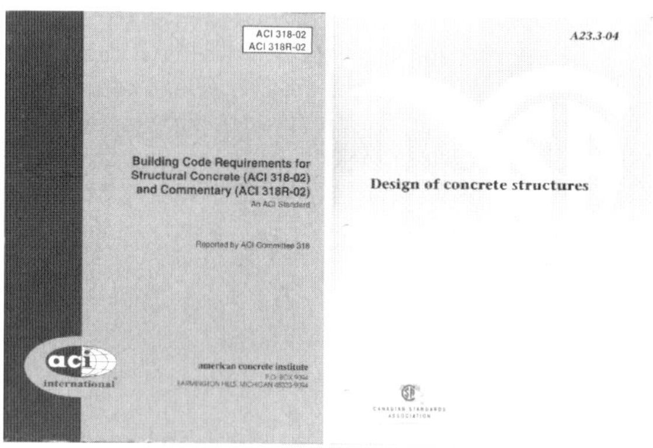

Figure 1: ACI and CSA Design Codes

The recently updated and reissued ACI-318-05 and CSA A23.3-04 Design Codes for Concrete (Figure 1) represent the culmination of extensive research involving all aspects of the behaviour of reinforced concrete. However, the shear design provisions in the ACI code remain based on traditional empirical relationships developed over 40 years ago, and do not reflect the vast improvements in understanding of the shear behaviour of reinforced concrete that have been gained over that time. In particular, there are significant concerns that the ACI shear design provisions are unconservative when applied to large beams and one-way slabs constructed without stirrups. As such, they must be replaced with rational, theoretically-sound design provisions that can predict the shear behaviour of these brittle, complex structural elements.

A particular aspect of the shear behaviour of reinforced concrete that is deserving of additional attention is the effect of the maximum aggregate size on the shear response of reinforced concrete sections. This is particularly true for reinforced concrete beams constructed without stirrups, since aggregate interlock is the dominant mechanism of shear transfer in these element types. Increasing the size of the coarse aggregate produces rougher cracks that are better able to transfer shear stresses. Likewise, reducing the maximum aggregate size decreases the shear strength of a concrete section. Furthermore, the use of high

M. Pandey et al. (eds), Advances in Engineering Structures, Mechanics & Construction, 153–164.

strength concrete or low-density aggregate can result in fracturing of the coarse aggregate particles as cracks form, thereby producing smoother cracks with a greatly reduced aggregate interlock capacity.

The Modified Compression Field Theory (MCFT) (Vecchio and Collins, 1985) employs equilibrium, compatability and experimentally verified stress-strain relationships to model the shear behaviour of cracked concrete. A fundamental relationship in the MCFT relates the shear stress on a crack surface due to aggregate interlock to the crack's width, the maximum aggregate size and the concrete strength. The aggregate effect was first codified when a general method of shear design was derived based on the MCFT and implemented in the AASHTO-LRFD bridge design guidelines. In 1994 the general method of shear design was implemented in the CSA concrete design code for buildings. Recently, an updated and simplified version of the general method has been developed (Bentz et al., 2005) and implemented in the 2004 CSA design code. The new general method, referred to as the Simplified Modified Compression Field Theory (SMCFT) has been found to be simpler than the original general method with, in many cases, improved predictive capabilities (Sherwood et al., 2005a).

As might be expected, there are certain areas of considerable disagreement between modern shear design methods based on the MCFT and the ACI method. The fundamental question that must be asked, therefore, is: "Are the existing ACI shear design methods sufficiently safe, such that a reworking of the provisions is not necessary". The corollary to this question is: "Has our understanding of the fundamental behaviour of reinforced concrete in shear advanced to such a stage such that modern shear design methods represent a clear improvement over traditional design methods?" The purpose of this paper is to explore the answers to these questions by studying the behaviour of reinforced concrete beams in shear, focusing on the role played by the coarse aggregate. A significant experimental program will be presented in which eighteen shear critical concrete beams constructed with different maximum aggregate sizes were tested to failure. The SMCFT design method will be discussed, and its predictive capabilities will be compared to those of the empirical ACI shear design method.

ACI Method of Shear Design

Morsch (1909) was amongst the first to research the behaviour of reinforced concrete in shear. At the beginning of the 20th century he developed the well-known 45° truss model, whereby shear was visualized to be transferred through the web of a cracked concrete member through a field of diagonal compression in the concrete and tension in transverse reinforcement. To produce an expression for the shear strength of a concrete section, he assumed that shear cracks that formed did so at an angle, θ, of 45°:

$$v = \frac{V}{b_w jd} = \frac{A_v f_v}{b_w s} \tag{1}$$

Reflecting the design philosophy at the time, f_v was taken to be the safe working stress in the stirrups. While Morsch knew from observations that failure shear cracks did not necessarily form at 45°, he saw no way to calculate the angle of what he termed secondary inclined cracks.

The 45° truss model entered use in various design methods and still forms the basis for the ACI expression for the shear resistance provided by stirrups. (The current ACI expression has simplified the equation by replacing the term jd with d.) As its use became more widespread, however, it was criticized for being overly conservative. In particular, the model assumed that only transverse reinforcement is effective at carrying shear, thereby predicting that a section without stirrups or bent-up bars would have no shear strength whatsoever. Clearly this is not the case. Extensive research efforts were undertaken in order to ascertain the so-called "concrete contribution" to shear resistance, which was eventually set at an empirically derived safe working shear stress of v_c=0.03f'_c. For the first time, the shear resistance of a reinforced concrete section was divided up into two components: a concrete contribution (V_c) and a web reinforcement contribution (V_s) predicted by the 45° truss model:

$$V = V_c + V_s \tag{2}$$

This method was used to design numerous concrete structures in the post-war construction boom of the 1950s and early 1960s. In 1955, however, a considerable portion of the roof of the Wilkins Air Force Warehouse in Selby, Ohio collapsed. The collapsed portions of the beams supporting the roof had been designed without stirrups, assuming that they could safely resist a working shear stress of 0.6MPa (90psi = 0.03 x 3000psi specified concrete strength). However, failure occurred at a shear stress of approximately 0.5MPa (70psi), corresponding to only about 80% of the safe service load on the roof. It therefore became apparent that unsafe designs could result from what had previously been considered to be a safe, conservative method.

As a result of the warehouse collapse, extensive research was undertaken to derive a better expression for V_c. In 1962, these efforts resulted in what was believed to be a simple, conservative expression for the failure shear based on a purely empirical curve-fit through 194 experimental data points (ACI Committee 326, 1962). This well-known expression (Equation 3) entered design use through incorporation into the 1963 American Concrete Institute Design Code, and has remained essentially unchanged since that time:

$$V_c = 0.167\sqrt{f_c'}\,b_w d \quad \text{(MPa units)} \tag{3a}$$

$$V_c = 2\sqrt{f_c'}\,b_w d \quad \text{(psi units)} \tag{3b}$$

The Size Effect in Shear

Beams

At the time Equation (3) was developed, it was not understood that the failure shear stress for members constructed without web reinforcement decreases as the member depth increases in a phenomenon known as the "size effect." Unfortunately, the average height of the specimens tested to develop Equation (3) was 340mm. As a result, the original researchers did not notice a size effect in their tests, and the ACI expression predicts a continuous and linear increase in shear capacity as the beam depth increases. Equation (3), while originally intended to be a conservative estimate, is therefore unconservative for deeper members constructed without web reinforcement because it can not predict this size effect. The failure of the warehouse beams was attributed to an unexpected tensile force in the beams, while it is far more likely that it was simply a result of the size effect (Lubell et al., 2004).

The MCFT predicts that the size effect in shear is related to the crack spacing in the web and the crack widths. As discussed by Sherwood et al. (2004): "…the larger crack widths that occur in larger members reduce aggregate interlock. Crack widths increase nearly linearly both with the tensile strain in the reinforcement and with the spacing between cracks…for the same reinforcement strain, doubling the depth of the beam will double the crack widths at mid-depth. To maintain beam action, a shear stress equal to about $V/b_w d$ must be transmitted across these cracks. The shear stress that can be transmitted across such cracks, however, decreases as the crack width increases and as the maximum aggregate size decreases." Thus, the limiting stress that can be transferred across cracks due to aggregate interlock in deep members is reached at a lower shear stress than in equivalent small beams. Once the limiting stress is reached, equilibrium cannot be maintained, and failure occurs.

The use of at least the minimum quantity of stirrups will largely eliminate the size effect by allowing more closely spaced cracks to form and by preventing the loss of aggregate interlock as the cracks widen. Web reinforcement consisting of additional longitudinal steel placed in layers along the height of a beam will also reduce the size effect. In this case, the size effect is related to the vertical spacing of the layers of reinforcement, rather than the overall height of the beam.

One-Way Slabs

To avoid the risk of a brittle shear failure, the ACI code requires that beams be constructed with stirrups if the factored shear force V_f exceed $\frac{1}{2}\phi V_c$. This requirement is relaxed for one-way slabs, however, as the code requires stirrups for these elements only when V_f exceeds the full value of ϕV_c. Yet it has been shown in an extensive companion research program that the web width (b_w) has no effect on the failure shear stress of reinforced concrete flexural members (Sherwood et al., 2005b). As such, deep one-way transfer slabs in high-rise construction are particularly vulnerable to the size effect if designed using the ACI code, as the designer may wish to proportion the slab such that no stirrups are required.

Aggregate Effects

Early attempts in the 1950s to develop rational theories of reinforced concrete in shear neglected the role played by aggregate interlock. Both implicit and explicit in these early theories was the assumption that all the vertical shear force in concrete sections without transverse reinforcement is carried in the uncracked concrete compression zone. This represented a reasonable first approximation of the complex behaviour of these element types. As research progressed, however, a belief gradually emerged that significant shear stress may, in fact, be transferred through the cracked web of a reinforced concrete beam. Fenwick and Pauley (1968) definitively showed this to be the case. Through direct measurement on subassemblies, it was possible to conclude that at least 60% of the vertical shear is carried by aggregate interlock at flexural cracks, with the remaining proportion being carried in the compression zone and through dowel forces.

Despite the successes of early classic studies on aggregate interlock, their results have been forgotten or otherwise neglected by many prominent modern shear researchers. Tureyn and Frosch (2003), for example, have formulated an expression for V_c based on the explicit assumption that all of the vertical shear force is carried in the compression zone, describing the assumption as a "reasonable approximation." Many others have taken similar approaches, particularly those using fracture mechanics principles (for example, Bazant and Yu, 2005). The fact that the importance of aggregate interlock is not appreciated has slowed the implementation of theoretically-sound design methods for shear. In particular, the size effect in shear can not be adequately accounted for unless aggregate interlock is explicitly considered.

A Modern Method of Shear Design

A considerable step forward in shear design methods was the development of a general method of shear design based on the MCFT. Design methods based on the MCFT have a firm theoretical base and are not derived by empirical curve fits to experimental data. As such, MCFT-based shear provisions are able to predict the behaviour of reinforced concrete elements in shear where no experimental data is available. Particular strengths of MCFT-based shear design methods include the ability to accurately predict the size and aggregate effects.

The recently developed SMCFT (Bentz et al., 2005) is based on the methods in the AASHTO-LRFD and the 1994 CSA Standards, but has been considerably simplified. Simple expressions have been developed for β, the crack angle, θ and the longitudinal strain in the web, ε_x, thereby eliminating the need to iterate to solve for these values.

The SMCFT employs the following relationship to determine the shear resistance of a concrete section:

$$V = V_c + V_s = \beta\sqrt{f_c'}\,b_w d_v + \frac{A_v f_y}{s} d_v \cot\theta \qquad (4)$$

The term β in Equation (4) is a parameter that models the ability of cracked concrete to transfer shear. It is a function of 1) the longitudinal strain at the mid-depth of the web, ε_x, 2) the crack spacing at the mid-depth of the web and 3) the maximum coarse aggregate size, a_g. It is calculated using an expression that consists of a strain effect term and a size effect term:

$$\beta = \frac{0.40}{(1+1500\varepsilon_x)} \cdot \frac{1300}{(1000+s_{ze})} = (\text{strain effect term}) \cdot (\text{size effect term}) \qquad (5)$$

The longitudinal strain at the mid-depth of a beam web is conservatively assumed to be equal to one-half the strain in the longitudinal tensile reinforcing steel. For sections that are neither prestressed nor subjected to axial loads, ε_x is calculated by:

$$\varepsilon_x = \frac{M_f/d_v + V_f}{2E_sA_s} \tag{6}$$

The effect of the crack spacing at the beam mid-depth is accounted for by use of a crack spacing parameter, s_z. This crack spacing parameter is equal to the smaller of either the flexural lever arm (d_v=0.9d or 0.72h, whichever is smaller) or the maximum distance between layers of longitudinal crack control steel distributed along the height of the web. To be effective, the area of the crack control steel in a particular layer must be greater than $0.003b_ws_z$.

The term s_{ze} is referred to as an "equivalent crack spacing factor" and has been developed to model the effects of different maximum aggregate size on the shear strength of concrete sections by modifying the crack spacing parameter. For concrete sections with less than the minimum quantity of transverse reinforcement and constructed with a maximum aggregate size of 20mm, s_{ze} is equal to s_z. For concrete with a maximum aggregate size other than 20mm, s_{ze} is calculated as follows:

$$s_{ze} = \frac{35s_z}{15 + a_g} \geq 0.85s_z \tag{7}$$

To account for aggregate fracturing at high concrete strengths, an effective maximum aggregate size is calculated by linearly reducing a_g to zero as f'_c increases from 60 to 70MPa. The term a_g is equal to zero if f'_c is greater than 70MPa. The square root of the concrete strength is limited to a maximum of 8MPa.

Since specimens with transverse reinforcement do not exhibit a size effect, s_{ze} is set equal to 300mm for specimens with at least the minimum quantity of stirrups as per Equation (8). This has the effect of reducing the size effect term to 1.

$$\frac{A_vf_y}{b_ws} = 0.06\sqrt{f'_c} \tag{8}$$

The angle of inclination of the cracks at the beam mid-depth, θ, is calculated by the following equation:

$$\theta = (29° + 7000\varepsilon_x)(0.88 + s_{ze}/2500) \leq 75° \tag{9}$$

Experimental Program

The preceding discussion has shown that considerable differences exist between the ACI shear design provisions and the SMCFT. In particular, the ACI method can not account for either the size effect or the aggregate effect. To investigate these differences, a series of ten large-scale and eight one-fifth scale-model shear-critical reinforced concrete beams were constructed and loaded to failure in the Mark Huggins Laboratory in the Department of Civil Engineering at the University of Toronto.

As summarized in Table 1 and Figure 2, the large beams measured 1510mm tall x 300mm wide x 9000mm long. Nine of the beams were reinforced with five No. 30 rebars in tension at an effective depth of 1400mm and with two No. 20 rebars for compression reinforcement. These beams were constructed without stirrups. The tenth large beam (specimen SB-10-H-S) was reinforced with eight No. 30 rebars in tension, two No. 20 rebars in compression, and approximately the minimum quantity of stirrups required by Equation (8) for the f'_c on the day of test. The stirrups were in the form of single legs of 9.5mm (3/8'') diameter rebars on alternating sides of the beam, spaced at 235mm. The one-fifth scale model beams measured 330mm tall x 122mm wide x 1800mm long. The reinforcement in nine of the small beams consisted of four 9.5mm diameter rebars at an effective depth of 280mm. The tenth scale-model beam (SSB-10-H-S) was reinforced in shear with 5mm smooth bars placed at 160mm on alternating sides.

Table 1: Summary of Experimental Program and Observations

Test Specimen	Specimen Properties					Experimental Observations				ACI Method		SMCFT Method		
	ρ_s (%)	$\frac{A_v f_y}{b_w s}$	f'_c (1) (MPa)	s_{ze} (mm)	$a_{g,eff}$ (mm)	P_{exp} (kN)	V_{exp} (2) (kN)	V_{exp} (3) (kN)	Δ_{ult} (mm)	V_{ACI} (kN)	$\frac{V_{exp}}{V_{ACI}}$	V_{SMCFT} (kN)	$\frac{V_{exp}}{V_{SMCFT}}$	ε_x (mm/m)
SB-10-N-1	0.83	0	38.4	1764	10	499	264	277	9.0	435	0.64	239	1.10	0.56
SB-10-N-2	0.83	0	40.3	1764	10	454	241	254	8.8	445	0.57	243	0.99	0.57
SB-10-H-1	0.83	0	73.6	2940	0	449	239	252	6.6	582	0.43	223	1.07	0.53
SB-10-H-S	1.33	0.50	71.2	300	0	1388	708	721	27.5	802	0.90	729	0.97	1.04
SB-20-N-1	0.83	0	31.4	1260	20	499	264	277	9.7	393	0.70	256	1.03	0.60
SB-20-N-2	0.83	0	33.2	1260	20	500	264	277	9.9	404	0.69	261	1.01	0.61
SB-40-N-1	0.83	0	28.1	1071	40	453	241	254	8.2	372	0.68	262	0.92	0.61
SB-40-N-2	0.83	0	28.5	1071	40	545	287	300	10.2	374	0.80	263	1.09	0.62
SB-50-N-1	0.83	0	41.0	1071	50	512	270	283	9.2	449	0.63	298	0.91	0.69
SB-50-N-2a	0.83	0	40.1	1071	50	565	297	310	9.4	444	0.70	295	1.01	0.69
SB-50-N-2b	0.83	0	40.1	1071	50	614	321	334	11.1	444	0.75	295	1.09	0.69
										Average:	0.68		1.02	
										Coefficient of Variation:	17.8%		6.6%	
SSB-10-N-1	0.83	0	41.9	353	10	72.7	36.6	36.8	3.5	36.9	1.00	32.5	1.13	0.90
SSB-10-N-2	0.83	0	41.9	353	10	76.1	38.3	38.5	3.9	36.9	1.04	32.5	1.18	0.90
SSB-10-H-1	0.83	0	77.3	588	0	74.9	37.7	37.9	3.2	47.3	0.80	33.6	1.12	0.93
SSB-10-H-S	1.34	0.50	77.3	300	0	132	66.3	66.5	7.7	67.3	0.99	59.6	1.11	1.03
SSB-20-N-1	0.83	0	39.2	252	20	77.7	39.1	39.3	3.5	35.7	1.10	33.4	1.17	0.93
SSB-20-N-2	0.83	0	38.1	252	20	75.9	38.2	38.4	3.5	35.2	1.09	33.1	1.15	0.92
SSB-40-N-1	0.83	0	29.1	214	40	83.3	41.9	42.1	6.2	30.8	1.37	31.0	1.35	0.86
SSB-40-N-2	0.83	0	29.1	214	40	69.3	34.9	35.1	3.9	30.8	1.14	31.0	1.13	0.86
										Average:	1.07		1.17	
										Coefficient of Variation:	15.1%		6.7%	

Notes:
(1) day of test
(2) Calculated at d_v=0.9d from face of loading plate, incl. self-weight
(3) Calculated at d from face of support, incl. self-weight

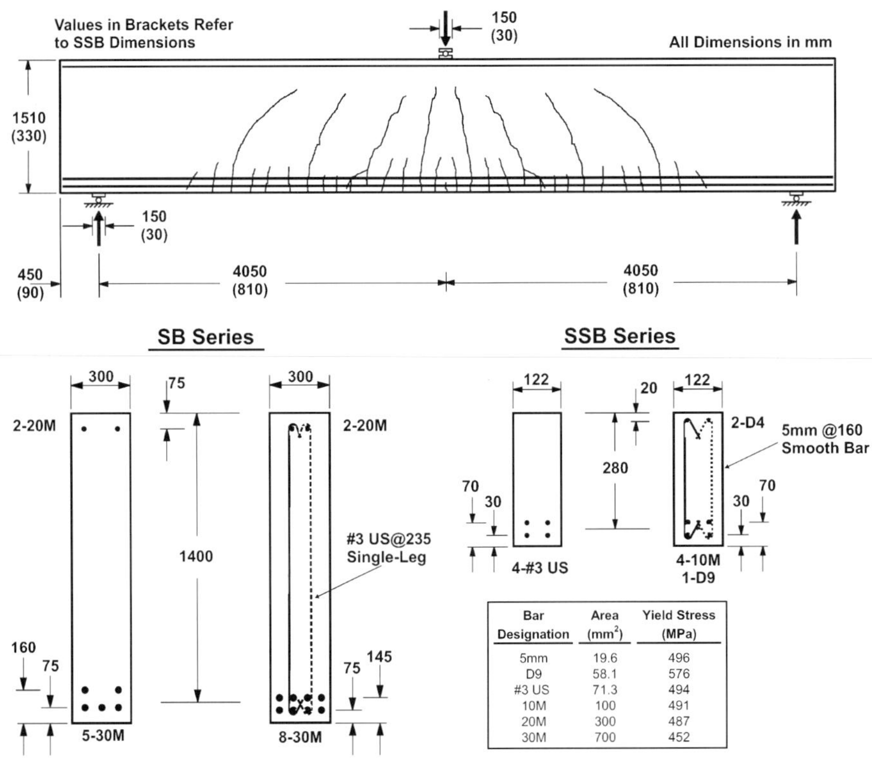

Bar Designation	Area (mm^2)	Yield Stress (MPa)
5mm	19.6	496
D9	58.1	576
#3 US	71.3	494
10M	100	491
20M	300	487
30M	700	452

Figure 2: Specimen Design

The N series of beams were constructed with normal strength concrete and with four different maximum aggregate sizes (10, 20, 40 and 50mm). Two duplicate specimens were cast for each aggregate size. The H-series of beams were constructed with high-strength concrete and a maximum aggregate size of 10mm. The concrete used was commercially available from a local ready-mix supplier. The aggregate was strong crushed limestone shipped from a quarry on Manitoulin Island and conforming to CSA grading requirements.

The beams were moist-cured for five days, after which they were removed from their formwork. The beams were loaded to failure in three-point bending at an a/d ratio of 2.89. The large beams were tested in a 4500kN force-controlled Baldwin Test Frame, and the small beams were tested in a 1000kN displacement controlled MTS actuator (Figure 3). Upon reaching 85% of the monotonic failure load of the duplicate specimen, the applied load for beams SB-10-N-2 and SB-20-N-2 was cycled 20 times from 225kN to this load. This process was repeated at 90% and 95% of the monotonic failure load. After Specimen SB-50-N-2 failed on the east side of the beam (SB-50-N-2a), this side was clamped together with a series of externally installed Dywidag bars, and the beam was reloaded until failure occurred on the west end (SB-50-N-2b).

Figure 3: Specimen Test Setup

Experimental Results

Experimental results are summarized in Table 1 and typical load deflection curves are presented in Figure 4. All beams failed in shear prior to reaching their flexural capacity. The large beams without stirrups exhibited extremely brittle behaviour, failing at ratios of mid-span deflection to span of 1/750 or less. Failure was sudden and was preceded by relatively little cracking. The large beam with stirrups (SB-10-H-S) failed after a significant amount of cracking and deflection. Prior to failure, the failure shear crack reached a width of 4mm, and the Δ/L ratio was 1/300. The use of minimum stirrups and additional longitudinal steel increased the shear capacity by 2.9 times over the equivalent specimen without stirrups, providing a dramatic and beneficial effect on the shear strength of the section. The small beams all exhibited slightly greater ductility. The small beam with stirrups was the most ductile of all the specimens with a Δ/L ratio of 1/200. Like the equivalent large specimen, this beam was extensively cracked prior to failure. However, the use of minimum stirrups and additional longitudinal steel increased the shear capacity by only 1.8 times over the equivalent beam without stirrups. Thus, stirrups are far more effective for deep members than for shallow members.

The results shown in Figure 4 also indicate that the failure load generally increased for increasing aggregate size. The most likely explanation for this result is the increased surface roughness of the failure shear cracks caused by the larger aggregate. That is, failure was initiated at a higher shear stress in the beams with large aggregate due to enhanced aggregate interlock capacity at the cracks. Note that Specimen SB-10-H-1, constructed with high-strength concrete, had the lowest peak load of all the large beams tested. This was due to reduced aggregate interlock caused by aggregate fracturing. In this case the effective aggregate size was 0mm.

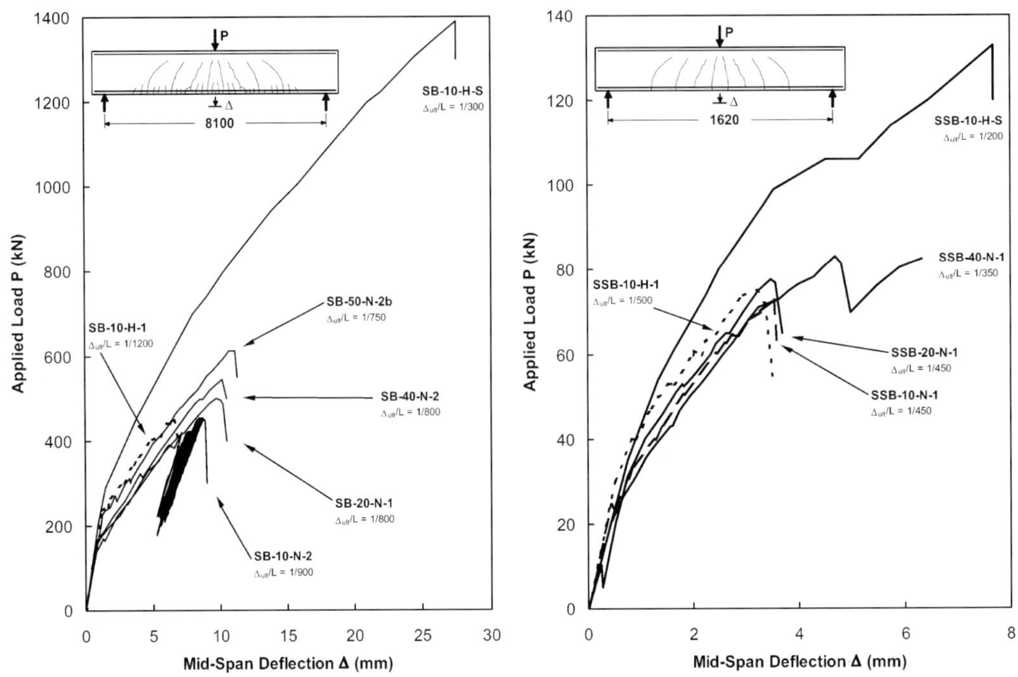

Figure 4: Typical Load-Deflection Curves of Test Specimens

The typical progression of failure in a large beam without stirrups is illustrated in Figure 5. These photos are of the shear critical region on the east side of Specimen SB-40-N-1 and were taken with a high-speed digital camera. The cracks have been digitally enhanced for clarity. In Figure 5a, the progression of cracks at 98% of the peak load is shown, and it can be clearly seen that a dominant flexural-shear crack has formed. The cause of the size effect is clearly demonstrated, in that the average longitudinal spacing of the cracks increases from 150mm at the level of the steel to 900mm at the beam mid-depth. In order to maintain a linear strain profile, these cracks are wider near the beam mid-depth than they are at the level of the steel. As the load is increased to the peak load, the dominant flexural shear crack extends slightly (5b). After the peak load has been reached, the dominant shear crack extends towards the loading point and widens (5c, 5d). After the dominant crack extends towards the load point, it also extends back towards the support point (5e) due to dowel forces in the longitudinal steel. At final failure, shown in Figure 5f, the failure crack is very wide and the bottom cover has been ripped from the bottom steel.

A Comparison of Methods

The failure shear predictions generated by the ACI Design Code and the SMCFT are summarized in Table 1. To account for the beam self-weight, it is appropriate, when using the ACI shear design method, to express the failure shear stress as the shear stress located at a distance d from the face of the support. When using the SMCFT, it is appropriate to calculate the failure shear at a distance d_v from the face of the loading plate. While the shear due to self-weight is slightly reduced, calculating the failure shear at this location takes into account the far more dominant effect of the moment on the longitudinal strain in the web, ε_x. The SMCFT produced safe and accurate failure shear predictions. The average ratio of experimental to predicted failure shear was 1.17 for the small beams, with a coefficient of variation of

6.7%. The average ratio of experimental to predicted failure shear was 1.02 for the large beams, with a coefficient of variation of only 6.2%. Comparison with the ACI predictions indicates that the ACI method has a lower experimental/predicted value for the small beams. Note, however, that the SMCFT is consistently conservative for the small beams over the entire range of aggregate sizes, resulting in a lower coefficient of variation. The increased scatter evident in the ACI predictions of the small beams shear strengths is indicative of an inability to properly account for the effects of the maximum aggregate size.

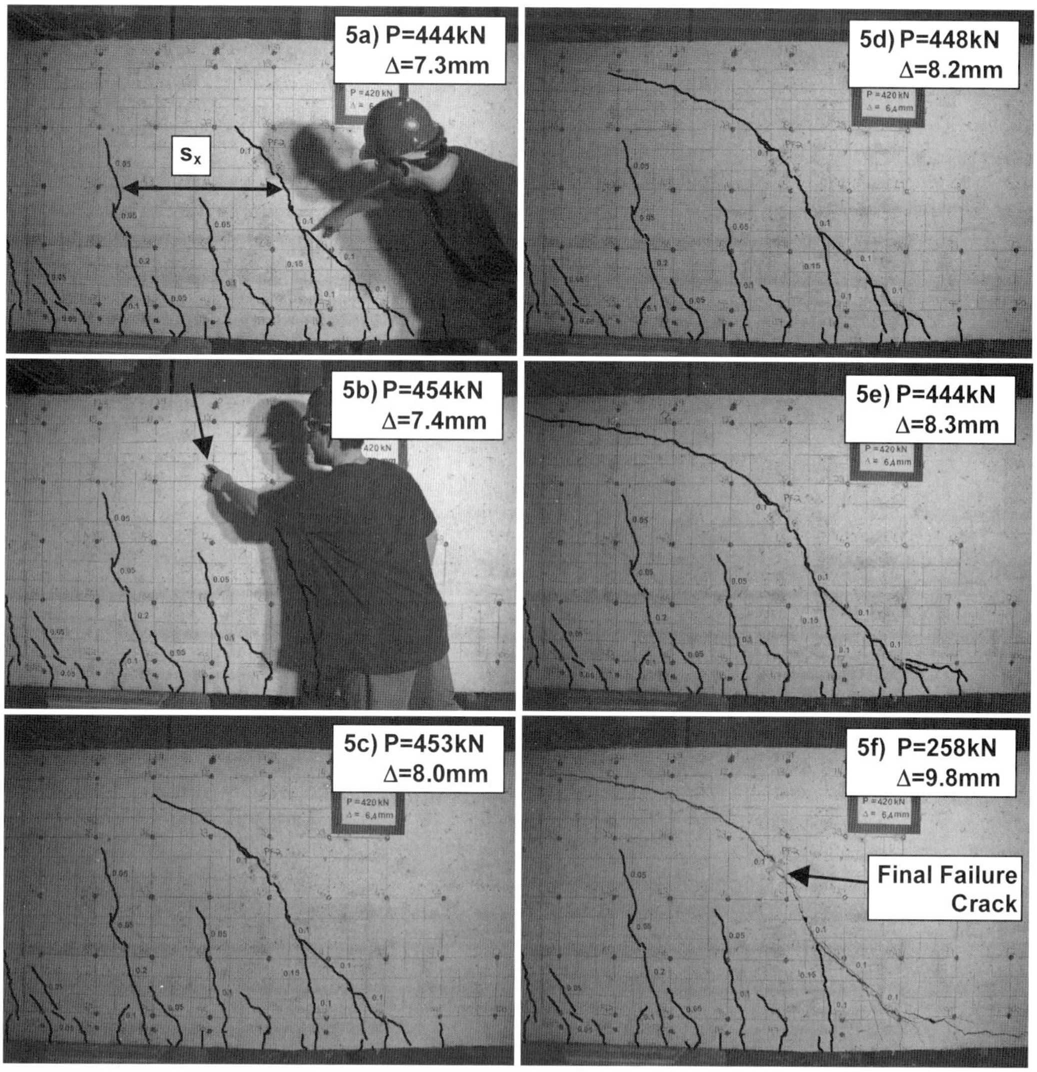

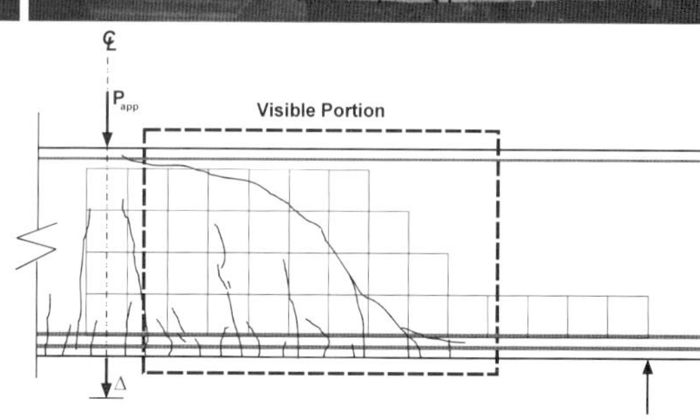

Figure 5: Progression of Failure Crack in Specimen SB-40-N-1

The ACI predicted failure shears for the large beams without stirrups ranged from 43% to 80% of the experimental failure shears. Clearly the ACI design method produces grossly unconservative predictions for large beams constructed without stirrups. Since the size effect is eliminated by the use of minimum stirrups, the ACI method produced an acceptable prediction of the failure shear stress of specimen SB-10-H-S. Despite the inability to account for aggregate effects, the ACI method produced otherwise excellent predictions of the failure shears of the scale model beams. This is to be expected, as the height of these beams was almost exactly the same as the average height of the beams tested to derive Equation (3).

Since the concrete strengths for all of the tested specimens varied, some thought is required to asses the true effect of the aggregate size. It is possible to normalize the experimental results by their concrete strengths and recalculate the experimental failure loads for the average f'_c of all the specimens (Sherwood, 2006). The normalized and recalculated experimental failure shears of the beams without stirrups are plotted in Figure 6 as a function of the aggregate size. The averages of duplicate specimens are shown, along with the experimental range at each aggregate size. This figure clearly and explicitly demonstrates that the shear strength increases as the aggregate size increases, but ceases to increase beyond an aggregate size of 25mm. The SMCFT successfully predicts the effects of the maximum aggregate size.

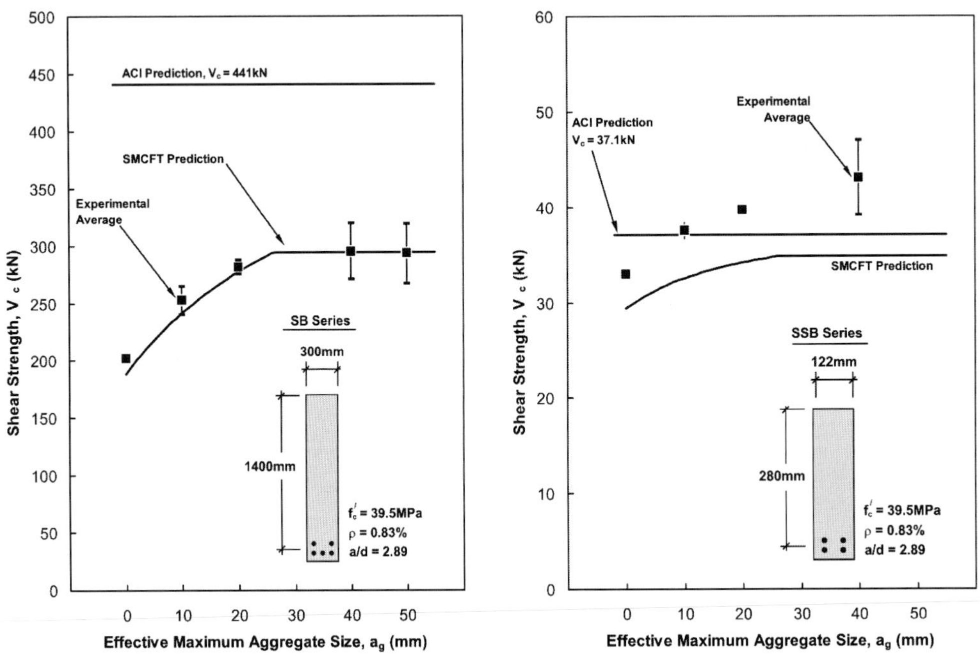

Figure 6: ACI and SMCFT Prediction of the Shear Strength of the Test Specimens

Since there is no difference between the shear behaviour of beams and one-way slabs, the large specimens can be thought of as 300mm sections taken from a wide, deep slab. If this slab was a large transfer element in high-rise construction, a likely ratio of dead load to live load would be 3:1 (Sherwood et al., 2005b), resulting in a safe service load of 60% of the failure load. As such, specimens SB-10-N-2 and SB-10-H-1 failed below the safe service load predicted by the ACI code. If a one-way transfer slab similar to these specimens designed by the ACI code were in use in a real structure, there would be a risk of failure under service loads, with little to no warning of impending collapse. The ACI method, however, is perfectly acceptable when designing shallow one-way slabs.

The Size Effect Factor

A unique aspect of the SMCFT is the introduction of the effective crack spacing parameter, s_{ze} (Equation 7). It is implemented into the expression for V_c through the size effect factor, $1300/(1000+s_{ze})$.

A wide range of effective crack spacings (from 214mm to 2940mm) was investigated in this study by varying only two variables: the effective depth and the maximum aggregate size. A graph showing the SMCFT predictions of the size effect factors for all of the beams tested is presented in Figure 7. Also shown are average experimental values of duplicate tests. The experimental points for the beams with stirrups are shown in white. The values of $V_{c,exp}$ for these two beams were calculated by subtracting the steel contribution calculated by Equation (4) from the total experimental failure shear force. This figure shows that the size effect factor can accurately model the shear strength of concrete sections over a wide range of both depths and aggregate sizes.

The effective crack spacing of 2940mm for Specimen SB-10-H-1 represents the highest value for s_{ze} ever tested. Basing the 2004 CSA shear design method on a rational theory rather than empirical relationships allows it to accurately predict the shear behaviour of a reinforced concrete beam in which an experimental variable was set outside the range of previous experimental data. It is also worthwhile to note that the experimental results for the beams with stirrups follow the trend of the equivalent beams without stirrups. It is therefore appropriate to use an s_{ze} value of 300mm for beams containing at least the minimum quantity of stirrups.

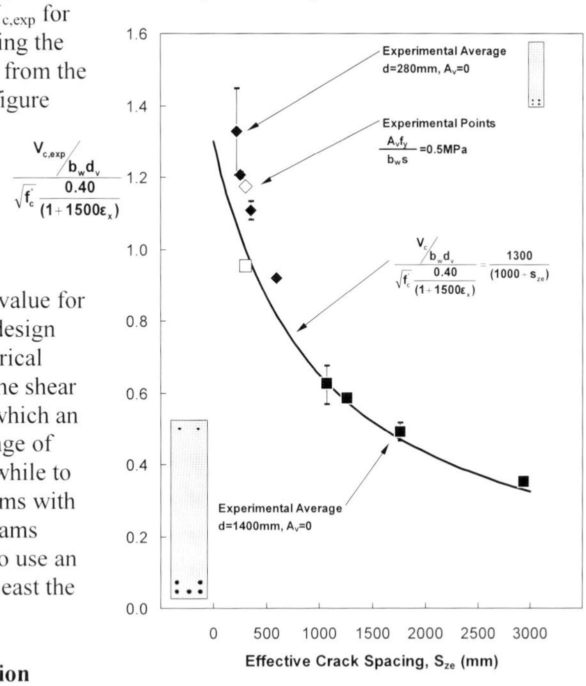

Figure 7: SMCFT Prediction of the Size Effect Term

Proportion of Shear Carried in Compression Zone

In order to definitively demonstrate the importance of aggregate interlock, the compression zone at the head of dominant shear cracks in several beams was instrumented with two columns of longitudinal concrete strain gauges. This was done with the intention of measuring the shear force carried by the compression zone.

As shown in Figure 8, the flexural strains measured by columns of gauges spaced at 80mm on the 365mm deep compression zone of Beam SB-10-N-2 allowed for the calculation of adjacent flexural stress profiles. Equilibrium requires that a horizontal complimentary shear stress of $v = \Delta C/(b_w \times 80)$ act at every level of the compression zone. Equilibrium further requires that the vertical shear stress equals the horizontal shear stress at every level of the compression zone. This allows for the vertical shear stress distribution to be calculated, as shown in Figure 8. The total shear force acting in the compression zone is therefore equal to $V = \int_A v \cdot dA$. For the shear stress distribution shown in the figure, the shear force carried in the compression zone is 50kN, representing 24% of the total shear force acting at the section. Clearly, then, 76% of the remaining shear must be carried by aggregate interlocking along the dominant shear crack and dowel forces. However, the shear crack is vertical at the level of the steel, indicating that the dowel forces transferred by the longitudinal reinforcement are extremely small.

Concluding Remarks

The ACI design code is dangerously unconservative when applied to the shear design of large concrete beams and one-way slabs constructed without stirrups. The ACI method is capable of accurately predicting V_c of members without stirrups only if their height is similar to the original experimental set used to calibrate the equation. To provide an adequate level of safety, the current ACI equation for V_c must be replaced with an expression that can account for the size effect in shear.

The primary mechanism of shear transfer in slender beams constructed without stirrups is aggregate interlock. It is incorrect to assume that the entire shear force is transferred in the compression zone. Clearly, then, any shear design provision that claims to be theoretically sound must be based on a theory that recognizes the critical importance of aggregate interlock in shear transfer. In this regard, the SMCFT represents a clear improvement over the ACI method.

The risks of basing design code provisions on empirical relationships rather than sound theory have been demonstrated in this paper. Developers of structural design codes owe it to the profession to base provisions wherever possible on rational theory rather than risky empiricism.

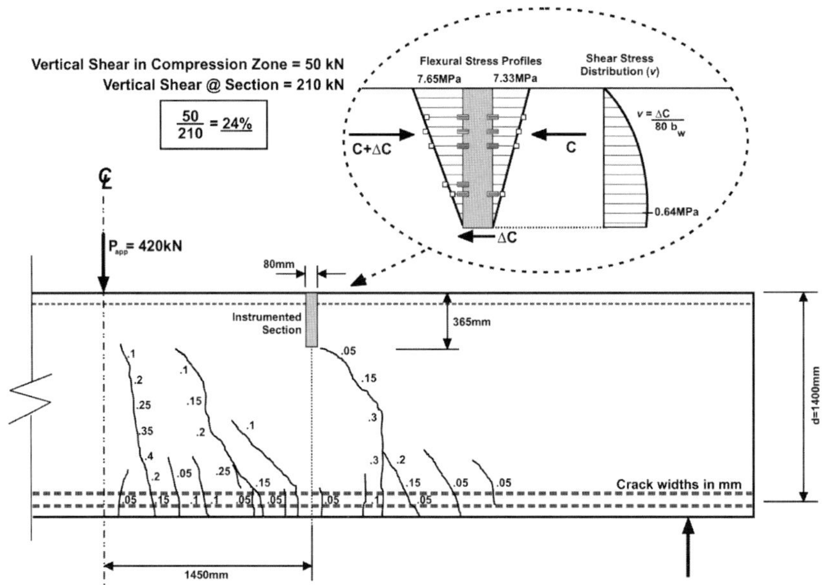

Figure 8: Calculation of Shear Carried in Uncracked Compression Zone, Specimen SB-10-N-2

References

ACI Committee 318 (2005) *Building Code Requirements for Structural Concrete (ACI 318-05) and Commentary*, American Concrete Institute, Farmington Hills, MI, USA

ACI-ASCE Committee 326 (1962) "Shear and Diagonal Tension," *ACI J., Proc.*, **59**: 1-30, 277-344, 352-396

Bazant, Z., Yu, Q. (2005) "Designing Against Size Effect on Shear Strength of Reinforced Concrete Beams Without Stirrups," *ASCE J. Struct. Engrg*, **131**(12) "I: Formulation" 1877-1885, "II: Verification and Calibration" 1886-1897

Bentz, E.C., F.J. Vecchio and Collins, M.P. (2005) "The Simplified MCFT for Calculating the Shear Strength of Reinforced Concrete Elements, *ACI Struct. J.*, submitted April 2005, 45pp.

CSA Committee A23.3. 2004. *Design of Concrete Structures (CSA-A23.3-04)*, Canadian Standards Association, Rexdale, ON

Fenwick, R.C. and Pauley, T. (1968) "Mechanisms of Shear Resistance of Concrete Beams," *ASCE Struct. J.*, **94**(ST10): 2235-2250

Lubell, A., Sherwood, T., Bentz, E. and Collins, M.P. (2004) "Safe Shear Design of Large, Wide Beams," *Concrete International*, January, **26**: 66-78

Mörsch, E. 1909. Concrete-Steel Construction. McGraw Hill Inc., New York, NY, USA

Sherwood, E.G., Bentz, E. and Collins, M.P. (2005a) "The Behaviour of Large, Lightly Reinforced Concrete Beams and One-Way Slabs," CSCE Annual Conference, Toronto, Ontario, June 2-4, 2005, Paper GC-167

Sherwood, E.G., Lubell, A., Bentz, E. and Collins, M.P. (2005b) "One-Way Shear Strength of Thick Slabs," *ACI Struct. J.*, 40pp., submitted July 2005

Sherwood, E.G. (2006) *Behaviour and Design of Large, Lightly-Reinforced Concrete Flexural Elements Subjected to Shear*, Ph.D. Thesis, University of Toronto, *in progress*

Tureyn, A.K. and Frosch, R.J. (2003) "Concrete Shear Strength –Another Perspective," *ACI Struct. J.*, **100**(5): 609-615

Vecchio, F and Collins, M.P. (1985) "The Modified Compression-Field Theory for Reinforced Concrete Elements Subjected to Shear," *ACI Journal*, **83**(2): 219-231

FRP REPAIR OF CORROSION-DAMAGED CONCRETE BEAMS – WATERLOO EXPERIENCE

Khaled A. Soudki
Canada Research Chair, Department of Civil Engineering, University of Waterloo, Waterloo, Ontario, Canada

Abstract: Corrosion of steel reinforcement is one of the main durability problems facing reinforced concrete infrastructures worldwide. This paper gives an overview on a seven year research program conducted at the University of Waterloo, sponsored by ISIS (Intelligent Sensing for Innovative Structures) Canada, to examine the viability of using fibre reinforced polymer (FRP) composites as a repair and strengthening method for corroded reinforced concrete structures. The majority of the research was carried out in the laboratory utilizing large-scale members. The results revealed that FRP repair successfully confined the corrosion cracking and improved the structural performance of corroded beams. Analytical models were developed to validate the experimental data. The FRP repair system was implemented in Fall 2005 to address corrosion damage in a bridge in the Region of Waterloo.

Introduction

Corrosion of reinforcing steel causes many structures in adverse environments to experience unacceptable loss in serviceability or safety far earlier than anticipated and thus need replacement, rehabilitation, or strengthening. As steel corrodes, there is a corresponding drop in the cross-sectional area. The corrosion products occupy a larger volume than the original steel which exert substantial tensile forces on the surrounding concrete and causes it to crack and spall off. If corrosion cracking can be prevented or delayed, a certain degree of structural strength may be maintained in a corroding RC beam.

Fibre reinforced polymer (FRP) systems are promising alternatives for the rehabilitation of deteriorated and deficient concrete members. In addition to their high strength to weight ratio, durability in adverse environments and high fatigue strength, FRP sheets can be easily externally bonded to reinforced concrete slabs, beams, and columns (ACI Committee 440 1996).

A multi-phase research program was undertaken at the University of Waterloo (see Table 1) to investigate the viability of using externally bonded fiber reinforced polymer (FRP) laminates to rehabilitate corrosion-damaged reinforced concrete beams (Soudki, et al. 2006, Badawi and Soudki 2004a and 2004b; Craig and Soudki 2005, 2002; El Maaddawy et al 2004, 2005a and 2005b; Masoud et al 2005 and 2001, 2000; Soudki and Sherwood 2003 and 2001, 1998). Several reinforced concrete beams (20 small-scale, 24 medium-scale and 50 large-scale beams) with variable chloride levels (0 to 3%) were constructed. The test variables were: level of corrosion damage at the time of FRP repair, location of corrosion damage, effect of short-term static, long-term sustained loading and fatigue loads. The beams were repaired by externally epoxy bonding FRP laminates to the concrete surface bonded to the tension face, with the fibre orientation in the longitudinal direction followed by transverse laminates bonded to the tension face and up each side of the beam, with the fibre orientation in the transverse direction. Two types of FRPs were used: Glass (GFRP) sheets had an ultimate strength of 600 MPa, an elasticity modulus of 26 GPa, and an ultimate elongation of 2.24%. The Carbon (CFRP) sheets had an ultimate strength of 960 MPa, an elasticity modulus of 73 GPa, and an ultimate elongation of 1.33%. The tensile reinforcement of the specimens was subjected to accelerated corrosion by means of impressed current up to 15% mass loss. Strain gauges were used on the FRP laminates to quantify tensile strains induced by the corrosion process. Accelerated corrosion was applied using a constant impressed current of 150 $\mu A/cm^2$. The current was impressed through the main longitudinal rebars, which act as the anode while the stainless steel bar in each specimen acts as the cathode. Following the corrosion phase, the specimens were tested in flexure in a four-point bending regime. Details of the test program, test methods, and test results are found elsewhere.

M. Pandey et al. (eds), Advances in Engineering Structures, Mechanics & Construction, 165–173.

Table 1. Overall experimental program

Researcher	Loading	Member	Dimensions (mm)	Number
Sherwood	Static	Flexural beams	102x154x1200	16
Craig	Static	Bond-beams	152x254x2000	37
	Static	Pull-out	150x150x150	24
Masoud	Static/Fatigue	Flexural beams	152x254x3200	20
	Fatigue	Flexural beams	120x175x2000	8
El Maaddawy	Sustained	Flexural beams	152x254x3200	29
Badawi	Static	Shear critical	152x254x3200	15
Rteil	Fatigue	Flexural beams	152x254x2000	60
Al Hammoud	Fatigue	Bond-beams	152x254x2000	29

In the following, key findings from FRP repair for reinforced concrete beams subjected to corrosion damage, are given.

Effects of FRP Repair on Serviceability

Corrosion Cracking

Figure 1 shows typical average crack width versus mass loss for corroded specimens. The width of the longitudinal cracks was measured at discrete time periods throughout the accelerated corrosion process for all the corroded specimens. It is evident that the FRP repair process reduced the crack opening by about 88% at the end of corrosion process. This implies a significant enhancement in appearance of FRP repaired corroded specimens by reducing crack opening due to further corrosion.

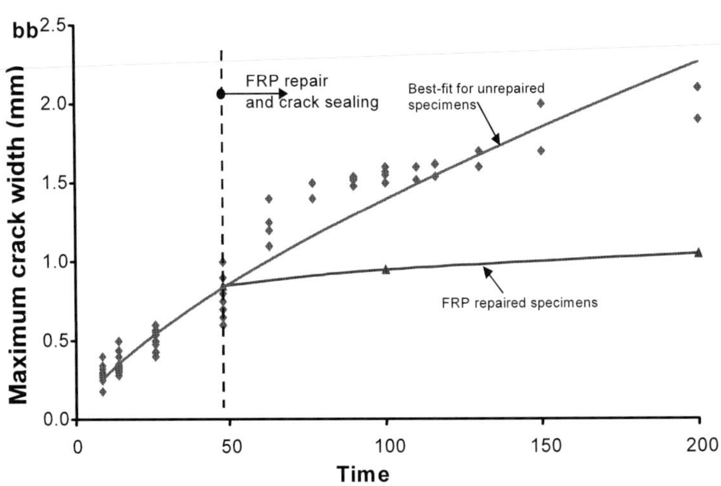

Figure 1. Crack width vs. time

Steel Mass Loss

Fig. 2 shows the average steel mass loss versus time relationship for various test specimens. A linear regression analysis for the steel mass loss results after repair showed that during the post-repair corrosion phase the steel mass loss rate in the specimens having the continuous-wrapping was on average about 32% lower than the level for the beams having the intermittent-wrapping. The presence of the sustained load during the post-repair corrosion phase increased the steel mass loss rate by about 9% and 12.5% for the specimens repaired with continuous and intermittent wrapping schemes, respectively. The specimens corroded under a sustained load had connected internal microcracks and external flexural cracks which increased the penetration of oxygen and moisture into the concrete and reduced the concrete resistivity and thus increased the steel mass loss rate to a level higher than that for the specimens corroded without load.

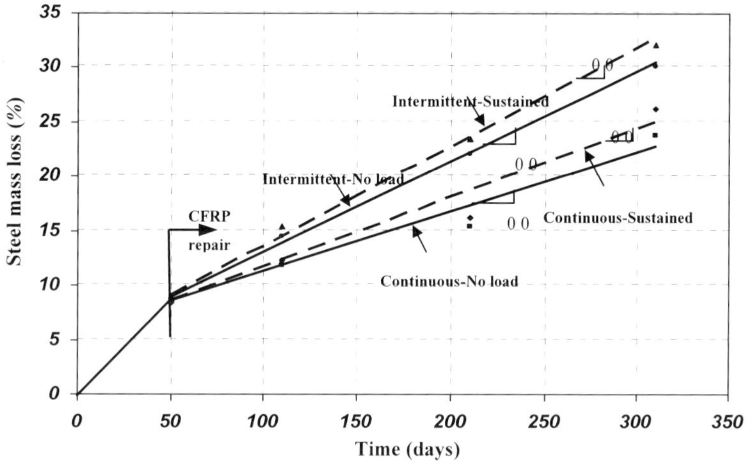

Figure 2. Steel mass loss versus time relationship

Bond Strength

Corrosion plays a devastating role in reducing the structural capacity of unconfined flexural members. CFRP confinement of the bond zone in bond-beam specimens serves to maintain steel-concrete bond interaction. The effect of CFRP on the confinement of corrosion-damaged members varies depending on whether the member has adequate bond or is bond deficient. For those members with inadequate bond length, the added CFRP confinement improved the performance of bond-deficient corroded members allowing them to outperform the unconfined specimen (Figure 3). It is important to understand the nature of failure of the CFRP confined specimens. Since no cracks were visible with the CFRP wrap in place, there are no indications of failures. Even under conditions of high ultimate bond stresses, the presence of low slip initiation bond stresses indicates that failure could potentially occur prematurely by bond pullout in the case of sustained loading or creep. Confinement was found to be more effective when applied prior to excessive corrosion of the specimens. Typically, small amounts of post-repair corrosion were found to have no effect or in some instances helped increase bond strength as a result of increased confining pressures. However, as the post-repair corrosion levels increased, the bond strength deteriorated. The overall structural performance of beams wrapped with CFRP was enhanced. However, caution and engineering judgement must be used in the application of this repair method since abrupt failure of the member due to bond pullout failure could occur without warning if repair is performed at high corrosion levels or if members were initially designed with inadequate bond. The confining wrap may increase the bond strength, but as with all repairs, this should not be used as a band-aid solution, and the cause of deterioration must be addressed to prevent further corrosion and deterioration.

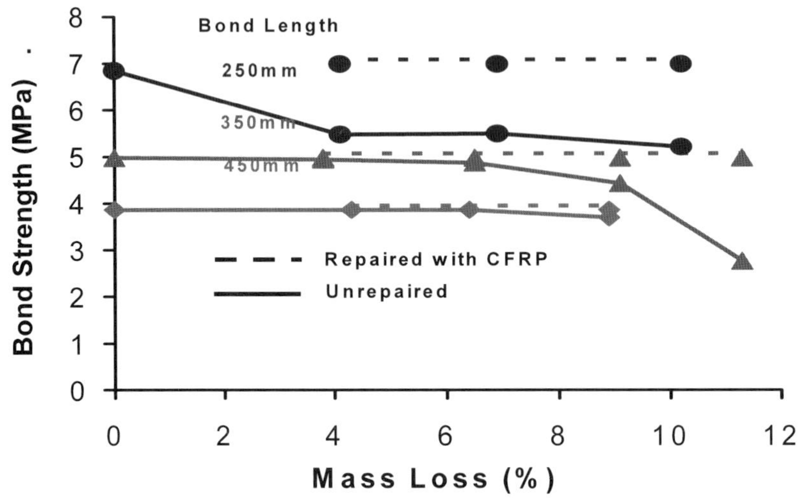

Figure 3. Bond strength vs. mass loss

Effects of FRP Repair on Static Strength

Figure 4 (a) and (b) show the normalized yield load vs. mass loss and normalized ultimate load vs. mass loss for the corroded-no load, corroded under load and CFRP repaired specimens. The normalized load is taken relative to the control specimen (at 0% mass loss). Corrosion of the steel reinforcement up to a 5.5% theoretical mass loss while the beam is unloaded resulted in a 10% and 9% reduction in the yield and the ultimate loads, respectively compared to those of the virgin uncorroded beam. The reduction in the yield and the ultimate capacity increased to 13% and 13.2%, respectively when the corrosion occurred under sustained load. A 12% theoretical mass loss of the steel reinforcement caused 16% and 11% reductions in the yield and the ultimate loads, respectively when corrosion occurred without load. Loading the beam with a 12% mass loss increased the reductions in yield and ultimate load to 22% and 17%, respectively compared to the virgin uncorroded specimen.

When CFRP external reinforcement was added there was an increase in the load carrying capacity. Compared with the uncorroded virgin beam, the uncorroded-strengthened beam exhibited a 30% and a 49% increase in the yield and the ultimate load, respectively. When a beam is corroded before the application of the CFRP laminates it doesn't reach the same strength as a strengthened uncorroded beam. The yield and the ultimate loads of the corroded-repaired beam at a 5.5% theoretical mass loss were 85% and 97%, respectively of that of the uncorroded-strengthened beam but they were higher than those of the virgin control beam by 11% and 43%, respectively. Increasing the level of corrosion damage to a 12% theoretical mass loss before repair reduced the yield strength to 77% of that of the uncorroded-strengthened beam while the ultimate strength remained at 97% of that of the uncorroded-strengthened beam. The yield load of the beam repaired after corrosion to a 12% theoretical mass loss was almost the same as that of the virgin control beam while the ultimate strength was still higher than that of the virgin beam by 43%. It is interesting to note that the ultimate strengths of the corroded-repaired beams were very close to that of the uncorroded-strengthened beam. This means that while corrosion of the steel reinforcement reduces the ability of the CFRP repair to increase the yield load its effect on the ultimate strength gained by repairing the beam with CFRP laminate is minimal.

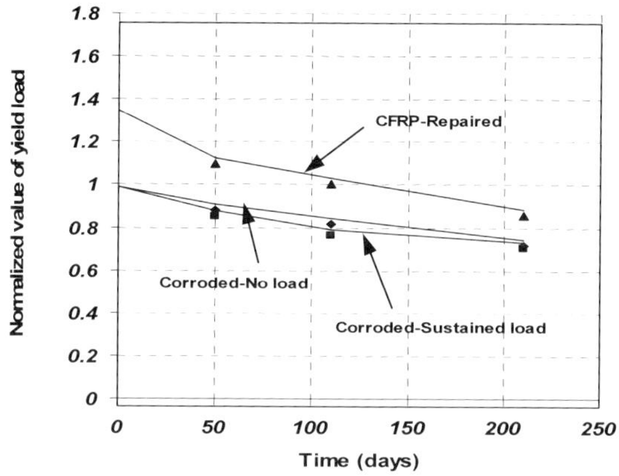

a) yield strength

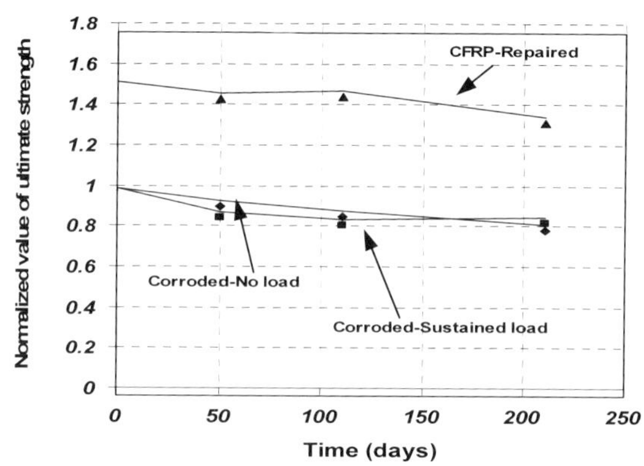

b) ultimate strength

Figure 4. Plots of normalized load vs. time

Effects of FRP Repair on Fatigue Strength

Figure 5 shows the fatigue life vs. load range for flexural and bond specimens. The life-load range variation of bond-beams is presented in Figure 5a. In set B-CD (with a 5% mass loss due to corrosion) the fatigue bond strength decreased by 23% and 15% compared to the control (uncorroded) beams at 3,000 cycles and 50,000 cycles respectively. The decrease was less at low load ranges (high fatigue lives) than at higher load ranges (lower fatigue lives). When CFRP sheets were added to strengthen the beams an increase of 15 % in fatigue bond strength compared to the fatigue strength of the corroded unstrengthened beams was achieved. This increase was about the same at all fatigue lives. It should be noted that CFRP sheets at long fatigue lives were able to restore the original uncorroded fatigue strength of the beams.

The variation of the fatigue life with the load range applied on flexure-beams in Figure 5b. The fatigue strength in set F-CD flexural beams was reduced by approximately 10% compared to those of set F-C. Repairing the 5% corroded beams with CFRP (set F-CR) increased the fatigue strength of the beams by 10% at 100,000 cycles and by 18% at 750,000 cycles compared to the corroded unrepaired beams (set F-CD). It should be noted that repairing with CFRP sheets restored the fatigue flexural strength of the corroded beams to a level equal to that of the original uncorroded beams.

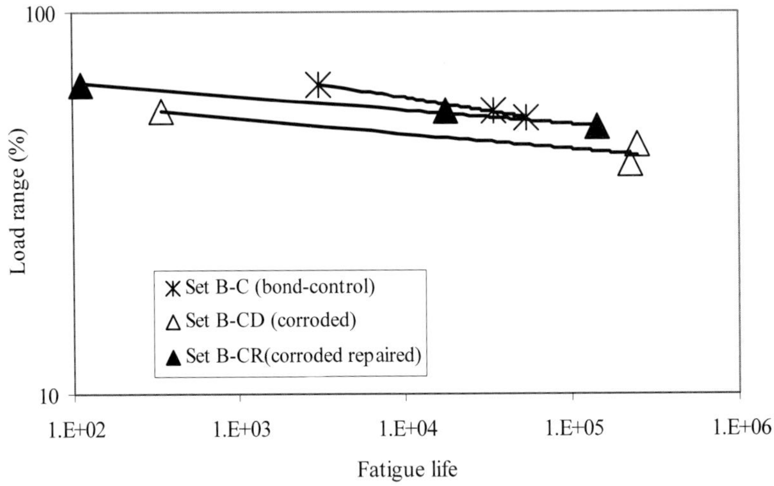

a)

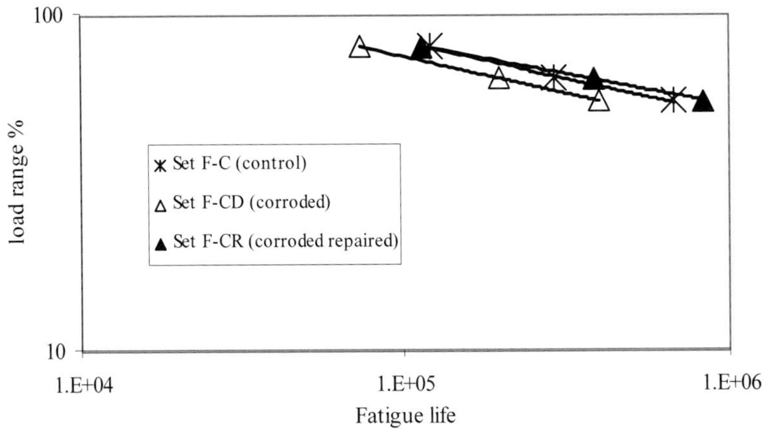

b)

Figure 5. Variation of the fatigue life with load ranges: a) bond beams, b) flexure beams

Analytical Modeling

A unified model that predicts the non-linear flexural response of corroded RC beams repaired with fibre reinforced polymer (FRP) sheets was developed. The model accounts for the effects of corrosion and FRP-wrapping on the transfer of load from the steel to the concrete between flexural cracks. The effects of the additional FRP reinforcement and the reduction in the steel area due to corrosion on the beam strength are predicted by the model. The beam cross section is discretized into finite layers while the beam span is modelled as a series of elements based on the mean crack spacing. The model was implemented into a new computer program. A comparison of the model's predictions with experimental results showed that the model accurately predicted response of both corroded and FRP-repaired beams. (Fig. 6).

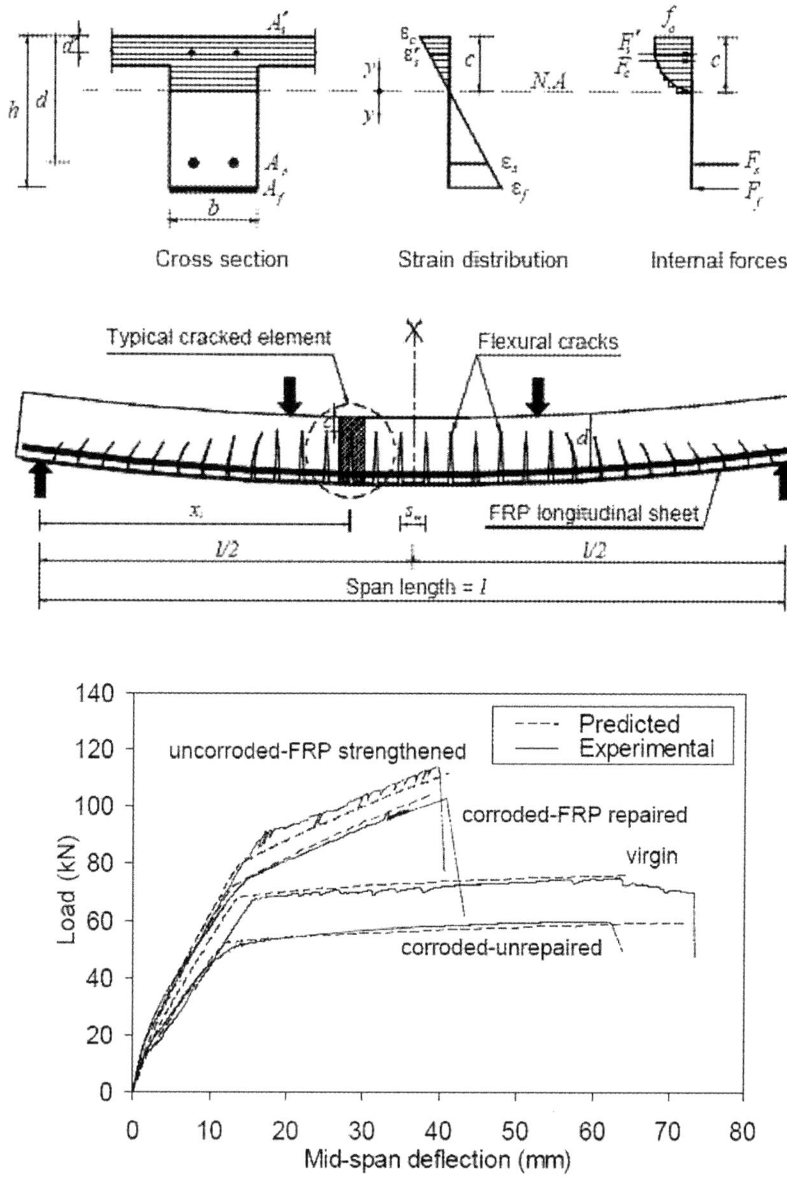

Figure 6. Analytical model and results

Field Implementation

In Fall 2005, the ISIS Waterloo group in collaboration with the C3 group successfully implemented the FRP repair onto a bridge girder in the region of Waterloo along with sensors to monitor the performance of the repair system. The FRP repair will be monitored over time to assess its effectiveness in the field. Figure 7 shows a photo of the FRP sheets on the girder.

Figure 7. Photo of CFRP repair on the bridge girder

Concluding Remarks

FRP composites are viable for the strengthening or repair of reinforced concrete beams that are experiencing steel reinforcement corrosion capable to maintain the structural integrity, serviceability and ultimate monotonic strength. The results in this paper provided important benchmark data, analytical modeling as well as field implementation of the FRP repair.

Acknowledgements

The author is a project leader in the Intelligent Sensing for Innovative Structures Network and wishes to acknowledge the support of the Network of Centres of Excellence Program and Natural Sciences and Engineering Research Council of Canada.

References

ACI Committee 440. (1996). State of the Art Report on Fiber Reinforced Plastic Reinforcement for Concrete Structures, ACI 440R-96, American Concrete Institute, Detroit, Michigan, 68 pp.

Badawi, M. and Soudki, K.A. 2004a. Control of Corrosion-Induced Damage in RC Beams using CFRP Laminates. *ASCE Journal of Composites for Construction*, Vol. 9, No. 2, pp. 195-201.

Badawi, M. and Soudki, K.A. 2004b. Flexural Performance of CFRP Repaired Corroded RC Beams. *ACI structural Journal*, submitted for review and possible publication, 33 pages.

Craig, B. and Soudki, K.A. 2005. Post-repair Performance of Bond Critical Corrosion Damaged Concrete Beams Repaired with CFRP. *ACI SP FRPRCS-7*, accepted May.

El Maaddawy, T., Soudki, K.A. and Topper T. 2004. Computer-based mathematical model for performance prediction of corroded beams repaired with fibre reinforced polymers. *ASCE Journal of Composites for Construction*, V 9, No 3, May, p 227-235.

El Maaddawy, T., Soudki, K.A., and Topper, T. 2005a. Analytical Model to Predict the Nonlinear Flexural Behavior of Corroded Reinforced Concrete Beams. *ACI Structural Journal*, Vol. 102, No. 4, pp. 550-560.

El Maadawy, T., Soudki, K.A., and Topper, T. 2005b. Long-term Performance of Corrosion-Damaged Reinforced Concrete Beams. *ACI Structural Journal*, accepted January, 35 pages.

Masoud, S., Soudki, K.A., and Topper, T. 2005. Modeling Concrete Cover Cracking and FRP Deformation due to Steel Reinforcement Corrosion. *Construction and Building Materials*, accepted August, 35 pages.

Masoud, S., Soudki, K.A., and Topper, T. 2001. Corroded RC Beams Strengthened with CFRP Sheets under Monotonic and Fatigue Loading. *ASCE Journal of Composites in Construction*, Vol. 5, No. 4, pp. 228-236.

Soudki, K.A. and Sherwood, E.G. 2003. Bond Behaviour of Corroded Steel Reinforcement in Concrete Wrapped with Carbon Fibre Reinforced Polymer Sheets. *ASCE Journal of Materials in Civil Engineering*, Vol. 15, No. 4, pp. 358-370.

Soudki, K.A. and Sherwood, T. 2000. Behaviour of Reinforced Concrete Beams Strengthened with CFRP Laminates Subjected to Corrosion Damage. *Canadian Journal of Civil Engineering*, Vol. 27, No. 5, pp 1005-1010.

Soudki, K.A., Rteil, A., Hammoud, R., Topper, T., 2006. "Fatigue Strength of FRP Repaired Corroded Beams," ISIS Special issue for the Canadian Journal of Civil Engineering (Invited paper).

COMPRESSIVE STRENGTH TESTING OF EARTHEN PLASTERS FOR STRAW BALE WALL APPLICATION

Brendon Taylor, Stephen Vardy and Colin MacDougall
Department of Civil Engineering, Queen's University, Kingston, Ontario, Canada

Abstract

Straw-bale construction is an emerging building method and many builders choose to plaster the straw bales with earthen plaster to reduce the embodied energy of the structure. A better understanding of the parameters affecting earthen plaster strength is essential for safe and effective use of this building technique. This study investigated the importance of initial plaster moisture content, drying time, clay content and, moisture content at the time of testing. Clayey silt soil, bagged ball clay and lime-cement are compared as plaster binders for straw-bale applications. Compressive testing was conducted on 50-mm plaster cubes and 100-mm by 200-mm plaster cylinders. It was found that as initial moisture content increased, strength and modulus of elasticity was unaffected for the earthen plaster. As the drying time increased between 10 days and 18 days, strength was unaffected but modulus of elasticity increased proportionally. As clay content increased, strength increased proportionally and stiffness was unaffected. As moisture content at the time of testing increased, both the strength and the stiffness decreased proportionally. Plaster made with soil was found to have greater strength than the plaster made with bagged clay or lime-cement plaster.

Introduction

In recent years the traditional practice of building plastered straw bale structures has seen a revival due to its economic and environmental benefits. As of 2004, over 50 permitted straw bale homes had been built in Ontario alone (OSBBC, 2004). This method of building originated in Nebraska in the late 19th century and a number of earthen rendered homes from the turn of the century are still lived in use. Load-bearing straw bale walls typically consist of a sandwich panel of stacked straw bales with plaster skins of Portland cement, lime, gypsum, clay, or a combination of these binders. Portland cement or lime-cement plasters are currently the most widely accepted, especially by building officials. However, the harmful environmental effects of cement and lime production have encouraged many environmentally conscious builders to consider earthen plasters.

Earthen plasters are typically mixed on-site and consist of local clay-rich soil, sand, water and chopped straw. They have been successfully used for centuries but are viewed with skepticism by many building officials. This is due, in part, to the lack of published research pertaining to the parameters that affect the strength of earthen plasters. Past research on earthen plasters has investigated parameters such as chopped straw content and sand content (Lerner et al. 2003; Ash et al. 2003). These tests have provided promising strength values as high as 2.00 MPa (Ash et al. 2003). These strength values are comparable to published values for Portland-cement plaster, ranging from 0.75 MPa to 1.98 MPa (Lerner et al. 2003; Vardy et al. 2005). However, some results are irreproducible due to a lack of proper soil analysis and there are many parameters yet to be investigated. A better understanding of how soil components and moisture content affect the strength of earthen plasters is essential to consistently building safe earthen rendered straw bale structures and allow for more widespread use of this environmentally friendly building material.

M. Pandey et al. (eds), Advances in Engineering Structures, Mechanics & Construction, 175–183.

Table 1. Test matrix describing batch parameters.

Batch	Drying Time	Initial M.C.	Drying Conditions	Sand/Soil	Binder Type/Source
Moisture Content:					
M1	14 days	0.126	lab	1.5	Clay / Soil
M2	14 days	0.132	lab	1.5	Clay / Soil
M3	14 days	0.134	lab	1.5	Clay / Soil
M4	14 days	0.144	lab	1.5	Clay / Soil
M5	14 days	0.146	lab	1.5	Clay / Soil
Drying Time:					
T1	10 days	0.14	lab	1.5	Clay / Soil
T2 (M3)	14 days	0.14	lab	1.5	Clay / Soil
T3	18 days	0.14	lab	1.5	Clay / Soil
Drying Conditions:					
C1	14 days	0.14	moist room (24h/48h)	1.5	Clay / Soil
C2 (M3)	14 days	0.14	lab	1.5	Clay / Soil
C3	14 days	0.14	drying oven (24h/48h)	1.5	Clay / Soil
Sand:Soil Ratio					
R1	14 days	0.14	lab	1.0	Clay / Soil
R2 (M3)	14 days	0.14	lab	1.5	Clay / Soil
R3	14 days	0.14	lab	3.0	Clay / Soil
Clay Source					
S1	14 days	0.14	lab	3.0	Clay / Bagged
S2 (M3)	14 days	0.14	lab	1.5	Clay / Soil
Lime-cement plaster					
Batch	**Drying Time**	**W/C.M.**	**Drying Conditions**		**Binder Type/Source**
P1	28 days	1.08	first 7 days moist, lab		Cement-lime / Bagged
P2	28 days	1.18	first 7 days moist, lab		Cement-lime / Bagged
P3	28 days	1.28	first 7 days moist, lab		Cement-lime / Bagged

Experimental Materials and Methods

Earthen plasters were mixed using either clayey-silt soil or commercially available bagged clay, masonry sand, and water. The soil was excavated to build a foundation on a residential building site in Haliburton, Ontario and subsequently used to plaster the walls of a load-bearing straw-bale building. The bagged clay was Ball Clay #123, a fine grained hydrous aluminum silicate clay with high unfired strength produced by the HC Spinks Clay Company.

The soil was broken up by hand and a mechanical sieve analysis for particles larger than 0.075 mm in diameter was conducted. Hydrometer tests in accordance with ASTM D 422 (2002) were performed for particles less than 0.075 mm.

Table 1 is the testing matrix, which shows the preparation parameters of each batch of plaster. Standard cube and cylinder specimens were prepared for each batch. Soil and sand for each mixture had a volume ratio of 1 : 1.5 (1 : 2.4 by mass), typical proportions for straw bale applications. The exceptions to this were batches R1 with soil to sand ratio of 1 : 1 and batch R3 with soil to sand ratio 1 : 3. Table 2 shows the percent by mass clay contents obtained in batches R1, R2 and R3 and the corresponding plaster strengths.

Table 2. Clay content for batches R1, R2 and, R3.

Batch	Sand/Soil (by Vol.)	% Clay (by mass)
R1	1	10.3
R2	1.5	7.9
R3	3	4.6

For plasters M1 … M5, the initial moisture contents (M.C.) were varied from 0.126 to 0.146 by mass. These moisture contents represent the workable limits for straw bale application. The drying time was varied for plasters T1, T2, and T3. Ten days is the earliest that the plaster hardens. Straw-bale builders generally regard 14 days as the time to reach adequate strength. The third drying time, 18 days, was chosen for linearity. Moisture content at time of testing was varied with batches C1, C2 and, C3. These batches were allowed to dry, and then subjected to extreme heat, moisture, or laboratory air. This is intended to simulate plaster subjected to hot, dry weather or a heavy rainfall, with the laboratory air acting as the control environment. Batch C1 was placed in a moisture room with 100% relative humidity prior to testing. Batch C2 dried in a laboratory environment. Batch C3 was placed in a drying oven at 110°C prior to testing. Cubes were left in the oven or moisture room for 24 hours and cylinders for 48 hours to ensure complete drying or moisture penetration.

Batch S1 was made with packaged clay purchased from a local earth-brick manufacturer. Portland-cement and lime plaster cylinders and cubes (P1, P2 and, P3) were cast in accordance with ASTM C 39 (1996) and ASTM C 109 (1998) respectively. The proportions of masonry sand, hydrated lime and Portland-cement were equal to 4.5 : 1.25 : 0.25 for all three batches.

To prepare the earthen plasters, the soil or bagged clay (batch S1 only) was massed and mixed thoroughly with water. The contents of the mixing bucket were allowed to soak for approximately two hours. The sand was then added to the mix.

The cubes and cylinders were allowed to dry in a controlled temperature room with a low-speed fan to speed the drying process. The cubes and cylinders were removed from the molds after 4 days. All specimens continued to dry in the laboratory for 10 more days, except for batches T1, T3, C1 and, C3. Specimens from batch T1 were tested after only 10 days and specimens from batch T3 were tested after 18 days. After 14 days, specimens from batch C1 and C3 were placed in the moisture room and drying oven respectively. Cubes were tested 24 hours later and cylinders were tested 48 hours later.

The cubes and cylinders were loaded until failure at a rate of 0.485 mm/min using an Instron Testing Machine. Due to the relatively low strength of the plaster, soft cork pads were used to cap the cylinders.

Results

The ideal soil for mixing plaster for straw-bale application is predominantly clay-sized particles because they act as a binder for other particles. The hydrometer test on the earthen soil indicates that it consists of 69% silt, 27% clay and 4% sand and can be classified as a clayey silt. The hydrometer test on the bagged ball clay revealed that it contained approximately 80% clay-sized particles and 20% silt-sized particles. Although only 80% of the mass is clay-sized, a larger proportion is likely clay minerals, specifically hydrous aluminum silicate. It is unknown whether or not the clay-sized particles in the soils are clay minerals.

Figure 1 is a typical stress-strain curve for an earthen plaster. The response is similar to that for concrete and cement-lime plaster. At stresses up to 40% of the ultimate stress (in this case about 35

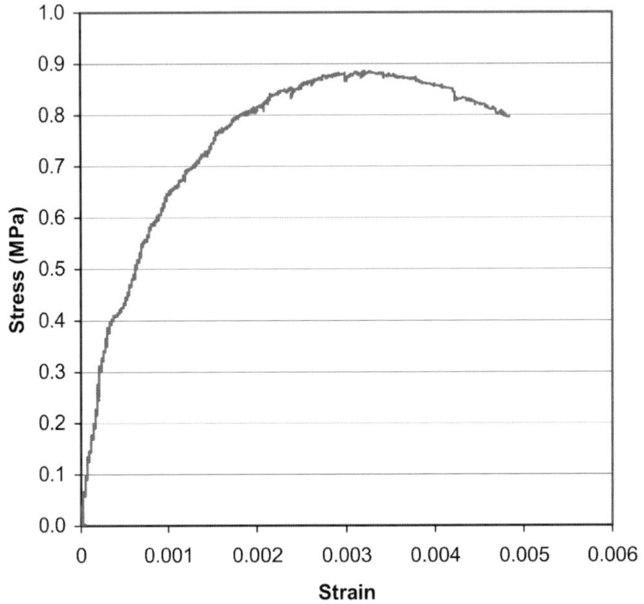

Fig. 1. Stress-strain response of cylinder A of batch C3 of earthen plaster.

MPa), the response is fairly linear. In accordance with ASTM C 469 (2002), the modulus of elasticity was taken as the slope of the stress-strain curve at 40% of the ultimate stress of the cylinder. Beyond 35 MPa, the response becomes non-linear. The ultimate stress, defined as the strength of the plaster, occurs at 0.89 MPa. The ultimate strain is 0.005.

Table 3 summarizes the average strength results for the cube and cylinder tests and modulus results obtained from the cylinder tests. The strength and modulus for each batch is the average value obtained from the three cubes or cylinders tested. Generally, there is little difference in the strength values obtained from cubes or cylinders for the earthen plasters. However, for the cement-lime plasters (P1, P2, P3), the cube strengths are significantly greater than the cylinder strengths.

Figure 2 shows the variation in strength with moisture content for the earthen plaster. There is little relationship between the initial moisture content and the strength of the plaster. As indicated by the data in Table 3, the modulus of the plaster also did not vary significantly with initial moisture content. This contrasts with cement-lime plasters, for which the ratio of water to cementitious materials was a critical parameter for the strength of the plaster (Vardy et al. 2005). The reason for the lack of sensitivity of earthen plasters to initial moisture content may result from the loss of water from the plasters, which was observed in these tests as visible shrinkage of the cubes and cylinders.

Figure 3 shows the plaster compressive strength obtained at various drying times. Clearly, there is no significant change in strength in the plaster between 10 and 18 days of drying. In contrast, Figure 4 shows the plaster modulus obtained at various drying times. The results indicate a significant increase in modulus between 14 and 18 days of drying. The results for batch T1, T2, and T3 in Table 3 indicate that the average modulus increases 2.4 times between 14 and 18 days of drying. Although the relationship is linear on this timescale, further investigation is necessary to determine plaster behaviour before 10 days and after 18 days of drying time.

Figure 5 shows the effect of increasing clay content on the plaster compressive strength. Batches R1, R2, and R3 as well as the bagged clay batch S1 are shown. For the earthen plasters, strength increases linearly with clay content. The bagged clay, however, does not follow the same trend,

Table 3. Summary of average compressive strengths and modulii of elasticity.

Batch	Initial Moisture Content	Cube Strength (MPa)	Cylinder Strength (MPa)	Modulus (MPa)
M1	0.126	1.5	1.4	1672
M2	0.132	1.2	1.3	1431
M3	0.134	1.1	1.2	2086
M4	0.144	1.1	1.3	1827
M5	0.146	1.0	1.2	1811
Batch	Drying Time (d)	Cube Strength (MPa)	Cylinder Strength (MPa)	Modulus (MPa)
T1	10	0.9	0.8	890
T2	14	1.0	1.1	758
T3	18	1.0	1.1	1848
Batch	Drying Environment	Cube Strength (MPa)	Cylinder Strength (MPa)	Modulus (MPa)
C1	drying oven (110 C)	1.8	2.1	2285
C2	laboratory	1.0	1.1	758
C3	moisture room (100%RH)	0.7	0.9	562
Batch	Sand/Soil by volume	Cube Strength (MPa)	Cylinder Strength (MPa)	Modulus (MPa)
R1	1.0	1.5	1.5	2500
R2	1.5	1.0	1.1	758
R3	3.0	0.7	0.8	1787
Batch	Clay Source	Cube Strength (MPa)	Cylinder Strength (MPa)	Modulus (MPa)
S1	commercial bagged clay	0.8	0.9	1731
S2	clayey silt soil	1.0	1.1	758
Batch	Water/Cementitious Mat.	Cube Strength (MPa)	Cylinder Strength (MPa)	Modulus (MPa)
P1	1.08	1.1	0.8	443
P2	1.18	1.1	0.7	839
P3	1.28	0.9	0.7	395

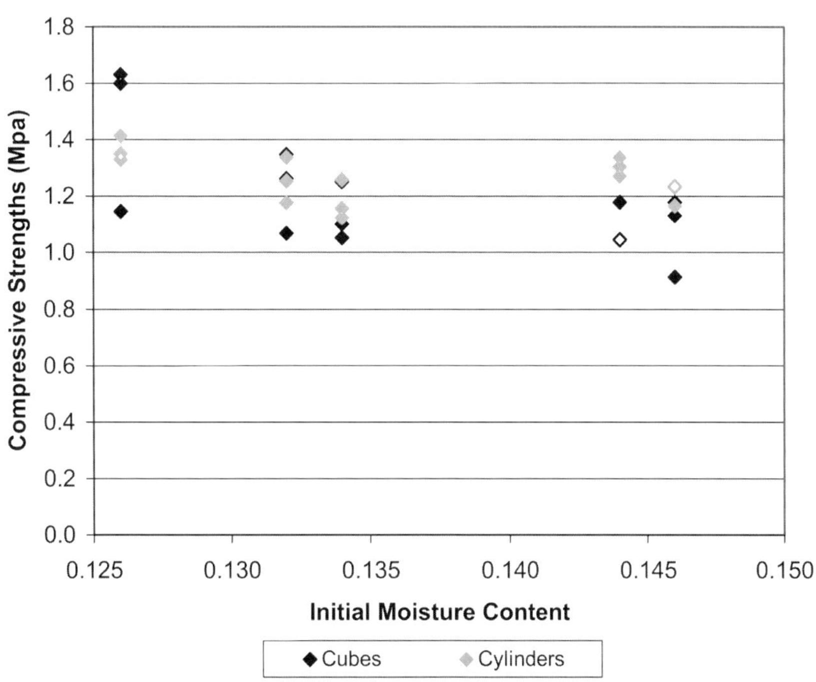

Fig. 2. Relationship between initial moisture content and compressive strength of earthen plaster.

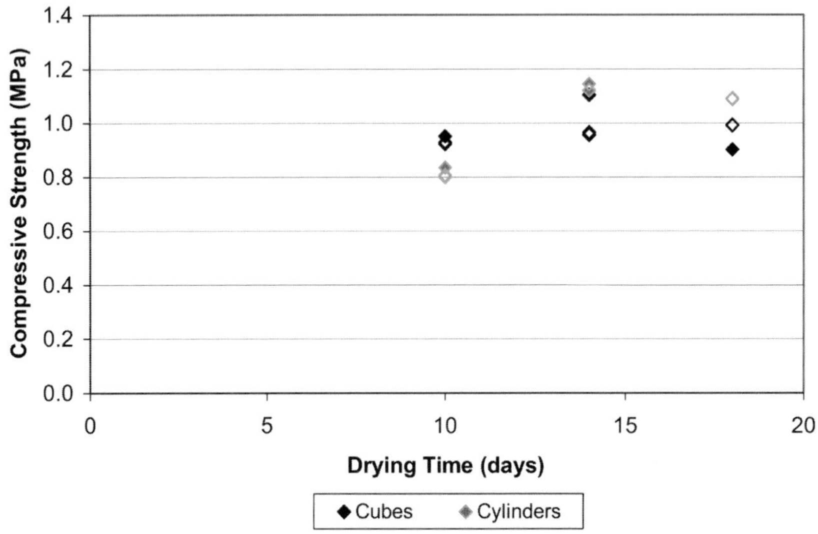

Fig. 3. Relationship between drying time and the compressive strength of earthen plaster.

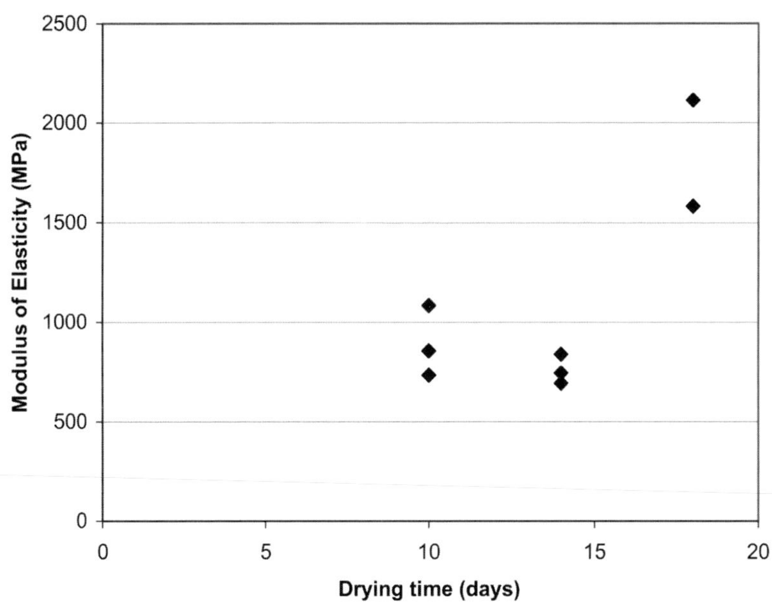

Fig. 4. Time dependency of the modulus of elasticity of earthen plaster.

and has a much lower strength than earthen plasters with similar clay content. The reason for these differences may be related to the vastly different gradations measured for the two types of soil. The bagged soil is 80% clay and 20% silt, while the earthen soil has 69% silt, 27% clay, and 4% sand. Further research is needed to identify the optimum soil gradation for earthen plasters. This has some important practical implications for builders using clay plasters, since the ideal is to use soil from the building site to reduce the energy needed to transport the building materials. Locally available soil may not have the optimum soil gradation to produce structural plasters.

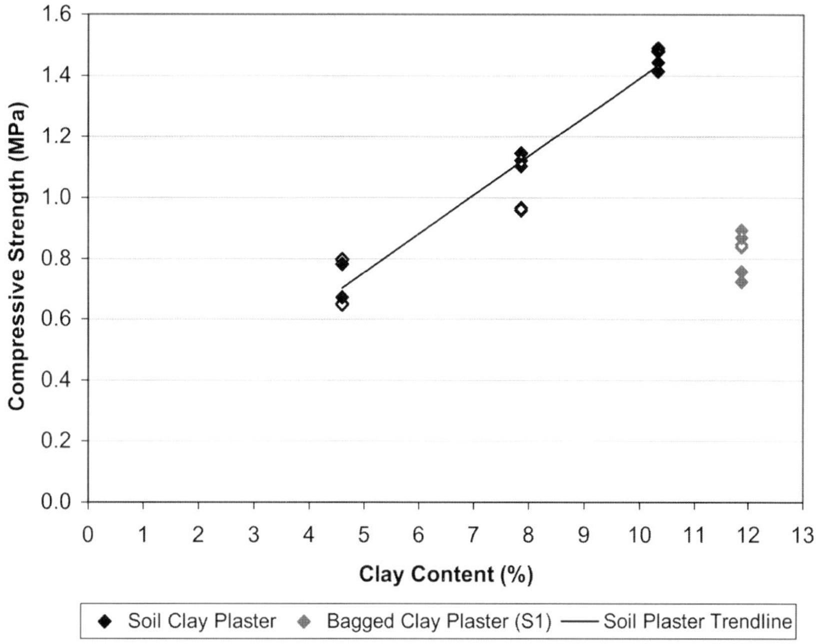

Fig. 5. Clay content and strength, including the bagged clay plaster batch (S1).

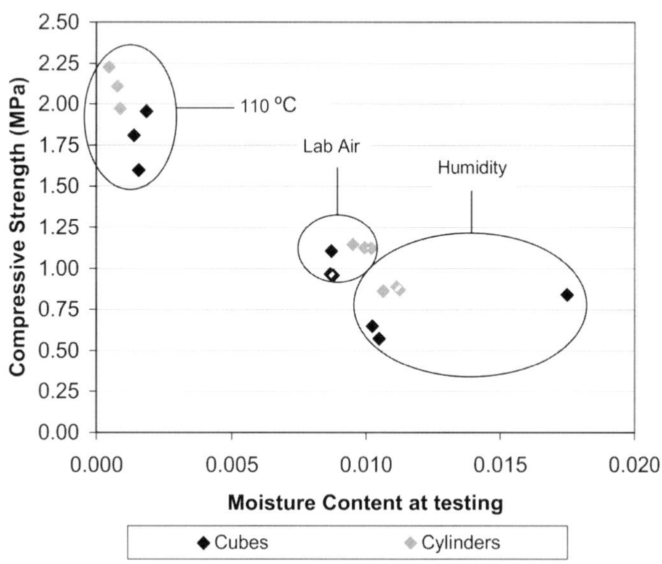

Fig. 6. Relationship between the moisture content at time of testing and strength.

Batches C1, C2, and C3 were dried in different environments prior to testing. This varied the moisture content of the plaster at the time of testing. The moisture content of each cube or cylinder was measured immediately after the compression test was complete. Figure 6 shows the drastic changes in moisture content and strength resulting from the different environments. Placing the plaster in a 110°C oven substantially reduced the moisture content and increased the strength.

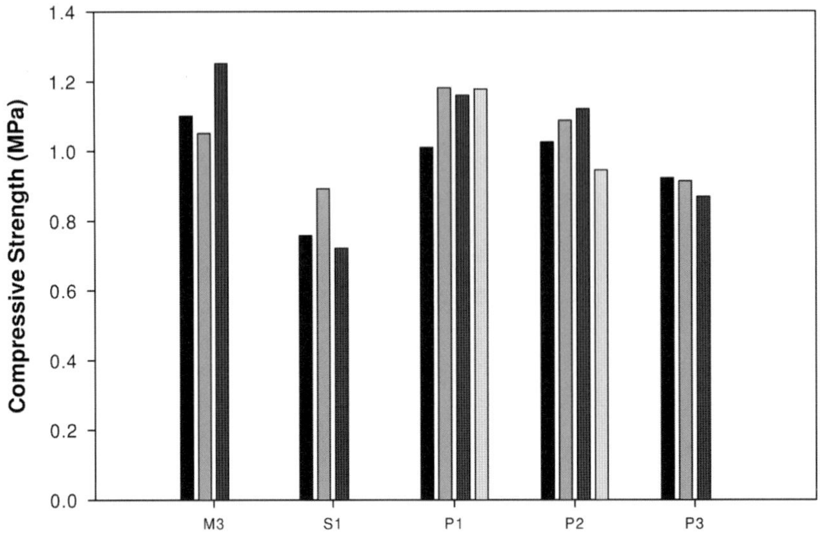

Fig. 7. Compressive strength of plaster made with various binder types.

Leaving the plaster in a humid environment increased the moisture content and resulted in lower strength as opposed to a lab environment. This can be compared with the initial moisture content results of Figure 2, which indicated no relationship between initial moisture content and strength. This suggests that the range of initial moisture contents tested for batches M1, ..., M5 did not result in significantly different final moisture contents. This is an area that needs further investigation. Furthermore, these results point to the critical importance of ensuring adequate moisture protection for the clay in building applications. In addition, the exposure of the plaster to hot temperatures during, for example, summer days, is likely to have a beneficial effect on the strength of the plaster.

Three batches of cement-lime plaster, P1, P2, and P3 were tested for comparison with the earthen and bagged clay plasters. The water to cementitious materials ratio (w/cm) ranged from 1.08 to 1.28. As the water to cementitious materials ratio increased, the compressive strength decreased. The results in Table 3 indicate that the average strength and elastic modulus increase as w/cm decrease, a trend noted by Vardy et al. (2005). Figure 7 compares the compressive cube strength of typical soil clay plaster (M3), bagged clay plaster (S1), and three batches of lime-cement plaster (P1, P2, P3). The average strength of the earthen plaster is slightly higher than the cement-lime plasters, and significantly greater than the bagged clay plaster. The average elastic modulus, given in Table 3, of the earthen plaster is 2086 MPa, which is significantly greater than the bagged clay plaster (1731 MPa) and the cement-lime plasters (395 MPa–839 MPa). It is encouraging that earthen plaster can equal and even surpass lime-cement plaster in strength since earthen plaster has only a small fraction of the embodied energy of lime-cement products.

Summary and Conclusions

This study investigated the effect of moisture content, drying time, drying conditions, and clay content on the strength and elastic modulus of an earthen plaster. The plaster was mixed using soil consisting of 69% silt, 27% clay and 4% sand. The earthen plasters were also compared to a plaster mixed using a commercially available clay, and cement-lime plasters. The specific conclusions of this work are:

(1) The stress-strain response of the earthen plaster was similar to that of concrete or cement-lime plaster.
(2) The initial moisture content of the earthen plaster had a negligible effect on the strength and elastic modulus. This is in contrast to structural concrete or cement-lime plasters.
(3) There is a negligible increase in compressive strength, but a significant increase in elastic modulus with 10 to 18 days of drying of an earthen plaster.
(4) Increased clay content significantly increased the strength of the earthen plaster.
(5) Placing the plaster in a 110°C oven substantially reduced the moisture content and increased the strength. Leaving the plaster in a humid environment increased the moisture content and resulted in lower strength as opposed to a lab environment.
(6) The earthen plaster had higher strength and elastic modulus than typical cement-lime plasters used for straw-bale construction. The earthen plaster also had higher strength and elastic modulus than the plaster mixed with commercial clay.

References

Ash, C., Aschheim, M. and Mar, D. (2003) "In-Plane Cyclic Test of Plastered Straw Bale Wall Assemblies, Ecobuild Network", www.ecobuildnetwork.org, accessed September 14, 2005.

ASTM C 109 (1998) "Standard Test Method for Compressive Strength of Hydraulic Cement Mortars (Using 2-in of 50-mm Cube Specimens)',, *Annual Book of ASTM Standards*, Section 4 – Construction, Volume 04.01 – Cement; lime; Gypsum, West Conshohocken, PA.

ASTM C 469 (2002) "Static Modulus of Elasticity and Poisson's Ratio of Concrete in Compression", *Annual Book of ASTM Standards*, Section 4 – Construction, Volume 04.02 – Concrete and Aggregates, West Conshohocken, PA.

ASTM D 422 (2002) "Standard Test Method for Particle-Size Analysis of Soils", *Annual Book of ASTM Standards*, Section 4 – Construction, Volume 04.08 – Soil and Rock, West Conshohocken, PA.

ASTM C 39 (2004) "Standard Test Method for Compressive Strength of Cylindrical Concrete Specimens", *Annual Book of ASTM Standards*, Section 4 – Construction, Volume 04.02 – Concrete and Aggregates, West Conshohocken, PA.

Lerner, K. and Donahue, K. (2003) "Structural Testing of Plasters for Straw Bale Construction", Ecological Building Network, www.ecobuildnetwork.org/pdfs/Plaster_tests.pdf, accessed September 14, 2005. Ontario Straw Bale Building Coalition – OSBBC (2004) "Straw Bale Building: An Information Package for Building Officials in Ontario", OSBBC, Spencerville, Ontario.

Vardy, S., Tipping, T. and MacDougall, C. (2005) "Compressive Testing and Analysis of a Typical Straw Wall Plaster", in *Proceedings of Engineering Sustainability 2005 Conference*, Pittsburgh, USA.

Steel Structures

GBT-BASED STRUCTURAL ANALYSIS OF THIN-WALLED MEMBERS: OVERVIEW, RECENT PROGRESS AND FUTURE DEVELOPMENTS

D. Camotim[1], N. Silvestre[1], R. Gonçalves[2] and P.B. Dinis[1]
[1] *Civil Engineering Department, ICIST/IST, Technical University of Lisbon, Av. Rovisco Pais, 1049-001 Lisboa, Portugal*
E-mail: dcamotim@civil.ist.utl.pt
[2] *EST Barreiro, Polytechnical Institute of Setúbal, Rua Stinville 14, 2830-114 Barreiro, Portugal*

Abstract

This paper provides an overview of the Generalised Beam Theory (GBT) fundamentals and reports on the novel formulations and applications recently developed at the TU Lisbon: the use of conventional GBT to derive analytical distortional buckling formulae and extensions to cover (i) the buckling behaviour of members with (i_1) branched, closed and closed/branched cross-sections and (i_2) made of orthotropic and elastic-plastic materials, and (ii) the vibration and post-buckling behaviours of elastic isotropic/orthotropic members. In order to illustrate the usefulness and potential of the new GBT formulations, a few numerical results are presented and briefly discussed. Finally, some (near) future developments are briefly mentioned.

1. Introduction

Generalised Beam Theory (GBT) was first proposed by Richard Schardt in 1966 and has, since then, fostered a vast amount of theoretical and applied research activity at the University of Darmstadt. However, due to the fact that practically all publications originating from this research group were available exclusively in German (including the book published by Schardt in 1989), GBT had virtually no impact for non-German-speaking researchers up until the early 90s. This situation was altered by J.M. Davies, who learnt about GBT in the mid 80s and, almost single-handed, disseminated it among the English-speaking technical and scientific communities together with his Ph.D. students Leach and Jiang, Davies employed GBT to perform in-depth investigations on the buckling behaviour of cold-formed steel members and, in particular, showed that this approach is a valid and often advantageous alternative to finite element or finite strip analyses (Davies, 1998, 2000). Moreover, it appears that Davies can also be credited with encouraging Schardt to start publishing in English (Schardt, 1994a,b).

Although GBT has recently attracted considerable attention from several researchers around the world (e.g., Rendek and Baláž, 2004; or Simão and Silva, 2004), it seems fair to say that the vast majority of the novel formulations and applications originated from the Technical University of Lisbon – this can be attested by the review paper recently published by the authors (Camotim et al., 2004), which summarises the work carried out prior to 2004. Therefore, the objective of this work is to provide a follow-up of that paper, by (i) reporting on the research activity concerning GBT undertaken in the last couple of years and (ii) addressing the developments expected for the foreseeable future. At this stage, it should be pointed out that, due to space limitations, it is only possible (i) to provide a brief overview of the new findings, (ii) to present a very small number of illustrative examples and (iii) to mention the main references, where the interested reader may find much more detailed accounts of all these topics dealt with in this paper.

M. Pandey et al. (eds), Advances in Engineering Structures, Mechanics & Construction, 187–204

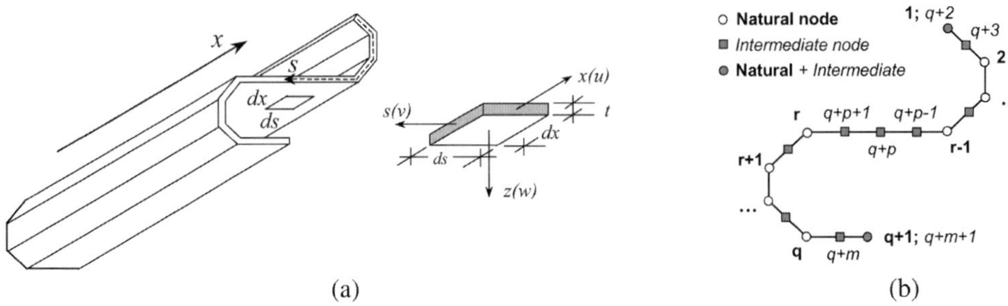

Figure 1. Thin-walled member (a) geometry, axes/displacements; (b) cross-section discretisation.

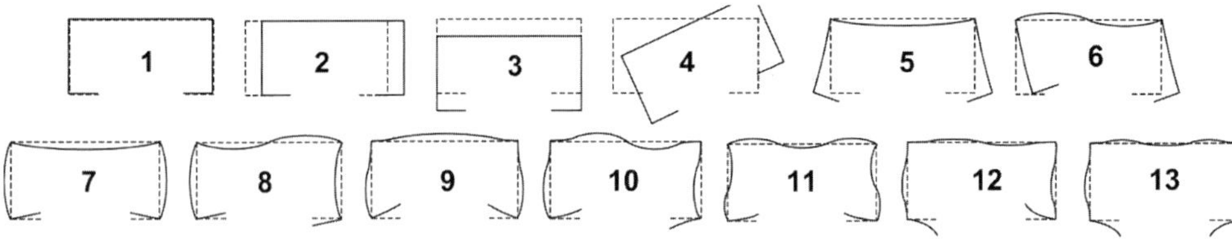

Figure 2. Lipped channel first thirteen deformation mode in-plane shapes.

1.1. *Conventional GBT*

In this paper, *conventional GBT* designates the formulation intended to perform stability (bifurcation) analyses of thin-walled members with unbranched (folded-plate) thin-walled members and made of linear elastic isotropic materials (e.g., most cold-formed steel profiles) – this designation stems from the fact that both (i) the vast majority of Schardt's publications and (ii) all of Davies's work concern members with these characteristics. Moreover, all the recently developed GBT formulations can be viewed, to a smaller or larger extent, as modifications or extensions of this conventional one.

A conventional GBT analysis involves (i) a *cross-section analysis*, leading to the GBT deformation modes and corresponding modal mechanical properties, and (ii) a *member linear stability analysis*, to obtain the member bifurcation stress resultants and associated buckling mode shapes (e.g., Davies, 1998; or Schardt, 1994a). In the case of the arbitrary q-walled member shown in Figure 1(a) and for the cross-section discretisation depicted in Figure 1(b) ($q + 1$ natural and m intermediate nodes), the performance of the cross-section analysis leads to the system of $q + m + 1$ GBT equilibrium equations (one per deformation mode)

$$EC_{ik}\phi_{k,xxxx} - GD_{ik}\phi_{k,xx} + EB_{ik}\phi_k + W^{\sigma}_{j.o}X_{jik}\phi_{k,xx} = 0, \tag{1}$$

where (i) $(\cdot)_{,x} \equiv d(\cdot)/dx$, (ii) $\phi_k(x)$ are modal amplitude functions, (iii) E, G are Young's and shear moduli, (iv) $W^{\sigma}_{j.o}$ are uniform (usually single-parameter) pre-buckling stress resultants and (v) the various matrix/tensor components are related to the cross-section stiffness (C_{ik}, D_{ik}, B_{ik}) and geometric effects (X_{jik}). Together with its boundary conditions, system (1) defines a standard eigenvalue problem, the solution of which (i) can be obtained by means of several standard methods (e.g., finite differences, finite elements or Galerkin's method) and (ii) yields the member bifurcation stress resultants and buckling modes. The latter are combinations of the GBT deformation modes, illustrated in Figure 2 for the case of a lipped channel cross-section ($q = 5$ and $m = 7$).

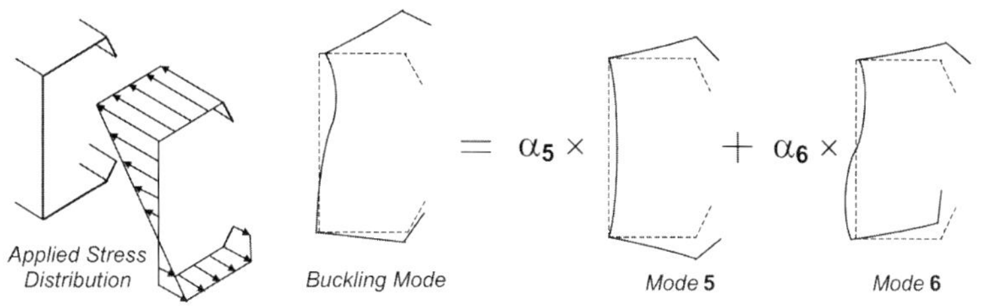

Figure 3. Lipped channel member buckling mode as a linear combination of modes **5** and **6**.

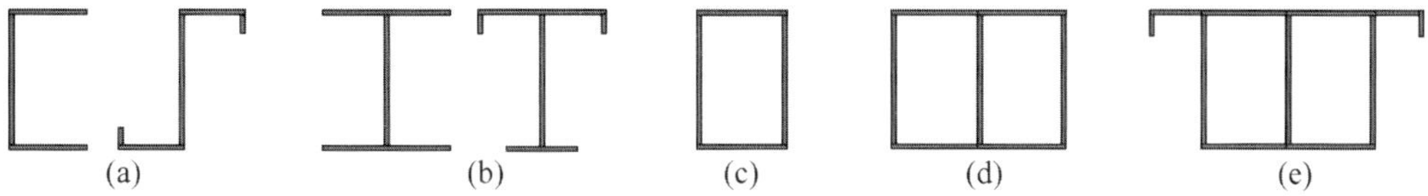

Figure 4. (a) Unbranched open, (b) branched open, (c) unbranched closed, (d) branched closed and (e) arbitrary branched open/closed thin-walled cross-sections.

1.2. *Recent Developments*

The novel GBT applications and formulations addressed in this paper concern (i) the development of *distortional buckling formulae* (using the conventional GBT), (ii) the *buckling analysis* of members (ii_1) with arbitrary cross-section shapes (branched and/or closed multi-cell) and (ii_2) made of isotropic/orthotropic elastic or isotropic elastic-plastic materials, (iii) the *vibration* analysis of isotropic/orthotropic elastic folded-plate members (loaded or unloaded) and (iv) the *post-buckling* analysis of isotropic/orthotropic elastic folded-plate members. This sequence is kept in the presentation and the various applications and/or formulations are grouped according to the type of structural analysis (buckling, vibration or post-buckling).

2. Buckling Analysis

2.1. *Distortional Buckling Formulae*

Since GBT allows for the possibility of performing approximate buckling analyses including any number of deformation modes, one may develop analytical formulae providing accurate distortional buckling stress estimates for lipped channel, Z-section and rack-section columns, beams and beam-columns with several support conditions (Silvestre and Camotim, 2004a,b). These formulae are obtained through *symbolic* GBT-based bifurcation analyses that include only either *one* (columns) or *two* (beams and beam-columns) cross-section distortional deformation modes: modes **5** (symmetric) and **6** (anti-symmetric) in Figure 3, where this procedure is illustrated for the case of lipped channel members. Extensive parametric studies showed that the buckling stress estimates provided by the GBT-based formulae (i) are consistently accurate and (ii) always compare favourably with the values yielded by formulae previously developed by other authors.

2.2. *Cross-Section Shape*

The cross-section of a thin-walled member is classified according to its mid-line, which may be (i) *open* or *closed* and (ii) *branched* or *unbranched* (see Figure 4). As mentioned earlier, the conventional GBT is valid only for folded-plate members, i.e., members with open unbranched cross-sections. Thus, in order to extend its application to all other cross-section shapes, one must modify

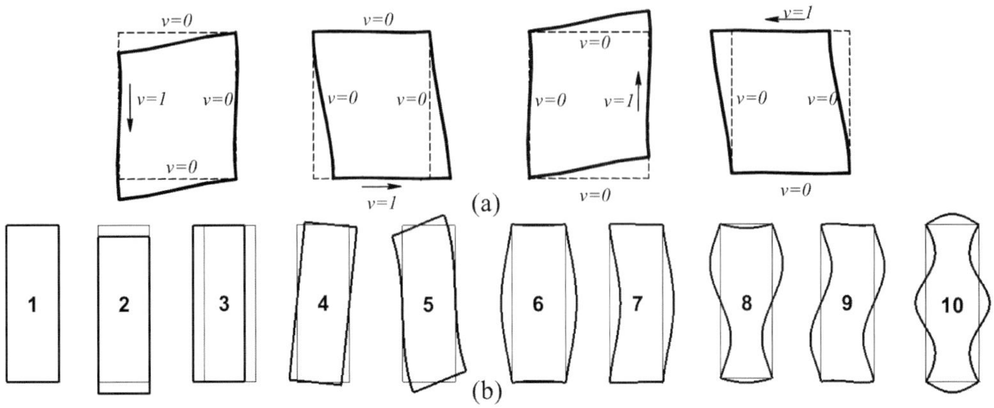

Figure 5. (a) Determination of the additional deformation modes accounting for shear deformation and (b) the 10 most relevant deformation modes of a narrow rectangular hollow cross-section.

the cross-section analysis procedure – since the member analysis remains unaltered, both system (1) and its boundary conditions retain their *forms*. This task has now been completed for any conceivable cross-section shape and comprised the following three stages:

(i) The first extension concerned (unbranched) single-cell closed cross-sections and involves the inclusion of four *additional* deformation modes, which (i_1) are related to the *shear deformation* of the cross-section mid-line and (i_2) stem from the imposition of unit transverse displacements in each wall, while preventing all the other ones, as shown in Figure 5(a) (Gonçalves and Camotim, 2004a). Moreover, one must incorporate the term $\int_s Gtv_i v_j ds$ in the analysis, to account for the virtual work associated with the shear strains. The 10 most relevant deformation modes of a narrow rectangular hollow section are displayed in Figure 5(b) and one notices that, besides the global (**1–4**) and local-plate (**6-10**) modes, similar to the ones yielded by the conventional GBT, a novel *distortion* (not *distortional*) mode **5** appears – together with mode **4**, it models the cross-section shear deformation (Gonçalves and Camotim, 2004b).

(ii) Next, a methodology that can handle arbitrarily branched open cross-sections was developed, thus overcoming difficulties related to (ii_1) the proper selection of the elementary warping and flexural functions and (ii_2) the solution of the statically indeterminate folded-plate problem (Dinis et al., 2006). One must view the cross-section as a combination of an unbranched sub-section and an ordered sequence of branches, which leads to the straightforward identification of the *dependent natural nodes*, i.e., the natural nodes where the warping displacements cannot be imposed (they must be *calculated*) – the number of such nodes is equal to $\sum(m_{wi} - 2)$, with the summation extending to all branching nodes and m_{wi} being the number of walls emerging from branching node i. These concepts are illustrated in Figure 6, where (ii_1) the cross-section depicted in Figure 6(a) has six dependent nodes (see Figure 6(e)) and (ii_2) two possible combinations of an unbranched sub-section and the corresponding branch sequence are displayed in Figures 6(b)–(d) (note that in both cases one must go up to second-order branches).

(iii) Finally, Gonçalves et al. (2006) have just developed the "definitive" formulation, in the sense that it is applicable to fully arbitrary cross-sections, namely those combining closed cells with open branches. A brief description of the steps and procedures involved in this formulation is presented next and illustrated through the cross-section depicted in Figure 7(a): an I-shaped section with closed cells separating the web from the unequal flanges – it has 13 walls, 2 closed cells and 12 natural nodes (6 branching ones):

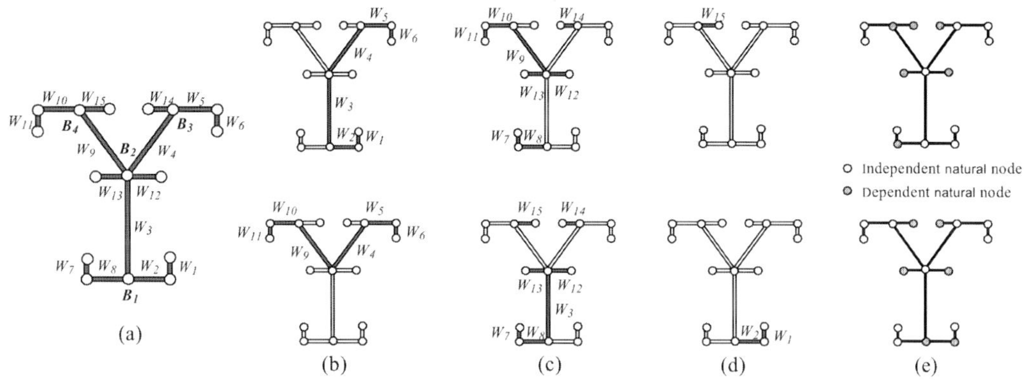

Figure 6. Illustrative branched section (a) geometry and two possible (b) unbranched sub-sections, (c) first-order branches, (d) second-order branches and (e) independent/dependent natural nodes.

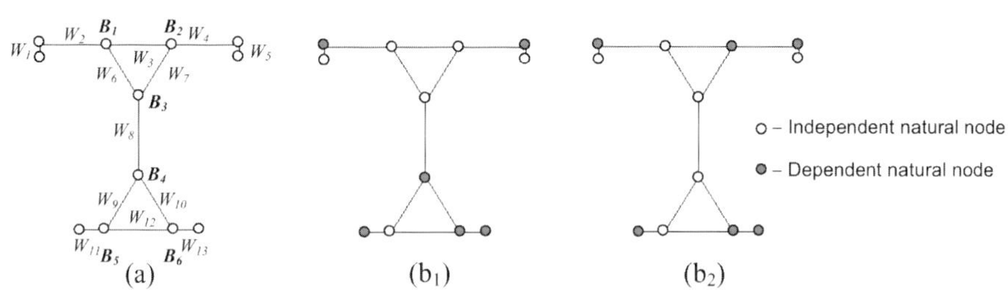

Figure 7. (a) Cross-section geometry and (b) dependent and independent natural nodes.

(iii.1) Choice of the dependent natural nodes – Figures 7(b_1)–(b_2) show two possible choices for the 6 dependent and 6 independent natural nodes to be considered in the analysis.

(iii.2) Determination of the "warping initial shape functions", by imposing elementary functions at each independent natural node and assuming that Vlassov's hypothesis holds in all walls, i.e., following the methodology developed by Dinis et al. (2006).

(iii.3) Determination of the "local-plate initial shape functions", by imposing elementary flexural functions at each intermediate node – 13 intermediate nodes were included in the illustrative example (mid-points of each internal wall and free ends of the external ones).

(iii.4) Identification of the conventional *deformation modes*, yielded by the simultaneous diagonalisation of the stiffness matrices $[C_{ik}]$ and $[B_{ik}]$. In the case of the illustrative example, one identifies 19 deformation modes, 11 of which are shown in Figure 8: 5 warping (**2**–**6**) and 6 local-plate (**13**–**18**) – note that modes **2**–**4** are "rigid-body" ones and mode **1** (axial extension) has been omitted.

(iii.5) Sequential imposition of unit membrane shear strains in each wall belonging to a closed cell, keeping all remaining walls free of those strains, and determination of the corresponding "initial shear shape functions". In the example, 6 unit shear strains are imposed (3 per closed cell), by combining u and v displacements – enforcing these unit strains may be a quite cumbersome task and some guidelines on how to carry it out can be found in the work by Gonçalves et al. (2006).

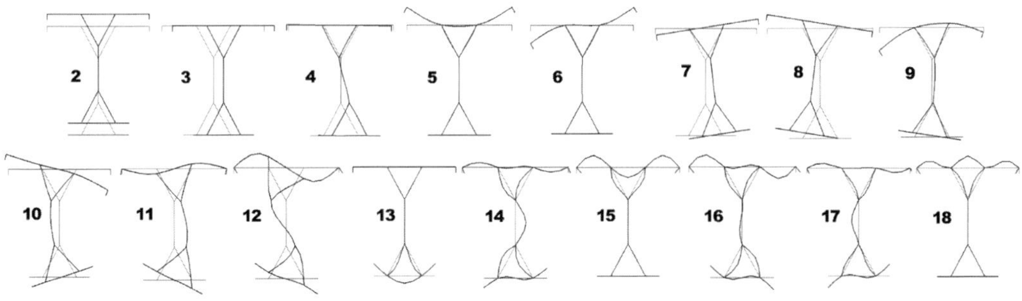

Figure 8. In-plane shapes of the 17 most relevant warping (**2–6**), shear (**7–12**) and local-plate (**13–18**) modes.

(iii.6) Identification of the *shear deformation modes*, again by simultaneously diagonalising matrices $[C_{ik}]$ and $[B_{ik}]$[1] – in this case, one obtains the 6 shear deformation modes (**7–12**) shown in Figure 8.

2.3. *Material Behaviour*

A GBT formulation applicable to members made of materials other than isotropic and linear elastic requires modifications that depend on the specific type of material behaviour.[2] This task has been carried out for (i) *orthotropic* linear elastic materials (e.g., laminated plate fibre-reinforced plastics – FRP) and (ii) isotropic *non-linear elastic-plastic* materials (e.g., stainless steel or aluminium). In the former case, first addressed about four years ago (Silvestre and Camotim, 2002), the layer (lamina) *plane-stress* constitutive law reads

$$\begin{Bmatrix} \sigma_{xx} \\ \sigma_{ss} \\ \sigma_{xs} \end{Bmatrix} = \begin{bmatrix} \bar{Q}_{11} & \bar{Q}_{12} & \bar{Q}_{13} \\ \bar{Q}_{12} & \bar{Q}_{22} & \bar{Q}_{23} \\ \bar{Q}_{13} & \bar{Q}_{23} & \bar{Q}_{33} \end{bmatrix} \begin{Bmatrix} \varepsilon_{xx} \\ \varepsilon_{ss} \\ \gamma_{xs} \end{Bmatrix}, \tag{2}$$

where $\bar{Q}_{ij}$ are "transformed reduced stiffness components" depending on the layer (i) fibre and plastic matrix material properties and (ii) fibre orientation (Jones, 1999). Moreover, the mechanical behaviour of a laminated plate member also varies with its layer properties and configuration (Silvestre and Camotim, 2002). In the most general case (arbitrary orthotropy, i.e., anisotropy), the GBT system (1) becomes

$$C_{ik}\phi_{k,xxxx} + H_{ik}\phi_{k,xxx} - D_{ik}\phi_{k,xx} + F_{ik}\phi_{k,x} + B_{ik}\phi_k - X_{jik}W^{\sigma}_{j.0}\phi_{k,xx} = 0 \tag{3}$$

and its boundary conditions must be modified accordingly. It is worth noting (i) the additional tensors H_{ik} and F_{ik}, accounting for material coupling effects between torsion and longitudinal/transversal flexure, and (ii) that, due to the layer-variation of the material properties, the various tensor components are now *mechanical properties* – material constants and geometrical characteristics fused together. Another aspect that deserves to be mentioned is the need to include in the analysis deformation modes that take into consideration the non-linearity of the warping displacement variation within the width of each wall and, therefore, are also associated with membrane shear strains, i.e., do not satisfy Vlassov's assumption (Silvestre and Camotim, 2004c; Silvestre, 2005). The warping

[1] Since the conventional and shear modes are not identified jointly, the overall matrices $[C_{ik}]$ and $[B_{ik}]$ are not diagonal – only their principal sub-matrices exhibit this property.

[2] Although most of the novel GBT formulations have not yet been applied to other than isotropic and linear elastic members, this is a straightforward (even if time-consuming) task that is planned for the near future.

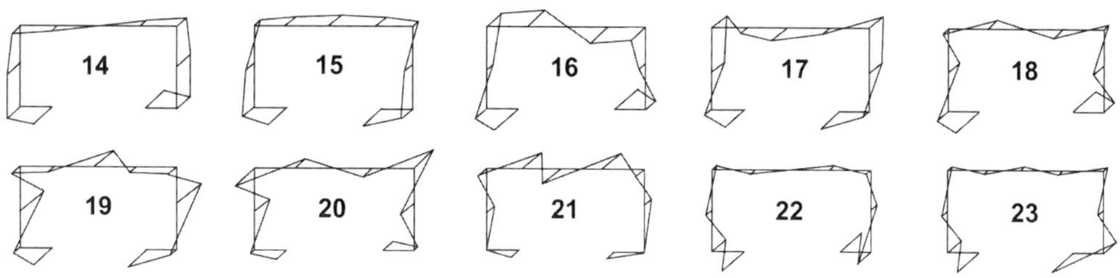

Figure 9. Lipped channel ten most relevant shear modes associated with non-linear warping.

configurations of this set of additional shear deformation modes are shown in Figure 9, for the case of a lipped channel cross-section.

Concerning the buckling analysis of isotropic non-linear elastic-plastic members, one begins by recalling that, if no strain reversal occurs along the fundamental equilibrium path, an elastic-plastic solid and its "hypoelastic comparison solid" have identical critical bifurcation stresses/loads (Hill, 1958). Thus, by (i) monitoring the evolution of the instantaneous moduli (on the fundamental path) and (ii) adopting *incremental* constitutive relations, it is possible to determine the member plastic bifurcation behaviour using non-linear elastic stability theory. For fundamental states with only longitudinal normal stresses, the plane-stress incremental constitutive relations read

$$\begin{Bmatrix} \dot{\sigma}^B_{xx} \\ \dot{\sigma}^B_{ss} \\ \dot{\sigma}^B_{xs} \end{Bmatrix} = \begin{bmatrix} \dot{E}_{11} & \dot{E}_{12} & 0 \\ \dot{E}_{21} & \dot{E}_{22} & 0 \\ 0 & 0 & \dot{G} \end{bmatrix} \begin{Bmatrix} \dot{\varepsilon}^B_{xx} \\ \dot{\varepsilon}^B_{ss} \\ \dot{\gamma}^B_{xs} \end{Bmatrix}, \qquad \dot{\sigma}^M_{xx} = E_T \dot{\varepsilon}^M_{xx}, \tag{4}$$

where (i) $\dot{\sigma}_{ij}$ and $\dot{\varepsilon}_{ij}$ are stress-rate and strain-rate components, (ii) $\dot{E}_{ij}$ and $\dot{G}$ are the instantaneous elastic and shear moduli, (iii) E_T is the longitudinal uniaxial tangent modulus and (iv) $(\cdot)^B$ and $(\cdot)^M$ are superscripts identifying bending and membrane terms. The material uniaxial stress-strain law is commonly described by Ramberg–Osgood type expressions (Rasmussen, 2003) and, due to the well-known "plate plastic buckling paradox" (Hutchinson, 1974), *both* J_2-deformation and J_2-flow small strain plasticity theories were included in the GBT formulation (Gonçalves and Camotim, 2004, 2005). After the incorporation of the instantaneous moduli, the hypoelastic bifurcation analysis leads to the system of *incremental* GBT equations

$$C_{ik}\dot{\phi}_{k,xxxx} - D_{ik}\dot{\phi}_{k,xx} + B_{ik}\dot{\phi}_k + X_{ik}\dot{\phi}_{k,xx} = 0 \tag{5}$$

and associated boundary conditions, where (i) functions $\dot{\phi}_k$ provide the deformation mode amplitude rates and (ii) all tensor components are *load-dependent* through the instantaneous moduli (Gonçalves and Camotim, 2005). Note that (5) is applicable to rather general (uniaxial-stress) loading conditions – the only restriction is that they must satisfy the basic hypothesis of Hill's "comparison solid" concept: the material behaviour may be assumed as hypoelastic in the close vicinity of the bifurcation point (Hill, 1958; Hutchinson, 1974).

3. Vibration Analysis

Given the well-know mathematical similarity between the stability (bifurcation) and *vibration* eigenvalue problems, the derivation of a GBT vibration formulation constitutes a relatively easy task: it suffices to replace the geometric effects by their *dynamic* counterparts, as done by Schardt and Heinz (1991) for isotropic linear elastic members, and by Silvestre and Camotim (2004d), for *orthotropic* linear elastic laminated plate members. Very recently, these authors proposed a novel combined formulation that makes it possible to analyse the vibration behaviour of *loaded* orthotropic members (Silvestre and Camotim, 2005a,b). In this case, the GBT system of equilibrium equations reads

$$(\mathcal{K}_{ik} + W_m^0 \mathcal{G}_{ik} - \omega^2 \mathcal{M}_{ik})\phi_k + (\mathcal{K}_{ij} - \omega^2 \mathcal{M}_{ij})\varphi_j = 0, \tag{6}$$

$$(\mathcal{K}_{hk} - \omega^2 \mathcal{M}_{hk})\phi_k + (\mathcal{K}_{hj} - \omega^2 \mathcal{M}_{hj})\varphi_j = 0, \tag{7}$$

and corresponds to the assembly of two subsystems, both including coupling components – while the first is associated with the *conventional* modes (see Figure 2 – amplitudes ϕ_k), the second is related to the *shear* modes (see Figure 9 – amplitudes φ_j). The system is expressed in terms of differential operators concerning *linear stiffness* ($\mathcal{K}_{ik}$, $\mathcal{K}_{ij}$, $\mathcal{K}_{hk}$, $\mathcal{K}_{hj}$), *geometric stiffness* ($\mathcal{G}_{ik}$) and mass ($\mathcal{M}_{ik}$, $\mathcal{M}_{ij}$, $\mathcal{M}_{hk}$, $\mathcal{M}_{hj}$) effects. The boundary conditions include generalised *normal* and *shear* stress resultants, involving terms that stem from (i) normal stress equilibrium and (ii) the variation of the shear stresses along the cross-section wall thickness. For composite members displaying cross-ply orthotropy (the case of the illustrative example presented in Section 6.2) the system (6)–(7) may be rewritten in matrix form as (Silvestre and Camotim, 2005a,b)

$$\begin{aligned}(\mathbf{C_c}\boldsymbol{\phi}_{,xxxx} - \mathbf{D_c}\boldsymbol{\phi}_{,xx} + \mathbf{B_c}\boldsymbol{\phi} + \mathbf{C_{cs}}\,\boldsymbol{\varphi}_{,xxx}) - \xi_B W_m^0(\mathbf{X}_{\mathbf{c}.m}\boldsymbol{\phi}_{,xx})\\ - \xi_V \omega^2(\mathbf{R_c}\boldsymbol{\phi} - \mathbf{Q_c}\boldsymbol{\phi}_{,xx} + \mathbf{Q_{cs}}\,\boldsymbol{\varphi}_{,x}) = \mathbf{0},\end{aligned} \tag{8}$$

$$(\mathbf{C_{sc}}\boldsymbol{\phi}_{,xxx} + \mathbf{C_s}\,\boldsymbol{\varphi}_{,xx} - \mathbf{D_s}\,\boldsymbol{\varphi}) - \xi_V\omega^2(\mathbf{Q_{sc}}\boldsymbol{\phi}_{,x} + \mathbf{Q_s}\,\boldsymbol{\varphi}) = \mathbf{0}, \tag{9}$$

where (i) the subscripts $(\cdot)_{\mathbf{c}}$, $(\cdot)_{\mathbf{s}}$ and $(\cdot)_{\mathbf{cs}}$ stand *conventional*, *shear* and coupling *conventional-shear* mode quantities, and (ii) the various dynamic stiffness matrices R_{ik} and Q_{ik} account for the effects of mass forces related to the *in* and *out-of-plane* cross-section translations, rotations and translation-rotation couplings. By making (i) $\xi_B = 1$ and $\xi_V = 0$, (ii) $\xi_B = 0$ and $\xi_V = 1$ or (iii) $\xi_B = \psi (0 < \psi \leq 1)$ and $\xi_V = 1$, this system defines linear eigenvalue problems associated with (i) buckling analyses, (ii) a free vibration analyses of load-free members and (iii) free vibration analyses of loaded members – in the last case, the loads W_m^0 are known *a priori* and ω^2 are the problem eigenvalues.

4. Post-Buckling Analysis

A previously developed and numerically implemented non-linear GBT formulation to analyse the post-buckling behaviour of initially imperfect isotropic linear elastic folded-plate members (Silvestre and Camotim, 2003) was recently extended to encompass orthotropic FRP composite members (Silvestre and Camotim, 2004e; Silvestre, 2005). This involved major modifications with respect to the conventional GBT, namely (i) to employ the stress-strain relations given in (2) and (ii) to consider the non-linear strain-displacement relations

$$\begin{aligned}\varepsilon_{xx} &= u_{,x} + \frac{1}{2}(v_{,x}^2 + w_{,x}^2) - z w_{,xx} - \bar{u}_{,x} - \frac{1}{2}(\bar{v}_{,x}^2 + \bar{w}_{,x}^2) + z\bar{w}_{,xx},\\ \varepsilon_{ss} &= v_{,s} + \frac{1}{2}(v_{,s}^2 + w_{,s}^2) - z w_{,ss} - \bar{v}_{,s} - \frac{1}{2}(\bar{v}_{,s}^2 + \bar{w}_{,s}^2) + z\bar{w}_{,ss},\\ \gamma_{xs} &= u_{,s} + v_{,x} + v_{,x}v_{,s} + w_{,x}w_{,s} - 2z w_{,xs} - \bar{u}_{,s} - \bar{v}_{,x} - \bar{v}_{,x}\bar{v}_{,s} - \bar{w}_{,x}\bar{w}_{,s} + 2z\bar{w}_{,xs},\end{aligned} \tag{10}$$

where the bars identify the terms associated with the initial imperfections. Moreover, one must also include an additional set of *transverse extension* deformation modes in the cross-section analysis,

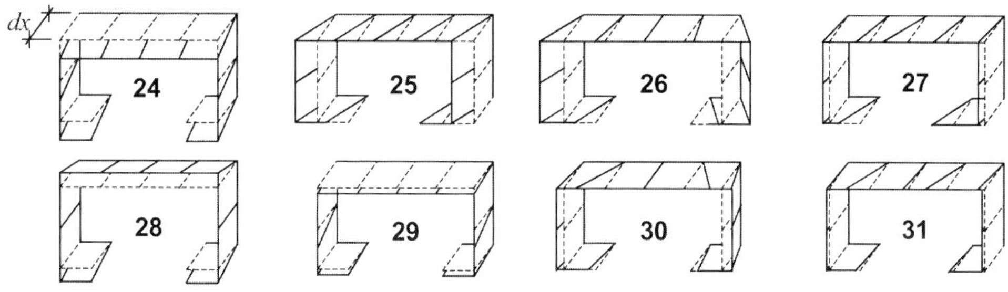

Figure 10. Lipped channel eight most relevant transverse extension modes.

which (i) account for the "bowing effect" stemming from the transverse bending of cross-section walls and (ii) do not comply with Vlassov's assumption – they stem from the sequential imposition of unit transverse and null warping displacements at both the natural and intermediate nodes. The configurations of the eight most relevant transverse extension deformation modes of a lipped channel cross-section are displayed in Figure 10.

After (i) incorporating (10) into the principle of virtual work, (ii) performing several operations, described in detail elsewhere (Silvestre and Camotim, 2003; Silvestre, 2005) and (iii) including the *special* cross-section analysis, one is led to the member equilibrium equations, written variationally as

$$\delta U_1 + \delta U_2 + \delta U_3 - \delta \bar{U}_1 - \delta \bar{U}_2 - \delta \bar{U}_3 + \delta \Pi_q + \delta \Pi_w = 0, \tag{11}$$

where (i) the strain energy terms $\delta U_1, \delta U_2$ and δU_3 are linear, quadratic and cubic functionals of the mode amplitude functions ϕ_i, (ii) their "bar counterparts" contain the imperfection amplitude functions $\bar{\phi}_i$ and (iii) the $\delta \Pi$ terms stand for the virtual work done by distributed or concentrated external loads.

5. Numerical Implementation

The solution of the buckling and vibration eigenvalue problems is obtained by means of either (i) Rayleigh–Ritz's or Galerkin's method, in the case of simply supported members, i.e., members with locally and globally pinned and free-to-warp end sections, or (ii) a GBT-based beam finite element formulation that uses Hermitean cubic polynomials to approximate the buckling/vibration modes, for members with other end support conditions (Camotim et al., 2004; Silvestre, 2005; Gonçalves and Camotim, 2005; Dinis et al., 2006). As for the solution of system (11), it requires the development of another (non-linear) GBT-based beam finite element (also using Hermite and Lagrange cubic polynomials to approximate $\phi_k(x)$ and $\varphi_j(x)$) and resorting to an incremental-iterative numerical technique to determine the member equilibrium path (Silvestre and Camotim, 2003; Silvestre, 2005). After the usual integrations, the FEM system of algebraic equations reads

$$(\mathbf{K}_0^{(e)} + \mathbf{K}_1^{(e)} + \mathbf{K}_2^{(e)})\mathbf{d}^{(e)} - (\mathbf{K}_0^{(e)} + \bar{\mathbf{K}}_1^{(e)} + \bar{\mathbf{K}}_2^{(e)})\bar{\mathbf{d}}^{(e)} = \mathbf{f}_\mathbf{e}^{(e)}, \tag{12}$$

where (i) $\mathbf{K}_p^{(e)}$ are linear ($p = 0$) and non-linear ($p = 1, 2$) *secant* stiffness matrices, (ii) $\mathbf{d}^{(e)}$, and $\bar{\mathbf{d}}^{(e)}$ are the vectors of generalised and initial imperfection nodal displacements and (iii) $\mathbf{f}_e^{(e)}$ is the external applied force vector. The well-known Newton–Raphson predictor-corrector iterative technique was employed to determine the member post-buckling equilibrium paths (e.g., Crisfield, 1991–1996).

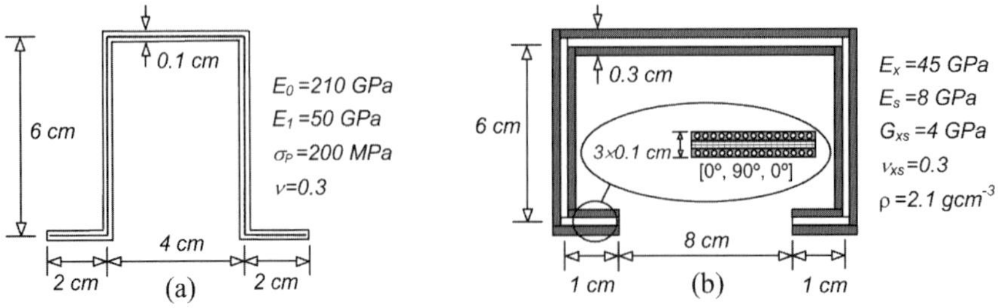

Figure 11. Geometrical and material properties: (a) buckling and (b) vibration and post-buckling.

6. Illustrative Examples

In order to illustrate the application and potential of the above GBT formulations, some numerical results are presented and briefly discussed next. Even if buckling, vibration and post-buckling results are included here, the space limitations make it impossible to cover more than a small fraction of the different (i) features outlined earlier and (ii) possibilities offered by the various GBT analyses. All the relevant geometrical and material properties of the thin-walled members analysed are given in Figures 11(a)–(b) – note that the elastic-plastic uniaxial stress-strain law is described by a bi-linear expression. For validation purposes, some GBT-based linear elastic results are compared with FEM-based values yielded by the codes ADINA (Bathe, 2003) or ABAQUS (HKS, 2002) and adopting fine shell element member discretisations.

6.1. *Buckling*

The results shown in Figures 12(a)–(b) and 13(a)–(b) concern the bifurcation behaviour of isotropic elastic and elastic-plastic hat-section beams (uniform major axis bending) with end sections that are locally and globally pinned and may warp freely (Gonçalves and Camotim, 2005). While Figure 12(a) shows the 9 most relevant deformation modes (out of 15 – the cross-section discretisation involved the following intermediate nodes: 3 in each web, 1 in the flange mid-point and 1 in each stiffener free end), Figure 12(b) depicts three GBT-based buckling curves, corresponding to elastic and elastic-plastic (J_2-flow and J_2-deformation) beams and providing the variation of the critical moment M_{cr} with the normalised length L/h (logarithmic scale). In addition, the white dots in Figure 12(b) stand for FEM-based elastic critical moments obtained using the code ADINA (Bathe, 2003).[3] As for Figures 13(a)–(b), they make it possible to compare the ADINA (perspective) and GBT-based (in-span cross-section) elastic critical buckling mode shapes of the beams indicated in Figure 12(b): lengths $L/h = 5$ (six-wave local-plate buckling modes $\mathbf{7}+\mathbf{9}$) and $L/h = 10$ (single-wave distortional-flexural-torsional buckling modes $\mathbf{3}+\mathbf{4}+\mathbf{6}$). Although an in-depth discussion of the results shown in these two figures is beyond the scope of this paper, the following general comments and remarks are appropriate:

(i) The linear elastic M_{cr} values yielded by the GBT-based analyses practically coincide with the ones obtained using ADINA. Moreover, one notices that the critical buckling mode nature varies with L/h as follows: (i) local-plate buckling (modes $\mathbf{7}+\mathbf{9}$) for short beams ($L/h \leq 8.8$), (ii) distortional-flexural-torsional buckling (modes $\mathbf{3} + \mathbf{4} + \mathbf{6}$) for intermediate beams ($8.8 \leq$

[3] No validation is presented for the elastic-plastic GBT-based results, as the authors know no commercial FEM code capable of calculating plastic bifurcation moments – e.g., neither ADINA (Bathe, 2003) nor ABAQUS (HKS, 2002) offer such possibility.

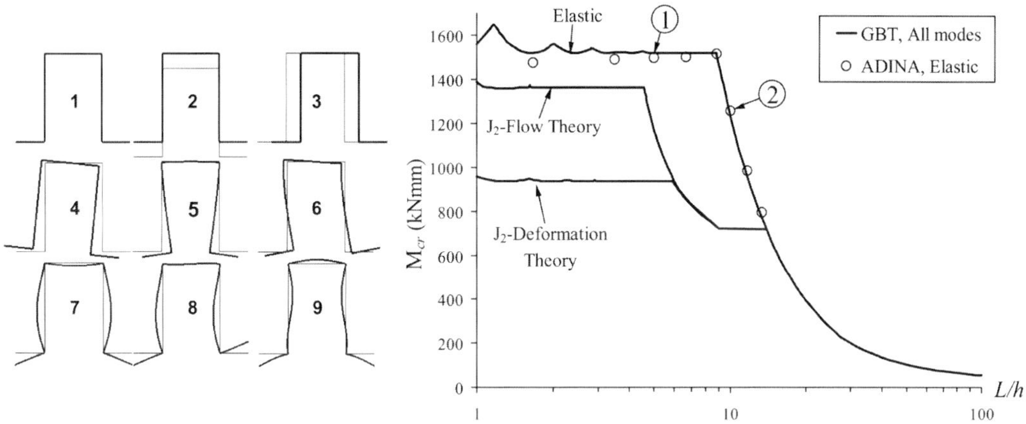

(a) (b)

Figure 12. (a) Hat-section first 9 deformation mode shapes and (b) beam elastic and elastic-plastic M_{cr} vs. L/h curves (GBT-based and ADINA results).

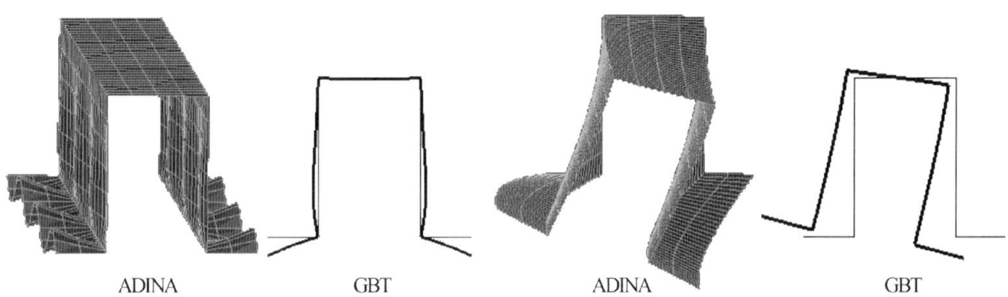

① – Local-Plate Mode ② – Distortional-Flexural-Torsional Mode

(a) (b)

Figure 13. ADINA and GBT-based elastic critical buckling mode shapes – beams with (a) $L/h = 5$ and (b) $L/h = 10$.

$L/h \leq 20$) and (iii) "classical" lateral-torsional buckling (modes **3** + **4**) for the longer beams ($L/h \geq 20$). Finally, note that the accuracy of the ADINA results deteriorate for $L/h < 4$ (local-plate buckling), due to the occurrence of stress concentrations in the shell element model (in the vicinity of the end sections) – obviously, they lower the M_{cr} values.

(ii) There is also a virtual coincidence between the elastic critical buckling mode shapes yielded by GBT and ADINA obviously, the former concern the most deformed beam cross-sections.

(iii) The elastic-plastic GBT-based results confirm the well-known fact that deformation theory leads to lower critical buckling loads than flow theory. This is particularly true for local-plate buckling, which involves almost exclusively transverse plate bending – recall that the transverse plate bending stiffness $t^3 \dot{E}_{22}/12$ decreases more rapidly for deformation theory (Gonçalves and Camotim, 2004b). The differences are much less relevant for distortional-flexural-torsional buckling and vanish for lateral-torsional buckling (the beams buckle elastically). Due to higher local-plate buckling stresses, the length interval related to critical distortional-flexural-torsional buckling is larger for flow theory than for deformation theory.

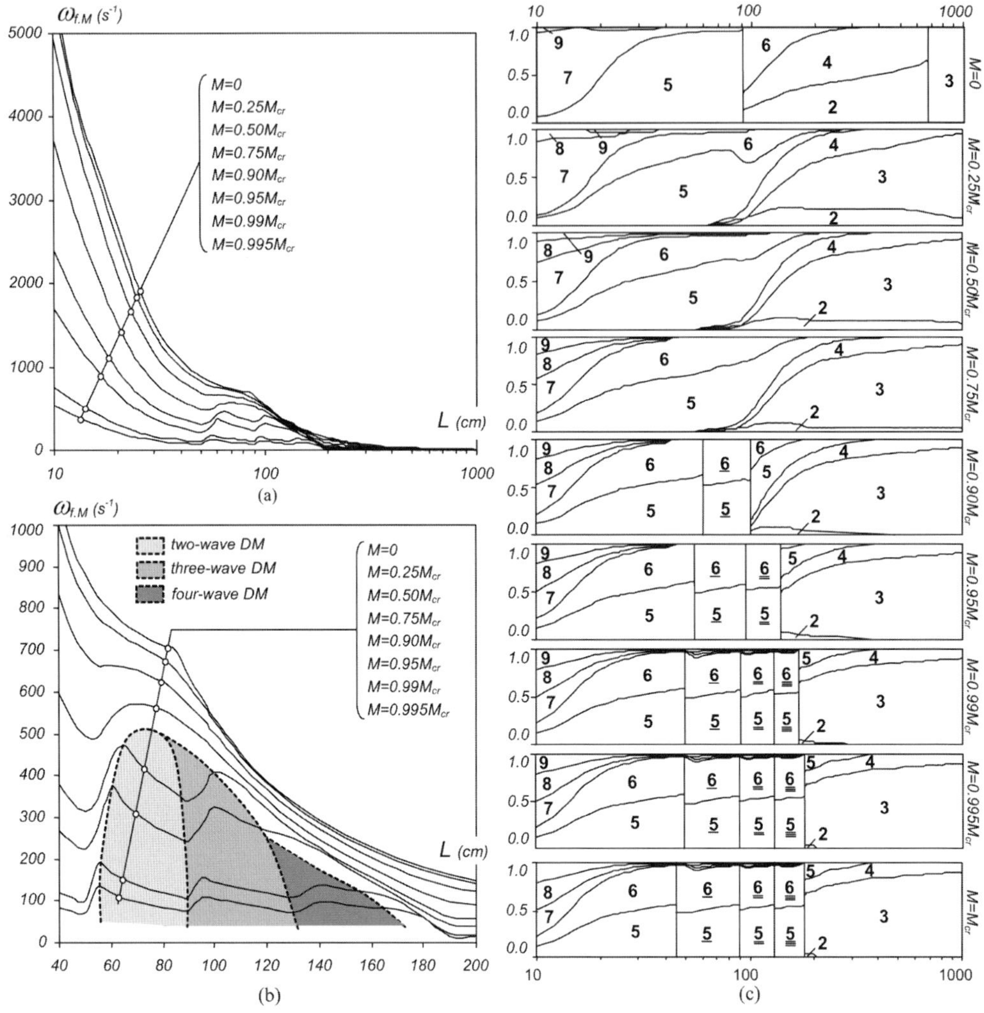

Figure 14. Variation of ω_f, for (a) $10 < L < 1000$ cm and (b) $40 < L < 200$ cm, and (c) modal participation diagram of the beam fundamental vibration mode with L and M/M_{cr}.

6.2. *Vibration*

The vibration results shown in Figures 14(a)–(c) concern laminated plate lipped channel beams (members under uniform major axis bending) with (i) walls formed by three equally thick orthotropic layers made of identical FRP materials (epoxy resin reinforced with e-glass fibres) and exhibiting a cross-ply configuration [0°, 90°, 0°], (ii) locally/globally pinned and free-to-warp end sections and (iii) uniform mass density $\rho = 2.1$ gcm^{-3} (Silvestre and Camotim, 2005a,b). The curves in Figure 14(a) show how the fundamental natural frequency $\omega_{f \cdot M}$ varies with the beam length L (logarithmic scale) and bending moment ratio M/M_{cr} – the upper curve is related to the load-free member ($M = 0$) and is always associated with single-wave vibration modes ($\omega_{f \cdot 0} = \omega_{1 \cdot 0}$).[4] The remaining seven curves correspond, in descending order, to increasing M/M_{cr} value – each

[4] Due to space limitations, the results concerning the buckling behaviour and load-free vibration behaviour of the beams cannot be presented here. The interested reader is referred to the recent works by Silvetre and Camotim (2005a,b).

curve is associated with a fixed percentage of M_{cr}, which varies with the beam length. Because M_{cr} corresponds to buckling modes exhibiting several waves for 55 cm $< L <$ 180 cm, it is interesting to look more closely at the curves $\omega_{f \cdot M}(L)$ in this length range – thus, Figure 14(b) shows these curves for 40 cm $< L <$ 200 cm. Moreover, the modal participation diagrams shown in Figure 14(c) supply additional information about the contribution of the GBT deformation modes (see Figure 2) to the fundamental vibration modes of beams with nine M/M_{cr} values.[5] Note that the presence of none, one, two or three bars underneath a mode number indicates a single, two, three or four-wave contribution to the vibration mode. The analysis of the results displayed in Figures 14(a)–(c) prompts following conclusions and/or comments:

(i) For $L <$ 55 cm or $L >$ 180 cm, the beams buckle in single-wave critical buckling modes (see footnote 4), which means that one has $\omega_{f \cdot M}(L) \equiv \omega_{1 \cdot M}(L)$ and all curves have fairly similar shapes. Moreover, it is worth pointing out that, although the beam fundamental vibration modes always exhibit a single wave, their shapes differ from those of the beam (i_1) critical buckling modes and (i_2) load-free fundamental vibration modes – indeed, the beam fundamental vibration mode shape "travels" between them as the ratio M/M_{cr} increases (see Figure 14(c)). For instance, note that the participation of mode **2**, which contributes significantly to the load-free member *flexural-torsional* vibration modes, continuously decreases as M/M_{cr} grows, until it vanishes for $M = M_{cr}$ – mode **2** is absent from the beam critical buckling mode (it appears in its pre-buckling path).

(ii) For 55 $< L <$ 180 cm, the beam critical buckling modes have more than one wave (see footnote 4) and the shapes of the curves $\omega_{f \cdot M}(L)$ become visibly different as the value of M/M_{cr} increases, as clearly illustrated in Figure 14(b) – these curves (ii_1) cease to decrease monotonically and, for large enough M/M_{cr} values, (ii_2) they are no longer "smooth", exhibiting sudden and quite pronounced slope reversals. Moreover, Figure 14(c) shows that the beam fundamental vibration mode wave number varies between one and the number of waves appearing in the beam critical buckling mode, as M/M_{cr} grows – for $M \geq 0.90M_{cr}$, the fundamental vibration mode successively exhibits 2, 3 and 4 waves (Figure 14(b) identifies very well the $L - M/M_{cr}$ combinations associated with each case). Apparently, the M/M_{cr} value that triggers a non-single wave number depends on the percentage difference between M_{cr} and $M_{b \cdot 1}$ (bifurcation moment leading to a single-wave buckling mode), i.e., a $M_{cr}/M_{b \cdot 1}$ decrease lowers the M/M_{cr} value corresponding to the change (Silvestre and Camotim, 2005a,b).

(iii) Figure 14(c) shows that a beam with L = 100 cm vibrates in (iii_1) a flexural-torsional-distortional mode (**2** + **4** + **6**) for $M = 0$, (iii_2) another flexural-torsional-distortional mode (**2** + **4** + **6** + **5**) for $0 < M/M_{cr} < 0.3$, (iii_3) a single-wave distortional mode (**5** + **6**) for $0.3 \leq M/M_{cr} < 0.8$ and (iii_4) similar two or three-wave distortional modes for $0.8 \leq M/M_{cr} < 0.9$ and $0.9 \leq M/M_{cr} \leq 1$, respectively. Since the combination of modes **5** and **6** attenuates the rotation of the tensioned flange-lip assembly and increases that of the compressed one (see Figure 2), it is possible to conclude that, for increasing M/M_{cr} values, the amplitudes of the tensioned and compressed flange-lip motions tend to decrease and increase, respectively. Therefore, for moderate-to-high M/M_{cr} values ($0.3 \leq M/M_{cr} < 1$), only the compressed flange-lip and the web upper half (due to compatibility) vibrate – they exhibit either 1, 2 or 3 waves (see the last two modes in Figure 15).

In order to validate the GBT-based results, some shell finite element analyses were carried out using the code ABAQUS (HKS, 2002) – they all involved beams with L = 100 cm and the results

[5] It was found that the shear modes associated with non-linear warping (see Figure 9) do not contribute to the beram fundamental vibration modes. Nevertheless, they were included in all analyses.

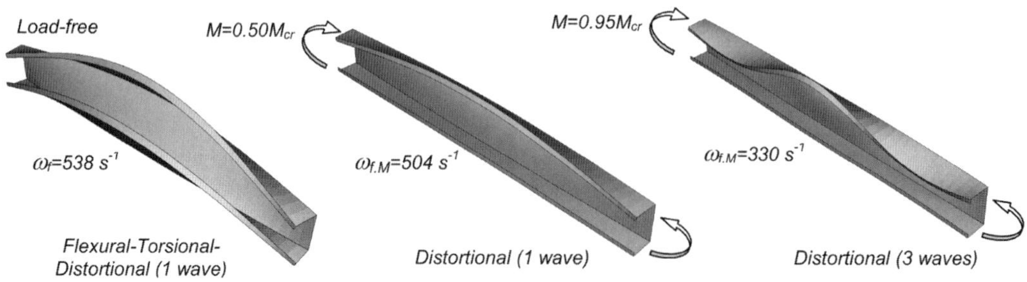

Figure 15. FEM-based beam fundamental frequencies and vibration mode shapes ($L = 100$ cm).

determined concerned $M/M_{b \cdot 1} = 0.271, 0.407, 0.488, 0.516, 0.537, 0.543$ (the last ratio means $M = M_{cr}$). The GBT and FEM-based $\omega_{f \cdot M}$ values read (i) $\omega_{f \cdot M} = 502, 463, 409, 325, 146, 0\ s^{-1}$ (GBT) and (ii) $\omega_{f \cdot M} = 508, 487, 463, 418, 315, 268\ s^{-1}$ (FEM), thus making it obvious that they become increasingly apart as $M/M_{b \cdot 1}$ grows. In particular, note that, for $M = M_{cr}$, GBT yields the (theoretically expected) *null* $\omega_{f \cdot M}$, while the FEM-based $\omega_{f \cdot M}$ value is about 50% of $\omega_{f \cdot 0} = 538\ s^{-1}$ (it only becomes null for $M = 1.0134 M_{cr}$). It was subsequently found that this rather surprising discrepancies stem from the fact that the ABAQUS shell FEA incorporate the stiffening effect due to the primary (first-order) bending deflections, which is not taken into account by the developed GBT formulation. This fact was overcome by performing the FEM vibration analyses in "adequately pre-cambered beams" (Silvestre and Camotim, 2005a,b), thus enabling a meaningful comparison between the two sets of $\omega_{f \cdot M}$ values. Figure 15 shows the FEM-based fundamental vibration mode shapes and frequency values for beams with $L = 100$ cm acted by $M = 0$; 0.5; $0.9M_{cr}$ – the FEM values $\omega_{f \cdot 0} = 530\ s^{-1}$, $\omega_{f \cdot 0.5} = 504\ s^{-1}$ and $\omega_{f \cdot 0.9} = 330\ s^{-1}$ agree very well with the GBT ones $\omega_{f \cdot 0} = 538\ s^{-1}$, $\omega_{f \cdot 0.5} = 502\ s^{-1}$ and $\omega_{f \cdot 0.9} = 325\ s^{-1}$ (differences of 1.5, 0.3 and 1.5%).

6.3. *Post-Buckling*

The results presented here concern the distortional post-buckling behaviour of two identical FRP laminated plate lipped channel columns (i) having locally/globally pinned and free-to-warp end sections, (ii) with the material properties and cross-section dimensions displayed in Figure 11(b) (identical to the ones considered in the vibration analyses presented in the previous subsection), (iii) containing critical-mode initial geometrical imperfections with amplitudes $v_0 = \pm 0.15 \cdot t$ (t is the wall thickness and v_0 is the outward/inward motion of the flange-lip corners at mid-span), (iv) with length $L = 40$ cm and (v) discretised into (v_1) 6 natural and 17 intermediate (5 in the web and flanges and 1 per lip) nodes and (v_2) 8 finite elements (Silvestre and Camotim, 2004e). It is worth noting that the adopted column geometry ensures that (i) bifurcation occurs in a single-wave distortional mode and that (ii) local-plate/distortional mode interaction effects are not relevant – the ratio between the minimum local-plate and distortional buckling loads is 1.6 (Silvestre and Camotim, 2004c).

Figure 16(a) shows the post-buckling equilibrium paths $\sigma/\sigma_{cr \cdot D}$ vs. v/t ($\sigma_{cr \cdot D} = 38.6$ MPa is the column critical buckling load and v is the additional outward/inward flange-lip motion) of the two columns, henceforth termed "outward" and "inward". Also included are the results yielded by the shell FEA performed in ABAQUS (HKS, 2002). Figure 16(b), on the other hand, displays diagrams that provide the contributions of the various GBT deformations modes (depicted in Figures 2, 9 and 10) to various column deformed configurations located along their equilibrium paths. Finally, Figure 16(c) provides the post-buckling evolution of the mid-span normal stress distribution along

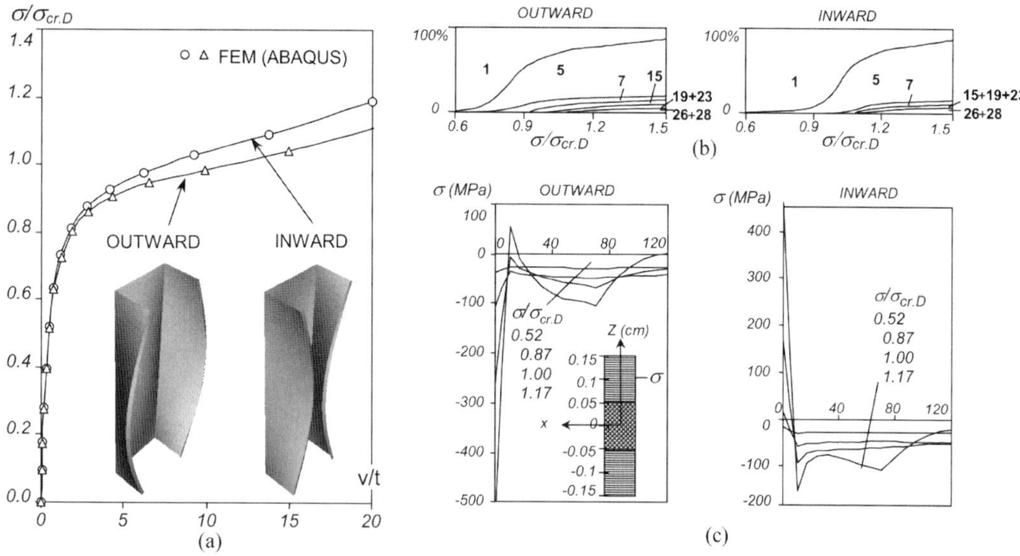

Figure 16. (a) Outward and inward column distortional post-buckling equilibrium paths $\sigma/\sigma_{cr \cdot D}$ vs. v/t, (b) diagrams providing the modal decomposition of the deformed configurations and (c) post-buckling evolution of the normal stress distribution σ ($x = 20$ cm; $z = 0.1$ cm; s) for the outward and inward columns.

the mid-lines of the outward and inward column outer layers (longitudinally aligned fibres).[6] After a close observation of the above figures, one is led to the following conclusions:

(i) First of all, one instantly recognises the important role played by the initial imperfection "sign". Indeed, the inward column post-buckling stiffness and strength are larger than their outward column counterparts by a non-negligible amount. Like in the case of the isotropic members (Silvestre and Camotim, 2003), this distortional post-buckling *asymmetry* stems mostly from the different contributions of the shear modes **15 + 19 + 23** to the outward and inward column deformed configurations.

(ii) Then, attention should be drawn to the virtual coincidence between the GBT-based equilibrium paths and the post-buckling results yielded by ABAQUS. The fact that the GBT analyses never involved more than 450 degrees of freedom provides a clear assessment of the high computational efficiency of this approach. The corresponding (and similarly accurate) FEM results required the discretisation of the thin-walled columns by means of very refined shell element meshes.

(iii) In the initial pre-buckling stages, the normal stress distribution is uniform, since mode **1** dominates. As post-buckling progresses and regardless of the v_0 sign (outward or inward motions), the normal stress distribution becomes non-linear in the web and flanges, mostly due to the shear modes **15 + 19 + 23**.

(iv) In the outward column post-buckling stages, the contributions of modes **5** and **15** account for the fact that the compressive stresses (iv_1) increase near the web-flange node and (iv_2) decrease in the vicinity of the flange-lip corner. Moreover, for $\sigma/\sigma_{cr \cdot D} \geq 1.0$, tensile stresses start to develop around the flange-lip node and rather high compressive stresses appear close to the lip free ends. Conversely, the inward column exhibits a compressive stress increase near the flange-lip corner (all the flanges are under compression) and high tensile stresses develop in

[6] Due to symmetry, only one half of the normal stress distributions are represented in Figure 16(c).

the neighbourhood of the lip free ends – however, these tensile stresses are lower than the compressive ones appearing in the outward column. This last fact is due to the relevant participation of mode **19**, reinforcing mode 5 in the outward column and opposing it in the inward one.

7. Conclusion and Future Developments

After an extremely brief overview of the concepts/procedures involved in a "conventional" GBT analysis, the paper reported on the GBT formulations/applications recently developed at the TU Lisbon, namely:

(i) The use of "conventional GBT" to derive approximate analytical formulae to estimate distortional buckling stresses in lipped channel, zed-section and rack-section cold-formed steel members.
(ii) Extensions of the conventional GBT to cover members with (ii_1) arbitrary open/closed branched cross-sections and (ii_2) made of orthotropic linear elastic and non-linear elastic-plastic materials.
(iii) A GBT formulation to analyse the vibration behaviour of loaded folded-plate members made of orthotropic linear elastic materials.
(iv) A geometrically non-linear GBT formulation to analyse the post-buckling behaviour of folded-plate members made of orthotropic linear elastic materials.

To illustrate the application and provide an idea of the capabilities of the GBT approach to solve structural analysis problems, a few numerical results were presented and very briefly commented. For validation purposes, some of these results were also compared with values yielded by shell finite element analyses, performed in the commercial codes ADINA and ABAQUS.

Because a very rich research activity on GBT formulations, implementations and applications is still under way at the TU Lisbon, further developments are to be expected in the near-to-intermediate future. For instance, the topics being currently investigated include:

(i) GBT formulations to analyse the buckling behaviour of thin-walled members made of orthotropic linear elastic or non-linear elastic-plastic materials that exhibit arbitrary closed/branched cross-sections.
(ii) GBT formulations to analyse the vibration behaviour of thin-walled members made of orthotropic linear elastic materials and exhibiting arbitrary closed/branched cross-sections.
(iii) GBT formulations to analyse the buckling and vibration behaviour of isotropic linear elastic thin-walled members exhibiting arbitrary closed/branched cross-sections and subjected to non-uniform internal force and moment diagrams (i.e., under stress gradients).
(iv) GBT formulations to analyse the first and second-order behaviour of isotropic elastic-plastic thin-walled members with unbranched open cross-sections.
(v) A GBT formulation to analyse the post-buckling behaviour of orthotropic linear elastic thin-walled members with open branched and arbitrary closed/branched cross-sections.
(vi) The influence of local-plate/distortional and local/global mode interaction effects on the elastic post-buckling behaviour of cold-formed steel members.
(vii) The development of GBT-based beam finite elements intended to enable the performance of elastic first-order and buckling analyses of plane and spatial frames made of thin-walled steel members.

References

Bathe, K.J./ADINA R&D Inc. (2003). *Adina System.*

Camotim, D., Silvestre, N., Gonçalves, R. and Dinis, P.B. (2004). GBT analysis of thin-walled members: new formulations and applications, in *Thin-Walled Structures: Recent Advances and Future Trends in Thin-Walled Structures Technology*, J. Loughlan (ed.), Canopus Publishing Ltd., Bath, pp. 137–168.

Crisfield, M. (1991–1996). *Nonlinear Finite Element Analysis of Solids and Structures: Essentials* (Vol. 1) and *Advanced Topics* (Vol. 2), John Wiley & Sons, Chichester.

Davies, J.M. (1998). Generalised beam theory (GBT) for coupled instability problems, in *Coupled Instability in Metal Structures: Theoretical and Design Aspects*, J. Rondal (ed.), Springer Verlag, Vienna, pp. 151–223.

Davies, J.M. (2000). Recent research advances in cold-formed steel structures, *Journal of Constructional Steel Research*, **55**(1–3), 267–288.

Dinis, P.B., Camotim, D. and Silvestre, N. (2006). GBT formulation to analyse the buckling behaviour of thin-walled members with arbitrarily 'branched' open cross-sections, *Thin-Walled Structures*, **44**(1), 20–38.

Gonçalves, R. and Camotim, D. (2004a). Buckling analysis of single and multi-cell closed thin-walled metal members using generalised beam theory, in *Proceedings of Fourth International Conference on Coupled Instabilities in Metal Structures* (CIMS'04, Rome, 27–29 September), pp. 119–130.

Gonçalves, R. and Camotim, D. (2004b). GBT local and global buckling analysis of aluminium and stainless steel columns, *Computers & Structures*, **82**(17–19), 1473–1484.

Gonçalves, R. and Camotim, D. (2005). Thin-walled member plastic bifurcation analysis using generalised beam theory, *Computers & Structures*, accepted for publication.

Gonçalves, R., Dinis, P.B. and Camotim, D. (2006). GBT linear and buckling analysis of thin-walled multi-cell box girders, in *Proceedings of SSRC Annual Stability Conference* (San Antonio, 8–11 February), pp. 329–352.

Hibbit, Karlsson and Sorensen Inc. (2002). *ABAQUS Standard* (Version 6.3).

Hill, R. (1958). A general theory of uniqueness and stability in elastic-plastic solids, *Journal of Mechanics and Physics of Solids*, **6**, 236–249.

Hutchinson, J.W. (1974), Plastic buckling, in *Advances in Applied Mechanics*, Vol. 14, C.S. Yih (ed.), Academic Press, New York, pp. 67–144.

Jones, R. (1999). *Mechanics of Composite Materials*, Taylor & Francis, Philadelphia.

Rasmussen, K.J.R. (2003). Full-range stress-strain curves for stainless steel alloys, *Journal of Constructional Steel Research*, **59**(1), 47–61.

Rendek, S. and Baláž, I. (2004). Distortion of thin-walled beams, *Thin-Walled Structures*, **42**(2), 255–277.

Schardt, R. (1966). Eine Erweiterung der technische Biegetheorie zur Berechnung prismatischer Faltwerke, *Stahlbau*, **35**, 161–171.

Schardt, R. (1989). *Verallgemeinerte technische Biegetheorie*, Springer Verlag, Berlin.

Schardt, R. (1994a). Generalised beam theory – an adequate method for coupled stability problems, *Thin-Walled Structures*, **19**(2–4), 161–180.

Schardt, R. (1994b). Lateral torsional and distortional buckling of channel and hat-sections, *Journal of Constructional Steel Research*, **31**(2–3), 243–265.

Schardt, R. and Heinz, D. (1991). Vibrations of thin-walled prismatic structures under simultaneous static load using generalized beam theory, in *Structural Dynamics*, W.B. Krätzig et al. (eds.), Balkema, Rotterdam, pp. 921–927.

Silvestre, N. (2005). *Generalised Beam Theory: New Formulations, Numerical Implementation and Applications*, Ph.D. Thesis in Civil Engineering, IST, Technical University of Lisbon [in Portuguese].

Silvestre, N. and Camotim, D. (2002). First and second-order generalised beam theory for arbitrary orthotropic materials, *Thin-Walled Structures*, **40**(9), 755–789 + 791–820.

Silvestre, N. and Camotim, D. (2003). Non-linear generalised beam theory for cold-formed steel members, *International Journal Structural Stability and Dynamics*, **3**(4), 461–490.

Silvestre, N. and Camotim, D. (2004a). Distortional buckling formulae for cold-formed steel C and Z-section members: Part I – derivation and Part II – validation and application, *Thin-Walled Structures*, **42**(11), 1567–1597 + 1599–1629.

Silvestre, N. and Camotim, D. (2004b). Distortional buckling formulae for cold-formed steel rack-section members, *Steel & Composite Structures*, **4**(1), 49–75.

Silvestre, N. and Camotim, D. (2004c). Influence of shear deformation on the local and global buckling behaviour of composite thin-walled members, in *Thin-Walled Structures: Advances in Research, Design and Manufacturing Technology* (ICTWS 2004, Loughborough, 22–24 June), J. Loughlan (ed.), Institute of Physics Publishing, Bristol, pp. 659–668.

Silvestre, N. and Camotim, D. (2004d). Generalised beam theory to analyse the vibration behaviour of orthotropic thin-walled members, in *Thin-Walled Structures: Advances in Research, Design and Manufacturing Technology* (ICTWS 2004, Loughborough, 22–24 June), J. Loughlan (ed.), Institute of Physics Publishing, Bristol, pp. 919–926.

Silvestre, N. and Camotim, D. (2004e). Generalised beam theory formulation to analyse the post-buckling behaviour of orthotropic laminated plate thin-walled members, in *Abstracts of 21st International Congress of Theoretical and Applied Mechanics* (ICTAM 04, Warsaw, 15–21 June), pp. 345–346.

Silvestre, N. and Camotim, D. (2005a). Local and global vibration behaviour of loaded folded-plate anisotropic members, in *Programme and Book of Abstracts of the Twelfth International Congress on Sound and Vibration* (Lisbon, 11–14 July), pp. 191–192 (full paper in CD-ROM Proceedings, Paper 604).

Silvestre, N. and Camotim, D. (2005b). GBT-based local and global vibration analysis of loaded composite thin-walled members, *International Journal of Structural Stability and Dynamics*, **6**(1), in press.

Simão, P. and Silva, L.S. (2004). A unified energy formulation for the stability analysis of open and closed thin-walled members in the framework of the generalized beam theory, *Thin-Walled Structures*, **42**(10), 1495–1517.

EVOLUTIONARY OPTIMIZATION OF BRACED STEEL FRAMEWORKS FOR TALL BUILDINGS USING A HYBRID OC-GA METHOD

C.-M. Chan and K.-M. Wong
Department of Civil Engineering,
The Hong Kong University of Science and Technology, Hong Kong, P.R. China

Abstract

Having many attractive advantages, genetic algorithms (GAs) have been applied to many design optimization problems. However, the practical application of GAs to realistic tall building design is still rather limited, since GAs require a large number of structural reanalyses and perform poorly in local searching. While the Optimality Criteria (OC) method can be applied effectively to the element sizing optimization of tall buildings, there is no guarantee that the OC method can always lead to the global optimum. In this paper, the so-called hybrid OC-GA method is presented to fully exploit the merits of both OC and GA for topology and element sizing optimization of braced tall steel frameworks. While the GA is particularly useful in the global exploration for optimal topologies, the OC technique serves as an efficient local optimizer for resizing elements of selected topologies. The effect of population size and the importance of the local OC search operator have been investigated. The applicability and efficiency of the hybrid OC-GA method were tested with two braced steel building examples. Results indicate that the incorporation of the OC operator into the GA has remarkably improved the efficiency and robustness of the evolutionary algorithm and thus make the hybrid method particularly useful for topology optimization of practical tall building structures involving a large number of structural elements and the use of numerous structural forms.

1. Introduction

For the past few decades, genetic algorithms (GAs) first developed by Holland (1975) have gained wide popularity and demonstrated their advantages over the conventional gradient-based optimization techniques. GAs are stochastic search methods, which mimic the principle of the survival of the fittest in natural selection. Due to their generality, GAs have been applied to a wide range of design problems especially those with discrete sizing variables, geometrical and topological variables. Unlike conventional optimization techniques, GAs are able to explore simultaneously the entire design space with a population of designs and therefore is capable of seeking for the global optimum. GAs can be applied directly and conveniently to structural design problems with both discrete and continuous design variables.

GAs can be regarded as a type of zero-order method, which requires numerous functional evaluations for achieving solution convergence. Consequently, as the scale and complexity of building structures increase, the required computational effort also increases, thus making GAs prohibitively difficult in solving practical large design problems. Another shortcoming of the GA approach is its lack of precision in searching for the definitive global optimum point. Since GAs are stochastic techniques, they are incapable of determining the precise global optimum and converge generally only to a near-optimum point.

One effective approach for the element sizing optimization of building structures has been based on the Optimality Criteria (OC) method, which has been shown to suit particularly well for tall building design with many design variables (Chan, 2001; Chan 2004). In the OC method, a set of necessary optimality criteria for the optimal design is first derived and a recursive algorithm is then applied to resize the element sizing design variables to indirectly satisfy the optimality criteria. For the lateral stiffness design of tall buildings, the OC method generally converges quite rapidly in a few design

M. Pandey et al. (eds), Advances in Engineering Structures, Mechanics & Construction, 205–214.

cycles indicating a weak dependence of the computational efficiency on the number of design variables. Although the OC method can be remarkably efficient in sizing optimization, they cannot be applied to topological optimization problems with addition and removal of discrete structural elements. Furthermore, there is no guarantee that the OC method can always lead to the global optimum (Kirsch 1993).

To overcome the problems associated with GAs while maintaining their merits, hybrid methods incorporating local search techniques into GAs, which provide great flexibility for hybridization, were proposed. Sakamoto and Oda (1993) proposed a hybrid method comprising a genetic algorithm and the generalized OC method to optimize both the layout and cross sectional area of simple trusses with nine nodes. Element sizing designs were refined locally by a proportional scaling in their proposed algorithm. Yeh (1999) inserted the fully stressed OC method to GAs for optimising element sizes of truss structures subject to stress and displacement constraints. The best design amongst a population size of 60 individuals in any generation was optimised by the fully stressed OC method. The results suggested that the hybrid GA method is more superior to a pure GA in terms of both quality of the optimal design and convergence behaviour. Chan et al. (2003) developed a hybrid method which combines GA with a rigorously derived OC technique. The method has the advantage over the conventional GAs and is capable of solving element sizing design problems of practical tall building structures in which the conventional OC method has encountered the problem of achieving solution convergence. Espinoza et al. (2005) demonstrated that a hybrid GA method with local search algorithm required significant reduction in the number of function evaluations for obtaining the optimal design when compared to simple GAs alone in solving a groundwater remediation problem. Fawaz et al. (2005) presented an evolutionary algorithm with a globally stochastic but locally heuristic search strategy. Considerably fewer computational operations have been found in the shape optimization of a simple 18-bar truss with stress constraints.

In this paper, the so-called hybrid OC-GA method is further extended to both topological and element sizing optimization of skeletal steel building frameworks. A local search operator based on a rigorously derived OC technique is developed and embedded in the framework of a GA. While the GA is used to explore the entire design space and generate improved topologies, the OC operator is applied as an efficient local optimizer for element resizing of selected topologies. The hybrid OC-GA method works in concert with the global GA searching method and the local OC optimizer to provide better optimal topological and element sizing design of building frameworks than that GA could provide alone. Different rates of the local OC search operator have been investigated for different population sizes with the aim of determining the most appropriate value of the probability of an OC operation for the topological and element sizing optimization of tall steel frameworks using the proposed hybrid OC-GA method.

2. General Design Problem Formulation for steel frameworks

Consider a general steel building framework having initially $i = 1,2,\ldots,N$ elements (or element fabrication groups), the minimum cost design the optimal topology and element sizing design in terms of material cost can be formulated as:

Minimize: $$W(t_i, A_i) = \sum_{i=1}^{N} w_i \cdot t_i \cdot A_i \qquad (1a)$$

Subject to: $$d_j = \frac{\delta_j - \delta_{j-1}}{h_j} \le d_j^U \qquad (j = 1,2,\ldots, M) \qquad (1b)$$

$$\left(\frac{\sigma_i}{\sigma_i^U} - 1\right) t_i \le 0 \qquad (i = 1,2,\ldots, N) \qquad (1c)$$

$$A_i \in \boldsymbol{A}_i = (a_{i1}, a_{i2}, \ldots, a_{in_i}) \qquad (i = 1,2,\ldots, N) \qquad (1d)$$

$$t_i = 0 \text{ or } 1 \qquad (i = 1,2,\ldots, N) \qquad (1e)$$

Eq. (1a) defines the material cost of the building framework in which w_i denotes the unit material cost per unit cross sectional area of element i and the design variables t_i, A_i represent the Boolean variable and the cross sectional area of the element, respectively. The Boolean variable t_i defines the presence or absence of the element i. If $t_i = 1$, then the element i exists; otherwise $t_i = 0$ and the element is removed from the structural model. The element sizing variable A_i is selected from a specified set of discrete commercial sections $\boldsymbol{A}_i$ as given in Eq, (1d), where n_i represents the number of available discrete sections for the ith element. Eq. (1b) defines the set of $j = 1,2,..,M$ serviceability lateral drift criteria, where δ_j, and δ_{j-1} are the lateral deflections of two adjacent floor levels j and j-1; h_j is the corresponding jth story height; d_j, and d_j^U are the drift ratio and its corresponding allowable limit. Eq. (1c) defines a set of element strength constraints, where σ_i, and σ_i^U are the individual element stress and its allowable limit. Since the design of tall buildings is predominately controlled by serviceability lateral stiffness requirements, element strength design constraints are secondary design considerations that only a small number of them are critical to a tall building design.

3. Hybrid OC-GA Method

3.1 Hybridization Strategy

In the proposed OC-GA method, a local search Optimality Criteria algorithm is incorporated into the GA process and thus called as an OC operator, which has the ability of performing element sizing optimization for a selected topology. The GA framework first starts with a randomly generated initial population and then produces new offspring designs with random changes in both the topology and element sizing by crossover and mutation. Unlike the genetic operators, the OC operator is a deterministic gradient based algorithm which improves a design with predefined topology subject to the specified structural design constraints. The OC technique is an efficient local search method which can resize rapidly the element sizes through the use of a recursive algorithm that satisfies a set of prescribed necessary optimality conditions. The balance between the global exploration of topology by the GA and the exploitation of efficient local element optimization by the OC is crucial to the success of achieving progressively improved designs whilst avoiding the occurrence of premature convergence. To achieve a satisfactory cooperation between GA and OC, the proposed hybridization strategy involves promoting frequent topology changes in the GA and precluding premature dominant designs generated by the OC at the early generations.

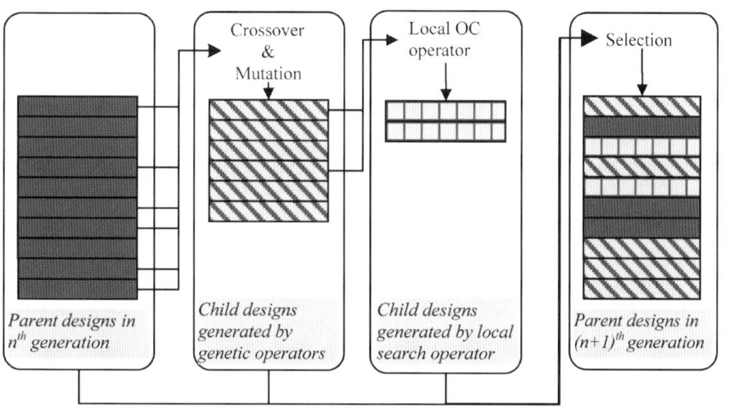

Figure 1. Schematic of the hybrid OC-GA method

An application of the OC operator can be illustrated in Fig.1. At the nth generation, child designs are first recombined by crossover and mutation from the parent population. The OC operator is then applied to a portion of the child designs stochastically to undertake the local search OC element sizing optimization. The rate of OC, p_{oc}, denoting the probability of a child design that takes on the OC operation, is used to control the application of the local search operator. If a randomly generated real number ranging from 0 to 1 is found to be smaller than a prescribed value of p_{oc}, the OC operator will then be invoked. In theory, the OC operator may be applied to every child design (i.e. when p_{oc}=1). However, it may become impractical to do so for realistic structures with a large population due to the excessive computation required. It is important that an appropriate value of the probability of OC be determined to strike a right balance between computational efficiency and the quality of the optimised designs. After the fitness of each design is evaluated, an enlarged sampling selection (Gen and Cheng, 1997) is employed such that the population size is temporarily enlarged to contain both the parent and offspring designs during the selection process. Based on their fitnesses, the offspring designs are set to compete with their parents and the surviving candidates then form the next parent population in the $(n+1)^{th}$ generation.

3.2 Local OC search operator

The Optimality Criteria (OC) approach has long been recognised as a highly efficient method for element sizing optimization of large-scale structures. The rigorously derived OC method is shown to be particular suitable for lateral stiffness optimization problems associated with tall buildings subject wind-induced serviceability design constraints (Chan, 2001; Chan, 2004). Generally speaking, the design of a tall building with a height-to-width aspect ratio larger than 5 is likely to be governed by wind-induced drift and motion perception serviceability design criteria. The rigorously derived OC method is herein adopted as a local search operator for the lateral drift design of tall skeletal steelworks.

To commence the rigorously derived OC method, the lateral drift constraints Eqs.(1c and 1d) must be formulated explicitly in terms of the design variables A_i. Using the principle of virtual work, the collective set of lateral top and interstory drift constraints can be expressed explicitly in terms of A_i:

$$d_j(A_i) = \sum_{i=1}^{N}\left(\frac{e_{ij}}{A_i} + e'_{ij}\right) \le d_j^U \qquad (j = 1, 2, .., M) \tag{2}$$

where e_{ij} and e'_{ij} are the respective virtual strain energy coefficient and its correction factor of the ith steel element of the structure associated with the jth drift constraint (Chan, 1997). Once a finite element analysis is carried out for a randomly selected child design with a given topology under the actual and virtual loading conditions, the internal element forces and moments are obtained and the element virtual stain energy coefficients are then readily calculated.

Upon establishing the design constraints into explicit functions, the constrained optimization problem can then be transformed into an unconstrained Lagrangian function which involves the objective function Eq.(1a) and the explicit drift constraints Eq.(2) associated with corresponding Lagrangian multipliers. Based on the stationary conditions derived from the Lagrangian function, the following recursive linear relations can be used to resize the active sizing variables A_i (Chan, 1997) :

$$A_i^{v+1} = A_i^{v} \cdot \left\{1 + \frac{1}{\eta}\left(\sum_{j=1}^{M}\frac{\lambda_j e_{ij}}{w_i A_i^2} - 1\right)_v\right\} \tag{3}$$

where λ_j denotes the Lagrangian multiplier for the corresponding j^{th} drift constraint, v represents the current iteration number; and η is a relaxation parameter. During the recursive resizing iteration process, any element found to reach its size bounds is deemed an inactive element having its size set at its corresponding size limit. Before Eq. (3) can be used to resize A_i, the Lagrangian multipliers λ_j must first be determined. Considering the sensitivity of the drift constraints due to the changes in the design variables, one can derive a set of M simultaneous equations to solve for M number of λ_j. Having the current design variables A_i^v, the corresponding λ_j^v values are readily determined by solving the simultaneous equations. Having the current values of λ_j^v, the new set of design variables A_i^{v+1} can then be obtained by the respective recursive relations Eq. (3). Therefore, the recursive applications of the simultaneous equations to find the λ_j^v and the resizing formula Eq. (3) to find the design variables constitute the OC algorithm (Chan, 1997). By successively applying the recursive OC algorithm until convergence, a local optimal solution for the design optimization problem is then found. The local optimum with improved element sizes will then complete for survival with all parent designs and other child designs generated from crossover and mutation in the selection process.

3.3 Overall design procedure

Fig. 2 shows the flowchart of the hybrid OC-GA method. The framework of a simple GA is adopted and an OC operator is added after the crossover and mutation operations. An initial population of designs is randomly generated to commence the design process. Structural analysis using SAP2000 (CSI, 2001) is then carried out on each individual design to determine the response performances. Once internal element forces and the drift responses of each individual design within a generation are found, fitness evaluations according to the design problem formulation Eq. (1) are conducted. In the fitness evaluation process, a penalty-based modified objective function is used. Any constraint violation found will reduce the fitness of a design and is reflected by a penalty function which is imposed on the objective function. The selection operator will determine survival candidate designs from the current generation and pass them to the next generation. The tournament selection with elitism mechanism is implemented in this hybrid method. If the convergence criteria are not satisfied, the genetic crossover and mutation operators together with the OC operation will then be applied to produce new designs. The hybrid OC-GA process can be repeated until the optimum solution is found.

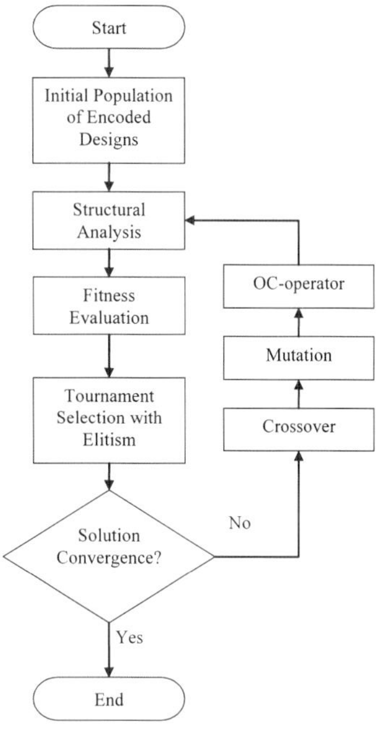

Figure 2. Flowchart of the hybrid OC-GA method

4. Illustrative examples

Simultaneous topological and element sizing optimization of a braced tall steel frameworks subject to lateral drift constraints is considered in two examples. Cross diagonal bracing elements are allowed to be added to the central bay of a 40-story, 3-bay planar building frame as shown in Fig. 3. The effectiveness of the hybrid OC-GA method on performing the optimal form searching as well as element sizing for the building framework was assessed by varying the different values of the probability of the OC operator and population sizes. The results produced by the hybrid method are compared with that generated by the GA method alone.

4.1 Example 1: Continuous addition of diagonal braces to a 40-story 3-bay framework

In this example, cross diagonals are assumed to hide from view behind the central lift core and are added only to the central bay of the 40-story 3-bay planar frame. These diagonal braces are added continuously from the ground level up to a specific story level denoted by a topological variable. The topological variable determines a story level which separates the frameworks along the height into two parts: stories below which are braced in the central bay; and stories above which use an unbraced rigid frame composed of only beams and columns. The optimal stopping level of the central bracing and optimal element sizes are sought by the hybrid OC-GA so as to minimize the total cost of the structure. The optimization involves searching for the most cost effective structural framework ranging from a pure rigid frame to a fully braced frame. Wind loads, as shown in Fig. 3, are derived based on the Hong Kong Wind Code (1983) using a general terrain wind profile. No gravity loading is applied to this framework example. Element strength design constraints and second-order $P-\Delta$ effects are not considered.

For the example framework having the cross diagonal braces as the topological and sizing variables and cross sectional sizes of the beams and columns as element sizing variables, the minimum cost function can be stated as follows:

$$W = \sum_{i=1}^{N_{column}+N_{beam}} \left(w_i \cdot A_i\right) + \sum_{j=1}^{N_{brace}} \left(t_j \cdot w_j \cdot A_j\right) \quad \text{where } t_j \begin{cases} = 0 & if\ s_{t_j} > T \\ = 1 & if\ s_{t_j} \le T \end{cases} \tag{4}$$

in which T defines the stopping level of the diagonal braces. The variable t_j denotes the on/off status of the jth pair of diagonal cross braces at j^{th} story. For the story levels above story T, the diagonal cross braces of these levels are removed from the central bay. However, for the story levels below or equal to story T, pairs of diagonal braces are added to the central bay of these levels and their associated costs are included in the cost function of the structure as $t_j = 1$. It should be noted that all beams are rigidly connected to columns such that the stability of the framework is always maintained whenever any diagonal brace is removed from the structure.

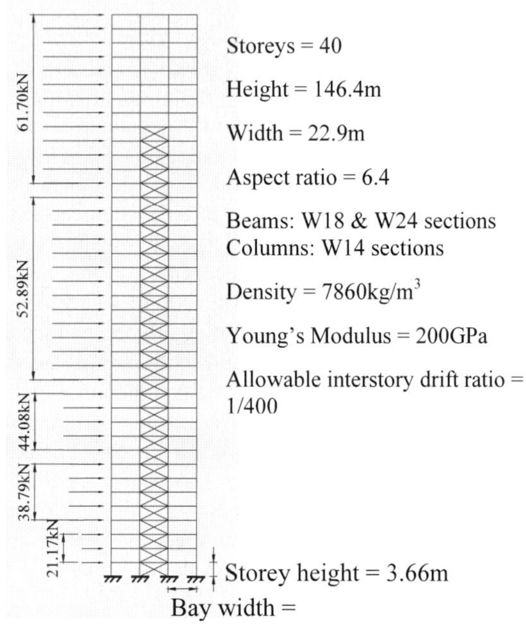

Figure 3. A 40-story 3-bay building frame under applied wind loads

All columns and diagonal braces are selected from among 36 discrete sections over the range of American AISC W14X22 to W14X730; beams are limited to W18 and W24 shapes selected from among 47 discrete sections over the range of W18X35 to W24X492. To account for symmetry and reversal of wind loads, exterior columns and beams are grouped together over two adjacent stories, as were interior ones. Diagonal braces are also grouped similarly to have the same size once over every adjacent two stories. The numbers of sizing variables for beams, columns and braces are 40, 40 and 20 respectively. Together with the topological variable T, the structure has a total of 101 design variables.

Two major parameters affecting the performance of the hybrid OC-GA, the population size and the probability of OC-operator, are studied in this example. Fixed population sizes of 10, 25, and 50 are used with a value of p_{oc} ranging from 0.0 (i.e., a pure GA without any OC operation) to 1.0 (i.e. all child designs undertake the OC operation).

Integer representation is adopted. Uniform crossover is applied with a probability of 80% such that 8 out of 10 of the parent designs are chosen to produce offspring designs. The mutation rates for the topological and element sizing variables are 20% and 5%, respectively. The reason for using a higher mutation rate for the topological variables is to increase the exploration of new creation of different forms of the braced frame. A quadratic penalty function is used for all design constraints. Binary tournament selection with elitism is employed in this example. The OC-GA algorithm is set to stop either when the maximum generation reaches 200 or when the same best-fit design is found for 20 consecutive generations. The optimization is carried out using a Pentium 4 3.0GHz computer with 512 MB memory.

Since the hybrid OC-GA is a stochastic algorithm, five independent runs are conducted with randomly generated initial designs. The average final structure weight produced by the hybrid OC-GA method for three population sizes is plotted with the different values of p_{oc} as shown in Fig. 4. For all runs wherein the probability of OC-operator is non-zero, the average final structure weights are found to be significantly over 20% less than those obtained by the pure GA (where p_{oc}=0).

With the use of the local OC operator, most of the final least-weight designs are found to satisfy almost all specified drift constraints, with a slight violation of less than 1% in the drift constraints being found in only a few final designs. For the case of the pure GA runs (where p_{oc} =0), over half of the final designs are found to be infeasible with a maximum violation of 10% in lateral drift constraints. Based on the results obtained for this example, the hybrid OC-GA method is able to produce more superior designs than the pure GA method.

Considering the results of all OC-GA runs using a value of p_{oc} ranging from 0.05 to 1.0, the average structural weights are fairly uniform among themselves with a small variation within 5% of the smallest value as shown in Figure 4,. This implies that under the current settings of the hybrid OC-GA method for this problem, the performance of the OC-GA algorithm was quite insensitive to the value of p_{oc}. When the population size is 10, the variation of the average structure weight against the p_{oc} is more fluctuating as compared with that of the population sizes of 25 and 50. It is evident that when a small population size is used, the likelihood of resulting in premature convergence to a local optimum becomes higher.

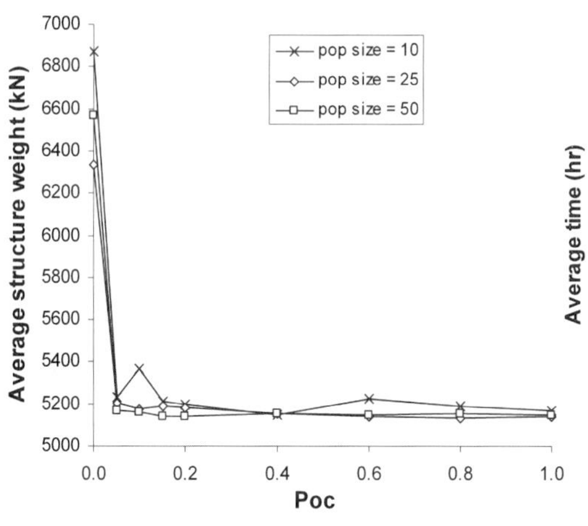

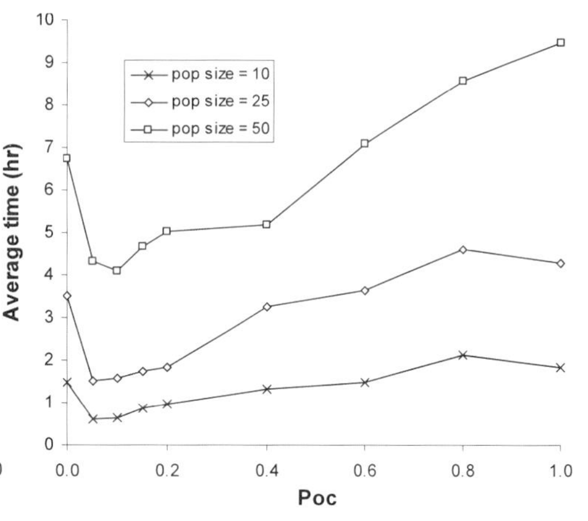

Figure 4. Effect of p_{oc} on average structure weight **Figure 5.** Effect of p_{oc} on average computer time

Since the values of the best-fit designs generated by the hybrid OC-GA method with different values of p_{oc} are similar, the computational time required for convergence becomes a crucial factor to determine the most appropriate value of p_{oc} to be applied for this example problem. Figure 5 shows the average computation time used for the runs with different values of p_{oc}. Apparently the least computation time required is found when the value of p_{oc} is around 0.05-0.1. It is evident that the use of a relatively small value of p_{oc} can cause a significant reduction in the computational effort required to produce a reasonable final design using the hybrid OC-GA method. However when p_{oc} is larger than 0.1, a gradual increase in the computational time is observed with an increasing p_{oc} as shown in Fig. 5. In general, experience indicates that the best value of p_{oc} for topological and element sizing optimization of building frameworks is found to be about 0.1, meaning that only 10% of the offspring designs are needed to undertake the local OC sizing optimization in order to achieve the best-fit designs by the OCGA method with the least computational effort.

4.2 Example 2: Design of a 40-story 3-bay framework allowing random addition or removal of diagonal bracings

The effectiveness of the hybrid OC-GA method is further investigated in this more complex example 2 design problem in which pairs of diagonal cross bracing members are randomly added to or removed from the central bay of any two adjacent floor levels of the same 40-story, 3-bay steel framework used in example 1. In addition to 80 element sizing design variables for columns and beams, 40 grouped topological and element sizing variables for braces were introduced to form a design problem having a total of 120 design variables. This design problem involves searching an optimal structure from 2^{20} possible topologies, where 2 represents the on/off topology choices and 20 represents the number of grouped stories. All optimization parameters used for this example are kept the same as that of example 1. Four different population sizes of 10, 25, 50 and 100 are considered with the use of different values of p_{oc} ranging from 0.05 to 1.0 for this example.

Fig. 6 shows the effect of the p_{oc} on the quality of the final designs obtained by the OC-GA method. The structure weights of all final designs of four different population sizes generated by the hybrid OC-GA method are given as point marks; the average values of these runs of each population size are joined by lines as given in Fig. 6. Similar to the first example, much superior designs are generally found by the hybrid OCGA when compared to that of the pure GA (where $p_{oc} = 0$). Regarding lateral drift performance of these runs, almost all final designs of totally 140 OC-GA runs with various values of p_{oc} are found feasible with only a few designs within the smallest population of 10 having slight 2% violation in drift. On the other hand, all the final designs generated by the pure GA (where $p_{oc}=0$) are found to be infeasible with some significant violations exceeding 100% in the specified lateral drift constraints.

As shown in Fig. 6, the larger the population size is used, the better the average results of the optimum designs are generally produced. However, the benefit of using a larger population size is apparently not significant particularly when $p_{oc} > 0.2$. It should be noted that the OC-GA method is able to produce steadily optimum designs with smaller spread of scattered final designs for the larger population sizes of 50 and 100. When the population size is relatively small (10 or 25), a larger spread of scattering of final designs is found as $p_{oc} < 0.2$. To improve the robustness and quality of the optimum designs, a larger value of p_{oc} (greater than 0.2) is needed to be applied for smaller population sizes of 10 and 25. When the population size is comparable to the similar order of the number of design variables (i.e. 50 or 100 for this example with 120 design variables), small values of p_{oc}, in the range of 0.05 to 0.2, is generally found to produce good quality designs.

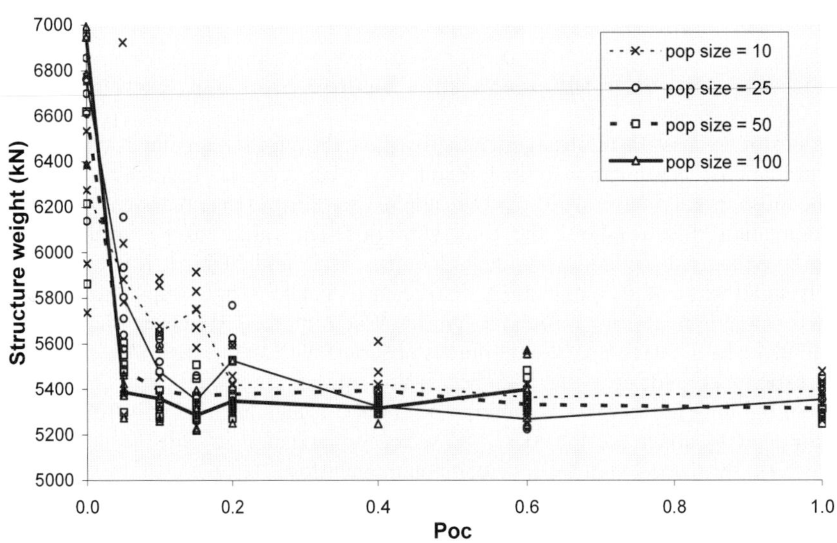

Figure 6. Effect of p_{oc} on structural weight

The scattering of the final design of all four population sizes diminishes with increasing p_{oc} values. This suggests that more frequent application of the OC-operator improves the quality of final designs and produce equally good results in terms of the objective function. Although much heavier designs are generally found by the pure GA method (where $p_{oc} = 0$), the final designs generated by GA are found all infeasible with significant drift violations.

For all the OC-GA runs, the design optimization process is terminated when the maximum generation reaches 200 or when almost the same best-fit design is maintained for 20 consecutive generations. The computational time required for solution convergence of each run is given in Fig. 7. For a given value of p_{oc}, the average computer time required is generally found to be linearly proportional to the population size. In other words, the time required for a population size of 100 is more or less two times that of a population size of 50. For all the OC-GA runs of a given population size, the average computer time required increases generally with an increasing value of p_{oc}. In general, the larger the population size, the smaller the value of p_{oc} should be adopted in terms of computational efforts. For this example, a relatively small value of p_{oc} ranging from 0.05 to 0.2 is found adequately sufficient to produce the best-fit designs with the least computational effort.

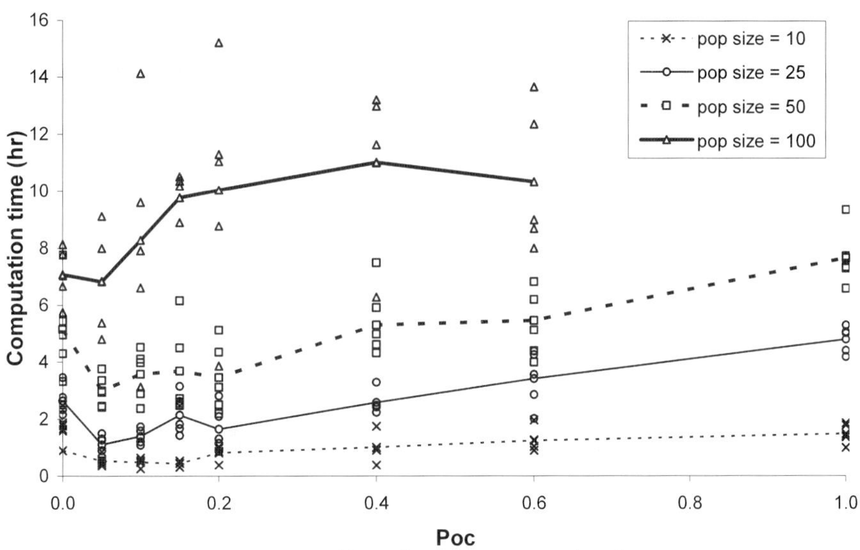

Figure 7. Effect of p_{oc} on computation time

5. Conclusion

In this paper, a novel hybrid OC-GA method incorporating a local search OC algorithm is presented for simultaneous topological and element sizing design optimization of tall steel building frameworks. The hybridization involves the exploration of topologies by the GA method and the refinement of element sizes by the local search OC algorithm. Two 40-story, 3-bay framework examples were tested to demonstrate the effectiveness of the hybrid OC-GA method on generating more superior optimal designs than conventional simple GAs. The incorporation of the OC operator into the GA process has remarkably improved the quality of the best-fit designs in terms of structural material consumption and computational efficiency. Results of the two examples indicate that robust and rapid solution convergence can be found by the hybrid OC-GA method. The use of a relatively small value of the probability of OC operator ranging from 0.05 to 0.2 is generally found to produce good quality best-fit designs with the least computational effort for a population size with a similar order of the number of design variables. The hybrid OC-GA method promises to become a useful tool for optimizing both the topology and element sizes of large-scale practical tall building structures.

Acknowledgement

The work described in this paper was substantially supported by a grant from the Research Grants Council of the Hong Kong Special Administrative Region, China (Project No. HKUST6302/04E).

References

Building Department of HKSAR. (1983).Code of practice on wind effects, Hong Kong.

Chan C.-M. (1997). How to optimize tall steel building frameworks, In: Guide to Structural Optimization. ASCE Manuals and Reports on Engineering Practices No.90, American Society of Civil Engineers, 165-195.

Chan, C.-M. (2001). Optimal Lateral Stiffness Design of Tall Buildings of Mixed Steel and Concrete Construction, Journal of Structural Design of Tall Buildings, 10(3), 155-177.

Chan, C.-M., Liu, P. and Wong, K.-M. (2003). Optimization of large scale tall buildings using hybrid genetic algorithms, Proc. 5th World Congress on Structural and Multidisciplinary Optim., Lido di Jesolo, Italy, May, 2003.

CSI (2001) SAP2000 (version 7.40) Analysis Reference, Computers and Structures, Inc., Berkeley, California.

Espinoza F. P., Minsker B. S., Goldberg D. E. (2005) Adaptive hybrid genetic algorithm for groundwater remediation design, Journal of Water Resources and Planning Management, Vol.131, 14-24.

Fawaz Z., Xu Y.G. & Behdinan (2005). 'Hybrid evolutionary algorithm and application to structural optimization', Structural and Multidisciplinary Optimization, Vol.30, 219-226.

Gen M., Cheng R. (1997) Genetic Algorithms and Engineering Design, New York: John Wiley & Sons, Inc.

Holland, J. H. (1975). Adaptation in Natural and Artificial Systems. Ann Arbor: University of Michigan Press.

Kirsch U. (1993) Structural Optimization: Fundamentals and Applications, Springer-Verlag, Berlin Heidelberg New York.

Sakamoto.J., and Oda, J. (1993). Technique of optimal layout design for truss structures using genetic algorithm, In Structures, Structural Dynamics, and Materials Conference, La Jolla, CA, AIAA paper #93-1582.

Yeh I.C.(1999). Hybrid genetic algorithms for optimization of truss structures ,Computer-Aided Civil and Infrastructure Engineering, Vol.14, No. 3, 199-206.

FUNDAMENTAL CHARACTERISTICS OF NEW HIGH MODULUS CFRP MATERIALS FOR STRENGTHENING STEEL BRIDGES AND STRUCTURES

Mina Dawood, Emmett Sumner and Sami Rizkalla
Department of Civil, Construction, and Environmental Engineering. North Carolina State University, USA, E-mail: sami_rizkalla@ncsu.edu

Abstract

Due to corrosion and the continuous demand to increase traffic loads, there is a great need for an effective, cost-efficient system which can be used for the repair and strengthening of steel highway bridge girders. Recently, research has been conducted to investigate the use of carbon fiber reinforced polymer (CFRP) materials to address this need. This paper describes the details of an experimental program which was conducted to investigate the fundamental behavior of steel-concrete composite bridge girders strengthened with new high modulus CFRP (HM CFRP) materials. The behavior of the beams under overloading conditions and fatigue loading conditions was studied as well as the possible presence of a shear-lag effect between the steel beam and the CFRP strengthening. A series of proposed flexural design guidelines are presented which can be used to establish the allowable live-load increase for a strengthened beam and to design the required HM CFRP strengthening.

Introduction

The use of fiber reinforced polymer (FRP) materials for the repair and strengthening of concrete structures has gained widespread acceptance. Due to the success of this technique, several researchers have investigated the use of externally bonded CFRP materials for the repair and retrofit of steel bridges and structures. A number of different approaches have been investigated to assess the effectiveness of using CFRP materials for the retrofit of steel bridge members including repair of overloaded girders (Sen et al., 2001), repair of naturally deteriorated girders (Mertz and Gillespie, 1996), strengthening of undamaged girders (Tavakkolizadeh and Saadatmanesh, 2003c) and repair of girders with simulated corrosion damage (Al-Saidy et al., 2004). Other research has been conducted to study the fatigue durability of CFRP strengthening systems (Miller et al., 2001).

Early research focused on the use of conventional CFRP materials to repair steel-concrete composite bridge girders which were damaged due to severe overloading conditions (Sen et al., 2001). The CFRP strengthening system helped to increase the yield load and post-elastic stiffness of the beams.

Researchers have also investigated the use of CFRP materials to repair naturally corroded bridge girders (Mertz and Gillespie, 1996). Installation of the CFRP materials restored the elastic stiffness and moment capacity of the girders to levels comparable to those of the undamaged girders.

In another study, three undamaged steel-concrete composite beams were strengthened with one, three and five layers of CFRP strips respectively (Tavakkolizadeh and Saadatmanesh, 2003c). The CFRP materials also increased the ultimate capacity of the strengthened beams by up to 76 percent, however, the increase of the elastic stiffness was minimal. In a companion study, the tension flange of three other steel-concrete composite beams were notched with a 1.3 mm wide notch at midspan to simulate 25, 50 and 100 percent loss of the tension flange due to corrosion (Tavakkolizadeh and Saadatmanesh, 2003b). The repair restored the elastic stiffness and ultimate capacity of the girders to levels comparable to the undamaged state and helped to reduce the measured residual deflections due to overloading.

Other researchers have simulated corrosion damage by removing a uniform portion of the tension flange along the entire length of the girders (Al-Saidy et al., 2004). The repair technique was capable of restoring the lost strength of the damaged beams to levels higher than those of the unstrengthened girders. However, only 50 percent of the lost stiffness of the beams was recovered.

M. Pandey et al. (eds), Advances in Engineering Structures, Mechanics & Construction, 215–226.

Investigations on the fatigue durability of steel beams strengthened with CFRP materials have been limited. Tavakkolizadeh and Saadatmanesh (2003a) demonstrated that externally bonded CFRP patches can be used to reduce crack propagation rates and increase the fatigue life of cracked steel members. Prestressed CFRP patches can also be installed to help promote crack-closure effects and further extend the fatigue life of cracked steel members (Bassetti et al., 2000). The fatigue durability of naturally corroded steel bridge girders which were repaired with CFRP materials has also been used investigated (Miller et al., 2001).

The majority of the previous research has focused on the use of conventional modulus CFRP materials for the repair of steel bridge members. While substantial strength increases have been achieved, typically large amounts of strengthening were required to achieve substantial increases of the elastic stiffness of the beams. This is due to the relatively low modulus of elasticity of the CFRP as compared to steel and also possibly due to the presence of shear-lag effects between the steel beam and the CFRP strengthening.

Recently, a strengthening system has been developed at North Carolina State University which uses new high modulus CFRP (HM CFRP) materials. These materials have a modulus of elasticity approximately two times greater than that of steel. The effectiveness of using HM CFRP materials to repair steel bridge girders was demonstrated by testing three large-scale steel-concrete composite beams strengthened with different configurations of HM CFRP materials (Schnerch, 2005). The elastic stiffness and ultimate moment capacity of the beams were increased by up to 36 percent and 45 percent respectively. The testing demonstrated that prestressing the HM CFRP strips prior to installation on the steel beam increased the efficiency of utilization of the CFRP and only half the amount of strengthening was required to achieve a comparable increase of stiffness as a beam strengthened with unstressed CFRP laminates.

This paper presents the details and relevant findings of an experimental investigation which was conducted to study the behavior of the strengthened beams under overloading conditions and fatigue loading conditions. Also, the possible presence of a shear-lag effect between the steel and the CFRP, which can limit the effectiveness of the strengthening system, was investigated in detail for a number of different loading conditions. Additional details about the research program are available in Dawood (2005). A series of guidelines are also proposed which can be used by practitioners to design the required HM CFRP strengthening for a given steel-concrete composite beam.

HM CFRP Strengthening System

The carbon fibers used in this study were the DIALEAD K63712 high modulus pitch based carbon fibers produced by Mitsubishi Chemical Functional Products Inc. The fibers were pultruded into 4mm thick, 100mm wide laminates by Epsilon Composite Inc., a French pultrusion company. The modulus of elasticity and ultimate strain of the strips which were reported by the manufacturer are 460 GPa and 0.00334 respectively. The laminates have a fiber volume fraction of 70 percent and were fabricated with a glass fiber peel ply on both faces to minimize the need for surface preparation of the composite material. The strips were bonded to the tension flange of the steel beams using the Spabond 345 two part epoxy adhesive with the fast hardener which is manufactured by SP Systems North America. This adhesive was selected from six adhesives in a previous phase of this research (Schnerch, 2005). The tension flange of the steel beam was grit blasted immediately prior to installation of the strengthening system to remove rust and mill scale from the surface. The surface was subsequently cleaned by air blowing and solvent wiping. After the CFRP strips were installed a wooden clamping system was applied for at least 12 hours until the adhesive had thoroughly set. The adhesive was allowed to cure for at least one week prior to testing.

Experimental Program

A total of seven steel-concrete composite beams were tested to study the effectiveness of the HM CFRP strengthening system. In the first phase of the research three beams were tested to study the behavior of the strengthening system under overloading conditions. The second phase consisted of three beams which were tested under fatigue loading conditions to investigate the durability of the strengthening system. The effect of the bonding procedure on the fatigue behavior of the beams was also examined. In the third phase, the seventh beam was tested to study the possible presence of shear-lag between the steel and the CFRP. The strengthened beams that were tested in the first and second phase of the research were also used to investigate the possible presence of a shear-lag phenomenon. The beams tested in all three phases of the experimental program consisted of scaled steel-concrete composite beams which are typical of most highway bridge construction. The typical cross-section of the tested beams is shown in Figure 1.

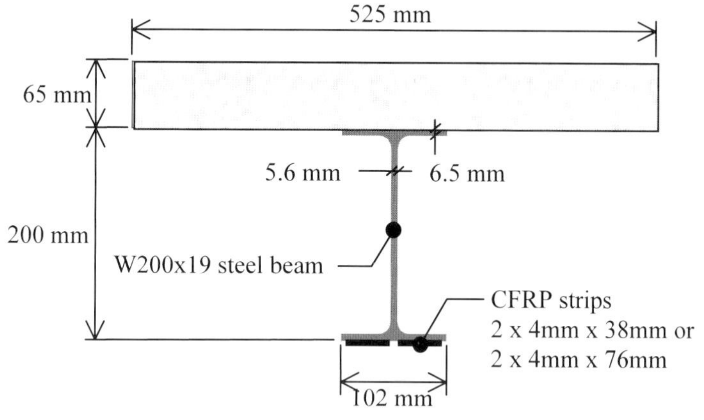

Figure 1: Cross-section of a typical test beam

The beams were strengthened with different levels of HM CFRP materials and tested in a four point bending configuration as shown in Figure 2.

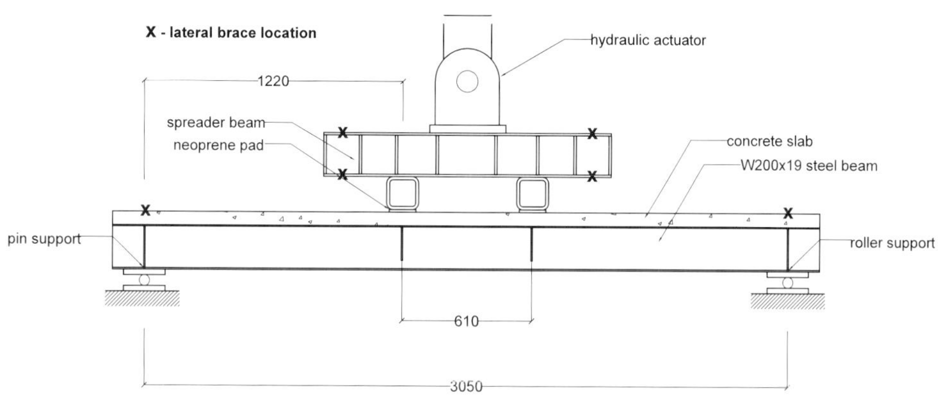

Figure 2: Beam test setup

The test matrix for the three phases of the experimental program is presented in Table 1. In the first phase, the behavior of the beams under overloading conditions was investigated. Two of the test beams were strengthened with two different levels of CFRP while the third remained unstrengthened to serve as a control beam for the overloading study. All three of the beams were unloaded and reloaded at various load levels to simulate the effect of severe overloading conditions.

The second phase was designed to study the fatigue durability of the strengthening system. Two different beams were strengthened with the same amount of CFRP materials, however, using different bonding techniques. The modified bonding technique involved increasing the final thickness of the cured adhesive and additionally including the use of a silane adhesion promoter. The third beam remained unstrengthened as a control beam for the fatigue study. All three beams were subjected to three million fatigue loading cycles with a frequency of 3 Hz. The minimum load in the loading cycle was selected as 30 percent of the calculated yield load of the unstrengthened beams to simulate the effect of the sustained dead-load for a typical bridge structure. The maximum load for the unstrengthened beam was selected as 60 percent of the calculated yield load to simulate the combined effect of dead-load and live-load. The maximum load for the two strengthened beams was selected as 60 percent of the calculated increased yield load of the strengthened beams to simulate the effect of a 20 percent increase of the allowable live-load level for a strengthened bridge.

The final beam was tested under monotonic loading conditions to investigate the possible presence of a shear-lag effect in the absence of more harsh loading conditions. The four strengthened beams which were tested in the first and second phases of the experimental program were also used to study the potential shear-lag effects.

Table 1: Test matrix for the three phases of the experimental program

Beam ID	**Reinforcement Ratio, ρ^***	**Adhesive Thickness, t_a**	**Concrete Strength, f_c'**	**Loading**
Overloading				
ST-CONT	0 percent	N/A	44 MPa	unload/reload
OVL-1	4.3 percent	0.1 mm	44 MPa	unload/reload
OVL-2	8.6 percent	0.1 mm	44 MPa	unload/reload
Fatigue				
FAT-CONT	0 percent	N/A	34 MPa	fatigue: P_{min}=50 kN, ΔP=50 kN
FAT-1	4.3 percent	0.1 mm	34 MPa	fatigue: P_{min}=50 kN, ΔP=60 kN
FAT-1b	4.3 percent	1.0 mm**	58 MPa	fatigue: P_{min}=50 kN, ΔP=60 kN
Shear-lag				
SHL	8.6 percent	0.1 mm	44 MPa	monotonic

*defined as the ratio of the cross-sectional area of the CFRP strengthening, accounting for the fiber volume fraction to the cross-sectional area of the steel beam
**included the use of a silane adhesion promoter

The tensile yield strength and modulus of elasticity of the steel beams were determined by coupon tests according to ASTM A370-02 as 380 MPa and 200,000 MPa respectively. The compressive strength of the concrete used for the concrete deck slabs for the seven test beams was determined from cylinder tests after 28 days in accordance with ASTM C39-03. The measured concrete cylinder strengths are presented in Table 1.

To accurately represent the actual behavior of a typical strengthened highway bridge, the two strengthened beams that were tested in the fatigue study were subjected to a sustained simulated dead-load prior to installation of the HM CFRP strengthening system. The simulated dead-load was applied using an independent loading apparatus as shown in Figure 3. Prior to installation of the CFRP, the simulated dead-load of 50 kN was applied by tightening nuts on a series of threaded rods as shown in the figure. The simulated dead-load was sustained on the beams while the CFRP strips were installed and during the curing process of the adhesive. After the adhesive cured, the load was transferred from the independent dead-load apparatus to the hydraulic actuator and the fatigue loading program was commenced.

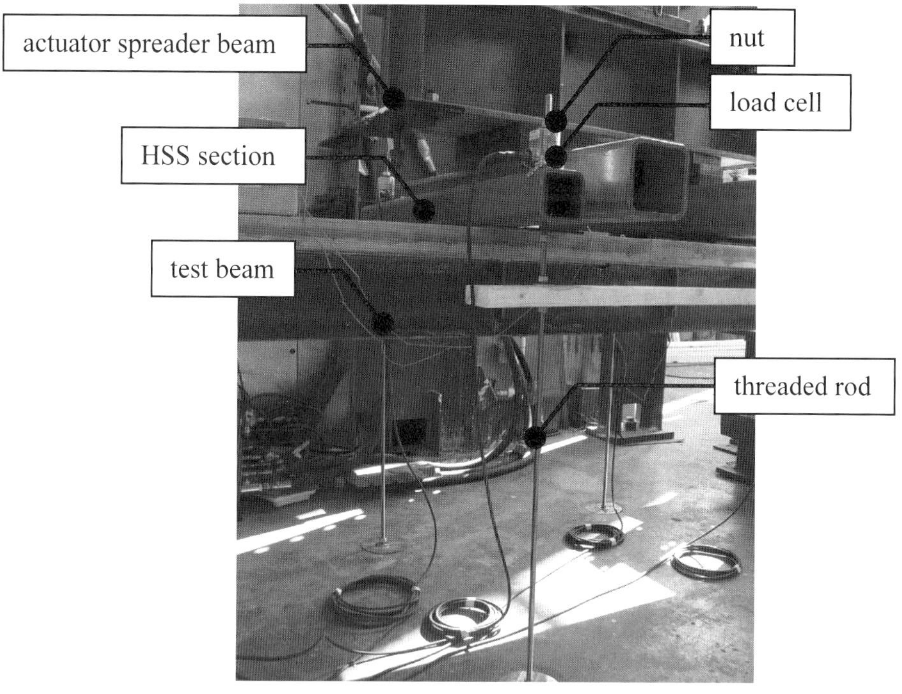

Figure 3: Independent dead-load apparatus

All of the test beams were instrumented to measure deflections at midspan and at the supports as well as to measure strains at various locations on the midspan cross-section of the beam. The measured strains were used to construct the strain profiles for the strengthened beams to investigate the possible presence of a shear-lag effect between the steel and the CFRP strengthening materials.

Experimental Results

This section presents the results of each of the three phases of the experimental program and discusses the relevant research findings. The three phases of the experimental program are presented separately in the following sections.

Findings of the Overloading Study

The load-deflection relationships of the three beams that were tested in the overloading study are presented in Figure 4. Beam ST-CONT remained unstrengthened to serve as a control beam for the overloading study while the remaining two beams were strengthened with different reinforcement ratios of CFRP. All three beams were unloaded and reloaded at various loading stages to simulate the effect of overloading conditions.

The load-deflection behavior of the three beams was essentially linear up to yielding of the steel. Unloading and reloading was essentially linear and exhibited minimal hysteresis. Prior to yielding of the steel, all three beams exhibited minimal residual deflections upon unloading. However, after yielding of the steel, the unstrengthened beam exhibited a significant increase of the measured residual deflection, as shown in Figure 4(a), while the two strengthened beams continued to exhibit minimal residual deflections up to rupture of the CFRP. After rupture of the CFRP occurred the behavior of the strengthened beams followed a similar trend to that of the unstrengthened beam.

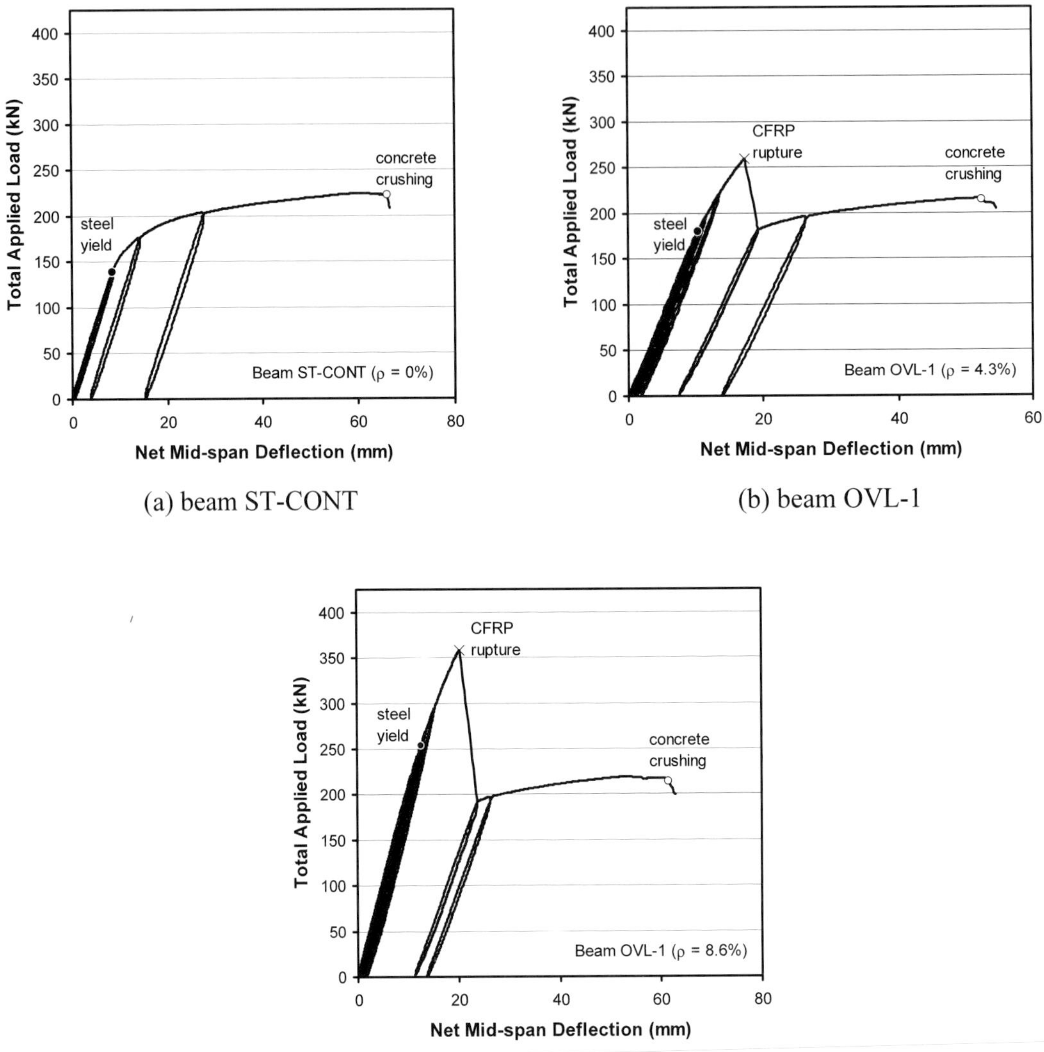

(c) beam OVL-2

Figure 4: Load-deflection behavior of the three overloading test beams

Table 2: Comparison of the overloading beams

Beam ID	Reinforcement Ratio	Stiffness Increase	Yield Load	Rupture Load	Crushing Load
ST-CONT	0 %	N/A	137 kN	N/A	222 kN
OVL-1	4.3 %	27 %	181 kN	259 kN	216 kN
OVL-2	8.6 %	46 %	253 kN	358 kN	216 kN

Table 2 presents the elastic stiffness increase, the yield load, the rupture load and the crushing load for the three beams tested in the overloading study.

The ultimate capacity of the unstrengthened beam, ST-CONT, was governed by crushing of the concrete while the ultimate capacity for the two strengthened beams, OVL-1 and OVL-2 was governed by rupture of the CFRP. The elastic stiffness, yield load and ultimate capacity of the beams were increased by 46 percent, 85 percent and 61 percent respectively using the higher reinforcement ratio. Inspection of Table 2 indicates that doubling the reinforcement ratio of the applied CFRP, from 4.3 percent to 8.6 percent, approximately doubled the elastic stiffness increase of the beams. Increasing the reinforcing ratio by two times also approximately tripled the increase of the measured yield load and ultimate capacity of the strengthened beams. This demonstrates that increasing the reinforcement ratio did not reduce the efficiency of utilization of the CFRP material. The improved performance of the strengthened beams indicates that it may be possible to increase the allowable live load level of a steel-concrete composite girder strengthened with HM CFRP materials. A proposed methodology for determining the allowable increase of the live load level is described later in this paper.

As mentioned previously, the presence of the HM CFRP materials helped to reduce the residual deflection of the strengthened beams due to the effect of overloading conditions. The residual deflections of the three beams that were tested in the overloading study were compared to evaluate the effectiveness of the HM CFRP strengthening system. Each of the three beams was unloaded at a load level of approximately 175 kN to simulate a severe overloading condition. The average measured strain at the steel tension flange of beams ST-CONT, OVL-1 and OVL-2 at the 175 kN load level were 2.5 ε_y, 1.0 ε_y and 0.6 ε_y respectively, where ε_y is the average yield strain of the steel determined from coupon tests. The beams were unloaded to a load level of 45 kN to simulate the sustained load acting on a structure, due to self-weight, after an overloading event. The plastic component of the residual deflection, after subtracting the initial measured elastic displacement of the beams at the 45 kN load level were compared. The measured residual deflection of beams OVL-1 and OVL-2 were respectively 5 times and 6.5 times lower than the measured residual deflection of beam ST-CONT. Consequently, under severe overloading conditions, an unstrengthened bridge girder may require repair or replacement while a strengthened bridge girder may remain serviceable.

Findings of the Fatigue Study

Three beams were tested in the fatigue study. Beam FAT-CONT remained unstrengthened to serve as a control beam for the fatigue study while beams FAT-1 and FAT-1b were strengthened with a reinforcement ratio of CFRP of 4.3 percent using two different bonding techniques. The two strengthened beams were tested with a 20 percent increase of the applied load range to simulate the effect of increasing the allowable live load for the strengthened beams. All three beams survived a three million-cycle fatigue loading course without exhibiting any indication of failure. Figures 5 (a) and (b) present the degradation of the stiffness and mean deflection respectively of the three beams throughout the three million-cycle loading course. Figure 5 presents the stiffnesses and mean deflections of the beams normalized with respect to the initial values at the beginning of the fatigue loading program.

All three beams exhibited a minimal degradation of the elastic stiffness of less than 5 percent throughout the three million fatigue loading cycles as shown in Figure 5(a). However, beam FAT-CONT exhibited a nearly 30 percent increase of the mean deflection due to the applied fatigue cycles as shown in Figure 5(b). This was likely due to the fatigue-creep behavior of the concrete deck slab. Both of the strengthened beams exhibited superior performance with an increase of only 10 percent in the measured mean deflection. The observed degradation of the two strengthened beams throughout the three million fatigue cycles was similar which indicates that the bonding technique did not affect the fatigue behavior of the strengthening system.

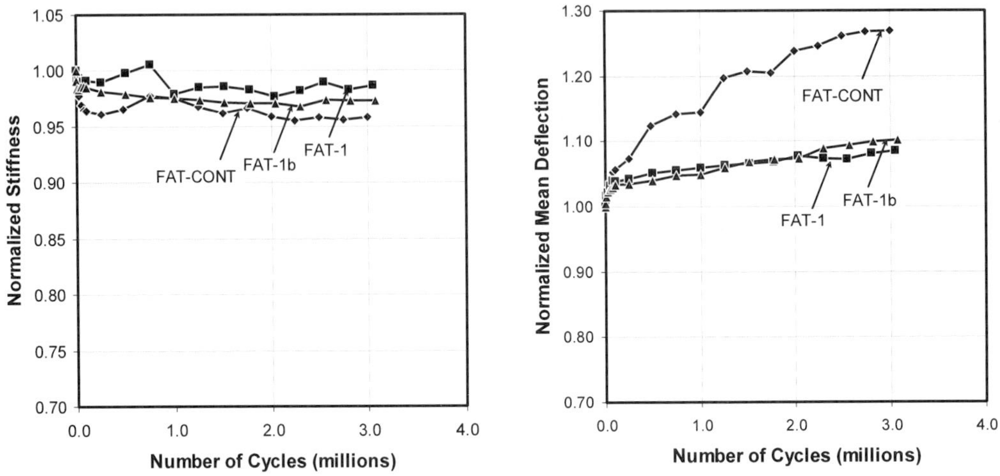

(a) stiffness (b) mean deflection

Figure 5: Degradation of (a) stiffness and (b) mean deflection for the fatigue beams

At the completion of the fatigue program, the three beams were loaded monotonically to failure. The load deflection behavior of the beams followed a similar trend to the observed load-deflection envelope of the three beams that were tested in the overloading study. The ultimate capacity of the two strengthened beams, FAT-1 and FAT-1b, was governed by rupture of the CFRP at a load of 250 kN. After rupture of the CFRP, the load deflection behavior of the two beams followed a similar trend to the load-deflection behavior of the unstrengthened beam, FAT-CONT. Crushing of the concrete for all three beams occurred at a measured load of between 200 kN and 215 kN. The findings of the fatigue study demonstrate the durability of the strengthening system when loaded at an increased live-load level. The allowable increase of the live load should be selected in accordance with the proposed design guidelines outlined later in this paper.

Findings of the Shear-lag Study

One additional beam, SHL, was tested to investigate the presence of a shear-lag effect between the steel beam and the CFRP strips under monotonic loading conditions. The beam was strengthened with a reinforcement-ratio of CFRP of 8.6 percent. The load-deflection behavior of beam SHL followed a similar trend to the load-deflection envelope of beam OVL-2, shown in Figure 4(c), which was strengthened with the same reinforcement ratio of CFRP.

The measured strain profiles at the midspan cross-section of beams SHL and OVL-2, immediately prior to rupture of the HM CFRP strips, are shown in Figure 6. The measured strain profiles for both beams indicate a slight discontinuity of the strain profile between the steel tension flange and the CFRP. However, the opposite sign of the discontinuity suggests that the behavior is not due to the presence of a shear lag effect. The measured discontinuity is likely due to the effect of residual stresses in the steel beam which formed during the manufacturing process or due to possible local instability or lateral movement of the tension flange of the steel beam. The presence of these effects was confirmed by independent strain measurements at different locations on the steel tension flange of the strengthened and unstrengthened test beams. The measured strain profiles for the remaining three strengthened test beams, OVL-1, FAT-1 and FAT-1b, were essentially linear and exhibited minimal discontinuities between the steel and the CFRP. The findings of the shear-lag study indicate that the effect of shear-lag is negligible and the plane sections remain plane assumption is appropriate for the analysis of the strengthened beams.

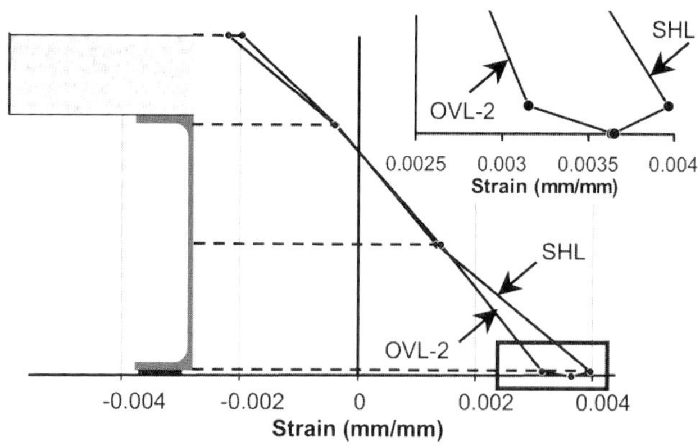

Figure 6: Strain profiles for beam SHL and OVL-2 (ρ = 8.6%)

Design Guidelines

Based on the findings of this research program a series of design guidelines have been developed which can be used by practitioners to design HM CFRP strengthening for steel-concrete composite beams (Schnerch et al., 2005). This section presents a summary of the proposed flexural design procedure including a discussion of the determination of the allowable increase of the live-load level for a strengthened beam.

The flexural analysis and design of a steel-concrete composite beam strengthened with HM CFRP materials are based on a non-linear moment-curvature analysis. The analysis satisfies the requirements of equilibrium and compatibility and neglects the effect of shear-lag between the steel and the CFRP. A non-linear material characteristic is used to represent the stress-strain behavior of the concrete and the steel while the CFRP is assumed to remain linear and elastic to failure.

The moment-curvature behavior of a given cross-section is determined based on the strain at the top level of the compression flange together with an assumed neutral axis depth. The cross-section is broken down into levels corresponding to the concrete deck, the longitudinal steel reinforcement of the concrete deck, the flanges and web of the steel beam, and the HM CFRP strips at the bottom of the cross-section. The strain, ε_x, at any distance, x, from the neutral axis of the section can be calculated using Equation (1) as,

$$\varepsilon_x = \frac{\varepsilon_c}{c} x \qquad \text{Equation (1)}$$

where ε_c is the strain at the top surface of the concrete deck and c is the assumed neutral axis depth. From the strain profile and the constitutive relationships of the materials, the corresponding stress profile for the beam can be established. The corresponding resultant forces for the different elements of the cross section can be calculated by integration of the stress profile using Equation (2),

$$F = b\int f(x)dx \qquad \text{Equation (2)}$$

where F is the calculated resultant force, b is the width of the element under consideration and f(x) is the constitutive relationship of the material. The assumed neutral axis depth is iterated until horizontal force equilibrium is satisfied. The corresponding moment can then be calculated by replacing the term f(x) in Equation (2) by the term x f(x) and re-evaluating the integrals. The curvature of the section can then be calculated from the strain at the top surface of the concrete deck, ε_c, and the appropriate neutral axis depth, c. This strain is subsequently increased to determine the next increment of curvature and the procedure is repeated. This procedure can be easily extended to predict the load deflection behavior of a beam with a given loading and support configuration by integration of the curvature profile using any commonly accepted method.

Based on the proposed moment-curvature analysis procedure, the allowable increase of live load for a steel-concrete composite beam strengthened with HM CFRP materials should be selected to satisfy three conditions. These three conditions are shown in Figure 7 with respect to the moment-curvature response of a typical steel-concrete composite beam section strengthened with HM CFRP materials. Due to the presence of the additional layer of HM CFRP material, the yield moment of the strengthened beam, $M_{Y,S}$, is greater than the yield moment of the unstrengthened beam, $M_{Y,US}$. This was verified by the findings of the overloading study. Based on the findings of the fatigue study, to maintain the fatigue life of the strengthened beam, the total applied moment acting on the strengthened section under service loading conditions, including the effect of dead load, M_D, and the increased live-load, M_L, should not exceed 60 percent of the increased yield moment of the strengthened section. To satisfy the strength limit state, the total factored moment based on the appropriate dead-load and live-load factors, α_D and α_L respectively, should not exceed the ultimate moment capacity of the strengthened section, $M_{U,S}$. Also, to ensure that the structure remains safe in the case of total loss of the strengthening system, the total applied moment, including the effect of dead-load and the increased live-load should not exceed the residual nominal moment capacity of the unstrengthened section, $M_{n,US}$.

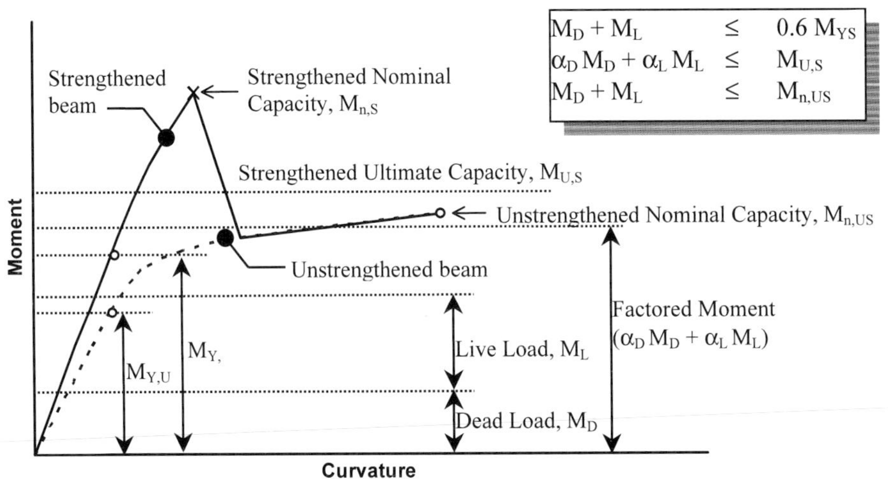

Figure 7: Load levels and moment-curvature behavior for a strengthened beam

While the nominal behavior of the member can be used to predict the behavior under service loading conditions, the design ultimate capacity should incorporate suitable reduction factors to ensure that the member remains safe. These reduction factors should account for the uncertainty of the HM CFRP material properties and should take into consideration the sudden, brittle failure which is typical of most HM CFRP strengthened steel-concrete composite beams. This guideline has adopted the approach outlined in ACI 440.2R-02 for the calculation of the ultimate capacity of concrete beams strengthened with externally bonded FRP materials. However, the approach has been modified to account for the inherent difference between the behavior of steel beams and concrete beams.

To account for the statistical uncertainty of the measured ultimate capacity of the HM CFRP materials, the mean strength of the CFRP reported by the manufacturer, $\bar{f}_{FRP,u}$, should not be used directly in calculating the ultimate capacity of the strengthened section. Rather, the average ultimate strength of the FRP should be reduced by 3 times the standard deviation, σ, as in Equation (3) (ACI 440.2R, 2002).

$$f^*_{FRP,u} = \bar{f}_{FRP,u} - 3\sigma \qquad \text{Equation (3)}$$

To account for possible environmental degradation of the CFRP materials throughout the lifetime of the strengthening, ACI 440.2R-02 recommends the use of an environmental reduction factor, C_E. For carbon fiber materials subjected to exterior exposure, which is typical for most bridge structures, a value of C_E of 0.85 should be used. Therefore, the design strength of the HM CFRP material can be calculated as

$$f_{FRP,u} = C_E f^*_{FRP,u} \qquad \text{Equation (4)}$$

The design ultimate strain of the CFRP material, $\varepsilon_{FRP,u}$ can be calculated by dividing the calculated design strength of the CFRP by the average elastic modulus, E_{FRP}, reported by the manufacturer.

The nominal moment capacity of the strengthened member, $M_{n,S}$, should be calculated using the proposed moment-curvature procedure and the design strength and ultimate strain of the CFRP. The nominal capacity of a steel-concrete composite beam strengthened with high modulus CFRP materials is typically governed by rupture of the CFRP materials. This type of failure occurs in a sudden, brittle manner without significant warning. To account for the brittle nature of failure, a strength reduction factor, ϕ, of 0.75 is recommended. This reduction factor is consistent with the reduction used in the AISC LRFD Specification (2001) for rupture type limit states. The design ultimate capacity of the strengthened beam, $M_{U,S}$ should be calculated as $\phi M_{n,S}$.

Conclusions

This paper presents the details of an experimental program which was conducted in three phases to investigate the fundamental behavior of steel-concrete composite beams strengthened using high modulus CFRP materials. Based on the findings of the first phase of the experimental program, it is evident that HM CFRP materials can be used to increase the elastic stiffness, yield load and ultimate capacity of steel-concrete composite beams which are typical of most highway bridge structures. Additionally, the presence of the CFRP helps to reduce the residual deflection due to overloading conditions which can help reduce or eliminate the need for future repair or replacement of the structure. The fatigue durability of the strengthening system was demonstrated in the second phase of the experimental program. Two beams were strengthened with a reinforcement ratio of CFRP of 4.3 percent and subjected to three million fatigue loading cycles with an increase of the simulated allowable live-load level of 20 percent as compared to an unstrengthened control beam. Both strengthened beams exhibited superior fatigue performance to the unstrengthened control beam with regards to the fatigue-creep behavior of the concrete deck. Further, the bonding technique did not appear to affect the fatigue behavior of the strengthening system. Based on the measured strain profile of the five strengthened beams which were investigated in the shear-lag study, the effect of shear-lag between the steel beam and the CFRP materials is minimal.

This paper also presents a proposed flexural design procedure, which is based on a non-linear moment-curvature analysis, which can be used to design the required HM CFRP strengthening for a steel-concrete composite beam. Based on the proposed guidelines, the allowable live load increase for a strengthened beam should satisfy three conditions. Particularly,

(i) $M_D + M_L \leq 0.6\, M_{YS}$
(ii) $\alpha_D M_D + \alpha_L M_L \leq M_{U,S}$
(iii) $M_D + M_L \leq M_{n,US}$

The findings of this research demonstrate that new externally bonded HM CFRP materials provide an effective, cost-efficient repair alternative for conventional steel-concrete highway bridge girders.

Acknowledgements

The authors would like to acknowledge the support provided by the National Science Foundation (NSF) Industry/University Cooperative Research Center (I/UCRC) for the Repair of Buildings and Bridges with Composites (RB^2C) and the support provided by Mitsubishi Chemical FP America Inc.

References

Al-Saidy, A.H., Klaiber, F.W. & Wipf, T.J. (2004). Repair of steel composite beams with carbon fiber-reinforced polymer plates. *Journal of Composites for Construction, 8 (2),* 163-172.

American Association of State Highway and Transportation Officials. (2002). *AASHTO LRFD bridge design specifications.* Washington, D.C.

American Concrete Institute. (2002). *ACI 440.2R-02 Guide for the design and construction of externally bonded FRP systems for strengthening concrete structures.*

American Institute of Steel Construction. (2001). *Manual of steel construction: load and resistance factor design. Third edition.*

Bassetti, A. Nussbaumer, A. & Hirt, M.A. (2000). Crack repair and fatigue life extension of riveted bridge members using composite materials. In Hosny, A.H. & Bakhoum, M.M. (Eds.), *Bridge engineering conference 2000 past achievements current practices future technologies* (pp. 227-237). Sharm El-Sheikh: Egyptian Society of Engineers.

Dawood, M. (2005). *Fundamental Behavior of Steel-Concrete Composite Beams Strengthened with High Modulus Carbon Fiber Reinforced Polymer (CFRP) Materials.* Master's Thesis, North Carolina State University, Raleigh, North Carolina.

Mertz, D.R. & Gillespie Jr., J. W. (1996). Rehabilitation of steel bridge girders through the application of advanced composite materials (Contract NCHRP-93-ID011). Washington, D.C.: Transportation Research Board.

Miller, T.C., Chajes, M.J., Mertz, D.R. & Hastings, J.N. (2001). Strengthening of a steel bridge girder using CFRP plates. *Journal of Bridge Engineering, 6 (6)*, 514-522.

Schnerch, D. (2005). *Strengthening of steel structures with high modulus carbon fiber reinforced polymer (CFRP) Materials.* Ph.D. dissertation, North Carolina State University, Raleigh, North Carolina.

Schnerch, D., Dawood, M. and Rizkalla, S. (2005). *Design Guidelines for the Use of HM Strips: Strengthening of Steel-Concrete Composite Bridges with High Modulus Carbon Fiber Reinforced Polymer (CFRP) Strips.* Available from http://www.ce.ncsu.edu/centers/rb2c/Events/RB2CJune05Mtg.htm.

Sen, R., Libby, L. & Mullins, G. (2001). Strengthening steel bridge sections using CFRP laminates. *Composites Part B: Engineering, 39,* 309-322.

Tavakkolizadeh, M. & Saadatmanesh, H. (2003a). Fatigue strength of steel girders reinforced with carbon fiber reinforced polymer patch. *Journal of Structural Engineering, 129 (2),* 186-196.

Tavakkolizadeh, M. & Saadatmanesh, H. (2003b). Repair of damaged steel-concrete composite girders using carbon fiber-reinforced polymer sheets. *Journal of Composites for Construction, 7 (4),* 311-322.

Tavakkolizadeh, M. & Saadatmanesh, H. (2003c). Strengthening of steel-concrete composite girders using carbon fiber reinforced polymer sheets. *Journal of Structural Engineering, 129 (1),* 30-40.

MODELING OF AXIALLY LOADED HSS SLENDER STEEL MEMBERS RETROFITTED WITH COMPOSITES

Amr Shaat, Waleed Safwat El Sayed Fahmy and Amir Fam
Department of Civil Engineering, Queen's University, Kingston, Ontario, Canada
E-mail: fam@civil.queens.ca

Abstract

This paper describes an analytical model developed to predict the behaviour of axially loaded slender members composed of steel hollow structural sections (HSS), retrofitted with carbon-fibre reinforced polymer (CFRP) composite sheets. A previous experimental study by the authors showed that gain in strength due to CFRP retrofitting was highly sensitive to the specimen's imperfection. As such, developing an analytical model was necessary to uncouple the effects of imperfection and number of CFRP layers. The model predicts the load versus axial and lateral displacements, and accounts for steel plasticity, the built-in through-thickness residual stresses, geometric non-linearity, including initial imperfection, and the contribution of CFRP sheets. The model was verified against results of the experimental program and showed reasonable agreement. The model was then used in a parametric study. The study demonstrated that retrofitting slender HSS columns using CFRP sheets increased both the axial strength and stiffness substantially.

Introduction

The progressive aging and deterioration, combined with the requirements for higher load capacity, results in an increase in the number of bridges which do not meet the current code standards. Conventional retrofitting techniques for steel members, in general, involve bolting or welding additional steel plates. These techniques have a number of shortcomings, including the added self weight of steel plates, the installation time, and the need for an elaborate and expensive shoring system. Fiber-reinforced polymers (FRP) are rapidly gaining acceptance as an effective material for retrofitting a wide range of structures, particularly due to their excellent corrosion resistance and fatigue properties (Hollaway and Cadei, 2002 and Shaat et al., 2004).

The majority of applications of these materials, so far, have been flexural members such as beams and bridge girders where bonding of FRP on the tension side is very effective in upgrading the structural performance of such members. Nevertheless, the failure mode of axially loaded steel members such as columns or truss members with medium to high slenderness ratios is generally governed by overall buckling, which is essentially a flexural problem. Therefore, bonding FRP sheets or strips of adequate stiffness in the longitudinal direction of slender members could be quite effective.

An experimental study was performed (Shaat and Fam, 2006) to investigate the effect of adhesively bonded carbon-FRP (CFRP) sheets on the behaviour of axially loaded HSS columns. The column's cross sectional shape is shown in Fig. 1(a) and the column length was 2380 mm, such that slenderness ratio was 68. In slender columns, where overall buckling governs, it is anticipated that the ultra-high modulus CFRP sheets would contribute to the flexural stiffness of the column, and at large deflections, may resist some tension on the outer surface. It was shown experimentally that CFRP sheets have indeed increased the columns' strengths by up to 23 percent. The study, however, revealed the sensitivity of axially loaded slender columns to their inherent geometric out-of-straightness and alignment (imperfections), which affects both the ultimate strength and stiffness of the specimens. Residual stresses are also an important factor, which can affect the behaviour of cold-formed sections. While the major parameter intended in the experimental investigation was the effect of number of CFRP layers, it is believed that geometric imperfections have also varied among the specimens. As such, no specific correlation could be established between the amount of strength gain and the amount of CFRP.

M. Pandey et al. (eds), Advances in Engineering Structures, Mechanics & Construction, 227–238.

A theoretical study including finite element analysis and analytical modeling is being developed by the authors to assess the exclusive contribution of CFRP sheets to the strength and stiffness of slender HSS steel members and understand their behaviour. The objective of this paper, which is the first phase in the analytical program, is to introduce a simplified analytical model to predict the load versus axial and lateral displacement responses of axially loaded HSS members retrofitted with CFRP. It is anticipated that further refinement of this approach will be needed in the future. The model is verified using the experimental results and is used to uncouple the effect of geometric imperfections from the effect of CFRP retrofitting. In the following sections, a brief summary of the experimental research program is given, followed by a detailed description of the analytical model. The model verification, using the experimental load-lateral displacement, load-axial displacement and load-axial strain responses, is then presented. Finally, the parametric study is presented.

Summary of Experimental Investigation

A brief summary of the experimental study is presented in this section, whereas more details can be found elsewhere (Shaat and Fam 2006). The experimental study included 5 axial compression tests, conducted on a standard 89 x 89 x 3.2 mm HSS section (Fig. 1(a)) with nominal yield strength (F_y) of 380 MPa. The length of the pin-ended members was 2380 mm, which corresponds to a slenderness ratio of 68. Ultra-high modulus unidirectional carbon fibre sheets were bonded to the specimens in the longitudinal direction. A single layer was 0.54 mm thick and had tensile strength and modulus of 510 MPa and 230 GPa, respectively. A layer of glass-FRP (GFRP) sheet was first installed directly on the steel surface before applying the CFRP layers to prevent direct contact between carbon fibres and steel, which could lead to galvanic corrosion. The GFRP lamina was 1.46 mm thick and had tensile strength and modulus of 855 MPa and 20.3 GPa, respectively. The stress-strain curves of steel, CFRP, and GFRP are illustrated in Fig. 1(b).

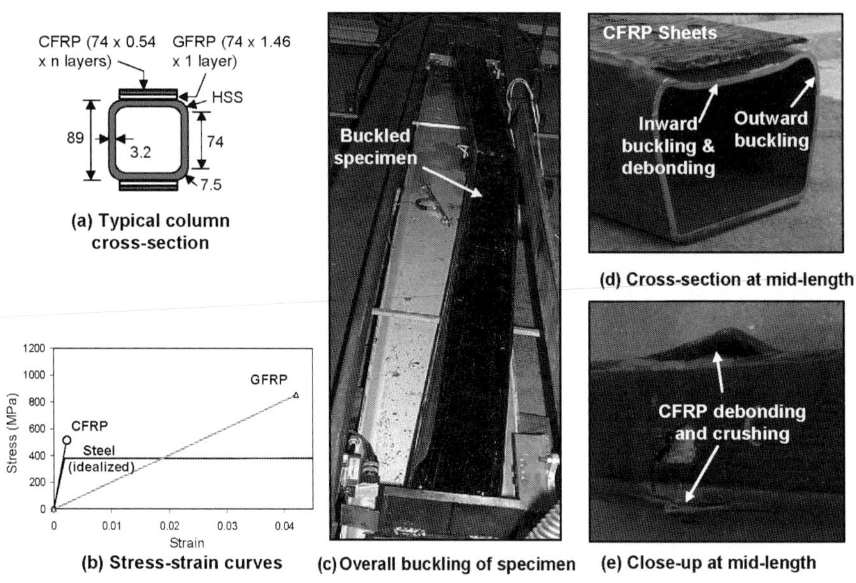

Fig. 1 Specimen cross-section, materials, test setup and failure mode

The tested specimens included a control (unretrofitted) specimen and three specimens retrofitted with one, three and five layers of CFRP, applied to two opposite sides in the plane of overall buckling. The fifth specimen was retrofitted with three layers, applied to all four sides of the specimen. The specimens were given identification codes. For example, 3L-2S indicates three CFRP layers applied on two opposite sides of the specimen.

The gain in axial strength of the FRP-retrofitted specimens ranged from 13 to 23 percent. The strength gains, however, did not correlate directly to the number of CFRP layers. As indicated earlier, this was attributed to the variability of geometric imperfections among the specimens, which is possibly due to a slight out of straightness of different values among the specimens, or minor misalignment within the test setup, or a combination of both. In all specimens, failure was mainly due to excessive overall bucking of the specimen, as shown in Fig. 1(c), followed by a secondary local buckling in the compression side, at or near mid length of the specimen. The local buckling took the form of inward buckling of the compression face and outward buckling of the two side faces, which was clearly revealed after testing by cutting the specimen, as shown in Fig. 1(d). For the FRP-retrofitted specimens, the secondary local buckling in the compression side was associated with a combined delamination and premature crushing of the FRP sheets. For specimen 3L-4S, retrofitted on four sides, the CFRP on the sides have also fractured due to the local bending associated with the outward buckling, as shown in Fig. 1(e).

Figure 2 shows the load versus axial strain of all specimens, at the two opposite sides, at mid-length. The figure shows that both sides are under compression, up to a certain load, where excessive buckling starts. At this load level, the strain readings at the outer surface revert to tension while strains at the inner surface show rapid increase in compression. The strain gauges on the compression side failed as a result of debonding and crushing of CFRP sheets. By carefully examining the strain readings, an average strain value of 0.0013 mm/mm can be defined as the strain at which CFRP failed in compression.

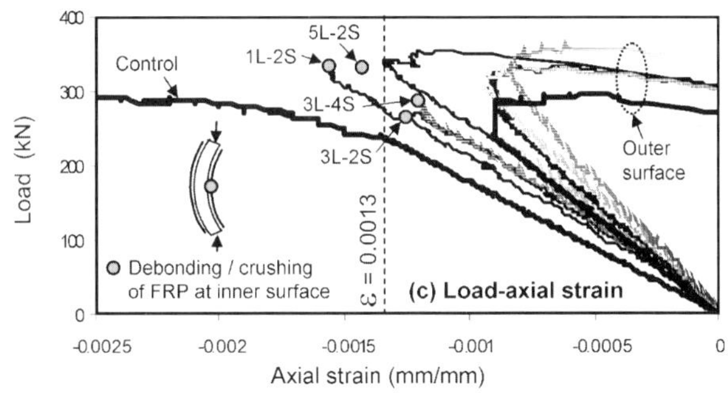

Fig. 2 Load-axial strain of experimental program

Analytical Model

In order to predict the load versus axial and lateral displacement responses of axially loaded slender HSS steel members retrofitted with CFRP sheets, a non-linear model has been developed. The model accounts for both material and geometric (second order effects) non-linearities as well as residual stresses. An incremental approach is used, where the concepts of equilibrium and strain compatibility are satisfied at each loading step. The stress-strain curve of steel is assumed to follow an elastic-perfectly plastic model, as shown in Fig. 1(b). On the other hand, FRP materials are assumed to behave linearly up to failure. The following sections provide detailed description of the model.

Residual Stresses in HSS Section

Residual stresses play an important role in the behaviour of steel structures and are normally induced in the manufacturing process. They typically result in a reduction of the flexural rigidity of slender members, and consequently, a lower buckling load may result (Weng, 1984). Although residual stresses are self-equilibrating, the cross sectional effective moment of inertia will be changed when parts of the section reach their yielding strength prior to other parts. An extensive experimental investigation of the residual stresses of hollow structural cold formed steel shapes was performed (Davison and Birkemoe, 1983 and Weng and Pekoz, 1990). The magnitudes of the measured residual stresses were found to vary, approximately, from 25 to 70 percent of the yield strength, depending on the manufacturing process.

Short columns are typically used in lieu of coupon tests to provide the average compressive stress-strain curves (Bjorhovde and Birkemoe, 1979). This type of tests demonstrates the overall column performance at very low slenderness ratio, in the absence of overall instability. The capacity of these columns is achieved when all fibres reach the yield stress and the corresponding load is defined as the yield load. Because of residual stresses, the short columns do not typically show a distinct yield point, but rather a gradual transition from the linear elastic behaviour to the fully plastic plateau, as a result of the gradual yielding. The magnitude of residual stresses F_{rs} can then be estimated by evaluating the difference between the proportional limit stress and the maximum stress levels. In this study, experimental short column tests have also been conducted by the authors on 175 mm long HSS columns of the same HSS sections used for the slender columns and the average load-strain curve is shown in Fig. 3. The behaviour shows a proportional limit load of 249 kN and a yield load of 410 kN. These two levels of load indicate that the magnitude of residual stress is in the order of 40 percent of the yield strength. In the proposed model, the through-thickness residual stress distribution will be idealized as shown in the schematic drawing in Fig. 3, as suggested by (Davison and Birkemoe, 1983) and by (Chan et al., 1991), where F_{rs} equals to the minimum value specified in literature as 0.25 F_y.

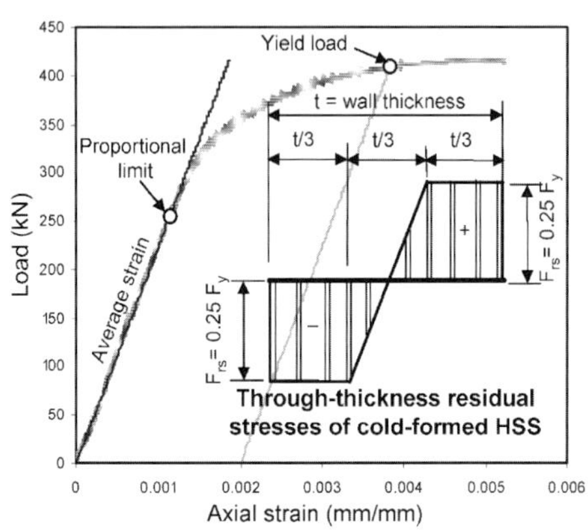

Fig. 3 Evaluation of residual stress magnitude

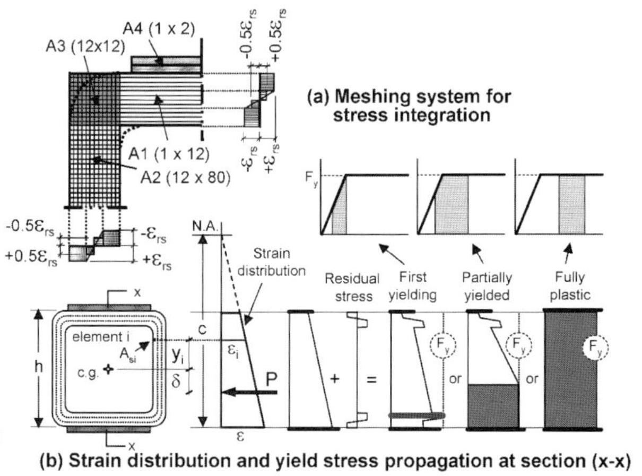

Fig. 4 Meshing system and yield propagation

In order to simulate the residual stress pattern, shown in Fig. 3, the wall thickness was first divided into three layers, as shown in Fig. 4(a). A pre-strain compressive value ε_{rs} of (-0.25 ε_y) was given to the inner third, while a tensile value of (+0.25 ε_y) was given to the outer third, where ε_y is the strain at yield, based on Fig. 1(b). The middle third was divided into two equal halves. The inner half was given a uniform value of (-0.125 ε_y), while the outer half was given a uniform value of (+0.125 ε_y), as shown in Fig. 4(a).

Meshing System

An element-by-element approach is adopted to integrate the stresses over the cross sectional areas of steel and FRP. The cross section was divided into four areas (A1 to A4), as shown in Fig. 4(a). The flat part of the flanges oriented normal to the plane of buckling (A1) was divided into 12 strips through the thickness, where the strain is constant across the width of each strip. The flat part of the flanges parallel to the plane of buckling (A2) was divided into 12 x 80 elements to capture the strain gradient along the depth of the section and also the residual stress distribution within the thickness. The conjunction between the flange and the web (A3) was idealized as a square and divided into 12 x 12 elements. Area A4 represents the FRP layers attached to area A1, and was divided into 1 x 2 elements. To model specimen 3L-4S with FRP sheets on four sides, an additional area of FRP, A5, attached to area A2, was also modeled. The distance between the center of each element and the centroid of the cross section is y_i, as shown in Fig. 4(b). It is assumed that perfect bond exists between the FRP sheets and steel surface and strains are linearly distributed along the depth of the section. The stresses at the centroid of each element are assumed constant within the element area.

Force Equilibrium and Moments
Figure 4(b) shows a cross-section at mid-height of a concentrically loaded slender compression member. Due to overall buckling, the axial force is eccentric with respect to the mid-height section. For a given strain gradient induced by the external eccentric axial load P, which is based on a strain level ε at the extreme compression side and neutral axis depth c, the strain ε_i in each element i located at a distance y_i from the centroid can be determined as follows:

$$\varepsilon_i = \left[1 - \frac{h}{2c} - \frac{y_i}{c} \right] \varepsilon \tag{1}$$

where h is the depth of the section. The strain ε_i is then added to the residual strain ε_{rs} to obtain the total strain ε_{i+rs} for the steel elements and check whether the element has yielded or not:

$$\varepsilon_{i+rs} = \varepsilon_i + \varepsilon_{rs} \tag{2}$$

Possible stress distributions at various stages of loading are shown in Fig. 4(b). The total axial load at this loading level, for a given ε and c, can be obtained by numerical integration of stresses over the cross section, for both the yielded and elastic elements as well as the FRP elements, as follows:

$$P = \sum_{elastic\ steel} \left(\varepsilon_i\, E_s\, A_{s_i} \right) + \sum_{plastic\ steel} \left(F_y A_{s_i} \right) + \sum_{FRP} \left(\varepsilon_i\, E_f A_{f_i} \right) \tag{3}$$

and the corresponding moment M is:

$$M = \sum_{elastic\ steel} \left(\varepsilon_i\, E_s\, A_{s_i} y_i \right) + \sum_{plastic\ steel} \left(F_y A_{s_i} y_i \right) + \sum_{FRP} \left(\varepsilon_i\, E_f A_{f_i} y_i \right) \tag{4}$$

where A_{s_i} and A_{f_i} are the areas of steel and FRP elements, respectively, and E_{si} and E_{fi} are Young's moduli of steel and FRP elements, respectively.

Lateral Displacement
Figure 5(a) shows a pin-ended slender compression member which is slightly curved initially, due to an imperfection e'. At any load level P, the total lateral displacements, measured from the vertical axis is δ. (Allen and Bulson, 1980) have shown that the shape of the deflected compression member can be represented by a Fourier series, which can be reduced to the following expression that relates the lateral displacement δ at a distance z along the member's length to the applied load P:

$$\delta = \frac{e'}{1 - \left(P / P_{cr} \right)} \sin\left(\frac{\pi z}{L} \right) \tag{5}$$

where L is the length of member and P_{cr} is the Euler buckling load and is given by:

$$P_{cr} = \frac{\pi^2 EI}{(kL)^2} \tag{6}$$

where EI is the flexural rigidity of a prismatic member, function of Young's modulus E and moment of inertia I of the member's cross section. The effective length factor k accounts for boundary conditions and is taken as unity for pin-ended members.

It should be noted that Eq. 5 is valid for lateral displacements of value up to 10% of the length $\left(\delta \le L/10 \right)$. It is also important to note that Equations 5 and 6 assume linear elastic behaviour of the material and that the residual stresses are not taken into consideration. In the following sections, methods are proposed to account for residual stresses, material non-linearity due to yielding, and the contribution of FRP.

Effective Moment of Inertia (I_{eff})
In order to account for gradual yielding of different parts of the cross section under the applied loading, the concept of "effective moment of inertia" is incorporated in this analysis. Contribution of any steel

element of area A_{s_i} to the flexural rigidity $(EI)_{s_i}$ is the product of the tangent modulus and the element's moment of inertia, as follows:

$$(EI)_{s_i} = E_t \, A_{s_i} \, y_i^2 \tag{7}$$

where E_t is the tangent modulus based on the stress-strain curve of steel. If the idealized elastic-plastic stress-strain curve of steel with Young's modulus E_s is used, then: for $\varepsilon_{i+rs} < \varepsilon_y$, $E_t = E_s$ and for $\varepsilon_{i+rs} \geq \varepsilon_y$, $E_t = 0$. This indicates that the flexural rigidity of the yielded parts becomes zero. Consequently, the effective bending stiffness $(EI)_{s_{eff}}$ of the entire section takes the following form:

$$(EI)_{s\,eff} = E_s \, . \sum_{elastic\ steel} \left(A_{s_i} \, y_i^2\right) \tag{8}$$

The effective moment of inertia $I_{s_{eff}}$ for the section can then be introduced in terms of the elastic parts only as follows:

$$I_{s\,eff} = \sum_{elastic\ steel} \left(A_{s_i} \, y_i^2\right) \tag{9}$$

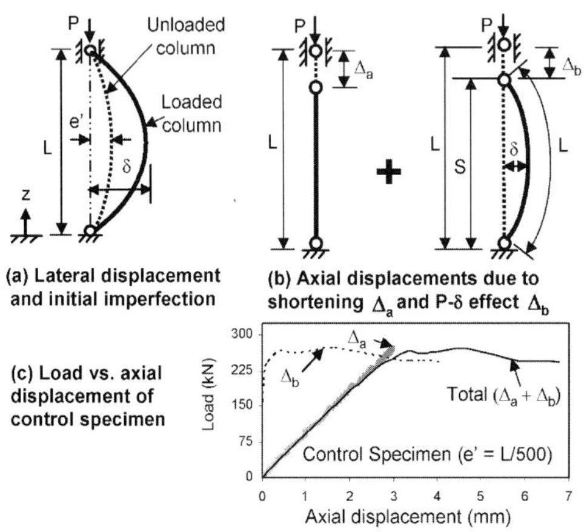

Fig. 5 Lateral and axial displacements of compression members

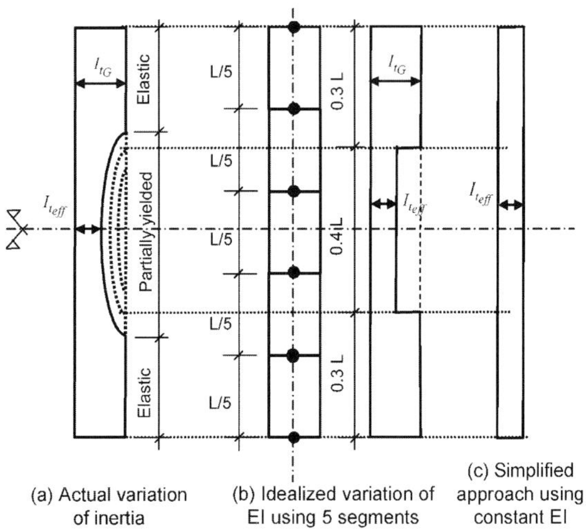

Fig. 6 Variation of stiffness (EI) along member's length due to yielding

First yielding will typically occur at the extreme fibres of the member's cross-section at mid-height. As the axial load and corresponding lateral deflection increase, yielding will spread within the cross-section and also in the longitudinal direction of the member, as shown in Figures 4(b) and 6(a). This indicates that the member will have a moment of inertia that varies with the applied load and also longitudinally within the yielded length as indicated by Eq. 9. As such, a more general expression for the Euler buckling load may be used in lieu of Eq. 6, which assumes a constant moment of inertia. The finite-difference method is used in this case (Ghali and Neville, 1989), where the member is divided into a number of segments of equal length λ and the equivalent concentrated elastic loads at each of the *m* internal nodes can be obtained. A series of simultaneous equations are then solved. The solution of these equations is an eigenvalue problem. An iteration procedure is utilized until a stable eigenvector $\{\delta\}$ is obtained. The buckling load P_{cr} can then be calculated from the largest eigenvalue γ as in Eq. 10. Figure 6(b) shows the idealized compression member and the variation of stiffness using five segments.

$$\gamma = \frac{12}{P_{cr} \, \lambda} \tag{10}$$

A conservative yet reliable simplification may be made by assuming that flexural stiffness of the critical section at mid-height (using Eq. 9) governs, and can then be assumed constant along the length of the member, as shown in Fig. 6(c). This will allow the use of the simple Euler buckling formula, as shown in Eq. 11.

$$P_{cr} = \frac{\pi^2 E_s I_{s\,eff}}{(kL)^2} \tag{11}$$

For HSS sections with FRP layers, the transformed effective moment of inertia $I_{t_{eff}}$ should be used in lieu of $I_{s_{eff}}$ in Eq. 11. $I_{t_{eff}}$ is calculated using the following equation:

$$I_{t_{eff}} = I_{s\,eff} + \sum_{FRP}\left[\frac{E_{f_i}}{E_s} I_{f_i}\right] \tag{12}$$

where $I_{f_i} = \left(A_{f_i}\, y_i^2\right)$. I_{f_i}, E_{f_i} and A_{f_i} are the moment of inertia, Young's modulus, and the area of intact FRP element *i*, respectively.

The lateral displacement of the member can now be calculated at any point along the member's length, at any load level, using Eq. 5.

Axial Displacement

The axial displacement Δ is the sum of two components, Δ_a and Δ_b, as shown in Fig. 5(b):

$$\Delta = \Delta_a + \Delta_b \tag{13}$$

where Δ_a and Δ_b are the displacements due to axial shortening and curvature arising from P-δ effect, respectively, and can be approximated as follows:

$$\Delta_a = \frac{PL}{E_s A_t} \tag{14}$$

where A_t is the transformed cross sectional area and is calculated as follows:

$$A_t = A_s + \sum_{FRP}\left[\frac{E_{f_i}}{E_s} A_{f_i}\right] \tag{15}$$

where, A_s is the cross sectional area of the HSS section

$$\Delta_b = L - S \tag{16}$$

where S is the chord length of the deformed shape of the compressed member (Fig. 5(b)), and is calculated based on a sine curve of arc length L and amplitude δ for the member's deflected shape.

Figure 5(c) shows the predicted load-axial displacement response of the control specimen. The figure shows that the contribution of the 'curvature' component Δ_b is only significant near and after the peak load, where the overall buckling occurs, whereas the axial shortening component Δ_a is dominant within the linear elastic range, before buckling.

Failure Criteria

The strain values in the steel cross section are incrementally increased, until the section reaches its full plastic capacity, as shown in Fig. 4(b). Elements with compressive residual stresses typically reach yielding before elements with tensile residual stresses. Eventually, all elements reach the yield strength. At this point, the effective moment of inertia of the steel cross section and the buckling load become zero, based on Equations 9 and 10. Consequently, a value for the lateral displacement δ can no longer be obtained using Eq. 5.

For FRP material in the compression side, the failure strains are limited by debonding and crushing at a strain value of 0.0013 mm/mm. After excessive overall buckling, the FRP on the outer surface may be subjected to some tensile strains, which are well below the ultimate value.

Generation of Full Responses

In order to obtain the load-lateral displacement (P-δ) response in compression, the procedure can be summarized as follows:

1. Assume a value of the extreme compressive strain ε and a neutral axis depth c (Fig. 4(b)).
2. For each element of the steel cross section, calculate its strain value ε_i, using Eq. 1, add the residual strain ε_{rs} as given by Eq. 2, and compare it to the yield strain value ε_y to check whether the element has yielded or not.
3. Calculate the corresponding stress of each element, using the steel stress-strain curve. For $\varepsilon_{i+rs} \geq \varepsilon_y$, the stress is limited to F_y.
4. Calculate the strain ε_i, using Eq. 1, and corresponding stress for each FRP element, based on linear stress-strain response. Compare the strain to ultimate values to check for failure of FRP.
5. Calculate the axial load P and bending moment M for the entire cross section, by using Equations 3 and 4, which are essentially a summary of steps 3 to 5 above.
6. Calculate the eccentricity $e = M/P$, which is induced by the non uniform stress distribution.
7. Calculate the cross sectional transformed effective inertia $I_{t_{eff}}$, excluding both the yielded steel elements and failed FRP elements, using Eq. 12.
8. Calculate the critical load P_{cr} using the conservative approach described by Eq. 11 (or alternatively using the more accurate approach by using Eq. 10). P_{cr} is used to calculate the lateral displacement at mid-length δ for a prescribed imperfection e' using Eq. 5.
9. Compare the eccentricity e obtained in step 6 with the lateral displacement δ, calculated in step 8. If the two values are different, assume a new value of the neutral axis depth c and repeat steps 2 to 8 until the two values are equal. This will provide one point on the load-lateral displacement curve.
10. Enter a higher value of strain ε in step 1 and repeat the process from steps 2 to 9 until the complete load-lateral displacement response is established.

In order to generate an approximate load-axial displacement P-Δ response of compression members, for an axial load P and corresponding lateral deflection δ, obtained earlier, the following procedure can be followed:

1. The axial shortening term Δ_a of the displacement is calculated using Eq. 14.
2. For a given lateral displacement δ, establish the deformed sine curve of a buckled member with mid-length amplitude of δ and an arc length of L.
3. Calculate the chord length of the sine curve S.
4. Calculate the 'curvature' component of the axial displacement Δ_b using Eq. 16.
5. The total axial displacement Δ is calculated using Eq. 13.
6. Repeat the previous steps for each load level P and its corresponding lateral displacement δ, until the complete P-Δ response is established.

Verification of Model and Parametric Study

The model described in the previous section has included several features, namely the P-δ effect (non-linear geometry), plasticity (yielding) of the steel section, the through-thickness residual stresses in the HSS section, the initial imperfection, and failure criteria of FRP in compression. In order to illustrate the significance of these features, the load-lateral displacement response of one specimen retrofitted with three CFRP layers, on two opposite sides, (3L-2S) has been predicted for five different cases. In case 1, Equations 5 and 6 have been used in their original form, assuming linear elastic materials (ignoring steel yielding and residual stresses). In case 2, plasticity of steel is considered, however, residual stresses are ignored. In case 3, both steel plasticity and residual stresses are accounted for but the failure criteria of FRP in compression is not applied (FRP is assumed fully intact in the compression side throughout the full response). In case 4, all the features of the model are applied, including the effect of variable inertia along the length, which is accounted for in calculating the buckling load, using the finite difference approach as shown in Fig. 6(b). Case 5 is similar to case 4, except that the simplified approach for calculating the buckling load is used, as shown in Fig. 6(c).

Figure 7 shows the experimental and analytical responses for the five cases. All predictions were made assuming initial imperfection ($e' = L/500$), which is the limit permitted by Canadian standards (CAN/CSA-S16-01). The effect of initial imperfection itself will be discussed later. Figure 7 clearly shows that ignoring the steel plasticity (case 1) would highly over estimate the axial strength. Ignoring the residual stresses (case 2) would also overestimate the load at which transition occurs from the elastic to plastic response. Also by assuming that FRP is fully effective in compression (case 3), the ultimate load is somewhat overestimated. It is, therefore, clear that cases 4 and 5 represent the most accurate predictions, using the full capabilities of the model. It is also clear that the simplified conservative approach used in case 5 is quite reasonable and would, therefore, be used in this paper for the predictions of all responses and the parametric study, next.

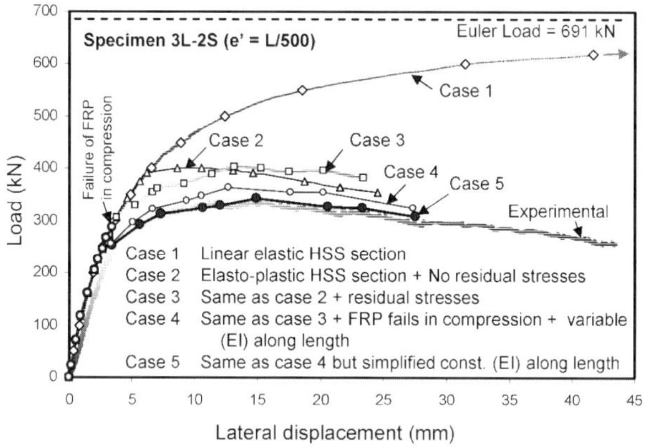

Fig. 7 Significance of various features of analytical model

The load versus lateral and axial displacements of the five specimens have been predicted and are presented in Figures 8(a) and 8(b), respectively. A summary of the predicted and experimental values is also presented in Table 1. The predictions are made for three different levels of initial imperfections, namely $L/300$, $L/500$, and $L/1000$. It is expected that most steel sections will practically have imperfection values less than $L/500$ (CAN/CSA-S16-01). The model shows reasonable agreement with test results for this common level of imperfections. However, for specimen 5L-2S, the behaviour was not accurately predicted since the maximum lateral displacement, measured experimentally, occurred near the quarter length point, rather than at mid-length. Also, for the same reason, this specimen did not show higher gain of strength compared to other specimens, with less number of CFRP layers. Figures 8(a) and 8(b) clearly emphasize the important effect of initial imperfection. The load-axial strain behaviour at two opposite sides of each specimen has also been predicted in Fig. 8(c), using the imperfection value that showed the best results for each respective specimen.

It has been shown that the model is capable of predicting the various responses of axially loaded slender HSS members retrofitted with CFRP sheets. It was also shown earlier, in the experimental results, that no clear correlation was established between the number of CFRP layers and strength gains, due to the variability of imperfections among the test specimens. In the following section, the model is used to study the sole effect of the number of CFRP layers, by fixing the level of imperfection. This study builds on the same specimens used to verify the model. The responses of control, 1L-2S, 3L-2S, and 5L-2S specimens are estimated twice, using two fixed values of imperfections, namely $L/300$ and $L/1000$, as shown in Fig. 9.

Effect of Number of CFRP Layers

Figure 9 (a and b) clearly shows that bonding CFRP sheets to slender HSS members can indeed increase their axial compressive strength. For example, the predicted percentage increases in strength of specimens with one, three and five CFRP layers are 11, 25, and 40 percent for specimens with $L/300$ imperfection and are 12, 26, and 40 percent, respectively, for specimens with $L/1000$ imperfection.

Also, the percentage increases in axial stiffness are 15, 31, and 48 percent for the *L*/300 specimens and are 13, 29, and 45 percent, for the *L*/1000 specimens. These findings show that the percentage increases in axial strength and axial stiffness, for the levels of imperfections studied, are independent of the initial imperfection value. The imperfection, however, affects the absolute values of ultimate loads.

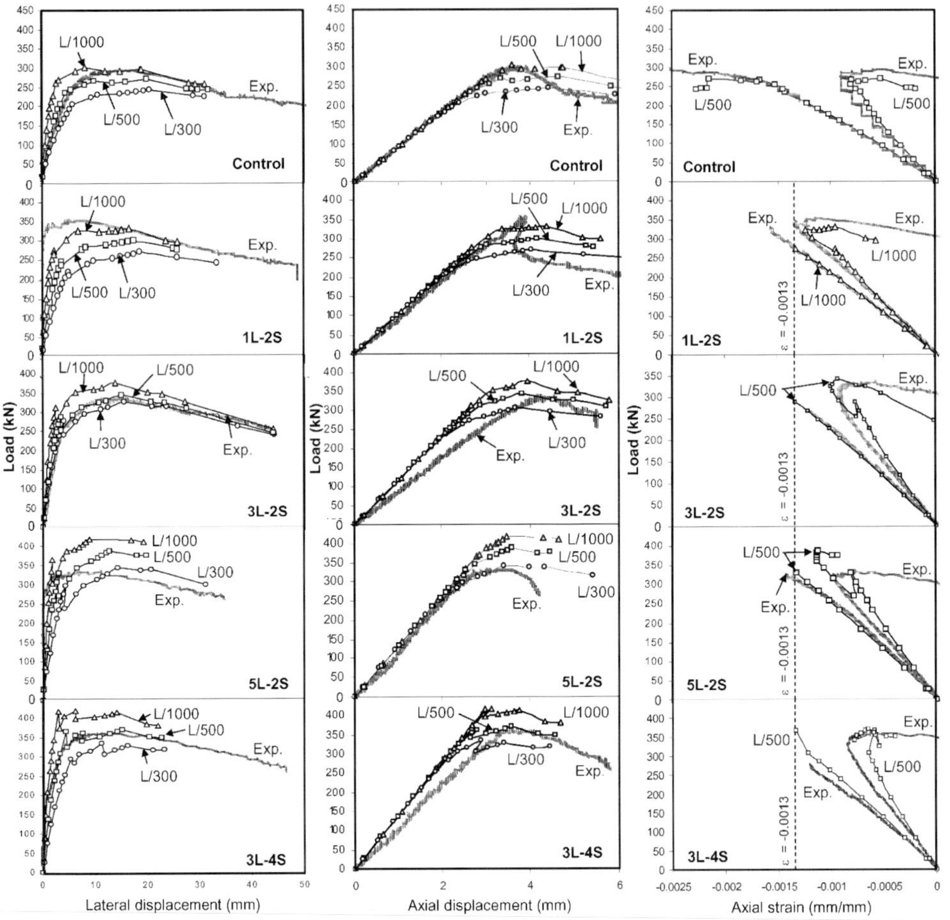

(a) Load vs. lateral displacement (b) Load vs. axial displacement (c) Load vs. axial strain

Fig. 8 Predicted responses of the slender HSS compression members

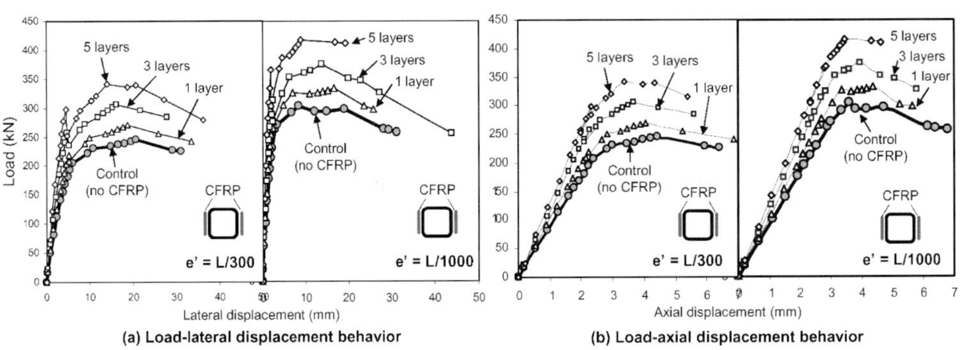

(a) Load-lateral displacement behavior (b) Load-axial displacement behavior

Fig. 9 Effect of number of CFRP layers on behavior of HSS compression members

Table 1. Comparison between experimental and analytical model results

Experimental		Analytical Model		(Analytical / Experimental) ratio
Specimen I.D.	P_{max} (kN)	e'	P_{max} (kN)	
Control	295	L/300	245	0.83
		L/500	271	0.92
		L/1000	297	1.01
1L-2S	355	L/300	271	0.76
		L/500	301	0.85
		L/1000	333	0.94
3L-2S	335	L/300	306	0.91
		L/500	344	1.03
		L/1000	375	1.12
5L-2S	332	L/300	344	1.04
		L/500	387	1.17
		L/1000	415	1.25
3L-4S	362	L/300	329	0.91
		L/500	371	1.02
		L/1000	415	1.15

Table 2. Percentage increases in axial strength and stiffness of retrofitted members

Imperfection value (e')	Specimen I.D.	P_{max} (kN.)	% Gain in strength	Axial Stiffness (kN/mm)	% Gain in stiffness
L/300	control	245	-	89	-
	1L-2S	271	11	102	15
	3L-2S	306	25	117	31
	5L-2S	344	40	132	48
L/500	control	271	-	91	-
	1L-2S	301	11	103	13
	3L-2S	344	27	117	29
	5L-2S	387	43	132	45
L/1000	control	297	-	91	-
	1L-2S	333	12	103	13
	3L-2S	375	26	117	29
	5L-2S	415	40	132	45

Conclusions

A non-linear model based on the concepts of equilibrium and strain compatibility has been developed to predict the axial load capacity of slender HSS compression members retrofitted by externally bonded CFRP sheets. The model is also capable of predicting the full load versus lateral and axial displacements. The member's initial imperfection, residual stresses, material and geometric nonlinearities are accounted for. The model was verified using experimental results and showed reasonable agreement. The effects of member's initial imperfection and number of CFRP layers are studied. The following conclusions can be drawn, based on the simplified conservative approach followed, and the range of imperfections studied in this paper:

1. Externally bonded CFRP sheets are effective in increasing the axial strength and stiffness of slender HSS compression members.
2. For a slender member retrofitted by a given CFRP reinforcement ratio, the member's initial imperfection has a pronounced effect on its axial strength but marginal effect on axial stiffness.
3. While initial imperfection affects the magnitude of member's axial strength, it does not affect the percentage increase in member's strength resulting from CFRP retrofitting, for the studied level of imperfections.
4. Ignoring residual stresses, steel plasticity, or premature delamination and crushing of CFRP in the inward side of a compression member undergoing overall buckling could highly overestimate the member's axial strength.

Future Work

The model will be further refined to account for the effect of the variable inertia throughout the length of the column by introducing larger number of segments.

Acknowledgement

The authors wish to acknowledge the financial support provided by the Natural Sciences and Engineering Research Council of Canada (NSERC).

References

1. Allen, H. G. and Bulson, P. S. (1980). "Background to Buckling." McGraw-Hill Book Company (UK) Limited.S.
2. Bjorhovde, R., and Birkemoe, P. C. (1979). "Limit States Design of HSS Columns." Canadian Journal of Civil Engineering, 6, 275-291.
3. Canadian Standards Association, CAN/CSA-S16-01, Limit states design of steel structures, Mississauga, Ontario.
4. Chan, S. L., Kitipornchai, S. and Al-Bermani, F. G. A. (1991). "Elasto-Plastic Analysis of Box-Columns including Local Buckling Effects." Journal of Structural Engineering. ASCE, 117(7): 1946-1962.
5. Davison, T. A., and Birkemoe, P. C. (1983) "Column Behaviour of Cold-Formed Hollow Structural Steel Shapes." Canadian Journal of Civil Engineering, 10, 125-141.
6. Ghali, A. and Neville, A. M. (1989). "Structural analysis: A unified classical and matrix approach" Chapman and Hall, London and New York.
7. Hollaway, L.C. and Cadei, J. (2002). "Progress in the Technique of Upgrading Metallic Structures with Advanced Polymer Composites", Progress in Structural Engineering and Materials, J. Wiley & Sons, 4(2):131-148.
8. Shaat, A., Schnerch, D., Fam, A., and Rizkalla, S. (2004) "Retrofit of steel structures using fiber-reinforced polymers (FRP): State-of-the-art.", Transportation Research Board (TRB) Annual Meeting, Washington, D.C., USA, CD-ROM (04-4063).
9. Shaat, A. and Fam, A. (2006) "Axial Loading Tests on CFRP-Retrofitted Short and Long HSS Steel Columns." Accepted for publication in the Canadian Journal of Civil Engineering.
10. Weng, C. C. (1984). "Cold-bending of thick steel plates at low R/t ratios." Master's thesis. Cornell University, at Ithaca, N.Y.
11. Weng, C. C., and Pekoz, T. (1990). "Residual Stresses in Cold-Formed Steel Members." Journal of Structural Engineering. ASCE, 116(6): 1611-1625.

QUANTIFYING INHERENT ROBUSTNESS IN STRUCTURAL STEEL FRAMING SYSTEMS

Christopher M. Foley[1], Kristine Martin[2] and Carl Schneeman[3]

[1]*Marquette University, Haggerty Hall 267, 1515 W. Wisconsin Ave., Milwaukee, WI 53233, USA, E-mail: chris.foley@marquette.edu*
[2]*Marquette University, Currently: Framatone ANP, Inc., Naperville, IL 60563, USA*
[3]*Marquette University, Currently: Walker Parking Consultants, Minneapolis, MN 55416, USA*

Introduction

It is well known that Vierendeel action in multi-story steel frames can be a source of inherent robustness and will provide a significant measure of general structural integrity in structural steel moment resisting framing systems. In situations where there is insufficient number of stories above a compromised column, or when simply-connected floor framing is assumed as in the case of most modern steel framing systems, the robustness inherent in the system remains to be fully understood. Grierson *et al.* (2005) points out those progressive collapse scenarios in steel structures that begin with upper-story columns becoming ineffective. The collapse sequence associated with these scenarios is a propagation of failures down the structural system to the ground level through debris loading accumulation.

Many analytical efforts to date only consider components found in the structural steel skeleton. Furthermore, the analytical models assumed that pin-connected beams and girders existed at interior columns and these analytical models did not support analysis considering ineffective interior columns, interior girders, exterior girders, or in-fill beams. If robustness in the structural steel framing system is to be quantified, the analyses must go beyond the simple removal of columns around the perimeter of the framework.

The objectives of this paper are to provide a brief overview of the methodologies that have been proposed and validated via experimental testing for quantifying the catenary and membrane mechanisms in concrete floor framing systems; and outline a new methodology for quantifying the membrane and catenary capacity in structural steel floor framing systems along with high-level provisions for ensuring structural integrity through the preservation of catenary and membrane action.

Membrane Action in Concrete Floor Systems

Researchers in the field of reinforced concrete have had a long history of attempting to understand the tensile behavior of structural concrete slab systems and proposing methodologies for quantifying the beneficial effects of catenary action and membrane action. Much of the research conducted in this regard has made its way into ACI 318 provisions for general structural integrity (Hawkins and Mitchell 1979; Mitchell and Cook 1984). Researchers studying the response of structural steel systems to fire have also begun in earnest to understand and capitalize on the inherent robustness present in steel framing systems that is contributed by the concrete deck (Allam *et al.* 2000; Bailey *et al.* 2000; Huang *et al.* 2003a; Huang *et al.* 2003b).

It has been long recognized that flat plate concrete floor systems have the potential to suffer from disproportionate collapse from a rather simplistic event: punching shear failure at interior and exterior columns (Hawkins and Mitchell 1979; Mitchell and Cook 1984). There was a series of systematic efforts carried out to develop design procedures that could limit the probability of a punching shear failure leading to progressive collapse. The first of such efforts was that conducted by Hawkins and Mitchell (1979). Mitchell and Cook (1984) enhanced this methodology to include procedures that allow quantifying the role of catenary action in the behavior of the floor system and its partnering with membrane action to mitigate progressive collapse in concrete floor system.

M. Pandey et al. (eds), Advances in Engineering Structures, Mechanics & Construction, 239–254.

When a concrete floor plate is loaded to the point of inelastic behavior, there is a tendency for the bottom fibers (assuming loading is from the top) to lengthen. This lengthening, however, is restrained by the concrete slab at the perimeter of the panel being loaded. Of course, steel beams in the systems considered in the present study will provide restraint to this outward movement. In the purely theoretical sense, the concrete slab will have a load versus vertical deflection response that exhibits snap through prior to the formation of membrane tension in the system. This *hanging net* effect cannot take place without significant vertical deformation in the floor system. In the hanging configuration, all sections through the floor plate are subjected to tensile forces and it is imperative that properly developed tension reinforcement exists in the slab and vertical support at the panel edges be maintained.

The beauty of the work of Hawkins and Mitchell (1979) is that the expressions for computing the membrane capacity of concrete floor panels are rather simplistic and include a significant amount of *engineering feel*. The fundamental assumption of the proposed methodology is that the deformed membrane between supports follows a circular shape. This makes the mathematics tractable and errors are minor when compared to the more correct catenary parabola. The basic slab system and membrane forces considered are schematically shown in Figure 1.

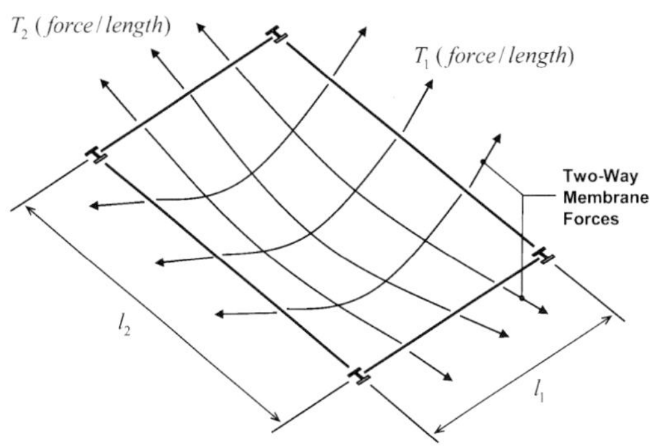

Figure 1 Two-Way Membrane Action in Reinforced Concrete Slab.

Two slab span directions are assumed: the first is defined as the short direction, l_1; and the second is termed the long direction, l_2. The reinforcement area on a per unit length basis in the short and long directions are A_{s1} and A_{s2}, respectively. The normal strains in the fibers of the membrane are assumed to be uniform over the membrane thickness and are functions of its curvature. Uniformly distributed loading over the surface of the membrane is assumed and positive loading is taken to be downward. Membrane tension forces (edge tensions) per unit length parallel to the short and long directions are T_1 and T_2, respectively. These forces are assumed to be in the direction tangent to the deformed membrane's mid-surface at the edges.

A typical structural mechanics solution procedure (*e.g.* imposition of vertical equilibrium, ensuring compatibility of deformations, and adherence to constitutive laws for the material) is employed to develop a relationship for the capacity of the tensile membrane that is a function of the edge tension, strain in the membrane (and therefore, vertical deflection) and the panel dimensions. When the panel dimensions differ (*i.e.* they are rectangular) the membrane capacity of the panel based upon the tensile reinforcement capacity at the edges can be written as (Hawkins and Mitchell 1979),

$$w_{edge} = \frac{2T_1 \sin\left(\sqrt{6\varepsilon_x}\right)}{l_1} + \frac{2T_2 \sin\left(\frac{l_1}{l_2}\sqrt{6\varepsilon_x}\right)}{l_2} \tag{1}$$

where: ε_x is the tensile strain in the membrane fibers parallel to the short direction, which is the dominant membrane direction. If the slab panel is square, there is no dominant direction. As l_2/l_1 increases, the slab panel begins to behave as a single direction membrane (*i.e.* a catenary).

Concrete slab systems quite often have different reinforcement patterns at the edges than that found in the *middle strip* areas within the panel span. As a result, if the mid-span reinforcement controls the tensile capacity of the membrane, the vertical load carrying capacity is (Hawkins and Mitchell 1979),

$$w_{pos} = 2\sqrt{6\varepsilon_x} \cdot \left[\frac{T_1'}{l_1} + T_2' \cdot \frac{l_1}{l_2^2} \right] \quad (2)$$

where: T_1' and T_2' are the tensile membrane forces per unit length within the mid-span (positive moment) regions of the panel parallel to the short- and long-directions, respectively.

The strain in the direction parallel to the short and long dimensions of the panel is related to one another as a result of the assumed circular shape of the membrane. If one knows the strain in the direction parallel to the short dimension, the strain in the direction parallel to the long dimension is computed using (Hawkins and Mitchell 1979),

$$\varepsilon_1 = \varepsilon_2 \cdot \left(\frac{l_2}{l_1} \right)^2 \quad (3)$$

Therefore, once the strains in the two directions are computed (short direction assumed, then long direction computed), the constitutive laws for the reinforcement can be used to determine the state of stress and then the tensile membrane forces on a per foot basis follow.

Once the strain in the direction parallel to the short direction is known, the maximum deflection within the panel can be computed using (Mitchell and Cook 1984),

$$\delta = \frac{3 l_1 \varepsilon_1}{2 \sin\left(\sqrt{6\varepsilon_1}\right)} \quad (4)$$

The vertical deflection is important when assessing the capacity of the membrane. Assuming end anchorage is present, the membrane is capable of carrying more loading in a highly deflected configuration for a fixed tensile force capacity. Therefore, if a large amount of loading is present and there is a fixed tensile capacity for the reinforcement in the membrane (assuming no rupturing of the reinforcement), then there is a tendency for the membrane to continue to deflect vertically to generate greater vertical components in the catenary forces. Therefore, the vertical deflection given by equation (4) can be used to determine if a slab panel will become debris loading for a panel below, or will impede modes of egress from the structure.

Mitchell and Cook (1984) provide an enhanced description of the post-failure response of concrete slab structures that is pertinent to situations considered in the present study. The response of a slab structure after initial failure depends upon the amount and details of the steel reinforcement, the vertical support conditions and the horizontal restraint conditions at the panel edges (Mitchell and Cook 1984). When the slab panel has vertical support surrounding its edges (*e.g.* steel beams at the perimeter of the panel), the slab is capable of providing its own in-plane compression ring restraint conditions at the perimeter. This compression ring helps to resist the horizontal component of the maximum tensile forces. If the edges of the panel are allowed to deform vertically, then this compression ring cannot form.

When "stiff" beams are present at the perimeter of the slab panel, the membrane action in the slab panel facilitates the slab system hanging off the perimeter beams. When an interior slab panel is considered, the adjacent regions of the floor system will help to restrain the edges of the overloaded panel. Edge or corner panels can develop the necessary compression ring behavior if the edges are supported by beams that have significant flexural stiffness when compared to the slab itself.

Membrane Action in Composite Deck Structural Steel Systems

Although structural steel floor framing systems are significantly different in many ways than that of a two-way flat plate or flat slab cast-in-place concrete system, there are enough similarities to justify using the theory and expressions developed by Hawkins and Mitchell (1979) and Mitchell and Cook (1984) in assessing the robustness of structural steel framing systems. It is felt that membrane and catenary action are indeed possible within the structural steel framing systems commonly found in buildings. More importantly, it is felt that this catenary and membrane behavior, to a large extent, is inherent in the systems typically constructed. The tension reinforcement present in these systems will need to be quantified and their anchorage discussed prior to detailed examination of ineffective supporting member scenarios.

In composite steel-concrete floor systems, there is typically welded-wire mesh and light gauge steel deck that can be utilized as tension reinforcement within the slab system should membrane and/or catenary action be needed. However, one must understand the usefulness of these components as reinforcing mechanisms in the slab system before one can count on this reinforcement as being inherent sources of membrane and catenary reinforcement for the floor system. The light-gauge steel deck is essentially a unidirectional spanning entity. In the direction parallel to the flutes in the deck, the steel deck is highly likely to be a very useful form of tension reinforcement for facilitating catenary action. However, in the direction orthogonal to the flutes, the steel deck likely has puddle welds or TEK screws that are unlikely to preserve tensile forces within the deck in this orthogonal direction. Furthermore, the fluted nature of the deck results in a tension force that has two distinct elevations at the floor deck soffit. This makes relying on the steel deck providing tensile membrane or catenary reinforcement in two directions very difficult. Therefore, the present analysis assumes that the steel deck provides one-way reinforcement within the floor framing system. It should be noted that if the steel deck panels are not continuous over the supporting beam, a force-transfer mechanism is questionable without edge beams providing vertical support.

The welded-wire fabric present in the floor system is also a source of membrane and catenary tension reinforcement. This steel fabric generally has a slightly elevated yield stress when compared to the usual mild-steel reinforcement. Furthermore, the spacing of the wires in the mesh can change with direction. This reinforcement will be assumed as sufficient to develop catenary and membrane forces if it is considered continuous through the panel perimeter and appropriately lapped.

In the steel building system considered in this study, a panel is defined as having two in-fill beams and two girders bounding a panel of concrete slab. In most cases, the perimeter of the slab panel will have puddle welds or even steel studs connecting the steel deck to the perimeter beams/girders. Furthermore, these perimeter members will have significantly greater flexural stiffness when compared that of the slab. As a result, the slab system can be assumed to develop compression ring anchorage if the perimeter beams remain in tact during a compromising event.

Ineffective Interior Column

Typical (economical) steel gravity floor framing systems implement simply-connected beams and girders. This system, theoretically, has no inherent robustness because if an interior column is lost, there are no moment resisting connections to help span the compromised column and a theoretical mechanism immediately forms. However, the scenario generated by an interior column being rendered ineffective results in activation of two-way membrane action in the composite-steel concrete floor framing system and two-way flexure/catenary grillage action in the structural steel framing. There is a synergy between these two component systems that has only recently been studied in relation to fire (Allam *et al.* 2000; Bailey *et al.* 2000; Burgess *et al.* 2001; Cai *et al.* 2002; Huang *et al.* 2000a; Huang *et al.* 2000b; Huang *et al.* 2003a; Huang *et al.* 2003b). In addition, the beams and girders do indeed have connections at their ends that support not only tension forces (Owens and Moore 1992), but bending moments as well (Astaneh-Asl *et al.* 1989b; Liu and Astaneh-Asl 2000a; Liu and Astaneh-Asl 2000b; Owens and Moore 1992; Wales and Rossow 1983).

In the present analysis, a deformation compatibility approach is used in conjunction with two separate static analyses: the first considering two-way membrane action in the slab; and the second considering two-way-grillage catenary/flexure action in the steel framing. These two analysis components are described in the schematics in Figures 2 and 3.

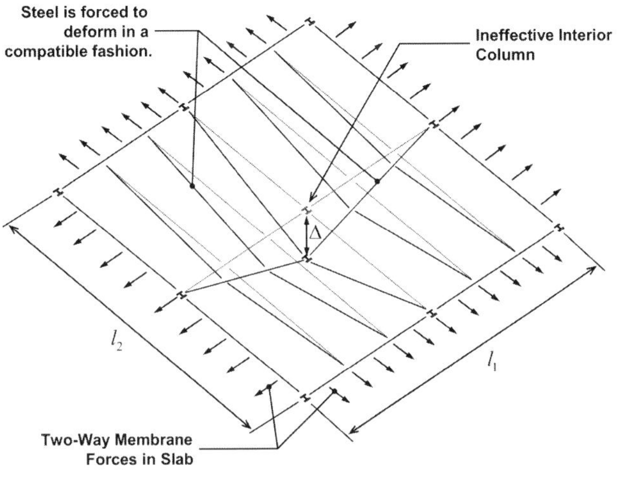

Figure 2 Two-Way Membrane Action Resulting from Ineffective Interior Column.

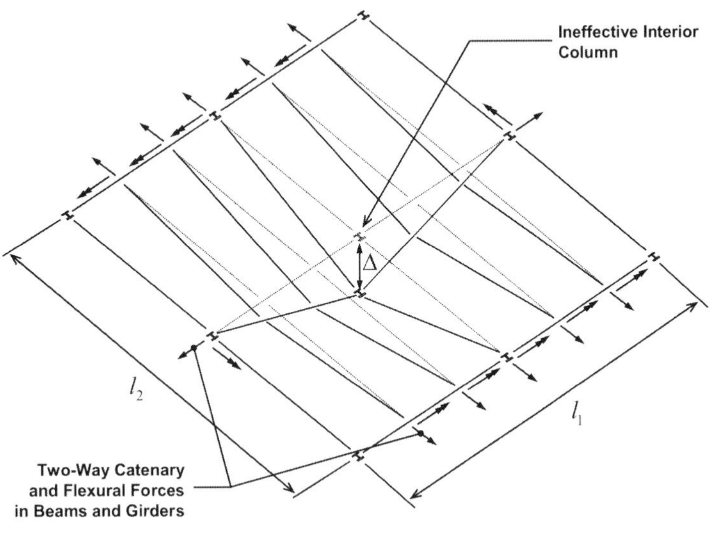

Figure 3 Two-Way Catenary/Flexure Action Resulting from Ineffective Interior Column.

As the interior column is rendered ineffective, the slab and grillage of steel members are forced to deform in a compatible manner and they both resist vertical deformation to the extent that their strength allows. The two-way membrane behavior in the slab is assumed to follow the theory described previously. Two way grillage (catenary/flexure) behavior in the steel framing can be computed using nonlinear structural analysis. These two theories can be used together to evaluate the robustness present in the typical interior 30-ft by 30-ft simple structural steel framing bay.

The framing connections that are assumed are considered flexible and are most commonly fabricated as double web angles (and sometimes referred to as web cleats). In order to assess the capabilities of double angle connections in facilitating the 3D grillage behavior, the web cleat moment capacity, tension capacity, and shear capacity needed to be determined. This process can be started by looking at the web cleat connection as being composed of bolt elements as shown in Figure 4.

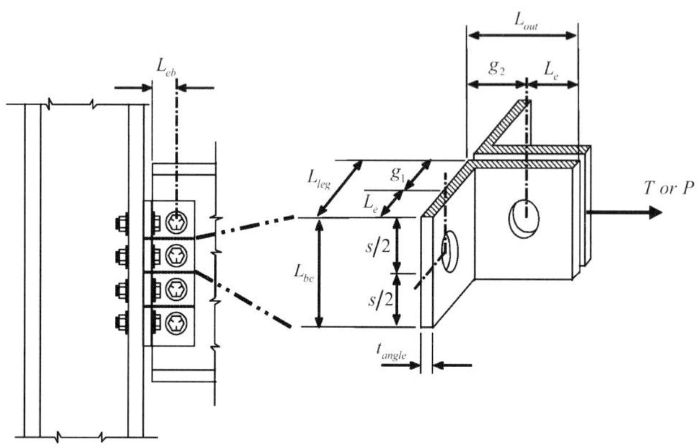

Figure 4 Web-Cleat to Bolt Element Transformation.

Researchers have been studying methodologies for determining pure-moment and tension capacities of bolted angle connections for quite some time (Astaneh-Asl *et al.* 1989a; Astaneh-Asl *et al.* 2002; Astaneh-Asl *et al.* 1989b; DeStefano and Astaneh-Asl 1991; DeStefano *et al.* 1991; DeStefano *et al.* 1994; Liu 2003; Liu and Astaneh-Asl 2000a; Liu and Astaneh-Asl 2000b; Shen and Astaneh-Asl 1999; Shen and Astaneh-Asl 2000; Wales and Rossow 1983).

The present study uses the approach of Shen and Astaneh-Asl (2000) and Liu and Astaneh-Asl (2000b) to develop nonlinear tension and compression behavior for bolt elements. These bolt elements can then be assembled to form web cleats whereupon moment-rotation behavior or tension/compression response of the connection can be developed. A trilinear tension-deformation response for the bolt element is derived using the procedure suggested by Liu and Astaneh-Asl (2000a); Liu and Astaneh-Asl (2000b); Shen and Astaneh-Asl (1999) and Shen and Astaneh-Asl (2000) with slight modifications. The compression and tension response models are shown in Figure 5.

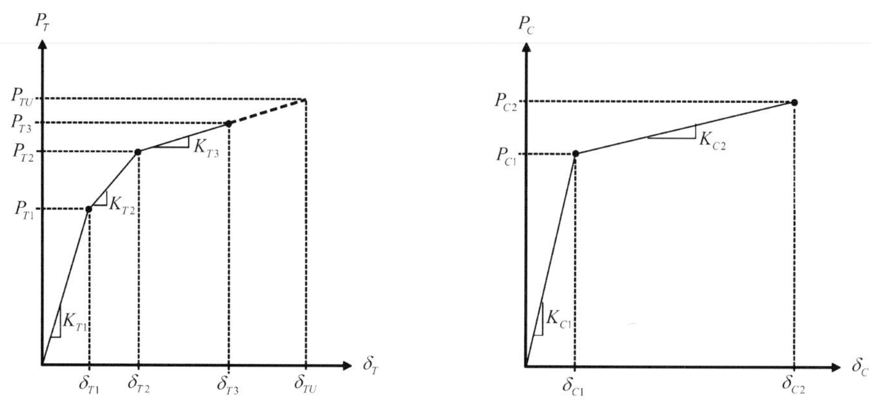

(a) (b)

Figure 5 Double Angle Bolt Element Response: (a) Tension and (b) Compression.

Three characteristic points on the tension response are generated using procedures recommended by Shen and Astaneh-Asl (2000) with slight modification. Point (P_{T1}, δ_{T1}) is defined using the yield moment in the legs of the angle. The initial stiffness, K_{T1}, is essentially the linear elastic stiffness of the bolt element considering bending of the legs perpendicular to the beam web and the axial extension of the leg parallel to the beam web. Point (P_{T2}, δ_{T2}) corresponds to the plastic mechanism capacity of the angle legs perpendicular to the beam web. The post-yield mechanism stiffness is defined as K_{T2}. The final point on the tension-deformation response is (P_{TU}, δ_{TU}). This point corresponds to the ultimate loading for the bolt element exclusive of bolt tension rupture or bolt shear rupture. It is defined through consideration of the angle legs perpendicular to the beam web forming catenary tension between the bolts in the support and the legs parallel to the beam web. The tension in the catenary at this ultimate loading is taken to be the loading corresponding to fracture on the net area through the angle leg perpendicular to the beam web. The final stiffness in the response is defined as K_{T3}.

The catenary tension force may or may not be able to form as a result of limits states being exceeded and therefore, a third point (P_{T3}, δ_{T3}) is defined. The loading, P_{T3}, is defined through consideration of the following bolt-element limit states;

- catenary tension fracture in the angle legs perpendicular to the beam web;
- tear-out bearing failure of the bolts in the beam web;
- tear-out bearing failure of the bolts in the angles;
- tension fracture of the bolts including prying action (Thornton 1985);
- tension fracture of the bolts excluding prying (superfluous);
- shear fracture of the bolts.

The yield point on the bolt element compression-deformation response (P_{C1}, δ_{C1}) is defined by considering three strength limit states;

- yield in the angle legs parallel to the beam web;
- yielding in the beam web;
- shear fracture of the bolts.

The ultimate loading capacity of the bolt element in compression is defined through consideration of the following strength limit states;

- crushing in the angle legs denoted by the ultimate tension stress being reached in the angle legs parallel to the beam web (conservative);
- crushing in the beam web denoted by the ultimate tension stress in the beam web being reached (again, conservative);
- 20% increase above the ultimate bolt shear stress magnitude.

The initial stiffness, K_{C1}, is defined using the smaller two stiffness magnitudes. The first is based upon web yielding and the second is based upon angle leg yielding. Post-yield stiffness is defined rather arbitrarily using a 0.5% multiplier to account for moderate strain hardening in the material on the way to crushing.

It should be noted that the behavior of the supporting element (*e.g.* a column flange, a column web, a girder flange) is omitted. This is likely very important, but the complexity incurred through consideration of this behavior would render the analysis proposed intractable. Expected yield and ultimate tensile stresses for the materials are used as recommended in the GSA guidelines (GSA 2003). Further details of the formulation and example computations can be found in Foley *et al.* (2006).

The tension and compression response for the bolt elements are shown in Figure 6 for the W18x35 and W21x68 wide flange shapes, respectively. These wide-flange shapes are consistent with the 3-story SAC-FEMA Boston building assumed as the analysis prototype (FEMA 2000b; Foley *et al.* 2006).

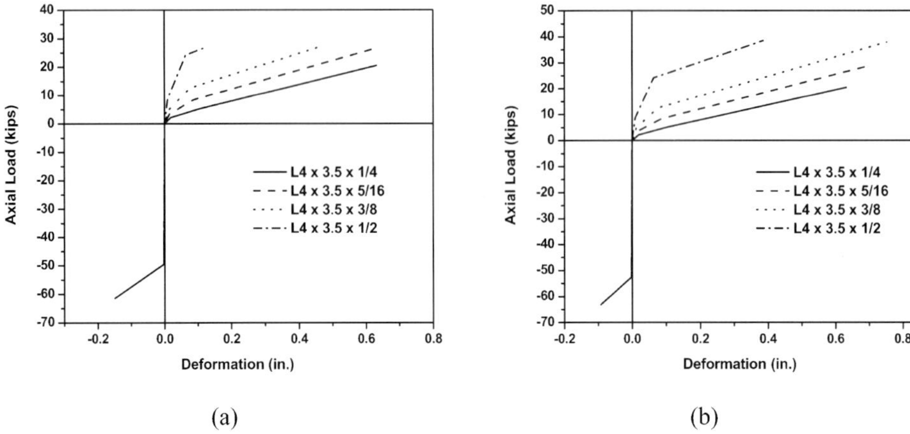

Figure 6 Bolt Element Tension and Compression Response for L4x3.5 Double Angles: (a) W18x35; and (b) W21x68.

The tension-deformation response varies considerably with beam shape and angle thickness. This is a byproduct of the varying limit states considered in the computations. For example, when thin angles are considered, the catenary tension action is allowed to form and rupture of the angle legs is the controlling limit state. However, as the angles get thicker, other limit states control the behavior. This is indicated by the "capping" of the tension forces in the 5/16, 3/8, and 1/2-inch angle thickness in the W18x35 beam shape and the 3/8 and 1/2-inch angle thickness with the W21x68 girder shape. The compression-deformation response is consistent indicating that the limit states controlling strength are consistent as well.

The bolt element ultimate strengths can be used to contribute to the determination of the tension capacity of the double angle connections through simple summation of the bolt element tension strengths in any given connection. The tensile capacity of the double-angle connection is determined through consideration of two additional limit states beyond those assumed for the bolt elements (Foley *et al* 2006):

- shear rupture of the bolts;
- tension fracture of the bolts including prying;
- block shear rupture in the angle legs parallel to the beam web;
- block shear rupture in the beam web;
- bearing tear-out failure in the angle legs parallel to the beam web;
- bearing tear-out failure in the beam web;
- catenary tension rupture in the angle legs perpendicular to the beam web.

The pure moment capacity of the double web-angle connection is determined using the bolt element tension- and compression-deformation response parameters described previously. The pure moment condition is defined by the deformation compatibility and internal equilibrium. The process for determining the pure moment capacity of the connection begins with defining the tension and compression response for each bolt element in the connection. A controlling state of deformation in the extreme tension angle, or extreme compression angle is assumed. These deformations are taken from the appropriate angle force-deformation curves. The connection rotation angle is then varied until the summation of all forces determined using the bolt element response curves sum to zero. This corresponds to the pure moment capacity of the connection. It should be noted that this process is iterative and the compression or tension deformation limit states may control the behavior. Details of the procedure can be found in (Foley *et al.* 2006).

The shear strength of the double angle connection given the beam shape chosen can be determined using the AISC Manual (AISC 2001). It should be noted that unfactored strengths were utilized and therefore, all manual-obtained strengths were divided by 0.75. The shear strengths for the double angles and beam shapes considered assume: $L_{ev} = 1.5''$; $L_{eh} = 1.5''$; $\phi = 1.0$; and 3/4" A325N bolts in STD holes. No consideration of expected strengths of the material in defining the shear strength was given.

The beams in the grillage are assumed to be W18x35's and the girders are W21x68's. From the AISC-LRFDM (AISC 2001), the W18 sections can support 3-5 rows of bolts, while the W21 sections can support 4-6 bolt rows with traditional spacing and end distances. Therefore, only these numbers of bolt rows were considered. Double-angle connections alone have a tensile capacity that ranges from 0.1-0.30 of the squash load of the cross-section (Foley *et al.* 2006). These are fairly significant tensile capacities (if taken as cumulative over all beam and girder members within the 3D system. The loading capacities are consistent with those found in testing by (Owens and Moore 1992). The moment capacities are very low, however (Foley *et al.* 2006). They range from 0.05 – 0.20 of the plastic moment capacity of the beam cross-section. This is consistent with the strength portion of the definition of a flexible connection (AISC 2005).

Bilinear moment-rotation response and axial load-extension response curves can be generated for the double angle connections using the bolt element response shown in Figure 5. Compression response characteristics are only used for defining moment rotation response. The connections in the grillage are not expected to go into compression in the ineffective column scenario considered. Tension secant stiffness for the bolt element, k_{BE}, is defined using point (δ_{T2}, P_{T2}) on the tension-deformation response. The tensile capacity of each bolt element in the double-angle connection then contributes to the tensile and moment capacity of the connection. The bilinear tension-deformation response of the bolt element is then characterized by the secant stiffness and the bolt element tensile capacity, P_{T3}.

The rotational and axial stiffness of the web-cleat connections are estimated using the magnitudes of the bolt element secant stiffness. In the case of axial tension, the axial stiffness of the double angle connection is simply the sum of the stiffness of each bolt element in the web cleat,

$$K_{\delta} = \sum_{i=1}^{n_b} k_{BE,i} \tag{5}$$

In general, if the bolt element stiffness, k_{BE}, is known and there is n_b bolt elements in the web cleat connection, the rotational stiffness can be computed as (Foley *et al.* 2006),

$$K_{\theta} = \sum_{i=1}^{n_b-1} i^2 \cdot \left(k_{BE} \cdot s^2\right) \tag{6}$$

where s is the pitch of the bolt elements (taken as a constant value of 3 inches).

The axial stiffness and flexural stiffness of the web cleat connections can be defined as a function of the axial rigidity and flexural rigidity of the connected member. This is mathematically defined as,

$$K_{\delta} = \alpha_{\delta} \cdot \frac{AE}{L} \tag{7}$$

$$K_{\theta} = \alpha_{\theta} \cdot \frac{EI}{L} \tag{8}$$

The rotational stiffness of the web-cleat connections are well below the stiffness limit corresponding to flexible connections (AISC 2005) given by $\alpha_{\theta} = 2$. The majority of the rotational stiffness multipliers are in the range; $0.05 \le \alpha_{\theta} \le 1.50$ (Foley *et al.* 2006). One exception is the 5 bolt arrangement in a W18x35 beam member. The axial stiffness multiplier for the majority of the connection arrangements lies in the range $0.10 \le \alpha_{\theta} \le 1.8$ With the 5-bolt connection in the W18x35 member giving $\alpha_{\delta} = 2.3$.

The analysis begins by computing the capacity of the concrete-steel composite slab system acting as a two-way membrane using equations (1) through (4). The steel deck is assumed to be 2VLI22 (Vulcraft 2005) and 40% of the cross-sectional area is assumed to be effective as tensile reinforcement (Foley *et al.*

2006). Welded-wire-mesh is assumed in the concrete deck: 6x6-W1.4xW1.4 (shrinkage and temperature reinforcement). When the interior column looses effectiveness, the concrete slab panel is 60-feet by 60-feet. The membrane capacity of the concrete slab-steel deck system is approximately 50-psf at 26.2 inches of vertical deflection at the center of the panel (Foley *et al.* 2006). This magnitude of vertical deflection corresponds to an approximate rotational demand of 0.073 radians, which is well below the limit of 0.21 radians (GSA 2003). It should also be noted that the rotation computed here is a total rotation (elastic plus plastic components). Therefore, the magnitude computed is conservative. The tension force in the steel deck running perpendicular to the in fill beams is approximately 566 lbs/in (Foley *et al.* 2006) along one edge of the panel.

The capacity of the steel grillage is then computed. A structural model for the steel floor framing system was developed for use in MASTAN2 (Zieman and McGuire 2000). A schematic of the analytical model is shown in Figure 7.

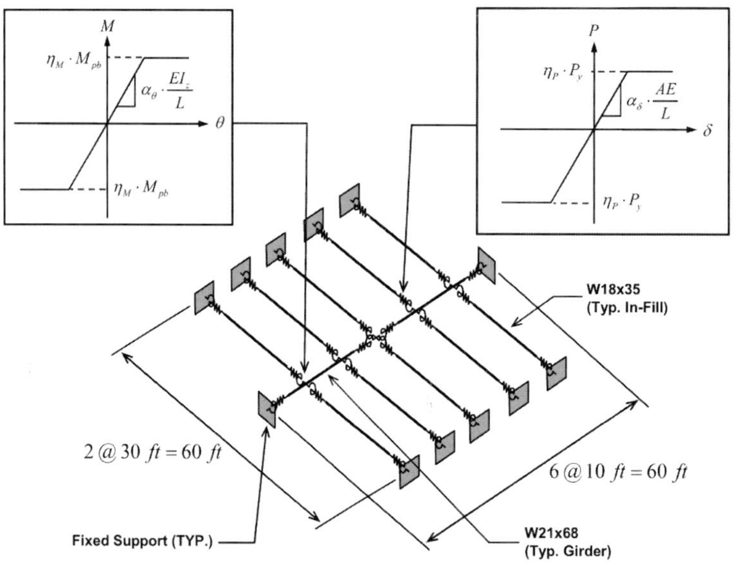

Figure 7 Steel Grillage Model Schematic (System 1) Illustrating Axial and Moment Connection Modeling for Nonlinear Analysis.

All members are modeled using multiple elements: in-fill beams are modeled using 10 elements and girders are modeled using 9 elements. The in-fill beams were modeled using 4 analytical segments. Two segments (*i.e.* 1/2 of the beam length) were centered on the beam mid-span. The end 1/4 lengths of beam were subdivided into 4 additional segments to facilitate connection modeling. Therefore, all in-fill beams contain end segments that are $1/16^{th}$ of their span. The end segments in the girders (at column supports and interior column location) were broken down into 4 segments yielding end connection segments of $1/12^{th}$ the girder span.

The end connections were modeled in the analytical segments of the beams and girders located immediately adjacent the fixed supports, the supporting girders, and the interior column. The connection rotational stiffness, K_θ, was input using the built-in capability. The connection moment capacity was interjected into the analytical model by adjusting the beam or girder's plastic moment capacity to $\eta_M \cdot Z_x$. The axial loading characteristics were included in a slightly different manner. MASTAN2 does not allow axial spring characteristics to be directly modeled. The cross-sectional areas of the beam or girder in the end connection segments were defined to be $\eta_P \cdot A_g$. This reduction in cross-sectional area also created implied linear spring stiffness in this isolated region of the beam equal to $\alpha_\delta \cdot AE/L = \eta_P \cdot AE/L$.

The method of modeling connections creates a "stub member" that has an axial capacity and a moment capacity that is the same as the connection intended. Three systems with varying connection characteristics were considered (Foley *et al.* 2006):

System 1	System 2	System 3
$\eta_M = 0.10 \quad \alpha_\theta = 0.50$	$\eta_M = 0.50 \quad \alpha_\theta = 5.0$	$\eta_M = 0.30 \quad \alpha_\theta = 2.0$
$\eta_P = 0.20 \quad \alpha_\delta = 0.20$	$\eta_P = 0.30 \quad \alpha_\delta = 0.30$	$\eta_P = 0.30 \quad \alpha_\delta = 0.30$

System 1 has strength and stiffness characteristics typical of web-cleat connections used in structural steel floor framing systems. System 2 has strength and stiffness characteristics typical of partially restrained beam-to-girder connections (Rex and Easterling 2002) and there is long-standing use of partially restrained girder-to-column connections. The axial strength and stiffness were increased slightly from that of System 1. A third system was considered. This system had a better balance between axial capacity and moment capacity than system 2. The axial strength and stiffness for the connections in system 3 were left the same as those in system 2. The bending strength and stiffness of the connections were reduced to a level slightly above that in System 1 and below that in system 2. The moment and axial strength characteristics are consistent with web cleat connections that are relatively thick (compared to the typical thickness used) and a number of bolt rows that fills up the beam and girder web (Foley *et al.* 2006).

MASTAN2 then uses these pieces of information to create an interaction (yield) surface of the form shown in Figure 8 for three systems considered.

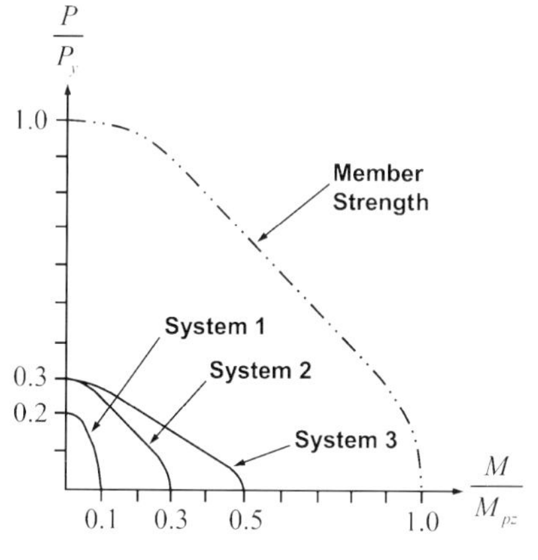

Figure 8 Member and Connection Interaction Surfaces for Connected Member and Three Grillage Systems (connection characteristics vary).

It should be noted that minor-axis bending is assumed to have a connection capacity that is equal to the minor axis moment capacity of the members and the connection stiffness in the minor-axis direction is infinite relative to the flexural rigidity of the connected beam (*i.e.* the connection is fully-restrained).

Each floor system is evaluated independently under the assumption that it carries its own loading. The slab system was determined previously to be capable of supporting approximately 50 psf through membrane action. The total unfactored live loading used for design of the system is: 80-psf dead loading; and 50 psf office occupancy live loading. The total point in time live loading that can be assumed present at the time a column is rendered ineffective can be computed as (GSA 2003);

$$q_{p.i.t.} = 1.0(80\ psf) + 0.25(50\ psf) = 93\ psf$$

The steel grillage will then be required to carry the following superimposed loading (with a deformation that is compatible with the slab membrane);

$$q_{grillage} = \beta_{dynam} \cdot (1.0D + 0.25L) - 50 = \beta_{dynam} \cdot (93psf) - 50psf$$

At pseudo-static loading levels $(\beta_{dynam} = 2.0)$ prescribed in the GSA Guidelines (GSA 2003), the grillage will need to support a uniformly distributed loading of 136 psf. However, this assumes that the supporting column is "vaporized". Furthermore, former studies (Liu *et al.* 2005; Marchand and Alfawakhiri 2004; Powell 2005) and relatively recent research (Foley *et al.* 2006) have shown that the multiplier commonly used to simulate dynamic loading can vary considerably. If the supporting column is not "vaporized", but simply compromised (*i.e.* it still has a fraction of its initial load capacity), then one might argue that the point-in-time loading alone needs to be carried $(\beta_{dynam} = 1.0)$ without dynamic multiplication. Therefore, in this case, the grillage must support 43 psf superimposed loading.

The MASTAN2 model shown in Figure 7 was analyzed using 2nd order inelastic analysis and a reference superimposed loading on the steel grillage of 108 psf. The load deformation response for the three systems is shown in Figure 9.

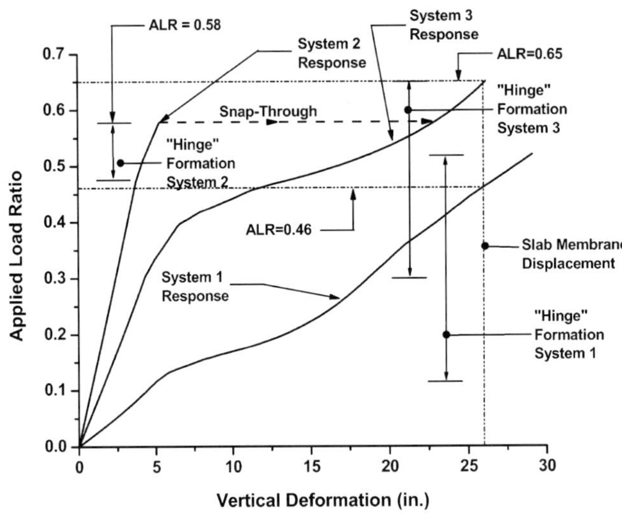

Figure 9 Load Deformation Response of Three Grillage Systems Considered.

The load deformation response of system 1 indicates that there is a very early transition from flexural behavior to catenary behavior in the grillage. The connection strengths and stiffness result in the cross-sections at the ends of the members reach the yield surfaces very early in the response and the large displacements result in catenary tension in the grillage forming. This transition is exhibited by the shallow yield plateau-like response and subsequent stiffening behavior. The applied load ratio that results in deformations compatible with the membrane displacement computed earlier (26 inches) is 0.46. This indicates that the capacity of the system (both slab and grillage) is;

$$q_{cap} = 0.46(108) + 50 \approx 100\ psf$$

Therefore, System 1 can definitely support the point in time live loading and there is some reserve for dynamic amplification: $\beta_{dynam} = 100/93 = 1.08$. If one were to assume that the system could continue to deflect without membrane reinforcement in the slab rupturing, or the anchorage of this reinforcement being compromised (*e.g.* deflection to approximately 30 inches), the membrane capacity would increase and the catenary capacity of the grillage could increase. This increase is shown in Figure 9 at $ALR = 0.52$. This would result in the system capacity moving upward to,

$$q_{cap} = 0.52(108) + 50 \approx 106\ psf$$

and the dynamic multiplier would naturally increase as well to $\beta_{dynam} = 106/93 = 1.14$. One should note that shrinkage and temperature welded wire fabric reinforcement was assumed as well as 22-gauge steel deck. Greater capacities can likely be attained if thicker deck is used and mild-steel reinforcement rather than welded wire mesh (Foley *et al.* 2006).

At 26 inches of vertical displacement, the total rotation over the beam and girder span of 30 feet was computed previously as approximately 0.07 radians. This is very close to the plastic rotational limit of 0.06 radians recommended for web-angle connections (FEMA 2000a). However, the present rotational demand is "total" and the plastic demand will likely align itself close to this limit. Therefore, the rotational demands at the level of loading considered are not likely to cause rupture of the connections.

The same reference loading was applied to the steel grillage of system 2. The load deformation response of the grillage system 2 is also shown in Figure 9. It is interesting to note that the catenary (stiffening) response is not present. The reason for this is that a plastic mechanism (flexural) forms at an applied load ratio of 0.58 with vertical deformation slightly less than 5 inches. This amount of vertical deformation is not sufficient to *activate* the geometric stiffness for the members in the floor system. In other words, analytically, catenary action is not allowed to form and the system numerically "fails". It is understood that there will be a *conversion* to catenary action once the mechanism forms, but the structural analysis is not able to consider this transformation because the tangent stiffness matrix of the system is singular at the instant this group of beam mechanisms forms.

The number of hinges that form in System 2 and the loading range over which they form is much less than that of System 1. One would like to have a system where there is a significant number of hinges forming so that full advantage of the structural indeterminacy and load redistribution is taken. When the hinges form over very short loading ranges, there is less redundancy and toughness in the *system*. The significantly smaller deformation in System 2 at the formation of the collapse mechanism would indicate that the grillage will form a bending moment collapse mechanism first with subsequent reliance on back-up capacity catenary action after significant vertical deformation.

Experimental rotations attained by Rex and Easterling (2002) for the partially-restrained beam-to-girder connections were reported to be on the order of 0.05 radians. If one were to rely on catenary action after the flexural mechanisms occurs, the vertical deformations in the system would likely rapidly increase to those found in the first system (approximately 26 inches). As a result, even though the flexural mechanism forms early at 5 inches of deformation there will need to be an additional 21 inches of deformation in the grillage needed to activate catenary action. As a result, the rotational demands on these connections are likely to be on the order of 0.07 radians. It is unclear if the PR beam-to-girder connection can support his level of rotational demand without fracture.

The axial stiffness and strength of the connections in system 3 are consistent with those of system 2 and therefore, it is expected that the catenary behavior will be the same in the two systems once it is activated. The load deformation response of system 3 is shown in Figure 9. After the formation of the flexural collapse mechanism in system 2, it is likely that the steel grillage will need to abruptly accumulate an additional 20 inches of deflection in order to reach the catenary tension stiffening that comes from the contribution of geometric stiffness. This behavior is analogous to snap-through behavior in arches and is schematically indicated in Figure 9. It should be noted that the response of system 3 indicates that system 2 will indeed be able to reach the same load carrying capacity of system 3, but it is not economically advantages to provide additional bending moment capacity and stiffness when there is no enhancement in load carrying capacity. Furthermore, dynamic snap-through behavior may have adverse ramifications with regard to system integrity and toughness.

System 3 is capable of supporting an applied load ratio of 0.65 at 26 inches of vertical deformation, which is compatible with the deformations needed for the slab system membrane to support 50 psf. This reveals that a relatively economical (simple-framing) system can support the following superimposed floor loading;

$$q_{cap} = 0.65(108) + 50 \approx 120\ psf$$

This magnitude of loading reveals that this system can allow for a dynamic amplification factor of $\beta_{dynam} = 120/93 = 1.30$.

Concluding Remarks

An introduction to methodologies available for computing the membrane tension capacity of concrete-steel composite slab systems was provided. A hypothetical event whereby an interior column in a typical structural steel framing system is rendered ineffective was outlined and a methodology for computing the load carrying capacity of the steel floor framing system after this event was described. Inelastic static structural analysis was conducted using bilinear connection characteristics typical of simple framing connections in steel systems. The analysis conducted included bilinear connection modeling for axial load-deformation response and moment-rotation response within the structural steel framing system.

A static nonlinear analysis of the typical 30-ft by 30-ft framing system that included nonlinear connection behavior consistent with that of web-cleat connections was conducted. The analysis indicated that while it is doubtful that the typical structural steel framing system could support the GSA-level dynamic loading estimates, one can say with certainty that the typical structural steel framing system can ***resist*** progressive collapse in the event an internal column is rendered ineffective. This statement is supported by the fact that the point-in-time loading of $1.0D + 0.25L$ can be supported through catenary and flexural action in the structural steel framing and membrane action in the composite concrete-steel deck system with moderate levels of dynamic amplification reserve.

The analysis conducted suggests that it is better to have smaller moment capacity and flexural stiffness for connections distributed throughout the floor framing system (as is typically found in structural steel interior framing arrangements). When the moment capacity is low, there is a smooth transition between the formation of the flexural mechanism and the catenary tension behavior that is essentially secondary after the initial collapse. If the moment capacity is too large, there will likely be snap-through-type behavior whereupon a significant magnitude of vertical displacement will rapidly take place prior to the formation of catenary action.

In general, a balance between flexure and catenary action in the steel grillage can be attained when the following axial and moment characteristics are met in regard to the connections at the ends of the beams and girders in the structural steel system;

$$M_{conn} \leq 0.30 M_{pb} \text{ and } K_\theta \leq 2\frac{EI}{L} \qquad P_{conn} \leq 0.3 P_y \text{ and } K_\delta \leq 0.3\frac{AE}{L}$$

It is interesting to note that this behavior is nearly approached for typical structural steel framing systems. Axial capacity, axial stiffness, and rotational stiffness characteristics needed are easily attained by providing web-cleat connections (double angle connections) that "fill up" the web of the connected beam or girder. The moment capacity recommendations may be able to be attained if better modeling of the connection limit states is performed. Foley *et al.* (2006) found that 1/2-inch thick web connection angles that fill up the web of the connected member can approach $0.2M_{pb}$.

The results of the study suggests that typical structural steel framing systems have significant levels of inherent robustness and general structural integrity without providing any special design effort. Furthermore, if slight increases in connection angle thickness and the number of bolt rows used in these connections are provided, the toughness of the system in response to abnormal events can be significantly enhanced.

References

AISC. (2001). *Manual of Steel Construction*, American Institute of Steel Construction, Chicago, IL,

AISC. (2005). *Load and Resistance Factor Design Specifications for Structural Steel Buildings*, American Institute of Steel Construction, Chicago, IL,

Allam, A., Burgess, I., and Plank, R.(2000) "Simple Investigations of Tensile Membrane Action in Composite Slabs in Fire." *International Conference on Steel Structures of the 2000's*, Istanbul, Turkey, pp. 327-332 (www.shef.ac.uk/fire-research/publications.html),

Astaneh-Asl, A., Call, S. M., and McMullin, K. M. (1989a). "Design of Single Plate Shear Connections." *Engineering Journal*, First Quarter, American Institute of Steel Construction, Chicago, IL, pp. 21-31.

Astaneh-Asl, A., Liu, J., and McMullin, K. M. (2002). "Behavior and Design of Single Plate Shear Connections." *Journal of Constructional Steel Research*, 58, 1121-1141.

Astaneh-Asl, A., Nader, M. N., and Malik, L. (1989b). "Cyclic Behavior of Double Angle Connections." *Journal of Structural Engineering*, 115(5), American Society of Civil Engineers, Reston, VA, pp. 1101-1118.

Bailey, C. G., White, D. S., and Moore, D. B. (2000). "The Tensile Membrane Action of Unrestrained Composite Slabs Simulated Under Fire Conditions." *Engineering Structures*, 22, 1583-1595.

Burgess, I. W., Huang, Z., and Plank, R. J.(2001) "Non-Linear Modelling of Steel and Composite Structures in Fire." *International Seminar on Steel Structures in Fire*, Shanghai, China, pp. 1-15 (www.shef.ac.uk/fire-research/publications.html),

Cai, J., Burgess, I. W., and Plank, R. J. (2002). *A Generalized Steel/Reinforced Concrete Beam-Column Element Model for Fire Conditions.* University of Sheffield - Department of Civil and Structural Engineering, Sheffield, U.K., www.shef.ac.uk/fire-research/publications.html.

DeStefano, M., and Astaneh-Asl, A. (1991). "Axial Force-Displacement Behavior of Steel Double Angles." *Journal of Constructional Steel Research*, 20, Elsevier Science, Ltd., 161-181.

DeStefano, M., Astaneh-Asl, A., DeLuca, A., and Ho, I.(1991) "Behavior and Modeling of Double Angle Connections Subjected to Axial Loads." *1991 Annual Technical Session - Inelastic Behavior and Design of Frames*, Chicago, IL, Structural Stability Research Council, pp. 323-334,

DeStefano, M., DeLuca, A., and Astaneh-Asl, A. (1994). "Modeling of Cyclic Moment-Rotation Response of Double-Angle Connections." *Journal of Structural Engineering*, 120(1), American Society of Civil Engineers, Reston, VA, pp. 212-229.

FEMA. (2000a). *State of the Art Report on Connection Performance (FEMA-355D)*, SAC Joint Venture and Federal Emergency Management Agency, Washington, D.C.,

FEMA. (2000b). *State of the Art Report on Systems Performance of Steel Moment Frames Subject to Earthquake Ground Shaking (FEMA-355C)*, SAC Joint Venture and Federal Emergency Management Agency, Washington, D.C.,

Foley, C. M., Martin, K., and Schneeman, C. (2006). *Robustness in Structural Steel Framing Systems, Draft Report - January 2006.* American Institute of Steel Construction, Chicago, IL,

Grierson, D. E., Xu, L., and Liu, Y. (2005). "Progressive-Failure Analysis of Buildings Subjected to Abnormal Loading." *Computer-Aided Civil and Infrastructure Engineering*, 20, Blackwell Publishing, Malden, MA, pp. 155-171.

GSA. (2003). "Progressive Collapse Analysis and Design Guidelines for New Federal Office Buildings and Major Modernization Projects." General Services Administration.

Hawkins, N. M., and Mitchell, D. (1979). "Progressive Collapse of Flat Plate Structures." *ACI Journal*, 76(8), American Concrete Institute, Detrioit, MI, pp. 775-808.

Huang, Z., Burgess, I. W., and Plank, R. J. (2000a). "Effective Stiffness Modelling of Composite Concrete Slabs in Fire." *Engineering Structures*, 22, 1133-1144.

Huang, Z., Burgess, I. W., and Plank, R. J. (2000b). "Three-Dimensional Analysis of Composite Steel-Framed Buildings in Fire." *Journal of Structural Engineering*, 126(3), 389-397.

Huang, Z., Burgess, I. W., and Plank, R. J. (2003a). "Modeling Membrane Action of Concrete Slabs in Composite Buildings in Fire. I: Theoretical Development." *Journal of Structural Engineering*, 129(8), 1093-1102.

Huang, Z., Burgess, I. W., and Plank, R. J. (2003b). "Modeling Membrane Action of Concrete Slabs in Composite Buildings in Fire. II: Validations." *Journal of Structural Engineering*, 129(8), 1103-1112.

Liu, J. (2003). "" University of California - Berkeley, Berkeley, CA.

Liu, J., and Astaneh-Asl, A. (2000a). "Cyclic Testing of Simple Connections Including Effects of Slab." *Journal of Structural Engineering*, 126(1), 32-39.

Liu, J., and Astaneh-Asl, A. (2000b). *Cyclic Tests on Simple Connections Including the Effects of the Slab.* Report No. SAC/BD-00/03, SAC Joint Venture,

Liu, R., Davison, B., and Tyas, A.(2005). "A Study of Progressive Collapse in Multi-Storey Steel Frames." *Proceedings of the 2005 Structures Congress and the 2005 Forensic Engineering Symposium*, New York, NY, American Society of Civil Engineers, pp. CD-ROM,

Marchand, K. A., and Alfawakhiri, F. (2004). *Blast and Progressive Collapse.* Facts for Steel Buildings Number 2, American Institute of Steel Construction, Chicago, IL,

Mitchell, D., and Cook, W. D. (1984). "Preventing Progressive Collapse of Slab Structures." *Journal of Structural Engineering*, 110(7), 1513-1532.

Owens, G. W., and Moore, D. B. (1992). "The Robustness of Simple Connections." *The Structural Engineer*, 70(3), Institution of Structural Engineers, London, U.K., pp. 37-46.

Powell, G. P.(2005) "Progressive Collapse: Case Studies Using Nonlinear Analysis." *Proceedings of the 2005 Structures Congress and the 2005 Forensic Engineering Symposium*, New York, NY, American Society of Civil Engineers, pp. CD-ROM,

Rex, C. O., and Easterling, W. S. (2002). "Partially Restrained Composite Beam-Girder Connections." *Journal of Constructional Steel Research*, 58, 1033-1060.

Shen, J., and Astaneh-Asl, A. (1999). "Hysteretic Behavior of Bolted-Angle Connections." *Journal of Constructional Steel Research*, 51, Elsevier Science Ltd., 201-218.

Shen, J., and Astaneh-Asl, A. (2000). "Hysteresis Model of Bolted Angle Connections." *Journal of Constructional Steel Research*, 54, Elsevier, 317-343.

Thornton, W. A. (1985). "Prying Action - A General Treatment." *Engineering Journal*, Second Quarter, American Institute of Steel Construction, Chicago, IL, pp. 67-75.

Vulcraft. (2005). *Vulcraft Steel Roof and Floor Deck Catalog*, Vulcraft - A Division of Nucor Corportation, www.vulcraft.com.

Wales, M. W., and Rossow, E. C. (1983). "Coupled Moment-Axial Force Behavior in Bolted Joints." *Journal of the Structural Division*, 109(5), American Society of Civil Engineers, Reston, VA, pp. 1250-1266.

Zieman, R. D., and McGuire, W. (2000). *MASTAN2 Ver. 1.1*, John Wiley & Sons, New York, NY.

STAINLESS STEEL REBAR FOR SEISMIC APPLICATIONS

Alberto Franchi[1], **Pietro Crespi**[1], **Aldo Bennani**[2] **and Marco Farinet**[2]
[1]*Politecnico di Milano, Milan, Italy*
E-mail: alberto.franchi@polimi.it, pietro.crespi@polimi.it
[2]*Cogne Acciai Speciali Aosta, Italy*
E-mail: aldo.bennani@cogne.com, marco.farinet@cogne.com

Abstract

This paper is meant to present the mechanical characteristics of austenitic stainless steel rebars under monotonic and cyclic loadings. Furthermore, some results concerning the experimental tests on column prototypes subjected to cyclic loadings are presented, with the intent of comparing with results obtained on analogous columns reinforced with standard high ductile carbon steel rebars.

Keywords: stainless steel, rebar, ductility, cyclic loading, seismic loading

1. Introduction

Stainless steel rebars are used primarily in reinforced concrete structures because of their corrosion resistance. It is well known from the literature (Park et al., 1982; Pipa and Carvalho, 1990; Monti and Nuti, 1992; Macchi et al., 1996; Franchi et al., 1996; Pantazopoulou, 1998; Riva and Franchi, 2001; Riva et al., 2001) that, for seismic applications, the reinforcing steel has to provide enough ductility, i.e. plastic elongation at ultimate stress, and enough strain hardening, measured by the ratio of ultimate stress over yield stress. At the same time, it is very well known from the literature, that austenitic stainless steel shows a remarkable degree of ductility and strain hardening. The question, therefore, comes out as follows: how does behave austenitic stainless steel reinforcement under seismic loading conditions?

2. Process Route

The melting shop used by Cogne Acciai Speciali (Aosta, Italy) for the production of standard stainless steel grades is equipped with an 80 Ton UHP furnace and an AOD converter with the same capacity (EC Report, 2005). The casting process carries out trough a continuous casting machine in square billet 160 mm. The billets are ground and inspected, in order to eliminate all the surface defects due to the casting operations. On line, the hot rolling process is performed in the wire rod and bar mill, consisting on a heating furnace, roughing stand and a continuous mill, capable of producing wire rod in coil from 5.5 to 32 mm and bars from 18 to 100 mm. The rolling mill provides to transform directly the billets into rebars, whose diameters can range between 14 to 50 mm; suitable heat treatment, sand blasting and picking process give the final surface aspect and mechanical characteristics to stainless steel rebars.

3. Tensile Tests under Monotonic Loading on AISI304 Ø16

Figure 2 shows the apparatus equipment for the quasi-static tensile tests under elongation control.

M. Pandey et al. (eds), Advances in Engineering Structures, Mechanics & Construction, 255–264.

Fig. 1. Rolling mill.

Fig 2. Elongation test apparatus.

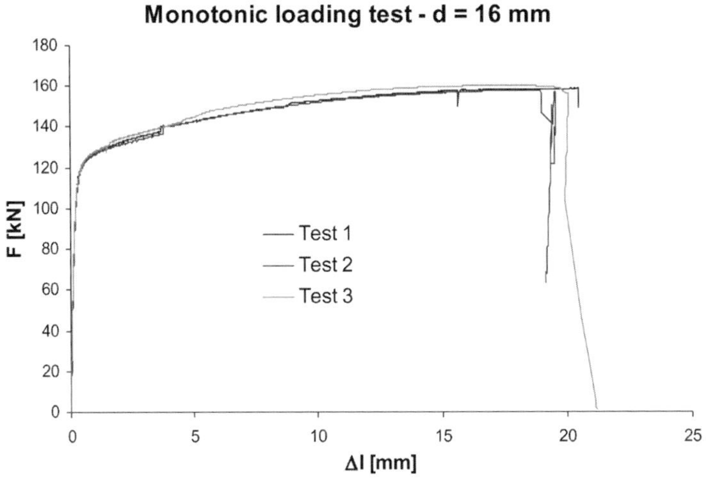

Fig. 3. Load-displacement of the tensile test.

The load-transducer displacement is reported for three analogous specimens in Figure 3. The end of the horizontal line of the graphs (about 20 mm) is due to the limited (20 mm) maximum elongation of the displacement transducer and not to the final stage of the stainless steel bar specimen.

3.1. *Results of the Monotonic Tensile Loading Test*

Figure 4 shows the stress-strain diagrams of the No. 3 stainless steel samples together with an optimized carbon "Tempcore" bar.

Experimental results concerning four major quantities, like yield stress-ultimate stress, ultimate over yield ratio, uniform elongation at maximum force are given in Table 1, in comparison with a 450 MPa carbon "Tempcore" steel rebar. It is evident that the AISI304 rebar is much superior with respect the carbon steel, even if this last is of the class 500 MPa instead of 450 MPa.

3.2. *Cyclic Loading on AISI304 Ø16* $(\Delta l/l = 1\%)$

The geometrical data of the test specimen are given in Figure 5. The test has been repeated for three specimens. A mark every 10 mm has been placed along a straight line of the external surface in order to measure the uniform maximum elongation after failure (A_{gt}). The distance between the two grips

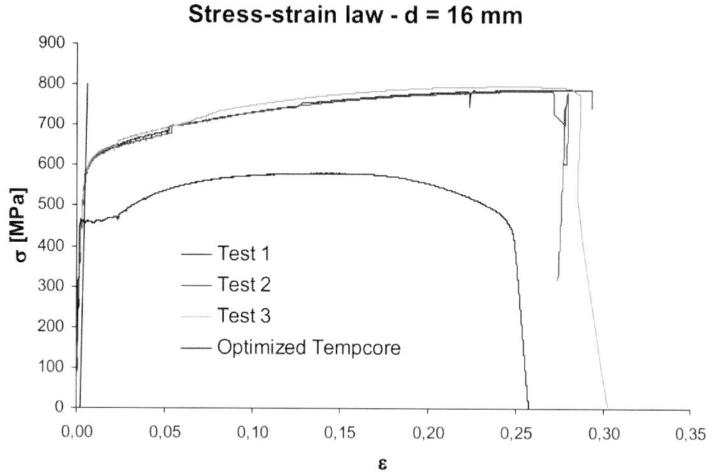

Fig. 4. Comparative results of the uniaxial tensile test between carbon and stainless steel rebar.

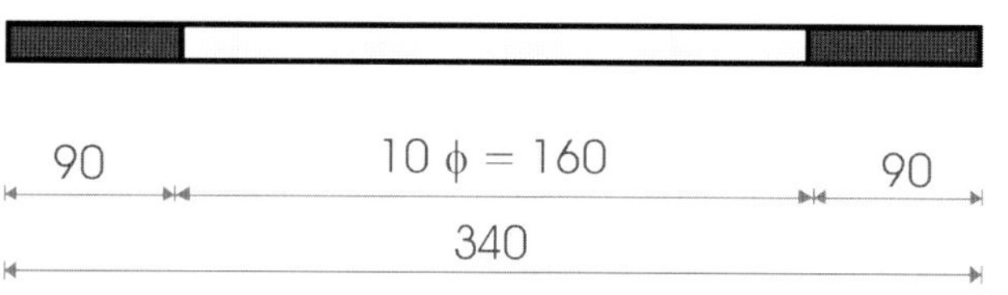

Fig. 5. Geometrical data of the specimens for the cyclic loading tests (dimensions in mm).

has been fixed to 10 times the diameter of the bar. The tests have been performed in displacement control with symmetric tension-compression cycles; frequency 1÷3 Hz; amplitude of the imposed displacement: $\Delta L = \pm 1.6$ mm (= 1% of 160 mm).

3.3. *Results of the Cyclic Loading*

Figure 6 shows the hysteretic stress-strain diagram for about 200 cycles.

Figure 7 shows the load-displacement cyclic diagram but limited to the first 10 cycles; analogous cycles, corresponding to the carbon "Tempcore" rebar, are inserted, for comparison purposes, in the same figure.

It is interesting to point out: AISI304 presents higher values of the force for a given elongation than the carbon steel; this is because of the higher yield point and of the higher strain hardening of the AISI304 with respect the "Tempcore". The decrease of the maximum resistance, especially

Table 1. Resistances and ductility values.

	R_e [MPa]	R_m [MPa]	R_m/R_e	A_{gt} [%]
Inox (Ø16) test 1	558.01	788.89	1.41	29.00
Inox (Ø16) test 1	554.66	785.10	1.42	26.75
Tempcore (Ø14)	472.78	584.95	1.24	14.39
Tempcore (Ø14)	478.36	599.03	1.25	12.89

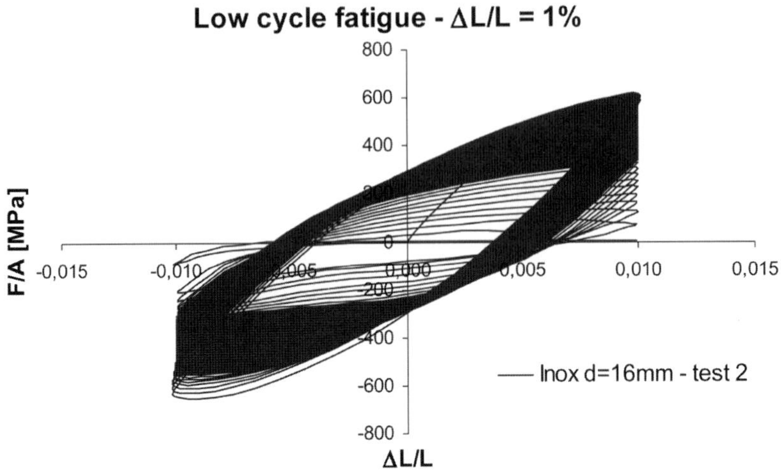

Fig. 6. Cyclic hysteretic load-displacement behaviour up to failure.

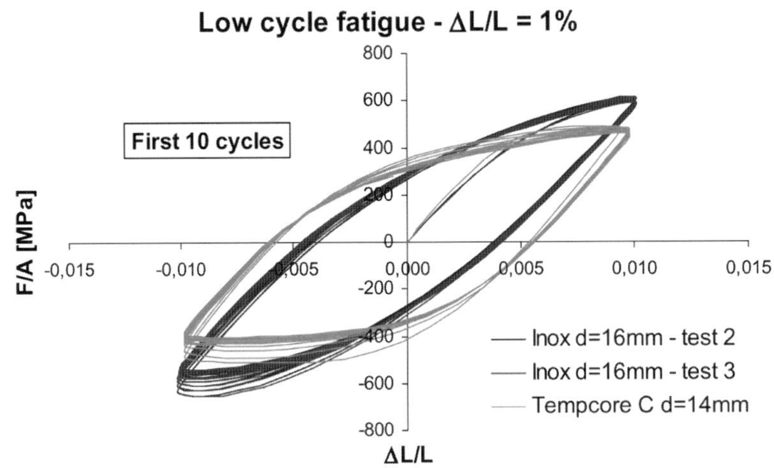

Fig. 7. Load-elongation hysteretic curves for the first 10 cycles.

in compression (i.e. in the negative side of the horizontal axis) is due to: (i) the crack initiation in the plastic hinge of the middle section and (ii) to the geometric effect, which is due to the fact that the pure compression mode becomes a bending-compression deformation mode. Figure 8 shows the decrease of the tension (greater values) and of the compression (smaller values) force along the cycles; it is evident how the carbon steel curve lies always below the lines of the AISI304. Low cycle fatigue resistance of AISI304 is almost twice (200÷250 cycles) as much as the carbon steel (100÷120 cycles).

Figure 9 gives total energy dissipation with the number of cycles, while Figure 10 gives the energy dissipated inside each cycle for all cycles.

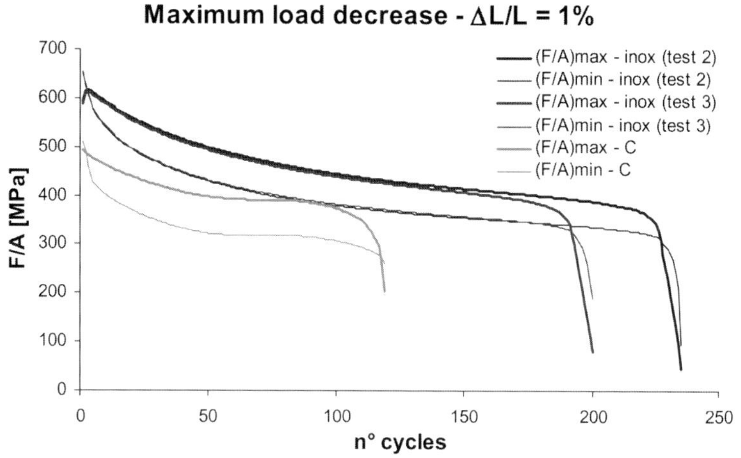

Fig. 8. Resistance decrease along the cycles.

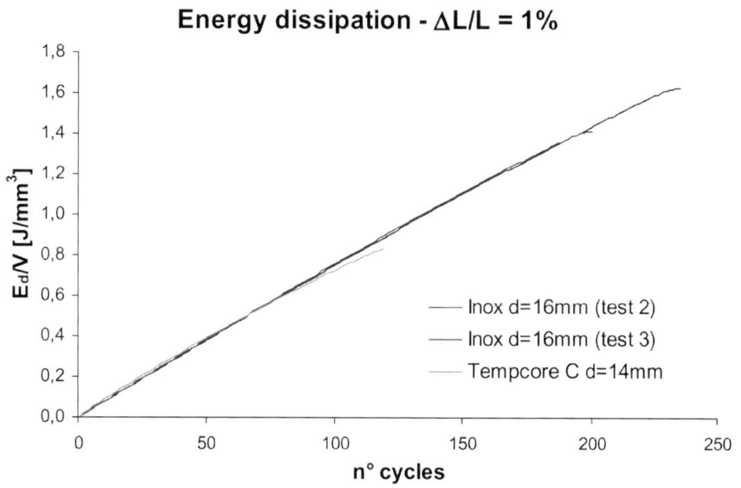

Fig. 9. Energy dissipation along the cycles.

4. Tests on R.C. Frames under Cyclic Loading

4.1. *The Prototype Design*

Figure 11 shows the steel reinforcement of the column and of the foundation. The steel reinforcement of the column consists of 4 Ø16 and stirrups Ø8 at 70 mm distance, inside the plastic hinge, made of AISI304.

4.2. *Test Rig*

The test apparatus consists of a stiff horizontal steel beam at which the foundation of the column is tight through two steel plates connected by 4 pre-tensioned steel bars. The r.c. foundation has been designed in such a way that it remains in the elastic domain while the plastic hinge will develop at the bottom part of the column. The top section of the column is connected to a horizontal actuator

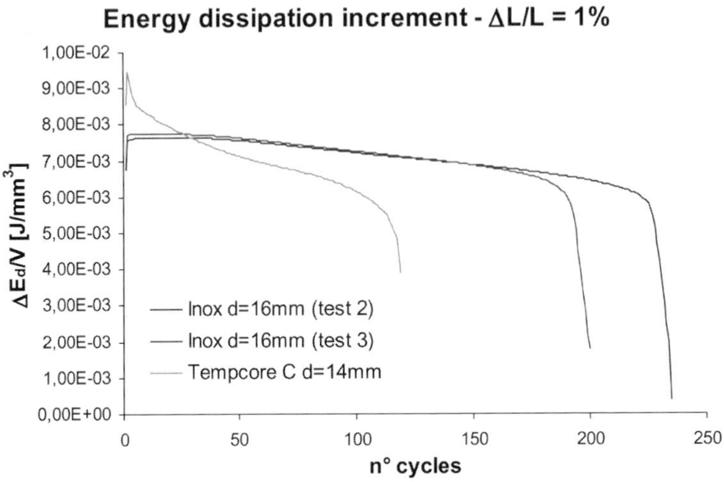

Fig. 10. Variation of energy dissipation in any cycle.

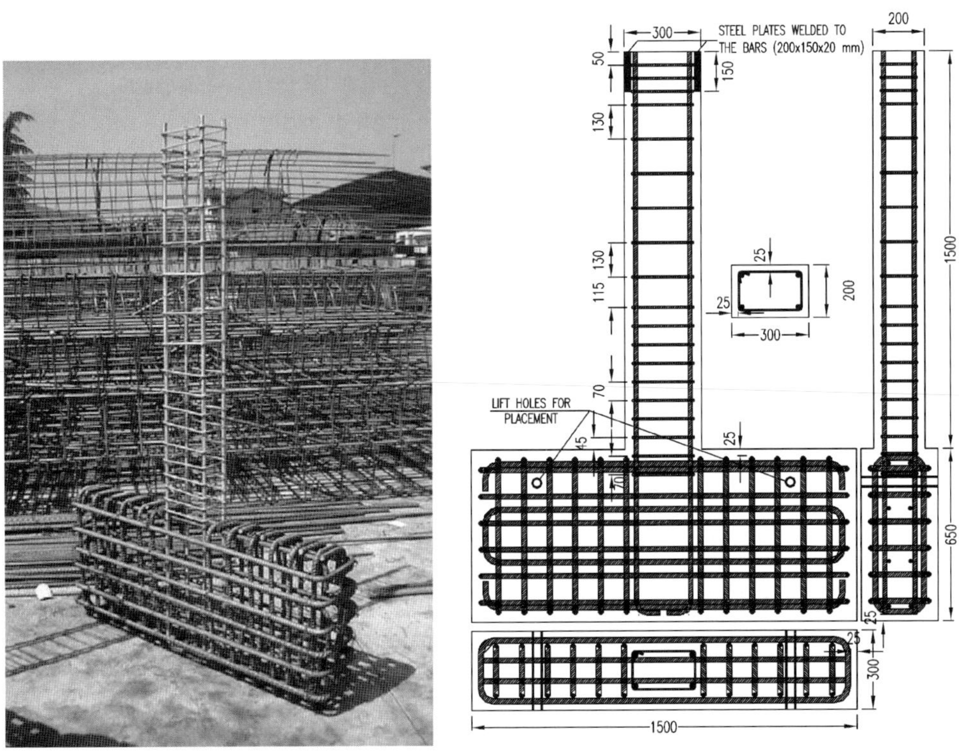

Fig. 11. Steel reinforcement of the column prototype.

Fig. 12. Test rig and the column prototype.

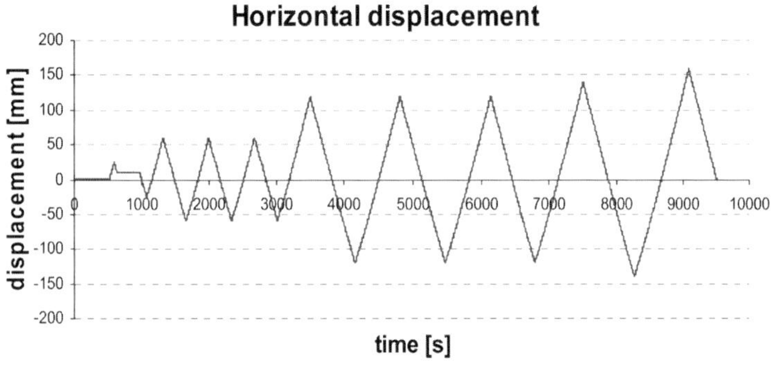

Fig. 13. Horizontal displacement of the top section of the column.

in displacement control. It moves cyclically, to the left and to the right with respect to the initial configuration, of a given horizontal displacement. The horizontal actuator is connected to a load cell, which gives the horizontal applied force, and a horizontal displacement transducer gives the horizontal displacement. Figure 12 shows also the vertical displacement transducers, located in the bottom of the column, which are necessary for evaluating the "plastic hinge" rotation.

4.3. *Loading History*

The tests are performed under displacement control, i.e. the horizontal displacement of the top of the column is imposed by the electro-mechanical displacement actuator and the corresponding horizontal force is measured by the load cell. The displacement time history is illustrated in Figure 13.

A first cycle is performed in order to evaluate δ_y, i.e. that displacement at which, for the first time, yield is reached in a point of the steel reinforcement. Then three cycles are performed by reaching $\pm 3\delta_y$ and another three cycles attaining $\pm 6\delta_y$. If the steel does not break, the steel could be defined as a high ductility steel reinforcement.

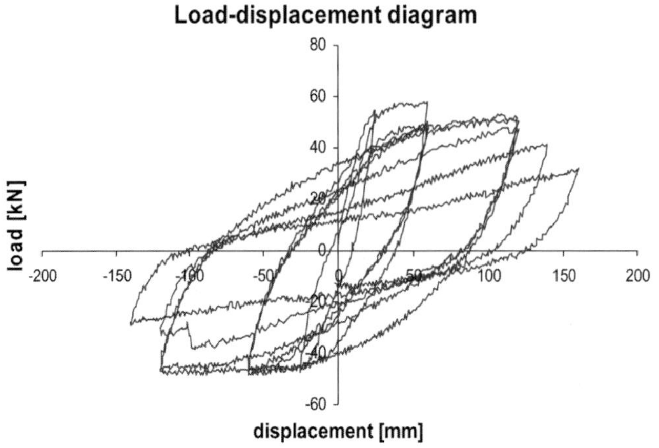

Fig. 14. Load-displacement curves of the top section of the column.

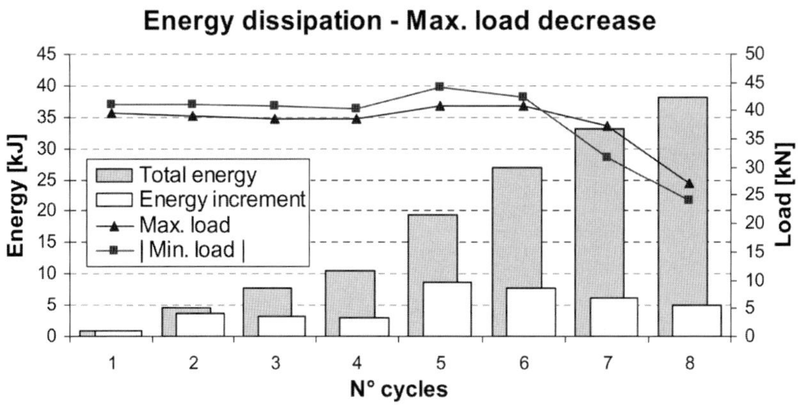

Fig. 15. Carbon steel energy dissipation.

4.4. *Results*

Figure 14 shows a load-displacement hysteretic cyclic diagram. It shows a decrease of the peak load during the cycles, mainly due to both a damage of the steel reinforcement and geometrical effects, i.e. the instability of compressed bars which deform in a flexural mode instead of a pure axial mode. The slope of the curves is a measure of the stiffness of the column and it presents two different situations: (i) an elastic phase, i.e. a phase where the steel bars load or unload in the elastic regime; (ii) an elastic plastic stage, where the bars are loaded in tension or in compression/bending in the plastic range; these slopes are continuously decreasing because of the continuous damage of the concrete material during the cycles.

The bar chart of Figures 15 and 16 shows, for each cycle, the maximum horizontal force and the cumulative dissipated energy, expressed as a function of the cycle number.

It is easy to verify that the maximum load carried out by the column reinforced by stainless steel is more than 20% the corresponding reinforced with carbon steel and, more important, the resistance stays constant during the cycles more than the columns reinforced with "carbon" steel. The total energy dissipated is almost twice in favour of the stainless steel.

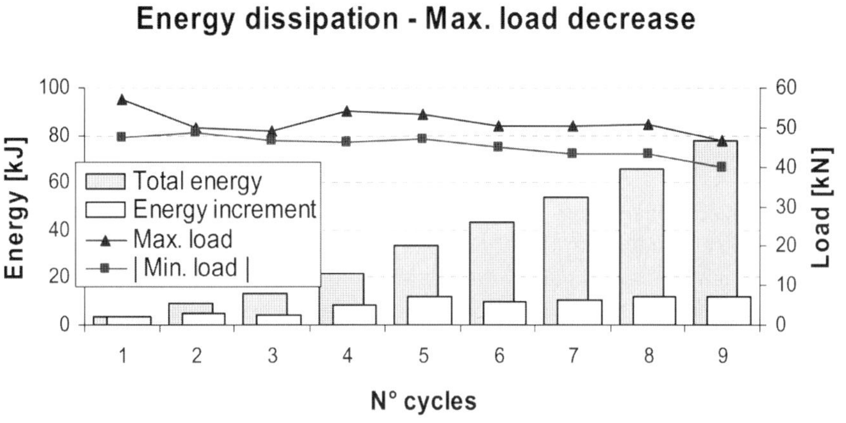

Fig. 16. Stainless steel energy dissipation.

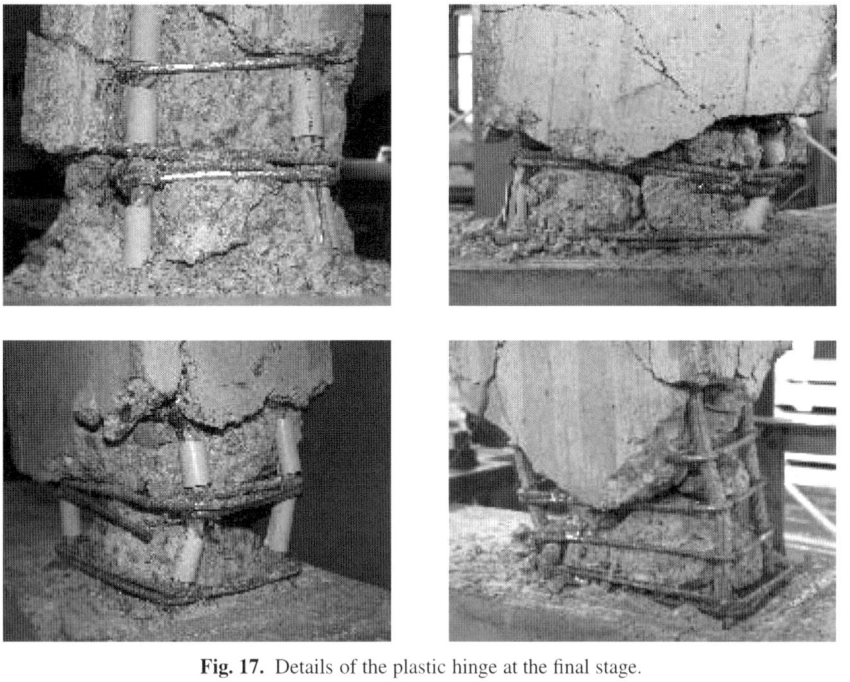

Fig. 17. Details of the plastic hinge at the final stage.

We can conclude that the same column, reinforced with stainless steel, will present after the same seismic action, probably, very less severe damages than the analogous column reinforced with carbon steel. Figure 17 shows pictures of the damaged "plastic hinge" at the final cycle of the loading history. It may be interesting to remark that some specimen present a plastic tube which prevent bond between the longitudinal bar and the concrete in the plastic hinge zone. The question was: does the bond influence the behaviour of the plastic hinge? The answer is: no. Interesting to observe in the right picture on the top of Figure 17 that the deformation of the "plastic hinge" is mainly due to shear; this happens when the concrete trust has collapsed. If the steel bar does not fail, then shear may be considered as the collapse mode of the column. The stainless steel greater strain hardening behaviour imposes a revision, in Eurocodes 2 and 8 (EN1992, 2004, EN1998, 2004), of the maximum shear force the reinforced concrete member has to sustain if maximum advantage of the stainless steel resistance has to be exploit. Otherwise, like shown in the left picture of the second row, a failure of a stirrup is very likely to happen.

5. Conclusions

AISI304 stainless steel rebars have been tested under quasi-static tensile load as well as under tension-compression cyclic loading. The class of resistance is a 500 MPa characteristic yield limit. The rebars have demonstrated unusual ductility level, both in the monotonic and the cyclic loading, as compared to traditional carbon steel rebars. Tests on column prototypes have shown an analogous ductile structural behaviour but at a higher horizontal force. This fact suggests the idea of using stainless steel in seismic areas for a more limited behaviour factor (larger horizontal forces) but using, at the same time, a limited steel area (because of the higher resistance of the stainless steel) and therefore resulting in expected limited damages to the structure after the earthquake.

Acknowledgement

The technical and financial support of Cogne Acciai Speciali of Aosta, Italy, is gratefully acknowledged.

References

EC, 2005, HIPER – Increase Infrastructure Reliability by Developing a Low Cost and High Performance Stainless Steel Rebar, Growth Programma, Final Report, 2005-12-09.

EC, SRD, Technical Steel Research, "Optimization of Ductility of Welded Steel Bars, Ribbed Coils and Mesh Fabric for Reinforced Concrete Elements under Severe Seismic Loads, Report EUR20506 EN.

EN1992-1-1, 2004, Eurocode 2: Design of Concrete Structures – Part 1: General Rules and Rules for Buildings.

EN1998, 2004, Eurocode 8: Design of Structures for Earthquake Resistance.

Franchi, A., Riva, P., Ronca, P., Roberti, R. and La Vecchia M., 1996, Failure Modalities of Reinforcement Bars in Reinforcing Concrete Elements under Cyclic Loading, *Studi e Ricerche*, Vol. 17, 157–187.

Macchi, G., Pinto, P. and Sanpaolesi, L., 1996, Ductility Requirements for Reinforcement under Eurocodes, *Structural Engineering International*, Vol. 4, 249–254.

Monti, G. and Nuti, C., 1992, Nonlinear Cyclic Behaviour of Reinforcing Bars Including Buckling, *Journal of Structural Engineering*, Vol. 118(12), 3268–3284.

Pantazopoulou, S.J., 1998, Detailing for Reinforcement Stability in R.C. Members, *ASCE Journal of Structural Engineering*, Vol. 124(6), June, 623–632.

Park, R., Priesley, M.J.N. and Gill, W.D., 1982, Ductility of Square Confined Concrete Columns, *Journal of the Structural Division, ASCE*, Vol. 108 (No. ST4), April, 929–950.

Pipa, M. and Carvalho, E.C., 1990, Experimental Evaluation of the Behaviour of Structures Designed for Two Ductility Levels, *European Earthquake Engineering*, Vol. 4(1).

Riva, P. and Franchi, A., 2001, Behaviour of Reinforced Concrete Walls with Welded Wire Mesh Subjected to Cyclic Loading, *ACI Struct. Journal*, May–June, 324–334.

Riva, P, Franchi, A. and Tabeni D., 2001, Welded Tempcore Reinforcement Behaviour for Seismic Applications, *Materials and Structures*, Vol. 34, May, 240–247.

SPREAD OF PLASTICITY: AN ADAPTIVE GRADUAL PLASTIC-HINGE APPROACH FOR STEEL FRAMES

Yanglin Gong
Department of Civil Engineering, Lakehead University, Thunder Bay, Ontario, Cnada P7B 5E1
E-mail: ygong@lakeheadu.ca

Abstract

This paper presents a new plastic-hinge method for inelastic analysis of steel frames. The proposed plastic-hinge model employed two parameters in the modeling. The first parameter involves mimicking the spread of plasticity through a section depth, while the second incorporates the spread of plasticity along a member length. Procedures to determine the key parameters are developed using moment-curvature-thrust relationship for steel beam-columns. The proposed analysis method is especially advantageous when modeling the spread of plasticity along a member length using various discretization schemes. Two numerical examples are performed to demonstrate the accuracy and simplicity of the method.

Keywords: steel frames, nonlinear analysis, plasticity, plastic-hinge, beam-columns

Introduction

Considerable studies have been undertaken on the inelastic analysis of steel frames in recent years (Chen and Toma 1994, Xu et al. 2005). The inelastic analysis methods are generally classified into two types: the distributed plasticity method and the plastic-hinge method. The distributed plasticity method discretizes frame members both along their length and through their cross section into many elements. The spread of plasticity is traced by the sequential yielding of the elements. This method is usually adopted to create benchmark solutions, as it is too computationally intensive and not suitable for practical design purposes. On the contrary, the plastic-hinge method usually involves using single or multiple elements to model a frame member, thus making it more efficient and the preferred method in engineering practice. The plastic-hinge method assumes that inelastic deformations are concentrated at plastic hinges at the end of elastic elements. The early studies used an elastic-plastic-hinge model, where the relationship between moment and curvature is linear up to the full plastic-moment of a section, after which the section becomes a perfect hinge. Though this approach is easy to implement, it often overestimates the ultimate strength of structural systems. More recently, refined- and quasi-plastic-hinge approaches (Liew et al. 1993, Attalla et al. 1994) with two-surface yielding criteria were proposed to account for the gradual plastification within steel members. Often, a model was constructed to simulate the gradual softening of plastic-hinges whose force point falls within the two yielding surfaces (Chen and Chan 1995, Hasan et al. 2002, Xu et al. 2005, Xu and Liu 2005).

This study proposes a new gradual plastic-hinge model for the inelastic analysis of planar steel frames. The analysis approach belongs to the domain of matrix displacement method. The plastic-hinge model is capable of mimicking the spread of plasticity both through the depth of a section and along the length of an element. The moment gradient of a frame member is directly taken into account in the plastic-hinge model. The model is unique and applicable to a general steel beam-column.

In this study, it is assumed that the cross sections are doubly symmetric and the stress-strain relation for steel material is elastic-perfectly-plastic. Only moment yielding is considered, while shear and axial yielding are ignored. Local plate, torsional, and lateral-torsional buckling are not considered.

Inelastic Beam-Column Model

A hybrid element (see Fig.1) is employed to model planar inelastic beam-columns. The element consists of two potential plastic-hinges at the ends of an elastic beam-column. Each plastic-hinge is modeled by a nonlinear zero-length rotational spring (which is called a hinge-spring or spring hereafter). For the elastic beam-column, E, I, and A correspond to Young's modulus, moment of

M. Pandey et al. (eds), Advances in Engineering Structures, Mechanics & Construction, 265–276.

inertia, and cross-sectional area of the element. All the plastification is assumed to be concentrated at end plastic-hinges.

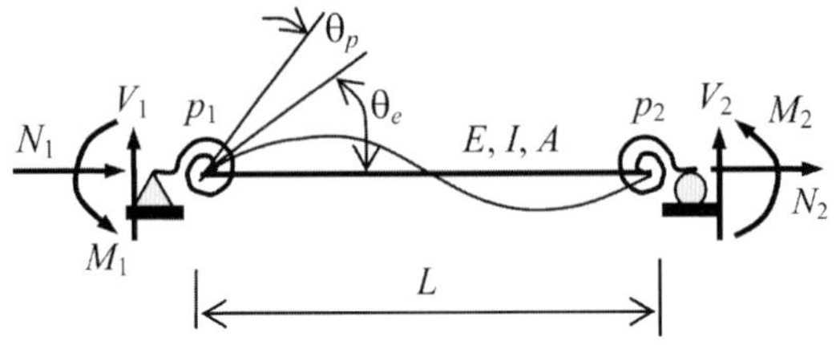

Fig. 1 Hybrid beam-column element and its end rotation

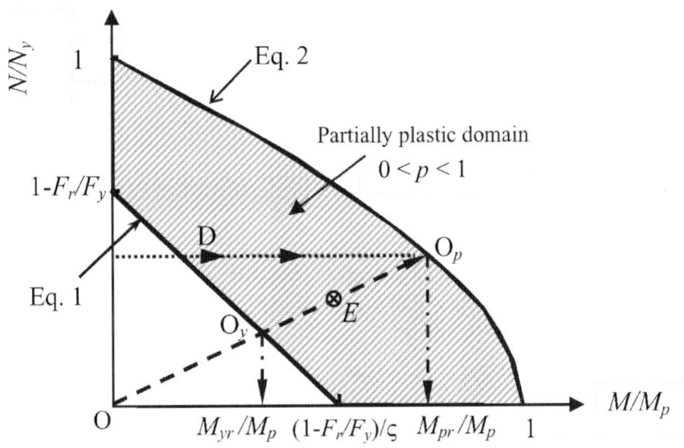

Fig. 2 Yielding criteria under combined axial force and bending moment

The interaction between the axial force N and moment M is considered using a two-surface criterion, as shown in Fig. 2. For the first-yield surface, the equation is

$$M/M_y + N/N_y = 1 - F_r/F_y \tag{1}$$

where: $M_y = F_y S$ is the first-yield moment of the section under a pure moment and S is the elastic modulus of the cross section; $N_y = AF_y$ is fully-plastic axial force capacity under a pure axial force; F_y is yielding stress of steel; and F_r is the peak residual stress in the flange of cross-section. In the meantime, the full-yield surface is expressed as

$$M/M_p + (N/N_y)^a = 1 \tag{2}$$

where $M_p = F_y Z$ is the plastic moment of the section under a pure moment and Z is the plastic modulus; and the exponent a depends on the shape of the section. For example, a=1.3 for a wide-flange section under strong axis bending (Duan and Chen 1990). Therefore, any section having a force point falling within the shaded area is partially yielded (see Fig. 2).

Adaptive Gradual Plastic-Hinge Model

To assign a plasticity-factor p (see Eq. 8 for definition) to a partially-yielded section (e.g., a section with force point E in Fig. 2) during a loading process, it is assumed that the ratio M/N for the section remains invariant when $M \neq 0$ (see line O-O_y-E-O_p in Fig. 2. O_y and O_p thus represent the first-yield and full-yield 'times' for the section, respectively.). Defining $\xi = (M_p/N_y)(N/M)$, the reduced first-yield moment M_{yr} for the section in post-elastic range under combined axial force and bending moment is

$$M_{yr} = M_y \left(1 - F_r / F_y\right) / \left(1 + \xi / \varsigma\right) \tag{3}$$

where $\varsigma = Z/S$ is shape factor. The reduced full-yield moment M_{pr} is found from Eq. 2 as

$$M_{pr} / M_p + \left(\xi M_{pr} / M_p\right)^a = 1 \tag{4}$$

Note that the assumption of N/M as invariant is not necessarily true. This assumption purely serves the purpose of evaluating the plasticity-factor for a partially yielded hinge.

The rotational stiffness of a hinge-spring degrades from infinity to zero as the moment M at the hinge section increases from M_{yr} to M_{pr}. This degradation is represented by a moment-plastic rotation curve (see Fig. 3). On this curve, point (0, M_{yr}) corresponds to the first-yield of the hinge, while point (θ_{pu}, M_{pr}) corresponds to the full-yield. This curve is expressed as (Xu et al. 2005)

$$\left(\frac{M - M_{yr}}{M_{pr} - M_{yr}}\right)^{\eta} + \left(\frac{\theta_{pu} - \theta_p}{\theta_{pu}}\right)^{\eta} = 1 \qquad (\theta_p < \theta_{pu}, M_{yr} < M < M_{pr}) \tag{5}$$

where θ_{pu} is the plastic rotation at which the hinge-spring has zero stiffness; and exponent η is a parameter dependent on the shape of the cross-section. Upon differentiating moment M with respect to plastic rotation θ_p in Eq. 5, the instantaneous rotational stiffness of the hinge-spring is found as

$$R_m = \frac{dM}{d\theta_p} = \frac{M_{pr} - M_{yr}}{\theta_{pu}} \left(1 - \frac{\theta_p}{\theta_{pu}}\right)^{\eta-1} \left[1 - \left(1 - \frac{\theta_p}{\theta_{pu}}\right)^{\eta}\right]^{\frac{1}{\eta}-1} \qquad (0 < \theta_p < \theta_{pu}) \tag{6}$$

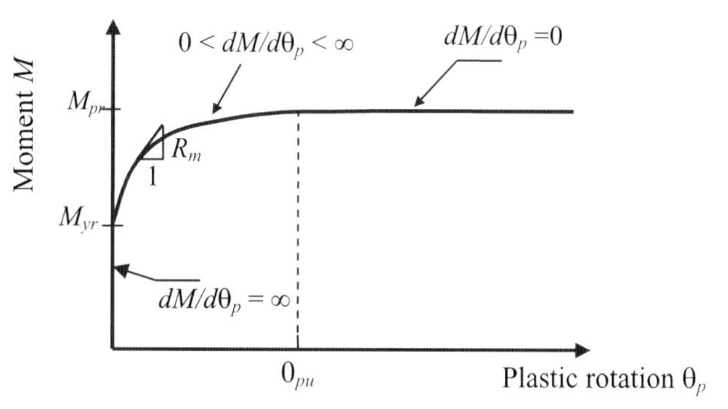

Fig. 3 Moment-plastic rotation curve of plastic-hinges

Equation 6 can be rewritten as

$$R_m = \frac{M_{pr} - M_{yr}}{\theta_{pu}} \left[1 - \left(\frac{M - M_{yr}}{M_{pr} - M_{yr}}\right)^{\eta}\right]^{\frac{\eta-1}{\eta}} \left(\frac{M - M_{yr}}{M_{pr} - M_{yr}}\right)^{1-\eta} \qquad (M_{yr} < M < M_{pr}) \tag{7}$$

Then, the plasticity-factor p for a plastic-hinge is defined as

$$p = \frac{\Delta\theta_e}{\Delta\theta_e + \Delta\theta_p} = \frac{1}{1 + 3EI/R_m L} \tag{8}$$

where: $\Delta\theta_e = \Delta M \bullet L/3EI$ is the incremental end rotation of the elastic beam-column (see Fig.1); ΔM is a moment increment; and $\Delta\theta_p = \Delta M/R_m$ is the incremental rotation of the spring. For a fully elastic section $R_m = \infty$ and $p=1$, for a fully plastic hinge $R_m=0$ and $p=0$, while for a partially plastic hinge $0<p<1$.

Upon introducing the plasticity-factor by Eq. 8, the stiffness matrix $\mathbf{K}_e$ for the hybrid element in Fig.1 is found as (Xu 2001, Hasan et al. 2002),

$$\mathbf{K}_e = \mathbf{S}\,\mathbf{C}_s + \mathbf{G}\,\mathbf{C}_g \tag{9}$$

where: $\mathbf{S}$ is the standard stiffness matrix for an elastic frame member; $\mathbf{C}_s$ is a correction matrix expressed in terms of plasticity-factors p (Hasan et al. 2002); $\mathbf{G}$ is the standard geometric stiffness matrix; and $\mathbf{C}_g$ is the corresponding correction matrix formulated as a function of p (Hasan et al. 2002).

A unique feature of the stiffness matrix $\mathbf{K}_e$ is that the stiffness of the springs R_m is not directly included in Eq. 9. Instead, the contribution of the hinge-springs to the element stiffness is incorporated through the plasticity-factors p. Such a treatment for R_m significantly improves the accuracy of the element model. To explain this, the R_m-M curves (see Eq. 7) and the corresponding p-M curves for an element of W200×46 section (CISC 2004) are drawn in Figs. 4 and 5, respectively. From Fig. 4, it is seen that spring stiffness R_m is extremely sensitive to moment M in the early post-elastic range. The stiffness R_m is infinite when $M=M_{yr}$, then R_m dramatically drops as M begins to exceed M_{yr}. R_m is also very sensitive to slight variations in parameter η. For instance, R_m=166323 kN-m/radian for $\eta=1.8$ under $M=1.03M_{yr}$, which is 2.5 times of that for $\eta=1.5$ under the same moment. This high sensitivity of R_m to M and η can be interpreted as a difficulty in modeling R_m numerically since some approximation in the R_m model is generally unavoidable. Therefore, if R_m is directly included in the stiffness matrix, considerable errors may be introduced. On the contrary, the plasticity-factor p has a relatively even degradation rate with the increasing of M, as shown in Fig. 5. Thus, some inaccuracy in R_m model does not cause much error in p value. Therefore plasticity-factor p serves as an 'error filter' in the hinge model.

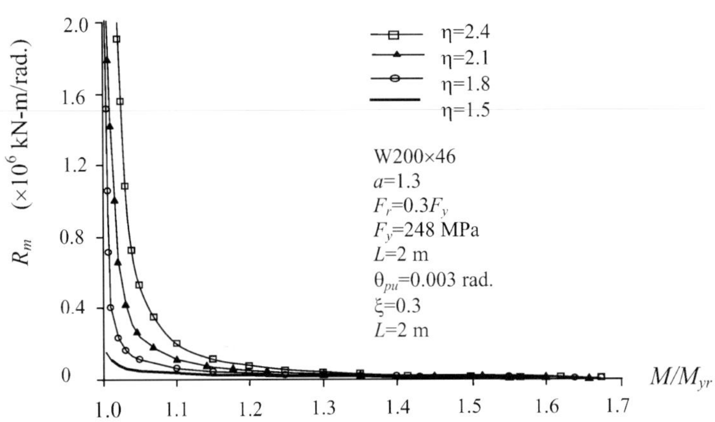

Fig. 4 Relations between moment and stiffness of hinge-spring

There are two key parameters, η and θ_{pu}, in the hinge-spring model Eq. 7. This study makes full use of these two parameters by employing η to simulate the spread of plasticity through the depth of a cross section and θ_{pu} to mimic the spread of plasticity along the length of an element.

It is obvious that parameter η has a significant impact on the degradation rate of the spring stiffness. For instance, the spring having η=1.8 degrades faster than the spring having η=2.1 (see Fig.5). In fact, this variation in degradation rate reflects the geometric difference among various cross sections. For example, an I-section degrades faster than a rectangular section under bending since an I-section has more materials allocated away from its neutral axis than a rectangular section has. Thus, the determination of η value is dependent on the shape of cross section. As it is shown in the numerical examples, η is taken as 1.8 for wide flange I-sections (and it was calibrated with experimental results). It appears that η=2.1 is reasonable for a rectangular section and η=2.4 for a solid circular section. By determining the η value in such a way one can simulate the spread of plasticity through a cross section.

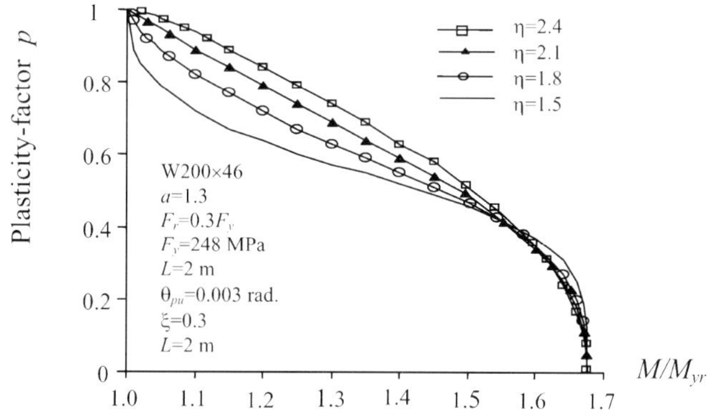

Fig. 5 Relations between moment and plasticity-factor

Before determining parameter θ_{pu}, it is instructive to examine how θ_{pu} affects the plasticity-factor of a hinge-spring. The relations between p and M for different θ_{pu} values are presented in Fig. 6 for the same element of W200×46 section. Figure 6 clearly illustrates that a larger θ_{pu} value will result in a softer hinge-spring. When θ_{pu} value is very small, the hinge-spring behavior is approaching that of a conventional elastic-plastic-hinge.

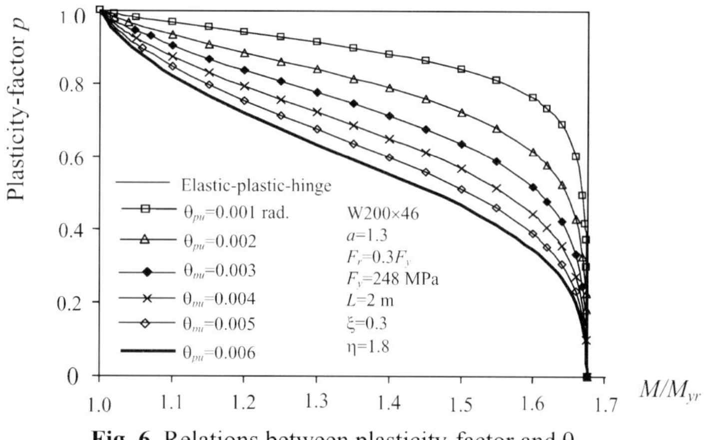

Fig. 6 Relations between plasticity-factor and θ_{pu}

To simulate the spread of plasticity along the length of a member, the parameter θ_{pu} must consider the distribution of bending moment along the length of the member (called moment gradient). In the following, the moment-curvature-thrust relations for beam-columns will be reviewed first, and then θ_{pu} value will be computed considering different moment gradients.

Moment-Curvature-Thrust Curves (M–ϕ–N) of Beam-Columns

The moment-curvature-thrust curves for general beam-columns were given by Chen (1971). A typical presentation of these curves is shown in Fig. 7, where the curvature is normalized by ϕ_y ($\phi_y=2\varepsilon_y/d$, where d=depth of a section, $\varepsilon_y=F_y/E$). According to Chen (1971), it needs two functions to represent M-ϕ relation in the post-elastic range.

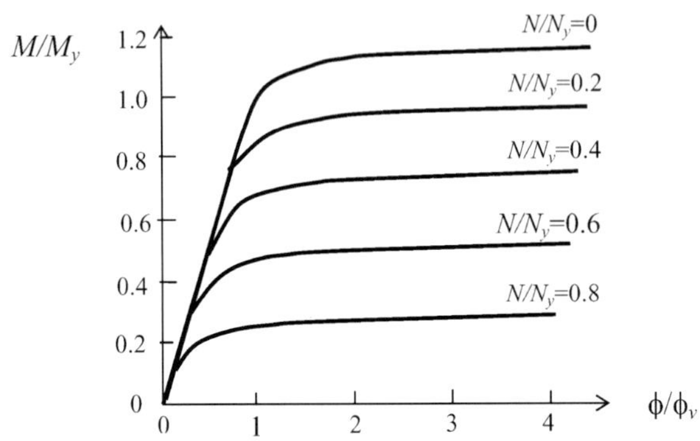

Fig. 7 Moment-curvature-thrust curves

In this study, the moment-curvature-thrust curves in Fig. 7 are expressed approximately as a unified form as shown in Fig. 8a. In the elastic range ($M \le M_{yc}$, where M_{yc} is the first-yield moment allowing for axial force), $\phi=M/EI$ (and $\phi_{yc}=M_{yc}/EI$ at the first-yield). After $M>M_{yc}$, M-ϕ relation becomes nonlinear, and $\phi=\phi_e+\phi_p$ (ϕ= total curvature, ϕ_e=elastic curvature, and ϕ_p=plastic curvature). The section is approximately fully yielded at curvature ϕ_{fu}, where ϕ_{fu} can be expressed as a constant ω_1 multiplied by ϕ_{pc} ($\phi_{pc}=M_{pc}/EI$, where M_{pc} is the full-yield moment allowing for axial force). Note that the determination of moments M_{yc} and M_{pc} does not involve the assumption of M/N being constant (hence different symbols are used). From Eqs. 1 and 2, we have

$$M_{yc} = M_y\left(1 - F_r/F_y - N/N_y\right) \tag{10}$$

$$M_{pc} = M_p\left(1-\left(N/N_y\right)^a\right) = \varsigma M_y\left(1-\left(N/N_y\right)^a\right) \tag{11}$$

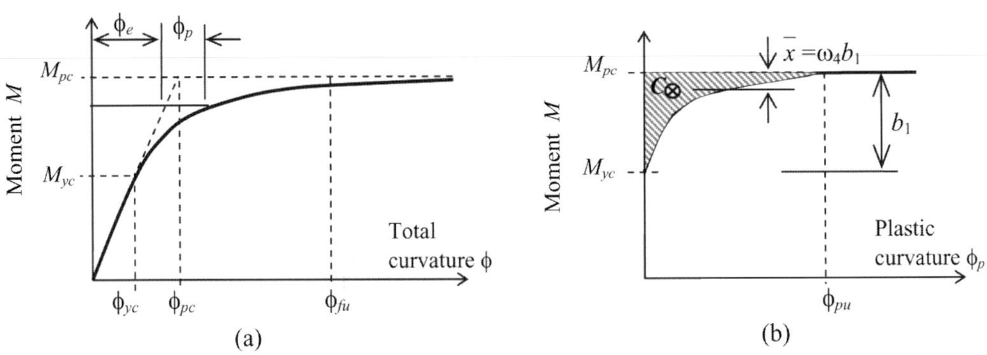

Fig. 8 Moment-curvature relationships: (a) total curvature (b) plastic curvature

For I-sections bending about strong axis, ω_1=1.9. The corresponding moment is found to have an average value of 96% of M_{pc} for $0 \le N/N_y \le 0.6$ (Chen 1971). The plastic curvature at the onset of full-yield is approximately equal to $\phi_{pu}=\phi_{fu}-\phi_{pc}=(\omega_1-1)\phi_{pc}=\omega_2\,\phi_{pc}$. For applying M-ϕ_p relation later in this study, the shaded area to the left of the curve (see Fig. 8b) can be found to be ($\omega_3 \times \phi_{pu} \times b_1$) and the centroid of the shaded area to the $M=M_{pc}$ line is $\bar{x}=\omega_4 b_1$. From Fig. 8a, we have

$$\phi_{pu} = \omega_2 \left(M_{pc}/EI\right) = \left(\omega_2/EI\right)\varsigma SF_y\left[1-\left(N/N_y\right)^a\right] = \left(2\omega_2\varepsilon_y\varsigma/d\right)\left[1-\left(N/N_y\right)^a\right] \tag{12}$$

where: $\varepsilon_y=F_y/E$, $I/S=d/2$, and d=depth of a cross section.

Computation of Parameter θ_{pu}

For the hybrid element, the total rotation θ at an end is comprised of two parts (see Fig. 1). The first part, θ_e, represents the end rotation of the elastic beam-column, while θ_p represents the plastic rotation of the hinge. Thus, at the full-yield $\theta=\theta_e+\theta_{pu}$. To simulate the spread of plasticity of an actual inelastic beam-column, the θ of the hybrid element must be identical to that of an actual beam-column.

Consider an actual inelastic beam-column under linear distributed first-order moments (see Fig. 9a). Define moment gradient $\kappa=\pm|M_2/M_1|$, where M_2 and M_1 are the bending moments at Ends 1 and 2, respectively, and $M_2 \le M_1$. Moment gradient κ is positive for double curvature and negative for single curvature ($-1\le\kappa\le1$). The lengths of the plastic zones at Ends 1 and 2 are L_1 and L_2, respectively. The corresponding curvature distribution along the member is illustrated in Fig. 9b. The plastic curvature begins at the section where moment is equal to M_{yc}. Figure 9c shows the distribution of plastic curvature alone at the onset of full-yield at End 1. Note that the identical relationship between M and ϕ_p at End 1 and in Fig. 8b. Amongst the total rotation End 1 undergoes, the portion of plastic rotation comes from the plastic curvatures along the length of the member. Assume that θ_{pu} of the hinge-spring is equal to the plastic rotation θ_p at End 1when the end is at the onset of full-yield.

The conjugate beam method is used to evaluate the plastic rotation at End 1 for the beam-column under a curvature loading shown in Fig. 9c. The conjugate beam is under the loading of plastic curvatures alone since only plastic rotation is of concern. The end rotation of the actual beam-column is equal to the shear at the same section of its conjugate beam.

For calculating θ_{pu}, moments M_{yc} and M_{pc} shall be computed from Eqs. 10 and 11, respectively. This can be seen clearly from the fact that the load path from the section with zero moment along the member length to End 1 follows line D-O_p in Fig. 2 (N remains unchanged along the member length).

It is desirable to correlate θ_{pu} to the moment gradient κ because plastic zones within a member are dependent on the moment distribution along the length (see Fig. 9a). In the following θ_{pu} values corresponding to four particular moment gradient κ are determined first. Then, the interpolation method is used to compute θ_{pu} value for any moment gradient κ.

For the convenience of deriving formulations for θ_{pu}, the ratio γ is defined as

$$\gamma = M_{yc}/M_{pc} = \frac{1-F_r/F_y - N/N_y}{\varsigma\left(1-\left(N/N_y\right)^a\right)} \tag{13}$$

Then, θ_{pu} values corresponding to four different moment gradient κ are found as follows:

(1) κ=1, double curvature. $M_1=M_2=M_{pc}$ and $\phi_p=\phi_{pu}$ at both end sections. The lengths of plastic zones are $L_1=L_2=0.5(1-\gamma)L$. The centroid of curvature loading at each end is $\bar{x}=\omega_4 L_1$, while the area of plastic curvature loading at each end is equal to ($\omega_3\times\phi_{pu}\times L_1$). Therefore, the shear force at End 1 of the conjugate beam under the plastic curvature loading is equal to

$$\theta_{pu} = \omega_3\phi_{pu}L_1\left(L-2\bar{x}\right)/L = \omega_2\omega_3\varsigma\varepsilon_y(1-\gamma)\left[1-\omega_4(1-\gamma)\right]\left[1-\left(N/N_y\right)^a\right]\left(L/d\right) \quad (\kappa=1) \tag{14}$$

(2) κ=0, single curvature. $M_1=M_{pc}$ and $\phi_p=\phi_{pu}$ at End 1, while M_2=0 and ϕ_p=0 at End 2. Plastic zones are $L_1=(1-\gamma)L$ at End 1 and L_2=0 at End 2. Hence, the shear force at End 1 of the conjugate beam under the plastic curvature loading is equal to

$$\theta_{pu} = \omega_3\phi_{pu}L_1\left(L-\bar{x}\right)/L = 2\omega_2\omega_3\varsigma\,\varepsilon_y(1-\gamma)[1-\omega_4(1-\gamma)]\left[1-\left(N/N_y\right)^a\right](L/d)\quad (\kappa{=}0) \tag{15}$$

(3) κ= –λ, single curvature. M_1=M_{pc} and ϕ_p= ϕ_{pu} at End 1, and M_2=M_{yc} and ϕ_p=0 at End 2. Then L_1=L at End 1 and L_2=0 at End 2. The shear force at End 1 of the conjugate beam under the plastic curvature loading is equal to

$$\theta_{pu} = \omega_3\phi_{pu}L\left(L-\bar{x}\right)/L = 2\omega_2\omega_3\left(1-\omega_4\right)\varsigma\,\varepsilon_y\left[1-\left(N/N_y\right)^a\right](L/d)\quad (\kappa{=}-\lambda) \tag{16}$$

(4) κ= –1, single curvature. M_1=M_2=M_{pc} and ϕ_p=ϕ_{pu} along the length of the member. The conjugate beam is under uniform plastic curvature loading. The shear force at the end of the conjugate beam is

$$\theta_{pu} = \phi_{pu}L/2 = \omega_2\varsigma\,\varepsilon_y\left[1-\left(N/N_y\right)^a\right](L/d)\quad (\kappa{=}-1) \tag{17}$$

Equations 14 to 17 are for a general beam-column. The factors ω_2, ω_3, ω_4, a, ς, and γ are dependent on the sectional properties of the beam-column.

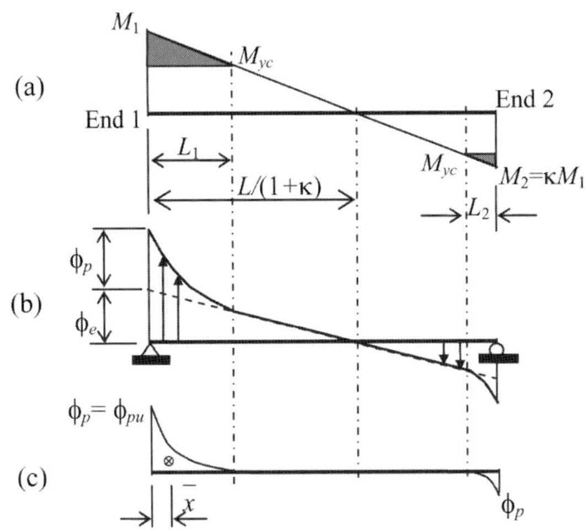

Fig. 9 Conjugate beam method for computing end rotation: (a) linear moment distribution and plastic zones (b) total curvature distribution and curvature loading for conjugate beam (c) plastic curvature distribution when End 1 is at the incipience of full-yield.

I-Sections

This section is to illustrate how to use Eqs. 14 through 17 to determine θ_{pu}-κ relation for wide flange I-sections bending about strong axis (i.e. W-sections in CISC 2004). For W-sections: it is found that ω_2=0.9, ω_3=0.308, and ω_4=0.21 using the moment-curvature-thrust curves given by Chen (1971); the average section shape factor ς =1.14; the peak residual stress is commonly taken to be 30% of yielding strength (i.e., F_r=0.3F_y).

It is found that θ_{pu} is not sensitive to N/N_y ratios when κ=1 and 0. Thus, Eqs. 16 and 17 are simplified respectively for W-sections as

$$\theta_{pu} = 0.116\,\varepsilon_y(L/d)\quad (\kappa{=}1) \tag{18}$$

$$\theta_{pu} = 0.233\,\varepsilon_y(L/d)\quad (\kappa{=}0) \tag{19}$$

Assuming a moderate axial force N=0.2N_y for a general beam-column, then $\varsigma\,[1-(N/N_y)^a] = 1.0$ in Eqs. 16 and 17, which are respectively simplified as

$$\theta_{pu} = 0.438\,\varepsilon_y (L/d) \quad (\kappa = -0.5) \tag{20}$$

$$\theta_{pu} = 0.9\,\varepsilon_y (L/d) \quad (\kappa = -1) \tag{21}$$

Based on Eqs. 18 through 21, a piecewise θ_{pu}-κ curve is established for W-sections in Fig. 10. Using the interpolation method, a unique θ_{pu} value is available corresponding to the κ value of an element. The moment gradient κ of an element at the first-yield of one hinge-spring is used to determine the θ_{pu} value for that spring. After the θ_{pu} value is computed for a particular hinge-spring, its value will remain unchanged during the rest of the loading history.

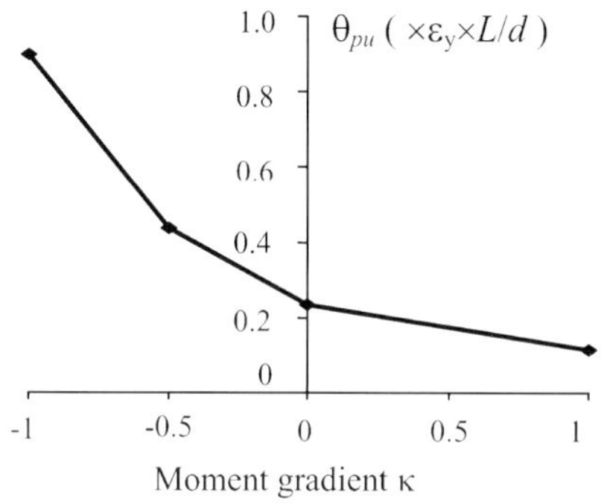

Fig. 10 Relation between moment gradient κ and θ_{pu} value for W-sections

Analysis Procedures

Incremental single-step method, as described in the following, is employed to conduct nonlinear analysis for steel frameworks.

1) Discretize frame members into elements. Initialize plasticity-factors to be p_1=p_2=1 for all elements.
2) Form element stiffness matrix $\mathbf{K}_e$ for each element, which involves using plasticity-factors from previous loading step to compute $\mathbf{C}_s$ and $\mathbf{C}_g$ (p=1 for the first loading step). Then assemble overall structure stiffness matrix $\mathbf{K}$. If the tangent stiffness matrix $\mathbf{K}$ is found to be singular, which indicates the structure collapses, terminate analysis.
3) Solve for incremental nodal displacements by $\Delta\mathbf{F}=\mathbf{K}\bullet\Delta\mathbf{u}$, where $\mathbf{u}$ and $\mathbf{F}$ are the vectors of overall nodal displacements and loads respectively. Calculate incremental deformations and forces for each element. Update the total overall nodal displacements and loads, and member internal forces by $\mathbf{u}=\sum\Delta\mathbf{u}$, $\mathbf{F}=\sum\Delta\mathbf{F}$, and $\mathbf{f}_j=\sum\Delta\mathbf{f}_j$, respectively, where $\mathbf{f}_j$ is the vector of member internal force.
4) Check the yielding status for each hinge-spring at the ends of each element. The combined actions of axial force and bending moment at a hinge-spring are, from Eqs. 1 and 2, denoted as $\beta_1=M/M_y+N/N_y$ and $\beta_2=M/M_p+(N/N_y)^a$. Then update p for each spring as below:
   (i) If $\beta_1 \le (1-F_r/F_y)$, the hinge-spring is still fully elastic with p=1.
   (ii) If $\beta_1>(1-F_r/F_y)$ and $\beta_2<1$, the hinge-spring is partially plastic. Calculate M_{yr} and M_{pr} from Eqs. 3 and 4, respectively. If the hinge-spring is found to be yielding for the first time, calculate its θ_{pu} value using θ_{pu}-κ relation. Compute the post-elastic rotational stiffness of the hinge-spring from Eq. 7 and the plasticity-factor p from Eq. 8.
   (iii) If $\beta_2 \ge 1$, the hinge-spring is fully plastic with p=0.
5) Go back to step 2.

Numerical Examples

The inelastic analysis is illustrated for two structures comprised of steel members of wide flange I-section. All the members are oriented with their webs in the loading plane. The exponent η in Eq. 5 is taken as 1.8. Young's modulus is E=2×10^5 MPa.

Example 1: A Simply-Supported Beam

The first example is a simply-supported beam (see Fig. 11), which was tested by Lay and Galambos (1964). The section is W130×28, with the measured properties being A=3606 mm^2, I=11.16×10^6 mm^4, S=168.8×10^3 mm^3, Z=190.1×10^3 mm^3, d=132 mm, F_y=247 MPa, and F_r= 19.8 MPa. $M_p = Z \times F_y$ = 47 kN-m, and load capacity P=71 kN. Equations 14 to 17 are used to establish θ_{pu}-κ relation for this beam. ω_2=0.9, ω_3=0.308 and ω_4=0.21.

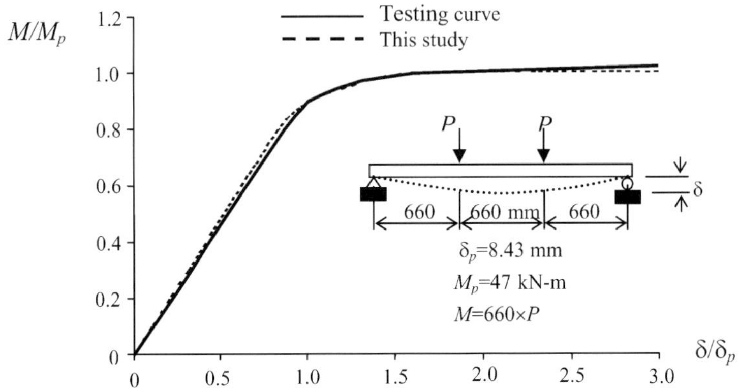

Fig. 11 Moment-deflection curve of the simply-supported beam

The middle portion of this beam is a region of constant moment. This entire region begins to yield when its moment exceeds the first-yield moment. Therefore, this beam is of particular interest for this study since the wide spread of plasticity both through the depth of the section and along the length of the beam took place. Such a structure with a large plastic zone is generally regarded as being difficult for a plastic-hinge method to model.

First, the beam was analysed by using four elements, as shown in Fig. 12b where the digit inside a rectangular box represents the element number. The middle portion is represented by two identical elements 2 and 3. The moment-deflection curve at midspan section is drawn in Fig. 11. Compared with the test curve, the errors of the theoretical analysis are within 5%. Next, the beam was re-analysed using a 3-element discretization scheme (see Fig. 12a). Table 1 records the θ_{pu} value for hinge-springs of different discretization schemes. The deflection at the one-third point at the incipience of collapse is also recorded in Table 1. The plasticity-factors under 90% of the collapse load are depicted in Fig. 12, where an oval represents a hinge-spring and the number inscribed in it is the corresponding plasticity-factor p.

Table 1 Analysis of the simply-supported beam using different discretization schemes

Discretization scheme	θ_{pu} for hinge-springs			Deflection at one-third point at collapse (mm)
	Element 1	Element 2	Element 3	
3-element	0.00068	0.00626	0.00068	11.723
4-element	0.00068	0.00313	0.00313	11.723

For the 3-element scheme, Table 1 shows that the θ_{pu} value for the hinge-springs of element 2 is 9.2 times as large as that for the hinge-springs of elements 1 and 3. Therefore, the hinge-springs of element 2 have a significantly smaller p value (i.e., p_1=p_2=0.18) than the hinge-springs of elements 1 (p_2=0.67) and 3 (p_1=0.67) as illustrated in Fig. 12a, even though all these hinge-springs are subjected to an identical moment. This substantially smaller p value for element 2 is translated into a much smaller stiffness for the middle portion of the beam, which is in keeping with the fact that the entire middle portion is a plastic zone. Thus, the parameter θ_{pu} is capable of imitating the spread of plasticity along the length of a member.

It is noted from Table 1 that the two analyses obtain identical deflection. It is further noted from Figs. 12a and 12b that the plasticity-factors are identical for these two discretization schemes. The two discretization schemes have identical beam stiffness at every loading step. This can be explained by inspecting Eqs. 7 and 8 and θ_{pu} values in the table. The θ_{pu} value for elements 2 and 3 of the 4-element scheme is half of that for element 2 of the 3-element scheme (θ_{pu} is proportional to element length L. The elements in the middle portion have $\kappa = -1$). Hence, the product $R_m L$ in Eq. 8 remains unchanged whether the middle portion of the beam is represented by one or two (or more) elements.

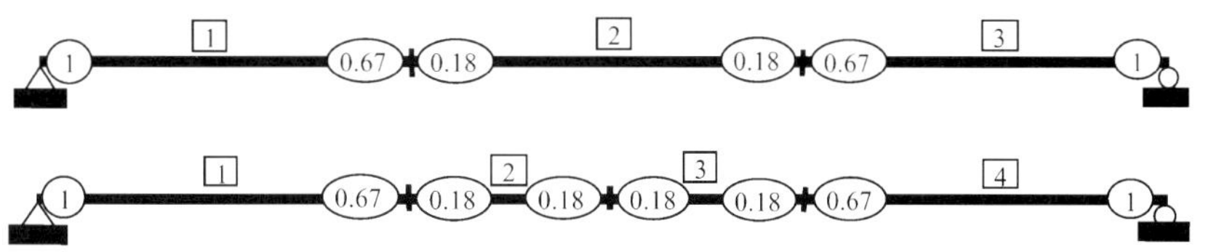

Fig.12 Plasticity-factors for two discretization schemes at $M=0.9M_p$: (a) 3-element, (b) 4-element

Example 2: A 2-Story Moment Frame

Consider the steel frame subjected to the gravity loads shown in Fig. 13. The structure supports specified loads of 110 kN/m on floor beams B1 and B2 and 51 kN/m on roof beams B3 and B4, where the factored loads are 1.4 times the corresponding specified loads. F_y=248 MPa and F_r=0.3F_y.

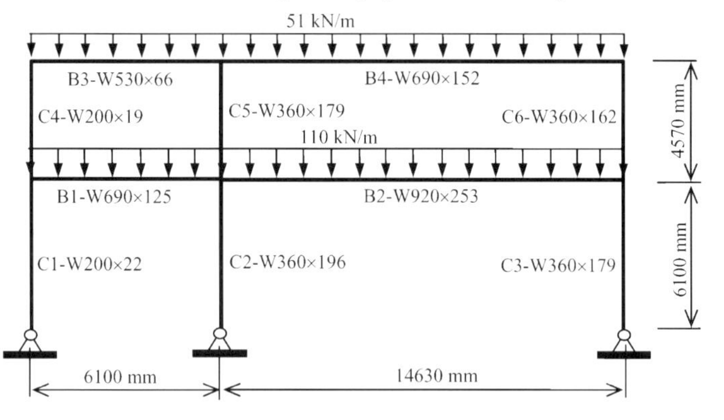

Fig. 13 Two-story moment frame

For the nonlinear analysis of this example, each column is represented by two elements while each beam is represented by four elements. The distributed loads are lumped at nodal points. Equations 18 to 21 are used to establish the θ_{pu}-κ relationship. The analysis terminated when the frame failed at load ratio λ=1.09. The lateral displacement of the top right corner of the frame during the loading history is described by the continuous line in Fig. 14. The analysis results are in close agreement with those from other researchers.

Conclusions

This paper presents a nonlinear analysis method for steel frames using a new plastic-hinge model. Two separate parameters are used in the hinge model to mimic the spread of plasticity both through section depth and along member length respectively. The proposed plastic-hinge model is general and applicable to all kinds of beam-columns, while specific numerical examples were conducted for steel frames with wide flange I-sections. The method is simple to implement since it only involves the modification of a conventional elastic matrix displacement procedure. Numerical examples demonstrated the accuracy of the proposed analysis method.

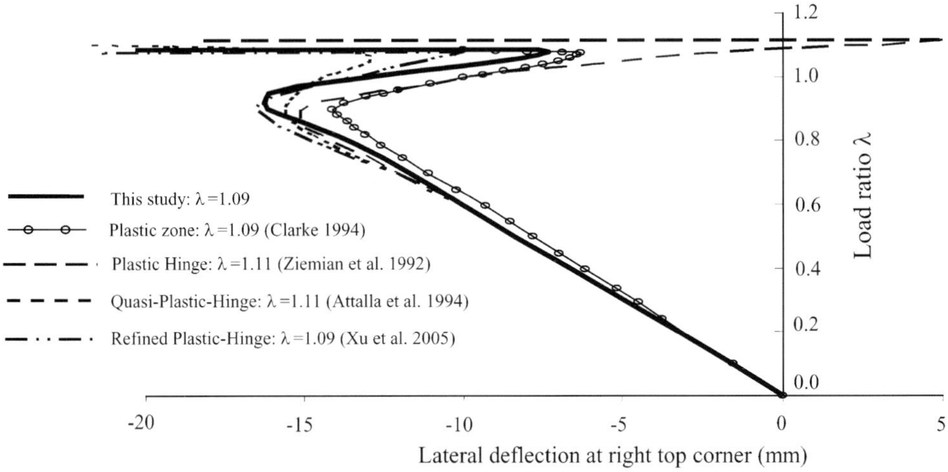

Fig. 14 Lateral deflection of the moment frame

References

Attalla, M.R., Deierlein, G.G., and McGuire, W. 1994. Spread of plasticity: quasi-plastic-hinge approach. Journal of Structural Engineering, ASCE, **120**(8): 2451-2473.

Chen, W.F. 1971. Further studies of inelastic beam-column problem. Journal of the Structural Division, ASCE, **97**(2): 529-544.

Chen, W.F, and Chan, S.L. 1995. Second-order inelastic analysis of steel frames using element with midspan and end springs. Journal of Structural Engineering, **121**(3): 530-541.

Chen, W.F., and Toma, S. 1994. Advanced analysis of steel frames. CRC Press, Boca Raton.

CISC. 2004. Handbook of steel construction, 8th Edition. Canadian Institute of Steel Construction.

Clarke, M.J. 1994. Chapter 6: plastic-zone analysis of frames. In: Chen, W.F., and Toma, S, editors. Advanced analysis of steel frames. CRC Press.

Duan, L., and Chen, W.F. 1990. A yield surface equation for doubly symmetrical sections. Engineering Structures, **12**(2): 114-119.

Hasan, R., Xu, L., and Grierson, D.E. 2002. Push-over analysis for performance-based seismic design. Computer and Structures, **80**: 2483–2493.

Lay, M.G., and Galambos, T.V. 1964. Tests on beam and column subassemblages. Frits Engineering Laboratory Report No. 278.10, Lehigh University.

Liew, J.Y.R., White, D.W., and Chen, W.F. 1993. Second-order refined plastic-hinge analysis for frame design. Part I. Journal of Structural Engineering, ASCE, **119**(11): 3196-3216.

Xu, L. 2001. Second-order analysis for semirigid steel frame design. Canadian Journal of Civil Engineering, **28**: 59-76.

Xu, L., and Liu, Y. 2005. Nonlinear analysis of inelastic steel frames. Proceedings of the 4th International Conference on Advances in Steel Structures, June 13-15, Shanghai, China.

Xu, L., Liu, Y., and Grierson, D.E. 2005. Nonlinear analysis of steel frameworks through direct modification of member stiffness properties. Advances in Engineering Software, **36**: 312-324.

Ziemian, R.D., McGuire, W., and Deierlein, G.G. 1992. Inelastic limit states design part I: planar frame structures. Journal of Structural Engineering, ASCE, **118**(9): 2532–49.

EMPIRICAL ANALYSIS OF MEMETIC ALGORITHMS FOR CONCEPTUAL DESIGN OF STEEL STRUCTURAL SYSTEMS IN TALL BUILDINGS

Rafal Kicinger and Tomasz Arciszewski
Department of Civil, Environmental and Infrastructure Engineering,
George Mason University,
Fairfax, VA, USA
E-mail: rkicinge@gmu.edu

Abstract

This paper discusses the results of extensive design experiments in which memetic algorithms were applied to optimize topologies of steel structural systems in tall buildings. In these experiments, evolutionary algorithms were employed to determine optimal configurations of structural members (topology optimization) while the optimal cross-sections of members (sizing optimization) were found using continuous/discrete optimization algorithm implemented in SODA. The impact of all major evolutionary computation parameters on the performance of memetic algorithms was investigated. Two classes of complex structural design problems were considered: design of a wind bracing system in a tall building and design of the entire steel structural system in a tall building. The total weight of the structural system was assumed as the optimality criterion with respect to which the designs were optimized while satisfying all design requirements specified by appropriate design codes.

In the conducted experiments various key EC parameters and their values were considered having the largest impact on the performance of memetic algorithms. It was discovered that the type of EA, the rate of mutation operator, and the size of parent population were critical for the success of structural optimization processes. Specifically, evolution strategies produced on average significantly better results than genetic algorithms for the design problems considered in the paper. Also, low mutation rates, i.e. 0.025, resulted in best performance of memetic algorithms. Furthermore, small parent population sizes were generally preferred to large populations. For the simpler problem of conceptual design of a wind bracing system, optimal results were produced even when the population with a single member was used. In the case of the second and more complex design problem slightly larger population sizes were required consisting of 5 members.

Results of a large number of design experiments allowed formulating initial recommendations regarding optimal parameter settings for memetic algorithms for structural design applications. The experiments also produced a body of structural design knowledge, both quantitative and qualitative in nature. They identified regions of the design spaces in which high-performance solutions can be found. They also defined the ranges of the total weight of structural systems associated with high-performance solutions for both classes of design problems. Furthermore, significant qualitative differences between high-performance solutions have been identified. The structural shaping patterns exhibited by high-performing designs ranged from crossed macrodiagonal patterns composed of X bracings to irregular patterns consisting of various types of bracings.

Keywords: Memetic algorithms, evolutionary computation, structural design, conceptual design, tall buildings

1. Introduction

Evolutionary computation (EC) is used for solving many complex problems in science and engineering. It allows conducting robust optimization and at the same time has modest requirements on the formulation of the problem to be solved. For example, it does not require continuous variables, differentiable objective functions, etc. Thus, it can be applied to many structural design problems, particularly those with discrete or symbolic variables, or objective functions with nonlinear and stochastic components as is frequently the case in conceptual design.

Parallel search conducted by a population (superset) of solutions is one of the key characteristics of evolutionary algorithms (EAs). It facilitates global exploration of the search space and helps EAs

M. Pandey et al. (eds), Advances in Engineering Structures, Mechanics & Construction, 277–288.

escape local optima. In design spaces, these local optima frequently correspond to already known design concepts. Parallel search, however, also has an adverse impact on the ability of EAs to efficiently refine near-optimal solutions. EAs are usually not as good as local search algorithms in converging to the optimal solution once the optimal region of the search space has been found.

Thus, several researchers proposed hybrid algorithms combining excellent global exploration characteristics of EAs and efficient refinement capabilities of local search algorithms (Hoeffler et al., 1973; Moscato, 1989). These hybrid algorithms are called memetic algorithms (MAs) but other names like hybrid EAs, or Lamarckian EAs, have also been used in the literature. Even though several applications of MAs to structural design have been reported, e.g., (Hoeffler et al., 1973; Quagliarella and Vicini, 1998; Sakamoto and Oda, 1993), none of them had such a broad empirical scope as the computational studies described in this paper.

This paper discusses the results of extensive design experiments in which MAs were applied to optimize topologies and cross-sections of steel structural systems in tall buildings. Two classes of complex structural design problems were considered: conceptual and detailed design of a wind bracing system in a tall building and conceptual and detailed design of the entire steel structural system in a tall building. In these experiments, evolutionary algorithms (EAs) were employed to determine optimal configurations of structural members (conceptual design) while the sizing optimization (detailed design) was conducted using continuous/discrete optimization algorithm implemented in SODA (Grierson, 1989). The impact of all major evolutionary computation (EC) parameters on the performance of design processes was investigated. The following parameters were tested: the type of an evolutionary algorithm, parent and offspring population sizes, the type of the generational model, crossover and mutation rates, and the length of design processes (number of fitness evaluations).

The remainder of the paper is organized as follows. First, a brief description of MAs is provided. Next, experimental settings used in the computational experiments are presented. Furthermore, results of extensive computational studies are reported and grouped with respect to experimental parameters being tested. Finally, research conclusions are provided, including recommendations for optimal experimental settings for structural design experiments utilizing MAs as well as directions of future research.

2. Memetic Algorithms

Memetic algorithms are hybrid algorithms in which EAs are integrated with a local search. In these algorithms, EAs are usually used for global exploration of the search space while fine-tuning of solutions is conducted by the local search. MAs have largely emerged in the late 1980s and early 1990s (Moscato, 1989) with the progress in the field of evolutionary computation and improved understanding of the dynamics of evolutionary algorithms. Some earlier studies on hybrid optimization methods utilizing EAs were also reported in the literature, e.g., a hybrid of evolution strategies (Rechenberg, 1973) and linear programming for topology optimization of trusses (Hoeffler et al., 1973).

The overall structure of a canonical MA is presented in Figure 1. It is a slightly modified version of a simple MA presented in (Hart et al., 2005a). Figure 1 shows that the differences between EAs and MAs occur at point 6. In the case of a MA, additional local search is conducted to improve (fine-tune) new individuals generated by an EA in steps 3-5.

This canonical algorithm structure can be instantiated in many different ways. Depending on the type of EA used (e.g., a genetic algorithm (GA) (Holland, 1975), evolution strategies (ES) (Rechenberg, 1973), etc.) and the type of local search (e.g., steepest ascent/descent, greedy search, etc.) different hybrids can be created. In this way, each component of the MA can be adjusted to a specific problem, or a class of problems. Current state-of-the-art review in this field can be found in (Hart et al., 2005b).

MAs have also been applied to solve several structural design problems. Hoeffler et al. (1973) combined linear programming and ES to optimize topologies of truss systems. Linear programming methods were applied to identify initial truss layout and ES were subsequently used to determine optimal positions of joints. This hybrid strategy resulted in reduced total weight of trusses when compared to results obtained using sole linear programming methods. Sakamoto and Oda (1993)

```
1. Initialize the population
2. Evaluate all members of the population
While the termination condition is not satisfied
{
    3. Select individual(s) in the population to be parent(s)
    4. Create new individuals by applying the variation operators to the copies of parent(s)
    5. Evaluate new individuals
    6. Improve individuals using local search
    7. Replace some/all of the individuals in the current population with the new individuals
}
```

Figure 1: Structure of a canonical memetic algorithm

applied a MA composed of a GA and the generalized optimality criterion method to optimize topologies and cross-sections of members in trusses. As before, minimum weight truss systems were sought subject to displacement constraints. At about the same time, Adeli and Cheng (1993) proposed a hybrid algorithm for the optimization of space truss structures. In this case, a GA combined with penalty-function method produced optimal solutions faster than a standard GA.

In the experiments reported in this paper, evolutionary algorithms were combined with a continuous/discrete optimization algorithm implemented in SODA (Grierson, 1989) to optimize topologies and cross-sections of members in steel structural systems of tall buildings. Extensive sensitivity analyses of evolutionary computation parameters were also conducted to determine their optimal settings for structural design experiments.

3. Experimental Design

3.1. Structural Systems in Tall Buildings

Computational experiments with MAs were conducted for two classes of complex structural design problems: design of a wind bracing system (Problem I) and design of an entire steel structural system in a tall building (Problem II). They are illustrated in Figure 2. In Problem I, an optimal configuration of wind bracing elements was sought while keeping the configurations of all other members the same, i.e., all beams, columns, and supports had the same topological configurations in all experiments. On the other hand, an optimal configuration of all structural members of the steel structure in a tall building was sought in Problem II (see Figure 2b). Furthermore, the first problem was further subdivided into the following 3 subproblems:

- Problem Ia - design of a wind bracing system composed of simple X bracings and no bracings (empty cells) only.
- Problem Ib – design of a wind bracing system composed of K bracings and no bracings (empty cells) only.
- Problem Ic – design of a wind bracing system composed of all 7 types of wind bracing elements shown in Figure 3a.

This was motivated by the fact that high-performance solutions for each of these subproblems are not only qualitatively but also quantitatively different (Kicinger, 2004). The impact of these differences on the performance of MAs was investigated. Figure 3b-c represent types of beam and support elements in conducted experiments, respectively. Columns were assumed the same (fixed joints) in all design experiments reported in this paper (for Problems I and II). Hence, there were not evolved by evolutionary algorithms but their cross-sections were optimized by the local search. The optimal cross-sections of structural members were selected from the catalog of standard shapes specified in (American Institute of Steel Construction, 1989).

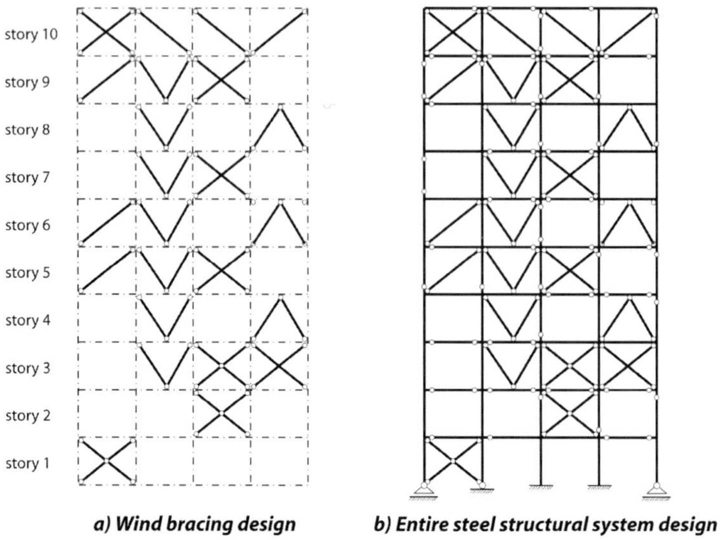

Figure 2: Classes of structural design problems considered in computational experiments

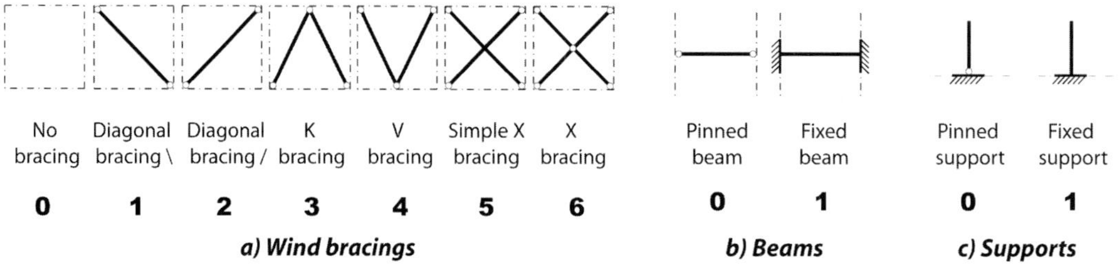

Figure 3: Types of structural members used in the design experiments

The parameters of the design problems and their values are presented in Table 1. It shows that 30-story buildings with 5 bays were considered. The 1st order structural analysis and the local optimization of cross-sections of members were conducted by SODA (Grierson, 1989). SODA is a commercial computer system developed for the analysis, design, and optimization of steel structural systems. Table 2 shows the magnitudes of dead, live, and wind loads assumed in these calculations.

Table 1: Parameters of the design problems

Domain Parameter	Value(s)
Number of stories	30
Number of bays	5
Bay width	20 feet (6.01 m)
Story height	14 feet (4.27 m)
Distance between transverse systems	20 feet (6.01 m)
Structural analysis method	1st order
Beams	pinned, fixed (Problem II)
Columns	fixed
Supports	pinned, fixed (Problem II)
Wind bracings	no, diagonal (/), diagonal (\), K, V, simple X, and X

Table 2: Magnitudes of dead, live, and wind loads

Load Parameter	Value(s)
Dead load magnitude	50 psf (2.39 kN/m2)
Live load magnitude:	
- building	100 psf (4.78 kN/m2)
- roof	30 psf (1.43 kN/m2)
Wind load:	
- Wind speed	100 mph (160.9 km/h)
- Wind importance factor	1.0
- Wind exposure category	C

3.2. Design Representations

In the reported experiments, steel structural systems were represented by linear genomes. For Problems Ia, Ib, and Ic the genomes were homogenous, i.e., they were composed of identical genes encoding wind bracing elements. For Problems Ia and Ib these genes had 2 values (binary representations) and for Problem Ic the genes had 7 values (integer-valued representation). Entire steel structural systems in tall buildings considered in Problem II were represented by nonhomogeneous genomes. They included 3 types of genes:

- encoding wind bracing elements with 7 possible values (see Figure 3a)
- encoding beam elements with binary values (see Figure 3b)
- encoding supports with binary values (see Figure 3c)

The length of the genome depended on the problem considered. For Problems Ia-Ic the genome was 150 genes long whereas for Problem II it consisted of 306 genes.

3.3. Evolutionary Computation Parameters

The major goal of the reported research is determination of optimal experimental settings for MAs applied to complex structural design problems. In order to achieve it, two classes of design problems were selected (described above) and an extensive evolutionary computation parameter search (sensitivity analysis) was conducted. It involved the following parameters and their values: the length of the design process (specified by the number of fitness evaluations), parent and offspring population sizes, the rate of mutation operator, and the rate of crossover operator.

The experiments were divided into two major groups depending on the termination criterion used in individual runs: short-term experiments (up to 1,000 fitness evaluations) and long-term experiments (up to 10,000 fitness evaluations). This distinction is important from the structural design point of view because evaluations of generated designs are usually very expensive (more than 99% of computational time).

Sensitivity analyses were conducted during short-term processes. The optimal combination of parameters' values found in the short-term processes was subsequently used in the long-term experiments. The performance analysis of MAs was conducted for both short- and long-term experiments. It included the following performance criteria:

- performance improvement of the best design at the end of the experiment compared to the best design from an initial population
- performance improvement of the average design at the end of the experiment compared to the average design from an initial population

Both improvements were measured by the reduction of the total weight of a structural system being designed.

An extensive parameter search was conducted during short-term experiments. For all combinations of parent and offspring population sizes shown in Table 3, a search for optimal rates of mutation and crossover was conducted. In each case, 12 combinations of mutation and crossover rates were considered, i.e. (mutation rate 0.025, crossover rate 0), (mutation rate 0.025, crossover rate 0.2), etc. The design processes were repeated 5 times for each combination of parameter values using a different value of a random seed each time.

Two types of memetic algorithms were tested: MA-GA and MA-ES. The former utilizes GAs while the latter uses ES. Both MAs were investigated in order to determine which produces better results when combined with the local search for the two classes of structural design problems. The initial population of parents was generated randomly for each experiment. As discussed above, the fitness of a design was determined by the total weight of the steel structural system calculated using the first-order structural analysis. Whenever an infeasible design was generated, it was assigned the fitness value of 0. In other words, the death penalty method was used to handle infeasible solutions (Coello Coello, 2002).

Table 3: Evolutionary computation parameters and their values

EC Parameter	Value(s)	EC Parameter	Value(s)
Type of MA	MA-ES, MA-GA	*Initialization method*	random
Pop. sizes (parent, offspring)	(1,5), (1,25), (5,25), (5,125) or (50,250) for MA-ES(μ+λ) (5,25), or (50,50) for MA-GA (5,25) for MA-ES(μ,λ)	*Fitness*	Total weight of the structural system (determined by the 1st-order analysis)
Generational model	Overlapping for MA-ES(μ+λ), Nonoverlapping for MA-ES(μ,λ) and MA-GA	*Constraint handling method*	death penalty (infeasible designs assigned 0 fitness)
Selection (parent, survival)	(uniform stoch., truncation) for MA-ES, (fitness prop., uniform stoch.) for MA-GA	*Termination criterion*	1,000 evaluations (short-term), or 10,000 evaluations (long-term)
Mutation rate	0.025, 0.1, 0.3, or 0.5	*Number of runs*	5 (in each experiment)
Uniform crossover rate	0, 0.2, or 0.5		

4. Experimental Results

This section reports the results of design experiments introduced in the previous section. They are grouped with respect to the parameter being investigated.

4.1. Optimal Rates of Mutation and Crossover Operators

Initial experiments focused on finding the optimal rates of mutation and crossover operators understood here as the rates which produced the best progress of MAs. An extensive parameter search was conducted to determine the optimal rates (see Table 3). Obtained results differed for various types of evolutionary algorithms. Typical results for MA-ES are presented in Figure 4 which shows the average best-so-far fitness values and 95% confidence intervals (vertical lines) calculated using Johnson's modified t test (Johnson, 1978) obtained in a series of design experiments with MA-ES(5+25) for Problem Ia. In these experiments, the rate of uniform crossover was equal to 0.2.

A clear pattern can be identified in Figure 4 regarding the impact of the mutation rate on the fitness of produced designs: lower mutation rates produce better fitness (i.e. lower fitness because it is a minimization problem) of designs produced. This pattern was observed in all design experiments involving MA-ES for various parent and offspring population sizes, and crossover rates, as illustrated in Figure 5. It shows average best-so-far fitness values and corresponding confidence intervals obtained at the end of short-term experiments for Problem Ia. Similar patterns were obtained for Problems Ib-c, and Problem II when ES was used.

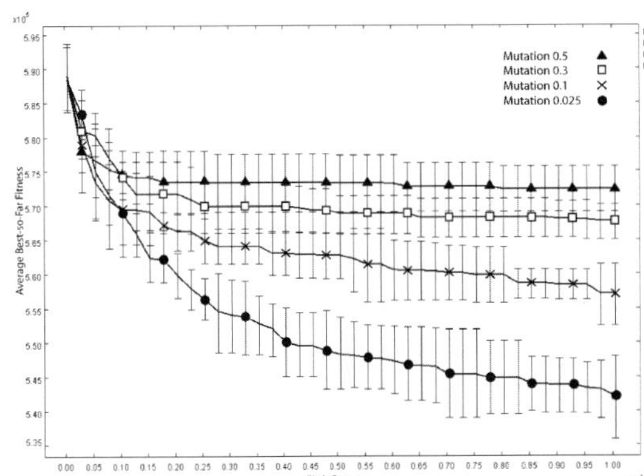

Figure 4: Progress of MA-ES for various mutation rates

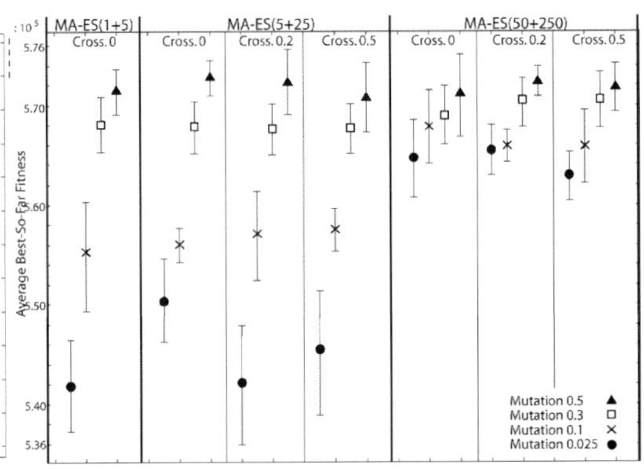

Figure 5: Impact of mutation rates on MA-ES

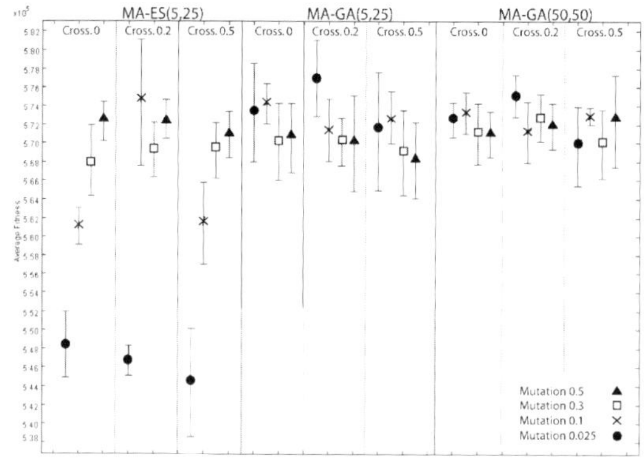

Figure 6: Comparison of the impact of mutation rates on MA-ES and MA-GAs

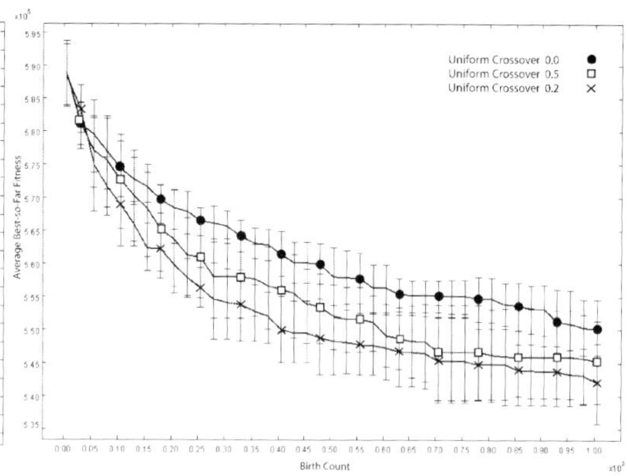

Figure 7: Impact of crossover rates on MA-ES

Different results and patterns were obtained when MA-GAs were employed to optimize topologies of steel structural systems. Figure 6 compares the results produced by two MAs, one employing MA-ES and one utilizing MA-GA. Here, the graphs produced by MA-ES(5,25) (left) are compared to the graphs produced by MA-GA(5,25) (center) and MA-GA(50,50) (right). It is clear that the results produced by MA-GAs are significantly different in terms of patterns produced: higher mutation rates produce better results, particularly when low crossover rates are used. In this case, however, the differences among the results produced by MA-GAs with various rates of mutation are small.

A search for the optimal rate of the crossover operator was conducted by analyzing the results of the design experiments in which various crossover rates were used but the mutation rate was kept the same. Figure 7 presents typical results obtained in the experiments in which MA-ES were used. It shows the average best-so-far fitness values and 95% confidence intervals obtained in the design experiments with MA-ES(5+25) and 3 different rates of crossover, i.e. 0.0, 0.2, and 0.5. The mutation rate was kept the same and equal to 0.025. Figure 7 shows that various crossover rates yielded only small differences in the fitness of produced designs. No clear pattern could be observed, as it was the case with the mutation operator. These observations were further confirmed by the results presented in Figure 8. It shows that there was no trend favoring specific crossover rates. On the contrary, in some cases the best results were achieved with no crossover at all and sometimes the best results were obtained when very high crossover rates are used, i.e. when the rate was equal to 0.5. Figure 8 also shows that even if there were differences among the fitness values obtained with various crossover rates, they were not significant (confidence intervals overlap in all cases). These results were consistent for all design problems reported in the paper. Similarly as in the case of MA-ES, MA-GAs do not exhibit any clear pattern in terms of preferred crossover rates. The graph showing these results was, however, omitted.

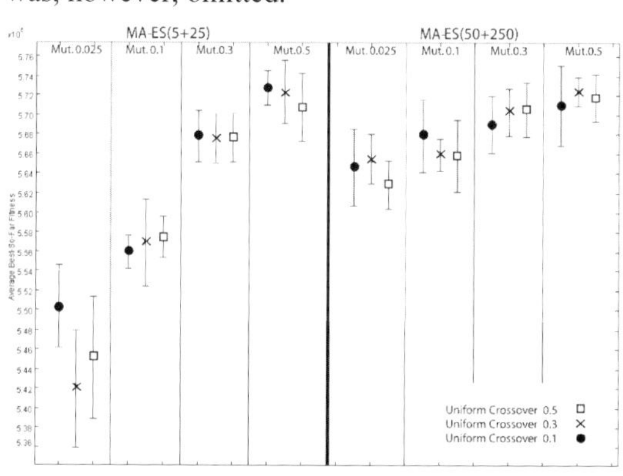

Figure 8: Impact of crossover rates on MA-ES

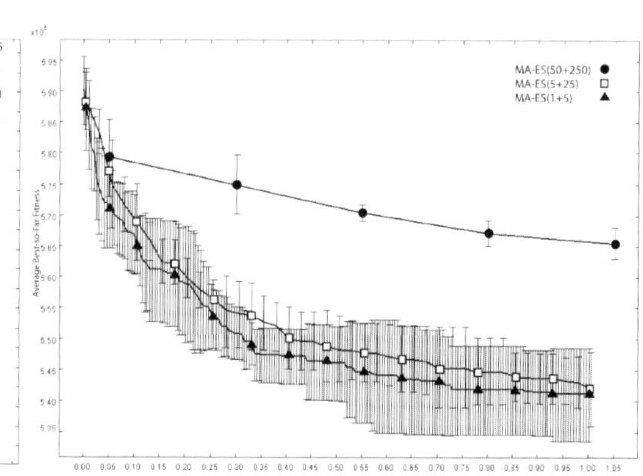

Figure 9: Impact of population sizes on MA-ES

Concluding, MA-ES produce the best results when low rates of mutation operator are used, e.g. 0.025. On the contrary, higher rates of mutation seem to be preferred by MA-GAs but the differences in the obtained results are not as significant as in the case of MA-ES. No such patterns were obtained for crossover rates.

4.2. Optimal Parent and Offspring Population Sizes

The next group of experiments focused on determining the optimal sizes of populations of parents and offspring for MAs. Three different combinations of sizes of parent and offspring populations were considered for MA-ES and two combinations for MA-GAs.

Typical results obtained with MA-ES for Problem I are presented in Figure 9. It shows the results of the design experiments in which three combinations of the parent and offspring population sizes were used, including MA-ES(1+5), MA-ES(5+25), and MA-ES(50+250). Mutation and crossover rates were kept the same in all experiments shown in Figure 9 and equal to 0.025 and 0.2, respectively.

It is clear that MA-ES using large population sizes, i.e. MA-ES(50+250), produced inferior results compared to the other two MA-ES with smaller population sizes. On the other hand, it also produced the smallest variance. The other two MA-ES with smaller population sizes achieved almost the same optimization progress in terms of the average best-so-far fitness of the produced designs. However, MA-ES(1+5), i.e. the 'greedy' MA-ES preserving only the single best individual to the next generation, exhibited much larger variance compared to MA-ES(5+25) which preserves the top 5 individuals to the next generation. Thus, in this case parallel search conducted by MA-ES(5+25) reduces the variance of the obtained results without decreasing the performance of the algorithm. On the other hand, when the size of populations is increased too much, e.g. as in MA-ES(50+250), the reduction of variance comes at a cost of a substantial decrease of the performance of the algorithm.

The outcomes were again different for MA-GAs. In both cases, i.e. for MA-GA(5,25) and MA-GA(50,50), the performance of the algorithm was almost identical. Figure 10 shows typical results of the design experiments involving MA-GA(5,25) and MA-GA(50,50). The specific results presented in this figure were produced by the two algorithms with the same mutation and crossover rates equal to 0.3 and 0.5, respectively.

The two best-so-far curves shown in Figure 10 are almost identical in their nature. The only difference between the two curves is the reduction of variance for the algorithm with larger population sizes, i.e. for MA-GA(50,50). Similar behavior was also observed for MA-ES (see Figure 9). Figure 11 shows that MA-ES produced exactly the same patterns for Problem II as for Problem I. Good optimization progress was obtained for small and medium population sizes, i.e., MA-ES(1+5) and MA-ES(5+25), while the best results in terms of progress and variance were achieved by MA-ES(5+25).

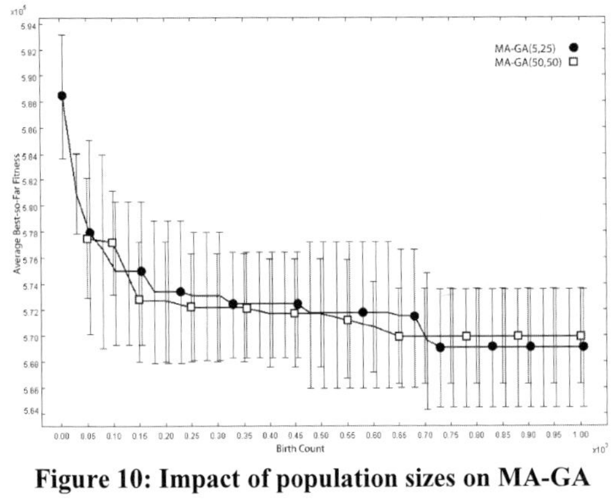

Figure 10: Impact of population sizes on MA-GA

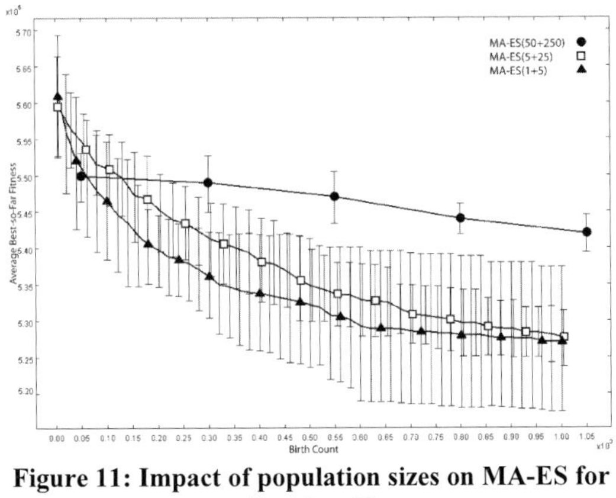

Figure 11: Impact of population sizes on MA-ES for Problem II

Concluding, small population sizes seem to be preferred by MA-ES for these problem domains. However, too small population sizes increase the variance of the obtained results. Good results in terms of both performance and variance were produced when moderate sizes of population sizes were employed, e.g. 5 in the case of the parent population and 25 in the case of the offspring population. The impact of the sizes of parent and offspring populations on the performance of MA-GAs seems to be negligible and related only to the reduction of variance of the obtained results. It didn't influence the actual performance of the algorithms in these problem domains.

4.3. Optimal Generational Model

The type of generational model in EAs determines whether or not best solutions found so far should be kept for the next generation. In the overlapping model, the best solution(s) from the parent population can compete for survival with solutions from the offspring population. On the other hand, in the nonoverlapping model only solutions from the offspring population can compete for survival.

In order to determine the impact of the type of the generational model (overlapping vs. nonoverlapping) on MAs, a series of design experiments involving two kinds of MA-ES was conducted, namely MA-ES(5+25) (overlapping) and MA-ES(5,25) (nonoverlapping). The experiments included a total of 24 design experiments (12 for each algorithm) utilizing all 12 combinations of mutation and crossover rates (see Table 3). Figure 12 shows typical results obtained in these experiments. Here, mutation and uniform crossover rates were equal to 0.025 and 0.2, respectively. Figure 12 shows that there are no significant differences between MA-ES(5,25) and MA-ES(5+25). This type of behavior was observed in all conducted experiments. In several cases MA-ES(5+25) slightly outperformed MA-ES(5,25) but in other cases it produced inferior results. The differences between the two generational models were, however, small both in terms of variance and fitness of the generated designs. Generally, it can be concluded that MA-ES with the overlapping and nonoverlapping generational model produce comparable results in these problem domains.

4.4. Optimal Type of Evolutionary Algorithm

The choice of the type of EA determines the properties of global exploration in MAs. In order to determine which EA produces better results in the investigated problem domains, a series of design experiments was performed. Figure 13 shows a comparison of the behavior of two algorithms, i.e., MA-ES and MA-GA, for Problem Ia. Two average best-so-far curves in the upper part of Figure 13 correspond to the best results obtained with MA-GAs with two combinations of parents and offspring population sizes, i.e. MA-GA(5,25) and MA-GA(50,50). In both cases the mutation rate was equal to 0.3 and crossover rate was equal to 0.5. The results produced by MA-GAs are compared to the average best-so-far performance produced by MA-ES with the overlapping (MA-ES(5+25)) and nonoverlapping (MA-ES(5,25)) generational models. In this case, the rates of mutation and crossover were equal to 0.025 and 0.2, respectively

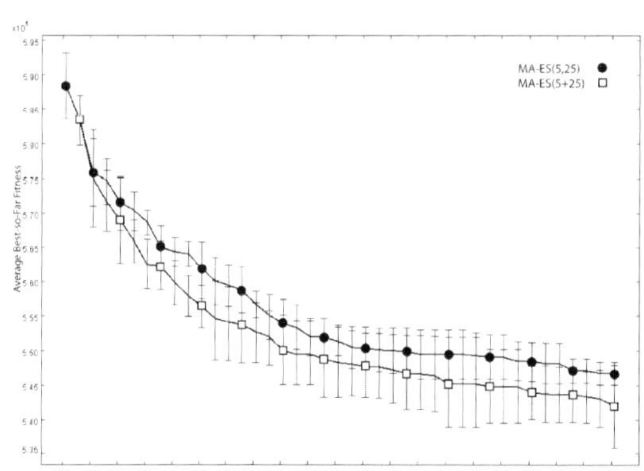

Figure 12: Impact of the generational model on MA

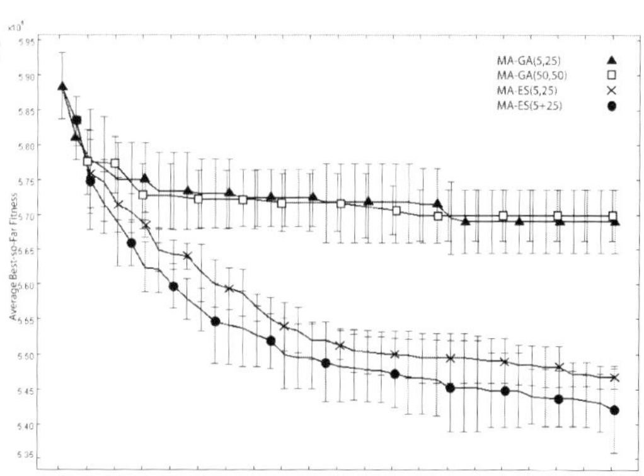

Figure 13: Impact of the type of EA on MA

Figure 13 clearly shows that MA-ES outperformed MA-GA in this problem domain. The average fitness value produced by MA-GA(5,25) after 1,000 evaluations was equal to 569,056 lbs. compared to 542,029 lbs. achieved by MA-ES(5+25). This corresponds to almost 5% better results, on average, produced by MA-ES. The performance improvement between an average design produced after 1,000 fitness evaluations and an average design in the initial population was equal to 19,434 lbs., or 3.3%, for MA-GA(5,25). On the other hand, for MA-ES(5+25) these values were equal to 46,461 lbs. and 7.9%, respectively.

Concluding, the results of the design experiments revealed that MA-ES performed better than MA-GAs in these problem domains. Hence, they were employed in the long-term design experiments reported in the next section.

4.5. Length of Evolutionary Processes

The length of an evolutionary process influences the computational effort required to run a design experiment. This issue is particularly relevant to structural engineering due to the fact that evaluations of structural designs are usually computationally expensive. In yet another series of design experiments with MAs, the impact of the length of design processes (measured by the total number of fitness evaluations) on the quality of produced designs was investigated.

Figure 14 shows the progress of the long-term experiments (10,000 fitness evaluations) with MA-ES for Problem Ic and compares it to the average fitness obtained after 1,000 evaluations (short-term experiment). The average performance improvement between the long-term processes and short-term processes was equal to about 21,900 lbs., or 4.3 percent. The difference between the average fitness after 10,000 fitness evaluations and the average fitness of the initial parents was equal to more than 55,500 lbs., or 10.2 percent. Similar results were produced by MA-ES for Problem II. Figure 15 shows a similar comparison of the long- and short-term experiments. In this case, the average fitness of the designs produced in the long-term experiments equaled 502,879 lbs. and was more than 21,500 lbs., or 4.1 percent, better than the average fitness obtained in the short-term experiments. The overall performance improvement in the long-term experiments was, on average, equal to more than 58,000 lbs., or 10.3 percent, compared to about 36,000 lbs., or 6.4 percent, achieved in the short-term optimization experiments.

The long-term experiments took ten times more computational time than the short-term experiments. Considering the fact that each evaluation took about 1 minute on a Pentium IV processor, the total time required for a single run of a long-term experiments was equal to almost 7 days compared to about 16 hours necessary for a short-term experiment. Thus, there is always a strong trade-off between the length of a design process and available computational resources. In this case, we can utilize the fact that after a certain number of fitness evaluations the best-so-far fitness curves tend to level-off (see Figure 15) and significant performance improvements are no longer obtained. This threshold, however, needs to be defined empirically for individual problems.

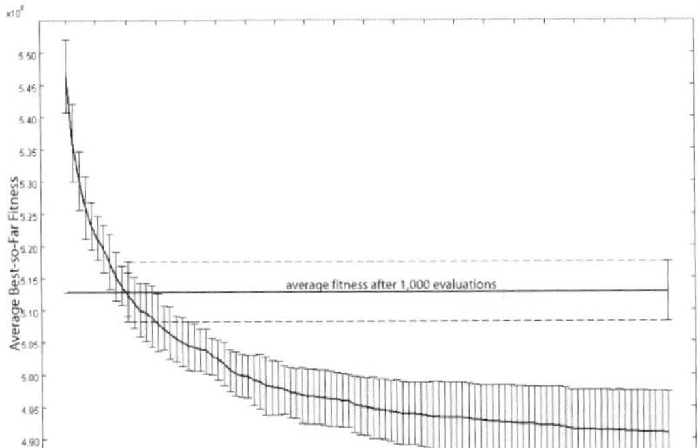

Figure 14: Impact of the length of design process on performance of MA-ES for Problem Ic

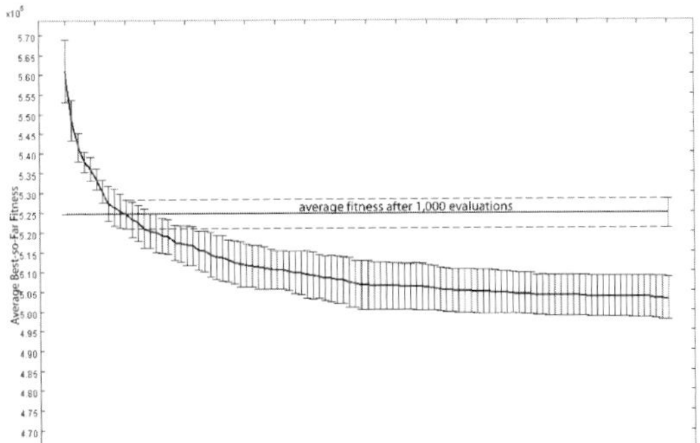

Figure 15: Impact of the length of design process on performance of MA-ES for Problem II

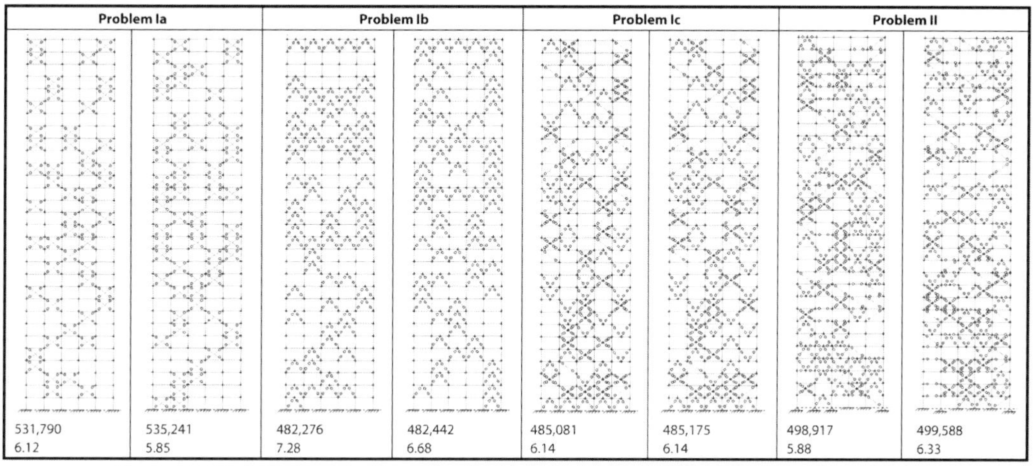

Figure 16:Best designs produced in design experiments with MAs for each problem

4.6. Optimal Designs

In extensive design experiments, (sub)optimal designs were produced for each class of structural problems and subproblems investigated in the paper. The final designs generated by MAs for each problem differed not only in the total weight but also in structural shaping patterns produced by configurations of wind bracings. The best designs produced in these experiments are shown in Figure 16. The values located below designs indicate their total weight in lbs. (top) and their maximum horizontal displacement (bottom) in inches.

Figure 16 shows significant differences in fitness (total weight) corresponding to best solutions found for each problem. For example, steel structural systems with the wind bracing system composed solely of X bracing elements (Problem Ia) have generally higher total weight by about 40,000-50,000 lbs. than best designs for other problems. On the other hand, they typically exhibit smaller horizontal displacements (better stiffness). They also show emerging structural shaping pattern of crossed macrodiagonal bracings in the lower and middle part of the structural system (see Figure 16). On the contrary, best solutions for Problems Ib-c and II exhibit random-looking configurations of wind bracing elements.

5. Conclusions

This paper reported results of a large number of design experiments with memetic algorithms. It also formulated initial recommendations regarding optimal parameter settings for these algorithms when applied to complex structural design problems. These heuristics can be used by researchers and practitioners to quickly set up their design experiments without conducting expensive parameter optimization.

In the reported studies, key evolutionary computation parameters and their values were identified. It was discovered that the type of evolutionary algorithm, the rate of mutation operator, and the size of the parent population were critical for the success of structural design processes. Specifically, evolution strategies produced on average significantly better results than genetic algorithms for the design problems considered in the paper. Also, low mutation rates, i.e. 0.025, resulted in better performance of memetic algorithms. Furthermore, small parent population sizes were generally preferred to large populations.

The experiments also produced significant body of structural design knowledge, both quantitative and qualitative in nature. Specifically, they identified regions of the design spaces in which high-performance solutions could be found. They also determined the ranges of the total weight of structural systems associated with high-performance solutions for both classes of design problems and showed that these ranges are significantly different. Finally, qualitative differences between high-performance solutions have been identified in terms of structural shaping patterns exhibited by configurations of wind bracing elements.

The research presented in this paper will be continued, including the extension of the scope of the empirical studies to other structural design problems. Also, other local search algorithms will be combined with evolutionary algorithms and applied to several structural engineering problems. Another promising direction of future research includes more advanced representations of structural systems and their impact on the performance of memetic algorithms.

References

Adeli, H., and Cheng, N. T. (1993). "Integrated genetic algorithm for optimization of space structures." *Journal of Aerospace Engineering*, 6(4), 315-328.

American Institute of Steel Construction. (1989). *AISC Manual of Steel Construction*, Ninth edition, AISC, Chicago, IL.

Coello Coello, C. A. (2002). "Theoretical and numerical constraint-handling techniques used with evolutionary algorithms: a survey of the state of the art." *Computer Methods in Applied Mechanics and Engineering*, 191, 1245-1287.

Grierson, D. E. (1989). "Computer-automated optimal design for structural steel frameworks." *Proceedings of the NATO ASI Conference on Optimization and Decision Support Systems in Civil Engineering*, B. H. V. Topping, ed., Edinburgh, UK, 327-354.

Hart, W. E., Krasnogor, N., and Smith, J. E. (2005a). "Memetic evolutionary algorithms." Recent advances in memetic algorithms, W. E. Hart, N. Krasnogor, and J. E. Smith, eds., Springer-Verlag, Berlin Heidelberg, 3-27.

Hart, W. E., Krasnogor, N., and Smith, J. E. (2005b). *Recent advances in memetic algorithms*, Springer-Verlag, Berlin Heidelberg.

Hoeffler, A., Leysner, U., and Weidermann, J. (1973). "Optimization of the layout of trusses combining strategies based on Mitchel's theorem and on biological principles of evolution." *Proceedings of the 2nd Symposium on Structural Optimization*, Milan, Italy.

Holland, J. H. (1975). *Adaptation in natural and artificial systems*, University of Michigan Press, Ann Arbor, Michigan.

Johnson, N. J. (1978). "Modified t tests and confidence intervals for asymmetrical populations." *Journal of the American Statistical Association*, 73(363), 536-544.

Kicinger, R. (2004). "Emergent Engineering Design: Design creativity and optimality inspired by nature," Ph.D. Dissertation, School of Information Technology and Engineering, George Mason University, Fairfax, VA, USA.

Moscato, P. (1989). "On evolution, search, optimization, genetic algorithms and martial arts: Towards memetic algorithms." *Caltech Concurrent Computation Program Report 826*, California Institute of Technology, Pasadena, CA, USA.

Quagliarella, D., and Vicini, A. (1998). "Coupling genetic algorithms and gradient based optimization techniques." Genetic algorithms and evolution strategies in engineering and computer science: recent advances and industrial applications, D. Quagliarella, J. Periaux, C. Poloni, and G. Winter, eds., John Wiley & Sons, Chichester, England.

Rechenberg, I. (1973). *Evolutionsstrategie; Optimierung technischer Systeme nach Prinzipien der biologischen Evolution*, Frommann-Holzboog, Stuttgart-Bad Cannstatt.

Sakamoto, J., and Oda, J. (1993). "Technique for optimal layout design for truss structures using genetic algorithms." *Proceedings of the 34th AIAA/ASCE/ASME/AHS Structural Dynamics and Material Conference AIAA/ASME Adaptive Structures Forum*, New York, NY, 2402-2408.

ELASTO-PLASTIC FINITE ELEMENT ANALYSIS OF PRYING OF TOP- AND SEAT-ANGLE CONNECTIONS

M. Komuro[1], N. Kishi[1] and A. Ahmed[2]

[1]*Department of Civil Engineering and Architecture, Muroran Institute of Technology, Muroran 050-8585 Japan, E-mail: komuro@news3.ce.muroran-it.ac.jp*

[2]*Department of Building, Civil and Environmental Engineering, Concordia University, Montreal, Quebec H3G 1M8, Canada*

Abstract

In order to precisely investigate an interaction between column flange and top angle's vertical leg and the effects including prying action of bolts on $M - \theta_r$ characteristics of top- and seat-angle connections, nonlinear finite element (FE) analyses for top- and seat-angle connections as one of the angle type connections were performed. In those analysis, contact model with small sliding option was applied between contacting pair surfaces of all connecting elements. Bolt pretension force was introduced in the initial step of analysis. Numerical results together with those estimated by using Kishi–Chen power model were compared with the experimental ones to examine those applicabilities. Parametric study was also performed by varying connection parameters, material properties of connection assemblages, and magnitude of bolt pretension. From this study, the following results are obtained: (1) proposed numerical analysis method can be applicable to estimate elasto-plastic nonlinear behavior of angle type connections; (2) pretension force of bolts has no effect on prying action at the initial and ultimate level of loading, but gives some effects a little at the intermediate loading level; (3) a prying action can be increased with decreasing the thickness of flange angle or increasing of gage distance from the angle heel to the centerline of bolt hole; and (4) the initial connection stiffness and ultimate moment capacity of the connections can be increased due to increasing of angle thickness, beam depth and bolt diameter, and decreasing of gage distance.

Keywords: prying force, FE analysis, top- and seat-angle connection, moment-rotation behavior

Introduction

It is widely known that the behavior of bolted connection is significantly influenced due to a prying action. Many researchers have attempted to investigate nature and influence of prying action on connection behavior and the supporting point for prying action of angle, etc. Fleischman (1988) and Chasten et al. (1989) had investigated the effect of prying action by tension bolts experimentally. It is confirmed by them that the interaction between top angle and column flange causes additional tension force in bolts due to prying action, which may depend on the factors such as: bending stiffnesses of bolt and top angle, and location of connecting bolts.

In this paper, in order to precisely investigate an interaction between column flange and top angle's vertical leg and the effects including bolt action on $M - \theta_r$ characteristics of top- and seat-angle connections, an elasto-plastic finite element (FE) analysis was performed by using ABAQUS code. The performance of the FE analysis was discussed by comparing with the experimental moment-rotation results conducted by Azizinamini et al. (1985) and Harper (1990). It is confirmed from the comparison that the nonlinear moment-rotation behavior of top- and seat-angle connections

M. Pandey et al. (eds), Advances in Engineering Structures, Mechanics & Construction, 289–301.

Table 1. Geometrical properties of top- and seat-angle connections.

<table>
<tr><th rowspan="2">FE model</th><th rowspan="2">Column section</th><th rowspan="2">Beam section</th><th colspan="5">Top and seat angles</th><th rowspan="2">Bolt diameter d_b (mm)</th></tr>
<tr><th>Angle section</th><th>l (mm)</th><th>g (mm)</th><th>r (mm)</th><th>q (mm)</th></tr>
<tr><td>A1, FE5, FE6, A1np</td><td>W12×96</td><td>W14×38</td><td>L6×4×$^3/_8$</td><td rowspan="2">203</td><td rowspan="2">64</td><td rowspan="2">140</td><td rowspan="2">89</td><td rowspan="6">22</td></tr>
<tr><td>A2, FE8, A2np, FE8np</td><td>W12×96</td><td>W14×38</td><td>L6×4×$^1/_2$</td></tr>
<tr><td>Test 3</td><td>W8×24</td><td>W8×21</td><td>L6×3½×$^3/_8$</td><td>152</td><td>51</td><td>89</td><td>70</td></tr>
<tr><td>FE1, FE7, FE1np, FE7np</td><td>W12×96</td><td>W14×38</td><td>L6×4×$^3/_4$</td><td rowspan="9">203</td><td>64</td><td rowspan="9">140</td><td rowspan="9">89</td></tr>
<tr><td>FE2, FE2np</td><td>W12×96</td><td>W14×38</td><td>L6×3½×$^3/_8$</td><td>51</td></tr>
<tr><td>FE3, FE3np</td><td>W12×96</td><td>W14×38</td><td>L6×6×$^3/_8$</td><td>114</td></tr>
<tr><td>FE4, FE4np</td><td>W12×96</td><td>W14×38</td><td>L6×4×$^1/_2$</td><td rowspan="4">63.5</td><td>19</td></tr>
<tr><td>FE9, FE9np</td><td>W12×96</td><td>W8×28</td><td>L6×4×$^3/_8$</td><td rowspan="5">22</td></tr>
<tr><td>FE10, FE10np</td><td>W12×96</td><td>W8×28</td><td>L6×4×$^1/_2$</td></tr>
<tr><td>FE11, FE11np</td><td>W12×96</td><td>W8×28</td><td>L6×4×$^3/_4$</td></tr>
<tr><td>FE12, FE12np</td><td>W12×96</td><td>W8×28</td><td>L6×3$^1/_2$×$^3/_8$</td><td>51</td></tr>
<tr><td>FE13, FE13np</td><td>W12×96</td><td>W8×28</td><td>L6×6×$^3/_8$</td><td>114</td></tr>
</table>

for large deformation region can be analyzed by using proposed FE analysis method taking pretension force of bolts, contact effects and sliding frictional effects of connecting elements into account. In addition, an applicability of Kishi–Chen power model was also verified by comparing with the experimental as well as the numerical results from FE analysis.

A parametric study was also conducted to investigate the effect of parameters on prying action and moment-rotation behavior of the connections. In this analysis, the dimensions of connection models, material properties of the connection components, and magnitude of the pretension forces surcharged to the bolts were taken as variables.

FE Analysis Method

Geometry of Connection Models

Dimensions of connection models for FE analysis are shown in Table 1 corresponding to the descriptions of column and beam section sizes, in which the connection parameters listed in Table 1 are shown in Figure 1. The other connection parameters are kept constant in all connection models and those are also shown in Figure 1. In this table, test specimens A1 and A2 were taken from the experiments of Azizinamini et al. (1985) and Test 3 was taken from Harper's thesis (1990). Specimens for numerical analysis are the same to those of test specimens A1, A2, and Test 3 and the specimens for numerical simulation in case varying some parameters of test specimen are also listed in Table 1. In those specimens 'np' means that pretension of bolts is ignored. Nominal name and geometrical dimensions for all column and beam sections considered here are determined following the AISC-LRFD specification (1994). As an example of mesh geometries of connection, a half model of A2 is shown in Figure 2. In this model, the total number of elements and nodes are 10,330 and

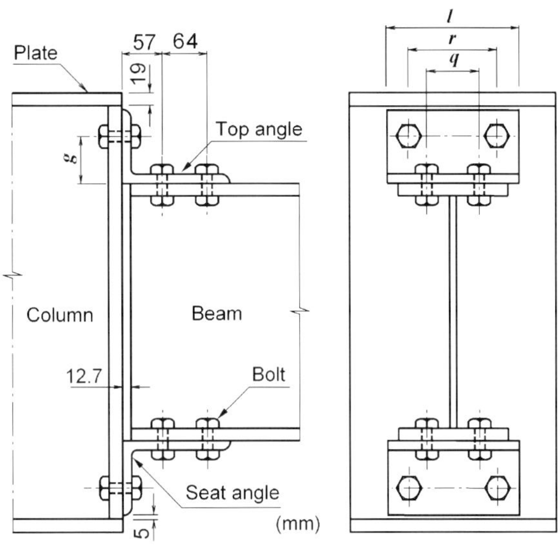

Fig. 1. Details of top- and seat-angle connection.

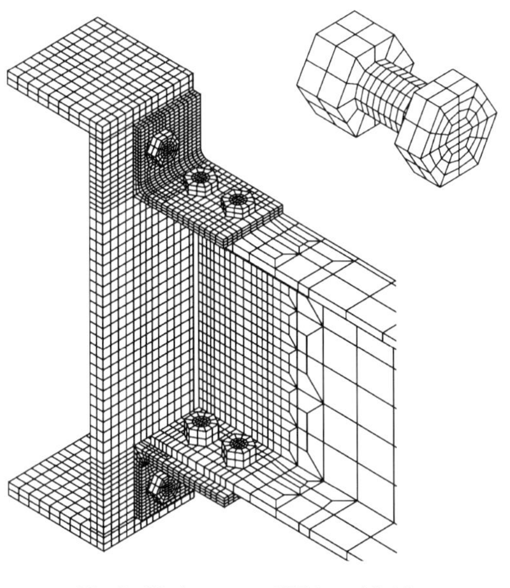

Fig. 2. Mesh pattern of FEA model A2.

17,984, respectively. All components including bolts are composed of eight node solid elements. Bolts in the mesh are divided into shank, head and nut elements to consider their individual effects on connection behavior. The diameter of bolt hole is set to be 1.6 mm (1/16 in) larger than that of bolt in accordance with Azizinamini et al.'s (1985) test data.

Mechanical Properties

Mechanical properties of angles and bolts used in numerical analysis are listed in Table 2. The mechanical properties of angles for all connection models except models FE7 and FE7np are the

Table 2. Mechanical properties used in FE analysis for top- and seat-angle connections.

Test or FEA model	Top and seat angles		Bolt	
	Yield stress σ_y (MPa)	Ultimate strength σ_u (MPa)	Yield stress $\sigma_{y,b}$ (MPa)	Ultimate strength $\sigma_{u,b}$ (MPa)
Test 3	297	517	635	830
A1, A2, FE1~FE6, FE9~FE13, A1np, A2np,FE1np~FE4np, FE9np~FE13np	365	550	635	830
FE7, FE7np	250	400		
FE8, FE8np	365	550	830	1,035

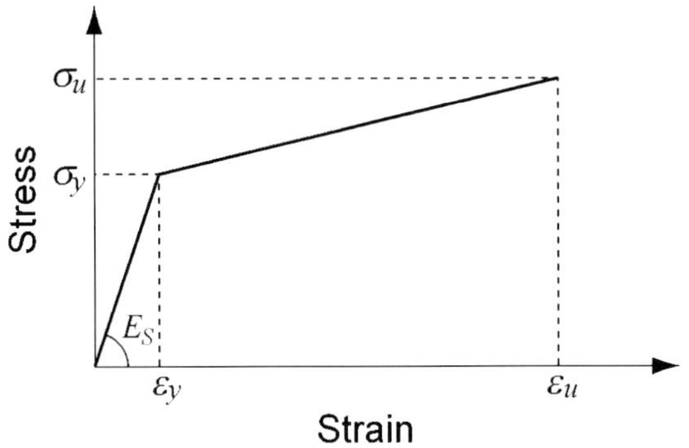

Fig. 3. Stress-strain relationship for steel used in this analysis.

same among them, which are taken from the test data for experiments conducted by Azizinamini et al. (1985). To investigate the effects of mechanical properties of angle on prying action, those for models FE7 and FE7np are assumed to be equal to the nominal values of A36 steel. Yield stress and ultimate strength of steel for beam, column and angles of all connection models are assumed to be 365 MPa and 550 MPa, respectively. Mechanical properties for bolts of all connection models except models FE8 and FE8np are assumed to be the nominal values of A325 bolts and those for models FE8 and FE8np are assumed to be equal to the nominal values of A490 bolts following the AISC-LRFD specifications because no coupon test results were reported yet. A bilinear elasto-plastic stress-strain relationship with isotropic hardening rule is assumed for all connection members taking Young's modulus of elasticity $E = 206$ GPa and Poisson's ratio $\nu = 0.3$ as shown in Figure 3. Here, ultimate strain ε_u is assumed to be 10% for bolt and 20% for another connection components.

Analytical Procedure

Numerical analyses of all connection models were performed using ABAQUS standards (1998), which was developed based on Finite Element Methodology. All components for each connection were modeled using first-order eight-node C3D8 solid elements. Here, pretension forces of bolts for all connection models except models A1np, A2np, FE1np through FE4np, FE7np through FE13np, FE5 and FE6 were prescribed up to 40% of the ultimate strength of bolt, and those for connection models FE5 and FE6 were prescribed up to 20% and 60%, respectively. The numerical results for

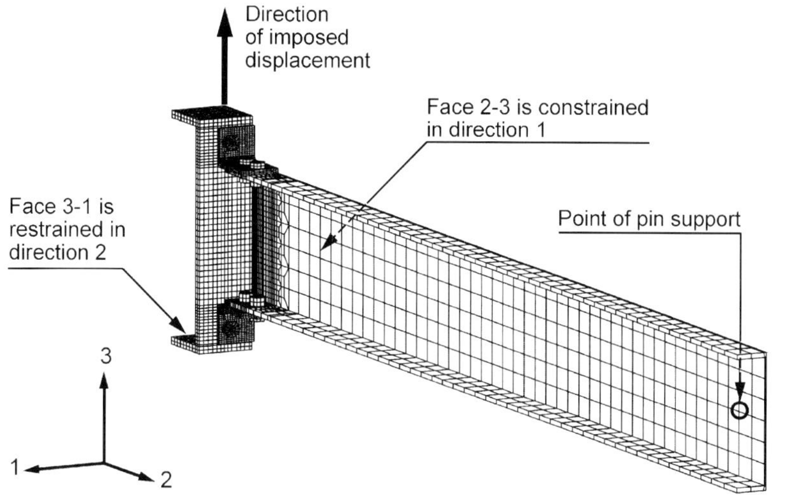

Fig. 4. Boundary conditions of FEA model of top- and seat-angle connections.

the models without pretension forces in bolts are used to estimate prying forces due to introducing pretension forces in bolts.

Numerical analysis was performed considering real experimental setup and loading method applied by Azizinamini et al. (1985), in which (1) two beams were symmetrically connected to the column flanges in a cruciform shape; (2) the ends of those beams were simply supported; and (3) the center of top plate of stub column was moved in upward direction as prescribed bending moment to be surcharged to the connection assemblages. Based on such experimental setup and considering structural symmetry, one-quarter model of connection composed of stub column, beam, top and seat angles and bolts was used for numerical analysis. Figure 4 shows a FE analysis model used for numerical analysis. The FE analyses considering pretension forces in bolts were performed in three loading steps. In the first step, a pressure equivalent to a prescribed pretension force is applied to the predefined section of bolt shank. As a result, the length of bolt shank at the pretension section changes by necessary amount to carry the prescribed load. In the second step, the prescribed load in bolt is replaced by changing the length of pretension section back into the initial length. In the third step, bending moment is introduced to the beam-to-column connection by employing vertical displacement at the middle section of plane 3–1 of the stub column (see Figure 4).

To precisely analyze the behavior of connecting members, contact model with small sliding option was applied for the contact surfaces between the vertical leg of angle and column flange, between the horizontal leg of angle and corresponding beam flange, and between the bolt and bolt hole elements. Moreover, to consider friction force occurring between sliding surfaces, Coulomb's frictional coefficient is assumed to be 0.1.

Analysis Results and Discussions

An Applicability of Proposed FE Analysis Method and Power Model

Initially, to assess an applicability of the proposed FE analysis method for simulating moment-rotation behavior of top- and seat-angle connections, numerical analyses for three connection models were performed and those numerical results together with those obtained using Kishi–Chen's three-parameter power model (1990) were compared with the experimental ones. The results of the first

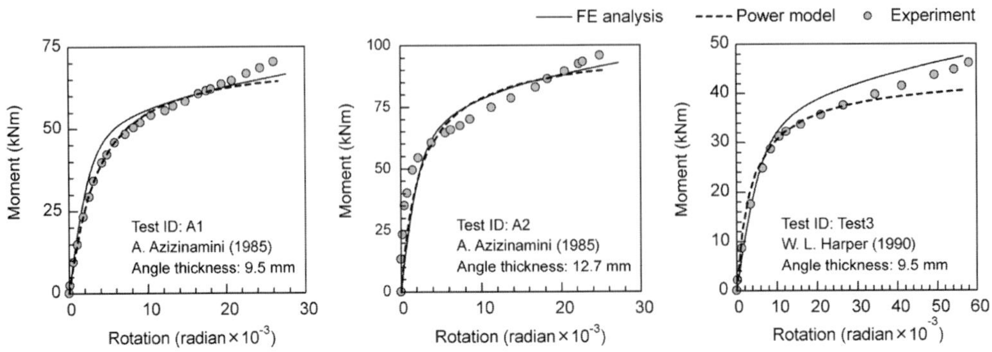

(a) A1 (b) A2 (c) Test 3

Fig. 5. Comparison of $M - \theta_r$ curves among FE analysis, power model and experiment.

two experiments (A1 and A2) reported by Azizinamini et al. (1985) and the other Test 3 reported by Harper (1990) were used for the comparisons. Following those experiments, a prescribed pretension force (40% of ultimate strength of bolt) was considered for all bolts of the connection. The comparisons of $M - \theta_r$ curves among analytical and experimental results are shown in Figure 5. It is observed that the curves obtained from FE analysis and power model prediction equation are almost similar to the experimental ones in the linear elastic and early plastic ranges. However, maximum moments from FE analysis for connection models A1 and A2, and from power model for all the three cases are smaller than those from test data. The differences among those three results in the plastic region may be caused by the following reasons:

(1) In the power model, angles are treated as perfectly elasto-plastic material and bolts are assumed to be perfectly rigid body, and the locations where plastic-hinges are formed in angles are logically fixed;
(2) There are some differences between the material properties used in FE analysis and those of test specimens.
(3) There may be experimental errors.

Even though analytical results differ a little from experimental ones, it is supposed that the proposed FE analysis method can be applied to more precisely investigate nonlinear connection behavior of top- and seat-angle connections considering pretension force introduced in bolts. In addition, it is also reasonable to conclude that the power model has a potential to predict $M - \theta_r$ curves of top- and seat-angle connections satisfactorily.

Stress-Deformation Behavior of Connection

Figure 6 shows the deformation configuration at ultimate state for model A2. From this figure, it is observed that although the horizontal maximum displacement is occurred at the heel of top angle, the vicinity of the bolt hole of top angle's vertical leg is deformed severely. Figure 7 shows the von Mises stress contour at ultimate state for model A2, in which at this time, applied connection moment is evaluated to be 103.5 kNm. From this figure, it is confirmed that the comparatively higher stresses were developed near the bolt hole and fillets of top angle. The bending moment evaluated using the reaction force developed at the beam end support can be converted into tension and compression forces in the connection, which are transferred to the column flange through the bolts fastening angles to the column flange. The bolts are elongated due to the bending-tension force applied to the

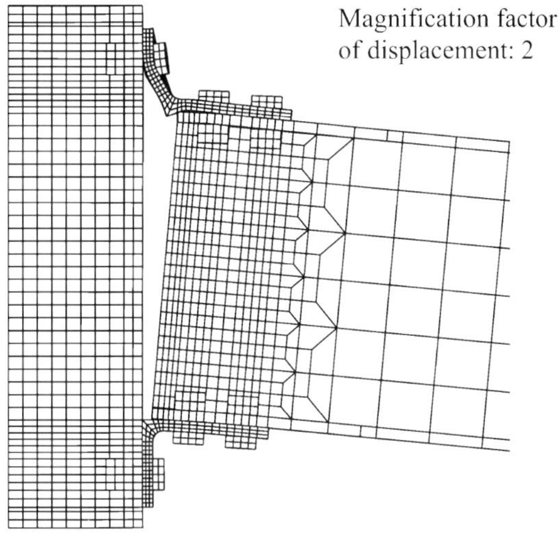

Fig. 6. Deformation configuration of connection model A2 at ultimate state.

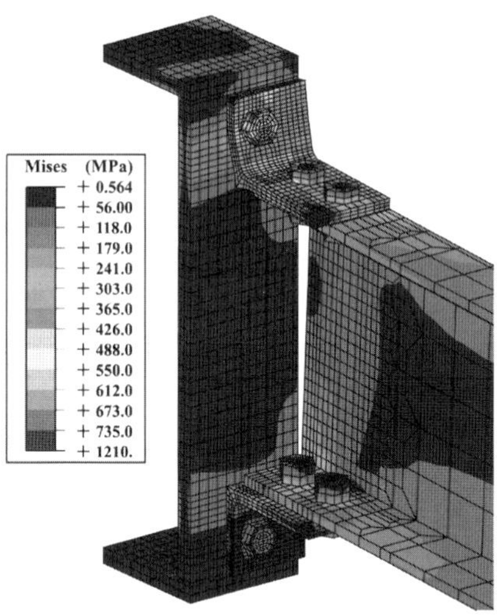

Fig. 7. Von Mises stress contour at ultimate state for model A2.

top angle, and the top angle's vertical leg behaves as a lever supported in the area from the upper portion of bolt hole through the top edge. As a result of interaction among top angle, column flange and bolts, a reaction pressure is developed in the area from the centerline of bolt hole through the top edge. To keep the forces acting in the top angle in equilibrium state, an equivalent additional tensile force corresponding to the reaction pressure is surcharged to the bolts. The reaction pressure developed on vertical leg can be substituted by a resultant, which is well known as prying force. So, tension bolts undergo not only pretension and bending-tension forces but also prying force.

Distribution of Forces in Tension Bolt

Figure 8(a) shows the distribution of internal forces in tension bolt for the case of model FE12np, in which initial pretension force introduced in bolts is ignored. The total tensile force surcharged to the

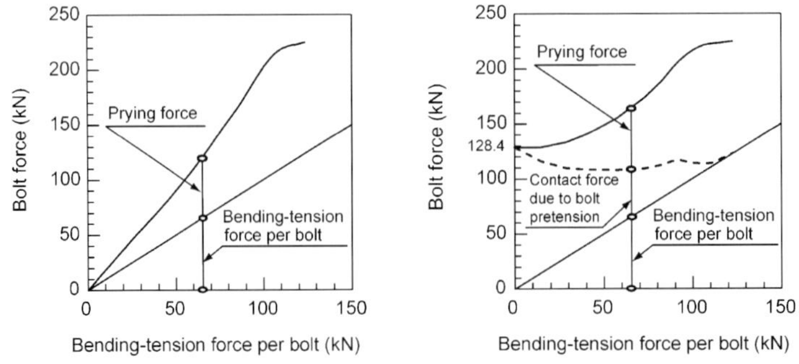

(a) Model FE12np (b) Model FE12

Fig. 8. Distribution of forces in tension bolt.

bolt for this case is evaluated as the summation of shear force applied to the top angle (hereinafter, shear force) which is equivalent to the tensile force applied to the upper flange of beam, and prying force because no pretension is applied to the bolts. From this figure, it can be observed that tensile force applied to the bolt is almost one and half time as much as shear force up to near the ultimate state. If the bending stiffness of top angle were infinite, tensile force applied to bolt would be equal to shear force applied to the angle.

It is apparent that prying force is also subjected to the tension bolts in case introducing pretension of model FE12 (Figure 8b). However, it is difficult to estimate the prying force subjected to tension bolts because bolt force consists of three components: shear force applying to top angle, contact force due to bolt pretension, and prying force. The shear force is determined as the transferred force from beam flange, whereas the other two forces can be determined as contact force that is the summation of contact force due to bolt pretension and prying force. In this study, to classify the contact force into the two forces, it is assumed that the distribution of prying force corresponding to shear force in case considering bolt pretension is similar to that for case ignoring pretension forces in bolts because it is confirmed that distribution of prying force in both cases considering with/without pretension forces in bolts is almost the same according to the results obtained from pre-analysis for investigating the influence of bolt pretension on prying action. Figure 8(b) shows the distribution of forces subjected in tension bolt of model FE12, which is the case by considering pretension forces in bolts. It is seen that three components of bolt force can be properly estimated. From this figure, it is observed that the contact force component caused by pre-tensioning of bolts is decreased nonlinearly with increasing of the shear force applied to the top angle's vertical leg.

Influence of Bolt Pretension on Prying Force and Connection Behavior

Figure 9 shows the comparisons of (a) distributions of bolt forces; (b) $M - \theta_r$ curves; and (c) initial connection stiffness among four connection models considering different levels of pretension forces in bolts. These models are A1np, FE5, A1, and FE6 for which initial pretension forces (C_0) are considered to be 0%, 20%, 40%, and 60% of the ultimate strength of bolt, respectively, and all the other properties are same among them. From Figure 9(a), it can be observed that the bolt tensile forces starting from different levels of pre-tensioning force get closer among them with increasing of bending tension force, and finally those reach at the same point. Thus, prying force near ultimate level of loading is not affected by the magnitude of pretension force. At the higher level of loading, contact force due to pre-tensioning of bolts is entirely neutralized, and only prying force is retained.

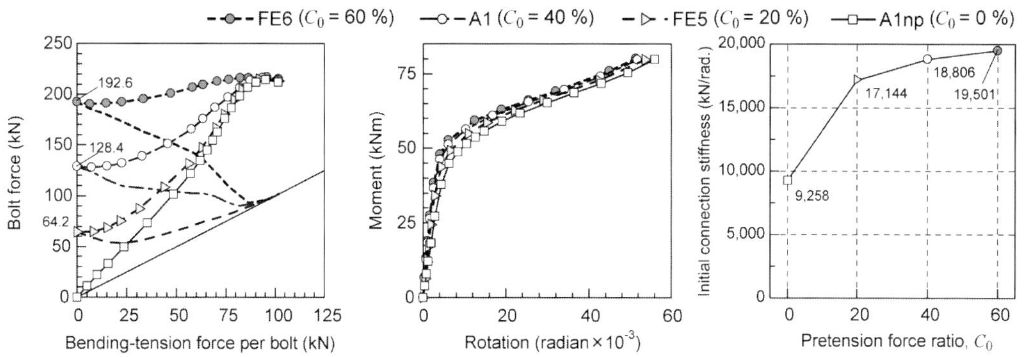

(a) Prying forces (b) M–θ_r curves (c) Initial connection stiffness

Fig. 9. Bolt pretension effect on prying force, $M - \theta_r$ curves, and initial connection stiffness.

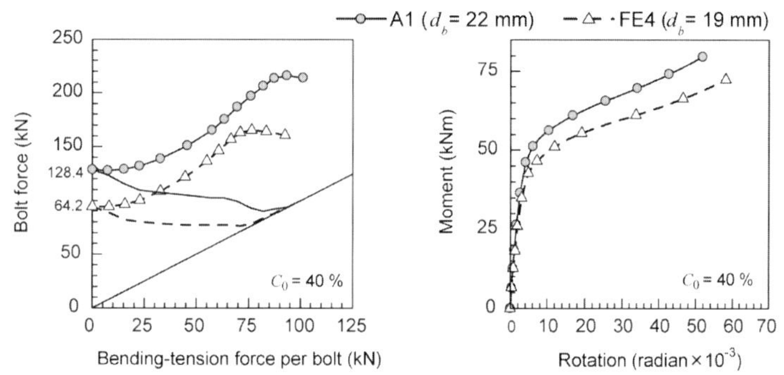

(a) Prying forces (b) M–θ_r curves

Fig. 10. Influence of bolt diameter on prying force and moment-rotation behavior.

From Figure 9(c), it can be observed that initial connection stiffness is largely increased by 85% for increasing pretension force from nil to snug-tightened level (20% of ultimate strength of bolt). Further increment of pretension forces in bolts is less effective in enhancing initial connection stiffness. When pretension force for bolt is increased from $C_0 = 20\%$ to 40 % and 60 %, initial connection stiffness is increased by 10% and 14%, respectively, with reference to the value at $C_0 = 20\%$. However, the ultimate moment capacity remains almost the same among the four cases (see Figure 9b).

Influence of Connection Parameters on Prying Force and $M - \theta_r$ Behavior

Bolt Diameter

The influence of bolt diameter on bolt force and $M - \theta_r$ curve of connection was investigated using numerical results obtained by inputting $d_b = 22$ mm and 19 mm for bolt diameter. Figure 10 shows the comparisons of distribution of prying force and $M - \theta_r$ curve of connection for the two cases. From Figure 10(a), it is observed that (1) at the beginning of loading, prying force develops very similar in both connection models; (2) but near the ultimate state, a prying force is increased when bolts with larger diameter is used.

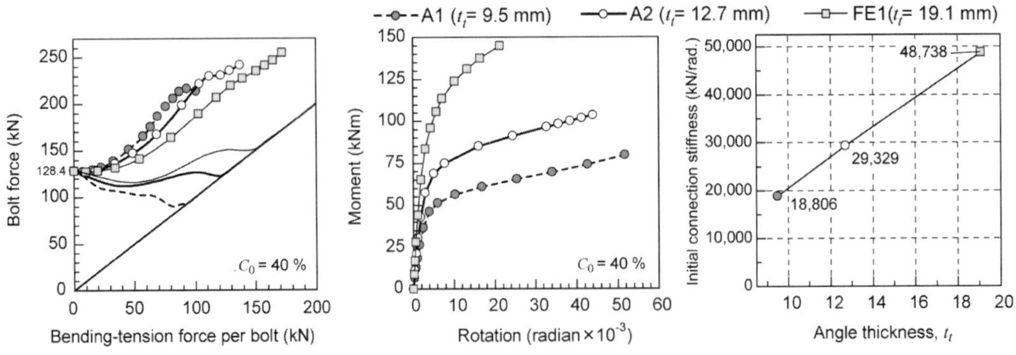

(a) Prying forces (b) M–θ_r curves (c) Initial connection stiffness

Fig. 11. Influence of flange angle thickness.

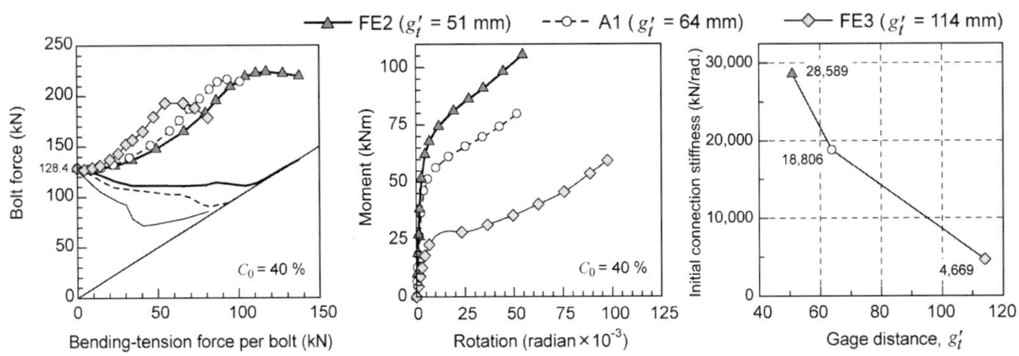

(a) Prying forces (b) M–θ_r curves (c) Initial connection stiffness

Fig. 12. Influence of gage distance from angle heel to the centerline of bolt hole.

From Figure 10(b), it is seen that even though the $M - \theta_r$ curves are almost similar to each other irrespective of the size of bolt diameter in the elastic region, the connection moment can be increased by increasing the size of bolt diameter in plastic region.

Angle Thickness

To investigate the effects of angle thickness on bolt force and $M - \theta_r$ curve of connection, numerical analysis were performed for connection model A1, A2, and FE1 in which those angle thicknesses (t_t) are assigned to be 9.5 mm, 12.7 mm, and 19.1 mm, respectively. Figure 11 shows the comparisons of bolt force, $M - \theta_r$ curve and initial connection stiffness among those three connection models. From Figure 11, it is seen that (1) the thinner the angle thickness is, the bigger the prying force is exerted and the smaller the connection moment is; (2) increasing angle thickness from 9.5 mm to 12.7 mm and 19.1 mm, ultimate connection moment is upgraded by 1.4 and 2.1 times more than that in case of 9.5 mm angle thickness, respectively; and (3) an initial connection stiffness is linearly increased corresponding to an increment of angle thickness.

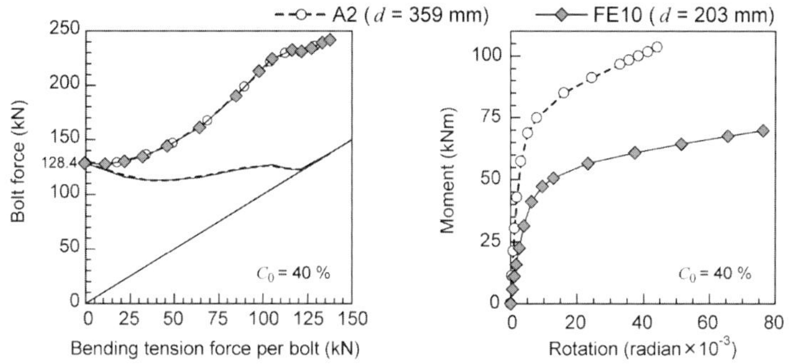

(a) Prying forces (b) M–θ_r curves

Fig. 13. Influence of beam height on prying force and moment-rotation behavior.

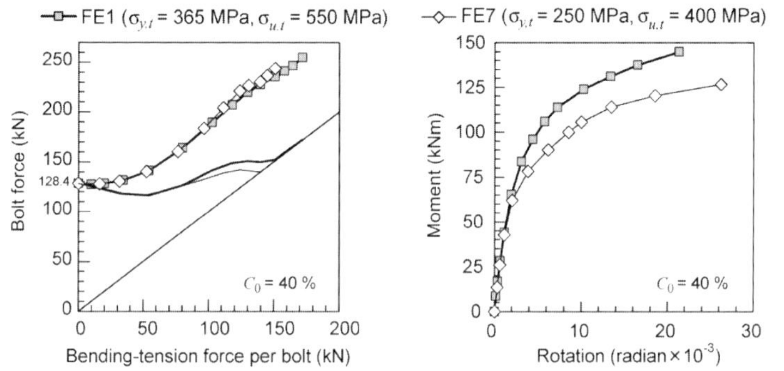

(a) Prying forces (b) M–θ_r curves

Fig. 14. Influences of mechanical properties of angle on prying force and moment-rotation behavior.

Gage Distance from Angle Heel to Centerline of Bolt Hole

Figure 12 shows the distributions of bolt force, $M - \theta_r$ curves, and initial connection stiffness in cases varying gage distance from top angle's vertical leg (g_t') as $g_t' = 51$ mm, 64 mm, and 114 mm, in which those are for models of FE1, A1, and FE3, respectively. From this figure, it is confirmed that decreasing the gage distance, (1) prying force is more increased; (2) a plastification of connection is early progressed and strain hardening effects can be expected in the large rotation area; and (3) initial connection stiffness is almost linearly increased.

Height of Beam Section

Figure 13 shows the distributions of bolt forces and $M - \theta_r$ curves in cases changing the beam height (d) as $d = 203$ mm and 359 mm, in which those are for models of A2 and FE10, respectively. From this figure, it is confirmed that the distribution of bolt force and prying force were never effected by beam height and connection moment is increased corresponding to the beam height only.

Mechanical Properties of Angle

Figure 14 shows the distributions of bolt force and $M - \theta_r$ curves in cases changing grade of angles as Azizinamini et al.'s test data and A36 steel, in which those are for models of FE1 and FE7, respectively. From this figure, it is confirmed that bolt force is not affected by mechanical properties

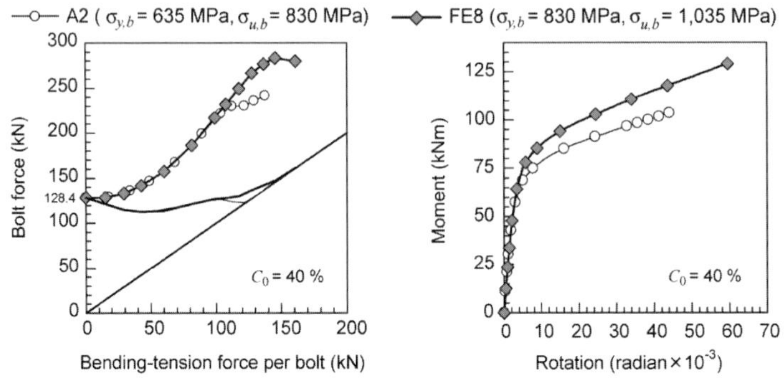

(a) Prying forces (b) M–θ_r curves

Fig. 15. Influences of mechanical properties of bolt on prying force and moment-rotation behavior.

of angle but connection moment is upgraded in the elasto-plastic region by using higher strength of angles.

Mechanical Properties of Bolt

Figure 15 shows the distributions of bolt force and $M - \theta_r$ curves, in cases changing grade of bolts as A325 and A490, in which those are for models of A2 and FE8, respectively. From this figure, it is observed that the bolt force is increased near ultimate state by taking higher grade bolts, connection moment capacity can be upgraded a little in the elasto-plastic region; and connection rotation also can be increased.

Conclusions

In this study, in order to investigate the interaction between column flange and top angle's vertical leg and these effects including bolt action on $M - \theta_r$ characteristics of top- and seat-angle connections, elasto-plastic finite element (FE) analysis was performed. Here, numerical analysis results together with the prediction by Kishi-Chen power model are compared with experimental ones to examine an applicability of the proposed analysis method and power model. In addition, to investigate the influence of connection parameters on prying force and moment-rotation behavior of top- and seat-angle connection, a parametric study was conducted by varying dimensions of connection models and magnitude of pretension force surcharged to the bolts. The following conclusions are obtained from this study:

(1) Proposed numerical analysis method can be applicable to estimate elasto-plastic nonlinear behavior of angle type connection;
(2) Power model is able to predict $M - \theta_r$ curves of the connection satisfactorily and can be applicable for nonlinear analysis of steel frames with semi-rigid connections;
(3) Pretension force of bolts has no effect on prying action at the initial and ultimate level of loading, but gives some effects a little at the intermediate loading level;
(4) A prying action can be increased with decreasing of thickness of flange angle or increasing gage distance from the angle heel to the centerline of bolt hole;
(5) Depth of beam section has no effect on prying action;
(6) Use of stiffer bolts or weaker angles in connection may increase the prying action; and

(7) Initial connection stiffness and ultimate moment capacity of the connections can be increased due to increasing of angle thickness, beam depth and bolt diameter, and decreasing of gage distance.

References

ABAQUS, 1998, *Standard User's Manual, Version 5.8*, Hibbitt Karlsson & Sorensen, Inc.

American Institute of Steel Construction (AISC), 1994, *Manual of Steel Construction, Load and Resistance Factored Design (LRFD)*, Vols I & II(2).

Azizinamini, A., Bradburn, J.H. and Radziminski, J.B., 1985, *Static and Cyclic Behavior of Semi-Rigid Steel Beam-Column Connections*, Structural Research Studies, Department of Civil Engineering, University of South Carolina, Columbia, SC.

Chasten, C.P., Fleischman, R.B., Driscoll, G.C. and Lu, L.W., 1989, Top-and-seat-angle connection and end-plate connections: Behavior and strength under monotonic and cyclic loading, *Proceedings of National Engineering Conference*, American Institute of Steel Construction, Chicago, Vol. 3, pp. 6-1–6-32.

Fleischman, R.B., 1988, Experimental and theoretical analysis of component behavior in top-and-seat-angle connections, ATLSS Project A3.1, Master's Thesis, Lehigh University, Pittsburgh, PA.

Harper, W.L., 1990, *Dynamic Response of Steel Frames with Semi-Rigid Connections*, Structural Research Studies, Department of Civil Engineering, University of South Carolina, Columbia, SC.

Kishi, N. and Chen, W.F., 1990, Moment-rotation relations of semi-rigid connections with angles, *Journal of Structural Engineering, ASCE*, 116(7), 1813–1834.

FULLY STRESSED SEISMIC DESIGN OF DAMPERS IN FRAMED STRUCTURES

Robert Levy and Oren Lavan
Faculty of Civil and Environmental Engineering,
Technion – Israel Institute of Technology, Haifa 32000, Israel
E-mail: cvrlevy@tx.technion.ac.il

Abstract

This paper presents an efficient and practical procedure for the optimal design of added damping in framed structures. The total added damping is minimized while inter-story performance indices for linear and nonlinear structures are chosen and restricted to allowable values under the excitation of an ensemble of realistic ground motion records. Optimality criteria are formulated based on fully stressed characteristics of the optimal solution and a simple analysis/redesign procedure is proposed for attaining optimal designs. Results of three examples presented compare well to those obtained using formal gradient based optimization.

Introduction

In the modern design of buildings to withstand strong earthquakes life safety is no longer the only concern but rather, performance whereby a prescribed level of damage is designed for. Hence, retrofitting of structures for a higher level of seismic protection may be needed. One of the means for achieving this enhancement to seismic performance is using supplemental damping which is the concern of this paper.

Various procedures for the design of added viscous damping, for *linear* behavior of damped 2D structures, were proposed by several researchers (for example Constantinou and Tadjbakhsh 1983; Zhang and Soong 1992; Fu and Kasai 1998; Inaudi *et al.* 1993; Gluck *et al.* 1996; Takewaki 1997; Lavan and Levy 2005a). Some of these procedures were extended to linear 3D structures (for example Wu *et al.* 1997; Takewaki *et al.* 1999). These methodologies, will usually require mathematics of stochastic processes, optimization methods, and/or variational mathematics – tools that are not that familiar to the practicing engineer. Garcia (2001) simplified the Sequential Search Algorithm, originally proposed by Zhang and Soong (1992) for stochastic models of the excitation, and made it appropriate for practical use. However, his method is restricted to linear structures under the excitation of a single deterministic record that does not guarantee optimal damping distribution of the dampers.

Procedures were proposed for the design of viscous dampers for *yielding* structures as well (Kim *et al.* 2003; Shen and Soong 1996; Lavan and Levy 2005b. The primary concern of the methodologies proposed by Kim *et al.* (2003), and by Shen and Soong (1996), was to estimate the total added damping needed rather than its distribution. The procedure proposed by Lavan and Levy (2005b) requires some nonlinear programming background and variational mathematics and may not be that easy to implement in the practicing design office. Allowable stress algorithms of the analysis/redesign procedures may be more suitable.

Allowable stress algorithms go back to the classical design of trusses, whereby the weight is minimized for a given allowable stress. Design problems of this type may be achieved iteratively using a two step algorithm in each iteration cycle. In the first step an analysis is performed for a given preliminary design, whereas in the second step the design is changed using a recurrence

M. Pandey et al. (eds), Advances in Engineering Structures, Mechanics & Construction, 303–315.

relationship which, for the truss problem is the ratio between the current stress and the allowable stress for each member. The algorithm possesses a fixed point (Levy 1991), i.e. fully stressedness and exhibits monotonic convergence properties. Convergence yields a *statically determinate fully stressed design, with members out of the design having strains smaller than the allowable*. This result appeared in the literature as early as 1900 (Cilley 1900). It was later shown (Levy 1985) that this design is a Karush–Kuhn–Tucker point and therefore, an optimal design. Algorithms closely related are the optimality criteria based algorithms as described by Khot *et al.* (1976), Venkayya (1978) and Rozvany (1989) to name only a few.

This paper presents a fully stressed design algorithm of the analysis/redesign type for the design of added viscous damping for linear as well as yielding structures for a given ensemble of realistic earthquake records and a specified target performance index. An algorithm of this kind is well liked by design engineers because its process is transparent and uses available and familiar dynamic time-history analysis programs that are common in civil engineering practice rather than unfamiliar mathematical optimization tools that need problem specific tailoring. Results were found to be in good agreement with optimal designs achieved using gradient based optimization. It should be noted in passing, that the fully stressed design described herein for viscous dampers has not been rigorously proven to be optimal yet.

Problem Formulation

This paper addresses the optimization problem of minimizing the added damping subject to constraints on local performance indices for framed structures excited by an ensemble of realistic ground motion records. The local performance indices *are interstory performance indices* for 2D frames, and interstory performance indices of the peripheral frames for 3D structures. Dampers are initially located at each story of the 2D frames or at each story of each peripheral frame in the 3D structures. As the optimization process progresses towards the optimum, however, some of the dampers will attain zero values.

The formulation of the optimization problem is comprised of the total added damping as an objective function, and an inequality constraint on the upper bound of each of the local performance indices which are computed based on the behavior of the structure, i.e., satisfying the equations of motion of the damped structure. The damping coefficients which are the design variables are required to be nonnegative.

The local performance indices are normalized by their allowable values such that a value of unity indicates that the local performance index is "fully stressed". Useful local performance indices are the maximal interstory drift, maximal interstory ductility, interstory hysteretic energy, combination of interstory ductility and interstory hysteretic energy such as the damage index of Park and Ang (1985), etc. or the maximal values of all of the above.

Linear Elastic Frames

A class of structures, usually regular ones, can be brought to behave elastically under an earthquake excitation, by the addition of a reasonable amount of damping. In these cases nonlinear analysis methodologies are not essential and some of the nonlinear performance indices are meaningless. Thus, linear tools are used.

Equations of motion: The equations of motion of a linear dynamic viscously damped system are given by:

$$\mathbf{M}\ddot{\mathbf{x}}(t) + [\mathbf{C} + \mathbf{C}_d(\mathbf{c}_d)] \cdot \dot{\mathbf{x}}(t) + \mathbf{K} \cdot \mathbf{x}(t) = -\mathbf{M} \cdot \mathbf{e} \cdot \mathbf{a}_g(t); \quad \mathbf{x}(0) = \mathbf{0}, \quad \dot{\mathbf{x}}(0) = \mathbf{0} \tag{1}$$

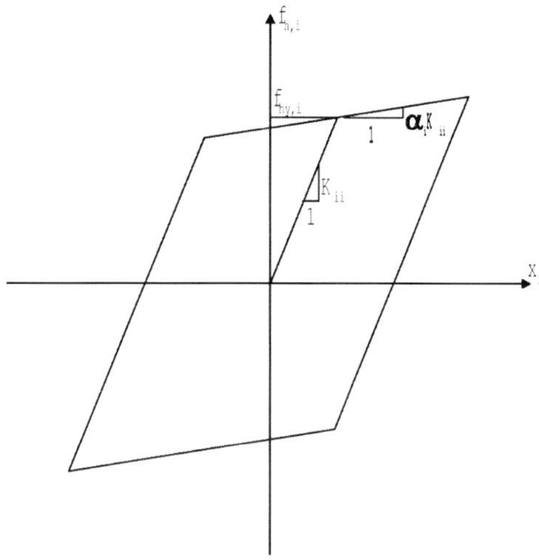

Fig. 1. Bilinear hysteresis diagram.

where $\mathbf{x}$ = the displacement vector of the degrees of freedom; $\mathbf{M}$ = mass matrix; $\mathbf{K}$ = stiffness matrix; $\mathbf{C}$ = inherent damping matrix; $\mathbf{c}_d$ = added damping vector; $\mathbf{C}_d(\mathbf{c}_d)$ = supplemental damping matrix; $\mathbf{e}$ = location matrix which defines location of the excitation, and $\mathbf{a}_g$ = vector of ground motion record. In the present work a damper is assigned to each story in the 2D frames, and at each story of the peripheral frames in the 3D structures.

Performance index: For linear structures, where the structure does not suffer structural damage, the maximal interstory drift becomes an important response parameter since it is a measure of nonstructural damage. Hence, the maximal interstory drift normalized by the allowed value, which is given as $pi_i = \max_t(|d_i(t)|/d_{all,i})$, is chosen as the local performance index for the 2D frames. Here $d_i(t)$ is the i-th story drift which is a linear function of $\mathbf{x}$ (such that $d_i(t) = \mathbf{L}_i\mathbf{x}(t)$ where $\mathbf{L}_i$ = transformation matrix), and $d_{all,i}$ is its allowable value. For the 3D structures a similar local performance index is used, but this time $d_i(t)$ is an interstory drift of a peripheral frame.

Optimization problem: The optimization problem is thus formulated as:

$$\text{minimize:} \quad J = \mathbf{c}_d^T \cdot \mathbf{1}$$

subject to:

$$\left.\begin{array}{l} pi = \max\limits_i \left(\max\limits_t (|\mathbf{L}_i \mathbf{x}(t)|/d_{all,i}) \right) \le 1.0 \\ \text{where } \mathbf{x}(t) \text{ satisfy the equations of motion} \\ \mathbf{M}\ddot{\mathbf{x}}(t) + [\mathbf{C} + \mathbf{C}_d(\mathbf{c}_d)] \cdot \dot{\mathbf{x}}(t) + \mathbf{K} \cdot \mathbf{x}(t) = -\mathbf{M} \cdot \mathbf{e} \cdot a_g(t); \\ \mathbf{x}(0) = \mathbf{0}, \dot{\mathbf{x}}(0) = \mathbf{0} \\ 0 \le \mathbf{c}_d \end{array}\right\} \forall a_g \in \text{whole ensemble.} \qquad (2)$$

Nonlinear Shear Frames

An addition of a reasonable amount of added damping may not be sufficient to result in elastic behavior of some classes of structures, usually irregular ones, under an earthquake excitation (Shen and Soong 1996; Uriz and Whittaker 2001). In these cases the damped structure develops plastic hinges. It is thus essential for the methodology to be based on nonlinear tools.

Equations of motion: The equations of motion of a nonlinear shear frame damped by linear viscous dampers are given by:

$$\mathbf{M}\ddot{\mathbf{x}}(t) + [\mathbf{C} + \mathbf{C}_d(\mathbf{c}_d)] \cdot \dot{\mathbf{x}}(t) + \mathbf{T} \cdot \mathbf{f}_h(t) = -\mathbf{M} \cdot \mathbf{e} \cdot a_g(t); \quad \mathbf{x}(0) = \mathbf{0}, \dot{\mathbf{x}}(0) = \mathbf{0} \tag{3}$$

$$\dot{\mathbf{f}}_h(t) = \mathbf{f}(\mathbf{L}\dot{\mathbf{x}}(t), \mathbf{f}_h(t)),$$

where $\mathbf{f}_h$ = interstory restoring force vector. In this work, a bi-linear hysteretic behavior is chosen (Figure 1); $\mathbf{T}$ =transformation matrix to transform $\mathbf{f}_h$ to the degrees of freedom coordinates, and $\mathbf{L}$ = matrix whose rows are $\mathbf{L}_i$. For a shear frame, the time derivative of the interstory restoring force of the i-th story, is a function of the i-th interstory restoring force and drift velocity only, i.e. $f_i = f_i(\dot{d}_i(t), f_{h,i}(t))$, where $f_{h,i}$ is the interstory restoring force of the i-th story.

Performance index: For nonlinear structures, where the structure is expected to suffer plastic deformations and dissipate energy by means of plastic behavior, the structural damage, that is measured by the damage index, becomes an important response parameter (see Williams and Sexsmith 1995 for a state of the art review on this topic), hence, it is chosen to represent the performance index. The story damage index is chosen to be energy based damage index due to its cumulative nature (Bannon *et al.* 1981). This damage index is chosen as the hysteretic energy dissipated by the restoring force divided by its allowed hysteretic energy. The total energy of the hysteretic component of the i-th story restoring force is given by $\int f_{h,i}(t)\dot{d}_i(t)dt$ and contains both its hysteretic and elastic energies. At the end of the ground motion, when the elastic displacements are small, the elastic energy is negligible, hence that integral represents the hysteretic energy dissipated by the i-th floor restoring force. Hence pi_i, the i-th element of $\mathbf{pi}$, is given by the expression:

$$pi_i(t_f) = \int_0^{t_f} f_{h,i}(t)\dot{d}_i(t)dt \Big/ \mathrm{H}_{\mathrm{all},i} \tag{4}$$

where $\mathrm{H}_{\mathrm{all},i}$ is the allowable hysteretic energy of the i-th floor which is a fraction of its hysteretic energy at failure. The hysteretic energy at failure is taken proportional to the elastic energy at yielding (Bannon *et al.* 1981).

Optimization problem: The optimization problem takes the following formulation:

$$\text{minimize:} \quad J = \mathbf{c}_d^T \cdot \mathbf{1}$$

$$\text{subject to:}$$

$$\left.\begin{aligned}
& pi = \max_i(pi_i(\dot{\mathbf{x}}(t), \mathbf{f}_h(t), t)|_{t_f}) \le 1.0 \\
& \text{where } \dot{\mathbf{x}}(t) \text{ and } \mathbf{f}_h(t) \text{ satisfy the equations of motion} \\
& \mathbf{M}\ddot{\mathbf{x}}(t) + [\mathbf{C} + \mathbf{C}_d(\mathbf{c}_d)] \cdot \dot{\mathbf{x}}(t) + \mathbf{T} \cdot \mathbf{f}_h(t) = -\mathbf{M} \cdot \mathbf{e} \cdot a_g(t); \\
& \mathbf{x}(0) = \mathbf{0}, \dot{\mathbf{x}}(0) = \mathbf{0} \\
& \dot{\mathbf{f}}_h(t) = \mathbf{f}(\dot{\mathbf{x}}(t), \mathbf{f}_h(t)) \\
& 0 \le \mathbf{c}_d
\end{aligned}\right\} \forall a_g \in \text{whole ensemble.} \tag{5}$$

Fully Stressed Design

Optimal seismic design of added damping for linear structures, as well as for yielding shear frames, has been achieved by the authors using formal nonlinear programming optimization techniques (Lavan and Levy 2005a, 2005b). Closer to the designer's heart, however are the analysis/redesign type techniques where an initial choice of the design variables is made (in this case damping coefficients, c_{d_i}) then, based on analysis results and a pre-defined recurrence relationship, the initial values are changed. A new analysis is made and the process continues until the designer is satisfied. Usually methods of this type converge very fast (up to five iterations for reasonable convergence of between 5–10% error in the examples of this paper). Moreover, in the classical FSD of trusses, for example, monotonic convergence is exhibited by the objective function (Spillers 1975). The designer is, thus, in full control and may stop at any iteration knowing that his results are the best up to that point.

Having studied the results achieved by the optimization methodologies mentioned above an analogy to the classical fully stressed design (FSD) of trusses seems to emerge. It is observed that, for 2D frames, the optimal design will attain nonzero values of c_{d_i} in stories for which the local performance index has reached the allowable, and zero values of c_{d_i} in stories for which the local performance index is less than the allowable. *In other words dampers are assigned only where the performance index is full.* Similarly, for the 3D structures the optimal design will attain nonzero values of c_{d_i} in stories of peripheral frames only, for which the local performance index has reached the allowable in at least one loading condition. In stories with no dampers the performance index is less than the allowable.

The recurrence relationship that is suggested in this work thus targets "fully stressedness" of the local performance index and is written as

$$c_{d_i}^{(k+1)} = c_{d_i}^{(k)} (pi_i^{(k)})^{1/q} \qquad (6)$$

where $c_{d_i}^{(k)}$ = the i-th component of the damping vector at the k-th iteration, q is a convergence parameter and $pi_i^{(k)}$ = the actual i-th component of the performance index at the k-th iteration (using $\mathbf{c}_d^{(k)}$ as a damping vector). Note that $c_{d_i}^{(k)}$ and and $pi_i^{(k)}$ refer to the same location i.e. same story of the same peripheral frame. In case the active set which is the subset of the ensemble that is considered at a certain stage, as will be explained later on, is comprised of more than one ground motion, $pi_i^{(k)}$ is taken as the envelope of $pi_i^{(k)}$ for the records within the "active" set.

The choice of q in (6) affects the efficiency of the method. For larger values of q the method is more stable, i.e. the method is more likely to converge, however the convergence is slower. For the linear problem $q = 0.5$ may be used whereas for the nonlinear case a value of $q = 5$ seems to be appropriate.

Design Methodology

The design is done by first identifying the "active" ground motion and attaining the design for that record. Once this design is achieved, the performance of the damped structure is evaluated for the remaining records in the ensemble. If the damping for this ground motion violates constraints of other records in the ensemble, the ground motion for which $\max_i(pi_i)$ receives the largest value is added until an appropriate solution for the whole ensemble is reached.

The methodology is thus comprised of four main stages. These stages will be described in detail subsequently and are summarized as:

Stage 1. Select the "active" ground motion.

Stage 2. Attain the design for the records within the active set using the analysis redesign approach.
Stage 3. Apply the remaining records in the ensemble on the current design for feasibility check and go to Stage 2 if stopping criteria are not met.
Stage 4. Stop.

Stage 1: Selection of the "active" ground motion
In general, the active ground motion is chosen by first sketching a certain response quantity of a single degree of freedom system, having the natural frequency of the structure, versus the damping coefficient, for all records in the ground motion ensemble. For the linear problem, the response quantity sketched is the maximal displacement whereas, for the nonlinear case it is the input energy (according to Uang and Bertero 1990). The record for which the response quantity takes the largest values, for a reasonable range of damping, is chosen as the active record.

Stage 2: Analysis redesign
Following the discussion on fully stressed design an analysis/redesign approach using (6) is adopted.

As will be seen from the examples, the choice of a starting point does not have a large effect on the methodology since the methodology converges very fast to the region of the final solution. In the examples to follow a uniform distributed damping contributing a predetermined percentage of critical damping to the first mode shape is used, hence:

$$c_{d_i}^{(1)} = 2 \cdot \xi_{d_1} \cdot \omega_1 \cdot \frac{\boldsymbol{\phi}^{(1)^T} \mathbf{M} \boldsymbol{\phi}^{(1)}}{\boldsymbol{\phi}^{(1)^T} \boldsymbol{\phi}^{(1)}}, \tag{7}$$

where ξ_{d_1} = predetermined damping ratio of the first mode; ω_1 = circular frequency of the first mode, and $\boldsymbol{\phi}^{(1)}$ = first mode shape.

The analysis redesign stage is stopped when the constraint error, $\max_i(pi_i)-1$ takes a small value and the changes in the objective function or in the damping vector for two subsequent iterations is small.

Stage 3: Feasibility check and stopping criteria
Once a design for the "active" ground motions is achieved in Stage 2, the performance of the damped structure for each of the remaining ground motions separately in the ensemble is evaluated using a time history analysis. If the design achieved in Stage 2 violates constraints of other records in the ensemble, i.e. $\max_i(pi_i) > 1$, the ground motion for which $\max_i(pi_i)$ receives the largest value is added to the active set.

In Examples 2 and 3 only one record is active. This record is easily tracked by the algorithm, and it is expected that the optimization scheme is likely to use, in general, only a few of the records and not whole ensembles. Therefore, the scheme becomes practical in the sense of the computational effort.

The methodology is terminated when no additional ground motion is added to the active set at this stage. If an additional ground motion is needed, then Stage 2 is repeated with the new active set.

Formal Optimization Methodology

The examples that follow were solved using the analysis/redesign technique that was introduced in the previous section. For the sake of comparison, these examples were also solved using a formal optimization methodology similar to the one introduced by Lavan and Levy (2005b). This methodology uses an appropriate first order optimization scheme that requires the derivation of the gradient of pi. The derivation of this gradient is done by first formulating the equations of motion and the performance indices in a differentiable equivalent *state space* formulation, i.e. as a differentiable set of first order differential equations, and then applying a variational approach for the gradient derivation.

Examples

Example 1 – linear 2 story shear frame

In order to demonstrate the proposed methodology, and the characteristics of the optimization problem the 2-story shear frame introduced by Lavan and Levy (2005a) is studied.

A 5% Rayleigh damping was assumed for the first and second modes. The constraint on the maximum drift given in (2) was set to 0.009m, which is 50% of the maximum drift of the bare frame. The two periods of the structure are 0.281 s and 0.115 s. The structure was excited by the record LA02 from the "LA 10% in 50 years" ground motions ensemble (Somerville *et al.* 1997), which is the N-S component of El-Centro 1940 scaled by a factor of 2.01 downloaded from "http://quiver.eerc.berkeley.edu:8080/studies/system/ground_motions.html". The mass, damping and stiffness matrices (displacements DOFs) to be used in (2) are:

$$\mathbf{M} = \begin{bmatrix} 25 & 0 \\ 0 & 25 \end{bmatrix} \text{ton}; \quad \mathbf{C} = \begin{bmatrix} 120.7 & -32.4 \\ -32.4 & 72.1 \end{bmatrix} \text{kN} \cdot \text{s/m};$$

$$\mathbf{K} = \begin{bmatrix} 62500 & -25000 \\ -25000 & 25000 \end{bmatrix} \text{kN/m}. \tag{8}$$

The contribution of the dampers to the damping matrix is:

$$\mathbf{C}_d = \begin{bmatrix} c_{d_1} + c_{d_2} & -c_{d_2} \\ -c_{d_2} & c_{d_2} \end{bmatrix} \text{kN} \cdot \text{s/m} \tag{9}$$

and

$$\mathbf{e} = \{1 \;\; 1\}^T. \tag{10}$$

The example was solved for a single record. Thus the "active" ground motion in Stage 1 is LA02. Applying the proposed analysis redesign procedure of Stage 2, with starting values obtained from (7) as $\mathbf{c}_d^{(1)} = 2795 \cdot \mathbf{1}$ kN·s/m and using $q = 0.5$ in (6), Figure 2 shows a contour map of the constraint, $\max_i(pi_i) \leq 1$, the objective function at the optimum value (straight line) and the iterative progress towards convergence of the analysis redesign.

The total added damping and the constraint's error ($\max_i(pi_i) - 1$) versus the iteration number are shown in Figure 3.

As can be seen from that figure, the convergence to the region of the optimum is quite fast (a constraint error of −0.35% and a total damping of 1543 occurred in 5 iterations), however, full convergence took 12 iterations. The final damping is $c_{d_1} = 1300.4$ kN·s/m and $c_{d_2} = 181.4$ kN·s/m. There are no remaining records to apply on the design (feasibility check of Stage 3) since only a single record is considered in this example. Hence, the design achieved is the final design. The value of the total added damping for the optimal design and the value of pi are compared in the first line of Table 1 with the ones achieved using the gradient based approach. As can be seen, the gradient based optimization leads to a, somewhat, lower value of the objective function, however, the violation of the constraint is a bit larger. It should be noted that the formal optimization technique yielded the same height distribution and practically equal values of the dampers (not shown).

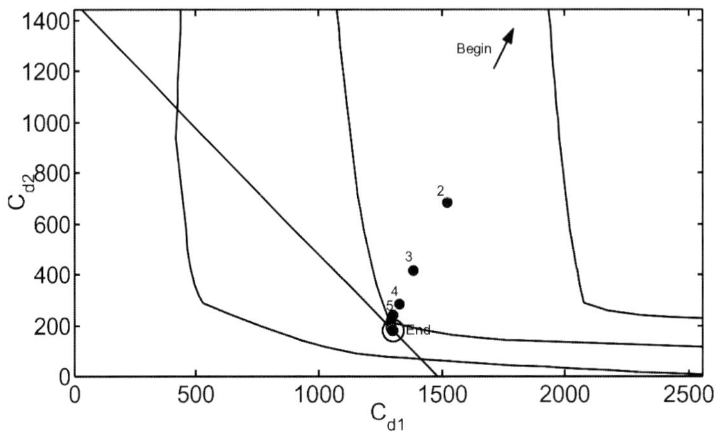

Fig. 2. Contour map of the constraint (curved lines), objective function at optimum (straight line) and iterative progress towards convergence (dots).

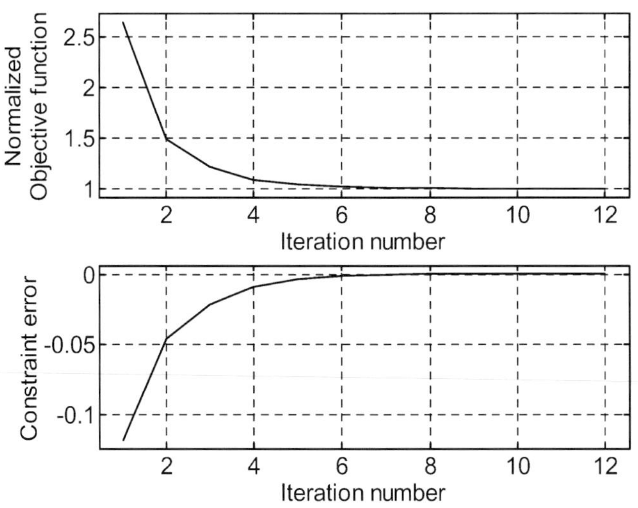

Fig. 3. Convergence to optimum.

Table 1. Optimal design values.

example	Analysis\Redesign		Gradient based optimization	
	Objective function	*pi*	Objective function	*pi*
1	1481.8 (101.94%)	1.0002	1453.6	1.0040
2	162005 (101.03%)	1.0000	160347.5	1.0065
3	418985.0 (100.00%)	1.0010	418972.0	1.0030

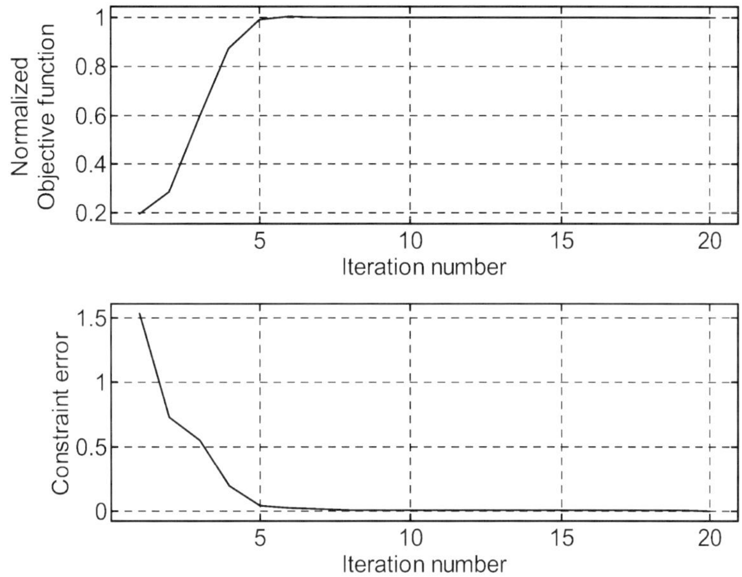

Fig. 4. Convergence to optimum.

Example 2 – linear 8 story 3 bay by 3 bay asymmetric framed structure

In order to demonstrate the applicability of the proposed methodology to 3D structures, the 8-story 3-bay by 3-bay asymmetric framed structure introduced by Tso and Yao (1994) is used. Inherent 5% Rayleigh damping in the first and second modes is assumed. The methodology was performed neglecting axial deformations, i.e. 3 degrees of freedom per floor were used (two horizontal displacements and torsional angle). The ground motion ensemble was chosen as the "LA 10% in 50 years" ensemble (Somerville *et al.* 1997), and the allowable drift at the peripheral frames was chosen as 1.0% of the story height. Stage 1 sketches the maximal displacement of a single degree of freedom system having the natural period of the frame (1.15 sec), versus the damping coefficient for each record in the whole ensemble. The record LA16 was chosen to start the process since its spectral displacement for all reasonable damping range had the largest value.

Applying the proposed analysis redesign procedure of Stage 2, with starting values obtained from (7) as $\mathbf{c}_d^{(1)} = 1000 \cdot \mathbf{1}$kN·s/m and using $q = 0.5$ in (6), Figure 4 shows the total added damping and the constraint's error ($\max_i(pi_i) - 1$) versus the iteration number for the record LA16.

As can be seen, the convergence to the region of the optimum is quite fast (a constraint error of 3.9% and a total damping of 160121 kN·s/m occurred in 5 iterations), however, full convergence took 20 iterations. The final damping and the components of **pi** for the damped frame are show in Figure 5.

Applying the remaining records in the ensemble on the design (Stage 3) indicated that there was no record that led to greater performance indices than that of the active record, LA16. Hence, the optimization process was terminated. A comparison of the total added damping and pi with these of the gradient based optimization solution is given in the second line of Table 1 with the same behavior as the previous example, i.e. the same distribution and practically equal values of the dampers where achieved using the formal optimization technique (not shown).

As can be seen (Figure 5), the optimal solution assigns damping in stories of peripheral frames that fully utilized their local performance index (stories number 2–5 of frame 1 and stories number

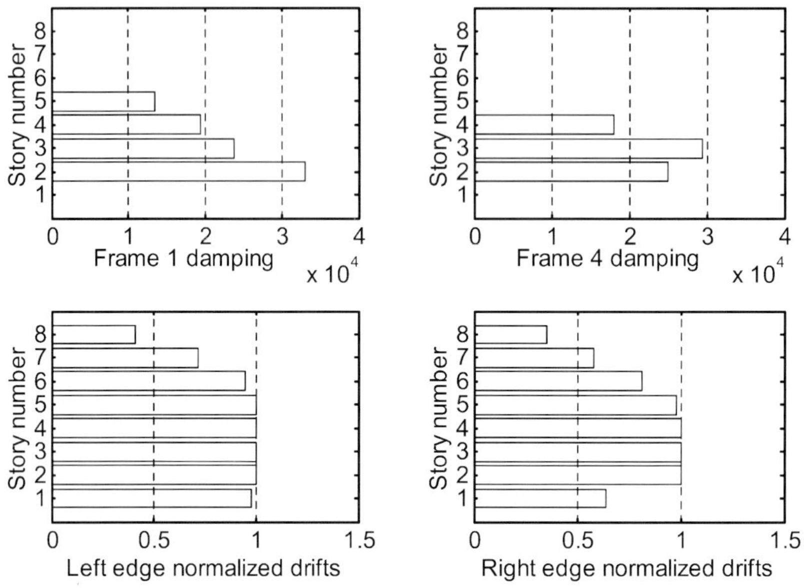

Fig. 5. (a) optimal damping of the damped structure for LA16: left edge (frame 1) and right edge (frame 4) and b) maximal drifts for LA16 (= envelope values): left edge (frame 1) and right edge (frame 4).

2–4 of frame 4), and no damping is assigned elsewhere.

Example 3 – yielding 10 story shear frame

In order to demonstrate the applicability of the proposed methodology to *yielding* shear frames under an ensemble of ground motion records, the 10-story shear frame introduced by Lavan and Levy (2005b) with inherent 2% Rayleigh damping in the first and second modes is used.

The fundamental period of the structure is 1.0 sec; $H_{\mathrm{all},i}$, which appears in (4), is taken as the elastic energy at yielding of the i-th story multiplied by $0.2 \times 16 = 3.2$. The secondary slope ratio for all floors was chosen as 0.02. The ground motion ensemble was chosen as the "SE 10% in 50 years" ensemble (Somerville *et al.* 1997). The record SE19 started the process (Stage 1) since its spectral input energy for the fundamental period of the structure had the largest value for a reasonable damping range (see Lavan and Levy, 2005b). A nonlinear analysis was performed on the bare frame for this record and revealed values much larger than 1 for the components pi_i. Applying the proposed analysis redesign procedure of Stage 2, with starting values obtained from (7) as $\mathbf{c}_d^{(1)} = 12095 \cdot \mathbf{1}$ kN·s/m and using $q = 5$ in (6), leads to the optimal damping and the components for the damped frame excited by SE19 are shown in Figures 6(a) and 6(b).

The total added damping and the constraint's error $(\max_i(pi_i) - 1)$ versus the iteration number are shown in Figure 7.

As can be seen from that figure full convergence took 12 iterations. Applying the remaining records in the ensemble on the design indicated that the design is feasible, thus, the design achieved is the final design. A comparison of the total added damping and pi with those of the gradient based optimization solution is given in the third line of Table 1. In this case of yielding structure, the same result observed in the linear cases is repeated, i.e. here too, the formal optimization technique yielded the same distribution and practically equal values of the dampers (not shown).

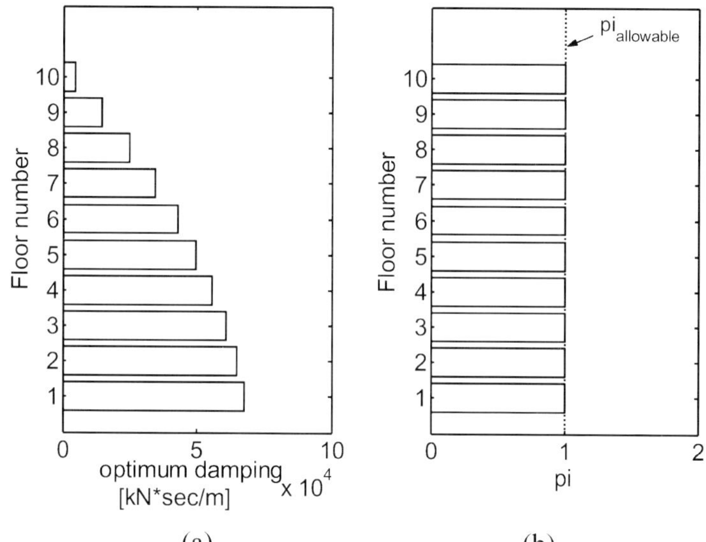

Fig. 6. (a) Design supplemental damping and (b) damage indices envelope for the designed damped frame excited by the SE 10% in 50 years ensemble.

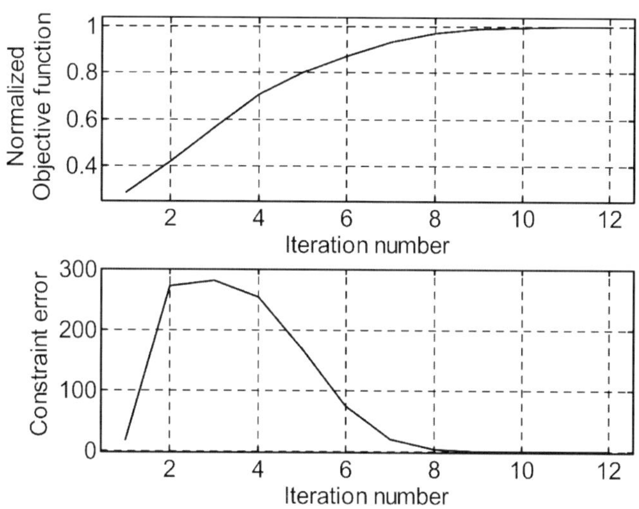

Fig. 7. Convergence to optimum.

Conclusions

A methodology for the optimal design of added viscous damping for an ensemble of realistic ground motion records with a constraint on the maximum drift for linear structures, and on the maximum energy based damage index for nonlinear shear frames, was presented.

The optimization methodology is based on an iterative procedure of the analysis/redesign type that is appropriate for engineering practice. This scheme seems to converge very fast to the region of the final design and is applicable to nonconvex problems.

The final solution coming from formal optimization is characterized by equal maximal drifts in the linear case, and equal maximal damage indices in the nonlinear case, for floors with assigned dampers, and lower maximal drifts/damage indices for floors with no assigned damping. This fully stressed result was targeted by the iterative procedure and is desired in structures due to the uniform distribution of damage (structural and nonstructural) throughout the structure.

References

Bannon H, Biggs JM, Irvine HM (1981) Seismic damage in reinforced concrete frames. The J. of the Structural Division 107 (ST9): 1713–1729.

Carr AJ (2001) RUAUMOKO – Program for inelastic dynamic analysis. Department of Civil Engineering, University of Canterbury, Christchurch.

Cilley FH (1900) The Exact Design of Statically Indeterminate Frameworks, An Exposition of its Possibility but Futility. Transactions, ASCE 43.

Constantinou MC, Tadjbakhsh IG (1983) Optimum design of a first story damping system. Computers and Structures 17(2): 305–310.

Fu Y, Kasai K (1998) Comparative study of frames using viscoelastic and viscous dampers. J. of Structural Engineering 124(5): 513–522.

Garcia DL (2001) A simple method for the design of optimal damper configurations in MDOF structures. Earthquake Spectra 17(3): 387–398.

Gluck N, Reinhorn AM, Gluck J, Levy R (1996) Design of Supplemental Dampers for Control of Structures. J. of Structural Engineering, ASCE 122 (12): 1394–1399.

Inaudi JA, Kelly JM, To CWS (1993) Statistical linearization method in the preliminary design of structures with energy dissipating devices. ATC-17-1: 509–520.

Khot NS, Berke L (1976) Structural Optimization Using Optimality Criteria Method. In New Directions in Optimum Structural Design (Ed. A. Atrek, R. H. Gallagher, K. M. Ragsdell and O. C. Zienkiewicz): 47–74.

Kim J, Choi H, Min KW (2003) Performance-based design of added viscous dampers using capacity spectrum method. J. of Earthquake Engineering 7(1): 1–24.

Lavan O, Levy R (2005a) Optimal design of supplemental viscous dampers for linear framed structures. Earthquake Engineering and Structural Dynamics (in press).

Lavan O, Levy R (2005b) Optimal design of supplemental viscous dampers for irregular shear frames in the presence of yielding. Earthquake Engineering and Structural Dynamics 34(8): 889–907.

Levy R (1985) On the Optimal Design of Trusses under One Loading. Quarterly of Applied Mathematics 43(2): 129–134.

Levy R (1991) Fixed Point Theory and Structural Optimization. Engineering Optimization 17(4): 251–261.

Miyamoto HK, Scholl RE (1996) Case study: seismic rehabilitation of non-ductile soft story concrete structure using viscous dampers. In: Proceedings of the Eleventh World Conference on Earthquake Engineering (held in Acapulco, Mexico), Elsevier Science Ltd, Paper No. 315.

Park YJ, Ang AHS (1985) Mechanistic seismic damage model for reinforced concrete. J. of Structural Engineering 111(4): 722–739.

Rozvany GIN (1989) Structural Design via optimality criteria: The Prager approach to structural optimization (Mechanics of Elastic and Inelastic Solids). Kluwer Academic Publishers.

Shen KL, Soong TT (1996) Design of energy dissipation devices based on concept of damage control. J. of Structural Engineering, ASCE 122(1): 76–82.

Somerville P, Smith N, Punyamurthula S, Sun J (1997) Development of ground motion time histories for Phase 2 of the FEMA/SAC steel project. Report No. SAC/BD-97/04.

Soong TT, Dargush GF (1997) Passive energy dissipation systems in structural engineering. John Wiley & Sons Ltd., Chichester.
Spillers WR (1975) Iterative structural design. North Holland Publishing Company, Amsterdam.
Takewaki I (1997) Optimal damper placement for minimum transfer function. Earthquake Engineering and Structural Dynamics 26(11): 1113-1124.
Takewaki I, Yoshitomi S, Uetani K, Tsuji M (1999) Non-monotonic optimal damper placement via steepest direction search. Earthquake Engineering and Structural Dynamics 28(6): 655–670.
Tso WK, Yao S (1994) Seismic load distribution in buildings with eccentric setback. Canadian Journal of Civil Engineering 21: 50–62.
Uang CM, Bertero VV (1990) Evaluation of seismic energy in structures, Earthquake Engineering and Structural Dynamics 19(1): 77–90.
Venkayya VB (1978) Structural Optimization: A Review and Some Recommendations. Int. J. Num. Meth. Eng. 13: 203–228.
Williams MS, Sexsmith RG (1995) Seismic damage indices for concrete structures: a state-of-theart review. Earthquake Spectra 11(2): 319–349.
Wu B, Ou JP, Soong TT (1997) Optimal placement of energy dissipation devices for threedimensional structures. Engineering Structures 19(2): 113–125.
Zhang RH, and Soong TT (1992) Seismic design of viscoelastic dampers for structural applications. J. of Structural Engineering, ASCE 118(5): 1375–1392.

INELASTIC ANALYSIS OF SEMIRIGID FRAMEWORKS

Y. Liu[1], L. Xu and D.E. Grierson
Civil Engineering, University of Waterloo, Canada

Abstract

This article presents an efficient method for inelastic analysis of semirigid planar steel frameworks. A compound element comprised of a plastic-hinge element and a semirigid connection element is located at member ends that may potentially undergo inelastic deformation. Nonlinear inelastic flexural behaviour is modeled by an empirical relation between moment and rotation for which the parameters are available from experimental results. A four-parameter model is employed to simulate the nonlinear moment-rotation behaviour of semirigid connections. The member stiffness matrix involving the compound element is expressed explicitly in terms of stiffness degradation factors that vary depending on the loading level. This permits direct account for the combined influence of inelastic and nonlinear connection behaviour on structure stiffness. A semirigid steel portal frame is analyzed to illustrate the proposed analysis method, and the results are compared with those obtained from experiments involving the same frame.

Keywords: semirigid connection, inelasticity, compound element

1. Introduction

Many studies have been devoted to developing practical methods of nonlinear analysis accounting for both semirigid connections and member plastic behaviour (Ziemian *et al* 1992; Chen *et al* 1994; Yau *et al* 1994; Chen *et al* 1996). However, little work has been done to investigate the interaction between the behaviour of semirigid connections and that of member plasticity. This study focuses on such interaction using the concept of a compound element, which is shown in Figure 1 and explained in detail in the following.

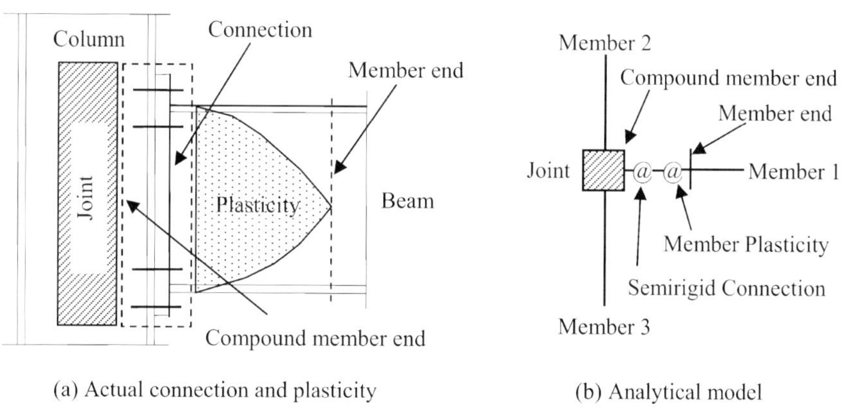

Figure 1. Connection and inelasticity model at a joint

Figure 1(a) is a typical beam-to-column connection joint involving member plasticity. Typically, the connection is semirigid and possibly includes bolts, welds and angles. The member plastic zone forms at the beam end due to concentrated internal loads. To facilitate nonlinear analysis, Figure 1(a)

[1] Corresponding author: Y27liu@uwaterloo.ca

M. Pandey et al. (eds), Advances in Engineering Structures, Mechanics & Construction, 317–328.

may be replaced by the analytical model shown in Figure 1(b), where one of the two springs represents the plasticity formed at the member end while the other spring represents the semirigid connection.

A method of analysis has been recently developed by the authors to deal with geometric and material nonlinearities (Grierson *et al* 2005, Xu *et al* 2005). The goal of this article is to extend this method to account for semirigid connections using a so-called compound-element approach. At each stage of the analysis, the combined-stiffness degradation due to semirigid connection and member plasticity behaviour is determined, and the corresponding tangent stiffness matrix for the structure is formed. The process ends when the specified external loads are completely applied on the structure or the limit loading state is reached.

2 Rotational Compound Element

This section uses the series theorem to develop a compound element representing the combined rotational stiffness behaviour of a semirigid connection together with a member-end plastic hinge. The determination of the stiffness of semirigid connections is discussed in detail, while that for member-end plasticity is adopted directly from the authors' previous research (Grierson *et al* 2005, Xu *et al* 2005).

2.1 Series Element Model

The series element model involves a semirigid connection spring, an inelastic spring and an elastic member all connected in series. The nature of the compound element is indicated in Figure 2, where parameters R_n, R_e, R_c, R_p denote the rotational stiffnesses of the compound member end, elastic member end, semirigid connection spring, and the member plasticity spring, respectively. Only end 1 of the member is considered, while end 2 may or may not have the same nature as end 1.

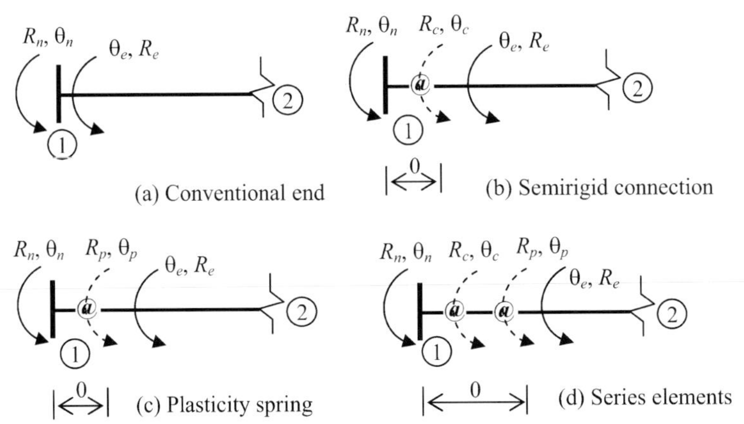

Figure 2. Compound-rotational element at the member end

The case in Figure 2(a) is conventionally used in structural analysis, where a beam-to-column connection at node 1 is assumed as either a pinned connection ($R_n = 0$) or fixed connection ($R_n = R_e$). This assumption makes analysis and design simple both for hand and computer-based analysis. However, if the effect of actual connections on structural responses must be considered, the model of the semirigid connection shown in Figure 2(b) is introduced in the analysis and design of structures. Another case is also popular in rigid-plastic analysis, where a plastic hinge is abruptly formed rather than gradually degrading from initial yield to full yield states. To improve the accuracy in this latter method, the inelastic model shown in Figure 2(c) is used to model the property of the gradual stiffness degradation due to the presence of plasticity. Finally, if both semirigid connection and plasticity behaviour may occur at the same time, the series-element model shown in Figure 2(d) should be introduced in analysis and design, as in the work of Yau and Chan (1994) where, however, the influences of plasticity and semirigid connections were considered separately. To facilitate structural analysis accounting for both connection and inelasticity stiffness degradations and their interactions, an integrated compound element is needed and is investigated in detail as follows.

The rotational deformation involving nonlinear connection and inelasticity behaviour indicated in Figure 2(d) is graphically represented in Figure 3(a) using two springs. It can be verified that the two series-connected springs may be substituted by the compound element shown in Figure 3(b) involving a single spring. The compound stiffness R that reflects the combined stiffnesses R_c and R_p can be derived as in the following (Liu 2005).

Provided that a moment M is applied at the joint in Figure 3(a), the relative rotations θ_c and θ_p are given by,

$$\theta_c = M / R_c ;\ \theta_p = M / R_p \tag{1}$$

Since the total relative rotation θ between the joint and the elastic member end is the summation of rotations induced by the connection and inelastic springs, from Eqs (1) the rotation θ can be expressed as,

$$\theta = \theta_c + \theta_p = M / R_c + M / R_p = M / R \tag{2}$$

where the compound rotational stiffness,

$$R = \frac{1}{1/R_c + 1/R_p} = \frac{R_c R_p}{R_c + R_p} \tag{3}$$

accounts for both connection and plasticity stiffness.

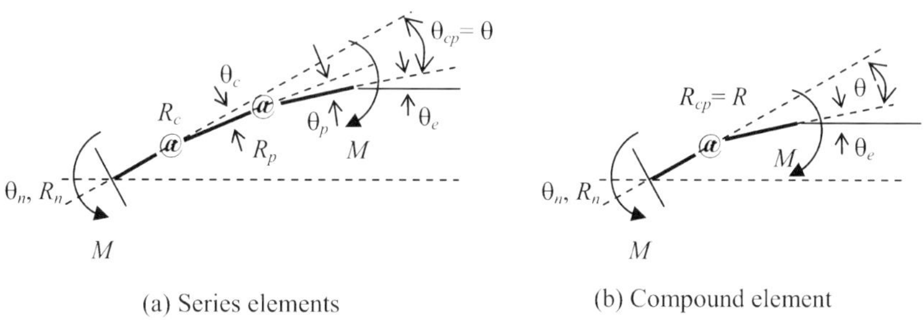

Figure 3. Simplified compound member-end model

2.2 Determining stiffness R_c

It remains to determine the stiffness of the compound element through Eq. (3). The inelastic stiffness R_p of the member is directly available in the literature (Grierson *et al* 2005). Thus, only the connection stiffness R_c in Eq. (3) need be established herein.

Several semirigid connection models have been proposed (Xu 1994), and of these models the four-parameter power model, originally proposed for modeling the post-elastic stress-strain relation (Richard *et al* 1975), is commonly used in analysis. Recently, this model has been further confirmed to be effective and accurate for predicting the behaviour of end-plate connections on the basis of experimental data for extended-end-plate and flush-end-plate connections (Kishi *et al* 2004). Thus, the following four-parameter model,

$$M = \frac{(R_{ce} - R_{cp})\theta_c}{\left\{1 + [(R_{ce} - R_{cp})\theta_c / M_0]^{\gamma}\right\}^{1/\gamma}} + R_{cp}\theta_c \tag{4}$$

is employed in this study to simulate the behaviour of semirigid connections. In Eq. (4), θ_c denotes the relative rotation of the semirigid connection, and the four parameters R_{ce}, R_{cp}, M_0, and γ are the initial rotation stiffness, strain-hardening/softening stiffness, reference moment, and shape parameter of the connection, respectively. The initial yield moment M_{cy} and corresponding rotation θ_{cy} determine the elastic stiffness $R_{ce}= M_{cy}/\theta_{cy}$, while the reference moment M_0, strain-hardening/softening stiffness R_{cp}, and rotation capacity θ_u determine the ultimate moment capacity to be,

$$M_u = M_0 - \theta_u R_{cp} \tag{5}$$

where the rotation capacity θ_u depends on the connection type and can be determined from the results of existing research (e.g., Bjorhovde *et al* 1990). The four parameters in Eq. (4) can be found for different types of connections from an existing database of experimental results (Xu 1994).

Differentiating Eq. (4) with respect to rotation θ_c determines the tangent stiffness of the connection to be (Richard *et al* 1975),

$$R_c = \frac{dM}{d\theta_c} = R_{cp} + \frac{R_{ce} - R_{cp}}{\left\{1 + [(R_{ce} - R_{cp})\theta_c / M_0]^\gamma\right\}^{1+1/\gamma}} \tag{6}$$

where R_{ce} is the elastic rotational stiffness at the initial condition $\theta_c = 0$, while R_{cp} is the strain-hardening/softening stiffness when rotation θ_c tends to infinity (for practical steel structures, θ_c can reach at most to the limiting rotation capacity of the connection when fracture occurs (Bjorhovde *et al* 1990).

It is seen from Eqs (4) and (6) that the four-parameter model reduces to a linear model with $R_c = R_{ce}$ when R_{cp} tends to R_{ce} , whereas a bilinear model is reached when the shape parameter γ approaches to infinity. If R_{cp} is set to zero (i.e., strain-hardening/softening is ignored), Eq. (4) reduces to the following three-parametric model that was previously suggested by Kishi and Chen (1987),

$$M = \frac{R_{ce}\theta_c}{[1 + (R_{ce}\theta_c / M_u)^\gamma]^{1/\gamma}} \tag{7}$$

where the reference moment M_0 is replaced by the ultimate moment M_u. Note that the rotation θ_c can be explicitly obtained from Eq. (7) as,

$$\theta_c = M / R_{ce}[1 \quad (M / M_u)^\gamma]^{1/\gamma} \tag{8}$$

Using an expression for post-elastic rotation previously derived by the authors (Grierson *et al* 2005, Xu *et al* 2005), and the connection rotation given by Eq. (8), the total relative rotation θ of the compound element can be explicitly expressed as,

$$\theta = \frac{M}{R_{ce}[1 - (M / M_u)^\gamma]^{1/\gamma}} + \phi_p \left\{ 1 - \left[1 - \left(\frac{M - M_y}{M_p - M_y} \right)^\alpha \right]^{1/\alpha} \right\} \qquad \frac{M_y}{M_p} \le \frac{M}{M_p} \le 1 \tag{9}$$

which represents the moment-rotation relationship of the compound element. The benefit of using the three-parameter model is that the rotation θ_c of the connection is directly obtained in the nonlinear analysis for the given moment *M*; the disadvantage is that the strain-hardening or softening nature of the connection is omitted. In contrast, strain hardening/softening is accounted for in the four-parameter model, but an iterative procedure is needed to find the relative rotation θ_c of the connection. Both of the connection models are considered for the verification analysis presented later.

2.3 Degradation Factors

The flexural stiffness degradation factor associated with the semirigid stiffness R_c is given by (Monforton *et al* 1963),

$$r_c = 1/(1 + 3EI / LR_c) \tag{10}$$

where EI/L is the flexural stiffness of the elastic member. The factor r_c is interpreted as the ratio of the end rotation of the elastic member to the combined rotation of the elastic member and the connection due to unit end-moment (Xu, 1994).

Similarly, the stiffness degradation factor associated with the inelastic stiffness R_p is given by (Grierson *et al* 2005),

$$r_p = 1/(1 + 3EI / LR_p) \tag{11}$$

where the factor r_p is interpreted as the ratio of the inelastic rotation M/R_p to the total elastic and inelastic rotation $M/R_p + ML/3EI$ under the action of bending moment M applied at the end connected to the compound element in the case when the far end of the elastic member is simply supported (Xu *et al* 2005).

To evaluate the combined stiffness effect, a stiffness degradation factor associated with the compound stiffness R is introduced and similarly expressed as $r = 1/(1+3EI/LR)$. The factor r is the ratio of the rotation of the compound element to the sum total of the rotation of the compound element plus the rotation of the elastic member when simply supported at the far end. From Eqs (3), (10) and (11), the compound stiffness degradation factor is found as,

$$r = \frac{1}{1 + 3EI / LR_c + 3EI / LR_p} = \frac{r_c r_p}{r_c + r_p - r_c r_p} \tag{12}$$

which maps $R \in [0, \infty]$ to $r \in [0, 1]$. From Eq. (12), the stiffness degradation factor for the compound element is a function of the degradation factors of the semirigid connection and member inelasticity, such that if any of these factors degrades to zero then the stiffness of the compound element will degrade to zero as well.

3 Characteristics of Compound Rotational Element

The behaviour of the compound rotational element is dependent upon the strength capacities of the connection and the connected beam members. For the current study, the effect of connected columns on the behaviour of the beam-column connection is ignored and the shear deformation of panel zones is not considered.

If considering only the effect of member plasticity, the moment-rotation relation in the post-elastic range is as shown in Figure 4(a), whereas when considering the effect of the semirigid connection alone the moment-rotation relationship is as shown in Figure 4(b). In Figure 4, M_u is the ultimate moment capacity of the connection, while M_y and M_p are the initial-yield and fully-plastic moment capacities of the connected member, respectively. Depending on the interaction of member inelasticity and nonlinear connection behaviour, three types of semirigid connections may be characterized by the compound element, as described in the following.

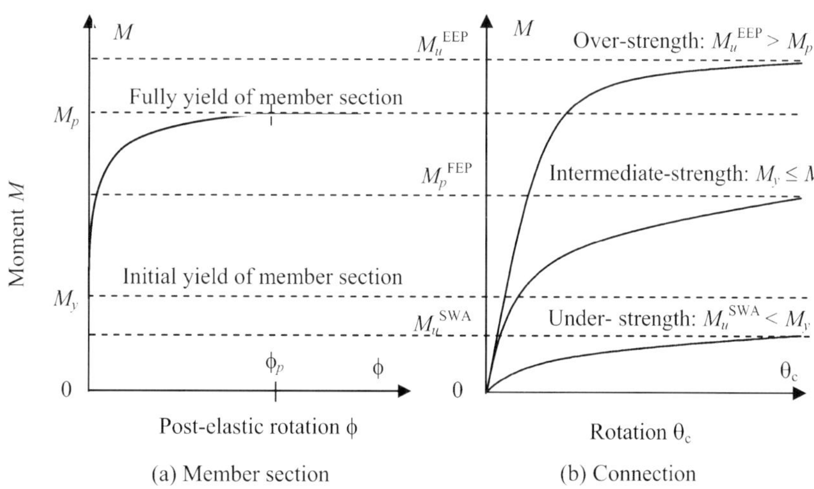

Figure 4. Stiffness-degradation relationship at a member end

(1) Under-strength connections: $M_u < M_y$

In this situation, the performance of the compound element is governed purely by the semirigid connection, and no inelasticity occurs in the vicinity of the member end. This may occur for single web-angle (SWA) connections with $M_u = M_u^{SWA}$. The combined moment-rotation behaviour of the compound element is shown in Figure 5, which is exactly the semirigid behaviour shown in Figure 4(b). This kind of connection is referred to as an Under-strength connection since the strength capacity of the compound element is less than the yield strength of the member. If M_u is small enough or the rotational constraint is ignored, this type of connection is categorized as a conventional simple or pinned connection (AISC 2001, CISC 2004).

(2) Intermediate-strength connections: $M_y \leq M_u \leq M_p$

In this second case, both the semirigid connection and member inelasticity govern the behaviour of the compound element, but the limit strength is determined by the nature of the connection. Such behaviour of the compound element may occur for a flush end-plate (FEP) connection. This type of connection is referred to as an intermediate-strength connection and corresponds to a partially restrained (AISC 2001) or semirigid (CISC 2004) connection.

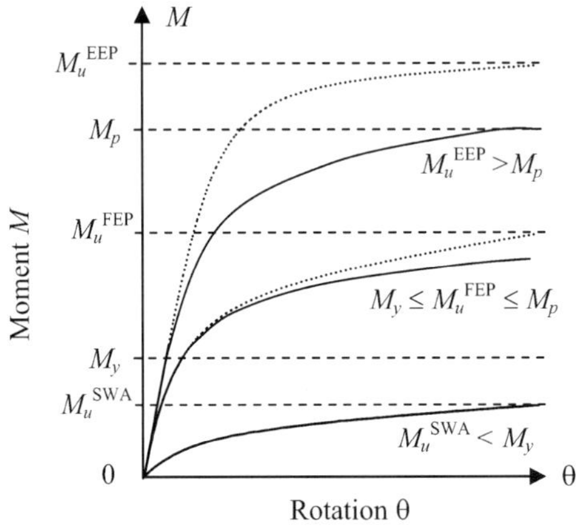

Figure 5. M-θ relations of compound

(3) Over-strength connections: $M_p < M_u$

Finally, when the ultimate moment of the connection is greater than the plastic moment of the connected member, and even though the connection influences the stiffness degradation of the compound element due to its nonlinear behavior, the member inelasticity dominates the behaviour of the compound element. An example for an extended end-plate (EEP) connection is illustrated in Figures 4 and 5 (where the dotted moment-rotation curve refers to the EEP connection considered alone). It can be seen from Figure 5 that the moment-rotation behaviour of the compound element (solid curve) is dominated by the plastic capacity of the member. This kind of connection is referred to as an over-strength connection and corresponds to a fully restrained (AISC 2001) or rigid (CISC 2004) connection (i.e., in the limit when the connection stiffness is considered infinite).

It can be seen from the foregoing discussion that a satisfactory design may be achieved if both the connection and the corresponding connected member have the same strength capacity, i.e., $M_p = M_u$. It is not good to use an over-strength connection, i.e., $M_u > M_p$, since the over-strength of the connection is not utilized in any way. Lower-strength connections may be used in circumstances where stiffeners can be added to avoid excessive deformation.

In the common case of a pinned connection, i.e., $r_c = 0$, the compound element has zero rotational stiffness regardless of the value of r_p. In this case, any connected member does not experience plastic behaviour. If $r_c = 1$, the compound element behaviour is determined by the inelastic behaviour of the member, i.e., $r = r_p$ in the case of a rigid connection. It is noted that when the plasticity factor is smaller than unity (e.g., $r_p = 0.7$), the r value of the compound element is close to the value of r_c (Liu 2005). This means that even if the member end has experienced some plasticity (e.g., 100-70 = 30%), the stiffness of the compound element is dominated by the connection. In other words, the level of inelasticity has insignificant effect on the stiffness degradation of the compound element. If the ultimate strength of a connection is close to, or lower than, the initial yield strength of the connected member, the influence of member plasticity on structural response may be ignored.

4 Structural Analysis

Once the stiffness degradation factor of a compound element is determined, as discussed in the previous sections, the structural analysis is readily conducted. This study focuses planar steel frameworks comprised of beam-column members with compact sections, for which plastic deformation is not precluded by local buckling (AISC 2001). The plastic bending, shearing or axial deformation (ϕ, γ or δ) of a member under the action of moment, shear or axial force (M, V or P) is concentrated at a member-end section (Xu *et al* 2005). Figure 6(a) shows a general member with Young's modulus E, shear modulus G, member length L, cross-section moment of inertia I, sectional area A, and equivalent shear area A_s. The parameters R_{pj}, T_{pj} and N_{pj} are respectively the post-elastic rotational bending, transverse shearing and normal axial stiffness of the member at the two end sections j =1, 2, while R_{cj}, T_{cj} and N_{cj} are respectively the rotational bending, transverse shearing and normal axial stiffness of the connections at the two end sections. Upon adopting a compound element at each member-end, the simplified model shown in Figure 6(b) is obtained, where the determination of the corresponding parameters is discussed in the following.

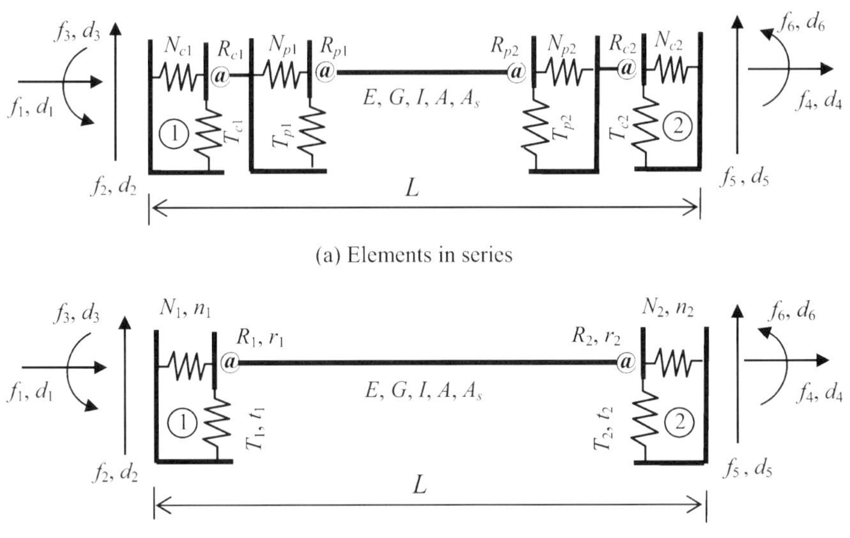

Figure 6. Beam-column-member model used in analysis

The evaluation of connection and member rotational stiffnesses R_{cj} and R_{pj} in Figure 6(a), and corresponding stiffness degradation factors r_{cj} and r_{pj} , has been discussed in detail in the previous sections. The member transverse shear and normal axial stiffnesses T_{pj} and N_{pj} were determined in previous research, where it was shown that the corresponding stiffness degradation factors t_{pj} and n_{pj} are given by (Grierson *et al* 2005, Xu *et al* 2005),

$$t_{pj} = 1/(1+3EI/L^3T_{pj})\,;\; n_{pj} = 1/(1+EA/LN_{pj}) \tag{13a, b}$$

which map T_{pj} or $N_{pj} \in [0, \infty]$ into t_{pj} or $n_{pj} \in [0, 1]$. Similarly, the transverse and normal stiffness degradation factors for the connection can be expressed as,

$$t_{cj} = 1/(1+3EI/L^3T_{cj})\,;\; n_{cj} = 1/(1+EA/LN_{cj}) \qquad (14a, b)$$

where T_{cj} and N_{cj} are the transverse shear and normal axial stiffnesses of the connection.

In this study, it is assumed that: 1) the transverse shear stiffness T_{cj} or normal axial stiffness N_{cj} of a connection is infinite when the materials are in the elastic range, and that the corresponding degradation factor t_{cj} or n_{cj} in Eqs (14) is unity; and 2) the stiffness T_{cj} or N_{cj} is zero when the materials are in the plastic range, and that the corresponding degradation factor t_{cj} or n_{cj} is zero. Such idealized perfectly elastic-plastic models are shown in Figure 7.

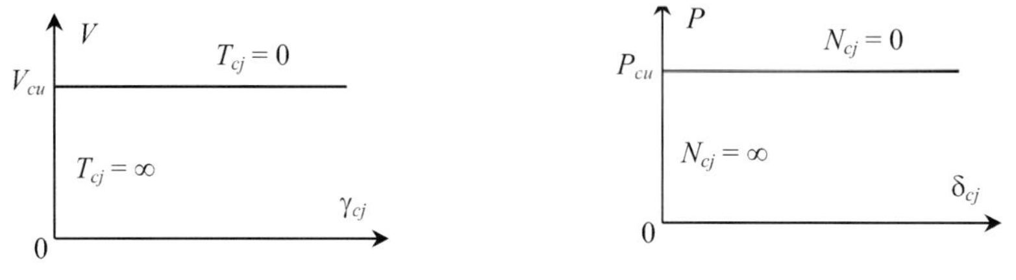

(a) Transverse case (b) Normal case

Figure 7. Idealized shear & axial force-displacement relations for connections

Upon determining the parameters characterizing the compound element, the general planar compound member shown in Figure 6(b) is generated, where f_i and d_i (i =1, 2,..., 6) are local-axis joint forces and deformations, respectively. The parameters r_j, t_j and n_j (j =1, 2) in Figure 6(b) are the so-called bending, shearing and axial stiffness degradation factors of the compound element. The factors r_j are calculated through Eq. (12), while t_j and n_j are similarly found as,

$$t_j = \frac{t_{cj}t_{pj}}{t_{cj}+t_{pj}-t_{cj}t_{pj}}\,;\qquad n_j = \frac{n_{cj}n_{pj}}{n_{cj}+n_{pj}-n_{cj}n_{pj}} \qquad (15a, b)$$

Based on the compound model shown in Figure 6(b), the local stiffness matrix $\boldsymbol{k}$ for each element is derived accounting for the effects of shear deformation and geometrical nonlinearity. The local element stiffness matrices are transformed into the global coordinate system and then assembled as the structure stiffness matrix $\boldsymbol{K}$. If $\boldsymbol{K}$ is nonsingular at the end of an incremental load step, the corresponding incremental nodal displacements $\Delta\boldsymbol{u}$ are solved for and the incremental member-end forces $\Delta\boldsymbol{f}$ and deformations $\Delta\boldsymbol{d}$ are found. After each load step i, the total nodal displacements $\boldsymbol{u} = \Sigma\Delta\boldsymbol{u}_i$ and member-end forces $\boldsymbol{f} = \Sigma\Delta\boldsymbol{f}_i$ and deformations $\boldsymbol{d} = \Sigma\Delta\boldsymbol{d}_i$ accumulated thus far over the load history are found. The initial-yield and full-yield conditions for each member-end section are checked to detect plastic behaviour, and the corresponding bending, shearing and axial stiffness degradation factors are found. The degraded stiffnesses R_c, T_c, N_c are determined based on the moments, shear and axial forces given by the analysis results at the current loading level. The degradation factors (r_p, t_p, n_p, r_c, t_c, n_c) are applied to modify the element stiffness matrices $\boldsymbol{k}$ and, hence, the structure stiffness matrix $\boldsymbol{K}$, before commencing the next load step. The incremental-load analysis procedure continues until either a specified load level $\boldsymbol{F}$ is reached or the structure stiffness matrix $\boldsymbol{K}$ becomes singular at a lower load level as a consequence of failure of part or all of the structure. (If the structure has not failed at load level $\boldsymbol{F}$, the analysis may be continued beyond that level until failure of the structure does occur.)

The final analysis results include the values of the bending, shearing and axial post-elastic compound stiffness degradation factors r, t and n indicating the extent of plastic and connection deformation in the beam-to-column connection regions. Further computational details are provided through the analysis example presented in the following section.

5 Portal Frame Example

To illustrate and verify the proposed method it is applied to the portal frame shown in Figure 8, for which experimental test results are available in the literature (Liew *et al* 1997). In the analysis, Young's modulus E = 200 GPa and shear rigidity G = 77 GPa. Residual stress in the members for bending and axial behaviour is taken as $\sigma_r = 0.3\sigma_y$, while for shearing behaviour $\tau_r = 0.05\tau_y$, where σ_y and τ_y are respectively the normal and shearing yield stresses for steel. From the published test data (Liew *et al* 1997), the properties of the beam are: cross-section area A = 4740 mm^2, moment of inertia $I = 5547\times10^4$ mm^4, plastic modulus $Z = 485\times10^3$ mm^3, normal yield stress σ_y = 345 MPa, and shear yield stress τ_y = 199 MPa (based on von Mises criterion). The properties for both of the columns are: area A = 7600 mm^2, moment of inertia $I = 6103\times10^4$ mm^4, plastic modulus $Z = 654\times10^3$ mm^3, and yield stresses σ_y = 336 MPa and τ_y = 194 MPa. Semirigid connections are modeled by the adopted four-parameter model (Richard *et al* 1975), for which the parameter values are obtained from the test results as described in the following. Upon applying a curve-fitting technique (Liu 2005) to the moment-rotation test results for the beam-to-column connection C1 (Liew *et al* 1997), the model parameters in Eq. (6) were determined to be M_0 = 79 kN-m, R_{ce} = 7202 kN-m/rad, R_{cp} = 144 kN-m/rad, and γ = 0.57; similarly, for the column-to-base connection C2, the model parameters were determined to be M_0 = 148 kN-m, R_{ce} = 24721 kN-m/rad, R_{cp} = 151 kN-m/rad, and γ = 0.78.

To match with the experimental test setup, the loads for the analysis procedure proposed by this study are monotonically increased up to the collapse load level by incrementally changing the magnitude of the load parameter H, while maintaining the fixed proportional coefficients on the horizontal and vertical loads shown in Figure 8. The beam is divided into three elements, while each column is taken as one element. The load-deflection behaviour of joint 6 was found by the analysis to be as given by Curve 1 in Figure 9(a). Also shown in Figure 9(a) are the test results (Liew *et al* 1997) and those found using a refined plastic hinge (PHINGE) analysis method (Chen *et al* 1996). It can be seen that at lower loading levels (H < 60 kN), the load-deflection results found by this study and PHINGE method are in good agreement with each other and those obtained in the test. At higher loading levels, the results of the current study are slightly less than those of the PHINGE method because the latter does not account for elastic shear deformation as herein. Note that the results obtained by both analysis methods are significantly less than the test results at higher load levels. The proposed method predicted structure collapse at load level H_c = 74 kN, which is close to the value of 77 kN predicted by the PHINGE method, but both of these values are considerably less than the 99 kN value found as the limit load by the experimental test.

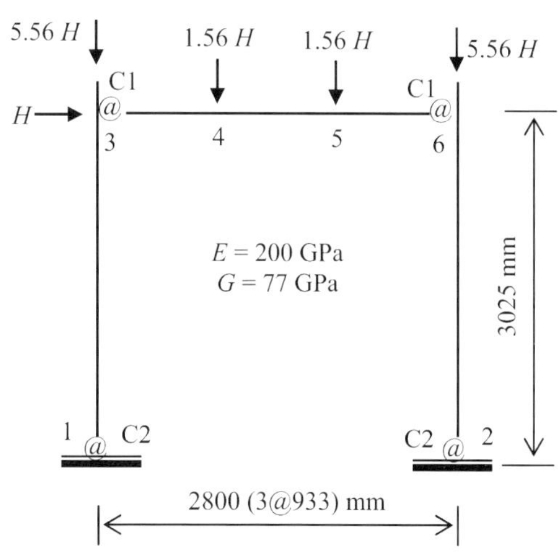

Figure 8. Tested semirigid portal frame

It likely that the above noted discrepancy between experimental and analytical results is a consequence of the analysis methods adopting connection behaviour data determined from separate pilot experiments (Liew *et al* 1997) that is different from that for the actual connections in the frame itself. In fact, a good prediction compared to the test results is found by adjusting the data for the beam-to-column connection C1 to be: M_0 = 111 kN-m, R_{ce} = 4629 kN-m/rad, R_{cp} = 1099 kN-m/rad, and $\gamma = 0.97$. The proposed analysis procedure then finds the load-deflection behaviour of joint 6 to be as defined by Curve 2 in Figure 9(a), which is observed to be in very good agreement with the experimental test results. The plastic behaviour shown in Figure 9(b) corresponds to Curve 2 in Figure 9(a). It is seen that the development of plasticity at member ends is not very significant. This is because the connections C1 and C2 respectively have ultimate moment capacities M_u= 133 kN-m and M_u= 151 kN-m, which are not that much greater than the yield moment capacities M_y= 100 kN-m and M_y= 134 kN-m of the beam and columns, respectively. In essence, the behaviour of the portal frame is governed by semi-rigid connection behaviour rather than member plasticity.

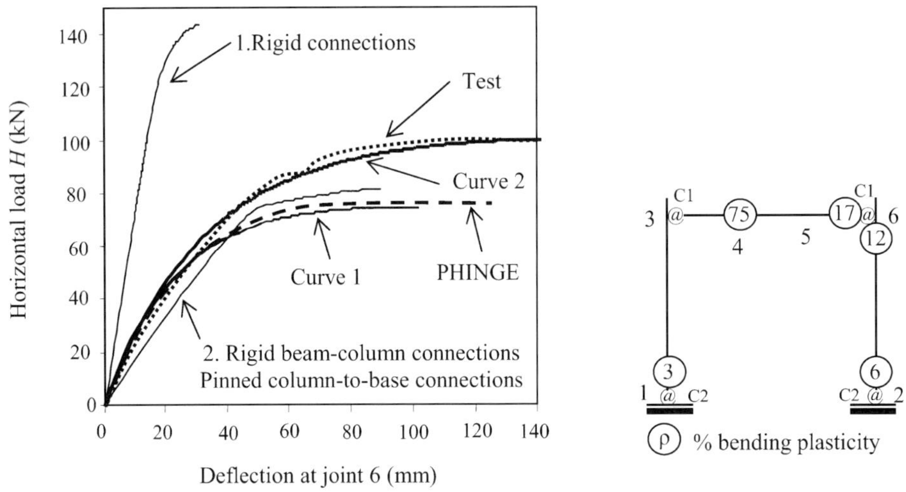

(a) Force-deflection curves (b) Plastic behaviour

Figure 9. Load-deflection and plasticity formation

Also shown in Figure 9(a) are two special cases where the portal frame was analyzed taking some or all of the connections to be rigid. For Case 1 when both the beam-to-column and beam-to-base connections were taken to be rigid, it can be observed from the corresponding load-deflection behaviour that the limit deflection is only about one-fifth of that found for the case of semi-rigid connections, at a limit load level H_f = 143.3 kN. For Case 2 when the beam-to-column connections were assumed rigid while the column-to-base connections were taken to be pinned, which is a conventional situation in design, the corresponding load-deflection behaviour is close to that when the connections are all semi-rigid, with a frame limit load capacity H_f = 82.4 kN. The post-elastic behaviour of the frame at the limit state for the two cases is shown in Figure 10. From Figure 10(a) for the case of all rigid connections, four plastic hinges (i.e., 100% plasticity) form in the beam and right column, while the left column base undergoes 52% plasticity under combined axial force and bending moment. The formation of the fourth plastic hinge at node 4 occurs when the horizontal load H_f = 143.3 kN. At the same time, the frame fails due to inelastic instability signaled by the horizontal displacement of node 6 becoming infinitely large (i.e., the corresponding stiffness coefficient tends to zero and causes the structure stiffness matrix to become singular). From Figure 10(b) for the case of beam-to-column rigid connections and column-to-base pinned connections, the beam experiences more serious plastic deformation than the columns. The formation of the plastic hinge at the right end of the beam occurs when the horizontal load reaches H= 77.7 kN. At limit load level H_f = 82.4 kN, the frame fails due to inelastic instability signaled by the horizontal displacement of node 6 becoming infinitely large (i.e., the same failure mode as for the rigid frame). Table 1 indicates the different degrees of member-end stiffness degradation for the scenarios when all connections are semi-rigid and the two cases where all or some of the connections are rigid.

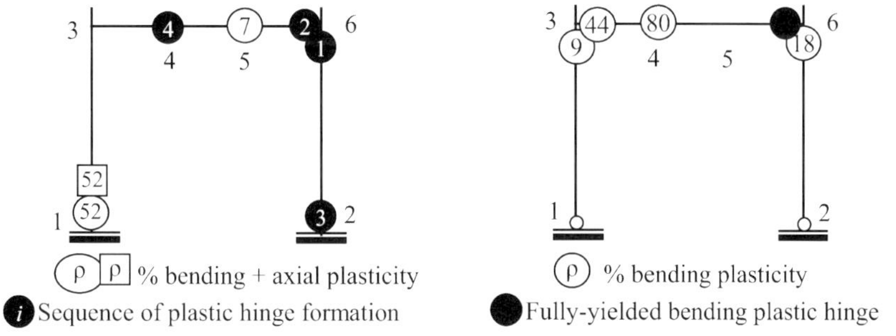

(a) Case 1:Rigid column base (b) Case 2: Pin-column base

Figure 10. Post-elastic behaviour of the rigid frame with different supports

Table 1. Stiffness degradation factors

Member	End	Semirigid				Rigid	
		Initial: r_{c0}	r_c	r_p	r	Case 1: r	Case 2: r
C13	E1	0.671	0.051	0.966	0.051	0.482	-
C26	E2	0.671	0.049	0.943	0.049	0.000	-
B34	E3	0.115	0.051	1.000	0.051	1.000	0.572
B56	E6	0.115	0.043	0.829	0.043	0.000	0.000

6 Summary and Conclusions

This article has presented a method of nonlinear analysis based on a compound-element that accounts for both member plasticity and semi-rigid connection behaviour. Depending on the properties of the compound element, connections can be categorized as being over-strength as for conventional fully-restrained or rigid connections, intermediate-strength as for partially-restrained or semi-rigid connections, and under-strength as for simple or pinned connections. Analysis results show that for an over-strength connection, both plasticity of the member and the semi-rigid nature of the connection affect nonlinear behaviour but the member strength dominates the failure state. An intermediate-strength connection has similar nonlinear behaviour but the connection strength dominates the failure state. For an under-strength connection, only the semi-rigid connection affects the nonlinear behaviour and the connection strength controls the failure state. Over-strength connections may be inappropriate for use in practice since their stiffness and strength are not fully utilized. A satisfactory design can be achieved if the connection and the connected member have about the same loading capacity.

A portal frame example demonstrated that the proposed method of nonlinear analysis can well predict the responses of structures with semirigid connections. Albeit, it was observed that the results are very dependent on properly modelling the behaviour of semi-rigid connections and that data from isolated connection tests may not be correct for assembled frameworks. The results show that the proposed method is effective and efficient for nonlinear analysis of steel frameworks taking into account flexural, shearing and axial stiffness degradation due to the combined action of member plasticity and semi-rigid connection behaviour.

Acknowledgements

This work forms part of the PhD research studies conducted by the first author under the supervision of the other two authors, and was funded by research grants from the Natural Science and Engineering Research Council of Canada.

References

AISC (American Institute of Steel Construction). (2001). *Manual of Steel Construction-Load and Resistance Factor Design* (LRFD), 3rd Edition, Chicago, Illinois.

Bjorhovde, R., Brozzeti, J., and Colson, A. (1990). "A classification system for beam to column connections." *J. Struc. Engrg*, ASCE, 116(11), 3059-3076.

CISC (Canadian Institution of Steel Construction). (2004). *Handbook of Steel Construction*. 8th Ed., Universal Offset Limited Alliston, Ontario, Canada.

Grierson, D.E., Xu, L., and Liu, Y. (2005). Progressive-failure analysis of buildings subjected to abnormal loading, *Journal of Computer-Aided Civil and Infrastructure Engineering*, **20**(3), 155-171.

Kishi, N. and Chen, W.F. (1987). *Moment-Rotation of Semi-Rigid Connections*. Structural Engineering Report, No. CE-STR-87-29, School of Civil Engineering, Purdue University, West Lafayette, Indiana.

Kishi, N., Komuro, M., and Chen, W.F. (2004). Four-parameter power model for M-θr curves of end-plate connections, ECCS/AISC Workshop Connections in Steel Structures V: Innovative Steel Connections, June, Amsterdam, The Netherlands.

Liew, J.Y.W., Yu, C.H., Ng, Y.H., and Shanmugam, N.E. (1997). Testing of semi-rigid unbraced frame for calibration of second-order inelastic analysis, *J. Construct. Steel Res*. **41**(2/3): 159-195.

Liu Y. (2006). Failure analysis of building structures under abnormal loads, *Ph.D. Thesis*, University of Waterloo, Ontario, Canada. (in progress)

Monforton, G.R. and Wu, T.S. (1963). Matrix analysis of semi-rigid connected frames, *J. Struct. Div.*, **89**(6):, 13-42.

Richard, R.M. and Abbott, B.J. (1975). Versatile elastic-plastic stress-strain formula, *J. Engrg. Mech. Div.*, ASCE, **101**(4): 511-515.

Xu, L. (1994). Optimal design of steel frameworks with semi-rigid connection. *Ph.D. Thesis*, Department of Civil Engineering, University of Waterloo, ON, Canada.

Xu, L., Liu, Y., and Grierson, D.E. (2005). Nonlinear analysis of steel frameworks through direct modification of member stiffness properties, *Advances in Engineering Software*, **36**(5): 312-324.

Yau, C.Y. and Chan, S.L. (1994). Inelastic and stability analysis of flexible connected steel frames by spring-in-series model, *J. Struct. Eng.*, ASCE, **120**(10): 2803-2819.

Ziemian, R.D., McGuire, W., and Deierlein, G.G. (1992). Inelastic limit states design Part I: Planar frame structures, *J. Struct. Engrg.*, ASCE, **118**(9): 2532-2549.

A NEW METHOD FOR ANALYSIS OF STRUCTURES INCLUDING NONLINEAR SEMIRIGID CONNECTIONS

Hamid Moharrami

Civil Eng. Dept., Faculty of Eng., Tarbiat Modarres University
Tehran, P.O. Box 14155-4838, Iran,
E-mail: hamid@modares.ac.ir and hamid.moharrami@gmail.com

Abstract

Semirigid connections show nonlinear behavior even due to small loadings. Therefore linear analysis is not a proper solution algorithm for structures that have such connections; rather a nonlinear analysis should be done. The conventional methods of nonlinear analysis of frames are inherently iterative, and their final results include some small order of approximation. They usually are done through modification of the stiffness matrix of structure and/or load vector.

In this paper, a new method of nonlinear analysis has been presented that contrary to iterative methods, it is non-iterative. It does the analysis in one step without change in the initial model and stiffness matrix of the structure or its load vector. Theoretically it does not include approximation and gives exact results. In this method to force internal moments follow their nonlinear moment-rotation curves, some virtual moments (that are primarily unknown) are imposed to the structure at semirigid connections. To find the unknown virtual moments, a quadratic programming problem is formulated and solved. After finding the values of virtual moments, employing superposition principle, exact nonlinear response of structure is obtained and internal forces and moments of members are calculated.

The method is capable to model semirigid connections with multilinear moment-curvature relations. The formulation of the problem for bilinear and trilinear moment-curvature relations has been presented here. Two examples are presented to demonstrate the robustness, capability and validity of the method.

Keywords: Semirigid connection, multilinear moment-rotation relation, nonlinear analysis, mathematical programming.

Introduction

In general connection of a beam to column can be categorized in three groups. The first group are rigid connections in which, theoretically saying, the angle between the two connected members does not change due to applied moments. The second group are hinged connections in which the connected members can have relative rotation without any resistance. In reality there is neither solid rigid connection, nor theoretical hinge connection, i.e. every rigid connection admits some rotation and every hinge connection tolerates some moment. Semi-rigid connections that have situation between the two groups constitute the third group. Their characteristics are determined by moment rotation relations that are usually nonlinear. To simplify nonlinear analysis, nonlinearity of materials is usually modelled as multilinear relations between stress and strain, the first part of which characterizes linear relations. The most simplified nonlinear relation is the bilinear elastic-plastic relation in which there is an elastic relation between stress and strain up to yield point.

If the structure is stressed up to this stress limit a linear analysis is sufficient for stress-strain calculations. However for further stress or strain a nonlinear analysis is necessary.

Literature is almost mature of nonlinear analysis methods. Crisfield (1991) and Owen and Hinton (1980) have cited good summaries of classical nonlinear analysis techniques. Among the major

M. Pandey et al. (eds), Advances in Engineering Structures, Mechanics & Construction, 329–342.

nonlinear analysis techniques the *Incremental Scheme, Initial Stiffness method, Newton-Raphson method* and combination of these methods can be mentioned

There are also some other techniques, that have been established for inelastic analysis of structures based on theorems of Structural Variation. Structural variation theory studies the effect of change of properties, or even removal, of a member on the entire structure. It takes advantage of linear analysis and sensitivity of structure to some self equilibrating unit loads that are applied at the end nodes of changing members. This technique has been applied to analysis of several types of inelastic skeletal structures including space trusses (Saka & Celik 1985), frames (Majid & Celik 1985) and grids (Saka 1997), etc. It has been also extended to nonlinear finite elements analysis (Abu Kassim & Topping 1985, and Saka 1991). Although this method takes advantage of initial stiffness matrix and does not require change in the stiffness matrix of structure during the analysis process, it is a historical and step by step method of analysis in which every step uses information from the previous step.

Nonlinear analysis of structures by mathematical programming is another field of research in this ground. De Donato (1977) presented fundamentals of this method for both holonomic (path independent) and nonholonomic material behaviours. In this method, it is assumed that displacement of nodes of an elastic-plastic structure comprises two parts namely elastic and plastic parts. Then, the problem of finding total displacement vector of a structure is formulated in the form of a quadratic programming (QP) problem with some complementarity yield constraints. These yield constraints state that individual members either are stressed within elastic limits and do not accept plastic deformations or they are stressed up to yield limit and, as a result, undergo some plastic deformations. The output of this sub-problem is linear and nonlinear deformation of structure. Despite its robustness, this method suffers from the considerable number of variables that enter in the QP sub-problem.

The goal of this research work is to bypass iterative techniques in the analysis of nonlinear structures and build up a method based on simple equilibrium relations to conduct analysis in one step. The idea of this technique has been initially examined by Moharrami et al (2000) for nonlinear analysis of structures including tension only and compression-only truss-type elements. Here in this paper it is extended to elastic-plastic flexural connections. This method holds simplicity of structural variation theorem, advantages and robustness of mathematical programming and precision of the results with less computation effort.

Formulation of the Solution Procedure

For simplicity of formulation of the problem, first the elastic-plastic model of nonlinear behaviour is considered and formulated. Then the formulation is extended to more complicated behaviours. In general the elastic-plastic behaviour of a semi-rigid connection is defined as in Fig.(1) in which moment in the connection is proportional to rotation up to certain limits. Beyond these limits, despite increase in rotation, moment remains constant. In the elastic-plastic semi-rigid connections the Moment-Rotation relation in each direction can be defined by two parameters M_{lim} and R_0 . R_0 is the stiffness of the connection and for rigid connections it tends to be infinity. It is zero for hinge connections and accordingly in this case M_{lim} is zero. To formulate the solution procedure, it is noted that the elastic-plastic behaviour of the connection can be mathematically written as :

$$M_{lim}^{-} \leq M \leq M_{lim}^{+} \qquad \textbf{(1)}$$

in which M_{lim}^{-} and M_{lim}^{+} are limits of negative and positive moments in the connection respectively. This is equal to the two following relations:

$$\begin{aligned} M_{lim}^{-} &\leq M^{-} \\ M^{+} &\leq M_{lim}^{+} \end{aligned} \qquad \textbf{(2)}$$

Noting that the Moment in the connection is either positive (M^{+}) or negative (M^{-}), the above relation can be written as follows:

$$M_{lim}^{-} \leq M^{+} - M^{-} \leq M_{lim}^{+} \qquad \textbf{(3)}$$

To formulate the new method, we first focus on a condition that one of the inequalities in Eq.(2) is governing. Later on the formulation will be extended for the case that both are concerned.

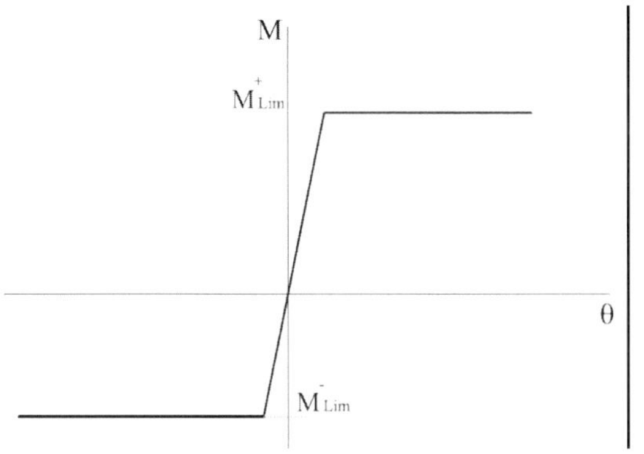

Fig.(1): Elastic-Plastic Behaviour

Elastic-Plastic Connections with Limited Capacity of Negative Moment

Consider a connection in which the second inequality in Eq.(2) is ignored, i.e. it is assumed that the behaviour of the connection is such that there is only a limit on negative moment but there is not any limit on positive moment, i.e. $M^+_{lim} = \infty$. In this case if a linear analysis is done and the value of moment exceeds the capacity of the connection it is necessary to add a positive moment to the connection such that after its distribution in the structure, it reduces the moment in the connection to the capacity limit M^-_{lim}. Let us assume that this unknown positive moment is X^+. To maintain the equilibrium and compatibility in the structure, it is necessary that this moment be accompanied by a moment at the other end and a pair of shear forces at the two ends of connected member.

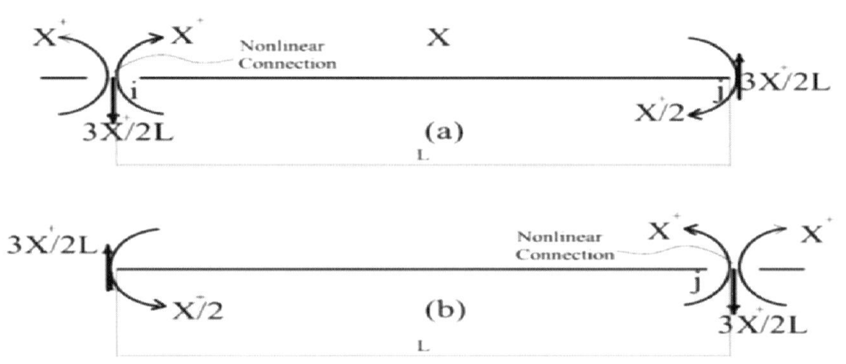

Fig(2): Application of Self-Equilibrated Positive Moment X^+ at two ends of a member and its related moment and shear forces. (a) Connection at **i** (b) Connection at **j**

Fig.(2) shows the application of self equilibrated moment and its accompanied moment and shear forces to a connection. From this point forward whenever it is talked about addition of a moment, it is meant a self equilibrated set of moments and shear forces.

Due to the applied moment X_i^+ in a connection or a joint named **i**, a moment $M_{ji} = \overline{m}_{ji} X_i^+$ will be produced in joint **j**. In this equation $\overline{m}_{ji}$ is the moment produced in joint/connection **j** due to unit moment in **i**. Considering that there are n such artificial moment X_i^+ on the structure, the resultant moment M_{ri} in a typical connection **i** can be obtained from the following equation:

$$M_{ri} = \hat{M}_i + X_i^+ + \sum_{j=1}^{n} \overline{m}_{ij} X_j^+ \quad \textbf{(4)}$$

In Eq.(4), $\hat{\bar{M}}$ is the moment obtained from linear elastic analysis. It should be pointed out that since in applying unit moments at the connection **i** at member **i-j** two external moments $M_i=1$ and $M_j=-.5$ are applied, these moments should be added to $\overline{m}_{ii}$ and $\overline{m}_{ji}$ values. This why the X_i^+ appears in Eq. (4). From this point forward, m_{ij} that comprises $M_i=1$ and $M_j=-.5$, is replaced for $\overline{m}_{ij}$ to include the effect of external unit moments and corresponding shear forces. Therefore Eq.(4) will be written as

$$M_{ri} = \hat{M}_i + \sum_{j=1}^{n} m_{ij} X_j^+ \quad \textbf{(5)}$$

Provided that X^+ are known, Eq. (5) can be used for calculation of moment in all joints/connections. Therefore It remains to find the values of X^+'s for all nonlinear connections. To that end substitute the M_{ri} from Eq. (5) into Eq. (2) for all nonlinear connections to have:

$$M_{ri} = \hat{M}_i + \sum_{j=1}^{n} m_{ij} X_j^+ \geq M_{lim}^- \qquad ,i = 1,2,\ldots,n \quad \textbf{(6)}$$

Eq. (6) provides a set of n inequalities that can be used for determination of n unknowns. However there is a special condition that should be observed during solution of this set of inequalities.
Considering that due to applied loads the value of M_{ri} from Eq.(5) does not exceed the moment capacity limit M_{lim}^-, then it would not be necessary to apply the unknown moment X_i^+, i.e. in this case $X_i^+=0$. Otherwise the artificial moment $X_i^+ \geq 0.$ should be applied to the joint/ connection to make $M_i = M_{lim}$. These mutual conditions which are in fact the complementary conditions for the set of inequalities (6) can be written in the following form:

$$X_i^+ (M_{ri\,ri}^- - M_{lim}^-) = 0. \qquad ;i = 1,2,\ldots,n \quad \textbf{(7)}$$

To solve the set of inequalities (6) with the conditions in Eq. (7), a Quadratic Programming (QP) problem as follows is established in which Eqs. (7) constitute its objective function and the set of inequalities in Eq. (6) comprise its constraints. It is noted that for arbitrary values of $X_i^+ \geq 0.$ that satisfies Eq. (6) the value of $X_i^+(M_{ri} - M_{lim})$ is always positive. Therefore minimization of the objective function reduces it to zero.

$$\begin{aligned} &\textit{Minimize} \ \sum_{i=1}^{n} X_i^+ (M_{ri} - M_{lim,i}^-) \\ &\textit{Subject to} \ \ M_{ri} = \hat{M}_i + \sum_{j=1}^{n} m_{ij} X_j^+ \geq M_{lim}^- \qquad ,i = 1,\ldots,n \end{aligned} \quad \textbf{(8)}$$

The solution of QP problem in Eq. (8) while reduces the objective function to zero provides the values of X_i^+'s. Then the Eq. (5) can be used to determine the moments in every joint of the structure regardless that there are or not any artificial moment.

Elastic-Plastic Connections with Limited Capacity of Positive Moment

In this type of connections it is assumed that positive moment is limited to M_{lim}^+ but it is not limited in negative direction. If we use the same unit moments that was used in the previous section, similar formulation can be derived except that to keep unique coordinate system there will be a negative sign in behind of X^-. In this case the sub-problem Eq. (8) will be as in Eq. (9)

$$\begin{aligned} &\text{Minimize} \sum_{i=1}^{n} X_i^-(M_{lim,i}^+ - M_{ri}) \\ &\text{Subject to} \quad M_{ri} = \hat{M}_i - \sum_{j=1}^{n} m_{ij} X_j^- \le M_{lim}^+ \qquad ,i=1,\ldots,n \end{aligned} \tag{9}$$

Elastic-Plastic Connections with Limited Capacity in both directions

The formulation of the problem in this case which provides the general elastic-plastic behaviour can be extracted from the formulation of previous sections. Noting that the artificial moment should be applied in either positive or negative direction, a general formula for evaluation of moment can be written as follows:

$$M_{ri} = \hat{M}_i + \sum_{j=1}^{n} m_{ij}(X_j^+ - X_j^-) \qquad ,i=1,\ldots,n \tag{10}$$

in which in any time one of the artificial moments can be present. i.e. in the presence of either of X_i^+ or X_i^-, the other one will have zero value. This can be mathematically stated as: $X_i^+ \times X_i^- = 0$. Substituting the general M_{ri} from Eq.(10) into Eq.(3) for M^+ -M^- and using a combination of objective functions in Eq.(8) and Eq.(9) and the complementary condition of $X_i^+ X_i^- = 0$ yields the following problem for the general elastic-plastic behaviour of a nonlinear connection.

$$\begin{aligned} &\text{Minimize} \sum_{i=1}^{n} [X_i^+(M_{ri} - M_{lim,i}^-) + X_i^-(M_{lim,i}^+ - M_{ri}) + X_i^+ X_i^-] \\ &\text{Subject to} \quad M_{lim,i}^- \le \hat{M}_i + \sum_{j=1}^{n} m_{ij}(X_j^+ - X_j^-) \le M_{lim,i}^+ \qquad ,i=1,\ldots,n \end{aligned} \tag{11}$$

Example 1.

To illustrate the solution procedure, and capabilities of the method a one-bay, one-story portal frame as shown in Fig.(3-a) is considered. In this example the moment-curvature relation of connections of beam to columns are assumed to be elastic-plastic in both positive and negative directions. All members are assumed to have the similar cross-sections. It is assumed that the frame will be loaded laterally up to 40 KN. The Moment capacity of connections in each direction has been intentionally assumed to be 85% of the members' plastic limit as shown in Fig (3-b).

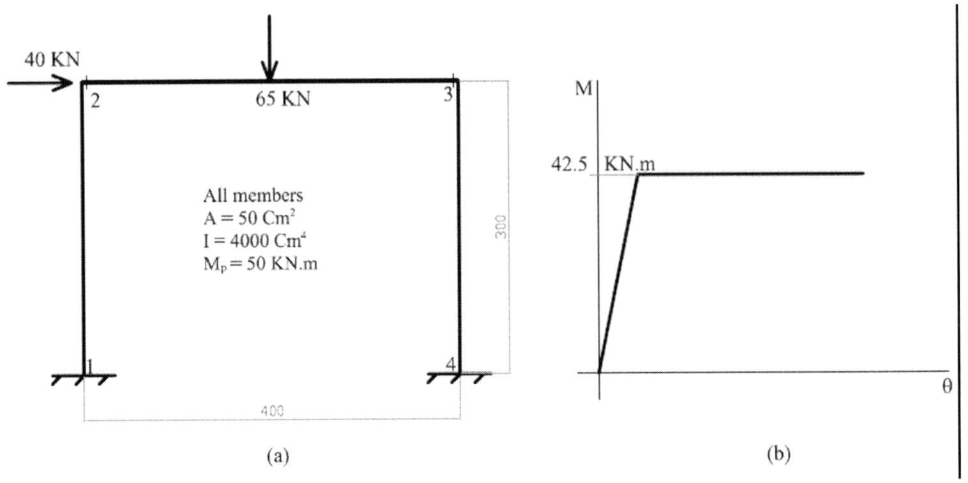

Fig.(3): Example 1, a) Properties of a Portal Frame, b)M- relations for connections 2 and 3

To have a comparison tool in hand, first the non-linear analysis of the frame will be done in a historical manner and then it is compared to the results of proposed solution procedure. Assuming that the structure behaves linearly, it is analysed under total lateral load (40-H & 65-V KN). The result of linear analysis has been reported in the first row of Table 1.

Table 1: Results of analysis of Example 1

Loading	M_1	M_2	Midspan	M_3	M_4
Total (Linear)	-23.892	0.961	41.456	-48.050	47.098
First Yield	-21.132	.850	36.668	-42.5	41.658
Load increment	-6.715	0.904	7.959	0	6.240
Total Nonlinear	-27.847	1.755	44.627	-42.5	47.898

Since value of M_3 =-48.05 exceeds the plastic limit of the connection (42.5 KN.m), the structure will experience inelastic behaviour under this intensity of load. The values of moments in all joints/ connections at the threshold of inelastic behaviour can be obtained by scaling the initial results by $42.5/48.050$. The results have been reported at second row of Table1. In the next stage, for the rest of loading (i.e. 4.620-H and 6.507-V), the frame model will contain a hinge at connection 3. The result of this load increment has been reported in the third row of the Table 1. The total moments due to nonlinear behaviour is the sum of second and third row of the Table and has been reported in the fourth row.

The analysis of the problem via proposed method requires analysis of the structure under unit moments that are applied in all elastic-plastic connections. The results of such analysis for unit loads at joints 2 and3 have been shown in Table 2.

Employing Eq. (10), the resultant moments at connections 2 and 3 can be written as follows:

$$M_{r2} = 0.961 + (1 - 0.6823)(X_2^+ - X_2^-) + (-0.5 + 0.5454)(X_3^+ - X_3^-)$$

$$M_{r3} = -48.050 + (-0.5 + 0.5454)(X_2^+ - X_2^-) + (1 - 0.6823)(X_3^+ - X_3^-)$$

Now according to Eq. (11) a quadratic programming problem is established as follows:

$$Minimize \sum_{i=1}^{2}[X_i^+(M_{ri}-(-42.5))+X_i^-(42.5-M_{ri})+X_i^+X_i^-]$$

$$Subject\ to\ \ M_{r2}=0.961+0.3177(X_2^+-X_2^-)+0.0454(X_3^+-X_3^-)\leq 42.5$$
$$M_{r2}=0.961+0.3177(X_2^+-X_2^-)+0.0454(X_3^+-X_3^-)\geq -42.5$$
$$M_{r3}=-48.050+0.0454(X_2^+-X_2^-)+0.3177(X_3^+-X_3^-)\leq 42.5$$
$$M_{r3}=-48.050+0.0454(X_2^+-X_2^-)+0.3177(X_3^+-X_3^-)\geq -42.5$$

Solution of this QP problem gives the following results:

$X_3^+=17.46783\ \ and\ \ X_2^+=X_2^-=X_3^-=0$

Table 2: Results of analysis of Example 1 for unit loads and Proposed Method.

Loading	M_1	M_2	Midspan	M_3	M_4
Unit moment at 2	+0.0458	1-0.6823	+0.1816	+0.5454-.5	-0.2264
Unit moment at 3	-0.2264	+0.5454-.5	+0.1816	1-0.6823	+0.0458
Total external loads	-23.892	0.961	41.456	-48.050	47.098
Unit moment at 3 × 17.46783	-3.955	0.793	3.172	5.55	0.800
Sum of the Two Above Rows	-27.847	1.754	44.628	-42.5	47.898

It remains to obtain internal forces in all sections using Eq. (10). The results of moment calculations have been shown in fifth and sixth rows of Table2. If the results of Table 1 and Table 2 are compared, it is observed that there is a very small difference between the two results which may have been because of numerical round off error in either of the algorithms/methods or both. As a conclusion, in this example it was shown that the proposed method is capable of solving the nonlinear problem and theoretically saying, it results in exact nonlinear solution.

Nonlinear Analysis with Tri-linear M-Y Curve

In this section the formulation of nonlinear analysis for tri-linear type of connection is presented. Fig. (4) shows such a typical relation. A tri-linear relation usually fits better on an M-Y curve. It can also model strain-hardening of a typical connection. This type of relation may be applicable for both negative and positive moment directions

As shown in Fig. (4) the tri-linear relation may be characterized by four parameters: M_1, M_{lim} ,R_0 and R_1. In which M_{lim} is the limit of bending moment capacity, R_0 and R_1 are initial and secondary flexural stiffness of the connection and M_1 is the first yield in the **M-Y** curve. Similar to the previous procedure for formulation of the problem, first the case of negative moment is considered.

Depending on the value of moment or rotation in a connection, the M-Y relation may coincide one of the three linear relations shown in Fig. (4).

If $\theta_i \leq \theta_{i1}$, then the connection behaves linearly. In this case: $M_r = \hat{M}_i$ and there will be no need for external virtual moment.

If $\theta_{i1} \leq \theta_i \leq \theta_{i2}$, the actual value of $\overline{M}_i^-$ for a given value of Y can be obtained from Eq. (12).

$$\overline{M}_i^- = M_{1i}+(\theta_i-\theta_{1i})*R_{1i} \quad \textbf{(12)}$$

From linear analysis we have: $\overset{\wedge}{\overline{M}}_i = R_{0i}\theta_i$ and $M_1^- = R_{0i}\theta_1$, therefore it looks possible to substitute for values of Y_i and Y_{1i} to find:

$$\overline{M_i^-} = M_{1i} + (\hat{M}_i^- - M_1^-) * \frac{R_{1i}}{R_0} \quad \textbf{(13)}$$

Eq. (13) stipulates that in the connection with reduced stiffness, only partial part of the moment in excess of M_1 is stored and some part of the excess moment should be distributed in the structure.

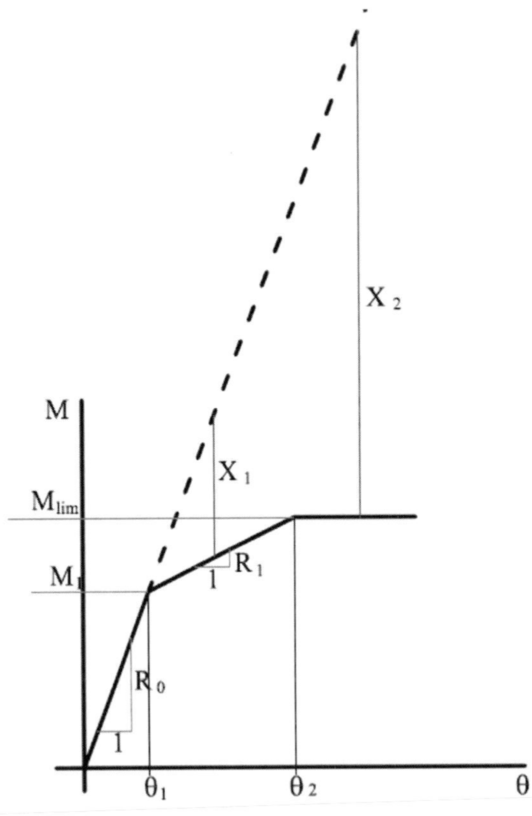

Fig.(4): Atypical Tri-linear M-Y relation for a connection

From the above explanation it is understood that the $(\hat{M}_i^- - M_1^-) * (1 - R_{1i} / R_0)$ part of the excess moment should be distributed in the structure. However any moment redistribution from connection into the structure not only affects the value of rotations of connection and corresponding moments in connected members but also affects the moment distribution in the structure which in turn affects again the value of excess moment in the connection. The study on this mutual moment redistribution between softened connection and the structure shows that there is a reciprocal relation as in Eq. (13) between remained moment in the connection and other parameters such as the moment distribution factor m_{ii} , the ratio of stiffness of member to the connection (EI/L)/R0 and the rate of stiffness softening R0/R1.

$$remained\ Moment\ in\ connection\ i = \gamma_i \times Moment\ \ in\ excess\ of\ M_1$$

$$\frac{1}{\gamma_i} = 1 + m_{ii}(\frac{R_0}{R_1} - 1)(\frac{4EI/L}{R_0}) \quad \textbf{(14)}$$

in which $\mathbf{m_{ii}}$ is the measured moment at connection **i** due to application of a unit moment at **i**. The statement in Eq. (14) can be mathematically written as the following equation for the softened range of behaviour of the connection.

$$M_{ri}^- = \hat{M}_i^- + \sum_{j=1}^{n} m_{ij}(X_{1j}^+ + X_{2j}^+ - (X_{1j}^- + X_{2j}^-)) \geq M_{1,i}^- + \Delta M_i * \gamma_i \tag{15}$$

The value of ΔM in the above equation includes all transferred moments from all virtually loaded connections onto the connection in question i.e.

$$\Delta M_i^- = \hat{M}_i^- - M_1^- + \sum_{j=1, j\neq i}^{n} m_{ij}(X_{1j}^+ + X_{2j}^+ - (X_{1j}^- + X_{2j}^-)) \tag{16}$$

Substituting for ΔM from Eq. (16) into Eq. (15) and rearranging for unknowns gives the yield condition for the case of stiffness softening in the connection as follows:

$$m_{ii}X_{1i}^+ + (1-\gamma_i)\sum_{j=1, j\neq i}^{n} m_{ij}(X_{1j}^+ + X_{2j}^+ - (X_{1j}^- + X_{2j}^-)) \geq (M_{1,i}^- - \hat{M}_i^-) * (1-\gamma_i) \tag{17}$$

Similarly when $\theta_i \geq \theta_{i2}$ we should add the external virtual moment X_{i2} to the connection to reduce the moment to M_{lim}. In this case the following relation can be written between resultant moment in joint **i** and limit moment:

$$M_{ri}^- = \hat{M}_i^- + \sum_{j=1}^{n} m_{ij}(X_{1j}^+ + X_{2j}^+ - (X_{1j}^- + X_{2j}^-)) \geq M_{lim}^- \tag{18}$$

It should be noted that, although X_{i1} and X_{i2} as shown in Fig.(4), are independent virtual moments, they can not exist simultaneously. Since it is not known a priori that how much is the θ_i, both of Eqs.(17&18) should be considered in the solution of the problem. In general the QP problem for calculation of unknown X's become as follows:

$$\begin{aligned}
&\text{Minimize} \sum_{i=1}^{n} \left\{ \begin{aligned} &X_{i1}^+[(\hat{M}_i^- - M_{1i}^-)(1-\gamma_i) + m_{ii}X_{1i}^+ + (1-\gamma_i)\sum_{j=1, j\neq i}^{n} m_{ij}((X_{1j}^+ - X_{1j}^-) + (X_{2j}^+ - X_{2j}^-))] \\ &+ X_{i2}^+[\hat{M}_i^- - M_{lim,i}^- + \sum_{j=1}^{n} m_{ij}((X_{1j}^+ - X_{1j}^-) + (X_{2j}^+ - X_{2j}^-))] \end{aligned} \right\} \\
&\text{Subject to}: \; m_{ii}X_{1i}^+ + (1-\gamma_i)\sum_{j=1, j\neq i}^{n} m_{ij}(X_{1j}^+ + X_{2j}^+ - (X_{1j}^- + X_{2j}^-)) \geq (M_{1,i}^- - \hat{M}_i^-) * (1-\gamma_i) \\
&\qquad \sum_{j=1}^{n} m_{ij}((X_{1j}^+ - X_{1j}^-) + (X_{2j}^+ - X_{2j}^-)) \geq M_{lim}^- - \hat{M}_i^- \qquad ; i = 1, \ldots, n
\end{aligned} \tag{19}$$

Similar QP problem can be established for reverse direction i.e. the case of a tri-linear connection in positive moment direction and unlimited linear relation in negative moment direction. If R_0, R_1, M_1 and M_{lim} properties in positive direction are assumed to be the same as their counterpart in negative direction, the following QP problem can be obtained. However one may use different properties for two different directions of moments.

$$\text{Minimize} \sum_{i=1}^{n} \left\{ \begin{aligned} &X_{i1}^-[(M_{1i}^+ - \hat{M}_i^+)(1-\gamma_i) + m_{ii}X_{1i}^- + (1-\gamma_i)\sum_{j=1, j\neq i}^{n} m_{ij}((X_{1j}^+ - X_{1j}^-) + (X_{2j}^+ - X_{2j}^-))] \\ &+ X_{i2}^-[M_{lim,i}^+ - \hat{M}_i^- + \sum_{j=1}^{n} m_{ij}((X_{1j}^+ - X_{1j}^-) + (X_{2j}^+ - X_{2j}^-))] \end{aligned} \right\}$$

$$\text{Subject to: } m_{ii}X_{1i}^{-}+(1-\gamma_i)\sum_{j=1,j\neq i}^{n} m_{ij}(X_{1j}^{+}+X_{2j}^{+}-(X_{1j}^{-}+X_{2j}^{-}))\geq(\hat{M}_i^{+}-M_{1,i}^{+})*(1-\gamma_i)$$

$$\sum_{j=1}^{n} m_{ij}((X_{1j}^{+}-X_{1j}^{-})+(X_{2j}^{+}-X_{2j}^{-}))\geq \hat{M}_i^{+}-M_{lim}^{+} \qquad ;i=1,\ldots,n \tag{20}$$

Combining equations (19) and (20) and noting that one of $X_{1i}^{+}, X_{1i}^{-}, X_{2i}^{+}$ *and* X_{2i}^{-} can possess nonzero value at a time, and the others are zero, yields the following QP sub-problem from which all unknown Virtual Moments can be evaluated.

$$\text{Minimize } \sum_{i=1}^{n}\left\{\begin{array}{l} X_{i1}^{+}[(\hat{M}_i^{-}-M_{1i}^{-})(1-\gamma_i)+m_{ii}X_{1i}^{+}+(1-\gamma_i)\sum_{j=1,j\neq i}^{n} m_{ij}((X_{1j}^{+}-X_{1j}^{-})+(X_{2j}^{+}-X_{2j}^{-}))] \\ +X_{i2}^{+}[\hat{M}_i^{-}-M_{lim,i}^{-}+\sum_{j=1}^{n} m_{ij}((X_{1j}^{+}-X_{1j}^{-})+(X_{2j}^{+}-X_{2j}^{-}))]+ \\ X_{i1}^{-}[(M_{1i}^{+}-\hat{M}_i^{+})(1-\gamma_i)+m_{ii}X_{1i}^{-}+(1-\gamma_i)\sum_{j=1,j\neq i}^{n} m_{ij}((X_{1j}^{+}-X_{1j}^{-})+(X_{2j}^{+}-X_{2j}^{-}))] \\ +X_{i2}^{-}[M_{lim,i}^{+}-\hat{M}_i^{-}+\sum_{j=1}^{n} m_{ij}((X_{1j}^{+}-X_{1j}^{-})+(X_{2j}^{+}-X_{2j}^{-}))]+(X_{i1}^{+}+X_{i2}^{+})\times(X_{i1}^{-}+X_{i2}^{-}) \end{array}\right\}$$

$$\text{Subject to: } m_{ii}X_{1i}^{+}+(1-\gamma_i)\sum_{j=1,j\neq i}^{n} m_{ij}(X_{1j}^{+}+X_{2j}^{+}-(X_{1j}^{-}+X_{2j}^{-}))\geq(M_{1,i}^{-}-\hat{M}_i^{-})*(1-\gamma_i)$$

$$\sum_{j=1}^{n} m_{ij}((X_{1j}^{+}-X_{1j}^{-})+(X_{2j}^{+}-X_{2j}^{-}))\geq M_{lim}^{-}-\hat{M}_i^{-} \tag{21}$$

$$m_{ii}X_{1i}^{-}+(1-\gamma_i)\sum_{j=1,j\neq i}^{n} m_{ij}(X_{1j}^{+}+X_{2j}^{+}-(X_{1j}^{-}+X_{2j}^{-}))\geq(\hat{M}_i^{+}-M_{1,i}^{+})*(1-\gamma_i)$$

$$\sum_{j=1}^{n} m_{ij}((X_{1j}^{+}-X_{1j}^{-})+(X_{2j}^{+}-X_{2j}^{-}))\geq \hat{M}_i^{+}-M_{lim}^{+} \qquad ;i=1,\ldots,n$$

Example 2

To illustrate the capability of the proposed method in the analysis, a portal frame with un-equal column lengths under vertical and lateral load as shown in Fig.(6-a) is considered. The M-Y relation for connections 2 and 3 is shown in Fig(6-b). It is assumed that the members can be loaded up to their plastic moment Mp. Beyond this moment, there will establish a plastic hinge that is assumed to be bilinear.

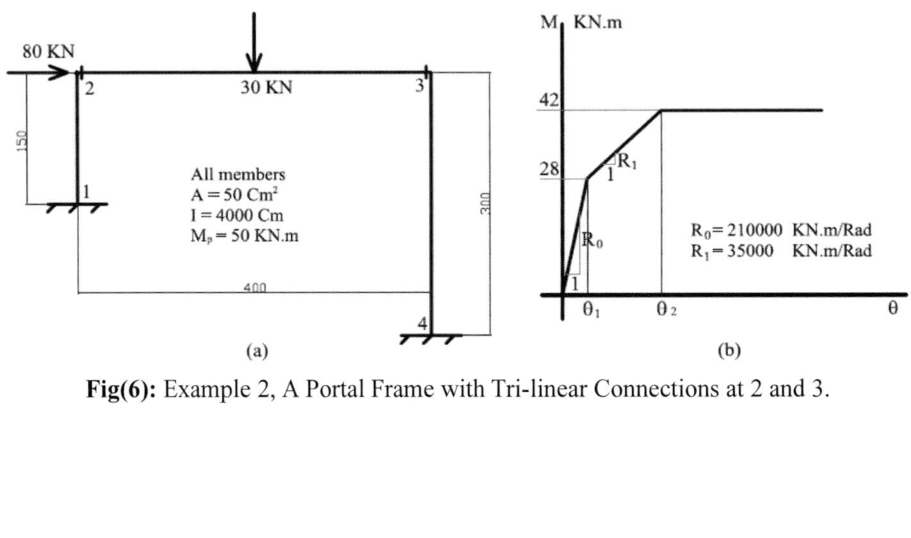

Fig(6): Example 2, A Portal Frame with Tri-linear Connections at 2 and 3.

It is to be noted that in general a connection is separate than a member and its role should be modeled as a torsional spring with rigidity of R_0 , R_1 and 0 at the three ranges of rotations. In this example the connections in 2 and 3 had to be modeled as in Figure 7-a. This required special analysis software. Instead the model in figure 7-b which can be analyzed by most of commercial softwares was used. In this model torsional rigidity of the members 2-6 and 3-7 are equal to the flexural stiffness of connection. Note that in nodes 1 and 4 there is no connection and the capacity of the structure at these nodes is determines by capacity of the members.

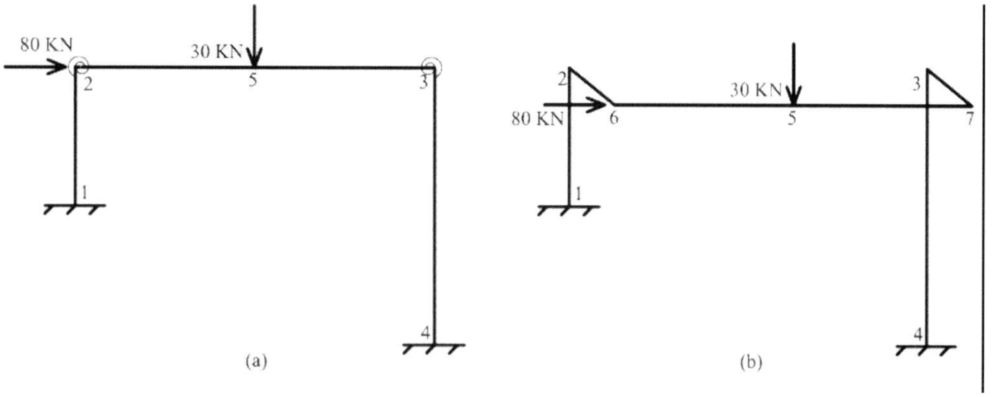

Fig.(7): Actual Analysis model (a) and its equivalent 3-D model (b)

To solve the problem we need to perform a linear elastic analysis for the applied loads and do some analysis for unit moments. The number of virtual unit moments depends on the number of potentially nonlinear points in the structure. Here we assume that all nodes will have nonlinear behavior. However the behavior of joints 1 and 4 will be assumed to be the same as elastic-plastic behavior of members. The moment at midspan does not seem to exceed its proportionate limit. As a result unit moment has not been applied there. Therefore there will be four virtual loadings. The results of aforementioned analyses have been reported in Table 3.

Table 3: Results of Analysis of Frame of Example 2 for Various Types of Loadings

Moment Position	Full External Load	Unit M at 1 in member 1-2	Unit M at 2 in member 2-3	Unit M at 3 in member 2-3	Unit M at 4 in member 3-4
Joint No. 1	-92.815	1.0-0.8503	0.12693	-0.10811	0.32734
Connection 2	7.726	0.5476-0.5	1.0-0.85089	0.4961-0.5	-0.00362
Midspan	28.256	0.00353	0.07260	0.07694	-0.03709
Connection 3	-11.213	-0.04054	0.4961-.5	1.0-0.84222	0.46202-0.5
Joint No. 4	27.206	+0.16367	-0.04828	-0.05064	1-0.31091

According to the results in Table 3, a QP sub-problem, can be established as follows. In the following QP problem a combination of Eq.(11) and Eq.(21) is used. This is because the joints 1 and 4 have bilinear behavior and obey the rules of Eq.(11) and joints 2 and 3 have tri-linear behavior and obey Eq.(21). In addition since in this simple problem we know the direction of applied virtual moments, it is evident that $X_1^- = X_{21}^+ = X_{22}^+ = X_{31}^- = X_{32}^- = X_4^+ = 0$. Therefore only those virtual moments that potentially are nonzero have been participated in the formulation. Because of this decision, only the complementary condition on $X_{i1}^{\pm} \times X_{i2}^{\pm} = 0$. has been considered.

$$\begin{aligned}
&Minimize\ \{ X_1^+[(\hat{M}_1^- - M_{lim,1}^-) + m_{11}X_1^+ - m_{12}(X_{12}^- + X_{22}^-) + m_{13}(X_{13}^+ + X_{23}^+) - m_{14}X_4^-) + \\
&X_{12}^-[(1-\gamma_2)(M_{1,2}^+ - \hat{M}_2^+) + m_{22}X_{12}^- - (1-\gamma)(m_{21}X_1^+ - m_{23}(X_{31}^+ + X_{32}^+) + m_{24}X_4^-)] + \\
&X_{22}^-[(M_{lim,2}^+ - \hat{M}_2^+) - [m_{21}X_1^+ - m_{22}X_{22}^- + m_{23}(X_{31}^+ + X_{32}^+) - m_{24}X_4^-)] + (X_{12}^- \times X_{22}^-) + \\
&X_{13}^+[(1-\gamma_3)(\hat{M}_3^- - M_{13}^-) + m_{33}X_{13}^+ + (1-\gamma_3)(m_{31}X_1^+ - m_{32}(X_{21}^- + X_{22}^-) + m_{34}X_4^-)] + \\
&X_{23}^+[(\hat{M}_3^- - M_{lim,3}^-) + m_{31}X_1^+ - m_{32}(X_{21}^- + X_{22}^-) + m_{33}X_{23}^+ - m_{34}X_4^-)] + (X_{13}^+ \times X_{23}^+) + \\
&X_4^-[(M_{lim,4}^+ - \hat{M}_4^{\,|}) - m_{41}X_1^+ + m_{42}(X_{21}^- + X_{22}^-) - m_{43}(X_{31}^+ + X_{32}^+) + m_{44}X_4^+]\}
\end{aligned}$$

$$\begin{aligned}
Subject\ to:\ & m_{11}X_1^+ - m_{12}(X_{12}^- + X_{22}^-) + m_{13}(X_{13}^+ + X_{23}^+) - m_{14}X_4^-) \ge M_{lim,1}^- - \hat{M}_1^- \\
& m_{22}X_{12}^- - (1-\gamma_2)(m_{21}X_1^+ + m_{23}(X_{13}^+ + X_{23}^+) - m_{24}X_4^-)) \ge (1-\gamma_2)(\hat{M}_2^+ - M_{12}^+) \\
& -m_{21}X_1^+ + m_{22}X_{22}^- - m_{23}(X_{13}^+ + X_{23}^+) + m_{24}X_4^- \ge \hat{M}_2^+ - M_{lim,2}^+ \\
& m_{33}X_{13}^+ + (1-\gamma_3)(m_{31}X_1^+ - m_{32}(X_{21}^- + X_{22}^-) - m_{34}X_4^- \ge (1-\gamma_3)(M_{13}^- - \hat{M}_3^-) \\
& m_{31}X_1^+ - m_{32}(X_{21}^- + X_{22}^-) + m_{33}X_{23}^+ - m_{34}X_4^- \ge M_{lim,3}^- - \hat{M}_3^- \\
& -m_{41}X_1^+ + m_{42}(X_{21}^- + X_{22}^-) - m_{43}(X_{31}^+ + X_{32}^+) + m_{44}X_4^+ \ge \hat{M}_4^+ - M_{lim,4}^+ \\
& X_{12}^- \times X_{22}^- = 0 \\
& X_{13}^+ \times X_{23}^+ = 0
\end{aligned}$$

In this sub-problem m_{ij} values are given in Table 3. Other parameters are as follows:

$M_{lim,1}^+ = -M_{lim,1}^- = M_{lim,4}^+ = -M_{lim,4}^- = 50 \quad KN.m$

$M_{1,2}^+ = -M_{1,2}^- = M_{1,3}^+ = -M_{1,3}^- = 28 \quad KN.m$

$M_{lim,2}^+ = -M_{lim,2}^- = M_{lim,3}^+ = -M_{lim,3}^- = 42 \quad KN.m$

$\gamma_2 = 0.788776$ *and* $\gamma_3 = 0.7593655$

Solution of the problem yields the following results:

$X_1^+ = 574.1396 \quad X_{21}^- = 54.6801 \quad X_{31}^+ = 10.6389 \quad X_4^- = 107.0615$

According to these results evaluation of internal forces in members is quite simple. Table 4 shows the calculation of final nonlinear response for moments.

It can be seen from the Table 4 that the position of vertical point load (Midspan) is still in its proportionate range and therefore, as it was foreseen, there is no need to apply external virtual moments. On the other hand except Node No. 1 it did not seem that other nodes can reach to nonlinear stage. But results show that Node No. 4 has reached to its plastic limit; and the moments at connections No. 2 & 3 are a little beyond the first yield of the connection. It is obvious that if the load is increased, the moments at connections 1 and 4 will remain constant and the moments of Midspan, Node 2 and Node 3 will increase.

Table 4: Calculation of nonlinear response of Example 2, Proposed method.

Effect of X_j at joints and connections	Values of Virtual Moments				Total Moment
Moment Position	X_1^+	X_{21}^-	X_{31}^+	X_4^-	$\hat{M}+X_1^++X_{21}^-+X_{31}^++X_4^-$
	574.1396	54.6801	10.6389	107.0615	
Node No. 1	85.9506	-6.9404	-1.1502	-35.0454	-50.0000
Connection No. 2	27.3277	-8.1531	-0.0415	3.8752	30.7340
Midspan	2.0254	-3.9699	0.8186	3.9705	31.1010
Connection No. 3	-23.2768	0.2134	1.6787	4.0658	-28.5319
Node No. 4	93.9691	2.6389	-0.5387	-73.7753	50.0000

Conclusion

In this paper, a new, non-iterative method of nonlinear analysis was proposed. In the proposed method to make internal moments follow their nonlinear moment-rotation curves, some virtual moments (that are primarily unknown) were imposed onto the structure at semirigid connections. To find the unknown virtual moments, a quadratic programming problem was formulated and solved. In the proposed method, the exact nonlinear response of structure including internal forces and moments of members are calculated by employing the values of virtual moments and using the superposition principle. Compared to the classical methods of nonlinear analysis, the following preferences can be mentioned for the proposed method.

- In the classical methods, the nonlinear analysis is done through iterative procedures which consist of modification of the stiffness matrix of structure and/or load vector, but in the proposed method the nonlinear analysis is performed in one step without change in the initial model and stiffness matrix of the structure or its load vector.
- the final result of iterative methods are somewhat dependant on the start point and the convergence criteria while in the proposed method it is obtained by solution of a Quadratic programming method and therefore not only it does not require any initial point, but also it does not require convergence criterion and mathematically saying it gives exact results.
- To increase the accuracy of the results in the classical methods it is necessary to decrease the incremental loading and tighten the convergence criterion. This increases the number of required analysis. However in the proposed method the results are exact and only preventing round off error in calculations increases the accuracy.

The method is capable to model semirigid connections with multilinear moment-curvature relations. The formulation of the problem for bilinear and trilinear moment-curvature relations was presented. Two examples were solved to demonstrate the robustness, capability and validity of the method. It was shown that the method not only can be used for nonlinear connections but also provided that the critical sections are pre-specified, it can do the plastic analysis in a flexural frame as well.

References.

Abu Kassim, A.M. and Topping, B.H.V. (1985). "The theorems of structural variation for linear and nonlinear finite element analysis", *Proceeding of the second international conference on Civil and Struc. Eng. Comp. CIVIL-COMP 85, Vol. 2, pp. 159- 168*

Crisfield M.A., (1991). "Nonlinear finite element analysis of solids and structures", Volume 1-Essentials. John Wiley & Sons, Chichester, U.K.

De Donato, O. (1977) "Fundamentals of elastic-plastic analysis", Engineering plasticity by mathematical programming, M.Z. Cohn and G. Maier, editors. Pergamon Press, New York, N.Y. pp. 325-349

Katsuki, S., Frangopol, D.M. and Ishikawa, N. (1993). "Holonomic elasto-plastic reliability analysis of truss systems. II: Applications." *Journal of Structural Engineering, ASCE, Vol. 119, No. 6 pp. 1792-1806*

Majid, K.I. and Celik, T., (1985). "The elastic-plastic analysis of frames by the theorems of structural variation", *International Journal for Numerical Methods in Engineering, Vol. 21 pp. 671-681*

Moharrami,H. and Reyazi M.M., (2000)"Analysis of structures including Compression-only and Tension-only members", *proceeding of European Congress on Computational Methods in Applied Science and Engineering (Eccomas 2000),* Barcelona, Spain.

Owen, D. R. J. and Hinton, E., (1980). "Finite elements in plasticity- theory and practise" Pineridge press Swansea.

Riyazi Mazloumi, M., (1999)" Analysis of structures including truss elements with limited axial strengths", M.Sc. thesis, Tarbiat Modarres University, Tehran, IR Iran.

Saka, M.P., (1997). "The theorems of Structural variation for Grillage systems" Innovation in Computer Methods for Civil and Structural Engineering, Edinburgh, Civil Comp Press, pp. 101 111

Saka, M.P., (1991). "Finite element application of the theorems of structural variation", *Journal of Computers and Structures, Vol. 41, No.3, pp. 519-530*

Saka, M.P., and Celik, T., (1985) "Nonlinear analysis of space trusses by the theorems of structural variation", *Proceeding of the second international conference on Civil and Struc. Eng. Comp. CIVIL-COMP 85, Vol. 2, pp. 153-158*

EVALUATION OF A FORTY-YEAR OLD T.Y. LIN DESIGNED ELLIPTICAL POST-TENSIONED CABLE-SUPPORTED ROOF

Randall W. Poston
Principal, WDP & Associates, Inc.
Austin, Texas, USA
E-mail: rposton@wdpa.com

Introduction

A structural evaluation of the roof and post-tensioning system and an assessment of the current integrity of the roof of the Norick Arena located on the State Fair Grounds in Oklahoma City, Oklahoma was conducted.

The roof of the Norick Arena is an elliptical post-tensioned concrete roof system with a major axis of about 400 ft. and a minor axis of about 320 ft. The original design was performed by T.Y. Lin and Associates using three-dimensional load-balancing for full dead load and live load. Although elliptical in shape, the post-tensioning tendons form an inverted cable net in which precast concrete panels were suspended and subsequently made monolithic by casting thickened grout beams to form a structural roof that is a relatively thin, monolithic shell membrane.

The roof system is supported by a reinforced concrete compression ring, which in turn is supported by rectangular reinforced concrete columns. Stub tendons are cast into the ring beam and then coupled to the main suspended roof tendons. The post-tensioning is an unbonded, wire-based, button-headed system covered with mastic and placed in a waterproof paper, which essentially acts as a duct. The major axis tendon ducts are 2-1/4 in. in diameter, whereas the minor axis tendon ducts are 1-1/4 in. in diameter. Typical details of the post-tensioning system are shown in Figure 1.

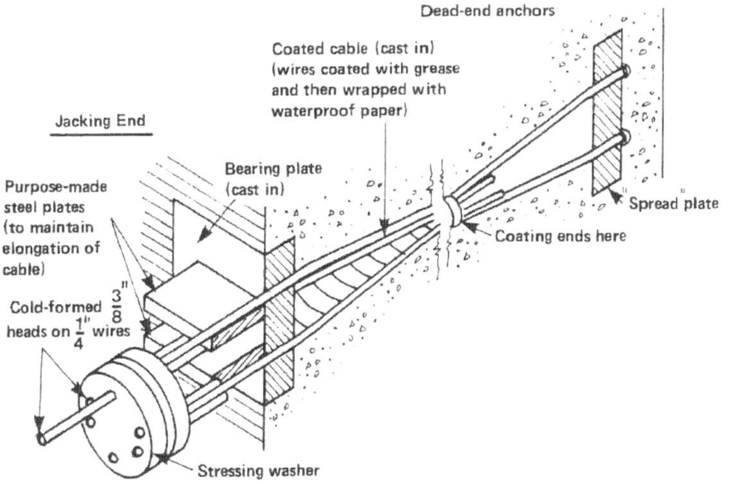

Figure 1. Post-tensioning system details from literature.

M. Pandey et al. (eds), Advances in Engineering Structures, Mechanics & Construction, 343–354.

The arena was constructed in 1965 using a four-stage post-tensioning process with the monolithic concrete membrane stressed at the third stage. The post-tensioning was primarily staged to mitigate cracking in the compression ring. The fourth stage post-tensioning was applied to balance about 12 psf of live load (Lin and Burns, 1981). Figure 2 is an overall view of the arena from ground.

Figure 2. General view of arena exterior.

This paper will summarize the investigation of the post-tension roof of the Norick Arena roof. Results from field investigation procedures and from structural analyses are presented in the context of assessment of current integrity.

Field Evaluation Procedures

The arena roof was evaluated using different nondestructive techniques consisting of surface penetrating radar (SPR) scanning, impact-echo (IE) testing and infrared thermography (IT). A geometry survey of the roof was also conducted, as well as a general visual inspection of the roof, compression ring beam and columns. Invasive probe openings were made at select post-tensioning components in order to assess the general condition of the post-tensioning system.

Geometry Survey

A differential elevation survey (see Figure 3) was conducted to determine the difference in elevation between the edge of the ring beam and various points near the center of the roof, including the center. The difference in the ring beam elevation and the PT at the center of the roof is the "sag" of the roof. Column deformation was determined using a laser level and a laser distance meter. These measurements of the roof sag and column deformation were made to compare to theoretical values computed from structural analyses.

Figure 3. Surveying to determine roof sag.

Invasive Probing and Visual Observations

There are no direct, nondestructive means to assess the condition of post-tensioning (PT) systems. The most direct way of examining tendon condition is to conduct selected probing at the anchorages, couplers, and other strategic locations along the tendon trajectory.

Visual observations were conducted at select probe openings of the post-tensioning system to document conditions. This included documentation of signs of corrosion of PT components, prestressing wire pitting, degradation of the "sisal craft" paper used for the tendon duct, and condition of the corrosion-protection mastic. Photographs were made to document conditions, and where appropriate, measurements were taken of pertinent details such as cover depth.

Nondestructive Testing

Various nondestructive testing (NDT) methods including surface penetrating radar (SPR), impact-echo (IE) and infrared thermography (IT) were utilized to assess various aspects of the concrete deck conditions. For brevity, only general results from the NDT will be presented.

Analysis

Structural System

The roof of the Norick Arena is shaped as an elliptic-paraboloid. The plan view of the roof is an ellipse with major and minor axes of about 400 ft. and 320 ft., respectively. The cross sections of the roof on planes parallel to the planes containing the major and minor axes of the ellipse are also parabolas. The deepest parabolas intersect at the center of the roof where the difference in elevation with respect to its perimeter was designed to be 17 ft. 3 in. from examination of the structural drawings. Figure 4 shows an overall plan view of the roof structural system and the reinforced concrete columns supporting it, and the section of the roof along the major axis of the ellipse. A concrete ring beam connects the columns and the structural roof. The PT tendons are anchored into the ring beam. Although the dimensions of the columns vary somewhat, the predominant dimensions are 18 in. by 58.5 in. with the strong axis parallel to the axis of the ring beam and a height of 62 ft. The ring beam is approximately 108 in. wide by an average 36 in. in depth.

The roof structural system consists of an arrangement of essentially "T-beams" spaced 10 ft. on center located on planes parallel to the planes containing both the major and minor axes of the elliptic-paraboloid. These beams consist of 9 ft. 7 in. square precast concrete panels of 3 in. thickness. The equivalent section used in the analysis is shown in Figure 6. The surface of the roof is covered by insulating and roofing material protecting it from weather. There were two roofing systems that had been applied over the 40-year history of the arena. The outermost layer of roofing insulation is a closed-cell foam and has a variable thickness ranging from about 4 to 6 in. It is somewhat thicker at the center of the roof. This layer prevented a direct visual evaluation of the current structural condition of the concrete surface of the roof other than at limited exposed areas of the probe openings.

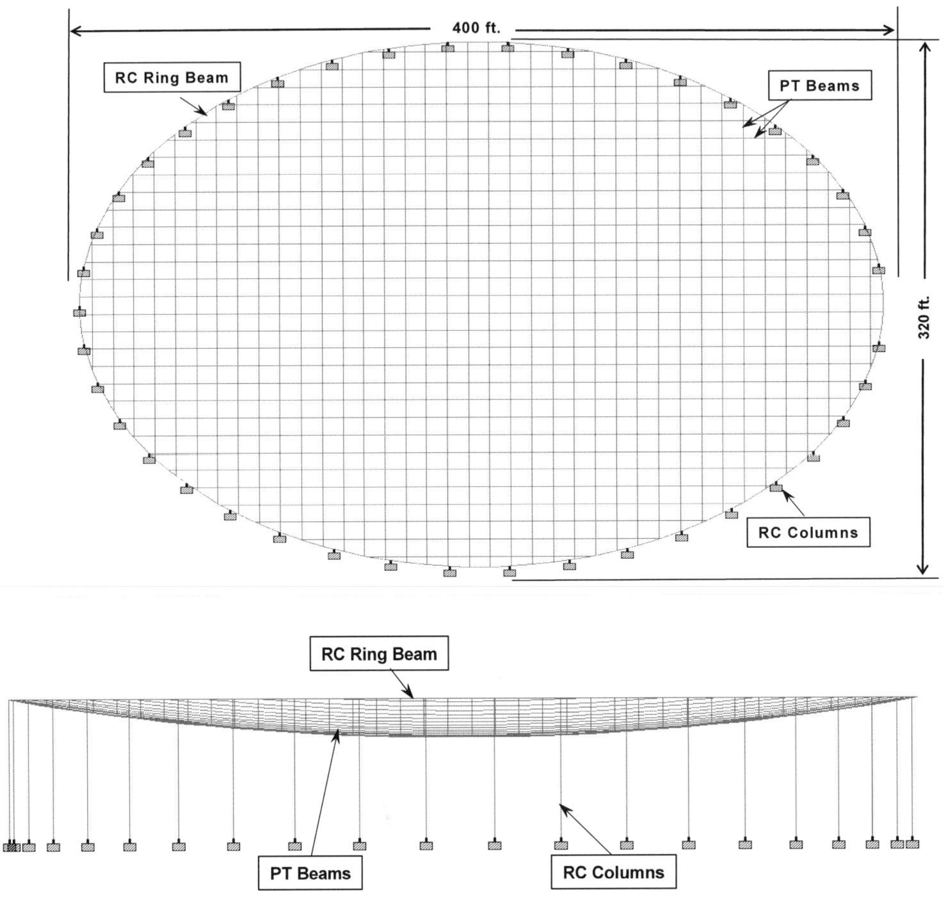

Figure 4. Roof plan view (top) and section along major axis (middle).

Material Properties

The material properties considered in the analysis were, f'_c, concrete compressive strength equal to 4000 psi. The prestressing steel wires that comprise the post-tensioning tendons had an ultimate strength, f_{pu}, equal to 240 ksi.

Post-tensioning Effects

The post-tensioning tendons were modeled using equivalent load effects. The effective anchorage force of each tendon after losses was determined to be 352 kips for the strong axis tendons and 224 kips for the minor axis tendons. The equivalent uniform uplift of the tendons was determined to be 68 psf by using the simplified load balancing model. This assumes that the effective stress, f_{se}, in the tendons considering time-dependent losses was equal to 0.6 f_{pu}, which is 144 ksi. Figure 5 shows a simplified load-balancing model with two "super tendons" along the major and minor axes to represent all post-tensioning tendons in the structural roof.

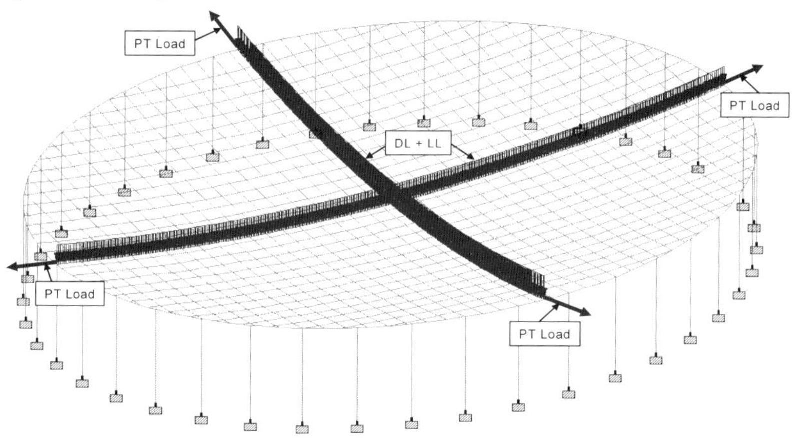

Figure 5. Equivalent force system of PT tendons.

The roof system was modeled with equivalent post-tensioned beams forming a 10 ft. waffle grid. The equivalent section that includes the precast panels is shown in Figure 6.

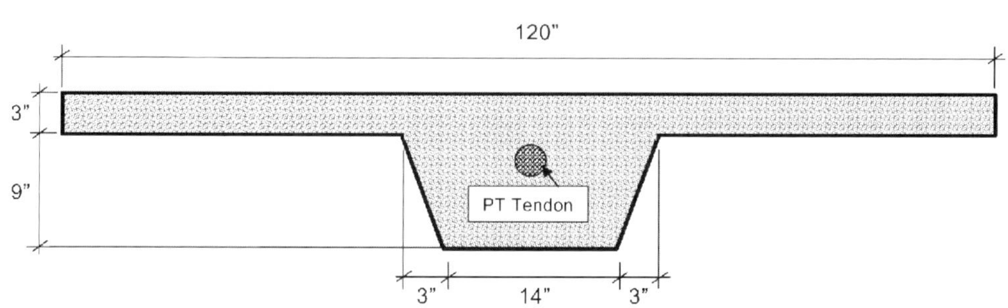

Figure 6. Equivalent roof post-tension beam cross section detail.

Loads

The roof system of the Norick Arena was analyzed for gravity loads. The dead load of the precast waffle deck section with its original built-up roofing was determined to be approximately 56 psf.

There are currently two types of roofing materials on the roof. There is an older built-up membrane and asphalt system covered by a newer sprayed-foam system. This sprayed-foam system is the principal weather-proofing. The thickness of the foam was measured to be about 4 in. as measured at select probe openings. At the center of the roof, the foam has been built up to about 6 to 10 in. to provide a slope for drainage. The average weight of this existing foam roofing material was determined to be 3 psf.

The snow load for the Norick Arena roof as determined from ASCE 7 (ASCE 2002) is 7 psf.

Load Cases

The structural condition of the post-tensioned concrete roof is important to assess if corrosion or other forms of deleterious phenomena have affected its integrity. In general, as a post-tensioned structure ages, its post-tensioning system experiences normal losses induced by relaxation of the steel tendons and by creep and shrinkage in the concrete. In addition, the concrete may experience cracking of some degree, which in turn affects its stiffness. Significant deterioration of the PT tendons due to corrosion would manifest as a loss of effective prestress and the roof's overall ability to carry load.

There were several variations of the structural model and load cases that were examined for comparison purposes with field measurements and to assess the sensitivity of various parameters on the roof behavior. Table 1 summarizes the various analysis cases that were examined.

Table 1. Summary of Analysis Cases.

Case No.	Description	Column Inertia	Long-term Modulus	DL (psf)	LL (psf)
1	Base case.	I_g	No	56	12
2	Considers cracked columns. Added 3 psf due to foam roof material.	$0.5\ I_g$	No	59	0
3	Considers time-dependent effects on concrete modulus.	$0.5\ I_g$	Yes	59	0
4	Determine LL to produce $6\sqrt{f_c^{'}}$ tension fiber stress.	$0.5\ I_g$	Yes	59	Varies 0 to 19
5	Determine effect of 10% loss of prestress due to deterioration.	$0.5\ I_g$	Yes	59	0
6	Determine effect of 25% loss of prestress due to deterioration.	$0.5\ I_g$	Yes	59	0
7	Determine effect of 50% loss of prestress due to deterioration.	$0.5\ I_g$	Yes	59	0

Base Case

This model assumed that the structural elements were uncracked, and therefore, assumed gross section properties.

Column Cracking

This model assumed that the columns supporting the roof were cracked based on an analysis of the cracking moment. Accordingly, the moment-of-inertia was assumed to be 0.5 I_{gross}.

Time-Dependent Effects

The time-dependent effects of creep and shrinkage in concrete were modeled using the ACI Committee 209 effective modulus approach (ACI 209, 1997) as follows:

$$E_{t=\infty} = \frac{E_{t=28days}}{1+\phi}$$

where $E_{t=\infty}$ = *long-term modulus of elasticity*

$$E_{t=28days} = 57000\sqrt{f_c'}$$

and ϕ = *time-factor; this case equal to 1.5*

Field Investigation Results

There were no visual observations made of the roof system that indicated that there has been a loss of integrity. There were observations made that selected columns have experienced some corrosion-related deterioration, but were not impacting the roof's integrity.

Visual Observations

Prestressing Wires.

Figure 7 shows a typical condition of the prestressing wires exposed at each tendon opening. The wires generally appeared to be in good condition with only evidence of minor surface corrosion. The dark color of the wires is discoloration principally from the mastic. This minor surface corrosion did not appear "fresh" as would have been evidenced by a bright orange corrosion product. There were no significant corrosion pits indicative of area loss.

The paper wrap, which acts as the unbonded PT duct, has experienced some deterioration as observed in Figure 7. The bitumen mastic that was used to protect the wires from corrosion was variable in thickness and coverage. However, as long as the roofing material prevents water intrusion, the deterioration of the paper duct and presence variable mastic is not considered overly significant.

Figure 7. Typical condition of prestressing wires.

Coupler.

The condition of the exposed coupler along the strong axis tendon is observed in Figure 8. There was little evidence of corrosion of the metal duct that was split open to observe the coupler, threaded rod or anchors. Also note that the "buttons" of each wire are intact and show no evidence of corrosion.

Figure 8. Condition of exposed coupler.

Anchorage.

Figure 9 shows the representative condition observed at the exposed anchorages of the PT tendons. The field investigation at the selected anchorages clearly indicated that there has been no significant corrosion of the prestressing wires, bearing plate or transfer plates.

Figure 9. Typical condition of exposed anchorage.

Geometry Survey

Based on the differential survey conducted, the roof sag as measured from the edge of the roof to the center was between approximately 17 ft. 4 in. and 17 ft. 8 in. The range is given since the exact thickness of the roof foam at these locations is not known. This measured value was compared to computed values from analysis under dead load only since at the time of the measurements there were no superimposed live loads.

NDT

Impact-echo testing revealed that the precast concrete panels were solid with no signs of corrosion-related deterioration. Infrared thermography indicated that the concrete deck surface was being adequately protected from moisture by the roofing.

Structural Analysis Results

The structural model for the Norick Arena was used to study the structural behavior of the structure under various load combinations and conditions. The implications of these results in the context of selected field measurements are presented. The theoretical sag at the middle is taken as 17 ft. 3 in. as determined from the original structural drawings. All deflection results are presented from this benchmark sag. Table 2 summarizes the results from the various analyses.

Case 1

This base model was considered in order to understand the expected behavior of the roof structure from the original design. This model considered a dead load of 56 psf, live load of 12 psf and the post-tensioning as fully effective. Gross section properties were considered for the column sections.

The analysis shows that the maximum downward deflection at the middle of the roof is about 2.1 in. below the benchmark sag. It was found that the columns developed bending moments in excess of the cracking moment.

Table 2. Summary of Analysis Results.

Case No.	Column Inertia	Long-term Modulus	DL (psf)	LL (psf)	Additional PT Losses from $f_{ps} = 0.6\ f_{pu}$ (%)	Center Deflection from Theoretical Sag (17 ft. 3 in.), Positive Downward (in)
1	I_g	No	56	12	0	2.1
2	0.5 I_g	No	59	0	0	1.3
3	0.5 I_g	Yes	59	0	0	3.2
4	0.5 I_g	Yes	59	Varies 0 to 19	0	Varies 3.2 to 7.5
5	0.5 I_g	Yes	59	0	10	4.2
6	0.5 I_g	Yes	59	0	25	5.6
7	0.5 I_g	Yes	59	0	50	8.1

Case 2

This case closely reflects Case 1, except that the column stiffness was reduced to 50% of that based on gross section properties to account for cracking. The field observations presented previously confirmed the presence of flexural cracks. Also, the dead load was increased to reflect the weight of the foam roof. The live load was removed to reflect conditions when field measurements were taken.

The field measurements indicate that the present roof sag is in the range of about 17 ft. 4 in. to 17 ft. 8 in. This is an increase in sag of 1 in. to 5 in. The computed value of Case 2 is in between these values.

Case 3

As part of the various analyses of the structure, this model considered long-term effects. This model is based on Case 2 with the addition of long-term effects. These effects are modeled by reducing the elastic modulus of the structural members by a factor corresponding to the long-term conditions expected on the structure. In this case, a factor of 0.4 was computed, indicating that for long–term conditions, the effective elastic modulus is about 40% of its original value. The structural analysis yields an increase in downward deflection of 3.2 in. The sag in the center of the roof increases by only about 2 in. considering time-dependent effects. The computed deflection is virtually in between the range of measured values. This strongly suggests that there has been no measurable loss in prestressing of the post-tensioning tendons.

Case 4

The ACI 318 Building Code (ACI 318, 2005) requires that a post-tensioned structure limit tension stress to $6\sqrt{f_c'}$ under full service loads. Case 4 was conducted to determine the level of live load acting on the structure needed to exceed the limiting tension stress. This value would be one possible index in determining the current load rating of the roof. Analyses showed this value to be about 19 psf. Under this loading and under similar conditions as in Case 3, the maximum downward deflection increases by 7.5 in. from the benchmark sag.

Cases 5 to 7

These analyses consider additional losses in the post-tensioning tendons due to some form of deterioration such as corrosion or other factors that might result in a partial loss of prestressing. These additional losses considered were 10%, 25% and 50%, respectively. These losses were applied to the same analysis model and loads of Case 3. The results are summarized in Figure 10. This graph shows that considering 0% losses in the post-tensioning, the deflection in the middle of the roof is about 3.2 in., whereas, 50% losses in the effective post-tensioning stress results in a deflection of 8.1 in. The analysis indicates that the center deflection of the roof is not overly sensitive to assumed losses in the post-tensioning from some form of deterioration. The increase in roof deflection is linear with the percentage of prestressing losses.

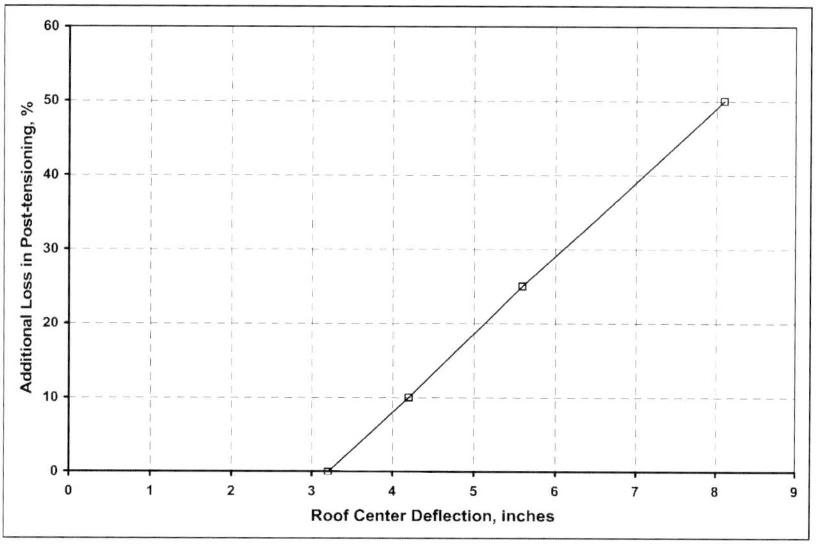

Figure 10. Deflection at middle of roof compared to assumed additional losses in post-tensioning.

Collapse Conditions

As a perspective of the conditions that would result in the "collapse" of the roof under its existing weight, a collapse analysis indicates that, on average, 65% of the post-tensioning would have to be lost from the effects of corrosion or as a result of some type of catastrophic event. The resulting increase in sag would be about 10 in.

Conclusions

A field evaluation and structural assessment was conducted for the post-tensioned concrete roof of the Norick Arena, which was designed by T.Y. Lin in 1965. There were no signs of significant deterioration that would indicate that the structural roof system has been compromised in its 40-year life thus far. It was determined that the roof will continue to safely carry the intended code-required loads as long as the roofing material is adequately maintained to mitigate ingress of water and as long as the roof drains are maintained.

The analysis indicates that the post-tensioning system was designed to balance a total load of about 68 psf. The present dead load of the roof is approximately 59 psf. Thus, the allowable additional live load from purely a load-balancing perspective is 9 psf. This exceeds the current snow load of 7 psf.

References

ACI Committee 209 (1997), "ACI 209R-92: Prediction of Creep, Shrinkage and Temperature Effects in Concrete", American Concrete Institute, Detroit, MI, 47 pp.

ACI Committee 318 (2005), "ACI 318-05: Building Code Requirements for Structural Concrete and ACI 318R-05: Commentary", American Concrete Institute, Detroit, MI, 430 pp.

ASCE (2002), "Minimum Design Loads for Buildings and Other Structures", American Society of Civil Engineers, Reston, VA, 376 pp.

Lin, T.Y. and Burns, N.H. (1981), "Design of Prestressed Concrete Structures", Third Edition, John Wiley & Sons, New York, NY, 646 pp.

STORY BASED DUCTILITY MODELS FOR DISPLACEMENT BASED DESIGN OF STEEL FRAMES

Mohammad Safi
Department of Civil Engineering, Power & Water University of Technology, Tehran, P.O. Box 16765-1719, Iran,
E-mail: msafi@pwit.ac.ir

Abstract

Estimation of ductility demand distribution through the height of the structure is a very hard task for seismic design engineers working on performance based design of buildings. In this paper a modified direct displacement based design procedure has been proposed. In this method the design force distribution among the height of the structure is obtained based on various ductility demand distributions derived from modal characteristics of the structure and mathematical formulations. The method has been applied to the moment steel frames in low, medium and high rise buildings and the results of various ductility distributions have been compared. The plastic mechanism has also been modeled and the efficiencies and deficiencies of each have been discussed through various numerical examples. The effect of yield mechanisms and ductility demand patterns for various building types on the equivalent SDOF parameters have been investigated compared to the time history analysis results to find the sensitive parameters.

Introduction

The purpose of Performance Based Design is to design the structure with sufficient and proportioned stiffness and strength in the structural members so as to develop inelastic action in the ductile designed members and to have appropriate over strength in the brittle members. Then the structure must be checked so that the demands do not exceed the existing capacities. This is best performed using a set of nonlinear dynamic analyses under earthquake with appropriate characters. Different design methods have been proposed based on performance criteria such as, Capacity Spectrum Method [1,2,3], N2 Method [4,5], Energy Based Methods and Displacement Based Design (DBD) Methods. In the last four decades the idea of DBD has been introduced and developed by different researchers started by introducing the concept of substitute structure [6]. This idea has been adopted for a direct displacement design of SDOF and MDOF reinforced concrete bridges [7,8,9]. Capacity Spectrum Method and the N2 Method have also been used to create other direct DBD procedures [1,2,3,4,9,10]. In all of these researches, seismic demand is specified as either a displacement spectrum or an acceleration-displacement response spectrum. Generally nonlinear inelastic behavior of a structural system can be accounted for either by an equivalent elastic response spectrum or an inelastic response spectrum. The former is associated with effective viscous damping and the latter is directly constructed based on relations between reduction factors and ductility.

In this paper the direct DBD method is briefly reviewed for multi story steel buildings. In this method the design force distribution among the height of the structure is obtained based on various ductility demand distributions derived from modal characteristics of the structure and mathematical formulations. The method has been applied to the steel braced frames with concentric and eccentric bracing systems in low, medium and high rise buildings. The plastic mechanism for each system has also been modeled and the efficiencies and deficiencies of each have been discussed through various numerical examples. The effect of yield mechanisms and ductility demand patterns for various

M. Pandey et al. (eds), Advances in Engineering Structures, Mechanics & Construction, 355–365.

building types on the equivalent SDOF parameters have been investigated compared to the time history analysis results to find the sensitive parameters. A design displacement spectrum has also been created for the parametric study based on the Iran earthquakes. Various factors affecting the dynamic response have also been provided in the procedure. It has been shown that this method is capable of predicting the response of braced frames especially high rise buildings in an efficient and robust manner.

Displacement Based Design of Steel Frames

Direct Displacement Based Design of multi story buildings is based on the generation of equivalent SDOF system or substitute structure concept. For this purpose, it is assumed that the structure vibrates in a pre-defined harmonic displaced shape. The base shears and the works developed by lateral external forces are also assumed the same for both equivalent and main structures [7-15]. Consider the relative displacement vector $\{\delta(h,t)\}$ for the multistory building with total height of H expressed in a decomposed form of displacement and time and assume a harmonic response with amplitude Δ for the system. We can write,

$$\{\delta(h,t)\} = \Delta.Sin(\omega.t).\{\Phi(h)\}\ ,\ 0 \le h \le H \tag{1}$$

which results in an acceleration vector $\{a(h,t)\}$ proportional to the assumed normalized displacement vector $\Phi(h)$ as follows,

$$\{a(h,t)\} = -\Delta.\omega^2.Sin(\omega.t).\{\Phi(h)\} = -\omega^2.\{\delta(h,t)\} \tag{2}$$

In order to obtain the equivalent system parameters, we define the normalized displacement vector $\{c(h,t)\}$ as,

$$\{c(h,t)\} = \frac{1}{\delta_{eff}}\{\delta(h,t)\} \tag{3}$$

where δ_{eff} is called the effective displacement. From equation 2 and 3 we may have,

$$c_i(h,t) = \frac{\delta_i(h,t)}{\delta_{eff}} = \frac{a_i(h,t)}{a_{eff}}, i = 1,2,\ldots,n \tag{4}$$

in which n stands for number of stories, a_{eff} is called the effective acceleration of the equivalent SDOF system and δ_i , a_i are the story displacement and acceleration respectively. Using equation 4, the base shear can now be determined in terms of the multi story structure and the equivalent system parameters as,

$$V_b = \sum_{i=1}^{n} f_i = \sum_{i=1}^{n} m_i.a_i = \left\{\sum_{i=1}^{n} m_i.c_i\right\}.a_{eff} = m_{eff}.a_{eff} \tag{5}$$

which leads to the definition for the effective mass as $m_{eff} = \sum_{i=1}^{n} m_i.c_i$. The lateral force at each level, f_i may also be determined using equations 4 and 5 as,

$$f_i = \frac{m_i.\delta_i}{\sum_{j=1}^{n} m_j.\delta_j} V_b \tag{6}$$

Equating the external works for the two systems, $V_b.\delta_{eff} = \sum_{i=1}^{n} f_i.\delta_i$ and using equation 6, we can obtain the definition for effective displacement as,

$$\delta_{eff} = \frac{\sum_{i=1}^{n} m_i.\delta_i^2}{\sum_{i=1}^{n} m_i.\delta_i} \tag{7}$$

The effective stiffness of the substitute SDOF system may also be obtained by entering the effective displacement into the displacement response spectrum with appropriate damping value and then substituting the obtained effective period and effective mass from equation 5 into the following equation,

$$K_{eff} = \frac{4\pi^2 m_{eff}}{T_{eff}^2} \tag{8}$$

The effect of story ductility may be considered substituting δ_i defined in equation 4 with the following relationship,

$$\delta_i = \mu_i.\delta_{yi} \tag{9}$$

where μ_i is the story ductility demand and δ_{yi} is the story yield displacement. The problem is now how to determine these two parameters. The story yield displacement δ_{yi} may be obtained by defining the story yield mechanism and has been discussed later in section 3. Determination of the ductility distribution through the height of the structure has also been discussed in section 3. Finally for detail design of the structure the base shear is obtained as $V_b = K_{eff}.\delta_{eff}$ and then the story forces f_i are computed using equation 6. Then the capacity design of the structure can be started considering the ductility capacities. This capacity-designed structure may then be verified using the time history or static push over analyses.

Design Displacement Spectrum

In this paper, a displacement response spectrum has been obtained through a deterministic procedure based on acceleration data for Iran earthquakes. These accelerograms were selected form more than 2000 records for different stations and earthquakes in Iran. The near field records were omitted and the accelerograms with medium to high magnitude (minimum 5 degrees in Richter scale) were selected. Using an Artificial Neural Network simulator (a committee neural simulator including competitive and back error propagating networks), prepared by the authors [17], the records were categorized according to their shapes (duration, sequence of peaks and their amplitude) by the competitive network to four categories. Each category represented a soil type thus the design displacement spectra for each soil type was obtained. The spectrum with 5 percent damping for soil type C (or II according to Iranian seismic code) has been presented in Fig. 1. A four-degree polynomial function has been matched to the data with a 0.98 standard deviation as shown in the figure. This equation was used to calculate the effective displacement in the numerical examples.

Definition of Effective Damping

Assuming a single displacement cycle based on the ultimate displacement the following well-known relationship between ξ_{eff} and the ductility demand μ for Elastic-Perfectly Plastic (EPP) behavior is

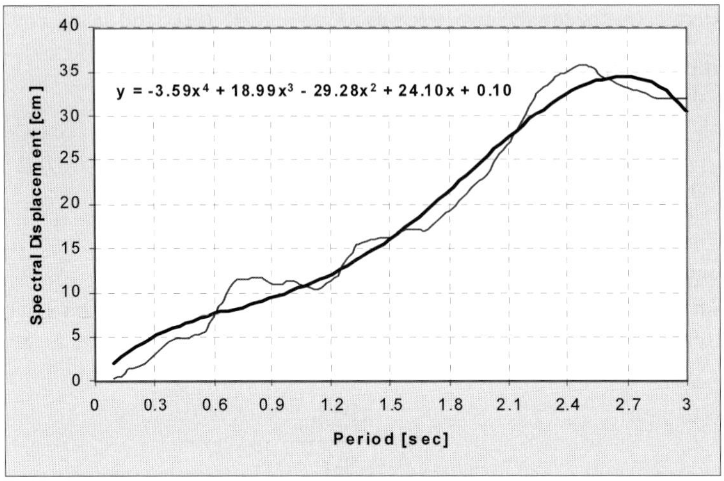

Figure 1: The displacement spectrum based on filtered Iran earthquakes and the design curve

obtained (Equivalent Energy Method [16]),

$$\xi_{eff} = \frac{2}{\pi}\left(1 - \frac{1}{\mu}\right) + \xi_{elastic} \tag{10}$$

where $\xi_{elastic}$ stands for the damping of the elastic structure. The equivalent viscous damping for bilinear systems with strain hardening ratio α and ductility μ may also be determined using the following equation [5],

$$\xi_{eff} = \frac{2}{\pi}\left(\frac{(1-\alpha).(\mu-1)}{\mu - \alpha\mu + \alpha\mu^2}\right) + \xi_{elastic} \tag{11}$$

The effective damping obtained above which is greater than the elastic viscous damping due to the hysteretic behavior is then used to get the effective period from the displacement spectrum. The presented spectrum shown in fig. 1 may be modified for other damping values using the EC8 [18] factor $\sqrt{7/(2+\xi)}$. Greater value of damping results in a greater effective period in the displacement spectrum and thus less base shear for the design.

Ductility Demand Models

In recent years many researchers have tried to find engineering procedures to do this task but each had some limitations. Studies have shown that Push Over Analysis is limited to low to medium rise buildings and is not recommended for high rise or flexible structures in which higher mode effects are not negligible and may govern the dynamic response [19]. Other existing methods such as the Capacity Spectrum Method and the N2 Method have similar limitations. In the first part of this section, simple plastic models for the story or local ductility of braced steel building with concentric and eccentric bracing systems have been proposed. In the second part, various functions for ductility demand distribution over the height of the structure have been introduced and then they have been verified by comparing to dynamic analysis results through various numerical examples.

Story Plastic Mechanism Models
In the presented DBD method, the story mechanisms are presumed to occur due to the yielding of the lateral resisting system. Therefore, the failure mechanism of the stories that depends on the structural geometry, yield strains and lateral load resisting system may be determined. In the case of a capacity-designed structure, the lateral mechanism can be predicted with good approximation. For steel braced frames considered in this study, the mechanisms are assumed to form by yielding of braces in concentric systems and by flexural yielding of link beams in eccentric systems. In these models one column lift is assumed for each story, which means that no rotational column plastic hinge is produced.

For concentric X bracing systems as shown in fig. 2-a the lateral story displacement δ_i can be written as a function of story shear V_i, brace span L_s, brace length L_b, brace sectional area A_b and elastic section modulus E as,

$$\delta_i = \frac{V_i.L_b^3}{E.A_b.L_s^2} \tag{12}$$

Using equation 12 and considering the relation $L_s / L_b = V_{yi} / \sigma_y.A_b$ from figure 2, the story yield displacement δ_{yi} in a concentric braced frame can obtained as,

$$\delta_{yi} = \frac{V_{yi}.L_b^3.\varepsilon_y}{\sigma_y.A_b.L_s^2} = \frac{L_b^2.\varepsilon_y}{L_s} \tag{13}$$

which does not depend on the brace sectional area. Similarly, for the Chevron bracing shown in fig. 2-b, the same equations may be obtained as follows,

$$\delta_i = \frac{2V_i.L_b^3}{E.A_b.L_s^2} \tag{14}$$

$$\delta_{yi} = \frac{2V_{yi}.L_b^3.\varepsilon_y}{f_y.A_b.L_s^2} = \frac{2L_b^2.\varepsilon_y}{L_s} \tag{15}$$

It is again clear that for Chevron bracing the displacement shape in yielding case of the structure is independent of the brace sectional properties. Thus for the concentric systems, an equal span and story height result in a linear deformed shape of the structure over the height at yield condition.

For eccentric bracing systems (fig. 2-c), the lateral displacement depends on plastic rotation capacity of the link beam, θ_p, and can be expressed through the following equations,

$$\delta_{yi} = H_i.Sin(\theta) - l_l.(1 - Cos(\theta_p - \theta)) \tag{17}$$

$$\theta = \pi - \beta - ACos\left[\frac{L_s^2 - l_l^2 + 2l_l.L_s.Sin(\beta - \theta_p)}{2L_s.L_b}\right] \tag{18}$$

where θ_p is defined as a function of section plastic moment M_p and plastic stiffness K_p as,

$$\theta_p = \frac{M_p}{K_p} = \frac{V_i.H_i}{2K_p} \tag{19}$$

For small length link beams, the shear mechanisms are formed while in moderate length link beams moment plastic hinges are generally formed. The latter case has been considered here. The value of θ_p is generally related to the connection and link beam details.

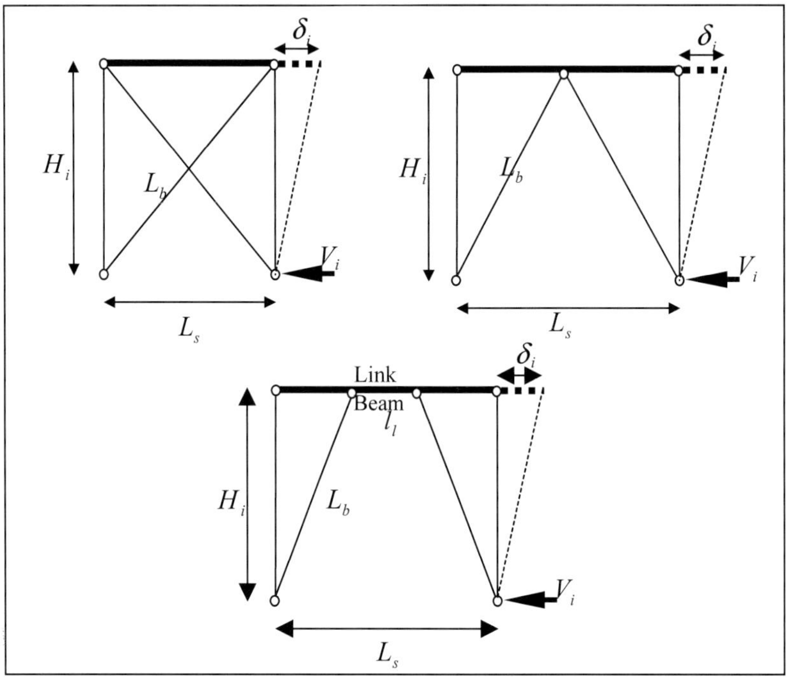

Figure 2: Structural models for (a) X bracing, (b) chevron bracing and (c) eccentric bracing systems

Ductility Demand Distribution Patterns

In the conventional DBD in which the ductility is assumed uniform over the height, the effective stiffness would not change with the ductility. If the ductility is distributed according to the ductile design of braces over the height for example based on the elastic modal vibration of the structure, the brace characteristics will interfere with the lateral displacement. The increase in the resulted ductility in comparison with the uniform distribution, will cause reduction in the effective mass and results in an increase in the effective period due to the rise in effective displacement and thus reduce the resulted effective stiffness of the substitute SDOF structure. In the presented DBD method, the lateral displaced shape of the structure is modified using multi-modal, polynomial and exponential distributions of ductility over the height of the structure in order to take into account higher mode effects and combined shear and flexural lateral deformations. Higher mode effect cause considerable changes in the dynamic response of large period or flexible structures such as tall buildings. Besides, the ductile behavior of the building also results in considerable increases in the system period. This issue has been discussed in the parametric study. In low rise buildings the shear behavior often governs the response and in medium rise buildings a combined shear and flexural deformation is normally expected.

For modal distribution of ductility it is assumed that the distribution is approximately conformed to the some first mode shapes which have a more than 98 percents of the system mass. For most of the structures, the first three modes of a cantilever with known equations may be assumed or alternatively an elastic modal analysis of the structure can be performed and then a mode combination procedure followed. However in order to avoid high strain demand in members in the structural design of the braced or wall buildings, the participation of higher modes must be limited. This issue is generally considered in the capacity design of the structure so that the mass portion of the first mode does not decrease to values less than 70 percents or alternatively reduce the number of modes that own the 98 percents of the system mass. In the presented study, as an alternative approach and in order to conform to the capacity design criteria, the effect of higher modes on the ductility demand distribution in the stories below the effective height have been neglected,

$$H_f = \frac{\sum_{i=1}^{n} f_i . H_i}{V_b} \tag{20}$$

The second selected pattern is an exponential function with parameter a as follows,

$$\phi_{\mu[EXP]} = \mu_{\max} \cdot \frac{1 - EXP(-a\, h/H)}{1 - EXP(-a)} \tag{21}$$

The third function is a simple polynomial function with parameter b as,

$$\phi_{\mu[P]} = \mu_{\max} \cdot \left(\frac{h}{H} \right)^b \tag{22}$$

The comparison of various ductility patterns using the mentioned functions has been plotted in fig. 3. The final displaced shape is obtained by multiplying the initial profile obtained from the mechanism models, introduced in the previous section, by ductility demand distribution. The maximum ductility capacity of the each story (defined as the ratio of maximum displacement capacity and yield displacement of the story) must also be determined. This parameter defines the limit state or the performance point. For concentric and chevron braced frames this is governed by the capacity of the brace to beam connection and for eccentric braces the plastic rotation capacity of the link beam defines the ductility. Some modifications for the effects of strain hardening and cumulative damage can also be assessed in the DBD method that has been discussed in the next section.

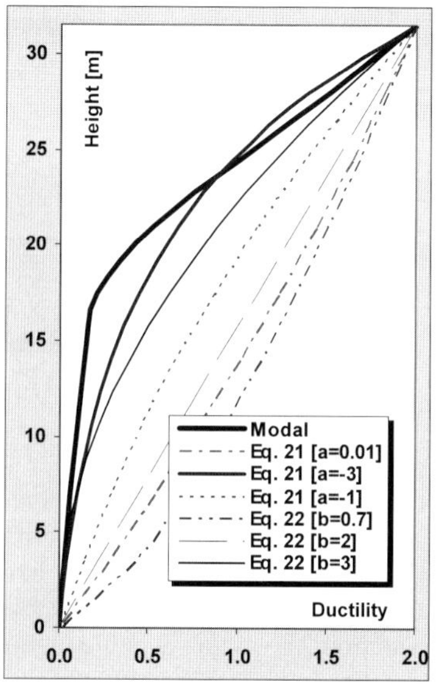

Figure 3: Various ductility distribution patterns considered in the study

Numerical Analyses

In this section, the presented modifications have been verified through some numerical examples including three, nine and twenty story braced frames. The geometric data of the models are presented in table 1 and fig. 5. The steel properties are, yield stress $\sigma_y = 245MPa$, initial elastic modulus $E = 2.10e5MPa$ with a bilinear nonlinear behavior with five percent strain hardening or second modulus $E_s = \alpha.E, \alpha = 0.05$. The dead load is assumed to be 3.9 kPa. and the reduced live load 1.4 kPa at floor levels. At roof level these values are assumed to be 3.2 kPa. and 1.0 kPa. respectively. The assumed data may be sufficient for DBD, but for the nonlinear push over and time history analyses the detail design of the members must also be available. This has been performed using the capacity design procedures and was performed using SAP2000 [21] commercial program. The modified strength reduction factors for MDOF structures taking into account the reductions due to structural over strength have been calculated based on the equations proposed in [23]. The effects of cumulative damage may also be considered using the idea presented in [5] that has been discussed subsequently. Nonlinear dynamic analyses have been performed using DRAIN2DX [22] program using three selected earthquake record which were compatible with the obtained response spectrum shown in fig. 1.

Table 1: Geometric data for numerical examples (Dimensions in meter)

n story	H_1	$(n-1)*H_i$	L_{s1}	$No*L_{span1}$	L_{s2}	$No*L_{span2}$	l_l
3 story	3.5	2*3.5	2(4.0)	2*4	2(3.5)	2*4	1.5
9 story	4	8*3.5	2(4.2)	2*4.2	2(3.5)	2*4	1.5
20 story	4	19*3.5	2(4.2)	3*4.2	2(3.5)	3*4	1.5

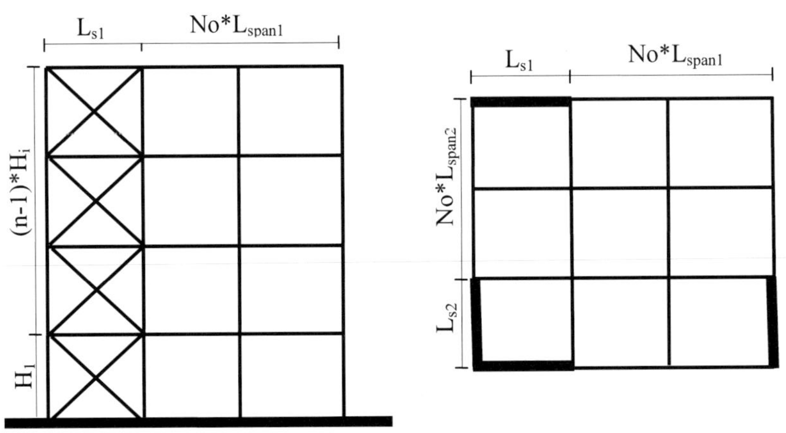

Figure 5: Geometric data for all numerical examples

Figure 6 shows the effect of the ductility demand distribution on story force and story drift using modal ductility distribution and equations 21 and 22 for a nine story building with eccentric bracing. The maximum ductility is assumed to be 2 according to fig. 8. It has been shown that the lateral story force has not been so sensitive to the ductility pattern but the displacement and drift directly change with the ductility in the DBD and equation 21 with a=-3 and equation 22 with b=3 may be acceptable comparing to the dynamic analysis results. In table 2 the effect of ductility pattern on DBD parameters (Effective parameters on the equivalent SDOF structure) has also been presented. As shown the effective damping, total base shear, mass and effective height ratio which may be assumed as the representative for lateral force distribution are not sensitive to the ductility distribution over height and just depend on the maximum ductility value. In figure 7 the period of the first mode from eigen value analysis has been compared to the periods obtained from linear and nonlinear time history analyses showing the effects of vertical loads and nonlinear behavior and also the effective periods from DBD method taking into account the effect of column deformation for the last story based ductility model.

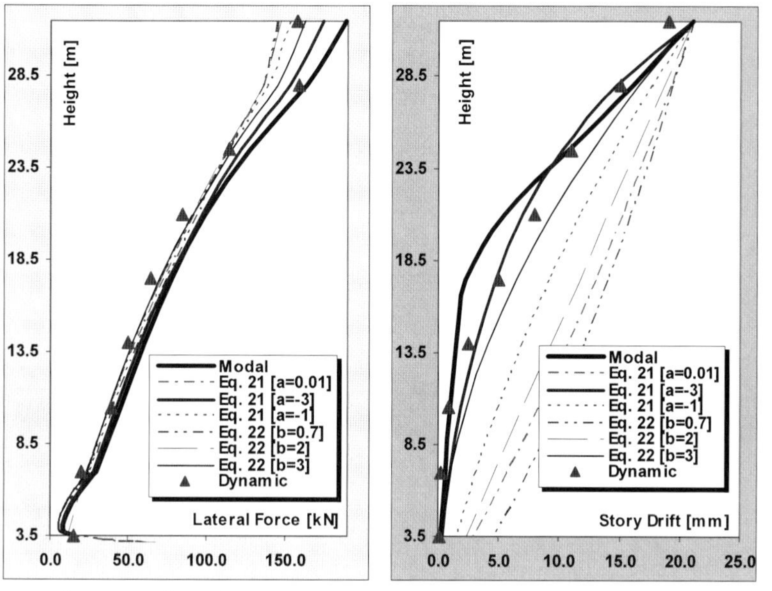

Figure 6: Effect of ductility demand distribution on story force and drift

Table 2: Effect of ductility demand distribution on DBD parameters

Effective Parameters	Units	Modal Pattern	Exponential (Equation 21)			Polynomial (Equation 22)		
			a=0.01	a=-3	a=-1	b=0.7	b=2	b=3
Effective Period	sec	0.85	1.30	0.92	1.14	1.35	1.25	1.01
Effective Damping	%	4.36	4.79	4.42	4.65	4.81	4.74	4.53
Mass Ratio	_	0.71	0.73	0.71	0.72	0.74	0.72	0.70
Effective Stiffness	N/mm	13601	5986	11783	7622	5658	6427	9550
Effective Displacement	mm	92	134	97	116	140	128	104
Effective Height Ratio	_	0.73	0.73	0.73	0.73	0.72	0.73	0.74
Total Base Shear	kN	798	674	756	683	685	674	705

Concluding Remarks

Displacement based procedures can directly lead the designer to the key design parameters such as interstory drifts and displacements. If the ductility is distributed according to the ductile design of braces over the height for example based on the elastic modal vibration of the structure, the brace characteristics will interfere with the lateral displacement. The results obtained from these studies have been summarized as follows,

- The modal ductility demand distribution and the resulted force distributions give acceptable results for tall buildings compared to the dynamic analyses. However in order to avoid high strain demand in members in the structural design of the braced buildings, the participation of higher modes must be limited. This issue is generally considered in the capacity design of the structure so that the mass portion of the first mode does not decrease to values less than 70 percents or alternatively reduce the number of modes that own the 98 percents of the system mass.

- As expected the ductility demand in eccentric systems had the highest values having other specifications constant. This ductility was highly dependent to the plastic rotations in the mid length link beams. It has also been shown that the lateral story force has not been so sensitive to the ductility pattern but the displacement and drift directly change with the ductility. The effective damping, total base shear, mass and effective height ratio which may be assumed as the representative for lateral force distribution are not sensitive to the ductility distribution over height and just depend on the maximum ductility value.
- As the relationship between member strains or member plastic rotations with the inter-story drifts are determined, both local and global performance criteria may be used for such a design method.

As shown by various examples, various effects such as column deformation effects, torsional effects, higher mode effects, $P-\Delta$ effects and low cycle fatigue effects may easily be assessed by the use of equivalent procedures in the direct DBD.

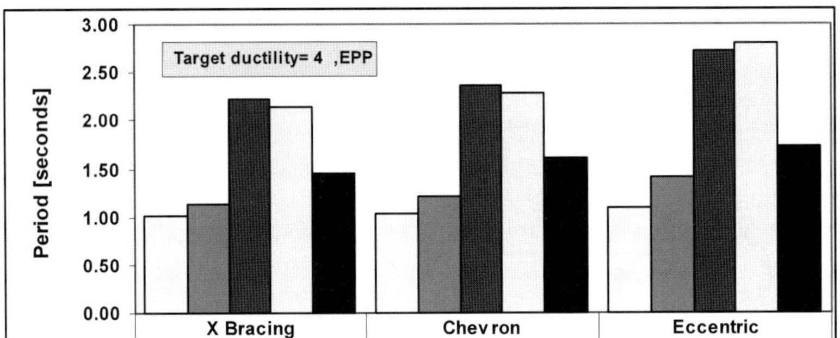

	X Bracing	Chevron	Eccentric
EVA	1.02	1.04	1.09
THA-L	1.14	1.21	1.41
THA-NL	2.22	2.37	2.72
DBD-CDC	2.15	2.29	2.81
DBD-CDN	1.45	1.61	1.73

Figure 7: Effect of column deformation, vertical load and nonlinearity on natural and effective periods for (a) nine and (b) twenty story buildings [Symbols: EVA: Eigen value analysis, THA: Time history analysis, L: Linear, NL: Nonlinear, CDC: Column deformation considered, CDN: Column deformation neglected]

Acknowledgment

The supports by the Moshanir Power Engineering Consultant Company are gratefully acknowledged.

References

Fajfar P. Capacity spectrum method based on inelastic demand spectra. Earthquake Engrg. Struct. Dynamics, pp Vol. 28, No. 9, pp 79–93, 1999.

Chopra AK, Goel RK., Capacity-demand-diagram methods for estimating seismic deformation of inelastic structures: SDOF systems, Report No. PEER-1999-02, Univ. of California, Berkeley, 1999.

Freeman SA., Development and use of capacity spectrum method, Proceedings of the 6th US National Conf. on Earthquake Engrg., Seattle, EERI, Oakland, California, 1998.

Fajfar P., A nonlinear analysis method for performance based seismic design, Earthquake spectra, Vol. 16, No. 3, pp 573-592, 2000.

Fajfar P., Equivalent ductility factors taking into account low-cycle fatigue, Earthquake Eng. Struct. Dynamics, Vol. 21, No. 9, pp 837–848, 1992

Shibata A, Sozen MA., Substitute-structure method for seismic design in RCs, Journal of the Structural Division ASCE, Vol. 102, pp 1-18, 1976.

Kowalsky MJ, Priestley MJN, MacRae GA., DBD, a methodology for seismic design applied to SDOF RC structures, Report No. SSRP-94/16, Univ. of California, San Diego, La Jolla, California, 1994.

Calvi GM, Kingsley GR., Displacement-based seismic design of MDOF bridge structures, Earthquake Engrg. Struct. Dynamics, Vol. 24, No. 9, pp1247–1266, 1995.

Qi X, Moehle JP., Displacement design approach for RC structures subjected to earthquakes, Report No. UCB/EERC-91/02. Berkeley, University of California, 1991.

Panagiotakos TB, Fardis MN., Deformation-controlled earthquake-resistant design of RC buildings, J. Earthquake Engrg., Vol. 3, No. 4, pp 495–518, 1999.

Priestly MJN, Kowalsky MJ, Ranzo G, Benzoni G., Preliminary development of direct DBD for MDOF systems, Proc. of the 65th Annual Convention, SEAOC, Maui, Hawaii, 1996.

Calvi G.M. & Pavese A., DBD of building structures, European seismic design practice, Belkema, Rotterdam, 1995.

Medhekar M.S. & Kennedy D.J.L., Displacement seismic design of buildings- Theory, Int. Jour. of Engrg. Structures, Vol.22, pp201-209, 2000.

Medhekar M.S. & Kennedy D.J.L., Displacement seismic design of buildings- Application, Int. Jour. of Engrg. Structures, Vol.22, pp210-221, 2000.

Chandler A.M. & Mendis P.A., Performance of RC frames using force and displacement seismic assessment methods, Int. Jour. of Engng. Structures, Vol.22, 2000.

Chandler AM, Tsangaris M., Lam NTK, Wilson JL, Edwards M., Hutchinson GL., Seismic performance of RC structures using displacement based principles, 11ECEE, Balkema, Rotterdam, 1998.

M.Tehranizadeh, M.Safi and S.A. Alavinasab, Calculation of Design Spectra for Iran Using Intelligent Classification of Earthquake Records, Submitted to JSEE.

European Seismic Standard, EC8, 1994.

Krawinkler H, Seneviratna GDPK., Pros and cons of a Push Over Analysis of seismic performance evaluation, Engrg. Struct., Vol.20, No.(4-6), pp452–464, 1998.

Paz M., Dynamic of structures,Mc Graw Hill inc., 1996.

SAP2000, Structural Analysis Program, Computers & Structures Inc., California, USA, 1999.

Kannan AE, Powell GH., Drain-2D: a general-purpose computer program for dynamic analysis of inelastic plane structures, Report no. EERC 73-6, Univ. of California, Berkeley, California, 1973.

Miranda E., Bertero V.V., Evaluation of strength reduction factors, Earthquake Spectra, Vol. 10, No. 2, pp 357-379, 1994.

Chandler AM, Lam NTK, PBD in earthquake engineering: a multidisciplinary review, Engrg. Struct., Vol.23, pp1525–1543, 2001.

Rosenblueth E, Herrera I., On a kind of hysteretic damping, Journal of Engineering Mechanics Division ASCE, Vol.90, pp37– 48, 1964.

Paulay T., A simple displacement compatibility based seismic design strategy for RC buildings, 12WCEE, New Zealand, 2000.

Paulay T., DBD approach to earthquake induced torsion in ductile buildings, Eng. Struct., Vol.19, No.9, pp699–707, 1997.

Prestandard for Seismic Rehabilitation of Buildings, Federal Emergency Management Agency, FEMA356, 2000.

"PLASTIQUE" - A COMPUTER PROGRAM FOR 3D INELASTIC ANALYSIS OF MULTI-STOREY BUILDINGS

Vlasis K. Koumousis, Eleni N. Chatzi and **Savvas P. Triantafillou**

Institute of Structural Analysis and Aseismic Research
National Technical University of Athens
NTUA, Zografou Campus GR-15773, Athens, Greece
E-mail: vkoum@central.ntua.gr

Abstract

The code "Plastique", suitable for the inelastic dynamic analysis of R/C structures and its theoretical background are presented. Every structural entity is represented by a single nonlinear element through the implementation of a macro modeling approach. Three different types of 2D-macro elements are formulated namely; beams, columns and shear walls. The structural behavior of each element is evaluated using a flexibility formulation based on both element edge regions that follow a distributed plasticity law. A fiber model is used to define the monotonic strength envelope at each section. The hysteretic behavior of the structural elements is monitored by a smooth hysteretic model of Bouc-Wen type. This model is capable to express the stiffness degradation, strength deterioration and pinching phenomena which are observed in R/C elements under cyclic loading. Plane frames consisting of combinations of plane elements are linked at the levels of floors via diaphragms to assemble the 3D mathematical model of the structure. Solutions are obtained by direct integration of the equations of motion, while an iterative procedure is implemented to satisfy equilibrium at every time step. Finally, a damage analysis is performed using an appropriate damage model. The numerical examples presented herein the reveal the features of the proposed analysis scheme.

Keywords: inelastic analysis, time history, Bouc-Wen, damage index

1 Introduction

A problem of major importance in structural engineering deals with the response of R/C structures subjected to dynamic loading. For load factored linear elastic analysis, suggested by the codes, the results are quite satisfactory, but do not reveal the characteristics of the true behaviour of the structure. However, if inelastic response is taken into account, more refined models are needed as to achieve a realistic behaviour. In recent years, significant research has been carried out in order to overcome the difficulties arising in such an analysis. Difficulties emanate not only from the inherent complexity of R/C structures, but also from the uncertainties related to terms such as dynamical loading, material nonlinearity and hysteresis.

Macro-modeling of structures has been one of the main methods introduced to simulate these complex phenomena. In macro-modeling simulation, the field of knowledge concerning the actual behaviour of reinforced concrete is incorporated in the structure using an element–based approach. In such a way, the well established, from matrix structural analysis, beam element is enriched with a moment curvature envelope describing the behaviour of both end sections, a hysteretic law and a relevant yield penetration rule for the beam. By introducing such an elasto-plastic element, one is able to simulate the gradual shift of the mechanical properties of the element as it passes from the elastic to the inelastic region of its response. The overall behaviour of the structure is assessed using a proper damage index.

M. Pandey et al. (eds), Advances in Engineering Structures, Mechanics & Construction, 367–378.

An aspect of utmost importance, for a non linear analysis, is the hysteretic rule needed to model the cyclic response of the structure. Over the last twenty years, significant development has occurred in the so-called phenomenological approach of hysteresis. Beginning with Bouc's original formulation (1967, 1969, 1971) of the single degree degrading hysteresis model with pinching, many modifications have been subsequently introduced, such as the Bouc-Wen model (1976, 1980), the Baber-Noori model (1985, 1986) and the Reinhorn model (1996). These hysteresis models –also known as smooth hysteretic models- are capable of simulating a number of different types of loops using a single smooth hysteretic function affected by a set of user-defined parameters. In doing so, one can easily model the three main phenomena describing the cyclic response of R/C elements namely; stiffness degradation, strength deterioration and pinching behaviour due to bond-slip effects.

Following these rules, many computer programs have been developed, capable to perform a non-linear structural analysis such as DRAIN-2D (Kanaan and Powell,1973), SARCF (Chung et al.,1998; Gomez et al.,1990), IDARC (Park et al., 1978;Kunnath et al., 1992) and ANSR (Oughourlian and Powell,1982). The "*Plastique*" code presented herein, although maintains the elastoplastic behaviour within the 2D plane frames, works with a 3D stiffness of the entire structure based on diaphragmatic action.

2 Material Properties

Material properties are defined through certain conventional stress-strain curves both for unconfined concrete and reinforcing steel. In the former a parabolic stress-strain relationship with a softening branch is used, while in the latter a bilinear stress-strain diagram with hardening is implemented. The aforementioned stress – strain curves are depicted in Figure 1.

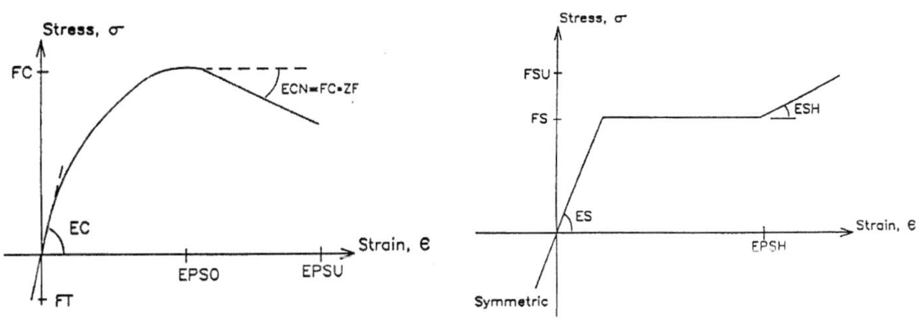

Figure 1. Stress –Strain diagrams a) for unconfined concrete and b) for reinforcing steel

3 Element Modeling

Three different types of two dimensional structural elements namely; beams, columns and shear walls are adequate to model the response of multi-storey buildings. By combining such elements one assembles multi-storey plane frames, which are linked together through diaphragms at the floor levels to create a 3D model of the structure.

3.1 Beam Element

Beam elements are considered as flexural elements with shear deformations with no axial deformations as beams belong to inextensional diaphragms and rigid zones at the ends to account for the stiffness increase at the joint, if needed. The element stiffness matrix varies throughout the analysis due to plasticity effects. In order to simulate such effects a hysteretic law and a spread plasticity model are introduced. The hysteretic model is formulated based on an initial moment-curvature relationship which represents the backbone skeleton curve. Such skeleton curves must be defined for each edge section of all elements. These curves can be either user defined or can be computed using a fiber model and certain given properties of the section under consideration such as geometry, concrete and reinforcement properties.

3.2 Column Element

The column element formulation is identical to that of a beam, as columns are treated as 2D elements that participate separately to intersecting 2D frames. Plasticity zones are developed at the end regions of columns and axial deformations are taken into account.

3.3 Shear Wall Element

Shear wall elements are modelled combining an axial linear-elastic spring together with a nonlinear shear and a nonlinear flexural spring in series. In this case not only a moment-curvature skeleton curve is required, but a shear force-shear deformation curve as well, which are either user defined or computed through a fiber model.

4 Spread Plasticity Model

Inelastic deformations vary along element's length. Consequently an element will also exhibit different flexibility characteristics. In order to formulate the elastoplastic flexibility matrix of such an element a spread plasticity model is introduced. This model can formulate an element's flexibility matrix by taking into account the current stiffness (i.e. flexural stiffness concerning beams and columns and also shear stiffness for walls) at each end section, a corresponding yield penetration length and/or the elastic core stiffness depending on the values of the yield penetration lengths.

5 Yield Penetration Model

A yield penetration model is used to compute the yield penetration lengths at the end sections of an element as shown in Figure 2. Yield penetrations parameters α_A and α_B specify the portion of the element where the acting moment is greater than the corresponding yield moment of the section. For simplicity, a linear moment distribution is assumed even in the case of the presence of distributed loads. However one can subdivide each structural element into a number of elements in order to capture closely a parabolic variation.

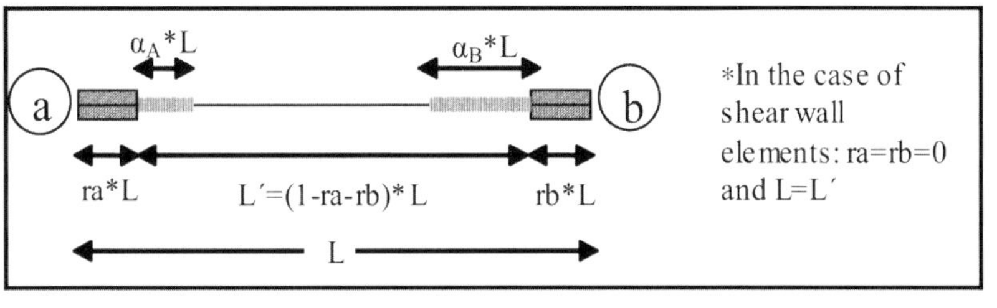

Figure 2. Yield Penetration Parameters

As long as the current moment distribution is defined, a set of geometrical oriented rules is used in order to define both the yield penetration lengths and the stiffness of the central part of the element. The yield penetration parameters are checked with the previous maximum penetration lengths, while the plastic regions of the element are considered only to expand and not to contract.

6 The Hysteretic Model

The smooth hysteretic model presented herein is a variation of the model originally proposed by Bouc (1967) and modified by several others (Wen 1976; Baber Noori 1985). The model was developed in the context of moment-curvature relationships of beam-columns. Therefore the stress variable is here referred to as "moment" (M) and the strain variable as "curvature" (φ). In the case of shear-walls the hysteretic loop can be described in terms of a shear force-shear deformation relationship.

The use of such a hysteretic constitutive law is necessary for the effective simulation of the behavior of R/C structures under cyclic loading, since often structures that undergo inelastic deformations and cyclic behavior weaken and lose some of their stiffness and strength. Moreover, gaps tend to develop due to cracking causing the material to become discontinuous. The Bouc-Wen Hysteretic Model is capable of simulating stiffness degradation, strength deterioration and progressive pinching effects.

The model can be visualized as a linear and a nonlinear element in parallel, as shown in Figure 3. The relation between generalized moments and curvatures is given by:

$$M(t) = M_y \left[\alpha \frac{\phi(t)}{\phi_y} + (1-\alpha) z(t) \right] \tag{1}$$

where M_y is the yield moment; φ_y is the yield curvature; α is the ratio of the post-yield to the initial elastic stiffness and z(t) is the hysteretic component defined below.

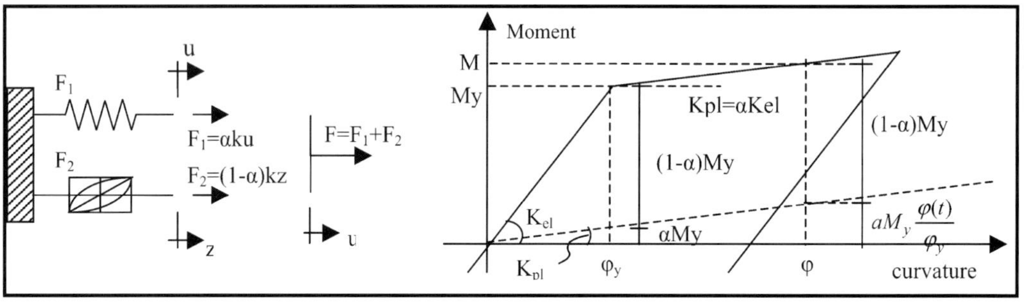

Figure 3. Bouc-Wen Hysteretic Model

The nondimensional hysteretic function z(t) is the solution of the following non-linear differential equation:

$$\dot{z}(t) = f(\dot{\phi}(t), z(t)) \frac{1}{\phi_y} \text{ or alternatively } \frac{dz}{d\phi} = K_z \frac{1}{\phi_y} \text{ where } K_z = \left[A - B \frac{1+sign(d\phi)}{2} \left(\frac{|z(t)|+z(t)}{2} \right)^{n_B} - C \frac{1+sign(d\phi)}{2} \left(\frac{|z(t)|-z(t)}{2} \right)^{n_C} - D \frac{1-sign(d\phi)}{2} \left(\frac{|z(t)|+z(t)}{2} \right)^{n_D} - E \frac{1-sign(d\phi)}{2} \left(\frac{|z(t)|-z(t)}{2} \right)^{n_E} \right] \tag{2}$$

In the above expression A, B, C, D & E are constants which control the shape of the hysteretic loop for each direction of loading, while the exponents n_B, n_C, n_D & n_E govern the transition from the elastic to the plastic state. Small values of n_i lead to a smooth transition, however as n_i increases the transition becomes sharper tending to a perfectly bilinear behavior in the limit ($n \to \infty$).

The program defaults are:

$$A = 1, \quad C = D = 0 \;\&\; B = \frac{1}{b^{n_B}}, \quad E = \frac{1}{e^{n_E}} \quad \text{where } e = \frac{-M_y^-}{M_y^+}, \quad b=1 \text{ and } n_B = n_E = n \tag{3}$$

The parameters C, D control the gradient of the hysteretic loop after unloading occurs. The assignment of null values for both, results to unloading stiffness equal to that of the elastic branch. Also, the model is capable of simulating non symmetrical yielding, so if the positive yield moment is regarded as a reference point, the resulting values for B and E are those presented in equation (3). The hysteretic parameter Kz is then limited in the range of 0 to 1, while the hysteretic function z varies from $-|M_y^- / M_y^+|$ to 1.

Finally, the flexural stiffness can be expressed as:

$$K = EI = \frac{dM}{d\phi} = M_y\left[\alpha\frac{1}{\phi_y} + (1-\alpha)\frac{dz}{d\phi}\right] = M_y\left[\alpha\frac{1}{\phi_y} + (1-\alpha)K_z\frac{1}{\phi_y}\right] = EI_0\left[\alpha + (1-\alpha)K_z\right] \quad (4)$$

6.1 Hysteretic behavior Variations

a) Stiffness Degradation

The stiffness degradation that occurs due to cyclic loading is taken into account by introducing the parameter η into the differential equation:

$$\frac{dz}{d\phi} = \frac{K_z}{\eta}\frac{1}{\phi_y} \rightarrow K = EI_0\left[\alpha + (1-\alpha)\frac{K_z}{\eta}\right] \quad where \quad \eta = 1.0 + S_k\frac{\mu_{\max} + \mu}{2} \quad (5)$$

The parameter η depends on the current, $\mu = \phi/\phi_y$, and maximum achieved plasticity, $\mu_{\max} = \phi_{\max}/\phi_y$. S_k is a constant which controls the rate of stiffness decay. Common values for S_k are 0.1 and 0.05.

b) Strength Deterioration

The strength deterioration is simulated by multiplying the yield moment M_y with a degrading parameter S_β:

$$M(t) = S_\beta M_y\left[\alpha\frac{\phi(t)}{\phi_y} + (1-\alpha)z(t)\right] \quad (6)$$

The parameter S_β depends on the damage of the section which is quantified by the Damage Index DI:

$$S_\beta = 1 - S_d DI \quad \text{where} \quad DI = \frac{\mu_{\max} - 1}{\mu_c - 1}\frac{1}{\left(1 - \frac{S_{p1}\int dE_{diss}}{4E_{mon}}\right)^{S_{p2}}} \quad (7)$$

In the above expression S_d, S_{p1}, S_{p2} are constants controlling the amount of strength deterioration; μ_c is the maximum plasticity that can be reached, $\mu_c = \phi_u/\phi_y$; $\int dE_{diss}$ is the energy dissipated before unloading occurs and finally E_{mon} is the amount of energy absorbed during a monotonic loading until failure as shown in **Figure 4**.

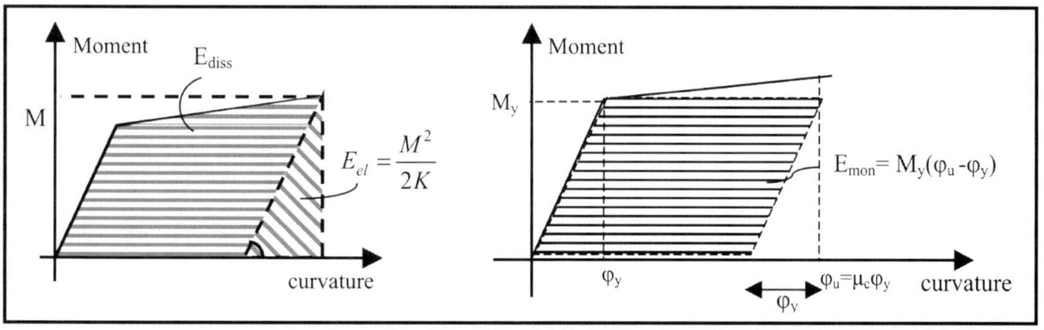

Figure 4. Dissipated Energy (Ediss) and Monotonic Energy (Emon).

c) Pinching or Slip

Pinching of hysteretic loops due to shear cracking and bond slip of the reinforcement is commonly observed in reinforced concrete structures during cyclic loading. This phenomenon is taken into account by introducing in the expression of *Kz* the "slip length" parameter a, and a function *f (z)*:

$$K_z^{pin} = \frac{K_z}{1 + \mathrm{a} \cdot f(z) \cdot K_z} \tag{8}$$

In the above equation K_z^{pin} is the hysteretic parameter affected by pinching, Kz is the original expression of the hysteretic parameter obtained from (3) and parameters a, f(z) are given by:

$$\mathrm{a} = A_s(\mu' - 1) \quad \& \quad \mathrm{f(z)=exp}\left[-\frac{(\mathrm{z}-\mathrm{z_m})^2}{z_s^2}\right] \tag{9}$$

where A_s is a control parameter which may be linked to the size of crack opening or reinforcement slip or both; μ' is the normalized curvature attained at the load reversal prior the current loading circle; z_s is the range where slip occurs. A non zero value of the parameter z_m will shift the effective slip region so that it is symmetric about $z=z_m$.

7 Method of Analysis

The computer program "*Plastique*" which is described herein, is capable to perform the different types of analysis, namely; Push-over analysis, Quasi-static analysis, Non-linear dynamic analysis and Eigenvalue analysis

The first two analysis types, although significantly simplified, can lead to valuable conclusions concerning the behavior of the structure and the possible collapse mechanism. The applied procedure can be described in brief as follows. In the case of 2-D analysis the structure is assumed to consist of a finite number of nodes interconnected by a finite number of elements. The types of elements have been described in section 3. In the case of 3-D analysis the structure is assumed to consist of the aforementioned 2-D frames, assuming a rigid diaphragm assemblage of their horizontal dof's per floor slab. Loads may be applied at the nodes or along the elements. In both cases though, they are transformed to nodal loads.

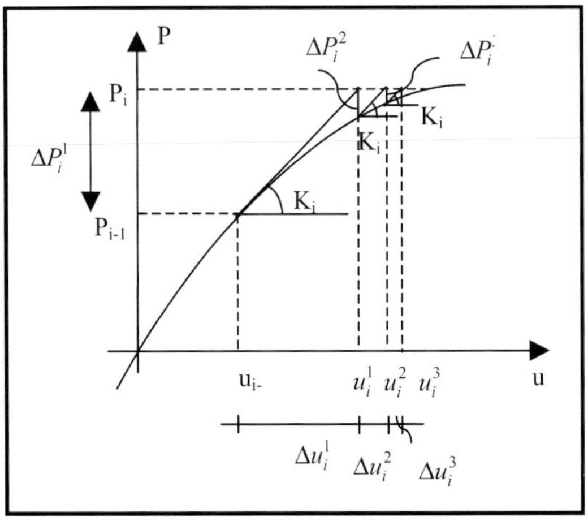

Figure 5. Modified Newton Raphson Method

After the formation of the stiffness matrix the equilibrium equations are solved by an efficient algorithm based on the Gaussian elimination method. The structure stiffness is stored in a banded form

to optimize the use of core storage and during
arithmetic operations are avoided. An incremental method is applied for all types of analysis. The specified loads are divided to sufficiently smaller sub-loads, in order to simulate more efficiently the stress redistribution which occurs due to the non linear behaviour of the structure. An iterative process (modified Newton Raphson Method) is incorporated in each load step so that a higher level of accuracy can be achieved, as shown in Figure 5.

The member forces are computed for each load increment and the tangent stiffness matrix is updated to account for changes in any stiffness coefficient of the element. A spread plasticity model is used, as described in section 4, in order to simulate the changes in the flexibility in each element. In the case of the dynamic analysis the Newmark Method is used for the direct integration of the equations of motion.

The equation of motion to be solved at any stage of the analysis is written as:

$$[M]\{\ddot{U}\}+[C]\{\dot{U}\}+\{P_{\text{int}}\}=\{P_{ext}\}=-[M]\{S\}\{\ddot{U}_g\} \tag{10}$$

where [M] is the mass matrix; [C] is the damping matrix; $[P_{int}]$ is the internal load vector of the structure; $[P_{ext}]$ is the external load vector of the structure; [S] is a modal influence vector; [U] is the structure displacement vector and $\{\ddot{U}_g\}$ is the ground acceleration vector.

The above system of equations is solved using the constant acceleration method, according to which equation (10) can be rewritten for time t+Δt and iteration k as:

$$[M]^{t+\Delta t}\cdot\{\ddot{U}\}^{(k)}+[C]^{t+\Delta t}\cdot\left\{\dot{U}^{(k)}\right\}+{}^{t+\Delta t}\{P_{\text{int}}\}^{(k-1)}+{}^{t}[K_t]^{t+\Delta t}\cdot\{\Delta U\}^{(k)}={}^{t+\Delta t}\{P_{ext}\} \tag{11}$$

The hysteretic model applied is the Bouc Wen - Baber Noori model which, as mentioned previously, is able to simulate R/C behaviour such as stiffness degradation, strength deterioration and pinching. At every step of the analysis, the Damage Indices of the elements and the structure are calculated providing an evaluation not only of the inflicted damage, but also of the structure's residual strength and capacity to withstand further loading. The Damage Index for each section is given by relation (7). The Damage Index for each element is computed as the maximum Damage Index of its sections' and finally the Damage Index of the whole structure can be obtained from the following expressions:

$$DI=\sum_i \frac{E_{tot,i}}{\sum_i E_{tot,i}}(DI)_i \tag{12}$$

where $(DI)_i$ is the Damage Index for each element and $E_{tot,i}$ is the total amount of absorbed energy per element.

8 Numerical Examples

8.1 Example 1

This example demonstrates the usage of the different analysis and design options of "*Plastique*". The structure used is a one bay single srorey 2D frame shown in Figure 6.

A monotonic push-over analysis can provide the collapse mechanism of the particular structure as shown in Figure 7.

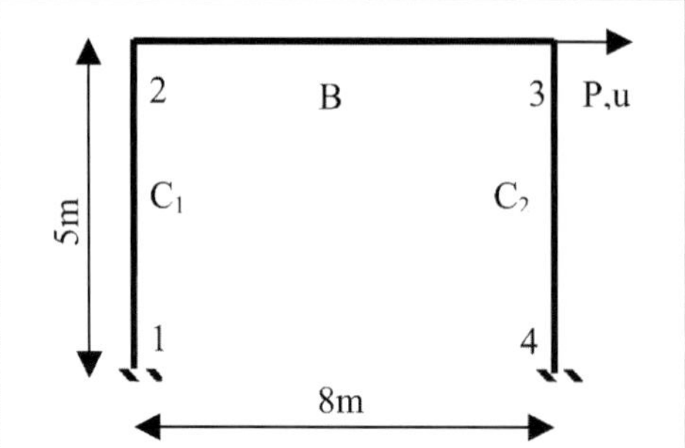

Figure 6. Structure -2 Column Frame

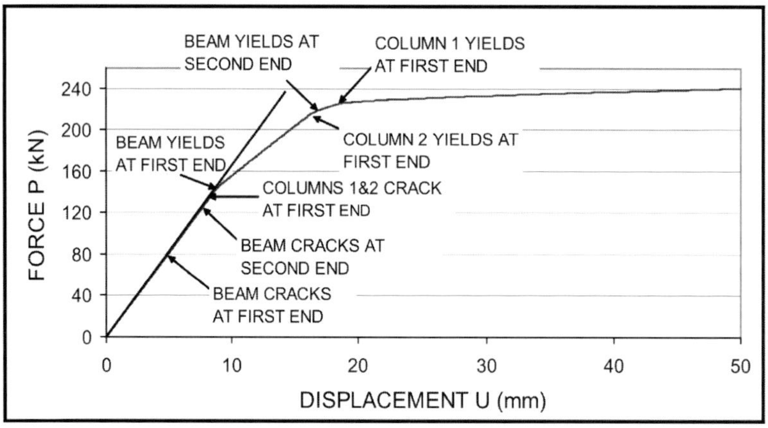

Figure 7. Possible Collapse Mechanism

In addition, a dynamic analysis is carried out. The accelerogram of the Northridge earthquake, normalized at 0.50g, was chosen for the particular analysis (Figure 8).

The resulting displacement time history, the force displacement hysteretic loop, as well as the final Damage Indices (DI) and yield penetration lengths are presented in the following figures (Figures 9 to 11).

In Figure 10, one can notice the gradual shift in the column section's stiffness due to stiffness and strength degradation effects, and the gradual shift of plastic deformation at a number of steps until the rest at a deformed state.

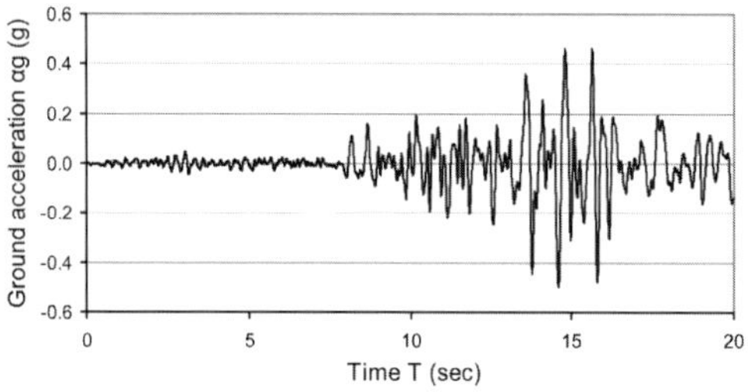

Figure 8. Northridge accelerogram

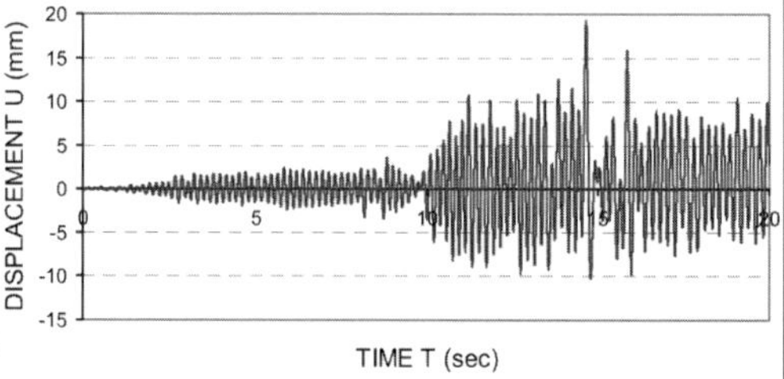

Figure 9. First floor displacement time history

The damaged state of the structure at the end of the analysis is depicted in Figure 11. The reduced sections are a graphical representation of the inelasticity distribution along each member's length. The damage index offers a measure of the accumulated damage at each member.

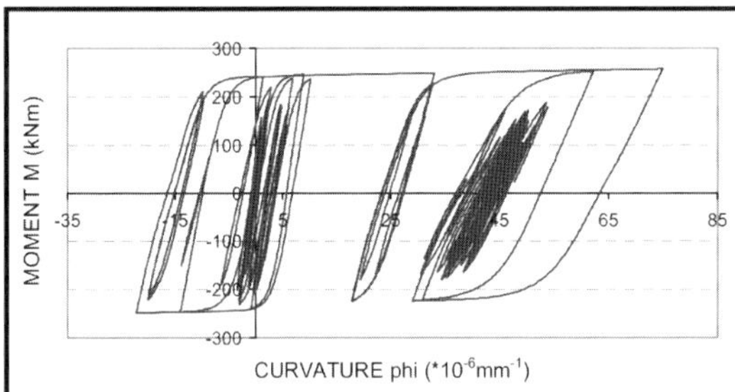

Figure 10. Moment-Curvature Hysteretic loop

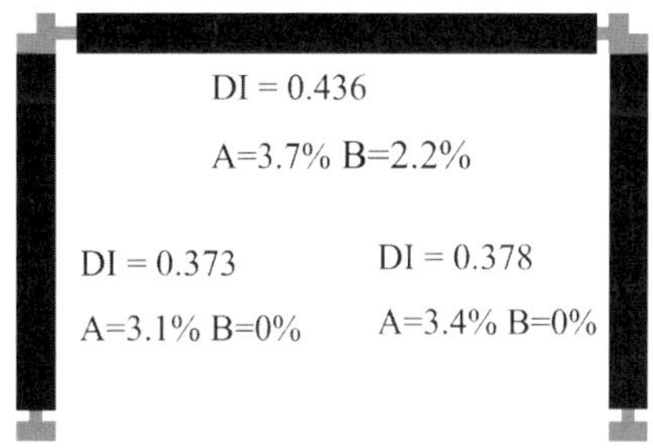

Figure 11. Final DI and yield penetration lengths

8.2 Example 2

As another example the behavior of a three storey building subjected to dynamical loading of time varying frequency content is presented. The Tabas (Iran – 0.68g) earthquake was chosen for this particular analysis and the method of wavelets was applied for the construction of the corresponding time-frequency spectrum. The analysis demonstrates the response of the structure to a short of "sliding" resonance between the first eigen frequency of the structure and the primary frequency of the excitation, that both are shifted as time progresses. The structure due to plastification and the excitation due to a time – frequency variation that often occurs in strong ground motion excitations.

The original Tabas accelerogram is presented in Figure 12 accompanied by its corresponding time-frequency spectrum (Figure 13).

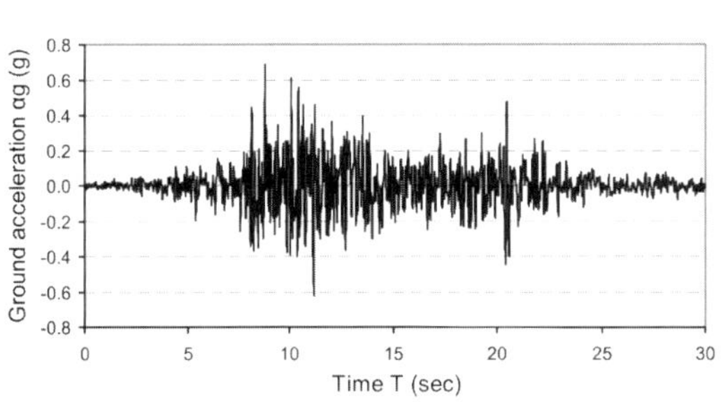

Figure 12. Tabas accelerogram

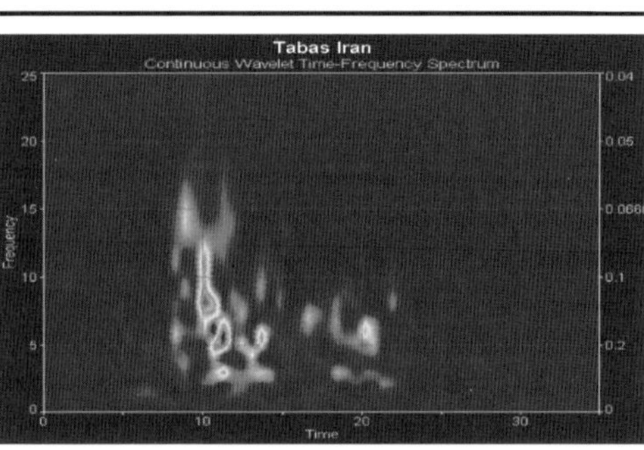

Figure 13. Time-Frequency spectrum

From the time-frequency spectrum one can identify the three peaks presented in Table 0. Since Max_2 and Max_3 have similar values and arise in relatively close time interval they can be substituted by an "equivalent" spectral frequency presented in the last row.

	Time t (sec)	Frequency F (Hz)
Max_1	11.1	8.12
Max_2	12	5.47
Max_3	14	5.55
$Max_{2,3}$ equivalent	13	5.50

Table 0: Frequency Peaks

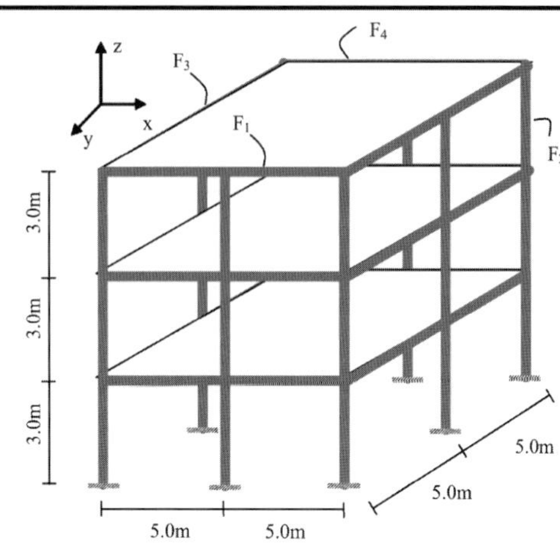

Figure 14. Three storey building

The 3D structure studied in the present example is depicted in Figure 14. A convenient set of geometrical/mechanical properties and hysteretic parameters was used in order to achieve an eigen frequency variation for the primary mode which would start from a value of 8.15Hz for a duration of 11.1 sec approaching 5.55 Hz around 13 sec. A dynamic analysis was performed using the Tabas accelerogram as excitation in the x direction The structure's response is presented in Figure 15, where it can be observed that after t = 13.5 sec the top floor displacement increases rapidly, leading the structure finally to failure at t = 14.4 sec.

By careful examination of the results it turns out that most of the critical sections of the structural elements yield and fail within the time interval of 11 to 14.4 sec. Similar evidence is provided by examining the variation of the primary frequency of the building, that remains steady at approximately 8.15Hz for the first 12.5 sec, decreasing rapidly in steps due to the successive degradation phenomena to the value of 5.55Hz at around 13.5 sec (Figure 15).

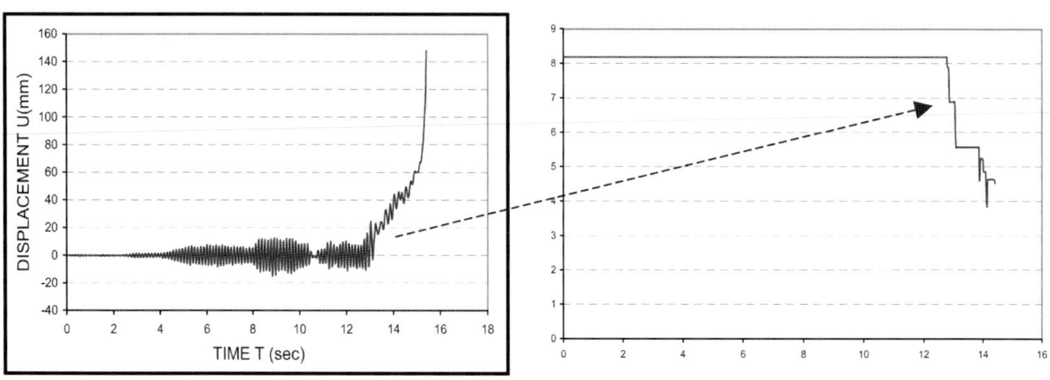

Figure 15. a) Top floor displacement time history- b) Frequency Variation.

In order to verify whether the failure is due to the previously described pattern of successive resonance a non linear analysis is performed on the previous structure after filtering out the second highest spectral frequency of the excitation (of approximately 5.552Hz), i.e by considering an excitation without the time – frequency variation. For this case although the structure suffers a severe damage does not collapse as it remains steady at a top storey displacement of the order of 50 mm (Figure 16).

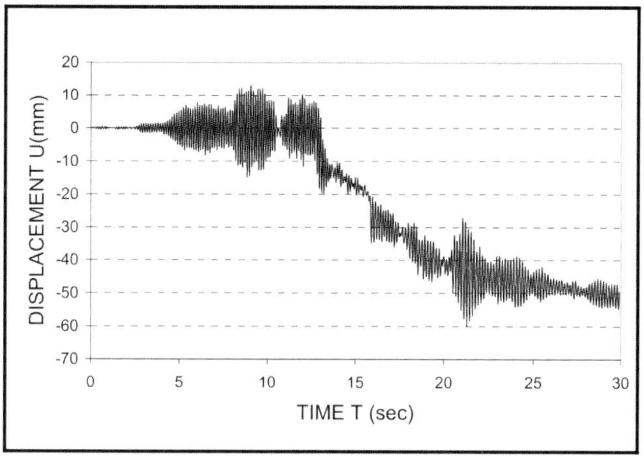

Figure 16. Top story displacement time history after filtering the accelerogram.

By examining the status of elements and its evolution in time a different development in the failure mechanism is observed mainly for t>13 sec, which is due to the absence of the filtered frequency.

9 Conclusions

A new computer program, *"Plastique"*, for the inelastic analysis of structures is presented. A macro modeling approach is implemented, combined with a spread plasticity and a yield penetration model to account for the inelastic phenomena on structural response. By introducing a smooth Bouc Wen type hysteretic model a close to reality simulation of R/C cyclic response is achieved. An iterative procedure is used in order to solve the equilibrium equations. The program proves to be a versatile tool offering different analysis options. In comparison to other similar nonlinear analysis programs, *"Plastique"* is capable of performing a full 3D inelastic analysis, using the same model for every plan orientation of the seismic excitation.

Inelastic analysis reveals a more realistic response of the structure, demanding though a more accurate description of the structural properties and their variations. Seismic hazard analysis becomes even more crucial, as compared to an elastic analysis, while time-frequency distribution of the frequency content of an earthquake, as demonstrated in the second example, can be decisive on the fate of the structure.

References

Klaus-Jürgen Bathe, "Finite Element Procedures".

E. N. Chatzi & S. P. Triantafillou. Diploma Thesis, (2004), «Inelastic analysis of multistory buildings with a Bouc-Wen type hysteretic model», NTUA.

Anil K. Chopra, "Dynamics of Structures", Prentice Hall Inc.

Robert D. Cook, David S. Malkus, and Michael E. Plesha, "Concepts and applications of Finite Element Analysis", Third Edition.

S. Dobson, M. Noori, Z. Hou, M. Dimentberg, and T. Baber, (1997), "Modelling and random vibration analysis of SDOF systems with asymmetric hysteresis".

Sashi K. Kunnath, John B. Mander, and Lee Fang, (1996), "Parameter identification for degrading and pinched hysteretic structural concrete systems".

Young J. Park, Andrei M. Reinhorn, and Sashi K. Kunnath, (1987), "IDARC : Inelastic damage analysis of reinforced concrete frame – shear-wall structures", Technical Report NCEER-87-0008, State University of New York at Buffalo.

J. S. Przemieniecki, "Theory of Matrix Structural Analysis".

R. E. Valles, A. M. Reinhorn, S. K. Kunnath, C. Li, and A. Madan (1996), "IDARC 2D Version 4.0 : A Computer Program for the Inelastic Damage Analysis of Buildings", Technical Report NCEER-96-0010, State University of New York at Buffalo.

Cold Formed Steel

ANALYTICAL PREDICTIONS OF STRENGTH AND DEFLECTION OF LIGHT GAUGE STEEL FRAME / WOOD PANEL SHEAR WALLS

C.Y. Chen, A.F. Okasha and C.A. Rogers
Department of Civil Engineering & Applied Mechanics, McGill University, Montreal, Canada
E-mail: colin.rogers@mcgill.ca

Abstract

It is anticipated that the construction of buildings that incorporate light gauge steel frame/wood panel shear walls as primary lateral load resisting elements will increase across Canada in coming years. At present, a codified method for the prediction of shear wall strength and stiffness is not available in Canada. For this reason an investigation of various analytical prediction methods was completed. The racking strength and stiffness of steel frame/wood panel shear walls have been shown to be highly dependent on the behaviour of the sheathing connections. An experimental program involving over 200 small-scale tests was first carried out to establish the performance of steel stud to wood sheathing connections. This information was then utilized in a comparison of five existing analytical/mechanics based methods to predict the strength and deflection of wood framed shear walls. These existing analytical methods were adapted for use with the steel framed walls. A comparison of the predicted strength and deflection values was then made with the results of full-scale shear wall tests. Based on the comparison between test and predicted shear wall response the elastic models presented by Källsner & Lam were recommended for use to predict the lateral resistance and deflection of light gauge steel frame/wood panel shear walls under monotonic and cyclic loading. At the same time, the shear capacity and initial stiffness as measured from tests of single sheathing connections with an edge distance of 25 mm, and which were evaluated using an equivalent energy approach, were recommended as the input connection parameters for both the strength and deflection models.

Introduction

A typical light gauge steel frame / wood panel shear wall is composed of cold-formed steel studs and tracks that are connected with self-drilling/tapping screws to either plywood or OSB sheathing. Guidelines for the design of these shear walls do not exist in current Canadian codes. A design method for the calculation of in-plane shear stiffness and strength has been proposed for use with the 2005 NBCC *(NRCC, 2005) (Branston, 2004)*, however this method is reliant on the results of full-scale testing. Eventually, with the proven applicability of an analytical model, researchers could extend the results of small-scale connection tests to aid in the design of full size shear walls. For this reason a study was carried out to evaluate the possibility of using existing analytical methods, which were originally developed for the design of wood framed shear walls, to predict the in-plane shear stiffness and deflection of steel frame / wood panel shear walls. Methods by Källsner & Lam *(1995)*, Easley *et al.*, *(1982)* and McCutcheon *(1985)* were adapted for use with steel frame shear walls. Since all of these methods are based, to a large extent, on the stiffness and strength properties of the individual sheathing connections in a wall, tests were carried out by Okasha & Rogers *(2004)* to identify the connection properties. With the results of the sheathing connection tests Chen *(2004)* then compared the analytical predictions with the results of shear wall tests.

Objective and Scope

The objective of this research was to recommend an analytical method, which is based on a mechanics approach, to predict steel frame / wood panel shear wall strength and deflection. This involved the evaluation of five existing methods, developed for the analysis of wood framed shear walls, in

M. Pandey et al. (eds), Advances in Engineering Structures, Mechanics & Construction, 381–391.

comparison with the results of shear wall experiments. The details of a simplified strength and deflection model are provided in this paper. Test programs on single-storey shear walls and individual sheathing connections are summarized. In addition, a comparison between the test results and the predictions of an analytical approach is presented for the monotonic and reversed cyclic shear wall tests carried out by Boudreault *(2005)*, Branston *(2004)* and Chen *(2004)*.

Shear Wall Test Program

The results of a series of 109 shear walls (16 configurations) subjected to lateral in-plane loading were used in this study (Table 1). The testing consisted of three different size single-storey wall specimens: 610×2440 mm, 1220×2440 mm and 2440×2440 mm, which were composed of light gauge steel C-studs (92.1×41.3×12.7 mm) screw connected to steel tracks (92.1×31.8 mm) (Fig. 1). Both the studs and tracks were fabricated of ASTM A653 *(2002)* steel with a nominal grade and thickness of 230 MPa and 1.12 mm, respectively. Three types of wood sheathing (12.7 mm DFP *(CSA O121, 1978)*, 12.7 mm CSP *(CSA O151, 1978)*, 11 mm OSB *(CSA O325, 1992)*) were connected to one side of each test wall with screws placed at a spacing of 76, 102 or 152 mm. Simpson Strong-Tie S/HD10 hold-downs were used to connect the chord studs to the test frame. A specially constructed shear wall test frame was utilized to allow for the application of a lateral in-plane load to the top of the wall, and to provide lateral support to limit out-of-plane movement *(Boudreault, 2005; Branston, 2004; Branston et al., 2004; Chen, 2004)*.

Figure 1. Typical 1220×2440 mm shear wall test specimens.

Monotonic and reversed cyclic tests were carried out using the CUREE protocol for ordinary ground motions *(Krawinkler et al., 2000; ASTM E2126, 2005)*. In most cases, 6 specimens (3 monotonic and 3 reversed cyclic) were tested per wall configuration. Subsequently, the results of each shear wall test were evaluated using a codified version of the equivalent energy elastic – plastic (EEEP) approach for calculating the design parameters of light framed shear walls *(ASTM E2126, 2005)*. It was decided that the EEEP model best represented the behaviour of light gauge steel frame / wood panel shear walls subjected to both monotonic and reversed cyclic loading (Fig. 2) *(Branston, 2004)*. The model results in an idealized load-deflection curve, of a simple bilinear shape, that can be easily defined and constructed, yet still provides a realistic depiction of the data obtained from laboratory testing. Moreover, the EEEP model recognizes the post-peak deformation capacity by taking into account the energy dissipated by the test specimen up to failure. In the case of each reversed cyclic test a backbone curve was first constructed for the resistance *vs.* deflection hysteresis. This backbone curve and the resistance *vs.* deflection curve for monotonic specimens were then used to create EEEP curves based on the equivalent energy approach.

Table 1. Matrix of shear wall tests.

Configuration	Wall Dimensions (mm)	Sheathing Type (mm)	Fastener Schedule (mm)	Researcher
1, 2, 3, 4	1220 × 2440	CSP 12.5	102 / 305	Boudreault *(2005)*
5,6	1220 × 2440	DFP 12.5	102 / 305	"
7, 8	1220 × 2440	CSP 12.5	152 / 305	Branston *(2004)*
9, 10	1220 × 2440	CSP 12.5	76 / 305	"
11, 12	1220 ×2440	DFP 12.5	152 / 305	"
13, 14	1220 × 2440	DFP 12.5	76 / 305	"
15, 16	610 × 2440	CSP 12.5	152 / 305	Chen *(2004)*
17, 18	610 × 2440	CSP 12.5	102 / 305	"
19, 20	610 × 2440	OSB 11	152 / 305	"
21, 22	1220 × 2440	OSB 11	152 / 305	Branston *(2004)*
23, 24	1220 × 2440	OSB 11	102 / 305	"
25, 26	1220 × 2440	OSB 11	76 / 305	"
27, 28	610 × 2440	OSB 11	102 / 305	Chen *(2004)*
29, 30	2440 × 2440	CSP 12.5	152 / 305	"
31, 32	2440 × 2440	CSP 12.5	102 / 305	"
33, 34	2440 × 2440	CSP 12.5	76 / 305	"

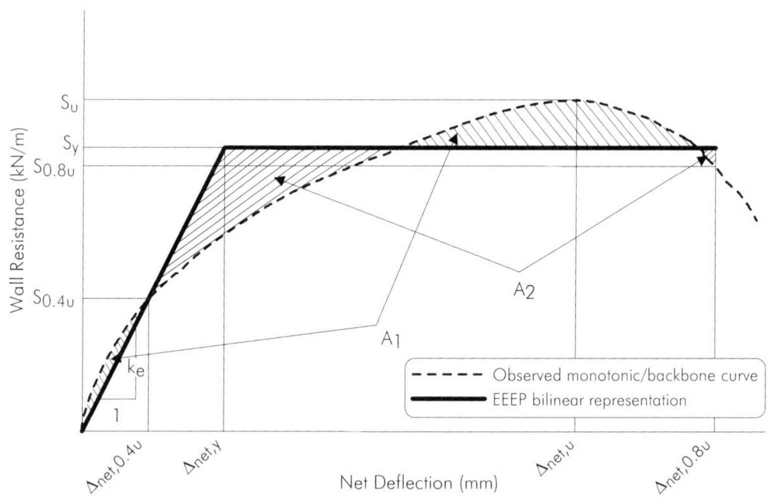

Figure 2. Equivalent energy elastic-plastic (EEEP) model *(Branston, 2004)*.

Sheathing Connection Test Program

The experimental program included 216 connection specimens that were subjected to tension monotonic and reversed cyclic loading. The monotonic tests were run at a rate of 2 mm/min, and the corresponding cyclic tests were performed by using the same CUREE test protocol as incorporated into the full-scale wall test program. The tests were conducted in accordance with the ASTM Standard D1761 *(1995)*. The test specimen matrix was selected in such a way to include the effect of the variation of wood sheathing type (DFP, CSP and OSB), thickness (9.5, 11, 12.5 & 15.5 mm) as well as orientation with respect to the grain and fastener edge distance (6, 9.5, 12.5, 16 & 25 mm). Moreover the effect of different steel stud thickness (0.84, 1.11, 1.37 & 1.73 mm) and strength (Grade 230 & 345) were included *(Okasha & Rogers, 2004)*. The primary aim of this research project was to provide information on connection performance to facilitate the predictions of the behaviour of steel frame / wood panel shear walls. Tests were carried out on individual screw sheathing to stud connections, as shown in Figure 3. The screw type, as well as sheathing and steel matched that used in the construction of the full-scale shear wall tests, the results of which were used to compare with the analytical predictions.

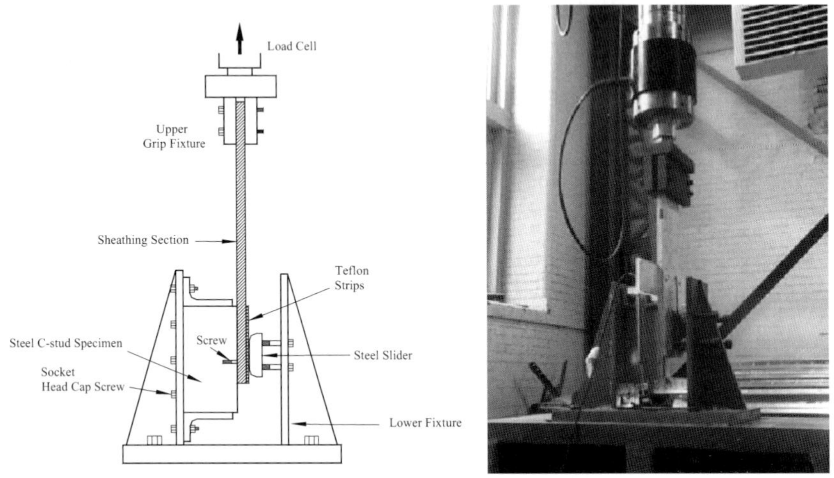

Figure 3. Sheathing connection test set-up.

Table 2. Parameters for prediction of monotonic tests *(Okasha & Rogers, 2004)*.

Specimen	Max. Load (kN)	EEEP Yield Load (kN)	k_e (0.4 max load) (kN/mm)	k_s (max load) (kN/mm)
CSP12.5-PR M	1.376	1.192	2.383	0.210
CSP25-PR M	1.740	1.487	1.513	0.197
OSB12.5-PR M	1.754	1.487	2.683	0.376
OSB25-PR M	1.955	1.643	1.168	0.194
DFP25-PR M	2.860	2.367	1.513	0.242

Table 3. Measured parameters of cyclic tests *(Okasha & Rogers, 2004)*.

Specimen	Max. Load (kN)	EEEP Yield Load (kN)	k_e (0.4 max load) (kN/mm)	k_s (max load) (kN/mm)
CSP25-PR C	2.228	2.024	0.691	0.209
OSB25-PR C	2.152	1.956	0.927	0.242
DFP25-PR C	3.186	2.785	0.735	0.311

The parameters listed in Tables 2 were adopted for the prediction of lateral resistance and deflection of full-scale shear walls under monotonic loading. Two failure modes were observed during the connection tests; screw pulled through the sheathing, and bearing / plug shear failure of the sheathing. These failure modes were in accordance with what was observed in the full-scale tests. The monotonic load capacity in connections loaded perpendicular-to-grain was higher (average 15%) than that in connections loaded parallel-to-grain. The results of the lower capacity parallel-to-grain specimens were therefore selected in order to establish the connection properties required for the monotonic wall analyses. The connection shear capacities listed in Table 3 for the cyclic tests were much higher than those obtained for the monotonic tests (Table 2). This was not in accordance with the full-scale test results, for which the shear strength of the cyclic tests was lower or close to that measured for the corresponding monotonic tests *(Chen, 2004)*. There are two possible explanations for this phenomenon; the loading speed for the cyclic connection tests was much faster than that used for the monotonic tests which resulted in a strain rate effect, and the large variation in the material composition of the wood panels, which may have a more noticeable effect when only single connections are tested compared with full size walls. In order to obtain a reasonable prediction of the wall behaviour, the load capacities for the monotonic tests were utilized for the shear walls under cyclic loading (Table 4). For the cyclic tests, the stiffness was taken as the average of the absolute values of the positive and negative results parallel-to-grain connection specimens. This average value takes into account the possible change in connection stiffness as the load in the wall changes direction.

Table 4. Parameters for prediction of cyclic tests *(Okasha & Rogers, 2004)*.

Specimen	Max. Load (kN)	EEEP Yield Load (kN)	k_e (0.4 max load) (kN/mm)	k_s (max load) (kN/mm)
CSP25-PR C	1.740	1.487	0.702	0.197
OSB25-PR C	1.955	1.643	0.926	0.230
DFP25-PR C	2.860	2.367	0.793	0.301

Analytical Method

A review of various analytical approaches that can be relied on to determine the strength and deflection of wood framed shear walls has been presented by Chen *(2004)*. These models were used to predict the lateral load capacity, which was defined as the yield shear strength, S_y, and the corresponding deflection of the wall $\Delta_{net,y}$ (Fig. 2). In this paper a summary of a strength and deflection model is presented. Figure 4 shows the assumed deformations and force distribution of a typical light gauge steel frame / wood panel shear wall. The lateral load at the top of the wall produces a moment and a horizontal force on the wall bottom. If the hold-downs are designed to fully transfer the tension force into the support through the end studs, the vertical forces acting on the end studs are balanced by the shear flow along the screw lines on the end studs, which is produced by sheathing rotation relative to the steel frame. The shear flow causes the axial forces in the end studs to distribute triangularly, with the maximum forces at the bottom of the end studs *(Stewart, 1987)*. With respect to the top track, if the screw spacing along the top edge of the sheathing and the spacing for anchors to the load beam are both small enough to assume the applied force is uniform, then no axial force exists. Similar for the bottom track, the applied force can be considered uniform if the screw and shear anchor spacing is small. The interior studs at the centreline of a panel or at the joint of two panels with the same width are assumed to carry no axial forces due to lateral loads causing in-plane shearing of the wall. The interior studs also provide out-of-plane support to stiffen the sheathing panel against shear buckling. The studs at the panel joints act as splices between adjacent wood panels; hence the design of the back-to-back studs needs to incorporate the shear force due to the opposite rotation of the two adjacent panels. Triangularly distributed forces also act perpendicular to the axes of the studs and tracks attached to the edges of panels, due to the relative displacements between studs and panels. In a capacity based design approach the size of the steel frame members is selected such that the frame itself does not fail. Given this information, and for simplification purposes, the frame members can be assumed to be rigid in the analytical models.

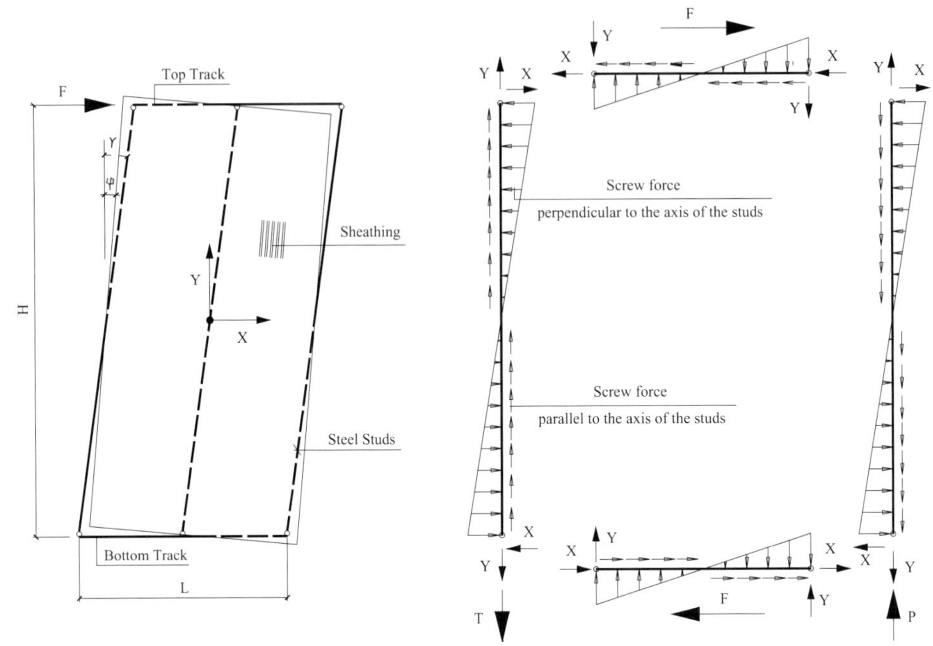

Figure 4. Deformations and force distribution in rigid framing members.

Simplified Strength Model

The racking performance of light gauge steel frame / wood panel shear walls is similar to that of wood framed shear walls. It is assumed that when a shear wall is subjected to lateral loading, the steel frame distorts as a parallelogram in which the top and bottom tracks maintain a horizontal position. The screws along the perimeter of a panel rotate about the flange of the studs; however, no obvious rotation of the screws connected to the interior studs occurs. The steel frame member connections act as hinges, which means that no lateral resistance develops in the frame itself. Rather, the lateral load is resisted by the composite action of the wood panels and steel framing through their relative rotation. The external work applied to the shear wall was assumed to be absorbed by the rotation of the screws.

In order to develop a model which can be used to predict the shear capacity of a light gauge steel frame / wood panel shear wall, some secondary behavioural characteristics need to be neglected or simplified. The following assumptions, which are similar to what was proposed by Källsner & Lam *(1995)* are applied in the model:

i) Deformation of the studs and tracks does not occur. These steel members are hinged to each other.
ii) The panels are rigid in their own plane and adjacent panels have no contact or overlap with each other.
iii) The relative displacements between the sheathing and framing are small compared with the panel size. The wood and steel also do not separate from each other during loading.
iv) No relative displacement exists between the centre of the sheathing panel and the corresponding centroid of the steel frame.
v) No horizontal panel joints exist in the same storey. Although in engineering practice, such joints are allowed, no tests with such configuration were included in this research.
vi) The shear wall is fully anchored onto the support or lower storey.
vii) The external work done by the racking loads is completely absorbed by the distortion of the sheathing-to-frame connections.
viii) The sheathing-to-frame connections have the same capacity in all directions.

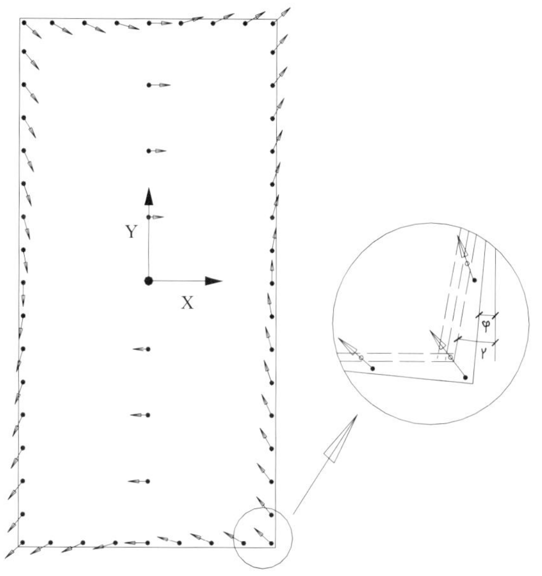

Figure 5. Assumed force distribution in sheathing connections.

The displacement of the sheathing relative to the steel frame can be viewed in Figure 4. All of the studs have rotated about their bottom ends through the angle γ, while the sheathing panel has rotated as a rigid body to an angle φ. In the simplified model these two rotations, which are taken as independent variables, result in the force distribution of the sheathing-to-frame connections as shown in Figure 5. Based on the assumed force distribution the shear capacity of the wall segment, $S_{y,wall}$, can be expressed as shown in Equation 1. The shear capacity is dependent on two factors, the first being the wall configuration including the connection pattern and the second the shear capacity per connection. This equation was originally presented by Källsner & Lam *(1995)* in their elastic model for wood framed shear walls.

$$S_{y,wall} = \frac{S_{y,conn}}{H \bullet \sqrt{\left(\frac{x_{\max}}{\sum_{i=1}^{N} x_i^2}\right)^2 + \left(\frac{y_{\max}}{\sum_{i=1}^{N} y_i^2}\right)^2}} \qquad \text{(Eq. 1)}$$

Where $S_{y,conn}$ is the shear strength of the individual sheathing connector (Tables 2&4), H is the height of the wall, x and y are the position of the fasteners (Fig. 5) and N is the number of sheathing connections.

Comparison of Predicted and Tested Shear Wall Capacity

Comparisons between the shear wall capacity measured during the laboratory testing and that predicted using the analytical model were performed. The intent was for the model to predict the shear wall capacity $S_{y,wall}$ at the level of the yield shear strength, S_y. At the same time, in order to verify that the elastic model by Källsner & Lam provided the most reasonable solution to predict the shear wall capacity, other models that were based on different assumptions and which have been applied in the prediction of wood frame shear walls, were also contained in the comparisons. These models, which were presented by Chen *(2004)*, include Källsner's & Lam's lower and upper plastic models, as well as models by Easley and McCutcheon. In this paper, only the test-to-predicted shear capacity results for the elastic Källsner & Lam model have been listed (Table 5). Each ratio and the associated statistical information represent the 16 wall configurations and a total of 103 individual shear wall test specimens.

Table 5. Full-scale shear wall test-to-predicted shear capacity *(Chen, 2004)*.

Monotonic Loading Cases	Källsner & Lam Elastic Model		
	Ratio	SD	COV
EEEP 12.5	1.270	0.181	0.143
EEEP 25	1.050	0.122	0.116
Max. Load 12.5	1.093	0.162	0.148
Max. Load 25	0.918	0.117	0.128
Cyclic Loading Cases	Källsner & Lam Elastic Model		
	Ratio	SD	COV
EEEP 25	1.012	0.129	0.127
Max. Load 25	0.885	0.122	0.137

In order to specify the connection test data that would best predict the shear wall capacity accurately, four cases were considered for the model for monotonic loading, *i.e.* EEEP 12.5, EEEP 25, Max. Load 12.5 and Max. Load 25; and two for cyclic loading, *i.e.* EEEP 25 and Max. Load 25. The shear capacity per connection, $S_{y,conn}$, was represented by the EEEP yield capacity or the maximum shear load; at the same time, two edge distances were considered, namely 12.5 mm and 25 mm. These cases correspond to the connection strength data listed in Tables 2 & 4. Based on the monotonic test predictions, the cases with an edge distance of 12.5 mm were not included in the predictions of cyclic tests. All of these cases were considered because it was not known which one would best represent the behaviour of the connection in the prediction of the performance of a full-size shear wall. The prediction using each model under each case was then compared with the average shear capacity, $S_{y,wall}$, of the tested full-scale walls with the matching sheathing configuration. An average shear capacity was obtained from the three or more tests that were performed for each wall configuration. Due to the large amount of the data, only the combined test-to-predicted ratios are listed in Table 5. As can be seen, the predicted shear capacity obtained using the test data from connections with a 25 mm edge distance and where the EEEP method was relied on to obtain the connection strength are the most accurate. A more detailed discussion of the comparison is provided by Chen *(2004)*.

Simplified Deflection Model

In the simplified deflection model, the same assumptions adopted for the strength model were made, except that the sheathing panels were not considered to be rigid. In contrast, the sheathing panels were defined as isotropic and deformable in terms of material properties; hence, the shear strain was assumed to be uniform over a whole panel. The purpose of the deflection model was to predict the deflection of the wall, $\Delta_{net,y}$, corresponding to the yield strength, S_y (or yield load capacity $S_{y,wall}$). The total displacement of the steel frame can be determined by considering the rotation of the frame and the bottom slippage on the support (Eq. 2).

$$\Delta_{walltop} = \frac{1}{k} FH^2 \bullet \left(\frac{1}{\sum_{i=1}^{N} x_i^2} + \frac{1}{\sum_{i=1}^{N} y_i^2} \right) + \frac{F}{GLt} + \left[\left(\frac{\Delta_{baseslip1} + \Delta_{baseslip2}}{2} \right) \right] + \left[\left(\Delta_{uplift1} - \Delta_{uplift2} \right) \times \frac{H}{L} \right] \quad \text{(Eq. 2)}$$

Where k is the stiffness of the individual sheathing connector (Tables 2&4), F is the applied shear force obtained from Eq. 1, H and L are the dimensions of the wall, G is the shear modulus of the wood sheathing, t is the thickness of the wood sheathing, Δ represents the base slip and uplift displacements, x and y are the position of the fasteners and N is the number of sheathing connections..

The final two components in Eq. 2 can be affected by many factors, such as the type of frame-to-support connections/anchorage, the shear modulus of these connections, the friction between a wall and its support, extension and slippage of the hold-down connections and the deformation of the steel frame. Although these factors have an impact on the behaviour of a tested wall, their inclusion would overly complicate the model, and hence they were not considered. Therefore, the deflection model used herein to predict the net lateral deflection was as shown in Eq. 3; which is similar to that presented by Källsner & Lam *(1995)* for predicting the deflection of wood framed shear walls.

$$\Delta_{net} = \frac{1}{k} FH^2 \bullet \left(\frac{1}{\sum_{i=1}^{N} x_i^2} + \frac{1}{\sum_{i=1}^{N} y_i^2} \right) + \frac{F}{GLt} \quad \text{(Eq. 3)}$$

Comparison of Predicted and Tested Shear Wall Deflection

Comparisons between the deflections measured during testing and the predictions made with Eq. 3 were performed to verify the accuracy of the model introduced above. Meanwhile, in order to verify that Källsner's & Lam's elastic model is more appropriate for the prediction of shear wall deflection than other models, Easley's model and McCutcheon's model were included in the comparison. The results of the Källsner & Lam model are presented in Table 6; the remaining test-to-predicted deflection ratios were tabulated by Chen *(2004)*. The same connection property cases, as described in the comparison of strength models, were incorporated in the deflection models. However, additional combinations were necessary because the deflected position of the wall depends on the estimated force, which was determined with Eq. 1 and the different connection strength values (Tables 2&4). Each combination of the listed deflection models and loading cases included all 16 wall configurations; a total of 103 individual shear wall test specimens. Only the combined ratio of the full-scale test-to-predicted deflection in each combination is listed in Table 6. More detailed information is provided by Chen *(2004)*.

For the most part, the prediction of lateral deflections was not as accurate as that of the lateral shear wall resistance. This can likely be attributed to the strong nonlinear behaviour of the sheathing-to-frame connections, as well as the overall nonlinear load *vs.* resistance performance of the shear walls (Fig. 2). The predictions based on the initial stiffness, k_e, of a connection tend to underestimate the lateral deflection under monotonic loading. In contrast, the predictions that incorporate the stiffness based on the ultimate load, k_s, can significantly overestimate the wall deflection. Källsner's & Lam's elastic model can provide a reasonable estimate of the shear wall deflection if the EEEP 25 load level and initial stiffness, k_e, connection properties are used. It should be noted, however, that the base slip and uplift of the wall have been ignored in the calculation of deflection. A variation in the type of holddowns and anchor bolts used for the full-scale shear wall tests could result in a change to the actual deflection of a wall.

Table 6. Full-scale shear wall test-to-predicted shear deflection *(Chen, 2004)*.

Monotonic Loading Cases	Källsner & Lam Elastic Model		
	Ratio	SD	COV
EEEP 12.5 & k_e	2.236	0.380	0.170
EEEP 25 & k_e	1.441	0.303	0.210
Max. Load 12.5 & k_e	1.923	0.339	0.176
Max. Load 12.5 & k_s	0.431	0.075	0.174
Max. Load 25 & k_e	1.263	0.301	0.238
Max. Load 25 & k_s	0.307	0.096	0.313
Cyclic Loading Cases	Källsner & Lam Elastic Model		
	Ratio	SD	COV
EEEP 25 & k_e	0.886	0.124	0.139
Max. Load 25 & k_e	0.775	0.133	0.171
Max. Load 25 & k_s	0.293	0.092	0.313

Conclusions and Recommendations

A number of analytical models, which had originally been developed for wood framed shear walls, were evaluated with respect to their ability to predict the strength and deflection of a series of single-storey light gauge steel frame / wood panel shear wall test specimens *(Boudreault, 2005; Branston, 2004; Chen, 2004)*. A simple analytical model based on the work of Källsner & Lam *(1995)* was presented and a comparison with the test results was completed. This was made possible by the sheathing connection tests that were carried out by Okasha & Rogers *(2004)*. The lateral shear yield resistance and the deflection of full-scale walls can be effectively predicted with the model if appropriate connection test data is available. Good agreement was obtained between the predicted and test strength values in both monotonic and cyclic cases using the Källsner & Lam approach. The deflection model only showed satisfactory agreement with the test results. The yield strength and initial stiffness per connection with 25 mm edge distance can be relied on to predict the yield lateral resistance and deflection of full-scale shear walls, if both the connection and full-scale test results are analyzed using the EEEP methods. In order to better predict the strength and deflection of the full-scale shear walls, the conditions for connection tests need to be kept consistent with those in full-scale tests, such as the loading speed and edge distance. In future connection tests, it is suggested that the edge distance parallel to the loading direction be the same as the perimeter screw spacing in the full-scale shear wall tests. As well, in order to obtain a better prediction of the full-scale tests, each connection test specimen should contain at least three screws to account for the variation of sheathing material.

Acknowledgments

The authors would like to acknowledge the support provided by the Natural Sciences and Engineering Research Council, the Canada Foundation for Innovation, the Canadian Sheet Steel Building

Institute, Bailey Metal Products Ltd., the Structural Board Association, Grant Forest Products Inc., Simpson Strong-Tie Co. Inc., Grabber Construction Products, ITW Buildex and the Canam Group Inc..

References

ASTM A653, 2002. Standard Specification for Steel Sheet, Zinc-Coated (Galvanized) or Zinc-Iron Allow-Coated (Galvannealed) by the Hot-Dip Process, West Conshohocken, PA, USA.

ASTM E2126, 2005. Standard Test Methods for Cyclic (Reversed) Load Test for Shear Resistance of Framed Walls for Buildings, West Conshohocken, PA, USA.

ASTM D1761, 1995. Standard Test Method for Mechanical Fasteners in Wood, West Conshohocken, PA, USA.

Boudreault, F.A., 2005. Seismic Analysis of Steel Frame / Wood Panel Shear Walls, Master's Thesis, Dept. of Civil Eng. and Applied Mechanics, McGill University, Montreal, QC, Canada.

Branston, A.E., 2004. Development of a Design Methodology for Steel Frame / Wood Panel Shear Walls, Master's Thesis, Dept. of Civil Eng. and Applied Mechanics, McGill University, Montreal, QC, Canada.

Branston, A.E., Boudreault, F.A., Chen, C.Y., Rogers, C.A., 2004. Light Gauge Steel Frame / Wood Panel Shear Wall Test Data : Summer 2003, Research Report, Dept. of Civil Eng. and Applied Mechanics, McGill University, Montreal, QC, Canada.

Canadian Standards Association, O121, 1978. Douglas Fir Plywood, Mississauga, ON, Canada.

Canadian Standards Association, O151, 1978. Canadian Softwood Plywood, Mississauga, ON, Canada.

Canadian Standards Association, O325, 1992. Construction Sheathing, Mississauga, ON, Canada.

Chen, C.Y., 2004. Testing and Performance of Steel Frame / Wood Panel Shear Walls, Master's Thesis, Dept. of Civil Eng. and Applied Mechanics, McGill University, Montreal, QC, Canada.

Easley, J. T, Foomani, M., Dodds, R. H., 1982. Formulas for Wood Shear Walls, *ASCE, J. of Struc. Div.* 108(ST11), 2460-2478.

Källsner, B., Lam F., 1995. Diaphragms and Shear Walls". STEP Lectures: Holzbauwerke nach Eurocode 5-Grundlagen, Entwicklungen, Ergaenzungen, Fachverlag Holz, Duesseldorf, Germany, 15/1-15/19.

Krawinkler, H., Parisi, F., Ibarra, L., Ayoub, A., Medina, R., 2000. Development of a Testing Protocol for Woodframe Structures, Report W-02, CUREE/Caltech Woodframe Project, Richmond, CA, USA.

McCutcheon, W. J., 1985. Racking deformation in Wood Shear Walls, *ASCE, J. of Struct. Eng.*, 111 (2), 257-269.

National Research Council of Canada, 2005. National Building Code of Canada 2005, Ottawa, ON, Canada.

Okasha, A.F., Rogers, C.A., 2004. Evaluation of Connection Performance for Steel Frame / Wood Panel Shear Walls, Research Report, Dept. of Civil Eng. and Applied Mechanics, McGill University, Montreal, QC, Canada.

Stewart, W.G., 1987. The Seismic Design of Plywood Sheathed Shear Walls, PhD Thesis, Dept. of Civil Eng., University of Canterbury, New Zealand.

INNOVATIVE COLD-FORMED STEEL I-JOIST

R.A. LaBoube
Department of Civil Engineering, University of Missouri-Rolla, Rolla, MO 65409, USA

Introduction

The use of cold-formed steel framing members as a method of both commercial and residential construction has increased markedly in recent years. Cold-formed steel members are used as the main framing members in low-rise and mid-rise commercial structures as well as for single and multiple family construction in the United States.

The predominant member of choice for the floor joist has been the C-section (Figure 1). The Csection possesses steel's many attributes to include high strength-to-weight-ratio; rapid construction times; high levels of dimensional stability; and favorable "green construction". Although the C-section's attributes are admirable, it does not lend itself to long spans and they are not utilities friendly. The standard web punching, typically 1.5 in by 4 in, does not readily accommodate mechanical, plumbing, and HVAC systems.

To provide a cold-formed steel member that addresses the short-comings of the standard Csection, research and development has been in progress to develop an I-shaped profile.

I-Shaped Joist Attributes

The I-shaped joist is a truss shaped profile that can more easily accommodate mechanical, plumbing, and HVAC systems (Figure 2). The joist would be fabricated by a continuous rollforming process with punched web openings and closed formed flanges that are mechanically clinched. The chords are pentagonal shaped that provide flat vertical surfaces for attachment to end closures, hangers, and

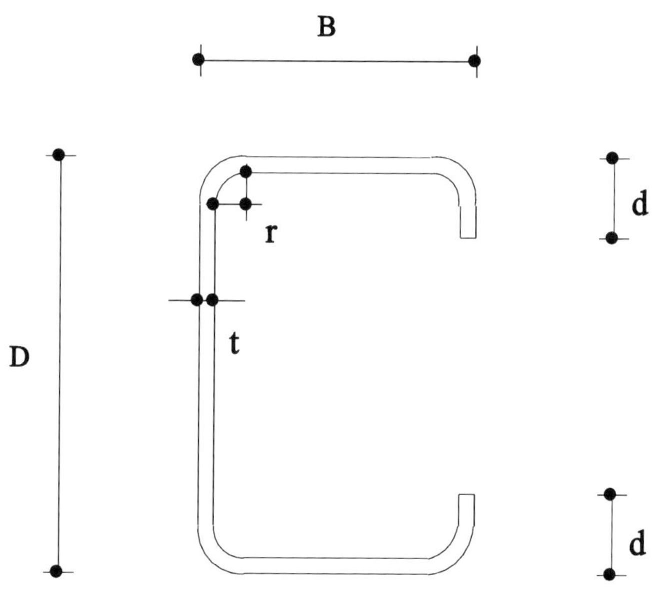

Fig. 1. Typical C-section joist.

M. Pandey et al. (eds), Advances in Engineering Structures, Mechanics & Construction, 393–399.

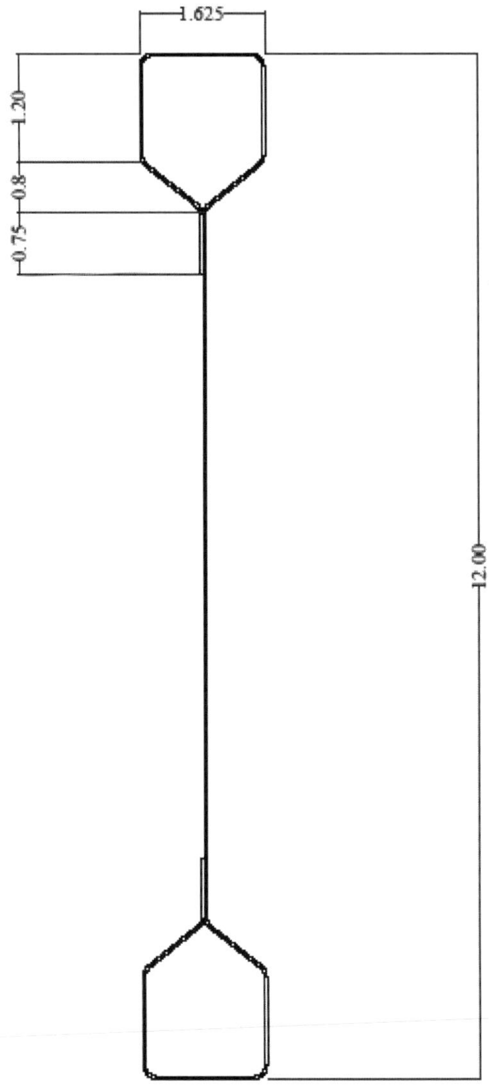

Fig. 2. Typical joist profile.

brackets (Figure 2). The trapezoidal shaped formed web openings along the length of the joist can accommodate large diameter utility passage within the floor cavity. The trapezoidal web openings are edge stiffened to provide for enhanced shear strength (Figure 3).

Experimental Study

The experimental study considered the varied strength issues related to the adequate performance of the joist to include clinching requirements for the closed chord, shear strength of the edge stiffened open web, and the flexure capacity of the closed chords.

Fig. 3. Reinforced web opening.

Fig. 4. Test rig figure.

Using a simulated uniform loading test rig that enabled load to be applied by pneumatic cylinders spaced 12 inches on center, the joists were tested to failure (Figure 4). The experimental studies were performed in the structural engineering laboratory in the Department of Civil Engineering at the University of Missouri-Rolla.

Table 1. Mechanical properties.

Sheet thickness (in)	Yield strength (ksi)	Tensile strength (ksi)	Elongation %
0.046	51.4	61.5	31.5
0.035	51.3	60.8	33.7

Table 2. Clinch connection test results.

Test No.	Clinch Type	Failure Load (lb/ft)
1	"A"	129
2	"R"	130
3	"A" & screws	149
4	"R" & screws	377

Clinch Requirements

The performance of the closed joist chord to achieve a bending compression flange failure is dependent upon the structural performance of the clinch connection. For this investigation, tests were performed on joists fabricated by using two different clinch geometries, Type A and Type R. The clinches were spaced at six inches on center. Figure 3 illustrates the clinch location within the cross section.

Two joist thicknesses were considered in this study, 0.046 in and 0.035 in. Table 1 summarizes the mechanical properties of the sheet steels used to fabricate the joists profile. Table 2 summarizes the test results.

Test specimen Nos. 1 and 2, had clinches spaced at six inches on center. For Test No. 1, failure was initiated by release of the clinch at an applied load of 90.6 lb/ft and subsequent failure of the compression web at a load of 129 lb/ft. For Test No. 2 the failure was sudden with simultaneous failure of the clinch and buckling of the compression web. Figure 5 shows a typical clinch failure.

To evaluate the performance of the joist with a closer spaced clinch pattern, self-drilling screws were added between the clinches. Thus, the connectors were spaced at three inches on center. Test Nos. 3 and 4 achieved an increase in strength which is indicative of the enhanced performance to be achieved by a closer spaced clinch connector. The failure mode for both tests was a buckling of the compression web, not a failure of the clinch connection.

Based on the test results, to achieve an adequate connection, a connector spacing of three inches is required. Although the closed joist chord did not achieve a bending compression flange failure, the tests demonstrated that clinch type A was superior to clinch type R when the connector spacing was reduced to three inches.

Flexure Capacity of Compression Chord

Two joist thicknesses were considered in this study, 0.046 in and 0.035 in. Table 1 summarizes the mechanical properties of the sheet steels used to fabricate the joists. To preclude failure of the

Fig. 5. Typical clinch failure.

Table 3. Flexure test results.

Test No.	Joist Thickness (in)	Joist Depth (in)	Joist Span (ft)	w_{test} (lb/ft)	w_{comp} (lb/ft)	w_{test}/w_{comp}
2-5	0.035	12	20	176.8	178.5	0.991
2-6	0.035	12	20	203.0	178.5	1.137
3-8	0.035	12	18	202.0	221.0	0.914
3-9	0.035	12	18	216.0	221.0	0.977
3-5	0.046	12	20	249.0	239.0	1.042
3-7	0.046	12	20	249.0	239.0	1.042

compression web, as occurred in the clinch study, the web was reinforced or embossed as shown by Figure 3. Clinches were of type "R" spaced three inches on center.

A total of six tests were performed in this study as summarized in Table 3. Figure 6 shows a typical joist failure. The tested capacity, w_{test}, was compared with the computed capacity, w_{comp}, where the computed capacity was based on the AISI specification (North, 2001). On the average, the flexure capacity of the joist was achieved for both thicknesses of joist. Also, the tests verified that the joist behavior is best modeled as a beam and not a truss.

Web Shear Capacity

A series of ten tests were performed to assess the affect of shear on the performance of the open-web I-joist. Two material thicknesses, 0.047 in and 0.060 in, were considered in the test program (Table 4). Clinches were of type "R" spaced three inches on center.

The load application was a simulated uniform loading condition (Figures 3 and 4) and for each test specimen the maximum applied uniform load, w_{test}, is listed in Table 4. The failure for each test

Fig. 6. Flexure failure.

Table 4. Shear test results.

Test No. (in)	Joist Thickness (in)	Joist Depth (ft)	Joist Span (lb/ft)	w_{test} (kips)	V_2 (kips)	V_n	V_2/V_n
4-1	0.035	18	20	242	2.26	2.12	1.07
4-2	0.35	18	20	241	2.25	2.12	1.06
4-3	0.60	18	20	300	2.80	2.84	0.99
4-4	0.60	18	20	308	2.87	2.84	1.01
4-6	0.35	16	20	275	2.57	2.12	1.21
4-7	0.35	16	19.67	275	2.54	2.12	1.20
4-8	0.35	16	19.67	269	2.49	2.12	1.17
4-9	0.35	14	19.67	237	2.19	2.12	1.03
4-10	0.60	14	19.67	306	2.83	2.84	1.00

specimen occurred in the flat portion of the web between the junction of the chord and web and the edge of the stiffened hole (Figure 7). This web depth for the section was taken as 0.75 in for all test specimens, as indicated on Figure 2. The failure occurred at the location of the first compression web diagonal. The shear was assumed to be linearly varying along the length of the span, thus the computed shear force at the first compression diagonal, V_2, is listed in Table 4.

The nominal shear strength of the web, V_n, was evaluated by using the AISI specification (North, 2001). The controlling shear limit state for all test specimens was web yielding because the web flat width was defined as 0.75 in for test specimens. This flat width defined the corresponding h/t ratios. Web yielding is defined by the following equation:

$$V_n = 0.6F_y ht, \tag{1}$$

Fig. 7. Typical shear failure.

where $F_y = 50.2$ ksi and $t = 0.047$ in or $F_y = 52.9$ ksi and $t = 0.060$ and $h = 0.75$ in. The accuracy of Equation (1) to estimate the shear capacity of the web is shown by the ratio of V_2/V_n in Table 4. For the test specimens in this study, the V_2/V_n ratio ranged from 0.99 to 1.21 with an average value of 1.08.

Summary and Conclusions

An experimental study was performed to consider the varied strength issues related to the adequate performance of an I-joist having an open web configuration. The strength issues include clinching requirements to achieve the closed chord profile, the shear strength of the edge stiffened open web, and the flexure capacity of the closed chords.

Based on the test program, two clinch types were studied and it was determined that clinch type "R" was more structurally effective. Also, a maximum spacing of three inches on center was required to preclude a premature failure of the clinch connection.

Both the shear and the flexural capacity of the I-joist can be adequately predicted by using the design provisions of the North American Specification for the Design of Cold-Formed Steel Structural Members (2001).

Reference

North American Specification for the Design of Cold-Formed Steel Structural Members, 2001, American Iron and Steel Institute, Washington, DC.

DETERMINATION OF THE LATERAL STRENGTH OF SHEAR WALL PANELS WITH COLD-FORMED STEEL FRAMES

Joel Martínez[1] **and Lei Xu**
Department of Civil Engineering, University of Waterloo
Waterloo, Ont., Canada N2L 3G1

Abstract

In current construction practice, lateral strengths of shear wall panels with cold formed steel framing are primarily determined by tests owing to the lack of analytical methods. Meanwhile, the use of numerical methods such as the finite element method has been limited to researchers investigating the behaviour of SWP. Moreover, the finite element method has rarely been employed in design practice to determine the lateral strength of shear wall panels because the modelling is cumbersome. Presented in this paper is an analytical method to determine the ultimate lateral strength of shear wall panels. The method accounts for the factors that affect the behaviour and the strength of shear wall panels, such as material properties and thickness of sheathing, sizes of the C-shape steel studs, spacing of fasteners, and so on. Lateral strengths obtained from the proposed method for sheathing wall panels were compared with those of recent experimental investigations. The results of the comparison demonstrate that the predicted lateral strengths are in good agreement with those of the tests. Therefore, the proposed method is recommended for engineering practice.

Keywords: lateral strength, shear wall panels, cold formed steel framing.

1. Introduction

In the search for new constructive methods and materials for low- and mid-rise residential buildings, where the quality of living, ease of construction, and cost-efficiency could be improved, cold formed steel (CFS) has been an attractive alternative to traditional materials such as timber. In cold formed steel framing construction, the shear wall panel (SWP) is one of the primary lateral load resisting systems, which has been extensively used in seismic applications in North America.

Typically, SWP in cold formed steel wall framing are constructed with vertically spaced and aligned C-shape cold formed steel studs. The ends of the studs are connected to bottom and top tracks of the wall. Sheathing may be present on either one or both sides of the wall with screw fasteners. In practice, studs are generally designed to support vertical loads, while sheathing is considered to resist lateral loads. However, the lateral strength of SWP cannot be determined alone by the strength of the sheathing, as the interaction among the sheathing, the studs, and the fasteners affect both the behaviour and lateral strength of SWP considerably. Generally, a SWP may experience both in-plane gravity and lateral loads as well as out-of-plane wind loads in the case of exterior walls. This study is focused on the evaluation of the lateral strength of SWP, which is the primary function of SWP. The strengths of SWP to resist in-plane gravity and out-of-plane loading are not considered.

To determine the behaviour and lateral strength of SWP, Serrette et al (1997, 2002) and Rogers et al (2004a) have carried out extensive experimental investigations. Gad et al (1999) conducted the

[1] Corresponding Author: j3martin@engmail.uwaterloo.ca

M. Pandey et al. (eds), Advances in Engineering Structures, Mechanics & Construction, 401–409.

finite element analysis on SWP, in which the studs and tracks were modelled by beam elements while sheathing was modelled with shell elements and sheathing-to-framing connections were modelled by nonlinear springs. Fulop and Dubina (2004b) conducted a series of tests and proposed a simplified model for determining the lateral strength of SWP based on replacing the sheathing with a pair of equivalent cross-bracing. A tri-linear force displacement relationship calibrated with test results was proposed for using the finite element method of analyzing SWP.

In practice, the lateral strengths of different SWP are available in the AISI design guideline (Brockenbrough, 1998) and standard (AISI, 2004). The values of the nominal lateral strengths of SWP presented in a tabulated form in the Lateral Design Standard of AISI (2004) are convenient to use and primarily determined on the basis of experimental tests, which provide an acceptable degree of confidence to the practitioners. However, as limited by the number of the tests being carried out, the freedom of selecting different sheathing materials, stud sizes and configurations of SWP is restricted as the tabulated values may not be applied or extended to SWP with different materials, configurations and construction details. Therefore, a reliable analytical method for determining the lateral strength of SWP is of importance to promoting cold formed steel framing technology.

Presented in this paper is an analytical method to determine the ultimate lateral strength of shear wall panels. This method accounts for the aspects that affect the behaviour and strengths of SWP associated with dimensions of the panel, material and cross-section properties of both sheathing and steel studs, and as well as the construction details such as the spacing of screw fasteners. The ultimate lateral strengths of different SWP are evaluated with the proposed method and the comparisons are made between the analytical results and the results obtained from recent experimental investigations (Serrette et al, 2002; Rogers et al 2004a; Fulop and Dubina 2004a).

2. Evaluation of the Ultimate Lateral Strength of SWP

2.1. Lateral Strength of SWP

The lateral strength of SWP is primarily contributed by the sheathing and framing studs and can expressed as,

$$P_R = P_{Sf} + P_F \qquad \text{Eq. (1)}$$

where P_{Sf} and P_F are lateral strengths associated with the sheathing and framing studs, respectively. In the case that the sheathing are provided in both sides of SWP, the lateral strength of the sheathing is given by

$$P_{Sf} = P_{Sf,1} + P_{Sf,2} \qquad \text{Eq. (2)}$$

where $P_{Sf,1}$ and $P_{Sf,2}$ are the lateral strength of sheathing presented on side 1 and 2 of the panel, respectively. In addition to the material and cross-section properties of sheathing, the ultimate lateral strength of sheathing are also highly affected by the characteristics and arrangement of sheathing-to-framing connections, which will be discussed in Section 2.2. The lateral strength of framing studs, P_F, can be determined as

$$P_F = K_F \Delta \qquad \text{Eq. (3)}$$

where K_F is the lateral stiffness of the framing studs and Δ is the lateral deflection the SWP impending the failure. Compared to that of the sheathing, the framing studs contribute little to the ultimate lateral strength of SWP, as the lateral stiffness of the studs is insignificant. Therefore, for the reason of simplicity, the elastic lateral stiffness of the framing studs is adopted as

$$K_F = \sum_{studs} \frac{3E_F I_F}{h^3} \qquad \text{Eq. (4)}$$

where E_F and I_F are the Young's modulus and the moment of inertia of the framing studs, respectively. h is the height of the panel.

Considering the compatibility of lateral deformation between sheathing and framing studs prior to the failure of the panel, the relationship between the sheathing strength and the lateral deformation of the panel is,

$$\Delta = \frac{P_{Sf}}{K_{Sf}} \qquad \text{Eq. (5)}$$

where K_{Sf} is the lateral stiffness of sheathing. Substituting Eq. (5) into Eq. (3) yields,

$$P_F = \frac{K_F}{K_{Sf}} P_{Sf} \qquad \text{Eq. (6)}$$

substituting Eq. (6) into Eq.(1), the lateral strength of SWP is

$$P_R = \left(1 + \frac{K_F}{K_{Sf}}\right) P_{Sf} \qquad \text{Eq. (7)}$$

2.2. Lateral Stiffness and Strength of Sheathing

To obtain the ultimate lateral strength of SWP as shown in Eq. (7), it remains to evaluate the lateral stiffness, K_{sf} and strength of sheathing, P_{sf}. Experimental investigations have shown that the predominant mode of failure in SWP is initiated at the sheathing-to-framing connections for the most common sheathing materials such as plywood, OSB, and gypsum board, etc. That is, the degradation of both lateral stiffness and strength of such panels are primarily due to the failure of the connections (Serrette et al, 2002; Rogers et al 2004b). Therefore, the failure of the connections has to be accounted for while evaluating both the lateral stiffness and strength of sheathing. To that end, the degraded sheathing stiffness may be calculated as

$$K_{Sf} = \frac{G_S A_C}{1.2h} + \frac{3E_S I_S}{h^3} \qquad \text{Eq. (8)}$$

where E_S and G_S are the Young's and shear modulus of sheathing, respectively; h is the height of the panel; and A_C is the reduced cross sectional area of the sheathing, defined as

$$A_C = t_S d_C n_C \qquad \text{Eq. (9)}$$

in which t_s is the thickness of the sheathing; d_C is the diameter of the screws, n_C is the number of screws along the cross section of the sheathing that is connected to the top collector member, and I_S is the moment of inertia of the reduced cross-section and is given by

$$I_S = n_C \frac{t_S d_C^3}{12} + 2t_S d_C \sum_{i=1}^{n_C/2} (i \cdot s_C)^2 \qquad \text{Eq. (10)}$$

where s_C is the screw spacing at the edge of the panel.

Considering the analogy between SWP and the eccentrically loaded bolted connection, in both cases the loads are applied eccentrically and the strength reductions are as a result of the failures of the connections or fasteners initiated at locations which are far from the centre of rotation. In this research, the inelastic method of evaluating strength of the eccentrically loaded bolted connection proposed by Brandt (1982) is employed and extended to evaluate the ultimate lateral strength of sheathing. Brandt's method involved an iterative process of locating the inelastic instantaneous center of rotation of the bolt group as shown in Figure 1; the ultimate strength of the connection is found when all of the forces (both internal and external) on the connection are in equilibrium. Extended from Brandt's method, the ultimate lateral strength of sheathing, P_{Sfi} (i=1, 2) can be evaluated as

$$P_{Sf,k} = C_u^{(n_x)} V_r \qquad (k = 1, 2) \qquad \text{Eq. (11)}$$

where n_x is the index of the last iteration, and V_r is the strength of a single sheathing-to-framing connection which is determined by the minimum value of the bearing resistance of the sheathing material, the shear resistance of the fastener, and the bearing resistance of the steel stud. The parameter $C_u^{(n_x)}$ is the ultimate strength reduction coefficient for the group of sheathing-to-framing connections and can be evaluated through the following procedure.

The coordinates of the inelastic instantaneous center shown in Figure 1 are given by

$$x_c = x_o + a_x \qquad y_c = y_o + a_y \qquad \text{Eq. (12)}$$

Where x_o and y_o are the coordinates of the elastic center, and a_x and a_y are the x and y components associated with the distance between the inelastic instantaneous center and elastic center of the bolt group and are given by

$$a_x = \frac{P_y J}{n_f M_o} \qquad a_y = -\frac{P_x J}{n_f M_o} \qquad \text{Eq. (13)}$$

where

$$J = \sum_{i=1}^{n_f} \left(x_{si}^2 + y_{si}^2 \right) \qquad \text{Eq. (14)}$$

$$M_o = -P_x e_{yo} + P_y e_{xo} \qquad \text{Eq. (15)}$$

in which: P_x and P_y are components in the x and y directions of the normalized external force, unitary force, P as shown in Figure 1; J is the polar moment of inertia of all the fasteners with respect to the elastic center (x_o, y_o); x_s and y_s are the coordinates of fastener i; n_f is the total number of fasteners within the panel; M_o is the moment produced by the components of the unitary force; and e_{xo} and e_{yo} are the load eccentricities with respect to the elastic center. It is assumed that the farthest fastener from the inelastic instantaneous center has the largest deformation (Brandt, 1982), for wood sheathing, which is taken as 10mm (0.39in) according to tests carried out by Okasha (2004). The deformation of fastener i (Δ_i) is linearly proportional to the distance (d_i) between the fastener and the inelastic instantaneous center as shown in Eq.(20). Furthermore, the normalized force of the fastener is a nonlinear function of the deformation of fasteners (Eq. 21).

Initially, the components of the distance between the inelastic instantaneous center and fastener i for the first iteration (denoted by the superscript in parentheses) can be obtained as

$$d_{xi}^{(1)} = x_{si} - x_c \qquad d_{yi}^{(1)} = y_{si} - y_c \qquad \text{Eq. (16)}$$

by setting the coordinates of the elastic center to be zero (x_c=0, y_c=0) and substituting Eq. (12) into Eq. (16), we have

$$d_{xi}^{(1)} = x_{si} - a_x \qquad d_{yi}^{(1)} = y_{si} - a_y \qquad \text{Eq. (17)}$$

Thus, the distance between the instantaneous center and fastener i is

$$d_i^{(1)} = \sqrt{d_{xi}^{2(1)} + d_{yi}^{2(1)}} \qquad \text{Eq. (18)}$$

The initial load eccentricities with respect to the inelastic instantaneous center are

$$e_{x_i}^{(1)} = e_{xo_i} - a_x \qquad e_{y_i}^{(1)} = e_{yo_i} - a_y \qquad \text{Eq. (19)}$$

The iterative process for determining the inelastic instantaneous center and ultimate shear strength reduction coefficient C_u is described in the following. For the j-th iteration,

the normalized deformation of fastener i, $\Delta_i^{(j)} = 0.39 d_i^{(j)} / d_{\max}^{(j)}$ Eq. (20)

where d_{max} is the distance between the inelastic instantaneous center to the farthest fastener.

The normalized force of fastener i, $\left(R_i^{(j)} / R_u^{(j)}\right) = \left(1 - e^{-10\Delta_i^{(j)}}\right)^{0.55}$ Eq. (21)

Moment of the fasteners normalized forces, $M^{(j)} = \sum_{i=1}^{n_f} \left(R_i^{(j)} / R_u^{(j)}\right) d_i^{(j)}$ Eq. (22)

Moment of the eccentric unitary force, $M_p^{(j)} = -P_x e_y^{(j)} + P_y e_x^{(j)}$ Eq. (23)

The normalized ultimate fastener force, $R_u^{(j)} = \dfrac{-M_p^{(j)}}{M^{(j)}}$ Eq. (24)

The normalized force components of fastener i,

$$R_{xi}^{(j)} = \left(-d_{y_i}^{(j)} / d_i^{(j)}\right)\left(R_i^{(j)} / R_u^{(j)}\right) R_u^{(j)} \qquad \text{Eq. (25)}$$

$$R_{y_i}^{(j)} = \left(d_{xi}^{(j)} / d_i^{(j)}\right)\left(R_i^{(j)} / R_u^{(j)}\right) R_u^{(j)} \qquad \text{Eq. (26)}$$

The normalized force of fastener i, $R_i^{(j)} = \sqrt{\left(R_{xi}^{(j)}\right)^2 + \left(R_{y_i}^{(j)}\right)^2}$ Eq. (27)

Components of the unbalanced forces, $F_x^{(j)} = P_x + \sum_{i=1}^{n_f} R_{xi}^{(j)}$ Eq. (28)

$$F_y^{(j)} = P_y + \sum_{i=1}^{n_f} R_{y_i}^{(j)} \qquad \text{Eq. (29)}$$

Displacement of the inelastic instantaneous center to its new position,

$$\delta_{ax}^{(j)} = \left(-F_y^{(j)} / n_f\right)\left(J / M_o\right) \qquad \text{Eq. (30)}$$

$$\delta_{ay}^{(j)} = \left(F_x^{(j)} / n_f\right)\left(J / M_o\right) \qquad \text{Eq. (31)}$$

Updated normalized load eccentricities for the next iteration,

$$e_{xi}^{(j+1)} = e_{xi}^{(j)} - \delta_{ax}^{(j)} \qquad \text{Eq. (32)}$$

$$e_{y_i}^{(j+1)} = e_{y_i}^{(j)} - \delta_{ay}^{(j)} \qquad \text{Eq. (33)}$$

Updated distance between the inelastic instantaneous center and fastener i,

$$d_{xi}^{(j+1)} = d_{xi}^{(j)} - \delta_{ax}^{(j)} \qquad \text{Eq. (34)}$$

$$d_{y_i}^{(j+1)} = d_{y_i}^{(j)} - \delta_{ay}^{(j)} \qquad \text{Eq. (35)}$$

$$d_i^{(j+1)} = \sqrt{\left(d_{xi}^{(j+1)}\right)^2 + \left(d_{y_i}^{(j+1)}\right)^2} \qquad \text{Eq. (36)}$$

The ultimate strength reduction coefficient, $C_u^{(j)} = \left|\dfrac{M^{(j)}}{M_p^{(j)}}\right|$ Eq. (37)

Repeat above procedure until, $\dfrac{C_u^{(j)} - C_u^{(j-1)}}{C_u^{(j)}} < \varepsilon$ Eq. (38)

where ε is a pre-assigned tolerance for convergence. The iterative process is terminated when the coefficient C_u will be invariant in further iterations, which indicates the equilibrium conditions are satisfied with respect to the updated location of inelastic instantaneous centre. As stated by Brandt (1982), and also found by this study, only a few iterations are required to obtain the ultimate strength reduction coefficient.

Having computed the ultimate strength reduction coefficient, the strength of the sheathing can be calculated based on Eq. (11) and the ultimate lateral strength of SWP can then be determined in accordance with Eq. (7).

3. Results comparison between analytical and experimental investigations

Experimental results (Serrette et al, 2002; Rogers et al 2004a; Fulop and Dubina 2004a) are used to validate the accuracy of the proposed analytical method of evaluating the ultimate lateral strength of SWP. The accuracy of the evaluated strengths are generally correlated with the material properties and the geometric dimensions of the components. As not all properties are reported in the foregoing literature, the material properties adopted in the evaluation may not be matched with those of the tested materials. In this study, both the material and geometric properties of steel studs were based on the values published by the Steel Stud Manufacturers Association (SSMA, 2001), and the material properties of sheathing being used in the calculation will be discussed in each individual case.

Rogers et al (2004) conducted a series of experimental investigations on SWP with three different sheathing materials, Oriented Strand Board (OSB), Douglas Fir Plywood (DFP), and Canadian Softwood Plywood (CSP). The cold formed steel studs were 92S41-1.12mm (358S158-44mils), spaced 610mm (24in) at the center, and double studs were placed at the chords. The sheathing was fastened with No. 8 screws (diameter = 4.06mm) on one-side of the panel. Screw spacing was 305mm (12in) in the field, and the edge spacing varied from 152mm (6in) to 76mm (3in). The length and height of the SWP were 1219mm (4ft) and 2438mm (8ft), respectively. The ultimate lateral strengths shown in Table 1 are the average values obtained from three specimens.

The following material properties are used to evaluate the ultimate lateral strengths of the foregoing SWP. The shear modulus of elasticity for OSB, DFP and CSP are 925MPa, 825MPa, and 497MPa, respectively (Okasha, 2004), while the modulus of elasticity associated with OSB (OSB, 1995), DFP and CSP (CANPLY, 2003) are 9917MPa, 10445MPa, and 7376MPa, respectively. The comparison between the analytical and test results is presented in Table 1.

Table 2 shows the comparison between the results of the proposed method and those of the experimental investigation conducted by Serrette et al (2002). The framing steel studs used in the two tests were 89S41 (350S158) with thicknesses of 1.37mm (54 mils) and 1.73mm (68 mils). The studs were spaced 610mm (24in) at the center, and double studs were placed at the chords. The sheathing material was OSB, and sheathing was presented on one side of the panel. The screw spacings on the edge and in the field of the panel were 51mm (2in) and 305mm (12in), respectively. The SWP dimensions were 1219mm (4ft) by 2438mm (8ft). The ultimate lateral strengths shown in Table 2 are the average values obtained from two specimens. The foregoing material properties associated with OSB sheathing are used to evaluate the strength of the SWP.

Presented in Table 3 is the comparison between the result of the proposed method and that of the experimental investigation conducted by Fulop and Dubina (2004a). The framing steel studs were 152S44-1.57mm (600S175-62mils) with 610 mm (24in) spacing. OSB sheathing was presented

Table 1. Comparison between analytical and tested results (Rogers et al, 2004a)

Shear Wall Panel description	Edge screw spacing (mm)	Lateral Strength (N/m)		Test / Predicted
		Test	Predicted	
OSB sheathing: 11 mm Steel stud 92S41-1.12mm Screw size: No. 8 Field screw spacing: 305mm	152	13257	12423	1.07
	102	19293	18181	1.06
	76	23550	23911	0.98
DFP sheathing: 12.5mm Steel stud 92S41-1.12mm Screw size: No. 8 Field screw spacing: 305mm	152	16010	15636	1.02
	101	23792	22920	1.04
	76	29721	30171	0.99
CSP sheathing: 12.5mm Stud 92S41-1.12mm Screw size: No. 8 Field screw spacing: 305mm	152	12752	12736	1.00
	102	16596	18521	0.90
	76	24880	24266	1.03

Table 2. Comparison between analytical and tested results (Serrette et al, 2002)

Shear Wall Panel description	Screw size (mm)	Stud Thickness (mm)	Lateral Strength (N/m)		Test / Predicted
			Test	Predicted	
OSB sheathing: 11mm Steel stud: 89S41 Screw spacing (mm) Edge: 51; Field: 305	No.8	1.37	34383	35373	0.97
	No.10	1.73	44964	42225	1.07

Table 3. Comparison between analytical and tested results (Fulop and Dubina, 2004a)

Shear Wall Panel description	Screw Size	Lateral Strength (N/m)		Test / Predicted
		Test	Predicted	
OSB sheathing: 10mm Steel stud: 152S44-1.57mm Screw spacing (mm) Edge: 102; Field: 305	d=4.8mm	21882	20120	1.09

on one side of the panel. The screw diameter was 4.6mm and the screw spacings were 102mm (4in) on the edge and 254mm (10in) in the field of the panel. Different from the foregoing two experimental investigations, the dimensions of the panel were 3600mm (≈12ft) by 2440mm (8ft). The ultimate lateral strengths shown in Table 3 are obtained from one specimen. As the material properties of sheathing were not available from the literature, the foregoing properties of OSB are employed in the analytical evaluation.

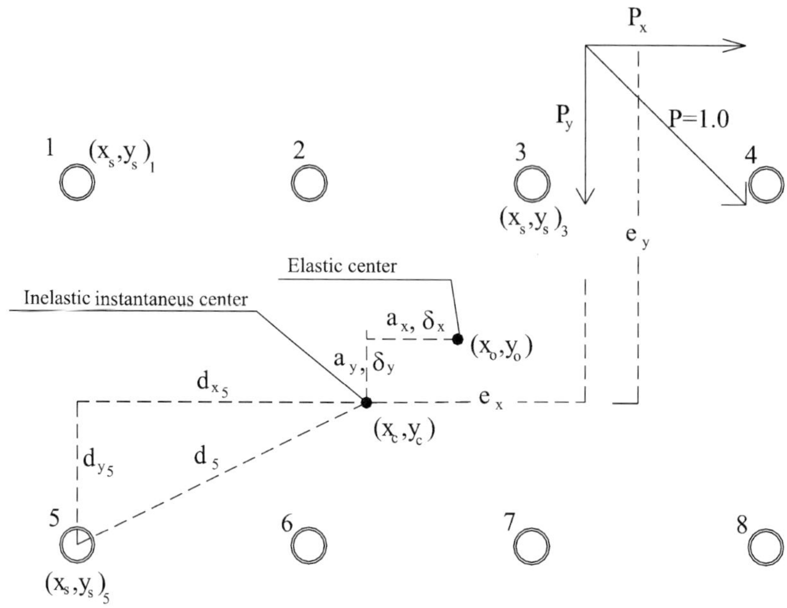

Figure 1. Fastener arrangement notation

4. Conclusions

The utilization of shear wall panels constructed with cold-formed steel and wood sheathing is becoming common practice for low- and mid-rise residential construction. However, analytical methods of evaluating the ultimate lateral strengths of the panels are needed to be developed in order to make cold formed steel systems more attractive to design practitioners. The method presented in this paper is comprehensive and can be used to evaluate the ultimate lateral strengths of SWP with different sheathing and framing materials, as well as panel dimensions and construction details such as fastener spacing. The comparisons made on the results obtained from the proposed method and the experimental investigations have shown good agreement between the evaluated and tested results. Therefore, the proposed method is recommended for engineering practice.

Acknowledgements

The first author would like to express his appreciation to the Mexican National Council for Science and Technology (Consejo Nacional de Ciencia y Tecnología) for its financial support. The authors are in debt to Dr. C. Rogers for providing experimental data.

References

AISI (2004). Standard for Cold-Formed Steel Framing-Lateral Design, *American Iron and Steel Institute*.

Brandt, G. D. (1982). Rapid Determination of Ultimate Strength of Eccentrically Loaded Bolt Groups, *Engineering Journal*, American Institute of Steel Construction.

Brockenbrough, R. L. & Associates (1998). Shear Wall Design Guide, *American Iron and Steel Institute,* Technical Data, RG-9804.

CANPLY (2003). Plywood Design Fundamentals, *Canadian Plywood Association.*

Fulop, L., & Dubina, D. (2004a). Performance of wall-stud cold-formed shear panels under monotic and cyclic loading Part I: Experimental research. *Thin-Walled Structures, 42*, 321-338.

Fulop, L., & Dubina, D. (2004b). Performance of wall-stud cold-formed shear panels under monotic and cyclic loading Part II: Numerical modelling and performance analysis. *Thin-Walled Structures, 42,* 339-349.

Gad, E. F., Chandler, A. M., Duffield, C.F., & Stark, G. (1999). Lateral Behaviour of Plasterboard-Clad residential Steel Frames. *Journal of Structural Engineering, ASCE*, January, 32-39.

Okasha, A. F. (2004). Performance of steel frame / wood sheathing screw connections subjected to monotonic and cyclic loading, *Department of Civil Engineering and Applied Mechanics*, *University of McGill*, Montreal, Master's Thesis.

OSB (1995). OSB Design Manual, Design Rated Oriented Strand Board, Structural Board Association.

Rogers, C. A., Branston, A. E., Boudreault, F. A., & Chen, C. Y. (2004a). Light gauge steel frame / wood panel shear wall test data: summer 2003, *Department of Civil Engineering and Applied Mechanics*, *University of McGill*, Montreal, Progress Report.

Rogers, C. A., Branston, A. E., Boudreault, F. A., & Chen, C. Y. (2004b). Steel Frame / Wood Panel Shear Walls: Preliminary Design Information for Use with the 2005 NBCC, *13th Word Conference on Earthquake Engineering.*

Serrette, R. L, Escalada, J., & Juadines, M. (1997). Static Racking Behavior of Plywood, OSB, gypsum, and Fiberbond Walls with Metal Framing, *Journal of Structural Engineering, ASCE*, August 1997, 1079-1086.

Serrette, R. L., Morgan, K. A., & Sorhouet, M. A. (2002). Performance of Cold-Formed Steel-Framed Shear Walls: Alternative Configurations. *Light Gauge Steel Research Group, Department of Civil Engineering, Santa Clara University*, Final Report LGSRG-06-02.

SSMA (2001). Product technical information. Steel Stud Manufacturers Association. ICBO ER-4943P, Retrieved January, 12, 2004, http://www.ssma.com

MODAL DECOMPOSITION FOR THIN-WALLED MEMBER STABILITY USING THE FINITE STRIP METHOD

B.W. Schafer
Johns Hopkins University, Department of Civil Engineering
Latrobe Hall 210, Baltimore, MD 21218, USA
E-mail: schafer@jhu.edu

S. Ádány
Budapest University of Technology and Economics, Department of Structural Mechanics
1111 Budapest, Műegyetem rkp. 3., Hungary
E-mail: sadany@epito.bme.hu

Abstract

This paper demonstrates how to decompose general stability solutions into useful subclasses of buckling modes through formal definition of the mechanical assumptions that underlie a class of buckling modes. For example, a thin-walled lipped channel column as typically used in cold-formed steel can have its buckling mode response decomposed into local, distortional, global, and other (transverse shear and extension) modes. The solution is performed by writing a series of constraint equations that are consistent with the mechanical assumptions of a given buckling class. The mechanical assumptions that defined the buckling classes were determined so as to be consistent with those used in Generalized Beam Theory (see e.g., Silvestre and Camotim 2002a,b) The resulting constraint equations may be used to constrain the solution before analysis, and thereby provide the opportunity to perform significant model reduction, or may be employed after the analysis to identify the buckling classes that participate in a given buckling mode. This paper shows the framework for this process in the context of the finite strip method (building off of Ádány and Schafer 2004, 2005a,b) and discusses some of the interesting outcomes that result from the application of this approach. Of particular interest, and discussed here, is the definition of global buckling modes, and the treatment of members with rounded corners – each of which provide certain challenges with respect to traditional definitions of the buckling classes. Examples are provided to illustrate the technique and challenges. The long-term goal of the work is to implement the procedures in general purpose finite element codes and thus enable modal decomposition to become a widely available tool for analyzing thin-walled member cross-section stability.

Introduction

Understanding cross-section stability of thin-walled members is critical to successful design. Inherently, or explicitly, design methods rely on prediction and separation of the cross-section stability modes. Different post-buckling behavior and strength are associated with each of the modes. For example, in the case of a cold-formed steel lipped channel column, typically three buckling modes are identified: local, distortional, and global. Local buckling modes are most typically handled by effective width methods which empirically include the potential for post-buckling reserve. Global buckling modes, such as flexural-torsional buckling, are handled through empirical column curves and are combined with the local post-buckling result in some fashion to account for local-global interaction. Distortional buckling is treated by modified effective width's or a modified column curve. Regardless of the design approaches employed, a fundamental first step is identification and prediction of the elastic cross-section stability modes.

Member Stability Solutions by the Finite Strip Method

Cross-section stability of open thin-walled members may be readily examined using the finite strip method. In a conventional stability solution the member is modeled as an inter-connected series of

M. Pandey et al. (eds), Advances in Engineering Structures, Mechanics & Construction, 411–422.

strips, as highlighted in **Figure 1**. The nodal lines of the strips have four degrees of freedom (DOF) each. In the local coordinates of the strip, the membrane DOF (u_i,v_i) follow plane stress assumptions, while the bending DOF (w_i,θ_i) follow thin plate bending assumptions. The membrane DOF are allowed to vary linearly in the transverse direction, while the bending DOF vary as a cubic (i.e., the typical beam shape function). Longitudinal deformation typically employs a trigonometric function, in its simplest form a single half-sine wave is enforced. The strips are transformed to global coordinates and assembled in a conventional manner to form the global stiffness matrices. A complete discussion of the finite strip method is available in Cheung and Tham (1998) and explicit details for construction of the elastic and geometric stiffness matrices as used in the open source program CUFSM are available in Schafer (1997).

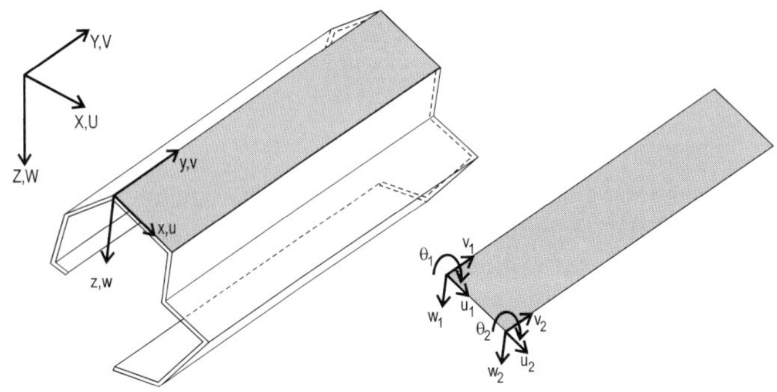

Figure 1. Coordinate systems and DOFs

The desired stability solution takes the form of an eigenvalue problem:

$$\mathbf{Kd} = \lambda \mathbf{K}_g \mathbf{d} \tag{1}$$

where **K** is the global elastic stiffness matrix, $\mathbf{K_g}$ is the global geometric stiffness matrix, λ is the buckling load multipliers, and **d** is the buckling mode shapes. **K** is dependent on cross-section geometry and material, while $\mathbf{K_g}$ is dependent on geometry and the applied longitudinal stress. The size of the eigenvalue problem is equal to 4*n*, where *n* is the number of nodal lines. Since both **K** and $\mathbf{K_g}$ vary as a function of length, the conventional approach is to sweep through all lengths of practical interest and construct the FSM buckling load multiplier vs. half-wavelength curve. Alternatively, one could fix the length and instead sweep through all sine wave "frequencies" of interest; this is in essence the approach in a conventional finite element method (FEM) stability analysis.

As an example of the information gained from a typical FSM analysis, consider the C-section of example (a) as shown in **Figure 2**. The developed FSM model has numerous internal nodal lines to ensure the accuracy of the local buckling solution. The model has a total of 21 nodal lines, and thus 84 independent DOF. The eigenvalue problem size is thus 84x84. Further, the eigen solution is performed at 59 different lengths to generate the results of **Figure 2**b. The analysis results demonstrate FSM's ability to capture all cross-section stability modes of interest, from local plate instabilities to global member instability, and FSM provides a means to perform an initial classification. Based on the half-wavelength, the presence of minima, and the observed cross-section deformations, out of 84 possible buckling modes examined at 59 different lengths, three classes of buckling modes are approximately defined: local, distortional, and global.

FSM greatly reduces the problem size below a conventional FEM (shell/plate element) stability analysis. However, the further conceptual reduction of the solution (e.g., our 84x84 eigen problem solved 59 times) down to three buckling classes: local, distortional, and global, is necessary for design. Without this reduction the possibilities remain too numerous, and it is impractical to provide engineers with reasonable guidelines on the post-buckling and collapse response of all the modes. As detailed in

Schafer and Adany (2005) exceptions exist where the FSM analysis is not sufficient to definitively identify the classes. If the buckling classes are defined properly from the start it should be possible to perform the solution directly for classes of modes; instead of indirectly as is done in a conventional FSM analysis. In this way, the conceptual reduction performed at the conclusion of an FSM analysis could become a mechanical reduction performed at the beginning of an analysis.

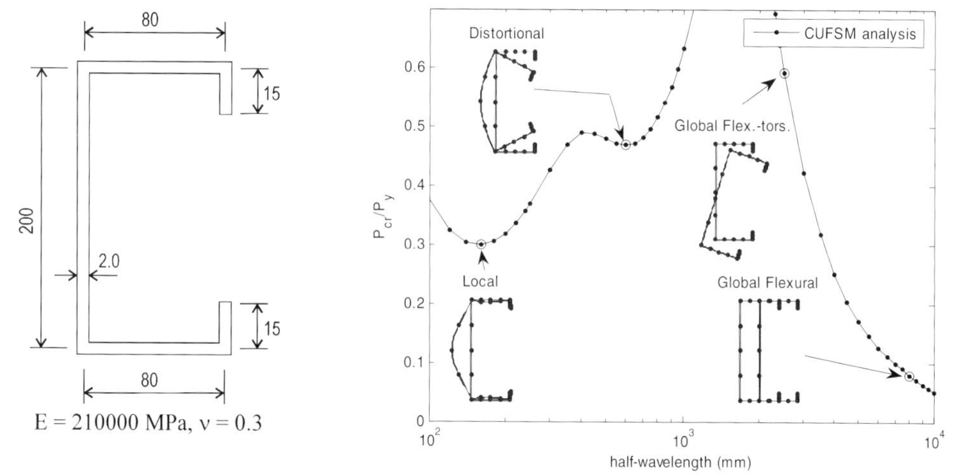

(a) example section (b) conventional finite strip method results

Figure 2. Finite strip analysis of a C-section with lips under pure compression, example (a)

Definition of Buckling Mode Classes

General

Generalized Beam Theory (GBT) is the only known method which is able to produce and isolate solutions for all the buckling modes: global, distortional and local. Further, modes identified via the GBT methodology are generally in accordance with commonly used definitions. Thus, our aim is to identify the critical assumptions in the GBT formulation that lead to the method's ability to isolate the modes. However, the goal is not to mimic the GBT formulation, but rather to construct the underlying GBT ideas as a series of distinct mechanical assumptions. These mechanical assumptions can then be enforced through constraints in more general methods such as FSM or FEM.

Mode Definitions

Based on an analysis of the GBT methodology, we put forth the following criteria as the critical mechanical assumptions necessary for defining the buckling modes. (Coord. system as in **Figure 1**.)

Criterion #1, membrane deformations:

- $\gamma_{xy} = 0$, membrane (in-plane) shear strains are zero,
- $\varepsilon_x = 0$, membrane transverse strains are zero, and
- $v = f(x)$, long. displacements are linear in x within an element/strip.

Criterion #2, longitudinal warping,

- $\varepsilon_y \neq 0$, long. strains/displacements are non-zero along the length.

Criterion #3, transverse flexure,

- $\kappa_y = 0$, no flexure in the transverse direction.

The buckling modes can then be defined as follows (see also Table 1).

- *Global* modes are those deformation patterns that satisfy all three criteria.
- *Distortional* modes are those deformation patterns that satisfy criteria #1 and #2, but do not satisfy criterion #3 (i.e., transverse flexure occurs).

- *Local* modes are those deformation patterns that satisfy criterion #1, but do not satisfy criterion #2 (i.e., no longitudinal warping occurs) while criterion #3 is irrelevant.
- *Other* modes are those deformations that do not satisfy criterion #1. Note, other modes do not exist in conventional GBT, but must exist in FSM due to the inclusion of membrane DOF.

Table 1. Mode classification table

	G modes	D modes	L modes	O modes
$\gamma_{xy} = 0$, $\varepsilon_x = 0$, v is linear	Yes	Yes	Yes	No
$\varepsilon_y \neq 0$	Yes	Yes	No	-
$\kappa_y = 0$	Yes	No	-	-

Modal Decomposition and the Constrained Finite Strip Method

General

The application of the above definitions implies that the appropriate deformation constraints, which are formulated in the three criteria, must be introduced into FSM. As a consequence, the original number of degrees of freedom (DOF) is necessarily reduced since the member can only deform in accordance with the strain conditions. Thus, our goal is to work out how the DOFs are reduced due to the various strain assumptions, and how this DOF reduction can practically be handled. Although the derivations are not extremely complicated, they are much longer than the limits provided by this paper. For this reason, here only a small example will be presented to demonstrate the method, which, at the same time, highlights all the important features of the more general derivations. Complete derivations can be found in (Ádány and Schafer 2005a).

Constraint Matrix Derivation

Consider the membrane deformation of a single finite strip, as shown in **Figure 1**. If the longitudinal distributions are assumed to be sinusoidal, as in the classical implementations of FSM, e.g., in CUFSM, the displacements can be expressed as a product of the assumed shape functions and the nodal displacements.

$$u(x,y) = \begin{bmatrix} \left(1-\frac{x}{b}\right) & \left(\frac{x}{b}\right) \end{bmatrix} \begin{bmatrix} u_1 \\ u_2 \end{bmatrix} \sin\frac{m\pi y}{a} \tag{2}$$

$$v(x,y) = \begin{bmatrix} \left(1-\frac{x}{b}\right) & \left(\frac{x}{b}\right) \end{bmatrix} \begin{bmatrix} v_1 \\ v_2 \end{bmatrix} \cos\frac{m\pi y}{a} \tag{3}$$

where u_1, u_2, v_1, v_2 are the transverse and longitudinal nodal displacements, m is the number of half-sine waves in the longitudinal direction, and a and b are the length and width of the strip, respectively.

Let us now introduce the criteria for zero transverse membrane strains:

$$\varepsilon_x = \frac{\partial u}{\partial x} = 0 \tag{4}$$

Substituting Eq. (2) into Eq. (4):

$$\varepsilon_x = \frac{\partial u}{\partial x} = \frac{-u_1 + u_2}{b}\sin\frac{m\pi y}{a} = 0 \tag{5}$$

and since the sine function is generally not equal to zero, u_1 and u_2 must be equal to each other in order to satisfy the equality.

This implies that the transverse displacements of the strip's two nodal lines must be identical, which is a natural consequence of the zero transverse strain assumption. In practice, the identical u

displacements prevent those deformations where the two longitudinal edges of the strip are not parallel, as illustrated in Figure 4.

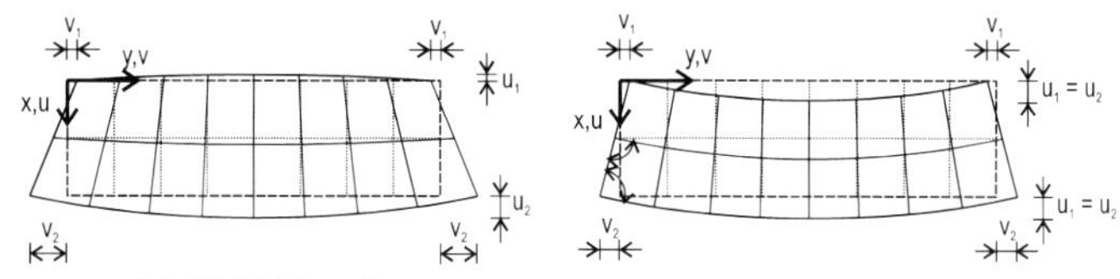

Figure 3. Effect of $\varepsilon_x = 0$ strain constraint on membrane deformations

The above derivation demonstrates that the introduction of a strain constraint reduces the number of DOFs, in this particular case from 4 to 3. Thus, we can define the new, reduced DOFs by u, v_1 and v_2, while the relationship of the original and reduced displacement vectors can be expressed as follows:

$$\begin{bmatrix} u_1 \\ v_1 \\ u_2 \\ v_2 \end{bmatrix} = \begin{bmatrix} 1 & 0 & 0 \\ 0 & 1 & 0 \\ 1 & 0 & 0 \\ 0 & 0 & 1 \end{bmatrix} \begin{bmatrix} u \\ v_1 \\ v_2 \end{bmatrix} \tag{6}$$

or in short:

$$\mathbf{d} = \mathbf{R}\mathbf{d}_r \tag{7}$$

where **R** is the constraint matrix, which is a representation of the introduced strain constraints.

Constraint Matrix

In case of the more general strain-displacement constraints, and more general cross-sections, the derivations are somewhat more complicated, but finally the associated constraint matrices (**R**) can be defined, as shown in Ádány and Schafer (2005a,b) for G and D modes and Ádány (2004) for L and O modes, and thus apply for all the criterion summarized in Table 1. Since a different **R** matrix may be constructed for each of the modal classes: G, D, L, and O, taken together they span the entire original nodal basis and represent a transformation of the solution from the original nodal basis to a basis where G, D, L, and O deformation fields are segregated.

The columns of the **R** constraint matrices are the deformation fields associated with the G, D, L, and O spaces. For the C-section of example (a), modelled only with nodal lines at the corners and the free edge, making for a 24 DOF model, the columns of the **R** matrix are provided graphically in **Figure 4**. **Figure 4**a and b provide the warping displacements and transverse displacements for the G and D modes. An important characteristic of these modes is that the transverse displacements are uniquely defined by the warping displacements. **Figure 4**c provides the transverse displacements for the L modes (note, no warping occurs in the L modes). As shown in the figure, these L modes appear to be in the nodal DOF basis, but they are not identical to the original FSM nodal basis, because they represent only the part of the nodal rotations that meet the L constraints – note nodal rotations occur in the G and D modes as well. Finally, **Figure 4**d and e provide the O modes, associated with shear and transverse extension. As discussed in detail in Ádány and Schafer (2005a,b) additional transformation inside the G, D, L, O spaces are possible and desirable. One attractive option is to use unit-member axial modes as detailed in Ádány and Schafer (2005b). In many applications we are interest in only one mode class, and in such a case the **R** matrix truly represents a constraint on the original DOF and can be applied to reduce the problem size.

Applications for the Constrained Finite Strip Method

Pure Mode Calculation

By the application of constraints associated with the strain assumptions of the various modes as described in Table 1, the problem of pure buckling mode calculation can be solved. Instead of solving the generalized eigenvalue problem of the member, as given in Eq. (1) one solves the reduced (or constrained) problem:

$$\mathbf{K_r d} = \lambda \mathbf{K_{gr} d} \tag{8}$$

where $\mathbf{K_r} = \mathbf{R}^T\mathbf{KR}$ and $\mathbf{K_{gr}} = \mathbf{R}^T\mathbf{K_g R}$ are the elastic and geometric stiffness matrix of the reduced DOF problem, respectively. **R** may be associated with any combination of the G, D, L, and O spaces or may include as little as one deformation field, for example D_1 of **Figure 4**, and thus reduce the problem to as little as one DOF.

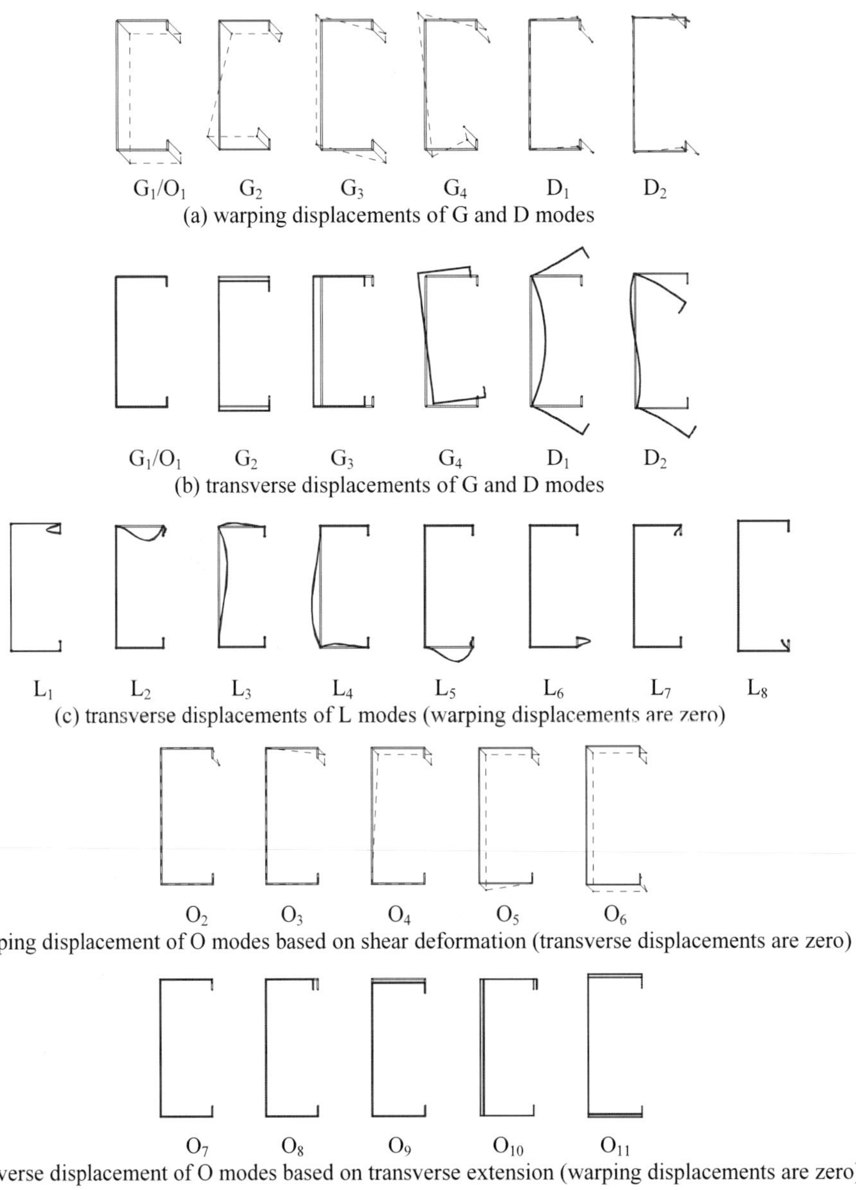

Figure 4. Deformation modes of example (a) (nodal lines at fold lines only, model has 24 DOF)

To demonstrate the pure mode calculation results consider again the analysis of example (a). In **Figure 5**a the conventional FSM analysis is compared to three separate analyses, one constrained to the L space, one constrained to the D space, and one constrained to the G space. The L space, spanned by the 8 deformation modes of **Figure 4**c requires solution of an 8x8 eigen problem. The D space, spanned by the last two deformation modes of **Figure 4**a and b requires solution of a 2x2 eigen problem, but actually the anti-symmetric D_2 mode does not contribute in this loading, so the solution can be simplified to a simple algebraic equation involving only the D_1 mode. The G space, spanned by the second through the fourth deformation modes of **Figure 4**a and b requires solution of a 3x3 eigen problem, and may be reduced to an algebraic equation for weak-axis flexural buckling and a 2x2 equation for flexural-torsional buckling. The constraint conditions successfully separate the modes.

The difference between the conventional FSM analysis and the D mode analysis is the most striking feature of the comparison. In **Figure 5**b the contribution of the L deformation modes and O deformation modes to this difference is demonstrated. Based on the mechanical definitions assumed here, the distortional minima in a conventional FSM analysis is not a pure mode; rather it includes interaction with additional mode classes; most notably, local buckling. Analysis of the combined LD space (9 DOF) provides a solution within ½% of the conventional FSM analysis (24 DOF). In this example the O modes must also be included (an additional 10 DOF) to close the error to zero. In other cross-sections distortional buckling may have a stronger interaction with global modes and thus the G space may be more important to include than the L space. No general conclusion can currently be made with regard to the significance of D modes which have interaction with L or G modes, but it is reasonable to assume that such a situation does impact the nature of the post-buckling response.

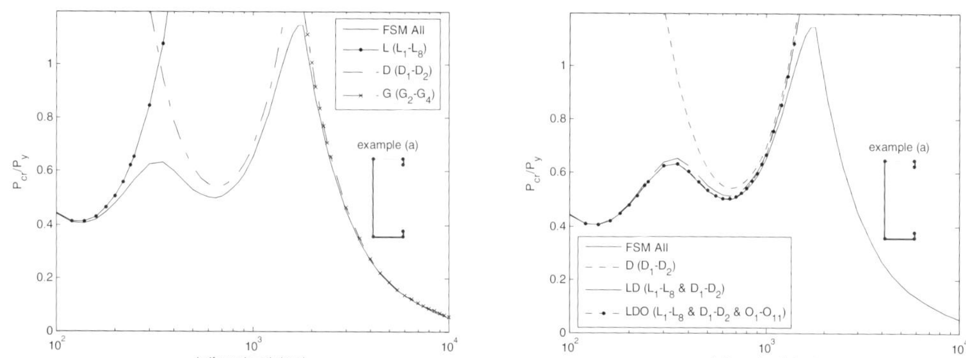

(a) FSM all mode analysis compared with constrained analyses in the L, D, and G mode classes

(b) FSM all mode analysis compared with constrained analysis in the D, D+L, and D+L+O mode classes

Figure 5. Comparison of conventional FSM analysis results with constrained models for example (a)

Mode Contribution Calculation

It is desirable to understand how the different pure modes of Eq. (8) for G, D, L, and O contribute in an all-mode or traditional FSM calculation, i.e. Eq. (1). This may be completed by transforming any displaced shape (buckled mode shape) into the eigenbasis created by the pure mode solution of Eq. (8). The eigenvectors **d** from the solution to Eq. (8) fully describe the pure mode solutions. Eq. (7) provides transformation from (or to) the pure mode space to the original DOF space.

The basis vectors must be normalized. Here, we select a normalization so that each base vector is associated with unit strain energy. In practice, let us consider again the eigenvalue problem of the member, as defined by Eq. (1). Any orthogonalized base vector satisfies Eq. (1), thus, we may write:

$$\mathbf{K}\mathbf{d}_o = \lambda \mathbf{K}_g \mathbf{d}_o \tag{9}$$

where $\mathbf{d}_o$ denotes the orthogonalized base vector. By pre-multiplying Eq. (9) with $\frac{1}{2}\mathbf{d}_o^T$:

$$\frac{1}{2}\mathbf{d}_o^T \mathbf{K}\mathbf{d}_o = \frac{1}{2}\lambda \mathbf{d}_o^T \mathbf{K}_g \mathbf{d}_o \tag{10}$$

where the left-hand side of the equation is the elastic strain energy, which by scaling the $\mathbf{d}_0$ vector can be set to unity. Any displacement vector can now be expressed as a linear combination of the basis vectors, by solving the linear matrix-equation as follows:

$$\mathbf{D}_0\mathbf{c} = \mathbf{d} \tag{11}$$

where $\mathbf{D}_0$ is a square matrix constructed from the orthonormal $\mathbf{d}_0$ base vectors so that each column of $\mathbf{D}_0$ would be a base vector; $\mathbf{d}$ is the given general displacement vector, while $\mathbf{c}$ is a vector containing the coefficients which are to be calculated. The contribution of any individual mode can be calculated as the ratio of the coefficient of that mode and the sum of all the coefficients, as follows:

$$|c_i| \Big/ \sum_{\text{all}} |c_i| \tag{12}$$

Similarly, the contribution of a mode class can be defined as:

$$\sum_{\text{mode}} |c_i| \Big/ \sum_{\text{all}} |c_i| \tag{13}$$

These definitions are preliminary, and imperfect. Currently the modal contributions defined in this manner are not unique. However, from a heuristic standpoint, they have value in allowing for an exploration of the various modal contributions.

Figure 6 provides modal contribution results for the cross-section of example (a), following Eq. (13). Generally the figure indicates the extent to which the G, D, L, or O classes contribute to the deformations at a given half-wavelength. For example, at the location of the distortional minimum (second minima in the curve of **Figure 6**a) D is the dominant class, but contributions from the other classes are observed. The defined contributions are imperfect, as they do not directly reflect the deformations impact on the buckling load. For example, at the local minima both D and O classes would appear to provide a significant modal contribution, but the analysis of **Figure 5**a indicates that the L class can provide the buckling load with only small error. While a robust modal contribution factor still remains a topic of future work, the value of such a metric is illustrated in **Figure 6**.

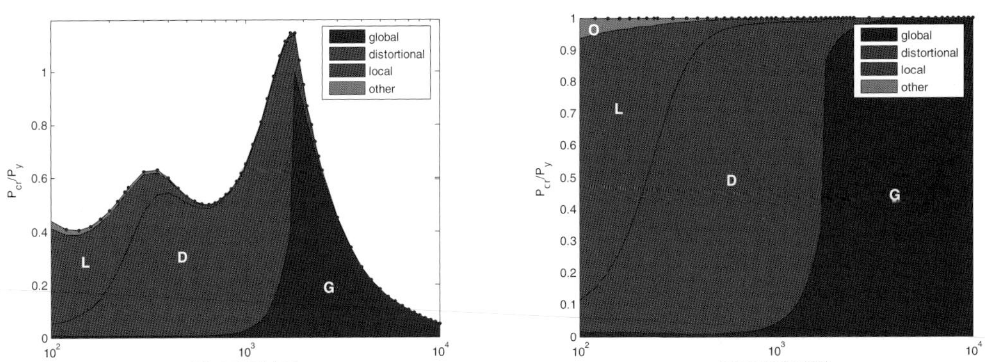

(a) under the conventional FSM analysis curve (b) as a simple portion of P_{cr}/P_y

Figure 6. Mode contribution/identification for example (a)

Discussion of Global Modes

A comparison of different solution methods for global modes highlight some important differences between constrained FSM, conventional FSM, and classically used analytical solutions. In **Figure 7**a the constrained FSM solution is compared to the classical analytical solution for flexural-torsional buckling (Timoshenko and Gere 1936). In the Figure, "Theory $G_1 - G_3$" represent the first three roots of the classical cubic equation that is solved for flexural-torsional buckling, and "$G_1 - G_3$" represent the first three eigenvalues of an FSM model constrained to only the deformations consistent with global modes (see **Figure 4**a and b). Constrained FSM gives higher critical forces. The difference is near 10% for any buckling lengths of practical importance. This difference is a direct consequence of the basic assumptions between beam and plate theory.

Constitutive Relations and Global Modes

In classical analytical solutions for global flexural buckling only the longitudinal normal stresses are considered, while the transverse normal stresses are assumed negligible. The problem is handled by a *beam model*, and only a one-dimensional constitutive relation, e.g.:

$$\begin{aligned}\sigma_x &= 0\\ \sigma_y &= E\varepsilon_y\\ \sigma_z &= 0\end{aligned} \tag{14}$$

Although rarely considered, transverse strains are not zero, even if they are small. FSM calculations are predicated on a *plate model*, and thus a two-dimensional constitutive relation, e.g.:

$$\begin{bmatrix}\sigma_x\\ \sigma_y\\ \tau_{xy}\end{bmatrix} = \begin{bmatrix}\dfrac{E}{1-\nu^2} & \dfrac{\nu E}{1-\nu^2} & 0\\ \dfrac{\nu E}{1-\nu^2} & \dfrac{E}{1-\nu^2} & 0\\ 0 & 0 & G\end{bmatrix}\begin{bmatrix}\varepsilon_x\\ \varepsilon_y\\ \gamma_{xy}\end{bmatrix} \tag{15}$$

It is assumed that the plates that form the cross-section are thin enough (compared to their width and length) that the Poisson effect should be considered. Thus, neither σ_x nor σ_y is negligible, while σ_z is implicitly assumed to be zero. If we calculate the pure global or distortional modes by applying the constraint matrix (**R**), we generate deformations that must satisfy the conditions that ε_x and γ_{xy} are zero. Thus, for constrained FSM the normal stresses can be expressed as follows:

$$\begin{aligned}\sigma_x &= \frac{\nu E}{1-\nu^2}\varepsilon_y\\ \sigma_y &= \frac{E}{1-\nu^2}\varepsilon_y\\ \sigma_z &= 0\end{aligned} \tag{16}$$

The difference between the beam model, Eq. (14), and the constrained FSM model, Eq. (16), is conspicuous. For longitudinal stresses the difference is equal to $1/(1-\nu^2)$. Considering that for steel the Poisson's ratio is approximately 0.3, the difference in the longitudinal stresses is approximately 10%. For flexural (global) buckling the only non-zero stress component (according to a beam model) is the longitudinal stress, consequently this difference in the longitudinal stresses directly transfers into a difference in the critical loads. The difference between the flexural buckling load calculated by the beam model and the constrained FSM is $1/(1-\nu^2)$ or 10% for $\nu=0.3$. Flexural-torsional global buckling involves shear stresses which are not affected by the Poisson's ratio, therefore, a smaller difference (i.e., < 10%) occurs between analytical and constrained FSM solutions.

Classical Flexural-Torsional Buckling Beam Model

Although the beam model is the classical model of structural mechanics, applied for centuries, and at least for decades even in buckling problems, its application for thin-walled members can be regarded as only an approximation. Two important approximations are involved: (i) transverse stresses are neglected and (ii) transverse plate flexure is neglected.

To asses the consequence of neglecting the transverse stresses, let us consider the flexural buckling of Example (a). Either the column buckles about its minor or major axis. The majority of the elastic strain energy develops from membrane strains/stresses. If the plates that make up the member are slender enough, the plane stress assumption is certainly more reasonable than the 1D stress assumption, which means that the transverse stresses (Poisson effect) cannot be neglected. Thus, from this aspect the beam model *under-estimates* the column rigidity. Transverse flexure, though small in extent, always takes place during flexural buckling. This may easily be demonstrated in a regular FSM or FEM analysis. It is obvious that transverse flexure provides additional flexibility to the column, consequently, from this aspect the beam model *over-estimates* the column rigidity. Thus, the beam model involves two competing approximations which nearly compensate each other so that the global buckling load provided by the beam model shows good agreement with more sophisticated models (e.g. FSM, see below) for most of practical cases, see **Figure** 7b.

Conventional FSM Model

A conventional FSM model (which is equivalent with a plate FEM model) considers both the Poisson effect and transverse plate flexure. If the analyzed plate elements are thin enough (which is always the case for cold-formed steel), and the applied finitization is sufficiently dense, both effects are considered in a correct way. Although these models still involve certain approximations (e.g. stresses/strains perpendicular to the plates are neglected), it is fair to say that FSM models calculate the elastic buckling of a thin-walled member with negligible error.

Constrained FSM Model

The constrained FSM model is a reduction of the regular FSM model, therefore it handles the plate elements as plates, i.e. together with the Poisson effect. However, the transverse strains as well as transverse flexure is completely eliminated, consequently a constrained FSM model behaves more rigidly than a regular FSM model. At the same time it differs from beam models, too, as already demonstrated via Eqs. (14) and (16), because it is not possible to simultaneously enforce zero transverse strains (which is a requirement from the basic GBT assumptions) and zero transverse stresses (which is the assumption of beam models). Since the constrained FSM satisfies the zero transverse strain condition, it gives larger rigidity, consequently higher critical forces than the beam models, as presented in **Figure** 7a.

Modified Constrained FSM

Unconstrained FSM provides the most rigorous solution, but does not allow decomposition nor identification of the buckling modes. Constrained FSM is clearly defined, but its lack of agreement with classical analytical solutions (the beam model) makes its practical use limited. Exact agreement between the beam model and constrained FSM is impossible. However, if we artificially assume that Poisson's ratio is zero, the constitutive equation for the constrained FSM model, Eq. (16) reduces to that of the beam model, Eq. (14). The two models are not identical, since transverse strains are zero in constrained FSM and non-zero in the beam model, but this difference has little effect, as **Figure** 7b shows.

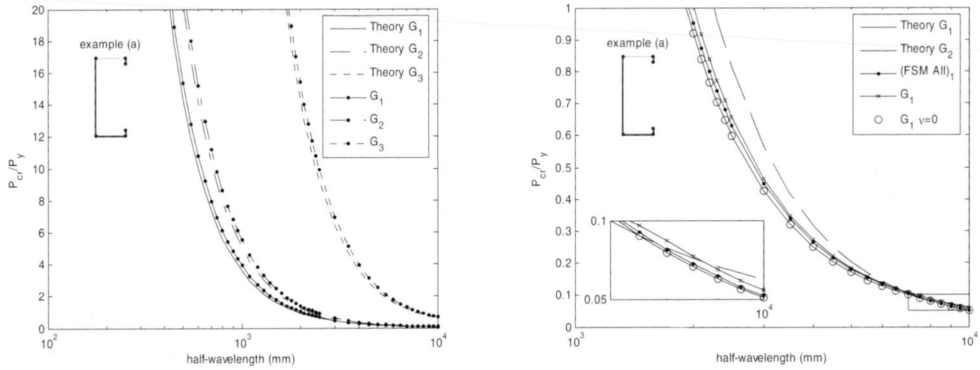

(a) analytical modes vs. pure global modes (b) comparison of global mode calculations

Figure 7. Comparison of global mode calculations for example (a)

The Impact of Corner Radii

In existing models with GBT the corners are always modeled as sharp, no corner radii is included. Comparison of FSM models with and without corner radii modeled, as shown in **Figure 8**a support the notion that the basic behavior is little influenced by small corner radii. A slight change in distortional buckling is observed, but this is still less than a few percent difference between the models.

However, from the standpoint of modal decomposition and the constrained FSM methods developed herein, the addition of corner radii has an important effect. **Figure 8**b illustrates the prediction of the local and distortional modes with three different models: an all mode, or conventional FSM analysis, and constrained models including only the D modes, and L modes. In addition, the figure also provides the predicted trasnverse displacements for the first minima (local buckling mode) predicted in the half-wavelength vs. buckling load plot. Even though the only change to the model is the addition of rounded corners, now the D modes do a better job of capturing the first minima than the L modes. Comparing this to the model with sharp corners of **Figure 5**a, the difference is striking. In fact, the predicted buckled shape via the L mode analysis is far too stiff when compared to a conventional (all mode) FSM analysis, and does not capture the appropriate deformations.

Comparing the transverse displacements in local buckling (**Figure 8**b); the L modes allow rotation only for the main nodes, and due to the angle changes around the corner all of the corner nodes are main nodes, thus the L modes effectively prevent the rotation of the corner as a global (rigid-body-like rotation). On the other hand, D modes allow transverse displacements, even though it is a special case of transverse flexure only, but since there are a many nodes in the vicinity of the corners, these transverse displacements result in a reasonable approximation of the real (all-mode) displacements. The addition of the corners also have a pronounced influence on the relationship between warping displacements and in-plane deformations. With the corners modeled explicitly, a gradient in the warping displacements through the corners allows a rotation to be engaged that closely mimics the corner rotations in the L modes of the sharp corner model. This may be best seen in the observed warping displacements, as given in **Figure 9**.

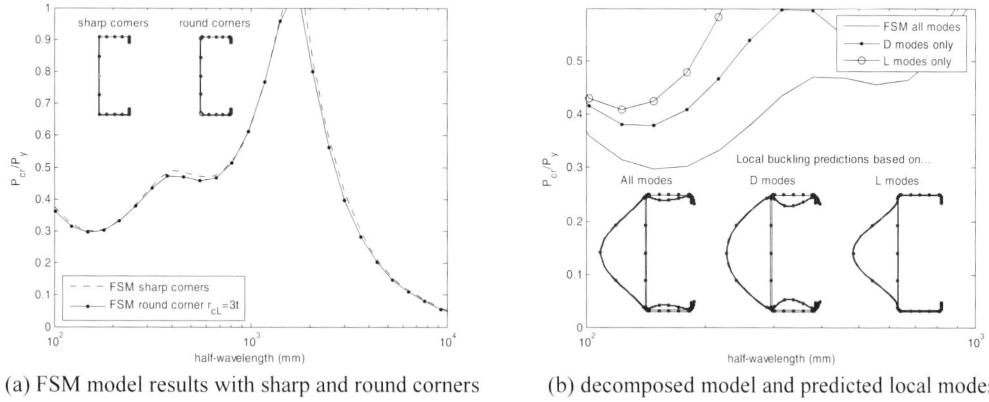

(a) FSM model results with sharp and round corners (b) decomposed model and predicted local modes

Figure 8. Impact of modeling cross-section with round corners on (a) FSM and (b) decomposition

The warping (longitudinal) displacements for the local buckling minima are provided in **Figure 9**. Comparison of **Figure 9**a and b, demonstrates that inclusion of corner radii alters the expected warping displacements. In a sharp corner model, the warping displacements only exist near the flange/lip juncture and have little impact on the actual result. With corner radii included in the model non-negligible warping also exists at the web/flange juncture; and further the warping distribution across the flange is approximately uniform. For constrained FSM models (**Figure 9**c and d) the D modes are unsuccessful in reproducing the actual (all mode) warping displacements, but the rounded corners allow warping at the web/flange juncture to engage bending in the web – and thus reasonably approximate the all mode local minima. Of course, the analysis consisting of L modes has, by definition, no warping, and in this case is a poor predictor.

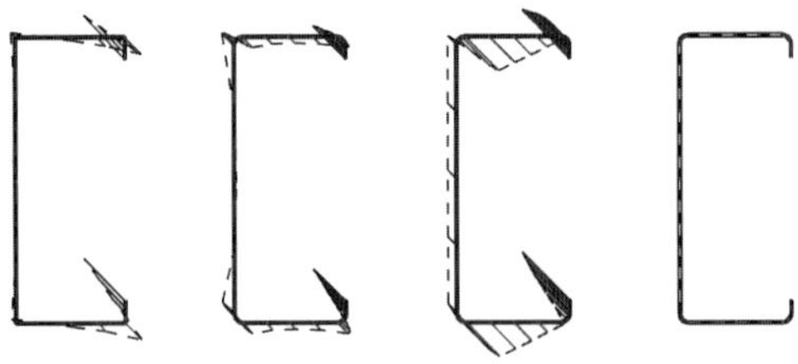

sharp corner round corner model
(a) all modes (b) all modes, (c) D modes (d) L modes

Figure 9. FSM predicted warping displacements for the local buckling minima

The constrained FSM method, based on GBT's mechanical assumptions, provides a unique means of decomposing a solution; however, it can not yet provide a general tool for modal classification and/or identification. The primary reason for this, is that reliance on heuristic definitions that describe the in-plane deformations; or simply point to minima in an FSM plot; are not themselves based on strict mechanical assumptions. Agreed upon definitions are needed in order to advance the field; however even simple issues such as models with corner radius included, demonstrate the challenges inherent in employing strict mechanical definitions. The corner radius models also suggest that further exploration of the relationship between criterion 3 regarding transverse flexure, and local plate bending are needed.

Conclusions

This paper provides an introduction to a new numerical method whereby general purpose finite element or finite strip methods can be constrained to deformation fields consistent with a particular class of buckling modes. The mechanical assumptions employed for this decomposition are based on Generalized Beam Theory, and implemented in a finite strip analysis. The decomposition (or constraining) of the conventional cross-section stability solution provides increased numerical efficiency and the potential to investigate cross-section stability behavior in a more detailed fashion. Some challenges exist when the approach is adopted; primarily related to the assumptions inherent in conventional solutions. For example, with global buckling modes it is shown that the constrained finite strip solutions are stiffer than analytic solutions due to the underlying constitutive relations. Models with and without corner radii present additional challenges for the mechanical definitions used to decompose the modes. Mechanics-based definitions of the cross-section stability modes do not agree well with current heuristic definitions when corner radii are present. Reasons for this difference are discussed in the paper. Modal decomposition brings the basic tools of Generalized Beam Theory for use in the more general finite strip method context, in doing so, both the advantages and limitations of such an approach are highlighted.

References

Ádány, S., Schafer, B.W. (2004). "Buckling mode classification of members with open thin-walled cross-sections." Fourth Int'l Conf. on Coupled Instabilities in Metal Structures, Rome, Italy, 27-29 Sept., 2004

Ádány, S., Schafer, B.W. (2005a). "Buckling mode decomposition of single-branched open cross-section members via finite strip method: derivation." Elsevier, *Thin-walled Structures*, (Submitted)

Ádány, S., Schafer, B.W. (2005b). "Buckling mode decomposition of single-branched open cross-section members via finite strip method: application and examples." Elsevier, *Thin-walled Structures*, (Submitted)

Cheung, Y.K., Tham, L.G. (1998). The Finite Strip Method. CRC Press.

Schafer, B.W. (1997). Cold-Formed Steel Behavior and Design: Analytical and Numerical Modeling of Elements and Members with Longitudinal Stiffeners. Ph.D. Disseration. Cornell University, Ithaca, NY.

Schafer, B.W., Ádány, S. (2005). "Understanding and classifying local, distortional and global buckling in open thin-walled members." Tech. Session and Mtg., Structural Stability Research Council. Montreal, Canada.

Silvestre, N., Camotim, D. (2002a). "First-order generalised beam theory for arbitrary orthotropic materials." *Thin-Walled Structures*, Elsevier, 40 (9) 755-789.

Silvestre, N., Camotim, D. (2002b). "Second-order generalised beam theory for arbitrary orthotropic materials." *Thin-Walled Structures*, Elsevier, 40 (9) 791-820.

STUDY ON SHEAR RESISTANCE OF COLD-FORMED STEEL STUD WALLS IN RESIDENTIAL STRUCTURE

Zhou Xuhong, Shi Yu, Zhou Tianhua, Liu Yongjian and Di Jin
Chang'an University, Xi'an, 710064, China
E-mail: zhouxuhong@126.com

Abstract

In this paper, tests and finite element analysis are used to study the shear resistance of cold-formed steel stud walls in low-rise residential structures. Firstly, the shear resistance of cold-formed steel stud walls under monotonic loading is tested. The test models, including walls with single-sided gypsum sheathing, walls with single-sided oriented strand board sheathing, and walls with gypsum sheathing on the back and oriented strand board on the face are made in full scale of engineering project. The test apparatus and test method and the failure process of specimens are introduced in detail. Then, the finite element analysis model of cold-formed steel stud walls considering geometric large deformation and materials nonlinear is presented to study their shear resistance. Walls were simulated as shell elements. The studs and tracks are simply connected. The screws connecting the sheathings to the frame are modeled by coupling methods. The solution method of equations is selected by ANSYS program automatically. Finite element analysis results in this paper are close to that of experiment. The results of test and finite element analysis show that sheathing materials influences the wall's shear resistance more greatly. The strength of steel has a less influence on the shear resistance of walls. As the decrease of stud spacing, height of wall and screw spacing at the perimeter, the walls' load ability increases obviously.

Keywords: cold-formed steel, assembled walls, shear resistance, experimental study, finite element analysis

1. Introduction

Assembled walls are the main load-bearing members of cold-formed steel residential buildings, and cold-formed steel stud walls are assembled by C-shaped steel studs (channel with lip flanges), U-shaped tracks (channel without lip flanges), gypsum board and oriented strand board (hereinafter OSB), which are connected by self-piercing or self-drilling screws(North American Steel Framing Alliance, 2000).

The shear resistance of cold-formed steel stud walls is associated with many factors, such as materials of studs and sheathing, screw spacing, height-width ration of wall, stud spacing and so on, so it is difficult to determine walls' shear resistance by theoretical calculation, but mainly by test method (American Iron and Steel Institute, 1998; Serrette and Ogunfunmi, 1996; Serrette et al., 1997). However, tests cannot totally reflect the influence of all the factors on the shear resistance of walls. Finite element analysis certified correct by test is an effective method to study the shear resistance of cold-formed steel stud walls (Xia et al., 2004; Emad et al., 1999). Many scholars at home and abroad only have studied the shear resistance for a certain kind of cold-formed steel stud walls because of a series of complex factors listed above. In order to provide design guidelines' tests and finite element analysis are presented to study the shear resistance of cold-formed steel stud walls in residential structures, and many factors influencing shear resistance are also analyzed in this paper.

M. Pandey et al. (eds), Advances in Engineering Structures, Mechanics & Construction, 423–435.

Table 1. Specimens of shear wall tests.

<table>
<tr><th>Denotation of wall specimens</th><th>Cold-formed steel member</th><th>Sheathing</th><th>Self-drilling fastener</th><th>Fastener interval</th></tr>
<tr><td>SSG</td><td rowspan="3">C89×44.5×12×1.0 studs (lipped channel section with web height 89mm, flange width 44.5mm,lip width 12mm and thickness 1mm), Spaced 600mm on center U92×40×1.0 tracks (channel section with web height 91mm, flange width 40mm and thickness 1mm) 50mm×1.0mm strap</td><td>gypsum sheathing on one side, with dimension 1.2m×3m and thickness 12 mm</td><td>ST4.2</td><td rowspan="3">150mm at the perimeter and 300mm in the middle field of board</td></tr>
<tr><td>SSO</td><td>OSB sheathing on one side ,with dimension 1.2m×2.44m and thickness 9 mm</td><td>ST4.8</td></tr>
<tr><td>DSGO</td><td>gypsum sheathing with thickness 12 mm on one side and OSB sheathing with thickness 9 mm on the other side</td><td>ST4.2 and ST4.8</td></tr>
</table>

2. Test Program

2.1. *Specimen Design*

The configurations of specimens are described in Table 1, and the walls with size of 3 m × 2.4 m (height by width) and 1.0 mm thick steel framing were sheathed with gypsum or OSB sheathing. But there was a vertical connecting seam on wall because every board only was half the width of wall, and there was a horizontal connecting seam on wall sheathed with OSB because the length of OSB sheathing was shorter than the length of wall, so a steel strap with 50 mm width and 1.0 mm thickness was fixed along the horizontal connecting seam to strengthen the wall. The studs in the middle of wall were single C-shaped cold-formed steel members, and the studs at both ends of the wall were two back-to-back C-shaped members connected by two lines of self-drilling screws. The 16 mm diameter uplift anchors were used to connect the walls to the test beam at the corner of wall, and the top and bottom tracks of the wall were fixed to the test beam by 12 mm diameter shear anchors. The yield strength of steel f_y was 320 N/mm^2, the tensile strength f_u was 379 N/mm^2, and the extension percentage of steel was 34%. Configuration of DSGO wall is illustrated in Figure 1.

2.2. *Test Apparatus*

The test apparatus is shown in Figure 2. Horizontal load was applied to wall through electro- hydraulic servo actuator, and lateral braces were also provided to limit out-of-plane movement of the test wall. The test process was operated by M2801 servo-control mechanism and computer. All the test dates were collected by a 7V08 date collector.

2.3. *Load Process and Failure Characteristics of Specimens*

Specimens were loaded to yield period by controlling 5∼7 loading steps. After being yield, specimens were loaded to failure by controlling displacement steps. Each loading stage stayed for about

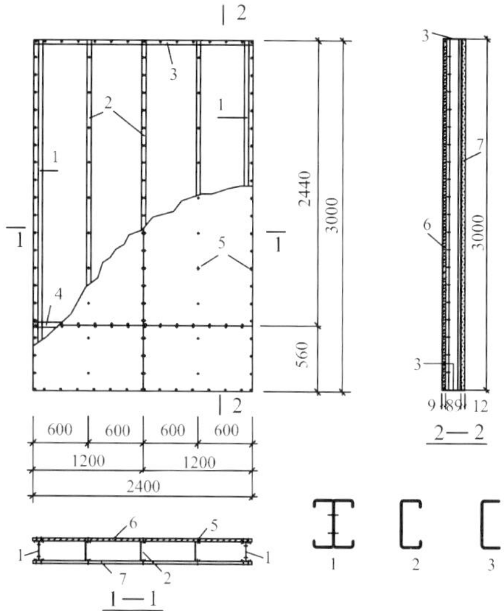

1. Back-to-back stud; 2. C-shaped stud; 3. Track; 4. Strap;
5. Self-drilling fastener; 6. Gypsum sheathing; 7. OSB sheathing

Fig. 1. Configuration of DSGO wall.

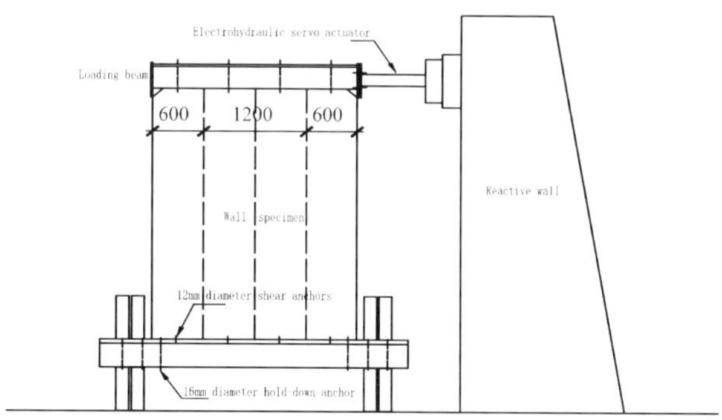

Fig. 2. Test apparatus.

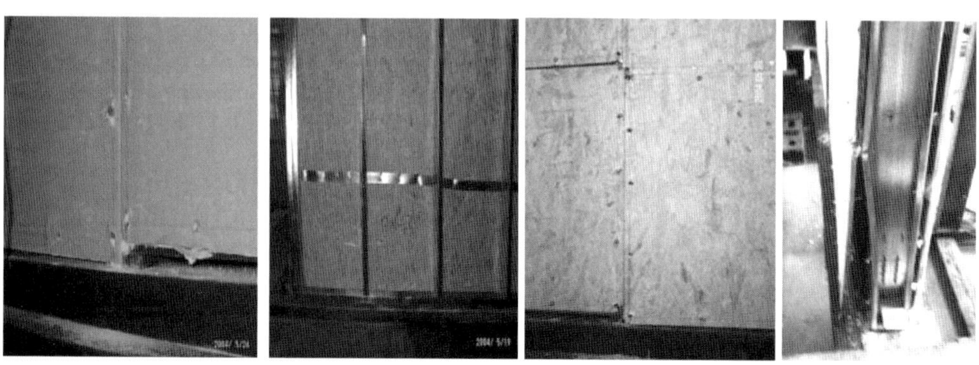

(a) Gypsum sheathing tearing at corner of wall (b) Buckling of strap (c) Slip of four panels of OSB sheathing (d) Local buckling of studs

Fig. 3. Failure modes of specimens.

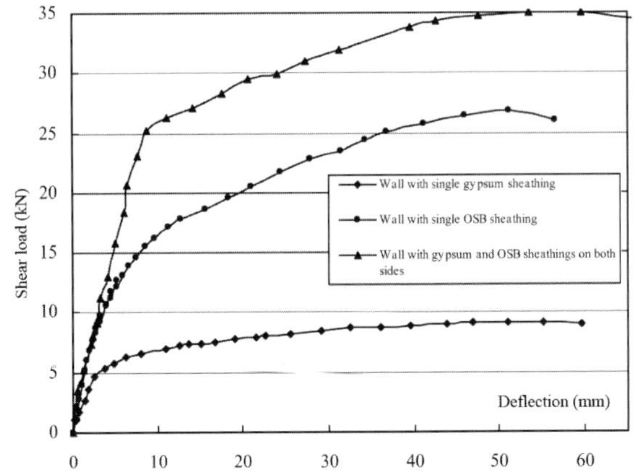

Fig. 4. Load-displacement curves for specimens.

3 minutes. The shear resistance and stiffness of SSG Specimen were lower. As a result, relative rotation was taken place between the two panels of gypsum sheathings, and their vertical connecting seam offset. The gypsum sheathing was tore at the corner of wall perimeter (shown in Figure 3a). The shear resistance strength and ductility of SSO wall were better than that of SSG wall, but the horizontal connecting seam in SSO wall influenced its bearing capacity and stiffness. With the load increasing, the steel strap buckled apparently (shown in Figure 3b), and the slip of the four panels of OSB sheathing was relatively great (shown in Figure 3c), which made the shear stiffness of wall reduced. The integrity, strength and stiffness of DSGO wall excelled that of SSG wall and SSO wall. The failure modes of three kinds of wall were similar. In general, when the failure of all wall specimens was occurred under shear load, most of the screw connecting sheathing and steel members around the wall were failed, and the stud on the end of wall usually locally buckled (shown in Figure 3d). But the gypsum sheathing or OSB sheathing did not drop off integrally due to the less damage of the screw connections in the middle field of wall, so we could conclude that the screws at the perimeter bore higher shear load than those in the middle field.

2.4. *Analysis of Test Results*

The shear load-displacement ($P - \Delta$) curves of all wall specimens are shown in Figure 4. Yield load of all walls could not been easily found from these curves, so all the characteristic loads including yield load were determined by the method prescribed in the Chinese *Specification of Testing Methods for Earthquake Resistant Building* (JGJ 101-96, 1996). The method for determining the characteristic loads of wall is illustrated in Figure 5. The horizontal line AB was drawn form the point A defining maximum load $P_{\max}$, then the secant OD was drawn intersecting line AB and curve OA at point D and C, respectively, when the area $ADCA$ was equal to the area $CFOC$. A vertical line was drawn form point D intersecting curve OA at point E. The shear load and the corresponding displacement of point E were the yield load P_y and the yield displacement Δ_y of wall respectively. With the displacement increasing, the load descend; when the load descend to $0.85P_{\max}$, the corresponding load and displacement were defined as the failure load P_u and the failure displacement Δ_u. The test results determined by the method described above are presented in Table 2.

The yield load P_y, maximum load $P_{\max}$ and failure load P_u of SSO wall were approximately 2.87, 2.94, 2.94 times those of SSG wall under the monotonic loading test. In other words, the load capacity of the wall with single-sided gypsum sheathing was 34∼34.8 % that of wall with single-sided OSB sheathing.

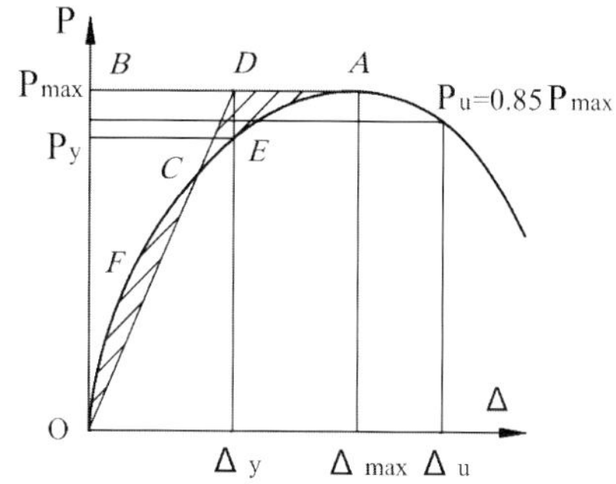

Fig. 5. Determination for special load point of specimens.

Table 2. Experimental results.

Wall type	P_Y (kN)	Δ_Y (mm)	P_{max} (kN)	Δ_{max} (mm)	P_u (kN)	Δ_u (mm)	Shear resistance (kN/m)
SSG	7.48	13.5	9.12	48	7.75	74.8	3.8
SSO	21.5	23.9	26.84	51.21	22.81	68	11.18
DSGO	29.12	19.6	34.99	59.62	29.72	74	14.58

The sum of yield load SSG wall and SSO wall was 28.98 kN, and the sum of maximum load of those two walls was 35.96 kN, the sum of failure load was 30.56 kN, which are close to the yield load 29.12 kN, the maximum load 34.99 kN and failure load 29.72 kN of DSGO wall respectively. Thus it can be seen that the total load capacity of two kinds of single-sided walls was close to that of double-faced walls with two corresponding kinds of sheathings.

3. Finite Element Analysis

3.1. Finite Element Analysis Model

Finite element analysis model was a mathematic representation of the practical structure, and the model and method of analysis should reflect the main performance of every member. The balanced equation of wall was listed as follows

$$[\mathbf{K}]\{\delta\} = [\mathbf{P}], \tag{1}$$

where $[\mathbf{K}]$ is the stiffness matrix of structure ($[\mathbf{K}] = \int_V [\mathbf{B}]^T [\mathbf{D}][\mathbf{B}]\, dV$), $[\mathbf{B}]$ stands for the geometric matrix, $[\mathbf{D}]$ stands for the material constitutive matrix, and V stands for the structure volume, $\{\delta\}$ is the joint displacement vector and $[\mathbf{P}]$ is the load vector.

$$\{\sigma\} = [\mathbf{D}]\{\varepsilon\}, \tag{2}$$

$$\{\varepsilon\} = [\mathbf{B}]\{\delta\}, \tag{3}$$

where $\{\sigma\}$ is the stress matrix and $\{\varepsilon\}$ is the stain matrix.

The stiffness matrix of structure $[\mathbf{K}]$ was not constant because the balanced equation considered the geometric and materials nonlinearity. Both the matrices $[\mathbf{D}]$ and $[\mathbf{B}]$ were related to the stress or strain.

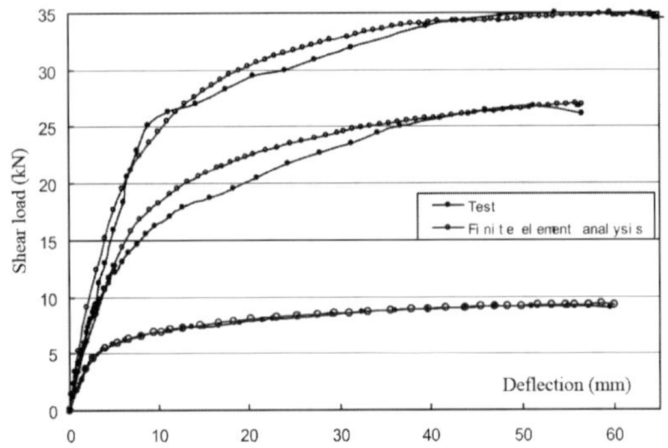

Fig. 10. Comparison of results between test and finite element analysis.

Table 4. The results of test and finite element analysis.

Wall type	Items	P_y (kN)	Δ_y (mm)	P_{max} (kN)	Δ_{max} (mm)	Shear resistance (kN/m)
SSG	test	7.48	13.5	9.12	48	3.8
	finite element analysis	7.71	16.59	9.29	59.79	3.87
SSO	test	21.5	23.9	26.84	51.21	11.18
	finite element analysis	22.46	20.33	26.98	55.15	11.24
DSGO	test	29.12	19.6	34.99	59.62	14.58
	finite element analysis	29.73	17.92	34.86	60.11	14.52

The finite element analysis results were close to those of tests (as shown in Figure 10). There were no obvious yield points in load-displacement curves in Figure 10. The characteristic points of curves were determined by the methods suggested in Chinese *Specification of Testing Methods for Earthquake Resistant Building* (JGJ 101-96, 1996). The comparisons of results between test and finite element analysis were listed in Table 4.

There was a clear error between walls' displacements at yield points or maximal load point measured through test and that calculated by finite element analysis, and the maximum error was 24.56%. But the errors on shear resistance of the SSG wall, SSO wall and DSGO wall were only 0.34% to 1.84%, and the errors on yield load of these three kinds of walls were only 2.1% to 6.03%. The finite element analysis results were close to those of the test. The shear resistance of SSO calculated by finite element analysis was 11.24 kN/m, which was only 3.1% lower than that of wall with the same configuration listed in Japan Iron and Steel Federation (2002). So the method of finite element analysis used in this paper was proved to be correct.

3.3. *Calculation Parameter Analysis of Cold-Formed Steel Stud Walls*

Based on the finite element analysis on the SSG wall, SSO wall and DSGO wall, a series of parameter analysis were done to study the influence of the steel strength, stud spacing, stud height, and screw spacing on the shear resistance of walls.

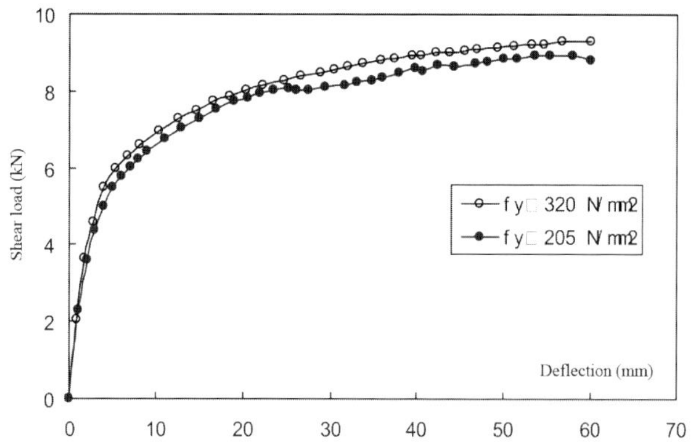

Fig. 11. Load-displacement curves of wall with single-sided gypsum sheathing on the face.

3.3.1. *The Influence of Steel Strength on the Shear Resistance of Walls*

The yield strength of steel of the wall with single-sided gypsum sheathing was 320 N/mm^2, and the maximum resistance of the wall was 9.29 kN. If the steel yield strength of this wall was changed to 205 N/mm^2, the maximum resistance calculated by finite element method was 8.95 kN. The shear resistance of the latter was only 3.65% lower than the former. The load-displacement curves of those two kinds of walls are shown in Figure 11. Obviously, the change of steel strength has little influence on the shear resistance of walls.

3.3.2. *The Influence of Studs Spacing on the Shear Resistance of Walls*

The studs of cold-formed steel stud walls usually spaced 400 mm and 600 mm. However, stud spacing of the test specimens and the finite element analysis models described above was 600 mm. Now the stud spacing of SSG wall, SSO wall and DSGO wall was adjusted to 400 mm, and the finite element analysis results of those walls are described in Table 5, the load-displacement curves are shown in Figure 12. When the stud spacing of SSG wall, SSO wall and DSGO wall was reduced from 600 mm to 400 mm, their shear resistance was increased by 14.47%, 24.11% and 29.96%, respectively. The stud spacing has obvious influence on the shear resistance of cold-formed steel stud walls, and this influence was strengthened with the increase of sheathing restriction.

Table 5. Finite element analysis results of walls.

Wall type	Stud spacing (mm)	P_{max} (kN)	Shear resistance (kN/m)
SSG	400	10.62	4.43
	600	9.29	3.87
SSO	400	33.47	13.95
	600	26.98	11.24
DSGO	400	45.29	18.87
	600	34.84	14.52

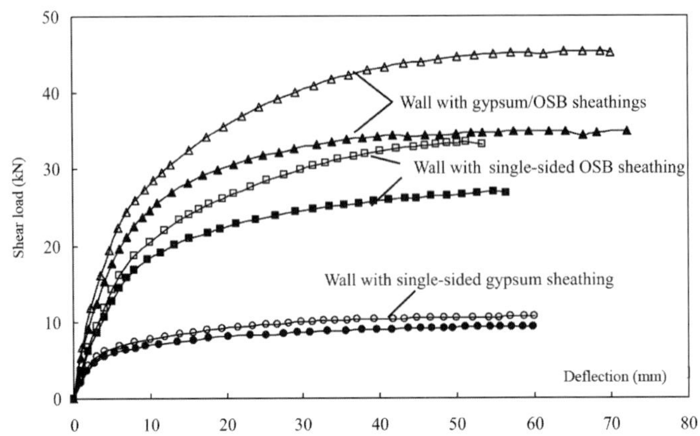

Fig. 12. Load-displacement curves of cold-formed steel stud walls.

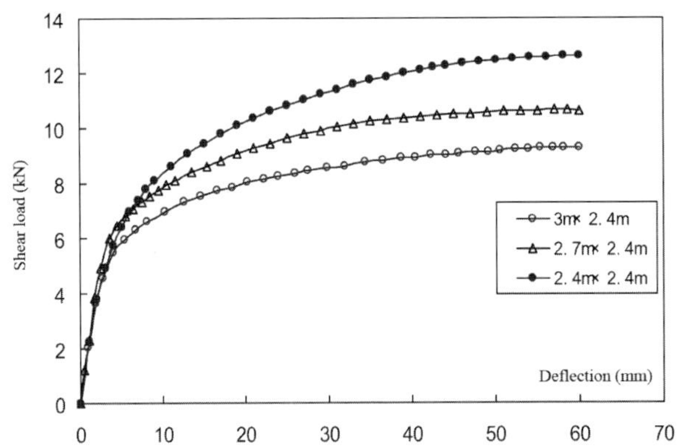

Fig. 13. Load-displacement curves of wall with single-sided gypsum sheathing on the face.

3.3.3. *The Influence of Height of Wall on the Shear Resistance of Walls*

Finite element analysis were presented on three kinds of walls with single-sided gypsum sheathing, whose width were all 2.4 m and the heights were 2.4 m, 2.7 m and 3 m, respectively. The results of finite element analysis are described in Table 6 and load-displacement curves are shown in Figure 13. When the height of walls was increased, the shear resistance of wall with dimension 3 m × 2.4 m was 12.84% lower than that of wall with dimension 2.7 m × 2.4 m, whose shear resistance was 15.59% lower than that of wall with dimension 2.4 m × 2.4 m. So the change of height has more influence on the shear resistance of walls.

3.3.4. *The Influence of Screw Spacing on the Shear Resistance of Walls*

The finite element analysis results of walls with different screw spacing are listed in Table 7. With the narrowing of screw spacing, the shear resistance of walls was increased. When the screw spacing of walls with single-sided gypsum sheathing, or with single-sided OSB sheathing or with gypsum sheathing on the back and OSB on the face, was adjusted form 150/150 (the screw spacing of wallboards was 150 mm at the perimeter and 150mm in the middle field) to 150/300, their shear resistance was improved by 3.36%, 2.85% and 1.93%, respectively. And if changed to 100/300, the shear resistance of walls was respectively improved by 6.98%, 1.33% and 10.54% compared with walls with screw spacing 150/300. So we can conclude that the narrowing of screw spacing in the

Table 6. Finite element analysis results of wall with single-sided gypsum sheathing on the face.

Dimension of walls (height × width)	Height-width ratio	P_{max} (kN)	Shear resistance (kN/m)
2.4 m × 2.4 m	1:1	12.63	5.26
2.7 m × 2.4 m	1.125:1	10.63	4.44
3 m × 2.4 m	1.25:1	9.29	3.87

Table 7. Finite element analysis results of walls.

Wall type	Screw spacing (mm)	P_{max} (kN)	Shear resistance (kN/m)
SSG	@100/300	9.93	4.14
	@150/300	9.29	3.87
	@150/150	9.60	4
SSO	@100/300	27.34	11.39
	@150/300	26.98	11.24
	@150/150	27.74	11.56
DSGO	@100/300	38.52	16.05
	@150/300	34.84	14.52
	@150/150	35.53	14.80

middle field lightly influenced walls' shear resistance when the screw spacing was less than 300 mm, but the walls' shear resistance was strengthened greatly with the decrease of screw spacing at the perimeter. This was in agreement with the phenomenon observed in the test.

3.4. *Design Suggestions*

The shear resistance of many kinds of walls was presented in Table 8 in order to guide actual engineering design. The results were got through the methods of test and finite element analysis introduced in this paper. The steel yield strength was 300 N/mm^2, the studs were C89 × 44.5 × 12 × 1, and the tracks were U92 × 40 × 1.

4. Conclusion

Tests and finite element analysis were presented in this paper to study the shear resistance of cold-formed steel stud walls in residential structures. We can draw the following conclusions:

(1) The material properties of the panel sheathing influence the shear resistance of cold-formed steel stud wall greatly. The shear load capcity of walls with single-sided gypsum sheathing is 34.8% of that of walls with single-sided OSB sheathing, and the total shear resistance of these two kinds of single-sided walls is close to that of wall with gypsum sheathing on one side and OSB sheathing on the other side.

Table 8. The shear resistance of cold-formed steel stud walls.

Dimension of wall (height×width) (m)	Stud spacing (mm)	Screw spacing (mm)	Sheathing	Shear resistance (kN/m)
3×2.4	600	150/300	12mm thick gypsum sheathing on one side	3.87
			9mm thick OSB sheathing on one side	11.24
			12mm thick OSB sheathing on one side	13.48
			12mm thick gypsum sheathing on one side and 9mm thick OSB sheathing with on the other side	14.52
			12mm thick gypsum sheathing on two sides, with thickness 12mm	7.31
			12mm thick OSB sheathing on both sides	25.1
			12mm thick gypsum sheathing on one side and 12mm thick OSB sheathing on the other side	17.11
		100/300	12mm thick gypsum sheathing on one side	4.14
			9mm thick OSB sheathing on one side	12.54
			12mm thick OSB sheathing on one side	15.01
			12mm thick gypsum sheathing on one side and 9mm thick OSB sheathing on the other side	15.89
			12mm thick gypsum sheathing on both sides	7.94
			12mm thick OSB sheathing on both sides	25.12
			12mm thick gypsum sheathing on one side and 12mm thick OSB sheathing on the other side	18.45
	400	150/300	12mm thick gypsum sheathing on one side	4.43
			9mm thick OSB sheathing on one side	13.95
			12mm thick OSB sheathing on one side	21.26
			12mm thick gypsum sheathing on one side and 9mm thick OSB sheathing on the other side	18.87
			12mm thick gypsum sheathing on both sides	8.26
			12mm thick OSB sheathing on both sides	42.72
			12mm thick gypsum sheathing on one side and 12mm thick OSB sheathing on the other side	27.85
		100/300	12mm thick gypsum sheathing on one side	4.82
			9mm thick OSB sheathing on one side	15.75
			12mm thick OSB sheathing on one side	21.98
			12mm thick gypsum sheathing on one side and 9mm thick OSB sheathing on the other side	20.57
			12mm thick gypsum sheathing on both sides	8.32
			12mm thick OSB sheathing on both sides	38.84
			12mm thick gypsum sheathing on one side and 12mm thick OSB sheathing on the other side	27.87
2.7×2.4	600	150/300	12mm thick gypsum sheathing on one side	3.91
			9mm thick OSB sheathing on one side	12.08
			12mm thick OSB sheathing on one side	14.30
		100/300	12mm thick gypsum sheathing on one side	4.24
			9mm thick OSB sheathing on one side	12.92
			12mm thick OSB sheathing on one side	14.67
2.4×2.4	600	150/300	12mm thick gypsum sheathing on one side	4.97
			9mm thick OSB sheathing on one side	13.70
			12mm thick OSB sheathing on one side	16.51
		100/300	12mm thick gypsum sheathing on one side	5.29
			9mm thick OSB sheathing on one side	14.33
			12mm thick OSB sheathing on one side	18.23

(2) The shear resistance of cold-formed steel stud walls is enhanced lightly with the increase of steel strength.
(3) The shear resistance of cold-formed steel stud walls is increased obviously when the stud spacing is reduced from 600 mm to 400 mm.
(4) With the decrease of wall height, the shear resistance of the wall is increased.
(5) With the narrowing of screw spacing at the perimeter, the shear resistance of cold-formed steel stud wall increases greatly.

Acknowledgement

The study is funded by the National Natural Science Foundation of China (50578013).

References

American Iron and Steel Institute (1998). *Shear Wall Design Guide.*

Emad, F.G., Adrian, M.C. and Colin, F.D. (1999). Lateral Behavior of Plasterboard-Clad Residential Steel Frames. *Journal of Structural Engineering*, 125(1), 32–39.

Japan Iron and Steel Federation (2002). *Design Guide for Thin Light-Gauge Steel Construction.* Gihodo Shuppan Co., Ltd., Japan.

JGJ 101-96 (1996). *Chinese Specification of Testing Methods for Earthquake Resistant Building.*

Kasal, B. and Leichti, R.J. (1992). Nonlinear Finite Element Model for Light-Frame Stud Walls. *Journal of Structural Engineering*, 118(11), 3122–3135.

North American Steel Framing Alliance (2000). *Prescriptive Method for Residential Cold-Formed Steel Framing*, Year 2000 Edition.

Serrette, R. and Ogunfunmi, K. (1996). Shear Resistance of Gypsum-Sheathed Light-Gauge Steel Stud Walls. *Journal of Structural Engineering*, 122(4), 383–389.

Serrette, R., Encalada, J., Juadines, M. and Nguyen, H. (1997). Static Racking Behavior of Plywood, OSB Sheathing, Gypsum, and Fiberboard Walls with Metal Framing. *Journal of Structural Engineering*, 123(8), 1079–1086.

Thomas, W.H. (2002). Concentrated Load Capacity and Stiffness of OSB: Calculation Versus Test. *Journal of Structural Engineering*, 128(ST7), 908–912.

Xia, B. and Dong, J. (2004). Finite Element Analysis of the Lateral Force Resistance of Light Gauge Steel Framed Compound Bearing Walls. *Building Structure Supplement*, No. 8, 334–337.

Zhou, T. and He, B. (2004). Test Report of the Shear Resistance of Cold-formed Steel Stud Wall. Xi'an University of Architecture & Technology.

Engineering Mechanics
and
Applied Mathematics

LEARNING, SELF-DIAGNOSIS AND MULTI-OBJECTIVE CONTROL OF AN ACTIVE TENSEGRITY STRUCTURE

Bernard Adam and Ian F.C. Smith

Ecole Polytechnique Fédérale de Lausanne (EPFL),
Applied Computing and Mechanics Laboratory,
Station 18, GC-G1-507,
CH-1015 Lausanne, Switzerland
E-mail: Ian.Smith@epfl.ch

Abstract

This paper presents a full-scale active tensegrity structure at EPFL and demonstrates how it can learn as well as carry out self-diagnosis and self-compensation. Tensegrities are generally flexible structures: small loads may lead to large displacements. We thus control slope by actively modifying the self-stress state between cables and struts. The structure benefits from past experience through case-based reasoning. It memorizes past control commands and adapts them in order to react to new applied loads up to forty times more rapidly than without this previous control information. Redundancy of this structure provides opportunities for "fault tolerant" behavior. The active control system can also be used to perform self-diagnosis and then to self-compensate local damage. For many cases of local damage, the structure remains capable of satisfying control goals. This paper also summarizes a multi-objective optimization method for control according to four criteria. In contrast with other applications involving multiple objectives, such as design where users prefer choices, this is a control task, thereby requiring identification of a single solution only. Also, the single dominant objective usually generates hundreds of possible solutions. Four objectives are evaluated firstly using Pareto optimality and then a unique solution is chosen through successive filtering of candidate solutions using a hierarchy of objectives. The combination of advanced computing techniques with structural control of serviceability criteria is providing many new possibilities for structural engineers. These results are expected lead toward more autonomous and self-adaptive structures that are able to evolve as their environment changes.

Introduction

Tensegrities are spatial and lightweight structures composed of compressed struts and tensioned cables that are stabilized by a self stress state. They are very flexible; small loads can induce large deflections. Serviceability control, performed by modifying the self stress state of the structure, has potential to create opportunities for using this type of structures in practical applications.

The active tensegrity structure built at EPFL contains five modules and covers a surface area of $15m^2$. It rests on three supports that altogether block six degrees of freedom in three dimensions. Two theses have thus far been completed using this structure by Fest (2002) and Domer (2003).

Each module consists of twenty-four stainless steel cables and six composite fiber bars that are connected to each other through thirteen joints. Compressed struts converge toward a central node and this constitutes the particularity of this design as inspired by the office of Passera & Pedretti, Lugano (Switzerland). The central node reduces the buckling length of compressed elements and leads to more slender struts (Fest et al, 2003).

The structure is equipped with three displacement sensors (nodes 37, 43, 48) and ten actuators: see Figures 1 and 2. This makes it possible to actively control the structure. The actuators are placed longitudinally in in-line pairs within each module and this makes it possible to modify the self stress state through modifying the strut length. Control commands (sequences of active strut contractions and elongations) are identified using stochastic search (Domer et al, 2006). This is one of the first large-

M. Pandey et al. (eds), Advances in Engineering Structures, Mechanics & Construction, 439–448.

scale active tensegrity structures that are able to satisfy a serviceability criterion. Djouadi et al. (1998), Skelton et al (2000) and Sultan (1999) studied tensegrity structure control only through numerical simulation.

Figure 1: EPFL tensegrity structure equipped with 10 actuators and 3 displacement sensors

The active control system can also be used for system identification if members break. Actuators make it possible to perturb the structure while sensors measure the response. Local damage does not necessary lead to a total collapse since the structure is redundant. The active control system is then used to self-compensate a broken element and satisfy the serviceability criterion considering a certain loss of carrying capacity.

Methodology

The goal is to control a serviceability criterion related to the structure in order to improve its usefulness. Our serviceability criterion consists of maintaining a constant slope of the upper surface of the structure. When a load alters the slope of the structure, we apply a control command. This modifies the self stress state and makes it possible to recover the initial slope.

However, finding an efficient control command is not an easy task due to high coupling between the elements and geometric non-linearity. Consequently we use a generate-analyze-verify process with stochastic search and case-based reasoning (Domer and Smith, 2005). The PGSL (Probabilistic Global Search Lausanne) stochastic search algorithm used in our case is a direct search algorithm developed at EPFL (Raphael and Smith, 2003). In order to take advantage of previous experience, altered configurations and corresponding control commands are stored in a case-base. When the structure is subjected to a load, the nearby configuration is retrieved from the case base and its corresponding control command is adapted to the new task. As cases are added in the case-base the average time necessary to identify a control command decreases (learning). Since the structure is able to improve performance progressively using past experience we considerer this to be a characterization of an intelligent structure.

In the case of local damage, localization is the "inverse problem" of determining a cause given an effect. It is possible to perturb the structure through micro-movements (±1mm) and measure its response through six indicators: RMS variation of the vertical displacements at the three measured nodes (37, 43, 48, see Figure 2) due to the micro-movement and slope variation in three directions due to the micro-movement. For this simple case, we used the algorithm described below. Measured indicator variations on the damaged physical structure are compared with numerically simulated indicator variations due to the same micro-movement and assuming a candidate local damage position. If measured and simulated variations vary in the same way, the assumption is kept. Otherwise the candidate is rejected and another local damage assumption is evaluated. The space of possible damage positions reduces iteratively with micro-movements until the damage is localized.

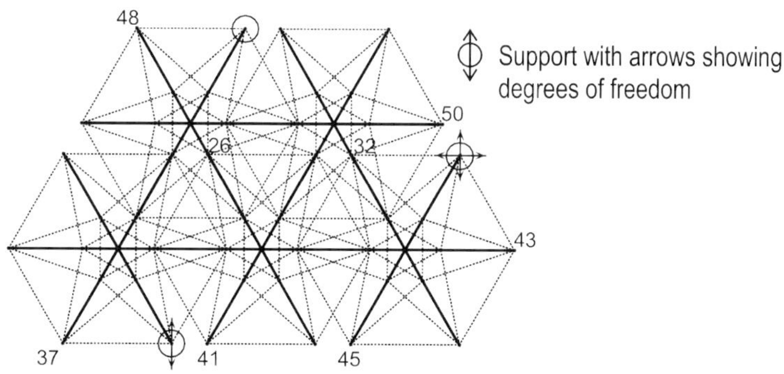

Figure 2: View of the structure from above with loaded nodes shown as numbers

Once the damage is localized, a self-compensating control command can be applied in order to satisfy the serviceability criterion. The structure can then be controlled despite a loss of carrying capacity. The principal constraint of this task is serviceability and not structural safety. Contrary to most traditional civil structures, the serviceability and structural safety of tensegrities are often conflicting objectives since measures that increase serviceability (for example, prestress) may lower structural safety through lower load-carrying capacity.

Results

Control commands are identified by numerically simulating the behavior of the structure using the dynamic relaxation method. They are then applied to the laboratory structure. Errors due to modeling and nonlinear behavior during the control command application lead to inaccuracies. For one applied point-load, the slope compensation varies between 79 and 100%, whereas for two point-loads the slope compensation varies between 65 and 100%. These second load cases are more difficult to compensate because of nonlinearity effects. If the compensation is not sufficient, it is possible to search a second control command using the first search final state as the second search initial state.

The case-based reasoning method makes it possible to create a structure that learns through using its own experience. Memorizing past altered configurations and the corresponding control command makes it possible to react faster to new applied loads by adapting past solutions. Control command search using case-based reasoning method is always faster then using stochastic search only. Moreover, the number of iterations needed to identify a control command decreases with increasing number of cases in the case-base, see the example in Figure 3. Improvements of up to 40 times have been observed. The structure does not learn regularly because of the stochastic nature of the process. Cases are retrieved from the case-base by comparing their notional “distance” to the actual task, considering altered slope value and active strut lengths.

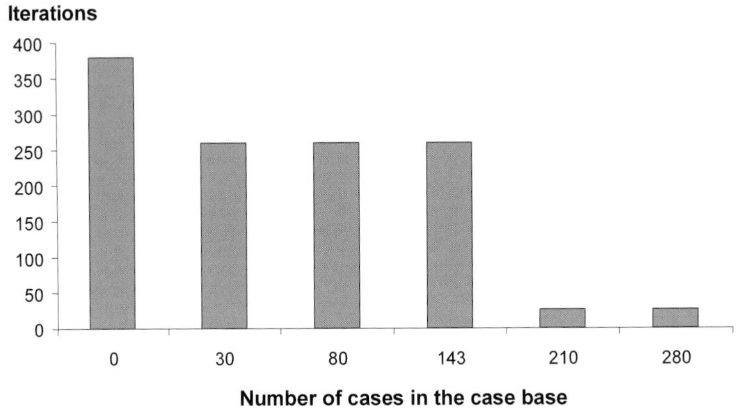

Figure 3: The structure learns as cases are added to the case-base (Domer and Smith, 2005)

The structure is perturbed by the control system to carry out self-diagnosis. This paper presents an initial numerical study. The goal of this study is to localize a broken lateral cable, knowing that the structure is not loaded. The initial example is that the broken cable is cable number 111. At the beginning, all thirty lateral cables in the five modules are present in the space of possible broken cables. The identification process assumes successively that each non rejected candidate has a broken lateral cable. The first micro-movement (strut 148: +1mm) makes it possible to eliminate twenty-two candidates since these situations involved behavior that was opposite in direction to that measured for cable 111. The second one (strut 60: +1mm) eliminates other four possible broken cables and then the third micro-movement (strut 60: -1mm and strut 148: +1mm) eliminates three more potentially broken cables and makes it possible to identify the broken lateral cable 111.

Micro-movements of active struts at the center of the structure eliminate candidates faster than micro-movements of active struts situated at the edge of the structure. Combining two micro-movements sometimes makes it possible to further eliminate candidates. It is not always possible to isolate one candidate with single micro-movements.

Once identified, this local damage can be compensated through applying a control command such that the structure still satisfies the serviceability criterion. Therefore the structure can still accept loads considering a loss of carrying capacity.

In this study point-loads are applied at nodes 37, 43 and 48. The maximum load in Newtons for which the structure can compensate the slope when the cable 111 is broken is presented at the second line of Table 1 (CC LD: Carrying Capacity with Local Damage). The maximum load for which the structure can compensate the slope when it is not damaged is shown at third line of Table 1 (CC NLD: Carrying Capacity with No Local Damage). The loss percentage of carrying capacity is indicated on the last line of Table 1. The loss of carrying capacity is understandably larger in the vicinity of the local damage (nodes 37 and 48 are close to cable 111). In these cases loads which should have passed through the broken element to reach the supports have to find another path.

Table 1. Loss of carrying capacity when the lateral cable 111 is broken

Node	37	43	48
CC LD [N]	350	440	310
CC NLD [N]	840	520	800
CC Loss %	58	15	61

Multi-Criteria Control

Many combinations of contractions and elongations of active struts can satisfy a single serviceability objective to an acceptable degree. This presents an opportunity to enhance control command search through use of additional objectives. Additional objectives should not significantly decrease control command quality with respect to the slope objective. Goals are to increase robustness of both the structure and the active control system in order to carry out multiple control events over service lives. The following four conflicting objectives are used to guide search:

- Slope: maintain top surface slope of the structure constant when subjected to loading,
- Stroke: maintain actuator jacks as close as possible to their midpoint,
- Stress: minimize stress of the most stressed element,
- Stiffness: maximize the stiffness of the structure.

The general form of a multi-objective optimization problem can be expressed as follows:

Minimize objective functions $\vec{f}(\vec{x})$

subject to inequality constraints $\vec{g}(\vec{x}) \leq 0$

and equality constraints $\vec{h}(\vec{x}) = 0$

where $\vec{x} \in R^n$, $\vec{f}(\vec{x}) \in \Re^k$, $\vec{g}(\vec{x}) \in \Re^m$, and $\vec{h}(\vec{x}) \in \Re^p$. Here, n represents the number of variables, k the number of objective functions, m the number of inequality constraints and p the number of equality constraints.

A Pareto filtering approach is employed in order to avoid the use of weight factors. In case of a multi-objective minimization task, a solution x^* is said to be Pareto optimal if there exists no feasible vector of decision variables x which would decrease some objective without causing a simultaneous increase in at least one other objective. This concept results in a set of solutions called the Pareto optimal set. The vectors x^* corresponding to the solutions included in the Pareto optimal set are called non-dominated (Pareto, 1896).

The multi-objective search method adapted to our tensegrity structure serviceability control task involves building a Pareto optimal solution set and selecting one solution (see Figure 4). The Pareto optimal solution set is identified according to the four objectives and the five constraints described above. Solution generation and Pareto filtering is carried out using the ParetoPGSL algorithm. Solutions are generated in order to minimize all objectives. Dominated solutions are rejected. ParetoPGSL stops after 1500 generated solutions since solution quality does not improve any further.

The selection strategy that is adopted hierarchically reduces the solution space until identification of a control command. It is developed in four steps and reflects the importance of the objectives. Control commands for which slope compensation is less than 95% are first rejected. In practical situations, slope compensation would be acceptable if its value was above this threshold. To keep objectivity with respect to the three remaining objectives, the remaining solutions are divided into thirds according to solution quality. The worst third of the solutions with respect to the stroke objective is rejected. The worst half of the remaining solutions with respect to the stress objective is then rejected. Finally, the best solution with respect to the stiffness objective is identified among solutions that are left. This becomes the control command that is applied to the structure. Therefore, each of the three objectives in the last three steps leads to rejection of the same number of solutions.

Control solutions describe the structural configuration when slopes are compensated. Sequences of application of control commands that transform the altered slope state to the compensated slope state involve verifying that no failure would happen during intermediate steps. The control command is divided into 1 mm steps. Strut contractions are placed at the beginning of the sequence and elongations at the end. In this way, energy is generally first taken out of the structure before it is added. Calculations are made using the dynamic relaxation method. The position of the structure is evaluated for each 0.1 mm of actuator travel. The sequence is then applied to the physical structure for experimental validation.

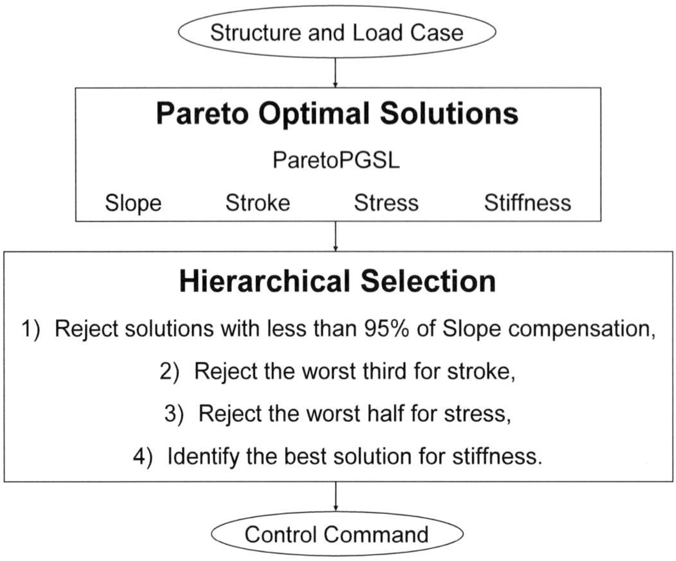

Figure 4: Multi-objective methodology: Hierarchical selection of Pareto optimal solutions

Results of Multi-Criteria Control

This methodology is tested for 24 load cases involving up to two vertical point loads from 391 N to 1209 N in magnitude (see Table 2). Loaded nodes are numbered according to Figure 2. One and two point load cases numbered from 1 to 24 are presented in Table 2.

Table 2. Load cases applied to the structure

Load case	Loaded	Load magnitude
1	26	625
2	26	900
3	26	1209
4	32	625
5	32	859
6	32	1092
7	37	391
8	37	550
9	37	700
10	48	391
11	48	550
12	48	700
13	6	1092
14	37 and 45	391
15	37 and 45	624
16	37 and 45	742
17	39 and 48	157
18	39 and 48	215
19	39 and 48	274
20	41 and 50	391
21	41 and 50	624
22	45 and 48	391
23	45 and 48	624
24	45 and 48	742

Examine load case 5: 859 N point load at node 32. Pareto optimal solutions are generated using the ParetoPGSL algorithm (see Figure 5). Solutions are presented in four dimensions with respect to the four objectives. The slope objective is shown on the vertical axis. Stroke and stress objectives are represented with the horizontal axis. The gray bar evaluates the stiffness objective. Values close to zero are considered best for all objectives.

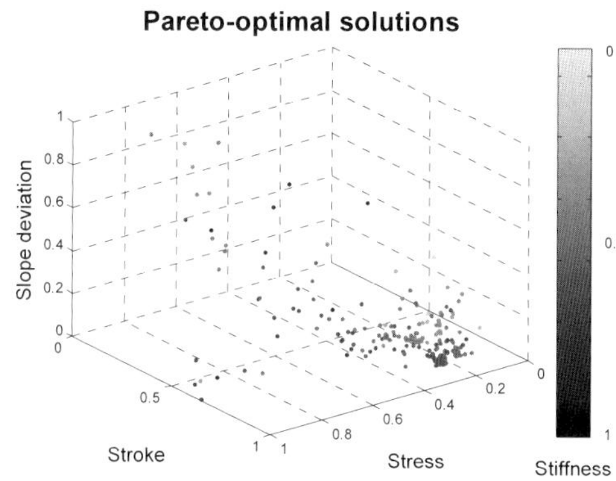

Figure 5: Pareto front respect to slope, stroke, stress and stiffness

The first step of the hierarchical selection strategy consists of rejecting all solutions for which slope compensation is less than 95% (see Figure 6).

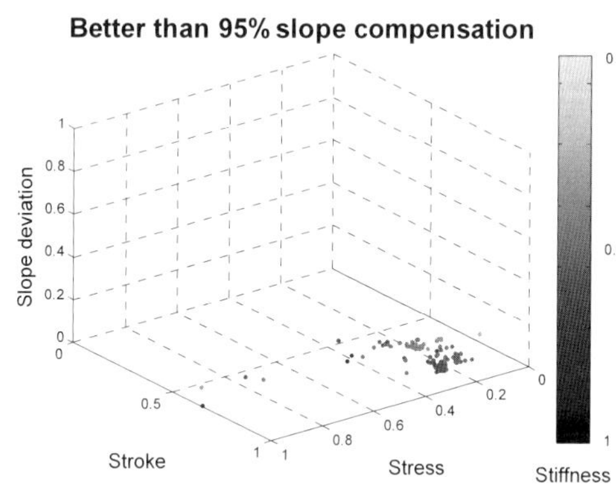

Figure 6: Solutions for which slope compensation is better than 95%

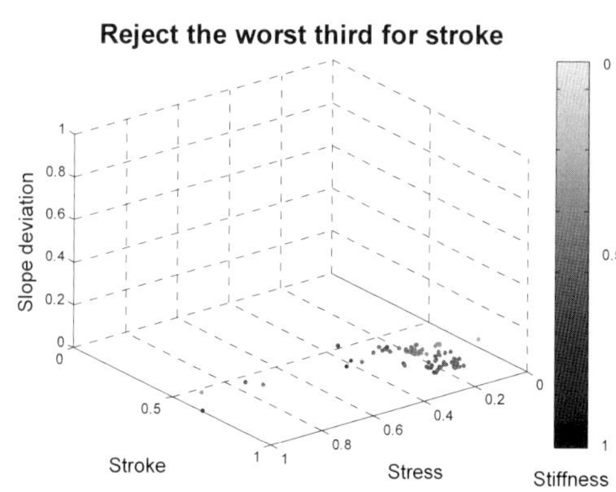

Figure 7: The worst third of the previous set with respect to stroke has now been rejected

The second step of the selection strategy involves dividing the remaining solution set into three parts according to stroke objective. The worst third is rejected (see Figure 7).

The third step of the selection strategy results in dividing the remaining solution set into two parts according to stress objective quality. The worst half is rejected (see Figure 8).

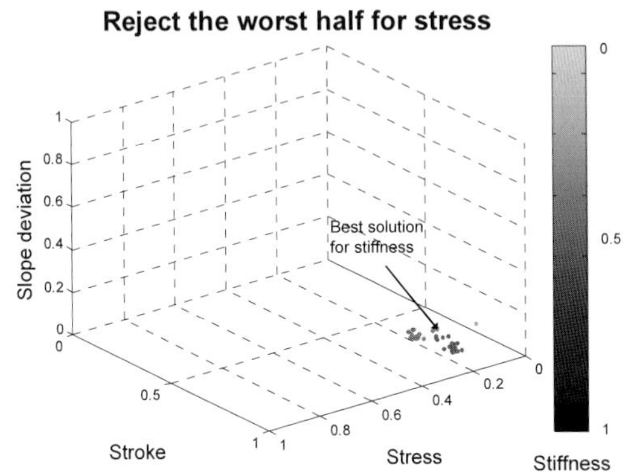

Figure: 8. The worst half of the previous set with respect to stress has now been rejected

The last step of the selection strategy consists of identifying the control command as the best solution with respect to stiffness objective. (see Figure 8). This solution represents the configuration of the structure when the slope is compensated. Sequence of application of the control command is then calculated to verify that no failure would happen and to observe slope evolution. The control command is applied to the loaded physical structure for experimental validation (see Figure 9). Slope evolution is plotted versus steps of 1mm of actuator travel.

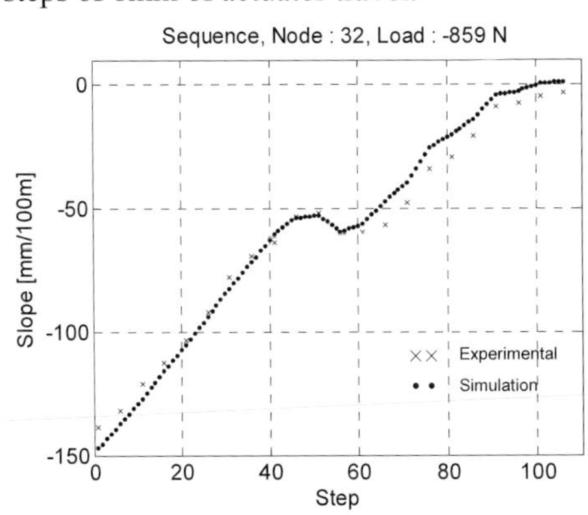

Figure 9: Experimental and numerical slope compensation sequence

Numerical simulation gives an altered slope of -147mm/100m and a compensated slope of 1mm/100m (99% compensated). Experimental testing gives an altered slope of 138mm/100m and a compensated slope of -4mm/100m (97% compensated). The average actuator travel is 1.5 mm. Stress

Table 3. Multiple load application scenario

Load event	Loaded	Load Magnitude
1	32	391
2	50	391
3	37	391
4	48	391
5	26	391
6	6	150

values are numerical only because the structure is not equipped with force sensors that would provide experimental data. Simulation and laboratory test results for slope are generally in good agreement.

Structural control for multiple load application events (Table 3) is presented in Figure 10. Slope evolution is plotted versus steps of 1mm of actuator travel. Structural behavior when control commands are identified using multi-objective search and single objective search are evaluated. Control commands are more rapidly effective when they are identified with multi-objective search. Single objective control command exhibit a more pronounced zig-zag profile that requires more steps to correct the slope. Multi-objective commands are useful to maintain robustness of both the structure and the control system whereas in single objective sequence no such maintenance can be assured. At the sixth control command multi-objective method makes it possible to compensate the slope whereas a single objective method leads to buckling of a strut.

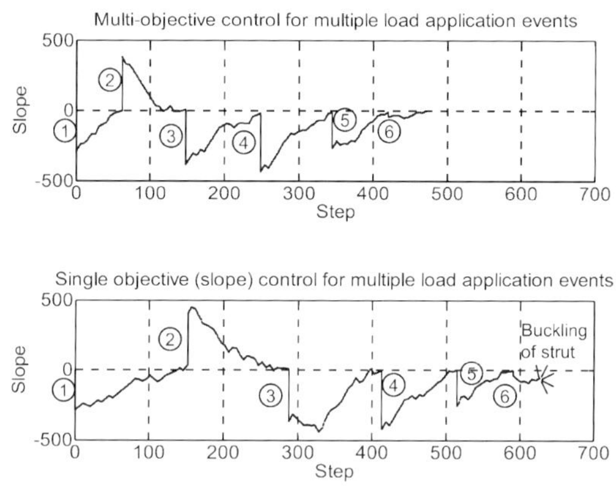

Figure 10: Multiple loads applied sequentially: multi-objective and slope-objective control commands

Conclusions

This paper presents a study of the application active control to a tensegrity structure so that opportunities for innovative applications can be identified. The following conclusions come out this research:

- When subjected to a load, a control command (sequence of strut contractions and elongations) makes it possible to modify the self stress state in order to maintain serviceability criteria.
- Use of previous control commands increases performance and creates an opportunity for the structure to learn.
- The active control system can also be used to perform self-diagnosis by comparing measured behavior of the damaged physical structure and simulated behavior resulting from an assumption of local damage location. These studies demonstrate good potential to localize a broken cable.
- The active control system can also be used to self-compensate a local damage and still satisfy the serviceability criterion. Simulation studies show opportunities for using damaged structures through considering loss of carrying capacity.
- In situations where satisfying a dominant objective results in many solutions, a Pareto approach together with hierarchical elimination of solutions is attractive, especially when tasks require single solutions such as during structural control.
- Multiple load application events are controlled more efficiently using multi-objective control.

These results lead toward more autonomous and self-adaptive structures that evolve in changing environments.

Acknowledgments

The authors would like to thank the Swiss National Foundation for supporting this work. Dr. E. Fest built the structure and the control system. Dr. B. Domer improved control and implemented the case-based reasoning study. Prof. B. Raphael provided support during programming of the control system. We are also grateful to Dr. K. Shea (University of Cambridge, UK), Passera & Pedretti SA (Lugano, Switzerland) and P. Gallay (EPFL, Switzerland) for their contributions.

References

Fest, E. (2002) "Une structure active de type tenségrité", Thèse no 2701, Ecole Polytechnique Fédérale de Lausanne.
Fest, E., Shea, K., Domer, B. and Smith, I.F.C. (2003) Adjustable tensegrity structures, *J of Structural Engineering*, Vol. 129, No 4, p. 515.
Domer, B. (2003). "Performance enhancement of active structures during service lives", Thèse no 2750, Ecole Polytechnique Fédérale de Lausanne.
Domer, B., Raphael, B., Shea, K. and Smith, I.F.C. (2003) "A study of two stochastic search methods for structural control", J of Computing in Civil Engineering, Vol 17, No 3, pp 132-141.
Domer, B., and Smith, I.F.C., (2005) "An Active Structure that Learns", J of Computing in Civil Engineering, vol. 19(1), pp. 16-24.
Djouadi, S., Motro, R., Pons, J.C., and Crosnier, B., (1998), "Active Control of Tensegrity Systems", Journal of Aerospace Engineering, Vol. 11, pp. 37-44.
Pareto, V., (1896), "Cours d'Economie Politique", vols I and II, Rouge: Lausanne, Switzerland.
Raphael, B. and Smith, I.F.C., (2003) "A direct stochastic algorithm for global search", J of Applied Mathematics and Computation, Vol 146, No 2-3, pp 729-758.
Skelton, R.E., et al., (2000), "An introduction to the mechanics of tensegrity structures", Handbook on mechanical systems design, CRC, Boca Raton, Fla.
Sultan, C., (1999) "Modeling, design and control of tensegrity structures with applications", PhD thesis, Purdue Univ., West Lafayette, Ind.

MICRO- AND MACRO-MECHANICAL APPROACHES FOR MODELLING OF POLYETHYLENE MATERIAL FOR PIPES

J.A. Alvarado-Contreras[1,3], H. Liu[1], M.A. Polak[1,*] and A. Penlidis[2]
[1]*Department of Civil Engineering, University of Waterloo, Waterloo, ON, Canada*
[2]*Department of Chemical Engineering, University of Waterloo, Waterloo, ON, Canada*
[3]*School of Mechanical Engineering, University of the Andes, Mérida, Venezuela*

Abstract

Two different approaches for modelling the mechanical behaviour of polyethylene materials are presented. In the first one, the emphasis is on the relationships between molecular features and mechanical properties. In the proposed model, the material is analyzed from a microscopic viewpoint and considered as an aggregate of crystals. The constitutive equation is expressed in a viscoplastic framework considering degradation at large deformations. For the second approach, the material response is considered to be nonlinear viscoelastic. A phenomenological approach is adopted, and attention is given on the formulation of a model that can be implemented for structural analysis of components such as pipes. In this part of the study, numerical and experimental data of creep for a medium-density polyethylene pipe material are presented. The efficacy of the micro- and macro-mechanical approaches is confirmed by experimental results.

1. Introduction

Polyethylene material has been widely used since the early 1940s. Its low cost, lightness, and chemical stability have favoured its popularity to substitute other materials in water, gas, and sewage disposal pipelines. During installation and utilization, pipes are typically subjected to complex loads and loading paths. This has motivated the interest, first, in understanding the relationships between molecular structure and end-use mechanical performance, and then, in improving the material properties. Many studies have concentrated on experimental and theoretical aspects of nonlinear history-dependent behaviour and on the deformation mechanisms at large deformations. The present work consists of two parts. The first one deals with the modelling of the mechanical response of polyethylene material from a microstructural viewpoint. The second part presents the macroscale approach to modelling the nonlinear behaviour of polyethylene based on responses obtained experimentally.

2. Structure of Polyethylene

Polymeric materials are made of macromolecules and appear in a variety of products. These macromolecules are long chains of thousands of repeating units, which give polymers many of their useful properties. In polyethylene, molecules are basically long chains of carbon and hydrogen atoms. The smallest unit (monomer) is ethylene (C_2H_4), whereas a typical polyethylene molecule contains from approximately 900 to 215,000 of these units. It has been shown that polyethylene possesses a structure with spherulitic morphology, as illustrated in Figure 1. Within this structure, molecules are capable of adopting two distinct arrangements, the amorphous and crystalline phases. Amorphous polyethylene is characterized by those segments where molecules have, on average, no favoured directions. On the order hand, crystalline polyethylene corresponds to segments of highly ordered molecules embedded within the amorphous phase. Polyethylene crystals form

* Corresponding author: M.A. Polak, E-mail: *polak@uwaterloo.ca*

M. Pandey et al. (eds), Advances in Engineering Structures, Mechanics & Construction, 449–458.

by folding the molecules alternately up and down and by arranging the straight segments between folds into a periodic array. The thickness of polyethylene crystals ranges from 50 to 250 Å (Å=10^{-10} m), and the lateral dimensions range from 1 to 50 × 10^{6} m. Only about 40 monomeric units are included in each of the straight segments between folds (Lin & Argon, 1994). The structure of a crystal is of the orthorhombic crystal class, as illustrated in Figure 1. The parameters a, b, and c are the lattice parameters, where the c-axis is parallel to the molecules.

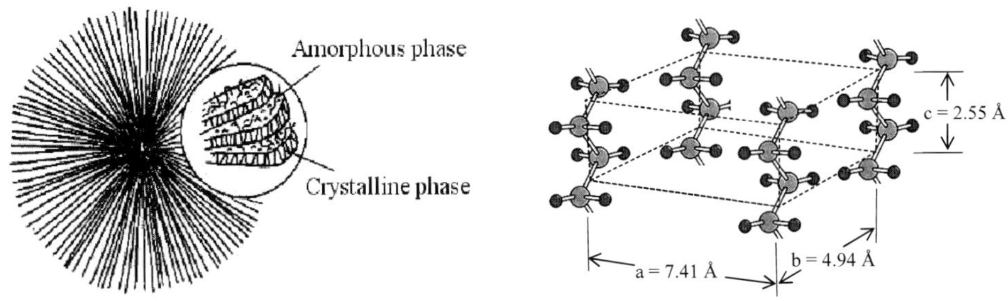

Figure 1. Schematic illustration of a polyethylene spherulitic structure and a crystal.

According to the American Society for Testing and Materials, polyethylene can be classified into four groups. Low-density polyethylene (LDPE) has a density ranging from 0.91 to 0.93 g/cm^3 and is predominantly amorphous. Medium-density polyethylene (MDPE) typically has average densities from 0.93 to 0.94 g/cm^3. High-density polyethylene (HDPE) has a density above 0.94 g/cm^3 with only a few branched chains. As a result, these molecules align into more compact arrangements with up to 90% crystalline regions. Linear low-density polyethylene (LLDPE) has longer and more branches than HDPE. Consequently, it has a higher density than LDPE but still a linear structure like HDPE.

3. Constitutive Micro-modelling

Although microstructure and mechanical properties of polymers have been the subject of many studies, it is only during the last two decades or so that deformation mechanisms have been properly described. The earlier models aid in the interpretation of structural observations under different loading conditions; however, they cannot predict the microstructural damage caused by deformation. The proposed model is expressed in the framework of viscoplasticity coupled with degradation at large deformations. This involves the concepts of damage mechanics considering the original microstructure, the particular irreversible rearrangements, and the deformation mechanisms. As mentioned above, molecules in polyethylene materials can adopt two basic arrangements. Nevertheless, to simplify the complex microstructure, the model regards the material exclusively as crystalline; the amorphous phase is ignored. Thus, a three-dimensional aggregate of randomly oriented and perfectly bonded crystals is used to describe the behaviour. In an aggregate, crystals are much larger than molecules, but smaller than material points. Initially, the material is assumed to be micro- and macro-scopically homogeneous.

The theoretical formulation of the micromechanical model has been presented in detail in Alvarado-Contreras et al. (2005). Consequently, we limit ourselves to a brief description of the most important features of the model, as summarized in Table 1. The model solves the stress-strain problems in two stages, one for the microscopic problem (local problem) and the other for coupling the micro- and macro-scopic problems (global problem).

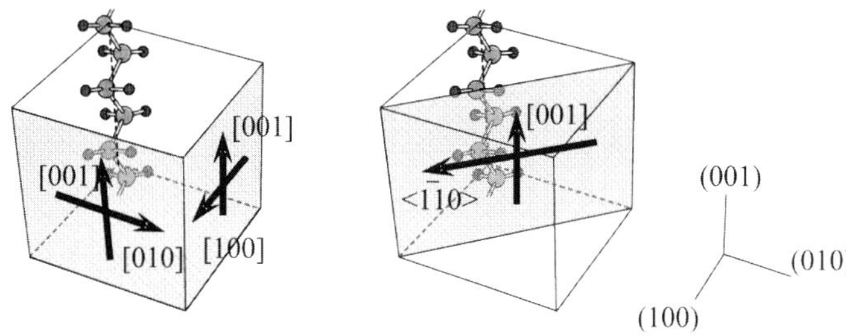

Figure 2. Slip systems in polyethylene crystals.

At the local problem, deformation mechanisms in single crystals are considered. These mechanisms are based on the theory of shear slip on crystallographic planes (Asaro & Rice, 1977). The material is initially isotropic and homogeneous, and the elastic deformations are ignored (realistic for large deformation analyses). The slip plane normal $\mathbf{n}^{\alpha}$ and slip direction $\mathbf{s}^{\alpha}$ define the slip systems (Figure 2). The second-order tensors $\mathbf{R}^{\alpha}$ and $\mathbf{A}^{\alpha}$ represent symmetric and skew-symmetric orientation tensors associated with the slip systems, as shown in Eq. 1. The deformation rate $\mathbf{D}$ and plastic spin $\mathbf{W}^{p}$ are consequences of the shear rates $\dot{\gamma}^{\alpha}$ on all slip systems, as indicated in Eq. 2 (Asaro, 1979). The constitutive equation connecting the microscopic deviatoric stress $\mathbf{S}$ and deformation rate $\mathbf{D}$ is expressed in Eq. 3. Here, the fourth-order compliance tensor M, based on a power-law model for the shear rates, is characterized by the inverse of the rate sensitivity n, the reference shear rate $\dot{\gamma}_0$, and the current critical shear strengths g^{α} (Eq. 4). Furthermore, Ω^{α} is a simple scalar damage variable associated with the atomic debonds of the slip planes, and it evolves according to the evolution law given in Eq. 5, where $\dot{\Omega}_0$ and m are the reference damage rate and rate exponent. Noting that damage and hardening processes may be considered independent of each other (Leimatre & Chaboche, 1994), the rate of hardening of a slip system evolves according to Eq. 6, where h_0 and c characterize the slip system hardening and saturation strength, respectively. To estimate the crystal orientation evolution, the local spin $\mathbf{W}$ is calculated as the sum of the rigid-body spin $\mathbf{W}^{*}$ and the plastic spin $\mathbf{W}^{p}$, where the plastic component is affected by an empirical release parameter Ψ which is a function of the material parameter ζ and slip system damage Ω^{α}, as specified in Eq. 7.

In general, microstructure and its changes are non-deterministic. The same is true for the fluctuations in stresses and strains. Thus, for the global problem, the macroscopic mechanical behaviour is calculated as a volume average of the corresponding crystal responses. First, let $\overline{\mathbf{D}}$ and $\overline{\mathbf{W}}$ be, respectively, the macroscopic deformation rate and spin applied to the crystal aggregate at some material point. The partition of the deformation rate among the crystals is calculated through Eq. 8, where P is a local fourth-order projection tensor that considers the inextensibility of the crystal molecules, and $\overline{\mathrm{P}}^{-1}$ represents the inverse of its volume average (Parks & Ahzi, 1990). Conversely, crystal spins, having only three independent components, simply equal the global spin, as in Eq. 9. Based on the above steps, the global response is estimated by averaging the stresses in all crystals. Then, if the homogenized reduced stress $\overline{\mathbf{S}}$ satisfies the boundary conditions to adequate tolerance, the solution is accepted. Otherwise, corrections based on the successive substitution scheme are performed, and the procedure is repeated using $\overline{\mathbf{S}}$ as a new trial stress. Once an estimate of $\overline{\mathbf{S}}$ has been found, it can be related to the global deformation rate $\overline{\mathbf{D}}$ throughout the relationship given in Eq. 10, where $\overline{\mathrm{M}}$ is the global average compliance tensor.

Table 1. Summary of the micro-mechanical model.

Local problem		
[1]	$\mathbf{R}^{\alpha} = \frac{1}{2}(\mathbf{n}^{\alpha} \otimes \mathbf{s}^{\alpha} + \mathbf{s}^{\alpha} \otimes \mathbf{n}^{\alpha})$ and $\mathbf{A}^{\alpha} = \frac{1}{2}(\mathbf{n}^{\alpha} \otimes \mathbf{s}^{\alpha} - \mathbf{s}^{\alpha} \otimes \mathbf{n}^{\alpha})$	Schmid tensors
[2]	$\mathbf{D} = \sum_{\alpha=1}^{8} \dot{\gamma}^{\alpha}\mathbf{R}^{\alpha}$ and $\mathbf{W}^{p} = \sum_{\alpha=1}^{8} \dot{\gamma}^{\alpha}\mathbf{A}^{\alpha}$	Deformation rate and plastic spin
[3]	$\mathbf{D} = \mathrm{M} : \mathbf{S}$	Local constitutive equation
[4]	$\mathrm{M} = \dot{\gamma}_0 \sum_{\alpha=1}^{8} \left(\frac{1}{(1-\Omega^{\alpha})g^{\alpha}} \right)^{n} \left| \mathbf{S} : \mathbf{R}^{\alpha} \right|^{n-1} \mathbf{R}^{\alpha} \otimes \mathbf{R}^{\alpha}$	Compliance tensor
[5]	$\frac{d\Omega^{\alpha}}{d\tau^{\alpha}} = \dot{\Omega}_0 \left| \frac{\mathbf{S} : \mathbf{R}^{\alpha}}{(1-\Omega^{\alpha})g^{\alpha}} \right|^{m}$	Damage evolution law
[6]	$\frac{d\dot{g}^{\alpha}}{d\dot{\gamma}^{\alpha}} = h_0 \mathrm{sech}^2\left(\frac{h_0}{c} \bar{\gamma} \right)$, where $\bar{\gamma} = \int_0^t \sum_{\alpha=1}^{8} \dot{\gamma}^{\alpha} dt$	Hardening evolution law
[7]	$\mathbf{W} = \mathbf{W}^{*} + \Psi \mathbf{W}^{p}$, where $\Psi = \tanh(\frac{1}{8}\zeta \sum_{\alpha=1}^{8} \Omega^{\alpha})$	Local spin
Global problem		
[8]	$\mathbf{D} = \mathrm{P} : \overline{\mathrm{P}}^{-1} : \overline{\mathbf{D}}$	Deformation rate partition
[9]	$\mathbf{W} = \overline{\mathbf{W}}$	Spin partition
[10]	$\overline{\mathbf{D}} = \overline{\mathrm{M}} : \overline{\mathbf{S}}$	Global constitutive equation

4. Numerical Results (Micro-mechanical Approach)

This section shows the capability of the model to represent the mechanical behaviour of high-density polyethylene based only on its crystalline microstructure. Calculations are carried out for an initially isotropic aggregate of 100 crystals in uniaxial tension. In the simulations, the deformation rate is prescribed, and the crystal stresses and orientation changes are determined. From them, the macroscopic stress state is calculated. Considering axisymmetric boundary conditions, a constant macroscopic strain rate of 10^{-3}/s and null macroscopic spin are applied. For each crystal, all eight slip systems are active and the initial resolved shear strengths are known, as listed in Table 2 (Parks & Ahzi, 1990). The material parameters are fitted from two tensile tests obtained from the literature (Hillmansen et al., 2000; G'Sell et al., 2002).

Figure 3 shows the normalized equivalent macroscopic stress $(\bar{\sigma}/\tau_0)$ versus the equivalent strain. The stress values are normalized using a τ_0 = 7.2MPa. This figure shows most of the features of the experimental responses are reproduced up to certain deformation values. The strengthening observed in the experimental tests is accounted for by adjusting the saturation strength c while the other parameters (see Table 3) remain unchanged. Two values for the saturation strength c in Eq. 6 were used, c = 0.75 and 5. As shown, the higher the c-value, the higher the stress rate (represented by the curve slope for a given equivalent deformation). Comparing, the calculated responses are stiffer than those observed in real polyethylene. This earlier strengthening is obvious at large deformations. As shown in Figure 3, an upturn in the stress-strain curve occurs beyond a strain of 1.3 for c = 0.75. This illustrates the importance of both the crystal lattice spin and the amorphous phase for large deformations; this last feature is ignored in the current formulation.

Figure 4 shows the evolution of the damage average for the eight slip systems for the case with c = 0.75. The solid and broken lines represent the damage evolution for the transverse and chain slip systems, respectively. The profiles depicted in the figure indicate that the largest damage values occur on the $(100)[001]$, $(110)[1\bar{1}0]$, and $(1\bar{1}0)[001]$ slip systems. All slip systems show three damage stages; decreasing, constant, and increasing. Thus, the model represents both the variations of damage gradients for the different slip systems as material is drawn as well as the sharp gradients near failure.

Table 2. Resolved shear strengths.

Slip-system	g^α [MPa]
(100) [001]	7.2
(010) [001]	7.2
$(110)\langle 001\rangle$	7.2
(100) [010]	7.92
(010) [100]	12.96
$\{110\}\langle 1\bar{1}0\rangle$	12.96

Table 3. Material parameters for the simulations.

n	$\dot{\gamma}_0$ [/s]	m	$\dot{\Omega}_0$	h_0 [MPa]	ζ
5	0.001	2	0.1	5.0	40.0

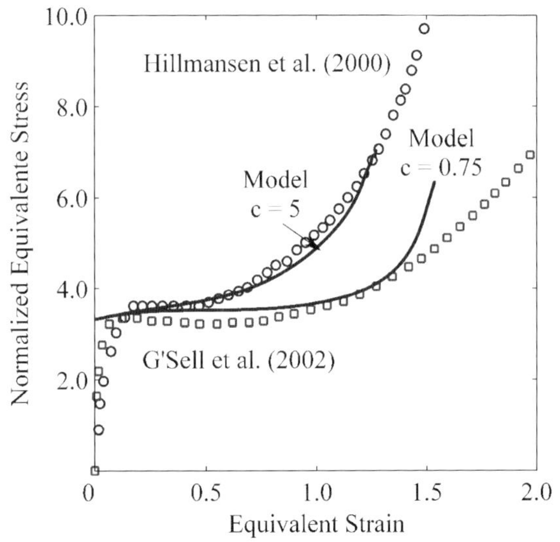

Figure 3. Normalized equivalent stress $(\bar{\sigma}/\tau_0)$ as a function of the equivalent strain for the two materials analyzed; $\bar{\sigma} = \sqrt{\frac{3}{2}\bar{\mathbf{S}}:\bar{\mathbf{S}}}$.

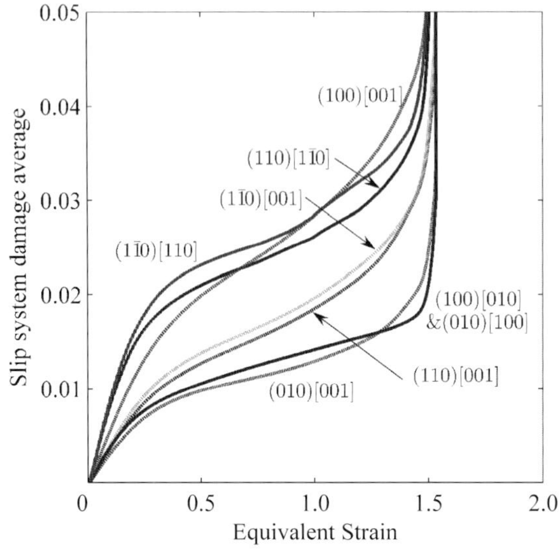

Figure 4. Evolution of damage average for the slip systems, taking c = 0.75.

5. Macro-mechanical Creep Response of Polyethylene

In viscoelasticity, the constitutive relationships are time dependent. Creep, time-relaxation, and loading rate effects are typical time-dependent responses. Creep tests are convenient for material testing and characterization. During a creep test, strain grows with time under a constant stress. The material deformation history can be described by a compliance (strain divided by stress)-time curve.

Creep tests on two high-density polyethylene (HDPE) materials were done in a previous study, and nonlinear viscoelastic models were generated for them (Liu & Polak, 2005). The models perform well in simulating creep behaviour at stresses below 10 MPa, where strains do not show drastic changes within the test durations (24 hours).

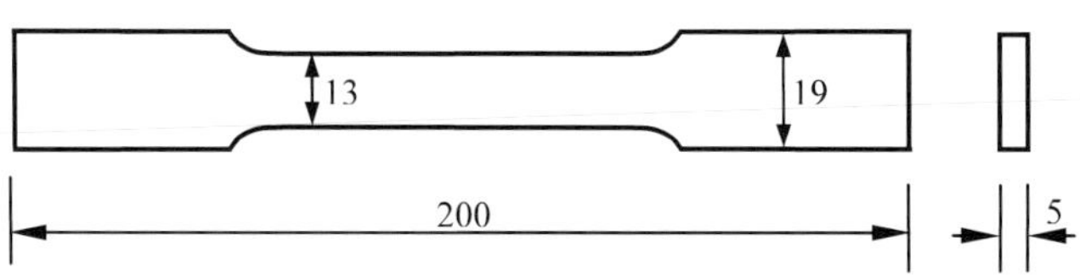

Figure 5. Material creep test specimen (units in mm).

In this study, further tests are presented for a medium-density polyethylene (MDPE) material. Tensile test specimens are cut off from a MDPE pipe, as shown in Figure 5. Creep test results for the MDPE samples are shown in Figure 6. During the test, load is applied by dead weights through a lever arm. Due to the small deformations, the (engineering) stress is assumed to be constant. A clip-on strain gauge with a data acquisition system is used to record the deformation history of the specimens. Displacements between the knife edges of the strain gauge are recorded, and the displacement-time history is converted to engineering strain-time history by dividing each displacement by the initial knife edges distance. The tests are carried out at room temperature (~22°C) for 24 hours. Figure 6(a) shows the creep strain-time curves from the tests on the MDPE pipe materials. The compliance curves at different stresses indicate strong viscoelastic nonlinearity, as shown in Figure 6(b).

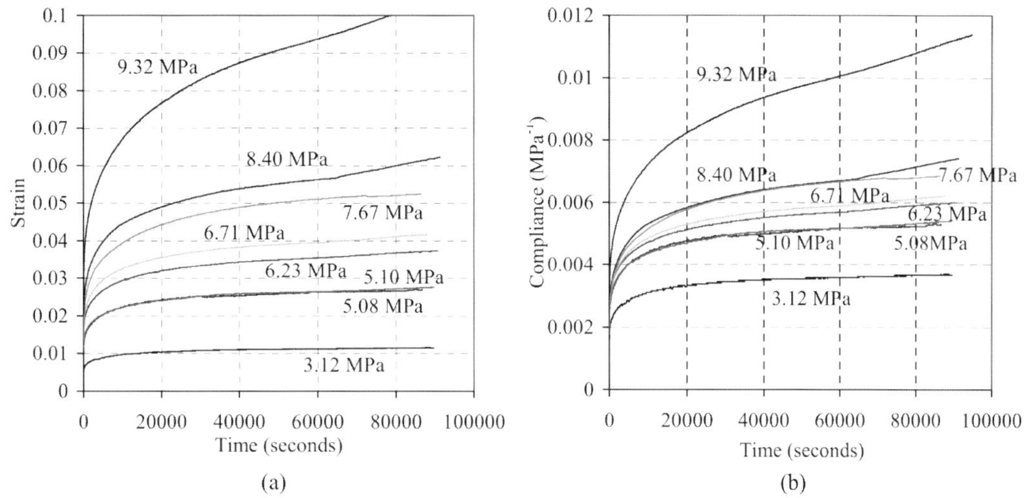

Figure 6. Creep tests for MDPE. (a) Strain-Time; (b) Compliance-Time.

6. Nonlinear Viscoelastic Modelling

Macro-modelling of polyethylene does not separate the crystalline and amorphous phase behaviour but describes the overall material behaviour in bulk. A nonlinear viscoelastic model is generated by adopting a linear interpolation approach to include stress influence on the material parameters. These parameters are taken to be piecewise linear functions instead of being prescribed functions. The model is presented as a matrix of material constants corresponding to the selected stresses. The parameters corresponding to stress levels between the stresses used for defining the model are found from a linear interpolation.

The proposed model is formulated from a multiple Kelvin element model, shown in Figure 7. The number of Kelvin elements for a specific material is determined in the calibration process. This makes the model more adaptable to material variations.

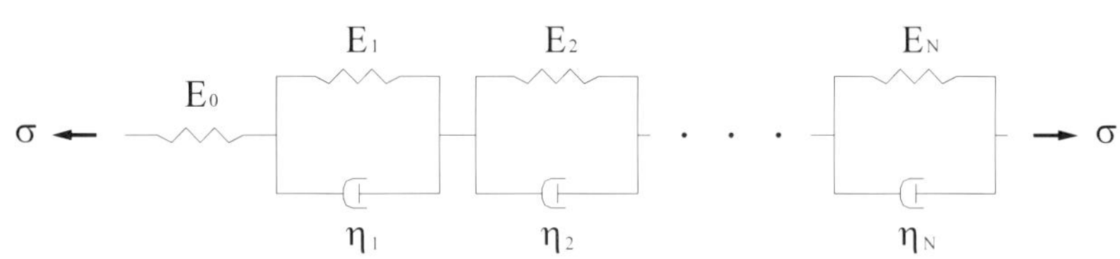

Figure 7. Multiple Kelvin element viscoelastic solid.

The constitutive equation of the nonlinear model is given by:

$$\varepsilon(t) = \int_0^t \left\{ \frac{1}{E_0(\sigma)} + \sum_{i=1}^{N} \frac{1}{E_i(\sigma)} \left[1 - \exp\left(-\frac{t-\tau}{\tau_i} \right) \right] \right\} \dot{\sigma}(\tau) d\tau \tag{11}$$

where E_0 and E_i are functions of stress, and $\tau_i = \eta_i / E_i$, known as relaxation times, remain constant. η_i are the Kelvin dashpot viscosities. The creep function is given by:

[12] $$\psi(t) = \frac{1}{E_0(\sigma)} + \sum_{i=1}^{N} \frac{1}{E_i(\sigma)} \left[1 - \exp\left(-\frac{t}{\tau_i} \right) \right]$$

Then, the strain response under a constant stress σ_i becomes:

[13] $$\varepsilon(t) = \sigma_i \left\{ \frac{1}{E_0(\sigma_i)} + \sum_{i=1}^{N} \frac{1}{E_i(\sigma_i)} \left[1 - \exp\left(-\frac{t}{\tau_i} \right) \right] \right\}$$

For a constant σ_i, the material parameters E_0 and E_i are calibrated from a given test and reduced to constants. For the development of the proposed model, several creep tests are performed at different stress levels. A linear least-squares fitting is used in the parameter calibration at each stress, and the number of Kelvin elements N is automatically chosen by the fitting criterion. A series of constant $E_0(\sigma_i)$ and $E_i(\sigma_i)$ values are generated corresponding to the creep stresses. Thus, the model can be presented as a table of E_0 and E_i values, as listed in Table 4.

In order to define the material behaviour at an arbitrary stress, the parameters are linearly interpolated from the parameters corresponding to stresses below and above the given value. Thus, E_0 and E_i are modelled as linear piecewise functions of stress. For modelling purposes, the instantaneous modulus $E_0(\sigma_i)$ and the coefficients of the exponential terms $x_i(\sigma_i) = 1/E_i(\sigma_i)$ (instead of $E_i(\sigma_i)$) are linearly interpolated for an arbitrary stress, σ. The interpolations are given by the following equations:

[14] $$E_0(\sigma) = E_0(\sigma_m) + \frac{\sigma - \sigma_m}{\sigma_n - \sigma_m} \left[E_0(\sigma_n) - E_0(\sigma_m) \right]$$

[15] $$x_i(\sigma) = x_i(\sigma_m) + \frac{\sigma - \sigma_m}{\sigma_n - \sigma_m} \left[x_i(\sigma_n) - x_i(\sigma_m) \right]$$

where σ_m and σ_n are the bounding stresses that define the interval in which σ is located. When the number of Kelvin elements is not equal for all the stress levels used to define the model, a zero value for the absent coefficients x_i can be used.

For numerical implementation, Eq. 11 can be discretized as:

[16] $$\varepsilon_k = \sum_{j=1}^{k} \left\{ \frac{1}{E_0} + \sum_{i=1}^{n} \frac{1}{E_i} \left[1 - \exp\left(-\frac{t_k - t_{j-1}}{\tau_i} \right) \right] \right\} (\sigma_j - \sigma_{j-1})$$

where t_j and t_k are times at discrete steps, and σ_j and ε_k are the corresponding stress and strain.

7. Numerical Results (Macro-mechanical Approach)

Five creep tests on the MDPE pipe material were used to develop the model, as in Table 4. The parameters of the nonlinear viscoelastic model (Eq. 11) are given in Table 4. Figures 8 and 9 compare experimental and simulated data. It is shown that the simulated strain data fit well. At stress levels other than those used for model development, the simulations match the experimental results as well, except when close to the highest stress (9.32 MPa) where the strain curve changes drastically and digresses from the main group strain (see Figure 9). Similar results have been found for HDPE materials at stresses higher than 10 MPa, as shown by Liu and Polak (2005).

Table 4. Nonlinear viscoelastic model for MDPE. Kelvin elements = 3.

		$\tau_1 = 500$	$\tau_2 = 10000$	$\tau_3 = 200000$
Stress (MPa)	E_0 (MPa)	E_1	E_2	E_3
3.12	640	1137.4169	1067.2127	1168.6089
5.10	470	804.3798	718.0750	588.7810
6.23	420	813.3631	668.5170	422.0754
8.40	410	690.8382	572.2448	224.6822

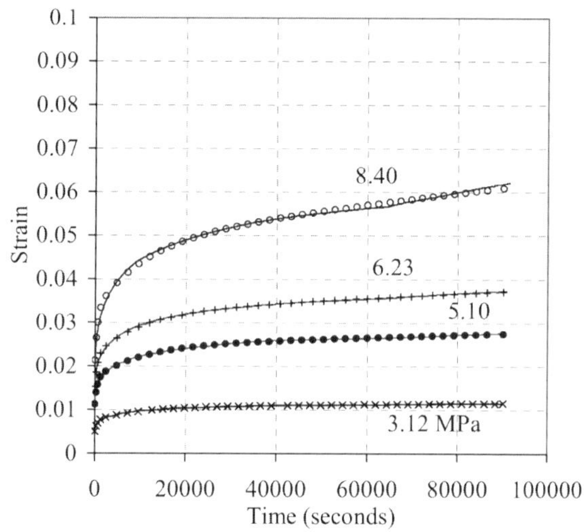

Figure 8. Creep tests vs. model simulations at modelling stresses.

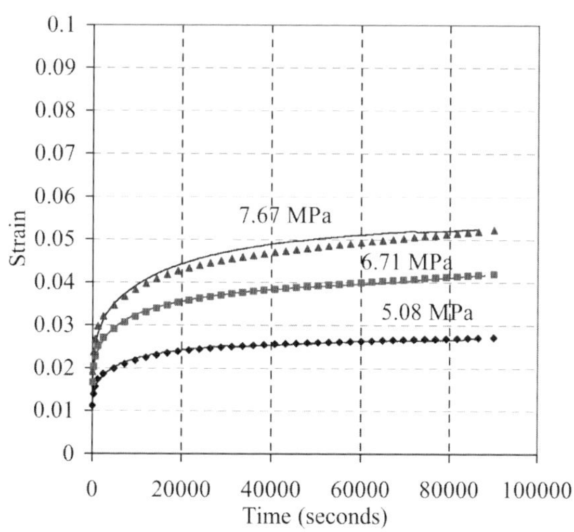

Figure 9. Creep tests vs. model simulations at verification stresses.

8. Conclusions

Two different approaches to study the mechanical behaviour of medium- and high-density polyethylene subjected to uniaxial tension are presented. From the results, the following conclusions can be drawn: 1) The described theories offer practical avenues of obtaining accurate representations of the behaviour of polyethylene from both the micro- and macro-scopic viewpoints; 2) For the microscopic approach, it is possible to couple the various deformation mechanisms and degradation processes taking place at the crystal level; 3) Material parameters are easily defined from experimental tensile tests; 4) The numerical stress-strain responses agree, up to certain deformations, with the experimental ones obtained by Hillmansen et al. (2000) and G'Sell et al. (2002); 5) For the macroscopic approach, the model simulates well the creep behaviour of medium- and high-density polyethylene; 6) The material parameters are obtained from creep tests directly on pipe material; 7) The model represents well the strong nonlinear viscoelasticity observed experimentally at intermediate stress levels; and 8) Both approaches could be applied to model other polymeric materials.

References

Alvarado-Contreras, J. A., Polak, M. A., and Penlidis, A. 2005. Damage modelling of polyethylene. *Proceedings of the 33d Annual General Conference of the Canadian Society for Civil Engineering*, Toronto, GC-185: 1-10.

Asaro, R. J. 1979. Geometrical effects in the inhomogeneous deformation of ductile single crystals. *Acta Metallurgica*, 27: 445-453.

Asaro, R. J. and Rice, J. R. 1977. Strain localization in ductile single crystals. *Journal of the Mechanics and Physics of Solids*, 25: 309-338.

G'Sell, C., Dahoun, A., Hiver, J. M., and Addiego, F. 2002. Compétition des mécanismes de cisaillement plastique et d'endommagement dans les polymers solides en traction uniaxiale, *Matériaux*, 1-5.

Hillmansen, S., Hobeika, S., Haward, R. N., and Leevers, P. S. 2000. The effect of strain rate, temperature, and molecular mass on the tensile deformation of polyethylene. *Polymer Engineering and Science*, 40: 481-489.

Lemaitre, J. and Chaboche, J. L. 1994. *Mechanics of Solid Materials*. Cambridge: Cambridge University Press.

Lin, L. and Argon, A. S. 1994. Review: Structure and plastic deformation of polyethylene. *Journal of Materials Science*, 29: 294-323.

Liu, H. and Polak, M. A. 2005. Nonlinear viscoelastic modelling of creep of polyethylene. *Proceedings of the 33d Annual General Conference of the Canadian Society for Civil Engineering*, Toronto, GC-140: 1-10.

Parks, D. M. and Ahzi, S. 1990. Polycrystalline plastic deformation and texture evolution for crystals lacking five independent slip systems. *Journal of the Mechanics and Physic of Solids*, 38: 701-724.

THE MECHANICS OF EARLY EMBRYO DEVELOPMENT: INSIGHTS FROM FINITE ELEMENT MODELING

Xiaoguang Chen and G. Wayne Brodland
Department of Civil Engineering, University of Waterloo, Waterloo, ON, Canada N2L 3G1
E-mail: brodland@uwaterloo.ca

Abstract

A finite element-based simulation of neurulation, a critical developmental event common to all vertebrates, is presented for an amphibian embryo. During this process, a sheet of tissue rolls up to form a tube, the precursor of the spinal cord and brain. Material property data for the simulation are based on the cellular fabric of the tissues and on tensile test data, and geometric data are obtained from three-dimensional reconstructions. A spatio-temporal correlation system is used to organize and correlate the data and to construct the finite element model. The simulations predict morphogenetic movements similar to those which occur in real embryos.

Keywords: mechanics of morphogenesis, finite element analysis; constitutive models; fabric evolution

Introduction

During early embryo development, structured sheets of cells undergo dramatic self-driven changes of shape in order to form organs and other crucial structures (Fig. 1) [Alberts *et al.* 1989; Gilbert, 2003]. When anomalies occur in these motions, serious and debilitating malformation birth defects, such as spina bifida, can result. In order to develop effective new strategies for preventing malformation defects it is necessary to understand the mechanics that underlie these motions.

To investigate the mechanics of "morphogenetic movements" is challenging because large strains occur in the tissues and their small scale makes it difficult to obtain accurate geometric and mechanical property data. For example, during formation of the neural tube (Fig. 1), the precursor of the spinal cord and brain, certain tissues experience strains as high as several hundred percent. Direct measurement of the tensile properties of the tissues involved in this process is difficult because they are less than half a millimeter in size and are extremely fragile.

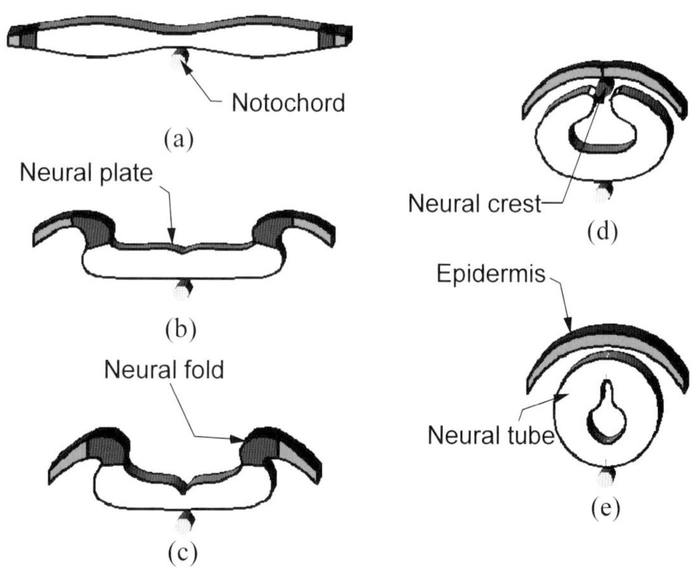

Figure 1. The process of neurulation

M. Pandey et al. (eds), Advances in Engineering Structures, Mechanics & Construction, 459–469.

Traditionally, embryological research has focused on identifying the molecular and genetic aspects of development [Huang and Ingber, 1999; Lecuit, 2003], and relatively little effort has been devoted to understanding the relevant mechanics. Recent work [Swartz, 2001; Stephanou, 2002; Chen, 2004; Robinson, 2004; Brodland, 2005; Moore et al, 2005] has shown, however, that mechanical processes must function in concert with chemical signal regulation and genetics to produce morphogenetic movements. In the present study, axolotls (*Ambystoma mexicanum*) embryos are used. Like a number of other amphibian embryos, they share important geometric similarities with human embryos and, for this reason, are often used as an animal model for neurulation.

This paper presents a whole-embryo finite element model of neurulation; the first of its kind. To capture the mechanical interactions that occur across cellular, tissue and whole-embryo scales, an advanced, multi-scale finite element approach is used. Cell-based simulations are used to construct a system of constitutive equations for embryonic tissues, and experimental data are used to determine the parameters in these equations. Images of live embryos and serial sections of fixed embryos provide the geometric data needed to complete the whole-embryo model.

Constitutive Model

In embryonic tissues, cells readily rearrange themselves, and, as a result, the stress-strain characteristics of embryonic tissues are quite different from those of mature tissues. Analytical studies and computer simulations show that the constitutive relationship is greatly affected by cell-level activities such as reshaping, rearrangement, and mitosis [Brodland *et al.*, 2000; Chen and Brodland, 2000; Brodland et al., 2005], a finding that is supported by experimental results [Brodland and Wiebe, 2004].

The model on which the analysis and simulations are based involves two primary assumptions:

1. Interfacial tension, γ, is assumed to be a primary driving force of cell-cell interactions. This tension is generated by circumferential microfilament bundles (CMBs), membrane-associated proteins, and cell membranes, and cell-cell adhesion generated by cell adhesion molecules (CAMs) and other mechanisms reduce the net contraction [Chen and Brodland, 2000].
2. Cell cytoplasm is assumed to be incompressible and characterized by an effective viscosity, μ.

To model cells with these characteristics using the finite element method, each n-sided cell is divided into n triangular elements (Fig. 2b) and rod-like elements are employed to model the interfacial tension. To account for the volume change that would occur in individual triangular elements due to motion of cytoplasm from one triangular element to another Poisson's ratio for the cytoplasm is set to zero. To keep the total volume in each cell constant, a volume constraint is applied to each cell.

Standard finite element approaches are used to determine element stiffness matrices and equivalent nodal loads and to assemble these results into a system of simultaneous equations [Chen and Brodland 2000],

$$\mathbf{C}\dot{\mathbf{u}} = \mathbf{f}\,, \tag{1}$$

where **C** is the damping matrix of the system and is derived from the triangular elements, **f** is the vector of driving forces produced by the rod-like elements carrying γ, and $\dot{\mathbf{u}}$ is the vector of nodal velocities. Using a forward difference scheme, this nonlinear equation can be rewritten as

$$\mathbf{C}\dot{\mathbf{u}} \approx \mathbf{C}\frac{\Delta\mathbf{u}}{\Delta t} = \frac{1}{\Delta t}\mathbf{C}\Delta\mathbf{u} = \mathbf{f}\,, \tag{2}$$

and

$$\frac{1}{\Delta t}\mathbf{C}(\mathbf{u}_{q+1} - \mathbf{u}_q) = \mathbf{f}_q\,. \tag{3}$$

Solution of these equations gives the time course of the resulting cell-cell interactions.

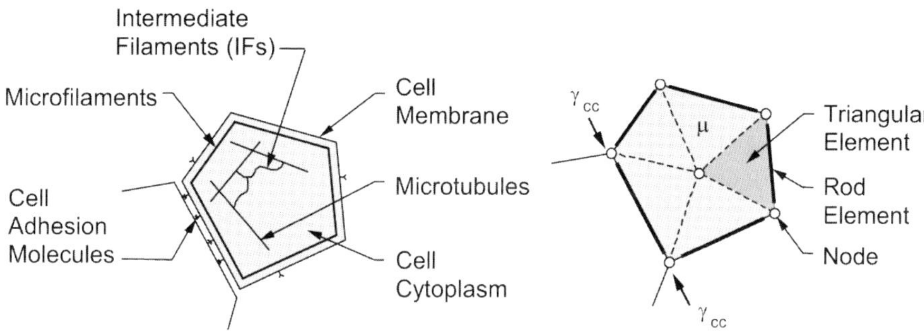

(a) A typical cell in plan view (b) The finite element model

Figure 2. Schematic of a cell and its corresponding finite element model. [Chen and Brodland, 2000]

Insights from finite element models of this kind have lead to the development of constitutive equations for "patches" of such cells (Fig. 3a). The cells in a patch are characterized by a composite cell (Fig. 3b) that captures average aspect ratio, κ, long-axis mean angle, α, and cell density, β [Brodland and Wiebe, 2004].

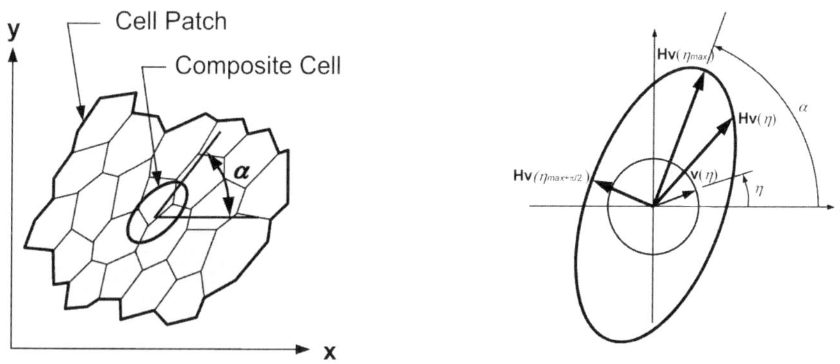

(a) (b)

Figure 3. A cell aggregate and its composite cell. [Brodland, 2004]

Brodland and Wiebe (2004) have demonstrated that the principal stresses in the sheet are generated by interfacial tensions according to Equations 4 and 5:

$$\sigma_1 = \frac{\rho\gamma\kappa}{2\pi}\int_0^{2\pi}\frac{\cos^2\theta}{\sqrt{\kappa\cos^2\theta+\dfrac{\sin^2\theta}{\kappa}}}d\theta \tag{4}$$

$$\sigma_2 = \frac{\rho\gamma}{2\pi\kappa}\int_0^{2\pi}\frac{\cos^2\theta}{\sqrt{\kappa\sin^2\theta+\dfrac{\cos^2\theta}{\kappa}}}d\theta \tag{5}$$

where stress σ_1 is in the α direction and stress σ_2 is in the direction normal to σ_1. These stresses result in an internal pressure,

$$p = \frac{\sigma_1+\sigma_2}{2}, \tag{6}$$

that acts together with the tension stresses.

Ultimately, the stress-strain characteristics of the tissue depend on the five parameters, γ, μ, κ, α and β [Brodland et al., 2005]. These equations can be incorporated into a tissue-level finite element model used to model an entire embryo (Fig. 4). Values for all parameters can be obtained from suitable images of real embryos and from mechanical property tests on tissue specimens. Tests have shown that these parameters vary with tissue type, location, orientation, and developmental stage [Brodland and Wiebe, 2004; Wiebe and Brodland, 2005].

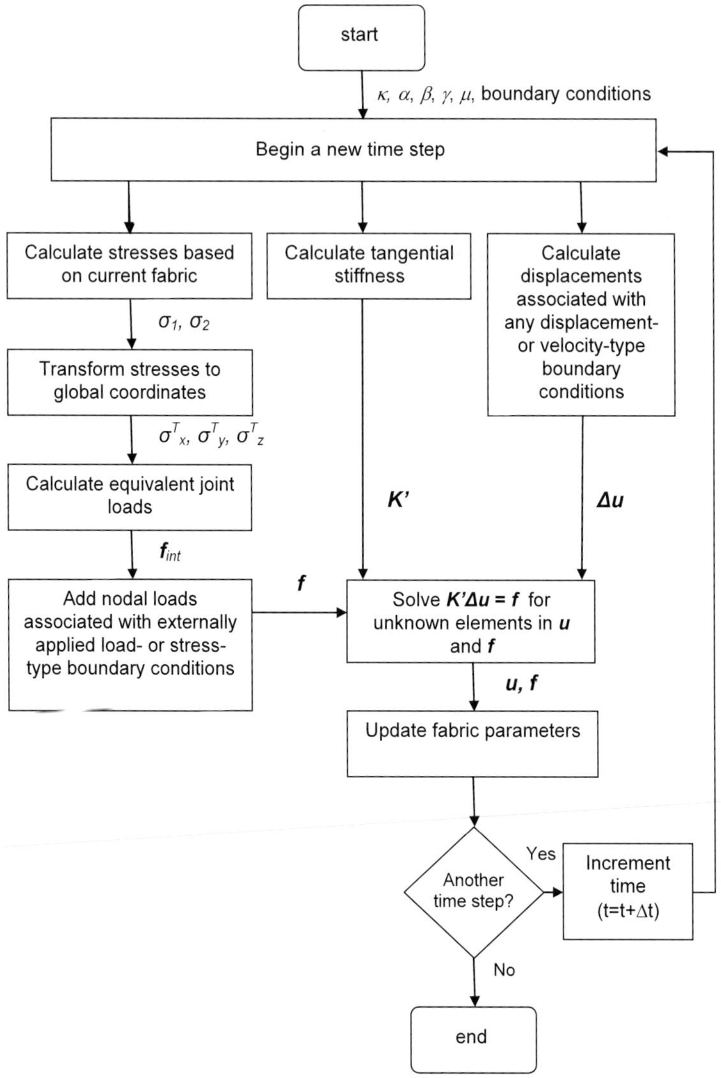

Figure 4. Finite element implementation of the constitutive equations.

Geometric Data

The geometry of an embryo is complex and varies with time. A complete geometric description, therefore, requires both thickness and shape data for each tissue layer and the fabric and material properties of each tissue, plus how these characteristics develop with time. The four primary steps involved in the collection of this "spatial-temporal data" are 1. determining embryo shape, 2. defining tissue fabric parameters, 3. determining embryo layer thickness, and 4. measuring material properties.

1. Determining embryo shape

To identify the specific three-dimensional (3D) shape of a live embryo is technically challenging, and our lab has developed an optical method for extracting live embryo 3D geometric information using robotic microscope images taken at several viewing angles (Fig. 5a). By establishing point correspondences among these views and employing suitable reconstruction algorithms [Brodland and Veldhuis 1998, Bootsma, 2005], we can calculate the spatial positions of a collection of surface points. These points can then be meshed and used to obtain a surface representation of the embryo (Fig. 5b). In-plane tissue motions and strain rates are determined using image processing methods that track the motion of groups of cells [Brodland and Veldhuis 1998; Bootsma, 2005].

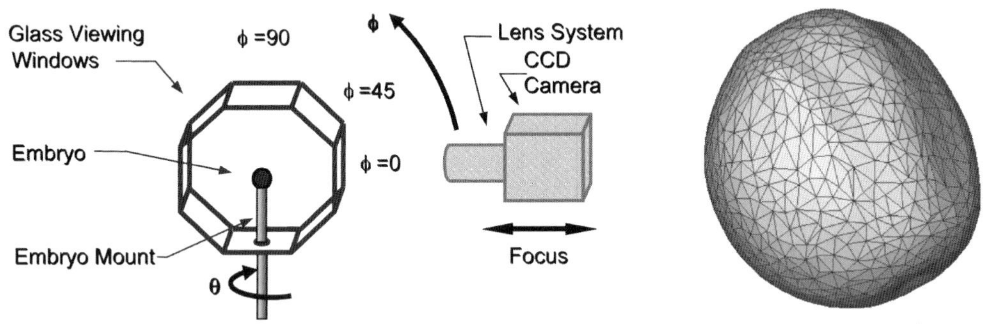

(a) Schematic diagram (b) A 3D reconstruction

Figure 5. The Robotic microscope and a reconstruction made using it.

2. Defining tissue fabric parameters

Tissue fabric parameters are extracted from the time-lapse images using specially designed image processing software. The software detects the boundaries of the cells and uses Fourier analysis to determine κ, α , and β values. An interpolation scheme is used to map fabric parameters onto the surface mesh.

3. Determining embryo layer thickness

At the beginning of neurulation, the embryo consists of three layers – the ectoderm, the mesoderm, and the endoderm. The thicknesses of each of these layers can be obtained from serial sections (Fig. 6) and merged with the surface reconstruction data. In practice, layer data are obtained from several embryos and averaged to improve data quality.

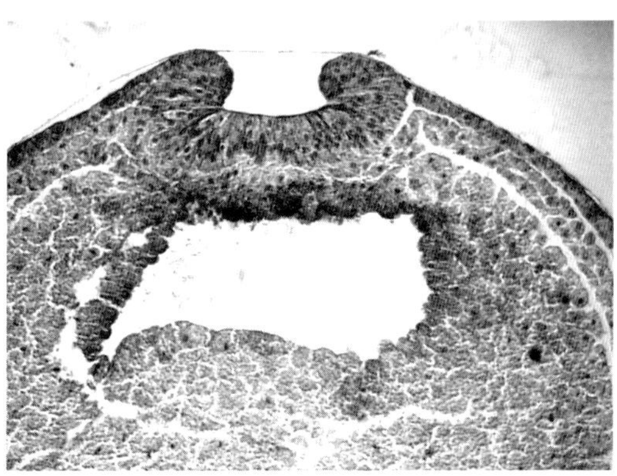

Figure 6. A transverse serial section

4. Measuring material properties

The tissues in an embryo are extremely fragile and, as a result, measurement of material properties is difficult. To obtain stress-strain data, a small specimen (300 by 500 μm) of tissue is attached to very thin wires. The specimen is stretched by drawing the wires apart under computer control. Applied force is calculated from measurements of wire bending, and real-time feedback is used to prevent strain reduction that would otherwise arise from the wire deflections. Companion, cell-level finite simulations are used to back calculate the γ and μ values [Wiebe and Brodland, 2005].

Feature Grid for Correlating Data

The spatial-temporal data collected in the above steps are obtained from a variety of sources, and each has a different natural coordinate system. To combine these data we devised a surface coordinate system, or feature grid. The grid and its attendant interpolation software serve as a vehicle through which element-specific data can be obtained for the whole-embryo finite element model.

Biologists generally divide the surface of a neurulation-stage embryo into three main regions: the neural plate region, the neural fold region, and the non-neural ectoderm region (Fig. 7). Potential feature grids devised for the neural plate are shown in Figure 8. The "parabolic" grid (Fig. 8a) was chosen for the neural plate region, and grid systems compatible with it were chosen for the other two regions.

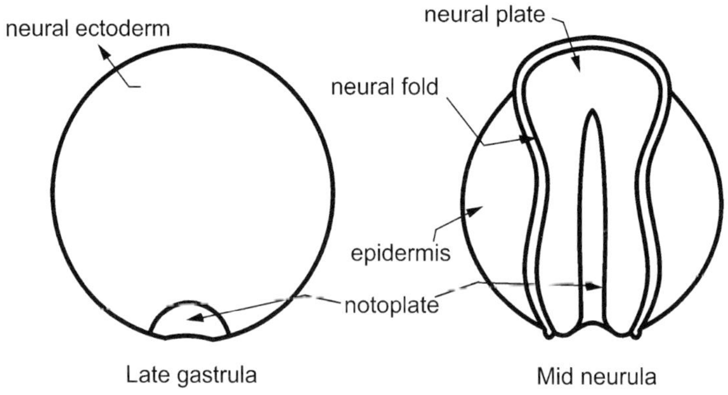

Figure 7. Embryonic region identification

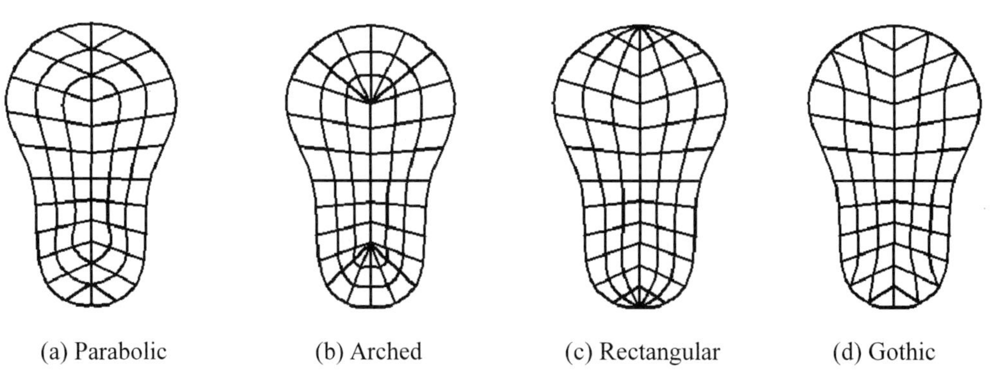

Figure 8. Parametric coordinate systems

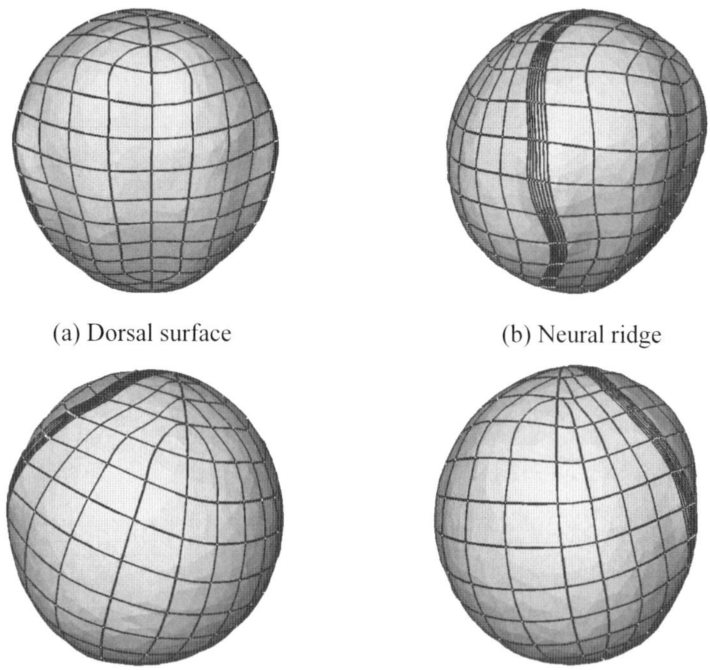

(a) Dorsal surface (b) Neural ridge

(c) Ventral region I (d) Ventral region II

Figure 9. Parametric coordinates shown from a variety of viewing angles

In the feature grid system, each surface point has unique u and v coordinate values, and the grid is mapped onto the 3D embryo surface reconstruction (Fig. 9) using a mathematical mapping (u, v) → (x, y, z). Mappings to other data sets are accomplished in a similar way. Details of the process [Bootsma, 2003] are beyond the scope of the present paper. All data are stored in both their raw form and in a feature grid form (Figure. 10).

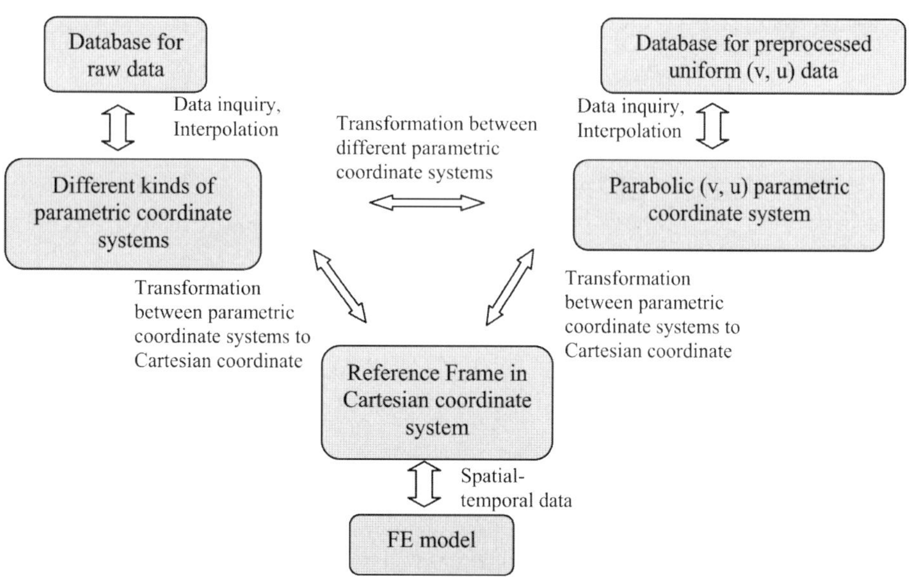

Figure 10. Schematic chart for inquiry system

The finite element mesh generator uses the feature grid as a basis for data inquiry and interpolation, and uses it to determine on an element-by-element basis, tissue geometry and thickness, cellular fabric and mechanical properties.

2D Simulations

To assess the accuracy of the collected data, the interpolation software and the finite element engine, a series of 2D simulations were conducted. Simulations of transverse aspects of neural tube development (Fig. 11) were found to be generally consistent with those obtained by Clausi and Brodland (1994). Which study's simulation best represents actual embryos is not yet known. The differences between the simulations are apparently due to differences in how the contraction forces are calculated: Clausi and Brodland used non-linear truss elements, whereas the current study uses cellular fabric.

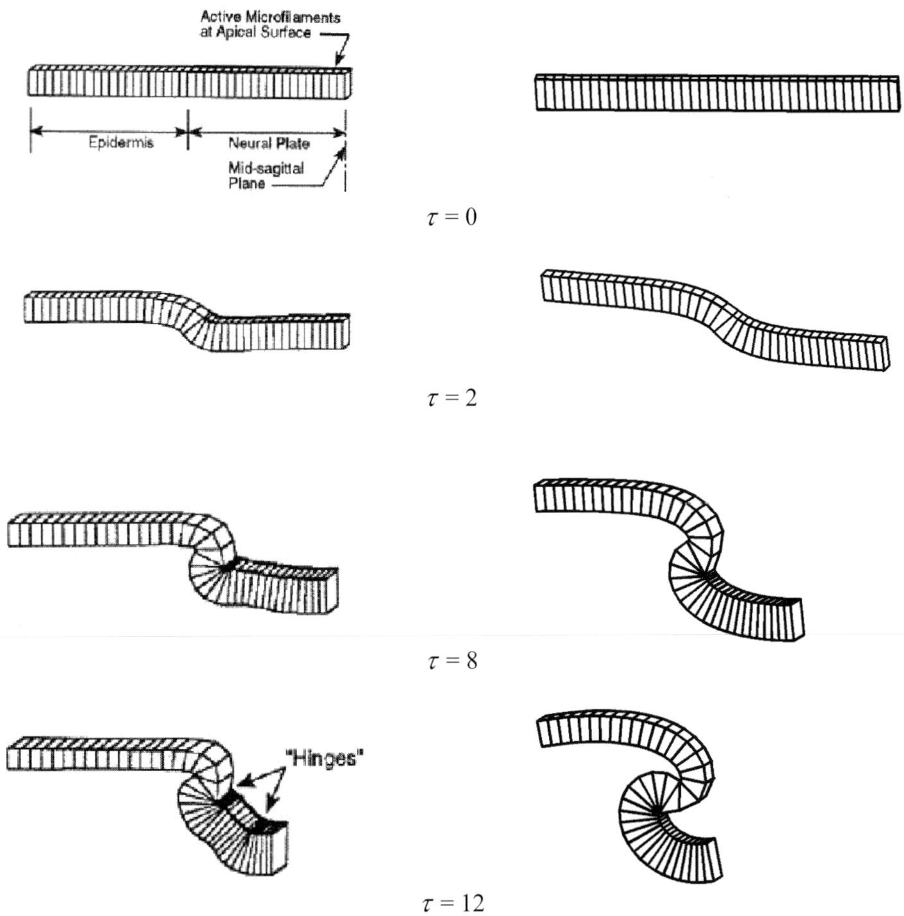

(a) Simulations conducted by Clausi and Brodland

(b) New simulations

Figure 11. Comparison of simulations of a transverse section of the embryo

3D Simulations

The initial configuration of a complete, three-dimensional finite element model, based on available geometric, fabric, and mechanical-property data, is shown in Fig. 12. Each finite element represents a triangular patch of approximately 150 cells. The finite element engine was used to determine how the tissues in the model embryo would interact with each other over time. The engine shows that the neural tube of the model embryo does indeed roll up (Fig. 12b-f) and that it does so in a manner that is similar to neural tube closure in real embryos.

Sensitivity analyses showed that morphogenetic movements are highly sensitive to details of the initial configuration and tissue mechanical properties, a finding consistent with previous 2D simulations [Clausi and Brodland, 1994].

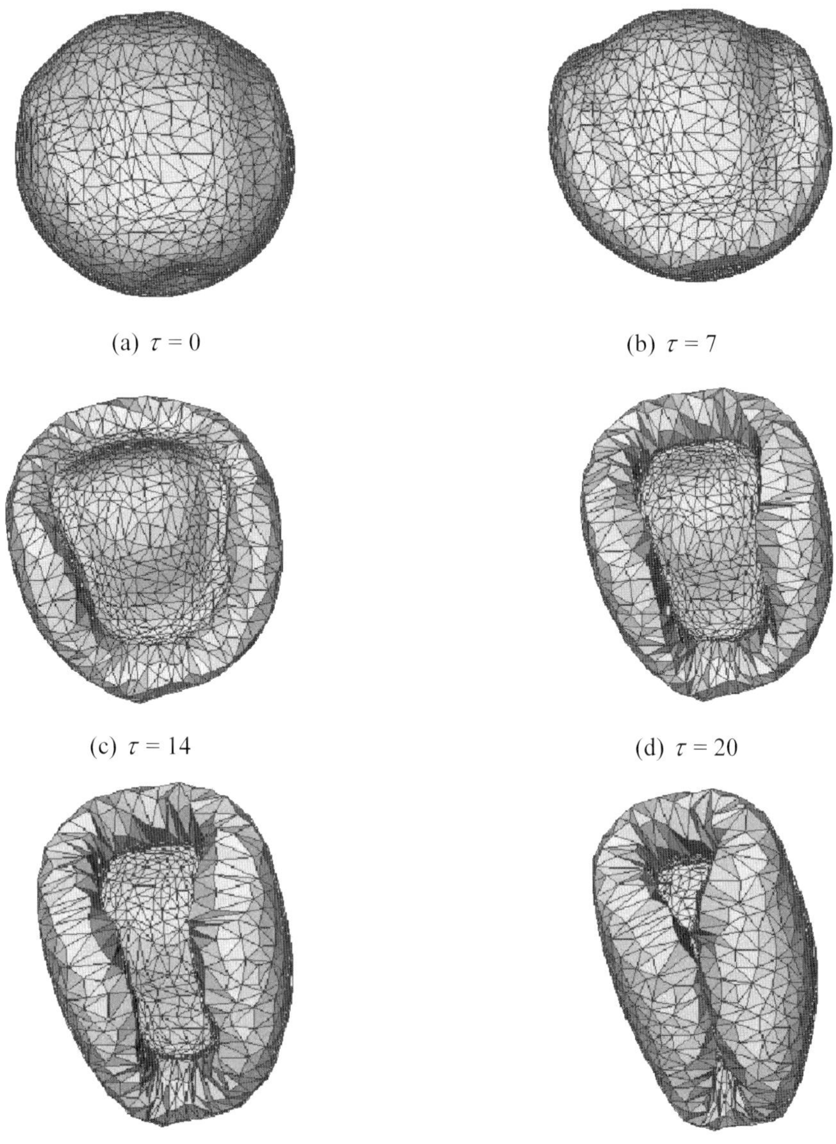

(a) $\tau = 0$ (b) $\tau = 7$

(c) $\tau = 14$ (d) $\tau = 20$

(e) $\tau = 24$ (f) $\tau = 30$

Figure 12. Whole-embryo simulations of neural tube rolling

Discussion and Conclusions

The finite element simulation presented here is the first published full-embryo model of neurulation. In time, it will be refined, as were its 2D predecessors, so that differences between the model and real embryos can be eliminated. The process of resolving these differences is an important scientific task because through it, deficits in understanding can be identified and relevant new experiments conceived.

The model demonstrates that realistic tissue motions are possible when a suitable cell-based constitutive equation is used. A key feature of the current constitutive model is that cells are able to flow past each other in-plane, a characteristic known to be important in real embryos. In previous (unpublished) attempts to model neurulation in 3 dimensions, cells were not free to rearrange in plane. This deficiency caused the tissues to be excessively stiff, especially with respect to in-plane shear, and it impeded the complex 3D deformations that must occur near the ends of the neural plate region. The present constitutive equation also made it possible for single finite elements to model multiple cells, making whole-embryo models computationally practical.

Recent experiments have shown that, during development, the fabric of the embryonic epithelia varies substantially with tissue type, location, and development stage. The constitutive model used here is able to successfully predict fabric evolution during *in vitro* tests, but additional statistical analysis software must be written before the accuracy of those predictions can be assessed in the context of neurulation. Such comparisons, however, are important to full model validation.

The constitutive model is structured so that, as the molecular pathways involved in tissue regulation are identified, their effects can be incorporated into the model. The model can hence serve as a bridge between gene expression and the morphogenetic movements of critical developmental events. This important integration of biology and mechanics is possible because the finite element method provides an open computational framework.

Simulations conducted to date show that tissue motions are highly sensitive to the geometry of the initial configuration and to tissue mechanical properties. This finding suggests that spina bifida and other neural tube defects might arise through a variety of subtle mechanical means. It also suggests that modest interventions might be sufficient to prevent neural tube defects. Identifying appropriate intervention methods is a critical goal for additional research, and the finite element model presented here holds promise as means to carry out preliminary evaluations of proposed interventions.

Acknowledgements

This research was funded by the Canadian Institutes of Health Research (CIHR). Animals were cared for in accordance with Canadian Council on Animal Care (CCAC) guidelines.

References

Alberts B, Bray D, Lewis J, Raff M, Roberts K, and Watson JD 1989, *Molecular Biology of the Cell 2nd Edition*, Garland Publishing, New York.

Bootsma G, 2003, Spatial data correlation, Cive 701 course project, Department of Civil Engineering, University of Waterloo.

Bootsma G.J. and Brodland G.W. 2005. Automated 3-D reconstruction of the surface of live early-stage amphibian embryos. *IEEE Trans Biomed Eng.* **52**:1407-1414.

Brodland G.W. and Clausi D.A. 1994. Embryonic tissue morphogenesis modeled by FEM. *J Biomech Eng.* **116**:146-155.

Brodland G.W. and Veldhuis J.H. 1998. Three-dimensional reconstruction of live embryos using robotic macroscope images. *IEEE Trans Biomed Eng.* **45**:1173-1181.

Brodland G.W and Chen H.H, 2000, The mechanics of cell sorting and envelopment, *J. Biomech.* **33**, 845–851.

Brodland G.W. 2003. New information from cell aggregate compression tests and its implications for theories of cell sorting. *Biorheology*. **40**:273-277.

Brodland G.W., Wiebe C.J., 2004. Mechanical effects of cell anisotropy on epithelia. *Computer Methods in Biomechanics and Biomedical Engineering* **7**(2), 91-99.

Brodland G.W., 2005, Do Lamellipodia Have the Mechanical Capacity to Drive Convergent Extension? *International Journal of Developmental Biology*

Brodland G.W., Chen, D.I-L., Veldhuis J.H., 2005, A cell-based constitutive model for embryonic epithelia and other planar aggregates of biological cells, *International Journal of Plasticity* (ready to print).

Chen C.S, Tan J, Tien J 2004. Mechanotransduction at cell-matrix and cell-cell contacts. Annu Rev Biomed Eng. **6**:275-302.

Chen H.H and Brodland G.W 2000, Cell-level finite element studies of viscous cells in planar aggregates, *ASME J. Biomech. Eng.* **122**, 394–401.

Huang S, Ingber D.E, 1999, The structural and mechanical complexity of cell-growth control, *Nat Cell Biol* **1**, 131-138.

Gilbert S.F., 2003, Development Biology, Seventh Edition, Sinauer Associates, Sunderland.

Lecuit T and Pilot F. 2003. Developmental control of cell morphogenesis: a focus on membrane growth. *Nat Cell Biol.* **5**:103-108.

Moore K. L., 1983 *Before We Are Born*, W.B. Saunders Company.

Moore K.A., Polte T, Huang S, Shi B, Alsberg E, Sunday M.E, Ingber D.E., 2005, Control of basement membrane remodeling and epithelial branching morphogenesis in embryonic lung by Rho and Cytoskeletal tension, *Developmental Dynamics* **232**, 268-281.

Robinson D.N. and Spudich J.A. 2004 Mechanics and regulation of cytokinesis. *Curr Opin Cell Biol.* **16**:182-188.

Stephanou A and Tracqui P. 2002. Cytomechanics of cell deformations and migration: from models to experiments. *C R Biol*. **325**:295-308.

Swartz M.A., Tschumperlin D.J., Kamm R.D., Drazen J.M. 2001. Mechanical stress is communicated between different cell types to elicit matrix remodeling. *Proc Natl Acad Sci*. **98**:6180-6185.

Wiebe C. and Brodland G.W. 2005. Tensile properties of embryonic epithelia measured using a novel instrument. *J Biomech*. **38**:2087-2094.

APPLICATIONS OF RANDOM FIELD MODELS IN STOCHASTIC STRUCTURAL MECHANICS

Christian Bucher
Institute of Structural Mechanics,
Bauhaus-University Weimar, Germany

Abstract

Stochastic structural mechanics deals with the analysis of random phenomena occurring in structural systems or components. There are two major categories of structural uncertainties which involve spatial correlation and which consequently require the treatment as random fields. These are:

- Material properties such as modulus of elasticity or strength,
- Geometrical properties such as shape or thickness

of structural components. The outcome of the stochastic structural analyses is significantly affected by the appropriate treatment of the random properties in the context of the Finite Element method.

The paper will provide an overview of random field representation as appropriate for the stochastic finite element method (cf. Matthies and Bucher, 1999). This includes integral representation models as well a point representation models. In addition, conditional random fields as required in the presence of pointwise deterministic information (e.g. from measurements) are introduced.

Example applications illustrate these concepts and discuss the numerical implications of random field modeling. These applications involve static and dynamic problems which arise in system identification (Macke and Bucher, 2000; Bucher et al., 2003) as well as dynamic stability issues due to geometrical imperfections of shells (Most et al., 2004).

1. Random Fields

A *random field* $H(\mathbf{x})$ is a real-valued random variable whose statistics (mean value, standard deviation, etc.) may be different for each value of $\mathbf{x}$ (Matthies et al., 1997; Matthies and Bucher, 1999), i.e.,

$$H \in \mathbb{R}; \quad \mathbf{x} = [x_1, x_2, \ldots x_n]^T \in \mathcal{D} \subset \mathbb{R}^n. \tag{1}$$

The *mean value function* is defined as

$$\bar{H}(\mathbf{x}) = \mathbf{E}[H(\mathbf{x})], \tag{2}$$

whereby the expectation operator $\mathbf{E}$ is to be taken at a fixed location $\mathbf{x}$ across the ensemble, i.e., over all possible realizations $H(\mathbf{x}, \omega)$ of the random field (see Figure 1).

The spatial correlation, i.e., the fact that we observe a specific dependency structure of random field values $H(\mathbf{x})$ and $H(\mathbf{y})$ taken at different locations $\mathbf{x}$ and $\mathbf{y}$ is described by the *auto-covariance function*

$$C_{HH}(\mathbf{x}, \mathbf{y}) = \mathbf{E}[\{H(\mathbf{x}) - \bar{H}(\mathbf{x})\}\{H(\mathbf{y}) - \bar{H}(\mathbf{y})\}]. \tag{3}$$

With respect to the form of the auto-covariance function we can classify the random fields. A random field $H(\mathbf{x})$ is called *weakly homogeneous* if

$$\bar{H}(\mathbf{x}) = \text{const.} \quad \forall \mathbf{x} \in \mathcal{D}; \quad C_{HH}(\mathbf{x}, \mathbf{x} + \boldsymbol{\xi}) = C_{HH}(\boldsymbol{\xi}) \quad \forall \mathbf{x} \in \mathcal{D}. \tag{4}$$

M. Pandey et al. (eds), Advances in Engineering Structures, Mechanics & Construction, 471–484.

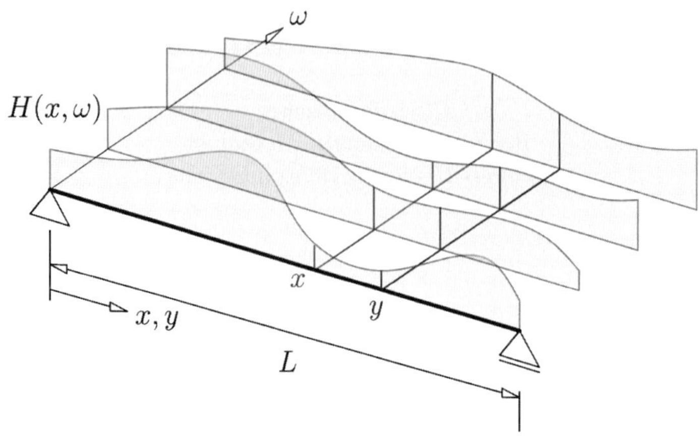

Fig. 1. Ensemble of realizations of one-dimensional random field.

This property is equivalent to the stationarity of a random process. If the covariance function depends on the distance only (not on the direction), i.e.,

$$C_{HH}(\mathbf{x}, \mathbf{x}+\boldsymbol{\xi}) = C_{HH}(\|\boldsymbol{\xi}\|) \quad \forall \mathbf{x} \in \mathcal{D}, \tag{5}$$

then a homogeneous random field $H(\mathbf{x})$ is called *isotropic*. For numerical computations it is useful to represent a continuous random field $H(\mathbf{x})$ in terms of discrete random variables $c_k;\ k = 1 \ldots \infty$ (Ghanem and Spanos, 1991, Brenner and Bucher, 1995):

$$H(\mathbf{x}) = \sum_{k=1}^{\infty} c_k \phi_k(\mathbf{x}), \quad \mathbf{x} \in \mathcal{D} \subset \mathbb{R}^n;\ c_k, \phi_k \in \mathbb{R}. \tag{6}$$

The functions $\phi_k(\mathbf{x})$ are deterministic spatial shape functions which are usually chosen to represent an orthonormal basis on $\mathcal{D}$. The random coefficients c_k can be made uncorrelated, which is an extension of orthogonality into the random variable case.

This representation is usually called *Karhunen–Loève Expansion*. It is based on the following decomposition of the covariance function:

$$C_{HH}(\mathbf{x}, \mathbf{y}) = \sum_{k=1}^{\infty} \lambda_k \phi_k(\mathbf{x}) \phi_k(\mathbf{y}), \tag{7}$$

in which λ_k and $\phi_k(\mathbf{x})$ are the eigenvalues and eigenfunctions, respectively. These are solutions to the integral equation

$$\int_{\mathcal{D}} C_{HH}(\mathbf{x}, \mathbf{y}) \phi_k(\mathbf{x}) \mathrm{d}\mathbf{x} = \lambda_k \phi_k(\mathbf{y}). \tag{8}$$

Mathematically, Equation (8) is an integral equation of the second kind.

In most Finite-Element applications the random field $H(\mathbf{x})$ is discretized right from the start as

$$H_i = H(\mathbf{x_i}); \quad i = 1 \ldots N. \tag{9}$$

A spectral representation for the discretized random field is then obtained by

$$H_i = \sum_{k=1}^{N} \phi_k(\mathbf{x_i}) c_k = \sum_{k=1}^{N} \phi_{ik} c_k. \tag{10}$$

Obviously, this is a matrix-vector multiplication

$$\mathbf{H} = \mathbf{\Phi c}. \tag{11}$$

The orthogonality condition for the columns of $\mathbf{\Phi}$ becomes

$$\mathbf{\Phi}^T \mathbf{\Phi} = \mathbf{I} \tag{12}$$

and the covariance matrix of the components of the coefficient vector $\mathbf{c}$ is

$$\mathbf{C_{cc}} = \text{diag}(\sigma_{c_k}^2). \tag{13}$$

Both conditions can be met if the columns $\boldsymbol{\phi}_k$ of the matrix $\mathbf{\Phi}$ solve the following eigenvalue problem:

$$\mathbf{C_{HH}}\boldsymbol{\phi}_k = \sigma_{c_k}^2 \boldsymbol{\phi}_k; \quad k = 1 \ldots N. \tag{14}$$

Statistically, the Karhunen–Loève expansion is equivalent to a representation of the random field by means of a Principal Component Analysis (PCA).

There are engineering applications in which the values of a structural property are known (e.g. from measurements) in certain selected locations. In geotechnical applications this may be a specific soil property which can be determined through bore holes. Between these locations, however, a random variability is assumed. The strategy to deal with this relies on a regression approach. First we assume that the structural property under consideration without any measurements can be modeled by a zero mean random field $H(\mathbf{x})$. This field is modified into $\hat{H}(\mathbf{x})$ by taking into account the additional knowledge.

Assume that the values of the random field $H(\mathbf{x})$ are known at the locations $\mathbf{x}_k$, $k = 1 \ldots m$. We then write a stochastic interpolation for the conditional random field:

$$\hat{H}(\mathbf{x}_i) = a(\mathbf{x}) + \sum_{k=1}^{m} b_k(\mathbf{x}) H(\mathbf{x}_k) \tag{15}$$

in which $a(\mathbf{x})$ and $b_k(\mathbf{x})$ are random interpolating functions whose statistics have yet to be determined. They are chosen to make the mean value of the difference between the random field and the conditional field zero, i.e. $\mathsf{E}[\hat{H}(\mathbf{x}) - H(\mathbf{x})] = 0$ and to minimize the variance of the difference, i.e. $\mathsf{E}[(\hat{H}(\mathbf{x}) - H(\mathbf{x}))^2] \rightarrow \text{Min}$.

Carrying out the analysis we obtain an expression for the mean value of the conditional random field.

$$\bar{\hat{H}}(\mathbf{x}) = \left[\, \mathbf{C}_{HH}(\mathbf{x}, \mathbf{x}_1) \;\; \mathbf{C}_{HH}(\mathbf{x}, \mathbf{x}_2) \;\; \ldots \;\; \mathbf{C}_{HH}(\mathbf{x}, \mathbf{x}_m) \,\right] \mathbf{C}_{HH}^{-1} \begin{bmatrix} H(\mathbf{x}_1) \\ H(\mathbf{x}_2) \\ \vdots \\ H(\mathbf{x}_m) \end{bmatrix}. \tag{16}$$

In this equation, the matrix $\mathbf{C}_{HH}$ denotes the covariance matrix of the random field $H(\mathbf{x})$ at the locations of the measurements. The covariance matrix of the conditional random field is given by

$$\hat{\mathbf{C}}(\mathbf{x}, \mathbf{y}) = \mathbf{C}(\mathbf{x}, \mathbf{y}) - \left[\, \mathbf{C}_{HH}(\mathbf{x}, \mathbf{x}_1) \;\; \mathbf{C}_{HH}(\mathbf{x}, \mathbf{x}_2) \;\; \ldots \;\; \mathbf{C}_{HH}(\mathbf{x}, \mathbf{x}_m) \,\right] \mathbf{C}_{HH}^{-1} \begin{bmatrix} \mathbf{C}_{HH}(\mathbf{y}, \mathbf{x}_1) \\ \mathbf{C}_{HH}(\mathbf{y}, \mathbf{x}_2) \\ \vdots \\ \mathbf{C}_{HH}(\mathbf{y}, \mathbf{x}_m) \end{bmatrix}. \tag{17}$$

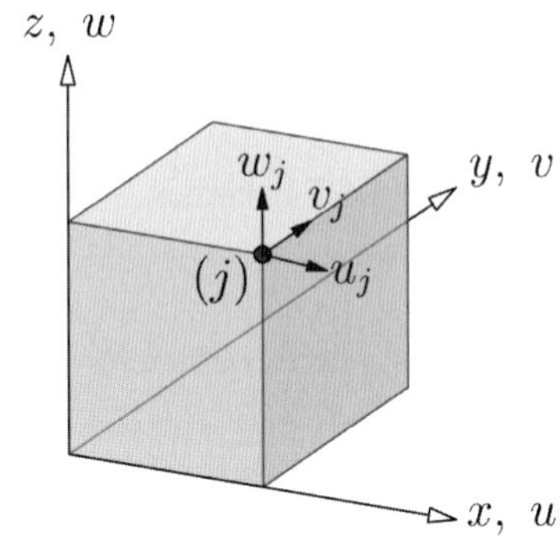

Fig. 2. Volume element with nodal displacements.

2. Stochastic Finite Element Formulation

As discussed earlier, the element stiffness matrix $\mathbf{K}^e$ relates the nodal forces $\mathbf{F}^e$ to the nodal displacements $\mathbf{U}^e$:

$$\mathbf{F}^e = \mathbf{K}^e\mathbf{U}^e \tag{18}$$

in which, for the element with m^e nodes (j); $j = 1 \dots m^e$ as sketched in Figure 2:

$$\mathbf{F}^e = \left[(f_{xj}, f_{yj}, f_{zj}); j = 1 \dots m^e\right]^T; \quad \mathbf{U}^e = \left[(u_j, v_j, w_j; j = 1 \dots m^e\right]^T. \tag{19}$$

Based on the principle of virtual work, the element stiffness matrix for a linear material law (assuming geometrical linearity as well) is obtained as

$$\mathbf{K}^e = \int_{V^e} \mathbf{B}^T(x, y)\mathbf{D}(x, y)\mathbf{B}(x, y)dV^e. \tag{20}$$

Typically, the strain interpolation matrix $\mathbf{B}(x, y)$ is chosen in polynomial form, i.e.

$$\mathbf{B}(x, y) = \sum_{k+l \le r}\sum \mathbf{B}_{kl}x^k y^l; \quad k, l, r \ge 0. \tag{21}$$

In this equation, $\mathbf{B}_{kl}$ are constant matrices. In fact, for the CST element shown above there is only one such matrix, i.e. $\mathbf{B}_{00}$. Assuming that the system randomness is described by a random elastic modulus $E(x, y)$, the elasticity matrix $\mathbf{D}(x, y)$ can be written as

$$\mathbf{D}(x, y) = \mathbf{D}_0 E(x, y). \tag{22}$$

Using the polynomial form of $\mathbf{B}(x, y)$, the element stiffness matrix finally becomes

$$\mathbf{K}^e = \sum_{k+l \le r}\sum \sum_{m+n \le r}\sum \mathbf{B}_{kl}^T\mathbf{D}_0\mathbf{B}_{mn} \int_{V^e} E(x, y)x^k y^l x^m y^n \,\mathrm{d}V^e. \tag{23}$$

The last term in this equation is a so-called *weighted integral* of the random field $E(x, y)$.

$$X_{klmn}^e = \int_{V^e} E(x, y)x^k y^l x^m y^n \,\mathrm{d}V^e. \tag{24}$$

Using this representation, it is possible to achieve a description of the random variation of the element stiffness matrix in terms of the mean values and the covariance matrix of the weighted integrals.

Due to the numerical rather than analytical integration procedures as utilized in FE analysis, this weighted integral is represented by linear combinations of the values R_j; $j = 1 \dots n$ of the random field at discrete integration points.

The global stiffness matrix is then assembled by applying standard FE techniques into the form

$$\mathbf{K} = \sum_{j=1}^{n} \mathbf{K}_j R_j, \tag{25}$$

which can be used as a starting point for a perturbation analysis with respect to the discretized random field R_j.

The general situation of a SFE analysis in nonlinear dynamics generally requires the solution of the following matrix-vector equation

$$\mathbf{M}\ddot{\mathbf{x}} + \mathbf{C}\dot{\mathbf{x}} + \mathbf{r}(\mathbf{x}) = \mathbf{f}(t). \tag{26}$$

In Equation (26), $\mathbf{M}$ is the mass matrix, $\mathbf{C}$ is the damping matrix, $\mathbf{x}$ denotes the vector of nodal displacements, $\mathbf{r}(\mathbf{x})$ is the vector of restoring forces depending nonlinearly on the nodal displacements, and $\mathbf{f}(t)$ is the applied load.

Within the FE concept, the restoring force vector $\mathbf{r}(\mathbf{x})$ is assembled from corresponding element forces, e.g. based on the principle of virtual work in the form

$$\mathbf{r}(\mathbf{x})\delta\mathbf{x}_e = \int_{V_e} \sigma_e(\epsilon)\delta\epsilon_e dV_e. \tag{27}$$

In Equation (27), the subscript e refers to a particular element. Obviously, this equation implies that the randomness of the material properties immediately affects the calculated restoring forces due to the integration over the volume of the element. Consequently any randomness of these material data will be reflected in the restoring forces as well as the tangential stiffness matrix $\mathbf{K}_T$ derived from them.

Utilizing a linearization approach at element level, classical SFE-methods as outlined above can be applied.

3. Perturbation Approach

For static problems, the linear finite element equations are

$$\mathbf{Ku} = \mathbf{F}, \tag{28}$$

where $\mathbf{K}$ is the stiffness matrix, $\mathbf{u}$ is the nodal displacement vector and $\mathbf{F}$ is the nodal force vector. In the following we will assume that only the stiffness matrix involves randomness, i.e. the Young's modulus $E(\mathbf{x})$ is described as

$$E(\mathbf{x}) = \bar{E}(1 + \varepsilon f(\mathbf{x})) \tag{29}$$

with $\bar{E}$ and ε as the mean and the coefficient of variation of the Young's modulus, respectively, and $f(\mathbf{x})$ as a zero-mean unit random field. Utilizing the stochastic finite element method the random field is discretized by a set of n zero-mean unit random variables R_j with covariances $Cov[R_j, R_k]$ $(j, k = 1, \dots, n)$. In case of small to moderate coefficients of variation ε of the Young's modulus a first-order perturbation approach suffices to render accurate second-moment results. Therewith, the nodal displacements can be written as

$$\mathbf{u} = \bar{\mathbf{u}} + \varepsilon \bar{\mathbf{K}}^{-1} \sum_{j=1}^{n} \left. \frac{\partial \mathbf{K}}{\partial R_j} \right|_{\mu_j} R_j \bar{\mathbf{u}}, \tag{30}$$

whereby $\bar{\mathbf{K}}$ and $\bar{\mathbf{u}}$ are the stiffness matrix and the nodal displacement vector evaluated at the means $\mu_j = 0$ of the random variables R_j. Taking expectations, the mean and the covariance of the nodal displacements are given by

$$E[\mathbf{u}] = \bar{\mathbf{u}} \tag{31}$$

and

$$Cov[\mathbf{u}, \mathbf{u}] = \varepsilon^2 \bar{\mathbf{K}}^{-1} \sum_{j=1}^{n} \sum_{k=1}^{n} \left. \frac{\partial \mathbf{K}}{\partial R_j} \right|_{\mu_j} \bar{\mathbf{u}} \bar{\mathbf{u}}^T \left. \frac{\partial \mathbf{K}}{\partial R_k} \right|_{\mu_k}^T Cov[R_j, R_k] \bar{\mathbf{K}}^{-T}, \tag{32}$$

respectively.

To include measured dynamic responses in the random field description – and therewith the nodal displacements – the generalized eigenvalue problem

$$(-\mathbf{M}\lambda_i + \mathbf{K})\mathbf{e}_i = \mathbf{0} \tag{33}$$

has to be solved, with $\mathbf{M}$ denoting the (deterministic) stiffness matrix, and λ_i and $\mathbf{e}_i$ the i-th eigenvalue and eigenvector, respectively. Expressing the eigenvalues by their Rayleigh coefficients

$$\lambda_i = \frac{\mathbf{e}_i^T \mathbf{K} \mathbf{e}_i}{\mathbf{e}_i^T \mathbf{M} \mathbf{e}_i} \tag{34}$$

and applying again a first-order perturbation approach, the random eigenvalues λ_i can be expressed as a linear combination of the random variables R_j, i.e.

$$\lambda_i = \bar{\lambda}_i + \varepsilon \sum_{j=1}^{n} \gamma_{ij} R_j = \bar{\lambda}_i + \varepsilon \sum_{j=1}^{n} \frac{1}{\bar{\mathbf{e}}_i^T \mathbf{M} \bar{\mathbf{e}}_i} \bar{\mathbf{e}}_i^T \left. \frac{\partial \mathbf{K}}{\partial R_j} \right|_{\mu_j} \bar{\mathbf{e}}_i R_j, \tag{35}$$

where $\bar{\mathbf{e}}_i$ are the eigenvectors evaluated at $\mu_j = 0$. Given m identified frequencies ω_i ($\lambda_i = \omega_i^2$), this information can be included in the random field description by transforming the random variables R_j such that they are conditional on the set of measurements $\mathbf{s} = [s_1, s_2, \ldots, s_m]$, whereby $s_i = \lambda_i - \bar{\lambda}_i$. Consequently, also the nodal displacements are conditional on the measurements $\mathbf{s}$, i.e. the conditional mean and covariance are

$$E[\mathbf{u}|\mathbf{s}] = \bar{\mathbf{u}} + \varepsilon \bar{\mathbf{K}}^{-1} \sum_{j=1}^{n} \left. \frac{\partial \mathbf{K}}{\partial R_j} \right|_{\mu_j} E[R_j|\mathbf{s}] \bar{\mathbf{u}} \tag{36}$$

and

$$Cov[\mathbf{u}, \mathbf{u}|\mathbf{s}] = \varepsilon^2 \bar{\mathbf{K}}^{-1} \sum_{j=1}^{n} \sum_{k=1}^{n} \left. \frac{\partial \mathbf{K}}{\partial R_j} \right|_{\mu_j} \bar{\mathbf{u}} \bar{\mathbf{u}}^T \left. \frac{\partial \mathbf{K}}{\partial R_k} \right|_{\mu_k}^T Cov[R_j, R_k|\mathbf{s}] \bar{\mathbf{K}}^{-T}. \tag{37}$$

4. Random Fields Conditional on Dynamic Measurements

4.1. *Formulation*

As can be seen from Equation (35), the random eigenvalues can be expressed as a linear combination of the random variables R_j with coefficients γ_{ij}. Constructing a new set of random variables $\mathbf{Q}$ by the transformation

$$\mathbf{Q} = \left[\boldsymbol{\gamma}_1, \boldsymbol{\gamma}_2, \ldots, \boldsymbol{\gamma}_n\right]^T \mathbf{R} = \boldsymbol{\Gamma}^T \mathbf{R}, \tag{38}$$

whereby the transformation vectors $\boldsymbol{\gamma}_i$ are defined as $\boldsymbol{\gamma}_i = [\gamma_{i1}, \gamma_{i2}, \ldots, \gamma_{in}]$, the mean and covariance of $\mathbf{Q}$ are

$$E\,[\mathbf{Q}] = \boldsymbol{\Gamma}^T E\,[\mathbf{R}] \tag{39}$$

and

$$Cov\,[\mathbf{Q}, \mathbf{Q}] = \boldsymbol{\Gamma}^T Cov\,[\mathbf{R}, \mathbf{R}]\,\boldsymbol{\Gamma}, \tag{40}$$

respectively. Given a set of m measurements $\mathbf{s}$, these measurements can be interpreted as a realization of the random vector $\mathbf{S} = [Q_1, Q_2, \ldots, Q_m]$. To include the measurement information in the statistical description of the random field, a conditional random field is defined with mean

$$E[Q_j|\mathbf{s}] = E[Q_j] + Cov[Q_j, \mathbf{S}]\big(Cov[\mathbf{S}, \mathbf{S}]\big)^{-1}\big(\mathbf{s} - E[\mathbf{S}]\big) \tag{41}$$

and covariance

$$Cov[Q_j, Q_k|\mathbf{s}] = Cov[Q_j, Q_k] - Cov[Q_j, \mathbf{S}]\big(Cov[\mathbf{S}, \mathbf{S}]\big)^{-1} Cov[\mathbf{S}, Q_k] \tag{42}$$

that possesses the desired properties (Ditlevsen, 1991). In other words, the random field described by Equations (41) and (42) is conditional on complete agreement with the measurements $\mathbf{s}$ made. The conditional mean and covariance of the original random variables $\mathbf{R}$ – as required by Equations (36) and (37) – are given by

$$E\,[\mathbf{R}|\mathbf{s}] = \boldsymbol{\Gamma}^{-T} E\,[\mathbf{Q}|\mathbf{s}] \tag{43}$$

and

$$Cov\,[\mathbf{R}, \mathbf{R}|\mathbf{s}] = \boldsymbol{\Gamma}^{-T} Cov\,[\mathbf{Q}, \mathbf{Q}|\mathbf{s}]\,\boldsymbol{\Gamma}^{-1}, \tag{44}$$

respectively.

4.2. *Numerical Example: Plate Structure*

As a numerical example a plate structure of size $4l$ by $2l$ ($l = 1.25$ m) with thickness $t = 0.25$ m is investigated (see Figure 3). Young's modulus $E = 30.0$ GPa, Poisson's ratio $\nu = 0.16$, correlation length $l_c = 1.0$ m, coefficient of variation $\varepsilon = 0.1$, mass density $\varrho = 2500$ kg/m^3, $p(x, y) = 1.6$ MPa.

It is assumed that a set of natural frequencies ω_i is measured (cf. Table 1) for one particular realization of the random field as shown in Figure 4. The unconditional mean and standard deviation of the vertical deflection under the given static load are shown in Figures 5 and 6, respectively. By including the measurements of the two lowest natural frequencies, the standard deviation of the deflection is considerably reduced as shown in Figure 7. As shown in Macke and Bucher (2000), the

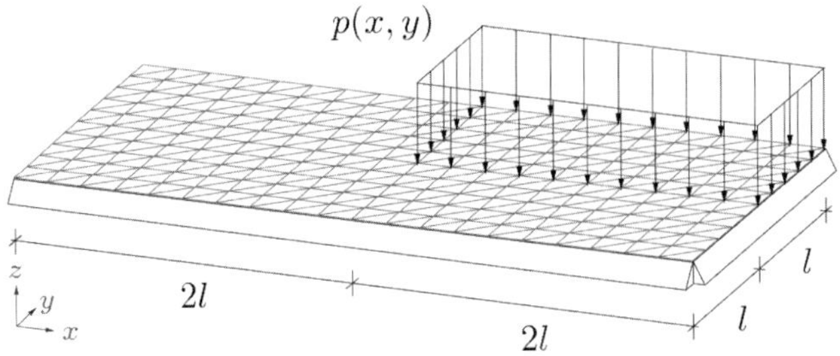

Fig. 3. Plate structure.

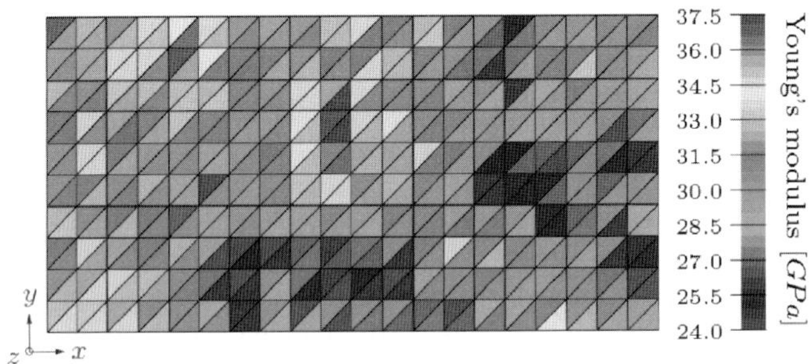

Fig. 4. Realization of random field.

Table 1. Mean and measured frequencies.

i	$\bar{\omega}_i$	ω_i
1	499	504
2	797	796
3	1290	1294
4	1686	1694
5	1976	1979
6	1976	1993
7	2457	2472

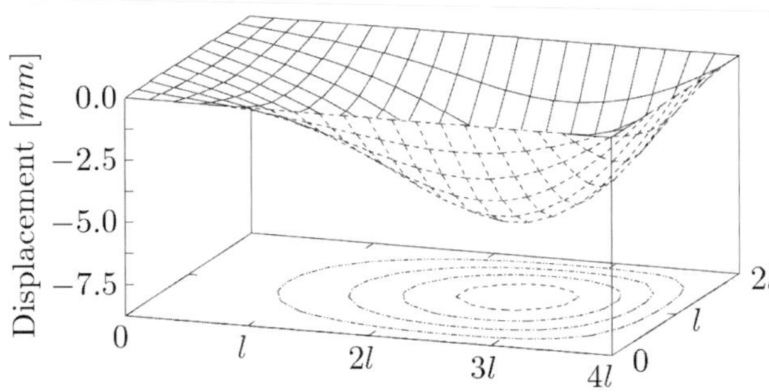

Fig. 5. Mean displacement $\bar{\mathbf{u}}$.

effect of including additional frequency measurement on further reduction of the standard deviation is very small.

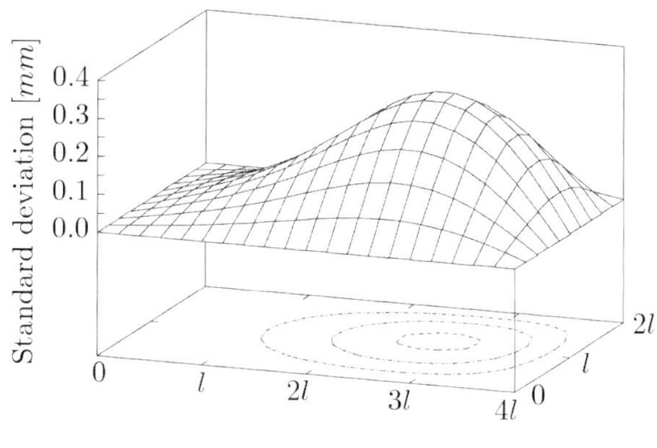

Fig. 6. Initial standard deviation.

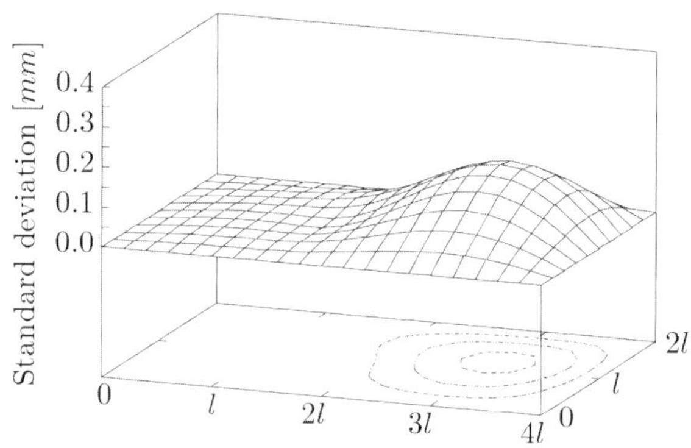

Fig. 7. Conditional standard deviation.

5. Dynamic Stability Analysis

5.1. *Nonlinear Stability Analysis*

The equation of motion of a geometrically nonlinear structural model is given by

$$\mathbf{M}\ddot{\mathbf{x}} + \mathbf{r}(\mathbf{x}, \dot{\mathbf{x}}) = \mathbf{f}(t). \tag{45}$$

For neighboring trajectories, the tangential equation of motion may be utilized to describe temporal evolution of the difference $\mathbf{y}(t)$

$$\mathbf{M}\ddot{\mathbf{y}} + \mathbf{C}\dot{\mathbf{y}} + \mathbf{K}\mathbf{y} = \mathbf{0}. \tag{46}$$

To analyze the dynamic stability behaviour of nonlinear systems an integration of Equation (45) is necessary until stochastic stationarity is reached. In each time step, the tangential stiffness matrix $\mathbf{K}$ has to be determined. With this kind of analysis a criterion for sample stability is developed. In order to speed up explicit time integration, this equation can be projected into a subspace of dimension m as spanned by the eigenvectors of the undamped system corresponding to the m smallest natural frequencies (Bucher, 2001). These eigenvectors are the solutions to

$$\left(\mathbf{K}\left(\mathbf{x}_{\text{stat}}\right) - \omega_i^2\mathbf{M}\right)\mathbf{\Phi} = 0; \quad i = 1 \ldots m \tag{47}$$

In this equation, $\mathbf{x}_{\text{stat}}$ is chosen to be the displacement solution of Equation (45) under static loading conditions. The mode shapes are assumed to be mass normalized. A transformation $\mathbf{x} = \mathbf{\Phi}\,\mathbf{v}$ and a multiplication of Equation (45) with $\mathbf{\Phi}^T$ represents a projection of the differential equation of motion for the reference solution into the subspace of dimension m as spanned by the eigenvectors:

$$\ddot{\mathbf{v}} + \mathbf{\Phi}^T \mathbf{r}(\mathbf{x}, \dot{\mathbf{x}}) = \mathbf{\Phi}^T \mathbf{f}. \tag{48}$$

The integration of Equation (48) by the central difference method (Bathe, 1996) requires a minimal time step.

The time integration in the subspace and the computing of the restoring forces on the full system causes the following problem: If the start displacement or velocity vector of the time integration is not zero, for example by static loading, the projection of this vectors into the subspace is an optimiziation problem caused by the higher number of variables in the full space. By using a least square approach:

$$\mathbf{v} = \mathbf{\Phi}^{-1}\mathbf{x}; \quad \mathbf{\Phi}^{-1} = \left(\mathbf{\Phi}^T \mathbf{\Phi}\right) \mathbf{\Phi}^T \tag{49}$$

this projection is optimal approximated, but not suitable for a subspace spanned by a small number of eigenvectors. A possibility to handle this, is to start the time integration in the subspace with a displacement and velocity vector equal to zero. The start vectors have to be saved in the full system and the restoring force vector has to be computed by addition of the start and the time integration vectors:

$$\mathbf{r}(\mathbf{x}, \dot{\mathbf{x}}) = \mathbf{r}(\mathbf{x}_{\text{start}} + \mathbf{\Phi v}, \dot{\mathbf{x}}_{\text{start}} + \mathbf{\Phi}\dot{\mathbf{v}});$$

$$\mathbf{v}(t = 0) = \dot{\mathbf{v}}(t = 0) = \mathbf{0}. \tag{50}$$

In the investigated cases the start vector $\mathbf{x}_{\text{start}}$ is the static displacement vector, the start velocities are assumed to be vanished.

To analyze the stability behaviour of the reference solution $\mathbf{x}_0(t)$, the long-term behavior of the neighboring motion (Equation (46)) is investigated. To reduce the dimension of the equation system, this equation can be projected into the same or a smaller subspace as Equation (48). Transformed into the state space description we obtain:

$$\dot{\mathbf{z}} = \begin{bmatrix} \mathbf{0} & \mathbf{I} \\ -\mathbf{\Phi}^T \mathbf{K} \mathbf{\Phi} & -\mathbf{\Phi}^T \mathbf{C} \mathbf{\Phi} \end{bmatrix} \mathbf{z} = \mathbf{A}[\mathbf{x}_0(t)]\mathbf{z} \tag{51}$$

From this equation, the Lyapunov exponent λ can be determined by a limiting process:

$$\lambda(\mathbf{x}_0, \mathbf{s}) = \lim_{t \to \infty} \frac{1}{t} \log ||\mathbf{\Theta}(\mathbf{x}_0, t)\mathbf{s}|| \tag{52}$$

in which $\mathbf{s}$ is an arbitrary unit vector. In Equation (52), $\mathbf{\Theta}(\mathbf{x}_0, t)$ is the transition matrix from time 0 to t associated with Equation (51). Based on the multiplicative ergodic theorem (e.g. Arnold and Imkeller, 1994) the Lyapunov exponent can also be calculated as an expected value:

$$\lambda(\mathbf{x}_0, \mathbf{s}) = E\left[\frac{d}{dt} \log ||\mathbf{\Theta}(\mathbf{x}_0, t)\mathbf{s}||\right]. \tag{53}$$

In the current investigation, the norm $||\mathbf{\Theta}(\mathbf{x}_0, t)\mathbf{s}||$ is expressed in terms of

$$\|\mathbf{\Theta}(\mathbf{x}_0, t)\mathbf{s}\| \leq \|\mathbf{\Theta}(\mathbf{x}_0, t)\| \cdot \|\mathbf{s}\| = \|\mathbf{\Theta}(\mathbf{x}_0, t)\|. \tag{54}$$

Finally, this result is used in calculating the Lyapunov exponent according to Equation (52) by using a matrix norm equal to the eigenvalue $\mu_{\max}$ of $\mathbf{\Theta}(\mathbf{x}_0, t)$ with the maximum absolute value. The time domain t has to be taken large enough that the Lyapunov exponent convergences to a stationary value. For the statistical estimation of the convergence of the Lyapunov exponent, Equation (53) is suitable.

5.2. *Linear Stability Analysis*

The Lyapunov exponent for the stability of the second moments of a linearized reference solution can be determined by the Itô analysis. The nonlinear stiffness matrix in Equation (46) can be expanded into an asymptotic series with respect to a static loading condition. Under the assumption that the fluctuating part is small enough this series can be truncated after the linear term:

$$\mathbf{M}\ddot{\mathbf{y}} + \mathbf{C}\dot{\mathbf{y}} + \left(\mathbf{K}\left(\mathbf{x}_{\text{stat}}\right) + \mathbf{f}(t)\mathbf{K}_1\right)\mathbf{y} = \mathbf{0}. \tag{55}$$

This equation of motion is projected into a subspace of dimension m and then transformed into its state space description analogous to Equation (51):

$$\dot{\mathbf{z}} = \left[\mathbf{A} + \mathbf{B}\mathbf{f}(t)\right]\mathbf{z}, \tag{56}$$

where the coefficient matrices **A** and **B** are constant. The fluctuating part of the loading function is assumed to be Gaussian white noise. Then Equation (56) represents a first order stochastic differential equation. For this system the Lyapunov exponent λ_2 for the second moments can be easily derived by applying the Itô calculus (e.g. Soong and Grigoriu, 1992; Lin and Cai, 1995).

The Lyapunov exponents for almost sure stability can be approximated for linear SDOF-systems analytically (Lin and Cai, 1995):

$$\lambda = -\zeta_0\omega_0 + \frac{\pi S_{ff}\omega_0^2}{4}, \tag{57}$$

where ω_0 is the eigenfrequency, ζ_0 is the modal damping ratio and S_{ff} is the power spectral density of the white noise excitation. The Lyapunov exponent λ_2 for the second moments can be calculated with

$$\lambda_2 = -2\zeta_0\omega_0 + \pi S_{ff}\omega_0^2. \tag{58}$$

By exploiting this, the Lyapunov exponent for the samples can be appproximated from the second moment exponent according to

$$\lambda = \frac{\lambda_2}{4} - \frac{\zeta_0\omega_0}{2}. \tag{59}$$

This equation can also be applied on MDOF-systems, it should be mentioned that the term $-\zeta_0\omega_0$ corresponds then to the Lyapunov exponent of the system without random parametric excitation.

5.3. *Reliability Investigation of a Shell Structure*

A cylindrical panel was considered, which is mentioned e.g. in Krätzig (1989) and Schorling and Bucher (1999). The assumed structure is shown in Figure 8. The geometrical and the material proberties were given as: radius $R = 83.33$ m, the half width and heigth $a = 5$ m, the thickness $h = 0.1$ m, the Young's modulus $E = 3.4 \times 10^{10}$ N/m^2, the mass density $\rho = 3400$ kg/m^3 and the Poisson's ratio $\mu = 0.2$. The constant load factor is $P = 1000$ N/m.

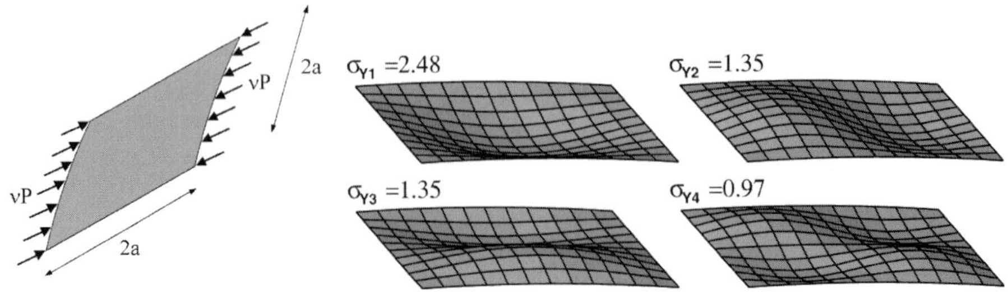

Fig. 8. Nonlinear cylindrical shell structure with associated weighted imperfection shapes.

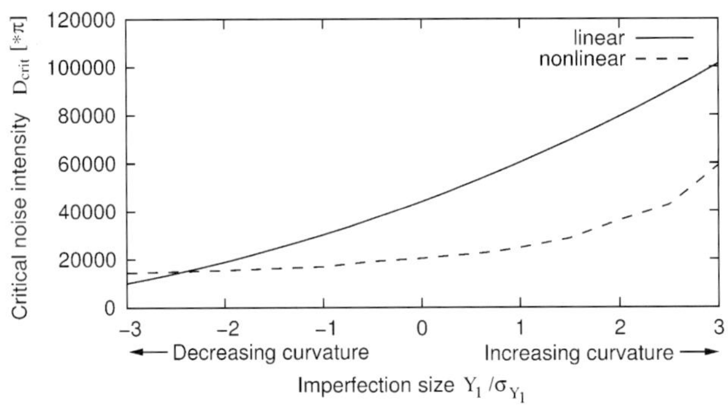

Fig. 9. Stability boundaries vs. imperfection size.

The structure is discretized with 7×7 nodes and meshed with geometrically nonlinear 9-node shell elements. At a static load factor of $\nu_{\text{crit}} = 16825$ the structure reaches an unstable state (Krätzig, 1989: $\nu_{\text{crit}} = 15120$; Schorling and Bucher, 1999: $\nu_{\text{crit}} = 16200$). The static load is assumed to be $P_0 = 0.85\nu_{\text{crit}}P$. The fluctuating load is considered as $P_{\text{fluct}} = \ell f(t)P$, where $f(t)$ is the unit white noise process and ℓ is the load factor. The damping is assumed as modal damping with the damping ratio $\zeta_k = 0.02$ for all modes.

The geometrical imperfections are considered in terms of radial deviations from the perfect panel surface and are modelled as a conditional Gaussian random field. The mean is assumed as zero and the standard deviation as $\sigma = 10^{-3}$ m. The correlation length of the exponential correlation function is considered with $l_H = 10$ m. The imperfection shapes are obtained by the decomposition of the covariance matrix according to Equation (14). The first four imperfection shapes are shown in Figure 8 as well. The corresponding standard deviations σ_{Yi} in uncorrelated normal space are indicated in the figure. The first shape is very similar to the buckling shape.

The structure was investigated by using the Itô analysis and it was found that only the first imperfection shape has a major influence on the stability behaviour. The critical noise intensity of the perfect system was obtained as $D_{0_{\text{crit}}} = 92000\pi$ with the linear and $D_{0_{\text{crit}}} = 20000\pi$ with the nonlinear method by averaging 20 simulations with 10^5 time steps. The nonlinear analysis uses a modal subspace spanned by 12 of the 213 eigenmodes with a critical time step of $\Delta t = 6.3 \cdot 10^{-3}$ s. The investigation of the first imperfection shape obtained by nonlinear analysis show observable deviations from the linear results. This points out that the nonlinearities of this structure have a higher influence as compared to the previous example. The obtained stability boundaries depending on the imperfection size are displayed in Figure 9 for the linear and the nonlinear analysis.

The failure probability for this one dimensional problem can be obtained analytical from the stability boundaries and is shown in Figure 10 depending on the noise intensity for both methods. It

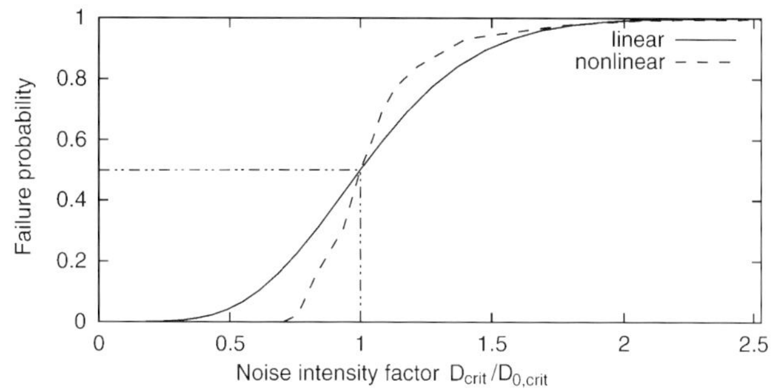

Fig. 10. Failure probabilities for both methods.

Table 2. Critical static and dynamic loads.

Model	ν_{crit}	$D_{0_{crit}}$ $\nu = 0.85\nu_{crit}$	$D_{0_{crit}}$ $\nu = 0.85\nu_{crit,7\times7}$
7×7	16825	91736π	91736π
11×13	15968	43946π	32116π
25×25	15744	32353π	13854π

is to be seen in the picture, that a sufficient approximation of the nonlinear probability graph is not possible with the linear method.

Furthermore the discretisation influence on the stability boundaries was investigated on the perfect panel. Additional to the 7×7 node model, systems modeled with 11×13 (Schorling and Bucher, 1999) and 25×25 nodes and meshed with geometrically nonlinear 9-node shell elements were analyzed. The critical static buckling loads are shown in Table 2. The dynamic stability boundaries are obtained by using the linear Itô analysis under considering the static load first with $0.85\nu_{crit}$ of the same model. This leads to different static loads. To obtain the stability boundaries by a constant static load this load was assumed as $0.85\nu_{crit}$ of the 7×7 node model. The results are shown additional in Table 2. It is to be seen that the influence of the discretization on the critical noise intensity is much higher than on the static buckling load.

The nonlinear method was not applicable for the 11×13 and 25×25 node models, caused by the hugh numerical effort. Simulations with the 11×13 node model by using a modal reduction from 657 eigenmodes to 33 eigenmodes (the critical time step is then $3.4 \cdot 10^{-3}$ s) did not lead to sufficient results. The Lyapunov exponent did not converge to a stationary value, caused be the to short time window of the simulations, limited by the available computer capacities.

References

Arnold, L. and P. Imkeller (1994). Fürstenberg–Khasminskii formulas for Lyapunov exponents via anticipative calculus. Technical Report Report Nr. 317, Institut für dynamische Systeme, University of Bremen.

Bathe, K.-J. (1996). *Finite Element Procedures*. Englewood Cliffs: Prentice Hall.

Brenner, C. E. and C. Bucher (1995). A contribution to the SFE-based reliability assessment of nonlinear structures under dynamic loading. *Probabilistic Engineering Mechanics* **10**, 265–273.

Bucher, C. (2001). Stabilization of explicit time integration by modal reduction. In W. A. Wall, K.-U. Bletzinger, and K. Schweizerhof (Eds.), *Proceedings, Trends in Computational Mechancics*. Barcelona: CINME.

Bucher, C., O. Huth, and M. Macke (2003). Accuracy of system identification in the presence of random fields. In A. DerKiureghian, S. Madanat, and J. Pestana (Eds.), *Applications of Statistics and Probability in Civil Engineering*, pp. 427–433. Millpress.

Ditlevsen, O. (1991). Random field interpolation between point by point measures properties. In *Proceedings of 1. Int. Conference on Computational Stochastic Mechanics*, pp. 801–812. Computational Mechanics Publications.

Ghanem, R. and P. D. Spanos (1991). *Stochastic Finite Elements – A Spectral Approach*. New York/Berlin/Heidelberg: Springer.

Krätzig, W. (1989). Eine einheitliche statische und dynamische Stabilitätstheorie für Pfadverfolgungsalgorithmen in der numerischen Festkörpermechanik. *Z. angew. Math. Mech.* **69-7**, 203–213.

Lin, Y.-K. and G.-Q. Cai (1995). *Probabilistic Structural Dynamics*. New York: McGraw-Hill.

Macke, M. and C. Bucher (2000). Conditional random fields for finite elements based on dynamic response. In M. Deville and R. Owens (Eds.), *Proc. 16th IMACS World Congress on Scientific Computation, Applied Mathematics and Simulation*, Lausanne, Switzerland, August 21–25, 2000.

Matthies, H. G. and C. Bucher (1999). Finite Elements for Stochastic Media Problems. *Comput. Methods Appl. Mech. Engrg.* **168**, 3–17.

Matthies, H. G., C. E. Brenner, C. G. Bucher, and C. G. Soares (1997). Uncertainties in Probabilistic Numerical Analysis of Structures and Solids – Stochastic Finite Elements. *Struct. Safety* **19**, 283–336.

Most, T., C. Bucher, and Y. Schorling (2004). Dynamic stability analysis of nonlinear structures with geometrical imperfections under random loading. *Journal of Sound and Vibration* **1–2**(276), 381–400.

Schorling, Y. and C. Bucher (1999). Stochastic stability of structures with random imperfections. In B. F. J. Spencer and E. A. Johnson (Eds.), *Stochastic Structural Dynamics*, pp. 343–348. Rotterdam/Brookfield: Balkema.

Soong, T.-T. and M. Grigoriu (1992). *Random Vibrations of Mechanical and Structural Systems*. Englewood Cliffs: Prentice Hall.

RELATIVE NDT EVALUATION OF THE SIDE WALLS OF A BRICK CHANNEL

Giovanni Cascante[1], Homayoun Najjaran[2] and Paola Ronca[3]

[1]*Department of Civil Engineering, University of Waterloo, ON, Canada*
[2]*Institute for Research in Construction, National Research Council Canada, ON, Canada*
[3]*Department of Civil Engineering, Polytechnic of Milan, Milan, Italy*

Abstract

The network of channels called Navigli was built in the surrounding country and downtown Milan, Italy. Wave propagation measurements were taken along the channel for a total length of 80 m. Each test consisted of the simultaneous measurement of the response of the wall with 15 transducers to an impulse load. The relative condition of the wall is evaluated by considering three main wave characteristics: group velocity, phase velocity, and attenuation coefficient. A fuzzy logic model is developed to make a relative evaluation of the condition of the sidewall.

Introduction

The evaluation of the condition of existing structures is a key element in the maintenance and safety plans of any municipality. This evaluation should be based on realistic measurements of different variables such as geometry and material strength. These variables are commonly measured with destructive tests. However, these tests are expensive and sometimes impossible to perform at regular spacings. Conversely, nondestructive tests (NDT) are fast and economical, and can be used in-situ without major disruption to normal operations. Wave velocity and attenuation depend on the stiffness and mass of the medium, thus variations in the medium properties should be inverted from the change in wave parameters.

The Navigli channel was mainly built during the XVI century. The current civic administration is facing the difficulty of saving what remains of these complex hydraulic and historic structures. Typical cross-sections of the wall are shown in Figure 1. The restoration of the Navigli should be performed without causing mayor disturbance to this historic structure. Therefore, the use of non-destructive methods for the condition assessment of the channel is an ideal and cost-effective solution. The pulse velocity method is widely used for the assessment of the quality of concrete; this assessment is based on empirical correlations between wave velocity and strength (Popovics and Rose 1994, Ronca 1993). Geophysical methods commonly used for site characterization are seismic reflection or refraction, spectral analysis of surface waves, seismic crosshole and downhole, ground penetrating radar, resistivity, and electromagnetic survey. On the other hand, the spectral analysis of surface waves (SASW) method has been proposed for the evaluation of shear wave profiles in layered media (Nazarian et al. 1988, Rix et al. 2002).

This paper summarizes the results of the nondestructive testing programme performed on the sidewalls of the Navigli channel. Wave propagation measurements were taken at two different elevations on the sidewall (bottom and top lines). All the results presented in this paper correspond to the bottom line only. The response of the wall was measured at 161 locations; the transducers (accelerometers) were spaced at 0.5 m, the total length of wall tested was 80m, and the total number of tests performed was 86. Each test consisted on the simultaneous measurement of the response of the wall on 15 transducers, thus each test covered a section of 7.5 m in length. The relative condition of the wall is evaluated by considering three main wave characteristics: group velocity, phase velocity, and attenuation coefficient. The sections that showed low values of group and phase velocities and high values of the attenuation coefficient are identified as the weaker sections. In the following sections, the theory of the techniques used for data analysis is presented, and then the experimental setup and the testing programme are described. Finally, the main results and conclusions from the testing programme are outline.

M. Pandey et al. (eds), Advances in Engineering Structures, Mechanics & Construction, 485–492.

Background

The data collected from seismic tests can be analyzed in both time and frequency domains. In the time domain, group or pulse velocities are evaluated; whereas, phase velocities and the attenuation coefficients are evaluated in the frequency domain. Wave velocity is related to the strength, stiffness, state of stress, and density of the medium. On the other hand, wave attenuation is an indicator of fractures, cementation, de-cementation, and compaction.

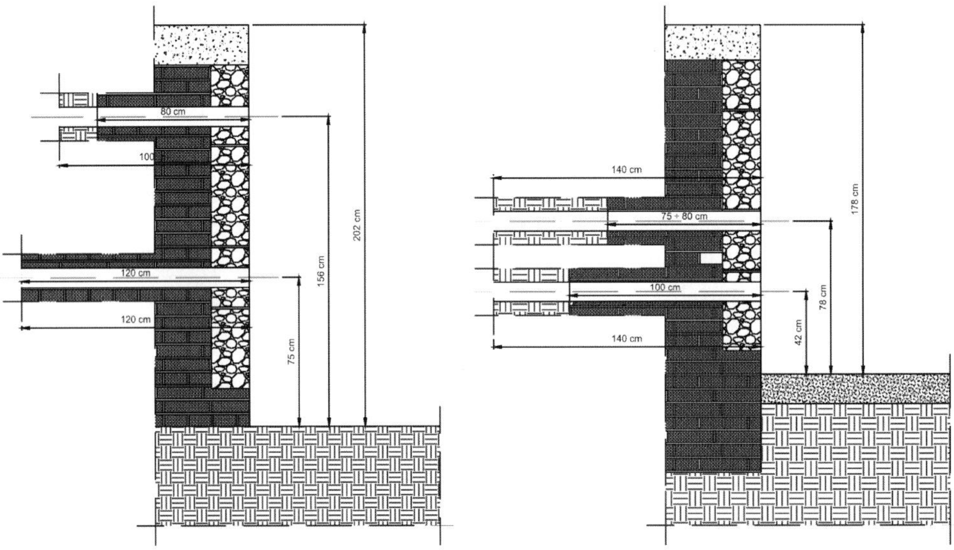

Figure 1. Typical cross-sections of the sidewall

The propagation of low-strain mechanical waves is a perturbation phenomenon that assesses the state of materials without causing permanent effects. When the wavelength is larger than the internal scale of the material, such as the brick size in a masonry wall, the wave velocity and attenuation coefficient can be defined for an equivalent continuum. The group shear and compressional wave velocities V_S and V_P are (Graff 1975):

$$V_S = \sqrt{\frac{G}{\rho}} \qquad V_P = \sqrt{\frac{M}{\rho}}$$

where G is shear modulus, M is constraint modulus and ρ is mass density of the medium. The velocities of compressional waves and shear waves are related through Poisson's ratio of the medium. If the medium is homogeneous, all the frequency components of a pulse travel at the same speed V_S or V_P. However if the medium is dispersive, the propagation of each frequency component has a different velocity. The velocity of the pulse is referred to as group velocity; whereas, the velocity of each frequency component is called phase velocity.

For low excitation frequencies, two thirds of the energy introduced into a medium from a circular rigid plate converts to surface waves (Richart et al. 1970). In conventional seismic surveys, surface waves are not used because they mask the reflection and refraction events from body waves. The use of surface waves for shallow applications has been improving since the development of the spectral analysis of surface waves (SASW) method. The depth of penetration of surface waves is proportional to the pulse wavelength. Surface waves practically penetrate to a depth of two times their wavelength. However for material characterization, their effective depth of penetration is approximately one-third their wavelength because the maximum displacements induced by surface waves take place close to the surface. The phase velocity of Rayleigh waves (R-wave) velocity V_{ph} at any frequency f is related to distance Δx and the phase difference $\Delta\phi$ between the receivers by:

$$V_{ph} = 2\pi f \frac{\Delta x}{\Delta \phi} \quad [1]$$

The wavelength λ as a function of frequency and phase velocity is given by

$$\lambda(f) = \frac{V_{ph}}{f} \quad [2]$$

As seismic waves propagate, their amplitude decreases with distance due to geometric spreading of the wave front and the intrinsic attenuation of the medium. Geometric attenuation is the result of the increasing surface area of the wave front as it propagates outward. In an ideal material, the amplitude of the wave front decreases only due to geometric spreading because the intrinsic attenuation is zero. However, in real materials, part of the elastic energy is absorbed. Intrinsic attenuation can be expressed as

$$\frac{A_2}{A_1} = e^{-\alpha\left(r_2 - r_1\right)} \quad [3]$$

where α is the attenuation coefficient, the wave amplitudes at distances r_1 and r_2 from the source are A_1 and A_2, respectively.

The two-dimensional Fourier transform (2D-FFT) has been used to compute dispersion curves of multimode signals (Zerwer et al. 2001). The arrangement of multiple-channel measurements into a matrix permits the use of the 2D-FFT, which generates a matrix of complex numbers. The contour plot of the magnitude of the 2D-FFT renders a plot of frequency versus wavenumber $k = 2\pi/\lambda$. Different waves are identified in the contour plot as a sequence of peaks. Peaks associated with non-dispersive waves plot as straight lines that pass through the origin; whereas, peaks corresponding to dispersive waves plot as curved lines with non-zero intercepts. Positive wavenumbers represent waves traveling in the forward direction (away from the source); whereas, negative wavenumbers represent waves traveling in the backward direction (towards the source).

Fuzzy-based methods have increasingly been used in a variety of civil and infrastructure-engineering problems from the evaluation of concrete and steel structures to water and wastewater applications (Liang et al. 2001; Najjaran et al. 2004). Fuzzy logic provides a language with semantics to translate qualitative knowledge into numerical reasoning. In many problems, the available information about the likelihoods of various items is vaguely known or assessed; hence, the information in terms of either measured data or expert knowledge is imprecise to justify the use of deterministic numbers. The strength of fuzzy logic is that it provides a rational and systematic approach to decision making through the integration of descriptive knowledge (e.g., very high, high, very low, low) and numerical data into a fuzzy model and uses approximate reasoning algorithms to propagate the uncertainties throughout the decision process. A fuzzy model contains three distinguished features: fuzzy numbers instead of, or in addition to numerical variables; relations between the variables in terms of IF-THEN rules; and an inference mechanism. A fuzzy number describes the relationship between an uncertain quantity x and a membership function $\mu(x) \in [0,1]$. A rule base determines the relationships between the inputs and outputs of a system using linguistic antecedent and consequent propositions in a set of IF-THEN rules. The inference mechanism uses approximate reasoning algorithms and the relationships to infer the outputs for given inputs.

Experimental methodology

The sidewall of the Navigli channel was instrumented at 2 m from the base of the channel. Steel plates at 0.5 m spacing were attached to the wall with epoxy resin. Measurements were obtained from 161 points (total length L= 80 m). The wall was divided into 11 sections: ten sections of 7.5 m and the last section of 5 m long. The length of the sections was limited by the number of channels of the data acquisition system (15 channels). The accelerometers were glued to the steel plates to conduct different multichannel surface wave tests (MASW). For each MASW test, 15 traces were recorded; each trace contained 2400 data points. After testing one section, the accelerometers were moved to the next section. A one-pound hammer was used as a source to make an impact at three locations, i.e., right, center and left of the array of accelerometers. Depending on the attenuation of the wall, the source was located at least at two different horizontal distances from the first receiver.

The equipment used in all the tests (Figure 2) consisted of a digital oscilloscope (HP 35610A), four filter-amplifiers (Krohn-Hite 3984), impulse hammer (Dyatran), 15 accelerometers (PCB and Dytran, frequency range 1 Hz to 5 kHz), and a laptop-based data acquisition system of 16 bit resolutions and 1 MHz maximum sampling rate (Iotech, Wavebook E16). The sampling rate used was 62.5 kHz per channel; thus, a maximum frequency of 6.25 kHz can be recognized from the time signals and a maximum frequency of 31 kHz can be identified from the Fourier spectra. The time window was 38 ms; the frequency resolution in the spectra was 26 Hz. The time responses are 2D-Fourier transformed to determine spatial and temporal frequencies from the contour plots in the frequency-wavenumber space.

Results

Each time signal is corrected by the amplification and sensitivity factors used in the field for each accelerometer. The corrected data is used to calculate the attenuation factor. However, the data is also normalized to the maximum response of each trace to enhance the interpretation of time traces and Fourier spectra. Time signals are significantly affected by noise after the eighth receiver because of the attenuation properties of the sidewall. Therefore, the responses of the accelerometers located in the first four meters of the array are given more weight in the data analysis. The arrival times are used to compute the corresponding group velocities (V_P and V_R). The arrival of compressional waves is weak because of the low amplitude of the generated p-waves and the attenuation properties of the sidewall.

The Fourier spectra indicate that after the first receiver the energy is mostly distributed between 250 Hz and 2,500 Hz; these frequencies correspond to wavelengths equal to λ=1.45 m and λ=0.30 m respectively. Therefore, the measured properties of the sidewall are representative of depths of 10 cm to 50 cm. Frequencies between 1 kHz and 4.5 kHz attenuate faster than frequencies below 1 kHz. Phase wave velocities change with distance because of the different conditions of the sidewall. Phase velocities are computed by curve fitting the change in phase angle with distance (Eq. 1) for each frequency component. A decrease of phase velocities with frequency indicates that the condition of the inside sidewall (0.50 m to 1.0 m) is better than the condition at the surface (0.0 to 0.25 m). Phase velocities give an indication of the relative condition of the sidewall with depth; however, the results should be interpreted carefully because wave reflections influence significantly the results. Cracks and voids reflect and diffract the wave front, thus relatively higher wave attenuation is expected in weaker sections than in sound sections. The attenuation or absorption coefficient (α, Eq. 3) can be computed in terms of the maximum response in time, frequency, or the area of the Fourier spectra. The attenuation information confirms previous observations that after the first 3.5 m of the array signals are drastically affected by the noise and wave reflections because the wave amplitude does not continue to decrease with distance in the second half of the array.

Two-dimensional Fourier transforms indicate wave velocities (V_R and V_P) that are closed to the measured group velocities. Energy peaks for negative wavenumbers indicate spatial aliasing for frequencies higher than 1300 Hz. The spatial aliasing is produced by the selected receiver spacing (Δx= 0.5 m). Spatial aliasing can be reduced by using a smaller spacing; however, this solution requires an increased number of measurements and thus more time for testing and data processing. The surface-wave group velocity (V_R), range of phase velocities (V_{ph1} to V_{ph2}), attenuation coefficient (α), and the frequencies of higher spectral energy (f_1) for all sections are summarized in Table 1. The average values for these variables are shown at the bottom of the table. Results for the right-hand side and the left-hand side of the sections are denoted by the letters R and L, respectively. The condition of the sidewall could change in a distance of few meters; therefore, it is possible to have consecutive sections with completely different conditions (e.g. sections 1L and 2R, Table 1).

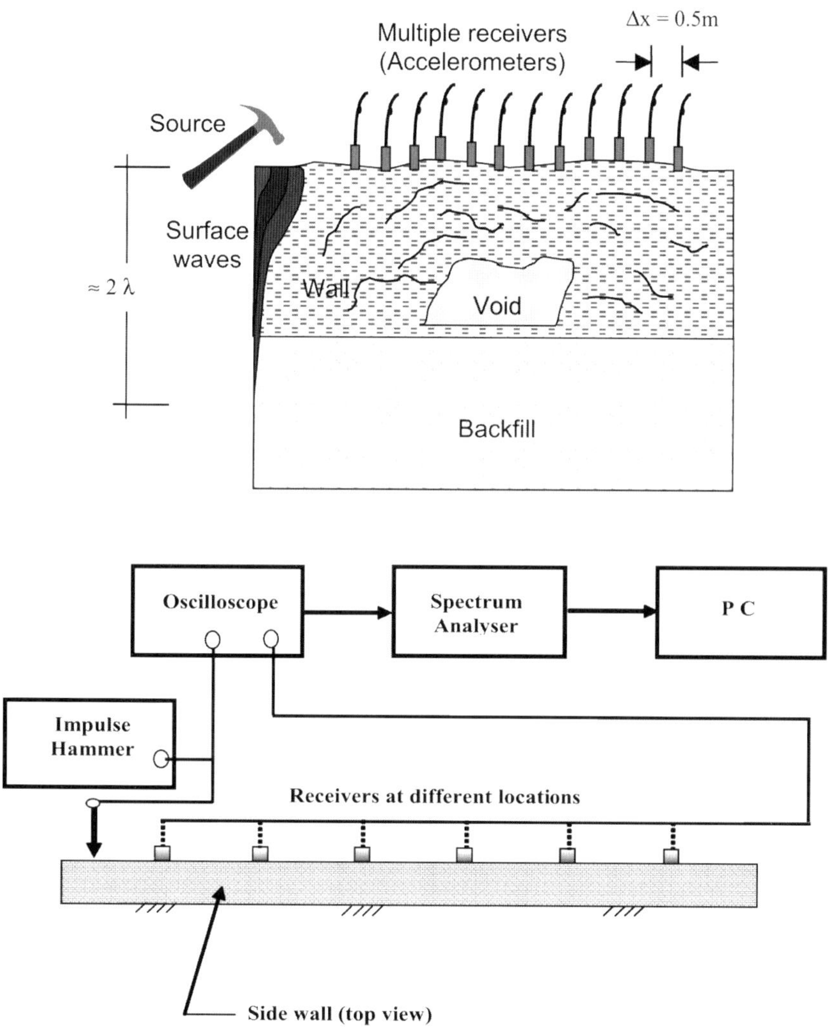

Figure 2. Test configuration of a multiple-channel seismic measurement

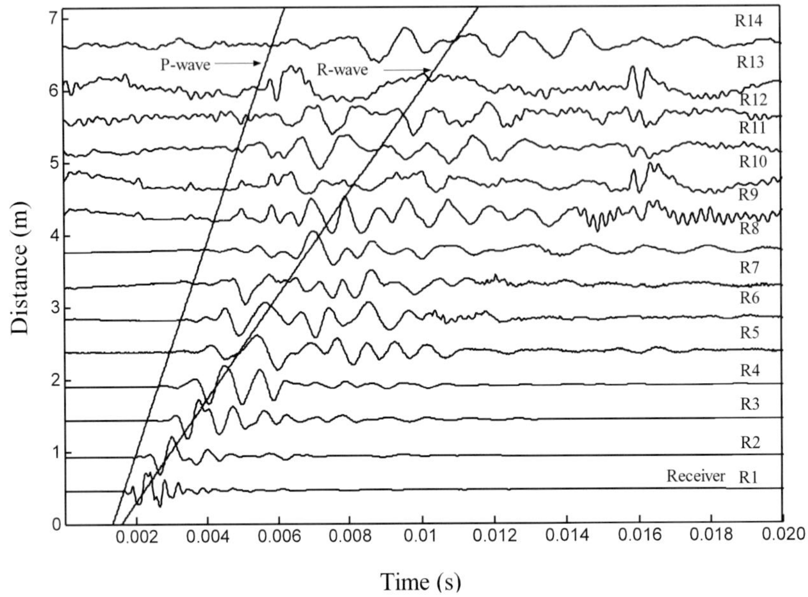

Figure 3. Typical time responses (Section 1). Solid straight lines indicate the arrival of compressional waves (p-waves) and surface waves (r-waves)

A fuzzy model is developed to assess the relative structural condition of different sections of the brick sidewall using seismic wave properties. The fuzzy model summarizes the results of nondestructive measurements in a quantitative parameter named condition index (CI). The most straightforward fuzzy modeling method is the direct approach in which expert knowledge is used to specify input and output variables (e.g. group velocity, phase velocity, wave attenuation, and condition index). Specific steps are: (a) to label the partitions of the input and output variables with linguistic terms (e.g. low, medium, high); (b) to define a set of linguistic rules (IF-THEN) that represent the relationships between the variables; (c) to select an appropriate reasoning method; and (d) to verify the model. A condition index CI = 0 indicates a section in relatively very good condition; whereas, CI = 1 refers to a damaged section. The input variables are partitioned into three fuzzy numbers low (L), medium (M), and high (H). The output is expressed using five partitions: very good (VG), good (G), moderate (M), deteriorated (DT), and damaged (DG). A fuzzy expert system (Najjaran et al. 2004) is used to develop the fuzzy model and evaluate the condition indices for the different sections. The results are summarized in the last two columns of Table 1. The condition index shows that Sections 1L, 6R, 8R, 9L, 10R, 10L, 11R and 11L are the relatively weaker sections. Visual inspection of these sections revealed that surface conditions are fine for most of them; however, voids and cracks are evident close to the line of measurement. In addition, low velocity and high attenuation could be the result of internal cracks and weaker conditions of the brick and mortar that could not be evident on the surface.

Conclusions

MASW tests were performed on 11 sections of the Navigli channel to assess the relative condition of the sidewall. The minimum and maximum group shear wave velocities were Vs=667 m/s and V_R=1200 m/s whereas the minimum and maximum shear wave phase velocities were V_{ph}= 500 m/s and V_{ph}=1140 m/s. The frequency with higher energy varied from 500 Hz to 1600 Hz; thus the penetration of surface waves was enough to reach the back surface of the sidewall (approximately at 1m from the surface of the wall).

A fuzzy model is developed to compute a condition index as a function of the group velocity, the maximum and minimum phase velocities, the frequency with higher energy in the Fourier spectrum, and the attenuation coefficient of each section. The condition index shows that Sections 1L, 6R, 8R, 9L, 10R, 10L, 11R and 11L are the weaker or less stiff sections of the sidewall. Visual inspection of these sections revealed that surface conditions are fine for most of them. However, voids and cracks are evident near the array of transducers. In addition, low velocity and high attenuation could be the result of more severe defects such as internal fractures and weaker conditions of the brick and mortar that could not be evident on the surface.

Acknowledgements

The authors express their gratitude to the regional and civic administrations of the city of Milan, Italy, and the technicians of the Laboratory of the Politecnico di Milan, Structural Engineering Department for their assistance. Special thanks to Andrea Bonizzi, who helped in the data collection and preparation of figures. This research is part of a study on nondestructive testing of geomaterials that is partially supported by the Earth and Environment Technology (ETech) of the Ontario Center of Excellence, the Natural Sciences and Engineering Research Council of Canada (NSERC), and the National Research Council Canada (NRC). Their support for this project is very much appreciated.

Table 1. Field data and calculated condition index (CI)

Section	V_R m/s	V_{ph1} m/s	V_{ph2} m/s	f_1 Hz	α	Condition Index	Condition State
1R	N/A	N/A	N/A	N/A	N/A	N/A	N/A
1L	754	800	600	700	0.55	0.917	**Damaged**
2R	1040	800	600	800	0.20	0.000	**Very Good**
2L	870	870	540	588	0.25	0.582	**Moderate**
3R	1000	1140	850	550	0.45	0.500	**Moderate**
3L	1030	880	700	1000	0.30	0.250	**Good**
4R	950	625	400	910	0.42	0.429	**Moderate**
4L	930	1000	600	900	0.23	0.250	**Good**
5R	930	730	430	1150	0.33	0.437	**Moderate**
5L	930	930	600	1100	0.33	0.311	**Good**
6R	767	750	600	650	0.56	0.917	**Damaged**
6L	1200	1000	600	1600	0.34	0.155	**Good**
7R	820	870	790	900	0.26	0.424	**Moderate**
7L	770	810	720	810	0.28	0.500	**Moderate**
8R	787	800	700	850	0.38	0.625	**Deteriorated**
8L	880	836	766	1150	0.40	0.465	**Moderate**
9R	837	800	600	1016	0.34	0.478	**Moderate**
9L	910	700	300	650	0.34	0.558	**Moderate**
10R	783	500	400	500	0.38	0.750	**Deteriorated**
10L	837	500	400	500	0.38	0.652	**Deteriorated**
11R	667	580	400	650	0.65	0.917	**Damaged**
11L	745	745	500	660	0.20	0.750	**Deteriorated**
Average	878	802	601	840	0.36	0.500	**Moderate**

Two analytical crack tip analyses for plastic solids involving work hardening nonlinear material are due to Hutchinson (1968) and Rice and Rosenberg (1968). They obtained singular stress field near crack tip using finite deformation theory of plasticity. They assume a stress field, known as HRR solution, consistent with singular strain energy density. Just like the LEFM solution, HRR solution is not valid at the tip because singular stress it predicts at the crack tip is incompatible with the limitation on stresses imposed by plasticity. HRR solution is not valid at large distance from the tip either. A finite element analysis by Mcmeeking et al. identifies the area over which HRR solution applies.

Singh et al. (1994) obtained stress, strain and displacement field around a crack in an infinite, isotropic and linear elastic plate. They used a non-classical small deformation theory. The classical theory assumes the displacement to be small such that replacing the deformed position of a particle by its initial un-deformed position is likely to induce negligible error in the solution. Singh et al. used the reverse argument that in an analysis of a problem involving small displacement, it is equally justified to use the deformed position of a particle, rather than its un-deformed position. Further, the consistency of analysis requires the use of deformed geometry if the boundary value problem is formulated in terms of true stress and true traction. On this basis, Singh et al. (1994) obtained a solution for the entire plate. They obtained the geometry of the deformed crack surface as part of their solution. In this presentation, we use their methodology to obtain stress and deformation field in an elastic-plastic plate. Solution uses deformed geometry and linearized stress-strain behavior.

2. Background

Consider a plate with a centrally located crack subject to far field tensile stress S. for the purpose of analysis; choose a Cartesian coordinate system with origin at the crack center, x-axis along the crack and y-axis perpendicular to it. Suppose the crack extends from $x = -a_0$ to $x = a_0$ on $y = 0$. Let

$$z = a_0 \cosh \xi, \quad \xi = \alpha + i\beta, \quad i = \sqrt{-1}$$

with $a = 0$ describing the initial geometry of a crack in elliptical coordinates. It is a special case of general transformation

$$z = x + iy = z(\xi)$$

that relates Cartesian coordinates (x, y) to elliptical coordinates (α, β). In the new system, crack surface $\alpha = 0$ extends from $\beta = 0$ to $\beta = 2\pi$. Because of the symmetry, it is sufficient to consider the upper right quadrant $(x \geq 0, y \geq 0)$ or $(\alpha \geq 0, 0 \leq \beta \leq \pi/2)$ of the plate.

The plate under external tensile stress normal to the crack is expected to develop a non-uniform stress field that can be obtained from solving the equation of equilibrium

$$\frac{\partial \sigma_{xx}}{\partial x} + \frac{\partial \sigma_{xy}}{\partial y} = 0, \quad \frac{\partial \sigma_{xy}}{\partial x} + \frac{\partial \sigma_{yy}}{\partial y} = 0. \tag{1}$$

These equations are transformed to complex plane then combined and expressed in the form

$$\frac{\partial}{\partial z}(\sigma_{xx} + \sigma_{yy}) + \frac{\partial}{\partial z}(\sigma_{xx} - \sigma_{yy} - 2i\sigma_{xy}) = 0.$$

It has a solution in terms of a stress function Φ such that

$$\sigma_{xx} + \sigma_{yy} = \frac{\partial^2 \Phi}{\partial z \partial \bar{z}},$$

$$\sigma_{xx} - \sigma_{yy} + 2i\sigma_{xy} = -\frac{\partial^2 \Phi}{\partial \bar{z} \partial \bar{z}}. \tag{2}$$

The main task now is to find a suitable stress function. The task becomes easier if the material behavior is linear and isotropic. For such a solid,

$$\frac{\partial u_x}{\partial x} + \frac{\partial u_y}{\partial y} = \frac{\partial u}{\partial z} + \frac{\partial \bar{u}}{\partial \bar{z}} = \frac{1 - \upsilon - 2k\upsilon^2}{E}\left(\frac{\partial^2 \Phi}{\partial z \partial \bar{z}}\right),$$

$$\frac{\partial u_x}{\partial x} - \frac{\partial u_y}{\partial y} + i\left(\frac{\partial u_y}{\partial x} + \frac{\partial u_x}{\partial y}\right) = 2\frac{\partial u}{\partial \bar{z}} = -\frac{1+\upsilon}{E}\left(\frac{\partial^2 \Phi}{\partial \bar{z} \partial \bar{z}}\right),$$

where u_x and u_y are the x- and y-components of displacement u $(= u_x + iu_y)$; $k = 0$ or 1 depending on whether the plate is in plane stress or plane strain. The material is linear if the mechanical parameters E and υ are constant. Otherwise, the material is considered non-linear. The above equations show that for a linear solid, the function Φ must be real and bi-harmonic, and hence $\partial^2\Phi/\partial z \partial \bar{z}$ must be real and harmonic. For a linear solid therefore, choose

$$\sigma_{xx} + \sigma_{uu} = \frac{\partial^2 \Phi}{\partial z \partial \bar{z}} = \frac{\partial f}{\partial z} + \overline{\left(\frac{\partial f}{\partial z}\right)}, \tag{3}$$

where f is an analytic function. The above equation can be integrated to yield

$$\frac{\partial \Phi}{\partial \bar{z}} = f + z\overline{\left(\frac{\partial f}{\partial z}\right)} + \bar{g}, \tag{4}$$

where g is another analytic function.

Suppose a particular choice of analytic function f and g solves a given boundary value problem for an isotropic linear solid. The same analytic functions can also be used to solve the problem of another isotropic linear solid if it is subject to the same boundary conditions. This feature can be exploited even for non-linear material provided its non-linear stress strain behavior can be replaced by piece-wise linear approximation. In the simplest case, non-linear material response can be replaced by bi-linear stress strain behavior involving mechanical parameters (υ, E) and (υ_n, E_n). Suppose the first segment I represents elastic behavior and the second segment n involves plastic deformation characterized by the plastic modulus E_p and the ratio $\upsilon_p = 1/2$ such that

$$\frac{1}{E_n} = \frac{1}{E} + \frac{1}{E_p}, \quad \frac{\upsilon_n}{E_n} = \frac{\upsilon}{E} + \frac{1}{2E_p}.$$

Once the analytic functions f and g are known or have been found, stress field can be obtained from (1). The displacement field in the elastic domain I is obtained from

$$u = z - Z = C + \frac{8f}{E} - 2\frac{1+\upsilon}{E}\frac{\partial \Phi}{\partial \bar{z}},$$

where C is a constant and z is the deformed position of a particle that initially, in the un-deformed configuration, occupied a position Z. The relation

$$u = z - Z = C_n + \frac{8f}{E_n} - 2\frac{1+\upsilon_n}{E_n}\frac{\partial \Phi}{\partial \bar{z}}$$

yields displacement in the plastic domain. The constants C and C_n, and the functions f and g must be chosen to ensure continuity of displacement and stress across the elastic plastic interface.

In the case of perfectly plastic material, the yield stress is constant; the stress strain curve is linear and horizontal (parallel to the strain axis) with slope $E_n = 0$. Since the stress strain curve is linear in the plastic domain, the sum $\sigma_{xx} + \sigma_{xx}$ can still be considered analytic. However, it is necessary only to choose an analytic function g for determination of the stress field in plastic domain. The yield criterion provides a second condition that can be used for this purpose. It is necessary only to choose an analytic function that determines $\sigma_{xx} + \sigma_{xx}$ such that the stress field it generates in conjunction with the yield criterion leaves the crack surface free traction. It is also necessary to maintain continuity between the stress fields of the elastic and plastic solutions across the common boundary.

3. Stress and Strain Field around Crack

To show how the proposal works, assume that the stress field in the elastic domain has been obtained from LEFM analysis. Irwin uses the stress field on the crack line $y = 0$ in the form

$$\sigma_{11} = \sigma_{22} = \frac{K_I}{\sqrt{2\pi r}}, \qquad \sigma_{12} = 0,$$

where K_I $(= S\sqrt{\pi a})$ is the stress intensity factor and r is the distance from the crack tip and S is the far-field stress. Consider plane stress, Mises yield criterion and use the effective stress

$$\sigma_e = \sqrt{\sigma_{11}^2 + \sigma_{22}^2 - \sigma_{11}\sigma_{12}}.$$

Suppose the material yields at $r = r_Y$ where the effective stress equals the yield stress σ_Y. $K_I = S\sqrt{2\pi r_Y}$. The load transmitted across the surface $y = 0$ in the plastic zone is $P = 2\sigma_Y r_Y$. If the condition $\sigma_{11} = \sigma_{22}$ holds along the crack line even in the plastic zone, the equilibrium condition predicts the plastic zone size of

$$l_p = \frac{P}{\sigma_Y} = 2r_Y = \left(\frac{S}{\sigma_Y}\right)^2 a. \tag{5}$$

To accommodate the increased plastic zone, one of the three conditions must hold: (1) the crack tip is displaced towards the center of the crack, (2) the elastic-plastic boundary moves away from the crack center, or (3) or a combination of the above two occurs. In Irwin's scheme, condition (2) is assumed to hold. That is, the elastic-plastic boundary is assumed to move such that the stress field of LEFM begins at a distance of $2r_Y$ from the crack tip. In order for stress field (2) to hold, the crack is assumed to be located at $a_{\text{eff}} = a + r_Y$ and the crack opens to a blunt configuration at the tip. The crack tip opening displacement (CTOD) is calculated from the formula

$$\frac{x^2}{a_{\text{eff}}^2} + \frac{y^2}{b^2} = 1$$

of the elastic solution. Thus CTOD at $x = a$ is

$$2\delta = 2\frac{b}{a + r_Y}\sqrt{2ar_Y + r_Y^2}.$$

If $r_Y = 1/2(S/\sigma_Y)^2 a \ll a$, the Singh solution for b along with the assumption that $S \ll E$ yields

$$2\delta = 4\frac{S}{E}\frac{S}{\sigma_Y}a.$$

Note that the method of evaluating crack opening displacement involves the use of elastic solution even though a part of the material undergoes plastic deformation. Moreover, the crack opening displacement of elastic solution indicates that the material particles initially on the crack surface ($\alpha = 0$ or $-a_0 < x < a_0$, $y = 0$) has opened into ($\alpha > 0$), otherwise there will be no crack opening. Obviously, if the boundary conditions are expressed in terms of true stress, the traction boundary conditions must be satisfied not on the un-deformed ($\alpha = 0$) surface but on the deformed ($\alpha > 0$) surface it opens into. This shows blunting and requires σ_{11}, vanishing at the crack tip. Further, in the case of plane stress, $\sigma_{22} = \sigma_Y$. This information can be used to select an analytic function f for the plastic zone as follows.

4. Stresses in Plastic Zone

In view of the fact that in plane stress, $\sigma_{11} = 0$, $\sigma_{22} = \sigma_Y$ at the blunted crack tip and $\sigma_{11} + \sigma_{22}$ must be harmonic, choose

$$\sigma_{11} + \sigma_{22} = \sigma_Y(\sin p(z - x_0) + \sin p(\bar{z} - x_0)).$$

where p is a constant and $z = x + iy$. Rewrite the above equation in the form

$$\sigma_{11} + \sigma_{22} = 2\sigma_Y \sin p(x - x_0) \cosh py. \tag{6}$$

On x-axis, $y = 0$ and $\sigma_{11} + \sigma_{22} = 2\sigma_Y \sin p(x - x_0)$. For a Mises solid therefore,

$$\begin{aligned}\sigma_{22} &= \sigma_Y \left\{ \sin p(x - x_0) + \frac{1}{\sqrt{3}} \cos p(x - x_0) \right\}, \\ \sigma_{11} &= \sigma_Y \left\{ \sin p(x - x_0) - \frac{1}{\sqrt{3}} \cos p(x - x_0) \right\}.\end{aligned}$$

To satisfy $\sigma_{11} = 0$ at the crack tip $x = x_t$, choose p such that $p(x_t - x_0) = \pi/6$. Note that on the extended crack-line in the plastic zone, $\sigma_{11} \neq \sigma_{22}$ except at the point where $p(x - x_0) = \pi/2$.

To reconcile the chosen plastic field with LEFM field in the elastic domain, choose $\sigma_{11} = \sigma_{22} = \sigma_Y$ at the yield point ($x = x_Y = r_Y$, $y = 0$). This is possible provided $p(x_Y - x_) = \pi/2$, or $x_Y = 3x_t - 2x_0$. Therefore, the plastic zone is $l_p = x_Y - x_t = 2(x_t - x_0) = \pi/3p$. To find its value, evaluate $P = \int_{x_t}^{x_Y} \sigma_{22} ds$ and equate it with $P = 2\sigma_Y r_Y$ obtained from LEFM. The result is $p = 1/(x_Y\sqrt{3})$ and the plastic zone size is

$$l_p = \frac{\pi}{\sqrt{3}} r_Y = \frac{\pi}{2\sqrt{3}} \left(\frac{S}{\sigma_Y} \right)^2 a$$

with p, l_p and x_Y known, it is easy to evaluate $x_0 = x_Y - \pi/2p$ and $x_t = x_Y - l_p$.

To find the crack opening in the plastic observe that on the crack surface $x = x_c$, $y = y_c$, $\sigma_{11} = \sigma_{22} = \sigma_Y$ and hence the equation of the deformed crack is

$$\sin p(x_c - x_0) \cosh py_c = 0.5. \tag{7}$$

Since the value of p is already known, the above equation can be used to find the deformed crack surface in the plastic domain. The surface in the elastic domain is obtained from the solution of Singh et al. and for $S \ll E$, it can be expressed in the form

$$x_e = a \cos \beta, \quad y_e = \frac{2S}{E} \sin \beta.$$

5. Crack Geometry

Consider the symmetry requirements on the deformed crack surface. Symmetry requires that $x = 0$ at $\beta = \pi/2$, and $y = 0$ at $\beta = 0$. The constraint imposed by continuity is that the elastic and have the same value at the interface $(\alpha = \alpha_f, \beta = \beta_Y)$ between two domains as a consequence, the points on the elastic segment of the deformed crack are found to have coordinates

$$x_1 = \left(a_0 - \frac{cS}{E}\cosh\alpha_f\right)\cos\beta,$$

$$y_1 = cS(\sinh\alpha_f + 2\cosh\alpha_f)\left(\frac{\sin\beta - \sin\beta_Y}{E} + \frac{\sin\beta_Y}{E_n}\right).$$

For an elastic solid, $E = E_n$ and hence

$$y_1 = cS(\sinh\alpha_f + 2\cosh\alpha_f)\left(\frac{\sin\beta}{E}\right) = c\sinh\alpha_f\sin\beta.$$

The crack opening parameter is therefore related to Young's modulus E and far-field stress S. In fact

$$\tan\alpha_f = \frac{2S}{E - S}.$$

However, this equation is not available if the behavior is elastic plastic.

To satisfy the continuity of stress across the elastic-plastic interface, we assume the same value for crack opening parameter on both elastic and plastic segments. Hence, the deformed position of the points on the plastic segment is

$$x_n = \left(a_0 - \frac{cS}{E}\cosh\alpha_f\right)\cos\beta + \left(a_0 - \frac{cS}{E_n}\cosh\alpha_f\right)(\cos\beta - \cos\beta_Y),$$

$$y_n = c\sinh\alpha_f\sin\beta = \frac{cS}{E_n}(\sinh\alpha_f + 2\cosh\alpha_f)\sin\beta.$$

The second of the above equation yield

$$\tan\alpha_f = \frac{2S}{E_n - S}.$$

Therefore, the value

$$\alpha_f = 0.5\ln\left(\frac{E_n + S}{E_n - 3S}\right)$$

of the crack opening parameter depends on the modulus E_n of the material at the tip and the far-field stress S.

For further generalization, divide the non-linear stress-strain curve into N segments. At the same time, divide the crack surface also in N domains such that points in k obey the stress-strain rule of the segment k, and $\beta = \beta_k$ is the interface between domains $k - 1$ and k.

Therefore, the deformed position of this interface is

$$x_k = \sum_{i=1}^{k}\left(a_0 - \frac{cS}{E_i}\cosh\alpha_f\right)(\cos\beta_i - \cos\beta_{i-1}),$$

$$y_k = \sum_{i=1}^{k} cS(\sinh\alpha_f + 2\cosh\alpha_f)\left(\frac{\sin\beta_{N-i} - \sin\beta_{N-i+1}}{E_{N-i+1}}\right).$$

For the purpose of book-keeping, β_N and β_0 in this equation corresponds, respectively, to the tip ($\beta = 0$) and the crown ($\beta = \pi/2$) and E_i is the modulus of the material in the ith segment. It is of course possible to convert the summed terms into integral forms by letting the number of segments become infinitely large such that, in the limit, each domain shrinks to a point and

$$\sum_{i=1}^{k}\left(a_0 - \frac{cS}{E_i}\cosh\alpha_f\right)(\cos\beta_i - \cos\beta_{i-1}) = -\int_{\pi/2}^{\beta}\left(a_0 - \frac{cS}{E_t}\cosh\alpha_f\right)\sin\beta d\beta,$$

$$\sum_{i=1}^{k} cS(\sinh\alpha_f + 2\cosh\alpha_f)\left(\frac{\sin\beta_{N-i} - \sin\beta_{N-i+1}}{E_{N-i+1}}\right)$$

$$= \int_0^{\beta}(\sinh\alpha_f + 2\cosh\alpha_f)\frac{cS\cos\beta}{E_t}d\beta,$$

where E_t is the tangent modulus. The right-hand side can be integrated provided a relationship between the tangent modulus and β can be established. But such a relation does exist. Recall the first of the two expressions for stress field in (5). Since the crack surface is free of traction, stress normal to it must vanish. Thus, in plane stress, $\sigma_{yy} + \sigma_{xx} = \sigma_t$ is the only non-zero stress, and Equation $(5)_3$ can be rearranged in the form

$$\cos 2\beta = \cos 2\alpha_f - \frac{\sinh 2\alpha_f(e^{2\alpha f} + 1)}{e^{2\alpha_f} + \sigma_t/S}.$$

This equation relates β on the crack surface to the tangential stress σ_t which in turn is related to the tangential modulus via the stress strain law. In that case, the stress-strain law need not be linearized. However, it must be emphasized that the purpose of linearization was to make the stress function biharmonic and allow the use of analytic function f and g in the analysis. It is of course possible to assume a form for stress and strain fields on some other ground in such a manner that the stress field is an approximate solution of the equations of equilibrium. This is the course adopted by Hutchinson (1968) and Rice and Rosenberg (1968). In our case, we have opted in favor of approximating the stress-strain law but, at the same time, choosing a stress field that satisfies the equations of equilibrium.

6. Results and Discussion

Let us first examine if the use of deformed configuration in the analysis has any influence on stresses. Consider the basic assumptions. Both linear elastic fracture mechanics and the classical infinitesimal strain theory assume small deformation. Therefore, displacement z–Z is considered small and hence $z \cong Z$ in lieu of deformed position z on the basis that this choice is expected to introduce negligible error. In view of this argument, the two theories use $\alpha_f = 0$ for the crack line, and assume the crack tip at $\alpha = \beta = 0$. Or, in terms of cylindrical polar coordinates with origin at the crack tip, the two theories use $\theta = \pi$ for the crack surface, and assume the crack-tip at $r = 0$. The two stress fields $(5)_3$ and $(5)_4$ both display singular behavior in the limit as $\alpha, \beta \to 0$ or as the crack-tip is approached. Both linear elastic fracture mechanics and HRR solutions exhibit such singularity (Barenblatt, 1962; Hutchinson, 1968; Irwin, 1957).

We adopt another view point. Let us assume small deformation and consider the displacement z–Z to be small such that $z \cong Z$. In that case, we propose to use the deformed position z of the particle in place of its un-deformed position Z in solving a boundary value problem. This particular choice can be justified on the basis of the argument employed in the classical theory that replacing one by the other is expected to introduce negligible error. There is yet another reason in favor of the present choice. If the boundary value problem is formulated in terms of true stress and true traction, and if the analysis is to be consistent with the formulation, solution must depend on the deformed geometry. Note that true stress and true traction are defined in terms of load over unit current or deformed area. The use of undeformed geometry in boundary value problems formulated in terms of true stress is only a convenient approximation. It should not be considered a requirement that must be imposed.

Suppose the crack opens under and the crack opening parameter α_f has non-zero value, implying crack tip blunting. The tip is initially at ($\alpha = 0$, $\beta = 0$) and is displaced under deformation to ($\alpha = \alpha_f$, $\beta = 0$). The stress field in (5) is no longer singular even though the stresses near the tip are still high. Stress distribution depends on α_f, which can be used as a parameter that controls stresses. Under sufficiently high stresses, the crack tip and the area around it yield and undergo plastic deformation. The extent of plasticity along the crack surface can be determined from $(5)_3$.

The problem involving elastic behavior was solved by Singh et al. (1994). They obtained the value

$$\alpha_e = 0.5 \ln\left(\frac{E+S}{E-3S}\right)$$

for the crack opening parameter. With $\alpha_f = \alpha_e$ in Equations $(5)_3$ and $(5)_4$, stresses can be obtained from (6).

Equation $(5)_1$ with $C = 0$ and $u = c\cosh(\alpha_e + i\beta) - a_0 \cos\beta$ yields the location

$$x = c\cosh\alpha_e \cos\beta = \frac{E}{E+S} a_0 \cos\beta,$$

$$y = c\sinh\alpha_e \sin\beta = \frac{2S}{E-S} a_0 \sin\beta$$

of points on the deformed crack surface.

For a bilinear solid, the crack tip opening and consequent blunting is governed by the modulus at the tip. The parameter

$$\alpha_n = 0.5 \ln\left(\frac{E_n+S}{E_n-3S}\right)$$

controls stresses in the plastic domain and determines the deformed shape of the crack opening at the tip. Since the tangential stress along the crack surface must remain continuous across elastic-plastic boundary, the parameter α must have a common value α_b at that point. The parameter α_e that determines the shape of the elastic part of the crack is no longer linked to Young's modulus E. It is therefore possible to assume $\alpha_f = \alpha_e = \alpha_n = \alpha_b$. In that case, the parameter α_n controls the stress field. The choice also ensures the continuity of stresses across the elastic-plastic boundary.

Since $\sigma_{yy} = E_n$ at the crack tip, a solid with lower modulus entails larger crack opening parameter, more blunting and lower stress at the tip. However, this solution is not applicable if the yield stress dominates stress field of the plastic domain.

In the case of ideal plastic behavior, there is no strain hardening and therefore $E_n = E$. It is inconceivable that Young's modulus will control stress field in the plastic domain. It is more likely to be controlled by the yield stress σ_Y which is expected to play an important role in stress analysis in the plastic zone.

Let us assume that the plate is in a state of plane stress and obeys Mises yield criterion. Consider Mises effective stress $\sigma_e = \sigma_Y$ and rewrite it in the form

$$3\frac{(\sigma_{yy} - \sigma_{xx} + 2i\sigma_{xy})}{2}\frac{(\sigma_{yy} - \sigma_{xx} - 2i\sigma_{xy})}{2} = \left(\sigma_Y^2 - \left(\frac{\sigma_{yy} + \sigma_{xx}}{2}\right)^2\right).$$

To satisfy the above equation, choose

$$\sigma_{yy} + \sigma_{xx} = 2\sigma_Y \sin t,$$

$$\sigma_{yy} - \sigma_{xx} + 2i\sigma_{xy} = \frac{2\sigma_Y}{\sqrt{3}} \cos t e^{i\varphi}.$$

Note that we use linearized stress-strain curve in the case of work hardening material. The curve is linear even for non-hardening material. Since the sum $\sigma_{yy} + \sigma_{xx}$ known to be harmonic for linear solids, choose an analytic function $f_p(z)$ in the plastic domain such that

$$\sigma_{yy} + \sigma_{xx} = (f_p' + \bar{f}_p').$$

Therefore,

$$\text{Re } f_p' = \sigma_Y \sin t,$$

where Re f_p' is the real part of $\partial f_p/\partial z$.

Since the crack surface is traction free, the only non-zero stress on it is tangential. It can be identified with the effective stress. In other words, $\sigma_{yy} + \sigma_{xx} = \sigma_Y$ or $t = \pi/6$, in the plastic domain of the crack surface.

On the surface $y = 0$, the shear stress must vanish because of symmetry. Therefore, x- and y-surfaces on the extended crack line are principal planes on which the principal stresses are

$$\sigma_2 = \sigma_Y(\sin t + \cos t/\sqrt{3}),$$

$$\sigma_1 = \sigma_Y(\sin t - \cos t/\sqrt{3}).$$

Some general conclusions regarding the stress distribution for $x \geq \alpha$ on the line $y = 0$ can be drawn even if the exact form of the function f_p' is unknown. For example, at the crack tip, $\sigma_1 = 0$, $\sigma_2 = \sigma_Y$, hence Re $f_p = \sigma_Y/2$. Thereafter, both principal stresses increase with Re f_p' until at Re $f_p' = \sqrt{3/4}\,\sigma_Y$, $\sigma_1 = \sigma_Y/\sqrt{3}$ and σ_2 attains a maximum value of $2\sigma_Y/\sqrt{3}$. Subsequently, σ_2 decreases while σ_1 continues to increase until the elastic plastic boundary is reached.

In view of the piece-wise linear stress-strain behavior assumed in the analysis, the analytic function f, or rather its derivative $\partial f/\partial z$ of the elastic domain can be used in the plastic domain as well. However, the analytic function g has no role in the plastic domain in which the yield criterion must be used for evaluating deviatoric stresses. It means that an additional term must be added to $\partial f/\partial z$ of the elastic domain to obtain the corresponding function of the plastic domain. Accordingly, choose

$$f_p' = A_1 + A_2 \coth\xi + A_3/\sinh\xi.$$

This choice immediately leads to

$$\sigma_{yy} + \sigma_{xx} = 8A_1 + 4\frac{A_2 \sinh(\xi + \bar{\xi}) + 2A_3(\sinh\xi + \sinh\bar{\xi})}{\sinh\xi \sinh\bar{\xi}}.$$

A yield criterion must be used to obtain deviatoric stress components. For a Mises solid in plane stress for example, the yield criterion can be rearranged to obtain

$$\frac{(\sigma_{yy}-\sigma_{xx}+2i\sigma_{xy})}{2}\,\frac{(\sigma_{yy}-\sigma_{xx}-2i\sigma_{xy})}{2}=\frac{1}{3}\left(\sigma_e^2-\left(\frac{\sigma_{yy}+\sigma_{xx}}{2}\right)^2\right).$$

To find the values of the constants A_1, A_2 and A_3, it is necessary to impose the condition that the stresses must be continuous across the elastic-plastic interface. Moreover, for a non-hardening material, stress along the crack surface in the plastic domain must remain at the yield value σ_Y. The second condition can be satisfied easily by choosing $A_1=\sigma_Y/8$ and

$$A_2\cosh\alpha_f+A_3\cos\beta=0.$$

Both α_f and β must therefore vary in the plastic domain. At the elastic-plastic interface, $\alpha_f=\alpha_e$ and $\beta=\beta_Y$ and hence

$$A_3=-\frac{\cosh\alpha_e}{\cos\beta}A_2.$$

The value $\alpha=\alpha_t$ at the crack tip can be obtained from

$$\cosh\alpha_t=-\frac{A_3}{A_2}=\frac{\cosh\alpha_e}{\cos\beta_Y}.$$

To find the value of A_2, use the expression for stresses on the line $\beta=0$ on which

$$\sigma_{yy}+\sigma_{xx}=\sigma_Y+8\frac{A_2\cosh\alpha+A_3}{\sinh\varepsilon}=\sigma_Y+8A_2\frac{\cosh\alpha-\cosh\alpha_t}{\cosh\alpha}.$$

For continuity across the elastic-plastic boundary $\alpha=\alpha_Y$, the above equation must yield a value equal to that obtained from the elastic solution. Therefore,

$$\sigma_Y+8A_2\frac{\cosh\alpha_Y-\cosh\alpha_t}{\sinh\alpha_Y}=S(-e^{2\alpha_e}+(e^{2\alpha_e}+1)\coth\alpha_Y).$$

The above equation can be solved for

$$8A_2=\frac{S(-e^{2\alpha_e}+(e^{2\alpha_e}+1)\coth\alpha_Y)-\sigma_Y}{\cosh\alpha_Y-\cosh\alpha_t}\sinh\alpha_Y.$$

Hence on $\beta=0$,

$$\sigma_{yy}+\sigma_{xx}=\sigma_Y+\left(\frac{S(-e^{2\alpha_e}+(e^{2\alpha_e}+1)\coth\alpha_Y)-\sigma_Y}{\cosh\alpha_Y-\cosh\alpha_t}-\sinh\alpha_Y\right)\frac{\cosh\alpha-\cosh\alpha-t}{\cosh\alpha}.$$

The difference in stresses is obtained with the help of the yield criterion and it can be expressed in the form

$$\sigma_{yy}-\sigma_{xx}=\sqrt{\frac{4\sigma_Y^2-(\sigma_{yy}+\sigma_{xx})^2}{3}}.$$

The above two equations can be used to find stresses σ_{xx} and σ_{yy} between the crack tip and the yield point on $\beta=0$.

Appendix

Uniaxial stress-strain equation for a bilinear solid can be expressed in the form

$$\varepsilon = \frac{\sigma_Y}{E} + \frac{\sigma - \sigma_Y}{E_n},$$

where E is Young's modulus (slope of the stress strain line of the first segment), E_n is the slope of the second line and σ_Y is the translation stress between the two linear segments. It can be identified with the yield stress. The transverse strain is

$$\varepsilon_{\mathrm{tr}} = -\upsilon \frac{\sigma_Y}{E} - \upsilon_n \frac{\sigma - \sigma_Y}{E_n},$$

where υ_n is the Poisson ratio for the second segment.

Suppose the strain in the second segment consists of a sum of elastic and plastic deformation such that

$$\frac{1}{E_n} = \frac{1}{E} + \frac{1}{E_{pn}}.$$

Therefore, $\varepsilon = (\sigma/E) + (\sigma - \sigma_Y)/E_{pn}$. Assuming plastic incompressibility, we obtain

$$\varepsilon_{\mathrm{tr}} = -\upsilon \frac{\sigma_Y}{E} - \upsilon_n \frac{\sigma - \sigma_Y}{E_n} = -\upsilon \frac{\sigma}{E} - \frac{\sigma - \sigma_Y}{2E_{pn}}.$$

For a work hardening solid in plastic deformation, $\sigma > \sigma_Y$ and hence the above equation can be rearranged to yield

$$\frac{\upsilon_n}{E_n} = \frac{\upsilon}{E} + \frac{1}{2E_{pn}}.$$

Suppose the crack surface open in the form used in LEFM solution or as predicted by the solution of Singh et al. (1994).

Suppose a crack $\alpha = 0$ opens under external load into a surface $\alpha = \alpha_f$. The presence of crack is likely to induce non-homogeneity in the stress field. Hence, to solve for stresses, choose

$$f = cA \cos\xi + cB \sin\xi + C = c(A + B)\cosh\xi - cBe^{-\xi} + C,$$

where A, B, C and c are constants. Since the crack surface remains free of traction,

$$\frac{\partial \Phi}{\partial \bar{z}} = f + z\overline{\left(\frac{\partial f}{\partial z}\right)} + \bar{g} + C = A \text{ constant on } \alpha = \alpha_f.$$

To satisfy the above condition, choose

$$\bar{g} = -2cA\cosh(2\alpha_f - \bar{\xi}) - cB\frac{\cosh 2\alpha_f}{\sinh\bar{\xi}}.$$

In view of these choices,

$$\frac{\partial \Phi}{\partial \bar{z}} = 2cA\{\cosh\xi - \cosh(2\alpha_f - \bar{\xi})\} + cB\frac{\cosh 2\alpha - \cosh 2\alpha_f}{\sinh\bar{\xi}} + C.$$

Stress field can now be obtained as follows:

$$\sum_s \equiv \sigma_{yy} + \sigma_{xx} = 4\frac{\partial^2 \Phi}{\partial z \partial \bar{z}} = 8A + 8B\frac{\sinh 2\alpha}{\cosh 2\alpha - \cos 2\beta},$$

$$\sum_d \equiv \sigma_{yy} - \sigma_{xx} - 2i\sigma_{xy} = \frac{\partial^2 \Phi}{\partial \bar{z} \partial \bar{z}} = 8A\frac{\sinh(2\alpha_f - \bar{\xi}}{\sinh \bar{\xi}} + 4B\frac{\cosh 2\alpha_f \cosh \bar{\xi} - \cosh \xi}{(\sinh \bar{\xi})^3}.$$

To evaluate the constants A and B, observe that the above stress field must be consistent with the applied far-field stress at large distances from the origin. In other words, they must satisfy the condition that both $\sum_s$ and $\sum_d \to S$ as $\alpha \to \infty$. These conditions lead to

$$A = -Se^{2\alpha_f}/8, \quad B = S(e^{2\alpha_f} + 1)/8.$$

When these constants are substituted in (2), (3) and (4), they lead to

$$f = \frac{cS}{8}\{-e^{2\alpha_f} \cosh \xi + (e^{2\alpha_f} + 1) \sinh \xi\} + C,$$

$$\frac{\partial \Phi}{\partial \bar{z}} = \frac{cS}{8}\left\{-2e^{2\alpha_f}(\cosh \xi - \cosh(2\alpha_f - \bar{\xi})) + (e^{2\alpha_f} + 1)\frac{\cosh 2\alpha - \cosh 2\alpha_f}{\sinh \bar{\xi}}\right\},$$

$$\sum_s \equiv \frac{\sigma_{yy} + \sigma_{xx}}{S} = -2e^{2\alpha_f} + (2e^{2\alpha_f} + 1)\frac{\sinh 2\alpha}{\cosh 2\alpha - \cos 2\beta},$$

$$\sum_d \equiv \frac{\sigma_{yy} - \sigma_{xx} - 2i\sigma_{xy}}{S} = -2e^{2\alpha_f}\frac{\sinh(2\alpha_f - \bar{\xi})}{\sinh \bar{\xi}} + (2e^{2\alpha_f} + 1)\frac{\cosh 2\alpha_f \cosh \bar{\xi} - \cosh \xi}{(\sinh \bar{\xi})^3}.$$

To find stress components, it is necessary to evaluate the real and imaginary parts, Re $\sum_d$ and Im $\sum_d$ of $\sum_d$. Subsequently, stresses can be obtained from

$$\sigma_{xx} = S\frac{\sum_s - \text{Re} \sum_d}{2}, \quad \sigma_{yy} = S\frac{\sum_s + \text{Re} \sum_d}{2}, \quad \sigma_{xy} = -S\frac{\text{Im} \sum_d}{2}.$$

The last two equations in (5) suggest that for a given applied load S, stresses in the plate depend only on the crack opening parameter α_f. It is therefore reasonable to assume that the parameter controls the stress field around the crack. To find its value, consider the displacement of points on the crack surface.

References

Barenblatt, G.I., 1962, "The mathematical theory of equilibrium cracks in brittle fracture", *Adv. Applied Mech.*, 7, 55–129.

Hutchinson, J.W., 1968, "Singular behavior at the end of a tensile crack in a hardening material", *Journal of the Mechanics and Physics of Solids*, 16, 13–31.

Irwin, G.R., 1957, "Analysis of stresses and strains near the end of a crack traversing a plate", *Journal of Applied Mechanic*, 24, 361–364.

Rice, J.R. and Rosenberg, G.F., 1968, "Plane strain deformation near a crack tip in a power-law hardening material", *Journal of the Mechanics and Physics of Solids*, 16, 1–12.

Singh, M.N.K., Dubey, R.N. and Glinka, G., 1994, "Notch and Crack analysis as a moving boundary problem", *Engineering Fracture Mechanics*, 135, 479–492.

MICRO/MESO-ANALYSIS OF POLYMER COMPOSITES WITH DAMAGE EVOLUTION

Fernand Ellyin, Zihui Xia and Yunfa Zhang

Department of Mechanical Engineering, University of Alberta
Edmonton, Alberta, Canada T6G 2G8
fernand.ellyin@ualberta.ca

Abstract

The micro/meso-mechanical approach for composites is commonly based on the analysis of a representative volume element or a so-called repeated unit cell (RUC). Through analysis of the RUC model one can predict not only macroscopic mechanical properties but also microscopic damage initiation and its propagation in composites. In this paper, we present an overview of our contributions in the following three essential areas in the micro/meso-mechanical analyses: (1) a unified form of periodic boundary conditions for the RUC modelling; (2) a nonlinear viscoelastic constitutive model for polymer matrix materials; and (3) a post-damage constitutive model based on the concept of smeared crack. Application examples combining the above three topics are presented, in which three types of glass/epoxy laminates are analyzed using finite element method. The predicted results are compared with experimental data and they are in good agreement.

Introduction

Composite materials are widely used in advanced structures in aeronautics, astronautics, automotives, marine, petrochemical and many other industries due to their superior properties over conventional industrial materials. In the past couple of decades many researchers have devoted considerable effort to evaluate macro-mechanical properties of composites by using micro/meso-mechanical modelling methods. The latter method provides overall behaviour of the composites from known properties of the reinforcing phase (particles, fibre or fibre yarns) and the matrix phase (polymers, metals) through analysis of a representative volume element (RVE) or a repeated unit cell (RUC) model, see Aboudi (1991), Nemat-Nasser and Hori (1993). Furthermore, damage in composites generally occurs in a microscopic scale and its effect is manifested progressively into meso- and macro-scales. Micro/meso-mechanical modelling can also be used to study damage initiation and growth in composites.

There are three essential prerequisites for successful micro/meso-mechanical analyses of composites, viz: (1) An appropriate RUC model must be selected and correct periodical boundary conditions should be applied to the RUC model. (2) The constitutive models must accurately describe the mechanical behaviour of the constituents, especially of the matrix material. (3) Appropriate damage criteria for different damage mechanisms and post-damage constitutive model should be introduced to simulate damage initiation and propagation in composites.

For many composite materials, such as fibrous laminates and textile composites, the microstructure can be envisioned as a periodic array of RUC. Therefore in the micro/meso-analysis based on a RUC, the equilibrium equation should be solved under appropriate periodic boundary conditions. Valid periodic boundary conditions should ensure the compatibility of the neighbouring RUCs, i.e. both displacement and traction continuity conditions should be met. However, there still exist ambiguities in application of correct periodic boundary conditions. For example, in cases of shear loading or multiaxial loading in which shear stress is involved, homogeneous boundary conditions (plane-remains-plane or uniform boundary tractions) were applied to the RUC in some previous publications. Many researchers, e.g. Sun and Vaidya (1996), Yuan et al. (1997), Xia et al. (2003a), have pointed out

M. Pandey et al. (eds), Advances in Engineering Structures, Mechanics & Construction, 505–516.

that the homogeneous boundary conditions will over- or under-estimate the effective modulus, since by applying homogeneous boundary conditions, the displacement and traction continuity conditions can not always be satisfied at the same time.

Since the micro/meso-mechanical analyses are based on properties of individual constituents of the composite, an accurate constitutive model of each constituent becomes a prerequisite in attempting to predict the composite behaviour. Although the reinforcing phases, such as fibres, ceramic particles behave elastically for most of their stress-strain range, the polymer matrices are viscoelastic or viscoelastic-viscoplastic materials. The analysis of Hashin (1966) demonstrated that the viscoelastic effect in a unidirectional fibre composite is significant for axial shear, transverse shear and transverse uniaxial stress, for which the influence of matrix is dominant. Comprehensive experimental studies on an epoxy polymer (Hu et al. 2003, Shen et al. 2004) have indicated that these materials exhibit complex time- and loading-history-dependent properties. The viscoelastic effects become even more pronounced under conditions of high temperature, sustained loading and/or high stress level. Therefore, it is necessary to develop an accurate constitutive model for the polymer matrix material.

Due to co-existence of multi-phase materials with quite different mechanical properties, damage mechanisms of composites become more complicated. The initiation and evolution of the damages are essentially in the microscopic scale. Therefore, micro/meso-mechanical approaches are desirable to capture these damage mechanisms. The polymer matrices usually have much lower strength and stiffness than the reinforcing phases. Some damage modes (matrix micro-cracking, fiber/matrix interfacial debonding) could occur at relatively low applied loads or even during the manufacturing process. However, certain degree of damages may be tolerated in composites before the structural failure modes (delamination, fiber fracture) occur. Therefore, it is essential for damage analysis of composites to be able to identify different damage modes, to predict damage evolution within each mode and between different modes. To this end, appropriate damage criteria and post-damage constitutive relations are required.

In this paper, our recent contributions in the aforementioned three areas for polymeric composites will be elucidated. A unified form of periodic boundary conditions for any multiaxial loading conditions will be presented first, and uniqueness of the solution by applying the unified periodic boundary conditions on the RUCs will be proved. For an epoxy polymer matrix a nonlinear viscoelastic constitutive model in differential form has been developed based on comprehensive multiaxial experimental test data. The nonlinear viscoelasticity is described through introduction of a modulus function and a time-scale factor. A post-damage constitutive model based on the concept of smeared crack has also been introduced to simulate the behaviour after damage initiation. It permits crack description in terms of stress-strain relations and stiffness reduction in particular orientations instead of the common element-death method in FEM codes. Application examples combining the above three essential topics will be presented, in which three types of glass/epoxy laminates (45° unidirectional, cross-ply, and ±45° angle-ply) will be analyzed using finite element method. The predicted results are compared with experimental data and it is shown that they are in good agreement.

Unified Periodic Boundary Conditions

Consider a large sample of periodic inhomogeneous body (composite) as shown in Fig. 1. The body has two length scales, a global (macroscopic) length scale, *D*, which is of the order of the size of the body, and a local (microscopic) length scale, *d*, which is proportional to the wavelength of the variation of the microstructure. Correspondingly

$$\delta = d / D << 1 \tag{1}$$

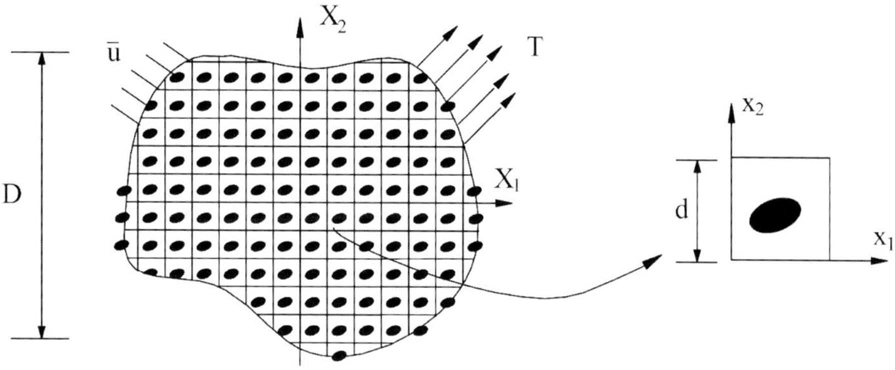

Fig. 1 RUC from a periodic composite.

Obviously, any function *f* in the body depends on two variables, *global (macroscopic) coordinates* X_i for the body and the *local (microscopic) coordinates* x_i for the unit cell. For small strain elasticity, the boundary value problem for the *composite body* can be defined as the following, from which the unknown field quantities, stress σ_{ij}, strain ε_{ij} and displacements u_i can be solved:

$$\sigma_{ij,j} = 0 \tag{2}$$

$$\sigma_{ij} = C_{ijkl}\varepsilon_{kl} \tag{3}$$

$$\varepsilon_{kl} = \frac{1}{2}(u_{k,l} + u_{l,k}) \tag{4}$$

$$\sigma_{ij}n_j = T_i \quad \text{on } \partial_\sigma, \qquad u_i = \bar{u}_i \quad \text{on } \partial_u \tag{5}$$

The boundary value problem has the feature that C_{ijkl} varies very rapidly within a short wavelength (order of d) on the global length scale X_i and therefore it is difficult to find a solution that solves the global problem and accounts for the local oscillation at the same time. For example, in a FEM solution, assuming roughly each unit cell should have several hundred elements to accurately capture the large variations due to the heterogeneity of the microstructure, then for the entire composite laminate or structure, the number of elements needed will increase by several orders. Hence, there is a motivation to seek a simplified solution. Since the composite body can be envisioned as a periodical array of the RUCs, it is adequate to obtain a solution based on an individual RUC. This implies that beyond a boundary layer of the composite body, each RUC in the composite has the same deformation mode and there is no separation or overlap between the neighbouring RUCs. That is, the stress and strain fields are periodic as the microstructure (Suquet, 1987).

Since the whole body, thus each unit cell is in balance, the equilibrium equation, Eqn. (2) and Eqns. (3)-(4) still apply in a RUC with volume *V*. However, the boundary conditions on the boundary of the RUC, ∂, should be properly determined. In the case of periodic media, the microscopic fields have to fulfill suitable periodicity conditions ensuring continuity of boundary displacements and tractions across adjacent cells. According to Suquet (1987), the displacement field for a periodic structure can be expressed as:

$$u_i(x_1, x_2, x_3) = \bar{\varepsilon}_{ij}x_j + u_i^*(x_1, x_2, x_3) \tag{6}$$

In the above, $\bar{\varepsilon}_{ij}$ is the global strain tensor of the periodic structure and the first term on the right side represents a linear distributed displacement field. The second term on the right side is a periodic function from one RUC to another. In addition, for a periodic RUC, the tractions on the opposite boundary surfaces should also meet the continuity condition, i.e.

$$\sigma_{ij}(P)n_j(P) = -\sigma_{ij}(Q)n_j(Q) \tag{7}$$

where P and Q are two periodic points (with the same in-plane coordinates) on the two opposite boundary surfaces, **n** is the unit outward normal vector to the surfaces, see Fig. 1. Note that the global strain $\bar{\varepsilon}_{ij}$ in Eqn. (6) and the corresponding global stress $\bar{\sigma}_{ij}$ can be defined as the averages over the RUC volume V:

$$\bar{\sigma}_{ij} = \frac{1}{V}\int_V \sigma_{ij}(x_1, x_2, x_3)\,dV \tag{8}$$

$$\bar{\varepsilon}_{ij} = \frac{1}{V}\int_V \varepsilon_{ij}(x_1, x_2, x_3)\,dV \tag{9}$$

where $\sigma_{ij}(x_1, x_2, x_3)$ and $\varepsilon_{ij}(x_1, x_2, x_3)$ are local (microscopic) stress and strain defined in the RUC.

Equations (6) and (7) are the periodic boundary conditions for a RUC. Together with the Eqns. (2), (3) and (4), we complete the boundary value problem for the RUC. This boundary value problem is well-posed, as shown e.g. in Suquet (1987). However, the periodic part of displacement, $u_i^*(x_1,x_2,x_3)$, in Eqn. (6) is usually unknown prior to the solution, thus it is not convenient to apply Eqn. (6) directly as displacement boundary conditions. In Xia et al. (2003a), a unified form of periodic boundary conditions for any multiaxial loading and suitable for finite element analysis is developed. For the sake of brevity only the conclusions are cited here. Interested readers are encouraged to consult the reference for further details and illustrative examples.

For any RUC, its boundary surfaces must always appear in parallel pairs, the displacements of two periodic points on a pair of parallel opposite boundary surfaces can be written as

$$u_i(P) = \bar{\varepsilon}_{ij}x_j(P) + u_i^*(P) \tag{10}$$

$$u_i(Q) = \bar{\varepsilon}_{ij}x_j(Q) + u_i^*(Q) \tag{11}$$

In the above "P" and "Q" identify the two corresponding points on a pair of two opposite parallel boundary surfaces of the RUC, see Fig. 1.

Note that $u_i^*(x_1,x_2,x_3)$ is a periodic function, i.e. its value is the same for two corresponding points at the two parallel boundaries (periodicity), therefore, the difference between the above two equations is

$$u_i(P) - u_i(Q) = \bar{\varepsilon}_{ij}[x_j(P) - x_j(Q)] = \bar{\varepsilon}_{ij}\Delta x_j \tag{12}$$

Since $\Delta x_j = x_j(P) - x_j(Q)$ are constants for each pair of the parallel boundary surfaces, hence, for a specified $\bar{\varepsilon}_{ij}$, the right side becomes constants. Equation (12) is a special type of displacement boundary conditions. Instead of giving known values of boundary displacements, it specifies the displacement-differences between two periodic points at opposite boundaries. Obviously, the application of Eqn. (12) will guarantee the continuity of displacement field, i.e. the neighboring RUCs cannot separate or encroach into each other at the boundaries after the deformation.

Furthermore, it has been proved in Xia et al. (2006) that if a RUC is analyzed by using a displacement-based finite element method, the application of only Eqn. (12) can guarantee the uniqueness of the solution and thus Eqn. (7) are automatically satisfied. For the sake of brevity only the conclusions are cited here.

Theorem: In a displacement-based FEM analysis, by applying the unified displacement-difference periodic boundary conditions on a RUC, a unique solution is obtained.

Lemma 1 For a fixed periodic structure, different RUC's may be defined, however, by applying the unified displacement-difference periodic boundary conditions, Eqn. (12), in the displacement-based FEM analysis, the solution will be independent of the choice of the RUCs.

Lemma 2 The solution obtained by applying the unified displacement-difference periodic boundary conditions, Eqn. (12), in the displacement-based FEM analysis, will also meet the traction continuity conditions, Eqn. (7).

It is noted that the unified periodic boundary conditions, Eqn. (12), can be easily applied in a FEM as the nodal displacement constraint equations. It is also noted that the proposed unified periodic boundary conditions are in the form of global strains. In the case of given global stresses, or a combination of the global stresses and strains, a proper proportion between the global strains must be applied. The required proportion can be determined without any difficulty through an iterative procedure, see the examples to follow.

One can also note that the derivation and proof procedures for the proposed unified periodic boundary conditions are not dependent on the properties of the constituent materials of the composites. Therefore, they can be applied to nonlinear micro/meso-mechanical analyses of the composites under any combination of multiaxial loads.

Nonlinear Viscoelastic Constitutive Model for Epoxy Polymers

Based on comprehensive experimental data (Hu et al. 2003, Shen et al. 2004) a new nonlinear viscoelastic model has been recently developed by the authors (Xia et al. 2003b, Xia et al. 2005). Here only a brief description will be given.

For the uniaxial stress state, the model can be represented by a finite number of nonlinear 'Kelvin-Voigt type' elements and a linear spring element, connected in series (Fig. 2). The constitutive equations, generalized to the multiaxial stress state, are sumarized below:

$$\{\dot{\varepsilon}_t\} = \{\dot{\varepsilon}_e\} + \{\dot{\varepsilon}_c\} \tag{13}$$

$$\{\dot{\varepsilon}_c\} = \sum_{i=1}^{n} \left(\frac{[A]}{E_i \tau_i} \{\sigma\} - \frac{1}{\tau_i} \{\varepsilon_{ci}\} \right) \tag{14}$$

$$\{\dot{\sigma}\} = E[A]^{-1} \{\dot{\varepsilon}_e\} \tag{15}$$

In the above, $\{\dot{\varepsilon}_t\}$, $\{\dot{\varepsilon}_e\}$, $\{\dot{\varepsilon}_c\}$, $\{\dot{\sigma}\}$ are the total strain-rate, elastic strain-rate, creep strain-rate, and stress-rate vectors (each contains six components, respectively). E is an elastic modulus which is assumed to be constant and $[A]$ is a matrix related to the value of Poisson's ratio, defined by

$$[A]=\begin{bmatrix} 1 & -\upsilon & -\upsilon & 0 & 0 & 0 \\ -\upsilon & 1 & -\upsilon & 0 & 0 & 0 \\ -\upsilon & -\upsilon & 1 & 0 & 0 & 0 \\ 0 & 0 & 0 & 1+\upsilon & 0 & 0 \\ 0 & 0 & 0 & 0 & 1+\upsilon & 0 \\ 0 & 0 & 0 & 0 & 0 & 1+\upsilon \end{bmatrix} \tag{16}$$

In Eqn. (14) $\tau_i = \eta_i / E_i$ $(i = 1, 2, \cdots, n)$ denotes the retardation time, E_i is the spring stiffness and η_i is the dashpot viscosity for the *i*-th 'Kelvin-Voigt type' element, respectively. Based on Eqn. (14), the retardation time τ_i has a damped exponential character as in an exponential-type function. Its value determines the time duration after which contribution from the individual 'Kelvin-Voigt type' element becomes negligible. Therefore, the number of the 'Kelvin-Voigt type' elements adopted in the constitutive equation depends on the required time range. For simplicity, we introduce a time scale factor α, and assume that

$$\tau_i = (\alpha)^{i-1} \tau_1 \tag{17}$$

In this way all τ_i are related through the scale factor α. A time span of order of *n* would be covered, if *n* 'Kelvin-Voigt type' elements were chosen and the value of α is taken to be 10.

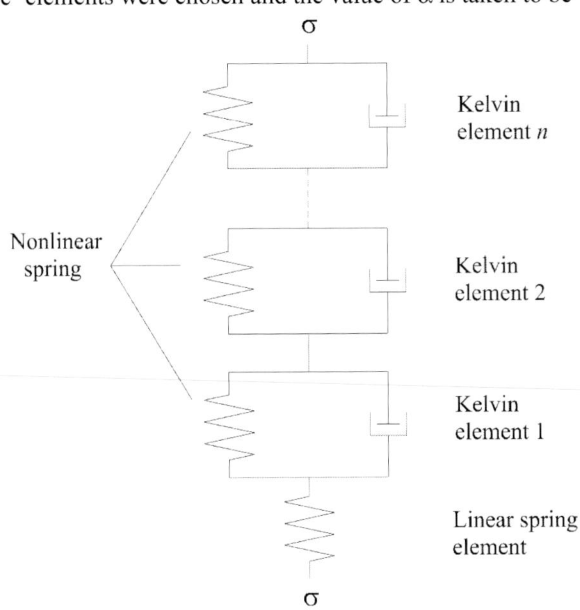

Fig. 2 A uniaxial visoelastic model represented by a finite series of 'Kelvin-Voigt type' elements coupled with an elastic spring.

The description of the nonlinear behaviour in the current model is achieved by letting E_i's be functions of the current equivalent stress, σ_{eq}. Furthermore, a single function form for all E_i's is assumed, i.e.

$$E_i = E_1(\sigma_{eq}) \tag{18}$$

with

$$\sigma_{eq} = \frac{(R-1)I_1 + \sqrt{(R-1)^2 I_1^2 + 12RJ_2}}{2R} \tag{19}$$

where $I_1 = \sigma_1 + \sigma_2 + \sigma_3$ is the first invariant of the stress tensor, $J_2 = s_{ij}s_{ij}/2$ is the second invariant of the deviatoric stress and R is the ratio of compressive to tensile 'yield stress'. Note that when $R = 1$, then Eqn. (19) reduces to the von Mises equivalent stress, $\sigma_{eq} = \sqrt{3J_2}$.

The five constants ($E, \upsilon, \alpha, \tau_1$, R) and the functional form of $E_1(\sigma_{eq})$ can be determined from uniaxial creep curves tested at different stress levels by a simple procedure described in Xia et al. (2003b).

A distinct feature of this constitutive model is its capacity to distinguish between loading and unloading cases by introducing a stress memory surface and a corresponding switch rule. The stress memory surface is defined as:

$$f_m^{\sigma}(\sigma_{ij}) - R_{mem}^2 = \frac{3}{2}s_{ij}s_{ij} - R_{mem}^2 = 0 \tag{20}$$

where $s_{ij} = \sigma_{ij} - \frac{1}{3}\sigma_{kk}\delta_{ij}$ are the deviatoric stress components. The radius of the memory surface, R_{mem}, is determined by the maximum von Mises stress level experienced by the material during its previous loading history, i.e. $R_{mem} = \sqrt{(3s_{ij}s_{ij}/2)_{\max}}$. Therefore, for a monotonic loading from a stress free state, the memory surface will expand isotropically with the increasing stress level. If σ_{ij}^t is the current stress point, $d\sigma_{ij}^t$ is the stress increment at time t, and $(\frac{\partial f}{\partial \sigma_{ij}})_{s_{ij}=s_{ij}^t}$ represents the direction of the normal to the memory surface at the current stress point, then the criterion to distinguish the loading/unloading cases is defined as follows:

- if the current stress point is on the memory surface and $(\frac{\partial f}{\partial \sigma_{ij}})_{\sigma_{ij}^t} \cdot d\sigma_{ij}^t \geq 0$, this signifies a loading case;
- if the current stress point is on the memory surface and $(\frac{\partial f}{\partial \sigma_{ij}})_{\sigma_{ij}^t} \cdot d\sigma_{ij}^t < 0$, then a switch from loading to unloading occurs;
- if the current stress point is inside the memory surface, i.e. $f_m^{\sigma}(\sigma_{ij}^t) - R_{mem}^2 < 0$, it is then an unloading case.

For the loading case the spring stiffness of the 'Kelvin-Voigt type' elements is defined as a function of the equivalent stress, $E_i = E_1(\sigma_{eq})$. For the unloading case, it is assumed that E_i remains the same during the entire unloading process, and its value is that of the E_i at which the switch from loading to unloading took place.

Post-Damage Constitutive Model Based on the Concept of Smeared Crack

Upon increasing the applied load, micro-cracks will develop in the matrix. These cracks cause reduction in stiffness of the laminate. In contrast to a predefined single dominant crack in isotropic materials, the orientation and location or even numbers of cracks in a laminate is unknown, thus this makes it difficult to deal with such cracks through the classical fracture mechanics approach. Instead, the so-called 'smeared crack' approach will be used. In this approach the reduction of load bearing capacity induced by a crack is described by stress-strain softening relationship. In this manner the discontinuity caused by a crack is smeared out, and this procedure can be implemented into a FEM code.

Once a crack is formed, it is assumed that it cannot transfer normal and shear stresses across the crack surfaces, i.e. σ_1, σ_{12} and $\sigma_{13} \to 0$. The subscript 1 denotes the Cartesian axis perpendicular to the crack plane while 2 and 3 are in the crack plane. However, the ability to transfer the other stress components is not affected by the crack formation. Let the stress and strain vectors in the local (crack) coordinate system be designated by $\{\sigma\}^{cr}$ and $\{\varepsilon\}^{cr}$, respectively. Thus, the post-damage constitutive model in the crack coordinate system is

$$\{\Delta\sigma\}^{cr} = E_t[D]\{\Delta\varepsilon\}^{cr} - \chi[B]\{\sigma\}^{cr} \tag{21}$$

or written in its full form:

$$\begin{pmatrix} \Delta\sigma_1 \\ \Delta\sigma_2 \\ \Delta\sigma_3 \\ \Delta\sigma_{12} \\ \Delta\sigma_{23} \\ \Delta\sigma_{31} \end{pmatrix}^{cr} = E_t \begin{pmatrix} \beta Z_1 & 0 & 0 & 0 & 0 & 0 \\ 0 & Z_1 & Z_2 & 0 & 0 & 0 \\ 0 & Z_2 & Z_1 & 0 & 0 & 0 \\ 0 & 0 & 0 & \beta Z_3 & 0 & 0 \\ 0 & 0 & 0 & 0 & Z_3 & 0 \\ 0 & 0 & 0 & 0 & 0 & \beta Z_3 \end{pmatrix} \begin{pmatrix} \Delta\varepsilon_1 \\ \Delta\varepsilon_2 \\ \Delta\varepsilon_3 \\ \Delta\gamma_{12} \\ \Delta\gamma_{23} \\ \Delta\gamma_{31} \end{pmatrix}^{cr} - \chi \begin{pmatrix} 1 & 0 & 0 & 0 & 0 & 0 \\ 0 & 0 & 0 & 0 & 0 & 0 \\ 0 & 0 & 0 & 0 & 0 & 0 \\ 0 & 0 & 0 & 1 & 0 & 0 \\ 0 & 0 & 0 & 0 & 0 & 0 \\ 0 & 0 & 0 & 0 & 0 & 1 \end{pmatrix} \begin{pmatrix} \sigma_1 \\ \sigma_2 \\ \sigma_3 \\ \sigma_{12} \\ \sigma_{23} \\ \sigma_{31} \end{pmatrix}^{cr} \tag{22}$$

In the above,

$$Z_1 = \frac{1-\upsilon}{(1+\upsilon)(1-2\upsilon)}, \quad Z_2 = \frac{\upsilon}{(1+\upsilon)(1-2\upsilon)}, \quad Z_3 = \frac{1}{2(1+\upsilon)}, \tag{23}$$

E_t is the modulus of the epoxy under uniaxial tensile loading at the instant of damage initiation, β is a small number which represents the loss of the stiffness in these three particular stress directions and the constant χ allows the three stress components to decrease to a near zero value in a sufficiently short time duration.

Application Examples

The response of three types of glass fiber/epoxy polymer matrix laminates subject to a tensile loading are presented herein, see Ellyin and Kujawski (1995), Zhang et al. (2005). These are: (1) a unidirectional laminate under 45° off-axis loading; (2) a [0°/90°]ns cross-ply laminate under transverse loading, and (3) a [±45°]ns angle-ply laminate under tensile loading, see Fig. 3(a), (b), (c), respectively. The 45° off-axis tension on the unidirectional laminate (Fig. 3a) and the tensile load on the $[\pm 45°]_{ns}$ angle-ply laminate (Fig 3c) result in combined normal and shear traction boundary conditions as shown in figures with

$$\bar{\sigma}_x = \bar{\sigma}_y = \bar{\tau}_{xy} = 0.5\bar{\sigma} \tag{24}$$

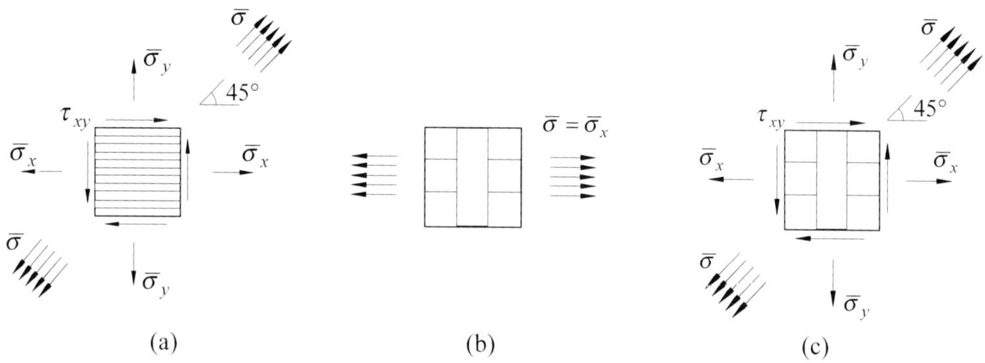

Fig. 3 (a) a unidirectional laminate under 45° off-axis loading; (b) a cross-ply laminate under transverse loading; (c) a [±45°]ns angle-ply laminate under tensile loading.

The RUC for the unidirectional laminate is a unit cube containing a cylindrical fiber and the RUCs for both $[0°/90°]_{ns}$ cross-ply laminate and $[\pm 45°]_{ns}$ angle-ply laminate are the same consisting of two cubes with the two cylindrical fibers at an angle of 90° (Fig. 4a, b).

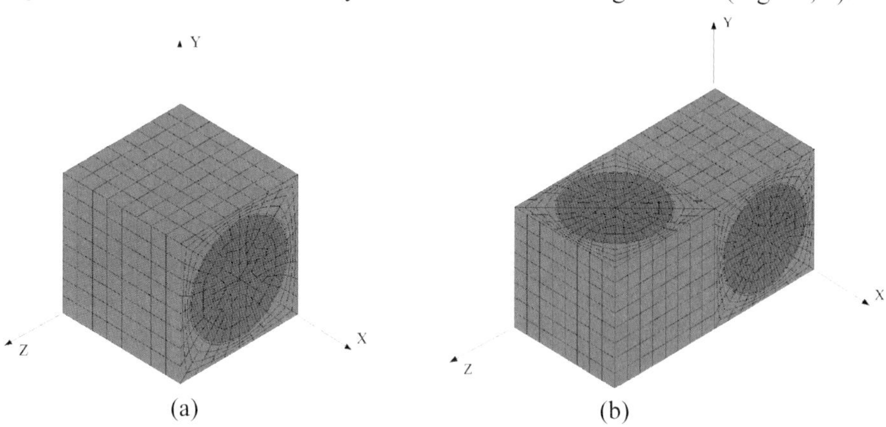

Fig. 4 Meshed RUCs for (a) unidirectional laminate and for (b) cross-ply laminate and [±45°]ns angle-ply laminate.

The finite element code ADINA was used to conduct the numerical analysis. The nonlinear viscoelastic constitutive model of the epoxy matrix and the post-damage constitutive model were implemented into the code through its user-defined subroutine. It is to be noted that in Eqn. (24) the multiaxial loads are the global stress components applied to the RUCs. The applied periodic boundary conditions, Eqn. (12), are in the form of global strain components. An iterative procedure is required to ensure proper proportion of the increments of the global strain components are applied so that Eqn. (24) is satisfied at each time step. The iteration procedure is as follows.

(i) For each time step, Δt, we apply a set of trial global strain increments, $\Delta\bar{\varepsilon}_{ij}$.

(ii) The solution gives the stress distribution in the RUC, so the global stress components are the average values over the volume V of the RUC (see Eqn. 8.)
(iii) Equation (24) is checked and, if it is satisfied (within a certain error limit), then one proceeds to the next time step. If not, new values of $\Delta\bar{\varepsilon}_{ij}$ are obtained and the steps (i) to (iii) are repeated. For a small time step, it could be assumed that the increments of $\Delta\bar{\varepsilon}_{ij}$ are proportional to the corresponding increment of average stress components, then the new values of $\Delta\bar{\varepsilon}_{ij}$ can be estimated.

Two damage mechanisms were considered in the analyses, i.e. matrix cracking and fibre breaking. For the epoxy polymer matrix, a maximum principal strain criterion was adopted, i.e. if $\varepsilon_1 \geq \varepsilon_{cr} = 4.8\%$, the matrix element was assumed to be damaged and subsequently the post-damage constitutive relation would be used. (The manufacturer did not provide the failure strain for the neat resin, the value assumed here was taken from another resin system). For the glass fiber, the average axial strain of the fibre was monitored and if it exceeded a prescribed maximum value, i.e. $\bar{\varepsilon}_{fa} \geq \varepsilon_{f\max} = 2.3\%$ then the fiber was assumed to have fractured and the calculation was then terminated. The above critical strain values were taken from experimental data and the fiber volume fraction of the laminates was assumed to be 52.5%.

Prediction of Stress-Strain Curves for the Three Types of Laminates

Figures 5 and 6 display the predicted global stress-strain curves of the three types of laminates and the comparison with the test results. All the calculations use the same set of material constants and the same set of constants used in the post-damage constitutive model. The test specimens were made of "Scotchply 1003" prepregs of 3M Company. The test data of cross-ply laminates were taken from the technical data of the 3M Company and that of the [±45°]ns angle-ply laminates and unidirectional laminates (UDC) under 45° off-axis loading were from Ellyin and Kujawski (1995). It is seen that the predictions of the drastically different responses of the three types of laminates are in good agreement with the experimental data.

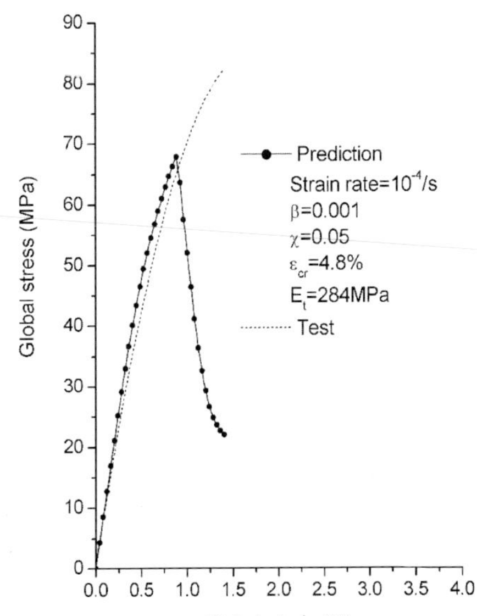

Fig. 5 Predicted stress-strain curve of a UDC under 45° off-axis tensile loading.

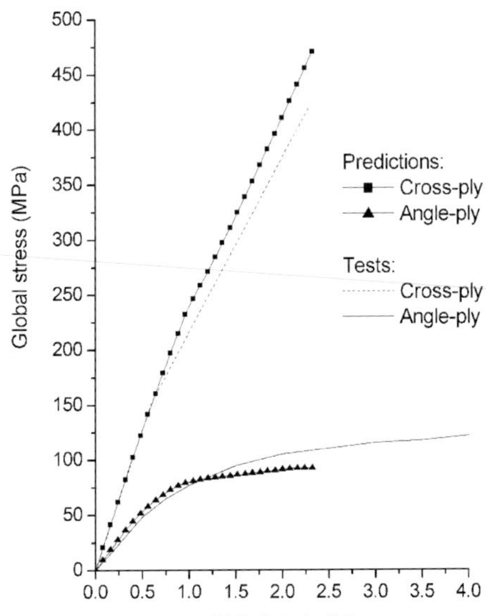

Fig. 6 Predicted stress-strain curves of cross-ply and angle-ply laminates.

In the case of the unidirectional laminate, it is noted that the predicted trend is in good agreement with the test results, Fig. 5. The predicted initial stiffness is 10.1 GPa, and the maximum load is 68 MPa, while the corresponding test results are 10.1 GPa and 82 MPa.

The effect of viscoelastic behavior of the matrix is manifested by the nonlinearity of the stress-strain curve, which is noticeable once the stress exceeds 40 MPa (about 0.5% strain). Since damage has not yet occurred at this load level (for the unidirectional laminate, damage initiates at the peak of the stress-strain curve), therefore this nonlinearity is mainly caused by the viscoelasticity of the epoxy matrix.

For the cross-ply and angle-ply laminates the test and predicted results are shown in Figure 6. For the cross-ply laminate, a bilinear stress-strain curve is predicted in which the two 'moduli' are approximately 25.5 GPa and 17.4 GPa. The corresponding test values are 25.5 GPa and 15.6 GPa, respectively. The knee between the two straight lines corresponds to the load level at which the transverse cracking of matrix occurs in the laminate. And the final failure of the specimen is due to the fracture of fibers in the 0° plies. From the technical data of the 'Scotchply', the test value of tensile strength of cross-ply laminate is 480 MPa, and the present prediction of 470 MPa is very close to that of the test. Note, however, that for the other two laminates, no fiber fracture occurs within the strain range of the present calculations.

The unique nonlinear stress-strain curve of the [±45°]ns laminate is also well predicted. For example, the predicted initial stiffness of 10.1 GPa agreed very well with the test value of 10.1 GPa. In contrast to the unidirectional laminate under 45° off-axis loading which failed at a relatively low global strain of 0.9%, at the same strain level, the [±45°]ns laminate is capable of carrying the applied load albeit at a reduced stiffness. However, prior to the 'yield' point, the stress-strain curve also manifested a nonlinear response. Since the damage has not yet occurred at this load level (50 MPa), this nonlinearity is mainly caused by the viscoelasticity of the epoxy matrix. Note that the nonlinearity of the stress-strain curve has different causes at different strain levels: at lower strain levels, it is mainly caused by the viscoelasticity of the matrix, while at higher strain values, the onset of damage and its evolution is the main contributor to the nonlinear response. Finally, it should be noted that the simulation is carried out up to about 2.3% applied global strain. Thereafter, it is difficult to continue the simulation, since the local deformation is very large and the present FEM model is based on the small deformation formulation.

Conclusions

To perform an effective micro/meso-mechanical analysis of composite materials and to obtain reliable predictions, the following three prerequisites are essential:

(1) Correct periodic boundary conditions must be applied to the repeated unit cells (RUCs).
- A unified form of periodic boundary conditions for RUCs of composites has been presented.
- Application of the proposed periodic boundary conditions can guarantee both the displacement and traction continuity. The solution is also independent on the choice of RUCs.

(2) The constitutive models must accurately represent the constituents' behaviours.
- A nonlinear viscoelastic constitutive model has been developed for epoxy polymers.
- The model is capable of predicting complicated time- and loading-history-dependent response of the polymer matrix materials.

(3) Proper damage criteria and post-damage constitutive model must be included in the analyses to simulate different damage modes in composite materials.
- A post-damage constitutive model based on smeared crack concept has been developed.

- Initiation and propagation of matrix cracking in composites are well simulated by using this post-damage constitutive model.

Acknowledgements

The Work presented here is supported, in part, by grants to F.E and Z.X. from the Natural Sciences and Engineering Research Council (NSERC) of Canada.

References

Aboudi, J., (1991). *Mechanics of Composite Materials, A Unified Micromechanical Approach.* Elsevier Science Publishers, Amsterdam.

Ellyin, F., Kujawski, D. (1995). Tensile and fatigue behaviour of glass fibre/epoxy laminates. *Construction and Building Materials*, **9**, 425-430.

Hashin, Z. (1966). Viscoelastic fiber reinforced materials. *AIAA Journal*, **4**, 1141-1147.

Hu, Y., Xia, Z, Ellyin, F. (2003). Deformation behavior of an epoxy resin subjected to multiaxial loadings, part I: experimental investigations. *Polymer Engineering & Science*, **43**, 721-733.

Nemat-Nasser S., Hori M. (1993). *Micromechanics: Overall Properties of Heterogeneous Materials*. Elsevier Science Publishers, Amsterdam.

Shen, X., Xia, Z., Ellyin, F. (2004). Cyclic deformation behavior of an epoxy polymer, part I: experimental investigation. *Polymer Engineering & Science*, **44**, 2240-2246.

Sun, C. T., Vaidya, R. S. (1996). Prediction of composite properties from a representative volume element. *Composites Science and Technology*, **56**, 171-179.

Suquet, P. (1987). Elements of homogenization theory for inelastic solid mechanics. In: Sanchez-Palencia, E., Zaoui, A., (eds.). *Homogenization Techniques for Composite Media.* Springer-Verlag, Berlin, 194-275.

Xia, Z., Zhang, Y., Ellyin, F. (2003a). A unified periodical boundary condition for representative volume elements of composites and applications. *International Journal of Solids and Structures*, **40**, 1907-1921.

Xia, Z, Hu, Y., Ellyin, F. (2003b). Deformation behavior of an epoxy resin subjected to multiaxial loadings, part II: constitutive modeling and predictions. *Polymer Engineering & Science*, **43**, 734-748.

Xia, Z., Shen, X., Ellyin, F. (2005). Cyclic deformation behavior of an epoxy polymer, part II: prediction of constitutive model. *Polymer Engineering & Science*, **45**, 103-113.

Xia, Z., Zhou, C., Yong. Q., Wang, X. (2006). On selection of repeated unit cell model and application of unified periodic boundary conditions in micro-mechanical analysis of Composites. *International Journal of Solids and Structures*, **43**, 266-278.

Yuan, F. G., Pagano, N. J., Cai, X. (1997). Elastic moduli of brittle matrix composites with interfacial debonding. *International Journal of Solids and Structures*, **34**, 177-201.

Zhang, Y., Xia, Z., Ellyin, F. (2005). Viscoelastic and damage analyses of fiber reinforced polymer laminates by micro/meso-mechanical modeling. *Journal of Composite Materials*, **39**, 2001-2022.

3M Minnesota Mining & Manufacturing CO., Technical Data of 'Scotchply' Reinforced Plastic Type 1002 & 1003, St. Paul, Minnesota, USA.

GALERKIN METHOD FOR STOCHASTIC ALGEBRAIC EQUATIONS AND PLATES ON RANDOM ELASTIC FOUNDATION

Mircea Grigoriu
Cornell University, Ithaca, NY 14853, USA
E-mail: mdg12@cornell.edu

Abstract

A new perspective is presented on the Galerkin solution for linear stochastic algebraic equations, that is, linear algebraic equations with random coefficients. It is shown that (1) a stochastic algebraic equation has an optimal Galerkin solution, that is, a Galerkin solution that is best in the mean square sense, and (2) the optimal Galerkin solution is equal to the conditional expectation of the exact solution with respect to a σ-field coarser than the σ-field relative to which this solution is measurable. Galerkin solutions that are not optimal are called sub-optimal. Both optimal and sub-optimal Galerkin solutions are defined and constructed. Optimal and sub-optimal Galerkin solutions are used to calculate statistics of the displacement of a simply supported plate sitting on a random elastic foundation. The accuracy of these Galerkin solutions is assessed by Monte Carlo simulation.

1. Introduction

Ordinary or partial differential equations with random coefficients, input and/or boundary conditions, referred to as stochastic differential equations, are used to formulate a broad range of mechanics problems. It is common to approximate the solution of stochastic differential equations by that of algebraic equations with random coefficients, called stochastic algebraic equations. There are no general and efficient methods for finding the probability law of the solution of a stochastic algebraic equation. Taylor series, perturbation, Neumann series, decomposition, equivalent linearization, iteration, and other approximate techniques can be used to solve these equations (Deb et al. 2001; Ghanem and Spanos, 1991; Grigoriu, 2002: section 8.3.1). Monte Carlo simulation is the only available method capable of providing estimates for the probability law of the solution of general stochastic equations, but can be numerically prohibitive if applied to solve realistic problems.

Our objectives are to (1) present an alternative interpretation of the Galerkin method for solving stochastic algebraic equations and (2) apply this method to calculate statistics for the displacement of a plate supported by a random elastic foundation. It is shown that there is an optimal Galerkin solution for a stochastic algebraic equation, that is, a Galerkin solution minimizing the mean square error. The optimal Galerkin solution is equal to the conditional expectation of the exact solution of a stochastic algebraic equation with respect to a σ-field coarser than the σ-field with respect to which this solution is measurable. Galerkin solutions that are not optimal are said to be sub-optimal. Second-moment properties, distributions, and other statistics are developed for both optimal and sub-optimal Galerkin solutions. Statistics are calculated for optimal and sub-optimal Galerkin solutions for a simply supported plate sitting on a random elastic foundation. The accuracy of the Galerkin solutions is evaluated by Monte Carlo simulation.

2. Stochastic Algebraic Equations

Consider the stochastic algebraic equation

M. Pandey et al. (eds), Advances in Engineering Structures, Mechanics & Construction, 517–525.

$$\big(\boldsymbol{a}+\boldsymbol{r}(\boldsymbol{Z})\big)\,\boldsymbol{X}=\boldsymbol{Y}, \tag{1}$$

where $\boldsymbol{a}$ denotes an (d,d) real-valued deterministic matrix, $\boldsymbol{Z}$ and $\boldsymbol{Y}$ are $\mathbb{R}^n$-valued and $\mathbb{R}^d$-valued random variables defined on some probability spaces $(\Omega_1,\mathcal{F}_1,P_1)$ and $(\Omega_2,\mathcal{F}_2,P_2)$, respectively, that may or may not coincide, and $\boldsymbol{r}$ is an (d,d) real-valued matrix whose entries depend on $\boldsymbol{Z}$. It is assumed that (1) the random variables $\boldsymbol{Y}$ and $\boldsymbol{Z}$ are independent and are in L_2, (2) the function $\boldsymbol{r}(\cdot)$ is measurable, so that $\boldsymbol{r}(\boldsymbol{Z})$ and $\boldsymbol{a}+\boldsymbol{r}(\boldsymbol{Z})$ are random variables on $(\Omega_1,\mathcal{F}_1,P_1)$, and (3) the operator $\boldsymbol{a}+\boldsymbol{r}(\boldsymbol{Z})$ is bounded almost surely (a.s.), that is, $\boldsymbol{a}+\boldsymbol{r}(\boldsymbol{Z}(\omega_1))$ is bounded for all $\omega_1\in\Omega_1\setminus N_1$, where $N_1\in\mathcal{F}_1$ and $P_1(N_1)=0$.

If $\det(\boldsymbol{a}+\boldsymbol{r}(\boldsymbol{Z}))\neq 0$ a.s., then Equation (1) has the unique solution

$$\boldsymbol{X}(\omega_1,\omega_2)=\big(\boldsymbol{a}+\boldsymbol{r}(\boldsymbol{Z}(\omega_1))\big)^{-1}\,\boldsymbol{Y}(\omega_2)=\boldsymbol{\beta}\big(\boldsymbol{Z}(\omega_1)\big)\,\boldsymbol{Y}(\omega_2)\quad\text{a.s.}, \tag{2}$$

which is a random variable on the product probability space $(\Omega_1\times\Omega_2,\mathcal{F}_1\otimes\mathcal{F}_2,P_1\otimes P_2)$, where $\Omega_1\times\Omega_2$, $\mathcal{F}_1\otimes\mathcal{F}_2$, and $P_1\otimes P_2$ denote the product sample space, the σ-field generated by the measurable rectangles $\{A_1\times A_2\}$, $A_1\in\mathcal{F}_1$, $A_2\in\mathcal{F}_2$, and the extension of the set function $R(A_1\times A_2)=P_1(A_1)\,P_2(A_2)$ defined for $A_1\in\mathcal{F}_1$ and $A_2\in\mathcal{F}_2$ to $\mathcal{F}_1\otimes\mathcal{F}_2$, respectively.

3. Galerkin Method

The Galerkin solution of Equation (1) requires to discretize the stochastic dimension, that is, the random variables $(\boldsymbol{X},\boldsymbol{Y},\boldsymbol{Z})$. Polynomials chaos representations and partitions of the sample space $\Omega_1\times\Omega_2$ or of the range of the random variables $(\boldsymbol{Y},\boldsymbol{Z})$ can be used to discretize the stochastic dimension of Equation (1). Galerkin method based on polynomial chaos has been applied successfully to solve a broad range of stochastic problems (Ghanem and Spanos, 1991), although there are some theoretical aspects of the method that remain to be clarified. For example, the m.s. convergence of polynomial chaos representations for $\boldsymbol{Y}$; $\boldsymbol{Z}$ to $\boldsymbol{Y}$; $\boldsymbol{Z}$ does not guarantee the m.s. convergence of the corresponding representations for $\boldsymbol{X}$ to the exact solution $\boldsymbol{X}$. Also, moments of order 3 and higher of polynomial chaos representations may not converge to corresponding target moments (Field and Grigoriu, 2004). Galerkin method using partitions of the range of the random variables $(\boldsymbol{Y},\boldsymbol{Z})$ views the solution $\boldsymbol{X}$ as an unknown function of $(\boldsymbol{Y},\boldsymbol{Z})$, that can be approximated by polynomials or other functions depending on some unknown coefficients. The solution of Equation (1) is found by solving a deterministic version of this equation obtained by viewing $\boldsymbol{Z}$ as a parameter taking values in $\boldsymbol{Z}(\Omega_1)$. The measure on $\boldsymbol{Z}(\Omega_1)$ is the density of $\boldsymbol{Z}$ rather than the Lebesgue measure (Babuška et al. 2004).

The version of the Galerkin method considered here is based on partitions of the product sample space $\Omega_1\times\Omega_2$. Let

$$\big(\emptyset,\Omega_k\big)=\mathcal{G}_{k,1}\subset\cdots\subset\mathcal{G}_{k,i}\subset\cdots\subset\mathcal{G}_{k,n_k}=\mathcal{F}_k,\quad k=1,2, \tag{3}$$

be two sequences of sub-σ-fields on probability spaces $(\Omega_k,\mathcal{F}_k,P_k)$, $k=1,2$, that can be constructed from, for example, finite partitions of the sample spaces Ω_k, $k=1,2$. Let $\{\Lambda_q\}$, $q=1,\ldots,m$, and $\{\Gamma_r\}$, $r=1,\ldots,m'$, be measurable partitions of the sample spaces Ω_1 and Ω_2, respectively. The σ-fields generated by the sets $\{\Lambda_q\}$, $\{\Gamma_r\}$, and $\{\Lambda_q\times\Gamma_r\}$ are of type $\mathcal{G}_{1,i}$, $\mathcal{G}_{2,j}$, and $\mathcal{G}_{1,i}\otimes\mathcal{G}_{2,j}$ in Equation (3), respectively. In the reminder of this section we define optimal and sub-optimal Galerkin solutions and give some of their properties.

Property 1. *The optimal Galerkin solution corresponding to the information content of a sub-σ-field $\mathcal{G}_{1,i}\otimes\mathcal{G}_{2,j}$ is*

$$\boldsymbol{X}_{i,j}=E\big[\boldsymbol{X}\mid\mathcal{G}_{1,i}\otimes\mathcal{G}_{2,j}\big]=E_1\big[\boldsymbol{\beta}(\boldsymbol{Z})\mid\mathcal{G}_{1,i}\big]\,E_2\big[\boldsymbol{Y}\mid\mathcal{G}_{2,j}\big]\ \text{(a. s.)}, \tag{4}$$

where E_k and E denote expectations with respect to the probability measures P_k, $k = 1, 2$, and $P_1 \otimes P_2$, respectively.

Generally, $\boldsymbol{X} = (\boldsymbol{a} + \boldsymbol{r}(\boldsymbol{Z}))^{-1}\,\boldsymbol{Y} = \boldsymbol{\beta}(\boldsymbol{Z})\,\boldsymbol{Y}$ is not $\mathcal{G}_{1,i} \otimes \mathcal{G}_{2,j}$-measurable for $i < n_1$ and/or $j < n_2$, that is, it is not a random variable with respect to the σ-field $\mathcal{G}_{1,i} \otimes \mathcal{G}_{2,j}$. The first equality in Equation (4) follows from the fact that the conditional expectation is the best mean square estimator for $\boldsymbol{X}$ with respect to the information content of $\mathcal{G}_{1,i} \otimes \mathcal{G}_{2,j}$ (Grigoriu, 2002: section 2.17.2). The validity of the second equality in Equation (4) results from properties of the conditional expectation, properties of σ-fields on product spaces, Fubini's theorem, and a theorem by Dynkin (Resnick, 1998: section 2.2).

Consider the special case in which $\mathcal{G}_{2,n_2} = \mathcal{F}_2$. The corresponding optimal Galerkin solution is

$$\boldsymbol{X}_{i,n_2} = E\big[\boldsymbol{\beta}(\boldsymbol{Z}) \mid \mathcal{G}_{1,i}\big]\,\boldsymbol{Y} \tag{5}$$

since $E[\boldsymbol{Y} \mid \mathcal{G}_{2,n_2}] = \boldsymbol{Y}$ a.s. (Equation (4)). This solution is used extensively in applications (Deb et al., 2001). Once the conditional expectations $E[\boldsymbol{\beta}(\boldsymbol{Z}) \mid \mathcal{G}_{1,i}]$ have been calculated, Equation (5) can be used to calculate statistics of the exact solution $\boldsymbol{X}$ approximately.

Property 2. *The optimal Galerkin solution in Equation (4) ranges from the expectation of the exact solution to the exact solution depending on the information content of the sub-σ-fields $\mathcal{G}_{1,i}$ and $\mathcal{G}_{2,j}$.*

Consider sub-σ-fields of $\mathcal{F}_k$ containing limited or full information on the random variables $\boldsymbol{Y}$ and $\boldsymbol{Z}$. The corresponding optimal Galerkin solutions are (Equation (4))

$$\begin{aligned}
\boldsymbol{X}_{1,1} &= E_1\big[\boldsymbol{\beta}(\boldsymbol{Z})\big]\,E_2\big[\boldsymbol{Y}\big] = E\big[\boldsymbol{X}\big] \\
\boldsymbol{X}_{1,n_2} &= E_1\big[\boldsymbol{\beta}(\boldsymbol{Z})\big]\,\boldsymbol{Y} \\
\boldsymbol{X}_{n_1,1} &= \boldsymbol{\beta}(\boldsymbol{Z})\,E_2\big[\boldsymbol{Y}\big] \\
\boldsymbol{X}_{n_1,n_2} &= \boldsymbol{\beta}(\boldsymbol{Z})\,\boldsymbol{Y} = \boldsymbol{X},
\end{aligned} \tag{6}$$

where the above equalities hold almost surely with respect to the product probability measure $P_1 \otimes P_2$. The above results follow from properties of the conditional expectation (Grigoriu, 2002: section 2.7.2). For example, $E_1[\boldsymbol{\beta}(\boldsymbol{Z}) \mid \mathcal{G}_{1,i}]$ is equal to $E_1[\boldsymbol{\beta}(\boldsymbol{Z})]$ and $\boldsymbol{\beta}(\boldsymbol{Z})$ a.s. for $i = 1$ and $i = n_1$, respectively. Also, $\boldsymbol{X}_{n_1,n_2}$ is equal to the exact solution $\boldsymbol{X}$ a.s. since the sub-σ-fields $\mathcal{G}_{k,n_k}$ coincide with the σ-fields $\mathcal{F}_k$, so that we have $E_1[\boldsymbol{\beta}(\boldsymbol{Z}) \mid \mathcal{G}_{1,n_1}] = \boldsymbol{\beta}(\boldsymbol{Z})$ and $E_2[\boldsymbol{Y} \mid \mathcal{G}_{2,n_2}] = \boldsymbol{Y}$ a.s.

Property 3. *The second-moment properties of the optimal Galerkin solution are*

$$\begin{aligned}
\boldsymbol{\mu}_{i,j} &= E\big[\boldsymbol{X}_{i,j}\big] = E\big[\boldsymbol{X}\big] \\
\boldsymbol{\gamma}_{i,j} &= E\big[\big(\boldsymbol{X}_{i,j} - E[\boldsymbol{X}_{i,j}]\big)\big(\boldsymbol{X}_{i,j} - E[\boldsymbol{X}_{i,j}]\big)^T\big] \\
&= E_1\Big[E_1\big[\boldsymbol{\beta}(\boldsymbol{Z}) \mid \mathcal{G}_{1,i}\big]\,E_2\big[\hat{\boldsymbol{Y}}\,\hat{\boldsymbol{Y}}^T\big]\,E_1\big[\boldsymbol{\beta}(\boldsymbol{Z}) \mid \mathcal{G}_{1,i}\big]^T\Big] \\
&\quad + E_1\Big[\hat{\boldsymbol{\beta}}(\boldsymbol{Z})\,E_2\big[\boldsymbol{Y}\big]\,E_2\big[\boldsymbol{Y}\big]^T\,\hat{\boldsymbol{\beta}}(\boldsymbol{Z})^T\Big]
\end{aligned} \tag{7}$$

with the notation

$$\begin{aligned}
\hat{\boldsymbol{Y}} &= E_2\big[\boldsymbol{Y} \mid \mathcal{G}_{2,j}\big] - E_2\big[\boldsymbol{Y}\big] \\
\hat{\boldsymbol{\beta}}(\boldsymbol{Z}) &= E_1\big[\boldsymbol{\beta}(\boldsymbol{Z}) \mid \mathcal{G}_{1,i}\big] - E_1\big[\boldsymbol{\beta}(\boldsymbol{Z})\big].
\end{aligned} \tag{8}$$

The expectation of $X_{i,j}$ is $E_1\{E_1[\boldsymbol{\beta}(\boldsymbol{Z}) \mid \mathcal{G}_{1,i}]\} \, E_2\{E_2[\boldsymbol{Y} \mid \mathcal{G}_{2,j}]\}$, and these expectations are equal to $E_1[\boldsymbol{\beta}(\boldsymbol{Z})] \, E_2[\boldsymbol{Y}] = E[\boldsymbol{\beta}(\boldsymbol{Z}) \, \boldsymbol{Y}] = E[\boldsymbol{X}]$ by properties of the conditional expectation and the independence of $\boldsymbol{Z}$ and $\boldsymbol{Y}$. Hence, the optimal Galerkin solution $\boldsymbol{X}_{i,j}$ in Equation (4) is an unbiased approximation for the exact solution $\boldsymbol{X} = \boldsymbol{\beta}(\boldsymbol{Z}) \, \boldsymbol{Y}$.

We have $\boldsymbol{X}_{i,j} - E[\boldsymbol{X}_{i,j}] = E_1[\boldsymbol{\beta}(\boldsymbol{Z}) \mid \mathcal{G}_{1,i}] \, \hat{\boldsymbol{Y}} + \hat{\boldsymbol{\beta}}(\boldsymbol{Z}) \, E_2[\boldsymbol{Y}]$ with the notation in Equation (8). The definition of the covariance matrix $\boldsymbol{\gamma}_{i,j}$, the independence of $\boldsymbol{Z}$ and $\boldsymbol{Y}$, and properties of the conditional expectation give the second relation in Equation (7). If $\mathcal{G}_{2,j}$ coincides with $\mathcal{F}_2$, then $\hat{\boldsymbol{Y}}$ in Equation (8) becomes $\boldsymbol{Y} - E_2[\boldsymbol{Y}]$.

In the remainder of the paper we denote σ-fields of the type $\mathcal{G}_{1,i}$, $\mathcal{G}_{2,j}$, and $\mathcal{G}_{1,i} \otimes \mathcal{G}_{2,j}$ by $\mathcal{G}_1$, $\mathcal{G}_2$, and $\mathcal{G} = \mathcal{G}_1 \otimes \mathcal{G}_2$ for simplicity. Generally, we chose the σ-fields $\mathcal{G}_1$, $\mathcal{G}_2$, and $\mathcal{G}$ to be coarser then $\mathcal{F}_1$, $\mathcal{F}_2$, and $\mathcal{F}_1 \otimes \mathcal{F}_2$, that is, we have $\mathcal{G}_1 \subset \mathcal{F}_1$, $\mathcal{G}_2 \subset \mathcal{F}_2$, and $\mathcal{G} \subset \mathcal{F}_1 \otimes \mathcal{F}_2$. Accordingly, the optimal Galerkin solution with respect to $\mathcal{G}$ is (Equation (4))

$$\tilde{\boldsymbol{X}} = E\big[\boldsymbol{X} \mid \mathcal{G}\big] = E_1\big[\boldsymbol{\beta}(\boldsymbol{Z}) \mid \mathcal{G}_1\big] \, E_2\big[\boldsymbol{Y} \mid \mathcal{G}_2\big]. \tag{9}$$

We also consider Galerkin solutions that differ from $E[\boldsymbol{X} \mid \mathcal{G}]$. These solutions are referred to as sub-optimal Galerkin solutions, and are also denoted by $\tilde{\boldsymbol{X}}$. Generally, sub-optimal Galerkin solution are biased approximations of the exact solutions. The error of a sub-optimal Galerkin solution $\tilde{\boldsymbol{X}}$ is

$$\tilde{\boldsymbol{X}} - \boldsymbol{X} = \Big(E\big[\boldsymbol{X} \mid \mathcal{G}\big] - \boldsymbol{X}\Big) + \Big(\tilde{\boldsymbol{X}} - E\big[\boldsymbol{X} \mid \mathcal{G}\big]\Big). \tag{10}$$

The first term in Equation (10) is the error of the optimal Galerkin solution, and cannot be reduced for a given $\mathcal{G}$. The second term in Equation (10) corresponds to the difference between the optimal and a sub-optimal Galerkin solutions. This component of the error can be reduced by improving the sub-optimal solution.

Property 4. *If $\{\Lambda_q\}$, $q = 1, \ldots, m$, is a partition of Ω_1, $\mathcal{G}_1 = \sigma(\{\Lambda_q\})$, and $\mathcal{G}_2 = \mathcal{F}_2$, then the mean, the correlation, and the distribution of the corresponding optimal and sub-optimal Galerkin solutions $\tilde{\boldsymbol{X}}$ have the expressions:*

$$\begin{aligned}
E\big[\tilde{\boldsymbol{X}}\big] &= \sum_{q=1}^{m} E_2\big[\boldsymbol{W}_q\big] \, P_1\big(\Lambda_q\big) \\
E\big[\tilde{\boldsymbol{X}} \, \tilde{\boldsymbol{X}}^T\big] &= \sum_{q=1}^{m} E_2\big[\boldsymbol{W}_q \, \boldsymbol{W}_q^T\big] \, P_1\big(\Lambda_q\big) \\
P\big(\tilde{X}_{i_1} \le \xi_1, \ldots, \tilde{X}_{i_s} \le \xi_s\big) &= \sum_{q=1}^{m} P_2\big(W_{q,i_1} \le \xi_1, \ldots, W_{q,i_s} \le \xi_s\big) \, P_1\big(\Lambda_q\big),
\end{aligned} \tag{11}$$

where $\boldsymbol{W} = \boldsymbol{\alpha}_q \, \boldsymbol{Y}$ is a vector in $\mathbb{R}^d$ with coordinates $W_{q,j}$, $j = 1, \ldots, d$, and

$$\boldsymbol{\alpha}_q = \begin{cases} E\big[\boldsymbol{\beta}(\boldsymbol{Z}) \mid \Lambda_q\big], & \text{for optimal Galerkin solution} \\ \boldsymbol{\beta}_q = \boldsymbol{\beta}(z_q), & \text{for sub-optimal Galerkin solution.} \end{cases} \tag{12}$$

We have

$$\tilde{\boldsymbol{X}} = \Big(\sum_{q=1}^{m} \boldsymbol{\alpha}_q \, 1_{\Lambda_q}\Big), \quad \boldsymbol{Y} = \sum_{q=1}^{m} \boldsymbol{W}_q \, 1_{\Lambda_q} \tag{13}$$

by properties of the conditional expectation and Equation (5). The above sub-optimal Galerkin solution corresponds to an approximate representation of $\boldsymbol{Z}$ setting this variable constant and equal to z_q in each Λ_q, that is, $\boldsymbol{Z}$ is approximated by the simple random variable $\tilde{\boldsymbol{Z}} = \sum_{q=1}^{m} z_q \, 1_{\Lambda_q}$, where 1_{Λ_q} denotes the indicator function for Λ_q defined by $1_{\Lambda_q}(\omega_1) = 1$ and 0 for $\omega_1 \in \Lambda_q$ and $\omega_1 \notin \Lambda_q$, respectively.

The results in Equation (11) follow from Equation (4), by the linearity of the expectation operator and the law of total probability. If $\boldsymbol{Y} = \boldsymbol{z}$ is a deterministic vector, then $E_2[\boldsymbol{W}_q] = \boldsymbol{w}_q = \boldsymbol{\alpha}_q \, \boldsymbol{z}$, $E_2[\boldsymbol{W}_q \, \boldsymbol{W}_q^T] = \boldsymbol{w}_q \, \boldsymbol{w}_q^T$, and $P_2(W_{q,i_1} \le \xi_1, \ldots, W_{q,i_s} \le \xi_s) = 1\big(w_{q,i_1} \le \xi_1, \ldots, w_{q,i_s} \le \xi_s\big)$. If $\boldsymbol{Y}$ is a random vector, then $E_2[\boldsymbol{W}_q] = \boldsymbol{\alpha}_q \, E_2[\boldsymbol{Y}]$ and $E_2[\boldsymbol{W}_q \, \boldsymbol{W}_q^T] = \boldsymbol{\alpha}_q \, E_2[\boldsymbol{Y} \, \boldsymbol{Y}^T] \boldsymbol{\alpha}_q^T$ and the probabilities $P_2(W_{q,i_1} \le \xi_1, \ldots, W_{q,i_s} \le \xi_s)$ can be calculated by, for example, Monte Carlo simulation.

4. Plates on Random Elastic Foundation

Consider a simply supported rectangular plate with unit stiffness sitting on a linear elastic foundation with stiffness $K(x, y)$ and subjected to a spatially distributed load $Q(x, y)$. The displacement $W(x, y)$ of the plate is the solution of the partial differential equation

$$\Delta\Delta W(x, y) + K(x, y) \, W(x, y) = Q(x, y), \quad (x, y) \in D = (0, a) \times (0, b), \tag{14}$$

where $\Delta = \partial^2/\partial x^2 + \partial^2/\partial y^2$ denotes the Laplace operator. The plate displacement satisfies the conditions $W = 0$ on the boundary ∂D of D, $\partial^2 W/\partial x^2 = 0$ on $\{0\} \times (0, b)$ and $\{a\} \times (0, b)$, and $\partial^2 W/\partial y^2 = 0$ on $(0, a) \times \{0\}$ and $(0, a) \times \{b\}$. It is assumed that the foundation stiffness K is random and the applied load Q is deterministic.

The foundation stiffness is modeled by the homogeneous translation field

$$K(x, y) = \alpha_1 + (\alpha_2 - \alpha_1) \, \Phi\big(G(x, y)\big) = \alpha_1 + (\alpha_2 - \alpha_1) \, \Phi\big(\boldsymbol{\theta}(x, y) \, \boldsymbol{Z}\big), \tag{15}$$

where $0 < \alpha_1 < \alpha_2 < \infty$ are some constants,

$$\begin{aligned} G(x, y) &= \sum_{k=1}^{n/2} \sigma_k \left(A_k \cos(\boldsymbol{\nu}_k \cdot (x, y)) + B_k \sin(\boldsymbol{\nu}_k \cdot (x, y)) \right) \\ &= \boldsymbol{\theta}(x, y) \, \boldsymbol{Z}, \quad (x, y) \in D = (0, a) \times (0, b), \end{aligned} \tag{16}$$

is a homogeneous Gaussian field, $n/2 \ge 1$ is an integer,

$$\begin{aligned} \boldsymbol{\theta}(x, y) &= \big[\sigma_1 \cos(\boldsymbol{\nu}_1 \cdot (x, y)) \ldots \sigma_n \cos(\boldsymbol{\nu}_n \cdot (x, y)) \;\; \sigma_1 \sin(\boldsymbol{\nu}_1 \cdot (x, y)) \ldots \sigma_n \sin(\boldsymbol{\nu}_n \cdot (x, y))\big], \\ \boldsymbol{Z}^T &= \big[A_1 \ldots A_{n/2} \; B_1 \ldots B_{n/2}\big], \end{aligned} \tag{17}$$

$\boldsymbol{\nu}_k = (\nu_{k,x}, \nu_{k,y})$ are wave frequencies, $\nu_{k,x}, \nu_{k,y} > 0$ are some constants, $\boldsymbol{\nu}_k \cdot (x, y) = \nu_{k,x} \, x + \nu_{k,y} \, y$, $\sigma_k > 0$ are some constants such that $\sum_{k=1}^{n/2} \sigma_k^2 = 1$, and (A_k, B_k) are independent $N(0, 1)$ variables. We note that K depends on n independent $N(0, 1)$ random variables, the entries of $\boldsymbol{Z}$, and its marginal distribution is uniform in (α_1, α_2). The Gaussian field G has mean 0, variance 1, and covariance function

$$E[G(x, y) \, G(x', y')] = \sum_{k=1}^{n} \sigma_k^2 \cos\big(\boldsymbol{\nu}_k \cdot (x - x', y - y')\big) \tag{18}$$

so that it is homogeneous. The translation field K in Equation (15) with G in Equation (16) is homogeneous and its covariance and correlation functions can be calculated from the probability law of G. In applications the covariance function of K, rather than that of G, is given, and we need to find a Gaussian field G such that K in Equation (15) has the required properties. The existence of G and the determination of the second-moment properties of G, if it exists, are discussed in (Grigoriu, 1995: section 3.1).

4.1. *Galerkin Solution*

The solution of Equation (14) involves two steps. First, Equation (14) is approximated by a stochastic algebraic equation. Second, optimal and sub-optimal Galerkin solutions are developed for this stochastic algebraic equation and some of their statistics are calculated.

The stochastic differential equation for W in Equation (14) with K in Equation (15) depends on the random variable $\boldsymbol{Z}$ in the definition of the random field K. The finite difference approximation of Equation (14) at an interior node has the form

$$\begin{aligned}&\left(20+K_{i,j}\right)W_{i,j}+2\left(W_{i-1,j-1}+W_{i-1,j+1}+W_{i+1,j-1}+W_{i+1,j+1}\right)\\&\quad-8\left(W_{i,j-1}+W_{i,j+1}+W_{i-1,j}+W_{i+1,j}\right)\\&\quad+W_{i,j-2}+W_{i,j+2}+W_{i-2,j}+W_{i+2,j}=Q_{i,j},\end{aligned} \tag{19}$$

where $W_{i,j}$, $K_{i,j}$, and $Q_{i,j}$ denote the values of $W(x, y)$, $K(x, y)$, and $Q(x, y)$ and the node (i, j) of the finite difference mesh. The above relation written at all nodes with adequate modifications to account for boundary conditions yields a stochastic algebraic equation of the type in Equation (1) with coefficients depending on the random variables $\boldsymbol{Z}$ and solution $\boldsymbol{X}$ with entries the displacements $W_{i,j}$.

Since $\boldsymbol{Z}$ in Equation (17) is a standard Gaussian vector in $\mathbb{R}^n$, $n = 12$, it has the representation

$$\boldsymbol{Z} = \boldsymbol{U}\,R, \tag{20}$$

where $\boldsymbol{U}$ is uniformly distributed on the unit sphere $S_n(1)$ in $\mathbb{R}^n$ centered at the origin of this space and R is a real-valued random variable following a chi distribution with n degrees of freedom. The random variable R is independent of $\boldsymbol{U}$ and has the distribution

$$P_1\left(R \le r\right) = I\left(r^2/2, n/2\right), \tag{21}$$

where $I(x, q) = \int_0^x t^{q-1}\, e^{-t}\, dt/\Gamma(q)$ and $\Gamma(q)$ denote the incomplete and the complete gamma functions, respectively.

We begin the construction of a partition $\{\Lambda_q\}$ of Ω_1 by dividing $\mathbb{R}^n$ in rings of radii $0 = r_0 < r_1 < \ldots < r_{n_r} = \infty$ such that each ring has the same probability content, that is,

$$P_1\left(r_{u-1} < R \le r_u\right) = I\left(r_u^2/2, n/2\right) - I\left(r_{u-1}^2/2, n/2\right) = 1/n_r, \quad u = 1, \ldots, n_r. \tag{22}$$

We then divide each ring in subsets of equal volume, and these subsets define the partition $\{\Lambda_q\}$ of Ω_1 used to construct the optimal Galerkin solution. The resulting sets Λ_q have the same probability content, that is, $P_1(\Lambda_q) = 1/(n_r\, n_l)$ for all q, where $n_l \ge 1$ denotes the number of subsets of equal volume in each ring. The conditional expectations $E_1[\boldsymbol{\beta}(\boldsymbol{Z}) \mid \Lambda_q]$ in the definition of the optimal Galerkin solution can be estimated from

$$E_1\left[\boldsymbol{\beta}(\boldsymbol{Z}) \mid \Lambda_q\right] \simeq \frac{1}{n_s}\sum_{u=1}^{n_s} \boldsymbol{\beta}\left(\boldsymbol{Z}(\omega_{1,u})\right) 1_{\Lambda_q}(\omega_{1,u}) \tag{23}$$

where $\{\boldsymbol{Z}(\omega_{1,u})\}$, $u = 1, \ldots, n_s$, are n_s independent samples of $\boldsymbol{Z}$.

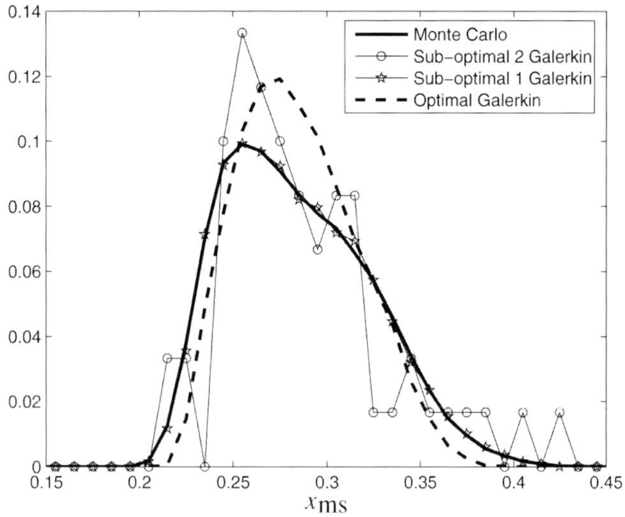

Fig. 1. Probability density function of X_{ms}.

4.2. *Numerical Results*

Numerical results are for $a = 60$, $b = 40$, a finite difference mesh with equal step of 10 in the x and y directions, $Q(x, y) = 1$ in D so that the load vector is $\boldsymbol{Y} = \boldsymbol{y} = \boldsymbol{1}$, and foundation stiffness K with parameters $\alpha_1 = 1$, $\alpha_2 = 7$, $n = 12$, $\sigma_k = 1/\sqrt{6}$, $\boldsymbol{\nu}_1 = -\boldsymbol{\nu}_4 = (1, 2)$, $\boldsymbol{\nu}_2 = -\boldsymbol{\nu}_5 = (2, 1)$, and $\boldsymbol{\nu}_3 = -\boldsymbol{\nu}_6 = (2, 2)$. The resulting stochastic algebraic equation depends on $n = 12$ independent $N(0, 1)$ variables, the entries of $\boldsymbol{Z}$, and has dimension $d = 15$, so that $\boldsymbol{X}$ and $\boldsymbol{Y}$ are vectors in $\mathbb{R}^{15}$.

We use for the optimal Galerkin solution $n_r = 5$ rings and partition of $S_n(1)$ as follows. The sphere $S_2(1)$ has been divided in 4 equal parts. The image of each of this parts in $S_3(1)$ has been divided in two equal parts. The division in two equal parts has been continued to partition the sphere $S_k(1)$, $k = 4, \ldots, 12$. The resulting number of partitions of $S_{12}(1)$ is 4096. Two partitions of Ω_1 have been used to construct sub-optimal Galerkin solutions, the partition $\{\Lambda_q\}$ defining the optimal Galerkin solution with $z_q \in \Lambda_q$ and a coarser partition. The corresponding solutions are referred to as sub-optimal 1 and sub-optimal 2 Galerkin solutions, respectively. The points z_q for the sub-optimal 2 Galerkin are the points of intersections of spheres of radii $(r_{u-1} + r_u)/2$, $u = 1, \ldots, n_r$, with the coordinates of $\mathbb{R}^n$. Each of these spheres intersects the coordinates of $\mathbb{R}^n$ at $2n$ points, so that there are $m = 2\,n\,n_r = 70$ points z_q and subsets Λ_q. The probability of the sets Λ_q is $P_1(\Lambda_q) = 1/(2\,n\,n_r)$ for all q's.

Let X_{ms} and X_{msn} denote the plate displacement at the mid span node of the finite difference mesh and at a node neighboring it, respectively. Figure 1 shows estimates of the probability density function of X_{ms} obtained by Monte Carlo simulation and optimal/sub-optimal Galerkin solutions. The estimate obtained by the sub-optimal 2 Galerkin solution exhibits relatively large fluctuations. The estimates of the density of X_{ms} by the other Galerkin solutions are similar to that obtained by direct Monte Carlo. The density of X_{ms} by the sub-optimal 1 Galerkin solution and Monte Carlo simulation nearly coincide. The Monte Carlo estimates have been calculated from 100,000 independent samples of $\boldsymbol{Z}$ and the Galerkin estimates have been calculated from Equations (11) and (12). Figure 2 shows estimates of the joint probability density of $(X_{\text{ms}}, X_{\text{msn}})$. The estimates of this density by optimal and sub-optimal 1 Galerkin solutions and by Monte Carlo are similar. The sub-optimal 2 Galerkin solution provides a less satisfactory approximation. As in the previous figure, the

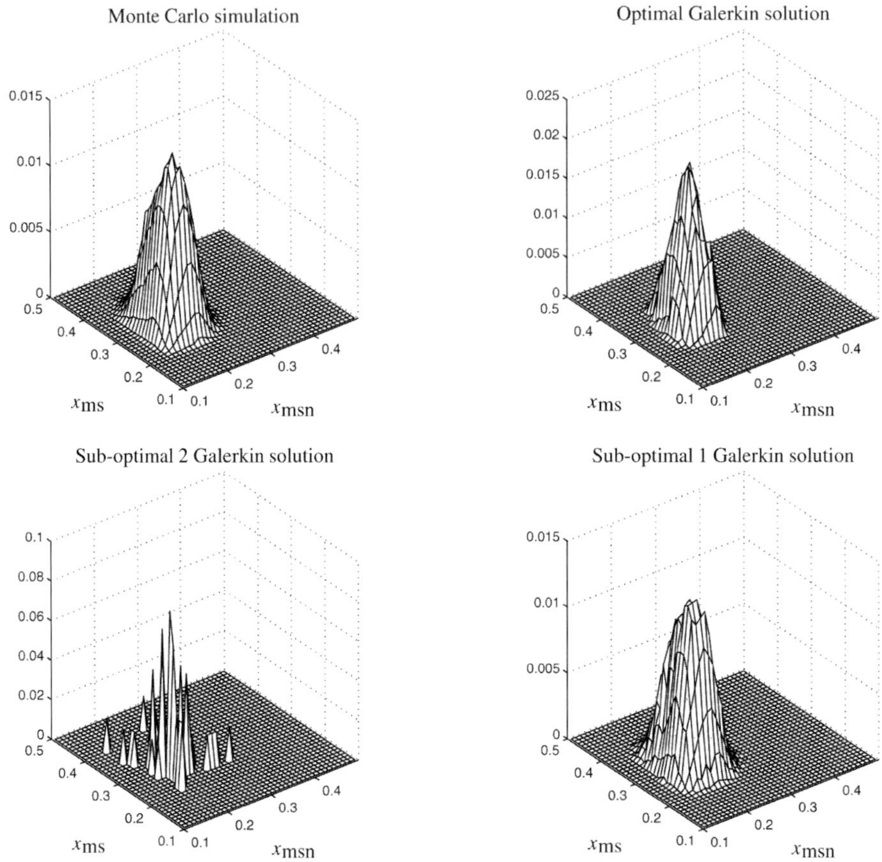

Fig. 2. Joint probability density function of $(X_{\mathrm{ms}}, X_{\mathrm{msn}})$

Monte Carlo estimates are based on 100,000 independent samples of $\boldsymbol{Z}$ and the Galerkin solutions have been calculated from Equations (11) and (12). The differences between the mean displacements by Galerkin solutions and Monte Carlo simulation are less then 0.0958%, 0.1267%, and 1.1056% for optimal, sub-optimal 1, and sub-optimal 2 Galerkin solutions, respectively. These differences and the plots in Figures 1 and 2 show that the optimal Galerkin solution is closer to the Monte Carlo result in the average, consistent with the fact that this solution is unbiased, but the sub-optimal 1 Galerkin solution provides a superior approximation for the probability law of the displacement field.

5. Conclusions

Galerkin solutions have been presented for a class of stochastic algebraic equations, that is, linear algebraic equations with random coefficients. The construction of the Galerkin solution is based on partitions of the sample space associated with the random parameters in the definition of a stochastic algebraic equation. It was shown that there is a Galerkin solution for an arbitrary stochastic algebraic equation that is optimal in the mean square sense. Moreover, the optimal Galerkin solution is equal to the conditional expectation of the exact solution of a stochastic differential equation taken with respect to a σ-field coarser than the σ-field relative to which this solution is measurable.

Galerkin solutions that are not optimal are referred to as sub-optimal. Algorithms are presented for the construction of both optimal and sub-optimal Galerkin solutions.

Optimal and sub-optimal Galerkin solutions have been used to calculate statistics for the displacement of the simply supported plate sitting on a random elastic foundation. The elastic foundation is modeled by a homogeneous translation random field with uniform marginal distribution. The accuracy of the Galerkin solutions depends on the partition of the sample space used in their definition, and was evaluated by Monte Carlo simulation.

References

Babuška, I. M., Temptone, R., and Zouraris, E. (2004). Galerkin finite element approximations of stochastic elliptic partial differential equations. *SIAM Journal of Numerical Analysis*, 42(2), 800–825.

Deb, M. K., Babuška, I. M., and Oden, J. T. (2001). Solution of stochastic partial differential equations using Galerkin finite element techniques. *Computational Methods in Applied Mechanics and Engineering*, 190, 6359–6372.

Field, Jr., R. V. and Grigoriu, M. (2004). On the accuracy of the polynomial chaos approximation. *Probabilistic Engineering Mechanics*, 19(1–2), 65–80.

Ghanem, R. G. and Spanos, P. D. (1991). *Stochastic Finite Elements: A Spectral Approach*. Springer-Verlag, New York.

Grigoriu, M. (1995). *Applied Non-Gaussian Processes: Examples, Theory, Simulation, Linear Random Vibration, and MATLAB Solutions*. Prentice Hall, Englewoods Cliffs, NJ.

Grigoriu, M. (2002). *Stochastic Calculus. Applications in Science and Engineering*. Birkhäuser, Boston.

Resnick, S. I. (1998). *A Probability Path*. Birkhäuser, Boston.

HYPERCUBE POINT CONCENTRATION SAMPLING TECHNIQUE

H.P. Hong
Department of Civil and Environmental Engineering, University of Western Ontario, N5A 6B9 Canada

Abstract

A new sampling technique referred to as the hypercube point concentration sampling technique is proposed. This sampling technique is based on the concepts of the Latin hypercube sampling technique and the point concentration method. In the proposed technique, first, the probability density function of the random variables is replaced by a sufficiently large number of probability concentrations with magnitudes and locations determined from the moments of the random variables. In other words, the probability density function is replaced by the probability mass function determined based on the point estimate method. The probability mass function is then used with the Latin hypercube sampling technique to obtain samples. For evaluating statistics of a complicated performance function of an engineering system, the proposed technique could be more efficient than the Latin hypercube sampling technique since for a given simulation cycle the required number of evaluations of the performance function in the former is less than that in the latter. The proposed sampling technique is illustrated through numerical examples.

Introduction

Many random variables are involved in an engineering system. Direct or simple Monte Carlo method and efficient reliability methods such as the first-order and second-order reliability methods (Madsen et al. 1986) can be employed to carry out probabilistic assessment of the system. The reliability methods are almost exclusively used to estimate the probability of failure while the Monte Carlo method are employed to calculate the statistics of the responses or the probability of failure of the system. Although the direct Monte Carlo method is simple to use, however it can be computationally intensive. To reduce the number of simulation cycles, more efficient simulation methods can be used (Iman and Conover 1980, Rubinstein 1981).

In this study, a new sampling technique is proposed. This new sampling technique is based on the Latin hypercube sampling (LHS) technique and the point estimate method (Rosenblueth 1975, 1981). In the LHS technique (Iman and Conover 1980), the domain of the random variables is partitioned into many mutually exclusive and collectively exhaustive hypercubes. The hypercubes that do not have common domain in subspaces are selected randomly; and a point within each of the selected hypercubes is chosen randomly for the analysis. For the proposed technique, the marginal probability density functions of the random variables are replaced by probability mass functions with a sufficiently large number of concentrations. The locations and their associated probability concentrations are obtained based on Rosenblueth's point estimate method (Rosenlueth 1975, 1981). The probability mass functions are then used with the Latin hypercube sampling technique to generate values of the random variables for the simulation analysis. The advantage of the new technique is that it requires less number of evaluations of the performance function of the system than the LHS technique. This is because several generated samples are likely to fall into a same concentration point. The proposed technique is described in detail in the following, and is illustrated by numerical examples.

M. Pandey et al. (eds), Advances in Engineering Structures, Mechanics & Construction, 527–536.

Background and discussions

Latin Hypercube sampling

Consider that the performance of an engineering system, Z, is a function of a set of s random variables $\boldsymbol{X}$, $\boldsymbol{X} = [X_1, \cdots, X_s]$,

$$Z = h(\boldsymbol{X}), \tag{1}$$

where $h(\bullet)$ is a deterministic function. The mathematical expectation of Z^k or the k-th moments of Z, $E(Z^k)$, is defined as,

$$E(Z^k) = \int_{\Omega} (h(\boldsymbol{x}))^k f_X(\boldsymbol{x}) d\boldsymbol{x}, \tag{2}$$

where Ω represents the domain of $\boldsymbol{X}$, and $f_X(\boldsymbol{x})$ is the joint probability density function of $\boldsymbol{X}$. The mean of Z, m_Z, equals $E(Z)$, and the variance of Z, σ_Z^2, equals $E(Z^2) - m_Z^2$. The probability that Z is less than or equal to a given value z_p, P_f, can be expressed as,

$$P_f = \int_{g \le 0} f_X(\boldsymbol{x}) d\boldsymbol{x}, \tag{3}$$

where $g \le 0$ represents the domain of Z-$z_p \le 0$.

Integrals in Eqs. (2) and (3) can be estimated using the simple simulation technique. Alternatively, the more efficient LHS technique can be employed (Iman and Conover 1980). According to this technique, for practical applications with independent random variables, the generation of Latin hypercube samples of size n could be carried out as follows. The domain of each random variable is divided into n mutually exclusive and collectively exhaustive intervals, and one value is selected randomly in each interval according to the probability distribution of the random variable. A value is randomly selected from the n values for each of the random variables to form the first Latin hypercube sample. The remaining n-1 values for each of the random variables are used to form the second Latin hypercube sample. That is, a value is randomly selected from the n-1 remaining values for each of the random variables to form the second Latin hypercube sample. This process is repeated until n Latin hypercube samples are obtained.

In particular, if the n mutually exclusive and collectively exhaustive intervals have equal probability, the k-th moments of Z is estimated as the average of the sum of the function $(h(\boldsymbol{x}))^k$ evaluated at each of the sample points; and the probability of failure P_f defined in Eq. (3) is approximated by the ratio of the number of sample points where $h(\boldsymbol{x})$-$z_p \le 0$ to n.

Rosenblueth's point estimate method

The point estimate method is developed to evaluate approximately the moments of Z based on the first few statistical moments of $\boldsymbol{X}$ (Rosenblueth 1975, 1981). The method does not require the knowledge of the probability distribution of $\boldsymbol{X}$ except their statistical moments such as the means, standard deviations, correlation coefficients and skewness coefficients. The method basically replaces the original probability density functions of random variables by probability concentrations with magnitudes of the concentrations and locations determined from the moments of the random variables. In particular, if Z is a function of only one random variable X, and two point concentration (i.e., two-point estimate method) is considered for the random variable X, the locations x_j and magnitudes p_j of the concentrations are given by (Rosenblueth 1975, 1981),

$$x_j = m_X + \xi_j \sigma_X, j = 1,2 \tag{4}$$

and,

$$p_j = (-1)^j \xi_{3-j} / \left(2\sqrt{1 + (\lambda_X / 2)^2} \right), \; j = 1,2, \tag{5}$$

where $\xi_j = \lambda_X / 2 + (-1)^{3-i} \sqrt{1 + (\lambda_X / 2)^2}$ and $j = 1,2$, and μ_X, σ_X and λ_X are, respectively, the mean, standard deviation, and coefficient of skewness of X. Using these concentrations the mean of Z, for example, can be approximated by,

$$E(Z) \approx \sum_{i=1}^{n} p_i h(x_i), \tag{6}$$

where $n = 2$. If Z is a polynomial of degree less than 4, Eq. (6) will provide an exact mean of Z.

In general, if n concentrations are considered, x_j and p_j, $j = 1,2,\cdots,n$, of the concentrations can be obtained by solving the following $2n$ equations,

$$\begin{bmatrix} 1 & 1 & \cdots & 1 \\ x_1 & x_2 & \cdots & x_n \\ \vdots & \vdots & \ddots & \vdots \\ (x_1)^{2n-1} & (x_2)^{2n-1} & \cdots & (x_n)^{2n-1} \end{bmatrix} \begin{bmatrix} p_1 \\ p_2 \\ \vdots \\ p_n \end{bmatrix} = \begin{bmatrix} m_0 \\ m_1 \\ \vdots \\ m_{2n-1} \end{bmatrix}, \tag{7}$$

where m_j is the j-th moment of X with respect to the origin. In such a case, the use of n concentrations matches the first $2n$-1 moments of the random variable X. Therefore, if Z is a polynomial of degree $\leq 2n$-1, the estimated mean of Z by using Eq. (6) with n concentrations is exact.

Discussions

Consider the case that Z, $Z = h(X)$, is a function of a random variable X. We discretize the domain of X into two mutually exclusive and collectively exhaustive intervals I_i, $i = 1,2$. The probability of the i-th interval p_{Ii} is given by $p_{Ii} = P(X \in I_i)$. According to the LHS technique we randomly select a sample x_{si}, $i = 1,2$, from I_i and calculate the expected value of Z using,

$$E(Z) \approx \sum_{i=1}^{n} p_{Ii} h(x_{si}), \tag{8}$$

where $n = 2$. This approximation, which uses only two samples, is unlikely to be accurate or satisfactory even for when $h(X)$ is a linear function of X. Clearly, we can overcome this by using Rosenblueth's two-point estimate method but with the following "sampling" interpretation so that we can extend it later. We replace the original probability distribution function of X according to Rosenblueth's two-point estimate method leading to x_j and p_j, $j = 1,2$, given by Eqs. (4) and (5). We divide the space into two intervals such that the cumulative probability distribution function for the i-th interval I_i varies from $\sum_{j=0}^{i-1} p_j$ to $\sum_{j=0}^{i} p_j$ where $p_0 = 0$, i=1 and 2. Therefore, the probability of the i-th interval I_i, p_{Ii}, equals p_i We randomly select a sample x_{si} from the interval I_i which is equal to the location of the i-th concentration x_i obtained according to Rosenblueth's two-point estimate method. Based on this sampling scheme the approximation to $E(Z)$ calculated by using Eq. (8) is identical to Eq. (6) since $p_{Ii},= p_i$, and $x_{si}, = x_i$. As already mentioned such an approximation is exact if $h(X)$ is a polynomial of degree less than 4. Therefore, one has judiciously selected the "sample" points for a sampling technique to become much more efficient. It is noteworthy that a similar observation can be made when n samples are employed.

The efficiency of using n point concentrations depends on the efficiency in solving the system of nonlinear equations (Eq. (7)) to find x_j and p_j, $j = 1,2,\cdots,n$. The well known approach to solve Eq. (7) (Erdelyi et al. 1953) is to first find the zeros (i.e., x_j, $j = 1,2,\cdots,n$) of the polynomial $\sum_{i=1}^{n} c_i x^i = 0$, where $c_n = 1$, and c_j, $j = 0,1,\cdots, n-1$, are obtained from,

$$M_{2n-2}\begin{bmatrix} c_0 & c_1 & \cdots & c_{n-1} \end{bmatrix}^T = -\begin{bmatrix} m_n & m_{n+1} & \cdots & m_{2n-1} \end{bmatrix}^T, \tag{9}$$

where $M_{2n-2} = \begin{bmatrix} m_0 & m_1 & \cdots & m_{n-1} \\ m_1 & m_2 & \cdots & m_n \\ \vdots & \vdots & \ddots & \vdots \\ m_{n-1} & m_n & \cdots & m_{2n-2} \end{bmatrix}$. The solution is then used to find p_i in Eq. (7).

The solution for x_j and p_j, does exist and p_j are larger than zero. This comes directly from a theorem in the orthogonal polynomials and quadrature formulas which state that (Erdelyi et al. 1953, Stroud and Secrest 1966):

For a non-negative weight function $f(x)$ (in our case it represents probability density function), if m_i exist and $\left|M_{2n}\right| \neq 0$, a unique sequence of orthogonal polynomial $\{q_i(x)\}$, $i = 1,\cdots, n$, (except normalization constants) can be constructed. $q_n(x)$ has n distinct roots (abscissas) which lie in the orthogonality interval. Using these n roots we can find the weights p_i such that,

$$\int_{\Omega} f(x)h(x)dx = \sum_{i=1}^{n} p_i h(x_i), \tag{10}$$

is exact if $h(x)$ is a polynomial of degree $\leq 2n-1$. Further, p_i are positive.

In other words, the above says that we can find the abscissas x_i and the positive weights p_i such that,

$$m_k = \int_{\Omega} x^k f(x)h(x)dx = \sum_{i=1}^{n} p_i x_i^k, \quad \text{or} \quad m_k = \sum_{i=1}^{n} p_i x_i^k \quad \text{for } k = 1,2,\cdots,2n-1, \tag{11}$$

which is equivalent to Eq. (7). Therefore, if some commonly used weighting functions can be transformed into the probability density functions by appropriate normalization constants, the obtained abscissas and weights for the Guassian quadrature formulas can be directly transformed into locations and probability concentrations in the point estimate method. There are three well known classical weighting functions associated with the Jacobi integration, the (generalized) Laguerre integration and the Hermite integration. With appropriate normalization constants, these weighting functions can be transformed into the beta distribution, the gamma distribution and the normal distribution. The correspondences between the abscissas and locations and between the weights and the probability concentrations for these cases are listed in Table 1. Note that the Legender integration is a particular case of the Jacobi integration. Values of the abscissas and weights for these quadrature formulas are tabulated in Stroud and Secrest (1966). They can also be calculated using the algorithms given in Press et al. (1992).

If Z, $Z = h(\boldsymbol{X})$, is a function of s random variables, we can generate n samples according to the LHS technique and evaluate $h(\boldsymbol{x})$ at these n sampling points in order to estimate the mean of $E(Z)$. The number of random variables s does not change the fact that we need to evaluate n times the function $h(\boldsymbol{x})$. However, if one uses the point estimate method with n concentrations for each random variable, one need to evaluate n^s times the function $h(\boldsymbol{x})$. This can be extremely large and make the point estimate method unattractive. For example, n^s equals 9765625 for $n = 5$ and $s = 10$. In such a case, one is better off by using the LHS technique or any other simulation techniques. It is noted that the point estimate schemes with less number of concentrations that have reported in the literature (see Hong (1998) and the references listed thereafter) may also be employed. However, none of these schemes can appropriately take into account the cross terms of order higher than 3.

In short, the above indicates that the point estimate method is very efficient if we are interested in a function of only one random variable, while the use of the LHS sampling technique is desirable if the number of random variables is large. In the following we propose a method that takes advantage of both of these methods.

Hypercube point concentration sampling

If Z is a function of s independent random variables, we replace each of the original probability distributions of the random variables by k probability concentrations whose locations and magnitudes are determined by the probability concentration method discussed in the previous section. Therefore, for the i-th random variable X_i, $i = 1,2,\cdots,s$, we have a "discrete" probability distribution function with locations and probability concentrations represented by $x_{i,j}$ and $p_{i,j}$, $j = 1,2,\cdots,k$. For k equal to 5, a schematic representation of the cumulative distribution function for the i-th random variable based on the above discretization is shown in Figure 1.

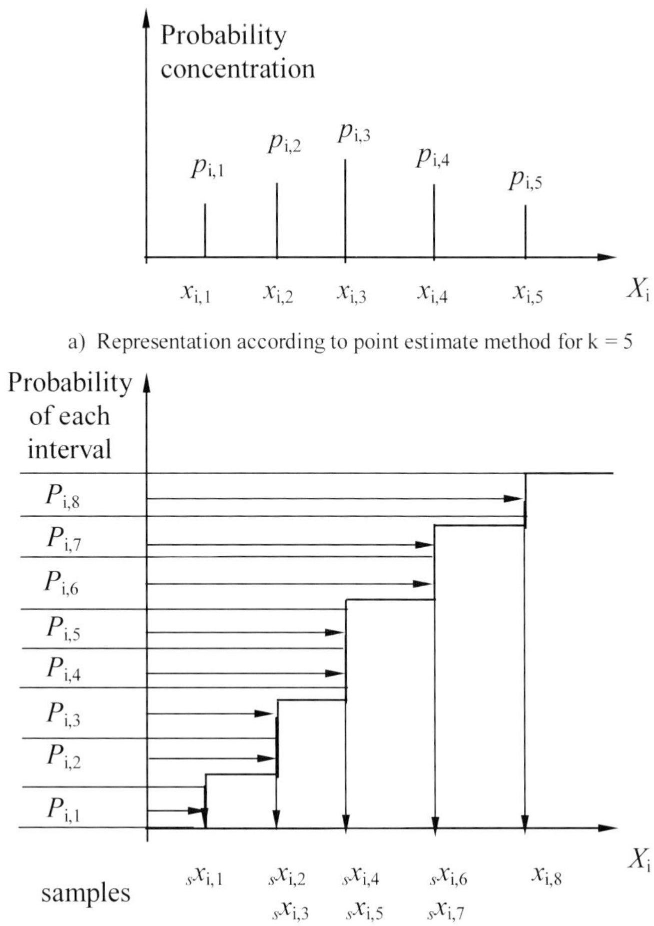

b) Partition and samples for n = 8

Figure 1. Schematic representations.

Now, for the i-th random variable we partition the range of the cumulative probability values (i.e., from 0 to 1) into mutually exclusive and collectively exhaustive n intervals each having a value $P_{i,m}$, $m = 1,2,\cdots,n$. This partition is based on the discrete representation of the probability distribution function defined by ($x_{i,j}$, $p_{i,j}$), $i=1,2,\cdots,s$, and $j=1,2,\cdots,k$. For example, for n equal to 8 and k = 5, this partition is illustrated in Figure 1 again for the i-th random variable. Since there are s random variables, the sample space is partitioned into n^s cells (hypercubes) each with probability $P_{1,m1}\times P_{2,m2}\times P_{3,m3}\times\cdots\times P_{s,ms}$ where $m1$ to ms can take values from 1 to n.

Given a random variable, for each of the partition, we randomly generate a sample. Let ${}_sx_{i,m}$ denote the randomly generated sample from the partition defined by $P_{i,m}$, $m=1,2,\cdots,n$, for the i-th random variable. The selected values ${}_sx_{i,m}$, $m=1,2,\cdots,n$, must coincide with one of the $x_{i,j}$, $j=1,2,\cdots,k$. Again, this is graphically illustrated in Figure 1 (in this particular case ${}_sx_{i,1}$ = $x_{i,1}$, ${}_sx_{i,2}$ = ${}_sx_{i,3}$ = $x_{i,1}$, ${}_sx_{i,4}$ = ${}_sx_{i,5}$ = $x_{i,3}$, ${}_sx_{i,6}$ = ${}_sx_{i,7}$ = $x_{i,4}$, ${}_sx_{i,8}$ = $x_{i,5}$). As in the LHS technique, we use these samples ${}_sx_{i,m}$, where $i=1,2,\cdots,s$ and $m=1,2,\cdots,n$, to form n samples $\boldsymbol{x}_j$, $j=1,2,\cdots,n$, in the s-dimension sample space. That is, the first sample $\boldsymbol{x}_1$ is formed by randomly selecting a value from the n values for each of the random variables. This results in $\boldsymbol{x}_1=\left({}_sx_{1,k1},{}_sx_{2,k2},\cdots,{}_sx_{s,ks}\right)$ and the probability associated with the cell from which the sample $\boldsymbol{x}_1$ was obtained, p_1, equals $P_{1,k1}\times P_{2,k2}\times P_{3,k3}\times\cdots\times P_{s,ks}$ where each of $k1$, $k2$, •••, ks, takes a values from 1 to n. The second sample $\boldsymbol{x}_2$ and the probability associated with the cell from which the sample $\boldsymbol{x}_2$ was obtained, p_2, are formed in the same way but based only on the remaining n-1 values for each of the random variables. This process is continued until $\boldsymbol{x}_n$ and p_n is formed. Since the samples obtained in this way represent corners of the Hypercubes formed by the Point Concentrations we will refer this sampling technique as the HPCS technique.

We can use these samples to estimate the expected value of Z, $E(Z)$ from the statistic, S, defined by

$$S=\sum_{j=1}^{n}n^{s-1}p_jh(\boldsymbol{x}_j),\qquad(12)$$

where p_j is the probability associated with the cell from which the sample $\boldsymbol{x}_j$ was obtained.

To show that the use of the above in estimating $E(Z)$ is adequate, one can shown, following a similar proof given by Iman and Conover (1980), that S is an unbiased estimator of $E(Z)$ when Z = $h(\boldsymbol{x})$ is a polynomial of degree less than $2k$-1 if k concentrations are employed to replace the probability distribution function.

Note that dependent random variables can be transformed into independent random variables by using the Rosenblatt transformation, and that if only the correlation coefficients between the dependent random variables are available (i.e., incomplete information), one could use the Nataf translation system to transform the correlated random variables into uncorrelated random variables (Madsen et al. 1986). For each of the independent random variables if its moments exist we can calculate $x_{i,j}$ and $p_{i,j}$ of the probability concentrations using Eq. (7). Alternatively, to avoid the evaluation of $x_{i,j}$ and $p_{i,j}$ we can transform the independent random variables into uniform, beta, exponential, gamma and/or normal variates since for these distributions $x_{i,j}$ and $p_{i,j}$ are readily available (see Table 1). In the transformed space, we can use the proposed HPCS technique to carry out the probabilistic analysis.

Illustrative numerical examples

Example 1. *Polynomial with two random variables.* Consider a simple example with Z defined by,

$$Z=\sum_{i=1}^{s}X_i^4+\sum_{i=1}^{s}\sum_{j=i+1}^{s}X_i^2X_j^2\qquad(13)$$

where $s = 2$, X_1 and X_2 are independent normally distributed random variables with a mean of zero and standard deviation of one. It can be shown that the exact mean equals 7.

Table 1. Relation between the locations, x_i, and probability concentrations, p_i, in the point estimate method and the abscissas, x_i', and weights, w_i, of well known integration formulas.

Guassian Integration formulas $\int_{\Omega_t} w(t)h(t)dt \approx \sum_{i=1}^{n} w_i h(t_i)$	Point estimate methods $\int_{\Omega_X} f(x)h(x)dx \approx \sum_{i=1}^{n} p_i h(x_i)$
Jacobi integration $w(t) = (1-t)^{\alpha}(1+t)^{\beta}$ $\Omega_t \in (-1,1)$	Beta distribution $f(x) = (b-x)^{\alpha}(a+x)^{\beta} / \left((b-a)^{\alpha+\beta+1} B(\alpha,\beta)\right)$, $\Omega_X \in (a,b)$ $p_i = w_i / \left((b-a)^{\alpha+\beta+1} B(\alpha,\beta)\right), x_i = \left((b-a)t_i + (b+a)\right)/2$
Legender integration $w(t) = 1$, $\Omega_t \in (-1,1)$	Uniform distribution $f(x) = 1/(b-a)$, $\Omega_X \in (a,b)$, $p_i = w_i / 2, x_i = \left((b-a)t_i + (b+a)\right)/2$
Generalized Laguerre integration $w(t) = t^{\alpha} e^{-t}$, $\Omega_t \in (0,\infty)$	Gamma distribution $f(x) = \lambda^{\alpha+1} x^{\alpha} e^{-\lambda x} / \Gamma(\alpha+1)$, $\Omega_X \in (0,\infty)$ $p_i = w_i / \Gamma(\alpha+1), x_i = t_i / \lambda$
Laguerre integration $w(t) = e^{-t}$, $\Omega_t \in (0,\infty)$,	Exponential distribution $f(x) = \lambda e^{-\lambda x}$, $\Omega_X \in (0,\infty)$, $p_i = w_i, x_i = t_i / \lambda$
Hermite integration $w(t) = e^{-t^2}$, $\Omega_t \in (-\infty,\infty)$	Normal distribution $f(x) = \frac{1}{\sqrt{2\pi}\sigma} \exp\left(-\frac{1}{2}\left(\frac{x-m}{\sigma}\right)^2\right)$, $\Omega_X \in (-\infty,\infty)$ $p_i = w_i / \sqrt{\pi}, x_i = m + t_i \sqrt{2}\sigma$

If we use the LHS technique, samples for a typical run with 100 cycles are shown in Figure 2. These samples are obtained by partitioning the domain of each random variable into 100 intervals each with the same probability. Now if we replace the original probability distribution function of each of the random variables by 5 concentrations obtained using the point estimate method, the samples for a typical run with 100 cycles obtained by using the HPCS technique are also shown in Figure 2. Note that by using the HPCS, one need to carry out only 13 evaluations for this particular run because many samples fall into the same point.

By repeating the above analysis many times, the average value and the standard deviation of the mean of Z are shown in Figure 3. The results shown in the figure indicate that the accuracy of the proposed HPCS technique is comparable to that of LHS the technique.

Example 2. *Polynomial with many random variables*. Consider Eq. (13) but with s equal to 5. X_i, $i = 1,2,\cdots,s$, are independent normally distributed variates with a mean of zero and standard deviation of one.

We replace the original probability distribution function of each of the random variables by k concentrations based on the point estimate method. For k equal to 3, 4 and 5, the joint probability distribution of the random variables is represented, respectively, by 243, 1024 and 3125 probability concentrations. Using n partitions with equal probability of $1/n$ for each of the random variables and carrying out simulation analysis using the HPCS, the obtained results are shown in Table 2a. Also shown in the table is the number of evaluations of $h(\mathbf{X})$, n_e, needed by the HPCS for the simulation runs carried out. Clearly, the results show that n_e is less than n

which represents the number of functional evaluations if the LHS technique is employed. The reduction in the computational effort is most significant if k is small and/or n large. Note that in all cases the obtained mean values are close to the exact value which is equal to 25.

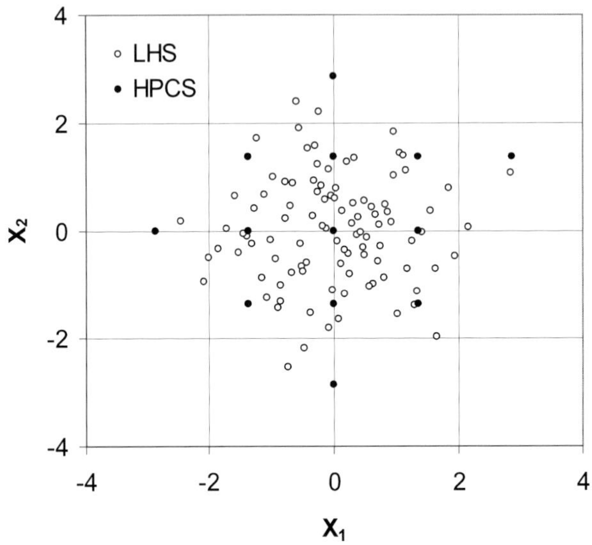

Figure 2. Illustration of sample points for the LHS technique and the HPCS technique.

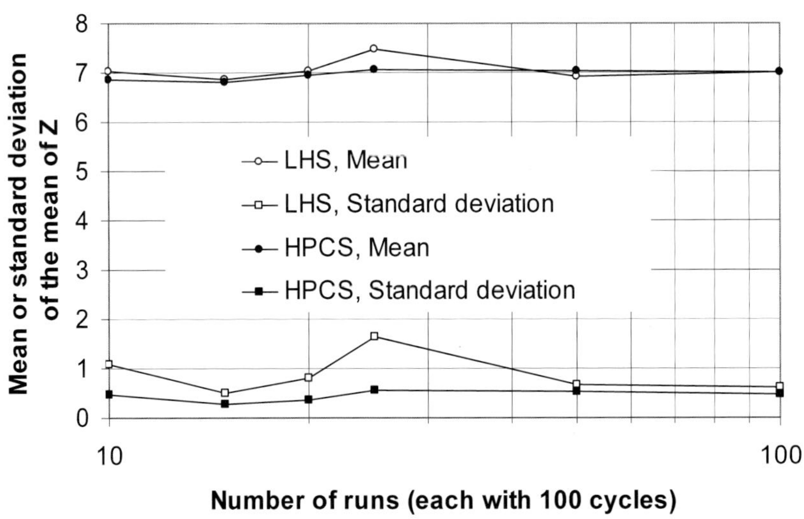

Figure 3. Mean and standard deviation of the mean of the performance function.

Table 2a. Results for Example 2 with $s = 5$.

	$k = 3$		$k = 4$		$k = 5$	
N	n_e	Estimated mean	n_e	Estimated mean	n_e	Estimated mean
100	55	24.66	68	24.47	73	24.45
500	115	24.98	168	24.76	205	24.76
1000	162	25.20	235	25.01	304	25.25
5000	224	24.96	408	24.98	594	24.97
10000	236	25.04	517	24.94	757	25.05

Table 2b. Results for Example 2 with $s = 10$.

	$k = 3$		$k = 4$		$k = 5$	
n	n_e	Estimated mean	n_e	Estimated mean	n_e	Estimated mean
100	94	76.14	98	73.13	100	74.64
500	434	74.30	484	74.63	489	73.92
1000	795	75.15	929	75.32	974	74.90
5000	2915	74.78	3783	74.81	4444	74.75
10000	4713	74.86	6530	74.71	8095	74.83

Repeat this analysis but with $s = 10$. The obtained results are shown in Table 2b. Comparison of the results shown in Tables 2a and 2b suggests that n_e for $s = 5$ is smaller than that for $s = 10$. This is expected since in the latter for k equal to 3, 4 and 5, the joint probability distribution function of the random variables is represented, respectively, by 59049, 1048576, and 9765625 concentrations.

Summary and conclusions

A new sampling technique is proposed in this study. This sampling technique is based on salient features of the Latin hypercube sampling technique and the point concentration method. In this technique, the original probability density function of a random variable is replaced by k probability concentrations determined from the point estimate method. These probability concentrations are then used with the Latin hypercube sampling technique to obtain samples. It is shown that by using this technique an unbiased estimator of the expectation of a performance function that is a polynomial of degree less than $2k$-1 can be obtained. For highly nonlinear functions, the proposed technique provides an approximate estimate and the error is due to terms of order higher than $2k$-1.

Illustrative numerical examples indicate that the proposed technique could be more efficient than the Latin hypercube sampling technique since the former could significantly reduce the required number of evaluations of the performance function. This is particularly important when the numerical evaluation of the performance function is computationally intensive. Also the numerical results suggest that the proposed technique provides relatively stable and accurate results even for highly nonlinear performance functions.

Acknowledgements

The financial support of the Natural Science and Engineering Research Council of Canada is gratefully acknowledged.

References

Erdelyi, A., Magnus, W., Oberhettinger, F. and Tricomi, F. G. (eds.) (1953) *Higher transcendental functions, Bateman manuscript project, Vol. II*, McGraw-Hill, New York.

Hong, H. P. (1996) Point-estimate moment-based reliability analysis, *Civil Engineering Systems*, Vol. 13, pp. 218-294.

Hong, H. P. (1998) An efficient point estimate method for probabilistic analysis, *Reliability Engineering & System Safety*, Vol. 59, No.3, pp. 261-267.

Iman, R. L. and Conover, W. J. (1980) Small sample sensitivity analysis techniques for computer models, with an application to risk assessment, *Commun. Statist. – Theor. Meth.* A9(17):1749-1842

MacGregor, J. G. (1997) *Reinforced concrete, mechanics and design*, Prentice-Hall, Upper Saddle River, NJ.

Madsen, H. O., Krenk, S. and Lind, N. C. (1986) *Methods of structural safety*, Prentice-Hall, Englewood Cliffs, NJ.

Press, W. H., Teukolsky, S. A., Vetterling, W. T. and Flannery, B. P. (1992) *Numerical recipes in FORTRAN: the art of scientific computing*, Cambridge University Press, New York.

Rosenblueth, E. (1981) Two-point estimates in probability, *Appl. Math. Modelling*, 5: 329-335.

Rosenblueth, E. (1975) Point estimation for probability moments. *Proc. Nat. Acad.. Sci.* U.S.A., **72**, 3812-3814.

Rubinstein, R. Y. (1981) *Simulation and Monte Carlo method*, John Wiley and Sons Inc., New York.

Stroud, A. H. and Secrest, D. (1966) *Gaussian quadrature formulas*, Prentice-Hall, Englewood Cliffs, NJ.

ASYMPTOTICALLY-CORRECT STRUCTURAL MODELLING OF THIN-WALLED ANISOTROPIC CLOSED CROSS-SECTION ROTATING SLENDER BEAMS

F. Khouli, F.F. Afagh and R.G. Langlois
Department of Mechanical and Aerospace Engineering,
Carleton University, Ottawa, Ontario, Canada
E-mail: fafagh@mae.carleton.ca

Abstract

An application of a comprehensive and compact methodology to obtain the asymptotically-correct stiffness matrix of anisotropic, thin-walled, closed cross-section, and rotating slender beams is presented. The Variational Asymptotic Method (VAM), which utilizes small geometrical parameters inherent to thin-walled slender beams, is used to obtain the displacement and strain fields, and the cross-sectional stiffness matrix without any ad hoc assumptions. The advantage of this approach is that the asymptotically-correct and populated 4×4 cross-sectional stiffness matrix provides all the necessary information about the elastic behavior of the rotating beam, thereby nullifying the need for refined beam theories that incorporate higher order deformation modes, like the Vlasov's mode. The implementation of the theory using MATLAB was validated against the Vartiational Asymptotic Beam Sectional Analysis (VABS) computer software, a two-dimensional finite element program that utilizes a more general approach to the VAM that is applicable to thick/thin-walled anisotropic cross-sections with arbitrary geometry. Sample applications of the theory to rotor blades are presented. The paper concludes with a discussion of how the presented material would be used directly in the dynamic modelling of rotating helicopter blades.

Introduction

The structural dynamics modelling of rotating composite blades closely follows the advances made in capturing the elastic behavior of anisotropic, slender, and rotating beams with arbitrary cross-sectional geometry. In the past, the highly coupled structural and dynamics aspects of the model along with its strong nonlinearity proved to be an unyielding obstacle, and engineers were forced to adopt various approximations that limited the scope and applications of their models (Volovoi et al., 2001), and were later proved to be less than stellar in predicting the behavior of the rotating blades (Hodges and Patil, 2005, Volovoi et al., 2001). Some modifications were added to these models, which improved their performance, but not to the levels required for today's advances in composite and highly flexible blades. However, these modified preliminary models are still used by the industry today despite existing limitations. One may attribute this to be the result of their simplicity in terms of their formulation, and the experience and insight into their functioning that has been accumulated over time. Performing 3D finite element analysis on the rotating composite blades is an expensive and an unfeasible option even with current computational capabilities, especially when the analysis is directed at devising vibration control strategies or studying the structural dynamics interaction of the blades with other rigid/flexible multibody systems.

Recent advances in the cross-sectional modelling of anisotropic composite beams with arbitrary geometry is a major triumph in overcoming the difficulties discussed above. It was found that for slender beams, asymptotical analysis of the 3D elastic energy can split the problem into a two-dimensional analysis over the cross-section and one-dimensional analysis along the span of the

M. Pandey et al. (eds), Advances in Engineering Structures, Mechanics & Construction, 537–547.

beam without any ad hoc assumptions. Utilizing other geometrical design aspects of the rotor blade, one can also arrive at a closed form solution of the stiffness properties of the cross-section. This significantly reduces the amount of effort the engineer has to spend on modelling the elastic behavior of the blade, since the asymptotically-correct development is found to be surprisingly compact for a problem that seemed to be impossibly complex from the earlier models' point of view.

The VAM was originally developed by Berdichevsky (1982) for elastic slender rods that have an inherent small dimensionless parameter, which is the slenderness ratio defined by the ratio of the characteristic dimension of the cross-section, 'a', to the elastic deformation wavelength, 'l'. Since only global elastic deformation modes that propagate along most of the beam span are of interest here, the deformation length is always of the order of the length of the beam. The theory was refined by Hodges and Cesnik (1994) and implemented in Variational Beam Sectional Analysis (VABS), a software package that utilizes the finite element method to obtain the elastic constants of any composite cross-sections with initial twist. VABS has been extensively validated against experimental data and results from other reliable 3D finite element software like ABAQUS and NASTRAN (Yu et al., 2002). Concurrently, Badir (in Berdichevsky et al., 1992) expanded the theory to thin-walled composite beams that have an additional small dimensionless parameter, which is the thinness ratio defined by the ratio of the thickness of the wall, 'h', to the characteristic dimension of the cross-section, 'a'. This allowed for a simple closed form solution for the stiffness constants, which has been used to model rotor blades with active materials (Cesnik and Shin, 1998, 2001a, 2001b). However, it was later found that Badir's work neglected the shell bending strain measure, which made it asymptotically-incorrect and produced results inconsistent with those produced by VABS for certain cross-sections. Hodges and Volovoi (2000) identified and corrected this flaw, and developed the asymptotically-correct theory for anisotropic thin-wall beams.

Rotor blades can be idealized as thin-walled closed cross-section beams while retaining high degree of fidelity. The asymptotically-correct theory has been successfully implemented using MATLAB to arrive at the elastic properties of any closed cross-section including helicopter rotor blades, which are at the centre of many research efforts of the Applied Dynamics Group at Carleton University. The versatility of the implementation allows it to obtain the elastic properties about any desired point in the plane of the cross-section like the elastic axis of the blade. Additionally, it allows for any desired material composition and distribution throughout the cross-section.

Theory

The Variational Asymptotic Method

A brief symbolic outline of the VAM and its general features is presented to complement the subsequent discussion. Let the 3D elastic energy of the beam be symbolically defined by the energy functional $\mathcal{F}$ with the small parameter that is for now called η such that (Cesnik, 1994):

$$\begin{aligned} \mathcal{F}(\Gamma,\eta) &= \mathcal{E}_1(\Upsilon,z_1)+\mathcal{E}_{\eta 0}(\Upsilon,z_1,\eta), \\ z_1 &= \begin{bmatrix} w_{11} & w_{12} & w_{13} \end{bmatrix}, \end{aligned} \tag{1}$$

where Γ is a 6×1 column matrix that represents the 3×3 symmetric Biot–Jaumann strain field, and Υ is a function of the axial coordinate along the span of the beam only (in this case it corresponds to Γ) (Danielson and Hodges, 1987).

The energy functional $\mathcal{F}$ is decomposed into two parts: $\mathcal{E}_1(\Upsilon,z_1)$ which contains all terms of order $\eta^0\equiv 1$ and $\mathcal{E}_{\eta 0}(\Upsilon,z_1,\eta)$ that contain all terms of order η^1 and higher with respect to this small parameter. The vector z_1 represents a perturbation in the classical 3D displacement field, which in reality is the in/out-of-plane warping functions to a first correction, that gives rise to low and high order terms as is apparent from its appearance in both parts of the energy functional. In order to find the first correction to the displacement field, z_1, the high-order component of the functional is discarded and then the functional is minimized with respect to z_1. The solution of the Euler minimization problem is not unique and the displacement field is four times redundant. The four rigid body modes have to be eliminated from z_1 (the warping field) over the surface area of the cross-section S, therefore, the following four constraints are imposed:

$$\begin{aligned} \min_{z_1} \mathcal{F} &= \min_{z_1} \mathcal{E}_1(\Upsilon, z_1), \\ \int_S \mathcal{C}_1(z_1)\,\mathrm{d}s = 0, \quad \int_S \mathcal{C}_2(z_1)\,\mathrm{d}s &= 0, \quad \int_S \mathcal{C}_3(z_1)\,\mathrm{d}s = 0, \quad \int_S \mathcal{C}_4(z_1)\,\mathrm{d}s = 0. \end{aligned} \tag{2}$$

When Equations (2) are solved over the cross-section they yield what is called the "zeroth-approximation" or the building block of the solution, z_1. This must not be confused with the order of the components of z_1 itself, which could be of some order of η, but rather it refers to it being obtained by minimizing the part of the energy that has zeroth order of η (i.e., η^0). In most cases, there is no closed form solution for z_1, and the problem is discretized over the cross-section with the constraints leading to a Sturm–Liouville problem followed by finite element calculations, which is the methodology of VABS (Cesnik et al., 1993, Cesnik and Hodges, 1997). The order of the components of z_1 is not known a priori but determined throughout the minimization procedure. The solution of Euler equations of the functional is symbolically written as:

$$z_1 = f_1(\Upsilon, \zeta_2, \zeta_3), \tag{3}$$

where ζ_2 and ζ_3 are the perpendicular axes defining the 2D Cartesian plane of the cross-section.

The displacement field is then perturbed again. Let the new perturbation be called z_2 such that:

$$z_2 = z_1 - f_1(\Upsilon, \zeta_2, \zeta_3). \tag{4}$$

The new perturbation is substituted back into the energy functional of Equation (1) to obtain:

$$\begin{aligned} \mathcal{F}(\Gamma, \eta) &= \mathcal{F}_1(\Upsilon) + \mathcal{E}_2(\Upsilon, z_2) + \mathcal{E}_{\eta 1}(\Upsilon, z_2, \eta), \\ z_2 &= \begin{bmatrix} w_{21} & w_{22} & w_{23} \end{bmatrix}. \end{aligned} \tag{5}$$

The function $\mathcal{F}_1(\Upsilon)$ represents all the terms that do not contain the new unknown, z_2. It is subscripted with 1 to indicate that it contains contributions from the first correction to the displacement field z_1. The function $\mathcal{E}_2$ contains the lowest-order terms involving z_2, while $\mathcal{E}_{\eta 1}$ contains all high-order terms.

Following the same procedure as before, the high order terms (i.e., $\mathcal{E}_{\eta 1}$) are discarded and the functional is minimized with respect to z_2 subject to the same constraints:

$$\begin{aligned} \min_{z_2} \mathcal{F} &= \mathcal{F}_1(\Upsilon) + min_{z_2} \mathcal{E}_2(\Upsilon, z_2) \\ \int_S \mathcal{C}_1(z_2)\,\mathrm{d}s = 0, \quad \int_S \mathcal{C}_2(z_2)\,\mathrm{d}s &= 0, \quad \int_S \mathcal{C}_3(z_2)\,\mathrm{d}s = 0, \quad \int_S \mathcal{C}_4(z_2)\,\mathrm{d}s = 0. \end{aligned} \tag{6}$$

Similarly,

$$z_2 = f_2(\Upsilon, \zeta_2, \zeta_3). \tag{7}$$

The process is repeated until the new perturbation yields no terms in the energy functional of order that is of the highest yielded by the previous perturbation, and at this point the displacement field is said to have converged. For example, assume that the perturbation z_k produced terms in the energy

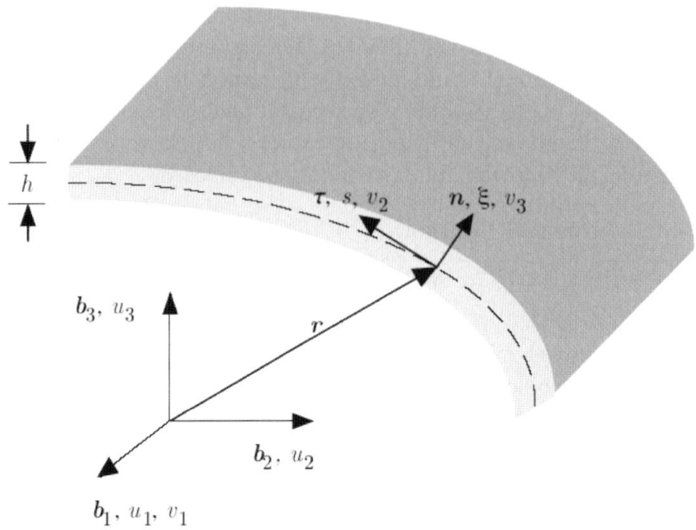

Fig. 1. Shell/thin-walled bar coordinate systems (Hodges and Volovoi, 2000).

functional with one being the highest order having an order of $\mathcal{O}(n)$. A further perturbation, z_{k+1}, produces terms that are of order $\mathcal{O}(n+1)$ and higher, then at this stage the iteration is terminated. In fact, one may not have to go as far as this in order to obtain the correct elastic behavior of the beam since η is a small parameter to start with. Terms of order $\mathcal{O}(\eta^0 \equiv 1)$ are the only ones needed to obtain the asymptotically-correct global elastic behavior of the beam.

The energy functional can then be written after the kth perturbation as:

$$\mathcal{F}(\Gamma, \eta) = \mathcal{F}_1(\Upsilon) + \mathcal{F}_2(\Upsilon) + \cdots + \mathcal{F}_k(\Upsilon) + \mathcal{E}_{k+1}(\Upsilon, z_{k+1}, \eta) + \mathcal{E}_{\eta k}(\Upsilon, z_{k+1}, \eta). \tag{8}$$

Alternatively, the energy functional can be expanded implicitly as an asymptotic series in the small parameter η:

$$\mathcal{F}(\Gamma, \eta) = \mathcal{O}(\eta^0) + \mathcal{O}(\eta^1) + \mathcal{O}(\eta^2) + \mathcal{O}(\eta^3) + \cdots + \mathcal{O}(\eta^k). \tag{9}$$

It must be evident by now that no ad hoc assumptions are made in order to arrive at the asymptotically-correct stiffness matrix that can be extracted from terms of order $\mathcal{O}(\eta^0)$. The process of dropping high order terms during the minimization or truncating them from the asymptotic expansion of the functional is equivalent to a correct and systematic neglect of insignificant terms in more conventional solution methods that rely on the Theory of Elasticity, equilibrium equations, and boundary conditions. But these conventional methods usually involve a multitude of partial differential equations, rendering the identification of these insignificant terms extremely difficult if not outright impossible.

The Asymptotically-Correct Stiffness Matrix

Hodges and Volovoi (2000, 2002) applied the VAM to thin-wall open/closed shells and strips depicted in Figure 1, and obtained the asymptotically-correct stiffness matrix for each case respectively. The intersection of the middle surface of the shell with the plane of the beam cross-section defines the contour s, which does not change along the span of the beam.

The energy functional of the composite beam in this case is the elastic shell energy per unit area written in terms of the six generalized shell strain measures:

$$\begin{aligned} 2\mathcal{U}_{\text{shell}} &= \psi^T Q \psi + 2\phi^T S \psi + \phi^T P \phi \\ \psi^T &= \begin{bmatrix} \gamma_{11} & h\rho_{11} & h\rho_{12} \end{bmatrix} \\ \phi^T &= \begin{bmatrix} 2\gamma_{12} & \gamma_{22} & h\rho_{22} \end{bmatrix}, \end{aligned} \tag{10}$$

where Q, S, and P are material property matrices that depend on the 2D elastic modulus tensor of the material, which is obtained from the reduced stiffness coefficients Q_{mn} of Classical Laminate Theory (Badir, 1992).

The six generalized shell strain measures are given in terms of the curvilinear displacements, which in turn can be expressed in terms of the cartesian displacements u_1, u_2, and u_3:

$$\begin{aligned} \gamma_{11} &= v_{1,1} & \rho_{11} &= v_{3,11}, \\ 2\gamma_{12} &= v_{1,2} + v_{2,1} & \rho_{12} &= v_{3,12} + \tfrac{1}{4R}(v_{1,2} - 3v_{2,1}), \\ \gamma_{22} &= v_{2,2} + \tfrac{v_3}{R} & \rho_{22} &= v_{3,22} - (\tfrac{v_2}{R})_{,2}, \end{aligned} \tag{11}$$

The subscripted comma indicates a differentiation with respect to the curvilinear coordinate subsequent to it, and R is the radius of curvature defined using the cartesian coordinates x_2 and x_3 along b_2 and b_3 respectively as $R = \dot{x}_2/\ddot{x}_3 = -\dot{x}_3/\ddot{x}_2$. The over-dot implies a differentiation with respect to the curvilinear coordinate s. The generalized strain measures in Equation (11) are substituted in the shell energy in Equation (10) and integrated over the contour to give the functional:

$$\oint 2\mathcal{U}_{\text{shell}}\, \mathrm{d}s \equiv \mathcal{F}(\psi, \phi). \tag{12}$$

It is reemphasized in this problem, that two small dimensionless parameters are present: the slenderness ratio $\eta_1 = a/l$, and the thinness ratio $\eta_2 = h/a$. All deformation modes are assumed to have the same order of magnitude $\mathcal{O}(\epsilon)$, and therefore they are all included in the development to make it general: extension, torsion, bending about the b_2 axis, and bending about the b_3 axis.

An asymptotic expansion of the shell energy functional in terms of the small parameters is sought in the form:

$$\begin{aligned} \oint 2\mathcal{U}_{\text{shell}} &= \mathcal{O}(\epsilon^2 \cdot \eta_1^0 \cdot \eta_2^0) + \mathcal{O}(\epsilon^2 \cdot \eta_1^1 \cdot \eta_2^0) + \\ &\quad + \mathcal{O}(\epsilon^2 \cdot \eta_1^0 \cdot \eta_2^1) + \mathcal{O}(\epsilon^2 \cdot \eta_1^1 \cdot \eta_2^1) + \cdots + \mathcal{O}(\epsilon^2 \cdot \eta_1^i \cdot \eta_2^j). \end{aligned} \tag{13}$$

Only the first term in Equation (13) is retained, since it is the dominant one, to give the elastic energy per unit length:

$$\mathcal{E} = \varepsilon^{\mathrm{T}} \overline{S} \varepsilon, \tag{14}$$

where $\varepsilon^{\mathrm{T}} = [u_1' \;\; \theta' \;\; -u_3'' \;\; u_2'']$, which is the classical linear strain measures vector.

The matrix $\overline{S}$, which has a closed form solution given in Hodges and Volovoi (2000, 2002), is the 4×4 asymptotically-correct stiffness matrix, which is beyond the Euler–Bernoulli Theory of bending and St. Venant's Theory of torsion in terms of its rigor. It takes into account all in-plane warping deformations since γ_{12} is never assumed to be zero throughout the development, in addition to the shell bending strain measure ρ_{22}. Given the natural boundary conditions on the tip of the anisotropic thin-wall beam, the correct elastic response can be obtained from the flexibility matrix, which is the inverse of the stiffness matrix:

$$\varepsilon = \overline{S}^{-1} \begin{bmatrix} F_1 \\ M_{\theta'} \\ M_2 \\ M_3 \end{bmatrix}. \tag{15}$$

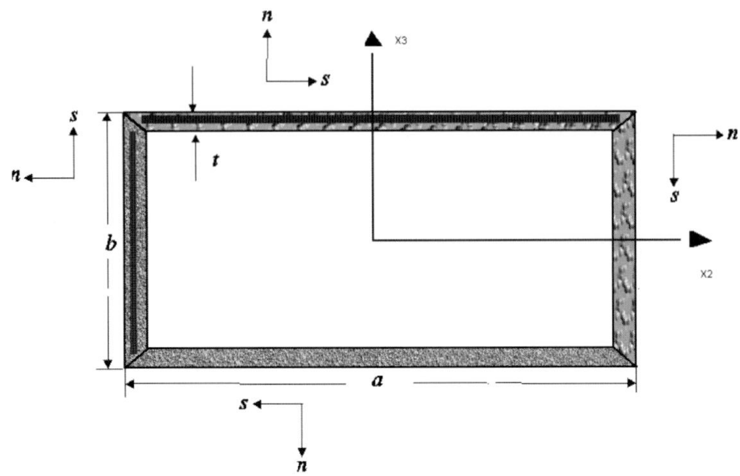

Fig. 2. Anisotropic single-cell box beam (Yu, 2005).

Table 1. The elements of the stiffness matrix obtained using the thin-walled anisotropic beam theory and VABS for Figure 2.

Stiffness Element	Thin-Walled	VABS (Popescu and Hodges, 2000)	Difference
S_{11}	1.2185×10^6 lb	1.2500×10^6 lb	2.52%
S_{12}	0.0522×10^6 lb – in	0.0521×10^6 lb – in	0.19%
S_{13}	0 lb – in	0 lb – in	Exact
S_{14}	0 lb – in	0 lb – in	Exact
S_{22}	0.1730×10^5 lb – in^2	0.1770×10^5 lb – in^2	2.26%
S_{23}	0 lb – in^2	0 lb – in^2	Exact
S_{24}	0 lb – in^2	0 lb – in^2	Exact
S_{33}	0.0508×10^6 lb – in^2	0.0543×10^6 lb – in^2	6.44%
S_{34}	0 lb – in^2	0 lb – in^2	Exact
S_{44}	0.1283×10^6 lb – in^2	0.1340×10^6 lb – in^2	4.25%

Implementation

Validation

The theory has been implemented in a MATLAB software module and validated against VABS. The first validation case was for a single-cell composite box beam VABS example shown in Figure 2.

The elements of the 4×4 stiffness matrix, $\overline{S}$, obtained from the implementation are compared to those produced by VABS in Table 1. Not all elements are shown since the stiffness matrix is symmetric.

The three-cell isotropic box beam shown in Figure 3 is another validation of the current implementation against VABS, where the stiffness matrix is now calculated about the lower-left corner of cross-section.

Similarly, The elements of the 4×4 stiffness matrix, $\overline{S}$, obtained from the implementation are compared to those produced by VABS in Table 2.

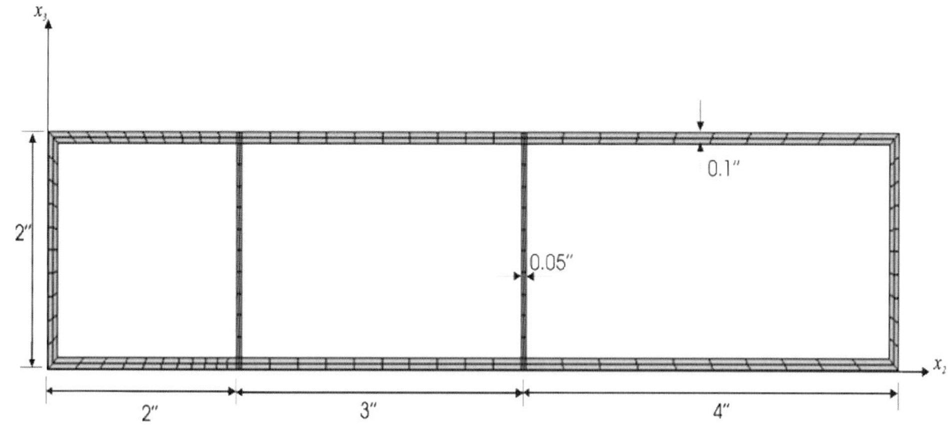

Fig. 3. Isotropic three-cells box beam (Yu, 2005).

Table 2. The elements of the stiffness matrix obtained using the thin-walled anisotropic beam theory and VABS for Figure 2.

Stiffness Element	Thin-Walled	VABS (Yu, 2005)	Difference
S_{11}	0.0611×10^{13} lb	0.0608×10^{13} lb	0.49%
S_{12}	0 lb − in	0 lb − in	Exact
S_{13}	0.0611×10^{13} lb − in	0.0608×10^{13} lb − in	0.49%
S_{14}	-0.2699×10^{13} lb − in	-0.2692×10^{13} lb − in	0.26%
S_{22}	0.0515×10^{13} lb − in^2	0.0540×10^{13} lb − in^2	4.63%
S_{23}	0 lb − in^2	0 lb − in^2	Exact
S_{24}	0 lb − in^2	0 lb − in^2	Exact
S_{33}	0.1073×10^{13} lb − in^2	0.1069×10^{13} lb − in^2	0.37%
S_{34}	-0.2699×10^{13} lb − in^2	-0.2692×10^{13} lb − in^2	0.26%
S_{44}	1.7091×10^{13} lb − in^2	1.7072×10^{13} lb − in^2	0.11%

The results presented in Tables 1 and 2 clearly validate the present implementation to model single/multi-cell thin-walled anisotropic cross-sections of rotor blades found on actual helicopters and turbine engines.

Applications

A two-cell anisotropic rotor blade with a SIKORSKY DBLN-526 Airfoil cross-section is the first example and it is depicted in Figure 4. Differences in the laminate design are highlighted by different line formats in Figure 4 including the webs, and they are given in Tables 3–5.

The solid-line region has elastic modula values that are the same as those in Table 3 but the wall thickness is 0.03 in, and the laminate stacking sequence is $[(40°/-40°/(30°/0°)_2)]$.

The stiffness matrix of this cross-section about the origin of the axes in Figure 4 was found to be:

$$\overline{S} = \begin{bmatrix} 0.4334 \times 10^{12}\ \text{lb} & 0.9970 \times 10^{7}\ \text{lb}-\text{in} & 0.3120 \times 10^{12}\ \text{lb}-\text{in} & -0.1725 \times 10^{12}\ \text{lb}-\text{in} \\ 0.9970 \times 10^{7} & 0.1094 \times 10^{9}\ \text{lb}-\text{in}^2 & 0.7177 \times 10^{7}\ \text{lb}-\text{in}^2 & -0.4463 \times 10^{8}\ \text{lb}-\text{in}^2 \\ 0.3120 \times 10^{12} & 0.7177 \times 10^{7} & 1.2372 \times 10^{12}\ \text{lb}-\text{in}^2 & -0.1792 \times 10^{10}\ \text{lb}-\text{in}^2 \\ -0.1725 \times 10^{12} & -0.4463 \times 10^{8} & -0.1792 \times 10^{10} & 1.3014 \times 10^{12}\ \text{lb}-\text{in}^2 \end{bmatrix}.$$

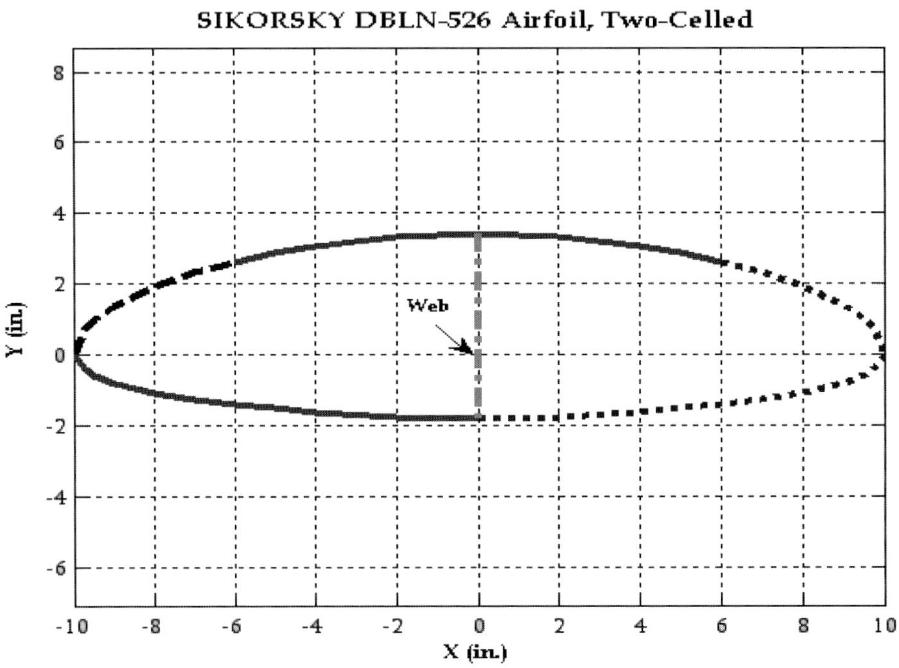

Fig. 4. Two-cell anisotropic SIKORSKY DBLN-526 Airfoil cross-section.

Table 3. Laminate design: dashed-line region in Figure 4.

Parameter	Value
wall thickness 'h'	0.01 in
laminate design	$[(0^\circ/90^\circ)]$, ply thickness = 0.005 in
E_l	20.59×10^6 psi
E_t	1.42×10^6 psi
G_{lt}	8.7×10^5 psi
G_{tn}	6.96×10^5 psi
$\nu_{lt} = \nu_{tn}$	0.42

Table 4. Laminate design: dotted-line region in Fig 4.

Parameter	Value
wall thickness 'h'	0.03 in
laminate design	$[45^\circ/60^\circ/-45^\circ/0^\circ/60^\circ/45^\circ]$, ply thickness = 0.005 in
E_l	0.11×10^{12} psi
E_t	0.02×10^{12} psi
G_{lt}	0.27×10^{11} psi
G_{tn}	0.96×10^{11} psi
ν_{lt}	0.37
ν_{tn}	0.42

Table 5. Laminate design: web region in Figure 4.

Parameter	Value
wall thickness 'h'	0.3 in
laminate design	[60°], ply thickness = 0.3 in
E_l	0.26×10^{12} psi
E_t	0.26×10^{12} psi
G_{lt}	0.1×10^{11} psi
G_{tn}	0.96×10^{11} psi
$\nu_{lt} = \nu_{tn}$	0.3

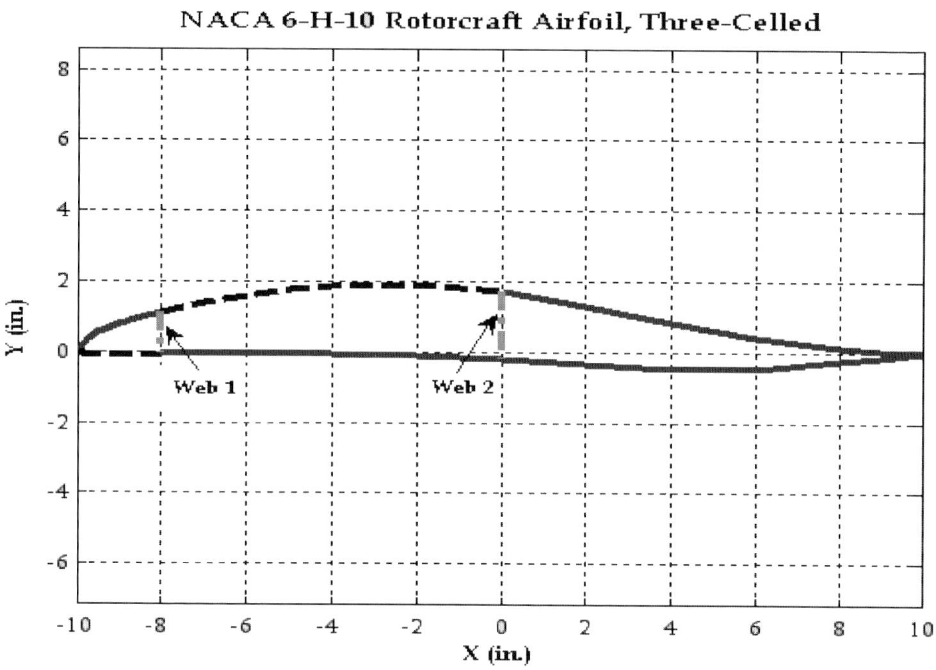

Fig. 5. Three-cells anisotropic NACA 6-H-10 Airfoil cross-section.

A second application example is the three-cells rotor blade cross-section shown in Figure 5.

The laminate design convention is the same as that for the previous example. The only exception is the second web, which has its ply fibre orientation at $-60°$ instead of $60°$ as in the first web. Similarly, the stiffness matrix of this cross-section about the origin of the axes in Figure 5 was found to be:

$$\overline{S} = \begin{bmatrix} 0.4018 \times 10^{11}\ \text{lb} & 0.7104 \times 10^{6}\ \text{lb} - \text{in} & 0.2683 \times 10^{11}\ \text{lb} - \text{in} & 1.2344 \times 10^{11}\ \text{lb} - \text{in} \\ 0.7104 \times 10^{6} & 0.2818 \times 10^{7}\ \text{lb} - \text{in}^2 & 0.1652 \times 10^{6}\ \text{lb} - \text{in}^2 & -0.4188 \times 10^{7}\ \text{lb} - \text{in}^2 \\ 0.2683 \times 10^{11} & 0.1652 \times 10^{6} & 0.2769 \times 10^{11}\ \text{lb} - \text{in}^2 & 0.6500 \times 10^{11}\ \text{lb} - \text{in}^2 \\ 1.2344 \times 10^{11} & -0.4188 \times 10^{7} & 0.6500 \times 10^{11} & 9.8795 \times 10^{11}\ \text{lb} - \text{in}^2 \end{bmatrix}.$$

Advantages

One must enumerate the advantages of this implementation in light of the fact that it will be utilized in tackling the dynamics of rotating blades for various research efforts in the Applied Dynamics Group at Carleton University. Relative to other specialized cross-sectional finite element packages like VABS and ABAQUS, the thin-walled asymptotic theory offers the following advantages:

- It provides a vast design space for parametric studies and easily interfaces with other disciplines like dynamics and control, which may necessitate keeping the information about the elastic deformation in a maximally compressed, yet correct, form.
- The meshing step and preparation of special input files are not required. The only input information required is the geometry of the airfoil (contour of the cross-section), which is readily available for any airfoil in simple and compact format, as well as the laminate design.
- As the thinness ratio increases, the analytical method becomes superior to VABS from the numerical point of view, since the finite element method results become unstable for high aspect ratio elements.
- It allows all but the essential variables to be eliminated offering a simple output and almost instantaneous execution time compared to more specialized software packages.

Conclusion

An implementation of a compact, yet comprehensive, asymptotically-correct anisotropic thin-wall theory to calculate the cross-sectional elastic constants of slender beams with airfoil-like cross-sections has been presented. It represents an essential step in the research efforts in the Applied Dynamics Group at Carleton University to model the structural dynamics of active rotor systems in maritime applications.

References

Badir, A.M. (1992). *Analysis of advanced thin-walled composite structures*. Ph.D., Georgia Institute of Technology.

Berdichevsky, V. (1982). On the energy of an elastic rod. *PMM*, 45, 518–529.

Berdichevsky, V., Armanios, E., and Badir, A. (1992). Theory of anisotropic thin-walled closed-cross-section beams. *Composites Engineering*, 2, 411–432.

Cesnik, C. and Hodges, D. (1997). VABS: A new concept for composite rotor blade cross-sectional modeling. *Journal of the American Helicopter Society*, 42(1), 27–38.

Cesnik, C. and Shin, S. (1998). Structural analysis for designing rotor blades with integral actuators. In *Proc. 39th AIAA Conf. on Structures, Structural Dynamics and Materials*, No. 2107, Long Beach, CA, AIAA.

Cesnik, C. and Shin, S. (2001a). On the modeling of integrally actuated helicopter blades. *International Journal of Solids and Structures*, 38, 1765–1789.

Cesnik, C. and Shin, S. (2001b). On the twist performance of a multiple-cell active helicopter blade. *Smart Materials and Structures*, 10, 53–61.

Cesnik, C., Sutyrin, V., and Hodges, D. (1993). A refined composite beam theory based on the variational-asymptotic method. In *Proceedings of the 34th Structures, Structural Dynamics, and Materials Conference*, No. 93-1616, La Jolla, CA, AIAA, pp. 2710–2720.

Cesnik, C.E.S. (1994). *Cross-sectional analysis of initially twisted and curved composite beams*. Ph.D., Georgia Institute of Technology.

Danielson, D. and Hodges, D. (1987). Nonlinear beam kinematics by decomposition of the rotation tensor. *Journal of Applied Mechanics*, 54, 258–262.

Hodges, D. and Patil, M. (2005). Correlation of geometrically-exact beam theory with the princeton data. *Journal of the American Helicopter Society*, 49(3), 357–360.

Hodges, D. and Volovoi, V. (2000). Theory of anisotropic thin-walled beams. *Journal of Applied Mechanics*, 67, 453–459.

Hodges, D. and Volovoi, V. (2002). Single- and multi-celled composite thin-walled beams. *AIAA Journal*, 40(5), 960–965.

Popescu, B. and Hodges, D. H. (2000). On asymptotically correct timoshenko-like anisotropic beam theory. *International Journal of Solids and Structures*, 37:535–558.

Volovoi, V., Hodges, D., Cesnik, C., and Popescu, B. (2001). Assessment of beam modeling methods for rotor blade application. *Mathematical and Computer Modelling*, 33, 1099–1112.

Yu, W. (2005). *VABS Manual and Examples*. Georgia Institute of Technology, Utah State University.

Yu, W., Volovoi, V., Hodges, D., and Hong, X. (2002). Validation of the variational asymptotic beam sectional analysis (VABS). *AIAA Journal*, 40(10), 2105–2113.

DYNAMIC SENSITIVITIES USING EFFICIENT REANALYSIS OF STRUCTURES[1]

U. Kirsch, M. Bogomolni and I. Sheinman
Department of Civil and Environmental Engineering,
Technion – Israel Institute of Technology, Haifa 32000, Israel
E-mail: kirsch@tx.technion.ac.il

1. Introduction

Design sensitivity analysis of structures deals with the calculation of the response derivatives with respect to the design variables. These derivatives, called the sensitivity coefficients, are used in the solution of various problems. In design optimization, the sensitivity coefficients are often required to select a search direction. These coefficients are used also in generating approximations for the response of a modified system. In addition, the sensitivities are required for assessing the effects of uncertainties in the structural properties on the system response. Calculation of the sensitivities involves much computational effort, particularly in large structural systems with many design variables. As a result, there has been much interest in efficient procedures for calculating the sensitivity coefficients. Developments in methods for sensitivity analysis are discussed in many studies (e.g. Haug et. al. 1986; Haftka and Adelman, 1989; Haftka and Gurdal, 1993; van Keulen et. al. in press). Methods of sensitivity analysis for discretized systems can be divided into the following classes:

a. Finite-difference methods, which are easy to implement but might involve numerous repeated analyses and high computational cost, particularly in problems with many design or response variables. In addition, finite-difference approximations might have accuracy problems. The efficiency can be improved by using fast reanalysis techniques.
b. Analytical methods, which provide exact solutions but might not be easy to implement in some problems such as shape optimization.
c. "Semi-analytical" methods, which are based on a compromise between finite-difference methods and analytical methods. These methods use finite-difference evaluation of the right-hand-side vector. They are easy to implement but might provide inaccurate results.

In general, the following factors are considered in choosing a suitable sensitivity analysis method for a specific application: the accuracy of the calculations, the computational effort involved and the ease-of-implementation. The implementation effort is weighted against the performance of the algorithms as reflected in their computational efficiency and accuracy. The quality of the results and efficiency of the calculations are usually two conflicting factors. That is, higher accuracy is often achieved at the expense of more computational effort.

Dynamic sensitivity analysis has been demonstrated by several authors. Using the mode superposition approach and assuming harmonic loading, the response sensitivities were evaluated by direct differentiation of the equations of motion in the generalized coordinates (Kramer and Grierson, 1989). In cases of earthquake loading the ground acceleration is usually given in discrete time steps, thus the loading is not given analytically. In several studies (Kim and Choi, 2000; van Keulen et. al, in press) the unconditionally stable implicit numerical equation was directly derived. It was found that the analysis equations and the sensitivity equations have the same left-hand side expression. Thus, it was possible to use the available factorized coefficient matrix. A numerical procedure was applied for calculation of the sensitivity of the response.

---

[1] This paper is a shortened version of the paper "Efficient design sensitivities of structures subjected to dynamic loading" by the authors, in press in the International Journal of Solid and Structures.

M. Pandey et al. (eds), Advances in Engineering Structures, Mechanics & Construction, 549–561.

Approximation concepts are often used to reduce the computational cost involved in repeated analysis of structures (Kirsch, 2002). However, most approximations that are adequate for structural reanalysis are not sufficiently accurate for sensitivity analysis. In this study, approximate reanalysis is used to improve the efficiency of dynamic sensitivity analysis by finite-differences. Given the results of exact analysis for an initial design, the displacements for various modified designs are evaluated efficiently by the recently developed Combined Approximations (CA) approach (Kirsch, 2002; 2003a). Originally, the approach was developed for linear static problems. Recently, accurate results were reported also for eigenproblem (Kirsch, 2003b; Kirsch and Bogomolni, 2004) and dynamic reanalysis problems (Kirsch et. al. submitted, in press).

Calculation of analytical derivatives using approximate analysis models have been demonstrated previously (Kirsch, 1994; Kirsch and Papalambros, 2001). It was found that accurate results can be achieved but, as noted earlier, analytical derivatives might not be easy to implement. It was demonstrated recently (Kirsch et. al. 2005, Bogomolni et. al. in press) that accurate derivatives can be achieved efficiently by CA and finite-differences for linear static problems and eigenproblems.

The present study deals with the design sensitivity analysis for discrete linear systems subjected to dynamic loading. The problem of dynamic analysis by mode superposition is first introduced, and the response derivatives with respect to design variables are presented as a combination of sensitivities of the eigenvectors and the generalized displacements. A procedure for reducing the number of differential equations that must be solved during the solution process is then proposed. Procedures intended to improve the accuracy of the approximations are developed, and efficient evaluation of the response derivatives by the combined approximations approach is presented. Numerical examples demonstrate the accuracy of the results.

2. Problem Formulation

2.1 Dynamic Analysis

Consider the equations of motion for a linear system subjected to dynamic forces

$$\mathbf{M}\ddot{\mathbf{r}} + \mathbf{C}\dot{\mathbf{r}} + \mathbf{K}\mathbf{r} = \mathbf{R} \tag{1}$$

where $\mathbf{M}$ is the mass matrix, $\mathbf{C}$ is the damping matrix, $\mathbf{K}$ is the stiffness matrix, $\mathbf{r}$ is the unknown displacement vector, and $\mathbf{R}$ is the load vector.

Considering mode superposition, we use the following transformation from the nodal displacements to the generalized displacements

$$\mathbf{r} = \sum_{k=1}^{p} \mathbf{\Phi}_k Z_k = \mathbf{\Phi}\mathbf{Z} \tag{2}$$

where p is the number of mode shapes considered (in general $p<<m$, where m is the number of degrees of freedom), $\mathbf{Z}$ is a vector of generalized displacements, and $\mathbf{\Phi}$ is the matrix of eigenvectors (mode shapes). The eigenvectors $\mathbf{\Phi}_k$ and eigenvalues $\lambda_k = \omega_k^2$ (ω_k are the circular frequencies) are obtained by solving the eigenproblem

$$\mathbf{K}\ \mathbf{\Phi}_k = \lambda_k \mathbf{M}\,\mathbf{\Phi}_k \qquad k = 1, \ldots, p \tag{3}$$

In the presentation that follows we assume damping such that classical modal analysis can be used. Substituting Eq. (2) into Eq. (1) and pre-multiplying the resulting equations by $\mathbf{\Phi}^T$, we obtain

$$\mathbf{I}\ddot{\mathbf{Z}} + \mathbf{\Lambda}\dot{\mathbf{Z}} + \mathbf{\Omega}^2\mathbf{Z} = \mathbf{P} \tag{4}$$

In these equations the right hand side vector in normalized coordinates is $\mathbf{P} = \mathbf{\Phi}^T\mathbf{R}$, and the mass matrix is an identity matrix $\mathbf{I}=\mathbf{\Phi}^T\mathbf{M}\,\mathbf{\Phi}$. The damping matrix and the stiffness matrix are diagonal low-order matrices

$$\mathbf{\Lambda}=\mathbf{\Phi}^T\mathbf{C}\,\mathbf{\Phi} \qquad \mathbf{\Omega}^2=\mathbf{\Phi}^T\mathbf{K}\,\mathbf{\Phi} \tag{5}$$

The elements on the diagonals of these matrices are given by $2\omega_k\zeta_k$, ω_k^2, respectively, $k = 1, \ldots, p$, ζ_k being the damping ratios. Thus, Eq. (4) consists of the p uncoupled equations

$$\ddot{Z}_k + 2\omega_k\zeta_k\dot{Z}_k + \omega_k^2 Z_k = P_k \qquad k = 1, \ldots, p \tag{6}$$

In many problems (e.g. earthquake loading) the load vector $\mathbf{R}$, and therefore the right hand side terms

$$P_k=\mathbf{\Phi}_k^T\mathbf{R} \tag{7}$$

are given as discrete values at each time step, and not by analytical functions.

In summary, computation of the dynamic response by modal analysis involves the following steps.

a. Determine the matrices **K**, **M**, and **C**.
b. Determine the p requested eigenpairs λ_k, $\mathbf{\Phi}_k$ by solving the eigenproblem of Eq. (3).
c. Compute the modal coordinates Z_k by solving Eqs. (6).
d. Compute the nodal displacements **r** by Eq. (2).
e. Calculate the element forces using the element stiffness properties.

2.2 Displacement Derivatives

The derivative expressions of the displacement vector **r** with respect to a design variable X_j, $\partial\mathbf{r}/\partial X_j$, are given by differentiating Eq. (2)

$$\frac{\partial\mathbf{r}}{\partial X_j} = \sum_{k=1}^{p}\left(\frac{\partial\mathbf{\Phi}_k}{\partial X_j}Z_k + \mathbf{\Phi}_k\frac{\partial Z_k}{\partial X_j}\right) \tag{8}$$

The derivatives $\partial\mathbf{\Phi}_k/\partial X_j$ can be evaluated efficiently by finite-differences using the CA approach, as will be shown later. Assuming that the damping ratios ζ_k are independent of the design variables (which is typical, for example, in civil engineering structures), we calculate $\partial Z_k/\partial X_j$ by differentiation of Eq. (6)

$$\frac{\partial\ddot{Z}_k}{\partial X_j} + 2\omega_k\zeta_k\frac{\partial\dot{Z}_k}{\partial X_j} + \omega_k^2\frac{\partial Z_k}{\partial X_j} = \frac{\partial P_k}{\partial X_j} - 2\frac{\partial\omega_k}{\partial X_j}\zeta_k\dot{Z}_k - \frac{\partial\omega_k^2}{\partial X_j}Z_k \tag{9}$$

Denoting

$$q_k = \partial Z_k/\partial X_j \qquad \dot{q}_k = \partial\dot{Z}_k/\partial X_j \qquad \ddot{q}_k = \partial\ddot{Z}_k/\partial X_j \tag{10}$$

and substituting Eqs. (7), (10) into Eq. (9) yields

$$\ddot{q}_k + 2\omega_k\zeta_k\dot{q}_k + \omega_k^2 q_k = \mathbf{\Phi}_k^T\frac{\partial\mathbf{R}}{\partial X_j} + \frac{\partial\mathbf{\Phi}_k^T}{\partial X_j}\mathbf{R} - 2\frac{\partial\omega_k}{\partial X_j}\zeta_k\dot{Z}_k - \frac{\partial\omega_k^2}{\partial X_j}Z_k \tag{11}$$

Note that the left hand sides of Eqs. (6) and (11) are similar, whereas the right hand sides are different. For identical initial conditions (e.g. $q_k = \dot{q}_k = 0$ for $t = 0$) this similarity can be used to reduce the number of differential equations that must be solved during the solution process.

In summary, given the eigenpairs and the response for a certain design and time, evaluation of the displacement derivatives involves the following steps.

a. Evaluate the derivatives of the eigenpairs ($\partial \mathbf{\Phi}_k/\partial X_j$ and $\partial \lambda_k/\partial X_j$).

b. Compute the right side of Eq. (11).

c. Compute the derivatives $q_k = \partial Z_k/\partial X_j$ by solving Eq. (11).

d. Evaluate the displacement derivatives $\partial \mathbf{r}/\partial X_j$ by Eq. (8).

Assuming a problem with p mode shapes and n design variables, the main computational effort is involved in the following two steps:

a. Solution of the pn differential equations (11).

b. Evaluation of pn derivatives of the eigenpairs ($\partial \mathbf{\Phi}_k/\partial X_j$ and $\partial \omega_k/\partial X_j$).

A procedure intended to reduce the number of differential equations to be solved during the solution process is proposed below. Efficient evaluation of the derivatives of the eigenpairs, using finite-difference and the CA approach, is presented later.

3. Reducing the Number of Differential Equations

Due to the linearity of Eq. (11), we can use superposition and divide it into the following 3 equations with identical initial conditions

$$\ddot{q}_k^{(i)} + 2\omega_k\zeta_k\dot{q}_k^{(i)} + \omega_k^2 q_k^{(i)} = F_k^{(i)} \qquad i = 1, 2, 3 \tag{12}$$

where

$$F_k^{(1)} = \mathbf{\Phi}_k^T \frac{\partial \mathbf{R}}{\partial X_j} + \frac{\partial \mathbf{\Phi}_k^T}{\partial X_j}\mathbf{R} \qquad F_k^{(2)} = -2\frac{\partial \omega_k}{\partial X_j}\zeta_k \dot{Z}_k \qquad F_k^{(3)} = -\frac{\partial \omega_k^2}{\partial X_j} Z_k \tag{13}$$

$$q_k = \sum_{i=1}^{3} q_k^{(i)} \qquad \dot{q}_k = \sum_{i=1}^{3} \dot{q}_k^{(i)} \qquad \ddot{q}_k = \sum_{i=1}^{3} \ddot{q}_k^{(i)} \tag{14}$$

Noting that the right hand sides of Eq. (6) and Eq. (12) for $i = 1$ are

$$P_k = \mathbf{\Phi}_k^T \mathbf{R} \tag{15}$$

$$F_k^{(1)} = \mathbf{\Phi}_k^T \frac{\partial \mathbf{R}}{\partial X_j} + \frac{\partial \mathbf{\Phi}_k^T}{\partial X_j}\mathbf{R} \tag{16}$$

and assuming that the load vector can be expressed in the form $\mathbf{R}(\mathbf{X}, t)=\mathbf{R}(\mathbf{X})\, g(t)$, then Eqs. (15), (16) describe similar functions in time with different amplitudes. For zero initial conditions (or, if we neglect the influence of the homogeneous solution), the ratio between the two displacement functions of Eqs. (6) and (12) is equal to the ratio between the right-hand side terms. Thus, given the solutions Z_k of Eq. (6) for all p modes, the solutions $q_k^{(1)}$ of Eq. (12) for $i = 1$ can be determined directly by

$$q_k^{(1)} = Z_k \frac{\mathbf{\Phi}_k^T \dfrac{\partial \mathbf{R}}{\partial X_j} + \dfrac{\partial \mathbf{\Phi}_k^T}{\partial X_j}\mathbf{R}}{\mathbf{\Phi}_k^T \mathbf{R}} \tag{17}$$

To find $q_k^{(2)}, q_k^{(3)}$, Eq. (12) must be solved for i=2 and i=3. For X_1 we have to solve the two equations

$$\ddot{q}_k^{(2)} + 2\omega_k\zeta_k\dot{q}_k^{(2)} + \omega_k^2 q_k^{(2)} = -2\frac{\partial\omega_k}{\partial X_1}\zeta_k\dot{Z}_k \tag{18}$$

$$\ddot{q}_k^{(3)} + 2\omega_k\zeta_k\dot{q}_k^{(3)} + \omega_k^2 q_k^{(3)} = -\frac{\partial\omega_k^2}{\partial X_1}Z_k \tag{19}$$

Given the solutions of Eqs. (18), (19) with respect to X_1, it is observed that the solutions for any other variable X_j can be determined directly by

$$q_k^{(2)}(X_j) = \frac{\dfrac{\partial\omega_k}{\partial X_j}}{\dfrac{\partial\omega_k}{\partial X_1}} q_k^{(2)}(X_1) \qquad\qquad q_k^{(3)}(X_j) = \frac{\dfrac{\partial\omega_k^2}{\partial X_j}}{\dfrac{\partial\omega_k^2}{\partial X_1}} q_k^{(3)}(X_1) \tag{20}$$

In the particular case where $\mathbf{\Phi}_k$ and $\mathbf{R}$ are orthogonal we obtain $P_k = \mathbf{\Phi}_k^T\mathbf{R} = 0$. From Eq. (6) we have $Z_k = \dot{Z}_k = \ddot{Z}_k = 0$, and from Eqs. (17) - (19) we find $q_k^{(1)} = q_k^{(2)} = q_k^{(3)} = 0$.

In summary, assuming a problem with p considered mode shapes and n design variables, the number of times that the differential equations (11) must be solved in order to perform sensitivity analysis is usually pn. Considering the procedure presented in this section and assuming that the solution of the analysis problem [Eq. (6)] is known, the number of times that the differential equations must be solved in order to perform sensitivity analysis is only $2p$ [Eqs. (18), (19)]. Thus, the ratio between the two numbers is $pn/2p=n/2$, which means a significant reduction in the computational cost. For example, for a problem with 10 design variables, the procedure presented requires about 20% of the effort involved in complete sensitivity analysis.

4. Derivatives of the Eigenpairs

4.1 Analytical Derivatives

For simplicity, we eliminate the subscripts k (mode shapes) and j (design variables). Thus, the eigenproblem of Eq. (3) is expressed as

$$\mathbf{K}\,\mathbf{\Phi} = \lambda\,\mathbf{M}\,\mathbf{\Phi} \tag{21}$$

The eigenvector is often normalized such that

$$\mathbf{\Phi}^T\mathbf{M}\,\mathbf{\Phi} = 1 \tag{22}$$

To evaluate the derivatives of the eigenpairs ($\partial\mathbf{\Phi}/\partial X$ and $\partial\lambda/\partial X$), we differentiate Eqs. (21), (22) with respect to a design variable X and rearrange to obtain

$$(\mathbf{K} - \lambda\mathbf{M})\frac{\partial\mathbf{\Phi}}{\partial X} - \frac{\partial\lambda}{\partial X}\mathbf{M}\mathbf{\Phi} = -(\frac{\partial\mathbf{K}}{\partial X} - \lambda\frac{\partial\mathbf{M}}{\partial X})\mathbf{\Phi} \tag{23}$$

$$\mathbf{\Phi}^T\mathbf{M}\frac{\partial\mathbf{\Phi}}{\partial X} = -\frac{1}{2}\mathbf{\Phi}^T\frac{\partial\mathbf{M}}{\partial X}\mathbf{\Phi} \tag{24}$$

or, in matrix form

$$\begin{bmatrix} \mathbf{K}-\lambda\mathbf{M} & -\mathbf{M}\mathbf{\Phi} \\ \mathbf{\Phi}^T\mathbf{M} & 0 \end{bmatrix} \begin{Bmatrix} \dfrac{\partial \mathbf{\Phi}}{\partial X} \\ \dfrac{\partial \lambda}{\partial X} \end{Bmatrix} = -\begin{Bmatrix} \left(\dfrac{\partial \mathbf{K}}{\partial X}-\lambda\dfrac{\partial \mathbf{M}}{\partial X}\right)\mathbf{\Phi} \\ \dfrac{1}{2}\mathbf{\Phi}^T\dfrac{\partial \mathbf{M}}{\partial X}\mathbf{\Phi} \end{Bmatrix} \tag{25}$$

In the solution of Eq. (25) care must be taken because the principal minor ($\mathbf{K}$-λ $\mathbf{M}$) is singular. In many cases we are interested only in the derivatives $\partial\lambda/\partial X$. These derivatives may be obtained by premultiplying Eq. (23) by $\mathbf{\Phi}^T$ and rearranging

$$\frac{\partial \lambda}{\partial X} = \frac{\mathbf{\Phi}^T(\dfrac{\partial \mathbf{K}}{\partial X}-\lambda\dfrac{\partial \mathbf{M}}{\partial X})\mathbf{\Phi}}{\mathbf{\Phi}^T\mathbf{M}\mathbf{\Phi}} \tag{26}$$

Note that this is only correct if the eigenvalue λ is distinct.

Several methods have been proposed to solve Eq. (25). In general, the solution involves much computational effort. Specifically, a matrix of the order (m+1), m being the number of degrees of freedom, must be factorized for each of the p considered mode shapes. In addition, the matrices $\partial\mathbf{K}/\partial X, \partial\mathbf{M}/\partial X$ must be calculated and forward and backward substitutions must be carried out for each design variable.

4.2 Finite-Difference Derivatives

In the forward-difference method, the derivatives are approximated from the exact displacements at the original point X and at the perturbed point X+δX by

$$\frac{\partial \mathbf{\Phi}}{\partial X} = \frac{\mathbf{\Phi}(X+\delta X)-\mathbf{\Phi}(X)}{\delta X} \tag{27}$$

where δX is a predetermined step-size. The accuracy can be improved by adopting the central-difference approximation, where the derivatives are computed from the exact displacements at the two points X- δX and X+ δX by

$$\frac{\partial \mathbf{\Phi}}{\partial X} = \frac{\mathbf{\Phi}(X+\delta X)-\mathbf{\Phi}(X-\delta X)}{2\delta X} \tag{28}$$

Finite-difference methods are the easiest to implement and therefore they are attractive in many applications. When $\mathbf{\Phi}(X)$ is known, application of Eq. (27) involves only one additional calculation of the displacements at X+ δX whereas Eq. (28) requires calculation at the two points X-δX and X+δX. For a problem with n design variables, finite difference derivative calculations require repetition of the analysis for n+1 [Eq. (27)] or 2n+1 [Eq. (28)] different design points. This procedure is usually not efficient compared to, for example, analytical and semi-analytical methods. An efficient solution procedure using the CA approach is described below.

As noted earlier, finite-difference approximations might have accuracy problems. The following two sources of errors should be considered whenever these approximations are used:

a. The truncation error, which is a result of neglecting terms in the Taylor series expansion of the perturbed response.

b. The condition error, which is the difference between the numerical evaluation of the function and its exact value. Examples for this type of error include round-off error in calculating $\partial\mathbf{\Phi}/\partial X$ from the original and perturbed values of $\mathbf{\Phi}$, and calculation of the response by approximate analysis. The latter can also be the result of a finite number of iterations being used within an iterative procedure.

These are two conflicting considerations. That is, a small step size δX will reduce the truncation error, but may increase the condition error. In some cases there may not be any step size which yields an acceptable error. Some considerations for choosing the forward-difference step-size are discussed elsewhere (Burton, 1992). In certain applications, truncation errors are not of major importance since it is often sufficient to find the average rate of change in the structural response and not necessarily the accurate local rate of change at a given point. Therefore, to eliminate round-off errors due to approximations it is recommended to increase the step-size.

It is well known that relatively small response values are not calculated as accurately as large response values (Haftka and Gurdal, 1993). The same applies to derivatives. Thus, it would be difficult to evaluate accurately small response derivatives by finite difference or other approximations.

5. Efficient Finite-Difference Derivatives

5.1 The Reduced Eigenproblem

Eigenproblem reanalysis by the CA method has been discussed in detail in previous studies (Kirsch, 2003b; Kirsch and Bogomolni, 2004). The solution procedure is briefly described in this section. We assume that the corresponding stiffness matrix $\mathbf{K}_0$ is given in the decomposed form

$$\mathbf{K}_0 = \mathbf{U}_0^T\mathbf{U}_0 \tag{29}$$

where $\mathbf{U}_0$ is an upper triangular matrix. The initial eigenpair $\boldsymbol{\Phi}_0$, λ_0 is obtained by solving the initial eigenproblem

$$\mathbf{K}\ \boldsymbol{\Phi}_0 = \lambda_0\, \mathbf{M}\, \boldsymbol{\Phi}_0 \tag{30}$$

Assume a perturbation δX in the design and corresponding changes $\delta\mathbf{K}$ in the stiffness matrix and $\delta\mathbf{M}$ in the mass matrix, respectively. The modified matrices are given by

$$\mathbf{K} = \mathbf{K}_0 + \delta\mathbf{K} \qquad \mathbf{M} = \mathbf{M}_0 + \delta\mathbf{M} \tag{31}$$

The object is to estimate efficiently and accurately the requested eigenpair $\boldsymbol{\Phi}$, λ without solving the complete set of modified equations

$$(\mathbf{K}_0 + \delta\mathbf{K})\, \boldsymbol{\Phi} = \lambda\, \mathbf{M}\, \boldsymbol{\Phi} \tag{32}$$

The solution process involves the following steps.

a. Calculate the modified matrices $\mathbf{K}$, $\mathbf{M}$ [Eqs. (31)].

b. Calculate the matrix of basis vectors $\mathbf{r}_B$

$$\mathbf{r}_B = [\mathbf{r}_1, \mathbf{r}_2, ..., \mathbf{r}_s] \tag{33}$$

where $\mathbf{r}_1$, $\mathbf{r}_2$, ..., $\mathbf{r}_s$ are the basis vectors, and s is much smaller than the number of degrees of freedom. For any requested eigenpair $\boldsymbol{\Phi}$, λ the basis vectors are determined separately, using the steps described in the next section.

c. Calculate the reduced matrices $\mathbf{K}_R$ and $\mathbf{M}_R$ by

$$\mathbf{K}_R = \mathbf{r}_B^T \mathbf{K} \mathbf{r}_B \qquad \mathbf{M}_R = \mathbf{r}_B^T \mathbf{M} \mathbf{r}_B \tag{34}$$

d. Solve the reduced $s \times s$ eigenproblem for the first eigenpair λ_1, $\mathbf{y}_1$

$$\mathbf{K}_R \mathbf{y}_1 = \lambda_1 \mathbf{M}_R \mathbf{y}_1 \tag{35}$$

where $\mathbf{y}_1$ is a vector of unknown coefficients

$$\mathbf{y}_1{}^T = \{y_1, y_2, \ldots, y_s\} \tag{36}$$

Various methods (e.g. inverse vector iteration) can be used for this purpose.

e. Evaluate the requested mode shape $\mathbf{\Phi}$ by

$$\mathbf{\Phi} = \mathrm{y}_1\mathbf{r}_1 + \mathrm{y}_2\mathbf{r}_2 + \ldots + \mathrm{y}_s\mathbf{r}_s = \mathbf{r}_B\,\mathbf{y}_1 \tag{37}$$

The requested eigenvalue is already given from Eq. (35) $\lambda = \lambda_1$.

It was found that high accuracy is often achieved with a very small number of basis vectors. In such cases the above solution procedure is most effective.

5.2 Improved Basis Vectors

The effectiveness of the solution approach depends, to a great extent, on the appropriate choice of the basis vectors. Proper selection of the basis vectors is perhaps the most important factor affecting the successful application of the method. It was found that the basis vectors determined by the method described in this section provide accurate results with a small computational effort.

The basis vectors for any requested eigenpair $\mathbf{\Phi}$, λ , are first calculated by the terms of the binomial series as follows (Kirsch et. al. submitted b). The first basis vector is selected as

$$\mathbf{r}_1 = \mathbf{K}_0^{-1}\ \mathbf{M}\,\mathbf{\Phi}_0 \tag{38}$$

Additional vectors are calculated by the terms of the binomial series

$$\mathbf{r}_k = -\mathbf{B}\,\mathbf{r}_{k-1} \tag{39}$$

where matrix $\mathbf{B}$ is given by

$$\mathbf{B} = \mathbf{K}_0^{-1}\delta\mathbf{K} \tag{40}$$

Calculation of each basis vector by Eq. (39) involves only forward and backward substitutions, since $\mathbf{K}_0$ is given in the decomposed form of Eq. (29) from thc initial analysis.

Substituting Eq. (40) into Eq. (39) yields

$$\mathbf{r}_k = -\mathbf{K}_0^{-1}\delta\mathbf{K}\,\mathbf{r}_{k-1} \tag{41}$$

It was found (Barthelemy et. al. 1988; Pedersen et. al. 1989) that the expression of Eq. (41) might cause inaccurate results in calculating sensitivities with respect to shape design variables. To improve the accuracy, it is possible to use the central-difference expression

$$\delta\overline{\mathbf{K}} = \mathbf{K}(X+\delta X) - \mathbf{K}(X-\delta X) \tag{42}$$

in Eq. (40), instead of the forward-difference expression [Eq. (31)]

$$\delta\mathbf{K} = \mathbf{K}(X+\delta X) - \mathbf{K}_0 \tag{43}$$

This modification may reduce significantly the number of basis vectors required to achieve sufficiently accurate results. In summary, the resulting expressions for calculating the basis vectors [instead of Eqs. (38) - (40)] are

$$\overline{\mathbf{r}}_1 = \mathbf{r}_1 = \mathbf{K}_0^{-1}\ \mathbf{M}\,\mathbf{\Phi}_0 \tag{44}$$

$$\overline{\mathbf{r}}_k = -\overline{\mathbf{B}}\ \overline{\mathbf{r}}_{k-1} \tag{45}$$

$$\overline{\mathbf{B}} = \mathbf{K}_0^{-1} \delta \overline{\mathbf{K}} \tag{46}$$

It should be noted that forward-difference derivatives [only one additional reanalysis for $(X+\delta X)$] can be used with the central difference expressions of Eqs. (44) - (46).

To improve the accuracy of the results for the higher mode shapes, we use Gram-Schmidt orthogonalizations of the approximate mode shapes and the basis vectors, with respect to the mass matrix (Bogomolni et. al. in press).

6 Numerical Example

Consider the fifty-story frame shown in Fig. 1. The number of degrees of freedom is 600, and the damping ratios for all modes are 0.05. The masses are assumed to be concentrated at the joints, and only horizontal inertia forces are considered. The inertia force is due to the frame self-weight and an additional concentrated mass of 50 ton in an internal joint and 25 ton in an external joint. The width of all elements is 0.5m, the depth of all columns is 1.0m and the depth of all beams is 0.8m. The modulus of elasticity is $3\text{x}10^7$ kNm^2. The loading is due to the ground acceleration of the El Centro earthquake, shown in Fig. 2. The object is to evaluate the sensitivities of the horizontal displacements at the 1st story and the 50th story with respect to the following four design variables;

X_1 – depth of the columns in the 1st story.
X_2 – depth of the beams in the 1st story.
X_3 – depth of the columns in the 50th story.
X_4 – depth of the beams in the 50th story.

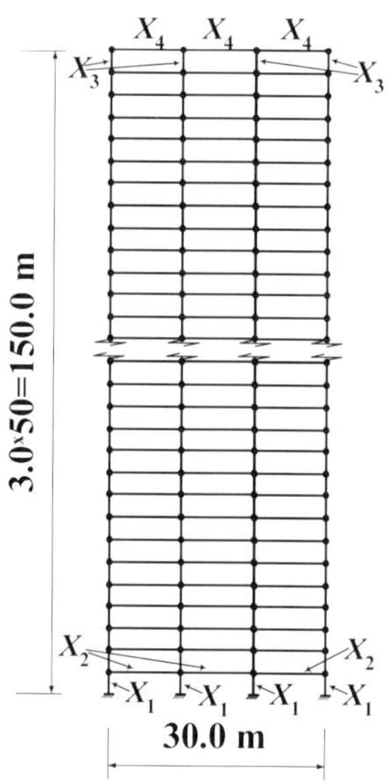

Fig. 1. Fifty-story frame

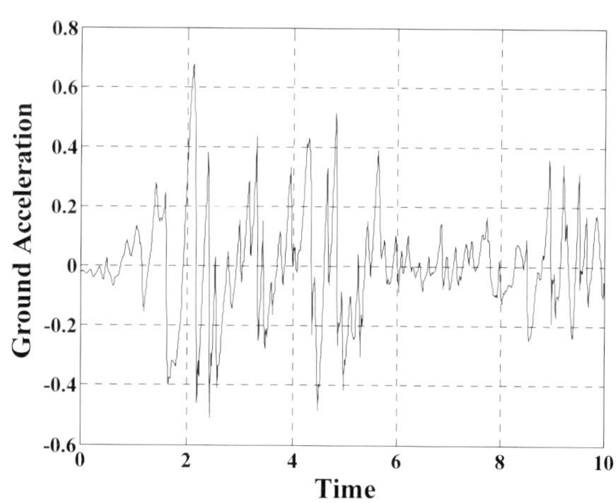

Fig. 2. El Centro Earthquake

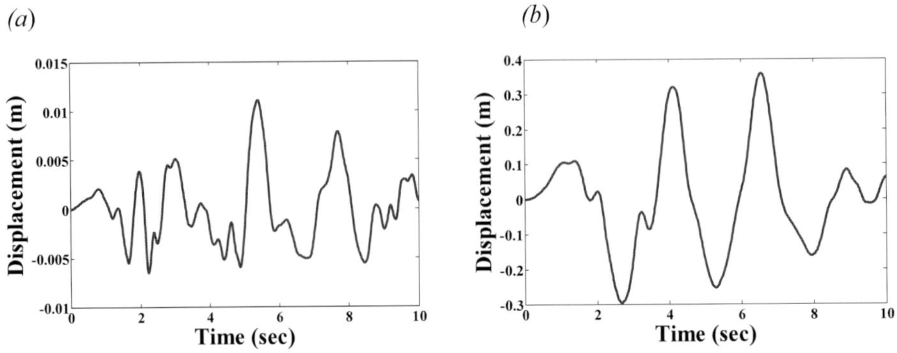

Fig. 3. Horizontal displacements: *(a)* 1st floor *(b)* 50th floor.

Table 1. Eigenvalue sensitivities, fifty-story frame

Sensitivity	Mode	FD(exact)	FD(CA2)
$\partial \lambda/\partial X_1$	1	0.0480	0.0480
	2	0.4723	0.4723
	3	1.4404	1.4404
	4	2.8998	2.8998
	5	4.9780	4.9780
	6	7.7050	7.7050
	7	11.232	11.232
	8	15.617	15.617
$\partial \lambda/\partial X_2$	1	0.0330	0.0330
	2	0.5834	0.5834
	3	1.7403	1.7403
	4	3.5348	3.5348
	5	5.9373	5.9373
	6	9.0852	9.0852
	7	12.876	12.876
	8	17.485	17.485
$\partial \lambda/\partial X_3$	1	-0.00315	-0.00315
	2	-0.02443	-0.02443
	3	-0.05199	-0.05198
	4	-0.04155	-0.04154
	5	0.06455	0.06457
	6	0.33255	0.33257
	7	0.85281	0.85282
	8	1.69010	1.69010
$\partial \lambda/\partial X_4$	1	-0.00774	-0.00779
	2	-0.06148	-0.06149
	3	-0.13726	-0.13723
	4	-0.12572	-0.12572
	5	0.13424	0.13425
	6	0.84037	0.84042
	7	2.28830	2.28840
	8	4.73610	4.73610

Choosing the time-step $\Delta t = 0.02$ sec. and considering the first 8 mode shapes, the results obtained by forward-difference derivatives using exact analysis formulation [FD(exact)] are compared with those

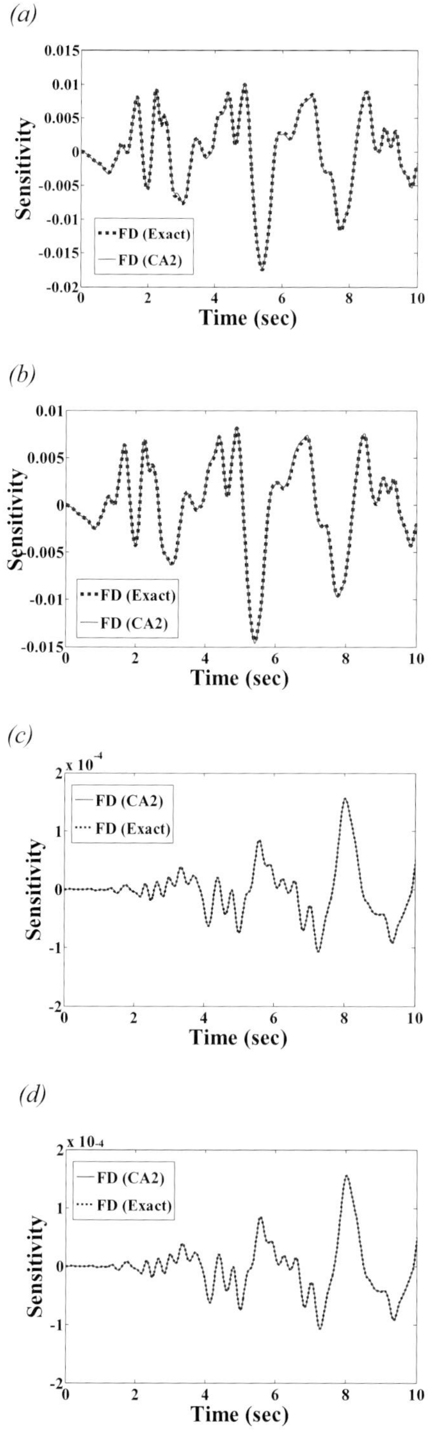

Fig. 4. Displacement- sensitivities, 1st floor, with respect to: (*a*) X_1 (*b*) X_2 (*c*) X_3 (*d*) X_4

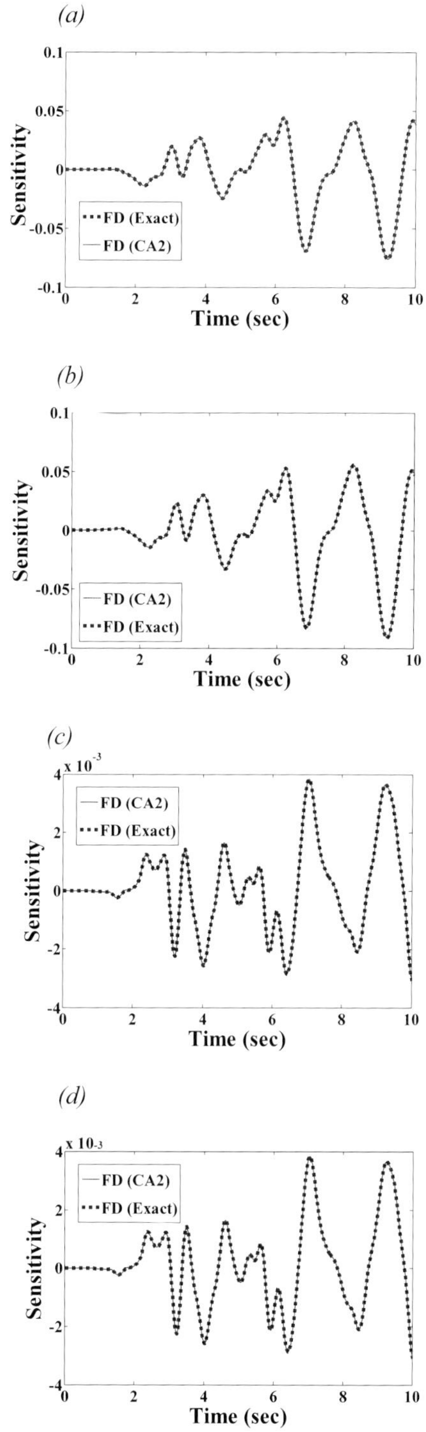

Fig. 5. Displacement- sensitivities, 50st floor, with respect to: (*a*) X_1 (*b*) X_2 (*c*) X_3 (*d*) X_4

achieved by the CA approach with only 2 basis vectors [FD(CA2)]. Table 1 shows the eigenvalue sensitivities, Fig. 3 shows the displacements, and Figs. 4, 5 show the displacement sensitivities of the 1st and the 50th stories. It is observed that high accuracy is achieved by the procedure presented.

Solving various frames with different numbers of degrees of freedom, it was found that in all cases only 2 basis vectors provide accurate sensitivities. This result is typical for small perturbations in a single design variable.

7 Conclusions

Calculation of response derivatives with respect to design variables often involves much computational effort, particularly in large structural systems subjected to dynamic loading. Approximation concepts, which are often used to reduce the computational cost involved in repeated analysis, are usually not sufficiently accurate for sensitivity analysis.

In this study efficient sensitivity analysis, using the recently developed combined approximations approach and finite-differences, is presented. Assuming modal analysis, a procedure intended to reduce the number of differential equations that must be solved during the solution process is proposed. Computational procedures intended to improve the accuracy of the approximations are developed, and efficient evaluation of the response derivatives by the combined approximations approach is presented. Numerical examples show that accurate results can be achieved efficiently. In general, sensitivity analysis by the CA method is used in problems of small perturbations in a single design variable. In such cases a very small number of basis vectors provide accurate results even for structures having large numbers of degrees of freedom.

References

Barthelemy, B., Chon, C.T., Haftka, R.T., 1988. Sensitivity approximation of static structural response. Finite Element in Analysis and Design 4, 249-265.

Bogomolni, M., Kirsch, U., Sheinman, I., In press. Efficient design sensitivities of structures subjected to dynamic loading.

Burton, R.R., 1992. Computing forward-difference derivatives in engineering optimization. Engineering Optimization 20, 205-224.

Haftka, R.T., Adelman, H.M., 1989. Recent developments in structural sensitivity analysis. Structural Optimization 1, 137-151.

Haftka, R.T., Gurdal, Z., 1993. Elements of Structural Optimization, Third Ed. Kluwer Academic Publishers, Dordrecht.

Haug, E.J., Choi, K.K., Komkov, V., 1986. Design sensitivity analysis of structural system, Academic press, New York.

Kim, N.H. Choi, K.K., 2000. Design sensitivity analysis and optimization of nonlinear transient dynamics, AIAA/USAF/NASA/ISSMO 8th Simposium on Multidisciplinary Analysis and Optimization. 6-8 Sept., Long Beach, CA.

Kirsch, U., 1994. Effective sensitivity analysis for structural optimization. Computer Methods in Applied Mechanics and Engineering 117, 143-156.

Kirsch, U., 2002. Design-Oriented Analysis of Structures, Kluwer Academic Publishers, Dordrecht.

Kirsch, U., 2003a. A unified reanalysis approach for structural analysis, design and optimization. Structural and Multidisciplinary Optimization 25, 67-85.

Kirsch, U., 2003b. Approximate vibration reanalysis of structures. AIAA Journal 41, 504-511.

Kirsch, U., Bogomolni, M., 2004. Procedures for approximate eigenproblem reanalysis of structures. International Journal for Numerical Methods in Engineering 60, 1969-1986.

Kirsch, U., Bogomolni, M., van Keulen, F., 2005. Efficient finite-difference design-sensitivities. AIAA Journal 43, 399-405.

Kirsch, U., Bogomolni, M., Sheinman, I., submitted. Efficient dynamic reanalysis of structures.

Kirsch, U., Bogomolni, M., Sheinman, I., In press. Nonlinear dynamic reanalysis by combined approximations.

Kirsch, U., Papalambros, P. Y., 2001. Accurate displacement derivatives for structural optimization using approximate reanalysis. Computer Methods in Applied Mechanics and Engineering 190, 3945-3956.

Kramer, G.J., Grierson, D.E., 1989. Computer automated design of structures under dynamic loads. Computers & Structures 32, 313-325,.

Pedersen, P., Cheng, G., Rasmussen, J., 1989. On accuracy problems for semi-analytical sensitivity analysis. Mechanics of Structures and Machines 17, 373-384.

Somerville, P., et al, 1997. Development of ground motion time histories for phase 2 of the FEMA/SAC steel project, Report No. SAC/BD-97/04.

van Keulen, F., Haftka, R.T. Kim, N.H., in press. Review of options for design sensitivity analysis. Part 1: linear systems.

SYMMETRIZATION OF SOME LINEAR CONSERVATIVE NONSELF-ADJOINT SYSTEMS

B.L. LY
AECL, Mississauga, Ontario, Canada

Abstract

We derive here equivalent self-adjoint systems for conservative systems of the second kind. Existence of the symmetrized systems confirms that certain conservative systems of the second kind behave as a true conservative system. In this way, study of stability can be carried out on the symmetrized system. In general, it is easier to study a self-adjoint system than a nonself-adjoint system. For the conservative system of the second kind, including the Pflüger column, we also presented a lower bound self-adjoint system. For a linear conservative gyroscopic system, we gave a zero parameter sufficient condition for instability and one for stability. The criteria depend only on the characteristics of the system. For a simple 2-DOF system, the present criteria yield the exact solutions.

1. Introduction

Two types of conservative nonself-adjoint systems are studied. One is the so-called conservative system of the second kind. The other one is a gyroscopic system.

Certain nonself-adjoint systems have only divergent type of instability, despite the presence of a polygenic force. Pflüger's column and Greenhill's shaft are two such systems. Leipholz (1974a, 1974b) called a true divergent nonself-adjoint system which has dynamic properties very similar to those of a self-adjoint system a conservative system of the second kind. He showed that such a system is self-adjoint with respect to an assigned self-adjoint operator, hence it is self-adjoint in a generalized sense. For such a system there exists a Lyapunov for predicting stability (Walker, 1972; Leipholz, 1974a) and a generalized Rayleigh quotient for determining the buckling load (Leipholz, 1974a). Inman and Olsen (1988) included velocity dependent forces in conservative systems of the second kind and proved the generalized self-adjointness and the existence of the eigenfunctions. In this way, the solution can be obtained by a modal analysis.

For certain asymmetric discrete systems, Inman (1983) demonstrated that there exists a similarity transformation that transforms the asymmetric system into an equivalent symmetric one, one that has the same eigenvalues.

Here we show for certain conservative systems of the second kind, an equivalent self-adjoint system can be derived. In this way, a conservative system of the second kind is symmetrized, similar to the symmetrization of an asymmetric discrete system. Existence of the symmetrized system confirms a conservative system of the second kind behaves like a self-adjoint system.

A gyroscopic system is nonself-adjoint because of the presence of a skew-symmetric operator. Walker (1991) presented a Lyapunov containing undetermined parameters for the study of the stability. By determining the parameters by trial and error, he was able to optimize the stability region. Because a zero-parameter stability criterion is easier to use, we derive a zero-parameter sufficient condition for instability and one for stability, all via symmetrization. For the example considered by Walker, the present method also yields the exact solutions.

M. Pandey et al. (eds), Advances in Engineering Structures, Mechanics & Construction, 563–570.

2. The Pflüger Column

Pflüger's column is a pinned column under a uniform tangential follower force. Walker (1972) included damping and non-uniform follower force. Here, we include guided end in addition to pinned end. Stability of Pflüger's column is governed by the differential equation

$$-\Omega^2 w + w'''' + pf(x)w'' = 0, \quad x \in [0, 1] \tag{1}$$

and the boundary conditions:

$$\begin{aligned} w = w'' = 0; &\quad \text{for a pinned end,} \\ w = w''' = 0; &\quad \text{for a guided end.} \end{aligned} \tag{2}$$

$w(x)$ is the lateral deflection, Ω is a frequency parameter, $p > 0$ is a load parameter, and $f(x)$ is a non-negative bounded function related to the distribution of the follower force. The operator $f\partial^2$ is nonself-adjoint with respect to the boundary conditions. The operator $-\Omega^2 + \partial^4 + pf(x)\partial^2$, however, is self-adjoint in a generalized sense (Leipholz, 1974a) with respect to the operator ∂^2 under the boundary conditions in Equation (2). The generalized self-adjointness implies

$$\int_0^1 (-\Omega^2 u + u'''' + pfu'')v''\,dx = \int (-\Omega^2 v + v'''' + pfv'')u''\,dx. \tag{3}$$

$u(x)$ and $v(x)$ are any two admissible functions satisfying all the boundary conditions for $w(x)$.

We will show that the Pflüger column is a conservative system by symmetrizing it. We will also show that Leipholz's generalized self-adjointness can be reduced to the classical self-adjointness.

Let us differentiate Equation (1) with respect to x and denote w' by $y(x)$. Then Equation (1) becomes

$$-\Omega^2 y + y'''' + p[f(x)y']' = 0. \tag{4}$$

The follower force now assumes the appearance of a unidirectional loading. Whether or not the operator in Equation (4) is self-adjoint also depends on the boundary conditions for $y(x)$.

At a pinned end of the column, $w'' = 0$ implies $y' = 0$. From Equation (1), $w = w'' = 0$ implies $w'''' = 0$, which in turn implies $y''' = 0$. As can be seen, a pinned end for w becomes a guided end for y. At a guided end of the column, $w' = 0$ implies $y = 0$, and $w''' = 0$ implies $y'' = 0$. It is seen that a guided end for w turns out to be a pinned end for y.

The system in Equation (4) is equivalent to the system in Equation (1), because Equation (4) and the associated boundary conditions are derived from Equation (1) via the transformation of variable: $y = w'$. Hence the eigensolutions of Equation (1) are also eigensolutions of Equation (4).

The operator in Equation (4) is self-adjoint. It can be shown in the usual manner that for admissible y and z, the following inner products hold:

$$\int_0^1 [-\Omega^2 y + y'''' + p(fy')']z\,dx = \int_0^1 [-\Omega^2 z + z'''' + p(fz')']y\,dx. \tag{5}$$

The expression in Equation (5) can also be derived from the generalized self-adjointness in Equation (3) by integration by parts to obtain

$$\int_0^1 [-\Omega^2 u' + u'''' + f(fu'')']v'\,dx = \int_0^1 [-\Omega^2 v' + v'''' + p(fv'')']u'\,dx \tag{6}$$

and then by denoting u' by y and v' by z. We have thus shown that the operator in Equation (4) is self-adjoint. So the Pflüger column can only have divergent type of instability.

The buckling load p_{cr} has a Rayleigh quotient:

$$p_{cr} = \inf_{y(x)} \frac{\int_0^1 (y'')^2 \, \mathrm{d}x}{\int_0^1 f(x)(y')^2 \, \mathrm{d}x}, \tag{7}$$

or in terms of w:

$$p_{cr} = \inf_{w(x)} \frac{\int_0^1 (w''')^2 \, \mathrm{d}x}{\int_0^1 f(x)(w'')^2 \, \mathrm{d}x}. \tag{8}$$

The foregoing derivation shows that the Pflüger column can be symmetrized into a self-adjoint system. Stability of the Pflüger column can be studied via its equivalent self-adjoint system. The relationship between the Pflüger column and its equivalent self-adjoint system is as follows.

Systems	Loading	End Conditions			
Pflüger's column	Tangential follower force	P-P	P-G	G-P	G-G
Equivalent self-adjoint	Unidirectional force	G-G	G-P	P-G	P-P

P stands for a pinned end, and G stands for a guided end. The equivalent self-adjoint system may be viewed as a continuum counterpart of the symmetrized discrete system. A G-G column has a rigid body translational mode. We will ignore it, as we are interested in the flexural modes only.

Note the static equation governing the bending moment in the Pflüger column, $M'' + f(x)M = 0$, is also self-adjoint in the classical sense.

As an example, to study the stability of a G-G column under a uniform tangential follower force, we can instead study a P-P column under its uniform self-weight. While the former is a nonself-adjoint problem, the latter is a well-known classical self-adjoint problem. The critical weight of the self-adjoint problem is 18.57, so is the buckling load of the nonself-adjoint problem.

It is interesting to note that for a divergent nonself-adjoint discrete system $Ax - \lambda Bx = 0$, where A is symmetric, positive definite, B is not symmetric, and λ is real, there exists a lower bound self-adjoint system $Ay - \sigma B^T A^{-1} By = 0$ such that $\sigma_1 \leq (\lambda_1)^2$ and

$$\sigma(x_1) = \frac{x_1^T A x_1}{x_1^T B^T A^{-1} B x_1} = \lambda_1^2.$$

Likewise, for Pflüger's column, a lower bound self-adjoint system exists:

$$u''(x) = \sigma \int_0^1 K(x, \xi) u''(\xi) \, \mathrm{d}\xi,$$

where $K(x, \xi) = G(x, \xi) f(x) f(\xi)$ is a symmetric, positive definite kernel, such that

$$\sigma_1 = \inf_{u(x)} \frac{\int_0^1 [u''(x)]^2 \, \mathrm{d}x}{\int_0^1 \int_0^1 K(x, \xi) u''(x) u''(\xi) \, \mathrm{d}x \, \mathrm{d}\xi} \leq (p_1)^2.$$

3. Greenhill's Shaft

Greenhill's shaft is a pin-ended bar in torsion. The system is not self-adjoint except when θ, the angle between the applied torque vector and the tangent to the end of the bar, is equal to 1/2.

Leipholz (1974a) studied a pure tangential torque ($\theta = 0$) and showed that the Greenhill shaft is a conservative system of the second kind. Walker (1973) considered the case $\theta \neq 1/2$ and included an axial compression and a damping force. He developed a Lyapunov functional for stability study. Here we examine a bar in a viscoelastic medium of low density under a pure tangential torque and a pure axial torque ($\theta = 1$), respectively. We will symmetrize the systems and improve the stability boundary obtained by Walker.

The linearized system (Bolotin, 1963) is described by the differential equation

$$\left\{\begin{matrix} \ddot{u}_1 \\ \ddot{u}_2 \end{matrix}\right\} + c\left\{\begin{matrix} \dot{u}_1 \\ \dot{u}_2 \end{matrix}\right\} + \left\{\begin{matrix} \partial^4 + p\partial^2 + k & L\partial^3 \\ -L\partial^3 & \partial^4 + p\partial^2 + k \end{matrix}\right\}\left\{\begin{matrix} u_1 \\ u_2 \end{matrix}\right\} = \left\{\begin{matrix} 0 \\ 0 \end{matrix}\right\} \tag{9}$$

$$u_1 = u_2 = 0$$

and the boundary conditions:

$$\left\{\begin{matrix} \partial^2 & \theta L\partial \\ -\theta L\partial & \partial^2 \end{matrix}\right\}\left\{\begin{matrix} u_1 \\ u_2 \end{matrix}\right\} = \left\{\begin{matrix} 0 \\ 0 \end{matrix}\right\} \tag{10}$$

with $\theta = 0$ or $\theta = 1$, respectively.

$u_1(x)$ and $u_2(x)$ are deflections in the two principal directions, L is the torque, p is the axial end compression, c and k represent the viscoelastic medium, and ∂ denotes $\partial/\partial x$.

Let us transform variables by denoting $z(x)\,\mathrm{e}^{rt} = u_1 + iu_2$, where $i = \sqrt{-1}$. Then the differential equation and the boundary conditions become, respectively,

$$(r^2 + cr + k)z + z'''' - iLz''' = 0 \tag{11}$$

and

$$z'' - i\theta L z' = 0. \tag{12}$$

For the case of a pure tangential torque ($\theta = 0$), let us denote $z'' = M$. Then $M = 0$ at either end, and

$$z(x) = \int_0^1 G(x, y)M(y)\,\mathrm{d}y, \tag{13}$$

where

$$G(x, y) = \begin{cases} x(y-1) & \text{for } 0 \leq x < y \\ y(x-1) & \text{for } y < x \leq 1 \end{cases}$$

is Green's influence function. Let $M = w(x)\,\mathrm{e}^{iLx/2}$. The expression in Equation (11) becomes

$$(r^2 + cr + k)\int_0^1 G(x, y)M(y)\,\mathrm{d}y + M'' + pM - iLM' = 0, \tag{14}$$

$$w(0) = w(1) = 0, \tag{15}$$

and

$$(r^2 + cr + k)\int_0^1 H(x, y)w(y)\,\mathrm{d}y + w'' + \left(p + \frac{L^2}{4}\right)w = 0. \tag{16}$$

$H(x, y) = G(x, y)\,\mathrm{e}^{-iL(x-y)/2}$ is a symmetric kernel. Greenhill's shaft is thus symmetrized.

For a given k and L, the critical load has a Rayleigh quotient

$$p_{cr} = -\frac{L^2}{4} + \inf_{w(x)} \frac{\int_0^1 (w')^2\,\mathrm{d}x - k\int_0^1\int_0^1 H(x, y)w(x)w(y)\,\mathrm{d}x\,\mathrm{d}y}{\int_0^1 w^2\,\mathrm{d}x}. \tag{17}$$

It is noted that

$$\int_0^1\int_0^1 H(x, y)w(x)w(y)\,\mathrm{d}x\,\mathrm{d}y = -\int_0^1 \overline{g'}\,g'\,\mathrm{d}x, \tag{18}$$

where $g(x)$ is the solution of

$$g''(x) = w(x)\,\mathrm{e}^{iLx/2}; \quad g(0) = g(1) = 0. \tag{19}$$

When $k = 0$, one can study the stability by considering, from Equation (16),

$$w'' + \left(p + \frac{L^2}{4}\right)w = 0; \quad w(0) = w(1) = 0, \tag{20}$$

which describes the free vibration of a string. So $p_{cr} = -(L^2/4) + \pi^2$ is the necessary and sufficient condition for stability when $k = 0$.

An estimate of p_{cr} in Equation (17) can be obtained by assuming $w(x) = \sin(\pi x)$, resulting in

$$p_{cr} = -\frac{L^2}{4} + \pi^2 + \frac{k\left[\left(\pi^2 - \frac{L^2}{4}\right)\left(\pi^4 - \frac{L^4}{16}\right) - 4\pi^2L^2\left(1 + \cos\frac{L}{2}\right)\right]}{\left(\pi^2 - \frac{L^2}{4}\right)^4}. \tag{21}$$

When $L = 0$, Equation (21) yields $p_{cr} = \pi^2 + (k/\pi^2)$. On the other hand, when $L = \pm 2\pi$, $p_{cr} = k/8((1/3) + (3/\pi^2))$. It is seen that $k > 0$ does increase the stability boundary, directly confirming Walker's conjecture about an increase in the upper bound on $L^2 + 4p$.

The case of a pure axial torque ($\theta = 0$) is considered next. Integrate Equation (11) twice with respect to x to obtain

$$(r^2 + cr + k)\int_0^1 G(x, y)z(y)\,\mathrm{d}y + z'' + pz - iLz' = a_1 + a_2x. \tag{22}$$

The constants of integration a_1 and a_2 are found to be zero, in view of the boundary conditions $z = 0$ and $z'' - iLz' = 0$. Consequently, Equation (22) becomes

$$(r^2 + cr + k)\int_0^1 G(x, y)z(y)\,\mathrm{d}y + z'' + pz - iLz' = 0. \tag{23}$$

The dependent variable needs to satisfy only the boundary conditions $z(0) = z(1) = 0$.

Equation (23) is the same as Equation (14). The symmetrized system in Equation (16) and the stability boundary in Equation (20) also holds true for Greenhill's shaft under an axial torque. Therefore, the shaft under an axial torque and that under a tangential torque behave in a similar manner and both have the same equivalent symmetrized system.

4. A Linear Conservative Gyroscopic System

The deflection of a conservative gyroscopic system being considered here satisfies the equation:

$$\ddot{x} + C\dot{x} + Kx = 0. \tag{24}$$

$C^T = -C$ and $K^T = K$. Walker (1991) gave a Lyapunov with undetermined parameters to study stability. We derive a zero-parameter sufficient condition for instability and one for stability.

A stable conservative gyroscopic system admits periodic solutions. Let $x = (u + iv)\,e^{i\omega t}$, where u and v are real vectors. After the real and the imaginary terms are collected separately, the following two equations are obtained:

$$\begin{aligned} &(K - \omega^2 I)u - \omega C v = 0, \\ &\omega C u + (K - \omega^2 I)v = 0. \end{aligned} \tag{25}$$

Elimination of, say, v from the above equations yields

$$[(K - \omega^2 I)C^{-1}(K - \omega^2 I) + \omega^2 C]u = 0. \tag{26}$$

An algebraic equation in ω^2 can be obtained by performing an inner product on Equation (26). Study of stability is easier from the behaviour of ω^2 as revealed by this algebraic equation.

To obtain a nontrivial algebraic equation, we will pre-multiply Equation (26) by $u^T C$, resulting in:

$$(u^T u)\omega^4 - [u^T(CKC^{-1} + K - C^2)u]\omega^2 + u^T CKC^{-1}Ku = 0, \tag{27}$$

which is the desired algebraic equation.

The system is stable (periodic) if ω^2 is positive. On the other hand, the system is unstable if ω^2 is negative or complex. The proposed sufficient conditions, one for stability and one for instability, follow Equation (27) immediately.

4.1. *Sufficient Condition for Instability*

The system is unstable if any of the following three conditions is satisfied:

$$u^T(CKC^{-1} + K - C^2)u < 0, \tag{28}$$

or

$$u^T CKC^{-1}Ku < 0, \tag{29}$$

or

$$\left[\frac{u^T(CKC^{-1} + K - C^2)u}{2u^T u}\right]^2 - \frac{u^T CKC^{-1}Ku}{u^T u} < 0. \tag{30}$$

The Cauchy–Schwartz inequality can be used to show that Equation (30) is implied by

$$u^T(C^{-1}KC + K - C^2)(CKC^{-1} + K - C^2)u - 4u^T CKC^1 Ku < 0. \tag{31}$$

With the identity $u^T Au = u^T A_S u$, where $A_s = A + A^T/2$, for any real matrix A, $A_S > 0$ (or $A_S < 0$) is sufficient for $u^T Au > 0$ (or $u^T Au < 0$). Therefore, a sufficient condition in terms of the operators for the system to be unstable is:

(A.1) $2(K - C^2) + C^{-1}KC + CKC^{-1} < 0$, or

(A.2) $CKC^{-1}K + KC^{-1}KC < 0$, or

(A.3) $(C^{-1}KC + K - C^2)(CKC^{-1} + K - C^2) - 2(CKC^{-1}K + KC^{-1}KC) < 0.$

The matrices in the above expressions are symmetric. Their negative-definiteness can be verified by using the Sylvester criteria. The union of the regions defined by Equations (A.1), (A.2) and (A.3) is a domain of instability.

For the two-degree-of-freedom system considered by Walker, the system is unstable if:

$$k_1 + k_2 + 16 < 0;$$

or

$$k_1 k_2 < 0;$$

or

$$(K_1 + k_2 + 16)^2 - 4k_1 k_2 < 0,$$

per (A.1), (A.2), and (A.3), respectively. The result is the same as the exact solution, because all the matrix products here are diagonal. The product of two 2×2 skew matrices is diagonal.

4.2. *Sufficient Condition for Stability*

The solution is stable (periodic) if all the following three conditions are satisfied:

$$u^T(CKC^{-1} + K - C^2)u > 0 \tag{32}$$

or

$$u^T CKC^{-1}Ku > 0 \tag{33}$$

or

$$\left[\frac{u^T(CKC^{-1} + K - C^2)u}{2u^T u}\right]^2 - \frac{u^T CKC^{-1}Ku}{u^T u} > 0. \tag{34}$$

An approximation will be made to simplify Equation (34). Let us require that Equation (34) be true for all real vectors, including w, which is the eigenvector in the eigenvalue problem

$$(CKC^{-1} + K - C^2)w = \lambda w. \tag{35}$$

Then Equation (34) becomes

$$w^T[(CKC^{-1} + K - C^2)^2 - 4CKC^{-1}K]w > 0. \tag{36}$$

Consequently, a sufficient condition for stability is:

(B.1) $2(K - C^2) + C^{-1}KC + CKC^{-1} > 0$, and

(B.2) $CKC^{-1}K + KC^{-1}KC > 0$, and

(B.3) $(C^{-1}KC + K - C^2)^2 + (CKC^{-1} + K - C^2)^2 - 4(CKC^{-1}K + KC^{-1}KC) > 0.$

The intersection of the regions in (B.1), (B.2) and (B.3) is a domain of stability.

According to the present criterion, the same 2-DOF system will be stable if all the conditions

$$k_1 + k_2 + 16 > 0;$$

and

$$k_1 k_2 > 0;$$

and

$$(k_1 + k_2 + 16)^2 - 4k_1 k_2 > 0,$$

are satisfied. Again, the result is the same as the exact solution.

5. Conclusions

Walker, Leipholz, and Inman and Olsen have studied conservative systems of the second kind. Our objective here is to obtain an equivalent self-adjoint system. Existence of the symmetrized systems confirms certain conservative systems of the second kind behave as a true conservative system. In this way, study of stability can be carried out on the symmetrized system. In general, it is easier to study a self-adjoint problem than a nonself-adjoint problem. For the conservative system of the second kind, including the Pflüger column, we also presented a lower bound self-adjoint system. For a linear conservative gyroscopic system, we gave a zero parameter sufficient condition for instability and one for stability. The criteria depend only on the characteristics of the system. For a simple 2-DOF system, the present criteria yield the exact solutions.

References

Bolotin, V.V., 1963, *Nonconservative Problems of the Theory of Elastic Stability*, Pergamon Press.

Inman, D.J., 1983, Dynamics of Asymmetric Nonconservative Systems, *ASME, Journal of Applied Mechanics*, 50(1), 199–203.

Inman, D.J. and Olsen, C.L., 1988, Dynamics of Symmetrizable Nonconservative Systems, *ASME, Journal of Applied Mechanics*, 55, 206–212.

Leipholz, H.H.E., 1974a, On Conservative Elastic Systems of the First and Second Kind, *Ingenieur-Archive*, 43, 255–271.

Leipholz, H.H.E., 1974b, On a Generalization of the Concept of Self-Adjointness and of Rayleigh's Quotient, *Mechanics Research Communication*, 1, 67–72.

Walker, J.A., 1972, Lyapunov Analysis of the Generalized Pflüger Problem, *Journal of Applied Mechanics*, Vol. 39, *Trans. ASME*, Vol. 94, Series E, pp. 935–938.

Walker, J.A., 1973, Stability of a Pin-Ended Bar in Torsion and Compression, *ASME, Journal of Applied Mechanics*, 40, 405–409.

Walker, J.A., 1991, Stability of Linear Conservative Gyroscopic Systems, *Journal of Applied Mechanics*, 58, 229–232.

INTERACTIVE KNOWLEDGE-BASED ASSISTANCE FOR CONCEPTUAL DESIGN OF BUILDING STRUCTURES

Rodrigo Mora, Hugues Rivard, Steve Parent and Claude Bédard
ETS, 1100 Notre-Dame Street West, Montreal, Canada, H3C 1K3
E-mail: hugues.rivard@etsmtl.ca

Abstract

During conceptual structural design the engineer proposes initial structural solutions to early architectural designs. At this stage, the decisions made by the engineer are based mostly on knowledge about structural behaviours and experience on the applicability of available construction technologies and materials to different design situations. This research proposes a knowledge-based computer approach to assist the engineer in proposing feasible structural solutions to the architect interactively. With this approach a structural solution is developed by the engineer from an overall description to a specific one through the progressive use of knowledge. A first prototype has been implemented and is being enhanced with a knowledge-base for design exploration. Therefore, an example of envisioned computer support is used to illustrate the capabilities of the proposed approach.

1. Introduction

Conceptual design is explorative in nature. During conceptual structural design, the engineer devises and compares feasible structural solutions to transfer loads to the ground safely and efficiently within a building architectural context (i.e. structural synthesis). The decisions made by the engineer are based mostly on knowledge about structural behaviours and experience on the applicability of available construction technologies and materials to different design situations. Due to the limited availability of resources, knowledge and experience minimize the need for time consuming analysis for decision making at this stage. However, the trade-off between knowledge and analysis depends on the building complexity (i.e. the more complex the building, the more analysis may be required to verify and evaluate proposed conceptual structural solutions).

Nowadays, advanced computer modeling tools are available to support structural system generation, analysis, and the integration to the architecture (Khemlani K, 2005). This kind of support is model-based since it relies on the geometric and data modeling capabilities of a building information model (BIM) that combines the building architecture with other disciplines. It has been demonstrated in practice (Solibri Inc., 2005) that explicit knowledge can be used in conjunction with BIM models in the form of requirements. These requirements constrain the model and maintain its consistency when changes take place. This type of knowledge support could be called passive since it validates or confirms design decisions that have already been made. However, these tools lack the knowledge required to assist the engineer to explore design alternatives and make decisions actively.

This paper proposes a knowledge-based approach that aims at providing interactive support for decision-making to help the engineer in the exploration of design alternatives and efficient generation of structural solutions. With this approach a structural solution is developed by the engineer from an abstract description to a specific one, through the progressive application of knowledge. Thus, knowledge is used by the engineer to refine conceptual structural design descriptions interactively. The approach is applicable to most typical buildings, such as office, apartment and institutional buildings of standard shape. "Sculptural" buildings such as the Guggenheim Museum in Bilbao, Spain are thus excluded. The paper is organized as follows: the next section summarizes relevant research in assisting structural design exploration. Then, the proposed approach for interactive knowledge-based

M. Pandey et al. (eds), Advances in Engineering Structures, Mechanics & Construction, 571–582.

support is presented, followed by the components used for its implementation. Next an example illustrates the advantages of the approach. An example of envisioned computer support is used to illustrate the capabilities of the proposed approach because a working prototype for knowledge-based support is currently being implemented.

2. Literature Review

Over the last three decades researchers have applied artificial intelligence (AI) techniques to assist engineers in exploring design alternatives over a vast array of possible solutions under constraints. Relevant techniques and examples are the following: expert systems (Maher 1988, Bédard and Ravi 1991), formal logic (Jain, Krawinkler and Law 1991, Eisfeld and Scherer 2003), grammars (Meyer 1995, Shea and Cagan 1998), case-based reasoning (CBR) systems (Bailey and Smith 1994, Kumar and Raphael 1997), evolutionary algorithms (Grierson and Khajehpour 2002, Sisk, Miles and Moore 2003, Rafiq, and Mathews and Bullock 2003) and hybrid systems that combine AI techniques such as a CBR system with a genetic algorithm (Soibelman and Peña-Mora 2000).

The impact of AI-based methods in design practice is negligible mainly because many of the proposed systems are standalone with no interactions with design representations currently employed in practice, such as building information models (BIM). In fact, only three of the above research projects (Meyer 1995, Bailey and Smith 1994 and Kumar and Raphael 1997) use architectural models with 3D geometry as input for structural synthesis. In the absence of such models, global gravity and lateral load transfer solutions can be explored to satisfy overall building characteristics and requirements. However, these solutions need actual architectural models to be substantiated and validated.

Another disadvantage of the above research systems that hinders their practical use is that the support provided is mainly automatic and the reasoning supported is monotonic (i.e. based on some inputs, these systems produce outputs that meet specified requirements). By contrast, a hierarchical decomposition/refinement (i.e. top-down) approach to conceptual design is followed in this research. This approach enables knowledge-based feedback to the engineer and engineer's interactions with an architectural-structural model at various decomposition/refinement levels. A similar approach has been proposed by Sacks et al. (2000), however, their approach automates design tasks that are exclusive to architects and engineers, such as positioning spaces and proposing structural layouts. In addition, it provides no interactions with building architectural models. The approach proposed in this paper is described in the next section.

3. Interactive Knowledge-Based Support for Conceptual Structural Design

A hierarchical decomposition/refinement approach to conceptual design is adopted in this research where different abstraction levels provide the main guidance for knowledge modeling. This approach is based on a top-down process model proposed by Rivard and Fenves (2000). To implement this approach the structural system is described as a hierarchy of entities where abstract functional entities, which are defined first, facilitate the definition of their constituent ones.

Figure 1 illustrates the conceptual structural design process. In Figure 1, activities are shown in rectangles, bold arrows pointing downwards indicate a sequence between activities, arrows pointing upwards indicate backtracking, and two horizontal parallel lines linking two activities indicate that these can be carried out in parallel. For clarity, in Figure 1 courier bold 10 point typeface is used to identify structural entities. As shown in Figure 1, the structural engineer first defines independent structural volumes holding self-contained structural skeletons that are assumed to behave as structural wholes. These volumes are in turn subdivided into smaller sub-volumes called structural zones that are introduced in order to allow definition of structural requirements that correspond to architectural functions (i.e. applied loads, allowed vertical supports and floor spans). Independent structural volumes are also decomposed into three structural subsystems, namely the horizontal, the vertical gravity, and the vertical lateral subsystems (the foundation subsystem is not considered in this research project). Each of these structural subsystems is further refined into structural assemblies (e.g. frame and floor assemblies), which are made out of structural elements and structural connections. The arrangement of structural elements and structural connections makes up the "physical structural system".

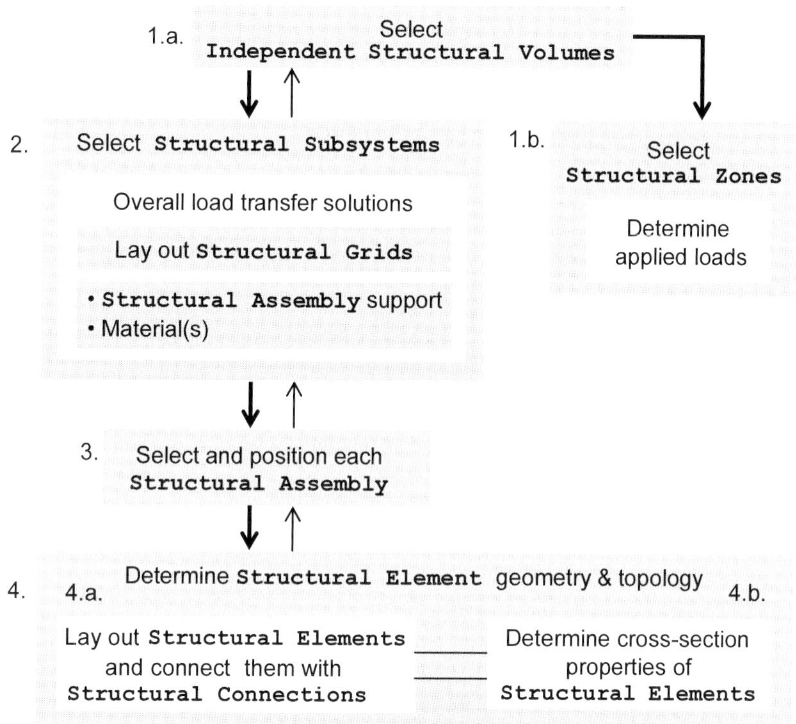

Figure 1. Simplified conceptual structural design

During activity number 2 in Figure 1 (i.e. select structural subsystems), the engineer defines overall load transfer solutions described in terms of supporting structural assemblies and corresponding material(s) and worked out based on tentative structural grids. An example of a structural solution at the subsystem level is the following: for a 9 by 12 structural grid, provide steel rigid frames for lateral support in the long building direction, steel braced frames for lateral support in the shortest direction, columns for vertical gravity support, and composite steel deck on W shape beams for horizontal gravity support. Structural grids determine tentative vertical supports (at gridline intersections), structural bays, likely floor framing directions, and floor spans.

Interactivity is intended between a structural engineer, a simplified model of the building architecture and the structural system, Architecture-Structure Model (ASM), simplified for conceptual design, and a structural design knowledge manager (DKM). During the synthesis process, an architectural model is made available first to the engineer. Then, with the progressive use of knowledge from the DKM the structural system is integrated to the architecture and the result is an integrated architecture-structure model (ASM). Table 1 summarizes the types of interactions that take place at each step of the process between the engineer, the ASM and the DKM. In Table 1 a pre-processing and a post-processing activity in the process are included (versus Figure 1). The pre-processing activity is an inspection of the architectural model, whereas the post-processing activity is the verification of the structural model.

Table 1. Interactivity table between the engineer, the ASM and the DKM

Engineer	ASM	DKM
Architectural Model Inspection		
Query – Look for potential structural problems, continuous load paths to the ground and constraints. Select - Select elements that may become structural	Display the architectural model Emphasize continuous physical elements from this model Highlight architectural grids (i.e. main functional dimensions) Display global dimensional/layout constraints	N/A
1.a. Select `Independent Structural Volumes (ISV)`		
Query - Verify building shape, occupancies, lengths and proportions. Select - Select ISV by grouping spaces.	Emphasize spaces Compute overall building dimensions and aspect ratios	Suggest seismic/expansion joints if applicable
1.b. Select `Structural Zones`		
Query - Check types of spaces and associated constraints Select - Select structural zones by grouping spaces	Emphasize spaces Show space occupancies Display space layout/dimensional constraints	Assign loads to each zone based on its occupancy
2. Select `Structural Subsystems`		
Query - Inspect the model globally Select - Select structural subsystems and materials • Structural assembly support • Material(s) • Lay out structural grids	Display overall building characteristics Display global architectural layout/dimensional constraints Emphasize architectural elements selected to become structural	Suggest structural subsystems and materials Rank overall structural solutions
3. Select and position `Structural Assemblies`		
Select - Select each structural assembly Verify – Validate the initial description from level 2 Specify - Position each assembly Lay out - May determine preferred floor framing directions	Display structural grids Display applied loads Display local architectural layout/dimensional constraints Emphasize architectural elements selected to become structural	Suggest feasible structural assemblies Rank structural assemblies
4. Determine `Structural Element` geometry and topology		
Verify- Anticipate problematic supporting conditions locally Lay out - May position special structural elements and supports locally	Emphasize openings and irregularities in assemblies Elaborate - Make selected architectural elements structural Compute element loads based on tributary areas	Elaborate - Lay out and connect primary structural elements (within gridlines) Elaborate – Lay out and connect secondary structural elements Refine – Select preliminary cross-section shape and size of structural members
Structural system verification		
Verification - Verify and support still unsupported members Verification - Verify critical members	Warn about lack of supports and show unsupported elements	N/A

As seen in Table 1 the main tasks performed by the engineer, the ASM and the DKM are the following: (1) the engineer queries the ASM model, selects entities, specifies, positions and lays out assemblies and elements, and verifies structural solutions. (2) The ASM model displays and emphasizes information accordingly, elaborates engineer's decisions, performs simple calculations on demand, and warns the engineer when supports are missing. And (3) the DKM suggests and ranks solutions, assigns loads, and elaborates and refines engineer's structural selections and layouts. Each activity performed by the engineer advances a structural solution and provides the course of action to enable the ASM and the DKM to perform subsequent tasks accordingly.

3.1. Knowledge-Based Exploration of Structural Alternatives

The knowledge-based exploration of structural alternatives takes place mostly at the abstraction levels of activities 2, 3, and 4 in Figure 1 and Table 1. At each subsequent level more information and knowledge is made available so that previously made decision can be validated and more accurate decisions can be made.

3.2 Select Structural Subsystems

At the structural subsystem level, overall structural solutions are studied by the engineer and described in terms of supporting assemblies and materials. Alternative structural grids are first proposed by the engineer respecting the architectural constraints from the ASM. These grids determine primary layouts for structural assemblies and elements and permit the subsequent validation of subsystem proposals by the engineer with the help of the computer. Then, for each structural subsystem, supporting assemblies are proposed by the engineer (with assistance from the DKM on demand). Next, the computer generates and places assemblies within the layout of the structural grids (i.e. structural frames) and from architectural floor slabs (i.e. for floor assemblies). Depending on the flexibility of the building architecture, the engineer and/or the DKM can propose alternative load transfer solutions based on the following factors: overall building characteristics (e.g. building location, seismic zone, type of building, size, number of stories, and construction area), building geometry (e.g. overall building dimensions and aspect ratios, main functional and bay dimensions) and predominant (i.e. global) architectural layout/dimensional constraints (i.e. that apply to the building as a whole). For a given structural layout, an approximate number of structural assemblies are obtained by the computer so that all feasible structural solutions can be ranked and evaluated by the DKM based on building requirements and preference factors. At this level, the engineer seeks to unify (if possible) the structural grids for the entire building. For buildings with multiple structural zones, structural subsystem descriptions can be detailed further. For example, in a building consisting of an apartment zone over an office zone over a parking zone, all the zones share the same vertical subsystems but each zone may require its own horizontal load transfer solution. Ensuring compatibility between the different assemblies selected can be achieved by the DKM using meta-knowledge heuristics. For example, a concrete rigid frame is compatible with a waffle slab.

3.3 Select and Position Structural Assemblies

Having the tentative structural bays determined at the subsystem level, at the structural assembly level (activity number 3) the engineer selects, specifies, positions, and validates each assembly with respect to local load conditions, supports and constraints given by the ASM. The DKM can also suggest feasible structural assemblies on demand. At this level, structural assemblies have not been populated yet with structural elements and connections. Structural assemblies already positioned at the subsystem level are modified, repositioned, or even removed by the engineer. Structural assemblies can also be added by the engineer or the DKM. For example, local space and/or storey factors (e.g. applied loads, availability of supports, layout constraints in spaces below) may even lead to partitioning a given floor assembly into two or more assemblies. At this level, structural assembly specifications consist of supporting element types organized by function, material, shape and size. This organization facilitates structural element grouping at the structural element level. For the typical bays, floor framing directions are specified and secondary element types, spacing and dimensions are determined. For evaluation, approximate assembly cost and weight can be obtained by the DKM for the bay dimensions given. At this level, ensuring element uniformity and compatibility within an assembly is cumbersome because of slight variations that may exist in architectural supports and constraints within and between stories (even within a single structural zone). ASM spatial verifications on adjacent bays combined with constructability knowledge from the DKM can be used to guarantee element uniformity and compatibility within a structural assembly.

3.4 Determine Structural Element Geometry and Topology

Before structural elements are generated by the computer, the engineer verifies each assembly locally to anticipate lack of supports or critically loaded bay sections. To assist the engineer, the ASM emphasizes openings and irregularities in assemblies. Then, the engineer may place local supports that will not be generated by the computer because they fall outside the overall structural layout. Next, the DKM and the ASM generate the physical structure. The ASM verifies that all structural elements are properly supported based on the specifications from the structural assembly to which these belong. If elements not properly supported exist, the ASM warns the engineer about the lack of supports. Alternative structural element's cross-section shape and dimensions can be proposed by the DKM based on element loading and support conditions, and considering element uniformity and compatibility within the floor assemblies to which these belong.

4. Implementation of the Proposed Approach

The implementation of the approach is based on an existing prototype for conceptual structural design which is called StAr (Structure-Architecture). StAr is a prototype system that assists engineers in the inspection of a 3D architectural model (e.g. while searching for continuous load paths to the ground) and the configuration of structural solutions. Assistance is based on geometrical reasoning algorithms (GRA), (Mora et al. 2006B) and an integrated architecture-structure representation model (ASM), (Mora et al. 2006A). The building architecture in the ASM representation model describes architectural entities such as stories, spaces and space aggregations, and space establishing elements such as walls, columns and slabs. Figure 2 illustrates an architectural model in StAr. The structural system is described in StAr as a hierarchy of entities to enable a top-down design approach, as discussed in section 3. Figure 3a illustrates core walls identified by StAr as continuous and selected by the engineer as structural, and Figure 3b presents the structural system generated by StAr. The geometric algorithms in StAr use the geometry and topology of the ASM model to construct new geometry and topology, and to verify the model. The algorithms are enhanced with embedded structural knowledge regarding layout and dimensional thresholds of applicability for structural assemblies made out of cast-in-place concrete. However, this knowledge is not sufficient for assisting engineers during conceptual design. StAr provides the kind of support described in the second column of Table 1, plus limited knowledge-based support (column 3) at levels 1.b and 4. Therefore, StAr is able to generate and verify a physical structure based on information obtained from precedent levels. However, no knowledge-based support is provided by StAr for exploration at levels 2, 3 and 4. This is the subject of this research. In addition, work is currently in progress to provide StAr with a graphical user interface (GUI) for inputs to replace the current interface with alphanumeric interactions with graphical outputs.

A structural design knowledge manager (DKM) is therefore being developed that gets architectural and/or partial structural information from the ASM directly or via GRA to assist the engineer to conceive, elaborate and refine structural solutions interactively. Once the engineer accepts a solution suggested by the DKM, it automatically updates (i.e. elaborates or refines) the partial ASM. Architectural requirements in the form of model constraints (e.g. floor depths, column-free spaces, etc.) from the ASM model are also considered by the DKM for decision-making.

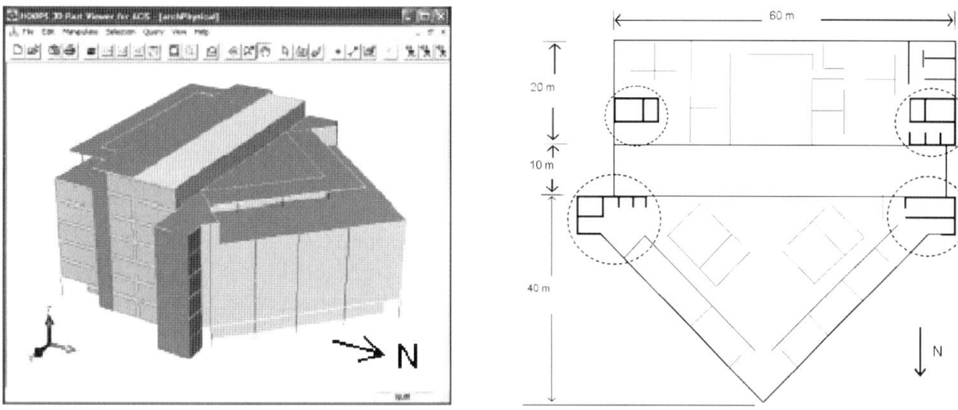

(a) 3D view of the architectural model (b) typical floor plan

Figure 2. Architectural model in StAr (after Mora et al. 2006B)

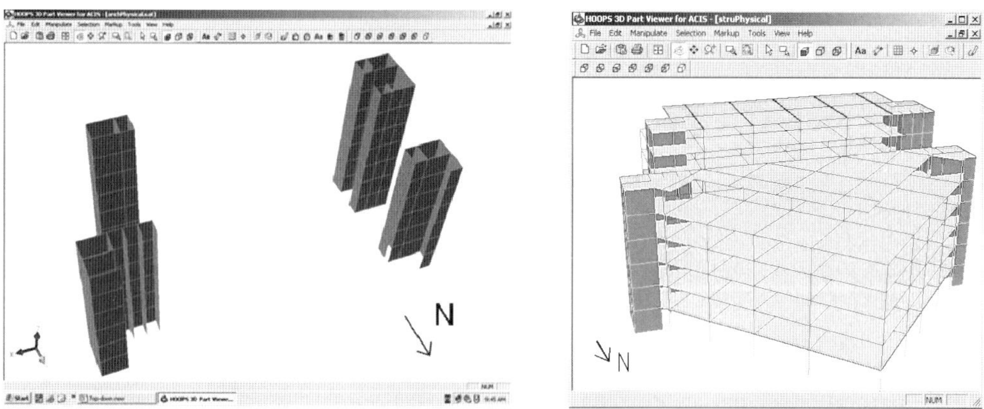

(a) Structural walls selected by the engineer using StAr (b) Structural system generated by StAr

Figure 3. Structural system in StAr (after Mora et al. 2006B)

The DKM encapsulates structural design knowledge by means of a set of technology nodes (Gomez 1998, and Fenves S.J., Rivard H., and Gomez N. 2000). The type of knowledge incorporated in the nodes is heuristic and considers available materials, construction technologies, constructability, cost and weight. A technology node represents the knowledge required to implement one design step (in the top-down hierarchy) utilizing a specific construction system or component. Nodes are organized into a hierarchy ranging from nodes dealing with abstract concepts (e.g. a structural subsystem) to those dealing with specific building entities (e.g. a reinforced concrete beam). The application of a technology node to a building entity from the ASM can be interpreted as making one decision about a design solution. Technology nodes support non-monotonic reasoning since they let the engineer retract any decision node and select another path in the technology tree.

Given that the StAr prototype is being enhanced with the design knowledge manager (DKM), the example that follows illustrates the envisioned support for interactive exploration of conceptual structural design solutions. In the example, the support already provided by StAr is indicated as well as the support envisioned by the DKM. DKM-StAr is used to indicate that some basic support is already provided by StAr but enhanced support is required from the DKM.

5. Example of Knowledge-Based Interactive Support for Conceptual Structural Design

As explained in section 1, the approach is applicable to most typical buildings, such as office, apartment and institutional buildings of standard (i.e. non-sculptural) shape. However, the example uses a deceptively simple office building with little architectural constraints in order to emphasize the knowledge-based interactive exploration of structural alternatives. Office buildings are characterized by having flexible space layouts which provides more room for structural layout exploration at the subsystem level. The building has a rectangular shape with 12-stories of offices, two parking levels underground. The building is located in an intermediate seismic zone. As shown in Figure 4 the building dimensions in plan are 48 m x 18 m. A 12 m x 6 m vertical circulation core is located at the center of the building. Column-free stories are preferred by the architect, storey heights are limited, and the façade must be as free as possible from structural elements. In Figure 4, space-organizing architectural grids are laid out and tentative column locations are proposed by the architect. However, the functional dimensions from the architecture are multiples of 3 m (e.g. 6 m, 9 m, 12 m, 18 m, etc). These dimensions are suitable to accommodate the parking spaces underground.

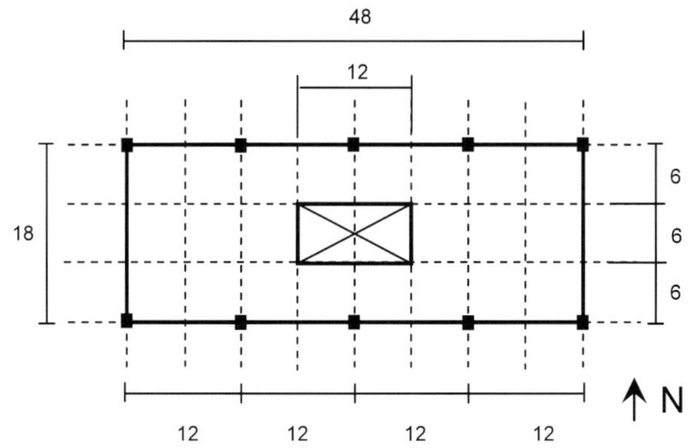

Figure 4. Floor plan of example office building (dimensions in meters)

After the inspection of the architectural model, the engineer selects the core walls to become structural. Then engineer defines one independent structural volume for the entire building. For this example no seismic or expansion joints are necessary as confirmed by the DKM. Next, the engineer selects four structural zones and the computer assigns applied loads to them accordingly: one for the parking stories (2.4 kN/m^2), a second zone grouping all the spaces in the ground floor (4.8 kN/m^2), a third zone grouping the office levels above the ground (2.4 kN/m^2), and a fourth zone being the equipment penthouse on top (3.6 kN/m^2). The current implementation of StAr assists the engineer in performing these initial tasks.

Next, the engineer explores overall load-transfer solutions (structural subsystems) as follows: vertical gravity loads will be transmitted primarily through column stacks and partly through the central core. Horizontal and lateral load transmission can be accomplished in several ways. For lateral loads the load transfer alternatives can be summarized: (1) rigid frames in one or both directions, (2) braced frames in one or both directions, (3) the central core only acting as a tube, (4) the building perimeter and the core acting as concentric tubes, and (5) a combination of the above. Through simple calculations the engineer verifies that the central core alone does not provide sufficient rigidity for overturning. Therefore, alternative number (3) is eliminated. Alternative number (4) is also eliminated because it involves overcrowding the façade with structural elements. In order to select an overall load transfer solution for the building, the engineer decides to explore three alternative structural layouts that s/he sketches using StAr (see Figure 5).

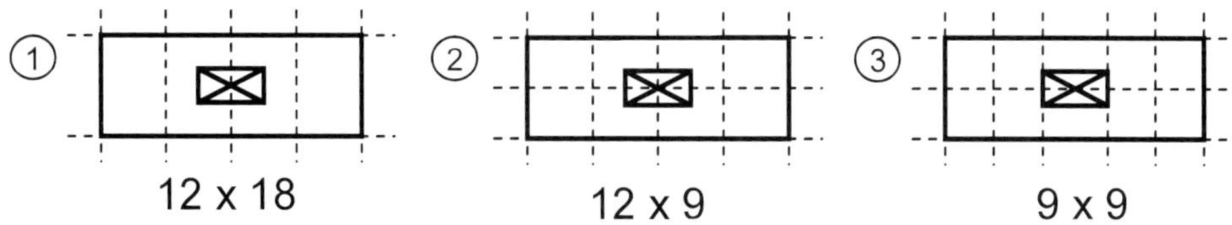

Figure 5. Alternative structural grids

For each structural layout the engineer proposes suitable load transfer solutions (using his/her own experience), which are validated by the DKM. Alternatively, the engineer may ask the DKM to suggest a feasible structural solution for each layout proposed. The overall load transfer solutions proposed by the engineer and validated by the DKM are described in Table 2. In Table 2, WF means wide flange (i.e. W shape beam). For all cases the engineer specifies that the building core also contributes to lateral support, and that vertical gravity load transfer is provided by column and wall stacks.

Table 2. Overall load transfer alternatives (i.e. at the subsystem level)

	# 1: 12 x 18	**# 2: 12 x 9**	**# 3: 9 x 9**
Material	steel	steel	concrete
Lateral support	braced frames	braced frames	rigid frames
Horizontal support	steel deck/open web	steel deck/open web/WF	two-way flat slab
Floor depth (mm)	1000	550	300

For the evaluation of the most suitable structural solution at the subsystem level, each solution should be ranked by the DKM considering several factors including estimated structural system cost and weight and the effect of the structure on the architectural cost, cost of the foundation, soil conditions, constructability, architectural requirements and constraints, and integration with mechanical electrical and plumbing (MEP) systems.

From a structural standpoint, alternatives number 2 and 3 are more stable because they provide more vertical supports for transferring the loads to the ground. However, these are also heavier structural solutions, alternative number 3 being the heaviest one. Alternatives number 2 and 3 are also more intrusive in spaces and facades than alternative one. However, they provide lower floor depths which are beneficial for building cost, constructability, and integration with MEP systems. Nevertheless, open web joists allow the passage of ducts and pipes through the joists. Alternatives number 2 and 3 distribute the load more uniformly over the foundation. However, heavier loads are also transmitted. The evaluation criteria can be customized to associate a weight for each factor and augmented to incorporate experiences from new projects.

With the rankings and data provided by the DKM, the engineer may try to persuade the architect to trade-off two columns inside spaces for lower building costs with shallower floor depths. It is assumed that the alternative number 2 is accepted by the architect with the condition that no braced frames should be placed. A modified alternative 2 is therefore selected with rigid frames instead of braced frames.

Then, the engineer proceeds to select and position each structural assembly individually. The DKM advises the engineer about the convenience of minimizing rigid connections with steel. For the vertical lateral subsystem, in the short direction the engineer specifies two interior rigid frames and in the long direction two rigid frames along the facades covering only the two mid-spans (see Figure 6). The remaining frames or frame sections are simple gravity. For the horizontal subsystem, a partial decision tree is described in Figure 7.

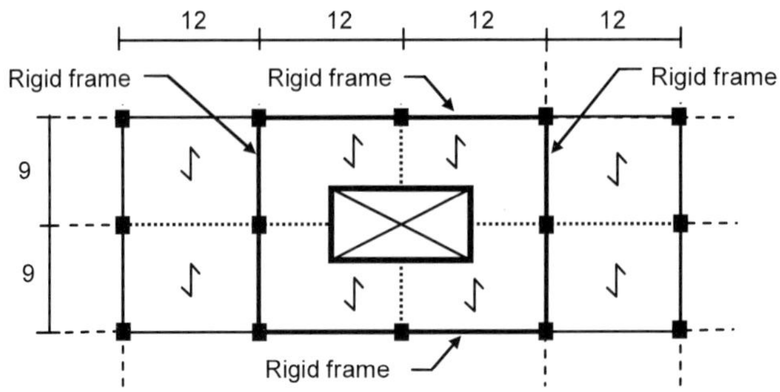

Figure 6. Floor assembly for alternative # 2 (dimensions in meters)

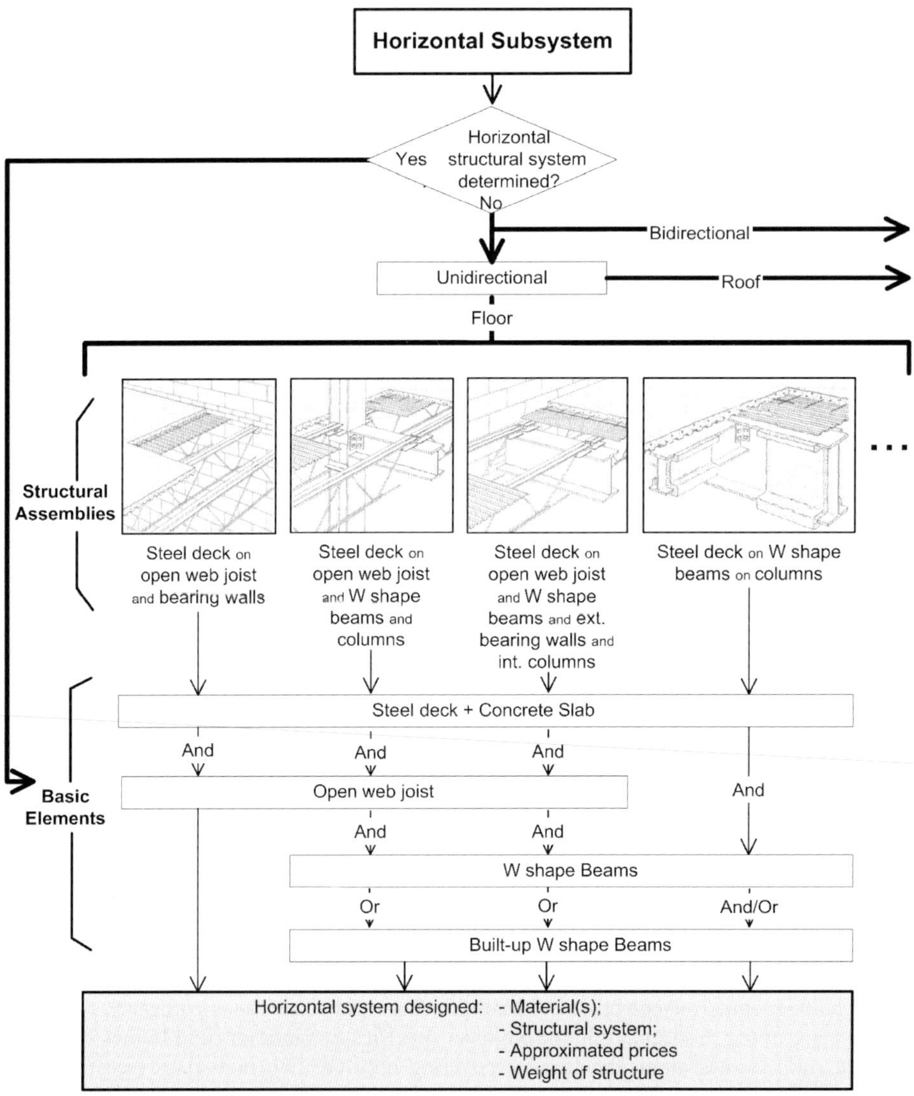

Figure 7. Partial decision tree for the horizontal subsystem (alternative # 2)

At the assembly level, the floor framing directions are determined by DKM-StAr (see Figure 6) and steel deck sections are identified, along with open web joist and W shape beam types. These are grouped by lengths (from the different bays) and the load that they carry. However, assemblies are yet populated with elements. At the element level, the engineer does not detect any openings or irregularities that may cause structural support problems and therefore no support verifications are required. Then, each floor assembly is populated with structural elements by DKM-StAr. Figure 8 illustrates a floor assembly for a typical floor (i.e. excluding the ground floor, the roof, and the first basement).

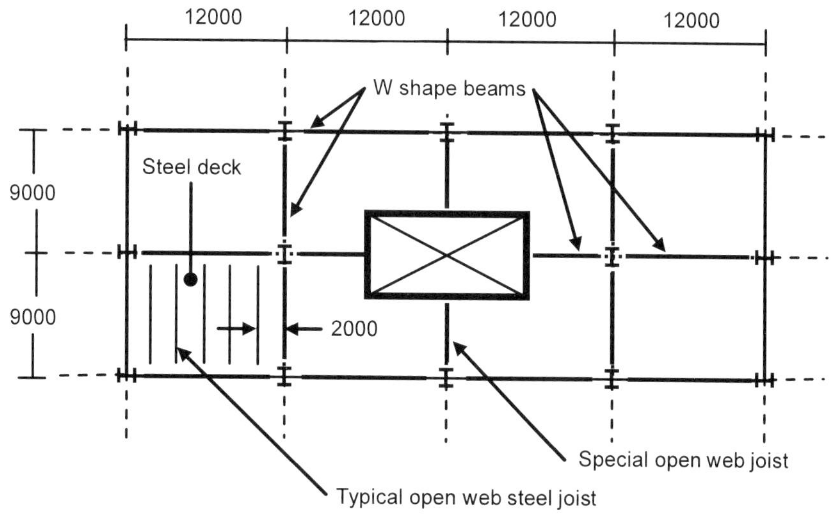

Figure 8. Floor assembly specified at the element level, alternative # 2 (dimensions in mm)

A fundamental difference between this approach and the ones proposed in section 2 is that here the architectural model is created by an architect and not by an architecturally constrained AI system, and alternative structural subsystems and layouts are proposed by the engineer and not by the computer. The computer only evaluates alternatives and suggests solutions on demand. Following this approach, the engineer can develop a structural a solution down to the assembly and element levels with knowledge-based assistance that corresponds to the level of detail required by the engineer.

6. Conclusions

An approach for knowledge-based interactive conceptual structural design has been proposed. The approach simplifies the conceptual design process by enabling the engineer to focus on the essential features only at each abstraction level while enabling quick structural synthesis. It has the following advantages over commercial applications for structural model generation: (1) it facilitates design exploration by proposing feasible design alternatives and enabling non-monotonic reasoning, (2) it constitutes a more efficient method for conceptual structural design because it simplifies the design problem by decomposition/refinement, (3) it enables more integrated design solutions because it uses structural design knowledge to evolve an architecturally constrained building information model, and (4) it facilitates decision-making and early architect-engineer negotiations by providing quantitative evaluation results. This research is work in progress. A knowledge-base is under development will be integrated to an improved StAr prototype.

Acknowledgments

The authors wish to acknowledge the financial support for this research from the Natural Science and Engineering Research Council of Canada (NSERC) and the Canada Research Chair Program (for the Chair in Computer-Aided Engineering for Sustainable Building Design).

References

Bailey S. and Smith I. (1994). "Case-based preliminary building design", ASCE J. of Computing in Civil Engineering, 8(4), 454-467.

Bédard C. and Ravi M. (1991). "Knowledge-based approach to overall configuration of multistory office buildings", ASCE J. of Computing in Civil Engineering, 5(4), 336-353.

Eisfeld M. and Scherer R. (2003). "Assisting conceptual design of building structures by an interactive description logic based planner", Advanced Engineering Informatics, 17(1), Elsevier, 41-57.

Fenves S.J., Rivard H., and Gomez N. (2000). "SEED-Config: a tool for conceptual structural design in a collaborative building design environment", AI in Engineering, 14(1), Elsevier, 233-247.

Gomez N. (1998). Conceptual structural design through knowledge hierarchies. PhD thesis, Department of Civil and Environmental Engineering, Carnegie Mellon University, Pittsburgh.

Grierson D.E. and Khajehpour S. (2002). "Method for Conceptual Design Applied to Office Buildings", ASCE J. of Computing in Civil Engineering, 16, 83-102.

Jain D., Krawinkler H., Law K. (1991). "Logic-based conceptual structural design of steel office buildings", Report No 49, Center for Integrated Facility Engineering, Stanford University.

Khemlani L. (2005). AECbytes product review: Autodesk Revit Structure, Internet URL: http://www.aecbytes.com/review/RevitStructure.htm.

Kumar B. and Raphael B. (1997). "CADREM: A Case-based system for conceptual structural design. Engineering with Computers"; 13(3), 153-164.

Maher M. (1988). "Expert systems for structural design", J. of Computing in Civil Engineering; 1(4), 270-283.

Meyer S. (1995). "A Description of the structural design of tall buildings through the grammar paradigm", Ph.D. Thesis, Dept. of Civil Engineering, Carnegie Mellon University, Pittsburgh.

Mora R., Rivard H., Bédard C. (2006A). "A computer representation to support conceptual structural design within a building architectural context", ASCE J. of Computing in Civil Engineering, to be published in the March issue.

Mora R., Bédard C., Rivard H. (2006B). "Geometric modeling and reasoning for the conceptual design of building structures", submitted for publication at the Journal of Advanced Engineering Informatics, Elsevier.

Rafiq M., Mathews J., Bullock G. (2003). "Conceptual building design – evolutionary approach", ASCE J. of Computing in Civil Engineering; 17(3), 150-158.

Rivard H. and Fenves S.J. (2000). "A representation for conceptual design of buildings", ASCE J. of Computing in Civil Engineering, 14(3), 151-159.

Sacks R., Warszawski A. and Kirsch U. (2000). "Structural Design in an automated building system", Automation in Construction, 10(3), Elsevier, 181-197.

Shea K. and Cagan J. (1998). "The design of novel roof trusses with shape annealing: assessing the ability of a computational method in aiding structural designers with varying design intent", Design Studies; 20, 3-23.

Sisk G., Miles J., Moore C. (2003). "Designer centered development of GA-based DSS for conceptual design of buildings", ASCE J. of Computing in Civil Engineering, 17(3), 159-166.

Soibelman L. and Peña-Mora F. (2000). "Distributed multi-reasoning mechanism to support conceptual structural design" ASCE J. of Computing in Civil Engineering; 126(6), 733-742.

Solibri Inc. (2005). Solibri Model Checker, Internet URL: http://www.solibri.com/services/public/main/main.php, last visited 6/10/2005.

HIGH PERFORMANCE COMPUTATIONAL MODELLING OF MICROSTRUCTURAL PHENOMENA IN POLYCRYSTALLINE METALS

K. Inal and K.W. Neale
Faculty of Engineering,
University of Sherbrooke,
Sherbrooke, Quebec, Canada J1K 2R1,
E-mail: Kenneth.Neale@USherbrooke.ca

Abstract

In this paper, applications of crystal plasticity theory to the numerical modelling of large strain plasticity phenomena are considered. In particular, instabilities and localized deformation phenomena for face-centred cubic (FCC) and body-centred cubic (BCC) polycrystals subjected to various deformation modes are investigated. In-house finite element analyses based on a rate-dependent crystal plasticity model have been developed to simulate the large strain behaviour for sheet specimens subjected to plane strain and plane stress deformation modes. In the formulation, the plastic deformation of an individual crystal is assumed to be due to crystallographic slip and simulations are performed using two approaches. In the first approach, each material point in the finite element analysis is considered to be a polycrystalline aggregate having a large number of FCC or BCC grains, and the Taylor theory of crystal plasticity is adopted to model the behaviour of the polycrystal. In the second approach, each grain is represented individually using one or more finite elements, and the constitutive response within each element is given by the single crystal constitutive model. Both approaches account for initial textures, as well as texture evolution during large plastic deformations. The numerical analyses incorporate parallel computing features. The results of simulations for the above-mentioned deformation modes are discussed, and in certain cases comparisons are made with experimental results for rolled aluminum sheet alloys and for draw quality steels.

Introduction

The mechanical properties of a polycrystalline metal depend on many attributes of its microstructure; consequently, considerable efforts have been devoted to the study of micromechanics. These studies indicate that, among the factors which result in the plastic deformation of single crystals and polycrystals, crystallographic slip occurring by the migration across the slip planes of atomic defects, termed dislocations, is the dominant one.

Crystallographic slip induces lattice rotations, which result in a non-random distribution of the crystal orientations in polycrystals. The textures developed during forming processes are macroscopic averages of such non-random orientations. Research indicates that texture occurs in many metal forming processes such as drawing, extrusion, rolling and sheet metal forming. These textures not only have profound effects on the mechanical and thermal properties of metals, but also have great influence on subsequent fabrication processes as well as on the quality of the products. Thus, it is obvious that accurate simulations of large strain phenomena should consider initial texture and its evolution, as well as the anisotropy induced by the evolution of microstructure and microscopic properties.

To model processes such as texture evolution and its influence on deformation-induced anisotropy, micromechanically based models of plastic behaviour are required. In particular, constitutive relations formulated on the concepts of crystal plasticity must be adopted. Since Taylor's pioneering work in 1938, the prediction of the deformation behaviour of polycrystalline solids from the response of their single crystal constituents has been the focus of many investigations. Thus, many crystal plasticity

M. Pandey et al. (eds), Advances in Engineering Structures, Mechanics & Construction, 583–593.

models have been proposed or modified to simulate the behaviour of polycrystalline metals during plastic deformation from the response of their single crystal constituents.

The mathematical modelling of material behaviour is a very effective way of reducing time and costs involved in optimizing manufacturing processes. Indeed, numerous complex forming operations have been simulated using numerical methods in order to predict critical parameters. Up to the 1980's, most applications involving numerical techniques such as the finite element method have been based on phenomenological constitutive models since microscopic models are significantly more demanding in terms of computational resources. However, the introduction of parallel computers has rendered metal forming modelling based on crystal plasticity feasible since they offer more computational power and storage than serial computer architectures. With proper parallelization techniques, realistic applications based on crystal plasticity can be performed on parallel supercomputers.

In this paper, applications of crystal plasticity theory to the numerical modelling of large strain plasticity phenomena are considered. Crystal plasticity theory is employed to model both FCC and BCC polycrystals where modelling of the polycrystalline aggregates is carried out at various scales. We first recapitulate the constitutive model. Then the parallel computing algorithms are briefly presented. In the last section we present two different applications where crystal plasticity theory is employed to simulate instabilities and localized deformation phenomena for FCC and BCC polycrystals subjected to plane strain tension and plane stress tension.

Constitutive Model

The polycrystal plasticity model formulated by Asaro and Needleman (1985) is employed in the analyses. Accordingly, the total deformation of a crystallite is taken to be the result of two distinct physical mechanisms: crystallographic slip due to dislocation motion on the active slip systems, and elastic lattice distortion. Within an FCC crystal, plastic deformation occurs by crystallographic slip on the 12{111}<110> slip systems. For a BCC crystal, crystallographic slip is assumed to occur on 24 slip systems, the 12{110}<111> + 12{112}<111> systems.

In the rate-sensitive crystal plasticity model employed, the elastic constitutive equation for each crystal is specified by:

$$\overset{\nabla}{\boldsymbol{\sigma}} = \boldsymbol{L}\boldsymbol{D} - \dot{\boldsymbol{\sigma}}^{0} - \boldsymbol{\sigma}\mathrm{tr}\boldsymbol{D} \tag{1}$$

where $\overset{\nabla}{\boldsymbol{\sigma}}$ is the Jaumann rate of Cauchy stress, $\boldsymbol{D}$ represents the strain-rate tensor and $\boldsymbol{L}$ is the tensor of elastic moduli. The term $\dot{\boldsymbol{\sigma}}^{0}$ is a viscoplastic type stress-rate that is determined by the slip rates on the slip systems of a FCC and BCC crystal. A detailed presentation of the crystal plasticity constitutive model can be found in Wu et al. (1997) and will not be repeated here.

The slip rates are governed by the power-law expression

$$\dot{\gamma}_{(\alpha)} = \dot{\gamma}_{(0)}\, sgn\,\tau_{(\alpha)} \left| \frac{\tau_{(\alpha)}}{g_{(\alpha)}} \right|^{1/m} \tag{2}$$

where $\dot{\gamma}_{(0)}$ is a reference shear rate taken to be the same for all the slip systems, $\tau_{(\alpha)}$ is the resolved shear stress on slip system α, $g_{(\alpha)}$ is its hardness and m is the strain-rate sensitivity index. The $g_{(\alpha)}$

characterize the current strain-hardened state of all the slip systems. The rate of increase of the function $g_{(\alpha)}$ is defined by the hardening law:

$$\dot{g}_{(\alpha)} = \sum_{\beta} h_{(\alpha\beta)} \left| \dot{\gamma}_{(\beta)} \right| \tag{3}$$

where $g_{(\alpha)}(0)$ is the initial hardness, taken to be a constant τ_0 for each slip system, and where the $h_{(\alpha\beta)}$ values are the hardening moduli. The form of the moduli is given by

$$h_{(\alpha\beta)} = q_{(\alpha\beta)} h_{(\beta)} \text{ (no sum on } \beta\text{)} \tag{4}$$

where $h_{(\beta)}$ is a single slip hardening rate and $q_{(\alpha\beta)}$ is the matrix describing the latent hardening behaviour of the crystallite.

The single slip hardening law employed in this investigation takes the following power-law form of the function $h_{(\beta)}$

$$h_{(\beta)} = h_0 \left(\frac{h_0 \gamma_a}{\tau_0 n} + 1 \right)^{n-1} \tag{5}$$

where h_0 is the system's initial hardening rate, n is the hardening exponent and γ_a is the accumulated slip.

Two different models are employed to obtain the response of a polycrystal comprised of many grains. In the Taylor model, the material response is obtained by invoking the Taylor assumption. Accordingly, at a material point representing a polycrystal of N grains, the deformation in each grain is taken to be identical to the macroscopic deformation of the continuum. Furthermore, the macroscopic values of all quantities, such as stresses, stress-rates and elastic moduli, are obtained by averaging their respective values over the total number of grains at the particular material point. In the FE/grain model an element of the finite element mesh represents a single crystal, and the constitutive response at a material point is given by the single crystal constitutive model. This approach enforces equilibrium and compatibility between grains throughout the polycrystalline aggregate in the weak finite element sense.

Parallel Computing

In general, Taylor-type polycrystal models are ideally suited for the parallelization of the computational procedures. Especially, when CPU time is considered, the simulations fall in the category of "embarrassingly parallel" (e.g., Sorensen et al. (1995)) applications, and they provide significant computational improvements. However, such "embarrassingly parallel" applications are strictly feasible only if the total program size fits within a single processor of the parallel computer. This is not the case for the simulations presented in this paper, and it was necessary to implement the polycrystal FE model in a data parallel form as described by Beaudoin et al. (1993) and Inal et al. (2002a, 2003).

The parallel computing algorithms employed in the simulations are designed to distribute data on the microscopic level (crystal data) over the processors of a distributed memory supercomputer. By this method, the global size of the simulation is distributed between the processors of the parallel computer. To illustrate this, consider a simulation with a total number N of crystals (Fig.1). The basic idea in the finite element formulation is that each material point is representing a polycrystal comprised of N crystals and the constitutive response is given through the Taylor polycrystal model (Fig. 1a). The global crystal data is distributed between the processors (Fig. 1b) such that each processor runs a part of the global program for $B = N/A$ crystals where A is the total number of processors used in the simulation. (Note that the processors read only the crystal data to which they are assigned and all arrays containing microscopic quantities have the maximum size of B instead of N.) Thus all processors compute microscopic arrays (for the set of crystals that they have assigned) independently. However, to compute the global stiffness matrix, the macroscopic values of stresses, stress rates and the moduli are required. These values are obtained by collective communication between the processors using the Message Passing Interface.

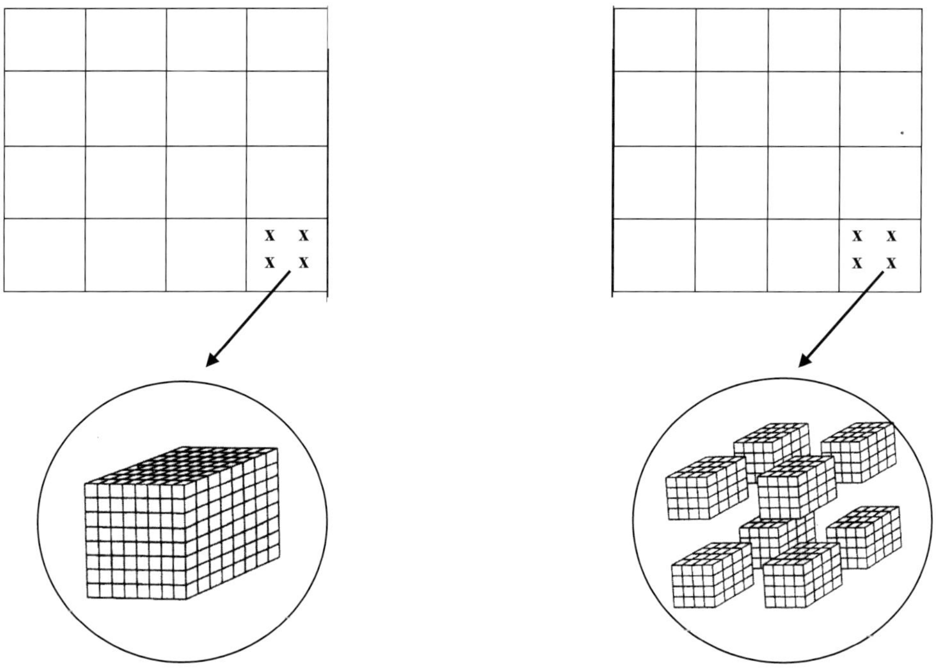

(a) (b)

Figure 1. (a) Polycrystal aggregate comprised of N crystals, (b) the distribution of this polycrystal aggregate between processors

The parallel computing algorithms which we have developed are essential for the simulations presented in this paper. These parallel sub-routines enable simulations with sufficiently fine meshes necessary to capture the key features of localized deformation for the aluminum alloy analysed.

Numerical Simulations

In this paper, crystal plasticity theory has been employed to simulate the effects of through-thickness texture gradients in the aluminum alloy AA5754 and the effects of strain paths in localized deformation in drawing quality (DQ) steel. It should be mentioned that the finite element meshes employed in both applications consist of four-node quadrilateral elements, made up of four 'crossed' constant strain triangular sub-elements. In presenting results, the quadrilateral is regarded as the basic element, and when reporting values of the field quantities the average value of the triangles is associated with the centroid of the quadrilateral.

The Effects of Through-Thickness Texture Gradients

It has already been mentioned that, to model processes such as texture evolution and its influence on deformation-induced anisotropy, models based on crystal plasticity have been employed in numerous studies. In these studies it has been assumed that the textures employed are representative of the entire volume. However, most forming processes do not produce materials with uniform spatial distributions of texture. Indeed, without sufficient care in the forming process, significant texture gradients develop (e.g., surface-to-midplane texture gradients in rolled materials, surface-to-core gradients in wires). Thus from a theoretical or practical point of view, it is important to investigate these texture gradient effects on plastic deformation properties.

In this section, a thin, orthotropic sheet specimen submitted to uniaxial tension is modelled (Fig. 2 where 40 x 56 elements are employed) under the assumption of plane strain conditions. The analyses assume no initial geometric imperfection. Localized deformation occurs as a result of the so-called "clamped" boundary conditions applied at the ends ($x_1 = \pm L_0$). With the tensile axis aligned in the x_1 direction, and x_3 being the direction normal to the sheet, the boundary conditions are

$$\begin{aligned} u_3 &= 0 \text{ along } x_1 = \pm L_0 \\ \dot{u}_1 &= \text{V (applied velocity) along } x_1 = L_0 \\ \dot{u}_1 &= \text{-V (applied velocity) along } x_1 = -L_0 \end{aligned} \tag{6}$$

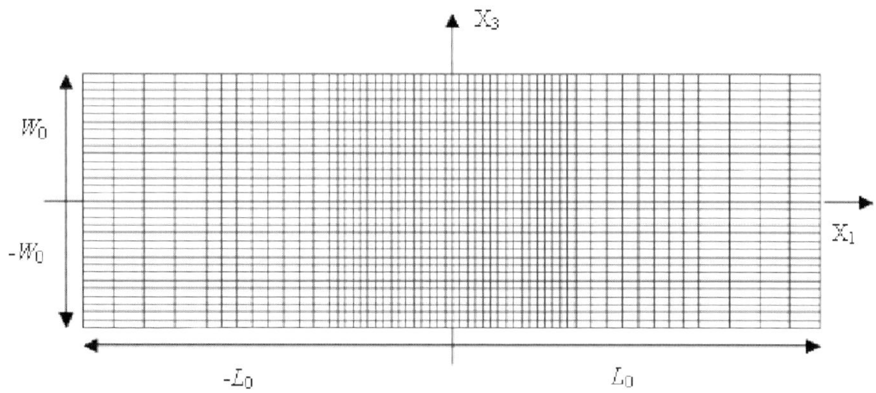

Figure 2. Finite element mesh used in the simulations

A set of discretized orientations of approximately 400 grains, measured at 7 different locations through the thickness of the aluminium alloy AA5754 is employed in the simulations (Figs. 3a-g). It can be seen that the initial textures become sharper towards the centre of the sheet. Thus, the degree of anisotropy increases towards the centre of the sheet. The values of the material properties used in the simulations are

$$\tau_0 = 95 \text{ MPa}, \; h_0/\tau_0 = 1.2, \; \tau_s/\tau_0 = 1.16, \; h_s/\tau_0 = 0, \; q = 1.0 \tag{7}$$

The slip system reference plastic shearing rate $\dot{\gamma}_0$ and the slip rate sensitivity parameter m are taken as $\dot{\gamma}_0 = 0.001\text{s}^{-1}$, and m=0.002, respectively with the crystal elastic constants taken as C_{11}=206 GPa, C_{12}=118 GPa and C_{44}=54 GPa.

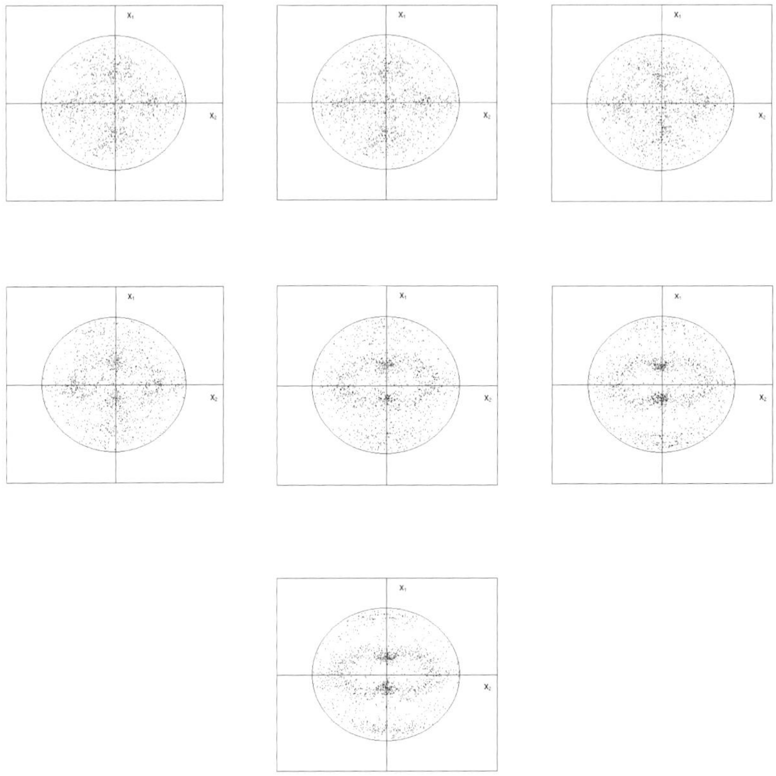

Figures 3a-g. Initial textures of the aluminum alloy AA5754 represented by {111} stereographic pole figures from the surface towards the centre of the sheet

A quantitative representation of shear band development is presented in Fig. 4 where contours of true strain (in the rolling direction) are plotted versus normalized elongation. It can be seen that at an elongation of U/L_0 = 0.065, a shear band passing through the centre of the specimen has already developed (Fig. 4a). With further stretching (U/L_0 = 0.07), even though strain has began to concentrate in this well defined shear band, a second shear band has developed perpendicular to the first one (Fig. 4b). The fully developed shear bands at U/L_0 = 0.15 are presented in Fig. 4c. Note that, although there are two fully developed shear bands intersecting at the centre of the specimen, the primary (first formed) shear band is sharper and wider than the secondary shear band. This pattern is due to the existing through-thickness texture gradients. Previous studies (Inal et al. (2002b, 2002c)) have indicated that when a single layer of texture was employed in the simulations (no texture gradients), multiple shear bands occurred simultaneously with the same intensities. It should also be mentioned that simulations of plane strain tension, where only a single layer of the initial textures (Figs. 3 a-g) was employed (no texture gradients), always predicted a single shear band.

A recent study by Inal et al. (2002b) has shown that when texture evolution is excluded from the analyses, localized deformation in the form of shear bands was not predicted during plane strain tension. To investigate the effect of through-thickness texture gradients on the predicted localisation modes (necking and/or shear banding), the simulation described above was performed once more, but with texture evolution excluded from the polycrystal model. Thus the stretching and rotation of the lattice vectors were excluded in the numerical analysis. Simulations have shown that, when through-thickness texture gradients are considered, even without texture evolution, localised deformation in the form of shear bands were predicted during plane strain tension (Fig. 5).

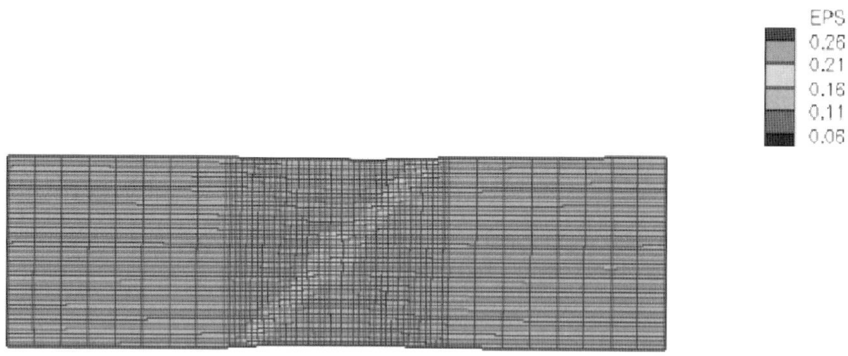

(a) $U/L_0 = 0.065$

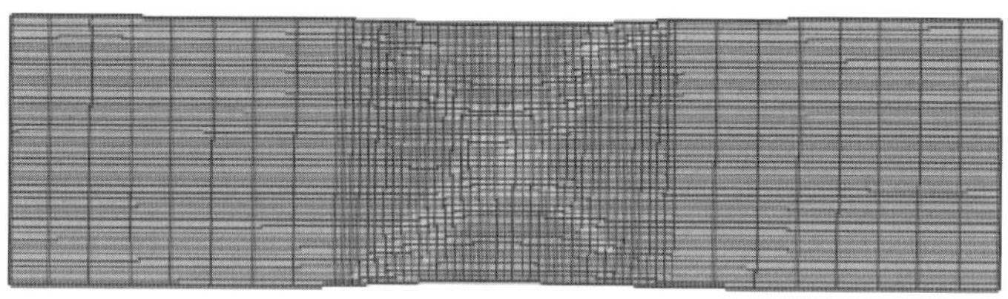

(b) $U/L_0 = 0.07$

(c) $U/L_0 = 0.15$

Figures 4a-c. True strain at various elongation levels

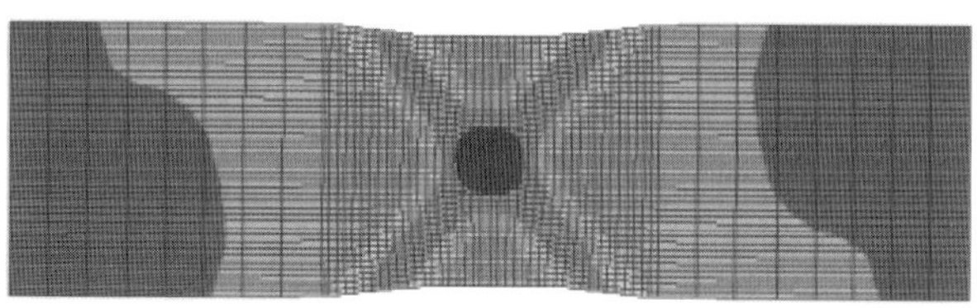

Figure 5. Deformed mesh at $U/L_0 = 0.12$ for the simulation without texture evolution

The Effects of Strain Paths on Localized Deformation in Drawing Quality (DQ) Sheet Steel

Many industrial processes require sheet metal to be subjected to several complex strain paths before the final product is manufactured. One such process is tube hydroforming, where tubed material is typically bent into a desired shape, placed within a die, then hyrdoformed to alter the tube cross-section. The strain path is complex in that the sheet material is first subjected to a near-plane strain path in the axial direction (i.e., bending), followed by a near-plane strain path in the circumferential direction (i.e., hydroforming). To simulate this process, sheet specimens were first pre-strained in the rolling direction (RD), then were rotated 90° clockwise and pulled along a second (orthogonal) path, such that the transverse direction (TD) was then aligned along the tensile axis (Fig. 6).

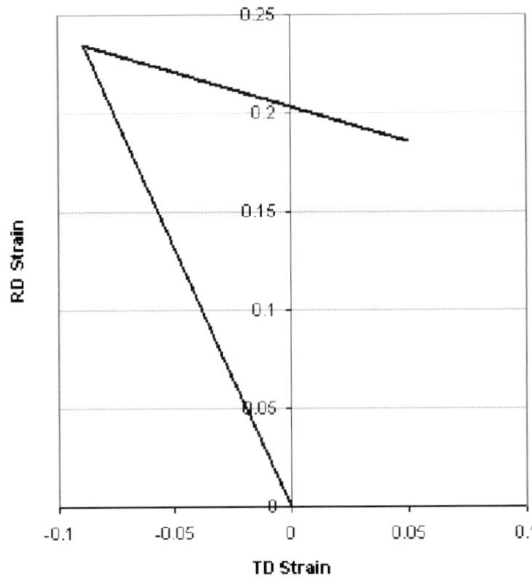

Figure 6. Experimentally observed strain paths for DQ sample

A so-called finite element (FE)/grain model together with a unit cell approach was employed to simulate the strain paths described above. In this model, each element of the finite element mesh represents a single crystal, and the constitutive response at a material point is that given by the single crystal constitutive model. A unit cell is defined as a globally small region of the sheet that contains all the essential micro-structural and textural features that characterize the sheet (e.g., Inal et al. (2005)). The sheet itself is subject to plane stress conditions (i.e., $\dot{\sigma}_{33} = 0$). Orientations within the measured texture data are randomly assigned in the mesh/unit cell. In other words, each element of the mesh represents an orientation from the measured texture. The loading imposed on the edges of the unit cell is assumed to be constant (Fig. 7), such that

$$\frac{\dot{\varepsilon}_{22}}{\dot{\varepsilon}_{11}} = \rho \tag{8}$$

where $\dot{\varepsilon}_{22}$ and $\dot{\varepsilon}_{11}$ are the (principal) logarithmic strain rates. The initial texture for the DQ steel represented by 400 grains/orientations is shown in Fig. 8. In this figure X_1 and X_2 correspond to the rolling and transverse direction of the sheet respectively. The values for the material parameters in the crystal plasticity analysis are $\dot{\gamma}_0 = 0.001\text{s}^{-1}$, m=0.05, $h_0 / \tau_0 = 28$, $\tau_0 = 54.5$ MPa, n= 0.18 and q=1.

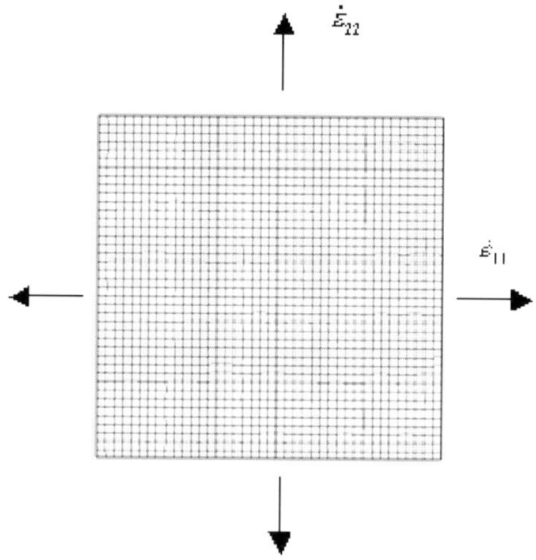

Figure 7. Schematic representation of a unit cell

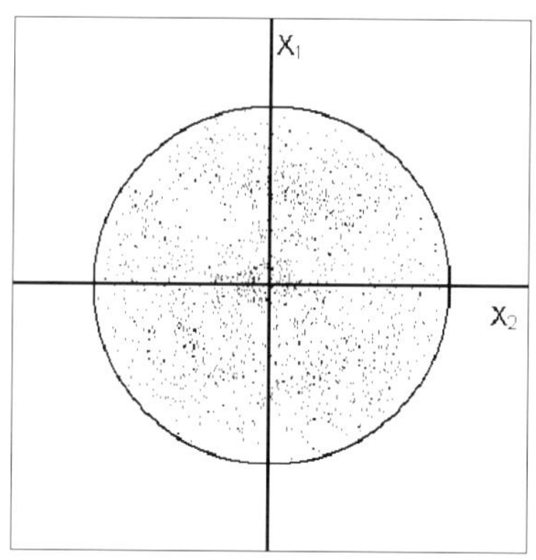

Figure 8. Initial texture represented in terms of {111} pole figure.

The strain paths observed in the experiments (Fig. 6) were imposed at the edges of the unit cell (Equation 8) and numerical simulations were performed with 1600 elements. Thus each grain in the initial texture is represented four times in our simulations. Figs. 9-10 present the measured and predicted textures after the strain paths defined in Fig. 6. It can be seen that the simulated texture is in good agreement with the measured texture. However, the simulated texture is slightly sharper than the measured one. This is probably due to the major drawback of the FE/grain model; i.e., the inevitable inhomogeneous spatial orientation distribution introduced numerically (since the initial texture is assigned randomly to the finite elements). To reduce the effect of spatial distribution of texture components in numerical simulations, the measured crystal orientations are randomly assigned N times to elements in the mesh. In our simulations N, was taken as 4. Usually the higher the N, the lower the effect of the inhomogeneous spatial orientation distribution introduced numerically. Thus, employing higher values on N will improve the overall predicted macroscopic and microscopic responses; predictions were slightly improved when N was taken as 8. However, employing electron backscattering diffraction (EBSD) data as input is the most efficient technique to minimize the effects of the spatial orientation distribution dependency.

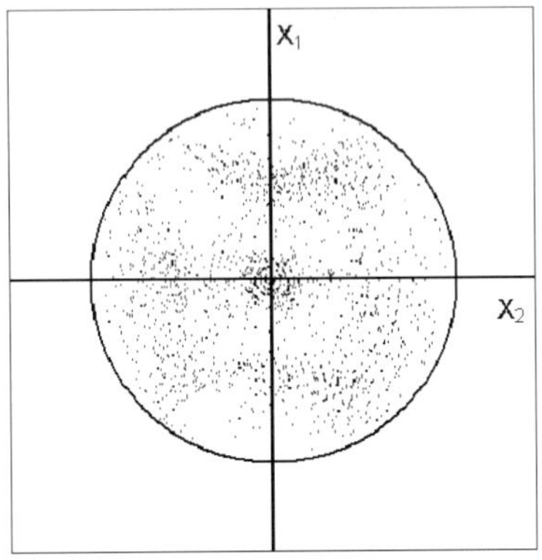

Figure 9. Experimentally measured texture

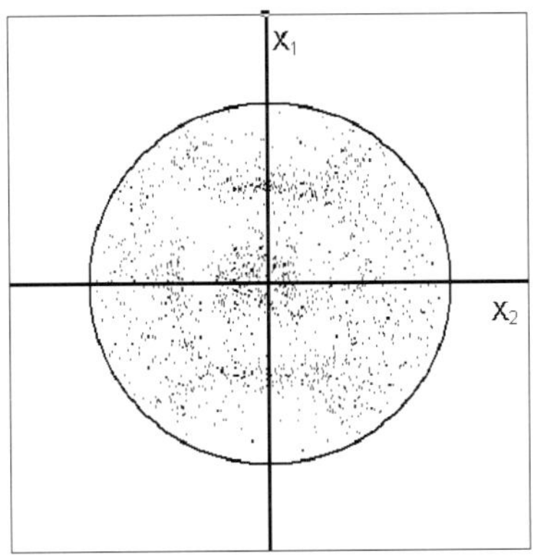

Figure 10. Simulated texture by FE/grain model

Conclusions

In this paper, the implementation of crystal plasticity constitutive relations in the numerical modelling of large strain phenomena was discussed. In particular, the effects of through-thickness texture gradient.

The common metals of industrial practice are polycrystalline aggregates which consist of single crystals or individual grains with lattice structures. As has already been discussed, the mechanical properties of a polycrystalline metal depend on many attributes of its microstructure. Thus, accurate modelling of large strain phenomena should include the effect of the initial microstructure and its evolution. Although phenomenological models are acceptable for many applications, they cannot explicitly include the basic physics of plastic deformation. However, large strain phenomena can be modelled more accurately based on crystal plasticity theories where the initial texture and its evolution, as well as the anisotropy induced by the evolution of microstructure and microscopic properties, are accounted for. Furthermore, with proper parallelization techniques and supercomputers, realistic applications with crystal plasticity theory are now feasible.

Numerical simulations have shown that both the Taylor model and the FE/grain model can be employed to simulate large strain plasticity phenomena in polycrystalline metals. It should be mentioned that the FE/grain model accounts for the grain morphologies (e.g., grain shapes and sizes), which cannot be modelled with the Taylor approach. However, the major drawback of the FE/grain model is the inevitable inhomogeneous spatial orientation distribution introduced numerically (since the initial texture is assigned randomly to the finite elements). By contrast, the spatial orientation distribution does not present this type of problem in simulations with the Taylor model since this model employs average values obtained from individual grains that do not depend on their relative locations.

Acknowledgements

This work was supported by the Natural Sciences and Engineering Research Council of Canada (NSERC). K.W. Neale holds the Canada Research Chair in Advanced Engineered Material Systems, and the support of this program is gratefully acknowledged.

References

Asaro, R.J., and Needleman, A. (1985): Texture development and strain hardening in rate dependent polycrystals. Acta Metallurgica, 33, 923-953.

Beaudoin, A.J., Mathur, K.K., Dawson, P.R., and Johnson, G.C. (1993): Three-dimensional deformation process simulation with explicit use of polycrystal plasticity models. International Journal of Plasticity, 9, 833-860.

Inal, K., Neale, K.W., Hamelin, C.J., Diak, B.J., and Pilkey, A.K. (2005): Numerical and experimental investigation of strain path effects on localized deformation in steel alloys. SAE Transactions, (in print).

Inal, K., Wu, P.D., and Neale, K.W. (2003): Parallel computing techniques for metal plasticity applications. High Performance Computing Systems and Applications, D. Sénéchal, Ed., NRC Research Press, Ottawa, Ontario, 133–137.

Inal, K., Neale, K.W., and Wu, P.D. (2002a): Parallel finite element algorithms for the analysis of multiscale plasticity problems. Applications of High-Performance Computing in Engineering VII, C.A. Brebbia, P. Meli and A. Zanasi, Eds., WIT Press, 75–84.

Inal, K., Wu, P.D., and Neale, K.W. (2002b): Instability and localized deformation in polycrystalline solids under plane-strain tension. International Journal of Solids and Structures, 39, 983-1002.

Inal, K., Wu, P.D., and Neale, K.W. (2002c): Finite element analysis of localization in FCC polycrystalline sheets under plane stress tension. International Journal of Solids and Structures, 39, 3469–3486.

Sorensen, N.J., and Andersen, B.S. (1995): A parallel finite element method for the analysis of crystalline solids. DCAMM Report, The Technical University of Denmark, 1-15.

Wu, P.D., Neale, K.W., and Van der Giessen, E. (1997): On crystal plasticity FLD analysis. Proceedings of the Royal Society of London, 453, 1831-1848.

INTERACTIVE VISUALISATION SYSTEMS FOR CONCEPTUAL BUILDING DESIGN: A PRACTICAL APPROACH

M.Y. Rafiq, M. Beck and I. Packham
University of Plymouth, UK
E-mail: mrafiq@plymouth.ac.uk

Abstract

At the conceptual stage of the design process where only a partial specification for a design is available and due to fuzzy nature of information at this stage it is difficult to program every design requirements. Experience has shown that evolutionary computation EC, (particularly the genetic algorithm) to be an effective decision support tool for conceptual design. To make EC useful in this stage of the design it needs strong human interaction and guidance to lead the search in discrete regions of the search space to explore and discover more appropriate design concepts. Humans are extremely good at perceptual evaluation of designs according to criteria that are extremely hard to program (Eckert et al., 1999). As a result, they can provide useful fitness evaluation for interactive evolutionary systems. They can also include personal preferences to lead the search and exploration to a preferred direction. This kind of interaction is extremely important to satisfy design/client requirements, particularly at the conceptual stage of the design process. This paper introduces a novel approach which demonstrates that interactive use of evolutionary computation, assisted by visualisation tools, leads to a human-led search. A system which support human-led search and it is based on an interactive visualisation clustered genetic algorithm, developed by Packham and co-workers (Packham, 2003; Packham and Denham, 2003; Packham et al., 2004; Rafiq et al., 2004), is introduced and its application on an example of a multi-disciplinary decision making process is demonstrated.

Introduction

For over a decade not only have techniques from the field of Evolutionary Computation, EC (particularly the Genetic Algorithm, GA), played a major role in design optimisation, but they are also seen as an analogy of design methodologies (Gero et al., 1997). More recently their role has been recognised as support tool for conceptual design (Goldberg, 2002).

The GA is typically a knowledge lean search process, and while this can be seen as one of its major strengths thus providing it with its wide applicability; it can be a hindrance when attempting to incorporate the richness of human knowledge, experience and insight into the search process.

This is particularly exemplified at the conceptual stage of the design process where: Firstly, only a partial specification for a design is available and due to fuzzy nature of information at this stage it is difficult to program every design requirements. It is argued that EC approaches need strong human interaction and guidance to lead the search in discrete regions of the search space to explore and discover more appropriate design concepts; Secondly due to the ill-defined nature of requirements the process itself is not one of search, but rather an exploration of the requirements and the potential solutions to those requirements (Gero, 1993; Maher et al., 1995).

However, with an ancestry in the artistic creations of Dawkins (1986) a new field of Interactive Evolutionary Computing (IEC) has begun to emerge and has been adapted to meet to a wide spectrum of application areas (Takagi, 2001).

M. Pandey et al. (eds), Advances in Engineering Structures, Mechanics & Construction, 595–606.

The authors believe that human-led IEC visualisation tools assisted by the search and exploration power of EC could be a powerful tool for human designers. To this end, the next section, discusses and presents a broad categorisation of visualisation and human iteration approaches adopted in EC systems. More detailed examples of visual and interactive systems as applied to conceptual building design will follow. Finally the paper will describe research on the Interactive and Visualisation and Clustering Genetic Algorithm (IVCGA).

Visual and Interactive Evolutionary Systems

Different approaches have been used to visualise the results of the EC search, and to make the EC process more interactive and user friendly. Lately research, discussed in this section, has been extended to allow users to interact with the evolutionary system to direct search in order to explore regions that are potentially more interesting.

The aim of this section is to discuss and review both the involvement of visualisation techniques for GA search, and the categories of interactive GA systems. Finally the major attributes of a visual interactive GA are discussed, and how such systems play a role in human led design.

Interactive Evolutionary Computation

Interactive Evolutionary Computation (IEC) is an umbrella term covering a range of techniques where the user interacts with the Evolutionary Computation search process to direct or modify the search based upon subjective preferences.

In a review of IEC, Takagi (2001) presents two definitions corresponding to a *narrow* and a *broad* view: In the narrow view IEC is seen as "the technology that EC optimises the target system based upon subjective human evaluation as fitness values for system outputs" (Takagi, 2001, p. 1275). This narrow view is more traditionally referred to as Interactive Genetic Algorithms (IGA). On the other hand, the broad definition is seen as "the technology that EC optimises the target system having an interactive human-machine interface" (Takagi, 2001, p. 1275), where the user modifies the GA parameters. A similar, but slightly broader, categorisation of IEC is also made by Parmee (2002).

In the IGA the user effectively replaces the fitness function. The classical example of this is the Biomorhps of Dawkins (1986), in this instance the user is presented with images of a series of candidate designs and selects a subset for reproduction. Their usage is normally in situations where it is not possible to provide a quantifiable fitness function, i.e. in the case of evaluating designing for their aesthetic content.

The approach adopted by IGA's has been successfully applied across a range of domains: Graphic art (e.g. Unemi, 2000; Todd and Latham, 1999); Industrial design (e.g. Graf, 1995); Face image generation (e.g. Takagi and Kishi, 1999); Speech processing (e.g. Watanabe and Takagi, 1995); Geological modelling (Wijns et al., 2003). The reader is referred to Takagi (2001) and Banzhaf (1997) for a more comprehensive review of applications.

While the IGA has only found limited usage in conceptual building design (Buelow 2002), it has been argued that such systems have a major role during conceptual design and "can fit naturally into human design thinking and industrial design practice" (Eckert et al., 1999).

Despite the obvious success of this approach it does suffer shortcomings, one of these being human fatigue. In a typical EC search the population size is usually one hundred and may evolve for a considerable number of generations. Human evaluation of this large number of individuals leads to not only inevitable fatigue, but also it is cognitively difficult for humans to rank or choose between this number of individual concepts generated by the system.

One approach to this problem is to use smaller population sizes across fewer generations. On more complex design problems with high dimensional search spaces a small population size may not cover the search space adequately but also exhibits rapid convergence.

To overcome the small population size issue, predictive evaluation techniques can be adopted. Here the user assigns a fitness score to a few selected individuals and the IGA predicts the fitness of the remaining individuals. This prediction can be either based on machine learning techniques (i.e. Neural Networks, Biles et al. 1996), or a Euclidian based similarity measure. The difficulty with Euclidian techniques is to develop a scale which mirrors the human evaluation scale.

The IGA may only have a limited role to play as an exploratory design tool in its fullest sense. If EC approaches are used as exploratory tools then it is necessary to allow the user to explore, or to move the search to, differing areas of the solution space which the user (i.e. expert designer) finds particularly interesting or due to a different emphasis on particular design requirements, and to have the ability to return to those areas for later investigation.

Currently it would seem that while IGA's are very effective in allowing a user to guide the search through the solution spaces, they do not readily allow the re-visiting of particular choice points. This ability to move the search, or focus at a specific region, is particularly important during conceptual design where the actual requirements themselves are being re-defined through the exploration process.

When the broader definition of Takagi (2001) is considered, again variants can be identified. One approach which uses human interaction in its fullest sense is the Human Based Genetic Algorithm (HBGA) of Kosorukoff (2001) and Kosorukoff and Goldberg (2002). The HBGA extends the IGA approach of the user driven selection, by allowing the user the option to perform and modify the GA operations of crossover and mutation. With the motivation that humans prefer to be creators rather than critics, in essence the evolutionary search process is now guided by human innovation. Moreover the HBGA is seen as a key member in collaborative problem solving and is playing a major role in the Free Knowledge Exchange (FKE) project and the DISCUS[1] project of Goldberg et al. (2003).

An alternative, and more traditional, approach is where the user is allowed to set or to modify GA parameters. Thus the work of Parmee et al. (2000) allows the user to set design objective preferences at the start of the GA search process, whilst the work of Mathews and Rafiq (Mathews, 2000; Rafiq et al., 2003; Rafiq et al., 2001) allows user modification of GA parameters during the GA run.

Visualising Evolutionary Computation

In terms of visualisation three main approaches can be identified: Firstly, tracing the ancestry of the evolving individuals or representing the genetic composition of individuals (Collins, 2002; Hart and Ross, 2001; Smith et al., 2002). The information presented in this approach is of interest to researches in Evolutionary Computation, rather than designers or users of such systems.

Secondly, the simplest attempt to introduce visualisation to help designers was that of a simple fitness plot and later on interaction with the fitness plot to backtrack the evolution process (Mathew, 2000), and in the multi-objective case, plotting the Pareto front or Pareto surface (Grierson and Khajehpour, 2002), puts this technique to good effect.

Thirdly, in high dimensional spaces it is difficult for the designer to see relationship between the design variables or the interaction with the objectives. The work of Abraham and Parmee (2004) uses a box-plot technique to display design variable – objective interaction, whilst Hayashida and Takagi (2000) used techniques to map high dimensional spaces to a 2-D mapped space.

[1] DISCUS: Distributed Innovation and Scalable Collaborations in Uncertain Settings.

In all cases discussed so far, visualisation is performed during or after the GA run has terminated and the user is not able to interact with the system either to refine the search or to move the search to a different area of the search space.

Closing Comments

This section discussed the ideas of interaction and visualisation as separate concepts and attempted a categorisation of each. In the following section a number of examples of illustrating how these concepts have been applied to Engineering Design problems are presented.

Following this, we present details of a unique approach which combines interaction and visualisation, the Interactive Visualisation and Clustering GA (IVCGA) developed by Packham et al (Packam, 2003; Packham and Denham, 2003; Packham et al., 2004; Rafiq et al., 2004)

Examples of Interactive and Visualisation Systems for Engineering Design

A comprehensive review by (Takagi 2001) lists a number of approaches to and applications of the IEC.

One of the first preliminary design systems devised by Pham and Yang (1993) allowed the user to view and evaluate solutions produced by the GA. Jo (1998) discovered that adding human interaction to his evolutionary design system allows domain knowledge to be incorporated online; solutions can be independently visualized in a space layout problem and the user was able to modify individual elements of the design. The interaction of a user has also been considered in a multi-objective environment: Fonesca and Fleming (1993) proposed a decision-maker (DM) that controls which objectives have more importance within a non-dominated set of solutions. They suggested the DM could be a human or an expert system. Horn (1997) pointed out that there are three different approaches to decision making in multi-criteria problems: make a multi-criteria decision before search, make a decision after search or integrate the search and decision making. The latter approach would appear to be the most powerful, incorporating iterative search and decision making.

Mathews and Rafiq (Mathews, 2000; Rafiq et al., 2003; Rafiq et al., 2001) developed a Conceptual Building Design (CBD) system using the Structured GA. CBD allows the user to manipulate the system more interactively, a powerful GUI was used which included a number of interactive dialogue boxes for effective user interaction. To allow better interaction with the system at run time, the design hierarchy was made available to the designer. The tree control, as shown in Figure 1, provides basic functionality for manipulating the nodes (different frame systems) and branches in a hierarchy (different design options). Using this facility allows the designer to include or exclude particular design options from the GA search, at the runtime. This facility is considered to be useful in a number of ways. It allows the user to force the GA search to follow a particular branch of the design hierarchy, which may not be considered by the GA as a best choice. This is also useful if the design brief requires a particular construction material to be used or a designer/client prefers a particular floor system, etc. to be considered.

In this limited interaction the system allows the designer to trace the design evolution process during a whole run of a GA operation. This is an important facility which adds transparency to the otherwise 'black box' GA operation. For example by clicking the mouse in a point on the graph, the corresponding details of the concept is shown in the second window. This facility was also made available during a genetic experiment while the GA was paused. An example of the use of this facility is presented in Figure 2.

The system also allows pausing the GA search at any time and changing either the GA parameters (e.g. mutation and crossover probabilities, etc.) or directing the search to specific branches of the design hierarchy.

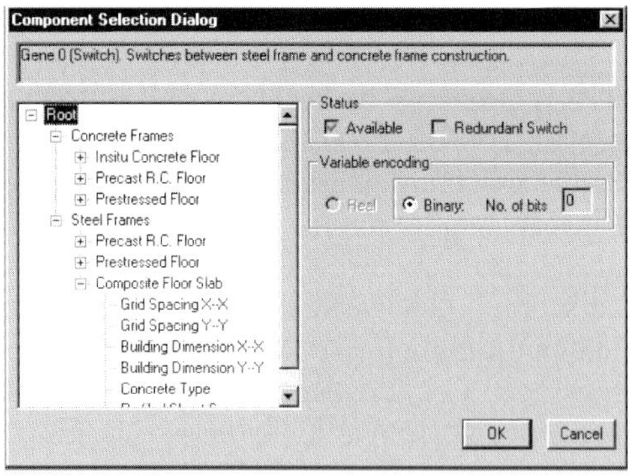

Fig. 1. Dialogue box representing the design hierarchy.

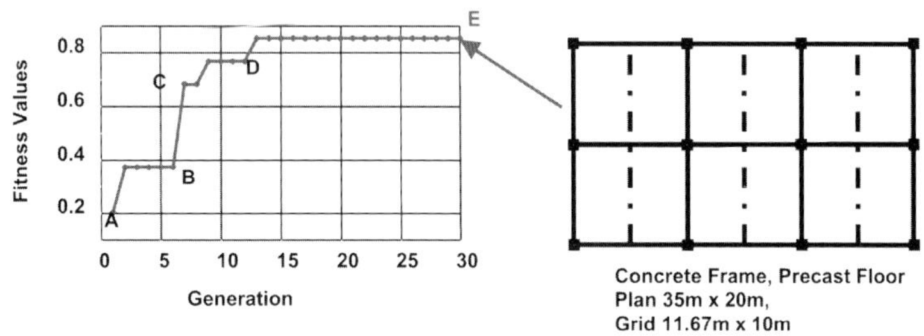

Fig. 2. Tracing design Evolution during a GA run.

Grierson and Khajehpour (2002) used Pareto optimisation techniques for multi-criteria conceptual design, involving genetic-based stochastic search and colour-filtered graphics. In this investigation, a large set of Pareto non-dominated designs were captured. They then used a computer colour filtering of the Pareto-optimal design set and created a large body of informative graphics that identify trade-off relationships between competing objective criteria, as well as design subsets having particular designer-specified attributes. A detailed illustration of the method for the cost-revenue conceptual design of high-rise office buildings, including several examples was presented. The work of Grierson and Khajehpour is an excellent example of knowledge discovery using a simple visualisation tool, and it demonstrates that by using visualisation tools effectively, it possible to discover the interrelationships between design parameters more readily.

Research in Plymouth Engineering Design Centre by Parmee and Bonham (1998) resulted in development of an Interactive Evolutionary Design System (IEDS) based on a system of iterative redefinition of variable and objective space by a designer. These ideas evolved from many years of research in using evolutionary computing to aid engineering design, starting from general ideas to locate and analyse robust regions of the search space. Development of cluster-oriented genetic algorithms (COGAs) allowed inclusion of user preferences between objectives to direct co-evolutionary search (Parmee et al., 2000). COGA extracts regions of good solutions from a GA run by filtering high performance solutions as the search progresses. More recently work on visualisation of COGA data has included the development of a novel technique called parallel coordinate box plots (Abraham and Parmee, 2004; Parmee and Abraham, 2004), that uses the parallel coordinate technique developed by Inselberg and Dimsdale (1994) to visualize many variables at once and compare the distribution of solutions between objectives using statistical analysis and discover overlap between various objectives.

The Interactive Visualisation and Clustering GA (IVCGA)

A major objective of the IVCGA, developed by Packham and colleagues (Packam, 2003; Packham and Denham, 2003; Packham et al., 2004; Rafiq et al., 2004), is to assist the designer through interaction to explore a range of feasible and innovative solution that best fit the design brief requirements. This process of exploration of search and solution spaces is more useful at the conceptual stage of the design process where design information is ill defined.

It is argued that such interaction is fundamental in real life design problems which are multi-dimensional, multi-criteria and multi-disciplinary in nature. Experts from different disciplines can interact individually with the system to explore their own areas of interest and evaluate the effect of changes that they propose on the overall design. It therefore, enhances the awareness of the project team on various aspects of the project which could lead to a mutually acceptable solution. This can lead to discovery of new knowledge about the specific regions of the search and solution spaces and more importantly understanding of collaborative design issues.

To this end, in the IVCGA the focus is on user interaction right from the start, allowing the user to freely interact with specific regions of the search space. Relating to the categories of visualisation and interaction discussed earlier, the IVCGA exhibits a number of unique features:

Firstly, in terms of visualisation the user is able to choose a number of high dimensional visualisation techniques which can help in understanding the relationship between design variables (i.e. 2D and 3D scatterplots, scatterplot matrix and parallel coordinates), additionally the data can be displayed in alternative coordinate systems, (i.e. 'Principal Components', 'Independent Components').

Secondly, a novel clustering technique based on kernel density estimation identifies the clusters in terms of the design variables. Clustering can be performed in alternative coordinate systems such as the principal components that reveal the 'natural' clusters in the data. Colour is used to identify different clusters.

Thirdly, in terms of interaction the user is in full control of the process and is either able to 'zoom in' in specific regions of the search space, i.e. on those regions containing specific colour coded clusters, enabling a more concentrated and focused search 'inside' such regions, or to extend the search 'outside' these regions to explore other possibilities which has not been currently discovered by the GA. It is both these cases which allow human experts to use their domain knowledge to guide the search to those areas of the search space which may meet differing, and possibly changing, design requirements. The nature of the IVCGA interaction places it inside the broader definition of IEC.

The Application of the IVCGA to Conceptual Building Design

Typically building design involves a number of individuals from a variety of disciplines, each with a differing emphasis on the design requirements. Thus, the objective of the study, presented in this section, is to illustrate how the IVCGA can be used in a multi-disciplinary design environment. The exercise has been conducted in close collaboration with an Architect, a Structural Engineer and a Heat and Ventilation Engineer, and it is noted that the differential requirements of each are:

- The Architect – Space, functionality and appearance and user comfort.
- The Structural Engineer – Building stability, strength and safety.
- The Heat and Ventilation Engineer – Mechanical, electrical, heating cooling and ventilation systems.

In this example, only two construction materials, concrete and steel, are included. Choice of suitable construction material could sometimes be a determining factor in selecting design options.

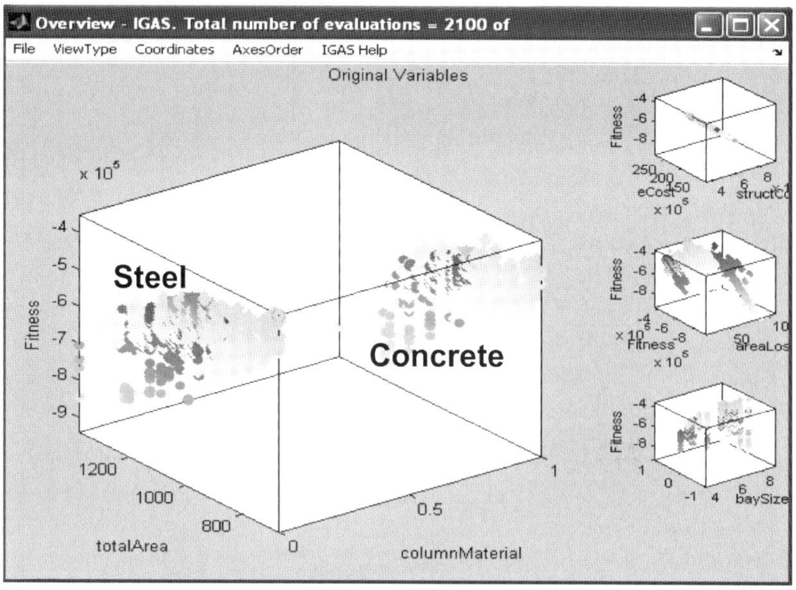

Fig. 3. Visualising solutions in Variable and Objective Spaces.

In Figure 3, concepts generated for each construction material (steel and concrete) is distinctly separated. This is a clear advantage of visualisation tools, which allow designers to investigate each alternative construction material independently and assess the suitability of each alternative against the design/client requirements. In most figures presented in this section the information is presented in the 'objective space' rather than the 'variable space'. This shows another distinct feature of the IVCGA is the ability to present data in either variable space or objective space. At times an objective space representation may be preferred to variable space when design variables are discrete variables.

To satisfy the architectural space requirements for this preliminary study, it is assumed that a major architectural requirement is the amount of available open space: A flexible floor area could be used by various clients with different functionality requirements throughout the life of the building. Therefore, there is a need to minimise the floor area lost due to columns and fire escapes. As the existence of columns in the floor makes the area around the columns restricted for use, it was decided to use a 1 m^2 unusable area for each column. In this study, a typical floor area of 1000 m^2 is assumed.

Figure 4 shows a population of designs generated after the first 20 generations of the GA. The frames on the right hand side show two alternative floor layouts in two different construction materials (steel and concrete). Both layouts satisfy the architectural requirement of minimum area loss. As the requirements suggests, fewer columns and other restrictive structural elements mean more letable floor area and hence more long term revenue.

The user is able to define which variables are plotted against each other, enabling designers to investigate any interrelationships between various design parameters, very quickly. The additional use of colour enables the user to emphasise the clusters that are important to the user whilst keeping other interesting data also available. For example, in Figure 4, concepts presented by cyan and magenta colours represent clusters of high fitness solutions.

Not only does the use of colour identify clusters of good designs, but also enables the designer to view individual solutions in the vicinity of these clusters in order to learn more about the interrelationship between design parameters. A particular strength of the IVCGA, is the ability to zoom in on particular areas of the solution space and to conduct a more focussed search in that area.

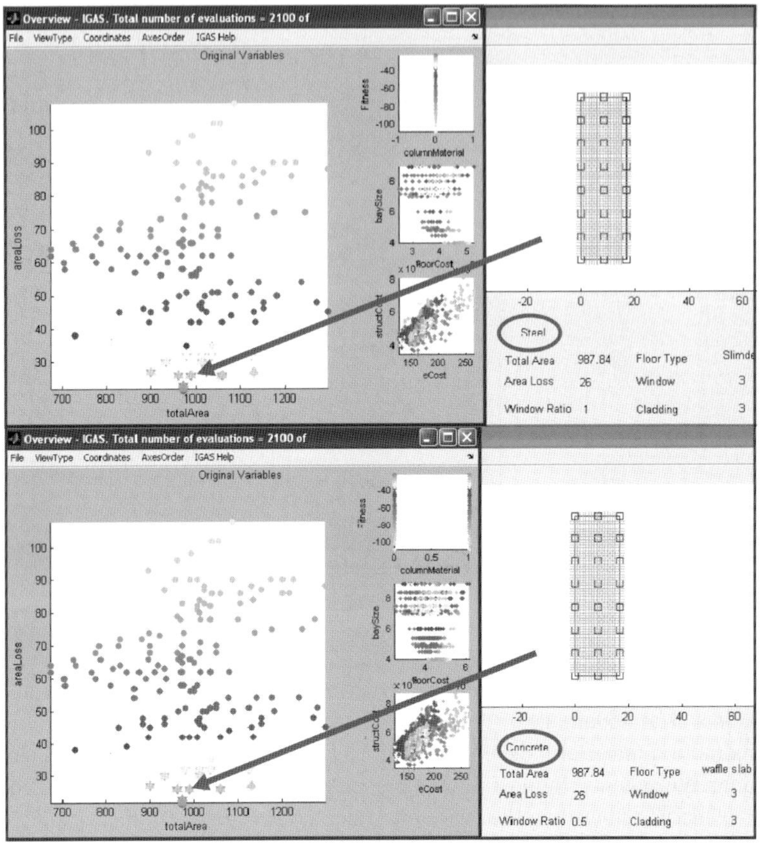

Fig. 4. Alternative concepts in two different construction materials.

To illustrate this, Figure 5 shows a 2D representation of the designs generated by the GA, based on the structural requirements. The left hand frame presents the entire population generated by the GA. No restrictions or filtering of information, to discourage the generation of unfeasible solutions, have been implemented. The rectangle on the left hand frame is interactively drawn by the designer, using the mouse, around the vicinity of 1000 m^2 required floor area, as was specified in the brief. The frame on the right hand side of Figure 5 shows only solutions which are enclosed inside this region. At this stage the system allows the designer to either view current designs generated by the GA within this region or conduct further runs the GA to populate this area with more designs. If this option is selected, the system automatically penalises any designs outside this region if generated by the GA operation. This feature of the system, which interactively and dynamically defines constraints to the problem, is a unique and very useful decision support tool.

A similar investigation could be conducted by each member of the design team to identify regions of suitable designs that best satisfy their individual disciplinary requirements. Finally, these areas are superimposed on each other to identify a mutually inclusive region that partially satisfies the requirements of all disciplines involved in the design. The frame at the left hand of Figure 6 presents all designs generated during separate runs of the GA. The cluster coloured copper represents all design that satisfies energy requirements. The green cluster represents Structural Engineer requirements and the cyan colour cluster represents the preferred region by the Architect. The red circle is the

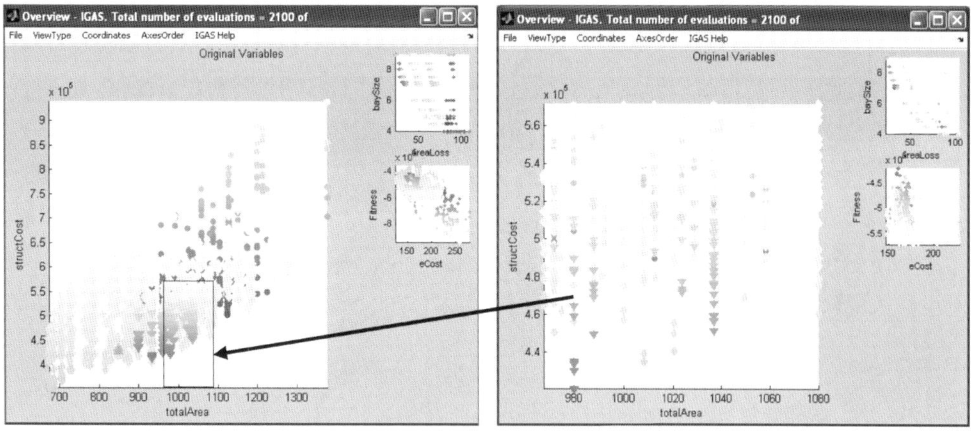

Fig. 5. 2D presentation of the most suitable region for the structural requirement.

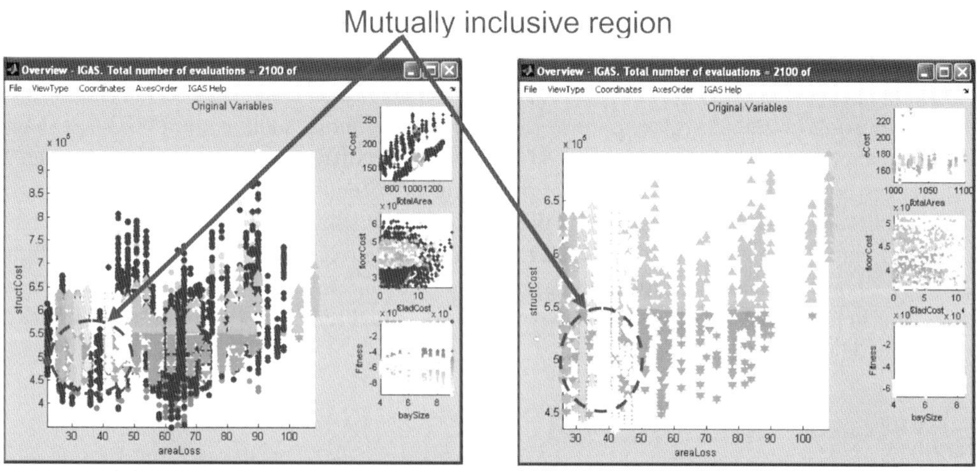

(a) All clusters **(a) Only 3 clusters**

Fig. 6. Mutually inclusive region for all three disciplines.

mutually inclusive region that partially satisfies the requirements of all parties involved. The frame at the right hand side of Figure 6 only shows clusters of suitable designs identified by the three disciplines independently.

Figure 7 present the above information in 2D and 3D scatter plots showing all design alternatives generated during separate runs of the GA. The cluster of designs coloured magenta represents the mutually inclusive region that partially satisfies the requirements of all parties involved.

Conclusions

The use of interactive visualisation undoubtedly gives confidence to the designer particularly at the conceptual stage of the design process by allowing her/him to conduct a human-led search in order to explore many possible solutions that satisfy design/client requirements. The IVCGA, is an effective decision support tool as it presents the designers with the opportunity to conduct search 'inside' and 'outside' the defined search and solution spaces and to evaluate the merit of each design generated,

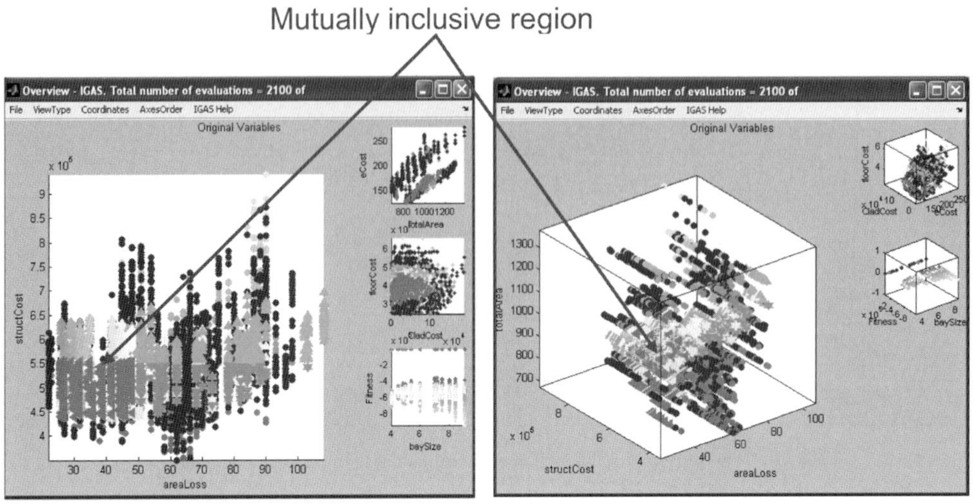

Fig. 7. Mutually inclusive region for all three disciplines in 2d and 3D views.

in term of existing or new design requirements. Visualisation tools, such as the IVCGA, enhances the multi-disciplinary design team members' understanding of the overall design issues and the consequences of the effect of changing one parameter or objectives on the overall design. Using visualisation tools it is possible to identify that region of the solution space that partially satisfies the requirements of all parties involved. Due to the fragmentation of activities between disciplines, agreeing on a compromise design framework presents difficulties in current multi disciplinary design practice. It is hoped that the use of tools such as IVCGA can facilitate better communication between parties involved in the design process.

References

Abraham, J.A.R. and Parmee, I.C. (2004) User-centric Evolutionary Design Systems – The Visualisation of Emerging Multi-Objective Design Information. In *Proceedings Xth International Conference on Computing in Civil and Building Engineering*, Weimar, Germany.

Banzhaf W. (1997) Interactive Evolution. In Bäck, T. (Ed.), *The Handbook of Evolutionary Computation*. Oxford: IOP Publishing and Oxford University Press.

Biles, J.A., Anderson, P.G. and Loggi, L.W. (1996) Neural Network Fitness Functions for a Musical IGA. In *IIA'96 / SOCO'96. International ICSC Symposia on Intelligent Industrial Automation and Soft Computing*, pp. 39–44.

Buelow, von P. (2002) Using Evolutionary Algorithms to Aid Designers of Architectural Structures Systems. In Bentley, P.J. and Corn, B.W. (Eds), *Creative Evolutionary Systems*. Morgan Kaufman.

Collins, T.D. (2002) Visualizing Evolutionary Computation. In Ghosh, A. and Tsutsui, S. (Eds), *Advances in Evolutionary Computation*. Springer Verlag.

Dawkins R. (1986) *The Blind Watchmaker*. New York: Norton.

Eckert, C., Kelly, I. and Stacy, M. (1999) Interactive Generative Systems for Conceptual Design: An Empirical Perspective. *Artificial Intelligence for Engineering Design Analysis and Manufacturing*, 13, 330–320.

Fonesca, C.M. and Fleming, P.J. (1993) Genetic Algorithms for Multiobjective Optimization: Formulation, Discussion and Generalization. In *Proceedings of the Fifth International Conference on Genetic Algoritms*. Morgan Kaufman, pp. 416–423.

Gero, J.S. (1993) Towards a Model of Exploration in Computer-Aided Design. In Gero, J.S. and Tyugu N. (Eds), *Formal Design Methods for Computer Aided Design*. Amsterdam: North-Holland.

Gero, J.S., Kazakov, V. and Schnier, T. (1997), Genetic Engineering and Design Problems. In Dasgupta, D. and Michalewicz, Z. (Eds), *Evolutionary Algorithms in Engineering Applications*. Berlin: Springer-Verlag, pp. 47–68.

Goldberg, D.E. (2002) *The Design of Innovation: Lessons from and for Competent Genetic Algorithms*. Boston, MA: Kluwer Academic Publishers.

Goldberg, D.E., Welge, M. and Llora, X. (2003) DISCUS: Distributed Innovation and Scalable Collaboration in Uncertain Settings, IlliGAL Technical Report 2003017. Illinois Genetic Algorithms Lab., University of Illinois at Urbana-Champaign.

Graf, J. (1995) Interactive Evolutionary Algorithms in Design. In *Proceedings International Conference on Artificial Neural Nets and Genetic Algorithms*, Ales, France, pp. 227–230.

Grierson, D.E. and Khajehpour, S. (2002) Method for Conceptual Design Applied to Office Buildings. *Journal of Computing in Civil Engineering*, 16(2), 83–103.

Hart, E. and Ross, P. (2001) GAVEL: A new tool for Genetic Algorithm Visualization. *IEEE Transactions on Evolutionary Computation*, 5(4), 335–348.

Hayashida, N. and Takagi, H. (2000) Visualized IEC: Interactive Evolutionary Computation with Multidimensional Data Visualization. In *Proceedings IECON 2000: IEEE International Conference on Industrial Electronics, Control and Instrumentation*, Nagoya, Japan, pp. 2738–2743.

Horn, J. (1997) Multicriteria Decision Making and Evolutionary Computation. In Bäck, T., Fogel, D.B. and Michalewicz, Z. (Eds), *Handbook of Evolutionary Computation*. Bristo, UK: Institute of Physics Publishing.

Inselberg, A. and Dimsdale, B. (1994) Multi-Dimensional Lines I: Representation & Multi-Dimensional Lines II: Proximity and Applications. *SIAM Journal of Applied Mathematics*, 54(2), 559–596.

Jo, J. (1998) Interactive Evolutionary Design System: Process and Knowledge Representation. In *Proceedings Fourth International Round-table Conference on Computational Models of Creative Design*, Heron Island, Australia, pp. 215–224.

Kosorukoff, A. (2001) Human-Based Genetic Algorithm. *IEEE Transactions on Systems, Man, and Cybernetics*, 5, 3464–3469.

Kosorukoff, A. and Goldberg, D. (2002) Evolutionary Computation as a Form of Organisation. In *Proceedings of the Genetic and Evolutionary Computation Conference (GECCO 2002)*. Morgan Kaufman, pp. 965–972.

Maher, L.M., Poon, J. and Boulanger, S. (1995) Formalising Design Exploration as Co-Evolution: A Combined Gene Approach. Preprints of the Second W\IPIF WG5.2 Workshop on Formal Design Methods for CAD.

Mathews, J.D. (2000) *Optimisation and Decision Support during the Conceptual Stage of Building Design: New Techniques Based on the Genetic Algorithm*, PhD Thesis, University of Plymouth, Plymouth, UK.

Packham, I.S.J. (2003) *An Interactive Visualisation System for Engineering Design Using Evolutionary Computing*, PhD Thesis, University of Plymouth, Plymouth, UK.

Packham, I.S.J. and Denham, S.L. (2003) Visualisation Methods for Supporting the Exploration of High Dimensional Problem Spaces in Engineering Design. In *Proceedings IEEE International Conference on Coordinated & Multiple Views in Exploratory Visualization (CMV2003)*, London, UK, pp. 2–13.

Packham, I.S.J., Rafiq, M.Y., Borthwick, M.F. and Denham, S.L. (2004) Interactive Visualisation for Decision Support and Evaluation of Robustness using Evolutionary Computing. In *Proceedings 11th International Workshop of the European Group for Intelligent Computing in Engineering (EG-ICE)*, Weimar, Germany.

Parmee, I. (2002) Towards Interactive Evolutionary Search and Exploration Systems. In *Proceedings Bird-of-a-Feather Workshop: Genetic and Evolutionary Conference*, New York.

Parmee, I.C. and Abraham, J.A.R. (2004) Supporting Implicit Learning via the Visualisation of COGA Multi-Objective Data. In *Proceedings IEEE Congress on Evolutionary Computation*, Portland, OR, USA, pp. 395–402.

Parmee, I.C. and Bonham, C.R. (1998) Supporting Innovative and Creative Design Using Interactive Designer/Evolutionary Computing Strategies. In *Proceedings Fourth International Round-Table Conference on Computational Models of Creative Design*, Heron Island, Australia, pp. 187–214.

Parmee, I.C., Cvetkovic, D., Watson, A.H. and Bonham, C.R. (2000) Multi-Objective Optimisation and Conceptual Engineering Design. *Journal of Evolutionary Computation*, 8(2), 197–222.

Pham, D.T. and Yang, Y. (1993) A Genetic Algorithm Based Preliminary Design System. *Institution of Mechanical Engineers, Part D: Journal of Automobile Engineering*, 207(2), 127–133.

Rafiq, M.Y., Mathews, J.D. and Jagodzinski, P. (2001) Interactive Role of the Human Computer Interaction in Design. In *Proceedings 8th Workshop of the European Group of the Structural Engineering Applications of Artificial Intelligence*, Loughborough, UK, pp. 31–42.

Rafiq, M.Y., Mathews, J.D. and Bullock, G.N. (2003) Conceptual Building Design – Evolutionary Approach. *Journal of Computing in Civil Engineering*, 17(3), 150–158.

Rafiq, M.Y., Packham, I.S.J., Easterbrook, D.J. and Denham, S.L. (2004) Biaxial Column Optimum Design with Better Representation of Search and Solution Spaces. *Journal of Computing in Civil Engineering*, to appear.

Smith, T., Bird, J. and Bullock, S. (2002) Beyond Fitness: Visualising Evolution – Workshop Overview. In *Proceedings of the 8th International Conference on the Simulation and Synthesis of Living Systems*, University of New South Wales, Sydney, Australia.

Takagi, H. (2001) Interactive Evolutionary Computation: Fusion of the Capabilities of EC Optimization and Human Evaluation. *Proceedings of the IEEE*, 89(9), 1275–1296.

Takagi, H. and Kishi, K. (1999) On-line Knowledge Embedding for an Interactive EC-Based Knowledge System. In *Proceedings 3rd International Conference on Knowledge-Based Intelligent Information Engineering Systems (KES-99)*, Adelaide, Australia, pp. 280–283.

Todd, S. and Latham, W. (1999) The Mutation and Growth of Art by Computers. In Bentley, P. (Ed.), *Evolutionary Design by Computers*. Morgan Kaufman, pp. 221–250.

Unemi, T. (2000) SBART 2.4: An IEC Tool for Creating 2D Images, Movies and Collages. In *Proceedings Workshop on Genetic Algorithms in Visual Art and Music*, Las Vegas, NV, USA, pp. 21–23.

Watanabe, T. and Takagi, H. (1995) Recovering System of the Distorted Speech Using Interactive Genetic Algorithms. In *Proceedings IEEE International Conference on Systems, Man and Cybernetics (SMC'95)*, Vancouver, Canada, pp. 684–689.

Wijns, C., Boschetti, F. and Moresi, L. (2003) Inversion in Geology by Interactive Evolutionary Computation. *Journal of Structural Geology*, 25(10), 1615–1621.

ASSESSING ALTERNATIVE PREFABRICATION METHODS: LOGISTICAL INFLUENCES

Robert Seaker[1] and Sang-Hoon Lee[2]
[1]*Department of Information and Logistics Tech., University of Houston, TX 77204, USA*
[2]*Department of Engineering Technology, University of Houston, TX 77204, USA*
E-mail: RFSeaker@central.uh.edu, slee15@uh.edu

Abstract

Any construction project that is completed on-time at the lowest total cost requires the consideration of logistics processes and economics. This study investigates the cost implications of moving and transforming materials in various materials network configurations associated with popular prefabrication construction methods. Efforts focus on the trade-offs that exist among contract-to-completion times, transportation costs, and assembly costs for the alternative construction methods. The findings suggest that the competitive advantage of prefabrication methods can be enhanced through an optimal combination of reduced construction times as well as the number, distance, and configurations of materials and sub-assembly shipments.

Introduction

The expression "time is money" appropriately applies to commercial initiatives, particularly construction projects. The duration of a construction project – from the time the contract is initiated to when the structure is completed –has a direct impact on the cost of the job. Bids from competing contractors, costs of construction loans, and earnings streams of business enterprises dependent on the finished structures are all affected by the time necessary to construct a building.

Prefabrication is a construction alternative that is primarily chosen because of its ability to reduce the total time for completing a project. Prefabricated construction refers to the utilization of sub-assembled structural components generally manufactured off-site (CII, 2002). From fully-manufactured buildings to those assembled from modularized, panelized, or even pre-cut components, the incremental degrees to which prefabrication can be employed is limitless. The reduction in construction times that these approaches provide can be mainly attributed to the degree of independence between site preparation and sub-assembly, more dependable job scheduling, and greater production efficiencies at the prefabrication sites as well as at the construction site (CII, 2002; Haas and Fagerlund, 2002; Kupitz and Goodjohn, 1991; Tatum et al., 1987).

While the reduction in construction time is perhaps the most visible and important benefit of prefabrication techniques, the costs associated with off-site activities pertaining to the movement of materials can impact the extent to which a particular level of prefabrication is economically feasible. The number, distance, and configurations of supply and sub-assembly shipments can add tens of thousands of dollars to projects of even a modest size. To gain further clarity regarding this issue, the effects of logistical activities associated with common prefabrication processes were investigated. The time and cost relationships that exist among these variables were identified, modeled, and analyzed. Efforts focused on the trade-offs that exist among contract-to-completion times, transportation costs, and assembly costs for alternative construction methods across multiple applications. Recommendations for strategies and configurations for improved economics are then offered.

M. Pandey et al. (eds), Advances in Engineering Structures, Mechanics & Construction, 607–614.

Logistics Processes and Variables

In the world of buying and selling consumer goods, logistics management involves the coordination of a variety of interrelated activities aimed at achieving optimal service and cost performance (Ballou, 2004; Stock and Lambert, 2001). Each activity can be classified as falling into one of three high-level functional areas: customer service, inventory control, and transportation management

Customer service activities involve identifying and locking in order fulfillment targets, ideally at the customer and stock-keeping unit level. These market-oriented metrics subsequently influence the service and cost objectives of the remaining two functional areas. Upon the identification of order fulfillment requirements, the levels of both inventory and transportation capabilities are collectively determined to support order fulfillment targets at the lowest total operational cost. While construction projects exhibit logistics variables similar to those found in conventional manufacturing processes, variations exist in how they are measured and applied.

For built to order (BTO) projects, the structural specifications, price, and completion time as stipulated in the contract is directly comparable to an order fulfillment target of a consumer goods supplier. This deliverable then would guide subsequent decisions regarding the amount and frequency of purchased inventory and the transportation activities required to support them. Within a construction environment, however, materials inventory are not recognized and measured in the same manner as consumer goods inventory. A commercial building is fixed asset and no alternative exists but to purchase a predetermined volume of materials as dictated by the building's specifications and the sequence of construction activities. Materials that are purchased earlier in the project will generate higher carrying costs than those purchased at later stages but the overall carrying cost effect would appear to be negligible.

The highest cost among the major logistics functions is most often attributed to transportation operations (Stock and Lambert, 2001). For most supply chain applications, the total cost from buying transportation services reflects the modes that are selected (i.e. air, ground, and water) as well as the distance and direction of shipments moving via those modes. Consistent with the adage "you get what you pay for", the per-pound freight charge for a small shipment that is rapidly transported over long distances is typically higher than that for a larger shipment that is transported at a slower speed over a shorter distance. Exceptions to this rule of thumb can be traced to the supply and demand of transportation equipment, non-linear ton-mile economics, and any additional services provided by the carrier.

Construction methods utilizing off-site prefabrication will naturally bear higher costs for transportation than conventional on-site methods. This can be attributed to any combination of the following: an increase in the total number of shipments, the distance and direction of shipments, and shipment configurations requiring higher-cost transportation capabilities.

In addition to transportation services, the facilities and manpower necessary for constructing a building will be reflected by the construction methods used. The nature of assembly processes and their locations in relation to those of their suppliers and to the building site will be important factors in determining the impact of logistics variables. Conventional, on-site "stick" construction will typically require less fixed overhead and less transportation activity than methods employing panelized or modularized components that are assembled off-site.

This paper focuses on the direct costs associated with the transportation alternatives pertaining to supply and assembly configurations of the various alternatives. Because of effects purported to be negligible in a close BTO operation, inventory carrying costs were not included in this analysis. Likewise, off-site overhead of the fabricators is considered a sunk cost and has not been incorporated into the model.

Method of Analysis

The core construction alternatives analyzed are off-site module prefabrication, off-site panel prefabrication, and traditional site-built construction. Each exhibit behaviors that can impact transportation needs and costs in various ways.

Guided by existing literature, a simple flow diagram was initially mapped to assist in the identification of common processes, locations, and logistical relationships required by alternative construction methods for the same project using the same suppliers. Figure 1 identifies the value-add tiers relevant to this investigation and the supply lanes that link them. It displays an aggregate supply lanes for shipments moving from origins A directly to construction site B as well as for shipments from the same origins to site construction site B via fabricator sites C.

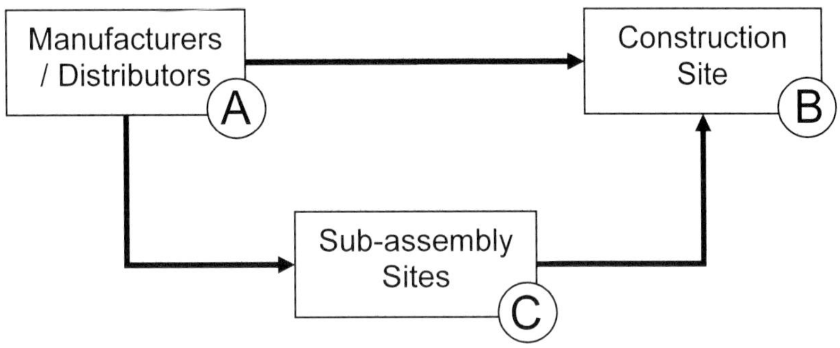

Figure 1: High level material flows for on-site and prefabricated construction alternatives

Traditional on-site construction efforts are supplied through lane A-B. Meanwhile, off-site sub-assembly efforts require that materials first move via lane A-C for sub-assembly and then move to the construction site via lane C-B. If supply lanes A-B and A-C are understood to be fairly similar based on the same bill of materials, shipment size, distance, and direction, then the combined time and costs effects of prefabrication activities at site C and the movement of materials via lane C-B is expected to be where significant effects exist in determining the most cost-effective method of construction.

Based on this framework for experimentation, relationships among the fixed and variable costs reflective of the resources and processes of actual construction operations were used to help build a more-detailed model. Comparative logistics scenarios for each alternative construction method were then created to better identify shared and isolated effects. The construction methods analyzed were:

a) No off-site prefabrication (traditional)
b) Panelized segments constructed off-site
c) Modular segments constructed offsite
d) Hybrids consisting of two or more of the above alternatives

Various levels and configurations of modularity require various transportation capabilities. Transportation alternatives for this study were limited to highway modes (e.g. 40' flatbed, 53' dry van) including oversized, permitted loads. The per-mile transportation rates based on shipment types that support the construction of a 3000 square foot structure are provided in Table 1.

Table 1: Shipment count and characteristics, by construction method, to supply materials for a 3000 square foot structure

Shipment Type	Average Shipment Weight	Number of Shipments	Rate Per Mile	Minimum Charge
No Pre-Fabrication, 53' Dry Van	37,500	4	$2.00	$300
Prefab Panels (20%-70% density loss), 53" Dry Van or Flatbed	20,000 - 30,000	5 - 7	$2.00	$300
Modules, Standard Load, Flatbed	16,000	10	$2.00	$300
Modules, Wide Load (8.5-12 ft.)	24,000	7	$7.00	$900
Modules, Wide Load (>12 ft.)	30,000	5	$15.00	$2,000

Per-mile transportation rates and minimum shipment charges were sampled from archival and web-based commercial sites on a basis. They were plotted and then averaged using line of best fit. Wide variations existed among the rates, particularly for local deliveries and oversized shipments. This could be attributed to the lack of uniformity in overhead allocation and equipment utilization among these specialized carriers. Additional validation was provided by two transportation professionals who reviewed the means and spread of the values.

For each scenario, the specifications of the finished structure as well as its site were held constant as time and total cost effects due to shipment volumes, configurations, and distances as well as opportunity costs of non-operating commercial enterprise were assessed.

Across multiple construction projects, huge variations in building specifications and transportation rates can exist. For example, modular construction can be used to construct small retail shops as well as multi-storied commercial structures such as the twenty-one floors of the Hilton Palacio del Rio hotel in San Antonio, Texas. Also, as earlier explained, oversized highway shipments can present large per-mile rate variations depending on shipment size, permits, among other factors. Therefore, key assumptions were required to help define and limit the scope. For each scenario,

a) Finished commercial structures are built under contract to the same specifications

b) Structures can be built using any construction method (i.e. modules, panels, kits, and traditional site-built) or any combination of methods.

c) All finished structures will bear a structural weight per square foot of 50 pounds and a construction cost to the customer of $100 per square foot. (averaged data from a variety of sources; e.g. Building Construction Cost Data, 2005)

Scenarios are differentiated by applying building sizes ranging from 3000 to 12000 square feet. Supply origins will first be held constant and then localized. Inventory carrying costs, construction loan interest, and off-site overhead were not included as variables.

Results

Interpretations of the data derived from the analysis are consistent with what is observed in a variety of supply chain environments. For example, manufactured products like potato chips that entail less density thus higher per-mile transportation costs than does its raw materials (i.e. potatoes) will put pressure on the manufacturing site to locate nearer the market. The data provided in Table 2 support this notion. Total costs for transportation services that support the supply of materials for finished structures of 3000 square-feet and 12,000 square-feet in size are provided. Each shipment type represents a sole application of that type to the project. That is, data for applications of mixed shipment types to the project are not listed.

The lanes identified by Figure 1 are also indicated. The two sizes define the range of sizes studied are used for the sake of efficiency in conveying the cost differences among shipment type (based on size and configuration), the miles traveled, and the total volume of materials moved. Transportation costs applying to finished structures of other sizes supplied over varying distances can be readily calculated by shipment type using this table because there are no multiplicative or exponential relationships among the variables as modeled. Calculations for projects of mixed technologies and/or shipment types could be approximated by weighed averages based of the percentage of usage or application among the variables.

Table 2: Total Transportation Charges per Lane for the Alternative Construction Methods (for 3,000 square-foot and 12,000 square-foot structures)

Shipment Type	3000 Square Feet				12000 Square Feet			
	25 miles	trans$ per sf	200 miles	trans$ per sf	25 miles	trans$ per sf	200 miles	trans$ per sf
A-B: No Pre-Fabrication	$1,200	$0.40	$1,600	$0.53	$4,800	$0.40	$6,400	$0.53
A-C: No Pre-Fabrication	$1,200	$0.40	$1,600	$0.53	$4,800	$0.40	$6,400	$0.53
C-B:								
No Pre-Fabrication	$1,200	$0.40	$1,600	$0.53	$4,800	$0.40	$6,400	$0.53
Prefab Panels (70% density loss)	$2,100	$0.70	$2,800	$0.93	$8,400	$0.70	$11,200	$0.93
Modules, Standard Load	$3,000	$1.00	$4,000	$1.33	$12,000	$1.00	$16,000	$1.33
Modules, Wide Load (8.5-12 ft.)	$6,300	$2.10	$9,800	$3.27	$25,200	$2.10	$39,200	$3.27
Modules, Wide Load (>12 ft.)	$10,000	$3.33	$15,000	$5.00	$40,000	$3.33	$60,000	$5.00

Table 2 also provides the transportation cost allocated per square-foot for the finished structure. This is helpful for determining how the type or mode of transportation used can impact the overall cost of the building. Based on the findings, the greatest deterrent to the geographic market expansion or proposing competitive bids as they apply to prefabrication technologies is the distance from the fabrication shop to the construction site and the type of transportation service used. Total cost differentials for the primary shipment types and distances traveled for a particular scenario are illustrated by Figure 2. At face value, large oversize loads seem to exhibit the highest total costs even with fewer total loads.

It is important to note that the values as they exist in Table 2 and Figure 2 should not be interpreted as either good or bad. Because trade-offs exist among the various operating assets and processes, the total cost of a particular construction method may be the lowest among competing alternatives even though the transportation cost by itself was the highest in the group. The cost for moving five oversized loads that were needed to assemble a 3000 square-foot building, for example, may have allowed for a particular prefabrication process that created a greater savings in construction costs and project time.

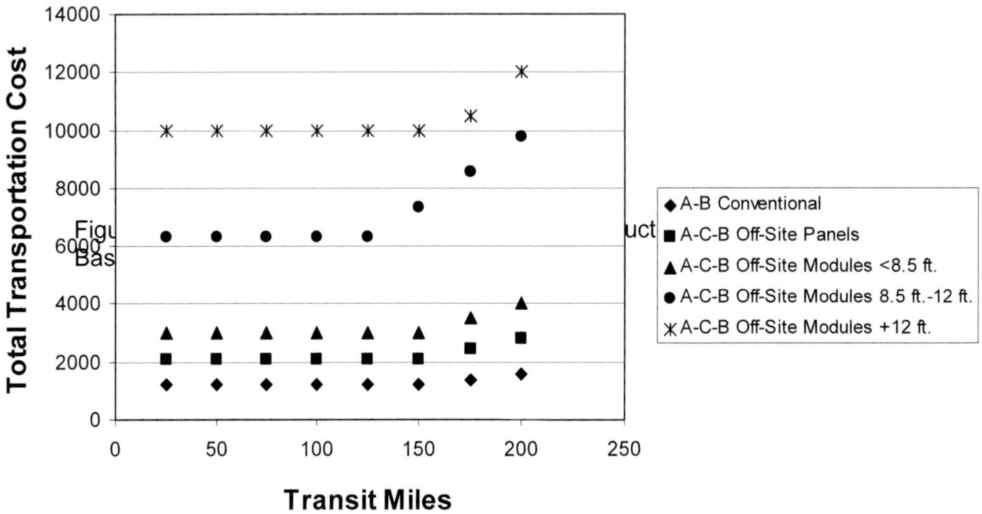

Figure 2: Total Transportation Charges per Mile for Alternative Construction Methods (for a 3,000 square-foot structure)

The rates, configurations, and distances for shipments in and out of a fabricator's facility can provide insight into sourcing, site location, and module design. In referencing Figure 1, total transportation costs from suppliers to the construction site have been calculated for the various construction methods based on the sole use of the shipment type noted. They pertain to a project involving a 3000 square-foot structure and supply legs of 25 miles and identified as:

a) A-B, Conventional On-site Construction: $1200

b) A-C-B, Panels with 70% density loss at site C ($1200 for four inbound loads moving 0 to 150 miles + $2100 for seven outbound loads moving 0 to 150 miles): $3300

c) A-C-B, Modules, no permits ($1200 for four inbound loads moving 0 to 150 miles + $3000 for ten outbound loads moving 0 to 150 miles): $4200

d) A-C-B, Modules, with permits ($1200 for four inbound loads moving 0 to 150 miles + $6300 for seven outbound loads moving 0 to 125 miles): $7500

e) A-C-B, Modules, with permits ($1200 for four inbound loads moving 0 to 150 miles + $10000 for five outbound loads moving 0 to 125 miles): $11200

Discussion

Modularity not only applies to the construction of a building. In manufacturing circles, using and integrating of common form factors or semi-finished components across multiple differentiated outputs has been highly successful for growing revenues and minimizing procurement and transformation costs (Zinn and Bowersox, 1988). Multiple automobile models assembled at the same plant, for example, may be built with common parts and sub-assemblies such as engines, chassis, and body panels. In any event, integrating standardized sub-assemblies into a process is intended to strike the balance among acceptable levels of customization, reduced delivery times, and cost containment (Waller et al., 2000). Similar effects are also expected when prefabrication construction techniques are applied. The results are truncated project times, greater production efficiencies, and acceptable levels of quality for a variety of finished structure that are accepted by the market.

Various logistics factors have been investigated regarding their impact on the feasibility of various prefabrication construction strategies. It was determined that building specifications, the relative value of the materials, as well as logical and tested procedures used for assembly did not allow enough opportunity to significantly reduce in-process inventory levels. Transportation cost factors, however, did play a critical role in determining the economic feasibility of prefabrication. The importance of transportation activities as they relate to the total cost of a supply chain have traditionally been marginalized in favor of asset utilization strategies such as flexible manufacturing, inventory minimization, and outsourcing. This is perplexing because total transportation expenses represent the highest cost of any supply chain network.

Construction time is also expected to decrease as the degree of off-site modularity is increased. This phenomenon will drive savings in construction loan interest and forgone operational earnings but based on the evidence derived by the research, other operating costs, particularly those emanating from transportation activities can off-set other savings. The level of transportation service required of a particular shipment will also help in determining the cost. For example, expedited delivery or oversized, permitted loads will require added resources and therefore, increase the rates. Based on the findings, oversized, permitted shipments may be the greatest operational threat to the growth in off-site modular construction. Increased distances for these shipments further exacerbate this effect.

Conclusion

Based on the results of this study, it is recommended that designers of modular construction methods incorporate lean manufacturing principles as they create solutions that are flexible enough to provide product variety, fast enough to offer marked reductions in construction times, and minimize total delivered costs to better compete with other construction alternatives.

A hybrid process that incorporates optimum assembly and logistics processes is envisioned. A combination module and panel solution, for example may add only a week to the project's duration but at a level of operational cost saving to make it worthwhile. Standardization of core materials for sub-assembled component and base modules that conform to conventional transportation equipment and services may be also included. Finally, if modular construction is less expensive when sub-assembly occurs closer to the market, then mobile prefabrication shops that source materials locally may offer the ultimate solution.

References

Ballou, R. H. (2004), Business Logistics Management: Planning, Organizing, and Controlling the Supply Chain, 5th edition, Upper Saddle River, NJ: Prentice-Hall.

Building Construction Cost Data (2005), 63rd edition, R.S. Means Company, Kingston, MA.

Construction Industry Institute (2002), "Prefabrication, preassembly, modularization, and offsite fabrication in industrial construction: A framework for design-making." *Research Summary 171-1* (July), Construction Industry Institute, Univ. of Texas at Austin, Austin, Tex.

Haas, C. T. and Fagerlund, W. R. (2002), "Preliminary Research on Prefabrication, Pre-assembly, Modularization, and Off-site Fabrication in Construction." *Research Report 171-11* (July), Construction Industry Institute, Univ. of Texas at Austin, Austin, Tex.

Kupitz, J. and Goodjohn, A. (1991), "Trends in Nuclear Power Reactor Design and Technology." Energy, Vol.16(1/2).

Stock, J. R. and Lambert , D. M. (2001), Strategic Logistics Management, 4th ed., Boston: McGraw-Hill Irwin.

Tatum, C. B., Vanegas, J. A., and Williams, J. M. (1987), "Constructability Improvement Using Prefabrication, Preassembly, and Modularization." *Source Document 25*, Construction Industry Institute, Univ. of Texas at Austin, Austin, Tex.

Waller, M. A., Dabholkar, P., and Gentry , J. J. (2000), "Postponement, Product Customization, and Market-Oriented Supply Chain Management", *Journal of Business Logistics*, Vol.21(2), pp. 133-160.

Zinn, W. and Bowersox , D. J. (1988), "Planning Physical Distribution with the Principle of Postponement", *Journal of Business Logistics*, Vol.9(2), pp. 117-136.

A STUDY OF QUASI-STATIC DELAMINATION IN SANDWICH STRUCTURES

Srinivasan Sridharan and Yupeng Li
Washington University in St. Louis, St. Louis, Missouri, 63130, USA
E-mail: ssrid@seas.wustl.edu

Abstract
Delamination of sandwich columns is studied using a relatively simple cohesive layer model. The model is described in some detail and is incorporated as a user supplied element (UEL) in a finite element package. The model is shown to predict accurately the test results of delamination of a facing sheet of a sandwich member. The accuracy of the model is seen to be superior to a model previously proposed by the authors, which predicts an earlier termination of crack growth. The UEL model is applied to a sandwich column investigated by earlier investigators – a column that is relatively stout (ratio of length (L) to depth (d) ≈ 7.3) and has stiff facing sheets (ratio of depth d to the thickness (h) of facings ≈ 15). The model is able to capture the onset of delamination buckling, sudden delamination growth at nearly constant compression, stable delamination growth and reaching of a limit point of the load carrying capacity. A slender sandwich column with relatively thin facings (L/d ≈ 15, d/h ≈ 40) is next considered. It indicates that overall bending tends to inhibit delamination growth under quasi-static loading as it tends to keep the delaminated surfaces in contact.

Introduction

An often encountered failure mode of sandwich structural components is the core facing debonding. Once such debonding occurs the integrity of the structure is compromised and a significant reduction in the stiffness and the load carrying capacity occurs. A sandwich member under compression is delamination-sensitive as the delaminated skin tends to buckle thus accentuating the risk of delamination growth.

Under quasi-static loading the extent of the growth of delamination depends upon the overall bending stiffness of sandwich beam as a whole as well as the bending stiffness of the delaminating facing sheet. As the structure bends in an overall sense, delamination tends to close notwithstanding the sense of bending and inhibit the growth of crack. The delamination growth may come to a standstill despite a continuing increase in the load carried. By the same token the delamination growth is inhibited in a sandwich column undergoing wrinkling - a mode of deformation composed of short waves. (The last point, however, is of academic interest only, as with the onset of wrinkling the total exhaustion of the load carrying capacity is not far.)

Previous experience (Sridharan, S., 2001 and El-Sayed, S. et al, 2002) has indicated that the delamination failure in sandwich members is principally in mode I (opening mode) with shear playing a negligibly small role. Thus the most significant parameter required for tracing the delamination growth is the critical value of strain energy release rate (SERR), viz. G_{Ic}. In order to trace the delamination growth without interference by the user, a cohesive layer model interposed between the facing and the core is employed. Authors have developed two types of the model (Li, Y. et al, 2005) which are used in conjunction with a widely used nonlinear finite element program (Abaqus, 2001):

(i) UMAT model: The cohesive layer is of finite thickness and represented by a layer of elements of a type available in a standard finite element package, but with *user supplied material* properties, hence called the UMAT model. The material property that is significant here is the relationship between incremental stress and incremental strain in the transverse direction.

(ii) UEL model: The cohesive layer has zero thickness initially and is represented by a set of *user supplied elements*. The user has to define the nodal forces and the current stiffness of the elements based on the relative displacement suffered by the element between its two separating surfaces.

M. Pandey et al. (eds), Advances in Engineering Structures, Mechanics & Construction, 615–626.

An effort is made to maintain utmost simplicity in the formulation of the model in either form. The paper also discusses briefly selection of the maximum cohesive stress of the material – which is crucial for capturing the crack initiation point precisely. It is seen, as in the case of laminated composites (Li, Y. et al, 2005) that the UMAT model ceases to be reliable as the crack advances and predicts a premature shutting down of crack growth.

Cohesive Layer Models

UMAT Model

This model employs a constitutive relationship in the form of a stress-strain relationship of the cohesive layer. The stresses and strains of the cohesive layer element are referred to axes which rotate with the element. Thus the effect of rigid body rotations is eliminated in the incremental strain-displacement relations. The model is defined in terms of the parameters describing the stress-strain relationships and the thickness of the cohesive layer. In the present context only the normal stress-strain relationship in the transverse direction is of pertinence, other relationships are assumed linear elastic. Poisson ratio is set to zero. For simplicity a nonlinear elastic relationship consisting of a linear elastic phase given by the modulus E_2 followed by a phase in which stress remains constant at σ_{max} (similar to "elastic-perfectly plastic" response) is assumed. The values of E_2 and σ_{max} must be representative of the core material through which the crack propagates. The other significant parameter is the thickness of the cohesive layer h_0, the initial thickness of the cohesive layer. Ideally h_0 is based on the observed dimensions of the process zone (Li, Y. et al, 2005). However the smallest possible dimension consistent with computational economy and reliability is chosen in the present work.

The cohesive material response calculations are performed in a module (UMAT) attached to the input program of Abaqus. A 4-noded plane strain/plane stress elements with reduced integration are selected for the cohesive layer for 2-D problems. Since reduced integration is used, a single integration point represents the whole element. The incremental strain ($\Delta\varepsilon_n$) is based on the current cohesive layer thickness h_c and so the total strain is approximately logarithmic[6]. As the nonlinear analysis proceeds, the value of G_I is monitored as in:

$$G_I = \sum \sigma_n h_c \Delta\varepsilon_n \tag{1}$$

Failure is deemed to occur as soon as the following fracture criterion is satisfied: $G_I \geq G_{Ic}$. The main program supplies the incremental strains which are based on the current (updated) dimensions and the user supplied material subroutine (UMAT) returns to the main program material stiffness matrix and the stresses at the integration point.

UEL Model

This cohesive layer model has zero initial thickness and is represented as a line in the undeformed state. Elements constituting the layer begin to acquire finite thicknesses as delamination occurs under mode I. The notion of strain is suspended and relative displacements of the nodes of the element are used to characterize the deformation and compute the nodal forces. Elements in model are 4-noded with 2 displacement degrees of freedom per node. The incremental deformation is characterized in terms of averaged normal relative displacement $\Delta\delta_1$ computed from the incremental nodal displacements. The stresses in the element are related to δ_1 and δ_2 respectively via nonlinear elastic relationships consisting of two phases (a linearly varying stress phase and a constant stress phase). δ_{10} and δ_{20} are the proportional limits of δ_1 and δ_2 and are prescribed taking into consideration of the stiffness of the material and the thickness of the process zone. They can be estimated as $\delta_{1o} \approx \frac{\sigma_{max}}{E_2}t$, $\delta_{2o} \approx \frac{\tau_{max}}{G_{12}}t$ where t represents process zone thickness; further δ_{10}, and δ_{20} must be chosen to be sufficiently small so that the elastic strain energy contribution is small compared to the cohesive strain energy in the layer. Nodal forces and the stiffness matrix are found in terms of the stress state in the element. These are transformed to the global axes and returned to the main program. The total strain energy stored per unit area in each mode in the element is given as a sum of incremental contributions as follows:

$$G_{Ic} = \sum \sigma_n \Delta\delta_1 \quad (2)$$

Failure is deemed to occur as soon as the following fracture criterion is satisfied: $G_I \geq G_{Ic}$.

Fig. 1 shows a typical element ABCD and its deformed configuration A'B'C'D'; the original coordinates and displacements are given by (x,y) and (u,v) respectively with appropriate subscripts. The main program supplies for each increment of loading the current nodal displacements at the end of the previous loading increment and their corresponding increments for each iteration with reference to the global coordinate system.

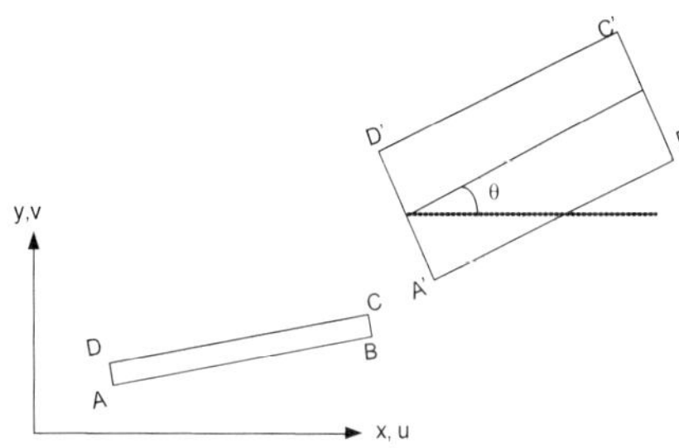

Fig. 1. Typical element ABCD and its deformed configuration A'B'C'D'

If θ is the averaged orientation of the element with reference to x-axis, then,

$$\tan(\theta) = \frac{y_2 - y_1}{x_2 - x_1} \quad (3)$$

Where:

$$y_2 = \frac{(y_C + v_C) + (y_B + v_B)}{2}; \quad y_1 = \frac{(y_D + v_D) + (y_A + v_A)}{2}$$
$$x_2 = \frac{(x_C + u_C) + (x_B + u_B)}{2}; \quad x_1 = \frac{(x_D + x_D) + (x_A + x_A)}{2} \quad (4)$$

The current relative displacements in the normal (δ_1) and tangential directions (δ_2) with respect to the current orientation of the element are then computed by transformation:

$$\delta_1 = (v_2 - v_1)\cos\theta - (u_2 - u_1)\sin\theta$$
$$\delta_2 = (v_2 - v_1)\sin\theta + (u_2 - u_1)\cos\theta \quad (5)$$

where:

$$u_1 = \frac{u_A + u_B}{2}; \quad u_2 = \frac{u_C + u_D}{2}$$
$$v_1 = \frac{v_A + v_B}{2}; \quad v_2 = \frac{v_C + v_D}{2} \quad (6)$$

Note A coincides with D and B with C in the original configuration. From the assumed stress-displacement relations, the normal and shear stresses carried by the element are determined. These in turn are employed to determine the nodal forces in terms of the element length. The tangential stiffness matrix for the element is set up by appropriate differentiation. The current nodal forces and the tangential stiffness matrix are then transformed to the global axes and returned to the main program. Note that the nodal forces are updated for every iteration.

Calibration and evaluation of performance of the models using a test case

Li performed tests (Li, X. et al, 2000) on a sandwich member of length 210mm completely restrained at its top face and loaded uniformly across its width at one end of the bottom facing. The specimen carried a pre-implanted delamination of 51mm. The geometry, the loading and the boundary conditions of the test specimen are indicated in Fig. 2. Of the three groups of specimens tested, only one group identified as "KKV sandwich", exhibited no crack kinking into the core. After the first micro-kink, the crack propagated in the core parallel to the interface about 0.5mm below the face/core interface. The specimens were loaded at varying, but small angles to the normal of the top face and the one loaded normal to face is selected for investigation here.

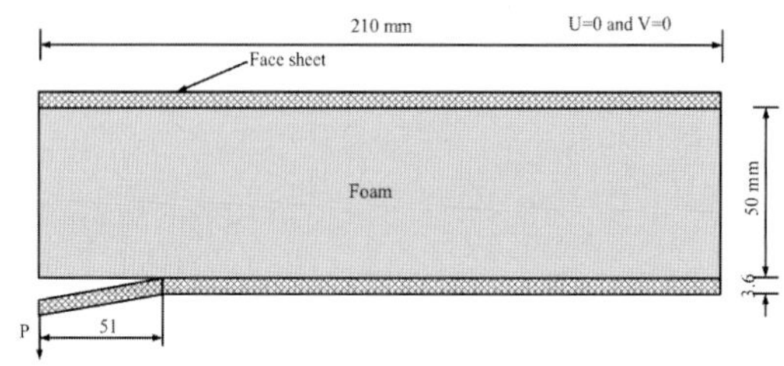

Fig. 2. Test configuration of Li's sandwich specimen

The facing sheets are made up of E-glass/vinyl ester while the core material is H200 Divinycell PVC foam. Table 1 lists the elastic properties of the materials. Here we use the averaged elastic properties as determined from compression and tension tests. (The most important property is the effective modulus in the longitudinal direction which determines the slope of the initial linear phase prior to crack initiation which varied from 23.2GPa in compression to 26.0GPa in tension - an average value of 24.6GPa is used for the comparison with experiment.).

Table 1: Material properties of Li's specimen

Material Property	Li's specimen	
	Face sheet	Core
Longitudinal Young's modulus, E_x (GPa)	24.6	0.165
Transverse Young's modulus, E_y (GPa)	11.8	0.263
Out-of-plane Young's modulus, E_z (GPa)	13.5	0.165
Poisson's ratio, υ_{xy}	0.31	0.32
Poisson's ratio, υ_{xz}	0.533	0.32
Poisson's ratio, υ_{yz}	0.446	0.32
In-plane shear modulus, G_{xy} (GPa)	3.78	0.0644
In-plane shear modulus, G_{xz} (GPa)	9.19	0.0644
In-plane shear modulus, G_{yz} (GPa)	3.74	0.0644
Critical strain energy release rate for mode I, G_{Ic} (J/m^2)	1270	

Selection of Parameters: General

Calculations have indicated that the SERR in the opening mode is dominant and thus the shear mode is virtually inactive. The critical strain energy release rate as measured by Li (Li, X. et al, 2000) is 1.270N/mm. The next in order of importance is the selection of maximum stress, σ_{max}. This determines the point at which delamination growth initiates. Consider for example the performance of the UEL model (Fig. 4) or the UMAT model (Fig. 6) for various levels of σ_{max}. The most striking feature of the results is that the unloading paths for a given model are parallel to each other; only the crack initiation point - which is almost coincident with the maximum load attained - is different. Thus crack propagation rate is virtually unaffected by the choice of σ_{max}. Similar observations have been made by the authors elsewhere (Li, Y. et al, 2005 and Alfano, G. et al, 2001).

The following considerations are germane to the selection of the σ_{max}:

(i) The maximum stress must be representative of the tensile strength of the material in which the crack propagates.

(ii) The more physically based cohesive stress-deformation laws involve a rising path till a maximum stress is attained and an unloading path taking the stress to zero when failure is deemed to be complete. On the contrary in the present study the cohesive stress is constant for the most part. If the crack opening displacement is the critical parameter indicating failure, it follows the σ_{max} in our model must be about one half of the actual strength of the material (used in the conventional models).

(iii) At crack initiation, the failure occurs at the crack tip by micro-cracking at an angle into the core as it were. Since we postulate crack growth parallel to the interface, the stress needed to initiate such a crack is probably greater than the actual tensile strength of the material.

The tensile strength of the core material - H200 Divinycell foam - is about 6MPa (Abot, J. L., 2000). The core material is likely to be reinforced due to the presence of a stiff adhesive (of strength 35MPa, Abot, J. L., 2000) bonding the facing sheet with the core for a small distance from the crack interface. It is difficult, therefore to assess precisely the strength of the material; suffice it to say together the effect of the adhesive and the crack kinking at crack initiation (item (iii) above) will put the effective strength significantly more than the actual strength of 6MPa. However in view of item (ii), it is thought appropriate to use the actual strength of 6MPa or values in that range for σ_{max}.

Performance of the UEL model

The finite element model consisted of 4-noded plane strain elements with reduced integration (CPE4R) throughout and the sizes of the element chosen are indicated in Fig. 3. The typical size of the element in the direction of the crack is 0.25mm and thus in one layer there were 840 elements end to end. (A trial run with a coarser mesh with 0.50mm size gave essentially the same results.) Loading was introduced by prescribing the deflection at the point of application of the load and computing the reaction thereof. A total deflection of 30mm was reached in 1000 increments.

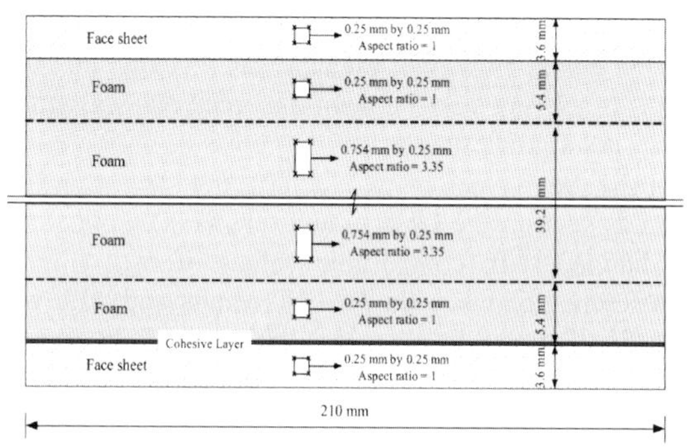

Fig. 3. Finite Element Configuration (not to scale)

The values of the maximum cohesive stress, σ_{max}, selected were 3, 6 and 9MPa respectively. Of these the first and the last are 50% less and 50% more respectively from the nominal strength of the core material, viz. 6MPa. The other parameters of the model: τ_{max} = 20.0Mpa, δ_{10} = 1.e-2, δ_{20} = 1.e-3, G_{Ic} = 1.270N/mm, G_{IIc} is set at an arbitrarily high value.

Li (Li, X., 2000) studied the problem experimentally by a sequence of loading up to incipient crack growth and unloading fully and reloading as before. Fig.4 displays the key points taken from these experimental results which the authors have taken the liberty of re-plotting ensuring the characteristics passed through the origin (elimination of zero error) but maintaining the slopes of the reloading characteristics and the loads recorded at the crack initiation at the end of each loading phase.

As already mentioned, the prediction of crack initiation point depends on the σ_{max} selected – the value of 3MPa is clearly too small for the corresponding behavior is too compliant in the vicinity of crack initiation and a value of 9MPa overestimates the load corresponding to crack initiation. Thus as far as foam core sandwich members are concerned , we may conclude that reasonable results can be obtained by selecting for σ_{max} the actual strength of the core for delamination predictions using the UEL model. A typical stress distribution around the crack tip is shown in Fig. 5 where it is seen that the model preserves the features of a sharp crack with a highly localized stress concentration with little damage upstream of the crack.

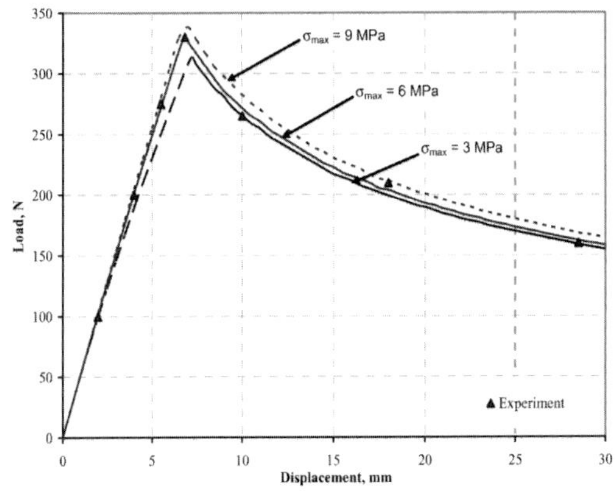

Fig. 4. Simulation results from UEL model vs. experimental results

Performance of the UMAT model

The finite element mesh configuration selected for the analysis is similar to the one used for the UEL model except for the mesh size in the longitudinal direction which is now 0.5mm. Two values of strength were investigated, viz. 7MPa and 9.4MPa which are marginally higher than those used for the UEL model. It was anticipated that the strength needed for the UMAT model must be higher than that employed for the UEL model, to compensate for the influence of compressive stress developing in the cohesive layer upstream of the crack. However, the analysis revealed that the cohesive zone (that part of the cohesive layer which is in tension starting from the crack tip) was sufficiently long (12 elements in length) so that this effect is not significant.

The results obtained for the two values of σ_{max} exhibit the same trend (Fig. 6), but differ in their prediction of crack initiation as seen in the UEL model results. The crack initiation results agree well with UEL model results obtained with comparable values of σ_{max} (Fig. 7). However there is an obvious and a significant difference between the UMAT results and UEL results which is highlighted in Fig. 7. It is seen that the UMAT model predicts that the crack ceases to grow after some growth and the load then begins to gradually increase with the crack opening displacement. There is no reason to suppose that such a prediction has any validity to it and in fact can be shown to be due to an intrinsic deficiency of the model.

Fig. 7 illustrates what happens when an initial crack length l_o higher than 51mm is assumed, say 71mm. The load-displacement characteristic has a smaller slope and crack initiation occurs at a point which lies on or close to the unloading characteristic associated with lo = 51mm. But the unloading characteristic associated with l_o = 71mm deviates from that associated with an initial crack length of 51mm and lies closer to that given by the experiment and UEL model. However this characteristic too gradually deviates from the correct result. It is clear therefore that the UMAT model, though capable of predicting crack initiation correctly gradually loses its ability to predict continued crack growth.

Fig. 8 shows the magnified image of stress distribution in the close vicinity of the crack tip when the crack opening displacement measured at the point of load application is 25mm, At this instant crack has progressed to a length of 101mm and the next crack tip element is on the point of failure.

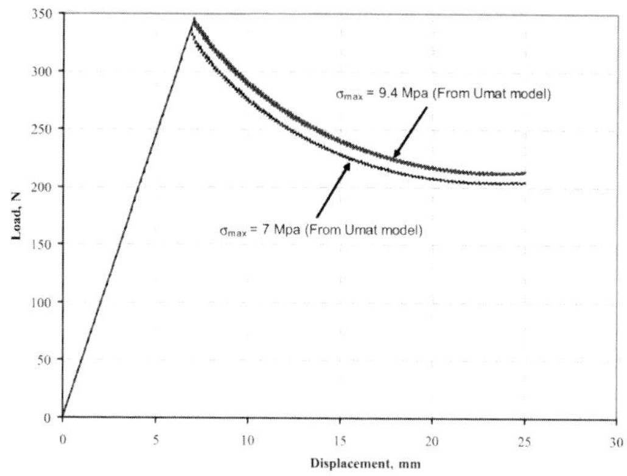

Fig. 6. Simulation results from UMAT model

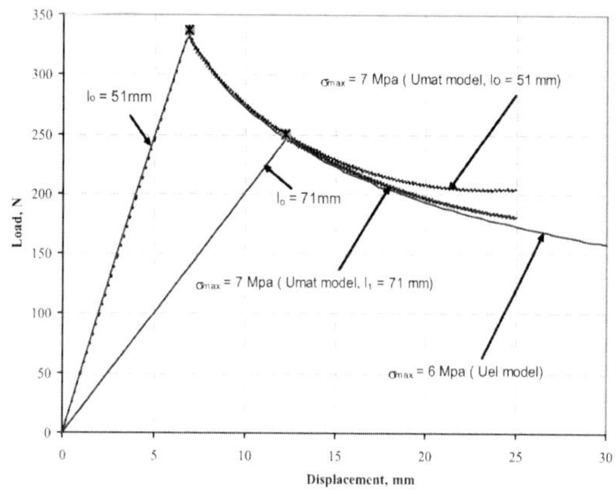

Fig. 7. Comparison between UEL results and UMAT results

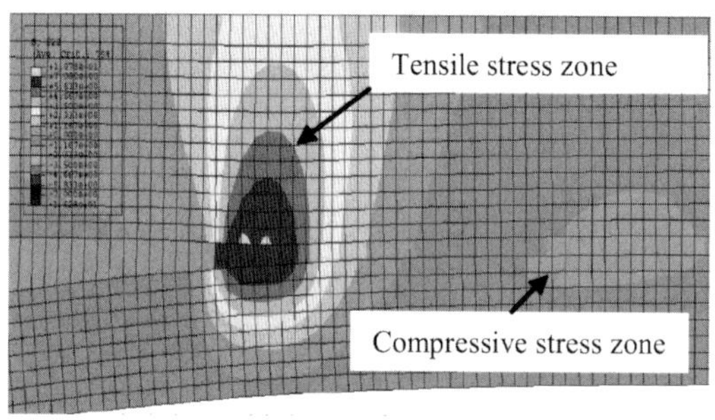

Fig. 8. Stress distribution around crack tip

Fig. 9 shows the magnified view of the deformation of the bottom surface of the cohesive elements close to the crack tip for the two cases: (i) The current crack length (l_{cr}) of 101mm is reached after a crack extension of 50mm from the initial crack length (l_o) of 51mm, and (ii) The initial crack length is taken as 101mm and the crack tip element is on the verge of failure ($l_{cr} = l_o = 101$mm). Note that the downward deflections are plotted above the x-axis in the figure.

There are two features that may be observed in these pictures:

(i) From Fig 8 , it is seen that there is a part of cohesive layer which carries tensile stress in the vicinity of the crack tip, but more significantly this stress gradually diminishes to zero as we travel upstream of the crack and changes to compression.

(ii) In case (i) the depth of the cohesive layer reduces as we travel upstream from the crack tip and a "neck" – a region of slightly reduced depth compared to neighboring regions on either side appears. The bottom surface of the cohesive elements is undulating at the crack tip. This is in contrast to the case (ii) which exhibits a smooth variation of the cohesive layer.

It is clear that item (ii) must be a result of what transpires during the failure of cohesive elements and the accompanying crack extension. As each element fails certain force is released on to the delaminated sheet which rotates so as to compress the cohesive layer downstream. Once a "neck" of certain threshold acuteness is formed, the crack growth is at first seriously inhibited and eventually stops.

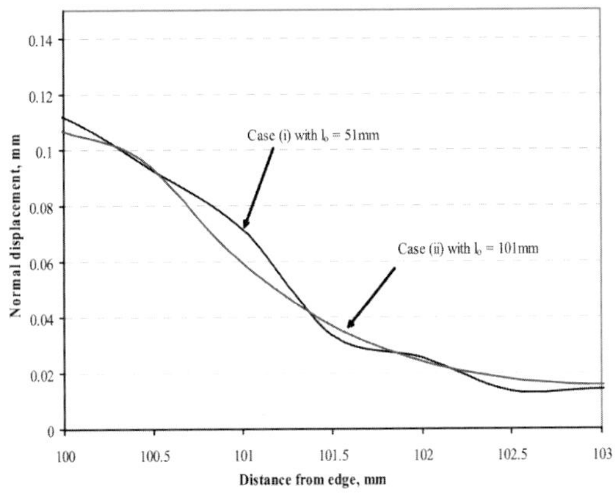

Fig. 9. Displacements along the bottom surface of the cohesive elements (when σ_{max} = 7Mpa)

These phenomena are fairly insignificant or virtually absent in the UEL model. Of these, item (i) can have an effect on crack initiation as the compression block upstream can send a wave of transverse compressive stress towards the crack tip thereby delaying the crack initiation. However, as already mentioned this effect appears to be not significant for sandwich members as the cohesive zone is of greater length than in laminated composites – of the order of 6-7mm. The second factor, too, kicks in much later, so much so the UMAT model can be used for a study of delamination in sandwich beams if one is not investigating large crack extensions.

Response of sandwich columns under compression

In this section we study the response of a compressed column. The compression carried by the column is small relative to the buckling load of the column. We note that the corresponding static problem has been studied by El-Sayed, S. et al, 2002 using a variant of the UMAT model which incorporates an unloading phase of the cohesive material response. The present study uses the UEL model for the problem and the salient features of the static response are briefly summarized. Two cases are considered; the first one is that of a relatively stout column with relatively thick facing sheets and the second is that of a slender column having very thin facing sheets.

Case I

Geometry, Materials and Boundary Conditions

The core and facing materials are exactly the same as the test specimen (Table 1) discussed in the last section and the cross-section of the member investigated is also the same (h = 50mm, t = 3.6mm). The total length of the column investigated is 420mm and it is considered clamped at either end. Because of symmetry involved – barring the unlikely scenario of sudden interference of antisymmetric modes of bifurcation- only half the column length of 210mm included between the line of symmetry on the left and the clamped end on the right is considered (Fig. 10). A rigid surface incapable of rotation and lateral translation is created to model the clamped end. A symmetrically located initial delamination of total length 102mm is deemed to exist between bottom facing sheet and the core. Fig. 10 shows the configuration and other pertinent details.

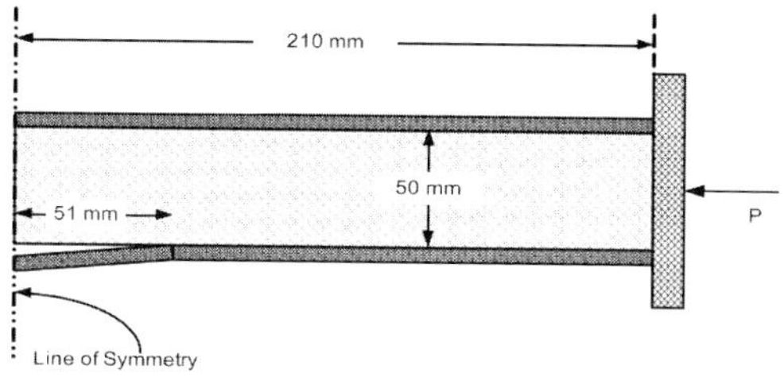

Fig. 10. Configuration of case I

Details of the Analysis

Finite Element mesh configuration is the same as in Fig. 3 though the length of the column and the cohesive layer elements are different. The size of the elements is 0.25mm in the longitudinal direction and plane strain elements (CPE4R) with reduced integration were employed in the analysis as before. Unlike the previous example, the column is treated as having unit width (1mm) and the loads are reported as per unit width. The load is inducted by prescribing the end-shortening (relative to the plane of symmetry). The critical value of SERR in the opening mode is once again taken as 1.27N/mm. The value of maximum stress, σ_{max} is taken as 10MPa and δo is taken as before equal to 0.01mm.

The model was perturbed slightly by incorporating imperfections in order to keep the delamination open and prompt the structure towards delamination buckling and growth. To this end, a linear stability analysis without the cohesive layer was conducted and the buckling loads and modes were obtained. The first three buckling loads were found to be 562N, 2079N and 3466N respectively. Of these the first (Fig. 11(a)) and the second (Fig. 11(b)) pertain to delamination buckling – the former keeps the delamination fully open and the latter partially closed. The last one (Fig. 11(c)) is an overall buckling mode. It is clear that the first alone is critical and therefore an imperfection in the form of this mode is incorporated in the model with a maximum deflection of 0.5mm at the center. Given the stoutness of the member (relative to the length, L/d =7.3) and the magnitude of the overall buckling load, it is clear that overall bending of the column plays a relatively minor role in the problem.

(a) First buckling mode (b) Second buckling mode (c) Third buckling mode

Fig. 11. The first three buckling modes

Buckling and Delamination

The column is compressed in a quasi-static manner, so that end-shortening increases from zero to a value at which the facings undergo an average strain of 1.5%. Nonlinearity sets in as soon as the delamination tends to buckle outward and at a load of 650N/mm (a longitudinal stress of 116MPa in the facing sheets) delamination begins to grow (Fig. 12). At first this is rather rapid and occurs with little increase in load (points A, B and C in Fig. 12), but soon settles down to a relatively modest rate and delamination occurs under increasing load thereafter. Finally at load of 1380N/mm (average stress of 246MPa), a limit point is reached and thereafter delamination growth is accompanied by shedding of the load by the member. At a load of 1294N/mm the delamination length recorded is 152.25 (the end-shortening = 3.15mm). Compressed to an end-shortening of say 3.5mm, almost complete delamination of the bottom facing occurs, but this would necessarily be accompanied by other forms of failure not considered here. Fig. 12 shows the variation of the load with end-shortening. The observed behavior is in very good agreement to that obtained by EL-Sayed, S., 2002 (not shown) though their model tends to be slightly stiffer at the more advanced delamination range. Also indicated in the figure are the delamination growths recorded at certain significant points in the history - in particular the values corresponding to end -shortening magnitudes of 0.8mm, 1.0mm and 1.2mm respectively. Fig. 13 shows the deformed shape when end shortening is 2mm.

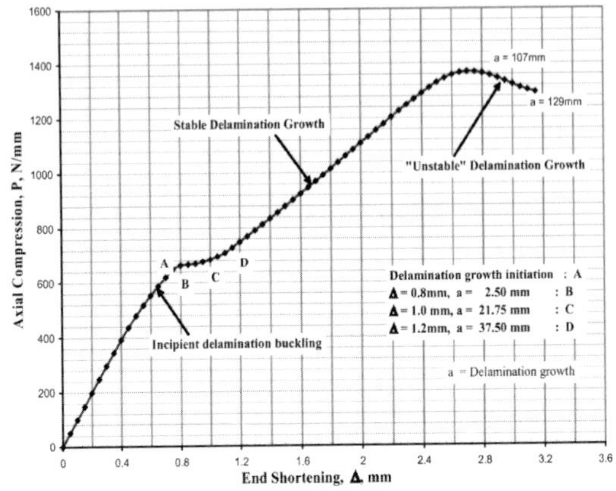

Fig. 12. Response of sandwich column under static loading, Case I

Fig. 13. Static response for case I: Deformed shape at an end shortening of 2mm

Case II

We next consider a sandwich column (Fig. 14) which is significantly slender in comparison to that presented in Case I. The column is clamped at its ends as before, the total length (L) now is 300mm and the total depth, d is 20mm. Thus the L/d is twice that in Case I. Further the thickness of the facing sheets, t is only 0.5mm each and thus d/t = 40 – again about thrice as much as in Case I. The motivation for studying such a case is to examine the influence of overall bending and wrinkling on the delamination growth. (The geometry is the same as that studied by El-Sayed, S., 2002). An initial delamination of the top facing sheet of 10mm at the center was considered (5mm on either side of symmetry).

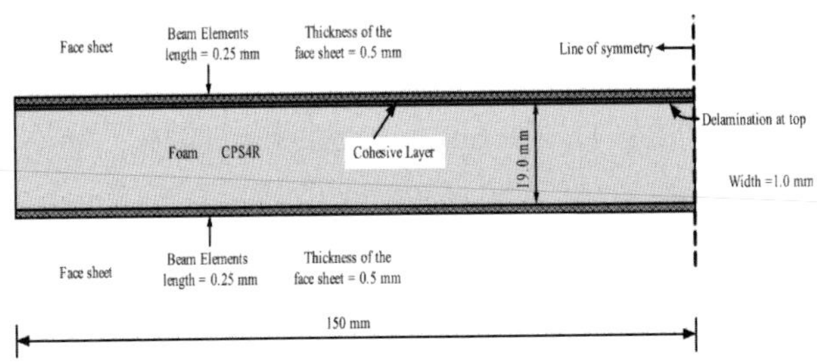

Fig. 14. Configuration of slender sandwich beam

Both the core and facing sheets are assumed to be isotropic, for simplicity with: E_f= 26900MPa, ν_f = 0.3 and E_c = 269MPa, ν_c = 0.3, where E and ν stand respectively for Young's modulus and Poisson ratio and the subscripts f and c stand for the facings and the core respectively. Two values of G_{Ic} are considered, viz. 1.270N/mm^2 and 0.635N/mm^2, with σ_{max} = 6MPa and δ_o = 0.01mm.

Finite Element Modeling

In view of the relatively small thickness of the facing sheets, modeling it with solid elements was not considered viable as the aspect ratio of the four noded elements needs to be maintained around unity. Therefore it was decided to use 2-noded shear-deformable beam elements to model the facing sheets. The core was modeled using 4-noded plane stress elements (CPS4R) with reduced integration.

The reference nodes of the beam elements of the top sheet were placed at the bottom of the beam section and those of the bottom sheet at the top of the section in order to ensure deformation compatibility with core elements. The element size along the longitudinal direction was maintained at 0.25mm and thus there were 600 elements in each layer of elements. The aspect ratios of the core elements were kept around unity. The width of the column was taken as 1mm as before.

Details of Static Nonlinear Analysis

As before, a linear stability analysis was performed to find the critical loads at which significant deformation will appear. The critical loads associated with delamination buckling and overall buckling were found to be 157.6N and 721.0N respectively. It will be seen as delamination buckling and growth occur the effective center of stiffness moves downwards and as a result some overall bending occurs at a fraction of the critical load associated with overall buckling predicted by linear stability analysis. This is clearly an example of interaction of delamination growth and overall buckling.

Load was introduced by prescribing end-shortening as before. The maximum end-shortening of 2mm is selected which is about 2.5 times that corresponding to delamination buckling. Small equal and opposite loads increasing from zero to 1N were applied at the bottom and top nodes respectively at the center of the column to prompt the delamination to remain open. End-shortening was applied in small increments, 1000 increments to reach a value of 2mm.

Features of Nonlinear Response

Consider first the case with G_{Ic} = 1.27N/mm. As the end-shortening increases to beyond the critical value, delamination occurs with about 12 elements (each 0.25mm long) failing one after another in quick succession. The load drops slightly but thereafter recovers to increase further. During this stage more elements fail with a total delamination growth of 9mm when the end-shortening of 2mm is reached. At this time the tangential stiffness of the structure has reduced significantly with the load beginning to level off. The behavior for G_{Ic} = 0.637N/mm is similar, except that delamination growth begins a little early, the load drop is more pronounced. And the total delamination growth is 18mm. Fig. 15 shows the axial compression vs. end shortening relationship.

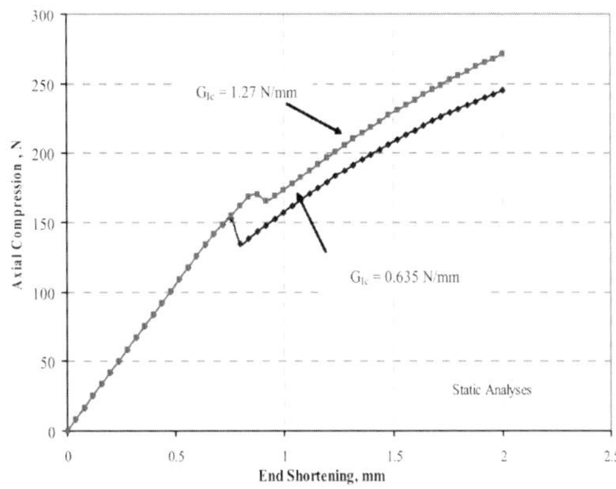

Fig. 15. Static Response of Sandwich Column, Case II

Notwithstanding the actual magnitudes of the loads involved, qualitatively there is a significant difference between the delamination behavior under static loading of the Cases I and II considered here (cf. Fig. 13 and Fig. 16). It is somewhat paradoxical to see that delamination growth is minimal for the present case of a slender column with thin facing sheets whereas it was extensive for the case of stouter column with thicker facing sheets. Careful examination of the deformed shape of the member as it evolves indicates that the delamination growth is influenced by overall bending of the entire column and highly localized rotation of the facing sheet that occurs at the crack tip. If the overall bending is such as to cause longitudinal tension at the facing sheet, delamination growth comes to a stand still; if it is such as to cause longitudinal compression in the delaminating facing sheet with a pronounced curvature, the sheet in the vicinity of the crack tip tends to rotate and come into contact with the core inhibiting delamination growth. Thus we may conclude the delamination growth is not a significant factor in a slender sandwich column having thin facings under static loading. There are other modes of failure, such as face sheet wrinkling and its interaction with overall bending which cause failure of the member (El-Sayed, S., 2002).

Fig. 16. Static Response, Case II: Deformed shape at end shortening = 2mm

Conclusions

The delamination phenomena in sandwich members are studied using cohesive layer models-designated as UMAT and UEL models. When applied to a test case, it was found that the UMAT model which has a finite thickness can predict initiation of delamination growth as well as rate of growth of delamination, but is unable to predict large delamination growth due to an inherent deficiency. The UEL model which has zero initial thickness is capable of tracing the entire delamination history. Apart from the critical value of strain energy release rate in the opening mode, the parameter that influences crack initiation is the strength of the cohesive model material; for sandwich delamination problems using the actual strength of the core material in conjunction with the proposed models gave satisfactory results.

Delamination was studied in a member carrying a prescribed end-shortening. Two cases are considered, the Case I of a column which was relatively stout with relatively thick facing sheets and Case II, a slender column with thin facing sheets. The overall bending of the member (in case II) had a significant influence in inhibiting the crack growth by virtue of contact between the facing sheet and the core in the quasi-static load application.

References

ABAQUS/Standard User's Manual, "UMAT: Define a Material's Mechanical Behavior," Hibbitt, Karlsson and Sorensen, Inc., Pawtucket, RI, Vol. 6.3, 2001, pp. 24.2.30.1-24.2.30.14.

ABAQUS/Standard User's Manual, "UEL: Define an Element," Hibbitt, Karlsson and Sorensen, Inc., Pawtucket, RI, Vol. 6.3, 2001, pp. 24.2.19.1-24.2.19.17.

Abot, J. L., "Fabrication, Testing and Analysis of Composite Sandwich Beams", Ph.D. Thesis, Northwestern University, Evanston, Illinois, 2000.

Alfano, G. and Crisfield, M. A., "Finite Element Interface Models for the Delamination Analysis of Laminated Composites: Mechanical and Computational Issues," *International Journal for Numerical Methods in Engineering*, Vol. 50, 2001, pp. 1701-1736.

El-Sayed, S. and Sridharan, S., "Cohesive Layer Models for Predicting Delamination Growth and Crack Kinking in Sandwich Structures," *International Journal of Fracture,* Vol. 117, No. 1, Sept. 2002, pp. 63-84.

Li, X., "Debonding fracture of foam core the tilted sandwich structure," Ph.D. Thesis, Florida Atlantic University, Boca Raton, Florida, 2000.

Li, Y. and Sridharan, S., "Performance of Two Distinct Cohesive Layer Models for Tracking Composite Delamination", International Journal of Fracture, in press, 2005.

Li, Y. and Sridharan, S., "Investigation of Delamination Caused by Impact Using a Cohesive Layer Model", AIAA Journal, Vol.43,No.10,2005, pp.2243-2250.

Sridharan, S., "Displacement-based mode separation of strain energy release rates for interfacial cracks in bi-material media", International Journal of Solids and Structures, Vol. 38, 2001, pp. 2787-6803.

ON THE MONTE CARLO SIMULATION OF MOMENT LYAPUNOV EXPONENTS

Wei-Chau Xie and Qinghua Huang
Department of Civil Engineering,
University of Waterloo, Canada
E-mail: xie@uwaterloo.ca, q2huang@engmail.waterloo.ca

Abstract

The moment Lyapunov exponents are important characteristic numbers for determining the dynamical stability of stochastic systems. Monte Carlo simulations are complement to the approximate analytical methods in the determination of the moment Lyapunov exponents. They also provide criteria on assessing how accurate the approximate analytical methods are. For stochastic dynamical systems described by Itô stochastic differential equations, the solutions are diffusion processes and their variances may increase with time of simulation. Due to the large variances of the solutions and round-off errors, bias errors in the simulation of momemt Lyapunov exponents are significant in the cases of improper numerical approaches. The improved estimation for some systems is presented in this paper.

1 Introduction

The moment Lyapunov exponents, which are define by

$$\Lambda(p) = \lim_{t\to\infty} \frac{1}{t} \log E\left[\|\mathbf{X}(t)\|^{p}\right], \tag{1}$$

characterize the moment stability of a stochastic dynamical system with state vector $\mathbf{X}(t)$, where $E[\cdot]$ denotes expectation and $\|\cdot\|$ denotes a suitable vector norm. The pth moment of the response of the system is asymptotically stable if $\Lambda(p) < 0$. Moreover, $\Lambda'(0)$ is equal to the largest Lyapunov exponent λ, which is defined by

$$\lambda = \lim_{t\to\infty} \frac{1}{t} \log \|\mathbf{X}(t)\| \tag{2}$$

and describes the almost-sure or sample stability of the system.

Even if the solution of a system is almost-sure stable, i.e. $\|\mathbf{X}(t)\| \to 0$ as $t \to \infty$ w.p.1 at a negative exponential rate, the chance that $\|\mathbf{X}(t)\|$ takes large values may lead to the instability of the pth moment. Hence, it is important to obtain the moment Lyapunov exponents such that the complete properties of dynamic stability of stochastic systems can be described.

Although moment Lyapunov exponents may be obtained by approximate analytical methods, such as stochastic averaging or perturbation, in some cases, Monte Carlo simulations have to be applied when it is difficult to do so. On the other hand, the accuracy of the approximate analytical methods needs to be verifed by numerical simulations.

A numerical algorithm for determining the moment Lyapunov exponents using Monte Carlo simulation developed by Xie (Xie, 2005) is described briefly below.

Consider a dynamical system whose state vector satisfies the following Itô stochastic differential equation

$$\mathrm{d}\mathbf{X}(t) = \mathbf{m}(\mathbf{X}, t)\mathrm{d}t + \boldsymbol{\sigma}(\mathbf{X}, t)\mathrm{d}\mathbf{W}(t), \tag{3}$$

M. Pandey et al. (eds), Advances in Engineering Structures, Mechanics & Construction, 627–636.

where $\mathbf{W}(t)$ is a vector of standard Weiner processes. Define

$$\rho_m(p) = \frac{E\left[\|\mathbf{X}(mT_N)\|^p\right]}{E\left[\|\mathbf{X}((m-1)T_N)\|^p\right]}, \qquad m = 1, 2, \cdots, M, \tag{4}$$

where $T_N = K\Delta$ is the time interval for normalization, Δ is the time step of iteration for an appropriate discrete scheme used to solve equation (3) numerically. In order to avoid data overflow or underflow, when system (3) is linear, at time mT_N the sth sample of response is normalized using

$$\mathbf{X}^s(mT_N) \triangleq \frac{\mathbf{X}^s(mT_N)}{\|\mathbf{X}^s(mT_N)\|} \tag{5}$$

before the iteration continues. Then for a large iteration time $T = MT_N$, by setting $\|\mathbf{X}^s(0)\| = 1$, the approximate moment Lyapunov exponents at time T are given by

$$\begin{aligned}\Lambda(p) &= \frac{1}{T}\log E\left[\|\mathbf{X}(T)\|^p\right] = \frac{1}{MT_N}\log E\left[\|\mathbf{X}(MT_N)\|^p\right] \\ &= \frac{1}{MT_N}\log\left\{\frac{E\left[\|\mathbf{X}(MT_N)\|^p\right]}{E\left[\|\mathbf{X}((M-1)T_N)\|^p\right]}\cdots\frac{E\left[\|\mathbf{X}(T_N)\|^p\right]}{E\left[\|\mathbf{X}(0)\|^p\right]}\right\} \\ &= \frac{1}{MT_N}\sum_{m=1}^{M}\log\rho_m(p).\end{aligned} \tag{6}$$

The expectation in (6) is determined by the sample average

$$E\left[\|\mathbf{X}(mT_N)\|^p\right] = \frac{1}{S}\sum_{s=1}^{S}\|\mathbf{X}_s(mT_N)\|^p, \tag{7}$$

where S is the total sample size for simulation.

However, notice that after the normalization operation $\|\mathbf{X}_s(mT_N)\| = 1$ is always satisfied according to equation (5). Actually equation (6) gives the approximate moment Lyapunov exponents at time T_N but not at time T, since $\rho_m(p)$ is always the expectation of response at time T_N in this case and

$$\Lambda(p) = \frac{1}{MT_N}\sum_{m=1}^{M}\log\rho_m(p) = \frac{1}{M}\sum_{m=1}^{M}\frac{1}{T_N}\log\rho_m(p). \tag{8}$$

One revision to correct the insufficiency of the above algorithm is to normalize the response by their expectation. For a linear system, by defining

$$\mathbf{Y}_m^s = \frac{\mathbf{X}^s(mT_N)}{E\left[\|\mathbf{X}((m-1)T_N)\|\right]}, \qquad \hat{\rho}_m^s = \|\mathbf{Y}_m^s\|, \tag{9}$$

it is easy to obtain the approximate moment Lyapunov exponents at time T through

$$\begin{aligned}\Lambda(p) &= \frac{1}{MT_N}\log\frac{E\left[\|\mathbf{X}(MT_N)\|^p\right]}{E\left[\|\mathbf{X}((M-1)T_N)\|\right]^p}\frac{E\left[\|\mathbf{X}((M-1)T_N)\|\right]^p}{E\left[\|\mathbf{X}((M-2)T_N)\|\right]^p}\cdots\frac{E\left[\|\mathbf{X}(T_N)\|\right]^p}{E\left[\|\mathbf{X}(0)\|\right]^p} \\ &= \frac{1}{MT_N}\left\{\log E\left[\left\|\frac{\mathbf{X}(MT_N)}{E\left[\|\mathbf{X}((M-1)T_N)\|\right]}\right\|^p\right] + \sum_{m=1}^{M-1}p\log E\left[\left\|\frac{\mathbf{X}(mT_N)}{E\left[\|\mathbf{X}((m-1)T_N)\|\right]}\right\|\right]\right\} \\ &= \frac{1}{MT_N}\left\{\log E\left[(\hat{\rho}_M)^p\right] + p\sum_{m=1}^{M-1}\log E\left[\hat{\rho}_m\right]\right\}.\end{aligned} \tag{10}$$

The solution of equation (3) is a diffusion process and its variance may increase significantly with time. Although equation (10) is exact theoretically when M is large enough, there are two main reasons which will lead to significant numerical errors. According to the Central Limit Theorem (Feller, 1965), for independent and identically distributed (i.i.d.) random variables $x_1, x_2, \cdots$ with the same mean μ and variance σ^2, the sample average $\bar{x} = \frac{1}{n}\sum_{i=1}^{n} x_i$ will tend to the normal distribution $\mathcal{N}(\mu, \sigma^2/n)$. This means that equation (7) will not give acceptable results of the expected values in the cases when the variance of response is so large that it is impossible to reduce the error of estimation with a finite sample size. On the other hand, due to the finite lengths of floating-point representations in computer, when two numbers are summed up, the smaller one will be neglected if the difference of their exponent bits exceeds the limit. Thus this truncated error in estimating the expectations will be dominant for simulations with large variances even if the chance that the responses take extremely large values is rare.

A simple example is the first-order linear stochastic system

$$\mathrm{d}x(t) = ax(t)\mathrm{d}t + \sigma x(t)\mathrm{d}W(t), \tag{11}$$

where a and σ are real constants. Its solution with the initial condition $x(0) = 1$ is

$$x(t) = e^{\left(a - \frac{1}{2}\sigma^2\right)t + \sigma W(t)}. \tag{12}$$

The moment Lyapunov exponents of this system are given by

$$\Lambda(p) = \frac{p}{2}\left[(p-1)\sigma^2 + 2a\right], \tag{13}$$

and the variance of norm is

$$\mathrm{Var}\left[x(t)\right] = e^{2at}(e^{\sigma^2 t} - 1). \tag{14}$$

It can be seen that the variance will increase exponentially with time when $\sigma \neq 0$.

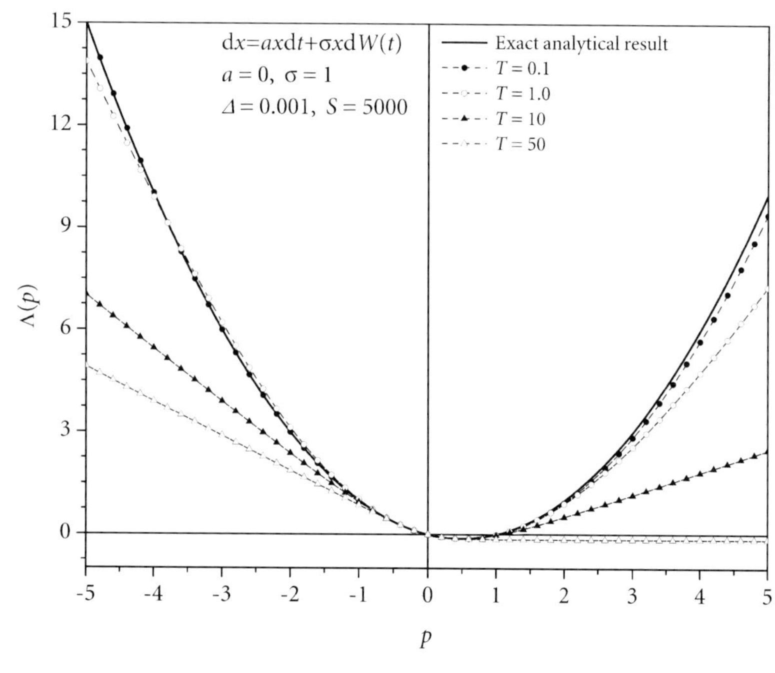

Figure 1: Moment Lyapunov exponents for different times of simulation

Figure 1 shows the numerical results using the explicit Euler scheme for different times T in the case $a=0$, $\sigma=1$. The time step for iteration is $\Delta=0.001$, the sample size is 5000 and equation (10) is used to determine the approximate moment Lyapunov exponents. It is obvious that the longer the time for simulation, the worse the results.

2 Estimation of the Expectation through the Logarithm of Norm

Due to the reasons discussed above, it is required to find a new algorithm to overcome the difficulty in estimating the expectation. Since the error is caused by large variance, it is clear that how to reduce the variance of response in order to obtain a good estimation of expectation using a finite number of samples is important.

Let $\{\mathbf{X}_i\}$ be i.i.d. random vectors with the same distribution as $\mathbf{X}(t)$, then

$$\begin{aligned}\log E\left[\|\mathbf{X}(t)\|^p\right] &= \frac{1}{n}\log\left(E\left[\|\mathbf{X}(t)\|^p\right]\right)^n = \frac{1}{n}\log\left(\prod_{i=1}^{n} E\left[\|\mathbf{X}_i\|^p\right]\right) \\ &= \frac{1}{n}\log E\left[\prod_{i=1}^{n}\|\mathbf{X}_i\|^p\right] = \frac{1}{n}\log E\left[\exp\left(p\sum_{i=1}^{n}\log\|\mathbf{X}_i\|\right)\right].\end{aligned} \tag{15}$$

Defining

$$S_n=\sum_{i=1}^{n}\log\|\mathbf{X}_i\|=\sum_{i=1}^{n}\rho_i, \qquad z_n=\frac{1}{n}S_n=\frac{1}{n}\sum_{i=1}^{n}\rho_i, \tag{16}$$

and letting

$$\mu=E\left[\log\|\mathbf{X}(t)\|\right]=E[\rho(t)], \qquad \sigma^2=\mathrm{Var}\left[\log\|\mathbf{X}(t)\|\right]=\mathrm{Var}\left[\rho(t)\right], \tag{17}$$

then

$$\log E\left[\|\mathbf{X}(t)\|^p\right]=\frac{1}{n}\log E\left[e^{npz_n}\right]. \tag{18}$$

With the notation

$$\xi_n=\frac{z_n-\mu}{\sigma/\sqrt{n}}, \tag{19}$$

equation (18) is converted to

$$\log E\left[\|\mathbf{X}(t)\|^p\right]=\frac{1}{n}\log E\left[e^{np\mu}\cdot e^{\sqrt{n}p\sigma\xi_n}\right]=p\mu+\frac{1}{n}\log E\left[e^{\sqrt{n}p\sigma\xi_n}\right]. \tag{20}$$

Let $F(\xi)$ be the distribution function of ξ_n, then $F(\xi)$ tends to $\mathcal{N}(0,1)$ as $n\to\infty$ according to the Central Limit Theorem, i.e. $F(\xi)\to\Phi(\xi)$ uniformly, where $\Phi(\xi)$ is the probability distribution function of the standard normal variable

$$\Phi(\xi)=\frac{1}{\sqrt{2\pi}}\int_{-\infty}^{\xi}e^{-\frac{1}{2}x^2}\,dx. \tag{21}$$

Using the Edgeworth expansion theorem for distribution (Gnedenko and Kolmogorov, 1954), $F(\xi)$ can be written as

$$F(\xi)=\Phi(\xi)+\sum_{k=3}^{\infty}c_k\Phi^{(k)}(\xi), \tag{22}$$

where the coefficients c_k are determined by the equivalence of moments on both sides.

The "tail effect" of distribution $F(\xi)$ is important since it is required to solve the expectation of $e^{\sqrt{n}p\sigma\xi_n}$ in equation (20). This means that accurate higher-order moments of ξ_n are needed in order to obtain a good approximation of $E\left[e^{\sqrt{n}p\sigma\xi_n}\right]$. However, it is rather difficult to do so for the solution of a general system (3) in practice. Therefore, in this paper, special systems are considered.

For a linear system with constant coefficients, i.e. the coefficient matrices in equation (3) take the form

$$\mathbf{m}(\mathbf{X},t)=\mathbf{m}\mathbf{X},\qquad \boldsymbol{\sigma}(\mathbf{X},t)=\boldsymbol{\sigma}\mathbf{X}, \tag{23}$$

where $\mathbf{m}$ and $\boldsymbol{\sigma}$ are constant matrice, it has been shown that the limiting distribution of $\rho(t)=\log\|\mathbf{X}(t)\|$ is normal as $t\to\infty$ if there is a constant h such that, for any vector $\mathbf{Y}$,

$$\langle\boldsymbol{\sigma}\mathbf{X},\mathbf{Y}\rangle=\mathbf{Y}^T\boldsymbol{\sigma}\mathbf{X}\geqslant h\,\|\mathbf{X}\|^2\,\|\mathbf{Y}\|^2 \tag{24}$$

is satisfied (Arnold, 1974, Khasminskii, 1980). In this case $F(\xi)$ will be normal since the distribution of sum of independent normal distributed random variables is also normal, i.e. $F(\xi)=\Phi(\xi)$. Thus

$$\begin{aligned}\log E\left[\|\mathbf{X}(t)\|^p\right]&=p\mu+\frac{1}{n}\log\int_{-\infty}^{\infty}e^{\sqrt{n}p\sigma\xi_n}\,\mathrm{d}\Phi(\xi)\\&=p\mu+\frac{1}{n}\log e^{\frac{1}{2}np^2\sigma^2}=p\mu+\frac{1}{2}p^2\sigma^2.\end{aligned} \tag{25}$$

Hence for the linear system with constant coefficients, by estimating the mean and variance of logarithm of norm, the moment Lyapunov exponents will be given by

$$\Lambda(p)=\frac{1}{T}\left(pE\left[\log\|\mathbf{X}(T)\|\right]+\frac{1}{2}p^2\mathrm{Var}\left[\log\|\mathbf{X}(T)\|\right]\right). \tag{26}$$

It is obvious that the variance of $\log\|\mathbf{X}(T)\|$ will be much less than the variance of $\|\mathbf{X}(T)\|$; therefore obtaining a good estimation through the sample average is possible.

3 Algorithm for Linear Systems with Constant Coefficients

According to equation (26), the algorithm of simulating the moment Lyapunov exponents for linear system with constant coefficients

$$\mathrm{d}\mathbf{X}(t)=\mathbf{m}\mathbf{X}(t)\mathrm{d}t+\boldsymbol{\sigma}\mathbf{X}(t)\mathrm{d}\mathbf{W}(t), \tag{27}$$

can be described below. The time step of iteration is Δ, time interval of normalization $T_N=K\Delta$, and the sample size is S.

Step 1. Set the initial conditions of state vector $\mathbf{X}(t)$ by

$$\|\mathbf{X}^s(0)\|=1,\qquad s=1,2,\cdots,S, \tag{28}$$

where $\|\mathbf{x}\|=\sqrt{\mathbf{x}^T\mathbf{x}}$.

Step 2. Between every normalization operation, i.e. $m=1,2,\cdots,M$, and for every sample, use appropriate discrete scheme for equation (27) to perform K iterations for the Monte Carlo simulation of $\mathbf{X}^s(t)$.

Step 3. Let

$$\rho^s(mT_N)=\log\|\mathbf{X}^s(mT_N)\|, \tag{29}$$

and then use equation (5) to normalize the value of $\mathbf{X}^s(mT_N)$ such that $\|\mathbf{X}^s(mT_N)\|=1$.

Step 4. Repeat steps 2 and 3 until $m=M$, i.e. $T=MT_N$. Then by the description in step 3,

$$\begin{aligned}\rho^s(T)=\log\|\mathbf{X}^s(MT_N)\|&=\log\frac{\|\mathbf{X}(MT_N)\|}{\|\mathbf{X}((M-1)T_N)\|}\cdots\frac{\|\mathbf{X}(T_N)\|}{\|\mathbf{X}(0)\|}\\&=\sum_{m=1}^{M}\log\frac{\|\mathbf{X}(mT_N)\|}{\|\mathbf{X}((m-1)T_N)\|}=\sum_{m=1}^{M}\rho^s(mT_N).\end{aligned} \tag{30}$$

Step 5. Use

$$
\begin{aligned}
E\left[\log\|\mathbf{X}(T)\|\right] &= E\left[\rho(T)\right] = \frac{1}{S}\sum_{s=1}^{S}\rho^{s}(T) = \bar{\rho}(T),\\
\mathrm{Var}\left[\log\|\mathbf{X}(T)\|\right] &= \mathrm{Var}\left[\rho(T)\right] = \frac{1}{S-1}\sum_{s=1}^{S}\left[\rho^{s}(T)^2 - \bar{\rho}(T)^2\right],
\end{aligned} \tag{31}
$$

to estimate the mean and variance of $\log\|\mathbf{X}(T)\|$.

Step 6. Use equation (26) to calculate the moment Lyapunov exponents for all values of p of interest.

4 Application on the Stability of a Viscoelastic System

Consider a single degree-of-freedom viscoelastic system excited by a zero mean wide-band stationary noise $\xi(t)$,

$$
\ddot{q}(t) + 2\varepsilon\beta\dot{q}(t) + \omega^2\left\{\left[1-\varepsilon^{1/2}\xi(t)\right]q(t) - \varepsilon\int_0^t h(t-s)q(s)\mathrm{d}s\right\} = 0, \tag{32}
$$

where $h(t)$ is the viscoelastic kernel function, the small parameter ε is introduced to denote that the damping, viscoelastic effect, and the amplitude of noise are small. The method of stochastic averaging (Khasminskii, 1966a,b) and the averaging method for integro-differential equations (Larinov, 1969) are applied to obtain the approximate analytical moment Lyapunov exponents.

By applying the transformation

$$
q(t) = a(t)\cos\Phi(t), \quad \dot{q}(t) = -\omega a(t)\sin\Phi(t), \quad \Phi(t) = \omega t + \varphi(t), \tag{33}
$$

equation (32) is converted to the form

$$
\begin{Bmatrix}\dot{a}(t)\\ \dot{\varphi}(t)\end{Bmatrix} = \varepsilon\mathbf{F}^{(1)}(a,\varphi,t) + \varepsilon^{1/2}\mathbf{F}^{(0)}(a,\varphi,\xi(t),t), \tag{34}
$$

$$
\begin{aligned}
\mathbf{F}^{(1)}(a,\varphi,t) &= \begin{Bmatrix} -2\beta a(t)\sin^2\Phi(t) - \omega\sin\Phi(t)\displaystyle\int_0^t h(t-s)a(s)\cos\Phi(s)\mathrm{d}s \\ -2\beta a(t)\sin\Phi(t)\cos\Phi(t) - \frac{1}{a(t)}\omega\cos\Phi(t)\displaystyle\int_0^t h(t-s)a(s)\cos\Phi(s)\mathrm{d}s \end{Bmatrix}\\
&= \begin{Bmatrix} F_1^{(1)}(a,\varphi,t)\\ F_2^{(1)}(a,\varphi,t)\end{Bmatrix},
\end{aligned} \tag{35}
$$

$$
\begin{aligned}
\mathbf{F}^{(0)}(a,\varphi,\xi(t),t) &= \begin{Bmatrix} -\frac{1}{2}\omega\xi(t)a(t)\sin 2\Phi(t) \\ -\frac{1}{2}\omega\xi(t)a(t)[1+\cos 2\Phi(t)] \end{Bmatrix}\\
&= \begin{Bmatrix} F_1^{(0)}(a,\varphi,\xi(t),t)\\ F_2^{(0)}(a,\varphi,\xi(t),t)\end{Bmatrix}.
\end{aligned} \tag{36}
$$

Since $\xi(t)$ is a wide-band noise, according to the method of stochastic averaging, system (34) can be approximated by the following averaged equations

$$
\mathrm{d}\begin{Bmatrix}\bar{a}(t)\\ \bar{\varphi}(t)\end{Bmatrix} = \varepsilon\begin{Bmatrix}\bar{m}_a\\ \bar{m}_\varphi\end{Bmatrix}\mathrm{d}t + \varepsilon^{1/2}\bar{\boldsymbol{\sigma}}\mathrm{d}\mathbf{W}(t), \tag{37}
$$

where

$$\bar{m}_a = \mathop{\mathcal{M}}_{t}\left\{F_1^{(1)}(a,\varphi,t) + \int_{-\infty}^{0} E\left[\frac{\partial F_1^{(0)}}{\partial a}F_{1\tau}^{(0)} + \frac{\partial F_1^{(0)}}{\partial \varphi}F_{2\tau}^{(0)}\right]\mathrm{d}\tau\right\},$$

$$\bar{m}_\varphi = \mathop{\mathcal{M}}_{t}\left\{F_2^{(1)}(a,\varphi,t) + \int_{-\infty}^{0} E\left[\frac{\partial F_2^{(0)}}{\partial a}F_{1\tau}^{(0)} + \frac{\partial F_2^{(0)}}{\partial \varphi}F_{2\tau}^{(0)}\right]\mathrm{d}\tau\right\},$$

$$\left[\bar{\boldsymbol{\sigma}}\bar{\boldsymbol{\sigma}}^T\right]_{ij} = \mathop{\mathcal{M}}_{t}\left\{\int_{-\infty}^{\infty} E\left[F_i^{(0)}F_{j\tau}^{(0)}\right]\mathrm{d}\tau\right\}, \quad i,j=1,2,$$

$$F_{j\tau}^{(0)} = F_j^{(0)}(a,\varphi,\xi(t+\tau),t+\tau), \quad j=1,2, \tag{38}$$

and

$$\mathop{\mathcal{M}}_{t}\{\cdot\} = \lim_{T\to\infty}\frac{1}{T}\int_0^T\{\cdot\}\,\mathrm{d}t \tag{39}$$

is the averaging operator.

When applying the averaging operation, $a(t)$ and $\varphi(t)$ are treated as constants and are replaced by $\bar{a}$ and $\bar{\varphi}$, respectively. After some calculations the corresponding terms in the averaged equations are

$$\bar{m}_a = \left[-\beta - \frac{1}{2}\omega\mathcal{H}^s(\omega) + \frac{3}{16}\omega^2 S(2\omega)\right]\bar{a}, \quad \bar{m}_\varphi = -\frac{1}{2}\omega\mathcal{H}^c(\omega) - \frac{1}{8}\omega^2\Psi(2\omega),$$

$$\left[\bar{\boldsymbol{\sigma}}\bar{\boldsymbol{\sigma}}^T\right]_{11} = b_{11} = \frac{1}{8}\omega^2 S(2\omega)\bar{a}^2, \quad \left[\bar{\boldsymbol{\sigma}}\bar{\boldsymbol{\sigma}}^T\right]_{12} = \left[\bar{\boldsymbol{\sigma}}\bar{\boldsymbol{\sigma}}^T\right]_{21} = 0,$$

$$\left[\bar{\boldsymbol{\sigma}}\bar{\boldsymbol{\sigma}}^T\right]_{22} = b_{22} = \frac{1}{8}\omega^2\left[2S(0) + S(2\omega)\right], \tag{40}$$

where

$$\mathcal{H}^s(\omega) = \int_0^\infty h(\tau)\sin\omega\tau\,\mathrm{d}\tau, \quad \mathcal{H}^c(\omega) = \int_0^\infty h(\tau)\cos\omega\tau\,\mathrm{d}\tau \tag{41}$$

are the sine and cosine transformations of the viscoelastic kernel function $h(t)$, $S(\omega)$ and $\Psi(\omega)$ are the cosine and sine power spectral density functions of the wide-band noise $\xi(t)$, respectively. Noting that transition density function for the solution of the averaged equation is the solution of the Fokker-Planck equation, which depends on the diffusion matrix $\bar{\boldsymbol{\sigma}}\bar{\boldsymbol{\sigma}}^T$ but not every single element $\bar{\sigma}_{ij}$, it can be set that

$$\bar{\sigma}_{12} = \bar{\sigma}_{21} = 0, \quad \bar{\sigma}_{11} = \sqrt{b_{11}} = \omega\bar{a}\sqrt{\frac{S(2\omega)}{8}}, \quad \bar{\sigma}_{22} = \sqrt{b_{22}} = \omega\sqrt{\frac{2S(0)+S(2\omega)}{8}}. \tag{42}$$

Letting $P = \bar{a}^p$ and applying Itô's Lemma to the first equation of (37) lead to the Itô differential equation for pth norm P,

$$\begin{aligned}\mathrm{d}P = \varepsilon&\left[p\left(-\beta - \frac{1}{2}\omega\mathcal{H}^s(\omega) + \frac{3}{16}\omega^2 S(2\omega)\right) + \frac{p(p-1)}{16}\omega^2 S(2\omega)\right]P\mathrm{d}t\\ &+ \varepsilon^{1/2}\omega p P\sqrt{\frac{S(2\omega)}{8}}\mathrm{d}W_1(t).\end{aligned} \tag{43}$$

Taking the expected value on both sides of eqation (43) results in

$$\mathrm{d}E[P] = \varepsilon\left[p\left(-\beta - \frac{1}{2}\omega\mathcal{H}^s(\omega) + \frac{3}{16}\omega^2 S(2\omega)\right) + \frac{p(p-1)}{16}\omega^2 S(2\omega)\right]E[P]\mathrm{d}t. \tag{44}$$

Hence the moment Lyapunov exponents are given by

$$\Lambda(p) = \lim_{t\to\infty} \frac{\log E[P]}{t} = \varepsilon \left[p \left(-\beta - \frac{1}{2}\omega \mathcal{H}^s(\omega) + \frac{3}{16}\omega^2 S(2\omega) \right) + \frac{p(p-1)}{16}\omega^2 S(2\omega) \right]. \tag{45}$$

It can be seen that the presence of viscoelasticity helps to stabilize the system, and the stronger the noise, the more unstable the system.

In oder to verify the accuracy of the approximate analytical results given by equation (45), the algorithm described in section 3 is used to simulate the moment Lyapunov exponents. The viscoelastic kernel function is assumed to be of the form

$$h(t) = \gamma e^{-\kappa t}, \tag{46}$$

and the wide-band noise is taken as the Gaussian white noise

$$\xi(t) = \sigma \dot{W}(t). \tag{47}$$

Then $S(2\omega) \equiv S_0 = \sigma^2$.

Let

$$x_1(t) = q(t), \quad x_2(t) = \dot{q}(t), \quad x_3(t) = \int_0^t h(t-s)q(s)\mathrm{d}s, \tag{48}$$

system (32) can be converted to the Itô differential equations

$$\mathrm{d}\begin{Bmatrix} x_1 \\ x_2 \\ x_3 \end{Bmatrix} = \begin{Bmatrix} x_2 \\ -\omega^2 x_1 - 2\varepsilon\beta x_2 + \varepsilon\omega^2 x_3 \\ \gamma x_1 - \kappa x_3 \end{Bmatrix} \mathrm{d}t + \begin{Bmatrix} 0 \\ -\varepsilon^{1/2}\sigma\omega^2 x_1 \\ 0 \end{Bmatrix} \mathrm{d}W(t). \tag{49}$$

Then the iteration equations for explicit Euler scheme are given by

$$\begin{aligned} x_1^{k+1} &= x_1^k + x_2^k \cdot \Delta, \\ x_2^{k+1} &= x_2^k + (-\omega^2 x_1^k - 2\varepsilon\beta x_2^k + \varepsilon\omega^2 x_3^k) \cdot \Delta - \varepsilon^{1/2}\sigma\omega^2 x_1^k \cdot \Delta W^k, \\ x_3^{k+1} &= x_3^k + (\gamma x_1^k - \kappa x_3^k) \cdot \Delta. \end{aligned} \tag{50}$$

Figure 2 shows typical results of the moment Lyapunov exponents for different values of σ, with the parameters taken as $\gamma = \kappa = \omega = 1.0$, $\varepsilon = 0.1$, $\beta = 0.05$. The sample size for estimating the expected value is $S = 5000$, time step $\Delta = 0.001$, and the total length of time of simulation is $T = 2000$, i.e. the number of iterations is $MK = 2 \times 10^6$. It can be seen that when σ is small, the approximate result from the averaging method agrees very well with the simulation result. When σ becomes larger, i.e. the noise becomes stronger, the discrepancy between the simulation and analytical results increases.

As a comparison, the simulation results using the revised algorithm in section 1 are plotted in Figure 3. It is clear that with the increase of σ, the algorithm in section 3 gives better results for $p < 0$.

5 Conclusion

For linear stochastic dynamical systems with constant coefficients, when the solutions have large variances, the algorithm used to simulate the moment Lyapunov exponents presented in this paper, which uses the mean and variance of logarithm of norm, gives better numerical approximation than the previous method, which uses the direct sample average of norm as the estimation of expectation. Effective algorithms for general stochastic systems are currently being developed.

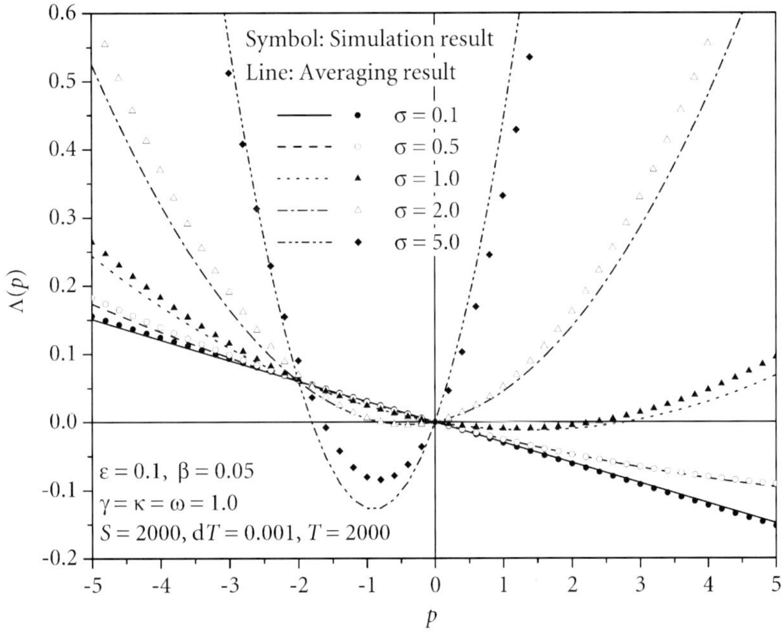

Figure 2: Moment Lyapunov exponents for different σ using algorithm in section 3

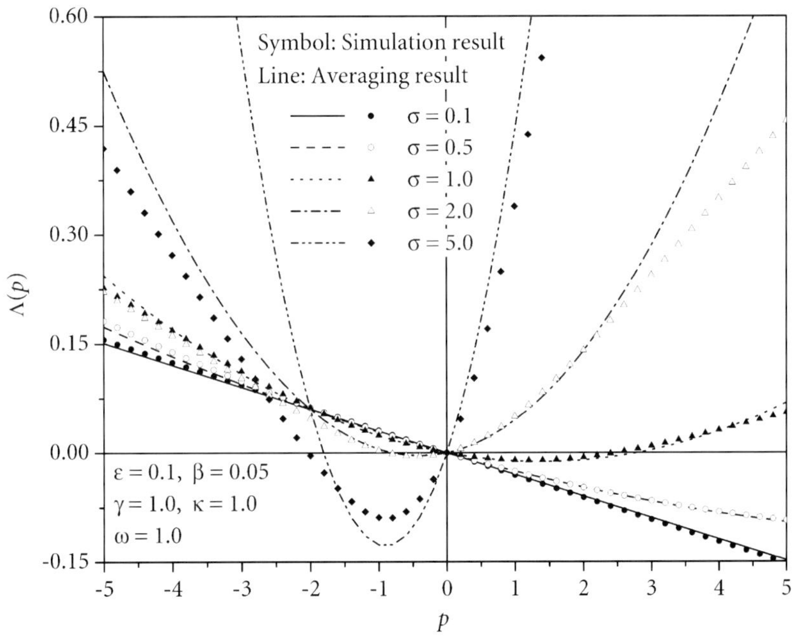

Figure 3: Moment Lyapunov exponents for different σ using algorithm in section 1

References

L. Arnold. *Stochastic Differential Equations: Theory and Applications*. John Wiley & Sons, Inc., New York, 1974.

W. Feller. *An Introduction to Probability Theory and Its Applications*, volume 2. John Wiley & Sons, Inc., New York, second edition, 1965.

B.V. Gnedenko and A.N. Kolmogorov. *Limit Distributions for Sums of Independent Random Variables*. Addison-Wesley Publishing Company, Inc., 1954. Translated from the Russian.

R.Z. Khasminskii. A limit theorem for the solutios of differential equations with random right-hand sides. *Theory of Probability and Its Applications*, 11(3):390–406, 1966a. English translation.

R.Z. Khasminskii. On stochastic processes defined by differential equations with a small parameter. *Theory of Probability and Its Applications*, 11(2):211–228, 1966b. English translation.

R.Z. Khasminskii. *Stochastic Stability of Differential Equations*. Kluwer Academic Publishers, Norwell, MA, 1980. English translation.

G.S. Larinov. Investigation of the vibration of relaxing systems by the averaging method. *Mechanics of Polymers*, 5:714–720, 1969. English translation.

W.C. Xie. Monte carlo simulation of moment Lyapunov exponents. *Journal of Applied Mechanics*, 72 (2):269–275, 2005.

Structural Optimization

MATHEMATICAL PROGRAMMING IN STRUCTURAL MECHANICS – THE PAST AND THE FUTURE

Adam Borkowski
Institute of Fundamental Technological Problems,
Polish Academy of Sciences,
Swietokrzyska 21, 00-049 Warsaw, Poland
E-mail: abork@ippt.gov.pl

1. Introduction

During the last phase of the Second World War huge quantities of goods were transported across Atlantic ocean from the United States and Canada to Europe. Diminishing the overall cost of this logistic task even by several percents meant sparing millions of dollars. This demand motivated US-authorities to allocate money for research and many mathematicians started to investigate a problem of transportation: how to organize the flow of goods between given locations in order to minimize the total cost of delivery. G.B. Dantzig proposed very efficient method of solving such problems – the simplex algorithm – and named the domain *Linear Programming(LP)*. He did not realize that the second term of this name will soon collide with the vast area of computer programming.

After Dantzig published his first paper on LP (Dantzig, 1948), this approach attracted much interest in the West. It remained unnoticed that similar results were obtained by L.V. Kantorovich in the Soviet Union already before the Second World War (Kantorovich, 1939). At the beginning of the 1950-ties a general theory of *Mathematical Programming (MP)* was developed, with major contribution given by H.W. Kuhn and A.W. Tucker (Kuhn, Tucker, 1951). The subject of this theory is a *Non-Linear Programming Problem (NLP-problem)*:

$$\min_{\mathbf{x}\geq\mathbf{0}} \{ f(\mathbf{x}) \mid g_i(\mathbf{x}) \leq 0, \mathbf{x} \in R^n, i = 1,2,\ldots,m \} \tag{1}$$

Here $\mathbf{x} \in R^n$ is a column matrix of unknowns, $f = f(\mathbf{x})$ is a cost function and $g_i = g_i(\mathbf{x})$ are constraints with $i = 1,2,\ldots,m$. Usually it is assumed that functions f and g_i are convex. In particular, the cost function may be quadratic and the constraints may be linear. This leads to a particular form of the NLP-problem called *Quadratic Programming Problem (QP-problem)*:

$$\min_{\mathbf{x}\geq\mathbf{0}} \{ \frac{1}{2}\mathbf{x}^{\mathrm{T}}\mathbf{D}\,\mathbf{x} + \mathbf{c}^{\mathrm{T}}\mathbf{x} \mid \mathbf{A}\,\mathbf{x} \geq \mathbf{b} \} \tag{2}$$

Matrices $\mathbf{A} \in R^{m\times n}$, $\mathbf{b} \in R^m$, $\mathbf{c} \in R^n$ and $\mathbf{D} \in R^{n\times n}$ are given. In order to assure convexity of the cost function, matrix $\mathbf{D}$ must be positive definite.

Finally, taking $\mathbf{D} = \mathbf{0}$ in (2), we obtain the simplest form of MP-problems, namely a *Linear Programming Problem (LP-problem)*:

$$\min_{\mathbf{x}\geq\mathbf{0}} \{ \mathbf{c}^{\mathrm{T}}\mathbf{x} \mid \mathbf{A}\mathbf{x} \geq \mathbf{b} \} \tag{3}$$

It turns out that each minimization problem in LP has its maximization counterpart:

$$\max_{\mathbf{y}\geq\mathbf{0}} \{ \mathbf{b}^{\mathrm{T}}\mathbf{y} \mid \mathbf{A}^{\mathrm{T}}\mathbf{y} \leq \mathbf{c} \} \tag{4}$$

Problems (3) and (4) are said to be mutually *dual* and the entries of $\mathbf{y} \in R^m$ are called *dual variables*. For the sake of simplicity, we quote MP-problems in their canonical form: the problems (1) to (4)

M. Pandey et al. (eds), Advances in Engineering Structures, Mechanics & Construction, 639–651.

contain only non-negative variables and inequality constraints. In general, free variables and equality constraints can be present as well.

At the beginning Linear Programming was used only in management and economics. The remnants of this period are still present in the terminology (the cost function, the shadow prices, etc.). A typical application of the model (4) would be maximizing the total production of a factory that uses different technological processes and different resources. The unknowns y_i are then the time slots allocated for each process, the entries of $\mathbf{c}$ represent given efficiency of each process, the entries of $\mathbf{A}$ tell us how much of each resource is consumed by a particular process and the entries of $\mathbf{b}$ describe available amount of each resource.

In parallel to things happening in Mathematical Programming, revolutionary changes occurred in Structural Analysis. Instead of relying on linear elasticity and on admissible stresses, a concept of safety factors against possible *ultimate states* was introduced. Again the Cold War precluded the exchange of ideas and the pioneering work of A.A. Gvozdev (Gvozdev, 1949) remained unknown in the Western hemisphere.

It was proved soon that the safety factor against plastic collapse does not depend on elastic properties of the structural material and that such factor can be found by maximizing the load multiplier over all statically admissible stress fields. For skeletal structures with a single dominant internal force (e.g. the axial force or the bending moment) a stress state $\mathbf{s}$ is statically admissible if each s_j remains less or equal to the yield stress s_{0j} and if $\mathbf{s}$ is equilibrated with a load $\mathbf{p}$. Assuming strains $\mathbf{q}$ and displacements $\mathbf{w}$ to remain small prior to the plastic collapse, we can write the equilibrium equation as $\mathbf{C}^{\mathrm{T}}\,\mathbf{s} = \mathbf{p}$, where $\mathbf{C}$ is the matrix of kinematics: $\mathbf{q} = \mathbf{C}\,\mathbf{w}$. Let loading be proportional: $\mathbf{p} = \mu\,\mathbf{p}_0$, where μ is an unknown multiplier and $\mathbf{p}_0$ is a given reference load. According to the static theorem, the ultimate value μ_* of the load factor can be found solving the problem:

$$\max_{\mu,\mathbf{s}} \{ \mu \mid \mathbf{s} \le \mathbf{s}_0, \mathbf{C}^{\mathrm{T}}\,\mathbf{s} - \mu\,\mathbf{p}_0 = \mathbf{0} \} \tag{5}$$

Looking at this model today we see at once that it is a LP-problem. However, the pioneers of the MP-based modeling of structural behavior had to overcome the barrier between economics and mechanics. Having accomplished that, they could enjoy the power of mathematics: the semantics of the production planning problems and the ultimate load problem is completely different but the formal structure of both problems is identical. We believe that the following papers, cited in the alphabetic order after the first author, were important in providing impetus to the MP-oriented approach: Biron & Hodge, 1968; Brown & Ang, 1964; Ceradini & Gavarini, 1965; Hodge, 1966; Koopman & Lance, 1965; Sacchi & Buzzi-Ferraris, 1966; Wolfensberger, 1964.

Special tribute should be given to two persons: Mircea Z. Cohn and Aleksandras Čyras. Already in 1956 Cohn published in Romania his first paper on the plastic structural analysis (Cohn, 1956). In 1972 an inspiring paper on the unified theory of plastic analysis appeared (Cohn et al., 1972). During the NATO Advanced Study Institute that took place in Waterloo in 1977 he was invited to deliver a keynote lecture (Cohn, 1979). The celebration of his 65th birthday in 1991 gathered over 60 contributors from 14 countries. The results of this meeting were published in a book edited by Cohn's former students D.E. Grierson, A. Franchi and P. Riva (Grierson et al., 1991). Their contribution to the considered domain is substantial (Franchi & Cohn, 1980), (Grierson & Gladwell, 1971), (Grierson, 1972), (Riva & Cohn, 1990).

Since Lithuania was a part of the Soviet Union at the time of his scientific carrier, Čyras was for a long time isolated from the Western scientific community. Most of his early papers were written in Russian (Čyras, 1963) and published in a local Lithuanian journal. His first paper in English, written with the present author, appeared in Poland in 1968 (Čyras & Borkauskas, 1968). Already well known in the Soviet Union and in other countries of the Eastern block, he was invited in 1974 by Wacław Olszak to present his results at the CISM-course (Čyras, 1974).

In 1969 Čyras published a book that contained many fundamental results on the applications of Linear Programming in the analysis and design of structures made of rigid-perfectly plastic material (Čyras, 1969). Two further books (Čyras, 1971), (Čyras et al., 1974) were also written in Russian language. The first English edition appeared in 1983 – this was the translation of the book (Čyras, 1982). In 2002 the 75th birthday of Aleksandras Čyras was celebrated in Vilnius. This motivated his former students R. Karkauskas and the present author to prepare new edition of the book (Čyras et al., 1974). Substantially updated and rewritten it appeared in 2004 (Čyras et al., 2004).

Being a founder and a long time Rector of the Institute of Civil Engineering in Vilnius (VISI), Čyras inspired many researchers to work on the MP-applications in Structural Analysis and Optimum Design. This group contributed substantially to progress in such areas as the ultimate state under constrained strains (Čyras & Čižas, 1966), the evaluation of displacements prior to collapse (Čyras & Baronas, 1971), the ultimate state of shells (Čyras & Karkauskas, 1971), (Čyras & Kalanta, 1974), the plastic shakedown problem (Čyras & Atkočiūnas, 1984). The present author took part in the development of general concept of the dual approach (Čyras & Borkauskas, 1969). In 1988 he published a book in Polish that was translated three years later into English (Borkowski, 1988).

The above overview is by no means complete. It reflects personal experience of the author who was involved in this fascinating scientific adventure.

2. Dual View of Mechanics

It seems that the first new insight brought by the MP-approach to Structural Mechanics was discovering the equivalence of kinematic and static formulations. In Mathematical Programming this property is known as duality: under certain premises each problem of constrained extremum has its dual and the values of cost functions for such problems attained at the solutions coincide. We already quoted the dual LP-problems (1) and (2). If $\mathbf{x}_*$ is the solution of (1) and if the constrained minimum $f'(\mathbf{x}_*) = \mathbf{c}^T\mathbf{x}_*$ is finite, then there exists a solution $\mathbf{y}_*$ of the dual problem (2) and the optimum values of the cost functions coincide: $f''(\mathbf{y}_*) = \mathbf{b}^T\mathbf{y}_* = f'(\mathbf{x}_*)$.

Let us expand slightly the QP-problem (4) by introducing additional variables $\mathbf{y} \in R^n$:

$$\min_{\mathbf{x}\geq\mathbf{0},\,\mathbf{y}\geq\mathbf{0}} \{\frac{1}{2}\mathbf{x}^T\mathbf{A}_{xx}\,\mathbf{x} - \frac{1}{2}\mathbf{y}^T\mathbf{A}_{yy}\,\mathbf{y} + \mathbf{b}_x^T\mathbf{x} \mid \mathbf{A}_{yx}\,\mathbf{x} + \mathbf{A}_{yy}\mathbf{y} + \mathbf{b}_y \leq \mathbf{0}\} \tag{6}$$

Here $\mathbf{b}_x \in R^m$, $\mathbf{b}_y \in R^n$, $\mathbf{A}_{xx} \in R^{m\times m}$, $\mathbf{A}_{yy} \in R^{n\times n}$, $\mathbf{A}_{yx} \in R^{n\times m}$. Moreover, $\mathbf{A}_{xx}$ is positive definite and $\mathbf{A}_{yy}$ is negative definite. The dual of (6) reads

$$\max_{\mathbf{x}\geq\mathbf{0},\,\mathbf{y}\geq\mathbf{0}} \{-\frac{1}{2}\mathbf{x}^T\mathbf{A}_{xx}\,\mathbf{x} + \frac{1}{2}\mathbf{y}^T\mathbf{A}_{yy}\,\mathbf{y} + \mathbf{b}_y^T\mathbf{y} \mid \mathbf{A}_{xx}\,\mathbf{x} + \mathbf{A}_{xy}\mathbf{y} + \mathbf{b}_x \geq \mathbf{0}\} \tag{7}$$

Duality has simple geometrical interpretation. Solving a pair of dual problems (6), (7) is equivalent to finding the *saddle point*

$$L(\mathbf{x}_*,\mathbf{y}_*) = \min_{\mathbf{x}\geq\mathbf{0}} \max_{\mathbf{y}\geq\mathbf{0}} L(\mathbf{x},\mathbf{y}) \tag{8}$$

of the *Lagrange function*

$$L(\mathbf{x},\mathbf{y}) = \frac{1}{2}\mathbf{x}^T\mathbf{A}_{xx}\,\mathbf{x} + \frac{1}{2}\mathbf{y}^T\mathbf{A}_{yy}\,\mathbf{y} + \mathbf{x}^T\mathbf{A}_{xy}\mathbf{y} + \mathbf{b}_x^T\mathbf{x} + \mathbf{b}_y^T\mathbf{y} \tag{9}$$

Such point can be reached in two ways. One can first establish the parabola that contains all maxima with respect to y-variables and then find the minimum on this curve. This sequence corresponds to the problem (6). Alternatively, one can begin with finding the parabola that corresponds to all minima with respect to the x-variables and then look for the maximum of this concave function. This way leads to the dual problem (7).

Point $(\mathbf{x}_*, \mathbf{y}_*)$ is the saddle point of L if it satisfies *Kuhn-Tucker conditions (KT-conditions)*:

$$\nabla \mathbf{L}_\mathbf{x} \geq \mathbf{0}\,, \nabla \mathbf{L}_\mathbf{y} \leq \mathbf{0}, \mathbf{x} \geq \mathbf{0}, \mathbf{y} \geq \mathbf{0} \tag{10}$$

$$\mathbf{x}^\mathrm{T} \nabla \mathbf{L}_\mathbf{x} = 0\,, \mathbf{y}^\mathrm{T} \nabla \mathbf{L}_\mathbf{y} = 0 \tag{11}$$

Here $\nabla \mathbf{L}_\mathbf{x} \in R^n$ and $\nabla \mathbf{L}_\mathbf{y} \in R^m$ are gradients of L with respect to $\mathbf{x}$ and $\mathbf{y}$. If all variables were free, then the KT-conditions would reduce to the common stationarity conditions:

$$\nabla \mathbf{L}_\mathbf{x} = \mathbf{0}\,, \nabla \mathbf{L}_\mathbf{y} = \mathbf{0} \tag{12}$$

or, explicitly, to the set of linear algebraic equations

$$\begin{aligned} \mathbf{A}_{\mathbf{xx}}\mathbf{x} + \mathbf{A}_{\mathbf{xy}}\mathbf{y} + \mathbf{b}_\mathbf{x} &= \mathbf{0} \\ \mathbf{A}_{\mathbf{yx}}\mathbf{x} + \mathbf{A}_{\mathbf{yy}}\mathbf{y} + \mathbf{b}_\mathbf{y} &= \mathbf{0} \end{aligned} \tag{13}$$

Note two features that distinguish this set: a) its matrix of coefficients is symmetric; b) the sub-matrices situated along the diagonal have special properties – $\mathbf{A}_{\mathbf{xx}}$ is positive definite, $\mathbf{A}_{\mathbf{yy}}$ is negative definite

It is easy to check by inspecting the Table 1 that the set of equations governing linear static analysis of elastic structures follows exactly the template (13). The goal of such analysis is to find displacements[1] $\mathbf{w}$, stresses $\mathbf{s}$ and reactions $\mathbf{r}$ of elastic structure caused by a given static load $\mathbf{p}_0$ and by a given kinematic load $\mathbf{w}_0$. The structure is represented by a common discrete model, where $\mathbf{E}$ is the matrix of elasticity, $\mathbf{C}$ is the matrix of compatibility, subscript p refers to the degrees of freedom with prescribed external forces and subscript w refers to the degrees of freedom with prescribed displacements.

It is seen from the Table 1 that displacements play the role of x-variables, whereas stresses and reactions correspond to y-variables in the model (13). The first two rows of the Table 1 contain the equilibrium equations. The third row comes from substituting strains $\mathbf{q} = \mathbf{C}\mathbf{w} = \mathbf{C}_\mathbf{p}\mathbf{w}_\mathbf{p} + \mathbf{C}_\mathbf{w}\mathbf{w}_\mathbf{w}$ into the constitutive equation $\mathbf{q} = \mathbf{E}^{-1}\mathbf{s}$. The last row merely says that $\mathbf{w}_\mathbf{w} = \mathbf{w}_0$. A matrix with zero entries can be treated either as positive semi-definite or as negative semi-definite. The inverse of matrix of elasticity is strictly positive definite. Hence, $-\mathbf{E}^{-1}$ is strictly negative definite.

Note that there are no inequalities in the Table 1 and that all variables are free with respect to sign. Hence, we don't need to take the KT-conditions into account.

[1] In the sequel we write „displacements", „strains", „stresses" and „loads" having in mind generalized variables taken usually in Structural Analysis.

Table 1. Governing equations of linear elastic analysis of structures

	$\mathbf{w}_p$	$\mathbf{w}_w$	$\mathbf{s}$	$\mathbf{r}$	1	
$\nabla \mathbf{L}_{w_p} =$	$\mathbf{0}$	$\mathbf{0}$	$\mathbf{C}_p^T$	$\mathbf{0}$	$\mathbf{p}_0$	$= \mathbf{0}$
$\nabla \mathbf{L}_{w_w} =$	$\mathbf{0}$	$\mathbf{0}$	$\mathbf{C}_w^T$	$-\mathbf{I}$	$\mathbf{0}$	$= \mathbf{0}$
$\nabla \mathbf{L}_s =$	$\mathbf{C}_p$	$\mathbf{C}_w$	$-\mathbf{E}^{-1}$	$\mathbf{0}$	$\mathbf{0}$	$= \mathbf{0}$
$\nabla \mathbf{L}_r =$	$\mathbf{0}$	$-\mathbf{I}$	$\mathbf{0}$	$\mathbf{0}$	$\mathbf{w}_0$	$= \mathbf{0}$

What do we gain by using the MP-based approach in elastic analysis? First, having filled the Table 1, we can derive easily the dual energy principles (compare the templates (6) and (7)):

a) kinematic principle –

$$\min_{\mathbf{s},\mathbf{w}} \{ \frac{1}{2} \mathbf{s}^T \mathbf{E}^{-1} \mathbf{s} - \mathbf{w}_p^T \mathbf{p}_0 \mid \mathbf{C}_p \mathbf{w}_p + \mathbf{C}_w \mathbf{w}_w - \mathbf{E}^{-1} \mathbf{s} = \mathbf{0}, \mathbf{w}_w = \mathbf{w}_0 \} \tag{14}$$

b) static principle –

$$\max_{\mathbf{s},\mathbf{r}} \{ -\frac{1}{2} \mathbf{s}^T \mathbf{E}^{-1} \mathbf{s} + \mathbf{w}_0^T \mathbf{r} \mid \mathbf{C}_p^T \mathbf{s} = \mathbf{p}_0, \mathbf{C}_w^T \mathbf{s} - \mathbf{r} = \mathbf{0} \} \tag{15}$$

Second, since the QP-problems (14), (15) contain no inequality constraints or non-negative variables, each of them can be reduced to a set of equations. This leads us very naturally to the fundamental computational tools of elastic analysis: the Stiffness (Force) Method and the Flexibility (Displacement) Method.

Third, the existence and uniqueness of solution for any given loading $\mathbf{p}_0, \mathbf{w}_0$ follows immediately from the convexity of problems (14), (15). Moreover, a generalization of the model (14), (15) to unilateral contact is straightforward. The replacement of $\mathbf{w}_w = \mathbf{w}_0$ in (14) by less restrictive condition $\mathbf{w}_w \geq \mathbf{w}_0$ induces sign constraint on $\mathbf{r}$ in the dual problem:

$$\min_{\mathbf{s},\mathbf{w}} \{ \frac{1}{2} \mathbf{s}^T \mathbf{E}^{-1} \mathbf{s} - \mathbf{w}_p^T \mathbf{p}_0 \mid \mathbf{C}_p \mathbf{w}_p + \mathbf{C}_w \mathbf{w}_w - \mathbf{E}^{-1} \mathbf{s} = \mathbf{0}, \mathbf{w}_w \geq \mathbf{w}_0 \} \tag{16}$$

$$\max_{\mathbf{s},\mathbf{r}} \{ -\frac{1}{2} \mathbf{s}^T \mathbf{E}^{-1} \mathbf{s} + \mathbf{w}_0^T \mathbf{r} \mid \mathbf{C}_p^T \mathbf{s} = \mathbf{p}_0, \mathbf{C}_w^T \mathbf{s} - \mathbf{r} = \mathbf{0}, \mathbf{r} \geq \mathbf{0} \} \tag{17}$$

Seemingly minor, this modification has dramatic consequences: a) the linearity of the problem is lost due to the KT-condition $\mathbf{r}^T (\mathbf{w}_0 - \mathbf{w}_w) = 0$; b) the energy principles (16), (17) can not be replaced by the sets of equations. Moreover, for certain loads $\mathbf{p}_0, \mathbf{w}_0$ the constraints of the problems (16), (17) might become contradictory. Thus the existence of solution is not warranted any more.

3. Expanding Area of Application

As shown by G. Maier in (Maier, 1970) Quadratic Programming allows us to model a broad range of piecewise-linear structural behaviors. The constitutive laws of such behaviors are given in Fig.2. An exhaustive overview of the QP-based approach, including both continuum and discrete models, can be found in (Borkowski, 2004). Let us recall, for the sake of brevity, only two cases – a cable-strut elastic structure and a structure made of strain-hardening material.

The entries of the stress matrix $\mathbf{s}$ for the cable-strut structure can be split into two sub-matrices: $\mathbf{s}_s$ represents axial forces in struts and $\mathbf{s}_c$ represents axial forces in cables. Obviously, cables work only in tension, while struts can be either extended or compressed. Table 2 shows the set of relations governing elastic behavior of the cable-strut system. In order to simplify things, we assume purely static loading.

The adjoint variable for the axial force in cable is its *slackness* – the axial strain taken with negative sign. Grouping the slackness unknowns into a column matrix $\mathbf{q}_c$, we consider it as a matrix of non-negative variables. Then, the entries of $\mathbf{s}_c$ remain formally unconstrained in sign: they will become non-negative in the solution due to the constraint $\nabla\mathbf{L}_{\mathbf{q}_c} \geq \mathbf{0}$.

Applying the templates (6), (7) to the Table 2, we obtain the following energy principles for the cable-strut system:

$$\underset{\mathbf{s}_s, \mathbf{w}, \mathbf{q}_c \geq \mathbf{0}}{min} \{\frac{1}{2}\mathbf{s}_s^T\mathbf{E}^{-1}\mathbf{s}_s - \mathbf{w}^T\mathbf{p}_0 \mid \mathbf{C}_s\mathbf{w} + \mathbf{C}_c\mathbf{w} - \mathbf{E}^{-1}\mathbf{s}_s = \mathbf{0}, \mathbf{C}_c + \mathbf{q}_c = \mathbf{0}\} \quad (18)$$

$$\underset{\mathbf{s}_s, \mathbf{s}_c}{max} \{-\frac{1}{2}\mathbf{s}_s\mathbf{E}^{-1}\mathbf{s}_s \mid \mathbf{C}_s^T\mathbf{s}_s + \mathbf{C}_c^T\mathbf{s}_c = \mathbf{p}_0, \ \mathbf{s}_c \geq \mathbf{0}\} \quad (19)$$

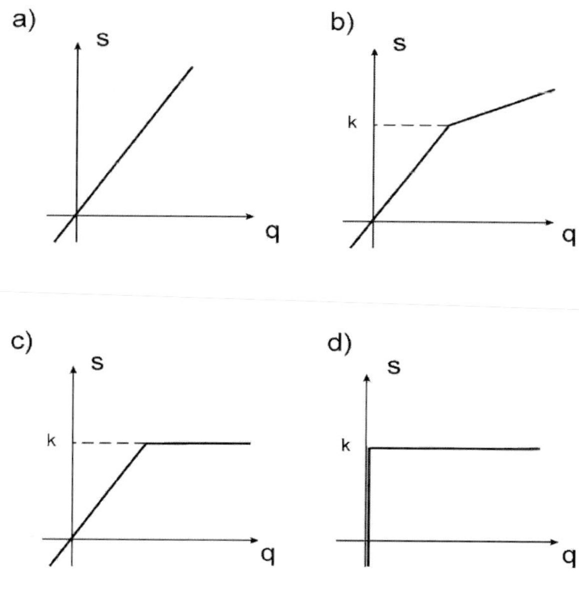

Fig. 2. Piecewise-linear behaviors of material: a) elastic; b) elastic-strain hardening; c) elastic-perfectly plastic; d) rigid-perfectly plastic.

Note that the existence of solution is not warranted: for a certain load $\mathbf{p}_0$ there might be no equilibrated stress state. Then the constraints of the QP-problem (19) become contradictory. On the other hand, if a solution exists, then it is unique due to the convexity of the problems (18), (19).

Table 2. Governing relations for a cable-strut structure

	$\mathbf{w}$	$\mathbf{q}_c$	$\mathbf{s}_s$	$\mathbf{s}_c$	1	
$\nabla \mathbf{L}_w =$	$\mathbf{0}$	$\mathbf{0}$	$\mathbf{C}_s^T$	$\mathbf{C}_c^T$	$\mathbf{p}_0$	$= \mathbf{0}$
$\nabla \mathbf{L}_{q_c} =$	$\mathbf{0}$	$\mathbf{0}$	$\mathbf{0}$	$\mathbf{I}$	$\mathbf{0}$	$\geq \mathbf{0}$
$\nabla \mathbf{L}_{s_s} =$	$\mathbf{C}_s$	$\mathbf{0}$	$-\mathbf{E}^{-1}$	$\mathbf{0}$	$\mathbf{0}$	$= \mathbf{0}$
$\nabla \mathbf{L}_{s_c} =$	$\mathbf{C}_c$	$\mathbf{I}$	$\mathbf{0}$	$\mathbf{0}$	$\mathbf{0}$	$= \mathbf{0}$
	$\mathbf{q}_c \geq \mathbf{0}$, $\mathbf{q}_c^T \nabla \mathbf{L}_c = 0$					

Similar procedure can be applied to a structure made of elastic-strain hardening material. A complete set of governing relations for such a structure is shown in Table 3. The first row of this table includes linearised yield condition $\mathbf{N}^T\mathbf{s} - \mathbf{H}\boldsymbol{\lambda} \leq \mathbf{k}_0$. A positive definite $(r \times r)$-matrix of hardening $\mathbf{H}$ is responsible for a shift of the yield planes caused by the plastic strains. The equation of equilibrium that relates the given load $\mathbf{p}_0$ to the unknown stress $\mathbf{s}$ can be recognized in the second row. The last row ensures the kinematic compatibility of strains $\mathbf{q} = \mathbf{q}_e + \mathbf{q}_p$ and displacements $\mathbf{w}$. Elastic strains obey the Hooke's law $\mathbf{q}_e = \mathbf{E}^{-1}\mathbf{s}$. Plastic strains are governed by the associated flow rule $\mathbf{q}_p = \mathbf{N}\,\boldsymbol{\lambda}$. Plastic multipliers represented by a column matrix $\boldsymbol{\lambda} \in R^r$ are supposed to be non-negative.

Table 3. Governing relations for a structure made of elastic-strain hardening material.

	$\boldsymbol{\lambda}$	$\mathbf{w}$	$\mathbf{s}$	1	
$\nabla \mathbf{L}_\lambda =$	$\mathbf{H}$	$\mathbf{0}$	$-\mathbf{N}^T$	$\mathbf{k}_0$	$\geq \mathbf{0}$
$\nabla \mathbf{L}_w =$	$\mathbf{0}$	$\mathbf{0}$	$\mathbf{C}^T$	$-\mathbf{p}_0$	$= \mathbf{0}$
$\nabla \mathbf{L}_s =$	$-\mathbf{N}$	$\mathbf{C}$	$-\mathbf{E}^{-1}$	$\mathbf{0}$	$= \mathbf{0}$
	$\boldsymbol{\lambda} \geq \mathbf{0}$, $\boldsymbol{\lambda}^T \nabla \mathbf{L}_\lambda = 0$				

Applying the templates (6), (7) to the Table 2, we obtain the following energy principles for elastic-strain hardening structures:

$$\min_{\mathbf{s},\mathbf{w},\boldsymbol{\lambda}\geq 0}\{\frac{1}{2}\boldsymbol{\lambda}^{T}\mathbf{H}\boldsymbol{\lambda}+\frac{1}{2}\mathbf{s}^{T}\mathbf{E}^{-1}\mathbf{s}+\mathbf{k}_{0}^{T}\boldsymbol{\lambda}-\mathbf{w}^{T}\mathbf{p}_{0}\mid \mathbf{C}\mathbf{w}-\mathbf{N}\boldsymbol{\lambda}-\mathbf{E}^{-1}\mathbf{s}=\mathbf{0}\} \tag{20}$$

$$\max_{\mathbf{s}}\{-\frac{1}{2}\boldsymbol{\lambda}^{T}\mathbf{H}\boldsymbol{\lambda}-\frac{1}{2}\mathbf{s}\,\mathbf{E}^{-1}\mathbf{s}\mid \mathbf{N}^{T}\mathbf{s}-\mathbf{H}\boldsymbol{\lambda}\leq\mathbf{k}_{0},\ \mathbf{C}^{T}\mathbf{s}=\mathbf{p}_{0}\} \tag{21}$$

Solutions of these QP-problems exist for any $\mathbf{p}_0$ since the admissible domain for stresses adjusts itself automatically to the loading. The convexity of the problems (20), (21) ensures uniqueness of the structural response. The only drawback of the model (20), (21) is its holonomic nature: a possible local unloading is not taken into account.

Assuming $\mathbf{H}=\mathbf{0}$, we obtain structure made of elastic-perfectly plastic material. The relevant dual QP-problems

$$\min_{\mathbf{s},\mathbf{w},\boldsymbol{\lambda}\geq 0}\{\frac{1}{2}\mathbf{s}^{T}\mathbf{E}^{-1}\mathbf{s}+\mathbf{k}_{0}^{T}\boldsymbol{\lambda}-\mathbf{w}^{T}\mathbf{p}_{0}\mid \mathbf{C}\mathbf{w}-\mathbf{N}\boldsymbol{\lambda}-\mathbf{E}^{-1}\mathbf{s}=\mathbf{0}\} \tag{22}$$

$$\max_{\mathbf{s}}\{\frac{1}{2}\mathbf{s}\,\mathbf{E}^{-1}\mathbf{s}\mid \mathbf{N}^{T}\mathbf{s}\leq\mathbf{k}_{0},\ \mathbf{C}^{T}\mathbf{s}=\mathbf{p}_{0}\} \tag{23}$$

still retain the uniqueness of solutions but the nice property of the existence of solution for any $\mathbf{p}_0$ is lost. The yield surface $\mathbf{N}^{T}\mathbf{s}=\mathbf{k}_0$ is now fixed and too high loading would cause the absence of statically admissible field of stresses.

If we would like to neglect elastic strains as well, the static energy principle (23) would loose its cost function (since there would be $\mathbf{E}^{-1}=\mathbf{0}$). This shows clearly that for a structure made of the rigid-perfectly plastic material the problem "find the response to the given load" is ill-posed. The right formulation is "find the load factor that corresponds to the state of plastic collapse". A complete set of relations for this formulation is given in Table 4.

Table 4. Governing relations for ultimate load factor.

	$\boldsymbol{\lambda}$	$\mathbf{w}$	$\mathbf{s}$	μ	1	
$\nabla\mathbf{L}_{\lambda}=$	$\mathbf{0}$	$\mathbf{0}$	$-\mathbf{N}^{T}$	$\mathbf{0}$	$\mathbf{k}_0$	$\geq\mathbf{0}$
$\nabla\mathbf{L}_{w}=$	$\mathbf{0}$	$\mathbf{0}$	$\mathbf{C}^{T}$	$-\mathbf{p}_0$	$\mathbf{0}$	$=\mathbf{0}$
$\nabla\mathbf{L}_{s}=$	$-\mathbf{N}$	$\mathbf{C}$	$\mathbf{0}$	$\mathbf{0}$	$\mathbf{0}$	$=\mathbf{0}$
$\nabla L_{\mu}=$	$\mathbf{0}$	$-\mathbf{p}_0^{T}$	$\mathbf{0}$	$\mathbf{0}$	1	$=0$
	$\boldsymbol{\lambda}\geq\mathbf{0},\ \boldsymbol{\lambda}^{T}\nabla\mathbf{L}_{\lambda}=0$					

The primal problem generated by this table is

$$\min_{\boldsymbol{\lambda},\mu}\{\mathbf{k}_{0}^{T}\boldsymbol{\lambda}\mid -\mathbf{N}\boldsymbol{\lambda}+\mathbf{C}\mathbf{w}=\mathbf{0},\ \mathbf{p}_{0}^{T}\mathbf{w}=1\} \tag{24}$$

and the dual one has been already given as Eq. (5). All sub-matrices located at the diagonal of the Table 4 have zero values. This leads to vanishing quadratic terms in the cost functions of the dual problems. On the other hand, we can not expect the solution to be unique in terms of stresses and/or collapse mechanisms, since linear functions are not strictly convex (concave).

4. Explicit and Implicit Optimization

Models based on Mathematical Programming give the user clear insight into the possible ways of structural optimization. Let us begin with the topological optimum design. Changing the layout of structural elements makes the entries of $\mathbf{C}$ variable. Hence, all relations where this matrix appears become non-linear. This circumstance explains why the topological formulation of the optimum design problem is still an open research area.

The search for optimum sizing of structural elements under given topology becomes much easier if we neglect elastic strains. Assuming additionally that a form of each element is given up to certain parameters, we can take the entries of $\mathbf{k}$ as unknowns, retaining the fixed matrices $\mathbf{C}$ and $\mathbf{N}$. Hence, basic relations of the rigid-perfectly plastic model remain linear. Note, that is not possible for elastic structures: in the sizing problem matrix $\mathbf{E}$ becomes variable which destroys linearity of the constitutive relations.

Let us assume that the unknown plastic modulae $\mathbf{k}$ are governed by a relatively small number of design variables the : $\mathbf{k} = \mathbf{G}^{\mathrm{T}}\mathbf{z}$, where $\mathbf{G}$ is $(g \times r)$-matrix of configuration. In order to maintain linearity of the problem, we adopt linear cost function $f(\mathbf{z}) = \mathbf{c}_0^{\mathrm{T}}\,\mathbf{z}$, where the entries of column matrix $\mathbf{c}_0 \in R^g$ are given cost coefficients. A problem to be solved reads: "given the structural layout find the optimum sizing $\mathbf{z}_*$ that minimizes f and assures given safety factor $\boldsymbol{\mu}_*$ against plastic collapse". In fact, we know an expected load carrying capacity $\mathbf{p}_* = \mu_*\mathbf{p}_0$ of the optimized structure, since both the safety factor $\boldsymbol{\mu}_*$ and the reference load $\mathbf{p}_0$ are given.

Table 5. Governing relations for optimum plastic design.

	$\boldsymbol{\lambda}$	w	s	z	1	
$\nabla\mathbf{L}_\lambda =$	$\mathbf{0}$	$\mathbf{0}$	$-\mathbf{N}^{\mathrm{T}}$	$\mathbf{G}^{\mathrm{T}}$	$\mathbf{0}$	$\geq \mathbf{0}$
$\nabla\mathbf{L}_w =$	$\mathbf{0}$	$\mathbf{0}$	$\mathbf{C}^{\mathrm{T}}$	$\mathbf{0}$	$-\mathbf{p}_*$	$= \mathbf{0}$
$\nabla\mathbf{L}_s =$	$-\mathbf{N}$	$\mathbf{C}$	$\mathbf{0}$	$\mathbf{0}$	$\mathbf{0}$	$= \mathbf{0}$
$\nabla\mathbf{L}_z =$	$\mathbf{G}$	$\mathbf{0}$	$\mathbf{0}$	$\mathbf{0}$	$-\mathbf{c}_0$	$\leq \mathbf{0}$
	$\boldsymbol{\lambda} \geq \mathbf{0}$, $\mathbf{z} \geq \mathbf{0}$, $\boldsymbol{\lambda}^{\mathrm{T}}\nabla\mathbf{L}_\lambda = 0$, $\mathbf{z}^{\mathrm{T}}\nabla\mathbf{L}_z = 0$					

Table 5 shows the internal structure of the problem of optimum plastic design. This problem is equivalent to the following pair of dual LP-problems:

$$\underset{\mathbf{s},\,\mathbf{z}\geq\mathbf{0}}{min}\{\mathbf{c}_0^{\mathrm{T}}\,\mathbf{z} \mid \mathbf{G}^{\mathrm{T}}\mathbf{z} - \mathbf{N}^{\mathrm{T}}\mathbf{s} \geq \mathbf{0},\ \mathbf{C}^{\mathrm{T}}\mathbf{s} = \mathbf{p}_*\} \tag{25}$$

$$\underset{\mathbf{w},\,\boldsymbol{\lambda}\geq\mathbf{0}}{max}\{\mathbf{p}_*^{\mathrm{T}}\,\mathbf{w} \mid -\mathbf{N}\,\boldsymbol{\lambda} + \mathbf{C}\,\mathbf{w} = \mathbf{0}, \mathbf{G}\,\boldsymbol{\lambda} \leq \mathbf{c}_0\} \tag{26}$$

Duality reveals an interesting role of the cost coefficients: according to the second constraint of the kinematic energy principle (26) these coefficients bound linear combinations of plastic multipliers. If we would take $\mathbf{G} = \mathbf{I}$ and $g = r$ which means that each plastic modulus is treated as an independent design variable, then the second constraint in (26) would reduce to the inequality $\boldsymbol{\lambda} \leq \mathbf{c}$. In those parts of optimum structure that undergo yielding this constraint must be satisfied as equality. Hence, assuming certain cost coefficients we in fact impose certain collapse mechanism on the optimum structure. It can be shown that this mechanism ensures uniform dissipation of energy over the structure.

Sizing is the case of explicitly formulated optimization problem. It is worth noting that the MP-approach allows us to uncover also certain possibilities of optimization hidden in the problems formulated from the analysis point of view. A good example of that is the problem of rigid-perfectly plastic structure brought to the state of plastic collapse by purely kinematic loading. Let us split again the degrees of freedom of the discrete structural model into the parts denoted by the indices p and w, as it was already done in the model (14), (15). Then the considered problem can be formulated in the following way: "given the structural layout and the distribution of plastic modulae $\mathbf{k}_0$ find the state $\mathbf{s}_*, \dot{\boldsymbol{\lambda}}_*, \dot{\mathbf{w}}_*$ of the structure brought to the plastic collapse by a given kinematic load $\mathbf{w}_0$"[2].

Table 6. Governing relations for kinematically induced plastic collapse.

	$\dot{\boldsymbol{\lambda}}$	$\dot{\mathbf{w}}_p$	$\dot{\mathbf{w}}_w$	$\mathbf{s}$	$\mathbf{r}$	1	
$\nabla L_\lambda =$	$\mathbf{0}$	$\mathbf{0}$	$\mathbf{0}$	$-\mathbf{N}^T$	$\mathbf{0}$	$\mathbf{k}_0$	$\geq \mathbf{0}$
$\nabla L_{w_p} =$	$\mathbf{0}$	$\mathbf{0}$	$\mathbf{0}$	$\mathbf{C}_p^T$	$\mathbf{0}$	$\mathbf{0}$	$= \mathbf{0}$
$\nabla L_{w_w} =$	$\mathbf{0}$	$\mathbf{0}$	$\mathbf{0}$	$\mathbf{C}_w^T$	$-\mathbf{I}$	$\mathbf{0}$	$= \mathbf{0}$
$\nabla L_s =$	$-\mathbf{N}$	$\mathbf{C}_p$	$\mathbf{C}_w$	$\mathbf{0}$	$\mathbf{0}$	$\mathbf{0}$	$= \mathbf{0}$
$\nabla L_r =$	$\mathbf{0}$	$\mathbf{0}$	$-\mathbf{I}$	$\mathbf{0}$	$\mathbf{0}$	$\mathbf{w}_0$	$= \mathbf{0}$
	$\dot{\boldsymbol{\lambda}} \geq \mathbf{0},\ \dot{\boldsymbol{\lambda}}^T \nabla L_\lambda = 0$						

Table 6 shows the governing relations for this problem. They correspond to the following pair of dual LP-problems:

$$\min_{\mathbf{w}_p, \mathbf{w}_w, \lambda \geq 0} \{ \mathbf{k}_0^T \dot{\boldsymbol{\lambda}} \mid -\mathbf{N}\dot{\boldsymbol{\lambda}} + \mathbf{C}_p \dot{\mathbf{w}}_p + \mathbf{C}_w \dot{\mathbf{w}}_w = \mathbf{0}, \dot{\mathbf{w}}_w = \mathbf{w}_0 \} \tag{27}$$

$$\max_{\mathbf{s}, \mathbf{r}} \{ \mathbf{w}_0^T \mathbf{r} \mid \mathbf{N}^T\mathbf{s} \geq \mathbf{k}_0,\ \mathbf{C}_p^T \mathbf{s} = \mathbf{0}, \mathbf{C}_p^T \mathbf{s} - \mathbf{r} = \mathbf{0} \} \tag{28}$$

The kinematic principle (27) says us that the collapse mechanism $\dot{\boldsymbol{\lambda}}_*, \dot{\mathbf{w}}_*$ corresponds to the minimum dissipated power $D = \mathbf{k}^T\dot{\boldsymbol{\lambda}}$. According to the static principle (28), the stresses $\mathbf{s}_*$ and the reactions $\mathbf{r}_*$ at the plastic collapse correspond to the maximum power of reactions done on the prescribed displace-

[2] For structures made from the rigid-perfectly plastic material the kinematic unknowns $\boldsymbol{\lambda}, \mathbf{w}$ should be replaced by their rates $\dot{\boldsymbol{\lambda}}, \dot{\mathbf{w}}$.

ment rates $\mathbf{w}_0$. Note that if the kinematic loading would be introduced in unilateral manner, which means replacing the last constraint in (27) by $\mathbf{w}_w \geq \mathbf{w}_0$, then the reactions would become sign constrained: $\mathbf{r} \geq \mathbf{0}$.

A. Čyras and his co-workers looked at the problem (28) from different perspective. They treated $\mathbf{p}$ as unknown loading and introduced linear *quality measure* of load $f = \mathbf{d}_0^T \mathbf{p}$. Here $\mathbf{d}_0 \in R^n$ is a given column matrix of weight factors. Then the following load optimization problem was formulated: "given the structural layout and the distribution of plastic modulae $\mathbf{k}_0$ find the ultimate load $\mathbf{p}_*$ that has the highest quality index f_*". Obviously, the dual problem revealed that the entries of $\mathbf{d}_0$ should be treated as prescribed displacement rates.

Approaching the problem from the kinematic side seems to be more natural. The implicit optimization of reaction forces that comes out via duality is probably more interesting in the continuum formulation. It can be shown then that by prescribing displacement rates on a part of the surface of the rigid-perfectly plastic body we obtain the distribution of surface tractions optimal in a certain sense (Borkowski, 2004).

5. Final Remarks

Computational complexity of MP-problems depends heavily upon their degree of non-linearity. It is quite easy to solve large LP-problems. Several simplex codes available on the market are able to solve problems with hundreds of thousands of variables and/or constraints. These codes usually use some version of sparse-matrix technique in order to cope with large matrices. Interestingly enough, the simplex algorithm, discovered over 50 years ago, is still the best solver.

QP-problems are more demanding and one can hardly expect to solve in reasonable time a problem with more than couple of hundreds variables and/or constraints. Despite huge effort spend on developing general purpose Non-Linear Programming solvers, the result is rather unsatisfactory. Most available codes work sufficiently well in the range of several tenths of variables and/or constraints.

Structural analysis and optimization taking into account elastic properties of the material is not reducible to Linear Programming. On the other hand, the efficiency of QP- and NLP-solvers is far below the efficiency of modern solvers of the sets of linear algebraic equations. This explains why expectations that Mathematical Programming will replace Linear Algebra in the domain of computing were not met.

On the other hand, the language of Mathematical Programming is excellent in teaching Structural Analysis and Structural Optimum Design. It discloses common background of the broad class of problems governed by geometrically linear kinematics, allows students to grasp the principal difference between bilaterally and unilaterally constrained problems, trains them in a good custom of looking at each problem from two perspectives – the kinematic one and the static one, simplifies checking of existence and uniqueness of solutions.

Obviously, a prerequisite of teaching the MP-based approach to the theory of structures is the prior knowledge of the Mathematical Programming by the students. A class on this subject should be taught during the first or second year of undergraduate studies, as a supplement to courses on Linear Algebra and Differential Calculus. The knowledge acquired on the MP-theory could be exploited in teaching not only Structural Analysis and Structural Optimum Design but also in the classes on other aspects of Civil Engineering (e.g. road planning, cost optimization, construction planning, etc.).

References

Biron, A. and Hodge, P.G., "Non-linear programming method for limit analysis of rotationally symmetric shells", *J. Non-Linear Mech.*, Vol. 3, 1968.

Borkowski A., "Analysis of skeletal structural systems in the plastic and elastic-plastic range", Elsevier, 1988.

Borkowski, A., "On dual approach to piecewise-linear elasto-plasticity", Part I: "Continuum models", pp. 337-351, Part II: "Discrete models", pp. 353-360, *Bulletin of the Polish Academy of Sciences, Technical Sciences*, Vol. 52, No. 4, 2004.

Brown, D. and Ang, A.H., "Structural optimization by nonlinear programming", *ASCE J. Struct. Div.*, 90, ST6, 1964.

Ceradini, G. and Gavarini, C., "Calcolo a rottura e programmazione lineare", *Giornale del Genio Civile*, 1965, gennaio-febraio.

Cohn, M.Z., "Fundamentals of the plastic structural analysis" (in Rumanian), *Industria Constructiilor* (Bucharest, Romania), V. 8, No. 11, 1956, pp. 655-666.

Cohn, M.Z., Ghosh, S. and Parimi, S., "Unified approach to the theory of plastic structures", *ASCE J. Eng. Div.*, 98, 1972, pp. 133-185.

Cohn, M.Z.: Introduction to engineering plasticity by mathematical programming, in: "Engineering plasticity by mathematical programming", in: *Proc. NATO Advanced Study Institute,* Pergamon Press, New York, 1979, Chapter 1, pp. 3-18.

Čyras, A. A., "Methods of linear programming in the analysis of elastic-plastic systems" (in Russian), Stroiizdat, Leningrad, 1969.

Čyras, A. A. and Borkauskas, A., "Die verallgemeinerte duale Aufgabe der Theorie des Grenzgleichgewichtes", *Bauplanung und Bautechnik*, Weimar, Vol. 23, No. 5, 1969, pp. 37-40,

Čyras, A. A. and Borkauskas, A., "Dual optimization problems in the theory of rigid-perfectly plastic solids" (in Russian), *Stroitielnaya Mech. I Rascchet Sooruzhenii* (Structural Mechanics and Analysis of Buildings), No. 4, 1969, pp. 5-10.

Čyras, A. A., "Optimization theory in the ultimate state analysis of deformable solids" (in Russian), Mintis, Vilnius, 1971.

Čyras, A. A. and Baronas, R., "Linear programming methods of displacement analysis in elastic-plastic frames, *Int. J. Num. Meth. Eng.*, Vol. 3, 1971, pp. 415-423.

Čyras, A. A. and Karkauskas, R., "Nonlinear analysis of rigid-plastic spherical shells" (in Russian), *Litovskii Mekhanicheskii Sbornik* (Lithuanian Mechanical Archives), No. 8, 1971, pp. 93-104.

Čyras, A. A. and Kalanta, S., "Optimal design of cylindrical shells by the finite-element technique", *Mechanics Research Communications*, CISM, Vol. 1, No. 3, 1974, p. 16.

Čyras, A. A., Borkauskas, A. E. and Karkauskas, R. P., "Optimum design of elastic-plastic structures – theory and methods" (in Russian), Stroiizdat, Leningrad, 1974.

Čyras, A. A., "Mathematical models for the analysis and optimization of elasto-plastic structures" (in Russian), Mokslas, Vilnius, 1982 (English translation: Butterworth, Toronto, 1983).

Čyras, A. A. and Atkočiūnas, J., "Mathematical model for the analysis of elastic-plastic structure under repeated variable loading", *Mechanics Research Communications*, CISM, Vol. 11, No. 5, 1984, pp. 353-360.

Čyras, A., Borkowski, A. and Karkauskas, R., "Theory and methods of optimization of rigid-plastic systems", Technika, Vilnius, 2004.

Čižas, A. and Čyras, A., "Analysis of elastic-plastic structures with constrained strains" (in Russian), *Litovskii Mekhanicheskii Sbornik* (Lithuanian Mechanical Archives), No. 1, 1967, pp. 102-114.

Dantzig, G. B., "Programming in a linear structure", *Comptroller*, VSAF, Washington, DC, 1948.

Franchi, A. and Cohn, M.Z., "Computer analysis of elastic-plastic structures, *Computer Methods in Appl. Mech. and Eng.*, Vol. 21, No. 3, 1980, pp. 271-294.

Grierson, D. E., and Gladwell, G. M. L., "Collapse Load Analysis Using Linear Programming", ASCE Proceedings, Journal of the Structural Division, Vol. 97, No. ST5, May 1971, pp 1561-1573.

Grierson D. E., "Deformation analysis of elastic-plastic frames", *J. Struct. Div. ASCE,* Vol. 98, 1972, pp. 2247-2267.

Grierson D. E., Franchi A. and Riva P. (Eds.), "Progress in structural engineering", Kluwer Academic Publishers, 1991.

Gvozdev, A. A., "Establishing load carrying capacity of structure by ultimate equilibrium method" (in Russian), Gostechizdat, Moscow, 1949.

Hodge, P.G., "Yield-point load determination by nonlinear programming", in: *Proc. XI Intern. Congr. Appl. Mech.,* Springer, 1966.

Kantorovich, L. V., "Mathematical methods in production management" (in Russian), Leningrad State University Publ., Leningrad, 1939.

Koopman, D.C. and Lance, R.H., "On linear programming and plastic limit analysis", *J. Mech. Phys. Solids*, Vol. 13, No. 12, 1965.

Kuhn, H.W., Tucker A.W., "Non-linear programming", in: *Proc. of the 2nd Berkeley Symp. of Math. Statistics and Probability*, Berkeley and Los Angeles, Univ. of California Press, 1951, pp. 481-492.

Maier G., "A matrix structural theory of piecewise-linear elasto-plasticity with interacting yield planes", *Meccanica*, Vol. 5, pp. 54–66, 1970.

Riva, P. and Cohn, M.Z., "Engineering approach to nonlinear analysis of concrete structures", *ASCE J. Struct. Eng.*, Vol. 116, No. 8, 1990, pp. 2162-2185.

Sacchi G. and Buzzi-Ferraris, G., "Sul criterio cinematico di calcolo a rottura di piastre inflesse mediante programmazione non lineare", *Rend. Ist. Lomb. Science e Lettere*, No. 101, 1966.

Wolfensberger, R., "Traglast und optimale Bemessung von Platten", Wildegg, 1964.

OPTIMIZATION METHOD OF PILE FOUNDATIONS

Anthony J. Hurd and Kevin Z. Truman
Department of Civil Engineering, Washington University in St. Louis, St. Louis, Missouri 63130, USA

Abstract

A computer-automated design and optimization process for pile foundations with rigid concrete slabs is presented. Optimality Criteria methodology is used to provide optimal pile designs. A three-dimensional optimization computer program has been developed that designs a foundation system with an optimal number of piles, geometric layout, pile orientation, batter, and size for a given structure subjected to multiple load cases. The optimization procedure controls displacements while reducing the overall weight of the pile foundation design. A new method for optimizing weightless variables, such as batter, was also created. Thus, the challenges of optimizing variables that indirectly affect the weight of the pile foundation can still be designed to create weight savings. In one example, the total volume of the steel piles is reduced from 61,920 in^3 to 49,570 in^3 by optimizing only the pile sizes. Furthermore, the weight is reduced again by simultaneously optimizing each pile group's size coupled with the weightless variable, batter.

1. Introduction

The purpose of this research is to create a computer-automated optimization process for large pile foundations. The U.S. Army Corps of Engineers (USACE) designs large-scale locks and dams that can easily contain thousands of piles, costing millions of dollars. The USACE currently uses pile analysis computer programs but none with optimal design. Therefore, the process for reducing the number and size of the piles is very time-consuming and uncertain, involving the tedious process of manual design, computer analysis, and redesign. This can easily take months and still result in a non-optimal final design. An automated computer optimization process would find an optimal design and take only minutes rather than months. The designer only enters an initial design, load cases, soil conditions, and constraints while the program alters the many design variables to create an optimal foundation design.

Work in pile foundation optimization was originally developed for the USACE by Hill in 1981, using a trial and error approach (Hill, 1981). His method involved first optimizing the batter and then finding an optimal pile spacing. The process finished by iteratively deleting the most and/or least stressed piles. However, this method is not numerically based, and it does not simultaneously optimize all the pile variables, so a true optimum solution is never found. Hoback and Truman used a numerical method to optimize pile designs of both rigid-slab foundations (Hoback et al., 1991) and flexible-slab foundations (Hoback et al., 1993). Their method utilized an Optimality Criteria method employed earlier by Cheng and Truman for structural frames (Cheng et al., 1983).

Optimality Criteria was chosen for this optimization for the following reasons: first, the Optimality Criteria method converges quickly with most examples converging in less than ten iterations. Second, Hoback and Truman were able to simultaneously optimize pile size, layout, batter, orientation, and number while still controlling all stress, strain, or displacement constraints.

The Optimality Criteria method works very well when optimizing a variable that *directly* affects the objective function, whether the objective function to minimize is weight, volume of steel, or cost. For

M. Pandey et al. (eds), Advances in Engineering Structures, Mechanics & Construction, 653–661.

example, pile size directly affects the weight of a design. However, some variables such as batter or geometric layout do not have direct effects on the weight of a design, but altering these weightless variables can clearly create large reductions in weight. Thus, a novel approach to optimizing these weightless variables was developed. Using a weightless scaling factor that acts as a pseudo-weight gradient, the Optimality Criteria method can still optimize these weightless variables. The challenges in choosing an appropriate weightless scaling factor were overcome by creating a unique way of gradually decreasing the scaling factor from iteration to iteration, approaching a more optimal design.

2. Optimality Criteria Method

The Optimality Criteria method requires an objective function to be minimized. Because the goal of this research is to minimize the cost of the pile foundation while still satisfying the constraints, the objective function is the total weight of the steel, W_T, in the piles which is directly related to cost:

$$W_T = \sum_{i=1}^{n} \rho L_i A_i \tag{2.1}$$

where:

	i	=	Pile number
	n	=	Number of piles
	ρ	=	Density of Steel
	L_i	=	Length of element i
	A_i	=	Area of element i

An unconstrained minimum weight of Equation 2.1 does not exist. Once displacement, stress, and/or strain constraints, h_j, are introduced to the problem, the Lagrangian function, L, can be written as

$$L = W_T + \sum_{j=1}^{m} \lambda_j h_j \tag{2.2}$$

where:

	j	=	Constraint number
	m	=	Number of constraints
	λ_j	=	Lagrange multiplier for constraint j

and when the constraint is displacement, u_j, with a maximum displacement in the jth direction of $\bar{u}_j$,

$$h_j = u_j - \bar{u}_j \leq 0 \tag{2.3}$$

A minimum of the Lagrangian will be located where the derivative with respect to each design variable, d, is equal to zero. Thus, a pile foundation design can be a local minimum weight if all the constraints, $h_j \leq 0$, are satisfied and there exist λ_j such that

$$\frac{\partial L}{\partial d_i} = \frac{\partial W_T}{\partial d_i} + \sum_{j=1}^{m} \lambda_j \frac{\partial h_j}{\partial d_i} = 0 \qquad i = 1, 2, \ldots n \tag{2.4}$$

where:

	d	=	Design variable (pile size, batter, etc.)
	i	=	Design variable number
	n	=	Number of design variables

It is important to note that the Lagrange multipliers, λ_j, cannot be negative because negative λ_j values would still allow the constraints to be satisfied while the weight of the piles increases. Thus,

$$\lambda_j \geq 0 \tag{2.5}$$

Equations (2.4) and (2.5) are the *Kuhn-Tucker conditions* (Kirsch, 1993). Equation (2.4) is rewritten to provide the *optimality criteria:*

$$T_i = \frac{-\left(\sum_{j=1}^{m} \lambda_j \frac{\partial h_j}{\partial d_i}\right)}{\frac{\partial W_T}{\partial d_i}} = 1 \qquad i = 1, 2, \dots n \tag{2.6}$$

If the optimality criteria, T_i, is less than one and the weight gradient, $\partial W_T/\partial d_i$, is positive, then the ith variable can be decreased. This is because the weight gradient is larger than the constraint gradient (the numerator), indicating a benefit in reducing the ith variable. The opposite is true if the weight gradient is negative. *Linear recurrence equations* are used to change the design variables, pushing the design toward a local minimum. The recurrence formula is based on the expanded power law, and T_i is used as the efficiency of each variable:

$$d_i^{k+1} = d_i^k + \frac{1}{r}(T_i - 1)d_i^k \qquad i = 1, 2, \dots n \tag{2.7}$$

where: k = Index of the iteration number
r = Convergence control parameter

The convergence control parameter, r, assures that the prediction of the design variable, d_i, for the next iteration does not go beyond the optimum. A reasonable value for r is 2, but $r > 1$.

Now, the only unknowns in the recurrence relation are the Lagrange multipliers, λ_j. The change in the active constraints will give us these values:

$$\Delta h_j = h_j^{k+1} - h_j^k = \sum_{i=1}^{n} \frac{\partial h_j}{\partial d_i} \underbrace{\left(d_i^{k+1} - d_i^k\right)}_{\Delta d_i} \qquad j = 1, 2, \dots m \tag{2.8}$$

Substituting Equation (2.7) for Δd and letting h_j^{k+1} go to zero as expected,

$$-h_j^k = \sum_{i=1}^{n} \frac{\partial h_j}{\partial d_i}\left[\frac{1}{r}(T_i - 1)d_i^k\right] \qquad j = 1, 2, \dots m \tag{2.9}$$

gives *m* linear equations with *m* unknown λ values. Any negative Lagrange multipliers, λ, require Equation (2.9) to be reevaluated with the corresponding constraint removed and recalculated to get the remaining λ values. If negative λ values still appear after recalculating, then this process will have to be repeated until all values are positive. After substituting the remaining positive Lagrange multipliers into Equation (2.7), the new design variables can be calculated, and the process is repeated to further improve the design variables.

Getting the partial derivative of the constraint with respect to the design variable, $\partial h_j/\partial d_i$, is rather simple when the constraint is displacement, u_j, as in Equation (2.3). Thus, with the goal to find $\partial u_j/\partial d_i$, begin with the pile stiffness equation,

$$[K]\{u\} = \{P\} \tag{2.10}$$

where: K = Global stiffness matrix (6x6)
u = Displacement vector (6x1)
P = Load vector (6x1)

The global stiffness is calculated by transforming and summing the local stiffness of each individual pile at one global point where we want to know and control the displacements. The local pile stiffness is calculated in the same manner used by the USACE and their rigid-pile analysis program, CPGA (CASE Task Group on Pile Foundations, 1983). Taking the derivative with respect to the design variable, d_i, we get:

$$\frac{\partial[K]}{\partial d_i}\{u\}+[K]\frac{\partial\{u\}}{\partial d_i}=\frac{\partial\{P\}}{\partial d_i}=0 \tag{2.11}$$

and rearranging terms gives:

$$\frac{\partial\{u\}}{\partial d_i}=-[K]^{-1}\frac{\partial[K]}{\partial d_i}\{u\} \tag{2.12}$$

3. Optimization of Weightless Design Variables

As seen in Equation (2.1), the weight of the pile foundation is only governed by the length and area of the piles. Thus, optimizing the size of the piles, and assuming an infinite selection of pile sizes, is rather straightforward because $\partial W_T/\partial d_i = \partial W_T/\partial A_i = \rho L_i$. The weight gradient, located in the denominator of the optimality criteria, Equation (2.6), is non-zero. However, for other design variables, such as batter (when the length is constant) or pile spacing, $\partial W_T/\partial d_i = 0$.

Even though the weight gradient of these topological design variables is zero, they still have an indirect effect on the weight. Unfortunately, this effect is immeasurable. To solve this problem, Hoback and Truman created a new optimality criterion that can model the behavior of topological variables (Hoback et al., 1993). First the optimality criteria for the weightless gradient is rewritten from Equation (2.4) with $\partial W_T/\partial d_i = 0$:

$$\frac{\partial L}{\partial d_i}=\sum_{j=1}^{m}\lambda_j\frac{\partial h_j}{\partial d_i}=0 \qquad i=1,2,\ldots n \tag{3.1}$$

Because the efficiency is one when the design variable is optimal, the weightless optimality criteria from Equation (2.6) is rewritten as:

$$T_i=\frac{-\sum_{j=1}^{m}\lambda_j\frac{\partial h_j}{\partial d_i}}{w}+1=0 \qquad i=1,2,\ldots n \tag{3.2}$$

The weightless scaling factor, w, replaces the weight gradient in the optimality criteria, effectively acting as a pseudo-weight gradient. Because the weightless optimality criteria is formulated to approach one at optimum, T_i can still be used in the same recurrence relationship, Equation (2.7).

The difficult step is finding an acceptable value for w. Hoback's method involved two steps. First, he scaled w so that the coefficients of the weightless Lagrange multipliers, λ_j, along the main diagonal of the linear equations were at least as large as the corresponding weighted coefficients. The second step involved increasing w until an estimated weight change converged. Although Hoback found this method successful, his work showed inconsistencies at finding a better, lower-weight design, and a new method has been developed.

After testing random optimization problems for the effects of the weightless scaling factor, w, on the optimization process, a pattern was discovered. First, a single value for w would not work for several reasons. If w is too large, the weightless variable would remain unchanged, for T_i will always be too close to one. The final solution would still optimize the weighted variables, but the final weight would be less than optimal without the optimization of the weightless variables. If w is too small, T_i will take on very large and very small values, throwing the weightless variable to their extremes, never reaching an optimum. This creates a very unstable optimization process that may never converge to a single design.

Figure 3.1 shows the typical results of simultaneously optimizing weighted variables and weightless variables. Weightless scaling factors of 10^{12}, 10^8, 10^6, and 10^3 are each tried from the same initial design point. When $w = 10^{12}$, the weightless variables were untouched because w was too high, and when $w = 10^3$, the optimization became unstable, resulting in sudden large increases and decreases in weight.

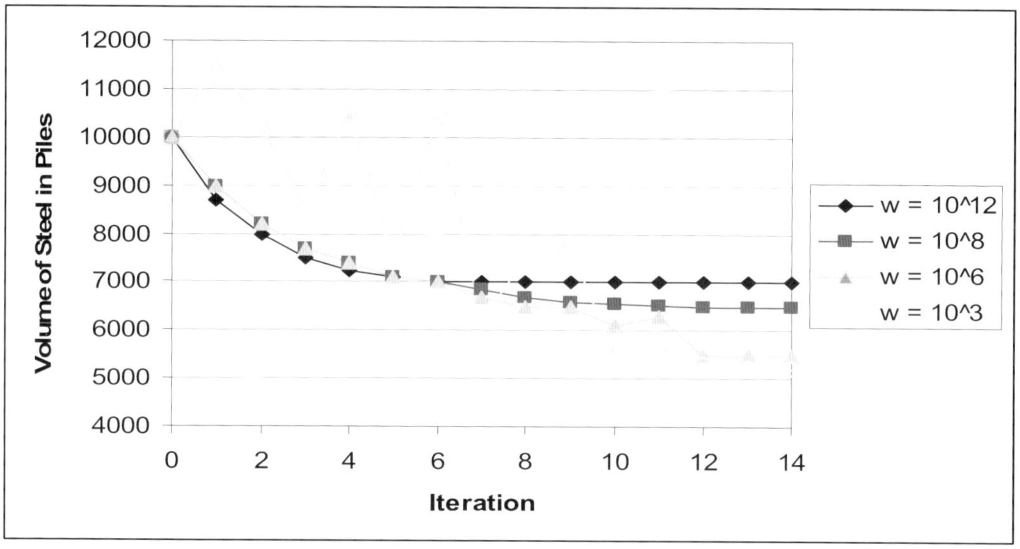

Figure 3.1. A typical optimization, including weightless variables, with different w values tested. Note that a gradually decreasing weightless scaling factor, beginning at 10^{12} and gradually reducing to 10^3, would end with a near-optimal design.

Because several random examples displayed similar results as figure 3.1, a new method for determining w was created. The weightless scaling factor, w, will begin at a very large value (such as 10^{12} in figure 3.1). Optimization will continue at that value of w until the weight converges (at iteration 4 in figure 3.1). For the next iteration, the square root, or another appropriate reduction factor, of w will be taken (thus, reducing w to 10^6). Then, this process will be repeated, continuing to reduce the weightless scaling factor and further decreasing the overall weight of the pile foundation design. Figure 3.1 shows how gradually decreasing w would result in a continuously decreasing weight if w were to decrease from 10^{12} to 10^6 to 10^3. Eventually, the weightless scaling factor will become too small, and the design will become unstable. At this point, the optimization is complete, and the lowest weight design is chosen.

4. Example

The following example illustrates the effortless weight reduction of a rigid-slab pile design, using the optimization methods described. First, the program will be run, optimizing only the pile *sizes*. Because pile size is a weighted variable, the optimization is rather straightforward since it does not require the use of the weightless scaling factor, w. Second, the optimization will include both pile size and the weightless variable, batter (weightless because the length will be held constant). For simplicity, all other variables will not be varied.

4.1 Optimization of a 2-D Pile Foundation, Optimizing Only Pile Size

The problem consists of a rigid slab foundation with 5-piles, arranged into 4 groups, as seen in figure 4.1. Group 1 initially has no batter but is only allowed to batter to the left while groups 2, 3, and 4 are battered to the right, initially at a value of 5. The loading, constraints, and site conditions are listed in figure 4.1.

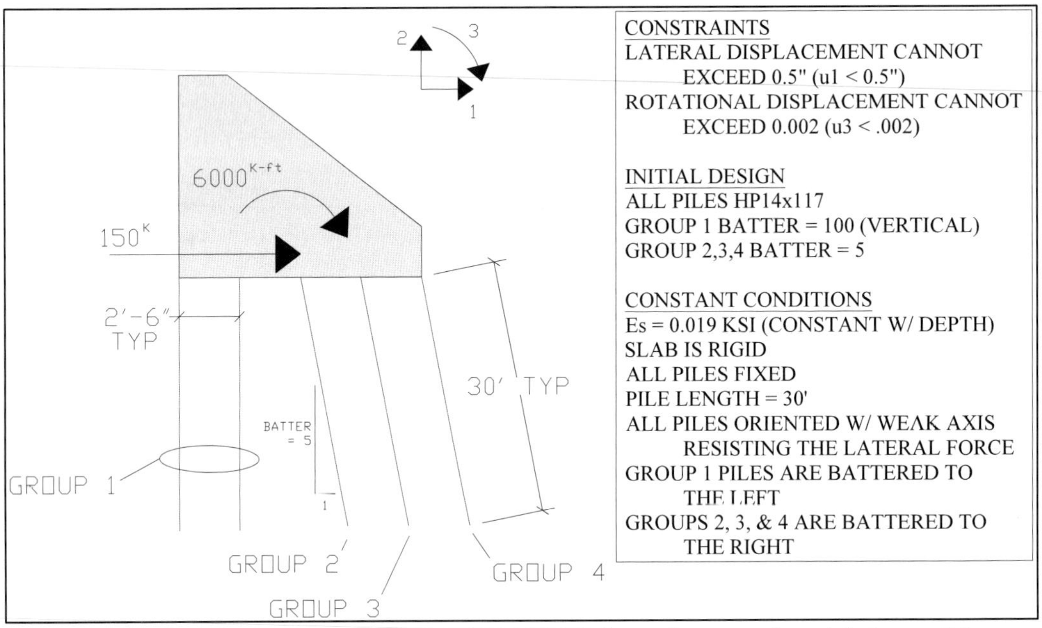

Figure 4.1. Example problem, showing constraints, initial design, and constant site and design conditions.

From analysis of the initial design and the prescribed loading, the displacements are already rather close to the maximum displacements. The lateral displacement is 0.46", and the rotational displacement is 0.0015 rad., both just under the constraints of 0.5" and 0.002 rad., respectively. The weight of the initial volume of steel in the five HP14x117 piles is **61,920 in^3**. With the displacements this close to their maximums, it is difficult to find a significantly lighter design by hand.

For this first optimization, only the pile sizes are optimized, keeping all other variables constant. After only four iterations, the design converged, and the total volume of the piles reduced to **49,570 in^3**. Even though the total weight of the piles decreased, the displacements remained under their constraints. The lateral displacement is now 0.49" and the rotational displacement is 0.0020 rad. The volume decrease is shown in Figure 4.2, and the original and optimized pile foundation design is compared in Table 4.1.

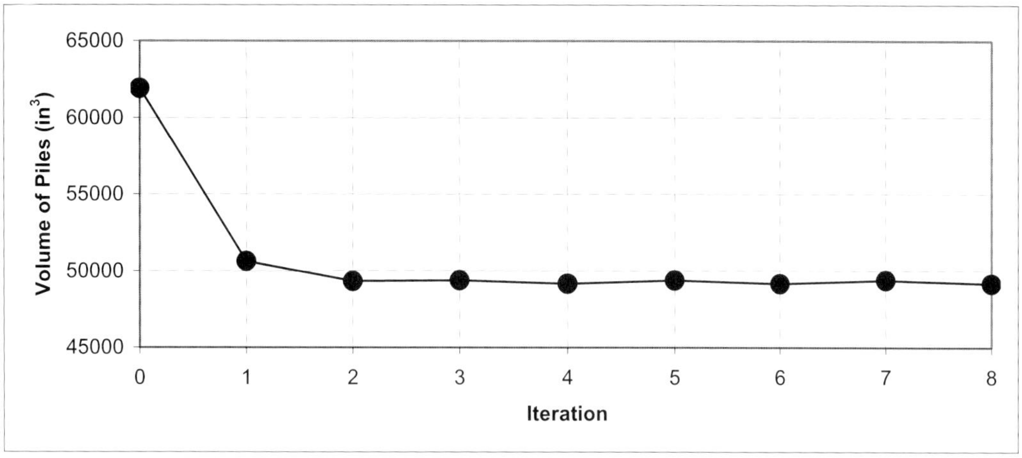

Figure 4.2. Convergence of optimization iterations.

		Original Design	*Optimized Design*
Pile Size:	*Group 1*	HP14x117	HP14x117
	Group 2	HP14x117	HP14x73
	Group 3	HP14x117	HP14x89
	Group 4	HP14x117	HP14x73
	Lateral Displacement	0.46"	0.49"
	Rotational Displacement	0.0015 rad.	0.0020 rad.
	Total Volume of Piles	**61,920 in^3**	**49,570 in^3**

Table 4.1. Original and Optimized Foundation Design

4.2 Introducing the Optimization of Batter

Optimizing the batter involves using the weightless scaling factor when pile length is held constant. The initial design will be the optimized design from example 4.1. From there the optimization is run, simultaneously optimizing pile size and batter. The batter's weightless scaling factor will begin at 10^{24} and will decrease at iterations that have only a small change in weight. The weight reduction is shown in the figure 4.3, and the corresponding weightless scaling factor, w, in each iteration can be seen in figure 4.4.

The weightless scaling factor begins very high, and is reduced gradually by taking it to the 2/3 power when the weight reduction from the previous iteration is low. Comparing Figures 4.3 and 4.4, w does not take on a weight reducing value until the fourth iteration ($w \approx 10^7$). This value proved to optimize batter while reducing the overall weight of the piles for several iterations. Finally, as the weight began to converge near iteration 11, w was further reduced, but the optimization became unstable at iteration 13. As explained earlier, when w becomes too low, the changes in the weightless variable (batter in this case) become too large, creating an unstable optimization. The final design, reached on iteration 12, changed the batters only slightly but was able to significantly alter the pile sizes to find an overall lighter pile design. The volume of steel was further reduced **from 49,570 in^3 to 46,580 in^3**.

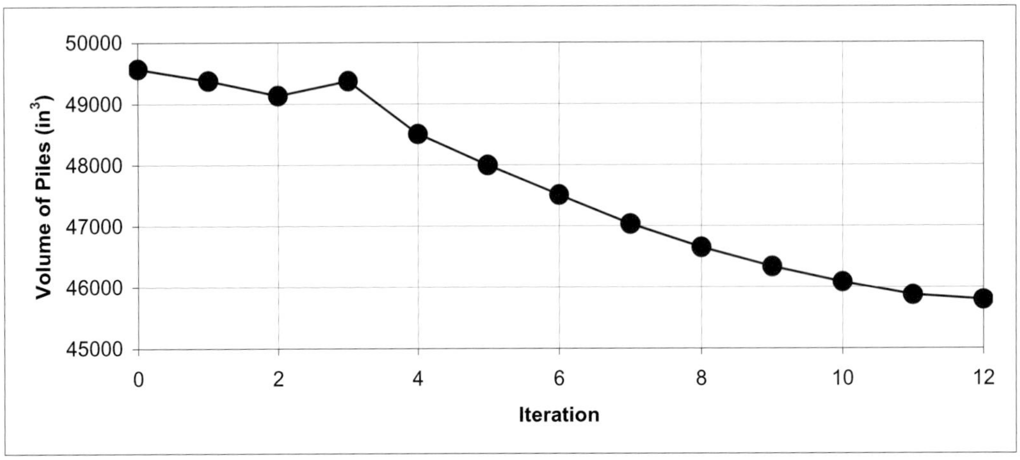

Figure 4.3. Convergence of optimization iterations after introducing the optimization of batter.

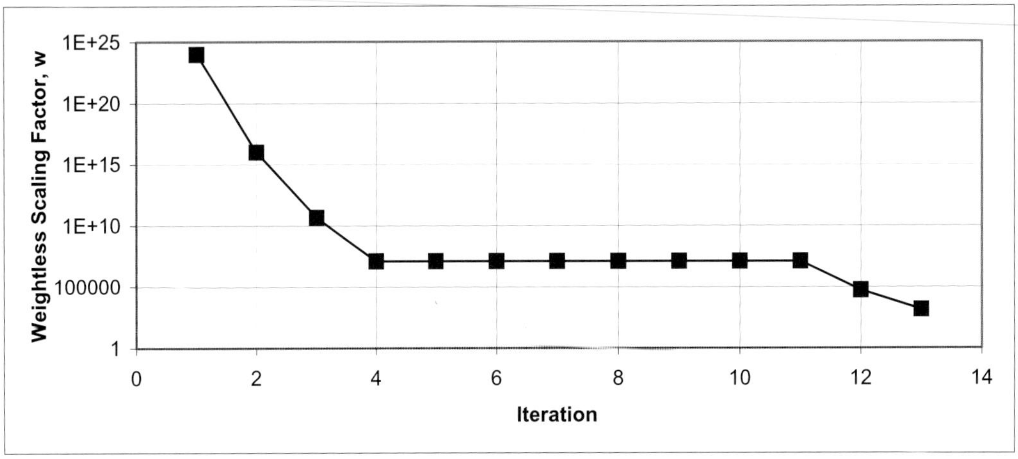

Figure 4.4. Change in the weightless scaling factor, w, with each iteration.

		Original Design	***Optimized Design***
Pile Size:	*Group 1*	HP14x117	HP14x89
	Group 2	HP14x73	HP14x73
	Group 3	HP14x89	HP14x73
	Group 4	HP14x73	HP14x117
Batter:	*Group 1*	100 (Vertical)	86
	Group 2	5	5.3
	Group 3	5	4.8
	Group 4	5	4.4
	Lateral Displacement	0.49"	0.46"
	Rotational Displacement	0.0020 rad.	0.0019 rad.
	Total Volume of Piles	**49,570 in^3**	**46,580 in^3**

Table 4.2. Original and Optimized Foundation Design after simultaneously optimizing pile size and batter

5. Conclusion

Optimality Criteria is an effective method at reducing the weight of steel in piles under rigid, concrete slabs. The method has successfully reduced the weight of piles in many real-life problems, both simple and complicated. Using weightless scaling factors provides a way to optimize zero-weight gradient variables such as batter or spacing. By gradually reducing this factor exponentially when weight reductions become sufficiently small, a near-optimal final design can be reached. However, it is important to note that the final design reached is only a *local* minimum, not a global minimum. Thus, the initial design plays an important role in determining the final pile layout. Because of the ease of use of this optimality criteria program, varying the initial design and rerunning the program with the same loads and constraints can result in several low-weight final designs, allowing the engineer to choose between several near-optimal designs.

Acknowledgements

This work has been performed with the support of the National Science Foundation through the NSF Graduate Teaching Fellows in K-12 Education program.

References

Case Task Group on Pile Foundations (1983). Basic Pile Group Behavior. Technical Report K-83-1, US Army Engineer Waterways Experiment Station, Vicksburg, MS.

Cheng, F. Y. & Truman, K. Z. (1983). Optimization Algorithm of 3-D Building Systems for Static and Seismic Loading. Modeling and Simulation in Engineering, 3, North-Holland, Amsterdam, 315-326.

Hartman, J. P., Jaeger, J. J., Jobst, J. J., & Martin, D. K. (1988). User's Guide: Pile Group Analysis (CPGA) Computer Program. Technical Report ITL-88, US Army Engineer Waterways Experiment Station, Vicksburg, MS.

Hill, J. L. (1981). User's Guide: Computer Program for Optimal Design and Analysis of Pile Foundations (PILEOPT). Instruction Report K-81-5, US Army Engineer Waterways Experiment Station, Vicksburg, MS.

Hoback, A. S., & Truman, K. Z. (1991). Optimization of the Layouts of Steel Piles Using Optimality Criteria. Dept. of Civil Engineering, Washington University, St. Louis, MO.

Hoback, A. S., & Truman, K. Z. (1993). Optimization of Steel Pile Foundations with Rigid or Flexible Concrete Slabs. Dept. of Civil Engineering, Washington University, St. Louis, MO.

Kirsch, U. (1993). Structural Optimization: Fundamentals and Applications. Springer-Verlag, Berlin.

US Army Corps of Engineers (1991). Design of Pile Foundations. Engineer Manual 1110-2-2906, Washington, DC.

GRADIENT BASED OPTIMIZATION OF ADDED VISCOUS DAMPING IN SEISMIC APPLICATIONS

Oren Lavan and Robert Levy
Faculty of Civil and Environmental Engineering, Technion – Israel Institute of Technology, Haifa 32000, Israel;
E-mail: cvrlevy@tx.technion.ac.il

Abstract

This paper presents a consistent approach for the optimal seismic design of added viscous damping in framed structures. The approach presented is appropriate for use in elastic as well as yielding frames. The sum of added damping is chosen as the objective function and the performance of the structure, under the excitation of an ensemble of deterministic ground motion records, is constrained. The performance of the structure is measured by the maximal inter-story drifts in both the linear and nonlinear cases. The nonlinear case however, uses an additional performance measure of the normalized hysteretic energy of the plastic hinges

Gradients of the performance measures are first derived to enable the use of an appropriate first order optimization scheme. Moreover, an efficient selection scheme enables the consideration of only a few records rather than the whole ensemble, hence making the optimization process efficient in terms of the computational effort.

Introduction

The problem of seismic retrofitting of existing structures has gained much attention lately due to the new *performance-based-design* approach, which allows engineers to design structures for a desired level of seismic performance. Installation of viscous dampers is an effective means for this seismic retrofitting, hence, the problem of optimal design of these dampers is of paramount importance. This problem was tackled by several researchers with limited results for the particular class of regular buildings (see for example Zhang and Soong 1992; Inaudi *et al.* 1993; Gluck *et al.* 1996; Takewaki 1997 to name only few). Since most existing buildings are irregular, available methodologies remain academic.

For an efficient and computationally effective solution of the optimization problem of dynamic systems subjected to time varying loads, first order schemes that require constraints' gradients are preferred. Zero order optimization schemes, e.g. genetic algorithms, require a large number of function and constraints evaluations, that is to say time history analyses, making them less attractive to use.

Several approaches for the gradient computation have been introduced in the literature. Hsieh and Arora (1985) derived the gradients of point-wise as well as integral type constraints for linear elastic systems by deriving the first variations of these constraints which depend on the variation on the displacements of the degrees of freedom. They further used a direct differentiation method of the equations of motion, and alternatively an adjoint variables method, to evaluate these variations on the displacements. Another approach for the gradient computation uses the finite difference method (see for example Falco *et al.*, 2004). Here the derivative of the constraint with respect to each design variable is approximated by the forward or backward finite difference approximation. This method actually requires an additional analysis for each design variable. Conte *et al.* (2003) distinguished two methods for computing the response sensitivities considering plastic behavior of the structure. The first method uses the differentiation of the response equations with respect to each of the design variables, and then discretizes the resulting response sensitivity equations in time. The second method discretizes the response equations in time, and then differentiates the resulting discrete response

M. Pandey et al. (eds), Advances in Engineering Structures, Mechanics & Construction, 663–671.

equations with respect to each of the design variables. It should be noted that both methods require an additional analysis for each design variable.

The present research proposes a gradient based approach for the optimal design of viscous dampers for the seismic retrofitting of existing, regular as well as irregular, structures. This approach uses a first order optimization scheme whose success lies in the ability to derive the gradients of the constraints with respect to the damping coefficients of the dampers. Thus, the main effort in this paper is the gradient derivation of constraints in linear as well as nonlinear dynamic optimization problems under earthquake excitations. The relatively small computational effort associated with their evaluation using the proposed scheme makes these gradients highly desirable.

Problem Formulation

The formulation of the optimization problem is comprised of the total added damping as an objective function, and an inequality constraint on the upper bound of each of the local performance indices which are computed based on the behavior of the structure, i.e., satisfying the equations of motion of the damped structure. These constraints are repeated for each ground motion record. The damping coefficients which are the design variables are required to be nonnegative and are assigned an upper bound.

Equations of motion

The general equations of motion of a yielding structure, retrofitted by added damping, and excited by an earthquake can be given by:

$$\begin{aligned}&\mathbf{M}\ddot{\mathbf{x}}(t)+\left[\mathbf{C}+\mathbf{C}_{\mathrm{d}}(\mathbf{c}_{\mathrm{d}})\right]\cdot\dot{\mathbf{x}}(t)+\mathbf{K}^{\alpha}\mathbf{x}(t)+\mathbf{B}_{\mathrm{fx}}\mathbf{f}_{\mathrm{h}}(t)=-\mathbf{M}\cdot\mathbf{e}\cdot\mathbf{a}_{\mathrm{g}}(t)\ ;\mathbf{x}(0)=\mathbf{0},\dot{\mathbf{x}}(0)=\mathbf{0}\\&\dot{\mathbf{f}}_{\mathrm{h}}(t)=\mathbf{f}(\dot{\mathbf{x}}(t),\mathbf{f}_{\mathrm{h}}(t))\ ;\mathbf{f}_{\mathrm{h}}(0)=\mathbf{0}\end{aligned} \tag{1}$$

where $\mathbf{x}$= displacements vector of the degrees of freedom (DOFs); $\mathbf{M}$= mass matrix; $\mathbf{C}$= inherent damping matrix; $\mathbf{c}_{\mathrm{d}}$= added damping vector; $\mathbf{C}_{\mathrm{d}}(\mathbf{c}_{\mathrm{d}})$= supplemental damping matrix; $\mathbf{K}^{\alpha}$=secondary stiffness matrix; $\mathbf{f}_{\mathrm{h}}(t)$=hysteretic forces/moments vector in local coordinates of the plastic hinges with zero secondary stiffness; $\mathbf{B}_{\mathrm{fx}}$=transformation matrix that transforms the restoring forces/moments from the local coordinates of the plastic hinges to the global coordinates of the DOFs; $\mathbf{e}$= excitation direction matrix with zero/one entries; $\mathbf{a}_{\mathrm{g}}(t)$=ground motion acceleration vector, and a dot represents differentiation with respect to the time.

Performance indices

Normalized hysteretic energy: Following Uang and Bertero, (1990), hysteretic energy accumulated at the plastic hinge *i*, (a measure of the structural damage in yielding frames), normalized by an allowable value, is given by:

$$E_{h,i}(t_f)=\left(\int_0^{t_f} f_{h,i}(t)\cdot v_i(t)dt\right)\Big/E_{h,i}^{all} \tag{2}$$

where t_f= the final time of the excitation\computation; $E_{h,i}(t_f)$=normalized hysteretic energy at the plastic hinge *i* at the time t_f; $f_{h,i}(t)$=hysteretic force/moment in the plastic hinge *i*; $v_i(t)$=velocity of the plastic hinge *i*, and $E_{h,i}^{all}$ =allowable value of the hysteretic energy at the plastic hinge *i* which is usually taken proportional to the elastic energy at yielding of the plastic hinge. In matrix notation, the hysteretic energy in the plastic hinges as depicted by Eq. 2 can be written as:

$$\mathbf{E}_{\mathrm{h}}(t_f)=\mathsf{D}^{-1}\left(\mathbf{E}_{\mathrm{h}}^{\mathrm{all}}\right)\cdot\int_0^{t_f}\mathsf{D}(\mathbf{f}_{\mathrm{h}}(t))\cdot(\mathbf{B}_{\mathrm{xf}}\dot{\mathbf{x}}(t))dt \tag{3}$$

where $\mathsf{D}(\mathbf{z})$= operator that forms a diagonal matrix whose diagonal elements are the elements of a given vector $\mathbf{z}$, and $\mathbf{B}_{\mathrm{xf}}$=transformation matrix that transforms the velocities from the global coordinates of the DOFs to the local coordinates of the plastic hinges.

Normalized maximal inter-story drifts can be written as:

$$\mathbf{d}_m=\max_t\left(\mathrm{abs}\left(\mathsf{D}^{-1}\left(\mathbf{d}^{\mathrm{all}}\right)\cdot\mathbf{H}_{\mathrm{x}}\mathbf{x}(t)\right)\right) \tag{4}$$

where $\mathbf{d}_m$=vector of normalized maximal inter-story drifts; $\mathbf{d}^{\mathrm{all}}$= vector of allowable maximal inter-story drifts; $\mathbf{H}_{\mathrm{x}}$=transformation matrix that transforms the displacements from the global coordinates of the DOFs to the coordinates of inter-story drifts, and the "abs" stands for the absolute function as it acts on each of the vector components separately.

Formal optimization problem
The formal optimization problem may now be written as:

mininimize: $J=\mathbf{c}_{\mathrm{d}}^{\mathrm{T}}\cdot\mathbf{1}$

subject to:

$$\left.\begin{aligned}
&\mathbf{E}_{\mathrm{h}}(t_f)=\mathsf{D}^{-1}\left(\mathbf{E}_{\mathrm{h}}^{\mathrm{all}}\right)\cdot\int_{t_o}^{t_f}\mathsf{D}(\mathbf{f}_{\mathrm{h}}(t))\cdot(\mathbf{B}_{\mathrm{xf}}\dot{\mathbf{x}}(t))dt\le\mathbf{1}\\
&\mathbf{d}_m=\max_t\left(\mathrm{abs}\left(\mathsf{D}^{-1}\left(\mathbf{d}^{\mathrm{all}}\right)\cdot\mathbf{H}_{\mathrm{x}}\mathbf{x}(t)\right)\right)\le\mathbf{1}\\
&\text{where } \mathbf{x}(t),\dot{\mathbf{x}}(t) \text{ and } \mathbf{f}_{\mathrm{h}}(t) \text{ satisfy the equations of motion}\\
&\mathbf{M}\ddot{\mathbf{x}}(t)+\left[\mathbf{C}+\mathbf{C}_{\mathrm{d}}(\mathbf{c}_{\mathrm{d}})\right]\cdot\dot{\mathbf{x}}(t)+\mathbf{K}^{\alpha}\mathbf{x}(t)+\mathbf{B}_{\mathrm{fx}}\mathbf{f}_{\mathrm{h}}(t)=-\mathbf{M}\cdot\mathbf{e}\cdot\mathbf{a}_{\mathrm{g}}(t)\,;\begin{cases}\mathbf{x}(0)=\mathbf{0}\\\dot{\mathbf{x}}(0)=\mathbf{0}\end{cases}\\
&\dot{\mathbf{f}}_{\mathrm{h}}(t)=\mathbf{f}(\dot{\mathbf{x}}(t),\mathbf{f}_{\mathrm{h}}(t))\,;\mathbf{f}_{\mathrm{h}}(0)=\mathbf{0}
\end{aligned}\right\}\forall\mathbf{a}_{\mathrm{g}}\in\text{ensemble } i \tag{5}$$

$$\mathbf{0}\le\mathbf{c}_{\mathrm{d}}\le\mathbf{c}_{\mathrm{d,max}}$$

where $\mathbf{1}$ = unity vector, and $\mathbf{c}_{\mathrm{d,max}}$=upper bound on $\mathbf{c}_{\mathrm{d}}$.

Optimization Scheme
Gradient derivation
The evaluation of the gradient of the objective function is trivial since this function depends on the design variables explicitly, and is given by $\nabla_{\mathbf{c}_{\mathrm{d}}}J=\mathbf{1}$. The evaluation of the gradients of the constraints, however, is not an easy matter since the constraints depend on the design variables through differential equations. It is achieved indirectly by formulating the problem in *state-space* notation and using optimal control theory.

The optimization problem (Eq. 5) is reformulated in terms of a single constraint on maximal values as:

$$\begin{aligned}&\text{mininimize:} \quad J = \mathbf{c}_{\mathrm{d}}^{\mathrm{T}} \cdot \mathbf{1} \\ &\text{subject to:} \\ &pi = \max\left(\max\left(\mathbf{E}_{\mathrm{h}}\left(t_f\right)\right), \max\left(\mathbf{d}_m\right)\right) \leq 1.0, \text{equations of motion and } \mathbf{0} \leq \mathbf{c}_{\mathrm{d}} \leq \mathbf{c}_{\mathrm{d,max}}\end{aligned} \tag{6}$$

where $pi =$ performance index.

A differentiable equivalent of the constraint: Before proceeding formally with the gradient formulation it is necessary, since use is made of variational approach, to replace the *max* function on *t* in Eq. 6 by a differentiable function.

Differentiable equivalent of $\mathbf{d}_m$. It is proposed to use a norm of the *p*-type differentiable function as an equivalent to $\mathbf{d}_m = \max\limits_t\left(\mathrm{abs}\left(\mathsf{D}^{-1}\left(\mathbf{d}^{\mathrm{all}}\right) \cdot \mathbf{H}_{\mathrm{X}}\mathbf{x}(t)\right)\right)$. Thus, $\mathbf{d}_m$ takes the form:

$$\mathbf{d}_m\left(t_f\right) = \left(\frac{1}{t_f}\int_0^{t_f}\left(\mathsf{D}^{-1}\left(\mathbf{d}^{\mathrm{all}}\right) \cdot \mathsf{D}\left(\mathbf{H}_{\mathrm{X}} \cdot \mathbf{x}(t)\right)\right)^p dt\right)^{\frac{1}{p}} \cdot \mathbf{1} = \left(\mathsf{D}\left(\mathbf{d}_{m,p}\left(t_f\right)\right)\right)^{\frac{1}{p}} \cdot \mathbf{1} \tag{7}$$

where $p =$ a large positive even number. It follows that:

$$\mathbf{d}_{m,p}\left(t_f\right) = \left(\mathsf{D}\left(\mathbf{d}_m\right)\right)^p \cdot \mathbf{1} = \frac{1}{t_f}\int_0^{t_f}\left(\mathsf{D}^{-1}\left(\mathbf{d}^{\mathrm{all}}\right) \cdot \mathsf{D}\left(\mathbf{H}_{\mathrm{X}} \cdot \mathbf{x}(t)\right)\right)^p dt \cdot \mathbf{1} \tag{8}$$

Differentiable equivalent of pi. The maximal component of a vector with non-negative entries, $\mathbf{z}$, can be evaluated using a differentiable weighted average of the form:

$$z_{\max} = \frac{\sum w_i^{\,q} z_i}{\sum w_i^{\,q}} \tag{9}$$

where w_i =weight of z_i, and $q =$ an index. When *q* is large, say *q*=*p*, and the components of $\mathbf{z}$ are used as their own weights, i.e. $w_i = z_i$, this weighted average approaches the value of the maximum component of $\mathbf{z}$. Since $\mathbf{E}_{\mathrm{h}}\left(t_f\right)$ and $\mathbf{d}_m$ are normalized quantities *pi* can be written as $pi = \max\left(\mathbf{E}_{\mathrm{h}}\left(t_f\right), \mathbf{d}_m\right)$ and reformulated Eq. 9 as:

$$pi = \frac{\mathbf{1}^{\mathrm{T}} \cdot \mathsf{D}^{q+1}\left(\mathbf{E}_{\mathrm{h}}\left(t_f\right)\right) \cdot \mathbf{1} + \mathbf{1}^{\mathrm{T}} \cdot \mathsf{D}^{q+1}\left(\mathbf{d}_m\right) \cdot \mathbf{1}}{\mathbf{1}^{\mathrm{T}} \cdot \mathsf{D}^{q}\left(\mathbf{E}_{\mathrm{h}}\left(t_f\right)\right) \cdot \mathbf{1} + \mathbf{1}^{\mathrm{T}} \cdot \mathsf{D}^{q}\left(\mathbf{d}_m\right) \cdot \mathbf{1}} \tag{10}$$

Substituting Eq. 7 yields:

$$pi = \frac{\mathbf{1}^{\mathrm{T}} \cdot \mathsf{D}^{q+1}\left(\mathbf{E}_{\mathrm{h}}\left(t_f\right)\right) \cdot \mathbf{1} + \mathbf{1}^{\mathrm{T}} \cdot \mathsf{D}^{\frac{q+1}{p}}\left(\mathbf{d}_{m,p}\left(t_f\right)\right) \cdot \mathbf{1}}{\mathbf{1}^{\mathrm{T}} \cdot \mathsf{D}^{q}\left(\mathbf{E}_{\mathrm{h}}\left(t_f\right)\right) \cdot \mathbf{1} + \mathbf{1}^{\mathrm{T}} \cdot \mathsf{D}^{\frac{q}{p}}\left(\mathbf{d}_{m,p}\left(t_f\right)\right) \cdot \mathbf{1}} \tag{11}$$

The *state space* formulation (a set of first order differential equations) of the optimization problem thus becomes:

$$\begin{aligned}
&\text{mininimize:} \quad J = \mathbf{c}_{\mathrm{d}}^{\mathrm{T}} \cdot \mathbf{1} \\
&\text{subject to:} \\
&\left.\begin{aligned}
&pi = \frac{\mathbf{1}^{\mathrm{T}} \cdot \mathsf{D}^{q+1}\left(\mathbf{E}_{\mathrm{h}}\left(t_f\right)\right) \cdot \mathbf{1} + \mathbf{1}^{\mathrm{T}} \cdot \mathsf{D}^{\frac{q+1}{p}}\left(\mathbf{d}_{m,p}\left(t_f\right)\right) \cdot \mathbf{1}}{\mathbf{1}^{\mathrm{T}} \cdot \mathsf{D}^{q}\left(\mathbf{E}_{\mathrm{h}}\left(t_f\right)\right) \cdot \mathbf{1} + \mathbf{1}^{\mathrm{T}} \cdot \mathsf{D}^{\frac{q}{p}}\left(\mathbf{d}_{m,p}\left(t_f\right)\right) \cdot \mathbf{1}} \le 1.0 \\
&\text{where} \\
&\dot{\mathbf{E}}_{\mathrm{h}}(t) = \mathsf{D}^{-1}\left(\mathbf{E}_{\mathrm{h}}^{\mathrm{all}}\right) \cdot \mathsf{D}\left(\mathbf{f}_{\mathrm{h}}(t)\right) \cdot \left(\mathbf{B}_{\mathrm{xf}} \mathbf{v}(t)\right) \;;\; \mathbf{E}_{\mathrm{h}}(0) = \mathbf{0} \\
&\dot{\mathbf{d}}_{m,p} = \frac{1}{t_f}\left(\mathsf{D}^{-1}\left(\mathbf{d}^{\mathrm{all}}\right) \cdot \mathsf{D}\left(\mathbf{H}_{\mathrm{x}} \cdot \mathbf{x}(t)\right)\right)^{p} \cdot \mathbf{1} \;;\; \mathbf{d}_{m,p}(0) = \mathbf{0} \\
&\text{and } \mathbf{x}(t), \mathbf{v}(t) \text{ and } \mathbf{f}_{\mathrm{h}}(t) \text{ satisfy the equations of motion} \\
&\dot{\mathbf{x}}(t) = \mathbf{v}(t) \;;\; \mathbf{x}(0) = \mathbf{0} \\
&\dot{\mathbf{v}}(t) = \mathbf{M}^{-1}\left(-\left[\mathbf{C} + \mathbf{C}_{\mathrm{d}}\left(\mathbf{c}_{\mathrm{d}}\right)\right] \cdot \mathbf{v}(t) - \mathbf{K}^{\alpha}\mathbf{x}(t) - \mathbf{B}_{\mathrm{fx}}\mathbf{f}_{\mathrm{h}}(t) - \mathbf{M} \cdot \mathbf{e} \cdot \mathbf{a}_{\mathrm{g}}(t)\right); \mathbf{v}(0) = \mathbf{0} \\
&\dot{\mathbf{f}}_{\mathrm{h}}(t) = \mathbf{f}\left(\mathbf{v}(t), \mathbf{f}_{\mathrm{h}}(t)\right) \;;\; \mathbf{f}_{\mathrm{h}}(0) = \mathbf{0}
\end{aligned}\right\} \forall \mathbf{a}_{\mathrm{g}} \in ensemble\ i \\
&\mathbf{0} \le \mathbf{c}_{\mathrm{d}} \le \mathbf{c}_{\mathrm{d,max}}
\end{aligned} \tag{12}$$

where $\mathbf{v}$ = the velocity vector.

Gradient derivation: Equation 12 has the general form of:

$$\begin{aligned}
&\text{minimize } f\left(\mathbf{c}_{\mathrm{d}}\right) \\
&\text{subject to:} \\
&g\left(\mathbf{y}\left(t_f\right)\right) - g_{\max} \le 0 \text{ where } \dot{\mathbf{y}}(t) = \mathbf{a}\left(\mathbf{y}(t), \mathbf{c}_{\mathrm{d}}, t\right); \mathbf{y}(0) = \mathbf{0}
\end{aligned} \tag{13}$$

where $\mathbf{y} = \left\{\mathbf{E}_{\mathrm{h}}^{\mathrm{T}} \quad \mathbf{d}_{m,p}^{\mathrm{T}} \quad \mathbf{x}^{\mathrm{T}} \quad \mathbf{v}^{\mathrm{T}} \quad \mathbf{f}_{\mathrm{h}}^{\mathrm{T}}\right\}^{\mathrm{T}}$. The gradient of $g(\mathbf{y}(t_f))$ is obtained from the following secondary optimization problem:

$$\begin{aligned}
&\text{minimize } g\left(\mathbf{y}\left(t_f\right)\right) \text{or} \int_0^{t_f} \frac{dg\left(\mathbf{y}(t)\right)}{dt} dt + g\left(\mathbf{y}(0)\right) \\
&\text{subject to:} \\
&\dot{\mathbf{y}}(t) = \mathbf{a}\left(\mathbf{y}(t), \mathbf{c}_{\mathrm{d}}, t\right); \mathbf{y}(0) = \mathbf{0}
\end{aligned} \tag{14}$$

The augmented function is given by

$$J_{\mathrm{a}} = \int_{t_o}^{t_f} \left\{ \frac{dg(\mathbf{y}(t))}{dt} + \boldsymbol{\lambda}^T(t)[\mathbf{a}(\mathbf{y}(t), \mathbf{c}_{\mathrm{d}}, t) - \dot{\mathbf{y}}(t)] \right\} dt + g(\mathbf{y}(0)) \tag{15}$$

where $\boldsymbol{\lambda} = \{\boldsymbol{\lambda}_{\mathrm{x}}^{\mathrm{T}} \; \boldsymbol{\lambda}_{\mathrm{v}}^{\mathrm{T}} \; \boldsymbol{\lambda}_{\mathrm{fh}}^{\mathrm{T}} \; \boldsymbol{\lambda}_{\mathrm{Eh}}^{\mathrm{T}} \; \boldsymbol{\lambda}_{\mathrm{xmp}}^{\mathrm{T}}\}^{\mathrm{T}}$. The variation of the augmented function, with t_f specified results in

$$\begin{aligned} \delta J_{\mathrm{a}}(\mathbf{c}_{\mathrm{d}}) = & \left[\left[\frac{\partial g(\mathbf{y}(t))}{\partial \mathbf{y}} - \boldsymbol{\lambda}(t)\right]^T \cdot \delta\mathbf{y}\right|_{t_f} + \int_{t_o}^{t_f} \left\{ \left[\frac{\partial}{\partial \mathbf{y}}\left(\boldsymbol{\lambda}^T(t) \cdot \mathbf{a}(\mathbf{y}(t), \mathbf{c}_{\mathrm{d}}, t)\right) + \frac{d\boldsymbol{\lambda}(t)}{dt}\right]^T \cdot \delta\mathbf{y}(t) \right. \\ & \left. + [\mathbf{a}(\mathbf{y}(t), \mathbf{c}_{\mathrm{d}}, t) - \dot{\mathbf{y}}(t)]^T \cdot \delta\boldsymbol{\lambda}(t) + \left[\frac{\partial\left(\boldsymbol{\lambda}^T(t) \cdot \mathbf{a}(\mathbf{y}(t), \mathbf{c}_{\mathrm{d}}, t)\right)}{\partial \mathbf{c}_{\mathrm{d}}}\right]^T \cdot \delta\mathbf{c}_{\mathrm{d}} \right\} dt \end{aligned} \tag{16}$$

Taking the first three variations as arbitrary results in the following three differential equations and boundary conditions to be satisfied:

$$\begin{aligned} \boldsymbol{\lambda}(t_f) &= \frac{\partial g(\mathbf{y}(t_f))}{\partial \mathbf{y}} \\ \frac{d\boldsymbol{\lambda}(t)}{dt} &= -\frac{\partial}{\partial \mathbf{y}}\left(\boldsymbol{\lambda}^T(t) \cdot \mathbf{a}(\mathbf{y}(t), \mathbf{c}_{\mathrm{d}}, t)\right) \\ \dot{\mathbf{y}}(t) &= \mathbf{a}(\mathbf{y}(t), \mathbf{c}_{\mathrm{d}}, t) \end{aligned} \tag{17}$$

The multiplier of the variation $\delta\mathbf{c}_d$ will yield the expression for the evaluation of the gradient $\nabla_{\mathbf{c}_d} g(\mathbf{y}(t_f))$. This expression becomes:

$$\delta J_{\mathrm{a}} = \frac{\partial J_{\mathrm{a}}}{\partial \mathbf{c}_{\mathrm{d}}} \cdot \delta\mathbf{c}_{\mathrm{d}} \Rightarrow \frac{\partial J_{\mathrm{a}}}{\partial \mathbf{c}_{\mathrm{d}}} = \int_{t_o}^{t_f} \frac{\partial\left(\boldsymbol{\lambda}^T(t) \cdot \mathbf{a}(\mathbf{y}(t), \mathbf{c}_{\mathrm{d}}, t)\right)}{\partial \mathbf{c}_{\mathrm{d}}} dt \tag{18}$$

which is the desired gradient since

$$\begin{aligned} \nabla_{\mathbf{c}_d} g(\mathbf{y}(t_{\mathrm{f}})) &= \frac{\partial}{\partial \mathbf{c}_{\mathrm{d}}}[g(\mathbf{y}(t_{\mathrm{f}}))] = \\ &= \frac{\partial}{\partial \mathbf{c}_{\mathrm{d}}}\left[\int_{t_o}^{t_f} \left\{ \left[\frac{dg}{dy}(\mathbf{y}(t))\right]^T \cdot \dot{\mathbf{y}} \right\} dt + g(\mathbf{y}(t_o))\right] \equiv \left.\frac{\partial J_{\mathrm{a}}}{\partial \mathbf{c}_{\mathrm{d}}}\right|_{\mathbf{a}(\mathbf{y}(t), \mathbf{c}_d, t) - \dot{\mathbf{y}}(t) = \mathbf{0}} \end{aligned} \tag{19}$$

The gradient of the constraint in Eq. 12, can now be evaluated from:

$$\nabla_{\mathbf{c}_d} pi(\mathbf{c}_{\mathrm{d}}) = -\int_0^{t_f} \left[\mathbf{v}^{\mathrm{T}}(t) \cdot \frac{\partial \mathbf{C}_{\mathrm{d}}^{\mathrm{T}}(\mathbf{c}_{\mathrm{d}})}{\partial \mathbf{c}_{\mathrm{d}}} \cdot \mathbf{M}^{-1} \cdot \boldsymbol{\lambda}_{\mathrm{v}}\right] dt \tag{20}$$

and the following set of differential equations and boundary conditions:

$$\begin{aligned}
&\boldsymbol{\lambda}_{\mathrm{x}}(t_f)=\mathbf{0};\boldsymbol{\lambda}_{\mathrm{v}}(t_f)=\mathbf{0};\boldsymbol{\lambda}_{\mathrm{fh}}(t_f)=\mathbf{0}\\
&\boldsymbol{\lambda}_{\mathrm{Eh}}(t_f)=\frac{1}{den^2}\left(den\cdot(q+1)\cdot\mathsf{D}^{q}\left(\mathbf{E}_{\mathrm{h}}\right)\cdot\mathbf{1}-num\cdot(q)\cdot\mathsf{D}^{q-1}\left(\mathbf{E}_{\mathrm{h}}\right)\cdot\mathbf{1}\right)\\
&\boldsymbol{\lambda}_{\mathrm{xmp}}(t_f)=\frac{1}{den^2}\left(den\cdot\left(\frac{q+1}{p}\right)\cdot\mathsf{D}^{\left(\frac{q+1}{p}-1\right)}\left(\mathbf{d}_{m,p}\right)\cdot\mathbf{1}-num\cdot\left(\frac{q}{p}\right)\cdot\mathsf{D}^{\left(\frac{q}{p}-1\right)}\left(\mathbf{d}_{m,p}\right)\cdot\mathbf{1}\right)
\end{aligned}\tag{21}$$

$$\begin{aligned}
&\dot{\boldsymbol{\lambda}}_{\mathrm{x}}(t)=\left(\mathbf{M}^{-1}\cdot\mathbf{K}^{\alpha}\right)^{\mathrm{T}}\cdot\boldsymbol{\lambda}_{\mathrm{v}}(t)+\left(-p\mathbf{H}_{\mathrm{x}}^{\mathrm{T}}\cdot\mathsf{D}^{p-1}(\mathbf{H}_{\mathrm{x}}\cdot\mathbf{x}(t))\mathsf{D}^{-p}\left(\mathbf{d}^{\mathrm{all}}\right)\cdot\boldsymbol{\lambda}_{\mathrm{xmp}}(t)\right)\\
&\dot{\boldsymbol{\lambda}}_{\mathrm{v}}(t)=-\boldsymbol{\lambda}_{\mathrm{x}}(t)+\left(\mathbf{M}^{-1}\cdot[\mathbf{C}+\mathbf{C}_{\mathrm{d}}(\mathbf{c}_{\mathrm{d}})]\right)^{\mathrm{T}}\cdot\boldsymbol{\lambda}_{\mathrm{v}}(t)-\left(\frac{\partial\mathbf{f}(\mathbf{v}(t),\mathbf{f}_{\mathrm{h}}(t))}{\partial\mathbf{v}}\right)^{\mathrm{T}}\cdot\boldsymbol{\lambda}_{\mathrm{fh}}(t)+\\
&\qquad+\left(-\mathbf{B}_{\mathrm{xf}}^{\mathrm{T}}\cdot\mathsf{D}(\mathbf{f}_{\mathrm{h}}(t))\cdot\mathsf{D}^{-1}\left(\mathbf{E}_{\mathrm{h}}^{\mathrm{all}}\right)\cdot\boldsymbol{\lambda}_{\mathrm{Eh}}(t)\right)\\
&\dot{\boldsymbol{\lambda}}_{\mathrm{fh}}(t)=\left(\mathbf{M}^{-1}\cdot\mathbf{B}_{\mathrm{fx}}\right)^{\mathrm{T}}\cdot\boldsymbol{\lambda}_{\mathrm{v}}(t)-\left(\frac{\partial\mathbf{f}(\mathbf{v}(t),\mathbf{f}_{\mathrm{h}}(t))}{\partial\mathbf{f}_{\mathrm{h}}}\right)^{\mathrm{T}}\cdot\boldsymbol{\lambda}_{\mathrm{fh}}(t)+\\
&\qquad+\left(-\mathsf{D}\left(\mathbf{B}_{\mathrm{xf}}\cdot\mathbf{v}(t)\right)\cdot\mathsf{D}^{-1}\left(\mathbf{E}_{\mathrm{h}}^{\mathrm{all}}\right)\cdot\boldsymbol{\lambda}_{\mathrm{Eh}}(t)\right)\\
&\dot{\boldsymbol{\lambda}}_{\mathrm{Eh}}(t)=\mathbf{0};\dot{\boldsymbol{\lambda}}_{\mathrm{xmp}}(t)=\mathbf{0}
\end{aligned}\tag{22}$$

$$\begin{aligned}
&\dot{\mathbf{x}}(t)=\mathbf{v}(t)\ ;\ \mathbf{x}(0)=\mathbf{0}\\
&\dot{\mathbf{v}}(t)=\mathbf{M}^{-1}\left(-[\mathbf{C}+\mathbf{C}_{\mathrm{d}}(\mathbf{c}_{\mathrm{d}})]\cdot\mathbf{v}(t)-\mathbf{K}^{\alpha}\mathbf{x}(t)-\mathbf{B}_{\mathrm{fx}}\mathbf{f}_{\mathrm{h}}(t)-\mathbf{M}\cdot\mathbf{e}\cdot\mathbf{a}_{\mathrm{g}}(t)\right)\ ;\ \mathbf{v}(0)=\mathbf{0}\\
&\dot{\mathbf{f}}_{\mathrm{h}}(t)=\mathbf{f}(\mathbf{v}(t),\mathbf{f}_{\mathrm{h}}(t))\ ;\mathbf{f}_{\mathrm{h}}(0)=\mathbf{0}\\
&\dot{\mathbf{E}}_{\mathrm{h}}(t)=\mathsf{D}^{-1}\left(\mathbf{E}_{\mathrm{h}}^{\mathrm{all}}\right)\cdot\mathsf{D}(\mathbf{f}_{\mathrm{h}}(t))\cdot(\mathbf{B}_{\mathrm{xf}}\mathbf{v}(t))\ ;\ \mathbf{E}_{\mathrm{h}}(0)=\mathbf{0}\\
&\dot{\mathbf{d}}_{m,p}=\left(\mathsf{D}^{-1}(\mathbf{d}_{all})\cdot\mathsf{D}(\mathbf{H}_{\mathrm{x}}\cdot\mathbf{x}(t))\right)^{p}\cdot\mathbf{1}\ ;\ \mathbf{x}_{m,p}(0)=\mathbf{0}
\end{aligned}\tag{23}$$

where $num=\mathbf{1}^{\mathrm{T}}\cdot\mathsf{D}^{q+1}\left(\mathbf{E}_{\mathrm{h}}(t_f)\right)\cdot\mathbf{1}+\mathbf{1}^{\mathrm{T}}\cdot\mathsf{D}^{\frac{q+1}{p}}\left(\mathbf{d}_{m,p}(t_f)\right)\cdot\mathbf{1}$;

$den=\mathbf{1}^{\mathrm{T}}\cdot\mathsf{D}^{q}\left(\mathbf{E}_{\mathrm{h}}(t_f)\right)\cdot\mathbf{1}+\mathbf{1}^{\mathrm{T}}\cdot\mathsf{D}^{\frac{q}{p}}\left(\mathbf{d}_{m,p}(t_f)\right)\cdot\mathbf{1}$.

Equations 23 return the equality constraints; Eqs. 21 and 22 give expressions for the evaluation of the Lagrange multipliers, $\boldsymbol{\lambda}_{\mathrm{x}}$, $\boldsymbol{\lambda}_{\mathrm{v}}$ and $\boldsymbol{\lambda}_{\mathrm{xmp}}$ which are needed for the evaluation of Eq. 20. Now since the elements of $\mathbf{C}_{\mathrm{d}}^{\mathrm{T}}(\mathbf{c}_{\mathrm{d}})$ are linear combinations of the elements of $\mathbf{c}_{\mathrm{d}}$, the differentiation of $\mathbf{C}_{\mathrm{d}}^{\mathrm{T}}(\mathbf{c}_{\mathrm{d}})$ with respect to $\mathbf{c}_{\mathrm{d,i}}$ (also needed in Eq. 20) is rather simple and easily programmed. The computation of the gradient for a single record is summarized as follows:

Step 1: Solve the equations of motion (Eq. 23).

Step 2: Solve the equations of the Lagrange multipliers (Eq. 21 with conditions for t_f in Eq. 22).

Step 3: Calculate the desired gradient (Eq. 20).

Optimization scheme

The gradients of the objective function and the constraints are needed at each iteration for first order optimization schemes. Thus the solution requires a time history analysis for each record (constraint) at every iteration cycle. In order to reduce the computational effort, optimization is first carried out for one "active" ground motion (loading condition), rather than for the whole ensemble. If the optimal solution for this ground motion violates other records in the ensemble, additional ground motions are added one at a time (*Stage 4* below). Following are the main steps in the methodology that are used for the optimization scheme.

***Step 1*: Select the "active" ground motion.** The record with the maximal displacement is selected to begin the process. It is evaluated from a SDOF with the 1st period of the undamped structure within the expected total damping ratio range.
***Step 2*: Compute an initial starting value for the damping vector.** The starting point is evaluated by first assuming a distribution of equal dampers for the damping vector. Then this damping vector is factored so as to satisfy $pi = 1.0$ where pi is computed from a time history analysis of the frame excited by the "active" ground motion of *Step1*.
***Step 3*: Solve the optimization problem for the active set of records**. An appropriate gradient based optimization scheme is used. The gradients are evaluated as described above. If more than one record is "active", say two, then the gradients are calculated separately for each record and the size of the problem doubles.
***Step 4*: Feasibility check.** A time history analysis is performed on the optimally damped structure for each of the remaining records in the ensemble, separately. One new ground motion is added to the active set only if its pi is largest and greater than 1.0. Then *Step 3* is repeated.
***Step 5*: Stop.**

Linear Frames –A Particular Case

A class of structures, usually regular, can be brought to behave elastically under an earthquake excitation, by the addition of a reasonable amount of added damping. In this case nonlinear analysis methodologies, which are computationally expensive, are not essential and some of the nonlinear performance indices, such as hysteretic energy, become meaningless. Since the optimization problem of linear structures is a particular case the same procedure presented earlier holds. Minor changes ease the computations. First, the equations of motion are reduced to their linear form by substituting $\mathbf{K}^{\alpha} = \mathbf{K}$ and $\mathbf{f}_{\mathrm{h}}(t) \equiv \mathbf{0}$ (or, equivalently, $\dot{\mathbf{f}}_{\mathrm{h}}(t) \equiv \mathbf{0}$; $\mathbf{f}_{\mathrm{h}}(t_o) = \mathbf{0}$). Substituting these relations into Eqs. 25 and 26 leads to simpler equations for the gradient computation as well. Then, the constraint on the hysteretic energy will be omitted since no hysteretic energy dissipates in the elastic range. This leads to simpler equations for the gradient computation as well since $\mathbf{E}_{\mathrm{h}}(t_f) = \mathbf{0}$. Inter-story drifts remain the only constraints. In 3D structures the inter-story drifts of the peripheral frames are used.

Fundamental Results

- *The optimal design of added damping in 2D frames, assuming linear behavior of the damped structure, and in 2D yielding frames is characterized by assigning damping only in stories that reached the allowable drift.*
- *The optimal design of added damping in 2D yielding shear frames is characterized by assigning damping only in stories that reached the allowable normalized hysteretic energy.*
- *The optimal design of added damping of 3D framed structures is characterized by assigning damping at the peripheral frames only, where the peripheral drift has reached the allowable.*

These observations are strikingly analogous to the classical "fully-stressed-design" behavior of optimal trusses reported in the sixties.

Conclusions

A gradient based methodology for the optimal design of added viscous damping for an ensemble of realistic ground motion records with constraints on the maximum inter-story drifts for linear frames, and additional constraints on maximum energy based local damage indices for nonlinear frames, was presented. This methodology is appropriate for use in linear, as well as nonlinear, frames.

The computational effort is appreciably reduced by first using one "active" ground motion record since experience shows that one or two records dominate the design.

The gradients of the constraints were derived so as to enable the use of an efficient first order optimization scheme for the solution of the optimization problem. The approach for the gradient derivation has several advantages over other approaches. It is appropriate for use when the equations of motion assume nonlinear plastic behavior as well, and it requires a relatively small computational effort, in the form of a single additional solution of a set of differential equations (that is, the equations for the Lagrange multipliers).

References

Conte, J. P., Vijalapura P. K., and M. Meghella 2003. Consistent finite-element response sensitivity analysis. *Journal of Engineering Mechanics*, ASCE Dec. 2003, 1380-1393.

Falco, S. A., S. M. B. Afonso, and L. E. Vaz. 2004. Analysis and optimal design of plates and shells under dynamic loads – I: finite element and sensitivity analysis. *Struct Multidisc Optim* 27, 189 –196.

Gluck, N., A. M. Reinhorn , J. Gluck, and R. Levy, 1996. Design of Supplemental Dampers for Control of Structures. *Journal of Structural Engineering*, ASCE 122 (12), 1394-1399.

Hsieh, C. C., J. S. Arora, 1985. Structural design sensitivity analysis with general boundary conditions: dynamic problem. *International Journal for Numerical Methods in Engineering* 21, 267-283.

Inaudi, J. A., J. M. Kelly, and C. W. S. To, 1993. Statistical linearization method in the preliminary design of structures with energy dissipating devices. ATC-17-1, 509-520.

Takewaki, I., 1997. Optimal damper placement for minimum transfer function. *Earthquake Engineering and Structural Dynamics* 26 (11), 1113-1124.

Uang, C. M., and V. V. Bertero, 1990. Evaluation of seismic energy in structures. *Earthquake Engineering and Structural Dynamics* 19 (1), 77-90.

Zhang, R. H., and T. T. Soong, 1992. Seismic design of viscoelastic dampers for structural applications. *Journal of Structural Engineering*, ASCE 118 (5), 1375-1392.

STRATEGIES FOR COMPUTATIONAL EFFICIENCY IN CONTINUUM STRUCTURAL TOPOLOGY OPTIMIZATION

Colby C. Swan and Salam F. Rahmatalla
Center for Computer-Aided Design, University of Iowa, Iowa City, Iowa 52242, USA
E-mail: colby-swan@uiowa.edu, srahmata@engineering.uiowa.edu

Abstract

A methodology of enhanced computational efficiency is presented for continuum topology optimization of sparse structural systems. Such systems are characterized by the structural material occupying only a small fraction of the structure's envelope volume. When modeled within a continuum mechanics and topology optimization framework such structures require models of very high refinement which is computationally very expensive. The methodology presented herein to deal with this issue is based on the idea of starting with a relatively coarse mesh of low refinement and employing a sequence of meshes featuring progressively greater degrees of uniform refinement. One starts by solving for an initial approximation to the final material layout on the coarse mesh. This design is then projected onto the next finer mesh in the sequence, and the material layout optimization process is continued. The material layout design from the second mesh can then be projected onto the third mesh for additional refinement, and so forth. The process terminates when an optimal design of sufficient sparsity, and sufficient mesh resolution is achieved. Within the proposed methodology, additional computational efficiency is realized by using a design-dependent analysis problem reduction technique. As one proceeds toward sparse optimal designs, very large regions of the structural model will be devoid of any structural material and hence can be excluded from the structural analysis problem resulting in great computational efficiency. The validity and performance characteristics of the proposed methodology are demonstrated on three different problems, two involving design of sparse structures for buckling stability, and the third involving design of a hinge-free gripper compliant mechanism.

Introduction

For nearly twenty years, beginning approximately with the work of Bendsøe and Kikuchi (1988), continuum structural topology optimization has been used to investigate optimal forms of structures and mechanical systems. During this time, myriad gains have been realized in understanding how different continuum topology formulations work in relation to an ever-widening circle of applications. Although there are many exceptions, continuum structural topology optimization methods have for the most part been applied to the design of structures and mechanical systems in two-dimensions. One would hope that the methods can eventually be applied to design of structures and mechanical systems in three-dimensions with the same ease that they currently enjoy in two dimensions. If this vision is to become reality, the potentially huge computational costs must be reduced to the extent possible while preserving the inherent generality and flexibility of continuum structural topology optimization. In many design applications, the structural will be very sparse in that the volume of structural material will occupy only a small fraction of the structural system's envelope volume. In such cases, continuum structural topology optimization can require highly refined finite element meshes to achieve convergent and interpretable material layout solutions and models that realistically estimate system performances (i.e. overall compliance; buckling stability; vibrational eigenvalues; etc).

One approach to achieving and dealing with fine meshes in continuum topology optimization [Maute et al. 1998] is to use adaptive mesh refinement to decrease the number of design variables and to seek smooth final topological forms. A similar approach is now extensively involved in finding optimal design forms using the evolutionary structural optimization method (ESO)[3]. In another approach, researchers enforced design symmetry during the optimization process by reducing the design space[4] by, or to remove the void nonstructural elements temporarily from the structural analysis, but reintroduce them if they needed [5,6]. The latter approach has been shown to be very effective in dealing with problems involving geometrical nonlinearity.

M. Pandey et al. (eds), Advances in Engineering Structures, Mechanics & Construction, 673–683.

The material usage constraint is another important factor and plays a considerable role in achieving low weight and certain performances when utilizing continuum topology optimization method. For example, most existing optimal large civil structures such as long span bridges are sparse in nature, where the real structural material occupies only a small percentage (less than 1%) of the structure's envelope volume. Therefore, in utilizing continuum structural topology optimization to obtain optimal design forms for such systems, it is crucial to impose stringent material usage constraint and implement very fine meshes in order to capture a realistic performance for such systems. The importance of such an approach has been demonstrated in a previous work [6] where it proved to be very effective when designing structural systems for buckling instability. Similarly, it has been shown in designing hinge-free compliant mechanisms to achieve considerable flexibility [7], the amount of structural material comprising the mechanism can be progressively reduced until the desired flexibility of the mechanism is achieved. It is also crucial toward this end to use stringent material usage constraint with very fine finite element meshes

This article presents a methodology for solving large-size sparse systems in continuum structural topology design framework based on sequential refinement and size reduction strategy in a new way that is conceptually simple and theoretically sound. In sequential refinement, the proposed methodology solved a preliminary problem involving relatively coarse meshes and moderate material usage constraint. The resulting optimal form from this stage, which comprises the solid structural material, is then mapped onto a finer mesh and with realistic material usage constraint. The new problem is then solved where a new topological form is obtained. The mesh refinement process is repeated until the final design converges to a realistic shape and performance with minimum error.

A size reduction strategy is implemented within each structural analysis, where the void nonstructural elements are removed temporarily from structural analysis but can come back quite easily and naturally if needed. The proposed size reduction technique has been tested on many linear and nonlinear systems involving geometrical nonlinearity and buckling instability and shown to be a very effective and powerful tool for reducing the computational costs, especially when dealing with sparse systems and very fine meshes.

It should be noted here that the current methodology is based on interpolation of nodal design variables using nodal basis functions[8] as opposed to element-based design variables. Although node-based design variables feature C^0 continuity, they must generally be used with perimeter constraints to achieve design convergence with mesh refinement.

Problem Formulations

Structural Model and Material Layout Description

The objective of continuum structural optimization is to find a layout of a structural material of specified properties in a defined spatial region that provides optimum structural performance. In order that the widest possible class of structural layouts can be considered, the methods in question must accommodate such generality. In this work, the spatial region that the candidate structural models can occupy is denoted Ω_s. To facilitate both description of the structural material layout in Ω_s and analysis of the performance associated with each layout considered, the domain is discretized into a relatively fine mesh of nodes and finite elements.

It is desired that at the end of the form-finding process, the structural region Ω_s will be decomposed into a collection of regions cumulatively denoted Ω_A that contain the structural material in question, and the remaining regions $\Omega_B = \Omega_s \setminus \Omega_A$ that are devoid of structural material. Since solution of the form-finding problem in this way is ill-posed, an alternative relaxed approach is usually employed, wherein it is assumed that an amorphous "mixture" of structural material A and a void material B exists throughout the structural region Ω_s. In each region of Ω_s, the nature of the mixture is characterized by a local volumetric density ϕ_A of structural material A. By permitting mixtures, the structural material A and a fictitious void material B are allowed to simultaneously occupy an infinitesimal neighborhood about each Lagrangian point $\mathbf{x} \in \Omega_s$. The volumetric density of structural material A at a fixed Lagrangian point $X \in \Omega_s$ is denoted by $\phi_A(X)$ and represents the fraction of an infinitesimal region surrounding point **X** occupied by material A. Natural constraints upon the volumetric densities are:

$$0 \leq \phi_A(X) \leq 1; \qquad 0 \leq \phi_B(X) \leq 1; \qquad \phi_A(X) + \phi_B(X) = 1. \tag{1}$$

Clearly, when $\phi_A(X) = 1$ the point **X** contains solid structural material, and when $\phi_A(X) = 0$ the point **X** is devoid of structural material. The last physical constraint of (1) states that the material volume fractions at **X** are not independent and so one need only be concerned with the layout of structural material A. The design of a structure is here considered to be the spatial distribution of the structural material A in Ω_s.

To describe the distribution of material A throughout Ω_s using a finite number of design parameters, the volumetric density at each of the NUMNP nodal point forms a set of NUMNP design variables. These are then interpolated over the space of all intermediate points in the structure using the nodal shape functions:

$$\varphi(\mathbf{X}) = \sum_{i=1}^{NUMNP} b_i N_i(\mathbf{X}) \qquad \forall\, \mathbf{X} \in \Omega_s \tag{2}$$

where b_i are the nodal volumetric density values associated with the structural material; and $N_i(\mathbf{X})$ are the nodal shape functions. This approach yields a C^0 continuous design variable field that is not susceptible to "checkerboarding" instabilities.

Given the finite element model of the structural region Ω_s, the structural loads and restraints (or supports) on this region are specified as the set of design loads. For each set of design loadings, and for each realization of the design vector $\mathbf{b} = \{b_1, b_2, K, b_{NUMNP}\}$, the response performance of the structure will be analyzed as a boundary value problem. From the computed response of the structure, the performance of the structure will be quantified, as will be the sensitivity of the performance to variations in the design variables.

Constitutive Mixing Rules

In the proposed design framework, each finite element comprising the spatial domain Ω_s of the structure will generally contain a spatially varying mixture of the structural and void materials. It is necessary to prescribe the stiffness (or elastic moduli) of such mixtures in terms of the stiffness characteristics of the solid material $\mathbf{C}_{solid}$, those of the fictitious void material $\mathbf{C}_{void}$, and the local volumetric density of the structural material $\varphi(\mathbf{X})$. Here, the well-known powerlaw formula (Bendsøe and Sigmund 1999) is used to accomplish this task, providing the local effective stiffness of the mixture $\mathbf{C}^*$ as:

$$\mathbf{C}^* = \varphi^p \mathbf{C}_{solid} + (1 - \varphi^p)\mathbf{C}_{void} \tag{3}$$

where typically the mixing rule parameter $p \in [1,4]$. With p=1, the Voigt rule of mixtures is obtained which does not penalize mixtures, but which does yield a convex formulation for compliance minimization problems (Swan and Kosaka, 1997) so that only one solution exists for the design problem. With p=4, mixtures are penalized in the final design, so that regions of Ωs tend to be either solid or void, but the optimization problem is not convex, and will admit a number of solutions that satisfy the first order optimality conditions.

Structural Analysis

For each design, a structural analysis problem is solved on the continuum domain Ω_s. In general terms, the structural analysis problem solved for each realization of the design vector **b** is the following: Find the displacement field $\mathbf{u}(\mathbf{X})$ $\Omega_s \rightarrow \Re^3$ such that the variational equilibrium problem is solved:

$$\int_{\Omega_s} \boldsymbol{\sigma} : \delta\boldsymbol{\varepsilon}\, d\Omega_s = \int_{\Gamma_s} \mathbf{h} \cdot \delta\mathbf{u}\, d\Gamma_s + \int_{\Omega_s} \rho\mathbf{g} \cdot \delta\mathbf{u}\, d\Omega_s \tag{4}$$

where $\boldsymbol{\sigma}(\mathbf{X})$ is the local stress field in the structure; $\mathbf{h}$ is a traction vector consistent with the design loads being applied to the structure; $\rho(\mathbf{X})$ is the local mass density of the structural material; $\mathbf{g}$ is the gravitational body force vector; $\delta\mathbf{u}$ is a kinematically admissible variational displacement field; and $\delta\boldsymbol{\varepsilon}$ is the corresponding variational strain field. In the structural model, the material features linear elastic behavior such that $\boldsymbol{\sigma} = \mathbf{C}^* : \boldsymbol{\varepsilon}$ where the effective elasticity tensor is design dependent and prescribed in accordance with Eq. (3). The matrix problem associated with variational equilibrium of the discrete finite element structural model, for which $\mathbf{u}(\mathbf{X}) = \sum_i N_i(\mathbf{X})\,\mathbf{u}_i$, is

$$\mathbf{0} = \mathbf{K}\cdot\mathbf{u} - \mathbf{f}^{\text{ext}} = \mathbf{f}^{\text{int}} - \mathbf{f}^{\text{ext}} \tag{5}$$

where:

$$\begin{aligned} \mathrm{K}_{jk}^{LM} &= \int_{\Omega_s} \mathrm{B}_{mj}^{L}\,\mathrm{C}^*_{\mathrm{mn}}\,\mathrm{B}_{\mathrm{nk}}^{\mathrm{M}}\,d\Omega_s \\ \mathbf{f}^{\text{int}} &= \mathbf{K}\cdot\mathbf{u} = \int_{\Omega_s} \mathbf{B}^T\boldsymbol{\sigma}\,d\Omega_s \\ \mathbf{f}^{ext} &= \int_{\Gamma_s} N\,\mathbf{h}\,d\Gamma_s + \int_{\Omega_s} N\,\rho\mathbf{g}\,d\Omega_s . \end{aligned} \tag{6}$$

In all of the above, N denote the nodal shape functions and $\mathbf{B}$ denote the standard strain-displacement matrices [*c.f.* Bathe, 1996]. The structural stiffness matrix $\mathbf{K}$ is positive definite due to the characteristics of the effective elasticity tensor $\mathbf{C}^*$, and this guarantees a unique solution to the structural analysis problem for each realization of the design $\mathbf{b}$.

Once the equilibrium solution to the problem of Eq. (5) is obtained, then the linearized geometrical stiffness matrix $\mathbf{G}$ can be computed based on the stress field $\boldsymbol{\sigma}$ in the structure:

$$G_{jk}^{LM} = \int_{\Omega_s} N_{,m}^{L}\,N_{,n}^{M}\,\sigma_{\mathrm{mn}}\delta_{jk}\,d\Omega_s \tag{7}$$

It is worth noting that G is not necessarily positive definite but rather depends heavily upon the nature of the stress field in the structure. A purely tensile stress field clearly makes G positive definite, although for any compressive stresses, G will not be positive definite.

Structural Performance Measures

As noted previously, structural topology design problems can be formulated in a number of alternative ways through utilization of assorted objective and constraint functions. Generally, the objective function measures the performance of the structure, and the constraint function limits the amount of structural material that can be used, although the roles can be reversed equally well. The significant aspects of using CSTO to design large-scale sparse structures can be demonstrated here using the linear elastic structural compliance performance measure and the critical load buckling factor.

Linear Elastic Structural Compliance

If a structure features a linear elastic response behavior, the resulting displacement field $\mathbf{u}$ in response to a set of applied external loads $\mathbf{f}^{ext}$ will be simply $\mathbf{u} = \mathbf{K}^{-1}\cdot\mathbf{f}^{\text{ext}}$ where $\mathbf{K}$ represents the stiffness matrix of the structure. For a given set of loads, the compliance $\Pi(\mathbf{b})$ of the structure is simply

$$\Pi(\mathbf{b}) = \frac{1}{2}\mathbf{f}^{\text{ext}}\cdot\mathbf{u}\ . \tag{8}$$

Structural concept designs $\mathbf{b}$ that are stiff with respect to the applied loads will have small compliance $\Pi(\mathbf{b})$, whereas structures that are not stiff with respect to the applied loads will have large compliance. To facilitate usage of gradient-based optimization solution techniques, it is necessary to compute the design derivatives of the compliance function. It can be shown that the design gradient of structural compliance is provided by the following expression:

$$\frac{d\Pi}{d\mathbf{b}} = -\frac{1}{2}\mathbf{u}\cdot\left(\frac{\partial\mathbf{K}}{\partial\mathbf{b}}\cdot\mathbf{u} - \frac{\partial\mathbf{f}^{\text{ext}}}{\partial\mathbf{b}}\right) \tag{9}$$

Linearized Buckling Performance Measure

Linearized buckling eigenvalue analysis proceeds as follows: A prescribed force loading $\mathbf{f}^{\text{ext}}$ is applied to the structure with its magnitude necessarily being less than that required to induce geometric instability in the structure. Once the resulting linear, elastostatic displacement solution $\mathbf{u} = \{u_i\} \in \boldsymbol{R}^N$ in response to the applied loading $\mathbf{f}^{\text{ext}}$ is obtained $\left(\mathbf{K}\cdot\mathbf{u} = \mathbf{f}^{\text{ext}}\right)$, where $\mathbf{K}$ is the linearized stiffness matrix, then the following eigenvalue problem is solved

$$\left[\mathbf{K}(\mathbf{b}) + \lambda\mathbf{G}(\mathbf{u},\mathbf{b})\right]\cdot\boldsymbol{\psi} = \mathbf{0} \tag{10}$$

In the preceding, $\mathbf{b} = \{\mathbf{b}_e\} \in \boldsymbol{R}^M$ is again the vector of design variables; $\mathbf{K}$ is the tangent stiffness operator; $\mathbf{G}(\mathbf{u},\mathbf{b})$ is the linearized geometric stiffness matrix; $\lambda = -(\boldsymbol{\psi}\cdot\mathbf{K}\cdot\boldsymbol{\psi})/(\boldsymbol{\psi}\cdot\mathbf{G}\cdot\boldsymbol{\psi})$ is an eigenvalue denoting the magnitude by which $\mathbf{f}^{\text{ext}}$ must be scaled to create instability in the structure, and $\boldsymbol{\psi}$ is a normalized eigenvector satisfying $\boldsymbol{\psi}\cdot\mathbf{K}\cdot\boldsymbol{\psi} = 1$. To avoid numerical difficulties in the solution of (10) stemming from the indefinite characteristics of $\mathbf{G}$, it is common (Bathe 1996) to solve a modified eigenvalue problem that deals with two positive definite matrices.

$$\left[(\mathbf{K}+\mathbf{G}) - \gamma\,\mathbf{K}\right]\cdot\boldsymbol{\psi} = \mathbf{0} \tag{11}$$

where

$$\gamma = \frac{\lambda - 1}{\lambda} \Leftrightarrow \lambda = \frac{1}{1-\gamma}. \tag{12}$$

In Eq. (11), the matrix $\mathbf{K}$ is positive definite irrespective of the loading applied to the structure, whereas the matrix ($\mathbf{K}$+$\mathbf{G}$) will only be positive definite when the magnitude of the loading applied to the structural model is less than the critical magnitude that creates instability in accordance with linearized buckling theory.

The design problem is formulated to maximize the calculated minimum-buckling load factor (λ), and accordingly the objective function f_E to be minimized for this problem would simply be the reciprocal of the lowest eigenvalue λ as follows.

$$f_E(\mathbf{u},\mathbf{b}) = \frac{1}{\min(\lambda)} \tag{13}$$

The optimization problem is thus stated to minimize the reciprocal of the first (or minimum) critical buckling load as follows

$$\min_{\mathbf{b},\mathbf{u}} f_E = \min_{\mathbf{b},\mathbf{u}}\left(\frac{1}{\lambda}\right) = \min_{\mathbf{b},\mathbf{u}}\left(-\max_{\|\boldsymbol{\psi}\|\neq 0}\frac{\boldsymbol{\psi}\cdot\mathbf{G}\cdot\boldsymbol{\psi}}{\boldsymbol{\psi}\cdot\mathbf{K}\cdot\boldsymbol{\psi}}\right) \tag{14}$$

subject to the normal bound constraints on the design variables (1), the linear structural equilibrium state equation (5), and a constraint on material resources.

The design gradient of the objective function can be expressed as:

$$\frac{\mathrm{d}f_E}{\mathrm{d}\mathbf{b}} = \frac{\partial f_E}{\partial\mathbf{b}} + \frac{\partial f_E}{\partial\mathbf{u}}\cdot\frac{\partial\mathbf{u}}{\partial\mathbf{b}} \tag{15}$$

To avoid explicit computation of the term $\frac{\partial\mathbf{u}}{\partial\mathbf{b}}$, adjoint design sensitivity analysis is employed by augmenting the objective function f_E with the equilibrium state equation as follows

$$\Xi = f_E + \mathbf{u}^a\cdot\mathbf{r} \tag{16}$$

where $\mathbf{u}^a$ is the adjoint displacement vector which functions as a matrix of Lagrange multipliers and determined by the solution of a linear adjoint problem. The design derivative of the augmented Lagrangian is then written as follows:

$$\frac{d\Xi}{d\mathbf{b}} = \left(\frac{\partial f_E}{\partial \mathbf{b}} + \mathbf{u}^a \cdot \frac{\partial \mathbf{r}}{\partial \mathbf{b}}\right) + \left[\frac{\partial f_E}{\partial \mathbf{u}} + \frac{\partial \mathbf{r}}{\partial \mathbf{u}} \cdot \mathbf{u}^a\right] \cdot \frac{\partial \mathbf{u}}{\partial \mathbf{b}} + \left(\mathbf{r} \cdot \frac{\partial \mathbf{u}^a}{\partial \mathbf{b}}\right) \tag{17}$$

The last term of Eq. (17) vanishes due to satisfaction of the equilibrium constraint (**r**=**0**), and the second term can be made to vanish by selecting the adjoint displacement vector to solve the following linear adjoint equality statement

$$\mathbf{K} \cdot \mathbf{u}^a = \psi \cdot \frac{\partial \mathbf{G}}{\partial \mathbf{u}} \cdot \psi \cdot \tag{18}$$

Since it can be shown that $d\Xi/d\mathbf{b} = df_E/d\mathbf{b}$, it follows that the design gradient expression for the objective function is

$$\frac{df_E}{d\mathbf{b}} = -\psi \cdot \left(\frac{\partial \mathbf{G}}{\partial \mathbf{b}} + \frac{1}{\lambda}\frac{\partial \mathbf{K}}{\partial \mathbf{b}}\right) \cdot \psi + \mathbf{u}^a \cdot \left(\frac{\partial \mathbf{K}}{\partial \mathbf{b}} \cdot \mathbf{u} - \frac{\partial \mathbf{f}^{ext}}{\partial \mathbf{b}}\right). \tag{19}$$

The preceding expression is valid only when the minimum eigenvalue is a simple, or non-repeated, eigenvalue. When the minimum eigenvalue is nonsimple, or repeated, the variation of the eigenvalue in design space is non-smooth, and direct usage of the expression in Eq. (19) is technically incorrect (Choi et al 1983; Seyranian *et al* 1994). Resolution of this issue is nontrivial, although it can be ameliorated somewhat by using small and variable move limits in the design optimization process. Despite this challenge, designs that successfully maximize the buckling stability of a structural system can nevertheless be obtained. Further details on formulation and solution of stability design problems using linearized buckling theory are provided in Rahmatalla and Swan[10].

Analysis Problem Size Reduction Technique

In continuum topology optimization of sparse structures with limited structural material usage and realistically large design loads, geometrical instabilities become a definite possibility. If modeled, geometrical instabilities in structural systems can result in finite deformations. The modeling of finite deformations in mixed solid-void grid-like meshes used in continuum structural topology optimization can result in excessive distortion of void or low-density elements that can in turn lead to numerical difficulties solving the structural analysis problem. Since the optimization process in continuum topology optimization typically removes structural material from low stress, or low-sensitivity areas, fairly substantial regions of low-density elements are common. As these elements are highly compliant, they contribute very little to structural stability, while being subject to excessive deformation that creates numerical difficulties. Therefore, it is sometimes advantageous to identify these large regions of void and low-density elements and to remove them, at least temporarily, from the structural analysis problem. An automated algorithm for identifying such regions and removing them from the structural analysis problem is presented and discussed below. It is worth noting that the procedure proposed and investigated here is reversible in that it permits low-density regions of the structure to return as high-density structural regions even after they have previously been removed from consideration during structural analysis.

The essence of the proposed analysis problem reduction technique can be captured in the three steps listed below:

1. All finite elements in the structural analysis model that are devoid of solid material, or nearly so, are identified as "void" elements. (Typically, in the examples presented below, if an element's volume fraction of solid material is less than or equal to .002, it is identified as "void".)

2. All nodes that are members only of "void" elements are identified as "prime" nodes. The degrees of freedom of such "prime" nodes are restrained, reducing the size of the analysis problem.
3. If only "prime" nodes comprise an element, that element is then denoted as a "prime" element. Such "prime" elements are then neglected in the structural analysis problem so that if they undergo excessive distortion it does not create any singularities in the system of finite element equations. It is worth noting, that "prime" elements are those that are surrounded by "void" elements.

A graphical description supporting the explanation of this technique for reducing the analysis problem is presented in Figure 1.

The matter of reducing the analysis problem by neglecting significant regions of void elements has been addressed in preceding works [9,10]. The current reduction techniques have proven to be both robust and efficient in all of the example problems presented in the section below. The techniques are especially powerful and effective when applied in design problems involving extremely sparse structures, since highly refined meshes are needed when very stringent material resource constraints are imposed. When a fine mesh is employed with a very limited amount of structural material, the proposed reduction techniques will allow for dramatic savings in computing effort.

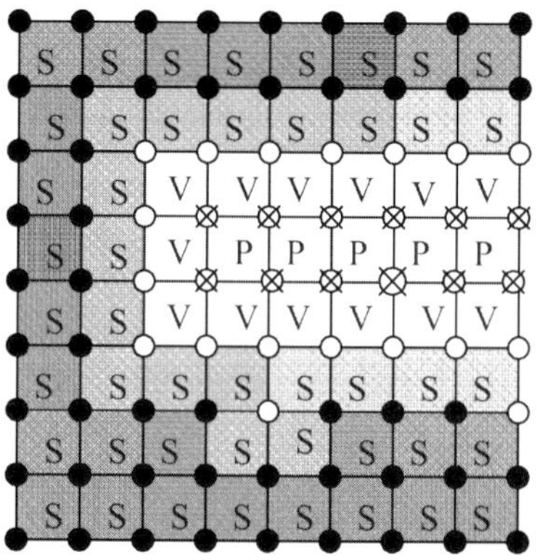

Figure 1. Schematic of partial mesh to illustrate analysis problem reduction technique. Nodes having vanishing design variable values are denoted with open circles; filled circles denote nodes associated with nonzero design values; nodes represented by open circles with X's are "prime" nodes whose degrees of freedom are restrained in the size reduction method. Elements designated with "S" are at least partially solid and those with "V" are devoid of material. Those designated with "P" are prime and need not be considered during structural analysis since all of their degrees of freedom are restrained.

Examples

The validity of the proposed methodology is demonstrated in this article on three problems,where the first two are benchmark linear stability design problems with extreme sparsity and the third involves design of a very sparse, hinge-free gripper compliant mechanism. In all problems presented, the void material is assumed to have a stiffness equal to that of structural material scaled down by a factor of 10^{-6}.

Stability Design of Sparse Structure under Fixed-Fixed End Conditions

In this problem, a point load is applied to the top central portion of a structural region with fixed supports at both lateral edges of the domain. A sparse, stable structural design is sought that carries the applied load back to the supports. This problem is solved here by optimizing the material layout such that the minimum buckling eigenvalue is maximized. Details on the formulation of such problems were presented in references 6 and 8. The problem has also been solved using different formulations and objective functions by Buhl *et al* (2000) and Gea and Luo (2001). In this specific example (Figure 2a), a point load of magnitude $1.0 \cdot 10^5$ is applied to the domain as shown which has relative dimensions of 100 by 50. The optimization problem is solved to find the constrained material layout under the applied loading that maximizes the minimum buckling eigenvalue (see Eq. 14).

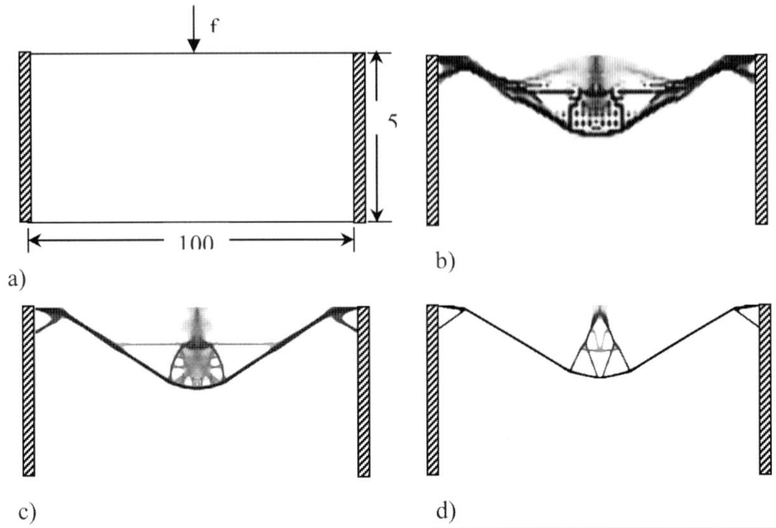

Fig. 2. Stability design of sparse structure for fixed-fixed end conditions. a) shows the design domain, support conditions, and loading; b) design obtained on a mesh of 100 x 50 bilinear elements with material usage constraint of 0.10; b) preceding design mapped onto mesh of 200 x 100 elements and further optimized while imposing a tightened material usage constraint of 0.05, and a perimeter constraint of 5000; d) preceding design mapped onto mesh of 400 x 200 elements and further optimized with material usage constraint of 0.03 and perimeter constraint of 3750.

The material layout optimization problem was first solved on a relatively coarse mesh of 100 by 50 bilinear quadrilateral continuum elements. The material layout shown in Figure 2b was obtained from a starting "design" of structural material completely occupying the entire the structural region, followed by imposition of a constraint on structural material: $V_{material} \leq 0.20 V_{domain}$ where V_{domain} represents the volume of the 100 by 50 structural domain. The design solution shown was obtained using a powerlaw parameter [p=1.75]. The material layout in Fig. 2b is both heavy and somewhat difficult to interpret. Consequently, the design shown was then projected onto a uniformly refined mesh of 200 by 100 bilinear continuum elements. With this refine mesh, a reduced material usage constraint of $V_{material} \leq 0.080 V_{domain}$ was imposed in addition to a perimeter constraint $P \leq \frac{1}{2}(l+h)$ where l and h are the lateral and vertical dimensions of the structural region in Fig. 2a. The material layout was then optimized on the refined mesh for 100 SLP (sequential linear programming) iterations with the result being as shown in Fig. 2c. To obtain a very clear, yet sparse, structural design, the design of Fig. 2c. was then mapped onto a mesh of 400 by 200 bilinear continuum elements. The material usage constraint was tightened to $V_{material} \leq 0.030 V_{domain}$, and a mixing rule powerlaw parameter of [p=4] was employed along with a tightened perimeter constraint $P \leq \frac{1}{3}(l+h)$. The final, optimal material layout design achieved is as shown in Fig. 2d. This final design shown in Fig. 2d. is similar those obtained in the previous study by the authors using this same problem[6].

Circular Domain Stability Design Problem

In this second test problem, the circle domain test problem introduced previously by the authors (Rahmatalla and Swan, 2003, 2004) is revisited. A downward acting point-load of magnitude $1.0 \cdot 10^9$ is applied at the center-node of the circular structural domain of radius R=50. In Fig. 3a, the coarsest mesh features 896 bilinear elements; the next mesh (3c) features 3584 elements; that of (3e) features 14,336 elements; and the finest (3g) features 57,344 elements. The design objective is to find the optimal form of structure that carries the design from the center back to the fixed boundaries. The design problem was first solved on the coarse mesh of 896 elements with a powerlaw value [p=1.75] and a material usage constraint $V_{material}/V_{structure} \leq 0.20$ and the resulting design is shown in Fig. 3b. This design was then mapped onto the mesh of Fig.3c and the layout optimization continued with [p=2.5] with a tightened material usage constraint of $V_{material}/V_{structure} \leq 0.05$. The resulting design (Fig. 3d) was then projected onto the mesh of (Fig. 3e), and the optimization process continued with [p=4.0] and an even tighter material usage constraint of $V_{material}/V_{structure} \leq 0.015$. The design obtained from Fig. 3f. was projected onto the finest mesh and further optimized with the same powerlaw value and material usage constraint. The buckling eigenvalue on the finest mesh and the material layout are in close agreement with those from the previous mesh, and in this sense have converged.

Size reduction was essential to these computations. As the refinement process utilized finer and finer meshes, the analysis problem size remained very modest in size. For example, even on the finest mesh of 57,344 elements, the reduced problem size involved only 4,464 degrees of freedom, and the actual structural and eigenvalue analysis were performed on an HP J-class workstation in less than 10 cpu-seconds.

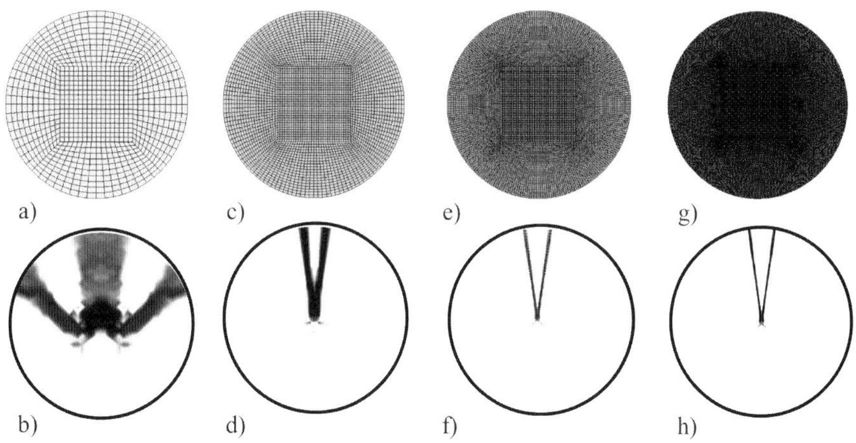

Figure 3. a) coarsest mesh of 896 elements; b) design with $V_{mat}/V_{structure}=0.20$ and $\lambda=2.17\cdot10^{3}$; c) mesh with 3,584 elements; d) material layout design with $V_{mat}/V_{structure}=0.05$ and $\lambda=1.20\cdot10^{2}$; e) mesh with 14,336 elements; f) material layout design with $V_{mat}/V_{structure}=0.015$ and $\lambda=1.00\cdot10^{1}$; g) mesh with 57,344 elements; h) material layout design with $V_{mat}/V_{structure}=0.015$ and $\lambda=1.05\cdot10^{1}$;

Sparse Compliant Mechanism Design Problem

In this final example, the importance of the proposed methodology is explored by considering the design of hinge-free gripper mechanism [(7)]. The design domain for this problem is shown in Fig. 4a. When a horizontal force is applied at the input port, the opposing output ports move vertically to pinch and thus grip a workpiece. The mechanism is designed with aluminum. In the first stage, the mechanism is designed utilizing a quadrilateral bilinear mesh of 5,000 elements and a material usage constraint of 12%. The resulting topology depicted in Fig.4b looks reasonable with no hinges in the resulting design; yet, such design is very stiff and very hard to use. Accordingly, a better design with higher flexibility can be achieved by using less material usage constraint; in this case, the material usage constraint is set at 3%. In order to capture a realistic performance with such a sparse system, the resulting topology is mapped onto a finer mesh of 22,500 elements and the resulting topology is shown in Fig.4c. In spite of this relatively fine mesh, it is still not capable of handling such a sparse system; therefore, it is very important to use a finer mesh. Thus, the resulting topology is then mapped onto a mesh of 90,000 elements, and the material usage constraint is kept at the same level of 3%; the resulting topology is shown in Fig. 4d. While this resulting topology looks much better and clearer than the previous one, more refinement is still needed in order to achieve a final design.

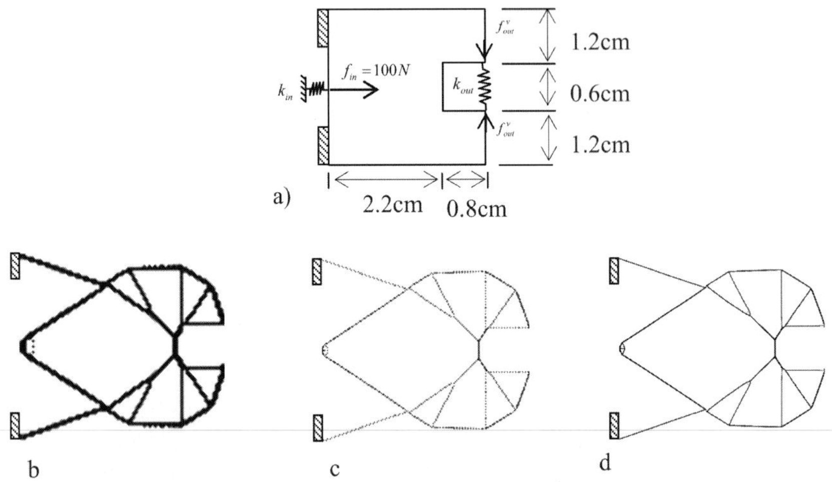

Figure 4. The gripper mechanism. a) the design domain and loading conditions; b) the resulting topology utilizing 5,000 quadrilateral bilinear elements with 12% material usage constraint; c) the resulting topology utilizing 22,500 finite elements and 3% material usage constraint; d) the resulting topology utilizing 90,000 elements and 3% material usage constraint.

Summary

In this work, a new methodology is introduced to solve large-size sparse systems in continuum topology optimization framework with relatively very low computational costs. The validity and performance of the proposed methodology has been demonstrated here in two examples involving linear and compliant mechanism design problems; however, the methodology has been successfully tested on numerous problems involving geometrical nonlinearity and buckling stability.

References

Bendsoe, M.P.; Kikuchi, N. 1988: Generating optimal topology in structural design using a homogenization method, *Comput. Meth. Appl. Mech. Engng*, 71, 197-224.

Bendsøe, M.P.; Sigmund, O. 1999: Material interpolations in topology optimization. *Archive of Applied Mechanics*, 69, 635-54.

Bruns, T. E.; Tortorelli, D. A. 2003: An element removal and reintroduction strategy for the topology optimization of structures and compliant mechanisms, *Int. J. Num. Meth. Engng*, 57(10), 1413-1430.

Buhl, T.; Pedersen, W.; Sigmund, O. 2000: Stiffness design of geometrically nonlinear structures using topology optimization, *Struct. Multidisc. Optim.*, 19, 93-104.

Bulman, S.; Sienz, J.; Hinton, E. 2001: Comparison between algorithms for structural topology optimization using a series of benchmark studies, *Computers and Structures*, 79(12), 1203-18.

Gea, H. C.; Luo, J. 2001: Topology optimization of structures with geometrical nonlinearities, *Computers and Structures*, 79, 1977-85.

Haber, R.B.; Jog, C.S.; Bendsøe, M.P. 1996: A new approach to variable-topology shape design using a constraint on perimeter. *Struct. Optim.* 11, 1–12.

Kosaka, I.; Swan, C.C. 1999: A symmetry reduction method for continuum structural topology optimization. *Computers and Structures*, 70(1), 47-61

Maute, K.; Schwarz, S.; Ramm, E. 1998: Adaptive topology optimization of elastoplastic structures, *Struct. Optim.* 15, 81-91.

Michell, A.G.M. 1904: The limits of economy in frame structures. *Philosophical Magazine* Sect. 6, 8(47), 589-597.

Rahmatalla, S.; Swan, C.C. 2003: Continuum topology optimization of buckling-sensitive structures, *AIAA J.* 41(5), 1180-1189.

Rahmatalla, S.; Swan, C.C. 2003: Form-finding of sparse structures using continuum topology optimization. *J. Struct. Engng.* 129(12), 1707-16.

Rahmatalla, S.; Swan, C.C. 2004: A Q4/Q4 continuum structural topology optimization implementation. *Struct. Multidis. Optim.* 27, 130-135.

Rahmatalla, S.F.; Swan, C.C., 2005: Sparse monolithic compliant mechanisms using continuum structural topology optimization, *Int. J. Num. Meth. Engng.*

Swan, C.C.; Kosaka, I. 1997: Voigt-Reuss topology optimization for structures with nonlinear material behaviors, *Int. J. Numer. Meth. Engng.*, 40, 3785-814.

DAMAGE DETECTION USING STATIC RESPONSE DATA AND OPTIMALITY CRITERION

Kevin Truman and Gus Terlaje
Department of Civil Engineering, Washington University in St. Louis, USA
E-mail: at1@cec.wustl.edu

Introduction

Damage of a structure is caused by many factors, some of which include earthquakes, wind, snow, increased live loads and simply the age of a structure in the form of fatigue or deterioration of its structural elements. Early detection of damage is vital in order to prevent the continuous degradation of the structure, ultimately leading to a catastrophic failure of the system. Traditional methods of detecting this damage are invasive and often require partial destruction of non-structural elements, such as wall panels and finish work. This detection process is executed blindly with little a priori knowledge of the possible location or extent of damage. Furthermore, damage often is not detected until it becomes so extensive that it is visible by building occupants. Structural health monitoring is employed to avoid unnecessary demolition through accurate prediction of damage location.

Background

Even though the area of structural health monitoring has been studied extensively in recent years, the majority of research and implementation has been based on damage detection using dynamic properties of the system. Using changes in vibration characteristics of a structure, damage may be detected and located (Salawu 1997). These vibrational characteristics are dependent on several system parameters such as; mass, stiffness, and damping, all of which require time and effort to obtain. Dynamic structural health monitoring has yielded significant results and therefore extensive research is justified, however requiring multiple system parameters creates inherent challenges in dynamic analysis. Because of the difficulties associated with dynamic analysis, a process using static properties merits exploration. Several methods have been developed using static responses to detect damage. Some methods combine both static and dynamic responses in order to locate damage. One such method predicts the location of damage by attempting to correlate expected and actual damage signs using first-order approximations for changes in static displacements and natural frequencies. Next a separate routine is used to determine the extent of damage at the predicted location (Wang et al.,2001). Although the results of such a procedure have been shown to be accurate for a planar truss and fixed-fixed beam, the effectiveness of the procedure for more complicated structures is not well known. Still other methods have been explored using solely static data. Such methods have investigated the correlation between measured displacements at various degrees of freedom and applied forces at other degrees of freedom in order to assemble the stiffness matrix of the system (Sanayei et al. 1991). Approximation techniques can then be used to obtain unknown displacement values. Such methods have accurately located and quantified damage for 2-D truss and beam element frames. A similar technique has been developed using unconstrained nonlinear optimization (Johnson et al. 2004). Displacements are incorporated into an error function which is dependent on only the cross sectional properties of each structural member. This method successfully detected damage in continuous beams and multi-bay, multi-story frames. Again, the effectiveness of this method is not known for other structural systems. The current theory will also present a method of detecting and quantifying damage using only static measurements in order to assemble the stiffness matrix of the structure. Once the damaged stiffness matrix is known, comparison with the healthy stiffness matrix can locate and quantify damage using the cross sectional properties of each structural element.

M. Pandey et al. (eds), Advances in Engineering Structures, Mechanics & Construction, 685–696.

Theory

Damage Detection

In order for a damage detection procedure to be practical to implement it must use easy to obtain information. Static response data, i.e. displacements resulting from static loads, can be obtained using electronic distance meters and can even be automated. Therefore it seems natural that static response data can serve well as a basis for a damage detection procedure. The measured displacements are a function of the loads applied to the structure and also the stiffness matrix of the structure. Damage must be quantified using the variables obtained from static response data.

What is damage?

Structural damage is usually associated with the inability of a structure to support loads. As a result, large displacements occur that normally would have been minimal. It is natural to associate the increased displacements of a structure with a decrease in overall stiffness. Therefore damage is defined as a reduction in stiffness of a structure. For this reason, the change in structural properties that comprise the stiffness matrix must be used as a measure for the location and severity of damage.

Structural Elements

The problem presented in this paper utilizes bending elements to construct the two dimensional frame structures. The stiffness matrix of a bending element is of the form:

$$K = \begin{bmatrix} \frac{12EI}{L^3} & \frac{6EI}{L^2} & \frac{-12EI}{L^3} & \frac{6EI}{L^2} \\ \frac{6EI}{L^2} & \frac{4EI}{L} & \frac{-6EI}{L^2} & \frac{2EI}{L} \\ \frac{-12EI}{L^3} & \frac{-6EI}{L^2} & \frac{12EI}{L^3} & \frac{-6EI}{L^2} \\ \frac{6EI}{L^2} & \frac{2EI}{L} & \frac{-6EI}{L^2} & \frac{4EI}{L} \end{bmatrix}. \qquad \textbf{Eq(1)}$$

By utilizing bending elements, the stiffness matrix is dependent on only the length, modulus of elasticity, and moment of inertia of each element. In reality the length of each element can be determined either from structural drawings or direct measurement of each member and the modulus of elasticity can be taken as the industry standard. Therefore, these values can be considered constant for both the healthy and damaged element. On the other hand, the moment of inertia of each element can change between the healthy and damaged member. By assembling the global stiffness matrix of the damaged structure and extracting the moment of inertia values, comparison with the healthy stiffness matrix values allows damage to be located and quantified.

Frame Assembly

A typical one-bay, one-story frame comprised of 3 bending elements is shown in Figure1(a). As shown, the frame has 3 degrees of freedom (DOF), two rotations and one lateral displacement. The damage detection procedure will require loads be applied at a subset of DOF and displacement measurements be taken at another subset of DOF. Although mathematically feasible, practical applications prevent concentrated moments be applied to a structure and additionally rotational displacements can be difficult to measure. Therefore, the frame shown in Figure1(a) is not a good model to use in actual damage detection routines. As a solution, rather than using a single bending element to connect each column span, two elements are used. This modification creates two additional degrees of freedom at the joint between each element, as shown in Figure1(b). As can be seen, now a vertical degree of freedom exists where loads can be applied and displacements can be measured. To reiterate, the additional DOF are not required for the mathematical formulation, but it does create a damage detection procedure that is more suitable to conversion to actual practice as it is less difficult to obtain vertical displacements than it is to obtain rotational displacements.

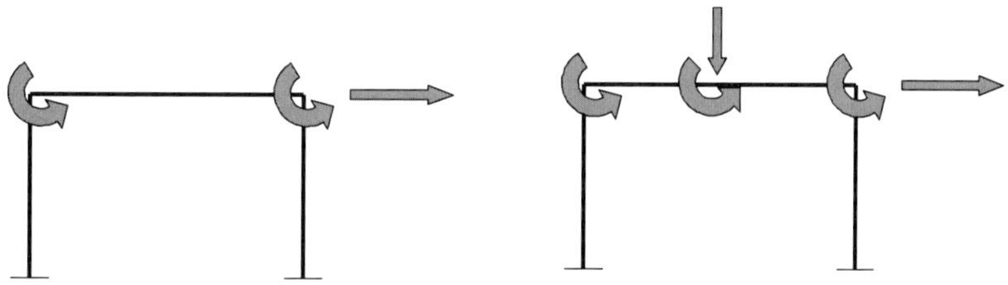

Figure 1(a,b). One-bay, One-story Frame Comprised of Bending Elements

Why use optimization?

As stated, the structures being considered in this paper are formed using bending elements, whose stiffness is a function of the moment of inertia of each structural member. Resulting is an engineering design problem with two main variables; a set of design variables (moment of inertia values) and a set of analysis values (measured displacements). Unfortunately, without an entire set of measured displacements at all DOF, there is not enough information to determine the design variables directly from the analysis values. As such, an optimization algorithm must be utilized to determine all unknown design variables. This can be accomplished by using constrained nonlinear optimization.

Fundamental to the optimization process is the force displacement relationship,

$$P = Ku\ , \qquad \textbf{Eq(2)}$$

where P is the static load vector applied to the structure, K is the global stiffness matrix, and u is the displacement vector of the structure. With knowledge of the healthy stiffness matrix of a structure, a set of displacements for all DOF can be calculated for any set of applied static loads. Next, by applying the same static loads to the present, damaged structure and measuring strategic displacements, insight can be gained as to which structural elements have reduced in stiffness. A general explanation of the optimization routine to gain this insight will be explained next.

Optimization

Because the moment of inertia of a damaged member can take on any positive real value, the structure contains continuous design variables and, in turn, can be formulated to have continuous constraint and objective functions. This engineering problem lends itself well to gradient-based optimization. Several gradient based optimization procedures have been developed over the years, but most follow a similar form. The typical constrained optimization problem has the form:

$$\begin{aligned} &Minimize \rightarrow \Phi = f(x_1, x_2, x_3, ..., x_n) \\ &s.t. \rightarrow g_i(x_1, x_2, x_3, ..., x_n) = 0 \end{aligned} \qquad \textbf{Eq(3a-b)}$$

The classical approach to solving constrained optimization is the method of Lagrange multipliers. This approach converts the constrained optimization problem into an unconstrained problem (Kirsch 1993). The objective function and constraint equations can be used to create another scalar valued function, known as the Lagrangian, which takes the form:

$$L = \Phi + \sum_{j=1}^{m} \lambda_j g_j . \qquad \textbf{Eq(4)}$$

where,

λ_j = lagrange multiplier for the jth constraint

By giving a set of necessary conditions to identify optimal points of equality constrained problems, each stationary point found by the optimization routine can be guaranteed to be a local minimum. These conditions, known as the Karush-Kuhn-Tucker conditions are as follows:

$$\nabla L = \nabla \Phi + \sum_{j=1}^{m} \lambda_j \nabla g_j = 0 \qquad \textbf{Eq(5)}$$

$$\lambda_j g_j = 0$$

The aforementioned conditions provide a powerful means to verify solutions. Next, the gradient of each constraint equation can be written as:

$$g_j(x + \Delta x) - g_j(x) = \sum_{i=1}^{n} \frac{dg_j}{dx_i} \Delta x_i \qquad \textbf{Eq(6)}$$

Note, Eq(5) and Eq(6) provide sufficient equations to solve for all lagrange multipliers.

Side Constraints

A useful concept in constrained optimization is the idea of side constraints. As an example, although mathematically feasible, a negative moment of inertia for a structural element is physically meaningless. As a result, the current problem must contain a subroutine within the optimization process that prevents design variables from taking on negative values. Side constraints restrict the range of values that each design variable can assume. Each iteration of the optimization process can check as to whether a design variable has left the feasible range. When such an event occurs, the algorithm can repeat the current iteration but artificially set the design variable to the maximum or minimum of the feasible range. In turn, the algorithm can adjust the gradient calculations to reflect the restriction on the design variable and the remaining design variables in the feasible range will change accordingly (Cheng et al.). An optimal solution can then be found with each design variable remaining within the feasible range.

Linking

Just as it may be desirable to restrict design variables to a certain range of values, it may also be desirable to force several design variables to take on the same value for every iteration of the optimization process. This is accomplished through linking. For example, the current formulation uses two bending elements to span a single bay. Although two elements are used, in reality this bay would be spanned by a single beam. In turn, the two separate elements should be forced to take on the same design value. Again, within the optimization algorithm it is possible to represent multiple design variables with a single variable yet still account for the change in both structural elements within the gradient calculations (Cheng et al.). Although both beam elements can be represented by a single variable, the change in cross sectional properties of each element must be accounted for as they both have an effect on the convergence to a local minimum of the objective function subject to the constraint equations.

Scale factor

To aid convergence of the optimization algorithm, it is beneficial to include a scale factor that will automatically scale all design variables to produce a structure that will satisfy at least one constraint equation exactly. To illustrate, the current problem will use displacement measurements as constraints. At the beginning of each iteration, the displacements will be calculated using the same load cases used to produce the measured displacements for the damaged structure. The most violated displacement will be used as the basis for the scale factor. For example, say at the beginning of the first iteration, where all design variables start at those of the healthy structure, the **calculated** displacement at the first degree of freedom is half as large as the **measured** displacement at that same degree of freedom. In order for the calculated and measured displacements to match, all design variables must be cut in half, thus reducing the stiffness of the structure by a factor of 2. As a result, the displacement constraint on the first DOF will automatically be satisfied. A scale factor is calculated and applied to the initial design variables of each iteration.

Damage Detection Problem

Objective function

In order to create an optimization routine that can locate and detect damage, an objective function must be found that achieves a minimum value at the damaged state of the structure. Without prior knowledge of the damage state it is difficult to create an objective function that reaches a minimum value at all possible damage states. In order to solve this challenge, it was hypothesized that the damage state of the structure would be that which produces the least damage compared to the healthy structure. Since damage has been defined as a reduction in stiffness of the structure, the objective function is written as to minimize the total reduction in stiffness. Although bending elements are used for each element, it was found that the reduction of the shear stiffness of each element produced the best results. As a result, the objective function is written as:

$$\Phi = \sum_{i=1}^{n} \frac{\Delta I_i}{L_i}. \qquad \textbf{Eq(7)}$$

where,

I_i = moment of inertia of the ith member
L_i = length of the ith member

Since each change in moment of inertia is divided by the length of each respective member, the function inherently weights the change in cross section properties according to the length of each member, thereby taking into account the affect each member has on the total stiffness of the structure.

Constraints

When loads are applied to the structure, naturally displacements occur at each degree of freedom. These displacements are a function of the applied loads, which are known, and the stiffness matrix of the damaged structure, which is not known. Using Eq(1), the displacements can be solved for in terms of the known and unknown values by:

$$u = K^{-1} * P. \qquad \textbf{Eq(8)}$$

If every displacement were measured this equation could easily be solved to find the damaged stiffness matrix. However, as stated earlier, the objective is to develop a damage detection procedure that is easy, accurate and practical. It is impractical to measure every displacement as one, it is time consuming, and two, often impossible due to obstructions. What remains is an expression with more unknowns than equations. Fortunately, strategic displacement measurements coupled with the correct objective function will provide a good basis that will determine the damaged stiffness matrix and therefore the cross-sectional properties of each member.

The number of constraints is equal to the number of measured displacements for all load cases. Using the measured displacements of the structure and Eq(8), equality constraints can be created that must be satisfied by the optimization procedure given by:

$$K^{-1} * P = u_{measured}. \qquad \textbf{Eq(9)}$$

where,

$u_{measured}$ = vector of measured displacements
K = global stiffness matrix
P = Load vector used to produce measured displacements

Next, in order to write the constraints in a form suitable for optimization, i.e. Eq(3b), Eq(9) can be written as:

$$(K^{-1} * P) - u_{measured} = 0. \qquad \textbf{Eq(10)}$$

Thus, Eq(10) provides a set of constraint equations that must be satisfied at the optimum set of design variables.

Side Constraints

In order to avoid convergence to a physically meaningless point, it was desired to include side constraints on each design variable. First, as explained earlier side constraints were applied to each design variable preventing negative values. Second, it is physically impossible for damage to create additional moment of inertia for each structural element, therefore an upper bound equal to the value of the healthy moment of inertia was placed on each design variable.

Even though complete damage or collapse of a member would reduce its moment of inertia to zero, it was not desirable to set the lower side constraint to zero in this process. This is because if the optimization process follows a path that forces one of the design variables to pass through zero, as will be seen, the recursive formula which predicts the design values for each subsequent iteration will be unable to change any design variable once it has been set to zero. To solve this apparent problem, the lower bound was set to a value that is small enough to be insignificant in reality but large enough to carry mathematical weight in the optimization process. Therefore, the lower side constraint was set to 5in^4 for every member.

Problem Statement

Using the above formulations the engineering problem becomes:

$$\begin{aligned} &\text{Min } \Phi = \sum_{i=1}^{n} \frac{\Delta I_i}{L_i} = \sum_{i=1}^{n} \frac{I_i - I_{i\_initial}}{L_i} \\ &\text{s.t. } \{u\} - \{u_{measured}\} = 0 \\ &\text{s.t. } 5 \le I_i \le I_{i\_healthy} \end{aligned} \qquad \textbf{Eq(11a-c)}$$

As before, the optimality criterion must be developed for the current engineering problem. To do this the Lagrangian must be written using the objective function and constraint equations. With the inclusion of lagrange multipliers for each constraint, the Lagrangian becomes:

$$L = \sum_{i=1}^{n} \frac{\Delta I_i}{L_i} + \sum_{j=1}^{m} \lambda_j (u_j - u_{j\_measured}), \qquad \textbf{Eq(12)}$$

where u_j are the calculated displacements of the healthy structure. From this, the necessary condition for optimality can be found by:

$$\nabla L = 0 \qquad \textbf{Eq(13)}$$

In turn, a local minimum of Eq(12) occurs when

$$\frac{dL}{dI_i} = -\frac{1}{L_i} + \sum_{j=1}^{m} \lambda_j \frac{du_j}{dI_i} = 0 \quad \text{(for } i\text{=1,2,...,n)} \qquad \textbf{Eq(14)}$$

Eq(14) can also be used as a barometer for how close each intermediate point within the optimization routine is to the optimal solution. Thus, a recursive formula can be created to determine the design variables for the next iteration based on the optimality criterion of the previous iteration (Cheng et al.). Manipulation of Eq(14) gives,

$$T_i = L_i \sum_{j=1}^{m} \lambda_j \frac{du_j}{dI_i}. \qquad \textbf{Eq(15)}$$

When the optimality criterion is satisfied T_i is equal to 1.

In order to solve for each lagrange multiplier, gradients of each constraint equation must be calculated in order to provide a system of equations sufficient to solve for all unknowns (Haug et al. 1979). From the definition of a derivative, the gradient of the constraint equations can be written as such:

$$\frac{g_j(I + \Delta I) - g_j(I)}{\Delta I} = \sum_{i=1}^{n} \frac{dg_j}{dI_i} \qquad \textbf{Eq(16)}$$

Simple manipulation of Eq(16) yields:

$$g_j(I + \Delta I) - g_j(I) = \sum_{i=1}^{n} \frac{dg_j}{dI_i} \Delta I_i \qquad \textbf{Eq(17)}$$

The ΔI term that appears in Eq(17) is determined using Eq(15). The change in I can be calculated using the recursive formula:

$$I_i^{k+1} = I_i^k + \frac{1}{r}(T_i - 1) I_i^k \qquad \textbf{Eq(18)}$$

Solving for ΔI yields:

$$\Delta I = I_i^{k+1} - I_i = \frac{1}{r}(T_i - 1)I_i^k \qquad \textbf{Eq(19)}$$

Substituting Eq(19) in to Eq(17) yields:

$$-g_j(I) = \sum_{i=1}^{n} \frac{dg_j}{dI_i} \frac{1}{r}(T_i - 1)I_i, \qquad \textbf{Eq(20)}$$

where r is a convergence control parameter that restricts the step size for each iteration (Cheng et al.). The control parameter is assumed to be 2 for all problems covered in this paper. Manipulation of Eq(20) yields:

$$rg_j - \sum_{i=1}^{m} \frac{dg_j}{dI_i} I_i = \sum_{i=1}^{m} -\frac{dg_j}{dI_i} T_i I_i \qquad \textbf{Eq(21)}$$

Eq(21) provides n equations for n unknown Lagrange multipliers, which can be used in Eq(15) and Eq(19) to determine the design values for the next iteration. The cycle is repeated until the optimality criterion is met.

Example

A two-bay frame is pictured in Figure 2. It is used to demonstrate the importance of the location of the applied loads in the damage detection process. Each structural element is assumed to be made of steel with modulus of elasticity 29,000 ksi and dimensions shown in Figure 2.

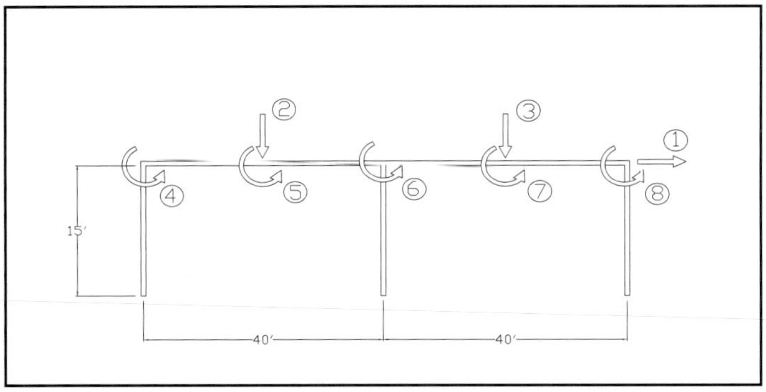

Figure 2. Two-bay Frame

Table 1 shows the initial and final design variables of each structural element. For this example, all elements will experience damage. All initial values are assumed to be known and will serve as the starting point for each member size. The final design values are not known, however, and will be determined by the optimization routine.

Table 1. Initial and Final Design Values

Member	$I_{initial}(in^4)$	$I_{final}(in^4)$
1	3500	1500
2	3500	1000
3	3500	1000
4	3500	1000
5	3500	1000
6	3500	700
7	3500	700

Case 1

For the first instance, 5 different load cases are used to measure displacements of the damaged structure. Loads are applied individually at each of the first 5 DOF and the resulting displacement at the same DOF is measured. The applied load and resulting displacement for each of the 5 DOF is shown in Table 2.

Table 2. Load Cases and Resulting Displacements for Case 1 & 2

Load Case	DOF Loaded	LOAD	DOF Measured	Displacement
1	1	10 k	1	0.10743 in
2	2	10 k	2	0.28251 in
3	3	10 k	3	0.30796 in
4	4	10 k-in	4	0.000017438 in/in
5	5	10 k-in	5	0.000012803 in/in

Each measured displacement is used as a constraint equation given by Eq(10). Table 3 shows the results of the optimization process using 5 displacements as constraint equations. As can be seen, the optimization routine converged to the damage state in 5 iterations. The first 7 columns of the last row are identical to the final design values shown in Table 1. It is also observed that all T_i values converged to 1. From the previous formulation (Eq.14-15) this signifies that the optimality criterion has been met and the current design variables are those that minimize the objective function and satisfy the constraint equations. The results are promising in that only 5 measured displacements were required for convergence to the correct damage state rather than all 8 measured displacements. However, two of the five measured displacements required not only rotation measurements but also concentrated moments to be applied, which can be difficult. In order for the damage detection procedure to be practical, these complications must be eliminated.

Table 3. Optimization Results for Case 1

Iteration	I_1	I_2	I_3	I_4	I_5	I_6	I_7	Obj	λ_1	λ_2	λ_3	λ_4	λ_5	T_1	T_2	T_3	T_4	T_5	T_6	T_7
1	3500	3500	3500	3500	3500	3500	3500	88.83	-45.864	-7.6083	-29.748	-23246	-920240	1.9243	1.2886	1.2886	1.2881	1.2881	0.6796	0.9914
2	1221	955.6	955.6	955.4	955.4	701.3	831.5	85.44	22.267	-6.6294	-27.925	-268290	-1E+06	1.3802	1.0788	1.0788	1.0795	1.0795	0.9998	0.7224
3	1453	993.3	993.3	993.4	993.4	701.2	716.1	84.16	6.2744	-10.394	-27.727	-238190	-1E+06	1.0627	1.0132	1.0132	1.013	1.013	0.9966	0.9562
4	1499	999.8	999.8	999.8	999.8	700	700.4	83.9	1.1466	-11.446	-27.548	-232980	-1E+06	1.0017	1.0004	1.0004	1.0003	1.0003	0.9999	0.9987
5	1500	1000	1000	1000	1000	700	700	83.89	0.9853	-11.478	-27.543	-232820	-1E+06	1	1	1	1	1	1	1

Case 2

If the number of displacement measurements is reduced to 4, the final design values are not close to the actual damage state. Case 2 uses the same initial and final design values and the same loading cases, however only the first 4 DOF measurements are used. The results of reducing the number of measured displacements to 4 are shown in Table 4. The structural members spanning each bay, namely I_2, I_3, I_4, and I_5, converged to values within 8% of their actual values and I_6 to within 2.1%. However, I_1 and I_7 were grossly misjudged with error in excess of 50% for I_1 and 70% for I_7.

Table 4. Optimization Results for Case 2

Iteration	I_1	I_2	I_3	I_4	I_5	I_6	I_7	Obj	λ_1	λ_2	λ_3	λ_4	T_1	T_2	T_3	T_4	T_5	T_6	T_7
1	3500	3500	3500	3500	3500	3500	3500	88.83	-167.66	-24.429	-25.749	224600	1.044	1.3504	1.3504	1.2011	1.2011	0.6952	1.5845
2	853.5	981.4	981.4	919.1	919.1	707.9	1079	86.16	-146.39	-27.873	-21.749	-37047	0.8936	1.085	1.085	1.0261	1.0261	1.0154	1.1011
3	808.1	1023	1023	931.1	931.1	713.3	1134	85.631	-147.38	-27.955	-21.3	-41909	0.9229	1.0127	1.0127	0.995	0.995	1.0014	1.0453
4	776.9	1030	1030	928.7	928.7	713.8	1159	85.625	-147.44	-28.131	-21.118	-42553	0.9534	1.0059	1.0059	0.9967	0.9967	1.0008	1.0259
5	758.9	1033	1033	927.2	927.2	714.1	1174	85.627	-147.39	-28.253	-21.013	-42938	0.9724	1.0035	1.0035	0.9981	0.9981	1.0005	1.0149
6	748.4	1034	1034	926.3	926.3	714.2	1183	85.628	-147.35	-28.325	-20.952	-43182	0.9839	1.002	1.002	0.9989	0.9989	1.0003	1.0086
7	742.3	1036	1036	925.8	925.8	714.3	1188	85.629	-147.32	-28.367	-20.917	-43329	0.9906	1.0012	1.0012	0.9994	0.9994	1.0002	1.0049
8	738.9	1036	1036	925.5	925.5	714.4	1191	85.629	-147.3	-28.392	-20.896	-43416	0.9946	1.0007	1.0007	0.9996	0.9996	1.0001	1.0028
9	736.9	1036	1036	925.3	925.3	714.4	1193	85.629	-147.29	-28.406	-20.885	-43467	0.9969	1.0004	1.0004	0.9998	0.9998	1.0001	1.0016
10	735.7	1037	1037	925.2	925.2	714.4	1194	85.629	-147.28	-28.414	-20.878	-43496	0.9982	1.0002	1.0002	0.9999	0.9999	1	1.0009

Case 3

At this point several alterations to the method can be made, two of which are to find an alternate objective function to be minimized or to try alternate loading cases from which to measure displacements. The former has been explored, using the minimization of strain energy, and minimizing the reduction in bending stiffness. However results have not shown to improve beyond Case 2. The latter alternative has proven to produce favorable results with a similar number of measured displacements. Table 5 shows an alternate loading condition with the resulting displacements using the final design values in Table 1.

Table 5. Load Cases and Resulting Displacements for Case 3

Load Case	DOF Loaded	LOAD	DOF Measured	Displacement
1	2	10 k	1	-0.024894 in
2	2	10 k	2	0.28251 in
3	2	10 k	3	-0.055233 in
4	2	10 k	4	0.00025174 in/in

For Case 3, a 10 kip load is applied at only degree of freedom 2 and displacements are measured at the first 4 DOF. Again the optimization routine is executed in an identical manner as the previous two cases and the results are shown in Table 6.

Table 6. Optimization Results for Case 3

Iteration	I_1	I_2	I_3	I_4	I_5	I_6	I_7	Obj	λ_1	λ_2	λ_3	λ_4	T_1	T_2	T_3	T_4	T_5	T_6	T_7
1	3500	3500	3500	3500	3500	3500	3500	100.54	-345.81	-19.46	207.68	-18825	3.1223	1.9186	1.9186	1.067	1.067	0.8183	1.8332
2	997.1	705.9	705.9	499.9	499.9	439.8	685.3	94.828	-924.66	-27.458	524.62	-36304	2.2241	1.5161	1.5161	1.4795	1.4795	1.2379	1.1959
3	1607	888.1	888.1	619.8	619.8	492.1	752.4	91.626	-1708.4	-16.893	790.29	-85194	1.3868	1.4052	1.4052	1.9392	1.9392	1.7721	1.1523
4	1691	941.4	941.4	802.9	802.9	601.2	713.7	85.433	-2055.6	-21.505	734.75	-144990	0.8756	1.1042	1.1042	1.3687	1.3687	1.2473	0.978
5	1586	990.5	990.5	950.9	950.9	675.6	705.8	84.161	-1889.3	-28.551	572.51	-159140	0.9451	1.0229	1.0229	1.0792	1.0792	1.0499	0.9987
6	1535	997	997	983.8	983.8	689.1	702	83.92	-1829.8	-30.415	533.78	-159880	0.988	1.0028	1.0028	1.0119	1.0119	1.0105	0.999
7	1525	998	998	989.2	989.2	692.5	701.3	83.893	-1818.7	-30.74	527.84	-159730	0.9977	1.0003	1.0003	1.0016	1.0016	1.0016	0.9998
8	1523	998.1	998.1	990	990	693	701.2	83.892	-1816.7	-30.793	526.99	-159660	0.9997	1	1	1.0002	1.0002	1.0002	1
9	1523	998.2	998.2	990.1	990.1	693.1	701.2	83.892	-1816.4	-30.799	526.9	-159650	1	1	1	1	1	1	1

The optimization routine did not identically match the actual damage state given in Table 1, however the found values are all within 2% of the actual damaged values. By adjusting the loading condition when measurements were taken, results improved compared to Case 2 and one rotational degree of freedom measurement and both concentrated moment loads were eliminated compared to Case 1. Utilizing a loading condition with only vertical loads also provides a process better suited for implementation on real structures.

Conclusion

As shown from examples presented, the damage detection process is highly dependent on the load cases used to produce the measured displacements utilized by the optimization routine. In order for the damage detection process to be easily implemented it is desired that only vertical loads and vertical displacements be required for the damage detection procedure. The example presented is just one example of the possible load cases that can be used to measure displacements. Future work will include the exploration of additional load cases coupled with strategic displacements measurements in order to determine the effectiveness with regards to damage detection.

The optimization routine lends itself well to such engineering problems as all design variables are continuous. Furthermore, the results show quick convergence to a local minimum as signified by the satisfaction of the optimality criterion. Since so few iterations are required computational effort is kept to a minimum. Minimizing computing effort will prove beneficial for damage detection in larger structures.

Acknowledgements

The research presented was funded by the U.S. Department of Education through the GAANN fellowship program.

References

Cheng, Franklin Y., and Truman, Kevin Z. Optimal Design of 3-D Reinforced Concrete and Steel Buildings Subjected to Static and Seismic Loads Including Code Provisions. *Civil Engineering Study Structural Series 85-20 Department of Civil Engineering University of Missouri-Rolla. Rolla, Missouri*. pp. 67-105.

Doebling, S.W., et al. (1998). A Review of Damage Identification Methods that Examine Changes in Dynamic Properties. *Shock and Vibration Digest, Vol. 30, No. 2*, pp. 91-105.

Haug, Edward J. and Arora, Jasbir S. (1979). *Applied Optimal Design: Mechanical and Structural Systems*. New York, New York: Wiley-Interscience. pp. 319-328.

Hjelmstad, K.D. and Shin, S. (1997). Damage Detection and Assessment of Structures from Static Response. *Journal of Engineering Mechanics, Vol. 123, No. 6*, pp. 568-576.

Johnson, L., Truman, K. (2004). Damage Detection Using Static Responses and Unconstrained, Nonlinear Optimization. *Masters Thesis. Department of Civil Engineering at Washington University in St. Louis.*

Kirsch, U. (1993). *Structural Optimization: Fundamentals and Applications.* Berlin, Germany: Springer-Verlag.

Salawu, O.S. (1997). Detection of Structural Damage Through Change in Frequency: A Review. *Engineering Structures, Vol. 19, No. 9*, pp. 718–723.

Sanayei, M. and Onipede, O. (1991). Damage Assessment of Structures Using Static Test Data. *AIAA Journal, Vol. 29, No. 7*, pp. 1174-1179.

Wang, X. et al. (2001). Structural Damage Identification Using Static Test Data and Changes in Frequencies. *Engineering Structures, Vol. 23, No. 6*, pp. 610-621.

Engineering Reliability

SYSTEM RELIABILITY ASSESSMENT OF STEEL GIRDER BRIDGES

Artur A. Czarnecki and Andrzej S. Nowak
University of Nebraska
W181 Nebraska Hall, Lincoln, NE 68588-0531, USA
E-mail: arturcz@umich.edu, anowak1@unl.edu

Abstract

Reliability can be considered as a rational evaluation criterion in assessment of bridge structures. The traditional element-based approach to bridge design and evaluation does not allow for consideration of interaction between the components that form a structural system and, therefore, it can be conservative. Safety of the structural system also depends on the degree of redundancy (load sharing) and ductility. As a result, it has been observed that the load carrying capacity of the whole structure can be much larger than what is determined by the design of individual components. Therefore, this paper is focused on the system behavior. The objective is to formulate a limit state function for the whole bridge, identify the load and resistance parameters, and develop an analysis procedure to assess the reliability of the bridge as a structural system. The major steps of the procedure include selection of representative structures, formulation of limit state functions, development of load and resistance models, development of the reliability analysis method, reliability analysis of selected bridges, and formulation of recommendations for practical bridge assessment. The live load is considered in form of a design truck. The analysis is performed for different values of span length, truck position (transverse and longitudinal), number of vehicles on the bridge (multiple presence), girder spacing, and stiffness of structural members (slab and girders). For each combination of these parameters, the bridge resistance is determined in terms of the weight of a truck (or trucks) causing an unacceptable deflection or instability of the considered bridge. The reliabilities are also calculated for individual components (girders) and compared to system reliabilities of the bridge. The resulting system reliability can serve as a tool in the development of a rational bridge design and evaluation procedure.

Keywords: Reliability index, System reliability, Steel girder bridges, Loads, Resistance

Introduction

There is a growing interest in the development and improvement of efficient procedures for the design, evaluation, repair and rehabilitation of bridge structures. Structural reliability can be considered as a rational measure of performance, as it is a function of uncertainties associated with loads and load carrying capacity (resistance). A traditional deterministic approach is based on the analysis of individual components. The capacity of the structure is reached when any of the components reaches its ultimate capacity. However, in practice, due to ductility and redundancy, there can be redistribution of load resulting in an increased load carrying capacity. The margin safety can be assessed more accurately using the system reliability approach.

Therefore, this paper deals with calculation of the reliability for the whole bridge, rather than individual components (girders). The main objective is to compare the reliability of components (girders) and system for steel girder bridges. The limit state function is formulated for the whole bridge, load and resistance parameters are identified, and the reliability analysis procedure is developed. The developed methodology is demonstrated on representative steel girder bridges. The recommendations are formulated for a practical bridge assessment.

M. Pandey et al. (eds), Advances in Engineering Structures, Mechanics & Construction, 699–710.

The ultimate limit states (moment and shear carrying capacities) are considered. The structural performance is determined as a function of load components, strength of material, and dimensions. The reliability analysis is performed for different values of span length, truck position (transverse and longitudinal), number of vehicles on the bridge (multiple presence), girder spacing, and stiffness of structural members (slab and girders). For each combination of these parameters, the bridge resistance (system resistance) is determined in terms of the gross vehicle weight, GVW, of two side-by-side trucks, with axle configuration of the design truck in AASHTO (2004). It is assumed that the ultimate capacity (maximum GVW) is reached when the resulting deflection exceeds the critical limit. The analysis is performed separately for each transverse position of two trucks.

Structural Analysis Model

The structural analysis is carried out using a nonlinear model. This allows for stress and load redistribution that can occur as the structure passes from an initial localized yielding in a member, to reaching the ultimate capacity of that member, and finally to a total bridge collapse. Moreover, the use of inelastic analysis allows to consider material nonlinearity, geometric nonlinearity and also boundary nonlinearity. In this study, the analysis was performed using ABAQUS, a commercially available finite element program. Material and other structural parameters are based on the collected information from the literature supplemented with engineering judgment.

For the purpose of finite element analysis, the geometry of bridge superstructure can be idealized in many different ways. For this study, a three-dimensional finite element method was applied to investigate the structural behavior of composite bridges. A concrete slab is modeled using isotropic, eight node solid elements, with three degrees of freedom at each node. The girder flanges and webs are modeled using three-dimensional, quadrilateral, four node shell elements with six degrees of freedom at each node. The reinforcement is represented by uniformly distributed layers of steel. The model assumes a complete connection between girders and concrete slab with no slip. Secondary elements were excluded from the analysis. The details of the model are given by Czarnecki and Nowak (2005a-c). An example of an FEM mesh for a bridge with four girders spaced at 3 m is shown in Figure 1.

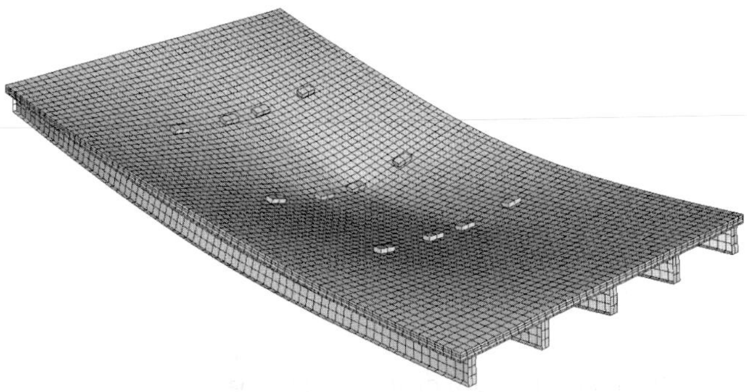

Figure 1 Deflected Shape, Finite Element Bridge Model

All investigated structures were designed as simply supported composite bridges in accordance with AASHTO LRFD (2004) Strength I limit state for flexure and shear. Span lengths ranging from 12 m to 42 m, with the intervals of 6 m, are considered. For each of the span lengths, three girder spacings are investigated: 1.8, 2.4, and 3 m. For all considered bridges, the longitudinal axis is assumed at right angle to the abutment. All bridges are designed as two-lane structures, and with deck slab thickness of 225 mm. Previous studies showed (Eamon and Nowak

2002 and 2004) that the effect of diaphragms on ultimate moment capacity is insignificant, therefore, the diaphragms are not considered in this study.

In the analysis, the load was applied in form of two side-by-side design trucks (AASHTO 2004). In the longitudinal direction, the trucks were positioned to generate the maximum bending moment. Different transverse positions were considered but the centerlines of the wheels of two adjacent trucks were placed no closer than 1.2 m.

Structural Reliability Analysis and Limit State Functions

The available reliability methods are presented in several publications (e.g. Nowak and Collins 2000) and, therefore, the details are not discussed in this paper. The methods vary with regard to accuracy, required input data, and computational effort. Most of the available procedures are suitable for a reliability analysis for individual components rather than structural systems. System reliability methods are more complex but they offer considerable advantages, and therefore they are the subject of the present paper.

In case of the component reliability, it is almost always possible to define a performance function g(X) of the basic random variables X such that $g > 0$ corresponds to a satisfactory performance and $g < 0$ corresponds to failure of a structure. On the other hand, the formulation of limit state function for the entire bridge is much more complex and requires a special approach. This is because of, among other things, the possible statistical dependence among the random variables, load redistribution after some members' failure, redundancy of the structure that is causing a load sharing.

So far, a number of approaches have been proposed that allow to define a limit state function, and consequently failure, for entire bridge. Zhou (1987) proposed that system failure occurs when two adjacent girders fail. Tabs and Nowak (1991) considered several girders must reach their ultimate capacity before the structure collapses. Ghosn and Moses (1998) defined the bridge resistance as the maximum gross vehicle load that is causing the formation of a collapse mechanism. Enright and Frangopol (1999) studied a number of system models for a five girder bridge. Estes and Frangopol (1999) assumed that failure occurs when three adjacent girders out of five fail. Liu and Moses (2001) considered a damaged steel girder bridge. It was assumed that damage can be caused by corrosion, collision, etc.

In this paper, however, it is assumed that the failure of a bridge is defined as a maximum load that the bridge can carry, or as 0.0075 of the span length deflection in any of the main members of the bridge, whichever governs. The deflection is calculated only due to live load, including static and dynamic components.

Bridge Load Models

The major load components for highway bridges are deal load, live load, dynamic load, environmental loads (temperature, wind, earthquake), and other loads (collision, braking). In this paper, however, only the first three are considered. Consideration of live load involves not only the weight of trucks, but also the distribution factor (fraction of the total truck load per girder), and truck position within the roadway (curb distance). The load models are based on the available statistical data, surveys, inspection reports, and analytical simulations. The load variation is described by cumulative distribution function, mean value or bias factor (ratio of mean to nominal value), and coefficient of variation.

Dead load, DL, is the gravity load due to self-weight of the structural and nonstructural components permanently attached to the bridge. Therefore, it includes the weight of girders, deck slab, wearing surface, barriers, sidewalks, and diaphragms, when applicable. The statistical parameters for dead load were selected from the available literature (Nowak 1999). Four components are considered: DL_1 – weight of factory made elements, DL_2 – weigh of cast-in-place concrete, DL_3 – weight of wearing surface (asphalt), and DL_4 – weight of miscellaneous items (e.g., railing, luminaries). All components of dead load are treated as normal random variables. For DL_1, the bias factor, $\lambda = 1.03$, and coefficient of variation, $V = 0.08$; for DL_2, $\lambda = 1.05$, and $V = 0.10$; for DL_4, $\lambda = 1.03 \sim 1.05$, and $V = 0.08 \sim 0.10$,; and for asphalt wearing surface it is assumed that the mean thickness is 75 mm and $V = 0.25$.

The live load model was developed in conjunction with calibration of the AASHTO LRFD Code (Nowak and Hong 1991; Nowak 1993). The statistical parameters (mean values, bias factors and coefficients of variation) are derived for the maximum lane moments and shears. The multiple presence of trucks is considered by using the observed frequencies of occurrence of two vehicles in the same lane or side-by-side. For a single-lane loaded case, the ratio of the mean maximum 75-year moment to AASHTO HL-93 design moment varies from 1.3 for shorter spans (10 m) to 1.2 for longer spans (50 m), while coefficient of variation, $V = 0.11$ for all spans. For the two-lane loaded case, bias factor for each truck varies from 1.2 for shorter spans (10 m) to 1.0 for longer spans (50 m), while coefficient of variation, $V = 0.11$ for all spans.

The basic load combination includes dead load and live load (static and dynamic). Live load is represented in form of a design truck as shown in Figure 2. It is assumed that the gross vehicle weight (GVW) is a random variable, but the axle spacing and percentage of the total load per axle remain constant. The transverse position of the truck within the roadway (curb distance) is also treated as a random variable. An example of the probability density function (PDF) of the curb distance is shown in Figure 3, for two traffic lanes. Each curve represents a curb distance for a line of wheels, spaced at 1.8m for a truck.

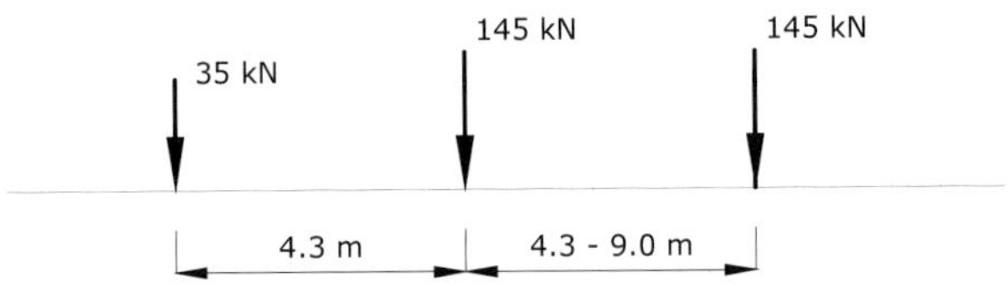

Figure 2 Design Truck

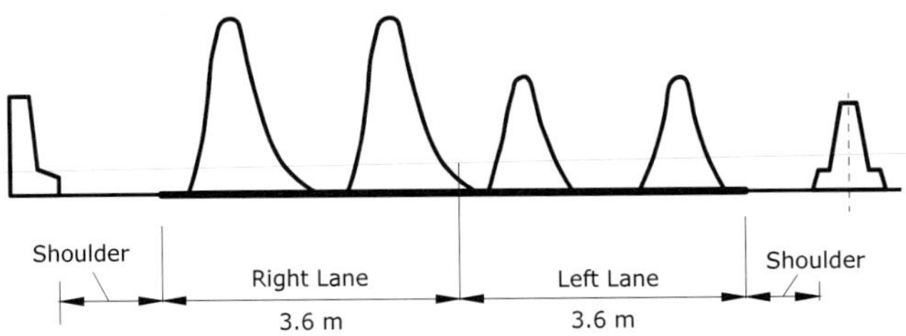

Figure 3 Probability Density Functions (PDF) of the Curb Distance. Each PDF Represents a Line of Truck Wheels (Tantawi 1986)

The variation in transverse traffic position is based on a survey on interstate highways in Southeastern Michigan. The PDF was approximated by a lognormal distribution with a coefficient of variation of 0.33. For a standard lane width of 3.63 m, the mean value of the distance from the lane edge to the centerline of the outermost vehicle wheel is equal to 0.91 m.

Dynamic load depends on roughness of the surface, dynamic properties of the bridge, and suspension system of the vehicle. Dynamic load factor is defined as the ratio of dynamic strain (or deflection) and static strain (deflection). Field tests conducted by Kim and Nowak (1997) and

Eom and Nowak (2001) showed that the dynamic load factor does not exceed 0.15 for a single truck and 0.10 for two heavily loaded trucks traveling side-by-side. Therefore, the mean dynamic load factor is conservatively taken as 0.10 with the coefficient of variation of 0.80.

Bridge Resistance Models

The load carrying capacity of a structure depends on its geometry (number of girders, girder spacing, and span length), connections, and most of all, the resistance of its components. Moreover, the resistance can be affected by uncertainties in strength of materials, dimensions and analysis. Therefore, the resistance of girders, R_{girder}, can be considered as a product of three factors representing strength of materials, dimensions and analysis, and consequently, as a product of random variables, it can be considered as lognormally distributed.

Girder resistance

The reliability analysis is performed for 54 composite girders. The statistical parameters of the load carrying capacity (resistance) for composite steel girders were derived by Nowak (1999). Ultimate moment and shear limit state equations are considered for interior girders. For moment carrying capacity, the bias factor and coefficient of variation are 1.12 and 0.10, respectively. For shear capacity, the bias factor and coefficient of variation are 1.14 and 0.105, respectively.

Bridge (system) resistance

The design of bridges is based on consideration of individual components, therefore, the performance of the whole structure can be underestimated because it does not account for redundancy and ductility. Failure of a component does not necessarily mean failure of the entire bridge. Therefore, bridge safety can be determined using system reliability approach that includes multiple failure path, load sharing and load redistribution after member failure. Consequently, the system reliability can be considered as a more accurate measure of safety. However, the system reliability computations are more difficult than the reliability analysis of a component, because there are many additional parameters.

The need for the system reliability analysis of bridge structures has long been recognized. There are many different modes of system failure. In this paper, the system resistance is considered in terms of the deflection of the main girders caused by live load. It is assumed that the ultimate limit state is reached when the maximum deflection of any girder exceeds 0.0075 of the span length.

The system resistance is considered in terms of the gross vehicle weight (GVW) of two side-by-side trucks with axle configuration of the design truck (AASHTO 2004), as shown in Figure 2. The calculations are performed for various transverse positions of the vehicles (within the roadway width). The incremental loading method is used. For each transverse truck position, the GVW is gradually increased until the deflection of one of the girders exceeds 0.0075 of the span length. The system resistance is defined as the GVW corresponding to this critical deflection. It can be different depending on the position of trucks within the roadway width. Each transverse position is associated with a certain probability of occurrence. Therefore, the system resistance, R_{system}, is equal to the expected value of the GVW, calculated for different transverse positions:

$$R_{system} = \sum_{i=1}^{n} p_i \cdot GVW_i \tag{1}$$

where GVW_i = gross vehicle weight of two trucks side-by-side, causing deflection equal to 0.0075 of the span length corresponding to i-th transverse position of trucks, p_i = probability of trucks occurring in the i-th position. Figure 4 shows a typical cross section of a 6-girder bridge and the truck positions considered in the system resistance analysis.

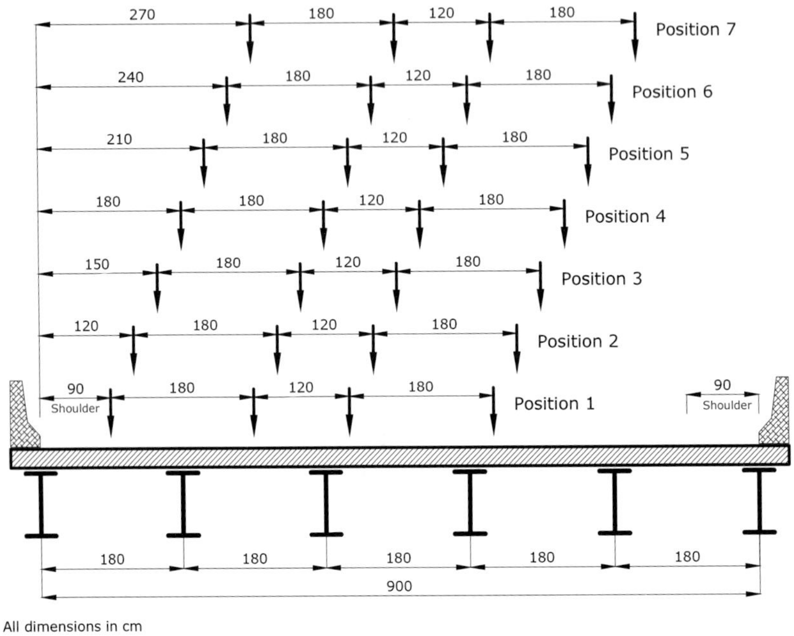

Figure 4 Truck Positions Considered in the System Resistance Analysis

An example of the deterministic load-deflection curve for a 30 m span composite steel girder bridge is shown in Figure 5.

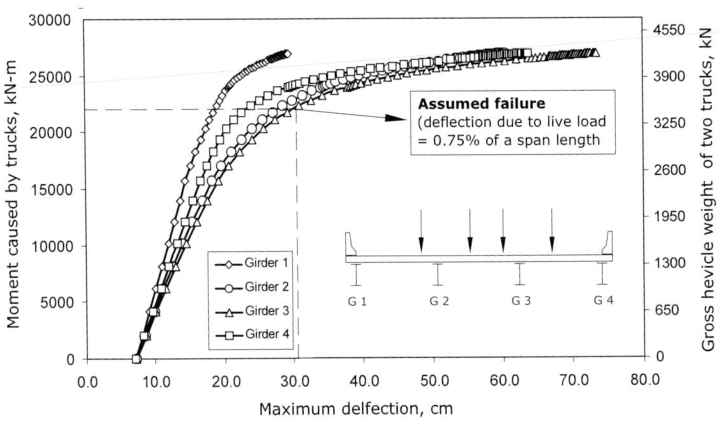

Figure 5 Deterministic Bridge Load-Deflection Relationships for Span Length of 30 m

Reliability Analysis for Selected Bridges

The reliability indices for individual girders were calculated using the procedure developed for calibration of the AASHTO LRFD Code (Nowak 1999). For each combination of span length and girder spacing, three design cases were considered: (1) with individual girders designed according to the code and thus providing reliability indices close to the target $\beta_T = 3.5$, (2) with under-designed girders, with the girder reliability indices close to $\beta_T = 2.0$, and (3) with over-designed girders and girder reliability indices close to $\beta_T = 4.5$. The ratio of the actual girder resistance and the minimum required resistance for the target β, is shown in Figures 6 and 7 for $\beta_T = 3.5$, for moments and shear forces, respectively. The ratios for shear are very large because the design is governed by the moment capacity, so the girders are over-designed with regard to shear. Therefore, the shear capacity was not considered in the system reliability analysis.

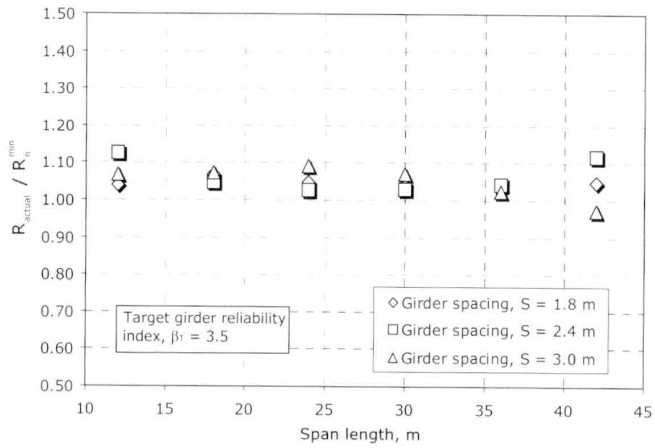

Figure 6 Ratios of Actual Resistance to Minimum Required Resistance for Target Girder Reliability Index $\beta_T = 3.5$, Moments

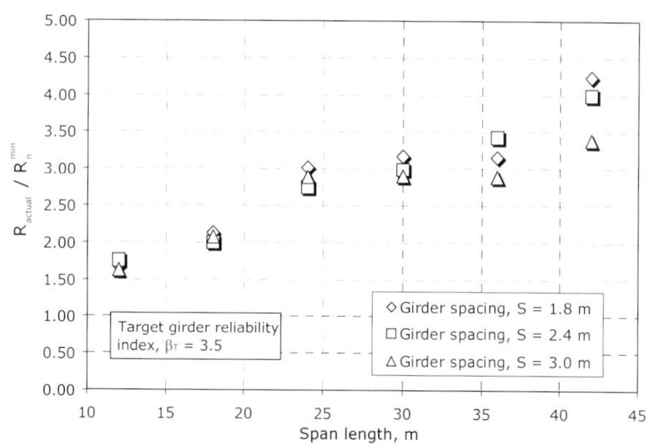

Figure 7 Ratios of Actual Resistance to Minimum Required Resistance for Target Girder Reliability Index $\beta_T = 3.5$, Shear Forces

In the system reliability analysis, the two main random variables are bridge system resistance and bridge live load. Both are represented in form of the gross vehicle weight (GVW) of two identical trucks placed side-by-side, each truck with axle configuration of the design truck

(AASHTO 2004). The system reliability index was calculated for various truck positions within the roadway width, and the final system reliability index was determined as an expected value using the weighting factors based on the curb distance distributions (Figure 3).

Figures 8a to 8c show the relationship between the girder reliability index and the corresponding system reliability index for three different girder spacings, and span length of 18, 30, and 42 m, respectively.

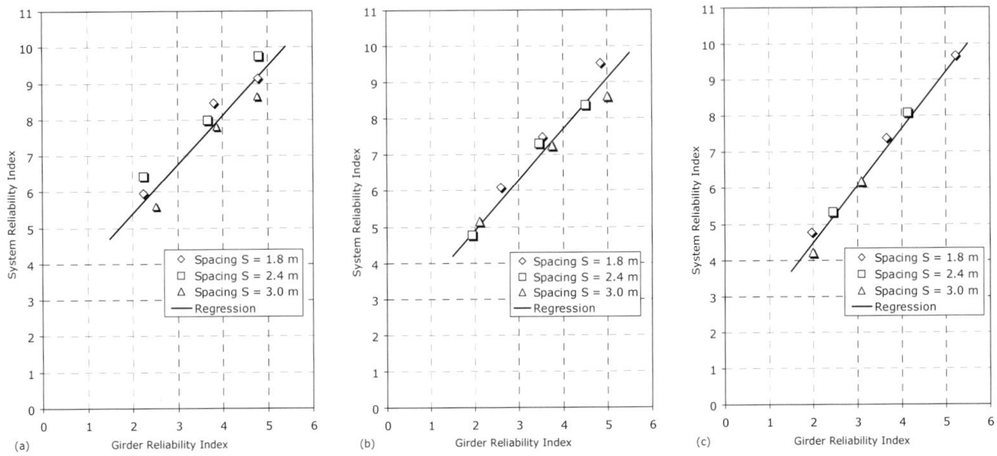

Figure 8 Girder Reliability Index versus System Reliability Index for Different Girder Spacings and Span Length of (a) 18 m, (b) 30 m, and (c) 42 m

The results indicate that the ratio of system reliability to girder reliability decreases with increasing girder reliability. This is because for over-designed bridges (β_T = 4.5), the ratio of system capacity to girder capacity is smaller than that for under-designed bridges (β_T = 2.0). Moreover, the system reliability indices increase with number of girders. This is due to the increased redundancy of the system. It is observed that not only does the ratio of system reliability to girder reliability decreases with increasing girder reliability, but it also decreases with the increase of span length or girder spacing.

Figures 9a to 9c show the girder and system reliability indices as a function of span length, for girder spacing of 3.0 m, for the three considered cases of the target girder reliability index, β_T, of 2.0, 3.5, and 4.5. Also shown are wide flange sizes for the selected main girders.

The next step was to investigate the effect of correlation between the resistances of individual girders represented by the yield stress, F_y, of structural steel. Four cases are considered: (1) no correlation, with coefficient of correlation $\rho = 0$, (2) full correlation with $\rho = 1$, (3) and (4) partial correlation, with different values of ρ depending on the number of girders. The correlation between girder strengths resulted in a reduction of the number of different random variables considered in system reliability analysis.

For a 6-girder bridge with girder spacing of 1.8 m, it was assumed that a partial correlation of $\rho = 0.33$ can be represented by the case of two adjacent girders having identical strength, and $\rho = 0.66$ is represented by four adjacent girders being identical. In addition, also considered were cases with randomly distributed correlated girders (two or four out of six). A similar approach was used in case of 5-girder bridges, with girder spacing of 2.4 m. However, as this time bridges had only 5 girders, so the global correlation between the strength of girders was $\rho = 0.40$ and $\rho = 0.60$ for cases when two and four girders were correlated, respectively. For 4-girder bridges, with girder spacing of 3.0 m, it was assumed that two or three girders were correlated resulting in the global correlation between the strength of girders of $\rho = 0.50$ and $\rho = 0.75$ for cases when two and three girders were correlated, respectively.

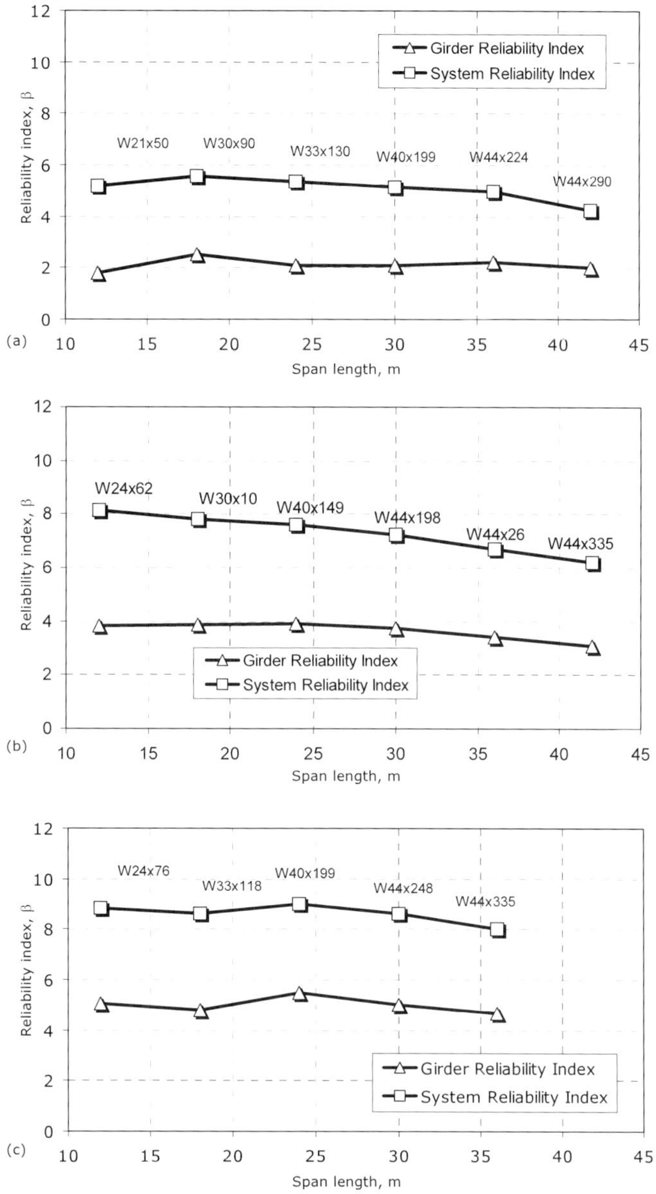

Figure 9 Girder Reliability Indices and Corresponding System Reliability Indices for Different Spans and Girder Spacing of 2.4 m, Target Girder Reliability Index of (a) β_T=2.0, (b) β_T=3.5, and (c) β_T=4.5

Examples of the effect of correlation on the system reliability indices for composite steel girder bridges are shown in Figure 10. The target reliability index for the girders is $\beta_T = 3.5$.

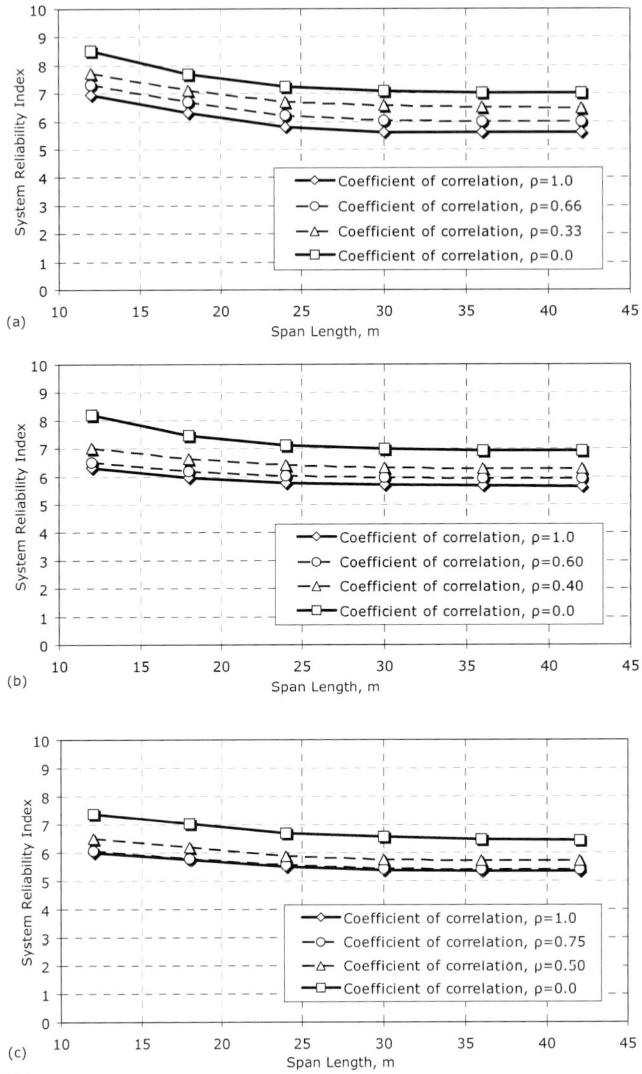

Figure 10 System Reliability Index for Different Degree of Correlation between Girder Resistance and Different Span Length, Girder Spacing of (a) 1.8 m, (b) 2.4 m, and (c) 3.0 m

Conclusions

This paper shows that reliability can be considered as a rational measure of structural performance in the design of bridges and assessment of exiting bridges. The reliability analysis procedure is demonstrated on a representative sample of composite steel girder bridges. Load and resistance parameters are identified, limit state function for the whole bridge was formulated, and reliability analysis is performed for girders and structural systems. It was found that the system reliability is considerably higher than the girder reliability, in particular in case of uncorrelated girder resistances. The difference between girder reliability and system reliability can be considered as a measure of bridge redundancy. The calculated reliability indices for the whole bridge, β_{system}, are compared with the reliability indices determined for individual girders, β_{girder}. It was observed that the ratio of β_{system} / β_{girder} decreases with increasing β_{girder}. It ranges from 1.6 for β_{girder} = 4.5 to 3.1 for β_{girder} = 2.0. It was also found that the correlation between girder resistances plays an important role and it can decrease β_{system} by 10 to 30% depending on the girder reliability, span length, and girder spacing. The paper shows that much can be gained by considering a bridge structure as a system.

References

AASHTO LRFD Bridge Design Specifications (2004), American Association of State Highway and Transportation Officials, Washington, D.C.

ABAQUS User's Manual-version 6.4 (1996), Hibbit, Karlsson & Sorenson, Inc., Pawtucket, Rhode Island.

Czarnecki, A. A., and Nowak, A. S. (2005a).“System Reliability Analysis for Steel Girder Bridges.” *Journal of Structural Engineering*, ASCE (submitted).

Czarnecki, A. A. and Nowak, A. S. (2005b). “Time-Variant Reliability Profiles for Steel Girder Bridges,” *Journal of Structural Safety* (submitted).

Czarnecki, A. A. and Nowak, A. S., (2005c). “Reliability-Based Evaluation of Steel Girder Brides,” *Bridge Engineering, Proceeding of the Institution of Civil Engineers* (submitted).

Eamon, C.D., and Nowak, A.S. (2002). “Effects of edge-stiffening elements and diaphragms on bridge resistance and load distribution.” *Journal of Bridge Engineering*, ASCE, 7(5), 258-266.

Eamon, C.D., and Nowak, A.S. (2004). “Effect of secondary elements on bridge structural system reliability considering moment capacity.” *Structural Safety*, 26(1), 29-47.

Enright, M.P., and Frangopol D.M.. (1999). “Reliability-based condition assessment of deteriorating concrete bridges considering load redistribution.” *Structural Safety*, 21, 159-195.

Eom, J. and Nowak, A.S. (2001) “Live Load Distribution for Steel Girder Bridges.” *Journal of Bridge Engineering*, ASCE, 6(6), 489-497.

Estes, A. and Frangopol, D.M. (1999). “Repair optimization of highway bridges using system reliability approach.” *Journal of Structural Engineering*, ASCE, 125(7), 766-775.

Ghosn, M., and Moses, F. (1998). “Redundancy in Highway Bridge Superstructures.” NCHRP Report 406. Transportation Research Board, Washington, D.C.

Kim, S-J. and Nowak, A.S. (1997). “Load Distribution and Impact Factors for I-Girder Bridges.” *Journal of Bridge Engineering*, ASCE, 2(3), 97-104.

Liu, Y. and Moses, F. (2001). “Bridge design with reserve and residual reliability constrains.” *Structural Safety*, 11(1), 29-42.

Nowak, A.S. (1993). "Live Load Model for Highway Bridges." *Structural Safety*, 13(1-2), 53-66.

Nowak, A.S. (1999). "Calibration of LRFD Bridge Design Code." NCHRP Report 368, Transportation Research Council, Washington, D.C.

Nowak, A.S. and Collins, K.R. (2000). *"Reliability of Structures"*, McGraw-Hill Companies, Inc., New York.

Nowak, A.S. and Hong, Y-K. (1991). "Bridge Live Load Models." *Journal of Structural Engineering*, ASCE, 117(9), 2757-2767.

Tabsh, S.W., and Nowak, A.S. (1991). "Reliability of highway girder bridges." *Journal of Structural Engineering*, ASCE, 117, 2372-2388.

Tantawi, H.M. (1986). "Ultimate Strength of Highway Girder Bridges." Ph.D. dissertation, Department of Civil and Environmental Engineering, University of Michigan, Ann Arbor, MI.

Zhou, J. (1987). "System Reliability Models for Highway Bridge Analysis." Ph.D. Dissertation, Department of Civil and Environmental Engineering, University of Michigan, Ann Arbor, MI.

RELIABILITY OF BILINEAR SDOF SYSTEMS SUBJECTED TO EARTHQUAKE LOADING

H.P. Hong[1] **and P. Hong**[2]
[1]*Department of Civil and Environmental Eng., University of Western Ontario, N5A 6B9 Canada*
[2]*Department of Civil Eng., Aeronautical and Industrial Institute of Nanchang, China*

Abstract: Probabilistic assessment of the ductility demand and reliability analysis were carried out for bilinear hysteretic SDOF systems. The assessment considered two sets of strong ground motion records, and was focused on the evaluation of the mean and the coefficient of variation of the ductility demand for a given value of the normalized yield strength. The results indicate that the ductility demand could be modeled as a Frechct (Extreme value type II) variate. Based on the obtained results, empirical equations were provided to predict the mean of the ductility demand for bilinear SDOF systems of different natural vibration periods, damping ratios, and ratios of the post yield stiffness to the initial stiffness. The numerical results show that the coefficient of variation (cov) of the ductility demand can go as high as to about 1.0 depending on the characteristics of the structure. Also, a simple approach was given to estimate the probability of incipient damage and the probability of incipient collapse using the developed probabilistic characterization of the ductility demand. The approach, which could be suitable for carrying out design code calibration analysis, is illustrated numerically.

Introduction

For a given strong ground motion, the peak responses of a linear or nonlinear single-degree-of-freedom (SDOF) system with and without strength degradation can be carried out using time-step integration methods. The obtained peak responses of SDOF systems can be employed in defining the linear elastic response spectrum and yield response spectra, and/or the ratios between peak linear elastic response and inelastic responses (Chopra 2000). These quantities are relevant for designing and assessing the safety of structures. Its use for the so-called displacement-based design has been discussed by many including Chopra and Geol (2000) and Borzi et al. (2001).

Let F_E denote the minimum strength required for a SDOF system to remain linear elastic during a ground motion, and $D_E(T_n,\xi)$ denote the peak linear elastic displacement where T_n and ξ are the natural vibration period and the damping ratio, respectively. If the strength of the structure is less than F_E, the system responds inelastically with yield displacement represented by $D_y(T_n,\xi,\mu)$ and peak inelastic displacement represented by $D_I(T_n,\xi,\mu)$, where μ represents the displacement ductility factor. Given a set of strong ground motion records, the yield reduction factor R_y, $R_y = D_E(T_n,\xi)/D_y(T_n,\xi,\mu)$, and the ratio R_μ, $R_\mu = D_I(T_n,\xi,\mu)/D_E(T_n,\xi)$, can be calculated. Note that $R_\mu = \mu/R_y$ which can be written as $R_\mu = \mu\phi$ where ϕ is defined as $1/R_y$ and is known as the normalized yield strength or the de-amplification factor. Note also that μ does not always increase monotonically as R_y decreases and more than one value of R_y could lead to the same ductility demand μ. By considering that for a given value of μ it is the largest yield strength, hence the largest ϕ (or smallest R_y) that is relevant for design, an iterative procedure that is described in detail in Chopra (2000) can be employed to evaluate the required ϕ (or R_y) for a given ductility factor μ. Note that the above is equivalent to say that given a value of ϕ, it is the maximum ductility demand, for all the normalized yield strength less than or equal to the specified value of ϕ, that is relevant for design. This view is adopted though out this study.

M. Pandey et al. (eds), Advances in Engineering Structures, Mechanics & Construction, 711–721.

Samples of R_y or R_μ obtained to meet specified target ductility level are employed to find statistics of the ratios R_y and/or R_μ and to develop empirical equations to predict the mean of the R_y and/or R_μ as functions of ductility demand μ (Veletsos and Newmark 1960, Krawinkler and Nassar 1992, Vidic et al. 1994, Miranda 2000 and Riddell et al. 2002, Hong and Jiang 2004). The means of R_y and/or R_μ are employed to scale the design response spectrum or peak linear elastic responses to obtain the design yield strength or the yield responses.

The evaluation of R_y or R_μ to meet specified target ductility factor is computationally intensive because the iteration mentioned previously. It is much more efficient to evaluate the ductility demand for a given value of the normalized yield strength because the latter does not require the iteration over the ductility factor. Also, it is noted that rather than develop empirical equations to predict the values of R_y, R_μ, or the normalized yield reduction factor ϕ for given values of μ, one may instead develop empirical equations to predict the ductility demand μ based on regression analysis conditioned on ϕ. A regression equation developed to predict the expected normalized yield strength is likely to differ from the one developed to predict the expected ductility (factor) demand. Perhaps, the former may be interpreted as a designer knows the ductility capacity of the structure to be designed and is interested in finding the minimum required yield strength; and the latter may be considered as a designer's task is to check a new design or evaluate an existing structure with a known yield strength level and is interested in finding what would be the ductility demand due to strong ground motions. Therefore, the latter that seems lacking in the literature is equally relevant as the former. The need for empirical equations to predict the ductility demand may be further justified based on that the uncertainty or variability associated with the ductility capacity is much greater than the yield strength (Nakashima 1997), hence, a designer could have better control on the yield strength level than on the ductility capacity, and a codification should be focused on incorporating the uncertainty in ductility capacity and ductility demand. Note that systematic assessment of the impact of uncertainty in ductility capacity on the structural reliability is not often investigated

In the following, statistics of the ductility demand are evaluated using two sets of strong ground motion records. The evaluation of samples of the ductility demand is carried out for given values of ϕ. This largely reduces the computing time since iterations over ϕ to find the ductility factor that matches a specified ductility level are not required. Also, empirical equations for the statistics of μ conditioned on the normalized yield strength are presented. In a few cases, comparison of these results to the ones obtained to meet specific ductility level is also given. The evaluation of the ductility demand presented in this study considers several damping ratios and the elastoplastic as well as bilinear hysteretic systems. A very simple method for assessing the reliability of bilinear system by using the developed empirical equations is presented. The method can be used to evaluate the probability of incipient collapse as well as the incipient of damage. Its use is illustrated by numerical examples.

Statistics, ratios and inelastic displacement

Records

Two sets of records considered in this study are for California earthquakes. The first set includes 230 components of records that were used by Miranda (2000) and were found in the database prepared by Silva (2001). This set was adopted by Hong and Jiang (2004) as well.

The second set of records adopted in the present study is the one used by Riddell et al. (2002) to represent earthquakes occurred in California. This set contains 44 records obtained for only two earthquakes, Northridge earthquake and Loma Prieta earthquake. 22 of these 44 components of records are common to the first set. The use of this set of records is aimed at gauging how sensitive are the estimated ratios mentioned previously to the selected records.

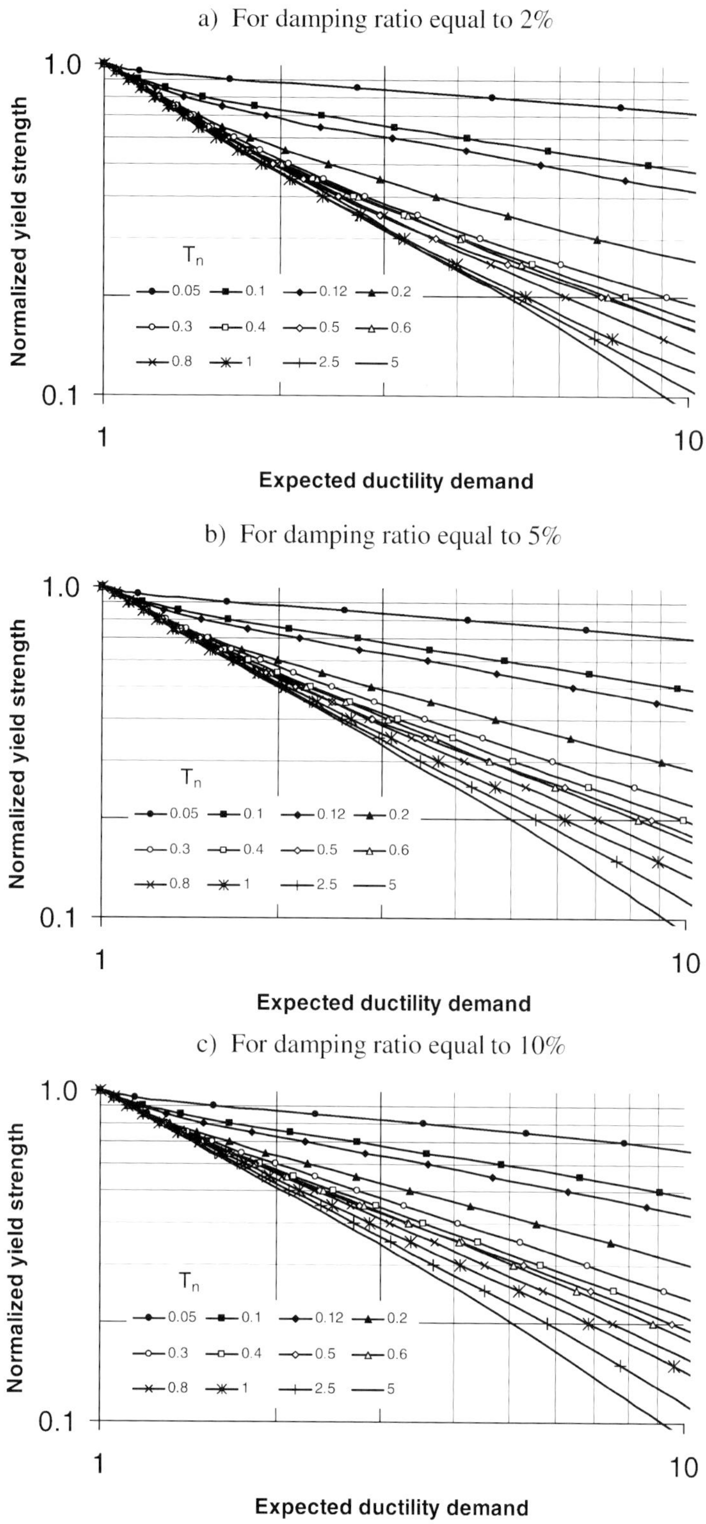

Figure 1. Expected ductility demand for elastic-perfectly-plastic SDOF system.

Results for elastocplastic systems

By using the first set of records and carrying out nonlinear dynamic analysis of elastic-perfectly plastic single-degree-of-freedoom (SDOF) system, the obtained mean of the ductility demand is presented in Figures 1a to 1c for several values of the natural vibration period and ξ = 2%, 5%, and 10%.

The results shown in the figure suggest that one may consider that the logarithmic of the expected ductility demand, m_μ, is a power function of ln(ϕ). This leads to

$$m_\mu = \exp\left((-\alpha_1 \ln\phi)^\beta\right), \tag{1}$$

where α_1 and β are parameters to be determined. The parameters for the models given in Eq. (1) may depend on the natural vibration period T_n and the damping ratio ξ. By minimizing the error ε defined by,

$$\varepsilon = \sum \left(m_{\mu i} - m_\mu(\phi)\right)^2, \tag{2}$$

the estimates of α_1 and β can be obtained. In Eq. (2), $m_{\mu i}$ is the mean of μ obtained from the samples such as those shown in Figure 4, $m_\mu(\phi)$ represents m_μ predicted using Eq. (1) for each given set of values of T_n and ξ.

If one is interested in obtaining simple empirical equations for predicting α_1 and β, the following fitted equations may be employed,

$$\alpha_1 = a_1 \exp\left(a_2 / T_n^{a_3}\right), \tag{3}$$

and,

$$\beta = \begin{cases} b_1 - b_2 \ln(T_n/0.2)/\ln(0.01/0.2) & 0.05 \le T_n < 0.2 \\ b_1 + (b_3 - b_1)\ln(T_n/0.2)/\ln(15/0.2) & 0.2 \le T_n < 5 \end{cases}, \tag{4}$$

where values of the parameters a_i and b_i, i = 1,2,3, are given in Table 1. An example of the predicted m_μ obtained by using the model given in Eq. (1) with α_1 and β calculated from Eqs. (3) and (4) is illustrated in Figure 2 for ξ equal to 5%. Comparison of the results shown in this figure and those presented in Figure 1b suggests that the empirical predicting model provides a good approximation to those given in Figure 1b.

Table 1. Parameters for Eqs. (3) and (4)

ξ	For Eq. (3)					For Eq. (4)				
		$\gamma = 0$	0.01	0.05	0.1		$\gamma = 0$	0.01	0.05	0.1
2%	a_1	0.893	0.866	0.813	0.786	b_1	1.22	1.14	1.14	1.14
	a_2	0.105	0.109	0.105	0.097	b_2	0.30	0.30	0.40	0.60
	a_3	0.972	0.952	0.941	0.928	b_3	0.82	0.82	0.82	0.82
5%	a_1	0.927	0.907	0.857	0.828	b_1	1.32	1.28	1.28	1.28
	a_2	0.162	0.158	0.148	0.139	b_2	0.30	0.40	0.70	1.00
	a_3	0.815	0.817	0.818	0.810	b_3	0.82	0.82	0.82	0.82
10%	a_1	0.973	0.952	0.876	0.858	b_1	1.42	1.38	1.38	1.38
	a_2	0.182	0.180	0.199	0.176	b_2	0.30	0.50	0.90	1.10
	a_3	0.752	0.755	0.697	0.711	b_3	0.82	0.82	0.82	0.82

The obtained cov of the ductility demand is illustrated in Figure 3 for ξ equal to 0.05. The results shown in the figure suggest that for the mean ductility demand less than about 10 (see Figure 1) the cov of μ increases as ϕ decreases. The cov of μ for relatively rigid structures is larger than that for the flexible structures, and decreases as the damping ratio increases. In almost all cases with a mean ductility demand less than 10, the cov of μ can be considered to be less than

1.0. Similar trends of the cov values were observed for the results obtained for ξ equal to 0.02 and 0.10.

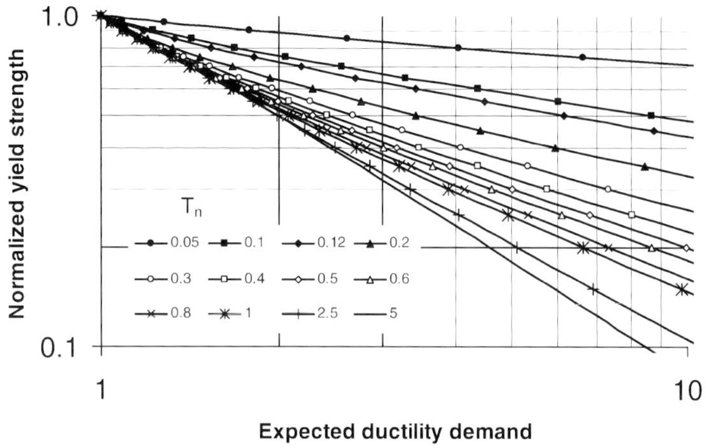

Figure 2. Predicted expected ductility demand using the model given in Eq. (1) for $\xi = 5\%$.

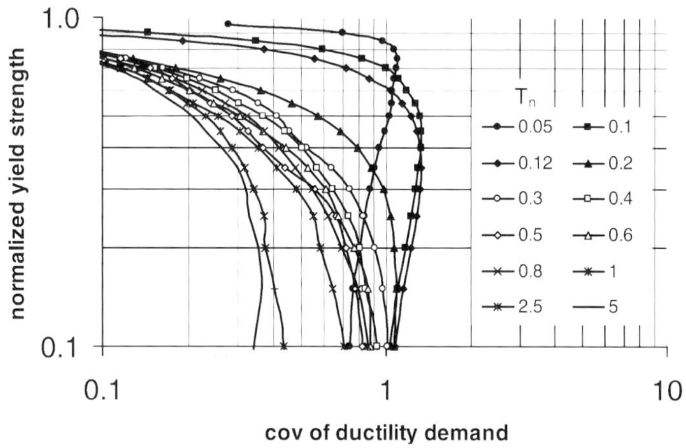

Figure 3. Coefficient of variation of the ductility demand for elastic-perfectly-plastic SDOF system with $\xi = 5\%$.

Now if the second set of records mentioned in the previous section is employed, the obtained mean and cov of the ductility demand for $\xi = 0.05$ are shown in Figures 4a and 4b. Comparison of the results shown in Figure 1b and Figure 4a and the results shown in Figure 3 and Figure 4b suggest that:

1) the difference between the predicted ductility demand obtained by using the first set of records and the second set of records is not very significant; and

2) The values of the cov of the ductility demand depend somewhat on the set of records used; however, the conclusion, that the cov of μ is less than about 1.0 for the mean of μ less than 10, is still adequate.

Note that no detailed analysis of the cov of μ was presented. This is because that the uncertainty in peak elastic displacement rather than that in the ductility demand is likely to play a dominant role in characterizing the uncertainty in the peak inelastic displacement since the cov of the peal elastic displacement usually ranges from 0.8 to higher than 10 for different sites in Canada.

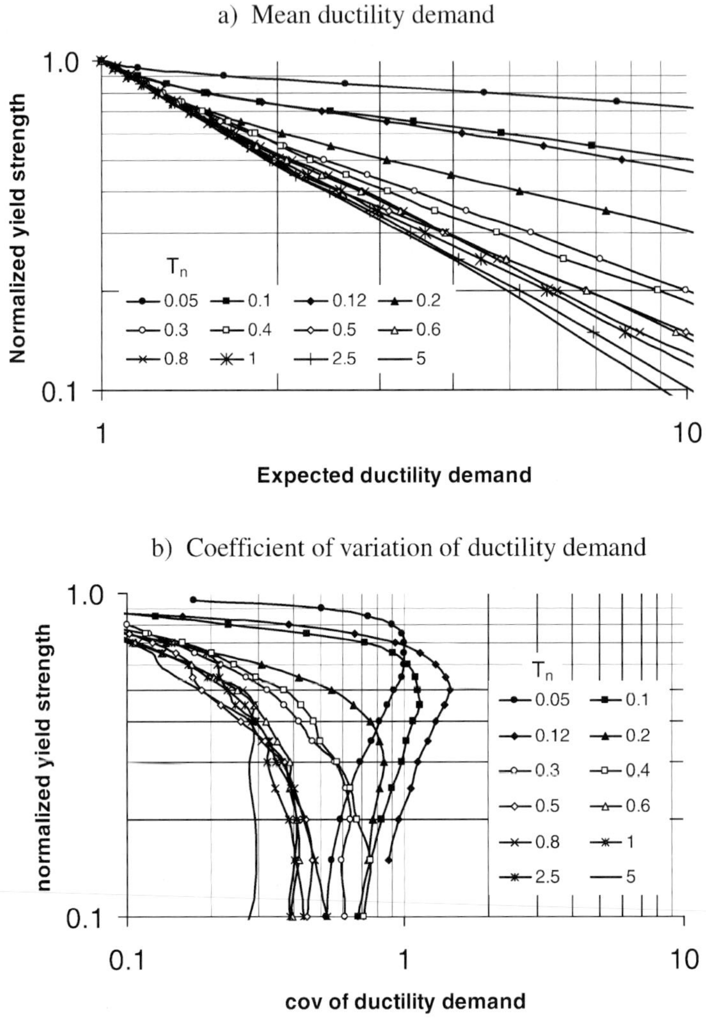

Figure 4. Statistics of ductility demand obtained using the second set of records for elastic-perfectly-plastic SDOF system with $\xi = 5\%$.

To investigate possible probabilistic distribution models for the ductility demand, we plot the samples of the ductility demand in the lognormal probability paper for a given value of ϕ. It was found that the samples slightly curved, therefore, the assumption that the ductility demand is lognormally distributed can be very convenient but may not be very adequate. However, if the samples are presented in the Frechet probability paper as illustrated in Figure 5, the ductility demand samples could be approximate by straight liners for each given values of ϕ. Therefore, the lognormal variate could be adopted to model the ductility demand but the use of Frechet distribution is preferred.

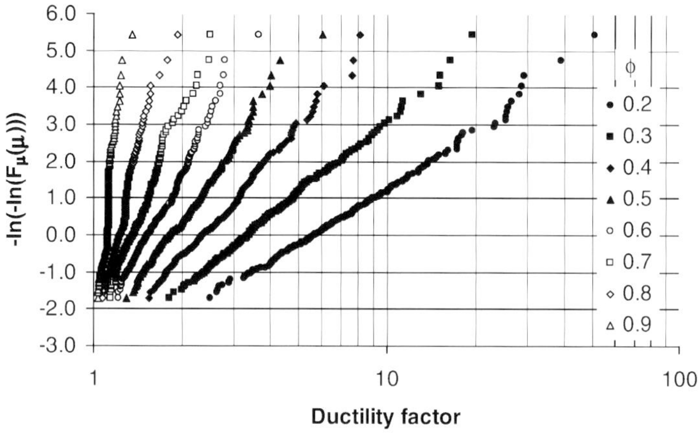

Figure 5. Frechet probability paper plot for the ductility demand samples

Results for Bilinear systems

To investigate the effect of the strain-hardening on the statistics of the ductility demand, bilinear hysteretic systems are considered in this section. Let γ denote the ratio of the post yield stiffness to the initial stiffness.

The analysis carried out for the elasto-perfectly-plastic SDOF systems is repeated for the bilinear system for the combinations of γ (= 0.01, 0.05 and 0.1) and ξ (= 0.01, 0.05 and 0.1) values. The obtained results are employed to find the values of α_1 and β for the model given in Eq. (1) as was done for elasto-perfectly-plastic SDOF systems. Eqs. (3) and (4) are then employed to fit these values leading to the parameters presented in Table 1.

Use of empirical equations in estimating reliabilities

Limit state functions and reliability analysis

The developed empirical equations can be employed for the probabilistic evaluation of the performance of a designed or an existing structural system, and for design code calibration. For the evaluation, only two levels, one for the incipient damage (incipient inelastic deformation) and the other for the incipient collapse (Wen 2001), will be considered. Let $D_R(T_n,\xi)$ denote the yield displacement capacity of the structural system, and ζ denote the ratio of $D_R(T_n,\xi)$ to $D_E(T_n,\xi)$. In this section, $D_E(T_n,\xi)$ represents the annual peak linear elastic displacement. Since the yield displacement capacity $D_R(T_n,\xi)$ less than the seismic demand $D_E(T_n,\xi)$ implies that the structure will at least sustain damage, and the maximum inelastic displacement (capacity) $\mu_R D_R(T_n,\xi)$ less than the maximum inelastic ductility demand implies that the structure will collapse, the limit state functions for these two performance levels are,

$$g_D = D_R(T_n,\xi)/D_E(T_n,\xi) - 1 = \zeta - 1, \tag{5}$$

and,

$$g_C = \left(\mu_R D_R(T_n,\xi)\right)/\left(\phi\mu(\phi)D_E(T_n,\xi)\right) - 1, \tag{6}$$

where g_D represents the limit state function of the incipient damage; g_C represents the limit state function of the incipient collapse, respectively; $\zeta = D_R(T_n,\xi)/D_E(T_n,\xi)$; and μ_R denotes the ductility capacity of the structural system. To emphasize that the ductility demand μ is a function of the normalized yield strength ϕ, the notation for the ductility demand μ in Eq.(6) is replaced by μ(ϕ). Damage occurs if g_D is less than zero and, collapse occurs if g_C is less than zero.

Note that if ζ is greater than unity (i.e., $g_D > 0$) implying that the yield displacement capacity of the structural system is greater than the linear elastic demand caused by the strong ground motions. Note also that since by definition $\zeta = D_R(T_n,\xi)/D_E(T_n,\xi)$, ζ represents the normalized yield strength (i.e., ζ = ϕ), Eq. (6) can be re-written as

$$g_C = \mu_R/\mu(\zeta) - 1. \tag{7}$$

The yield strength is usually considered to be normal or lognormally distributed with a cov of about 0.1 to 0.15 (Ellingwood et al. (1980), Nakashima (1997)). Therefore, the yield displacement capacity of the structure can be considered to be lognormally distributed with a cov of 0.15. The uncertainty associated with the ductility capacity μ_R is much more significant than that associated with the yield strength. According to Nakashima (1997), the cov of μ_R can vary from about 0.5 to 1.0. In this section it is considered that μ_R can be modeled as a lognormal variate with a cov with of 0.5. Further, it is considered that the peak linear elastic displacement $D_E(T_n,\xi)$ is lognormally distributed as well with a cov within 1 to 10. This can be justified based on the seismic hazard studies given by Adams and Halchuk (2003).

Based on these adopted probabilistic models, the evaluation of the probability of the incipient damage, P_D, is straight forward (Madsen et al. (1986)). It can be calculated from,

$$P_D = \text{Prob}(\zeta < 1) = F_\zeta(1), \tag{8a}$$

where

$$F_\zeta(\zeta) = \Phi\left(\left(\ln\zeta - \ln\left(m_\zeta/\sqrt{1+v_\zeta^2}\right)\right)/\sqrt{\ln(1+v_\zeta^2)}\right) \tag{8b}$$

where $m_\zeta = \left(1+v_E^2\right)m_R/m_E$, $v_\zeta^2 = (1+v_R^2)(1+v_E^2) - 1$, m_R and m_E denote the means of $D_R(T_n,\xi)$ and $D_E(T_n,\xi)$, respectively; v_R and v_E denote the cov of $D_R(T_n,\xi)$ and $D_E(T_n,\xi)$, and $\Phi(\bullet)$ is the standard normal probability distribution function. Note that ζ is lognromally distributed with mean m_ζ and cov of v_ζ since $D_R(T_n,\xi)$ and $D_E(T_n,\xi)$ are lognormally distributed.

The probability of the incipient collapse, P_C, $P_C = \text{Prob}(g_C \le 0)$, can be evaluated by recursively using the first-order reliability method (FORM) or the simulation techniques. In this study, the simulation technique is employed for the analysis. Note that $P_C = \text{Prob}(g_C \le 0)$ can be expressed as

$$P_C = \text{Prob}(\mu_R/\mu(\zeta) \le 1 | \zeta < 1)\text{Prob}(\zeta < 1) = \text{Prob}(\mu_R/\mu(\zeta) \le 1 | \zeta < 1)P_D, \tag{9}$$

since $\text{Prob}(\mu_R/\mu(\zeta) \le 1 | \zeta \ge 1) = 0$. The basic steps for evaluating $\text{Prob}(\mu_R/\mu(\zeta) \le 1 | \zeta < 1)$ by using simulation technique are:

1) Generate a sample of ζ according to the updated (or truncated) probability distribution function of ζ, $F_\zeta(\zeta)/P_D$
2) Find the mean of μ(ζ) using Eq. (1) with parameters defined in Eqs. (3) and (4), and the cov of μ(ζ) from figures similar to Figure 3;
3) Using the obtained value in Step 2) define the probability distribution of μ(ζ), which is considered to be Frechet distributed;
4) Generate samples of μ_R and μ(ζ) according to their probability distributions; and check if $\mu_R/\mu(\zeta)$ is less than or larger than unity;
5) Repeat Steps 1) to 5) to generate enough samples of $\mu_R/\mu(\zeta)$ and to count number of times that $\mu_R/\mu(\zeta)$ is less than one for estimating $\text{Prob}(\mu_R/\mu(\zeta) \le 1 | \zeta < 1)$.

Numerical examples

The formulations given in previous section are illustrated by simple numerical examples. For the analysis, it is assumed that the structure can be modeled as a SDOF system with natural vibration period equal to 0.5 (sec). Consider that the ratio between the yield displacement capacity of a bilinear hysteretic system m_R and the displacement corresponding to the design earthquake load for a linear elastic SDOF system, $D_{EN}(T_n,\xi)$, ϕ_D, is known. ϕ_D takes into account factors such as the ductility-related force reduction and the mean resistance is greater than the factored resistance. The yield displacement capacity is considered to be lognormally distributed with cov of 0.15. Further, consider that ϕ_D equals 0.5; the mean of the ductility capacity, $m_{\mu R}$, equals 4; the ratio of the post yield stiffness to the initial stiffness γ takes the value of 0, 0.01 or 0.05; and $D_{EN}(T_n,\xi)$ equals 475-year return period value of the peak linear elastic displacement demand $D_E(T_n,\xi)$ that is considered to be lognormally distributed. The cov of $D_E(T_n,\xi)$, v_E, that equals 0.8, 2 and 10 are considered. The small and large values of v_E were used to represent approximately the seismic hazard conditions for, respectively, the west and the east of Canada Adams and Halchuck (2003).

Based on these considerations, it can be shown that m_ζ is given by,

$$m_\zeta = \left(1+v_E^2\right)\frac{m_R}{m_E} = \left(1+v_E^2\right)\frac{\phi_D D_{EN}(T_n,\xi)}{m_E} = \phi_D\sqrt{1+v_E^2}\exp\left(\sqrt{\ln(1+v_E^2)}\Phi^{-1}\left(1-\frac{1}{475}\right)\right), \tag{10}$$

which is independent of m_E. In Eq. (10), $\Phi^{-1}(\bullet)$ denotes the inverse of the normal probability distribution function. Substituting Eq. (10) into Eq. (8) gives P_D equal to 3.44×10^{-2}, 1.10×10^{-2}, and 5.74×10^{-3} for v_E equal to 0.8, 2 and 10, respectively.

To simplify the evaluation of P_C, the numerical analysis was carried out by considering that the cov of $\mu(\zeta)$ is independent of ζ. The effect of this assumption on the estimated probability will be investigated by comparing the results obtained for the cov of $\mu(\zeta)$ equal to 0.4 and 0.8, which are shown in Table 2. The results shown in the table suggest that the probability of incipient collapse is not very sensitive to the assumed cov of the seismic demand (i.e., cov of $D_E(T_n,\xi)$, v_E) nor to the assumed cov of $\mu(\zeta)$. Also, it is noted that that the obtained probability of incipient collapse is insensitive to the considered post-yield stiffness. This may be explained by noting that a structure with T_n = 0.5 and for a normalized yield strength (i.e., ϕ_D) around 0.5 the expected ductility demand for $\gamma = 0$ does not differ significantly from that for $\gamma = 5\%$.

It should be noted that no attempt is made in this study to carry out a design code calibration excise (Madsen et al. (1986)). However, it is noteworthy that given the mean ductility capacity of structure $m_{\mu R}$ and the nominal or factored design earthquake demand $D_{EN}(T_n,\xi)$, the formulation and procedure given in this study can be used to calibrate the required resistance factor for the yield displacement (or strength) ϕ_D such that use of the factor in design will leads to the designed structures to meet a specified target reliability level.

Table 2. Estimated probability of incipient collapse

cov of $\mu(\zeta)$	γ	v_E		
		0.8	2	10
0.4	0	1.54E-03	1.15E-03	1.40E-03
	1%	1.52E-03	1.15E-03	1.33E-03
	5%	1.25E-03	1.02E-03	1.16E-03
0.8	0	2.41E-03	1.61E-03	1.54E-03
	1%	2.54E-03	1.50E-03	1.44E-03
	5%	2.17E-03	1.38E-03	1.34E-03

Conclusions

A statistical analysis of the ductility demand was carried out for bilinear hysteretic SDOF systems. The analysis results indicate that the ductility demand can be modeled as a Frechet (Extreme value type II) variate for given values of the normalized yield strength.

The mean of the ductility demand and the normalized yield strength when plotted in a logarithmic paper, follow approximately a straight line for SDOF systems having the same initial natural vibration period. This observation leads to a simple empirical equation in predicting the expected ductility demand. Model parameters for the proposed empirical predicting models were obtained for different natural vibration periods, damping ratios, and ratios of the post yield stiffness to the initial stiffness. The coefficient of variation (cov) of the ductility demand can go as high as to about 1 depending on the characteristics of the structure.

Using the developed probabilistic characterization of the ductility demand, a simple approach to estimate the probability of incipient damage or incipient collapse was given. Numerical results suggest that an accurate empirical predicting model for the cov of the ductility demand may not be necessary since sensitivity analysis results indicate that the variation of this cov on the probability of incipient collapse is not very significant.

Acknowledgements

The financial supports of the Natural Science and Engineering Research Council of Canada are gratefully acknowledged.

References

Adams J, Halchuk S. Fourth generation seismic hazard maps of Canada: values for over 650 Canadian localities intended for the 2005 National Building Code of Canada. Geol. Surv. of Canada, 2003, Open-File 4459.

Borzi B, Calvi GM, Elnashai AS, Faccioli E, Bommer JJ. Inelastic spectra for displacement-based seismic design, Soil Dynamics and Earthquake Engineering, 2001, Vol. 21:47-61.

Chopra AK, Goel RK. Evaluation of NSP to estimate seismic deformation: SDF systems, Journal of Structural Engineering, 2000, Vol. 126, No.4: 482-490.

Chopra AK. *Dynamics of Structures*, Prentice Hall, New Jersey, 2000.

Ellingwood B, Galambos TV, MacGregor JG, Cornell CA. Development of a probability based load criterion for American national standard A58. Nat. Bureau of Standards Special Pub. No. 577. 1980.

Hong HP, Jiang J. Ratio between inelastic and elastic responses with uncertain structural properties, Canadian Journal of Civil Engineering, 2004, Vol. 31, No. 4:703-711.

Krawinkler H, Nassar AA. Seismic based design on ductility and cumulative damage demands and capacities, *in Nonlinear seismic analysis and design of reinforced concrete buildings* (eds. Fajfar P, Krawinkler H.), Elsevier Science, New York, 1992.

Madsen HO, Krenk S, Lind NC. *Methods of structural safety*, Prentice-Hall, Englewood Cliffs, N. J. 1986.

Miranda E. Inelastic displacement ratios for structural on firm sites, Journal of Structural Engineering, ASCE, 2000, Vol. 126, No. 10:1150-1159.

Nakashima M. Uncertainties associated with ductility performance of steel building structures, in Seismic design methodology for the next generation of codes (eds. Fajfar P. and Krawinkler H.), Balkema, Rotterdam, 1997.

Riddell R, Garcia JE, Garces, E. Inelastic deformation response of SDOF systems subjected to earthquakes, Earthquake Engineering and Structural Dynamics, 2002, Vol. 31:515-538.

Silva W. PEER Strong motion database, Pacific Engineering, http://nisee.berkeley.edu/software_and_data/strong_motion/index.html, 2001.

Veletsos AS, Newmark NM. Effect of inelastic behavior on the response of simple systems to earthquake motions, Proc., 2nd World Conference on Earthquake Engineering, 1960, Vol. 2:895-912.
Vidic T, Fajfar P, Fischinger M. Consistent inelastic design spectra: strength and displacement, Earthquake Engineering and Structural dynamics, 1994, Vol. 23, No. 5: 507-521.
Wen YK. Reliability and performance-based design, Structural Safety, 2001, Vol. 23: 407-428.

RISK-BASED MAINTENANCE OPTIMIZATION OF AGING HIGHWAY BRIDGE DECKS

Zoubir Lounis
Institute for Research in Construction,
National Research Council, Ottawa, Ontario, Canada;
E-mail: Zoubir.Lounis@nrc.gc.ca

Abstract

This paper presents a practical approach for maintenance optimization of a network of aging highway bridge decks that integrates a stochastic deterioration model based on Bogdanoff's cumulative damage theory with an effective multi-objective optimization approach. The multi-objective maintenance optimization takes into account all relevant objectives, such as improving bridge deck condition, minimizing maintenance costs, and minimizing traffic disruption and associated user costs. The consideration of these three objectives enables to take full advantage of the available bridge inspection data and implicitly lead towards the minimization of the risk of failure due to bridge deck deterioration and maintenance activities. A multi-objective optimality index is proposed as an optimality criterion for priority ranking of the deficient bridge decks for maintenance. The obtained optimal maintenance project prioritization strategy achieves a satisfactory trade-off or compromise between the selected relevant and competing optimization objectives. The proposed approach is illustrated on a small network of ten bridge deck projects that are optimized for maintenance.

Introduction

The deterioration of reinforced concrete bridge decks due to reinforcement corrosion is recognized as the main cause of failure of bridge decks and it is estimated that one-third to one-half of the projected bridge rehabilitation costs in North America are related to bridge deck deterioration (Weyers 1998; Lounis and Mirza 2001). The corrosion of the reinforcing steel is caused by the chlorides (from deicing salts), which penetrate the concrete cover and destroy the protective passive film on the steel reinforcement. As the corrosion products accumulate, they generate high tensile stresses, which eventually lead to irreversible damage such as concrete cracking, delamination and spalling. Furthermore, corrosion leads to reduction of concrete and reinforcement cross sectional areas, loss of bond between steel and concrete, reduction in strength, and ductility. The effects of corrosion are compounded by other deterioration factors, such as initial damage (e.g. due to shrinkage cracking), increased traffic loads, freeze-thaw cycles, poor workmanship, and inadequate maintenance.

The extensive deterioration of highway bridge decks in North America and the limited funds allocated for their maintenance present considerable technological and economic challenges for bridge owners and managers, namely: (i) which bridge decks to maintain; (ii) when to maintain them; and (iii) how to maintain them (i.e. identify the most effective maintenance strategy, which could be a patch repair, overlay, partial depth replacement and overlay, cathodic protection, total replacement, etc.). The importance of a pro-active maintenance policy is confirmed by the study carried out by Dunker and Rabbat (1990) on the performance of highway bridges included in the U.S. *National Bridge Inventory (NBI)*, which showed that bridge deterioration varied considerably from state to state with the highest deterioration level being observed in the central and southeastern states, while the lowest deterioration was observed in the southwestern states. This considerable difference in structural deficiency between the different states was attributed to differences in design, construction, inspection, funding and most importantly maintenance policies.

Different approaches to maintenance optimization have been implemented in the different bridge management systems ranging from simplified economic models to advanced Markovian decision processes. In the literature on bridge management, the main optimization objective used for maintenance optimization is the minimization of the present value life cycle cost, which represents all the costs incurred throughout the life cycle of a bridge structure, including, the costs of design,

M. Pandey et al. (eds), Advances in Engineering Structures, Mechanics & Construction, 723–734.

construction, maintenance, repair, rehabilitation, replacement, demolition, and in some instances users' costs, and possibly costs of failure. Most of these systems are based on a single-objective optimization , namely, the minimization of the maintenance costs.

Given the importance and high consequences of failure of highway bridge structures, a risk-based maintenance management methodology can be more effective and objective as it enables the optimization of different types of structures and systems from different bridges within a network by considering not only their probability of failure but also their consequences of failure. The implementation of a quantitative risk-based bridge maintenance management, however, is very complex task due to the difficulties of assessing quantitatively the probability and the consequences of failure, especially for a large network of bridge structures.

The failure modes of bridge decks may include the loss of serviceability (e.g. excessive cracking, concrete delamination and spalling, and deformation), loss of functionality (e.g. poor traffic conditions, inadequate deck geometry/approach roadway alignment, limited clearance, etc.), and possibly the partial or total collapse. As mentioned earlier, the loss of serviceability and loss of functionality, although not catastrophic or life-threatening, are by far the most frequent failure modes for bridge decks and involve significant cumulative costs. The consequences of failure of highway bridges are multiple and may include loss of life, injury, excessive maintenance costs, user costs, traffic disruption, environmental impacts, etc. It is clear that some of these consequnces are incommensurable and cannot be evaluated in monetary terms.

To overcome the above difficulties, a multi-objective maintenance optimization approach that enables to consider all relevant criteria of a risk-based approach, namely the maximization of condition rating, minimization of maintenance costs, and minimization of user costs is presented in this paper. The proposed approach can address the requirements of many decision-makers (bridge owners/ managers/ engineers) seeking to satisfy implicitly and/or explicitly several objectives at the same time in planning the maintenance of a large network of aging bridge decks. The proposed approach enables a better evaluation of the effectiveness of maintenance strategies in terms of several criteria and determines the optimal solution that achieves the best trade-off between all criteria (including conflicting ones, such as condition and cost).

Methodology for Risk-Based Maintenance Optimization

The proposed methodology for network-level bridge deck maintenance optimization is based on the simultaneous satisfaction of three relevant and competing criteria, namely: (i) maximization of condition rating; (ii) minimization of maintenance costs; and (iii) minimization of user costs. The development of such a methodology requires the integration of three simple and practical decision support models, namely: (i) qualitative condition assessment model; (ii) qualitative deterioration prediction model; and (iii) multi-objective optimization model to determine the optimal maintenance strategy for a network of bridge decks. The proposed methodology is symbolically outlined in Fig.1. The qualitative condition assessment model enables to take full advantage of the available bridge inspection data. The deterioration of bridge decks is modeled using an appropriate stochastic process that captures the time-dependence and uncertainty of the deterioration mechanism. The deterioration prediction model is compatible with the existing condition assessment procedure, and is developed based on the historical field performance data collected during bridge inspections.

As mentioned earlier, the governing failure modes for bridge decks are the loss of serviceability and loss of functionality due to corrosion-induced damage. The consequences of bridge deck deterioration can range form a simple riding discomfort to a loss of life as a result of a traffic accident on the deteriorated deck or on a detour route (with a poor condition) due to the closure of one lane or the entire bridge during its maintenance or due to its excessive deterioration. As opposed to life cycle cost or cost-benefit criteria used in most bridge maintenance optimization studies, the use of a risk of failure as a criterion for maintenance optimization is more rational and relevant, however its

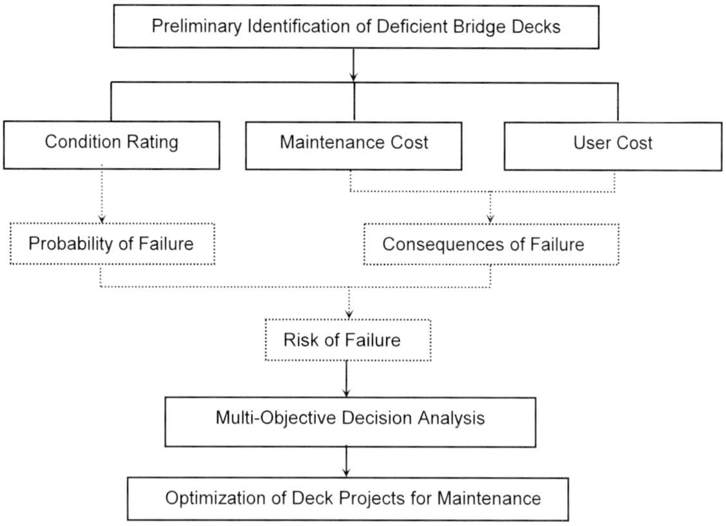

Fig. 1. Schematic of multi-objective maintenance optimization approach for bridge decks

implementation is not easy given the complexity of assessing the consequences of failure in monetary terms This means that monetary values need to be assigned for fatalities, injuries, and social costs which are not easily quantified, and various methods have been developed.

Given the difficulty of accepting the notion of placing any sort of value on human life, Starr (1969) evaluated the risk of death from various causes and identified two general categories for risk of death: (i) risk associated with voluntary activities in which the individual evaluates and adjusts his exposure to risk; and (ii) risk associated with involuntary activities, which are determined by regulations from governmental agencies. Starr (1969) indicated that the public typically was willing to accept voluntary risks 1,000 times greater than involuntary risks. Paté-Cornell (1994) proposed different ranges of acceptable levels of risks for the public and workers ranging from 10^{-8} to 10^{-3} per year.

The consideration of these three objectives within the multi-objective optimization framework is a practical approach for the solution of the risk minimization problem through the minimization of the probability of failure (by maximizing the condition rating) and minimization of the consequences of failure (by minimizing the maintenance costs and user costs), as indicated in Fig.1. The proposed approach overcomes the difficulties of assessing quantitatively the probability of failure and the consequences of failure for a large network of bridge decks. A detailed description of the models is given in the next sections.

Predicting Deterioration of Bridge Decks Using Bogdanoff's Cumulative Damage Model

Different types of models, including empirical, mechanistic, statistical, and logistic haven been proposed in the literature for the prediction of bridge deck deterioration (Golabi and Shepard 1997; Frangopol et al. 1997; Weyers 1998; Stewart and Rosowsky 1998; Lounis and Madanat 2002; Morcous et al. 2003). The prediction of the deterioration and service life of bridge decks is a difficult task due to the complexity of the mechanisms involved, such as the penetration of chlorides into concrete, onset of corrosion, damage initiation, damage accumulation, and coupling effects of corrosion and other deterioration factors (e.g. traffic load, initial damage, etc.). This deterioration prediction is further complicated by the considerable uncertainty in the governing parameters of the damage initiation and accumulation models, as well as in the uncertainty in the models themselves.

The fluctuations from mean performance and mean life are sufficiently large and cannot be ignored without serious consequences owing to the large fluctuations in the in-service environment, deck design, initial damage, etc. Hence, a probabilistic modeling of the damage accumulation and service life is required to achieve reliable results. Furthermore, these models should be developed by using the available data collected during the regular or special inspections of bridge structures in the inventory of highway agencies. These models can be updated as more data become readily available and/or by using data from long term in-service exposure, accelerated laboratory testing, or from the use of fundamental mechanistic models (Golabi and Shepard 1997; Lounis and Madanat 2002; Morcous et al. 2003).

Condition Assessment Model

In order to ensure the implementation of the proposed methodology, it is imperative to take full advantage of the bridge inspection data collected in the bridge inventory database of different highway agencies. In general, highway bridge decks (and other bridge components) are inspected every two years on average (FHWA 1995; MTO 1989). The inspector rates the condition of the bridge deck and assigns condition ratings for the deck. This condition rating consists of mapping the level of observed damage in the structure during visual inspections and non-destructive evaluation (or predicted using empirical or statistical methods) onto a discrete (1 to 9, 1 to 7, or 1 to 5) rating scale. Concrete bridge decks are inspected for cracking, spalling, delamination, potholing, rust staining, scaling, reinforcement corrosion, chloride contamination, and partial or full depth failures. In this paper, the condition rating (CR) of concrete bridge decks is based on the following seven-state rating scale, which reflects the different damage states associated with chloride-induced corrosion and is adapted from existing different condition assessment systems (Golabi and Shepard 1997; FHWA 1995; MTO 1989; Morcous *et al.* 2003). Table 1 provides a summary description of the adopted condition rating system.

Table 1. Condition rating system for concrete bridge decks

Condition Rating (CR)	Description
1	Excellent condition: no contamination; no corrosion; no repaired area.
2	Very good condition: minor cracks, no spalls or delaminations; chloride contaminated or repaired areas ≤ 2% (of total deck area).
3	Satisfactory condition: spalls or delaminations ≤ 2%; cracked, corroded, contaminated, or repaired area ≤10%.
4	Fair condition: spalls or delaminations ≤ 5%; cracked, corroded, contaminated, or repaired area ≤20%.
5	Poor condition: spalls or delaminations ≤ 10%; cracked, corroded, contaminated, or repaired area ≤25%.
6	Critical condition: spalls or delaminations ≤ 15%; cracked, corroded, contaminated, or repaired area ≥ 25%.
7	Failed condition (total loss of serviceability or functionality): extensive spalling, delamination, repaired areas ≥ 30%; maintenance required.

Such a condition assessment system is very practical and cost-effective (in terms of inspection/evaluation costs) for the long- and short-term analysis of maintenance needs for a network of hundreds or thousands of structures, which is the case for many highway agencies. However, for safety-critical elements and for high-risk structures, a detailed and a more rigorous condition assessment may be required.

Bogdanoff's Cumulative Damage Model

In this paper, a Bogdanoff's cumulative damage (CD)-based model is used to predict the future condition and service life of reinforced concrete bridge decks. The proposed model assumes a probabilistic evolutionary structure of the damage accumulation process. The condition of the bridge deck is discretized into a finite state space with seven (7) damage states. A basic element of the model is the concept of duty cycle, which is a repetitive period of operation in the life of a deck in which the damage accumulation is assumed non-negative (Bogdanoff 1978). A duty cycle is defined as one-year in which the deck is subjected to de-icing salts in winter, freeze-thaw cycles, traffic load (in addition to its own weight).

The probability distribution of damage after a duty cycle is assumed to depend only on the duty cycle itself and the damage accumulated at the start of the duty cycle; thus its is assumed independent of how the damage was accumulated at the start of the duty cycle. This represents the first-order type of stochastic process correlation underlying the Markovian process (Bogdanoff 1978). These assumptions lead to the fact that the damage process can be modeled as a discrete-time and discrete-state Markovian process (Bogdanoff 1978). The probabilistic evolution of damage is completely determined by the transition matrix for each duty cycle and initial damage state. The transition probability matrix for a duty cycle is given by:

$$\mathbf{P}=[p_{j,k}] \qquad j=1,2,\ldots\ldots,b;\text{ and } k=1,2,\ldots.,b \tag{1a}$$

with

$$p_{jk}= P(D_{t+1}=k \mid D_t=j) \tag{1b}$$

where p_{jk} represents the probability of the deck being in state k at the end of the duty cycle given it was in state j at the start of the duty cycle (with $j < k$ for non-maintained systems). Damage states 1 to 6 are transient states, whereas damage state 7, denoted "*state b*" is called an *absorbing state*, which is a state that cannot be vacated without a maintenance action.

Given the adopted condition rating scale and short duration of the duty cycle or transition time (1 year), the probability of deteriorating by more than one state (i.e. multiple damage states transitions) may be assumed negligible (Golabi and Shepard 1997; Lounis 2000; Morcous et al. 2003). Therefore, the transition matrix is greatly simplified and has only two elements per row, namely $p_{k,k}$ and $p_{k,k+1}$, which is referred to as the *"unit jump Markov chain"* (Bogdanoff 1978). Given the uncertainty in defining the end of life or failure criterion, it is possible to have different definitions of the absorbing state, depending on the requirements of the bridge owner, risk of failure, etc. The initial state of damage D_0 is identified by the vector $\mathbf{p}_0=[p_o(i)]_{i=1,b}$, where $p_o(i)$ is the probability of being in state i at time t=0. This initial damage may arise from poor materials, inadequate design and/or construction. It follows from Markov chain theory (Bogdanoff 1978; Ross 1996) that the damage state vector at time t, $\mathbf{p}_t$, is given by:

$$\mathbf{p}_t = \mathbf{p}_o\, \mathbf{P}_1\mathbf{P}_2\ldots\ldots.\mathbf{P}_t = [p_t(1)\;\; p_t(2) \qquad p_t(b)] \tag{2a}$$

where $\mathbf{P}_j$ is the transition matrix for the jth duty cycle, and $p_t(k)$ is the probability of being in state k at time t. If we assume that the duty cycles are all of constant magnitude throughout the deck lifetime, then the transition probability matrices are time-invariant and equal to $\mathbf{P}$, which yields a stationary stochastic process. Therefore, Eq. (2a) simplifies to:

$$\mathbf{p}_t = \mathbf{p}_o\, \mathbf{P}^t \tag{2b}$$

The above transition probability matrix is generated from the data collected during the inspections of the bridge decks. Contrary to lifetime models, the transition matrix and thus the proposed cumulative damage model can be developed from a limited set of data, which then can be further refined using the Bayesian updating approach (Golabi and Shepard 1997; Lounis and Madanat 2002). The probability that the deck be in damage state j at time t is given by:

$$P(D_t = j) = p_t(j) \tag{3a}$$

The cumulative distribution function of damage at time t, D_t, is defined by:

$$F_{Dt}(j) = P(D_t \leq j) = \sum_{k=1}^{j} p_t(k) \tag{3b}$$

The expected damage at time t, E[D(t)] is given by:

$$E[D(t)] = \sum_{j=1}^{b} jp_t(j) \tag{3c}$$

The service life (L) of the deck may be defined as the time to absorption at state b. For the case of an initial damage vector with $p_0(1)=1$, its cumulative distribution function F_L is given by (Bogdanoff 1978):

$$F_L(t) = P(L \leq t) = P_t(b) \qquad t=1, 2,\ldots, n \tag{4}$$

The expected service life E[L] is given by the mean time to absorption (Bogdanoff 1978), i.e.:

$$E(L) = \sum_{t=1}^{\infty} [1 - F_L(t)] \tag{5}$$

For the case of an initial damage vector with multiple nonzero elements, the cumulative distribution function of service life (or time to absorption) is given by (Bogdanoff 1978):

$$F_T(t) = P(T \leq t) = \sum_{k=1}^{b-1} p_0(k)F_{Tk}(t) \tag{6}$$

where $F_{Tk}(t)$is the cumulative distribution function of the time at which the damage state first enters the absorbing state, given the initial damage state is k.

The probabilistic prediction of accumulation of damage in the bridge deck using Eq.(2a) is illustrated in Fig.2, which indicates the evolution with time of the probability mass function of the damage. In Fig.2, it is seen that in the early stages of the deck life, the probability mass of the damage is near state 1, but with aging and damage accumulation, this probability mass shifts to higher damage states. Ultimately, if no maintenance is undertaken, all the probability mass accumulates in the absorbing state 7 or state "b".

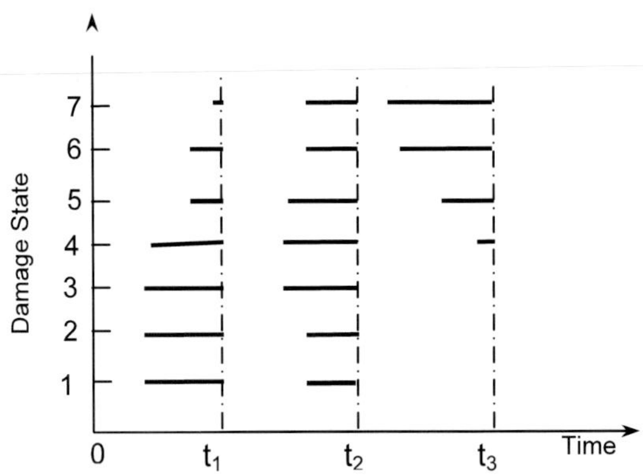

Fig. 2. Probabilistic evolution of bridge deck deterioration using Bogdanoff's CD model

Multi-objective Maintenance Optimization of Aging Bridge Decks

As discussed earlier, the actual maintenance optimization problem is multi-objective in nature as the bridge owner or manager seeks to satisfy simultaneously several objectives, such as the minimization of maintenance costs, improvement of safety, serviceability and functionality, minimization of maintenance time, minimization of traffic disruption, minimization of user costs, and minimization of risk of failure, etc. The solution of this maintenance management problem can be obtained using the techniques of multi-criteria or multi-objective optimization. Several approaches have been developed to solve multi-objective optimization problems, including multi-attribute utility theory (Von Neumann and Morgenstern 1947; Keeney and Raiffa 1976), weighted sum approach (Zadeh 1963), compromise programming, constraint approach, and sequential optimization (Koski 1984; Duckstein 1984; Lounis et al. 1993, 2000). In this paper, the compromise programming approach is used to solve the multi-objective maintenance optimization problem, and is presented in the next section.

Overview of Multi-objective Optimization Approach

For single-objective optimization problems, the notion of optimality is very well defined as the minimum or maximum value of some given objective function is sought. In multi-objective (or vector) optimization problems, the notion of optimality is not obvious because of the presence of multiple, incommensurable and conflicting objectives. In general, there is no single optimal (non-dominated or superior) solution that simultaneously yields a minimum (or maximum) for all objective functions. The Pareto optimality concept has been introduced as the solution to multi-objective optimization problems (Koski 1984; Duckstein 1984). A maintenance strategy $\mathbf{x}^*$ is said to be a Pareto optimum if and only if there exists no maintenance strategy in the feasible set of maintenance alternatives that may yield an improvement of some criterion without worsening at least one other criterion. The multi-objective maintenance optimization problem can be mathematically stated as follows:

Find: $$x^* = [x_1^*, x_2^*, \ldots\ldots, x_n^*] = optimum \tag{7a}$$

Such that: $$f(x) = [f_1(x), f_2(x) \ldots\ldots\ldots f_n(x)] = minimum \quad x \in \Omega \tag{7b}$$

and $$\sum_{j=1}^{n} C_{m,t}(x_j) \le C_{max} \tag{7c}$$

$$\Omega = \{x \in N : CR(x) \ge CR_{th}\} \tag{7d}$$

where: x^* = vector of optimum solutions; **f** = vector of optimization objectives (e.g. condition rating, maintenance cost, user cost, etc.); $C_{m,t}(x_j)$=maintenance cost of project x_j at time t ; C_{max}= available budget; **Ω**= subset of the bridge network that at time t contains deficient bridge decks having a condition rating (CR) above a specified threshold value (requiring maintenance); N= entire set of bridge deck projects within the network.

The concept of Pareto optimality mentioned above, may be stated mathematically as follows (Koski 1984; Lounis and Cohn 1993):

$$\mathbf{x}^* = \text{Pareto optimum} \tag{8a}$$

if $$f_i(\mathbf{x}) \le f_i(\mathbf{x}^*) \quad \text{for } i=1,2,\ldots,m \tag{8b}$$

and $$f_k(\mathbf{x}) < f_k(\mathbf{x}^*) \quad \text{for at least one } k \tag{8c}$$

In general, for a multi-objective optimization problem, there are several Pareto optima, and the problem is to select the solution that achieves the best compromise between all competing objectives. Such a solution is referred to as "*satisficing*" solution in the multi-objective optimization literature (Koski 1984; Lounis and Cohn 1995). The determination of this satisficing solution is discussed in the next section.

Decision-Making under Multiple and Conflicting Objectives

In compromise programming, the "*satisficing*" solution is defined as the solution that minimizes the distance from the set of Pareto optima to the so-called "ideal solution". This ideal solution is defined as the solution that yields minimum (or maximum) values for all criteria. Such a solution does not exist, but is introduced in compromise programming as a target or a goal to get close to, although impossible to reach. The criterion used in compromise programming is the minimization of the deviation from the ideal solution **f*** measured by the family of L_p metrics (Koski 1984; Lounis and Cohn 1993). In this paper, a multi-criteria optimality or multi-objective index, "MOI", is defined as the value of the weighted and normalized deviation from the ideal solution **f*** measured by the family of L_p metrics:

$$MOI_p(x) = \left[\sum_{i=1}^{m} w_i^p \left| \frac{f_i(x) - min\, f_i(x)}{max\, f_i(x) - min\, f_{i(x)}} \right|^p \right]^{1/p} \qquad (9)$$

This family of Lp metrics is a measure of the closeness of the satisficing solution to the ideal solution. The value of the weighting factors w_i of the optimization criteria f_i (i=1,...,m) depends primarily on the attitude of the decision-maker towards risk. The choice of p indicates the importance given to different deviations from the ideal solution. For example, if p=1, all deviations from the ideal solution are considered in direct proportion to their magnitudes, which corresponds to a group utility (Duckstein 1984). However, for p ≥2, a greater weight is associated with the larger deviations from the ideal solution, and L_2 represents the *Euclidian metric*. For p=∞, the largest deviation is the only one taken into account and is referred to as the *Chebyshev metric* or *mini-max criterion* and L_∞ corresponds to a purely individual utility (Duckstein 1984; Koski 1984; Lounis and Cohn 1995; Lounis and Vanier 2000). In this paper, both the *Euclidean* and the *Chebyshev* metrics are used to determine the multi-objective optimality index and corresponding satisficing solution.

Illustrative Example

The approach presented in this paper is applied for the maintenance optimization of 100 bridge deck projects from different bridges within a network of a highway agency. The objective here is to optimize the prioritization of the bridge deck projects for maintenance, considering simultaneously their condition rating, maintenance cost, and user costs, assuming a constraint on the allocated funding for the current year, which is assumed as $1.5 Million. The physical condition of the bridge decks is assessed using the 1-5 condition rating scale described earlier.

The deterioration of the bridge decks is predicted using the Bogdanoff's cumulative damage model described earlier, assuming a unit-jump and stationary deterioration model. A constant duty cycle is assumed throughout the service life of bridge decks that consists in one-year exposure to chloride-induced corrosion due to deicing salts, freeze-thaw cycles, and traffic loading. The transition probability matrix has the following elements (Lounis 2000): p_{11}=0.7, p_{22}=0.765, p_{33}=0.85, p_{44}=0.9, and p_{55}=0.98, p_{66}=0.98, and p_{77}=1. The probability mass function of the current network condition is shown in Fig.3(a), and is given by the following initial condition vector:

$$\mathbf{p}_0=[0.06\ \ 0.34\ \ 0.31\ \ 0.19\ \ 0.08\ \ 0.01\ \ 0.01] \qquad (10a)$$

Using Eq. 2(b), the predicted deterioration of the bridge deck network after 10, 20 and 30 years is shown in Fig. 3(b). For example, after 30 years, the probabilistic distribution of the condition of the deck network is given by the vector:

$$\mathbf{p}_{30}=[0.\ \ 0.\ \ 0.008\ \ 0.089\ \ 0.547\ \ 0.273\ \ 0.082] \qquad (10b)$$

Hence, after 30 years, no deck of the network is in damage state 1 or 2 (i.e. no damage or minor damage), and about 90% of the deck network is in damage state 5, 6 or 7 (i.e. poor to failed states).

For bridge decks with condition ratings (or damage states) 1 or 2, no action or some preventive maintenance may be needed. For decks in damage states 3 and 4, some intermediate maintenance actions may be required (patch repair, asphalt overlay). For the decks with damage states 5 or higher, the possible maintenance actions include deck overlay using latex-modified concrete or low-slump dense concrete, cathodic protection, partial or complete replacement of the deck. To illustrate the application of the approach, the maintenance optimization is carried on a group of bridge decks with damage states 5 or higher, i.e. the threshold condition rating CR_{th} is assumed equal to 5.

In this example, the maintenance alternatives are assumed optimized for the individual deficient bridge decks on the basis of life cycle cost minimization. The present value costs of the maintenance alternatives for the ten most damaged bridge decks are summarized in Table 2. The user costs are assumed to represent the sum of all costs incurred by the users during the maintenance activity, which include the delay costs, accident costs, and vehicle operating costs. These user costs depend primarily on the duration of the maintenance activity, average daily traffic, accident rate increase due to traffic detour or/and lane closure, and are summarized in Table 2.

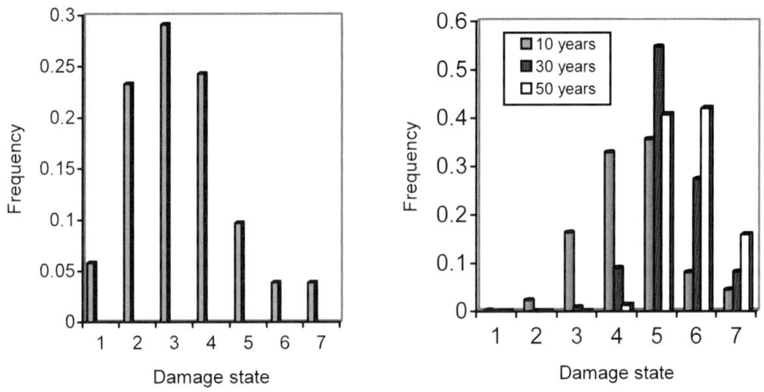

(a) Current network condition **(b) Network condition after 10, 30, and 50 years**

Fig.3. Current and future network condition predicted using Bogdanoff CD model

The normalized values (normalized with regard to the maximum value) of the three selected objective functions are shown in Fig. 4. Table 2 and Fig. 4 illustrate the conflicting nature of these criteria and the difficulty in prioritizing, as the project with the highest urgency in terms of condition rating (Project #1) is neither the same in terms of maintenance cost (Project #9) nor in terms of user cost (Project #4). If single objective-based optimizations are undertaken, the bridge deck projects will be prioritized for maintenance following these three different schemes, depending on the selected objective:

- Damage condition rating -based prioritization: the projects will be ranked in terms of decreasing condition rating, i.e. the project with the highest condition rating will given first priority, and end up with the project which exhausts the available budget;
- Maintenance cost-based prioritization: the projects will be ranked in terms of increasing cost, i.e. the project with the lowest maintenance cost will be given first priority, ending with the project at which the available budget is exhausted;
- User cost-based prioritization: the projects will be ranked in terms of decreasing user costs, i.e. the project with the highest user cost will be given first priority, ending with the project at which the available budget is exhausted.

The required total fund to address all maintenance needs for the 10 deck projects is $2.891 million, which is well in excess of the available budget of $1.5 million. From Table 2, the "ideal" (but non-existing) maintenance solution is associated with the following "ideal" objective vector $\mathbf{f}^*=[f_{1max}\ f_{2min}\ f_{3max}]^T= [7,\ 75000,\ 153000]^T$.

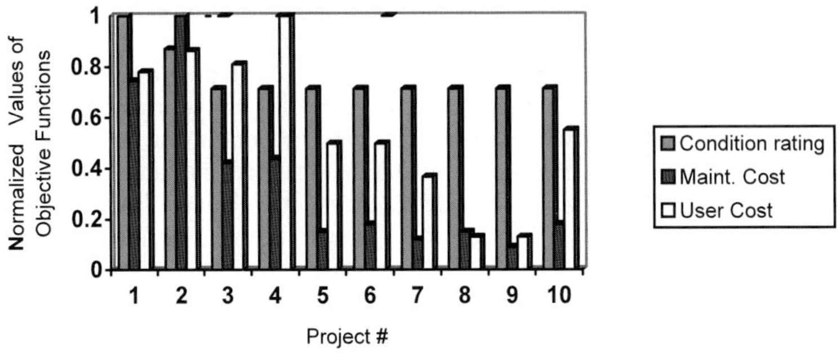

Fig. 4. Normalized values of objective functions for deck maintenance projects

Table 2. Multi-objective-based maintenance optimization of bridge decks

Bridge Deck Project #	Damage Condition Rating	Maintenance Costs ($1,000)	User Costs ($1,000)	Euclidean Metric MOI_2	Chebyshev Metric MOI_∞
1	7	620	120	1.260	1.000
2	6	832	132	1.130	1.000
3	5	350	124	0.424	0.363
4	5	364	153	0.382	0.382
5	5	125	76	0.583	0.579
6	5	150	76	0.587	0.579
7	5	100	56	0.730	0.729
8	5	125	20	1.002	1.000
9	5	75	20	1.000	1.000
10	5	150	84	0.528	0.519
	Average= 5.3	Σ=$2891,000	Σ=$861,000	MIN=0.382	MIN=0.363

Using Eq. (9), the values of the multi-objective optimality indices MOI_2 and MOI_∞, corresponding to the Euclidean and Chebyshev metrics, respectively, are determined for the bridge deck projects and are summarized in Table 2. Using the *min.* MOI_2 criterion, the "satisficing" solution is found to be Project # 4, however using the *min.* MOI_∞ criterion, the "satisficing" solution is found to be Project # 3. Fig. 5, however, illustrates the similarity between the rankings of projects for maintenance obtained using both the *min.* MOI_2 and *min.* MOI_∞ criteria for the other deck projects.

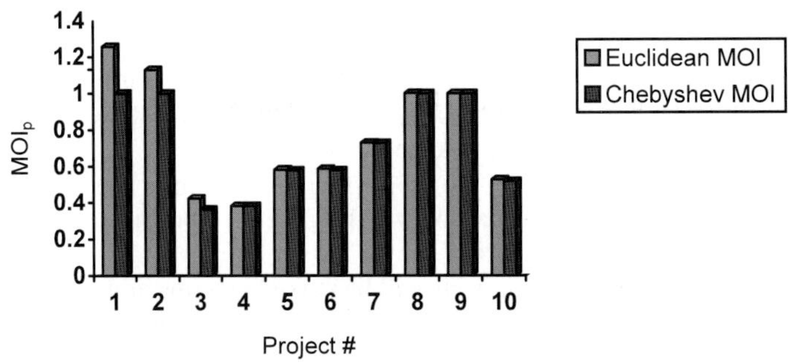

Fig. 5. Multiobjective-based maintenance prioritization of bridge decks

Considering now the budgetary constraint, the scheduling of projects maintenance will be as follows:

(i) Euclidean metric-based prioritization: Projects #4, #3, #10,#5, #6, #8, and #9, for a total cost of $1.439 million. The other projects may delayed until the next year; however, a detailed analysis may be required.

(ii) Chebyshev metric-based prioritization: Projects #3, #4, #10, #5, #6, #7, #8, and #9 for a total cost of $1.439 million.

From the above example, both the Euclidean and Chebyshev (or minimax) criteria for multi-objective optimization yield the same prioritization of bridge deck projects for maintenance. It should be pointed out that for both metrics, project #6 has always a higher priority than project #8, because the latter is not a "true" Pareto optimum, as it is dominated by solution (or project) #6. It is also possible to use the proposed approach by introducing different weighting factors on the different objectives, as discussed in the previous sections.

Conclusions

This paper illustrated that the bridge deck maintenance optimization problem can be formulated as a multi-objective optimization problem. The major merits of the approach are: (i) consideration of all possible (even conflicting) objective functions; (ii) ability to put more emphasis on the more relevant objectives (e.g. condition improvement); and (iii) rational decision-making regarding the selection of bridge projects for maintenance. The prioritization of the bridge decks is based on the satisfaction of several conflicting objectives simultaneously, including improving the physical condition, reducing the maintenance costs and user costs. The proposed multi-objective optimization approach provides a decision support tool for effective bridge management that enables decision-makers to select all relevant objectives in planning the maintenance of their bridge network. The development and integration of the proposed models for maintenance optimization will lead to an effective approach to bridge maintenance management, which optimizes the allocation of maintenance funds, as well as improves the risk management of bridge decks. The solutions obtained achieved a satisfactory trade-off between several competing criteria, including the maximization of the bridge deck condition, minimization of maintenance costs, and minimization of user costs. The proposed multi-objective optimality index can be used as an effective optimality criterion for the prioritization of deteriorated bridge decks for maintenance.

References

Bogdanoff, J.L.(1978). "A new cumulative damage model – Part 1." *J. Applied Mechanics*, ASME, 45(2), 246-250.

Duckstein, L. (1984). "Multi-objective optimization in structural design: The model choice problem." In *New directions in optimum structural design,* Atrek et al., eds., Wiley, 459-481.

Dunker, K.F., and Rabbat, B.G. 1990. "Highway bridge type and performance patterns." *J. Performance of Constructed Facilities*, ASCE, 4(3),161-173.

Golabi, K., and Shepard, R. (1997). "Pontis: A system for maintenance optimization and improvement of US bridge networks." *Interface,* 27, 71-88

Federal Highway Administration (FHWA). (1995). *Recording and coding guide for the structure inventory and appraisal of the nations' bridges.* Report FHWA-PD-96-001, Washington, D.C.

Frangopol, D.M., Lin, K.Y., and Estes, A.C. (1997). "Reliability of reinforced concrete girders under corrosion attack." *J. Struct. Engrg.*, ASCE,123(3), 286-297.

Keeney, R.L., and Raiffa, H. (1976). *Decisions with multiple objectives: Preferences and value tradeoffs*. J. Wiley & Sons, New York.

Koski, J. (1984). Multi-objective optimization in structural design. In *New directions in optimum structural design,* Atrek et al., eds., Wiley, New York, 484-503.

Lounis, Z., and Cohn, M.Z. (1993). "Multi-objective optimization of prestressed concrete structures." *J. Struct. Engrg.,* ASCE, 119(3), 794-808.

Lounis, Z., and Cohn, M.Z. (1995). An engineering approach to multi-criteria optimization of highway bridges. *J. Computer-Aided Civ. & Infrast. Engrg.,* 10(4), 233-238.

Lounis, Z. (2000)."Reliability-based service life prediction of aging concrete bridge decks." In *Life prediction and aging management*, Naus, D (ed.), RILEM publications, 965-972.

Lounis, Z., and Vanier, D.J. (2000). "A multi-objective and stochastic system for building maintenance management." *J. Computer-Aided Civ. & Infrast. Engrg.*,15, 320-329.

Lounis, Z., and Madanat, S.M. (2002). "Integrating statistical and mechanistic deterioration models for effective bridge management." *7th ASCE Conf. on Appl. of Adv. Tech. in Trans.*, 513-520.

Melchers, R.E. (1987). *Structural reliability: analysis and prediction*. Ellis Horwood, England.

Ministry of Transportation, Ontario (MTO). (1989).*Ontario Structure Inspection Manual*.Queen's Printer for Ontario.

Morcous, G., Lounis, Z., and Mirza, M.S. (2003). "Identification of environmental categories for Markovian deterioration models of bridge decks." *J. Bridge Engrg.*, ASCE, 8(6) 353-361.

Mori, Y., and Ellingwood, B. (1993). "Reliability-based service life assessment of aging concrete structures." *J. Struct. Engrg.,* ASCE, 119(5), 1600-1621.

Paté-Cornell, M.E.(1994). "Quantitative safety goals for risk management of industrial facilities." *J. Structural Safety*, 13, 145-157.

Starr, C. 1969. Social benefit vs. technical risk. *Science*, 165, 1232-1238.

Stewart, M. and Rosowsky, D.V. (1998). "Structural safety and serviceability of concrete bridges subject to corrosion." *J. Infrast. Systems*, ASCE, 4(4), 146-155.

Von Neumann, J. and Morgenstern, O. (1947). *Theory of games and economic behavior*. Princeton University Press, Princeton.

Weyers, R.E.(1998). "Service life model for concrete structures in chloride laden environments. *J. ACI Materials*, 95(4), 445-453.

Zadeh, L.A. 1963. Optimality and non-scalar valued performance criteria. *IEEE Transaction Aut. Cont*, 8(1), 59-60.

A COMPARISON OF PROBABILISTIC MODELS OF DETERIORATION FOR LIFE CYCLE MANAGEMENT OF STRUCTURES

M.D. Pandey and X.-X. Yuan
Department of Civil Engineering, University of Waterloo, Waterloo, Ontario, Canada

Abstract

The probabilistic modelling of deterioration in the time-dependent reliability analysis is a necessary step for developing a risk-based approach to the life cycle management of infrastructure systems. The decisions regarding the time and frequency of inspection, maintenance and replacement are confounded by sampling and temporal uncertainties associated with the deterioration of structural resistance. To account for these uncertainties, probabilistic models of deterioration have been developed under two broad categories, namely the random variable model and stochastic process model. The paper presents a conceptual exposition of these two models and highlights their profound implications to the age-based and condition-based preventive maintenances policies. The proposed stochastic gamma process model of deterioration is more versatile than the random rate model commonly used in the structural reliability literature.

Keywords: life cycle management, structural reliability, age-based replacement, condition-based maintenance, gamma process

1 Introduction

Time-dependent reliability analysis is necessary to develop optimum strategies for the life-cycle management of infrastructure systems that include roads, bridges, nuclear plants and transmission lines. The decisions regarding the time and frequency of inspection, maintenance and replacement are confounded by uncertainties associated with the deterioration of structural resistance. In general, the modelling of deterioration is influenced by sampling and temporal uncertainties. The sampling uncertainty refers to the variability of deterioration from sample to sample. The uncertainty inherent with the progression of deterioration over time is referred to as temporal uncertainty. The sampling uncertainty, an epistemic uncertainty, can be reduced by additional inspections. The temporal uncertainty on the other hand is aleatory in nature so that it cannot be eliminated completely by increasing inspections. An adequate consideration of temporal uncertainty is necessary for a credible and effective life cycle management of critical infrastructures.

The probabilistic models of deterioration can be classified in two broad categories, namely, random variable (RV) model and stochastic process model. In the RV model, parameters associated with an empirical deterioration law are randomized to reflect the sampling variability observed in a sample of deterioration data, such as the rate of deterioration (Hong 2000, Pandey 1998). The stochastic process model, such as the Markov chain or Gamma process, incorporates the temporal uncertainty associated with evolution of deterioration (Bogdanoff and Kozin 1985, Nicolai et al. 2004, van Noortwijk and Frangopol 2004). A key distinction between these two models is that a specific sample path is deterministic in RV model, but it remains uncertain in the stochastic process model.

The application of the random variable and stochastic process deterioration models have been hitherto reported, but a clear interpretation of conceptual distinctions between these two models and their impact on maintenance optimization problem have been lacking in the engineering literature. To address this issue, the paper evaluates the random variable and stochastic Gamma process models in a simplified setting of time-dependent structural reliability analysis. The two equivalent versions of the deterioration models are compared in terms of distributions of lifetime, deterioration magnitude and the life cycle cost. The paper presents an original exposition of the implications of deterioration models to the age-based and condition-based preventive maintenances policies.

M. Pandey et al. (eds), Advances in Engineering Structures, Mechanics & Construction, 735–746.

2 Models of Deterioration

2.1 Random Variable (RV) Deterioration Model

In case of a deteriorating system, the structural failure is defined as an event when the strength falls short of the applied stress. A corresponding limit state function is defined as

$$R(t) - s = [r_0 - X(t)] - s = \rho - X(t) \le 0 \tag{1}$$

where $R(t)$ denotes the deteriorating resistance at time t, s the load effect, r_0 the initial resistance, $X(t)$ the cumulative deterioration at time t, and $\rho = (r_0 - s) > 0$ is the available design margin or a failure threshold. For the sake of simplicity of the discussion, r_0, s and ρ are assumed as deterministic constants. Thus the failure is the event of cumulative deterioration $X(t)$ exceeding the threshold ρ.

The random variable (RV) model characterizes the randomness of the deterioration by a finite-dimension vector of time invariant random variables Θ as $X(t;\Theta)$. For example, consider a simple linear deterioration model as

$$X(t) = At \tag{2}$$

where A is the deterioration rate, which is typically randomized to reflect the variability in a large population of similar components. Given the probability distribution of random rate, $F_A(a)$, the distribution of the amount of deterioration, $X(t)$, is derived as $F_{X(t)}(x) = F_A(x/t)$. The mean, variance and coefficient of variation (COV) of $X(t)$ are expressed respectively as

$$\mu_{X(t)} = \mu_A t,\ \sigma^2_{X(t)} = \sigma^2_A t^2, \text{ and } \nu_{X(t)} = \frac{\sigma_{X(t)}}{\mu_{X(t)}} = \nu_A \tag{3}$$

According to the failure definition in Eq. (1), the cumulative probability distribution of the lifetime, T, can be written as

$$F_T(t) = P[T \le t] = P[\rho / A \le t] = P[A \ge (\rho / t)] = 1 - F_A[\rho / t] \tag{4}$$

Depending on the probability distribution of A, the lifetime distribution can be derived analytically or computed numerically.

Suppose the deterioration rate is a gamma distributed random variable with probability density function given as

$$f_A(a) = \frac{(a/\delta)^{\eta-1}}{\delta\,\Gamma(\eta)} e^{-a/\delta}, \text{ for } a \ge 0 \tag{5}$$

where η and δ are the shape and scale parameter, respectively. We denote the gamma density function by $ga(a;\eta,\delta)$ and its cumulative distribution function by $GA(a;\eta,\delta)$. From Eq.(2), the deterioration, $X(t)$, is also gamma distributed with density function $ga(x;\eta,\delta\, t)$. From Eq.(4), the lifetime ($T = \rho / A$) follows an inverted gamma distribution with the following density function:

$$f_T(t) = \frac{(\delta/\rho)}{\Gamma(\eta)} \left(\frac{\rho}{\delta\, t}\right)^{\eta+1} e^{-\rho/\delta t} \tag{6}$$

and the cumulative distribution function is written as

$$F_T(t) = 1 - \mathrm{GA}(\rho / t; \eta, \delta) \tag{7}$$

It can be shown that the moments of the lifetime are given as follows:

$$\mu_T = \frac{\rho}{\delta(\eta-1)},\ \sigma_T^2 = \frac{\rho^2}{\delta^2(\eta-1)^2(\eta-2)},\ \text{and } \nu_T = \frac{\sigma_T}{\mu_T} = \frac{1}{\sqrt{\eta-2}},\ \text{for } (\eta > 2) \tag{8}$$

Since the gamma distributed deterioration rate has mean $\delta\eta$ and COV $\sqrt{1/\eta}$, the moments of lifetime distribution can be related with that of the degradation rate as

$$\mu_T = \frac{\rho}{\mu_A(1-\nu_A^2)},\ \sigma_T^2 = \frac{\rho^2}{\mu_A^2(1-\nu_A^2)^2(1/\nu_A^2-2)} \text{ and } \nu_T = \frac{1}{\sqrt{1/\nu_A^2-2}} \tag{9}$$

2.2 *Gamma Process (GP) Deterioration Model*

The gamma process is a continuous-time Markov process with independent and gamma distributed increments (Abdel-Hameed 1975). The gamma process is a limit of a compound Poisson process with gamma distributed increments. The limit is reached when the Poisson rate approaches infinity in any finite time interval as the size of the increment tends proportionally to zero (Dufresne et al. 1991). In other words, the gamma process model implies that deterioration progresses with frequent occurrence of very small increment. A physical example such process is the flow accelerated corrosion in nuclear piping system.

In the stationary gamma process (GP) model, the cumulative deterioration $X(t)$ follows a gamma distribution $ga(x;\alpha\, t, \beta)$ with the shape parameter αt and the scale parameter β. The mean, variance and COV of $X(t)$ are given as

$$\mu_{X(t)} = \alpha\beta\, t,\ \sigma_{X(t)}^2 = \alpha\beta^2 t,\ \text{and } \nu_{X(t)} = \frac{1}{\sqrt{\alpha\, t}} \tag{10}$$

In this model, both mean and variance of the deterioration are linear in time. In RV model, the variance is quadratic in time as shown in Eq.(3). From Eq. (1), the cumulative lifetime distribution can be expressed as

$$F_T(t) = P[X(t) \geq \rho] = 1 - GA[\rho;\alpha t, \beta] \tag{11}$$

The probability density function, $f_T(t) = dF_T(t)/dt$, has no closed form expression, but it can be computed numerically. Similarly, the moments of the lifetime have to be evaluated by numerical integration.

3 Comparison of RV and GP Models

3.1 *Calibration of the Parameters of RV and GP Models*

In order to have a consistent comparison between the RV and GP models, a careful scheme is required for the calibration of the model parameters. In practical situations, typically a sample of lifetime data is available from the past failures records, and the mean (μ_T) and COV (ν_T) of the lifetime can be estimated from such a sample.

If we assume that the random lifetime sample is generated by a RV deterioration model, the parameters of random rate can be derived in terms of the mean and COV of the lifetime using Eq.(8) as

$$\eta = 2 + 1/\nu_T^2 \text{ and } \delta = \frac{\rho\nu_T^2}{\mu_T(1+\nu_T^2)} \tag{12}$$

If however the underlying degradation model is assumed to be a stochastic gamma process, the shape and scale parameters can be obtained by solving the following equations of moments:

$$\left.\begin{aligned} \mu_T &= \int_0^\infty P\{T \ge t\} dt = \int_0^\infty GA(\rho; \alpha\, t, \beta) dt \\ \mu_T^2 (1 + v_T^2) &= 2\int_0^\infty t P\{T \ge t\} dt = 2\int_0^\infty t GA(\rho; \alpha\, t, \beta) dt \end{aligned}\right\} \quad (13)$$

The two simultaneous equations must be solved numerically to compute the parameters α and β.

In summary, in the present calibration scheme, both RV and GP models are equivalent in a sense that they have identical mean and variance of the lifetime. The parameters of these two models are given in Tables 1 and 2.

Table 1: Parameters used in the model calibration

Mean Lifetime μ_T	***COV of Lifetime*** V_T	***Failure threshold*** ρ
50 units of time (fixed)	Varied from 0.1 to 0.9	100 units (fixed)

Table 2: Examples of RV and GP model parameters

Parameters		$v_T = 0.3$	$v_T = 0.6$	$v_T = 0.9$
RV Model	Shape (η)	13.1111	4.7778	3.2346
	Scale (δ)	0.1651	0.5294	0. 8950
GP Model	Shape (α)	0.2099	0.0408	0. 0078
	Scale (β)	10.01	64.63	1,453.6

3.2 *Comparison of Deterioration Distribution*

The evolution of deterioration is compared in the two equivalent RV and GP models. Figure 1 compares the mean rate of deterioration as a function of COV of the lifetime distribution (v_T) by fixing the mean lifetime and failure threshold as $\mu_T = 50$ and $\rho = 100$.

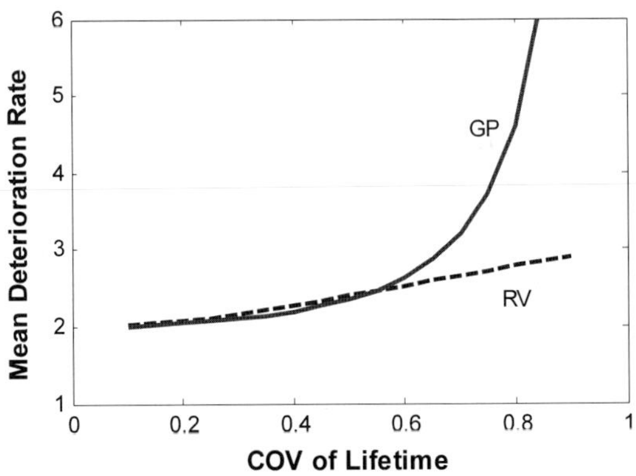

Figure 1: Comparison of the mean deterioration rate in equivalent RV and GP models ($\mu_T = 50$)

The mean deterioration rates in the RV and GP model are given as $\mu_A = \eta\delta$ and $\alpha\beta$, respectively. When $v_T < 0.6$, the mean rates in both models are almost identical. However, in cases of $v_T > 0.6$, the GP deterioration rate accelerates much faster than that in the RV model.

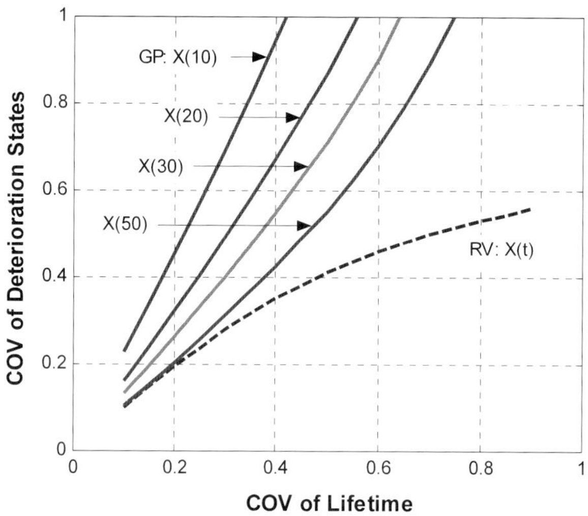

Figure 2: Coefficient of variation (COV) of deterioration states in equivalent RV and GP models

Figure 2 compares the COV of deterioration, $\nu_{X(t)}$, in the two models. It is a time-invariant and nonlinear function (Eq.3 and 9) of ν_T in the RV model. In contrast, $\nu_{X(t)}$ in GP model is a time-dependent parameter (Eq. 10), which is decreasing over time. Nevertheless, $\nu_{X(t)}$ of GP model is always greater than that of an equivalent RV model.

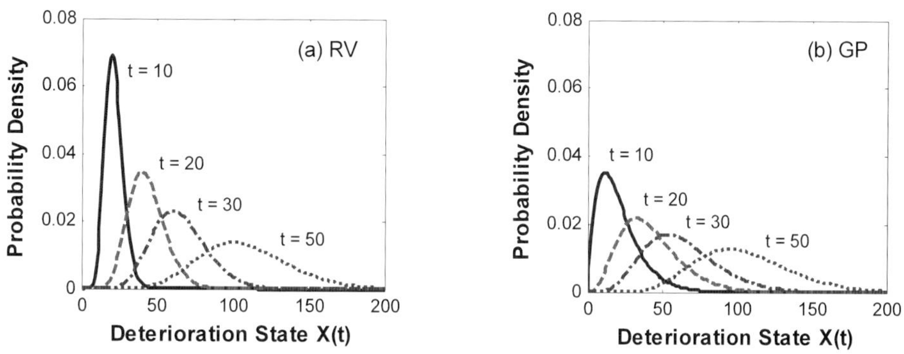

Figure 3: Probability density functions of deterioration $X(t)$ for $\nu_T = 0.3$

The probability density function (PDF) of $X(t)$ is given as $ga(x;\eta,\delta\, t)$ and $ga(x;\alpha\, t,\beta)$ in the RV and GP models, respectively. The evolution of the PDF with time is displayed in Figure 3 for $\nu_T = 0.3$. A key observation is that deterioration in GP model has greater variability than that in an equivalent RV model.

3.3 Comparison of Lifetime Distributions

The model calibration scheme is based on the method of moments that yields the parameters of RV and GP models such that the mean and COV of the lifetime distribution are identical in both models. In spite of the identical first two moments, the tails of lifetime distribution in RV and GP models can be remarkably different, as shown in Figure 4. Figure 4(a) shows that the lifetime distribution of the RV model is more skewed that that of the GP model. The difference between distribution tails can be seen more clearly through a comparison of survival functions in Figure 4(b). The GP model is more pessimistic than RV model about the prospect of survival due to uncertainty associated with evolution of deterioration that takes place in form of independent gamma distributed increments. In contrast, the deterioration in the RV model is fully correlated over the lifetime, which results in an overestimation of the survival probability.

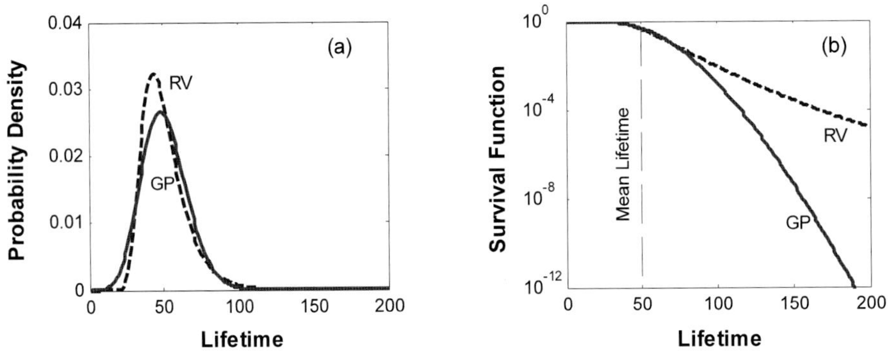

Figure 4: Comparison of lifetime distributions in equivalent RV and GP models for $\nu_T = 0.3$: (a) probability density, and (b) survival function

4 Age-Based Replacement Policy

The age-based replacement is the simplest policy for the renewal of aged fleet of structures and components. In this policy, a component is replaced when it reaches to a specific age (t_0) regardless of its condition. The component is of course replaced, if failure occurs before the replacement time, t_0.

Denote the total cost associated with all the consequences of a structural failure as C_F, and the cost of a preventive replacement as C_P. According to the renewal theory, the average cost per unit time in long term, also known as the mean cost rate K, can be computed as a function of the replacement age (Barlow and Proschan 1965):

$$K(t_0) = \frac{F_T(t_0)C_F + [1 - F_T(t_0)]C_P}{\int_0^{t_0}[1 - F_T(t)]dt} \tag{14}$$

It is easy to check that $K(t_o) \to C_F / \mu_T$ as $t_0 \to \infty$. Using Eq.(14), an optimal age of preventive replacement (t_0) can be found that would minimize the mean cost rate.

Since the calculation of the cost rate is sensitive to the lifetime distribution $F_T(t)$, it would be of interest to examine the impact of RV and GP model on the replacement policy. For an illustration, the cost data are assumed as $C_P = 10$ and $C_F = 50$. The basic data given in Tables 1 and 2 are used in the calculation of lifetime distribution. Figure 5 shows the variation of mean cost rate with the replacement age when the lifetime COV is $\nu_T = 0.3$. The optimal replacement age for both models is about the same, 29 units, but corresponding mean cost rate for the GP model ($K(29) = 0.45$ units) is greater than that of the RV model (0.38). The reason is that the GP model involves higher uncertainty than the RV model, as discussed in Section 3.

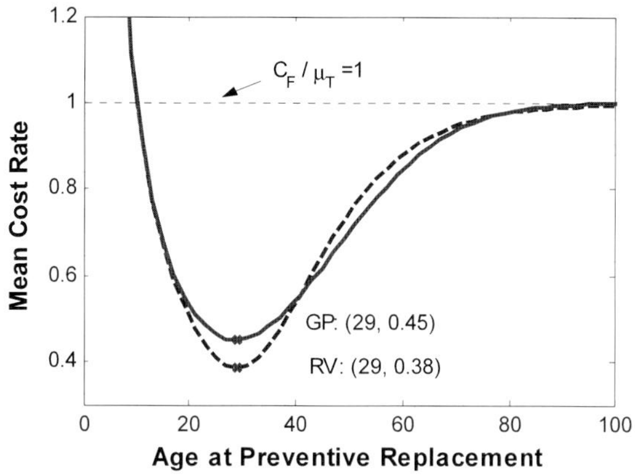

Figure 5: Mean cost rate as a function of replacement age in equivalent RV and GP models ($\nu_T = 0.3$)

A comprehensive comparison of optimal cost rate and replacement age results for other values of ν_T in the RV and GP models are shown in **Figure 6**. The results of optimal replacement age obtained from the two models are qualitatively different. In the RV model, the optimal replacement age decreases continuously as the lifetime COV ν_T increases. This observation is somewhat intuitive in the sense that when faced with increased uncertainty it is prudent to reduce the replacement age. The results of GP model exhibit two distinct trends. Initially, with increase in lifetime COV the replacement age decreases. However, as ν_T increases beyond 0.5, the trend reverses and the replacement age begins to increase. It means that in case of large uncertainty associated with component lifetime, the life-cycle cost can be optimized by extending the replacement age.

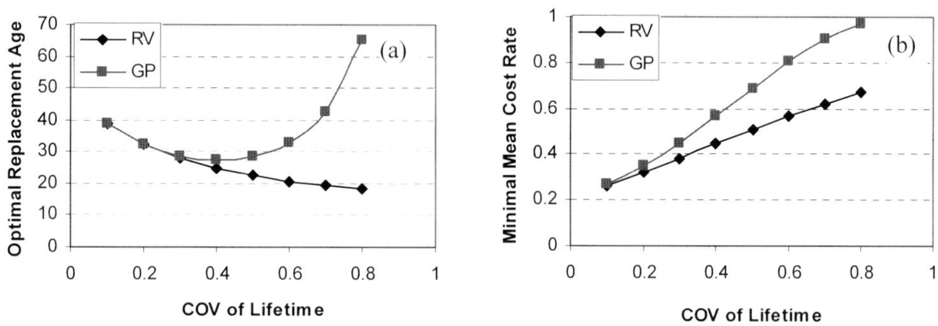

Figure 6: Comparison of age replacement policy in equivalent RV and GP models: (a) replacement age, and (b) mean cost rate

The comparison of the cost rate in Figure 6(b) shows that mean cost rate obtained from GP model is always higher than that of the RV model as a result of additional temporal uncertainty associated with GP model. The difference between the optimum cost rates increases with increase in lifetime uncertainty.

5 Condition Based Maintenance (CBM) Policy

5.1 The Strategy

The deterioration along a specific sample path is deterministic in the RV model, whereas it varies probabilistically in the GP model. In a linear RV model, one inspection determines the deterioration rate and it fixes the future deterioration path. An inspection in GP model reveals only its current state from which we can infer only the probability distribution of future deterioration. This distinction has profound implications to the optimization of condition-based maintenance strategies.

The condition based maintenance (CBM) strategy involves the periodic inspection of a structure at a fixed time interval t_I and cost C_I. We assume that the inspection is accurate such that the deterioration $X(t)$ can be measured with negligible error. The threshold for the preventive maintenance, $c\rho$ $(0 < c < 1)$, is a fraction of the failure threshold. The preventive maintenance (PM) results in complete renewal (as good as new) of the component. If $X(t_I) < c\rho$, no action is taken until the next inspection. A component is renewed with cost C_P when $c\rho < X(t_I) < \rho$. The structure would be immediately replaced upon failure at any time when $X(t) > \rho$, incurring a failure cost, C_F. Typically PM cost is much lower than the failure cost $(C_P < C_F)$.

The optimization of the condition-based maintenance means finding the inspection interval (t_I) and the PM ratio (c) that would minimize the mean cost rate. This in principle involves a two-dimensional optimization problem. In practical situations, however, the PM threshold $c\rho$ is known from experience or prescribed by operation standards or regulation. In such cases, the inspection interval is the only optimization variable. For illustration purpose, we fix the PM ratio as $c = 0.8$ and determine the optimal inspection interval and the associated mean cost rate for the RV and GP models.

5.2 RV Model for Condition Based Maintenance

The three possibilities that arise at any time of inspection are: do nothing, PM or renewal, as shown in the maintenance decision tree in Figure 7.

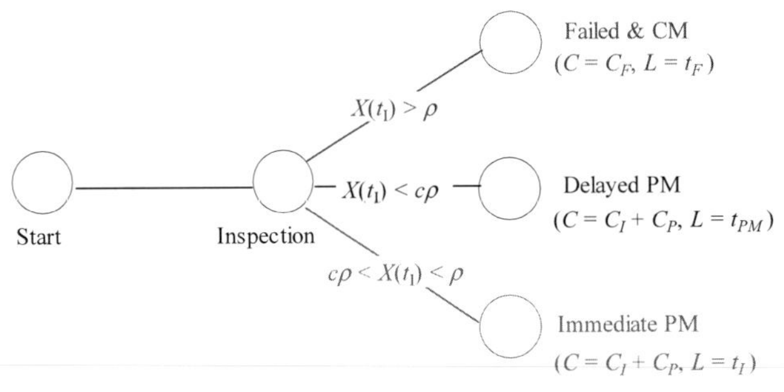

Figure 7: CBM decision tree for the RV model

At the time of first inspection t_I, if we observe $X(t_I) < c\rho$ we do nothing and the time of the future preventive maintenance can be predicted as $t_{PM} = c\rho t_I / X(t_I)$, since one inspection suffices to determine the sample path in the linear RV deterioration. This is referred to as delayed PM and note that no additional inspection is required before the time of replacement. The other two situations are straightforward. The PM is immediately conducted when $c\rho < X(t_I) < \rho$, and it is replaced when $X(t_I) > \rho$. A renewal is thus a delayed PM, immediate PM, or a CM.

The most important point is that the under assumption of RV deterioration model, only one inspection is required for the implementation of CBM strategy.

Using the maintenance decision tree and associated costs shown in Figure 7, it is easy to set up expressions for the mean renewal cycle cost, C, and length, L in terms of the time of first inspection interval t_I and PM ratio c. The renewal cycle cost is simply evaluated as

$$E[C(t_I,c)] = (C_I + C_P)P\{X(t_I) \le \rho\} + C_F P\{X(t_I) > \rho\} = (C_I + C_P - C_F)F_A(\rho/t_I) + C_F \quad (15)$$

Note that the mean cycle cost is independent of the PM ratio, c. The renewal cycle length is evaluated as

$$E[L(t_I,c)] = \int_0^{t_I} P\{X(t) < \rho\}dt + \int_{t_I}^{\infty} P\{X(t) < c\rho\}dt = \int_0^{t_I} F_A(\rho/t)dt + \int_{t_I}^{\infty} F_A(c\rho/t)dt \quad (16)$$

The derivation of Eq. (16) uses the mathematical identity for the expectation of a non-negative random variable, i.e., $E[T] = \int_0^{\infty} [1 - F_T(t)]dt$.

Eq. (15) and (16) can be easily evaluated given the parameters of the gamma distribution of the random rate A. According to renewal theory, the mean cost rate is given as

$$K(t_I,c) = \frac{E[C(t_I,c)]}{E[L(t_I,c)]} \quad (17)$$

When $c = 1$, the renewal cycle length equals the mean lifetime for any inspection interval. As the inspection interval $t_I \to \infty$, the CBM policy becomes equivalent to the age-based replacement policy.

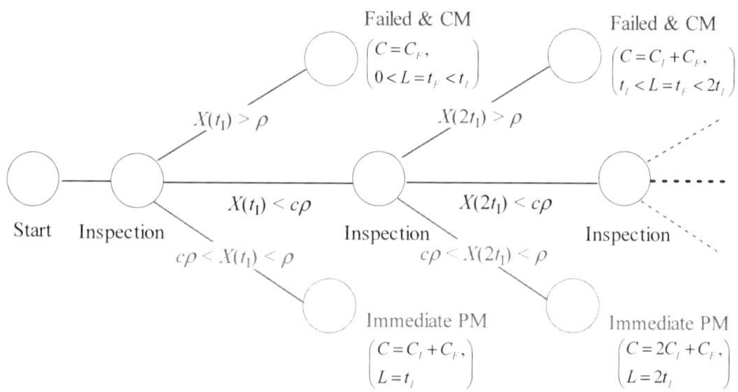

Figure 8: CBM decision tree for the GP model

5.3 *GP Model for Condition Based Maintenance*

The inspection and replacement scenarios in the GP model are more involved due to temporal uncertainty. The three possible situations arise at every inspection interval, namely, inspection and do nothing, preventive maintenance, or failure, as shown in the maintenance decision tree in Figure 8. In this case, a renewal is either an immediate PM or a corrected maintenance after unchecked failure. The renewal cycle cost is derived as (Park 1988, van Noortwijk et al. 1996)

$$E[C(t_I,c)] = \sum_{n=1}^{\infty} \left((nC_I + C_P)P\{\text{PM at } nt_I\} + [(n-1)C_I + C_F]P\{\text{Failed in } ((n-1)t_I, nt_I]\}\right) \quad (18)$$

The mean cycle length can be obtained as

$$E[L(t_I,c)] = \sum_{n=1}^{\infty} \left(nt_I P\{PM\ at\ nt_I\} + E[T_F \mid (n-1)t_I < T_F \le nt_I]\right) \quad (19)$$

Expressing each probability above in terms of the probabilities of the deterioration at corresponding times, we calculate numerically the mean cycle cost and mean cycle length using the following expressions (Park 1988):

$$E[C(t_I,c)] = C_P + (C_F - C_P)[1 + S(c\rho)] \\ \qquad - (C_F - C_I - C_P)[GA(\rho;\alpha\, t_I,\beta) + \int_0^{c\rho} s(x) GA(\rho - x;\alpha\, t_I,\beta)dx] \tag{20}$$

$$E[L(t_I,c)] = \int_0^{t_I} GA(\rho;\alpha\, t,\beta)dt + \int_0^{c\rho}\int_0^{t_I} s(x) GA(\rho - x;\alpha\, t,\beta)dtdx \tag{21}$$

where $S(x) = \sum_{n=1}^{\infty} GA(x; n\alpha\, t_I,\beta)$ and $s(x) = \sum_{n=1}^{\infty} ga(x; n\alpha\, t_I,\beta)$.

There are two limiting cases of the CBM policy. As the inspection interval t_I becomes large, the mean cost rate converges to C_F / μ_T. Secondly, as the PM ratio c approaches 1, t_I converges to the optimal replacement age similar to that in the age-based maintenance policy.

5.4 Numerical Examples

In the optimization of inspection interval with respect to the cost rate, the following cost data are used: $C_I = 1$, $C_P = 10$, $C_F = 50$ and the PM ratio $c = 0.8$. The parameters of the RV and GP models are the same as calibrated in Section 3.1.

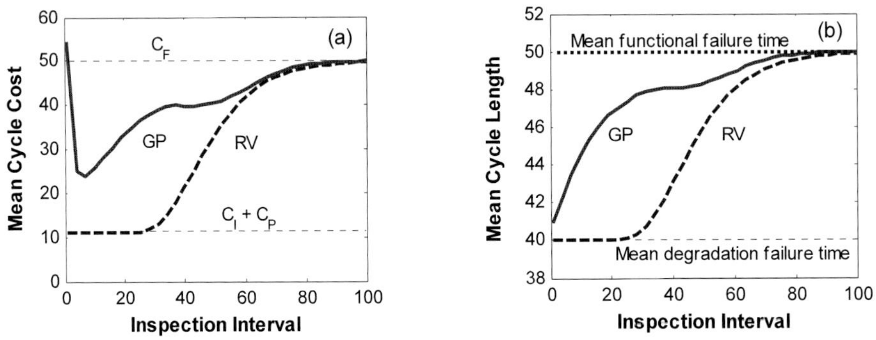

Figure 9: Results for equivalent RV and GP models ($\nu_T = 0.3$): Inspection interval versus (a) Mean cycle cost, and (b) mean cycle length

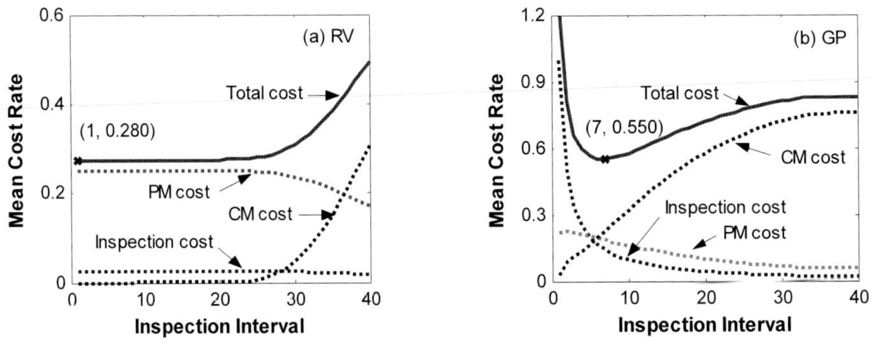

Figure 10: Mean cost rate as a function of inspection interval in equivalent RV and GP models ($\nu_T = 0.3$)

Figure 9 compares the mean cycle cost and mean cycle length obtained from the RV and GP models (for $\nu_T = 0.3$), whereas the mean cost rates, and its different component, are plotted in **Figure 10**. These results show remarkable differences between the two equivalent models of deterioration.

The cost curve for RV model has two plateaus, the lower one extends from $t_I = 0 - 30$ and the corresponding cost is $C_I + C_P$. The upper limit of the cost is C_F. Because of the lack of temporal uncertainty, the RV model favors a short inspection interval. The GP model shows a distinct optimum of inspection interval (t_{opt} = 7 units) in Figure 10(b). For a small inspection interval, the cycle cost is fairly high due to increase in cost associated with frequent inspections. This aspect is qualitatively different from the RV model, which associates a smaller cost with a short inspection interval. For a large interval, the GP cycle cost approaches to the failure cost as expected. In general, the GP cycle cost is higher than that for RV model due to the effect of temporal uncertainty.

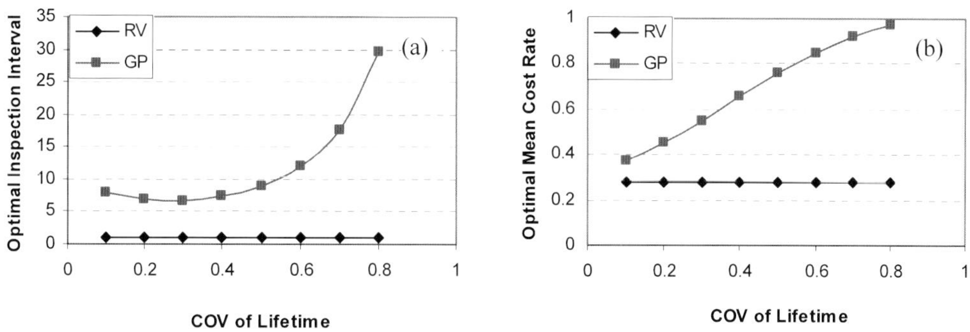

Figure 11: Comparison of DBM policy for RV and GP models: (a) inspection interval age, and (b) mean cost rate

The optimization results for RV and GP models for other values of the lifetime COV are compared in **Figure 11**. It is remarkable in the RV model that both the optimum cost rate and inspection interval are insensitive to the variance of lifetime distribution, which is due to the lack of consideration of temporal uncertainty. In contrast, the minimum cost rate in the GP model increases with lifetime uncertainty. The cost rate increases from 0.38 to 0.97 as the COV of lifetime (ν_T) is increased from 0.1 to 0.8. The optimal inspection interval increases from 7 to 30 units, as ν_T is increased from 0.1 to 0.8.

6 Conclusion

The purpose of infrastructure life cycle management is to mitigate the risk of structural failures caused by unchecked deterioration. The uncertainty associated with deterioration can be incorporated through probabilistic models that can be classified in two broad categories, namely the random variable (RV) model and stochastic process model. The paper introduces a versatile stochastic gamma process (GP) model and compares it with an equivalent random variable (RV) model with respect to distributions of lifetime, deterioration magnitude and the life cycle cost.

The paper underscores the points that a RV model cannot capture temporal variability associated with evolution of degradation. As a consequence, the deterioration along a specific sample path is deterministic in the RV model, whereas it varies probabilistically in the GP model. This distinction has profound implications to the optimization of age-based and condition-based maintenance strategies. The results presented in the paper show that the optimum cost and inspection interval obtained from stochastic process model are qualitatively different than that obtained from the RV model. The RV model appears to underestimate the life cycle cost due to lack of consideration of temporal uncertainty.

In summary, a careful consideration of the nature of uncertainties associated with deterioration is important for a meaningful time-dependent reliability analysis and life-cycle management of structures. If the deterioration process is affected by temporal uncertainty, it is mandatory to model it as a stochastic process.

Acknowledgements
The authors are grateful to the Sciences and Engineering Research of Canada (NSERC) and the University Network of Excellence in Nuclear Engineering (UNENE) for providing the financial support for this study.

References

Abdel-Hameed, M. 1975. A gamma wear process. *IEEE Transactions on Reliability*, 24(2), 152-153.

Barlow, R.E. and Proschan, F. 1965. *Mathematical Theory of Reliability*. John Wiley & Sons Inc., New York, NY.

Bogdanoff, J. L., Kozin, F. (1985). *Probabilistic Models of Cumulative Damage*. John Wiley & Sons, NY.

Dufresne, F., Gerber, H. U., Shiu, E. S. W. (1991). Risk theory with the gamma process. *ASTIN Bulletin*, 21(2): 177-192.

Hong, H. P. (2000). Assessment of reliability of aging reinforced concrete structures, *J. Struct. Eng. ASCE,* 126 (12): 1458-1465.

Nicolai, R. P., Budai, G., Dekker, R., Vreijling, M. (2004). Modeling the deterioration of the coating on steel structures: a comparison of methods. 2004 IEEE Int. Conf. on Systems, Man and Cybernetics.

Pandey, M. (1998). Probabilistic models for condition assessment of oil and gas pipelines. *Int. J. Non-Destructive Testing and Evaluation*, 31(5), 349-358.

Park, K. S. (1988). Optimal continuous-wear limit replacement under periodic inspections. *IEEE Trans. Reliability*, 37(1): 97-102.

Van Noortwijk, J. M., van Gelder, P. H. A. J. M. (1996). Optimal maintenance decisions for berm breakwaters. *Structural Safety*, 18(4): 293-309.

Van Noortwijk, J. M., Frangopol, D. M. (2004). Two probabilistic life-cycle maintenance models for deteriorating civil infrastructures. *Probabilistic Engineering Mechanics*, 19(4): 345-359.

TIME-VARIANT RELIABILITY ANALYSIS FOR SERIES SYSTEMS WITH LOG-NORMAL VECTOR RESPONSE

Sayan Gupta[1,*], **Pieter van Gelder**[1] **and Mahesh Pandey**[2]
[1]*Department of Civil Engineering, Technical University of Delft, The Netherlands*
[2]*Department of Civil Engineering, University of Waterloo, Canada*
**Corresponding Author: gupta.sayan@gmail.com*

Introduction

Failures in randomly vibrating structures are defined to occur if the response exceeds permissible thresholds, within specified time durations. Estimating the failure probability involves characterizing the probability of the exceedance of the structure response, which are modeled as random processes. An elegant approach for addressing this problem lies in expressing the failure probability in terms of the probability distribution function (PDF) of the extreme values associated with the response.

In structural series systems, failure of any of the individual components signals system failure. The system reliability is thus expressible in terms of the joint probability of exceedance of the component response processes. Often, the loads acting on the various components of a system have common source, and hence, the component responses, and in turn, their extremes, are mutually dependent. This emphasizes the need to characterize the joint PDF of these extreme values for estimating the system reliability.

A common approach in characterizing the extreme value distributions for random processes, is to study the associated first passage failures, based on the assumption that level crossings can be modeled as Poisson counting processes. The parameter of the counting process is related to the mean outcrossing rate, which in turn, can be estimated if the joint probability density function (pdf) of the process and its time derivative, is available (Rice, 1956). For Gaussian random processes, this is readily available and closed form expressions for the extreme value distributions have been developed (Lin, 1967; Nigam, 1983). This knowledge is, however, seldom available for non-Gaussian processes. A literature review on the various approximations developed for the extreme value distributions for scalar and vector non-Gaussian processes is available (Manohar and Gupta, 2005). Outcrossing rates of vector random processes have been studied in the context of problems in load combinations and in structural reliability. The focus of many of these problems have been in determining the probability of exceedance of the sum of the component processes, and the outcrossing event has been formulated as a scalar process outcrossing. Some of these results have been used in the geometrical approach (Leira, 1994, 2003) in the studies on development of multivariate extreme value distributions for vector Gaussian/non-Gaussian random processes. Multivariate extreme value distributions associated with a vector of Gaussian random processes have been developed (Gupta and Manohar, 2005), based on the principle that multi-point random processes can be used to model the level crossing statistics associated with the vector Gaussian processes. Similar principles have been applied in developing approximations for the multivariate extreme value distributions associated with a vector of non-Gaussian processes, obtained as nonlinear transformations of vector Gaussian processes (Gupta and van Gelder, 2005).

Here, we extend the above formulation to illustrate its usefulness in estimating the reliability of a randomly vibrating structural system, in series configuration. The response of the structural components have been modeled as a vector of mutually correlated log-normal loads and approximations have been developed for the joint extreme value distribution for the response of the structural components. This is of particular importance in the context of risk analysis of nuclear plants, where the dynamic loads arising from various load effects, are modeled as log-normal random processes.

M. Pandey et al. (eds), Advances in Engineering Structures, Mechanics & Construction, 747–759.

Problem Statement

We consider a linear structural system consisting of m components in series configuration. We assume that the structural system is excited by a n-dimensional vector of mutually correlated, stationary, log-normal loads $\{Y_k(t)\}_{k=1}^n$. The structure response of the jth component is given by

$$Z_j(t) = \tilde{g}_j[Y_1(t), \ldots, Y_n(t)] = g_j[X_1(t), \ldots, X_n(t)], \tag{1}$$

where $Y_j(t) = e^{X_j(t)}$, $(j = 1, \ldots, n)$ and $\tilde{g}_j[e^{X_1(t)}, \ldots, e^{X_n(t)}] = g_j[X_1(t), \ldots, X_n(t)]$. Here, $\{X_j(t)\}_{j=1}^n$ constitutes a vector of mutually correlated Gaussian random processes, $g_j[\cdot]$ is a deterministic nonlinear function which relates the random processes $X_j(t)$ to the component response $Z_j(t)$ and t is time. It is clear that $Z_j(t)$ is a non-Gaussian process whose probabilistic characteristics are difficult to estimate. A component failure is defined to occur when $Z_j(t)$ exceeds specified threshold levels and is given by

$$P_{f_j} = 1 - P[Z_j(t) \le \alpha_j; \forall t \in (0, T)], \tag{2}$$

where α_j denotes the threshold level, T is the duration of interest and $P[\cdot]$ is the probability measure. Equation (2) can be recast into the following time invariant format

$$P_{f_j} = 1 - P[Z_{m_j} \le \alpha_j] = 1 - P_{Z_{m_j}}(\alpha_j), \tag{3}$$

where $Z_{m_j} = \max_{0 \le t \le T} Z_j(t)$ is a random variable denoting the extreme value of $Z_j(t)$ in $[0, T]$ and $P_{Z_{m_j}}(\cdot)$ is the corresponding PDF. We assume that α_j, $(j = 1, \ldots, m)$ to be high and the spectral bandwidth ratio (Vanmarcke, 1972) of the processes $Z_j(t)$, $(j = 1, \ldots, m)$ to be such that the outcrossings of $Z_j(t)$ can be modeled as a Poisson point process. This leads to the following expression for $P_{Z_{m_j}}(\alpha_j)$:

$$P_{Z_{m_j}}(\alpha_j) = \exp[-\nu_j^+(\alpha_j)T]. \tag{4}$$

Here, $\nu_j^+(\alpha_j)$ is the mean outcrossing rate of $Z_j(t)$ across level α_j. An estimate of $\nu_j^+(\alpha_j)$ can be determined from the well known expression (Rice, 1956)

$$\nu_j^+(\alpha_j) = \int_0^\infty \dot{z} p_{Z_j \dot{Z}_j}(\alpha_j, \dot{z}; t, t) \mathrm{d}\dot{z}, \tag{5}$$

where $p_{Z_j \dot{Z}_j}(z, \dot{z}; t, t)$ is the joint pdf of the process $Z_j(t)$ and its instantaneous time derivative $\dot{Z}_j(t)$, at time t. A crucial step in this formulation lies in determining the joint pdf $p_{Z_j \dot{Z}_j}(z, \dot{z})$.

For a structural system comprising of m components, the structure is deemed to have failed if any of the constituent m components fail. Thus, the system failure, denoted by P_{f_s}, is expressed as

$$P_{f_s} = 1 - P[\cap_{j=1}^m \{Z_{m_j} \le \alpha_j\}] = 1 - P_{Z_{m_1} \ldots Z_{m_m}}(\alpha_1, \ldots, \alpha_m). \tag{6}$$

Here, $P_{Z_{m_1} \ldots Z_{m_m}}(\cdot)$ is the m-dimensional joint PDF for the vector of extreme value random variables $\{Z_{m_j}\}_{j=1}^m$. Assuming that the respective thresholds, α_j, corresponding to each component process $Z_j(t)$, are sufficiently high for the respective outcrossings to be rare, the level crossings, denoted by $\{N_j(\alpha_j)\}_{j=1}^m$, can be modeled as Poisson random variables. Since the different components have common source of excitations, $\{Z_j(t)\}_{j=1}^m$, and, in turn, $\{N_j(\alpha_j)\}_{j=1}^m$, are mutually correlated. Consequently, $\{Z_{m_j}\}_{j=1}^m$ are also expected to be mutually dependent. This implies the need for developing approximations for the joint multivariate PDF for the level crossings. Based on recent studies (Gupta and Manohar, 2005; Gupta and van Gelder, 2005) we construct the multivariate PDF for the extreme values of the vector of non-Gaussian random processes $\{Z_j(t)\}_{j=1}^m$. We first illustrate the proposed method for the case when $m = 2$ and then extend it to the more general multi-dimensional situation.

Bivariate Vector

We first consider the case when $m = 2$ and $Z_1(t)$ and $Z_2(t)$ constitute a vector of mutually dependent non-Gaussian random variables, given by

$$Z_1(t) = g[X_1(t), \ldots, X_n(t)], \tag{7}$$

$$Z_2(t) = h[X_1(t), \ldots, X_n(t)], \tag{8}$$

where $g[\cdot]$ and $h[\cdot]$ are deterministic nonlinear functions. Let $N_1(\alpha_1)$ and $N_2(\alpha_2)$ be the number of level crossings for $Z_1(t)$ and $Z_2(t)$, across thresholds α_1 and α_2, in time duration $[0, T]$. For high thresholds, $N_1(\alpha_1)$ and $N_2(\alpha_2)$ can be modeled as mutually dependent Poisson random variables. Introducing the transformations,

$$\begin{aligned} N_1(\alpha_1) &= U_1 + U_3, \\ N_2(\alpha_2) &= U_2 + U_3, \end{aligned} \tag{9}$$

where $\{U_j\}_{j=1}^3$ are mutually independent Poisson random variables with parameters $\{\lambda_j\}_{j=1}^3$, it can be shown that $N_1(\alpha_1)$ and $N_2(\alpha_2)$ are Poisson random variables with parameters $(\lambda_1 + \lambda_3)$ and $(\lambda_2 + \lambda_3)$ respectively and covariance equal to λ_3. This construct for multivariate Poisson random variables has been discussed in the literature (Johnson and Kotz, 1969). The parameters $\{\lambda_j\}_{j=1}^3$ are, as of yet, unknowns.

Taking expectation on both sides of Equation (9), it can be shown that

$$\begin{bmatrix} 1 & 0 & 1 \\ 0 & 1 & 1 \\ 0 & 0 & 1 \end{bmatrix} \begin{Bmatrix} \lambda_1 \\ \lambda_2 \\ \lambda_3 \end{Bmatrix} = \begin{Bmatrix} \langle N_1(\alpha_1)\rangle \\ \langle N_2(\alpha_2)\rangle \\ \mathrm{Cov}[N_1(\alpha_1), N_2(\alpha_2)] \end{Bmatrix}. \tag{10}$$

Here, $\mathrm{Cov}[N_1(\alpha_1), N_2(\alpha_2)] = \langle N_1(\alpha_1)N_2(\alpha_2)\rangle - \langle N_1(\alpha_1)\rangle - \langle N_2(\alpha_2)\rangle$ and if $Z_j(t)$ are stationary random processes,

$$\langle N_j(\alpha_j)\rangle = T \int_0^\infty \dot{z} p_{Z_j \dot{Z}_j}(\alpha_j, \dot{z}) \mathrm{d}\dot{z}, \quad j = 1, 2, \tag{11}$$

$$\langle N_1(\alpha_1)N_2(\alpha_2)\rangle = \int_{-T}^{T} (T - |\tau|) \left\{ \int_0^\infty \int_0^\infty \dot{z}_1 \dot{z}_2 p_{Z_1 Z_2 \dot{Z}_1 \dot{Z}_2}(\alpha_1, \alpha_2, \dot{z}_1, \dot{z}_2; \tau) \mathrm{d}\dot{z}_1 \mathrm{d}\dot{z}_2 \right\} \mathrm{d}\tau, \tag{12}$$

where $\tau = t_2 - t_1$. Details of the derivation for Equation (12) is available (Gupta and Manohar, 2005). Thus, a solution for $\{\lambda_j\}_{j=1}^3$ can be obtained from Equation (10). Furthermore, it has been shown (Gupta and Manohar, 2005) that the joint PDF for the extreme values are related to $\{\lambda_j\}_{j=1}^3$ through the relation

$$P_{Z_{m_1} Z_{m_2}}(\alpha_1, \alpha_2) = \exp\left[- \sum_{j=1}^{3} \lambda_j \right]. \tag{13}$$

A crucial step in this formulation, however, lies in evaluating the expressions $\langle N_1(\alpha_1)\rangle$, $\langle N_2(\alpha_2)\rangle$ and $\langle N_1(\alpha_1)N_2(\alpha_2)\rangle$, for which, a knowledge of the pdfs $p_{Z_1\dot{Z}_1}(\cdot)$, $p_{Z_2\dot{Z}_2}(\cdot)$ and $p_{Z_1Z_2\dot{Z}_1\dot{Z}_2}(\cdot)$, is essential. This, however, is seldom available, especially when $Z_j(t)$ are non-Gaussian. In the following section, a methodology has been presented for developing approximate models for these pdfs.

Approximation for Joint PDFs

Here, we first illustrate the development of the joint pdf, of the form, $p_{Z_1\dot{Z}_1}(\cdot)$. The method is based on the formulation developed by Naess (1985). Next, based on an earlier study (Gupta and van Gelder, 2005), we show the development of an approximation for $p_{Z_1Z_2\dot{Z}_1\dot{Z}_2}(\cdot)$.

Scalar Case

We rewrite $p_{Z_1\dot{Z}_1}(\cdot)$ in the form

$$p_{Z_1\dot{Z}_1}(z,\dot{z};t,t) = \int_{-\infty}^{\infty}\cdots\int_{-\infty}^{\infty} p_{X_2\ldots X_n Z_1\dot{Z}_1}(x_2,\ldots,x_n,z,\dot{z};t,t)\mathrm{d}x_2\ldots\mathrm{d}x_n, \quad (14)$$

where $p_{X_2\ldots X_n Z_1\dot{Z}_1}(\cdot)$ is the joint pdf of random variables $X_2,\ldots,X_n$, Z_1 and $\dot{Z}_1$, at time t. Using the standard technique of transformation of random variables, we seek the transformation between the joint pdf $p_{X_2\ldots X_n Z_1\dot{Z}_1}(\cdot)$ and $p_{X_1\ldots X_n\dot{Z}_1}(\cdot)$. In order to achieve this, we assume that at time t, Z_1 in Equation (7), is a function of X_1 with all other random variables being fixed. We assume that there are k solutions for X_1 for the equation $z_1 = g[X_1, x_2\ldots,x_n]$, for a given set of values for $Z_1 = z_1$, $X_2 = x_2,\ldots,X_n = x_n$. This leads to the expression

$$p_{X_2\ldots X_n Z_1\dot{Z}_1}(x_2,\ldots,x_n,z,\dot{z}) = \sum_{j=1}^{k}\left|\frac{\partial Z}{\partial X_1}\right|_j^{-1} p_{X_1\ldots X_n\dot{Z}_1}(x_1^{(j)},\ldots,x_n,\dot{z}). \quad (15)$$

Here, k depends on the form of the function $g[\cdot]$. The joint pdf $p_{X_1\ldots X_n\dot{Z}_1}(\cdot)$ can now be written as

$$p_{X_1\ldots X_n\dot{Z}_1}(x_1,\ldots,x_n,\dot{z}) = p_{\dot{Z}_1|X_1\ldots X_n}(\dot{z}|X_1 = x_1,\ldots,X_n = x_n)p_{X_1\ldots X_n}(x_1,\ldots,x_n). \quad (16)$$

Here, $p_{X_1\ldots X_n}(x_1,\ldots,x_n)$ is the n-dimensional joint Gaussian pdf and is completely specified if the mean and the covariance matrix of the vector Gaussian process is known. To determine the conditional pdf $p_{\dot{Z}_1|X_1\ldots X_n}(\cdot)$, we first write the time derivative of $\dot{Z}_1(t)$ from Equation (7), and when conditioned on $\{X_j = x_j\}_{j=1}^n$, is given by

$$\dot{Z}_1|_{\mathbf{X}} = \sum_{j=1}^{n}\left|\frac{\partial Z_1}{\partial X_j}\right|_{\mathbf{X}}\dot{X}_j = \sum_{j=1}^{n} g_j\dot{X}_j = \mathbf{G}\dot{\mathbf{X}}. \quad (17)$$

Here, $\mathbf{G} = [g_1,\ldots,g_n]$, $\dot{\mathbf{X}} = [\dot{X}_1,\ldots,\dot{X}_n]'$, the superscript ($'$) denoting transpose and $g_j = \partial Z_1/\partial X_j$, and when conditioned on $\mathbf{X}$, is a constant. $\dot{X}_j(t)$ are the time derivatives of $X_j(t)$ and are thus, zero-mean, stationary, Gaussian random process. Since $\dot{Z}_1|_{\mathbf{X}}$ is a linear sum of Gaussian random variables, $\dot{Z}_1|_{\mathbf{X}}$ is Gaussian, with parameters

$$\begin{aligned}\mu_{Z_1|\mathbf{X}} &= \mathbf{G}\langle\dot{\mathbf{X}}\rangle = 0,\\ \sigma^2_{\dot{Z}_1|\mathbf{X}} &= \mathbf{G}\langle\dot{\mathbf{X}}\dot{\mathbf{X}}^*\rangle\mathbf{G}' = \mathbf{G}\mathbf{C}_{\dot{\mathbf{X}}\dot{\mathbf{X}}}\mathbf{G}'.\end{aligned} \quad (18)$$

Here, $*$ denotes complex conjugation. Substituting Eqs. (14-16) to Equation (11), we get

$$\begin{aligned}\langle N_1(\alpha_1)\rangle &= T\sum_{j=1}^{k}\int \ddot{\Omega}_j \int |g_1^{(j)}|^{-1} \\ &\times \left\{\int_0^\infty \dot{z}p_{\dot{Z}_1|\mathbf{X}}(\dot{z};\mathbf{x}^{(j)},t)\mathrm{d}\dot{z}\right\} p_{X_1\ldots X_n}(x_1^{(j)},x_2,\ldots,x_n)\mathrm{d}x_2\ldots\mathrm{d}x_n. \end{aligned} \tag{19}$$

Here, Ω_j denotes the domain of integration determined by the permissible set of values $x_2,\ldots,x_n$ for each solution of $x_1^{(j)}$. Since $p_{\dot{Z}_1|\mathbf{X}}(\cdot)$ is Gaussian, it can be shown that (Naess, 1985)

$$\int_0^\infty \dot{z}p_{\dot{Z}_1|\mathbf{X}}(\dot{z};\mathbf{x},t)\mathrm{d}\dot{z} = \sigma_{\dot{Z}_1|\mathrm{x}}\Psi\left(\frac{\mu_{\dot{Z}_1|\mathrm{x}}}{\sigma_{\dot{Z}_1|\mathrm{x}}}\right), \tag{20}$$

where $\Psi(x) = \phi(x) + x\Phi(x)$, $\phi(x)$ and $\Phi(x)$ are respectively, the standard normal pdf and PDF. Without loss of generality, $\mathbf{X}$ can be assumed to be a vector of mutually independent Gaussian random variables. When $\mathbf{X}$ are correlated, appropriate linear transformations can be applied to make $\mathbf{X}$ mutually independent. These linear transformations, however, result in a new definition for the function $g[\cdot]$.

Equation (19) can now be expressed as

$$\langle N_1(\alpha_1)\rangle = T\sum_{j=1}^{k}\int \ddot{\Omega}_j \int f(x_1^{(j)},x_2,\ldots,x_n)p_{X_2\ldots X_n}(x_2,\ldots,x_n)\mathrm{d}x_2\ldots\mathrm{d}x_n, \tag{21}$$

where

$$f(x_1^{(j)},x_2,\ldots,x_n) = |g_1^{(j)}|^{-1}\sigma_{\dot{Z}_1|\mathrm{x}}\Psi\left(\frac{\mu_{\dot{Z}_1|\mathrm{x}}}{\sigma_{\dot{Z}_1|\mathrm{x}}}\right)p_{X_1}(x_1^{(j)}).$$

The difficulties involved in evaluating Equation (21) are: (a) in determining the domain of integration Ω_j, defined by the possible set of solutions for $X_1^{(j)}$, and (b) in evaluating the multidimensional integral. A recently developed numerical algorithm is used to overcome these difficulties. This has been discussed later in this paper.

Vector Case

We now focus on developing models for $p_{Z_1Z_2\dot{Z}_1\dot{Z}_2}(\cdot)$. As in the scalar case, we rewrite

$$\begin{aligned}&p_{Z_1Z_2\dot{Z}_1\dot{Z}_2}(z_1,z_2,\dot{z}_1,\dot{z}_2) \\ &\quad= \int_{-\infty}^{\infty}\cdots\int_{-\infty}^{\infty} p_{X_3\ldots X_nZ_1Z_2\dot{Z}_1\dot{Z}_2}(x_3,\ldots,x_n,z_1,z_2,\dot{z}_1,\dot{z}_2)\mathrm{d}x_3\ldots\mathrm{d}x_n,\end{aligned} \tag{22}$$

where the dimension of the integrals is $(n-2)$. The joint pdf $p_{X_3\ldots X_nZ_1Z_2\dot{Z}_1\dot{Z}_2}$ is rewritten as

$$p_{X_3\ldots X_nZ_1Z_2\dot{Z}_1\dot{Z}_2} = \sum_{j=1}^{k}|\mathbf{J}|_j^{-1}p_{X_1\ldots X_n\dot{Z}_1\dot{Z}_2}(x_1^{(j)},x_2^{(j)},x_3,\ldots,x_n,\dot{z}_1,\dot{z}_2), \tag{23}$$

where, for fixed values of $X_3,\ldots,X_n$, Z_1 and Z_2, there exist k solutions for X_1 and X_2, and $\mathbf{J}_j$ denotes the Jacobian matrix

$$\mathbf{J}_j = \begin{bmatrix}\partial Z_1/\partial X_1 & \partial Z_1/\partial X_2 \\ \partial Z_2/\partial X_1 & \partial Z_2/\partial X_2\end{bmatrix}, \tag{24}$$

evaluated at $(x_1^{(j)}, x_2^{(j)})$. As before, we now rewrite

$$p_{X_1\ldots X_n\dot{Z}_1\dot{Z}_2}(x_1^{(j)}, x_2^{(j)}, x_3, \ldots, x_n, \dot{z}_1, \dot{z}_2; t_1, t_2) = p_{\dot{Z}_1\dot{Z}_2|\mathbf{X}}(\dot{z}_1, \dot{z}_2|\mathbf{x}; t_1, t_2)p_{\mathbf{X}}(\mathbf{x}), \tag{25}$$

where $p_{\mathbf{X}}(\mathbf{x})$ is the n-dimensional Gaussian pdf. The time derivatives for $Z_1(t)$ and $Z_2(t)$, conditioned on $\mathbf{X}$, is expressed as

$$\dot{Z}_1(t_1) = \sum_{j=1}^{n} \left.\frac{\partial g}{\partial X_j}\right|_{\mathbf{X}} \dot{X}_j(t_1) = \mathbf{G}\dot{\mathbf{X}}(t_1), \tag{26}$$

$$\dot{Z}_2(t_2) = \sum_{j=1}^{n} \left.\frac{\partial h}{\partial X_j}\right|_{\mathbf{X}} \dot{X}_j(t_2) = \mathbf{H}\dot{\mathbf{X}}(t_2). \tag{27}$$

Here, $g_j = \partial g/\partial X_j$, $h_j = \partial h/\partial X_j$, evaluated at $\mathbf{X} = \mathbf{x}$ and $\mathbf{G} = [g_1, \ldots, g_n]$, $\mathbf{H} = [h_1, \ldots, h_n]$. Since $\mathbf{G}$ and $\mathbf{H}$ are constants and $\dot{\mathbf{X}}(t)$ constitutes a vector of zero-mean stationary, Gaussian random processes, $\dot{Z}_1(t)$ and $\dot{Z}_2(t)$, when conditioned on $\mathbf{X}$, are zero-mean, stationary Gaussian processes. The joint conditional pdf $p_{\dot{Z}_1\dot{Z}_2|\mathbf{X}}(\dot{z}_1, \dot{z}_2|\mathbf{x}; t_1, t_2)$ is therefore jointly Gaussian and is of the form

$$p_{\dot{Z}_1,\dot{Z}_2|\mathbf{X}}(\dot{z}_1, \dot{z}_2; t_1, t_2) = \frac{1}{4\pi^2|\Delta|^{0.5}} \exp\left[-\frac{1}{2}\mathbf{w}\Delta^{-1}\mathbf{w}'\right]. \tag{28}$$

Here, $\mathbf{w} = [\dot{z}_1, \dot{z}_2]'$, $\Delta \equiv \Delta(t_1, t_2) = \mathbf{T}\mathbf{C}_{\dot{\mathbf{X}}}(t_1, t_2)\mathbf{T}'$, the operator $|\cdot|$ denotes the determinant of a matrix, $\mathbf{T} = [\mathbf{G}, \mathbf{H}]'$ and $\mathbf{C}_{\dot{\mathbf{X}}}(t_1, t_2)$ is the covariance matrix $\langle \dot{\mathbf{X}}(t_1)\dot{\mathbf{X}}(t_2)^* \rangle$. Without loss of generality, it can be assumed that $\dot{\mathbf{X}}(t)$ constitutes a vector of mutually independent, stationary, Gaussian random processes. This leads to $\mathbf{C}_{\dot{\mathbf{X}}}(t_1, t_2) = \mathbf{C}_{\dot{\mathbf{X}}}(\tau)$ being a diagonal matrix, where $\tau = t_2 - t_1$.

Substituting Equations (22–28) into Equation (12), and rearranging the order of integrations, we get

$$\langle N_1(\alpha_1)N_2(\alpha_2)\rangle = \sum_{j=1}^{k} \int \ddot{\Omega}_j \int |\mathbf{J}|_j^{-1} \left\{ \int_{-T}^{T} (T - |\tau|)F(\tau)\mathrm{d}\tau \right\} \times p_{X_1}(x_1^{(j)})p_{X_2}(x_2^{(j)})p_{X_3\ldots X_n}(x_3, \ldots, x_n)\mathrm{d}x_3 \ldots \mathrm{d}x_n, \tag{29}$$

where

$$F(\tau) = \int_0^{\infty} \int_0^{\infty} \dot{z}_1\dot{z}_2 p_{\dot{Z}_1\dot{Z}_2|\mathbf{X}}(\dot{z}_1, \dot{z}_2; \tau)\mathrm{d}\dot{z}_1\mathrm{d}\dot{z}_2. \tag{30}$$

The above integral can be evaluated using symbolic software MAPLE or numerically evaluated. Subsequently, the inner integral in Equation (29), with respect to τ, is carried out numerically. The remaining $(n-2)$ dimensional integrals can be evaluated using the numerical algorithm described later in this paper.

Multivariate Vector

As has been shown earlier (Gupta and Manohar, 2005), the construct for bivariate vector of Poisson random variables can be easily generalized for the case $m > 2$. The number of mutually independent

Poisson random variables can be generalized to be given by $C_1^m + C_2^m$, where C_k^m denotes combination of m variables taken k at a time. Thus, for $m = 3$, consider six mutually independent Poisson random variables, $\{U_i\}_{i=1}^6$, with parameters $\{\lambda_i\}_{i=1}^6$ and define

$$\begin{aligned} N_1(\alpha_1) &= U_1 + U_4 + U_5 \\ N_2(\alpha_2) &= U_2 + U_4 + U_6 \\ N_3(\alpha_3) &= U_3 + U_5 + U_6 \end{aligned} \tag{31}$$

The equations relating $\{\lambda_i\}_{i=1}^6$ to the moments of $\{N_i\}_{i=1}^3$ can be shown to be given by

$$\begin{bmatrix} 1 & 0 & 0 & 1 & 1 & 0 \\ 0 & 1 & 0 & 1 & 0 & 1 \\ 0 & 0 & 1 & 0 & 1 & 1 \\ 0 & 0 & 0 & 1 & 0 & 0 \\ 0 & 0 & 0 & 0 & 1 & 0 \\ 0 & 0 & 0 & 0 & 0 & 1 \end{bmatrix} \begin{Bmatrix} \lambda_1 \\ \lambda_2 \\ \lambda_3 \\ \lambda_4 \\ \lambda_5 \\ \lambda_6 \end{Bmatrix} = \begin{Bmatrix} \langle N_1(\alpha_1)\rangle \\ \langle N_2(\alpha_2)\rangle \\ \langle N_3(\alpha_3)\rangle \\ \mathrm{Cov}[N_1(\alpha_1), N_2(\alpha_2)] \\ \mathrm{Cov}[N_1(\alpha_1), N_3(\alpha_3)] \\ \mathrm{Cov}[N_2(\alpha_2), N_3(\alpha_3)] \end{Bmatrix}. \tag{32}$$

It is to be noted that for $m > 2$, the formulation requires the evaluation of a set of integrals of the form in Equation (12) and at no stage does the order of the integrals becomes greater than that of Equation (12). In general, the number of such integrals that need to be evaluated is C_2^m.

Numerical Algorithm

A crucial step in the above formulation lies in evaluating integrals of the type as in Equation (21). Closed form solutions for the integrals are possible only for a limited class of problems. Here, we propose the use of Monte Carlo methods, in conjunction with importance sampling to increase the efficiency, for evaluating these integrals. The integrals in Equation (21) can be recast as

$$\begin{aligned} \mathcal{I}_j &= \int_{-\infty}^{\infty} I[q(\mathbf{X}) \leq 0] f(\mathbf{X}) \frac{p_{\mathbf{X}}(\mathbf{x})}{h_{\tilde{\mathbf{X}}}(\mathbf{x})} h_{\tilde{\mathbf{X}}}(\mathbf{x}) \mathrm{d}\mathbf{x} \\ &= \frac{1}{N} \sum_{j=1}^{N} I[q(\mathbf{X}_j) \leq 0] f(\mathbf{X}_j) \frac{p_{\mathbf{X}}(\mathbf{x}_j)}{h_{\tilde{\mathbf{X}}}(\mathbf{x}_j)}, \end{aligned} \tag{33}$$

where $h_{\tilde{\mathbf{X}}}(\cdot)$ is the importance sampling pdf and $I[\cdot]$ is an indicator function taking values of unity if $q(\mathbf{X}) \leq 0$, indicating that the sample lies within the domain of integration Ω_j, and zero otherwise. Since the problem is formulated into the standard normal space $\mathbf{X}$, $h_{\tilde{\mathbf{X}}}(\cdot)$ can be taken to be Gaussian with unit standard deviation and shifted mean. The difficulty, however, lies in determining where should $h_{\tilde{\mathbf{X}}}(\cdot)$ be centered. An inspection of Equation (33) reveals that the form of the integrals are similar to reliability integrals which are of the form

$$\mathcal{I}_j = \int_{-\infty}^{\infty} I[q(\mathbf{X}) \leq 0] p_{\mathbf{X}}(\mathbf{x}) \mathrm{d}\mathbf{x}. \tag{34}$$

This implies that for efficient computation of the integrals, the importance sampling pdf $h_{\tilde{\mathbf{X}}}(\cdot)$ may be centered around the design point for the function $q(\mathbf{X}) = 0$. If $q(\mathbf{X})$ is available in explicit form, first order reliability methods can be used to determine the design point. If $q(\mathbf{X})$ is not available explicitly, an adaptive importance sampling strategy can be adopted to determine the design point. In certain problems, the domain of integration, characterized by $q(\mathbf{X}) = 0$, may consist of multiple design points or multiple regions which contribute significantly to $\mathcal{I}_j$. This is especially true when

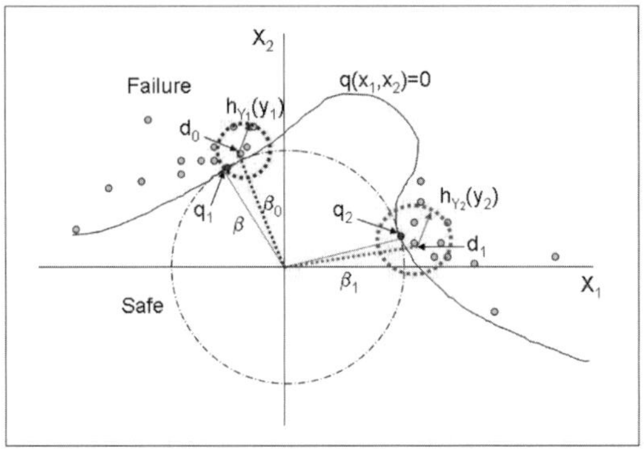

Fig. 1. Schematic diagram for numerical algorithm for evaluating multidimensional integrals; $g(x_1, x_2) = 0$ is the limit surface ion the $X_1 - X_2$ random variable space; $h_{Y_1}(y_1)$ and $h_{Y_2}(y_2)$ are the two importance sampling pdfs; two design points at distance β from the origin.

$q(\mathbf{X}) = 0$ is highly nonlinear, irregular or consists of disjointed regions. In these situations, it is necessary to construct a number of importance sampling functions, with each function centered at the various design points.

The steps for implementing the algorithm for numerical evaluation of integrals of the type in Equation (21), has been developed and discussed (Gupta and van Gelder, 2005). The sequential steps for implementing the algorithm is detailed below, with reference to the schematic diagram in Figure 1.

(1) Carry out pilot Monte Carlo simulations in the standard normal space. If there are too few samples in the failure domain, we carry out Monte Carlo simulations with a Gaussian importance sampling function with mean zero and a higher variance. On the other hand, if there are too few samples in the safe region, the variance of the importance sampling function is taken to be smaller. Repeat this step, till we have a reasonable number of samples in the failure and the safe regions. (2) We sort the samples lying in the failure domain according to their distance from the origin. (3) A Gaussian importance sampling pdf is constructed which is centered at the sample in the failure domain lying closest to the origin. Let this point be denoted by d_0 and its distance from the origin be denoted by β_0. (4) We check for samples in the failure domain, within a hyper-sphere of radius β_1, $\beta_1 - \beta_0 = \epsilon$, where ϵ is a positive number. (5) For samples lying within this hyper-sphere, we check for the sample d_1, which lie closest to the origin but is not located in the vicinity of d_0. This is checked by comparing the direction cosines of d_1 and d_0. (6) By comparing the direction cosines of all samples lying within the hyper-sphere of radius β_1, we can identify the number of design points. We construct importance sampling pdfs at each of these design points. If there exist no samples with direction cosines distinctly different from d_0, there is only one design point and a single importance sampling pdf is sufficient. (7) During importance sampling procedure corresponding to a design point, for each sample realization, we check if x_1 and $\dot{x}_1$ expressed in terms of the random variables $(Z_1, \ldots, Z_{2n-2})$ are real. The indicator function is assigned a value of unity if real, and zero otherwise. (8) An estimate of I_j is obtained from Equation (33).

Numerical Example

For illustrating the proposed formulation, we consider a simple structural system consisting of two components. The component responses, $Z_1(t)$ and $Z_2(t)$, are assumed to be lognormal random processes. We assume that $Z_1(t)$ and $Z_2(t)$ arise from a common source of load effects, and are expressed as

$$\begin{aligned} Z_1(t) &= \exp[X_1(t) + X_2(t) + X3(t)], \\ Z_2(t) &= \exp[X_1(t) - X_2(t) - X_3(t)], \end{aligned} \tag{35}$$

and hence, are mutually correlated. For the sake of simplicity, we assume that $\{X_j(t)\}_{j=1}^3$ are mutually independent, zero-mean, stationary, Gaussian random processes, with auto-correlation function given by

$$R_{jj}(\tau) = S_j^2 \exp[-\beta_j \tau^2], \quad j = 1, 2, 3, \tag{36}$$

where S_j and β_j are constants. The prescribed safety levels are assumed to be deterministic and constant over time. The time duration considered is $T = 10$ s.

First, we develop the marginal extreme value distributions for $Z_1(t)$ and $Z_2(t)$. Following Equation (21), we get

$$\langle N_j(\alpha_j)\rangle = T \int_{-\infty}^{\infty}\int_{-\infty}^{\infty} \sigma_{\dot{Z}_j|\mathbf{x}} \Psi\left(\frac{\mu_{\dot{Z}_j|\mathbf{x}}}{\sigma_{\dot{Z}_j|\mathbf{x}}}\right) \frac{p_{X_1}(\tilde{x}_1)}{\alpha_j} p_{X_2}(x_2) p_{X_3}(x_3) \mathrm{d}x_2 \mathrm{d}x_3, \tag{37}$$

where, for $z_j = \alpha_j$,

$$\begin{aligned} \tilde{x}_1 &= \ln[\alpha_j] - x_2 - x_3, && (38) \\ \mu_{\dot{z}_j|\mathbf{X}} &= \alpha_j(\langle \dot{X}_1\rangle + \langle \dot{X}_2\rangle + \langle \dot{X}_3\rangle) = 0, && (39) \\ \sigma^2_{\dot{z}|\mathbf{X}} &= \alpha_j^2\{\sigma^2_{\dot{X}_1} + \sigma^2_{\dot{X}_2} + \sigma^2_{\dot{X}_3}\} && (40) \end{aligned}$$

and $\{p_{X_j}(x_j)\}_{j=1}^3$ are Gaussian pdf with mean zero and standard deviation σ_j. It can be shown that Equation (37) can be further simplified to the form

$$\langle N_j(\alpha_j)\rangle = T\sigma_{\dot{Z}_j|\mathbf{x}} \Psi\Big(\frac{\mu_{\dot{Z}_j|\mathbf{x}}}{\sigma_{\dot{Z}_j|\mathbf{x}}}\Big) \frac{1}{\alpha_j} \langle p_{X_1}(\tilde{x}_1)\rangle. \tag{41}$$

The analytical predictions for the failure probability for $Z_1(t)$ and $Z_2(t)$, given by Equation (41), for various levels of α_1 and α_2 are compared with those obtained from full scale Monte Carlo simulations in Figures 2 and 3. respectively. The accuracy of the analytical predictions are observed to be acceptable.

Next, we construct the joint extreme value distribution for $Z_1(t)$ and $Z_2(t)$. For this example, Equation (29) can be written as

$$\langle N_1(\alpha_1)N_2(\alpha_2)\rangle = \int_{-\infty}^{\infty} \left[|\mathbf{J}|^{-1} \left\{ \int_{-T}^{T} (T - |\tau|)F(\tau)\mathrm{d}\tau \right\} p_{X_1}(\tilde{x}_1) p_{X_2}(\tilde{x}_2) \right] p_{X_3}(x_3)\mathrm{d}x_3, \tag{42}$$

where, for $Z_1 = \alpha_1$ and $Z_2 = \alpha_2$,

$$|\mathbf{J}| = \begin{bmatrix} g_1 & g_2 \\ h_1 & h_2 \end{bmatrix} = \begin{bmatrix} \alpha_1 & \alpha_1 \\ \alpha_2 & -\alpha_2 \end{bmatrix} = -2\alpha_1\alpha_2, \tag{43}$$

$$\begin{aligned} \tilde{x}_1 &= \frac{1}{2}\ln[\alpha_1\alpha_2], && (44) \\ \tilde{x}_2 &= \frac{1}{2}\ln[\frac{\alpha_1}{\alpha_2}] - x_3, && (45) \end{aligned}$$

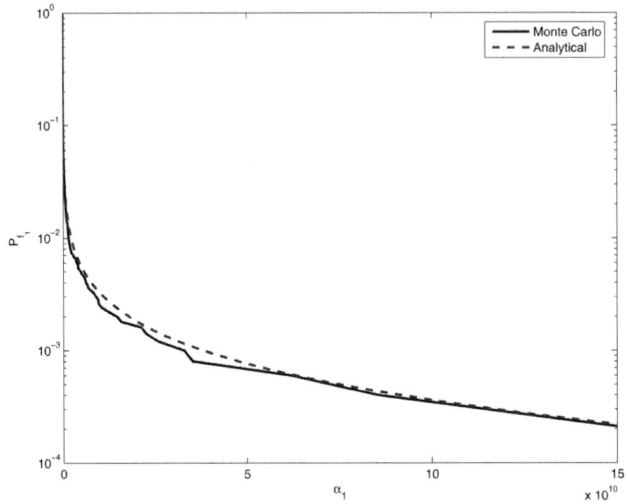

Fig. 2. Probability of exceedance, P_{f_1}, for $Z_1(t)$ across threshold level α_1.

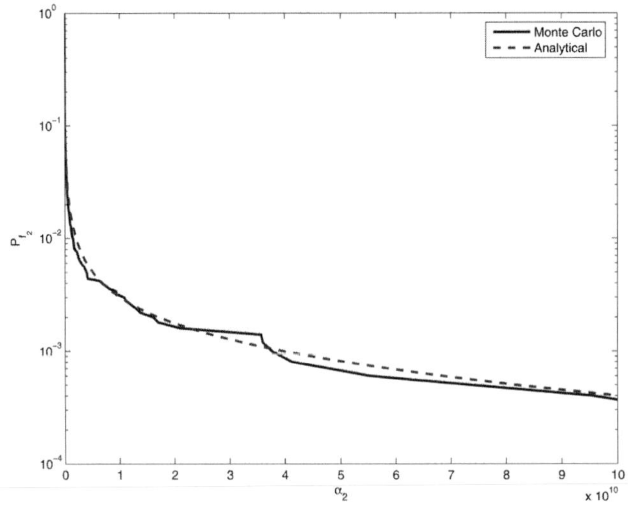

Fig. 3. Probability of exceedance, P_{f_2}, for $Z_2(t)$ across threshold level α_2.

and $F(\tau) = s_0(\tau)\{s_1(\tau) + s_2(\tau) + s_3(\tau) + s_4(\tau)\}$. Here,

$$
\begin{aligned}
s_0(\tau) &= 0.25(\pi c_{22})^{-3/2}, \\
s_1(\tau) &= \pi c_{12} c_{22}^{3/2}, \\
s_2(\tau) &= 2(c_{22}^2 c_{11} - c_{12}^2 c_{22})\sqrt{\pi c_{22}/(c_{11}c_{22} - c_{12}^2)}, \\
s_3(\tau) &= 2c_{12}^2(c_{22} - c_{12}^2/c_{11})\sqrt{\pi c_{22}/(c_{11}c_{22} - c_{12}^2)}, \\
s_4(\tau) &= 2c_{22}c_{12}\sqrt{\pi}\{2c_{12}/[\sqrt{c_{22}(c_{11}c_{22} - c_{12}^2)}(2 + 2c_{12}^2/(c_{11}c_{22} - c_{12}^2))]\} \\
&\quad + \tan^{-1}[c_{12}/\sqrt{c_{11}c_{22} - c_{12}^2}],
\end{aligned}
\tag{46}
$$

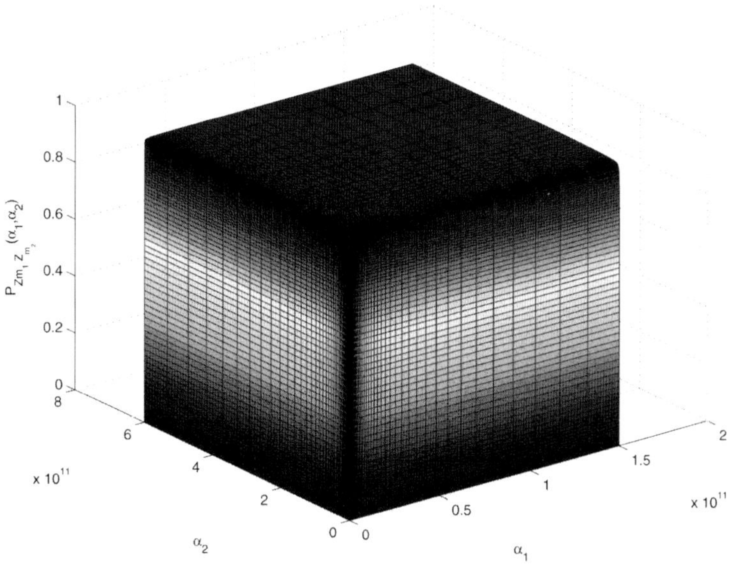

Fig. 4. Joint PDF for the extreme values of $Z_1(t)$ and $Z_2(t)$.

$$\begin{bmatrix} c_{11} & c_{12} \\ c_{12} & c_{22} \end{bmatrix} = \begin{bmatrix} g_1 & g_2 & g_3 \\ h_1 & h_2 & h_3 \end{bmatrix} \begin{bmatrix} \rho_{11}(\tau) & 0 & 0 \\ 0 & \rho_{22}(\tau) & 0 \\ 0 & 0 & \rho_{33}(\tau) \end{bmatrix} \begin{bmatrix} g_1 & h_1 \\ g_2 & h_2 \\ g_3 & h_3 \end{bmatrix}, \tag{47}$$

$g_1 = g_2 = g_3 = \alpha_1$ and $h_1 = \alpha_2$, $h_2 = h_3 = -\alpha_2$, and $\rho_{jj}(\tau) = \langle \dot{X}_j(t)\dot{X}_j(t+\tau)\rangle = -\partial^2 R_{jj}(\tau)/\partial\tau^2$, $(j = 1, 2, 3)$. The inner integral in Equation (42), given by

$$\mathcal{I} = \int_T^T (T - |\tau|)F(\tau)\mathrm{d}\tau, \tag{48}$$

can be evaluated numerically. This leads to the following simplified form for Equation (42):

$$\langle N_1(\alpha_1)N_2(\alpha_2)\rangle = -\frac{\mathcal{I}}{2\alpha_1\alpha_2} p_{X_1}(\tilde{x}_1)\langle p_{X_2}(\tilde{x}_2)\rangle. \tag{49}$$

The joint extreme value distribution function for $Z_1(t)$ and $Z_2(t)$ are computed analytically and is shown in Figure 4.

The marginal distribution for the exceedance probability, for various threshold levels of α_1, have been shown in Figure 5. In the same figure, the corresponding conditional exceedance probability, when conditioned on various threshold levels α_2, have also been shown. The significant levels of difference in the probability levels indicate the importance of the correlations that exist between the extreme values of $Z_1(t)$ and $Z_2(t)$. The results obtained from Monte Carlo simulations are also shown in the same figure and are observed to have close resemblance with the analytical predictions. The corresponding results for $Z_2(t)$ are shown in Figure 6.

The analytical predictions are compared with those obtained from Monte Carlo simulations carried out on an ensemble of 5000 samples of time histories for $Z_1(t)$ and $Z_2(t)$. These results have been shown in Figure 6.

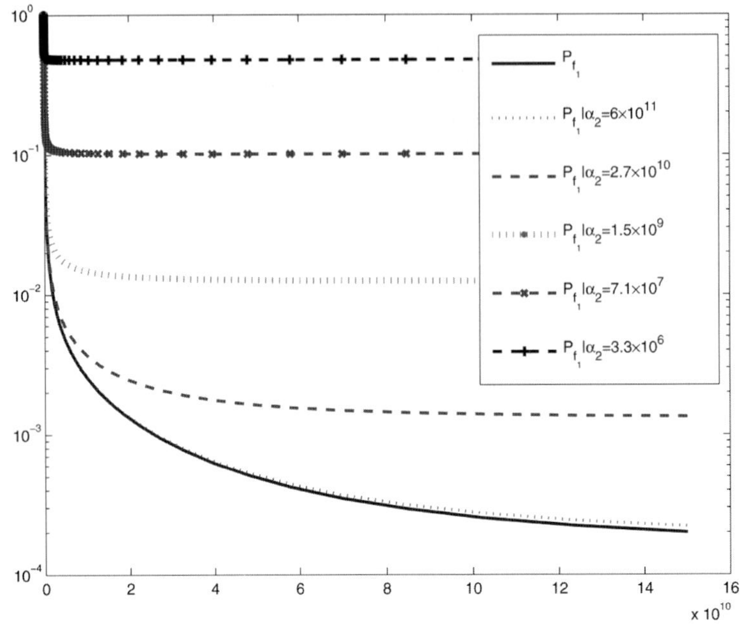

Fig. 5. Probability of exceedance for $Z_1(t)$; the markers indicate the estimates obtained from Monte Carlo simulations.

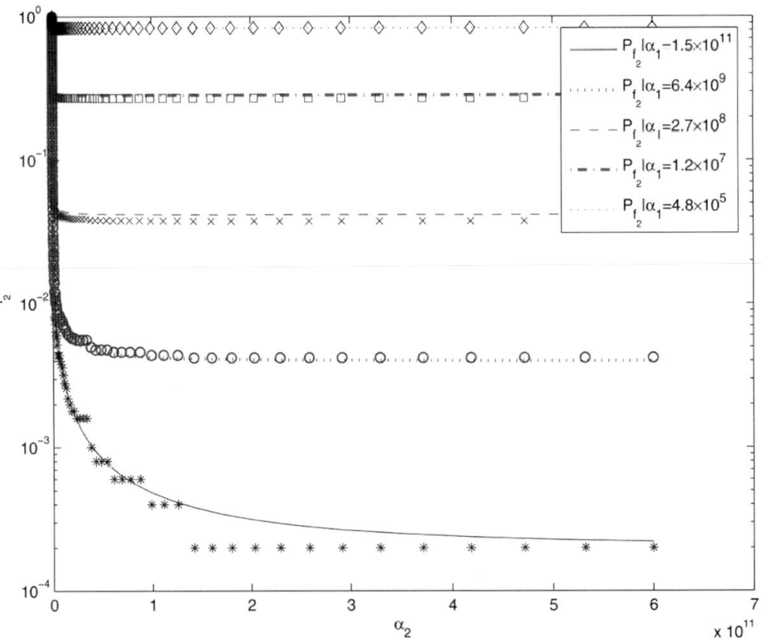

Fig. 6. Probability of exceedance for $Z_2(t)$; the markers indicate the estimates obtained from Monte Carlo simulations.

Concluding Remarks

A methodology has been developed for estimating the time variant reliability of a randomly vibrating series system, when the component response processes constitute a vector of mutually correlated log-normal random processes. A key feature in the development of the proposed method lies in the assumption, that for high thresholds, the number of level crossings of a non-Gaussian process can be modeled as a Poisson point process. The assumption of the outcrossings being Poisson distributed have been proved to be mathematically valid for Gaussian processes when the threshold approaches infinity (Cramer, 1966). However, it has been pointed out that for threshold levels of practical interest, this assumption results in errors whose size and effect depend on the bandwidth of the processes (Vanmarcke, 1972). While it can be heuristically argued that for high thresholds, the outcrossings of non-Gaussian processes can be viewed to be statistically independent and hence can be modeled as a Poisson point process, to the best of the authors' knowledge, studies on the validity of this assumption for non-Gaussian processes, do not exist in structural engineering literature. The multivariate extreme value distributions obtained by the proposed method, is thus expected to inherit the associated inaccuracies and limitations due to this assumption.

References

Cramer, H. (1966). On the intersections between the trajectories of a normal stationary stochastic process and a high level. *Arkive Mathematics*, 6, 337–349.

Gupta, S. and Manohar, C.S. (2005). Development of multivariate extreme value distributions in random vibration applications. *J of Eng. Mech.*, 131(7), 712–720.

Gupta, S. and van Gelder, P. (2005). Extreme value distributions for nonlinear transformations of vector Gaussian processes. *Probabilistic Engineering Mechanics*, under review.

Johnson, N.L. and Kotz, S. (1969). *Discrete Distributions*. John Wiley and Sons, New York.

Leira, B.J. (1994). Multivariate distributions of maxima and extremes for Gaussian vector processes. *Structural Safety*, 14, 247–265.

Leira, B.J. (2003). Extremes of Gaussian and non-Gaussian vector processes: A geometric approach. *Structural Safety*, 25, 401–422.

Lin, Y.K. (1967). *Probabilistic Theory of Structural Dynamics*. New York: McGraw-Hill.

Manohar, C.S. and Gupta, S. (2005). Modeling and evaluation of structural reliability: Current status and future directions. In *Recent Advances in Structural Engineering* (Eds. K.S. Jagadish and R.N. Iyenagar), University Press, pp. 90–187.

Naess, A. (1985). Prediction of extreems of stochastic processes in engineering applications with particular emphasis on analytical methods. Ph.D. Thesis. Norwegian Institute of Technology, Trondheim.

Nigam, N.C. (1983). *Introduction to Random Vibrations*. MIT Press, Massachussetts.

Rice, S.O. (1956). A mathematical analysis of noise. In *Selected Papers in Random Noise and Stochastic Processes* (Ed. N. Wax), Dover Publications, pp. 133–294.

Vanmarcke, E. (1972). Properties of spectral moments with applications to random vibrations. *J of Eng. Mech*, 98, 425–446.

Construction Management

IN-PIPE GROUND PENETRATING RADAR FOR NON-DESTRUCTIVE EVALUATION OF PVC LINED CONCRETE PIPE

Samuel T. Ariaratnam[1] and Noel Guercio[2]

[1] *DEWSC, Ira A. Fulton School of Engineering, Arizona State Univ., P.O. Box 870204, Tempe, AZ, 85287-0204, USA; E-mail: ariaratnam@asu.edu*
[2] *Project Engineering Consultants, 2310 W. Mission Lane, Suite 4, Phoenix, AZ, 85021, USA; E-mail: noel@pecaz.com*

Abstract

Underground utility services play an essential role in sustaining urban life. The majority of these utility services are delivered through pipeline networks, which are mostly buried underground and are interconnected through other urban systems to distribute or collect basic sustainable needs such as treated water, waste water, gas, communication, and power. Deterioration of underground infrastructure systems occurs due to ineffective maintenance management practices. Because new installation can be very costly and disruptive, the best course of action is to maintain the present infrastructure in a more effective way to maximize life span and prevent catastrophic failures. The accurate evaluation of current underground infrastructure must be done before any crucial decisions including lifecycle, rehabilitation and replacement intervals, and appropriate remedial methods can be made. Unfortunately, traditional technologies and management approaches have been limited by the use of insufficient data in the evaluation of the structural integrity of an aged infrastructure. This paper describes the testing, development, and application of a novel assessment technology, which combines in-pipe Ground Penetrating Radar (GPR) with Digital Scanning and Evaluation Technology (DSET) robotics to collect accurate information about the condition of the inside wall of concrete sewer pipes. A case study applying this innovative technology to sections of large diameter PVC-lined concrete pipe in the City of Phoenix is presented. The study and adoption of innovative pipeline assessment methods provide better information to improve the decision-making process, thereby making economical decisions to optimize resources in more efficient ways.

Introduction

The recent Infrastructure Report Card produced by the American Society of Civil Engineers (ASCE) assigned the U.S. infrastructure an overall average grade of D (ASCE, 2005). Within the overall infrastructure, both water and wastewater systems were giving grades of D-. This is alarming as the nation struggles to maintain the integrity of its buried network. The project cost for fixing this problem is estimated to be in the trillions of dollars. Municipalities are trying adopting fiscally responsible strategies for minimizing the impact of their failing water and wastewater systems. A situation analysis must first be performed through the review and assessment of Closed Circuit Television (CCTV) records of the current state.

Back in the 1950's the City of Phoenix, Arizona searched for a feasible solution to the problem of hydrogen-sulfide attacks on reinforced concrete sewer lines. Due to elevated temperatures in the Southwestern United States, Phoenix, along with other cities such as Los Angeles, Las Vegas, and Sacramento, turned to adopting a PVC liner to provide a protective barrier inside concrete pipe. The lined concrete pipe of choice for the City of Phoenix was Ameron's T-Lock™ pipe. T-Lock pipe is a

M. Pandey et al. (eds), Advances in Engineering Structures, Mechanics & Construction, 763–772.

reinforced concrete pipe lined with a polyethylene sheet. As illustrated in Figure 1, the liner is fixed to the concrete with imbedded "Tees" that run the length of the pipe section. This prevents the liner from detaching itself from the pipe. The degree of lining varies depending on what has been specified and is intended to protect the concrete from harmful sewer gases including hydrogen-sulfide.

The City of Phoenix wastewater collection system is comprised of approximately 6,400 km (4,000 miles) of pipeline ranging in diameter from 200 mm (8 in) to 2,250 mm (90 in) and including over 72,000 manholes. Of the Phoenix system, approximately 98 km (61.4 miles) are composed of lined concrete pipe. In addition to its own system, the City operates and maintains two systems from the Sub-Regional Operating Group (SROG). These systems convey wastewater from the Cities of Phoenix, Mesa, Tempe, Scottsdale, Glendale, Peoria and Tolleson to the 91st Avenue Wastewater Treatment Plant located at 91st Avenue and Southern Avenue. These two SROG-owned systems include the Southern Avenue Interceptor (SAI) and the 99th Avenue Interceptor. The SAI includes approximately 30 km (19.1 miles) of PVC lined concrete pipe and 99th Avenue system includes approximately 10 km (6.5 miles) of PVC lined concrete pipe.

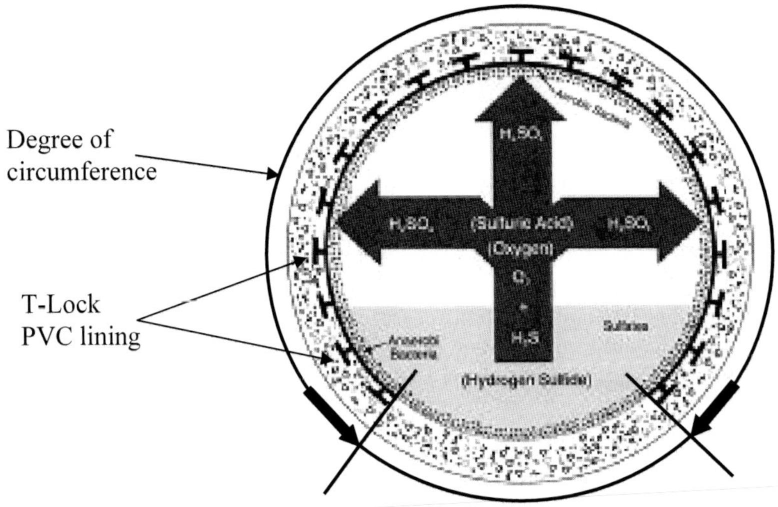

Figure 1. Cross-sectional view of reinforced concrete sewer pipe with T-Lock liner (Ameron 2001)

The City of Phoenix was the first municipality to initiate a major assessment program on the lined-concrete system. The program consists of a condition assessment of all lined concrete sewer pipe 750mm (30 in) in diameter and larger that are maintained and operated by the City, including those portions owned by the SROG. This also included the condition assessment of the associated manholes. The project was divided into two phases. Phase 1 included a condition assessment of all lined concrete sewer pipe constructed prior to 1990. Phase 2 included a condition assessment of all lined concrete sewer pipe constructed after 1989. Phase 2 also included an evaluation of the pipe erosion due to sediment and debris transportation within the sewage flow.

Sewer Inspection Methodology

The entire lined-concrete pipe system was assessed using a Closed Circuit Television (CCTV) for each manhole-to-manhole run. Each CCTV inspection was documented with a written field report and a viewable recording placed on electronic media. The inspection operations were normally completed during periods of relatively low system flow, which generally occurred during night time hours. As inspections were completed, the information was reviewed for entirety and clarity with additional field information requested as necessary. Each defect type found within a pipe run was quantified and assigned a rating based on the National Association of Sanitary Sewer Companies (NASSCO) Pipeline Assessment and Certification Program (PACP) standards and ratings to the resultant defect data. Table 1 presents the NASSCO grading system. It should be noted that the similar rating system used in Canada is through the North American Association of Pipeline Inspectors (NAAPI).

Table 1. NASSCO grading system

Defect Grade	Defect Condition	Rate of Pipeline Deterioration
5	Immediate Attention	Pipe has failed or will fail within the next 5 years
4	Poor	Pipe will probably fail in 5 to 10 years
3	Fair	Pipe may fail in 10 to 20 years
2	Good	Pipe is unlikely to fail in at least 20 years
1	Excellent	Failure is unlikely in the foreseeable future

By applying standardized defect codes to this assessment project, each similar defect will have comparable ratings to help in generating the condition assessment results. The codes also help to standardize the condition assessment procedure and convert the inspector's defect data into meaningful condition grading information.

Inspection Results

Various defects were observed during the CCTV visualization inspection. Condition assessment of the results found that a total of 537 out of 2,451 total defects (22%) were considered to be minor and given a condition grade of 1. Subsequently, these were omitted from the pipe segment ratings and the subsequent tables and figures. Common defects observed during the CCTV inspection included pinholes in the liner, blisters in the liner, detached weld strips, and holes in the liner. Several of these defects are illustrated in Figure 2. A breakdown of the Grade 2 through 5 defects is presented in Table 2.

Figure 2. Detached liner

Figure 3. Detached weldstrip

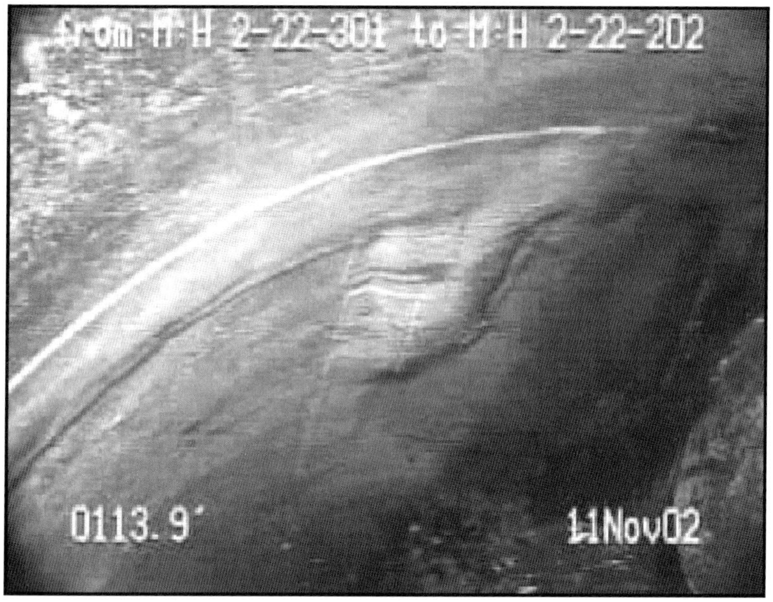

Figure 4. Blister in liner

Table 2. Breakdown of Phoenix defects

Defect Descriptions	Defect Code	Total # Found	% of Defects
Pinhole in Liner	LFPH	1038	54.2
Blister in Liner	LFB	278	14.5
Weld Strip Detached	LFWS	169	8.8
Hole in Liner	LFHR	127	6.6
	LFHL	99	5.2
	LFHC	31	1.6
Detached Liner	LFD	72	3.7
Wrinkled Liner	LFW	54	2.8
Grease Deposits	DAGS	23	1.2
Settled Deposits	DS	12	0.6
Obstructions	OBZ	5	0.3
Deposits, Attached	DAE	2	0.1
Visible Reinforcement	SRVC	1	0.1
Object Protruding	OBI	1	0.1
Construction Debris	OBN	1	0.1
Defective Tap	TFD	1	0.1
Total		**1914**	**100%**

Evolution of In-Pipe Ground Penetrating Radar

The defects discovered by the CCTV inspection were of the liner and not the actual concrete pipe. There was concern raised about the condition of the concrete pipe and reinforcements "behind" the liner. Subsequently, a need was recognized for utilizing a form of non-destructive method testing for evaluating the condition behind the liner. Morrison (2004) recommended the use of secondary assessment technologies during a pipe assessment program.

Several non-destructive evaluation methods were examined for suitability including: sonar; thermography; X-ray; laser profilometry; and ground penetrating radar. It was determined that ground penetrating radar (or GPR) was the most promising of the applications for looking behind the liner and into the concrete pipe. Furthermore, it was decided that it would be best if the GPR unit were placed on the CCTV robot and transported within the sewer line providing close contact with the lined pipe.

Field testing of a prototype unit was conducted in Tskuba, Japan in the fall of 2003 to determine whether readings could be made. Defects were manually created in the concrete behind sections of lined pipe. The first prototype, illustrated in Figure 5, performed well in the trials and was brought to Phoenix for a pilot project.

Figure 5. Prototype inspection robot with gpr unit

Due to the pipe size and potential flow conditions, it was determined to use a wheel robotic body rather than the track body used in the Japanese field tests. The second generation robot, illustrated in Figure 6, capture data behind the liner at pre-positioned liner defect positions along several pipeline sections. Unfortunately, a limitation was the fact that information could only be captured at the 12 o'clock position due to the fixed GPR unit.

The third generation robot, illustrated in Figure 7, solved any problems associated with clock face positioning through the use of multiple gpr units attached to two arms. This enables the unit to capture defect information anywhere along the 9 o'clock to 3 o'clock positions of a pipe section. An example of the corresponding ground penetration radar results is shown in Figure 8. The top results show the defect areas in relation to a folded view of the pipe, while the bottom results show a front view of the liner tears.

Figure 6. Second generation wheeled inspection robot with gpr

Figure 7. Third generation wheeled inspection robot is multiple gpr units

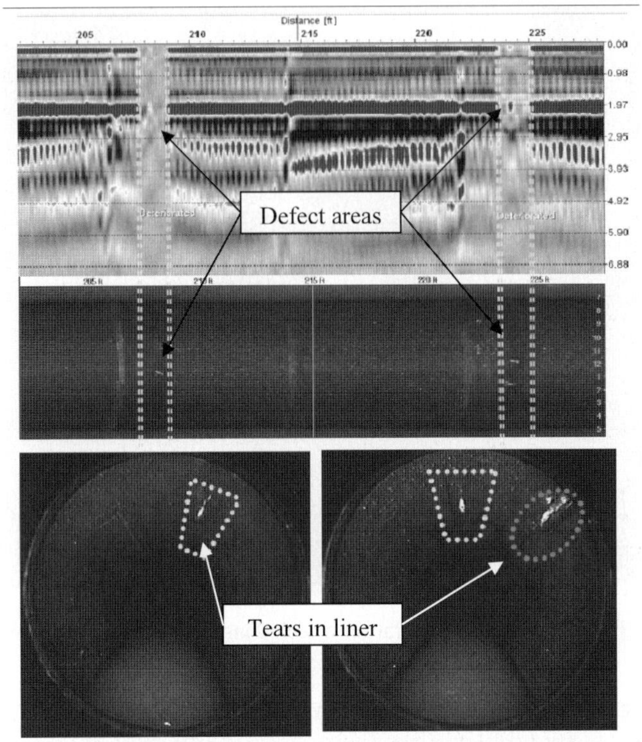

Figure 8. In-pipe ground penetrating radar results

Ground Penetrating Radar Technology

Ground penetrating radar is a wave-based electromagnetic geophysical method that fundamentally detects interfaces between materials possessing varying electrical (dielectric) characteristics. A typical digital GPR system consists of transmitting and receiving antenna elements, which emit outgoing electromagnetic pulses into the media being investigated and receive incoming reflections from this media; a central control unit which governs the characteristics of the transmitted electromagnetic waves, processes the received signal by amplifying and recording it and converting it to a digital format; and a color computer video display unit that receives the digital information from the control unit and, after processing the information through the GPR acquisition software, produces a graphical representation of the acquired data as a real-time two-dimensional continuous depth profile, which includes horizontal antenna position and vertical target depth and amplitude information. GPR operates by emitting electromagnetic radar impulses into a media (pipe) at a high repetition rate, from an antenna array towed through the pipeline. Reflections occur at interfaces of materials with differing electrical characteristics (dielectric permittivity). Reflections of various amplitudes (Figure 8) are produced at these interfaces and are detected by the receiving antenna element, depending on the incoming signal frequency, the magnitude of the difference in dielectric constants of the two materials. Additional information on GPR may be found in Holmes (2004).

Current state-of-the-art GPR systems have evolved into compact, reliable, user-friendly instruments able to be operated from start to finish by a well trained, technically experienced individual. Today's systems allow a single user to acquire and interpret large amounts of project data in a relatively short amount of time as well as the ability to transfer the information in real time.

Along with technological advancements in recent GPR systems comes an increase in available situations to apply the technology. Beyond the historically scientific applications in the geological, archaeological, or research realms, GPR has recently been increasingly applied for utilitarian purposes in the environmental and trenchless technology industries to locate and assess subsurface objects.

Destructive Verification through Core Samples

Core samples of pipe walls were obtained from two different locations to provide a visual verification of suspected problem areas detected by the ground penetrating radar. In both locations, a 5 m deep, 2.4 m wide, 7.6 m long trench was excavated at predetermined locations with a box shoring used for stabilization. A GPR survey was performed on the exterior surface of the top of the reinforced concrete pipe to determine the exact location to extract the coring sample. Two 75 mm diameter samples were recovered at each of the two locations and preserved for laboratory testing of compressive strength (Edwards and Nowaczyk, 2005). One sample was at a suspected defect location, while the other was in an unaffected section of the concrete pipe wall for comparison. Table 3 presents the results of the compressive strength tests performed to ASTM C39-01 standards.

The cores samples (Figure 9) verified the results found using the non-disruptive in-pipe ground penetrating radar unit transported on a robotic platform. Discoloration of the core samples and diminished compressive strength measures further demonstrate the problems that could occur if hydrogen-sulfide gases were to penetrate the liner and attack the concrete wall.

Table 3. Laboratory testing of concrete core compressive strength (ASTM C39-01)

Core Sample #	Measured Average Length of Recovered Concrete Core (mm)	Compressive Strength (psi)	Notes
1 (a)	102.87	9,770	Core recovered from relatively non-corroded 750mm pipe section
1 (b)	85.09	8,840	Core recovered from suspected corroded 750mm pipe section
2 (a)	106.81	8,020	Core recovered from relatively non-corroded 900mm pipe section
2 (b)	93.17	7,810	Core recovered from suspected corroded 900mm pipe section

Figure 9. Core samples from 750 mm concrete pipe section

Conclusions and Recommendations

As municipalities that previously adopted lined concrete sewer pipe engage in their respective inspection programs, it is critical that secondary inspection technologies be utilized to provide validation of the pipe wall condition "behind" the liner. This paper provides a description of the development and use of in-pipe ground penetrating radar (GPR) deployed on a robotic unit for the evaluation of PVC lined concrete pipe. The third generation unit employs multiple GPR units on two arms to facilitate the capture of data from any location above the pipe's spring line. This improved on the limitations of previous generation, which were restricted to only capturing information at the 12 o'clock position within a pipe.

Furthermore, coring samples were conducted to validate the results obtained from the GPR unit. Samples were extracted from two different pipe locations. For each location, two 75 mm diameter core samples were taken. One sample was at a suspected defect location, while the other was in an unaffected section of the concrete pipe wall for comparison. The samples from the defect section had lower compressive strength measures compared to the unaffected samples.

References

ASCE (2005). *Infrastructure Report Card 2005*. American Society of Civil Engineers, Reston, VA.

Ameron International (2001). *Amer-Plate® T-Lock® Price List [Brochure]*. Brea, CA.

Edwards, Mark D., and Nowaczyk, S.D. (2005). *Existing Sewer Line Geophysical Evaluation 41st Avenue and Peoria Avenue, Phoenix, Arizona Pecos Road and 40th Street, Phoenix, Arizona Report*. Prepared by Ninyo & Moore Consultants, Phoenix, AZ.

Holmes, Patrick (2004). "Emerging Methods for Utility Locates." *Proceedings of the No-Dig 2004 Conference*, New Orleans, LA, March 22-24, Paper B-4-01, pp. 1–7.

Morrison, Robert (2004). "Condition Assessment: Back to Basics." *Underground Infrastructure Management,* November/December, Peninsula, OH, pp. 51–55.

FREEFORM CONSTRUCTION APPLICATION RESEARCH

R.A. Buswell[1], R.C. Soar[1], A.G.F. Gibb[2] and A. Thorpe[2]
[1] *Wolfson School, Loughborough University, UK*
E-mail: r.a.buswell@lboro.ac.uk
[2] *Civil and Building Engineering, Loughborough University, UK*

Abstract

The level of automation technology and processes control, within the construction sector, faces unique challenges if it is to catch up with automotive and aerospace applications. The construction industry has problems relating to health and safety, environmental legislation and traditional methods of procurement. These are compounded by diminishing skills in the labour force. One way to address these issues is by increased automation and integration of design, modelling and process control. Digital Fabrication has demonstrated the feasibility of the integration of design and component production on a large scale. Freeform Construction builds on Digital Fabrication by integrating the control of final material deposition. This paper reports on recent meetings held with industrialists to gauge their perceptions of the technology and encourage discourse to identify both applications and opportunities for the wider research community. Examples of digital fabrication in construction are discussed. Freeform Construction is defined and potential applications are presented. An example of physical model generation from construction CAD software is described.

Keywords: construction automation, rapid manufacturing, additive manufacturing, freeform construction

Introduction

Freeform Construction is an initiative that considers the benefits of automated layer based, additive production methods for construction. These methods can be applied, conceptually at least, at any scale from desktop model to full scale building construction. This paper describes results from a recently completed project that investigated the potentials for such applications in conjunction with industry.

Issues for the Construction Industry

The construction industry has yet to realise the benefit technological step changes in process can bring, placing it behind industries such as aerospace, automotive and ship building. For most applications, the fundamental principles of construction have changed little for hundreds of years; material placement is still primarily a manual operation. Often the legal framework that enable the business of construction acts as a disincentive to trial new approaches. Competition for projects concentrate on first cost. The industry is conservative and innovations tend to be through incremental change. There is a growing skills shortage in the UK compounded by an aging population. Construction remains a hazardous environment. The industry is likely to face increasing pressures from developing environmental legislation, waste management, and energy conservation measures (Guthrie et al. 1999).

The UK government has been addressing these issues through a succession of initiatives, prompted by the Latham and Egan Reports (Egan 1998, Latham 1994) and ultimately by 'Constructing Excellence.'

M. Pandey et al. (eds), Advances in Engineering Structures, Mechanics & Construction, 773–780.

The drive is towards leaner, better Modern Methods of Construction. There is, however a need for a more radically different solutions. As human endeavour pushes further forward, construction will need to be able to respond to unique challenges in aggressive environments such as the North pole, desert, chemical contamination and off-world. The industry will need to respond to environmental issues with new materials and new solutions for buildings at 'end of life'.

Additive Processes

Construction can be described as a mixture of additive process and assembly of components. A big question is whether the layer by layer additive processes found in Rapid Prototyping and Rapid Manufacturing can inspire processes for larger scale automation useful to construction.

Traditionally, the realisation of a component design was costly because of the involvement of man hours for the production of a few bespoke parts. The development of computer aided modelling has allowed designs to be developed faster, particularly when combined with simulation and modelling software. These computer representations of solid 3D objects can be used to control additive manufacturing machines, producing a physical component. These machines reconstruct the 3D object by sequentially bonding '2D' layers of material. The technology's impact to existing manufacturing has been in the form of waste reduction when compared to machining processes and the elimination of tooling when compared to moulding or casting processes. A useful by-product of this approach is almost unlimited geometrical freedom and that moving parts can be constructed in a single build, negating the need for assembly. Additive manufacturing processes are seeding a revolution in customised, end-use parts, a*dding value* to parts (Wohlers 2004).

Additive manufacturing technologies have been used for construction applications, but have been largely restricted to concept modelling in architecture (Kalay, Skibniewski 2002). There are two notable exceptions: Pegna (1997) considered a layer deposition method suitable for traditional construction materials; Contour Crafting is the first automated layer manufacturing process that has been demonstrated to have the potential for generating large structures directly from digital data (Khoshnevis 2002).

Constructing from Digital Models

The ability to use solid 3D CAD models, with computer based analysis tools is generating impressive freeform architecture (Kolarevic 2003, Stacey, Beesley & Hui 2004). In recent years there have a number of buildings constructed using large scale manufacturing methods to create structural components that are assembled on site. Gehry's Zollhoff Towers is one.

Today Construction is using CAD/CAM to liberate architectural possibilities. We can control the shape of construction components directly from the digital design model. The next technological step is to control the deposition of the construction material directly from the building model. This is Freeform Construction.

Additive process can be applied at various scales from hand held items to whole buildings. As the application scale increases, it may become possible to manufacture modular, or volumetric components. Systems integration is a benefit at this scale. Services can be assembled inside the structure as it is built. Although building objects using additive manufacturing processes at this scale does not exist, processes such as 3D printing, are potentially scalable. The benefits of systems integration and structural optimisation are maximised if the building as a whole are produced by a machine.

Identifying Freeform Construction Applications

To generate applications of Freeform Construction technology that are grounded by industry, 17 practitioners and 6 academics, representing 13 organisations were canvassed for ideas during the course of a one day workshop. Engineers, architects, construction managers, academia, CAD software houses and construction innovations developers were represented. Two groups of 7-8 were selected such that each group was reasonably diverse in terms of the professions represented. In each of the groups, each person was asked to provide answers to the question:

'If you could have a 'freeform machine' today what would you use it for?'

Table 1 rationalises the responses. The discussion fell into broad categories that reflected the elements of a building and construction. Six technology categories also emerged. Potential applications tended to fall into those that can be delivered through improving computing technology and those that require a physical process and/or materials. Of those, some applications reflected what can be achieved through Rapid Manufacturing and CAD/CAM technologies today. Examples of others reflected what is being achieved in high end architecture though digital fabrication. More far reaching applications tended to fall into three further categories: new specialist processes that provide a solution for use in a conventional construction environment; those that would be particularly suited to off site fabrication using new technologies; and those applications that are only conceivable when new technologies are applied to the construction of the whole structure.

The sense of the group who contributed to Table 1 was that Freeform approaches could help in the manufacture of complex forms and in component and systems integration. One major issue identified, was that of responsibility. Where integrated components have two or more functions, i.e. a wall with heating water transport and electrical distribution, then who takes the responsibility for the design? If digital design information is used directly by the 'contractor' via a digitally controlled process, who is responsible for the construction; the designer, the contractor or those who built the machine?

The study demonstrated that there is a desire to improve how construction is procured. Technology can play a significant role and it is likely that the viability of new processes will be driven by specialist application. Major issues for large scale Freeform Construction, however, will be in proving the viability of the technology, establishing design process, contract requirements and supply chains to enable it.

An Example of Model Production

Table 1 identifies the application of model production. This is useful for visualisation, communication and conflict resolution. The application of architectural models is not new and Rapid Prototyping is a proven technology in this respect. For wide spread application in the industry, across architecture and engineering, the transition of information from designer and CAD to Prototyping machine must be simple. This initial investigation to identified some of the issues with using a model from a construction oriented CAD package to produce a physical model using a Rapid Prototyping process.

Virtually all Rapid Prototyping machines accept a data format called STL. This data format describes the object to be built using a single tessellated surface where the inside and outside faces of the surface is defined. What is 'inside' is deemed to be solid. This data model is 'sliced' into layers. The machine interprets which part of each layer is solid and which is not. The information controls the sequential reconstruction of the object in the physical world, a layer at a time, ground up. There is a continual cycle of operations, applying a new layer of material across the build platform and selectively turning the 'internal' parts of that layer solid using an activation process. In the case of the Selective Laser

Table 1. Classified Applications for Freeform Construction.

State of Technology Development	*Ongoing Improvements to Existing Technology*	*Widespread Use Today in Manufacturing*	*State of the Art Today on High End, Architectural Projects*	*Probable First Line of Technology Introduction*	*Probable First-Second line of Technology Introduction*	*More Speculative Technologies*
Building Element	**Modelling, Design & Computational Analysis Tools**	**Current Rapid and CAD/CAM Manufacturing**	**Current Digital Fabrication in Construction**	**Mix of Traditional & Hybrid Technologies**	**Technologies For Off Site Applications**	**Large Scale Whole Structure, On Site Approaches**
Design + Communication	Greater degree of client interaction in design of architectural space, i.e. hands on interaction with the model – a 'what you see is what you get' approach to construction.		Realisation of non-linear surface forms for buildings.			Generate new architectural possibilities in addition to improving personalised/ergonomic design – no more straight lines and corners.
Structure	Greater degree of numerical optimisation – minimisation of building material and improved building performance.	Contour crafting demonstrated to be capable of building large wall structures using CAD/CAM technology.	Realisation of support systems for highly curved surfaces.			Possibility of tension/compression structures.
Roofs & Canopies	Improved design for thermal insulation.		Shop drawings derived from 3D digital models.			Possibility of monolithic curved enclosures.
Floors, Ceilings and Walls	Improved design for thermal insulation				Integrating service systems into partitions and volumetric assemblies.	Integrating service systems into building fabric.

Table 1 Continued. Classified Applications for Freeform Construction.

Building Element	Modelling, Design & Computational Analysis Tools	Current Rapid and CAD/CAM Manufacturing	Current Digital Fabrication in Construction	Mix of Traditional & Hybrid Technologies	Technologies For Off Site Applications	Large Scale Whole Structure, On Site Approaches
Surfaces, Finishes & Decoration	Design possibilities for integrated storage, bespoke sanitary ware.	Reproduction of heritage architecture.	External surfaces - shop drawings derived from 3D digital models.	Embedding control systems in surfaces.	Possibility of seamless finishes.	Possibility of joint-less construction, precise surface property control. And greater variety of surface finishes.
Openings					Possibility of structural glazing systems.	Possibility of structural glazing systems.
Mechanical & Electrical Systems	Increased building modelling could generate optimal routes – minimising material use and in use losses i.e. improved efficiency.			Possibility of generating space/weight optimised service models for faster services installation in traditional construction.	Systems integration with fabric.	Systems integration with fabric. In addition, the model used to generate the building is the 'as built' information.
Fixings, Flashings and Junctions		Creation of bespoke façade system fixtures and fittings.		Improved weather protection at service penetrations.		Elimination of troublesome system interfaces.
Temporary Works			Examples of automated on site 'factory' environments that enclose the entire building while each floor is 'manufactured'. Generating moulds/formers for concrete works.	Forming water tight plant-rooms for early start on M+E works.		

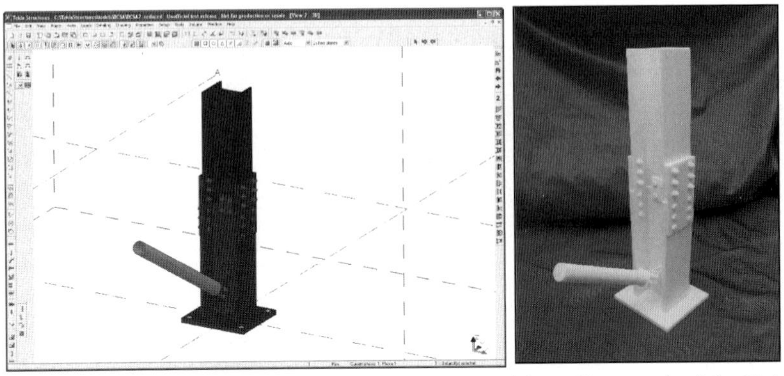

Figure 2. On the Left, the CAD Image of the Object and on the Right, a Photograph of the Finished Item.

Sintering, the solid parts of each layer of powdered material are melted, subsequently cooling and solidifying.

Where solid 3D CAD software is employed, conversion of digital model information to STL format suitable for Rapid Prototyping machines can be relatively trivial. There are two issues that are important for model production of building components. Although solid modelling is becoming more widely used, a great deal of design information is detailed in 2D CAD systems or are rendered 3D images using surface modelling. It is generally difficult, and for complicated systems practically impossible, to convert these models into good[1] STL files. The machine must be told what is and isn't solid and hence must have objects with perfectly joined edges, clearly defining volume. For example, it is clear to us that a pool ball is solid, but STL data would describe it as an infinitely thin surface that has a certain volume inside. If there is a hole in that surface, you can't define what is inside and what is out and hence, what is solid. Problems like this will cause a 'build' to fail. Surface modellers and 2D CAD software are not designed to describe volume and so can't be used to generate an accurate description of a surface that will instruct a machine to make a physical object.

The second issue for construction is that *assemblies* of components can be problematic. Construction system designs generally describe an assembly of individual components; steel frame systems are a good example. There are construction orientated CAD systems that use solid modelling, Tekla Structures is one example (Tekla 2005). This system models every component down to the nuts, bolts and washers and hence a model of a whole system, is described as an assembly of solid objects. If a model of the *complete* system is required, the data must be converted to *one* surface describing a *single* object. The issue is how to combine multiple objects in the assembly so that the desired model is described as one homogenous unit.

The left hand picture in Figure 2 details the CAD image of an example steel section modelled using Tekla Structures. This software models each component as a solid object but it does not have an integrated STL conversion function. To generate the STL file, the data was loaded into the solid modeller, Rhino 3D (Rhinoceros 2005), using a file format common to both (DXF). Using the automated functions within Rhino, the STL file generated was made up of many surfaces. Each surface defining the individual objects in the design. A model of the entire assembly was required. Using Magics, an STL file editing software package (Materialise 2006), each object surface was combined using an operation which merges one object with another, resulting in a single surface. The dimensions of the object were scaled to suit the size of the Rapid Prototyping machine and then built in nylon using a Selective Laser Sintering process. The final part is depicted on the right hand side of Figure 2.

[1]I.e. STL files that will procure a successful build on a Rapid Prototyping machine.

At present, if a model of a building, component or subassembly is required, it is likely that separate operations will be required to encode the design into solid modelling software in order to produce a STL file that is capable of instructing a machine to build the physical object. If an assembly of solid objects are used to describe the item to be built, additional operations will be required to generate a single object in STL. As software develops, it is likely that the latter will be achievable through prescribed software functions. If solid modelling becomes the standard way of describing building information in the future, the former issue will become obsolete.

Conclusions

There are many examples in construction where new technologies are moving away from the use of traditional methods of building procurement to those that can be designed and fabricated digitally. It is likely that this trend will grow. To enable the full benefits of construction automation, radically new process are required. The freeform approach is well placed to contribute. By engaging with industry, potential applications have been identified. These range from near market solutions to some that will take years, even decades to be exploited commercially.

Generally freeform methods can find ready application in geometrically complex building components or the integration of systems. Specialist applications seemed to be most probable locations for early development. One hurdle to overcome is that of information handling. Solid modelling as a real tool for construction is in it's infancy when compared to ship building or aerospace engineering. An additional factor is that to be successful, the digital building model information needs to be readily used by all members of the supply chain. This will mean greater use of CAD/CAM manufacturing technologies by suppliers. The modelling example, although fairly trivial, does demonstrate that there is a little way to go before the use and exchange of construction information with Rapid Prototyping type processes becomes trivial. Such issues will be exacerbated when generating full scale structures.

Acknowledgements

The Authors would like to thank the EPSRC for the funding of this project through the IMCRC at Loughborough University. We would like to thank BPB plc and Z Corporation for their support and in particular the contributions of Keith Humphrey and Tom Clay. We would also like to thank Clive Robinson and Tekla for the modelling example.

References

Egan, J. 1998, *Rethinking Construction*, Department of the Environment, London.

Guthrie, P., Coventry, S., Woolveridge, C., Hillier, S. & Collins, R. 1999, *The reclaimed and recycled construction materials handbook* CIRIA, London.

Kalay, Y.E. & Skibniewski, M.J. 2002, "Special Issue: Rapid Prototyping", *Automation in Construction,* vol. 11, no. 3.

Khoshnevis, B. 2002, "Automated Construction by Contour Crafting â€" Related Robotics and Information Sciences", *Automation in Construction Special Issue: The best of ISARC 2002,* vol. 13, no. 1, pp. 2-9.

Kolarevic, B. 2003, *Architecture in the Digital Age: Design and Manufacturing,* New York & London: Spon Press - Taylor & Francis Group.

Latham, M. 1994, *Constructing the Team*, HMSO, London.

Materialise 2006, 1/9/2006-last update, *Magics* [Homepage of Materialise], [Online]. Available: http://www.materialise.com/magics-rp/main_ENG.html [2006, 1/9] .

Pegna, J. 1997, "Exploratory investigation of solid freeform construction", *Automation in Construction,* vol. 5, no. 5, pp. 427-437.

Rhinoceros 2005, , *Rhinoceros* [Homepage of Rhinoceros], [Online]. Available: www.Rhino3d.com [2005, 12/10] .

Stacey, M., Beesley, P. & Hui, V. 2004, *Digital Fabricators,* 1st edn, University of Waterloo School of Architecture Press, Waterloo, Canada.

Tekla 2005, , *Tekla Corporation* [Homepage of Tekla], [Online]. Available: www.tekla.com [2005, 12/10] .

Wohlers, T. 2004, *Rapid Prototyping, Tooling & Manufacturing: State of the Industry*, Wohlers Associates, Colorado, USA.

RISK IDENTIFICATION AND RESPONSE METHODS: VIEWS OF LARGE SCALE CONTRACTORS WORKING IN DEVELOPING COUNTRIES

A. Samer Ezeldin and Wallied Orabi
Department of Construction Engineering
The American University in Cairo, Cairo, Egypt
E-mail: aezeldin@aucegypt.edu

Abstract

Risk management, as it relates to construction, is vital to the successful undertaking and completion of any construction project. One way to effectively manage project's risks is to develop more reliable means of identifying the most critical risks and the associated effective response methods. Research studies have extensively addressed this aspect of risk management. However, a small fraction of this research focused on identification of the critical risks encountering contractors working in the construction industry of developing countries, and few tackled identifying the risk mitigation measures employed in such an industry by domestic, international, and multinational contractors.

This paper presents a comprehensive methodology that addresses the risk identification and response methods for developing countries represented by Egypt. The paper is based mainly on the approaches used by large contractors either domestic or international.

The investigation, via a comprehensive questionnaire survey, tries to identify the most critical and significant risks that face the contractors working in the Egyptian construction industry and their associated effectively employed risk mitigation/elimination measures. Twenty-nine (29) construction project risks are classified into six (6) main categories according to their type and hundred and forty (140 risk) mitigation/elimination measures are introduced to overcome the impact of risks under each of these risk categories.

According to the collected data and the results of the statistical analysis procedures employed, the most critical risk encountered by the contractors working in the Egyptian construction industry are: 1)the financial inability of the client; 2)the improper management of construction projects; 3)inflation and interest rates; 4)in-house cash shortage; and 5)Foreign exchange and convertibility. 101 risk response methods were found to be effective from the 140 methods introduced. The most commonly used risk response method was the risk reduction technique.

Introduction

Risk is inherent in all human endeavors and the risk elements involved are diverse and varied (Odeyinka, 2000). Many researchers believe that the construction industry is one of the riskiest economy sectors because risks can affect productivity, performance, quality, and budget of a construction project (Ahmed et al., 2002; Kangari, 1995; Tah and Carr, 2000).

The common view of risk is that it is a negative event that results in loss, hazard, harm, and adverse consequences on the project (Mason, 1973; Moavenzadeh and Rossow, 1976). However, an increasing number of researchers are currently developing their definition of risk as an exposure to loss or gain from involvement in the construction process (Odeyinka, 2000; McKim, 1992; Healey, 1982; Perry and Hayes, 1985).

M. Pandey et al. (eds), Advances in Engineering Structures, Mechanics & Construction, 781–792.

Each party in a contractual relationship will perceive risks from their unique perspective. Owners, who are the ultimate beneficiaries of the contract, may only be considering the project from a construction industry share or production requirement perspective. Their greatest overall risk could reside in the ultimate product and not with the finished facility. Consultants' risks are not as extensive as the owner's or the contractor's and are, to a great extent, limited to design risks. A contractor's overall business risk can be thought of as a portfolio of risks made up of individual project risks. This portfolio changes frequently as some projects are completed and new projects are added. Unlike the designer, construction contractors work at higher risks created by the complexity of design and estimating total project costs (Smith and Bohn, 1999).

Risk management is becoming a crucial process to the success of construction projects because of such an industry that embraces many uncertainties. Furthermore, in developing countries the construction environment is much riskier due to the continuing evolution of laws and regulations, the high cost value of infrastructure projects that represent the majority of projects in such countries, and the economic and political instability.

Contractors working in these countries are in need of a simple but effective tool to help them properly identify the risks that they may encounter and the associated response methods to these risks. This study is focusing on Egypt, a developing country in the Middle East with an emerging reformed economical policy, a population of about 70 millions, and an increasing need of infrastructure and industrial projects.

The study has four main objectives:

1. Identify the most critical risks related to the Egyptian construction industry and their perceived criticality from the point of view of large scale contractors;
2. Identify the risk response methods employed by contractors working in the Egyptian construction industry and their effectiveness.
3. Compare the nature of construction risks encountered in the Egyptian construction to other international markets.
4. Develop a spreadsheet file that summarizes the findings of this research, which could help contractors in their preparation of effective risk management process for new projects.

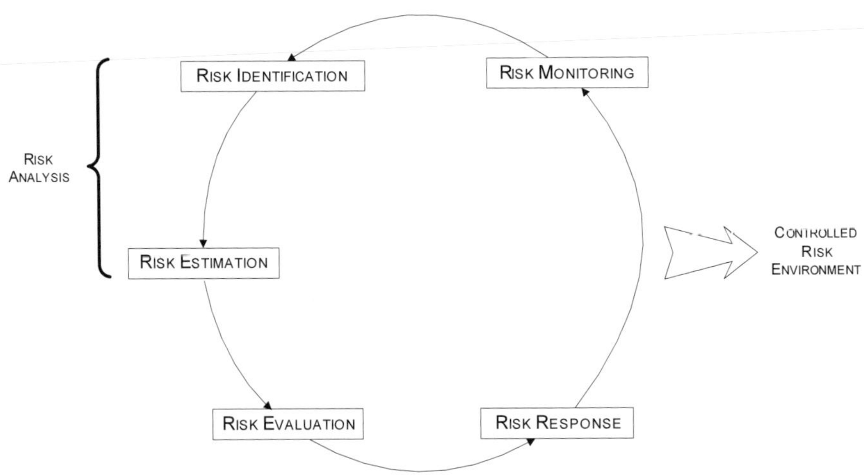

Figure 1: Controlled Risk Environment (after Baker et al., 1999)

Risk Management Process

Researchers, to a great extent, have agreed on the risk management process. However, some difficulties still exist on the terminology and steps' categorization. The five-step process introduced by Baker et al (1999) seems to have gained popularity because of its simplicity. The five-step process is captured in Figure 1.

Risk identification includes selection and classification of potential risks associated with a construction project. The primary basis for identifying risks is historical data, expert judgment, and adequate insight. Once identified, risks can be classified using a Hierarchical Risk Breakdown Structure (HRBS). Figure 2 shows an HRBS presented by Tah and Carr (2000) that allows risks to be separated into those that are related to internal sources and those that are prevalent in the external environment.

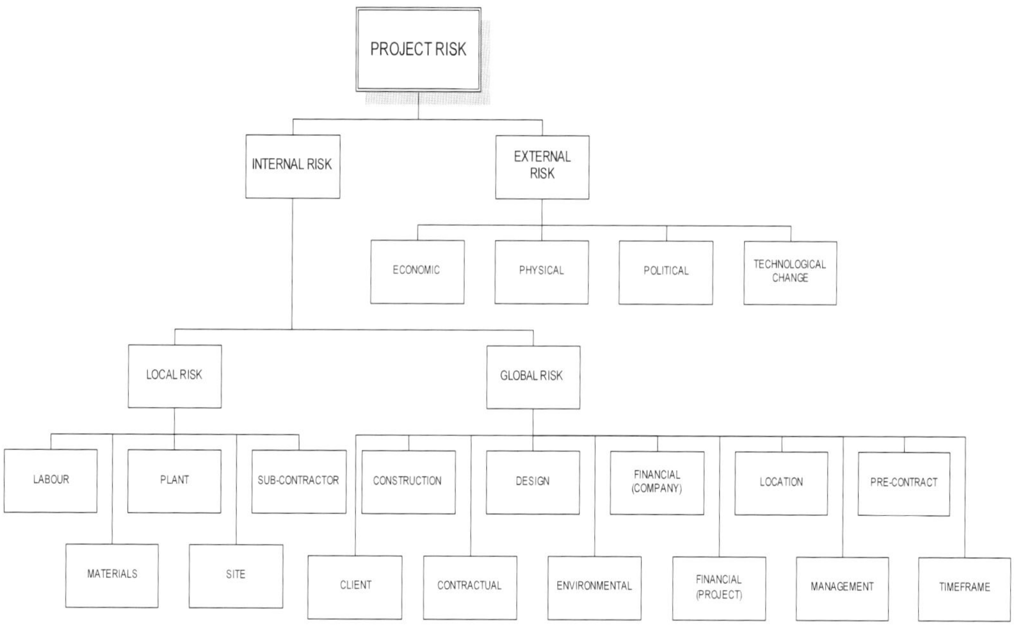

Figure 2: Hierarchical risk breakdown structure (after Tah and Carr, 2000)

Risk estimation or quantification is concerned with assessing the consequences of a certain risk. Risk estimating could include either quantitative or qualitative techniques.

Risk evaluation deals with the impact of risks. Evaluating the impact is based on the assessed consequences and the probability of occurrence of that risk. Kerzner (1998) introduced the following equation showing this concept:

$$Impact\ of\ risk = (likelihood\ of\ risk)\ x\ (consequence\ of\ risk) \qquad (1)$$

Risk response handles taking the appropriate actions to minimize or avoid the impact of risks. The risk response actions include the following: 1) risk avoidance; 2) risk sharing; 3) risk reduction; 4) risk transfer; 5) risk acceptance with contingency; and 6) risk acceptance without contingency.

Finally, risk monitoring aims to monitor the status of identified risks and ensure the proper implementation of planned responses and review their effectiveness.

Literature Review of Similar Studies

Wang el al (2000) conducted a study for evaluation and management of foreign exchange and revenue risks in China's BOT projects. The study was initiated by conducting unstructured interviews to filter the risks collected from the literature. This step was followed by a survey via a questionnaire in order to evaluate the criticality of risks. The study identified and proposed responses for about fifty (50) risks associated with BOT power projects in China.

Bing et al (1999) conducted a research to investigate the critical factors in International Construction Joint Ventures (ICJV) in East Asian countries and the most common and effective measures that the industry participants adopted for risk management. The research ranked a total of fifty-eight (58) risks categorized in three (3) main groups.

Smith and Bohn (1999) performed an investigation into the use of contingency in small construction firms. The main conclusion suggests that these firms assume proportionally greater business risk that suggested in the literature on contingency.

In their evaluation of Florida general contractors' risk management practices, Ahmed et al. (2002), via a questionnaire survey and in-depth interviews, evaluated the practices of risk analysis and management adopted by the general contractors in Florida construction industry and presented a comparison with four other US states considered to have highly profitable and modernized construction industry. The results revealed that in the Florida construction industry, risk analysis and management techniques are rarely used by the general contractors due to lack of knowledge coupled with doubts on the suitability of these techniques for the construction industry.

Research Methodology

In order to achieve the main objectives of this investigation, a specific research methodology has been followed.

The research was initiated by conducting literature review and performing unstructured interviews with expert practitioners in the construction industry. The data gathered from the literature and the practitioners were used to develop a comprehensive questionnaire. A questionnaire survey was administered to a carefully-selected group of qualified engineers in the construction industry. The data collected from the construction industry via the questionnaire survey was then processed by the means of statistical analysis for the purpose of generalizing its findings, as much as possible, to the entire Egyptian construction industry rather than the targeted sample. These findings were disseminated into a simple spreadsheet file developed to aid engineers working in the construction industry in preparing effective risk management processes for their new projects.

Figure 3 presents a schematic diagram of the methodology adopted in this research.

Questionnaire Development

The questionnaire consisted of three main parts over eleven (11) pages. The questionnaire content was finalized after conducting an extensive review of the relevant literature and consulting

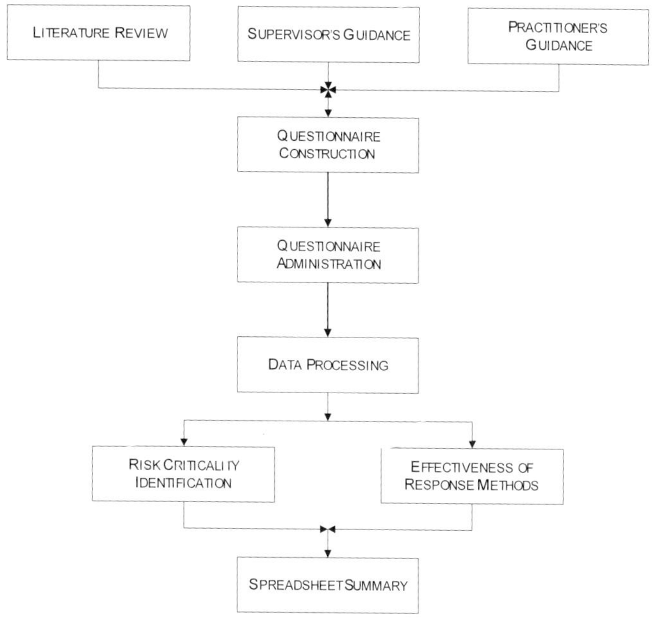

Figure 3: Schematic diagram of the research methodology

qualified and competent practitioners working in the field of construction. This questionnaire essentially contained a list of specific risks and a list of specific response methods from four response techniques: 1) retention; 2) reduction; 3) transfer; and 4) elimination.

The first part of the questionnaire included background information about the respondent, i.e. 1) name, 2) designation, 3) name of company he/she works for, 4) company nationality, 5) company ownership type, 6) types of projects acquired, 7) locations of project acquired, 8) experience, 9) contact information, and 10) asking respondent's permission for authorization of disclosure of his/her personal information for the sole purpose of this research. Some of the abovementioned information were labeled 'optional' to keep, as much as possible, the respondents privacy. The information labeled 'optional' was the name, company name, and contact information.

The second part of the questionnaire dealt with criticality of risks. In order to facilitate the process of identifying construction project risks from the perspective of the contractor, it was decided to classify risks according to their type. Hence, six main risk categories were identified: 1) financial and economic risks; 2) construction and design risks; 3) political/government risks; 4) client-generated risks; 5) subcontractors-generated risks; and 6) miscellaneous risks. A total of 29 different risks were grouped under these categories as shown in Table 1. The six subsections collectively included 29 close-ended questions using a 5-point Likert scale. An instruction of the indication of each point on the Likert scale was provided on top of the section. The scale points ranged from 1 (not critical at all) to 5 (exceptionally critical). Blank spaces were provided for the respondents, at their option, to list other key risks that were not listed in the questionnaire.

The third section of the questionnaire covered the effectiveness of risk mitigation measures. This section was divided into six subsections, one for each of the six risk categories. The six subsections collectively included 29 blocks each containing the response methods associated with the 29 risks. The response methods covered retention, reduction, transfer, and elimination

Table 1: Risks Included in the Questionnaire Survey

Risk Group	Risk Factors
[A] **Financial and Economic Risks**	– Foreign exchange and convertibility – Cash shortage – Inflation and interest rates – Competition.
[B] **Construction and Design Risks**	– Improper design – Improper project management – Constructability – Site safety – Low construction productivity – Defective work – Resourced unavailability – Defective material.
[C] **Political/Government Risks**	– Political instability – Change in laws – Corruption – Approvals and permits – Expropriation
[D] **Client-Generated Risks**	– Financial ability – Changing needs – Response to Claims – Possession of site
[E] **Subcontractors-Generated Risks**	– Technical qualifications – Financial ability – Variation of bids
[F] **Miscellaneous Risks**	– Differing site conditions – Physical damages – Force majeure – Partnership risks – Environmental protection

Table 2: Number of response methods per each risk category

	Elimination	Transfer	Reduction	Retention	Total / Category
[A] Financial	4	-	18	-	**22**
[B] Construction	6	7	22	5	**40**
[C] Political	5	3	18	2	**28**
[D] Client	5	-	6	6	**17**
[E] Subcontractors	1	3	10	-	**14**
[F] Miscellaneous	4	3	10	2	**19**
Total / Technique	**25**	**16**	**84**	**15**	**140**

techniques. The 29 blocks collectively included 140 close-ended questions. Table 2 displays the number of response methods used for each category of risks from each response technique. The complete listing of the 140 risk response methods are listed in Orabi (2003).

Questionnaire Administration

The members of the target population are the qualified contractors working in the Egyptian construction industry. A construction firm was deemed qualified if it was: i) a large scale domestic contractor that is recorded under class 'one' of building and complementary work categories according to the *Egyptian Federation of Construction Contractors*; or ii) an international or multinational contractor that is currently working in Egypt whether solely or under a partnership with another domestic or international contractor.

The respondent filling the questionnaire had to meet the following criteria: i)a total experience of at least ten years; ii)currently working or has been working, for at least five years, with a qualified contractor according to the criteria set before; and; iii) working in an area related to risk (i.e.: risk management, project management, project controls, tendering, … etc).

Based on these guidelines, the size of the target population of qualified contractors was found to be approximately 150 contractors. When applying a percentage of 10% and using an average of two replies from each contractor, the number of qualified replies should be around 30 responses. Such a number would satisfy the minimum requirement of statistical analysis.

Statistical analysis

The standard method of analyzing quantitative questionnaire data is by submitting them to various statistical procedures. These procedures are categorized under two broad categories: descriptive statistics and inferential statistics.

- *Descriptive Statistics* were used to summarize the data collected and present them in smaller space and easier to read and interpret format. The descriptive statistic procedures were performed on the different multi-item scales and each single item of the research questionnaire: Mean, Range, Mode, Median, Standard Deviation, and Skewness. The different risks and response methods were ranked according to their means in a descending order. This helped in achieving the first two objectives of this study since by this ranking process, the most critical risks facing contractors and the most effective response methods employed to mitigate the effect of risks were identified.
- *Inferential statistics* were used to measure the statistical significance of the results obtained from the descriptive statistics and if they were good enough to indicate a more generalized phenomenon. In other words, inferential statistics were used to generalize the results obtained from the statistical analysis performed on the sample data over the whole population. The inferential statistics procedure used in this research is the analysis of variance (ANOVA), which was performed to specify the variance of the results obtained between different company characteristics (i.e. company nationality and company ownership type). The significance tests are conducted at a 5% level of significance using an F distribution.

Analysis of Results

Company Nationality

Contracting firms of eight different nationalities have shown initial interest to participate in the survey, but only companies of seven different nationalities actually responded. Table 3 shows a breakdown of the response rate according to companies' nationality.

Table 3: Response rate breakdown by company nationality

Company Nationality	Initial interest	Actual response	Response rate (%)	Actual no. of replies	Replies / company
Egyptian	15	9	60	19	2.11
Lebanese	1	0	0	0	0.0
US	4	2	50	5	2.5
French	3	2	67	2	1.0
Greek	1	1	100	2	2.0
Japanese	2	2	100	2	1.0
Korean	1	1	100	1	1.0
Multinational	3	3	100	6	2.0
Total	**30**	**20**	**67**	**37**	**1.85**

Most critical Risks

The means of descriptive statistics were performed on the data collected from the respondents concerning the construction project risks to obtain the mean, minimum, maximum, standard deviation, and Skewness of each risk. The ranking of the 29 construction project risks are shown in Table 4 in a descending order according to their means. The criticality of the 29 construction project risks ranged from the lowest value of 2.50 to the highest value of 4.19.

Risk Responses Methods for the Most Critical Risks.

Of the originally listed one hundred and forty (140) response methods in the questionnaire, one hundred and one (101) were deemed effective to respond to risks in the Egyptian market according to the expert respondents. Only the responses for the top five (5) risks are included in this paper, Tables 5 to 9. The complete set of response methods to their respective risks can be found in Orabi (2003).

The most commonly used risk response technique was the risk reduction technique. Risk transfer and retention response techniques were found not suitable for eliminating and/or mitigating the impact of critical risks such as the financial and economic risks and few of the client-generated and subcontractors-generated risks.

Using contractual measures was a key risk response method that was perceived by all types of contractors as the most effective method to eliminate/mitigate the effect of most of the risks considered in this research.

In general, no apparent difference of the perception of effectiveness of the different risk response techniques studied between different company types whether based on nationalities or ownership types was noticed.

Comparison with other international Market

The third objective of this study was to compare the construction risks in the Egyptian market to other international markets. The study of Bing et al (1999) on China's construction market was chosen for that purpose. The top eighteen (18) risks of this investigation had a mean critical score of 3.0 or greater. The list of risks in the study of Bing et al. includes nine (9) of these risks. A closer comparison between the results of both studies indicates that each study identified five (5) risks in the top ten (10) list of the other study. In addition, the most critical risk in both studies was the same, namely; <u>*Client Financial Ability.*</u>

Table 4: The ranking of construction project risks according to their means

Rank	ID	Construction Project Risks	N	Min.	Max.	Mean	Std.	Skewness	
								Statistic	Std. Error
1	D1	Client's Financial Ability	37	2	5	4.19	.776	-.726	.388
2	B2	Improper Management	37	1	5	3.92	1.115	-.849	.388
3	A2	Inflation & Interest Rates	37	1	5	3.89	1.173	-.761	.388
4	A3	In-house Cash Shortage	37	2	5	3.81	.908	-.310	.388
5	A1	Foreign Exchange	37	1	5	3.68	1.292	-.575	.388
6	B1	Improper Design	37	1	5	3.49	1.044	-.194	.388
7	D3	Response to claims	37	1	5	3.49	1.017	-.296	.388
8	B7	Lack of Resources	36	1	5	3.33	1.146	-.468	.393
9	B6	Defective Work	36	2	5	3.31	.951	-.036	.393
10	E1	Subcontractors Technical Qualifications	36	1	5	3.28	1.003	-.064	.393
11	C5	Expropriation	37	1	5	3.22	1.669	-.097	.388
12	B8	Defective Material	37	1	5	3.19	1.050	-.400	.388
13	C4	Approvals & Permits	37	1	5	3.16	.928	-.119	.388
14	F4	Partnership Risks	34	1	5	3.12	.977	.166	.403
15	B5	Low Productivity	36	1	5	3.08	.967	-.174	.393
16	A4	Competition	36	1	5	3.03	1.134	-.057	.393
17	B4	Site Safety	37	1	5	3.03	1.093	.349	.388
18	F3	Force Majeure	37	1	5	3.00	1.374	.204	.388
19	C2	Change in Laws	37	1	5	2.97	1.142	.410	.388
20	E3	Subcontractors Variation of Bids	37	1	5	2.95	.998	.290	.388
21	C3	Corruption	37	1	5	2.95	1.026	.602	.388
22	E2	Subcontractors Financial Ability	37	1	5	2.89	1.125	-.024	.388
23	F1	Differing Site Conditions	36	1	5	2.89	1.036	-.257	.393
24	D2	Client's Changing Needs	36	1	5	2.89	1.008	.056	.393
25	D4	Possession of Site	37	1	5	2.84	1.167	.111	.388
26	B3	Constructability	36	1	5	2.72	.944	.818	.393
27	C1	Political Instability	37	1	5	2.68	1.107	.437	.388
28	F5	Environmental Protection	37	1	5	2.54	.931	.422	.388
29	F2	Physical Damages□	36	1	5	2.50	.811	.852	

Table 5: Risk Response Methods for Financial ability of the Client

ID	Response method description	Mean	Std.	Skew.
D11	Ensure that a reputable owner finances the project	3.84	0.834	.017
D12	Specify extension or compensation clauses in contract for payment	3.78	0.917	-.228
D13	Enter into fixed rate loan contract with lending banks	2.86	0.918	-.175
D14	Finance the project from in-house sources	2.58	1.079	.495

Table 6: Risk Response Methods for Improper Project Management

ID	Response method description	Mean	Std.	Skew.
B21	Hire competent project manager and project team	4.00	0.882	-.514
B23	Undertake sound pre-project planning to minimize management risks	3.49	0.961	.040
B25	Establish a fixed standard project management system for all projects to follow	3.22	1.031	.021
B24	Subcontract project management to a specialized organization	2.70	1.051	.036
B22	Centralize project management activities at head office	2.56	0.939	-.279

Table 7: Risk Response Methods for Foreign Exchange and Convertibility

ID	Response method description	Mean	Std.	Skew.
A14	Include compensation clauses for exchange rate in the contract	3.46	1.095	-.628
A12	Use dual-currency contracts with certain portion to be paid in local currency and others in foreign currency	3.22	1.158	-.108
A11	Obtain government guarantees of exchange rate and convertibility, e.g. fixed rate for long period	3.03	1.500	-.310
A13	Use transfer tools to hedge exchange rate , e.g. forward and swap	2.57	0.979	.587

Table 8: Risk Response Methods for Inflation and Interest Rates

ID	Response method description	Mean	Std.	Skew.
A21	Secure standby cash flow in advance	3.47	1.158	-.454
A22	Incorporate escalation clauses for interest, inflation rates and delays in the contract	3.41	1.117	-.128
A24	Ensure that a reputable owner finances the project	3.30	0.996	-.114
A28	Sign fixed or pre-defined prices with material suppliers	3.24	1.140	-.034
A23	Obtain payment and performance bonds from banks	3.24	1.090	.028
A27	Adopt as much as possible domestic product/labor to reduce cost	3.16	1.014	-.005
A25	Specify extension or compensation clauses in contract for payment	3.11	1.008	-.056
A26	Enter into fixed rate loan contract with lending banks	3.05	0.998	-.290

Table 9: Risk Response Methods for Cash Shortage

ID	Response method description	Mean	Std.	Skew.
A34	Ensure that a reputable owner finances the project	3.35	1.060	-.030
A32	Owner to secure standby financing (i.e. more than 100% financing commitments when needed)	3.35	1.274	-.116
A33	Obtain payment and performance bonds from banks	3.16	1.214	-.131
A31	Get Letter of Credit from government	3.05	1.353	-.174
A36	Specify extension or compensation clauses in contract for payment	2.78	1.124	.210
A35	Adopt alternatives to contract payment, e.g. land development rights, or resource swap	2.27	0.693	.110

The fourth objective of this research was to develop a simple spreadsheet file that summarizes the findings of this research. The findings of the study were collected and summarized in a Microsoft® Excel workbook. This workbook contains nine worksheets, the first worksheet is a welcome screen that gives the user two options: either (1) to view risks classified according to their nature or (2) to view risks ranked according to their criticality. The second worksheet lists the names of the six risk categories: (a) financial and economic risks; (b) construction and design risks; (c) political/government risks; (d) client-generated risks; (e) subcontractors-generated risks; and (f) miscellaneous risks. The third worksheet lists the 29 construction risks that were identified by this research in a descending order according to their criticality scores. The remaining six worksheets are for the six risk categories, one worksheet per each category and it includes the risks in this category and the response methods associated with each risk.

Conclusions

The following section presents an overview of this research and a summing up of its findings.

The most critical risks that the contractors working in the Egyptian construction industry experience in their projects were complied in a form of checklist. According to the results of the statistical analysis procedures employed, the most critical risks were found to be:

1. the financial inability of the clients;
2. the improper management of construction projects;
3. inflation and interest rates;
4. in-house cash shortage; and
5. Foreign exchange and convertibility.

According to the number of risk factors deemed critical, the international and multinational contractors perceive the Egyptian construction industry as being riskier than the domestic contractors do.

Under the four different utilized risk response techniques, 101 risk response methods were found to be effective from the 140 methods included in this research. Mainly, the most commonly used risk response technique was the risk reduction technique. The statistical analysis indicated no apparent difference of the perception of effectiveness of the different risk response techniques studied between different company types whether based on nationalities or ownership types was noticed.

References

Ahmed, S. M., Azhar, S., and Ahmad, I. (2002). Evaluation of Florida general contractors' risk management practices. Revista Ingenieria de Construccion (Construction Engineering Journal), 17(1), 4-10.

Baker, S., Ponniah, D., and Smith, S. (1999). Risk response techniques employed currently for major projects. Construction Management and Economics, 17, 205-213.

Bing, L., Tiong, R., Fan, W., and Chew, D. (1999). Risk management in international construction joint ventures. Journal of Construction Engineering and Management, ASCE, 125(4), 277-284.

Healey, J.R. (1982). Contingency funds evaluation. Transaction of American Association of Cost Engineers, B3.1-B3.4.

Kangari, R. (1995). Risk management perceptions and trends of US Construction. Journal of Construction Engineering and Management, ASCE, 121(4) 422-429.

Kerzner, H. (1998). Project management: a systems approach to planning, scheduling, and controlling. New York: John Wiley and Sons.

Mason, G.E. (1973). A quantitative risk management approach to the selection of a construction contract provisions. Ph.D. Thesis, Stanford University.

McKim, R.A. (1992). Systematic risk management approach for construction projects. Journal of Construction Engineering and Management, ASCE, 118(2), 414-415.

Moavenzadeh, F., and Rossow, J. (1976). Risk and risks analysis in construction management. Proceeding of the CIB W65, Symposium on Organization and Management of Construction, US National Academy of Science, Washington, DC.

Odeyinka, H. A. (2000). An evaluation of the use of insurance in managing construction risks. Construction Management and Economics, 18, 519-524.

Orabi, W. (2003), "Risk Identification and Responses Methods in the Egyptian Construction Industry: Views of Large Scale Contractors", M.Sc. Thesis, the American University in Cairo, 255 pp.

Perry, J. G., and Hayes, R. W. (1985). Risk and its management in construction projects. Proceedings, Conference of the Institution of Civil Engineers, Part 1, 78(June), 499-521.

Smith, G. R., and Bohn, C. M. (1999). Small to medium contractor contingency and assumption of risk. Journal of Construction Engineering and Management, ASCE, 125(2), 101-108.

Tah, J. H. M. and Carr, V. (2000). A proposal for construction project risk assessment using fuzzy logic. Construction Management and Economics, 18, 491-500.

Wang, S. Q., Tiong, R. L. K., Ting, S. K., and Ashley, D. (2000). Evaluation and management of foreign exchange and revenue risks in China's BOT projects. Construction Management and Economics, 18, 197-207.

BETTER VISUALIZATION FOR MULTIPLE-SITE INFRASTRUCTURE PROGRAMS USING GIS

Tarek Hegazy
Department of Civil Engineering, University of Waterloo, Waterloo, Ontario, Canada N2L 3G1
E-mail: tarek@uwaterloo.ca

Abstract

This paper introduces a model for optimizing and visualizing infrastructure maintenance programs of multiple-distributed sites. Two unique aspects of the model are discussed in the paper: the underlying Geographic Information System (GIS); and the powerful scheduling engine that optimizes execution plans. The GIS system stores and represents two main levels of information about the scattered sites involved in a construction/maintenance program. The first level pertains to pre-planning data such as resources, locations, optional estimates, and work constraints. This information is then used by the scheduling engine to generate an optimum schedule, and accordingly, a second layer of GIS information is generated containing activities' start and finish dates and the assigned crews. This layer of information is then used by the GIS system to visualize the crews' work assignments in a legible manner. An implementation program BAL is presented on an example application to illustrate the benefits of using GIS to support municipalities and owner/contractor organizations administering large number of infrastructure assets, such as buildings, highways, and bridges, etc.

Keywords: infrastructure, GIS, scheduling, optimization, resource management

Introduction

In recent years, interest in developing management systems for infrastructure networks, such as highways, bridges, airports, and water/sewer systems, etc., has grown rapidly to help sustain infrastructure services. Management systems concern mainly with prioritizing capital assets for maintenance and repair (M&R) purposes. The majority of management systems, however, provide little or no decision support during the execution of M&R programs, thus leading to cost overruns and delays. Among the biggest challenges facing the execution planning of M&R programs is the repetitive multiple-location nature of M&R programs and the difficulty in representing the large amount of information involved.

Currently, there exist many software tools for maintenance management. Some of these systems may provide traditional planning and scheduling features such as bar charts to schedule operations. While these systems are beneficial, they address some but not all aspects of infrastructure execution planning. They are not formulated to respect a given deadline and do not consider the distributed and repetitive nature of operations. In addition, they provide no decision support for cost optimization and do not legibly represent the large amount geographically dispersed data, particularly related to the assignment of resources among the sites. As such, the need has emerged for a new decision support tools that use GIS technology to support effective management of infrastructure M&R programs with multiple sites.

This paper provides details on the use of GIS technology to facilitate the development of a generalized model for optimizing and visualizing M&R programs that involve distributed sites. An example application is then used to illustrate implementation issues.

M. Pandey et al. (eds), Advances in Engineering Structures, Mechanics & Construction, 793–800.

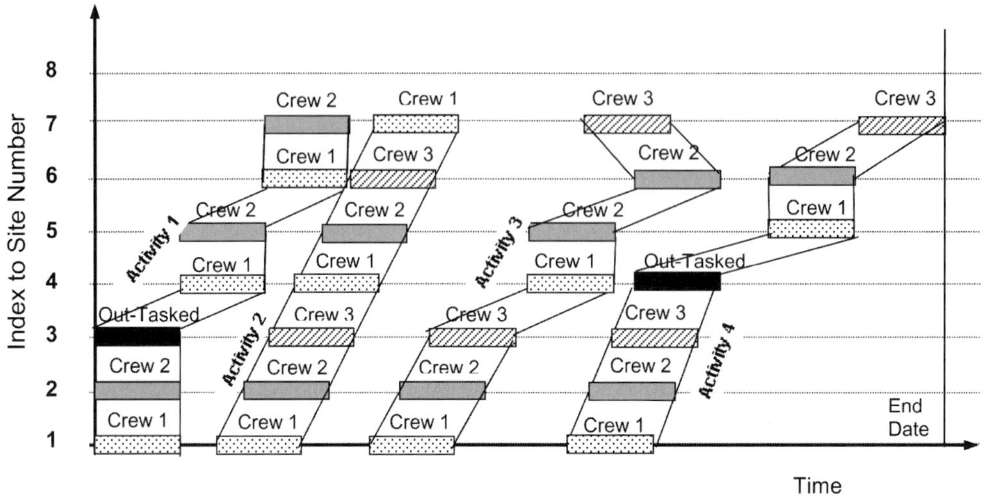

Fig. 1. Multiple-site schedule representation.

Geographic Information Systems (GIS)

Simply put, GIS is an information system that combines layers of information about a place to provide better information about that place. A GIS involves hardware, software, and procedures designed to support the capture, management, manipulation, analysis, modelling and display of spatially-referenced data.

A GIS is not an automated decision making system but a tool to query, analyze, and map data in support of the decision making process. For example, GIS can be used to help reach a decision about locating a new housing development in an area having minimal environmental impact and is also close to a population center. GIS presents information quickly, concisely, and clearly in the form of maps to facilitate efficient analysis of multiple decision scenarios. Presenting information in maps also well suits the powerful pattern recognition capability of human beings who can draw conclusions from maps much easier than from tables of rows and columns. Based on this description of GIS technology, it is clear that it has great potential to support execution planning decisions for multiple-site infrastructure M&R programs.

Distributed Scheduling Model

Recently, a new Distributed Scheduling Method, DSM, was introduced to facilitate the planning of resources in distributed and repetitive projects (Hegazy et al., 2004), and was implemented in a computer software program, called BAL. The DSM helps in scheduling multi-site M&R programs that are delivered using combination of in-house resources and outsourcing. The DSM model determines the optimum site execution order and assigns available resources to the various sites in a manner that maintains crew work continuity, and meets the cost, and resources constraints. To meet deadlines, speedy delivery options (often expensive) for various activities are stored and used in the scheduling process. One of the key features of the DSM is its legible representation of the large amount of schedule information, as shown in Figure 1. The figure shows time on the horizontal axis and site index on the vertical axis. All crew movements among sites are clearly shown. For example, crew 1 of the second activity will proceed to site 4 once it completes its work on site 1. Clearly, the time

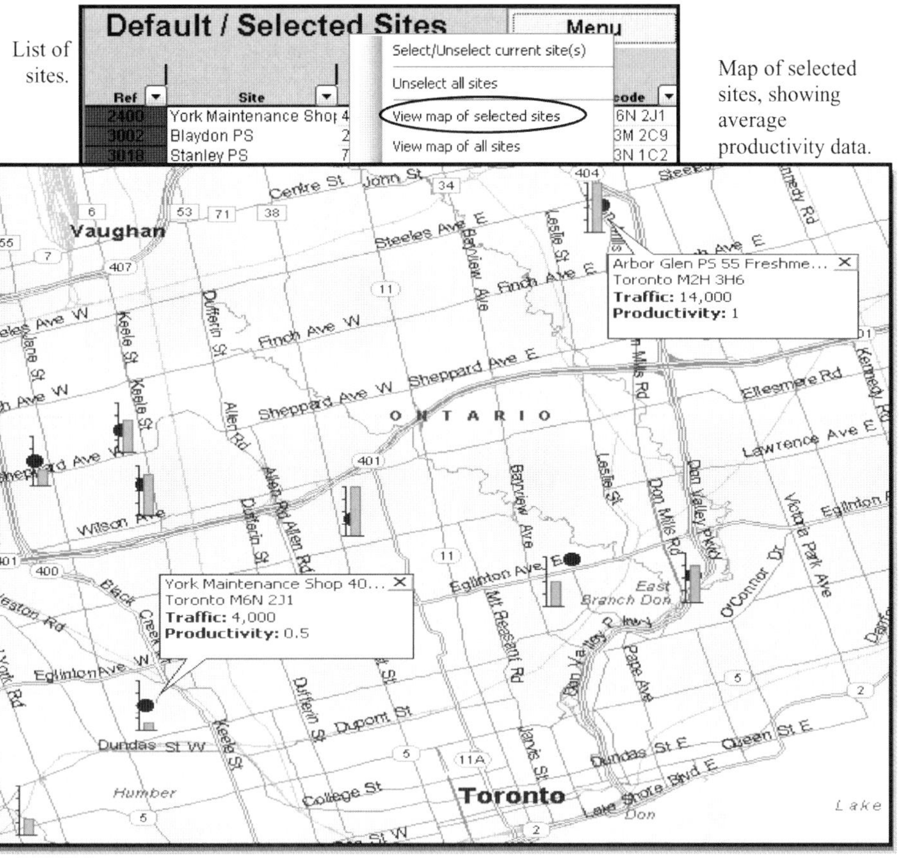

Fig. 2. Site locations and productivity factors.

and cost of moving resources from site 1 to site 4 depends on the distance between the sites and the speed of moving along possible travel routes. As such, the site order (vertical index) becomes an indicator of the sequence of operations and its consequent travel time and cost.

The DSM model is capable of generating schedules by manually changing the options for construction methods, number of crews, the site order, and the amount of interruption at various sites. However, with the large number of possibilities, even for a small network of sites, a cost optimization model becomes necessary to identify the optimum combination of these variables to meet schedule constraints. The optimization model in the DSM involves the setup of the objective function and optimization constraints. The objective function of the model is to minimize total construction cost. Along with proper ranges for the variables, two soft constraints are used: Project duration should be less than or equal to the deadline duration; and total aggregated amount of a given resource is less than or equal to the amount available. To handle the large-scale optimization involved, a non-traditional optimization technique, Genetic Algorithms, has been successfully used in the DSM.

GIS Support

To incorporate GIS features into BAL, a commercial GIS software, Microsoft MapPoint 2002, has been integrated into BAL through visual basic code. The GIS program was utilized to store

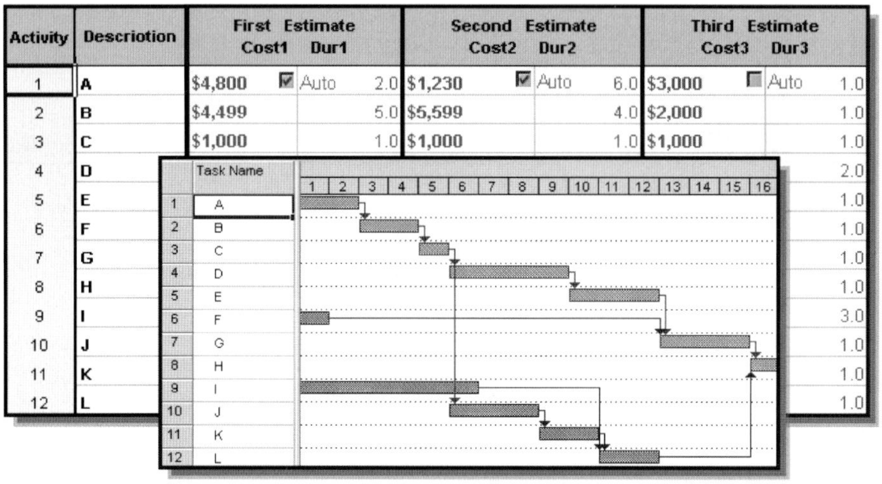

Fig. 3. Optional estimates and logical relationships.

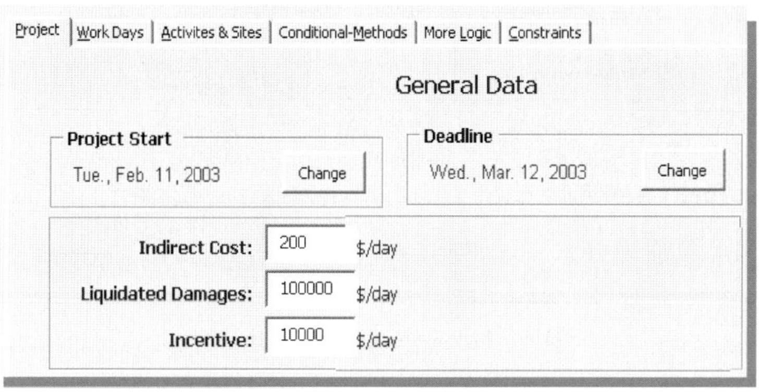

Fig. 4. Main input screen with general data.

sitespecific information, which lends itself well to mapping representation, to facilitate speedy and informed decisions for M&R programs. Site-specific pre-planning information includes location, local weather, land survey data, traffic volume, etc. Using this information, the GIS system automatically calculates the distances from one site to any other, considering the shortest travel routes. Once the distances are calculated, the GIS system then calculates the travel time from each site to any other, considering the speed limits specified for the highways or local roads along a route. These are then directly used to determine the time and cost to transport resources from one site to the other.

During the planning stage, BAL's scheduling engine stores the site-specific data of work quantities, available resources, construction methods, and the time/cost/other constraints. Using this information, the scheduling engine of BAL runs the Genetic Algorithm (GA) procedure and experiments with thousands of random solutions until an optimum is reached. For each random solution, the GA selects the number of crews to use, the site order, and a set of construction methods. In this process, the GIS system feeds the associated distances, moving time, and moving cost, to the scheduling engine. At the end of the GA procedure, an optimum work schedule is determined that respects the time, cost, and resource constraints. Afterwards, another layer of GIS information is generated; containing activities' start and finish dates at the various sites along with the assigned crews. This

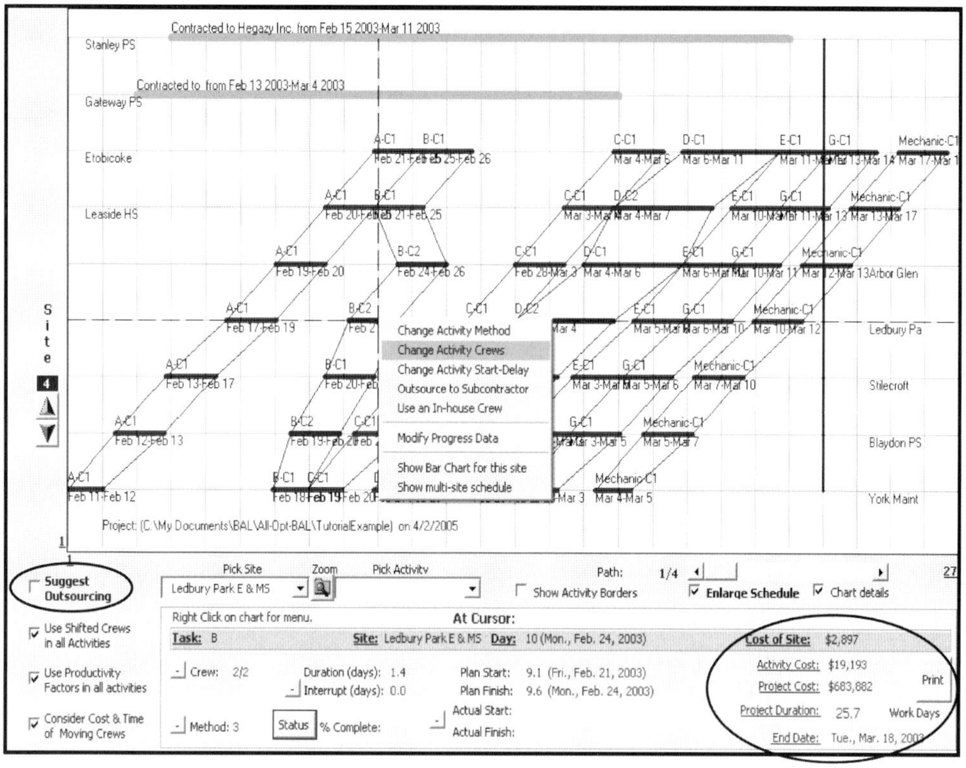

Fig. 5. Initial schedule.

layer of information is then used by a macro written on the GIS system to present the schedule and crew assignments legibly to all project participants.

Example Application

To illustrate the features of BAL and its GIS system, a simple example of an M&R program is presented. The program resembles a school board having a M&R program involving 9 schools. First, the nine schools were selected from a spreadsheet having all school locations at the Greater Toronto Area in Canada (Figure 2). Each school is defined by its address (recognizable by the GIS software). For each site in the list, various pre-planning information was stored, including traffic volume and local productivity factors (monthly values that depend on weather, site access, and other conditions). These productivity factors will be used in the scheduling process to adjust activities' durations, thus, providing realistic execution conditions. As shown in Figure 2, the GIS automatically reads the sites' information and shows a map of their locations and other related information. Various types of data representations are available, such as pie chart, column chart (shown) or sized circles. These data representations provide simple and legible options to compare the values (e.g., productivity factors) at the various sites.

In the present M&R example, 12 activities are involved in each typical site, as shown in Figure 3. The work quantities and three estimates that represent normal, overtime, or weekend work options are shown in the same figure. While the cheaper estimate is the default in generating an initial plan, the other options (faster but more expensive) may be necessary in case durations need to be

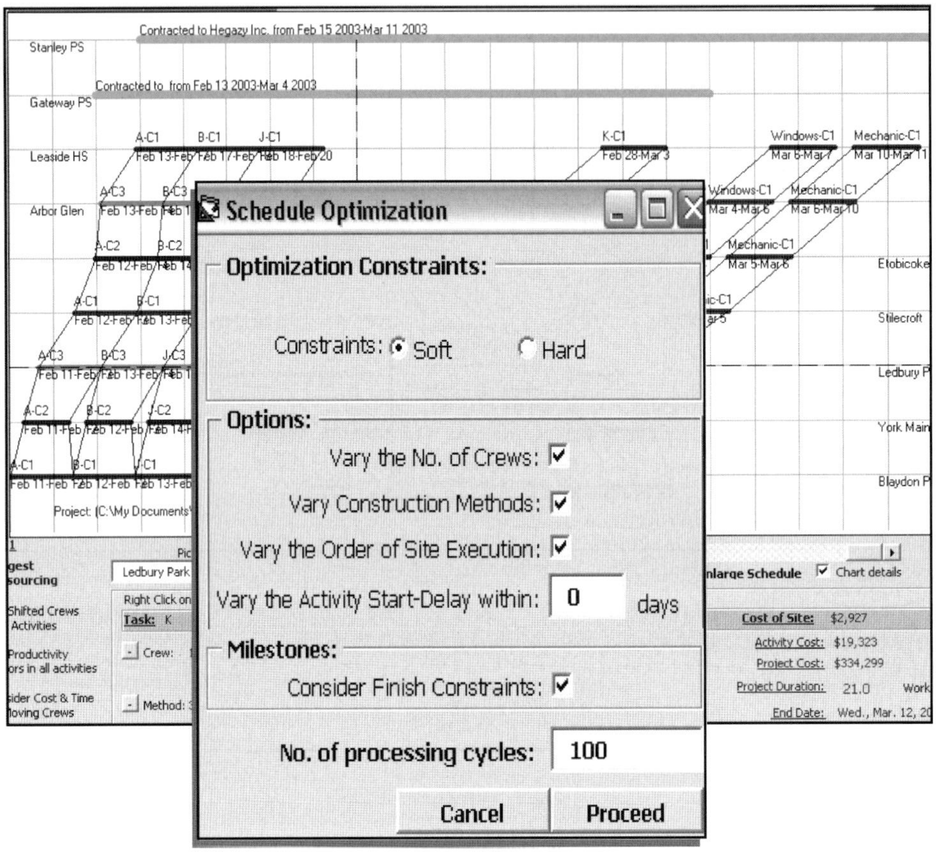

Fig. 6. Schedule optimization.

shortened to meet a strict deadline. In this case, finding the proper combinations of work options for the activities becomes part of the optimization feature.

Before producing a detailed schedule, the user needs to enter some general data such as start date (February 11, 2003, as shown in Figure 4) and the execution constraints such as deadline (March 12, 2003), incentive ($10,000/day), liquidated damages ($100,000), and any resource limits. As such, this example M&R program has only 22 working days (excluding weekends) to complete the 9 sites.

With the default option being to use in-house crews, the next step was to enter information regarding any special sites that were decided to be outsourced to contractors. In this example, two of the nine sites were pre-specified as outsourced. Once all site-specific information was specified, a detailed execution schedule was generated, as shown in Figure 5.

The initial schedule shows the various color-coded crews. The two contracted sites are shown at the top part of the schedule. It is noted that the initial site order is the one entered by the user in Figure 2. Based on that site order, project duration and cost (considering crew-moving time and cost) were automatically calculated, taking into account the site productivity factors. This resulted in a schedule of 26 days, as opposed to the 22-day deadline. Accordingly, the cost becomes $283,882, plus liquidated damages of $400,000, for a total of $683,882, as shown in Figure 5.

To improve the schedule, it is possible to manually change the crews, work options, and site order to try to meet the deadline and reduce cost (user modifiable options shown in Figure 5). Rather than

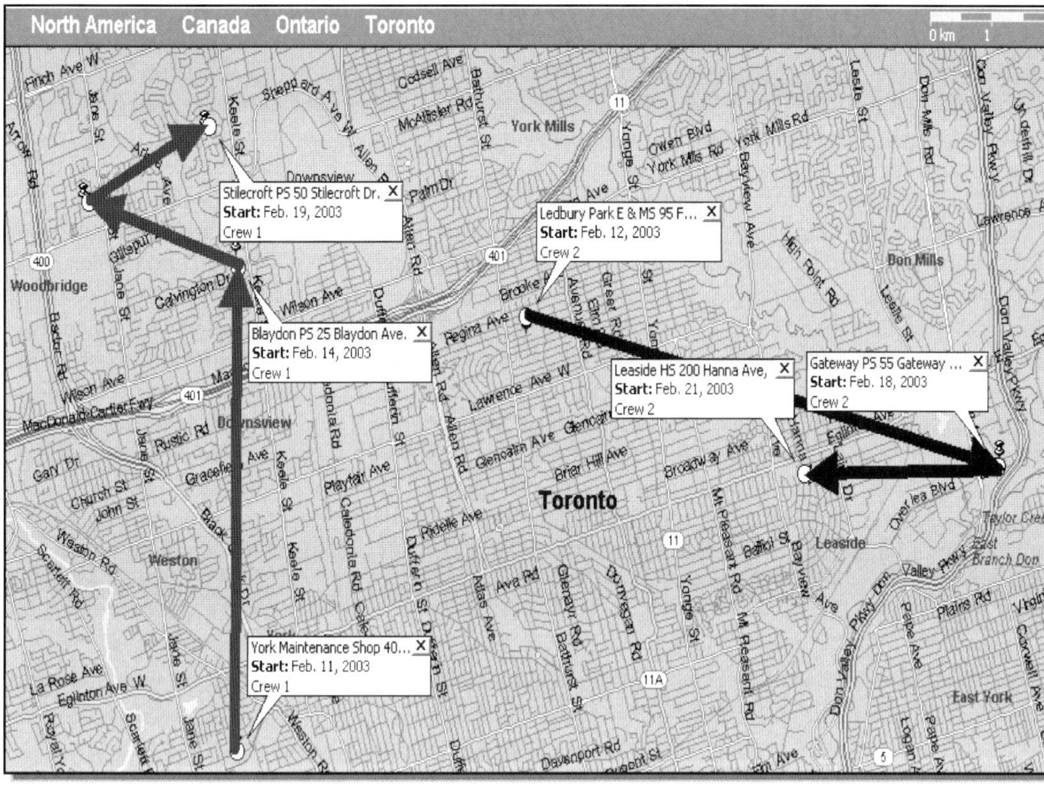

Fig. 7. Crew assignment map.

this manual adjustment process, it is possible to activate the optimization feature, and accordingly, a form for specifying optimization options appears (Figure 6) and allows the user full control over the optimization variables, to suit the project objectives. For this example, various optimization experiments were conducted. The optimum schedule is shown in Figure 6. Project duration of 21 days meets the deadline at a total cost of $334,299.

Once a satisfactory schedule was obtained, site-specific schedule data were automatically exported to the GIS system to provide legible maps of detailed movement of the various crews. An example map generated by the GIS system for the present example is shown in Figure 7. The figure shows two color-coded paths for two crews involved in an activity. The start time for each site is also shown on each site. As such, the map is simple to read and shows when each crew needs to start in which site. An alternative chart that shows detailed travel directions, which can also be generated from the GIS system.

Conclusions

This paper introduced a scheduling model and implementation software, BAL, for optimizing resource allocation in infrastructure projects with multiple-distributed locations. Two unique aspects of the program are discussed in this paper: (1) the powerful scheduling engine that optimizes the execution plan; and (2) the underlying Geographic Information System (GIS).

The GIS system stores and represents various levels of information about the scattered sites involved in a construction/maintenance program. Pre-planning information includes location, productivity factors, land survey data, and traffic volume, etc. Using this information, the GIS system automatically calculates the distances from one site to any other, considering the shortest travel routes. Based on the distances, the GIS system calculates the travel time from each site to any other, considering the speed limits specified for the highways or local roads along a route. These are then directly used to determine the time and cost to transport resources from one site to the other.

During the planning stage, BAL's scheduling engine stores the user-input data of available resources, construction methods, and the time/cost/other constraints. The scheduling engine then runs a Genetic Algorithm to optimize the number of crews to use, the site order, and the set of construction methods. Based on the optimized schedule, another layer of GIS information is generated; containing activities' start and finish dates at the various sites along with the assigned crews. This layer of information is then used to represent the work assignment in a legible manner to the various project participants. One of the outputs is a legible crew assignment map.

An example application was used to illustrate the benefits of using GIS to support schedule computation and better visualization of multiple-site multiple-crew execution plans. The proposed computer program is potentially usable by municipalities and owner/contractor organizations administering a large number of infrastructure assets, such as buildings, highways, and bridges, etc.

Reference

Hegazy, T., Elhakeem, A., and Elbeltagi, E. (2004) Distributed Scheduling Model for Infrastructure Networks. *Journal of Construction Engineering and Management, ASCE*, 130(2), 160–167.

GLASS-PLASTIC HYBRID CONSTRUCTION

J. Hildebrand and F. Werner
Bauhaus-Universität Weimar, Institute of Structural Engineering,
Dept. of Steel Structure, Marienstrasse 5, 99423 Weimar, Germany
E-Mail: joerg.hildebrand@bauing.uni-weimar.de, frank.werner@bauing.uni-weimar.de

Abstract

Transparent glass constructions are being used more and more often in the construction industry using up-to-date carrying systems. The progress in this field is determined by the methods to couple glass systems and transparent substructures in a durable manner while keeping the dimensions of the substructures as small as possible.
Experimental investigations have shown that carrying glass-hybrid beams are possible and that there are appropriate adhesive and surface pre-treatment substances available on the market to connect glass and plastic in an orderly manner.

1 Introduction

Maximum transparency as wished by architects is presently spoiled by necessary carrying elements made of steel, wood or concrete as well as by joint formation. A special problem which is to be solved here, is the connection of glass elements with each other as well as connecting them with the carrying substructure. (see Fig. 1 to 3).

Fig. 1: synagogue in Berlin

Fig. 2: substructure

Fig. 3: monastery Volkenroda

Two reasons cause these problems:

- There are no industrial methods available for the production of transparent cross-sections which have sufficient bending and longitudinal force carrying capabilities, e.g. double-T cross-sections.
- The connection of glass elements with each other and the connection with the substructure are at present strongly limited by the lack of approved and durability-tested transparent adhesive substances and scientifically tested dimensioning guidelines.

There exists a great number of partly new construction forms which try to realise efficient substructures that do not spoil transparency more than necessary while taking into account durability, heat protection, etc. (Dodd,2004). All of those have in common that generally the carrying parts are non-transparent. The safety and reliability of the constructions are realised by safety and laminated glasses and by local connection elements, mostly made of stainless steel. The deployed techniques do not

M. Pandey et al. (eds), Advances in Engineering Structures, Mechanics & Construction, 801–808.

permit the production of cross-sections which are effective with regard to bending and longitudinal force stresses and the subsequent tension and stability problems.
Simply agglutinating individual glass sheets to get rectangular cross-sections as carrying bar cross-sections does not lead to optimal cross-section forms. Furthermore, connecting bars with each other is problematic, so that a great number of discrete connection elements - mostly made of metal – has been used so far along with agglutinating. These discrete couplings, i.e. the stress transmissions mean a high safety risk especially when one takes into account long-term impacts. This is a result of the carrying properties of the material glass which is very sensitive with regard to local tensile stress, and through external influences, e.g. heat, stress concentrations can occur.
The deployment of highly ductile synthetic construction elements as coupling element between glass sheets and as connection element of cross-section parts of profiles or construction parts, respectively, opens completely new possibilities in the design of transparent construction parts (Bemm, 2003). In this investigated field there are no other findings known to the author as to the carrying behaviour, the possibility of producing adhesive connections, on the durability and robustness as well as on the dimensioning (Freytag, 2004; Hess, 2000).

2 State of Research

In recent years more and more investigations on the carrying capacity and on the stability of glass profiles with rectangular cross-section have been carried out since these were mostly used so far. These investigations concentrated on the problems of local stresses as well as on global influences (Hess, 2002). All of this leads to a significantly increasing deployment of glass constructions – this applies for carrying elements, too. This development is not giving impetus, though, with regard to qualitatively new forms of construction or structure.

The recent research on the field of constructive glass building can be divided into three main areas:

- Carrying behaviour of construction elements (Hess, 2000; Hess, 2002),
- Problems of the local carrying behaviour especially with regard to bearing and connection (Schuler et al, 2004; Bernard et al, 2004),
- Adhesive techniques (Wiesner, 2004; Müller, 2004).

Recent investigations try to further develop the given technical possibilities of glass as well as those of connection elements. This applies basically for (Weller et al, 2004):

- Laminated glass (Schober and Schneider, 2004),
- a variety of local mechanic holders (Veer et al, 2001),
- the attempt to replace silicate adhesives by new acrylic adhesives or similar things (Dilger et al, 2003).

These developments have their significance as to an increased safety, easier execution and an enlargement of the glass sheets in relation to the necessary supportive constructions. All of these investigations have in common that they basically try to improve efficiency and reliability using known paths (Bornemann et al., 2003; Welters and Dilger, 2003).

3 Idea

Even glass-plastic hybrid connections have been used very successfully for quite a while in the field of laminated glass technology. Connecting glass sheets with relatively thick interlayers of special synthetics result in sheets which have a high breaking resistance. This construction form is limited to elements such as shop windows, security walls in buildings, doors, etc. (Business, 2003). Substantial investigations showed that there is no approach to deploy these hybrid connections - in a new quality - to create carrying elements.
When designed correctly the plastic sheets are capable of transmitting normal as well as shearing stresses, even though the e-modulus of the plastic is significantly lower than the one of glass. The

glass sheets within the hybrid elements can be coupled shear proof over the plastic sheets when appropriate adhesive techniques are deployed. In doing so, one achieves cross-sections with a favourable longitudinal force and bending carrying capacity while the coupling element is not in danger of a brittle fracture like glass since it is made of ductile plastic. The transmission of shearing forces between bridge and flange occurs through the ductile plastic. (See Fig. 4)

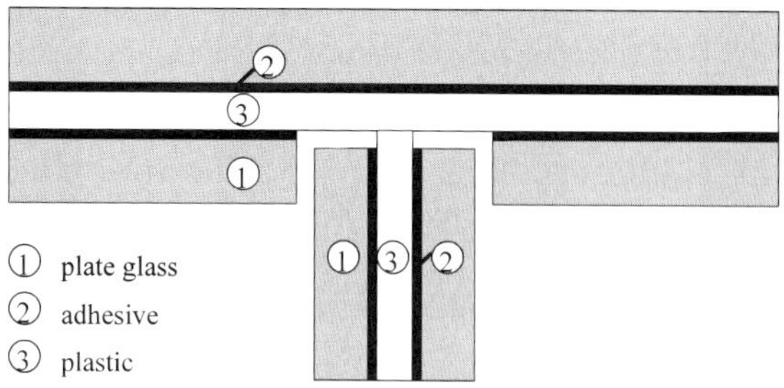

Fig. 4: principle of glass-plastic hybrid element

4 Investigation on the Adhesive

A problematic part in the production of glass-plastic hybrid connection is the large-area adhesion of glass sheets with plastic sheets. The goal is to produce connections which are durable, stress-relieved and optically clear. Especially the latter point poses a considerable technological challenge.
For the first experiments polycarbonates from the company Bayer (Makrolon GP 099) and from the company GE Plastics (Lexan PC) were chosen from the wide range of plastics (P). The used float glass belongs to the group of lime-natronsilicate glasses. The specimens consisted of a compound (3 mm glass – 3 mm plastic – 3 mm glass) with the dimensions 300 mm length and 80 mm width. With regard to the adhesives (A) and the surface pre-treatments (SPT) different combinations were investigated.
The aging cycle D3 „Heat, Coldness and Humidity" taken from the standard DIN EN ISO 9142 (Standard, 2004) represents the climate conditions occurring in the case of glass constructions best. A determined humidity was not set in the experiments, they were limited to temperature cycles in the following steps:

- Exposure (15 ± 1) h at a temperature $(+40 \pm 2)$ °C
- Change within (60 ± 20) min to a temperature (-20 ± 3) °C; exposure (2 ± 1) h
- Change within (80 ± 20) min to a temperature $(+70 \pm 2)$ °C; exposure (4 ± 1) h
- Change within (60 ± 20) min to a temperature $(+40 \pm 2)$ °C

The pictures in Table 1 show selected specimens after 25 cycles given above. All specimens showed large-area debondings; partly microcracks occurred in the Lexan PC which were availed by the surface pre-treatment. The wished compound effect could not be achieved. The experimental program was changed with regard to plastic, surface pre-treatment and adhesive.
In Table 2 shows specimens of another series of experiments with changed adhesives and surface pre-treatment methods. Using the Pyrosil method had a very positive influence on the compound effect between glass and plastic. In this process a pre-treatment of the glass surface as well as of the plastic surface was carried out. The Pyrosil method produces an increased surface energy and thus improves the bonding.
After 25 cycles no microcracks or debondings occurred in polycarbonate Makrolon GP 099 (see Table 2, specimen 3-3). In the case of polycarbonate Lexan PC (see Table 2, specimen 4-3) microcracks occurred and - at the edge – debondings.
In addition to an acrylic adhesive two other adhesives were deployed (Epoxy and PUR). Both specimens showed large-area debondings leading to the destruction of the glass in the case of specimen 6-3. Both adhesives are probably not adequate to produce a durable compound between glass and Lexan. The Pyrosil method had no influence on the compound effect.

Table 1: pictures of selected specimens glass-plastic-glass after 25 cycles

specimen 1-1 A	specimen 2-1 B	specimen 5-1 A
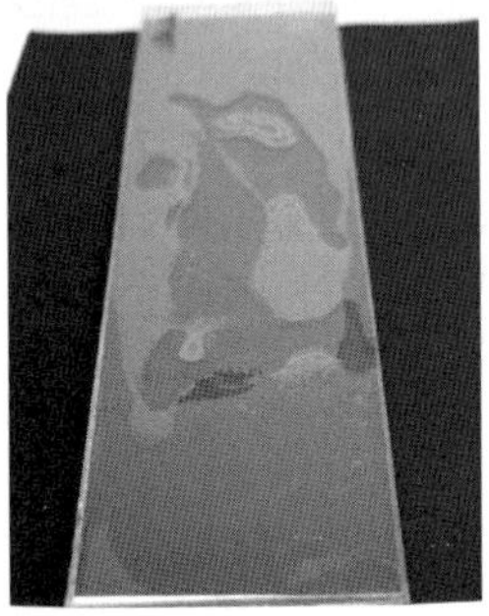	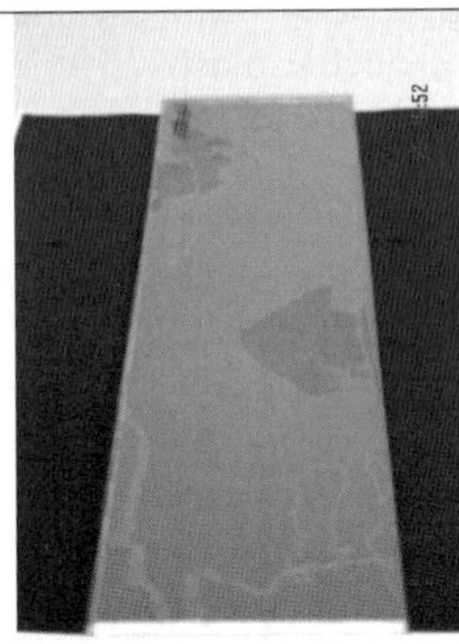	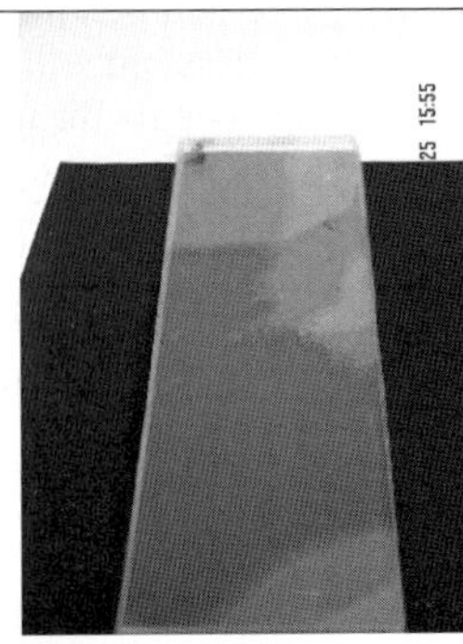
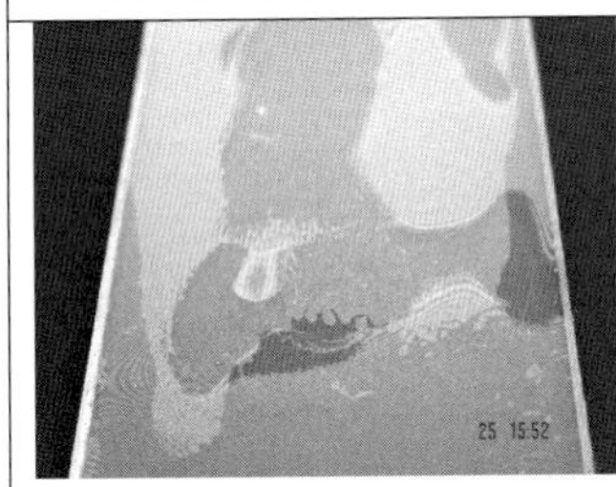	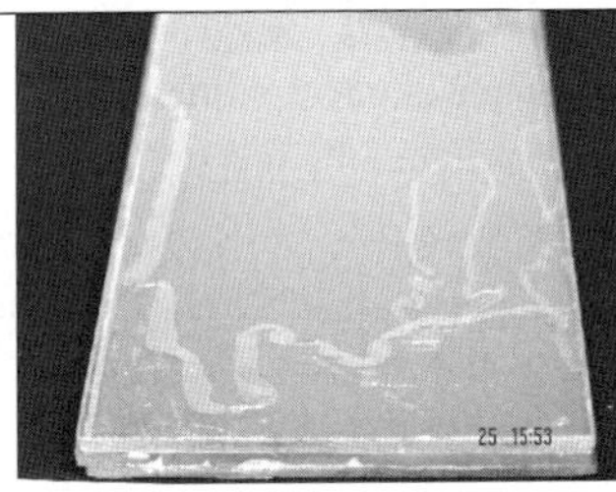	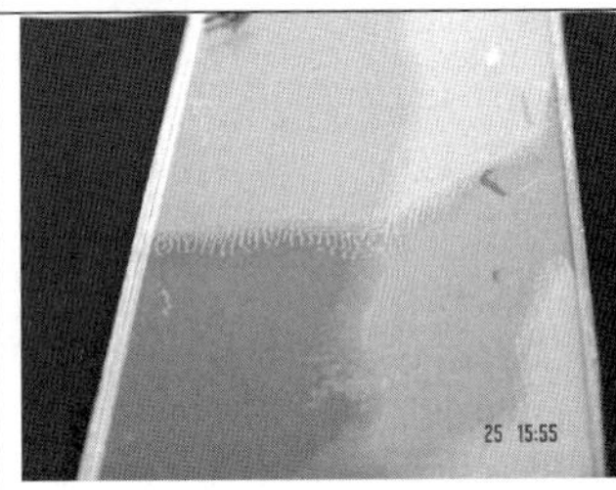
P: Lexan PC A: acrylic, uv-hardening SPT: alcohol	P: Lexan PC A: acrylic, uv-hardening SPT: plastic cleaner	P: Lexan PC A: silicone, uv-hardening SPT: alcohol

Table 2: pictures of selected specimens glass-plastic-glass after 25 cycles

specimen 3-3	specimen 4-3	specimen 6-3
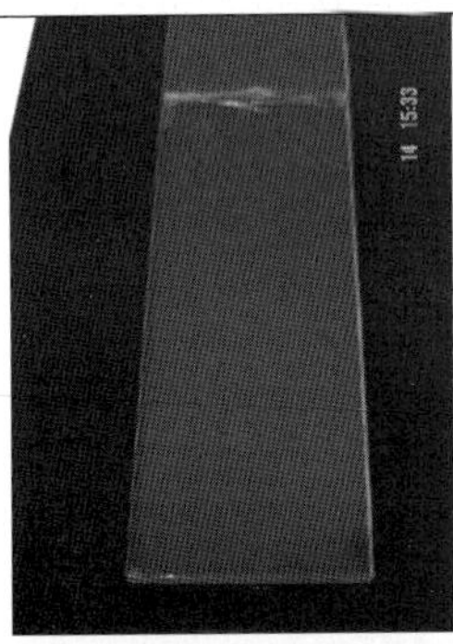		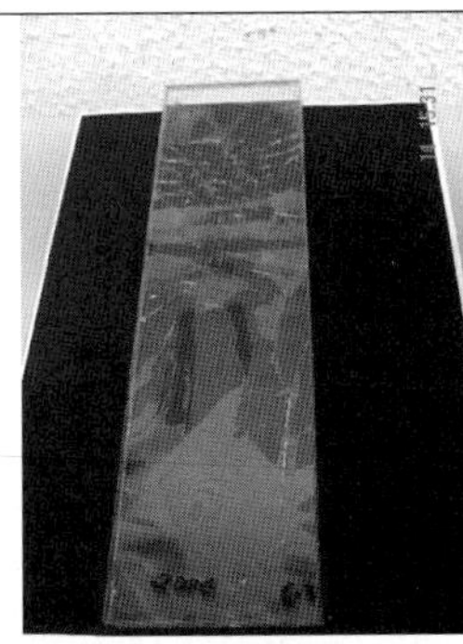
	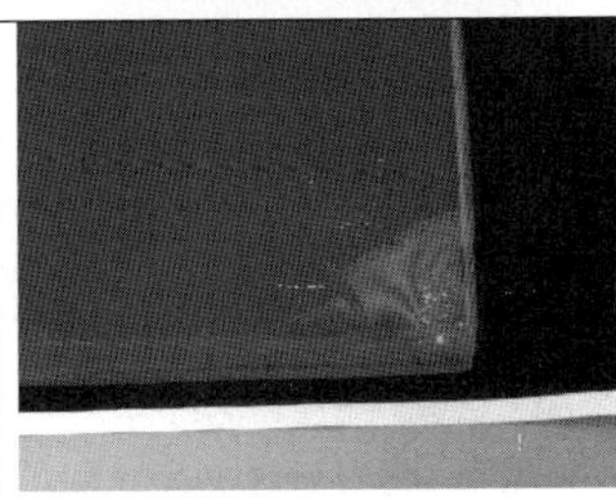	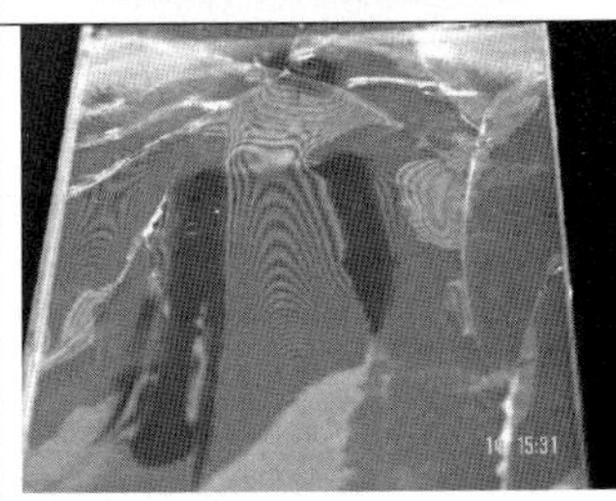
P: Makrolon GP 099 A: acrylic, uv-hardening SPT: Pyrosil	P: Lexan PC A: acrylic, uv-hardening SPT: Pyrosil	P: Lexan PC A: PUR SPT: Pyrosil

For the production of a bending element the following materials and methods were used:

- Makrolon GP 099 as plastic,
- Acryl 1K, uv-hardening as adhesive,
- Pyrosil method as surface pre-treatment.

5 Static Carrying Behaviour of a T-Beam

For a first investigation on the carrying behaviour of a glass-plastic hybrid element a T-beam of 1 m length was produced. The Figs. 5a and 5b show the position of the strain gauges (SG) on the beam which consisted of a 50 mm wide flange and a 71 mm high bridge. The glass and plastic sheets had a thickness of 3 mm.

In a 3-point bending test the stress change at the flange and at the bridge was investigated for a quasi-static stress. The load was increased in the interval 250 N, 500 N up to 1000 N and held 2 mins. A complete relief was carried out after each load. The concentrated load was brought to the centre of the beam by means of the construction shown in Fig. 6.

The load was brought in over a period of 50 sec. at 500 N and for 120 sec. at 1000 N to avoid effects of to fast stresses. The stress-time development is exemplarily shown for two steps. (see Figs. 7a and 7b). The load of 500 N produced a stress of 20 N/mm² (tensile) in the glass at the underside of the web. The minimal difference between both graphs can be explained by the different positions of the strain gauges. During the time period 50 sec. - 170 sec. a slight increase of the stress occurred in the glass at 500 N. This seems to indicate a force rearrangement from the plastic to the glass. After the relief there was no residual stress in the glass to be found. The maximal deflection in this test was at 0.75 mm.

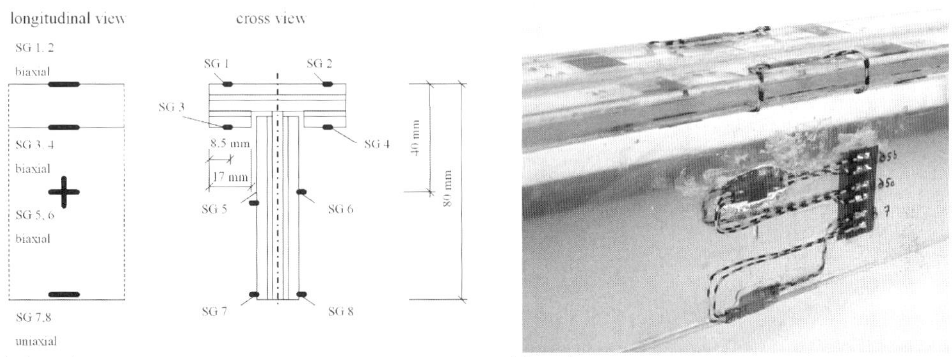

a) drawing b) web

Fig. 5: arrangement of the strain gauges (SG)

a) b)

Fig. 6: glass-plastic hybrid beam during the test

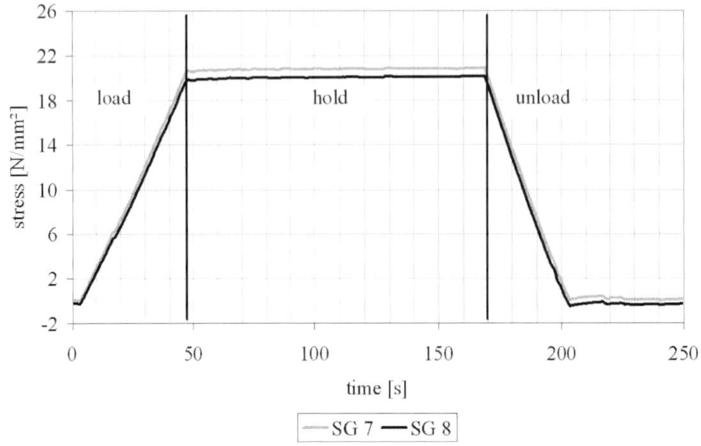

a) max. load: 500 N

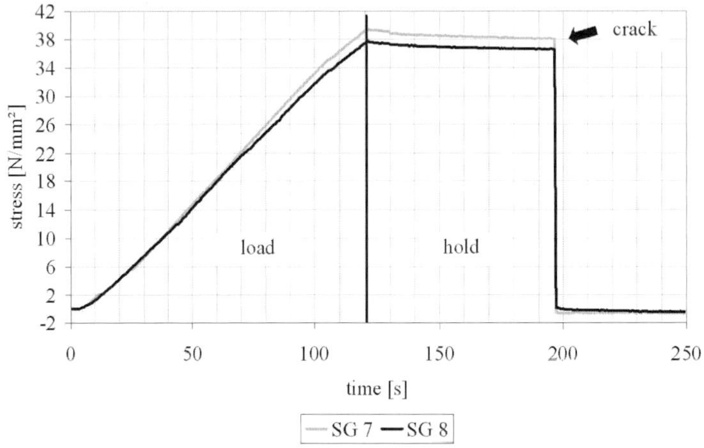

b) max. load: 1000 N

Fig. 7: stress-time development at the underside of the web

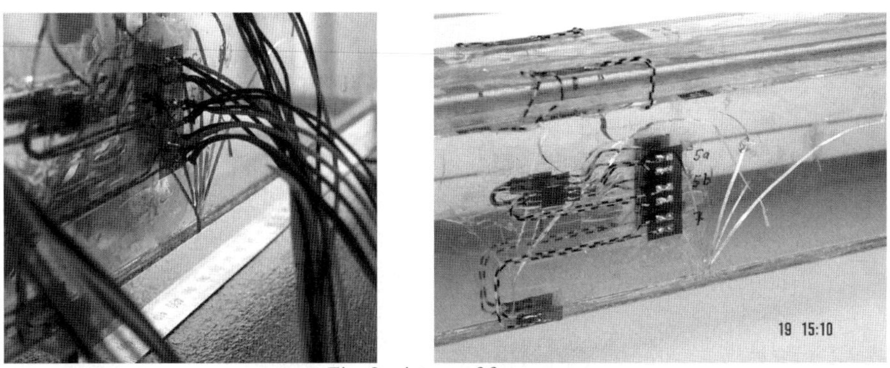

Fig. 8: picture of fracture

In Fig. 7b the max. carrying load of the T-beam is reached. At a stress of 38 N/mm² (tensile) a breaking of both glass sheets in the web occurred after 75 sec. A typical picture of fracture for float glass in the web was the result. (see Fig. 8) A maximum deflection of 2.0 mm could be measured. The residual carrying capacity of the plastic was sufficient to avoid the complete destruction of the beam.

6 Behaviour of the T-Beam at Temperature Changes

Apart from the resistance capacity against loads the temperature behaviour plays an essential role. A second T-beam was exposed to 25 temperature cycles according to standard DIN EN ISO 9142 – Aging Cycle D3. This resulted in a longitudinal in the centre of the 50 mm wide glass sheet of the flange. This crack ran 150 mm before the end of both beam sides. At the marked point in Fig. 10b the crack divides and cuts out a wedge-shaped piece of glass at the ends of the upper flange. The detailed image in Fig. 10a shows spots where very fine glass pieces can be seen which are the result of the steady opening and closing of the crack (probably formed in the first temperature cycles) because of the temperature change. This process could be confirmed by a FEM-analysis of both temperature conditions (20°C and +70°C). At a temperature of -20°C tensile stresses of such a magnitude occur at the upper side of the flange which let the glass break in the static experiment. When the T-beam is exposed to a temperature of +70°C a high tensile stress occurs at the underside of the upper glass sheet of the flange. This high tensile strength leads to a crack which opens and closes in the following cycles.
The deformation in the opposite direction as a consequence of both temperature conditions is shown in Fig. 9. The movement of the upper flange was a result of its unsymmetrical design, (upper glass sheet continuous, lower glass sheet interrupted), it was worsened by the different heat expansion coefficients of glass and plastic. When further developing glass-plastic hybrid elements one has to find constructive solutions that neutralise effects of such deformations in the opposite direction.

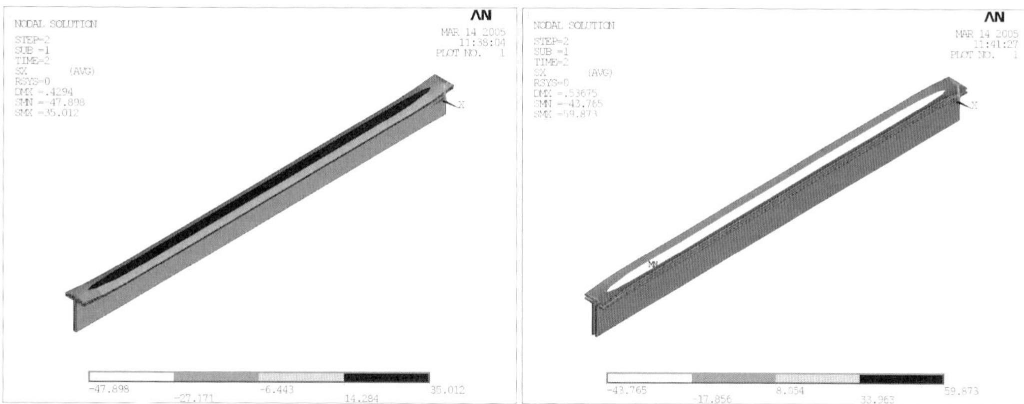

a) beam at temperature -20°C b) beam at temperature +70°C

Fig. 9: stress in cross direction of a hybrid beam at different temperatures

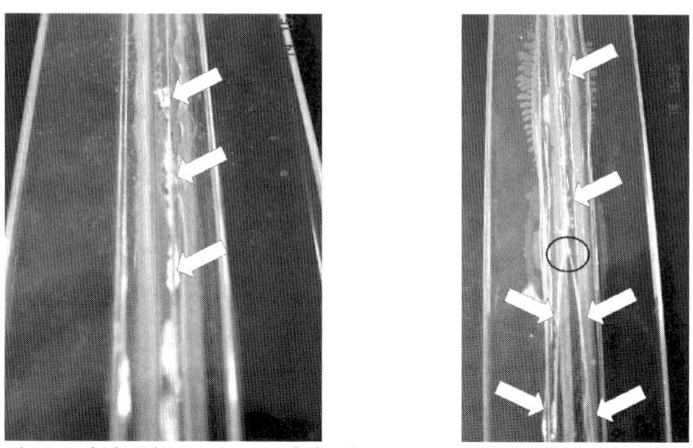

a) crack in the center of the beam b) crack at the end of the beam

Fig. 10: path of crack in the upper glass sheet of the flange

Acknowledgements

The department of steel structure, Bauhaus-Universität Weimar, thank the companies Glaskontor Erfurt, Rudolstädter Stahlbau, Innovative Klebtechnik Zimmermann, and the Institute for Joining Technology and Material Testing Jena (IFW Jena) for the support.

References

Bemm, M. (2003) Entwicklungen im Stahl-Glas-Bau und ihre besondere Bedeutung an historischen Bauten, 6.Informationstag des Instituts für Konstruktiven Ingenieurbau, Bauhaus Universität Weimar, Oktober 2003.

Bernard, F., Daudeville, L., and Gy, R. (2004) Load Bearing Capacity of Connections in Tempered Glass Structures, *Structural Engineering International*, 14(2), 107−110.

Bornemann, J., Welters, T., Dilger, K. and Schlimmer, M. (2003) Berechnung und Dimensionierung von Klebverbindungen mit der Methode der Finiten Elemente und experimentelle Überprüfung der Ergebnisse. DVS-Berichte Band 222, Verlag für Schweißen und verwandte Verfahren DVS-Verlag GmbH, Düsseldorf, pp. 74−79.

Business Magazine "European Products & Enterprises", edition Jan−Feb/2003, Invisible safety.

Dilger, K., Böhm, S., Welters, T., and Brackhage, K.-H. (2003) Entwicklung einer geregelten Einheit für das manuell geführte Klebstoffauftragen, *Schweißen & Schneiden*, 55(12), 664−671.

Dodd, G. (2004) Structural Glass Walls, Floors and Roofs, *Structural Engineering International*, 14(2), 88−91.

Freytag, B. (2004) Glass-Concrete Composite Technology, *Structural Engineering International*, 14(2), 111−117.

Hess, R. (2000) Glasträger – Forschungsbericht, Zürich: vdf Hochschulverlag AG an der ETH Zürich.

Hess, R. (2002) Konstruieren mit Glas, insbesondere unter Berücksichtigung der Sicherheit, Vortragsskript zum 21. Steinfurter Stahlbau-Seminar.

Müller, U (2004) Lastabtragende Klebungen, in *Glasbau2004*, Technische Universität Dresden, pp. 107−116.

Schober, H. and Schneider, J. (2004) Developments in Structural Glass and Glass Structures, *Structural Engineering International*, 14(2), 84−87.

Schuler, Chr., Bucak, Ö., Sackmann, V., Gräf, H., and Albrecht, G. (2004) Time and Temperature Dependent Mechanical Behaviour and Durability of Laminated Safety Glass, *Structural Engineering International*, 14(2), 80−83.

Standard DIN EN ISO 9142 (2004) Klebstoffe − Auswahlrichtlinien für Labor-Alterungsbedingungen zur Prüfung von Klebverbindungen (Adhesives – Selection guidelines for the laboratory − aging conditions for the test of adhesive joints), May.

Veer, F.A., Hobbelman, G.J., and van der Ploeg J.A. (2001) The Design of Innovative Nylon Joints to Connect Glass Beams, in *Stuctural Engineering, Mechanics and Computation, SEMC 2001*, pp. 447−454.

Weller, B. (2004) Forschung und Innovation im Glasbau an der TU Dresden, in *Glasbau2004*, Technische Universität Dresden, pp. 51−62.

Welters, T. and Dilger, K. (2003) Berechnung von Fügeverbindungen in der Klebtechnik durch Einsatz der Finite-Elemente-Berechnung. DVS-Berichte Band 225. Verlag für Schweißen und verwandte Verfahren DVS-Verlag GmbH, Düsseldorf , pp. 3−7.

Wiesner, S. (2004) Stand der Klebtechnik im Konstruktiven Glasbau, in *Glasbau2004*, Technische Universität Dresden, pp. 75−87.

DEVELOPMENT OF NEW BRIDGE RESTRAINER USING LAMINATED FIBER REINFORCED RUBBER

Nobutaka Ishikawa[1], **Yasushi Nishimoto**[1] **and Toru Ukishima**[2]
[1]*National Defense Academy, Yokosuka 238-0022, Japan*
E-mail: cgishikawa@m4.dion.ne.jp
[2]*Shibata Industrial Co. Ltd., Akashi 674-0082, Japan*

Abstract

This paper presents an experimental approach for the development of new bridge restrainer system using laminated fiber reinforced rubber (LFRR). After Kobe earthquake on January 1995, the design concept for the bridge restrainer has been revised so that the bridge should install the shock absorber which may be prevented from falling down due to earthquake shock. However, the shock absorbing system for the bridge restrainer has been required to satisfy the two performance requirements of high energy absorption and reduction of impact load. To this end, the laminated fiber reinforced rubber was developed to apply to the new bridge restrainer system as a shock absorber. In this study, the three kinds of tests of static compression, rapid speed loading and weight dropping impact for the LFRR specimen were first performed in order to investigate the efficiency of LFRR as a shock absorber. Then, the rubber-rolled pin was also developed as a new bridge restrainer system from the viewpoints of impact load reduction and high energy absorption.

Keywords: bridge restrainer, laminated fiber reinforced rubber, impact test, shock absorber

1. Introduction

As one of many fallout accidents of bridge girders occurred by the Kobe earthquake on January 1995 in Japan (Kawashima et al., 1997), the bridge restrainer system was damaged by impulsive loading in many places. Thus, the new bridge restrainer system has become needed to develop the shock absorbing device in order to prevent it from the failure if the equivalent earthquake occurs (Japan Road Association, 1996). In case of severe earthquake, a shock absorber is required to reduce the impact load which acts on the falling down prevention devices and also to absorb the high kinetic energy of girders. However, it is difficult to satisfy these two performance requirements at the same time. Although the high stiffness of a material is required in order to absorb the kinetic energy, it can not reduce the impact load. Therefore, a new shock absorber using the laminated fiber reinforced rubber (LFRR) is developed as shown in Figure 1 (Nishimoto et al., 2000, 2001).

In this study, the efficiency of LFRR is first investigated by performing three kinds of tests (static compression, rapid speed loading and weight dropping impact tests) from the viewpoints of two performance requirements of impact load reduction and high energy absorption. Second, the rubber-rolled pin using LFRR is developed as a new bridge restrainer by carrying out the impact

Fig. 1. LFRR.

M. Pandey et al. (eds), Advances in Engineering Structures, Mechanics & Construction, 809–821.

Table 1. Material properties of laminated fiber reinforced rubber.

	Fiber			Rubber	
	High tension (HMF)	Middle tension (MMF)	Low tension (LMF)	Hardness degrees of 50 (R50)	Hardness degrees of 65 (R65)
Material	6-nylon	6,6-nylon	Vinylon	Natural rubber	
Tensile strength	5292(N/3cm)	2646(N/3cm)	1764(N/3cm)	10.2(MPa)	20(MPa)
Elongation percentage at break	40(%)	25(%)	20(%)	600(%)	600(%)

Table 2. Test cases.

Specimen	Static compression test	Rapid speed loading test	Weight dropping test
R50	○	○	○
R65	○	○	○
LMF1	○		
LMF5	○	○	○
LMF25	○	○	○
LMF50	○		
MMF1	○		
MMF5	○	○	○
MMF25	○		
MMF50	○		
HMF1	○		
HMF5	○	○	○
HMF25	○		

test (Ishikawa et al., 1997). Finally, effects of LFRR and rubber-rolled pin are discussed from the viewpoints of mitigation of impact load and the energy absorption.

2. Effect of Laminated Fiber Reinforced Rubber (LFRR) as Shock Absorber

2.1. *Specimen and Test Cases of LFRR*

The shape of specimen of LFRR is a rectangular cross section with length of 150 mm, wide of 150 mm and depth of 50 mm. The LFRR is consisted of rubber with hardness degrees of 50, 65 and three kinds of fiber which are 6-Nylon (HMF), 6,6-Nylon (MMF) and Vinylon (LMF) with different tensile strength as shown in Table 1. Test cases are shown in Table 2 in which R50 and R65 stand for the rubber with hardness degree of 50 and 65, respectively and LMF5 means the laminated fiber reinforced rubber with five pieces of low tensile strength fiber.

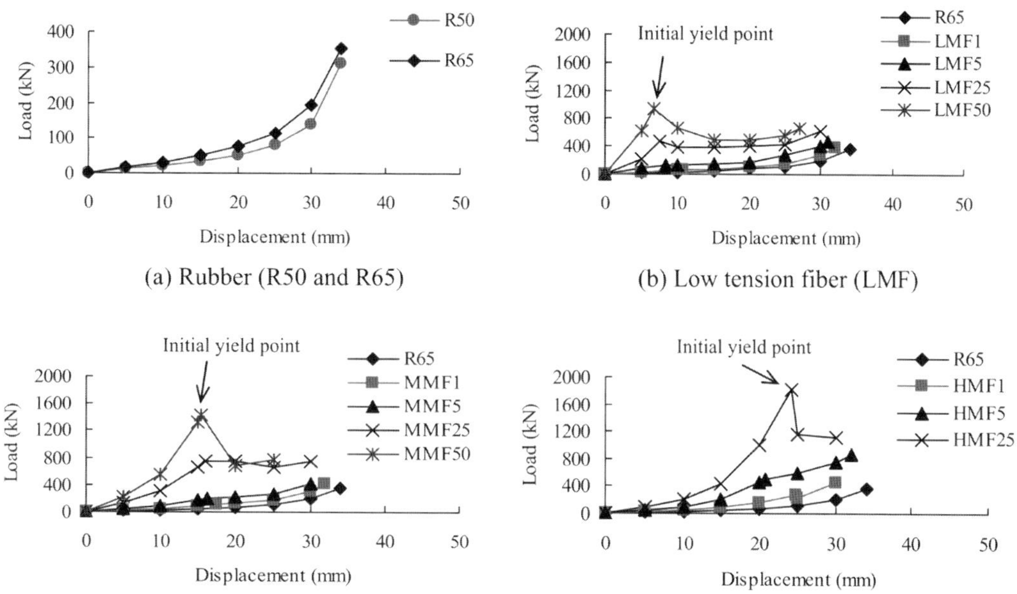

(a) Rubber (R50 and R65)

(b) Low tension fiber (LMF)

(c) Middle tension fiber (MMF)

(d) High tension fiber (HMF)

Fig. 2. Load ~ displacement curve by static compression test.

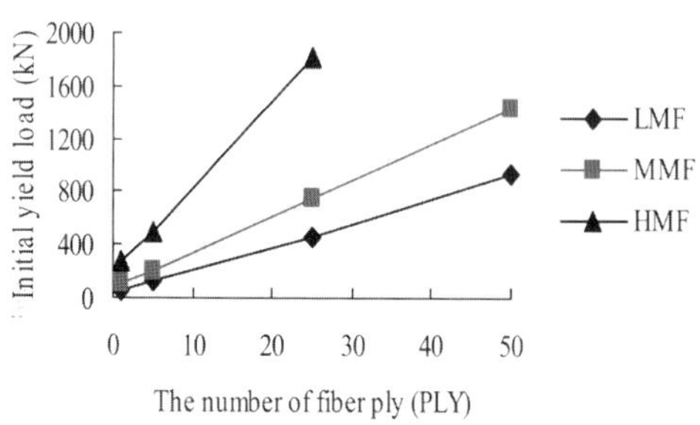

Fig. 3. Initial yield load ~ fiber ply relation.

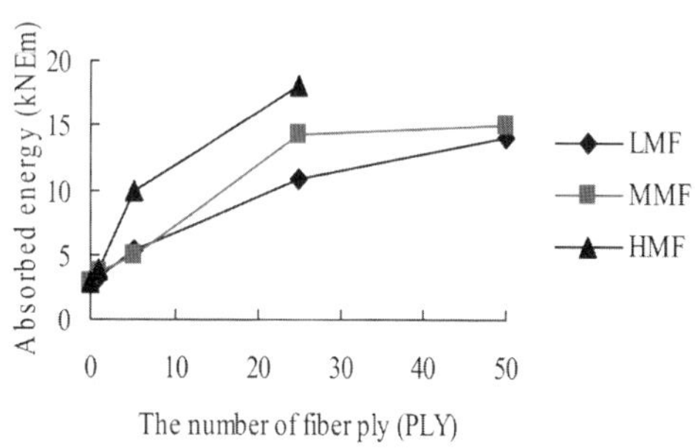

Fig. 4. Absorbed energy ~ fiber ply relation.

2.2. *Static Compression Test*

The static compression test was performed by using the 5,000 kN compression test machine.

Figure 2 shows the load-displacement relations by static compression test. The arrow in Figure 2 indicates the initial yield point at which the laminated fiber starts to break. It is found that the load-displacement curves of the LFRR show the elastic-plastic behavior and these are obviously different from those of the usual rubber. Figures 3 and 4 illustrate the initial yield load – the number of fiber ply relation and absorbed energy – the number of fiber ply relation, respectively. It can be seen that the initial yield load and the absorbed energy of LFRR increase with the increase of the number of fiber ply. Therefore, if we want to obtain the required initial yield load, we can select the number of fiber with an appropriate tensile strength. It is also noticed from Figure 4 that the absorbed energy (18 kN·m) of HMF (25 PLY) is about 6 times larger than that (3 kN·m) of usual rubber (0 PLY).

Fig. 5. Rapid speed loading machine.

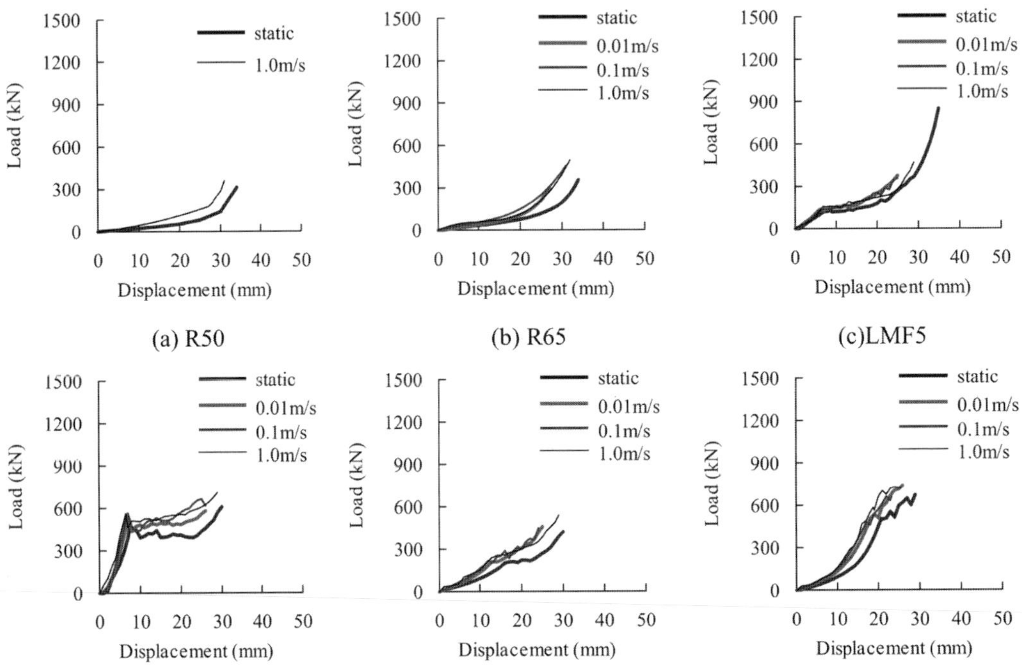

Fig. 6. Load ~ displacement curve by rapid speed loading test.

2.3. *Rapid Speed Loading Test*

The rapid speed loading test was executed by using the rapid loading machine as shown in Figure 5. The aim of this dynamic test is to examine the rate-effect of LFRR and the loading speed was set at 0.01 m/sec, 0.1 m/sec and 1.0 m/sec. The load was measured by the 1,000 kN load cell and the displacement was measured by the laser type displacement sensor.

Figure 6 shows the dynamic load-displacement relations including static compression test results. It is found that the loading speed has remarkable effect on the load-displacement relation in case of

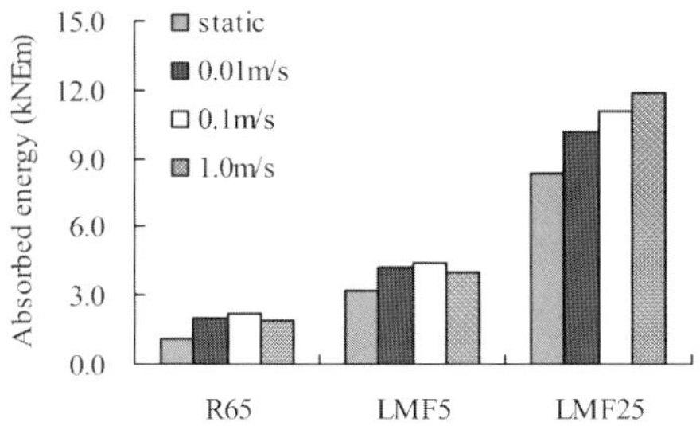

(a) Effect of fiber ply

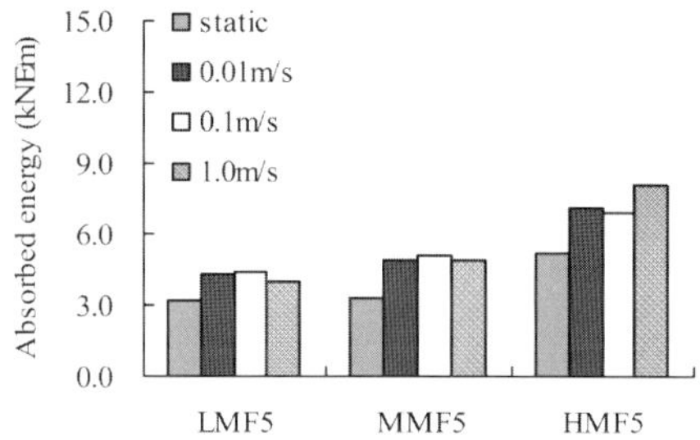

(b)Effect of fiber tensile strength

Fig. 7. Absorbed energy ~ loading speed relations.

Fig. 8. Weight dropping machine.

usual rubber when the displacement exceeds 10 mm. For example, the dynamic load is about 2.5 times larger than the static load at the displacement of 25 mm.

It is also noticed that from Figures 6c–6f that the initial yield load increases about 20%, but in plastic region after initial yield load, the dynamic load increases about 30–50% larger than the static one in case of LFRR. This may be caused that fiber breaks gradually and the LFRR becomes like the usual rubber. These facts mean that the usual rubber tends to be more sensitive in rate effect than LFRR.

Figure 7 represents the absorbed energy of LFRR and it is obvious that the absorbed energy of LMF25 in case of rapid speed loading (1.0 m/sec) becomes larger than those of R65, LMF5. It is also noticed that the absorbed energy increases as the fiber ply and fiber tensile strength increase.

2.4. *Weight Dropping Impact Test*

The weight dropping impact test was carried out by using the weight dropping machine as shown in Figure 8. The falling weight was set at $W = 4.0$ kN and the falling height was set at $H = 0.25$, 1.25 and 2.5 m ($V = 2.21$, 4.94 and 7.00 m/sec), that is, impact energy was set at $E = 1.0$, 5.0

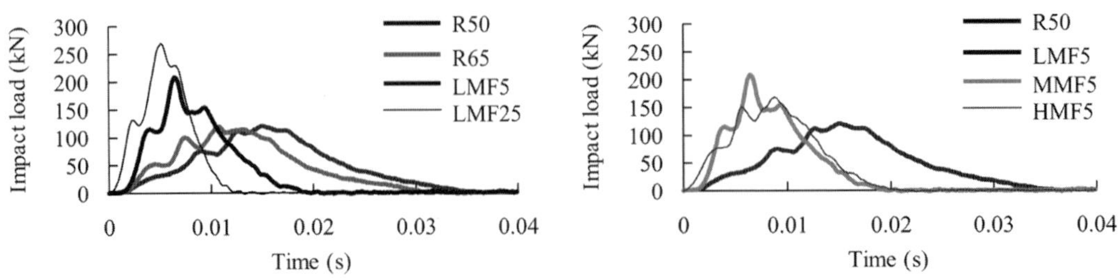

(a) Rubber (R50 and R60) and LMF (b) R50 , LMF5, MM5 and HMF5

Fig. 9. Impact load ~ time relation (input energy $E_i = 1.0$ kN·m).

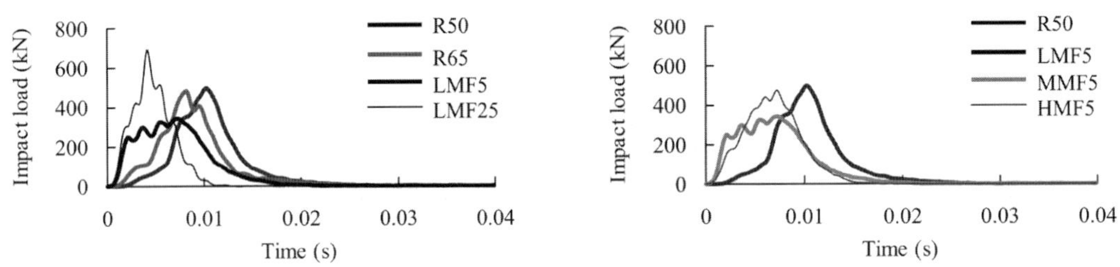

(a) Rubber (R50 and R60) and LMF (b) R50 , LMF5, MM5 and HMF5

Fig. 10. Impact load ~ time relation (input energy $E_i = 5.0$ kN·m).

and 10.0 kN·m. These mean that $E = 1.0$ kN·m is the energy that laminated fiber may never break, $E = 10.0$ kN·m is the energy that laminated fiber may break perfectly and $E = 5.0$ kN·m is the intermediate value between 1.0 and 10.0 kN·m. These energies are determined by the results of static compression test. The impact transmitted load is measured by using the 2,000 kN load cell attached at the steel plate under the specimen.

Figures 9, 10 and 11 illustrate the load-time relations obtained by weight dropping impact test. It can be seen from Figure 9 that the maximum impact transmitted load of LFRR is larger than the one of usual rubber and the impact duration time of LFRR is shorter than that of usual rubber in case of low input energy $E = 1.0$ kN·m. This may be the reason why the stiffness of LFRR in elastic region is larger than that of usual rubber. However, it should be noticed from Figures 10 and 11 that the maximum impact load of LFRR at large input energy becomes smaller than that of usual rubber.

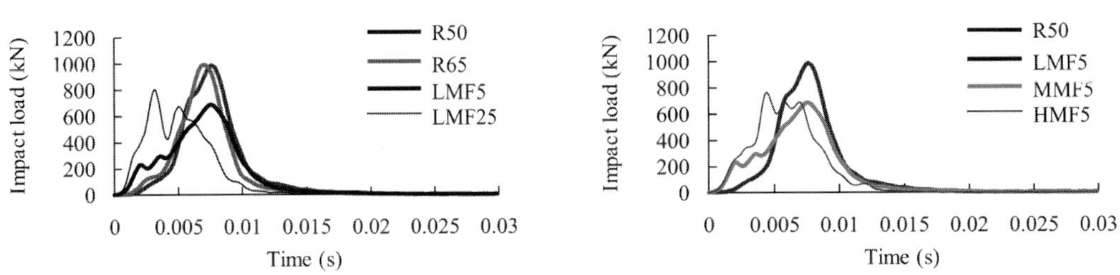

(a) Rubber (R50 and R60) and LMF (b) R50 , LMF5, MM5 and HMF5

Fig. 11. Impact load ~ time relation (input energy $E_i = 10.0$ kN·m).

Table 3. Maximum impact load.

	1.0kN・m	5.0kN・m	10.0kN・m
R50	120.6	497.7	987.7
R65	114.8	482.3	991.5
LMF5	203.8	326.3	687.1
LMF25	268.9	690.0	803.9
MMF5	167.5	414.4	670.9
HMF5	168.4	474.7	765.6

Table 4. Energy absorption ratio.

Specimen	Energy absorption ratio (%)	
	Ei=1.0kN · m	Ei=10.0kN · m
R65	72.1	85.0
LMF5	86.6	88.7
LMF25	88.9	90.4
MMF5	89.3	87.8
HMF5	86.5	85.5

The reason why the maximum load of LFRR is reduced may be due to that the LFRR can absorb the large kinetic energy by breaking the laminated fiber and decreasing the stiffness.

Table 3 shows the maximum impact load and the hatching figures mean that the laminated fiber breaks. For instance, the maximum loads of LMF5 in cases of $E = 5.0$ kN and 10.0 kN·m are 1/1.5 smaller those of usual rubber R65. Therefore, the LFRR is effective as a shock absorber rather than usual rubber from the viewpoint of the mitigation effect of impact load at high input energy.

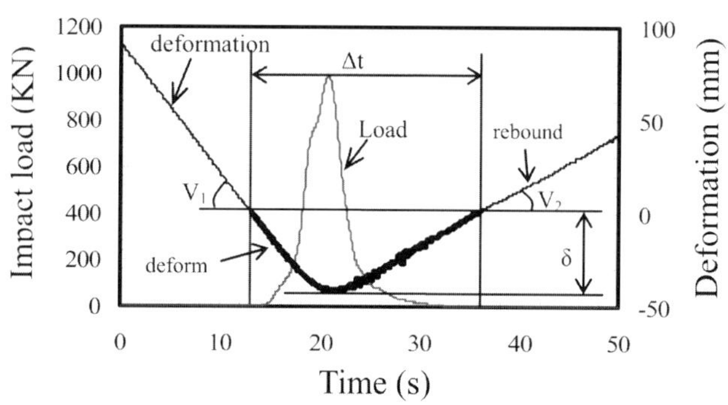

Fig. 12. Impact load, deformation-time relation. (R50) $E = 10.0$ kN·m.

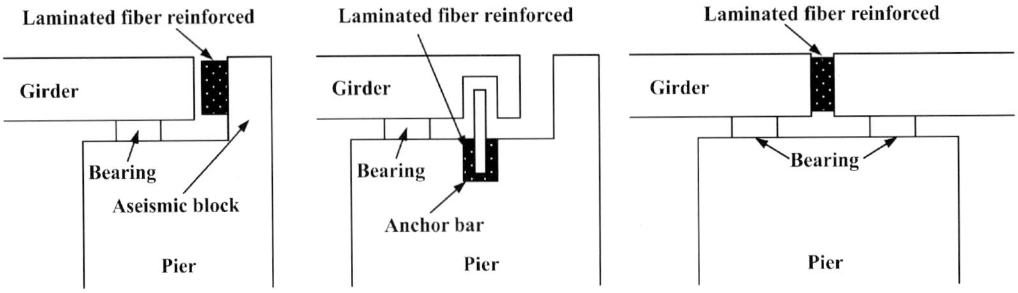

Fig. 13. Direct application of LFRR to bridge restrainer system.

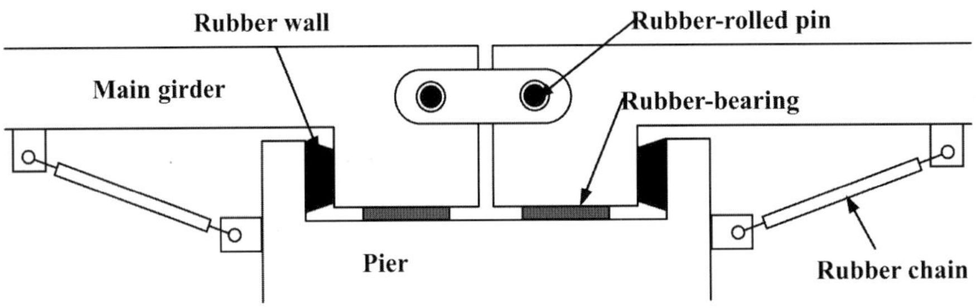

Fig. 14. Indirect application of LFRR to bridge restrainer system.

Table 4 expresses the energy absorption ratio (ΔE) which is defined as follows:

$$\Delta E = \frac{E - E'}{E} \times 100\%, \tag{1}$$

in which E and E' are the kinetic energy before and after collision, respectively. The velocity of weight is obtained by differentiating the traveling distance with respect to time as shown in Figure 12. It can be seen from Table 4 that the energy absorption ratio of LFRR is larger than that of natural rubber in every cases. Therefore, the LFRR can absorb the kinetic energy of weight, that is, the velocity of the weight (girder) after collision in case of LFRR becomes smaller than that of usual rubber. This means that the damage of girder using the LFRR may be smaller than that using usual rubber.

3. Application to New Bridge Restrainer System

The LFRR is directly applied to the shock absorber of aseismic block, anchor bar and pounding between girders as shock absorbers of new bridge restrainer system as shown in Figure 13.

However, it is difficult to prevent the fallout accident of girder by only one shock absorber in case of large scale earthquake. Therefore, it would be desirable that a bridge structure may be installed by combining various kinds of shock absorbing system such as rubber wall, rubber-bearing, rubber-pin and rubber chain as shown in Figure 14. For instance, the rubber chain device and rubber wall using LFRR might be useful as devices which can absorb large amount of energy for the huge earthquake.

Furthermore, the rubber-rolled pin would improve the existing bridge restrainer plate system as device which can mitigate the transmitting load to girders.

Herein, the effect of rubber-rolled pin is examined by performing the impact test from the viewpoints of mitigation of impact load and the energy absorption.

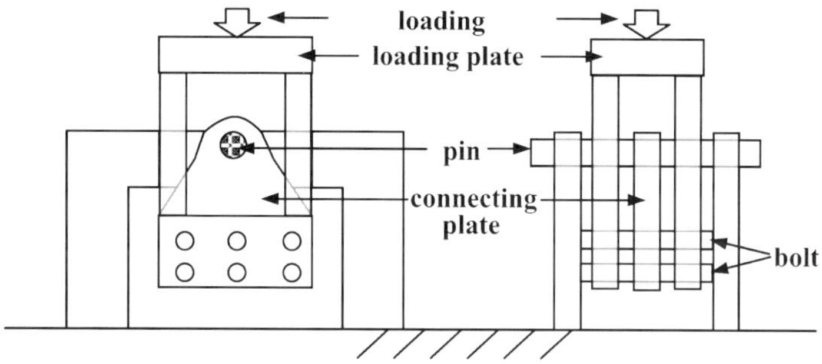

Fig. 15. Impact test of rubber-rolled pin.

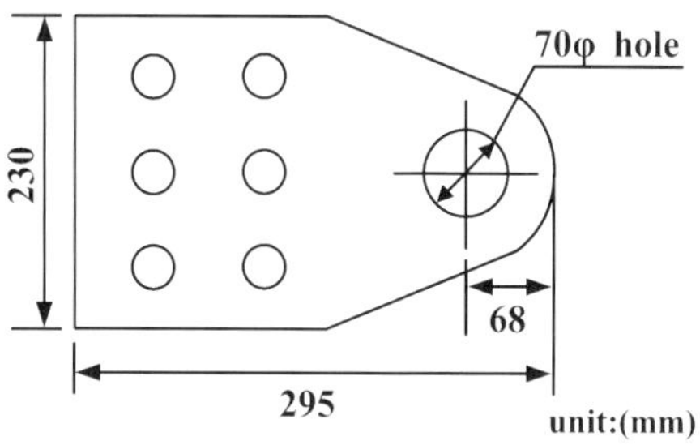

Fig. 16. Connecting plate.

3.1. *Impact Loading Test of Rubber-Rolled Pin*

The impact loading test using dropping weight machine was carried out by pushing down the loading plate which was linked with the connecting plate by bolts as shown in Figure 15. The steel pin and rubber-rolled pin pass through the connecting plate and cut them off by pushing down the loading plate. The connecting plate specimen (SS400, thickness 9 mm) is a half size of actual shape as shown in Figure 16 and its cross section is equal to the connecting plate which is used in the bridge of about 20 m span length. The pin specimens (SC35) were used as two types as shown in Table 5. Type A is the ordinary steel pin and type B is the rubber-rolled pin using LFRR in which they have the same total diameter of 60 mm. The load was measured by the load cell and displacement was measured by the laser type sensor, respectively. In impact test, the responding strains of connecting plate were measured by the 3-axial gauges which were arranged as shown in Figure 17.

The falling weight was set at $W = 1$ kN and falling height was set at $H = 10$–150 cm (impact velocity $V = 140$–540 cm/sec).

3.2. *Load Mitigation Effect of Rubber-Rolled Pin*

Figure 18 shows the circumferential strain-time relations at interval of 30° in the elastic range ($W = 1$ kN, $H = 10$ cm). It is found that the maximum strain (350 μ) at No. 7 ($\theta = 90°$) of type-B becomes about 1/4 smaller than the one (1350 μ)of type-A. It is also confirmed that the difference of the maximum strain at each position of type-B is smaller rather than the one of type-A (for example, the difference of No. 1 and No. 7 in type-B is about 40%, but it is about 75% in type-A). It is guessed that this phenomenon may be caused by the larger contact area between rubber-rolled pin and plate in type-B. Thus, type-B can disperse the transmitting load to connecting plate. Therefore, the rubber-

Table 5. Type of pin (unit:mm).

pin-type	type-A (steel-pin)	type-B (rubber-rolled pin)
cross section	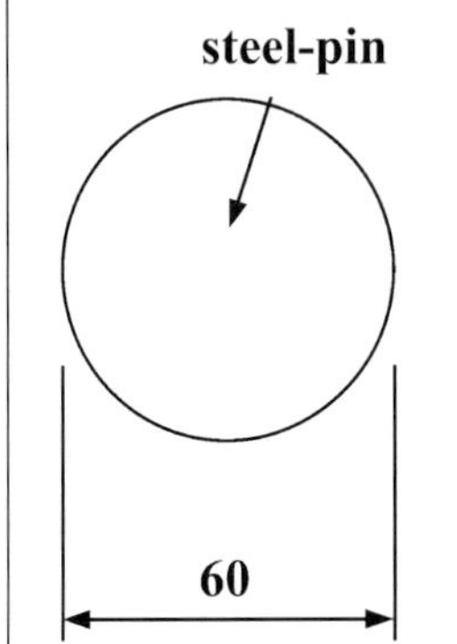	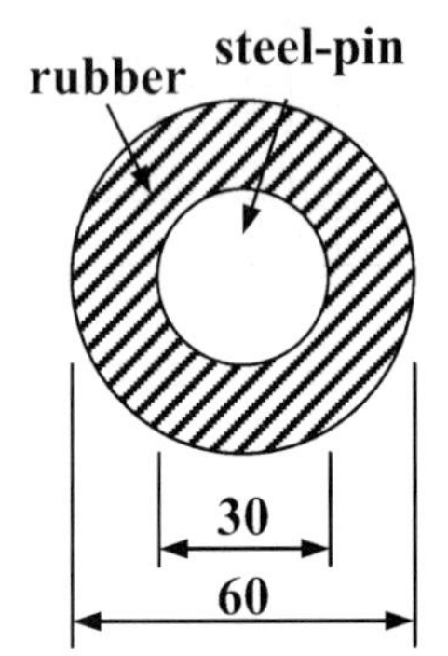

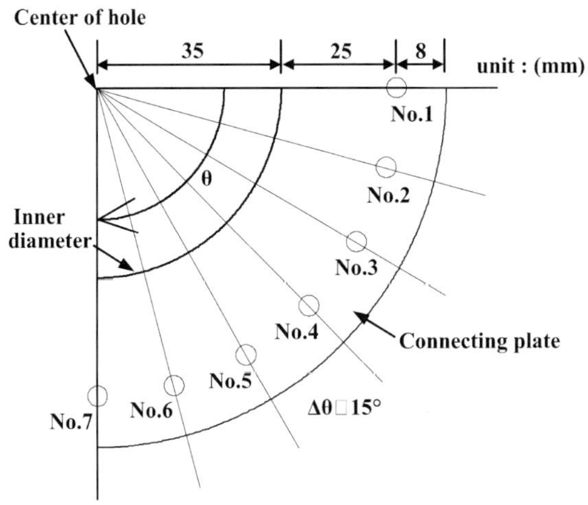

Fig. 17. Position of strain gauges.

rolled pin can expect not only the effect of mitigating impact load, but also the effect of dispersing impact load. This tendency was also confirmed in the radial components of strains.

Figure 19 shows the strain (at gauge No. 7) – time relations at $W = 1$ kN and $H = 10$ cm. It is apparent that the maximum strain of type-B is about 1/3 1/4 smaller than the one of type-A, and the responding period of radial strains of type-B is about twice as long as type-A. Therefore, type-B can prevent the stress concentration of connecting plate in elastic range.

3.3. *Energy Absorption of Rubber-Rolled Pin*

In the previous test, the load mitigation effect of rubber-rolled pin was examined within the elastic loading range. Next, the dropping height is increased from 10 cm to 150 cm and the elastic limit height was judged as the state of permanent strain remained obviously.

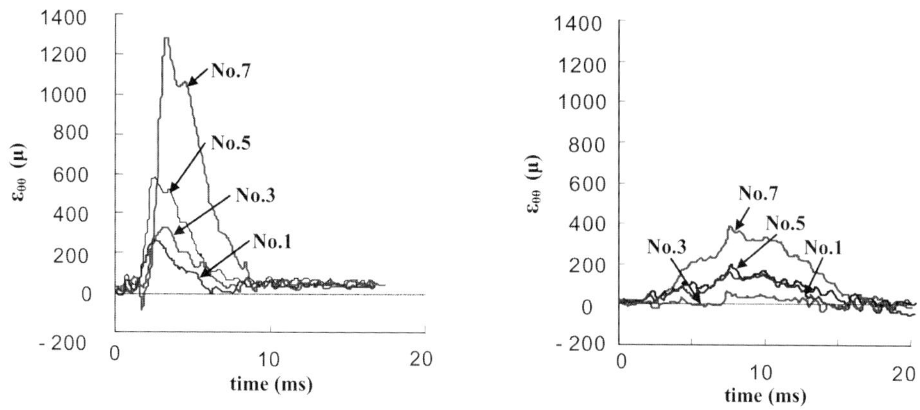

(a) Type-A (steel pin) **(b) Type-B (rubber-rolled pin)**

Fig. 18. Circumferential strain-time relations.

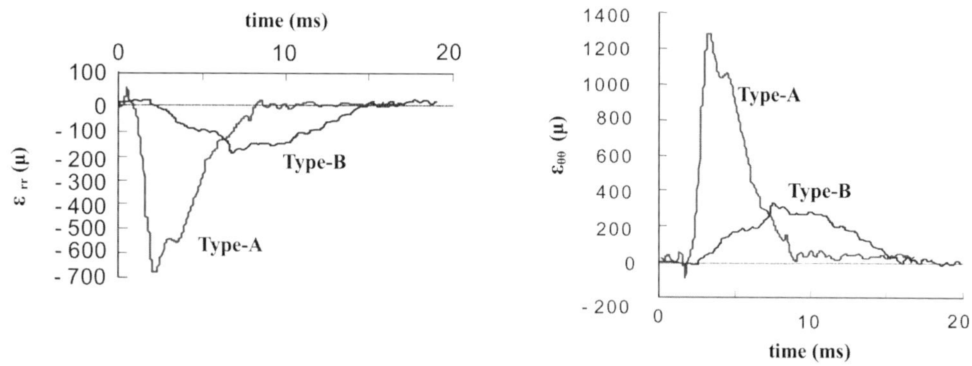

(a) radial strains **(b) circumferential strains**

Fig. 19. Strain (No. 7) – time relations at $W = 1$ kN, $H = 10$ cm.

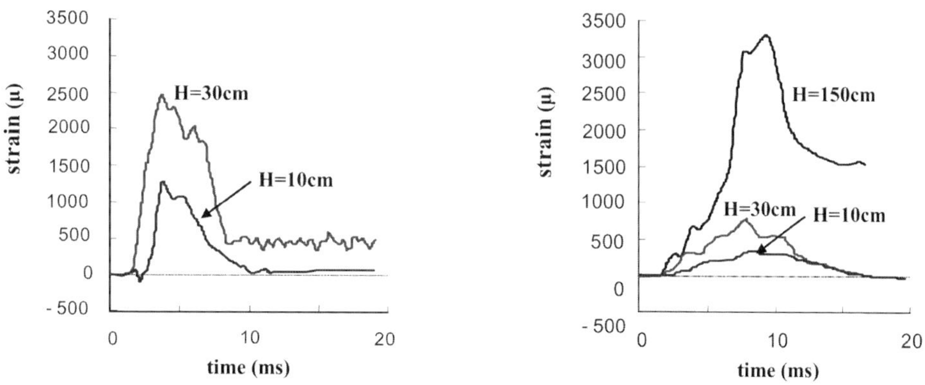

(a) Type-A **(b) Type-B**

Fig. 20. Strain (No. 7) – time relation at various falling height.

Figure 20 illustrates the strain (No. 7) – time relations by increasing the dropping height until the elastic limit state. It is found that the effect of mitigating the load at elastic limit level is similar to the previous elastic range, that is, the strain of 2500 μ at $H = 30$ cm of type-A is reduced to 800 μ (70% reduction) at $H = 30$ cm of type-B. It is also noticed that the elastic limit state of type-B ($H = 150$ cm) exhibits five times larger than the one of type-A ($H = 30$ cm). Therefore, the energy absorption of type-B has five times larger than the one of type-A, and therefore, type-B will be able to work under the elastic-plastic range of connecting plate for the considerable severe loading condition.

4. Conclusions

The following conclusions are drawn from this study.

4.1. *Development of LFRR*

(1) The maximum impact load of the LFRR in case of high kinetic energy is smaller than that of usual rubber, although its stiffness is large. Because, the LFRR can reduce the impact load by breaking the fiber in the rubber in case of high kinetic energy.
(2) The energy absorption ratio of LFRR is larger than the one of usual rubber. Therefore, the velocity after collision in case of LFRR becomes smaller than that of usual rubber, and such, the damage of bridge may be reduced for the earthquake shock.
(3) The LFRR is much better materials than usual rubber from the viewpoints of high absorbed energy and reducible effect of impact load at high input energy. Therefore, the LFRR is directly applied to the shock absorber of bridge restrainer system.

4.2. *Development of Rubber-Rolled Pin*

(1) The rubber-rolled pin exhibits obviously the mitigating effect of transmitting load to girders. This rate becomes smaller about 1/3–1/4 of the ordinary steel pin from the responding strains.
(2) The rubber-rolled pin has the transmitting and dispersing effects which would prevent the stress concentration of connecting plate in elastic and elastic-limit range.
(3) The energy absorption of rubber-rolled pin is about five times larger than the one of the ordinary steel pin. Thus, the rubber-rolled pin is advantageous as a bridge restrainer system for the severe loading condition.

References

Ishikawa, N., Sonoda, Y. and Hikosaka, H., 1997, Development of New Bridge Restrainer with Rubber-Rolled Pin for the Great Earthquake, *Earthquake Resistant Engineering Structures, Computational Mechanics Publications*, pp. 203–212.

Japan Road Association, 1996, Design Specification of Highway Bridge, Part V: Seismic Design [in Japanese].

Kawashima, K. and Unjoh, S., 1997, Impact of Hanshin/Awaji Earthquake on Seismic Design and Seismic Strengthening of Highway Bridge, *Journal of Structural Mechanics and Earthquake Engineering*, No. 556/I-38, pp. 1–30.

Nishimoto, Y., Kajita, Y., Ishikawa, N. and Nishikawa, S., 2000, An Experimental Study on Dynamic Properties of Laminated Fiber Reinforced Rubber as Shock Absorber of Bridge Restrainer System, *Journal of Structural Engineering*, Vol. 46A, pp. 1865–1874 [in Japanese].

Nishimoto, Y., Kajita, Y., Ishikawa, N. and Nishikawa, S., 2001, A Study on the Weight Dropping Impact Test and Prediction of the Impact Transmitted Load of Laminated Fiber Reinforced Rubber as a Shock Absorber for Bridge Restrainer System, *Journal of Structural Engineering*, Vol. 47A, pp. 1655–1664 [in Japanese].

COLLAPSE ANALYSIS OF REINFORCED CONCRETE SLABS: ARE THE UP AND DOWN ROADS ONE AND THE SAME?

David Johnson
School of the Built Environment, Nottingham Trent University,
Burton Street, Nottingham NG1 4BU, UK
E-mail: david.johnson@ntu.ac.uk

The road up and the road down are one and the same
(Heraclitus)

O ye'll tak' the high road and I'll tak' the low road,
And I'll be in Scotland afor ye
(Traditional Song)

Abstract

The comparative merits of hand and automated upper and lower bound techniques for the collapse load estimation of reinforced concrete slabs are examined. Examples, drawn from both theoretical and practical design work, are used to show that both hand and automated upper bound yield line techniques can produce significant, unsafe errors. Automated lower bound solutions, however, are shown to consistently provide safe estimates that are not unduly conservative, provided appropriate formulations are adopted. As long as the engineer is willing to dispense with the crutch of a yield line pattern, it is therefore contended that, whilst Heraclitus may be correct in that both the upper and the lower bound roads can lead to one and the same collapse load, the lower bound road gets you there, certainly more safely, and usually quicker, as the Traditional Song suggests.

Introduction

A unique solution for the collapse load of a reinforced concrete slab, as predicted by rigid-plastic theory, requires the simultaneous fulfilment of equilibrium, yield and mechanism conditions. This requirement is normally too onerous to be achieved directly and recourse is therefore commonly made to the use of either upper or lower bound solutions (Wood, 1961). Both approaches satisfy the equilibrium condition, but the upper bound approximation ignores the yield requirement, whilst the lower bound technique does not require the specification of a mechanism. The upper bound method can therefore postulate a collapse mechanism which incorporates "unsafe" violations of the slab yield requirement that moments do not exceed the fully plastic moment of the slab section. Lower bound solutions, on the other hand, can postulate moment distributions that are "safe" (in the sense that the plastic moment is nowhere exceeded), but which do not correspond to a collapse mechanism. For both formulations, either hand or automated approaches are possible and the principal features of a range of these will be outlined prior to a comparative study of the merits of the various techniques, as demonstrated by a range of examples.

M. Pandey et al. (eds), Advances in Engineering Structures, Mechanics & Construction, 823–831.

Upper Bound Techniques – The High Road

The traditional hand calculation approach to yield line analysis (Johansen, 1964) requires the initial specification of a potential mechanism for which a collapse load is calculated. This estimate will be an upper bound on the true collapse load. Subsequent trial mechanisms may then be investigated and the lowest collapse load found is taken as the exact value. A refinement is to investigate the effect of geometric variation of a given yield line arrangement, whilst maintaining the same basic topology. This refinement normally results in less significant reductions in the collapse load than is possible from the detection of a more critical mechanism and is therefore sometimes approximated by applying a 10-15% reduction in the collapse load. This reduction is assumed to cover this geometric refinement effect, together with subsidiary variations in collapse mechanisms, such as the presence of "corner levers" (Kennedy *et al*, 2003).

The yield line approach may be automated by casting it as a linear programming formulation (Munro *et al*, 1978). The slab system to be analysed is subdivided by a triangulated grid and the optimisation process then evaluates a critical mechanism, taking the edges of the triangles as potential yield line sections. The process does not necessarily identify either the correct form of mechanism or the exact geometric positioning of the yield lines, since it is limited by the stipulation that yield should only occur along the edges of the specified grid. To improve the accuracy of the estimated collapse load, a two part process has been suggested (Johnson, 1994), in which a "fine" grid is used to identify the likely critical mechanism and a "coarse" grid is then used with geometric optimisation to obtain an improved approximation to the positioning of the yield line pattern.

Lower Bound Techniques – The Low Road

In practice, using hand calculations, lower bound solutions are generally obtained by the "strip" method (Hillerborg, 1975). In its simplest form, the strip method uses lines of assumed zero shear force to allow the determination of equilibrium moment distributions that provide a "safe" solution if used to design the slab's reinforcement pattern. The basic methodology has been extended (Hillerborg, 1996) by a number of approximations and simplifications to allow the design of slabs with complex geometries, loadings, and boundary conditions and also to make it more convenient for practical application.

Lower bound solutions may also be generated by automated approaches, although these are closer to automated upper bound (yield line) techniques rather than to the strip system. As with the automated yield line system, triangulated nets are used, but, rather than taking the net edges as potential yield lines, prescribed distributions of bending moment (Krenk *et al*, 1994) within the triangular regions are presumed. If, as is common practice, the moment within each element is assumed to be a linear function of the values at its nodes, then a lower bound solution may be obtained by optimising the load subject to the nodal moment values not exceeding the slab's yield capacity.

Arrivals and Departures

When using either an upper or a lower bound technique, whether one arrives at the true collapse load or departs significantly from it depends on the approach selected and the slab under investigation. A number of illustrative examples will therefore be studied to show the comparative abilities of the different methods.

Rectangular slab simply supported on three sides

An automated yield line procedure presented by Shoemaker (Shoemaker, 1989) followed the hand calculation approach of examining a number of predetermined collapse modes, the form of which depended on the particular slab under consideration. In the case of a rectangular slab, simply supported on three sides and free on the fourth, the two collapse modes shown in Figure 1a were examined and the critical values of x and y were determined by a grid based search procedure. For a particular example, the solution obtained by this procedure was as shown in Figure 1b.

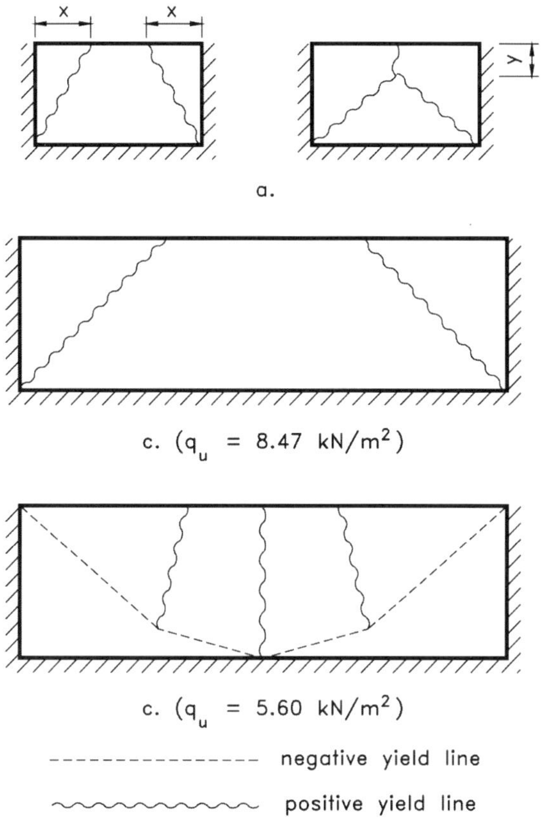

Figure 1: a. Possible modes b. Shoemaker solution c. Hillerborg solution

Subsequently, Hillerborg (Hillerborg, 1991) showed that, in the absence of top reinforcement, the mechanism shown in Figure 1c is critical and that the mechanism employed by Shoemaker had overestimated the collapse loads by at least 50%. This led Hillerborg to conclude that "*In my opinion, it is questionable whether yield line analysis is a suitable basis for the design of slabs, as the risk of getting an unsafe structure is too high.*"

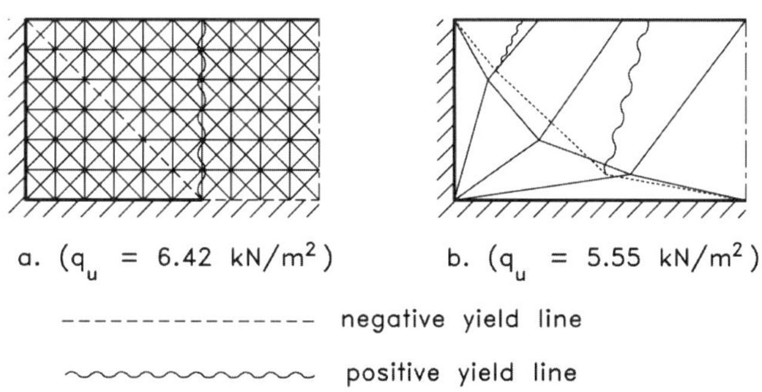

Figure 2: a. Fine mesh solution b. coarse mesh solution

The slab has also been analysed by the automated yield line procedure (Johnson, 1996). In this case, the "fine" net shown in Figure 2a, based on a half-slab model, was used to identify the nature of the collapse mode and the "coarse" net shown in Figure 2b was then used to optimise the geometry of the mode. As may be noted, this procedure produced a solution that corresponded closely to the one suggested by Hillerborg (Figure 1c). In this case, therefore, the automated yield line technique did successfully identify the appropriate mechanism. However, it might well be contended that the two stage solution is somewhat cumbersome since it is dependent on manual intervention to establish a coarse grid appropriate to the identified mode.

Practical design example

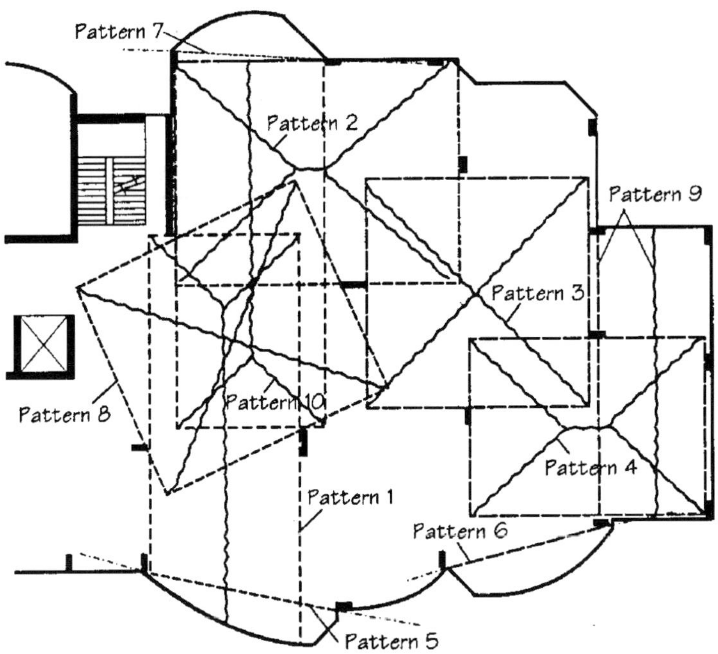

Figure 3: Part-plan of floor slab for block of flats and trial yield line patterns *(Kennedy et al, 2003)*

The part-plan of one storey of a seven-storey block of flats is shown in Figure 3 (Kennedy et al, 2003). The floor slab was initially analysed by hand using no less than ten possible yield line patterns, which are also shown in Figure 3. On the basis of this analysis, a required moment of resistance of 43.3 kNm/m was established for the slab, the critical mechanism being Pattern 2.

The analysis was subsequently checked by the use of the automated yield procedure, which produced the complex yield line pattern shown in Figure 4a, and an ultimate moment requirement of 47.2 kNm/m. The pattern demonstrates a basic problem with the use of the automated yield line method, namely that the form of the collapse mechanism is not readily obvious from Figure 4a. In these circumstances, reference to an associated contour plot of the collapse mode (Figure 4b) is often useful and, in this case, demonstrates that the collapse mode is of the "folded plate" nature. This collapse mode approximates to a combination of the Patterns 1 and 10 considered in the hand analysis (Figure 3) and these two patterns were, in fact, identified as the third and second most critical cases, respectively, by the hand analysis.

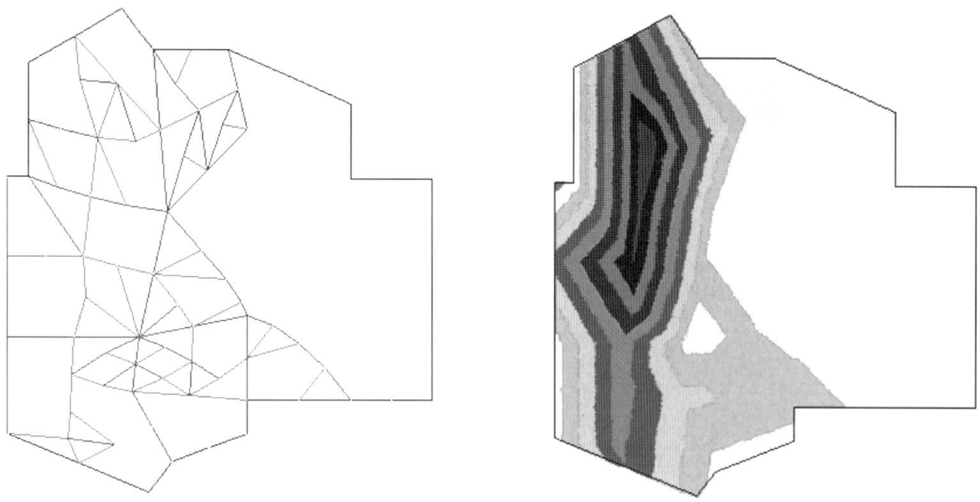

a. b.

Figure 4: a. Yield line pattern b. collapse mode contour plot

To obtain a bound on the true collapse load, an automated lower bound solution was also obtained and produced the collapse mode shown in Figure 5. As may be seen from Figure 5, the collapse mode is a variation on that given by the upper bound solution (Figure 4b) and the associated ultimate moment requirement of 64.1 kNm/m is a 36% increase on the 47.2 kNm/m obtained from the upper bound analysis. A subsequent upper bound analysis, based on a finer, more restricted net, gave an ultimate moment of 53.8 kNm/m, which corresponded with a value of 55.8 kNm/m derived from a hand analysis (Kennedy et al, 2003). These revised results, however, are still an underestimate of 15% from the lower bound prediction, suggesting that the safer lower bound should generally be preferred, unless it can be shown to produce unduly conservative results.

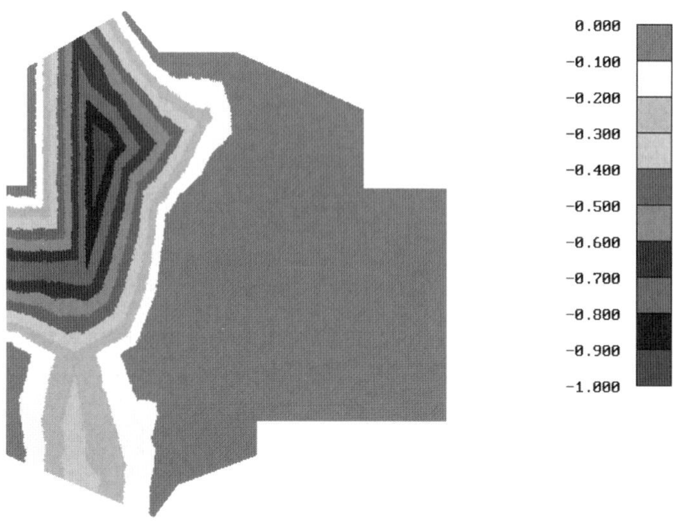

Figure 5: Lower bound solution collapse mode contour plot

Safety of the strip method

Hillerborg (Hillerborg, 1996) introduced a number of approximations and simplifications to the strip method in order to make it more convenient for practical application. Although it was the intention that the modifications should be conservative, no formal guarantees exist that this is the case and it is therefore conceivable that unsafe designs might be generated, although Hillerborg contends that *"the yield line theory of slabs designed according to the recommendations of the book never show that it is on the unsafe side, at least as far as I have found"*. To examine the validity of this claim, the automated yield line method, employing fine meshes to establish probable mechanisms and then coarse meshes to optimise the geometry of the detected mechanism, have been undertaken (Johnson, 2001) for a range of the examples considered by Hillerborg. The results of these analyses are summarised in Table 1.

Table 1: Automated yield line analyses summary

No.	Example	LF (fine) (coarse)
1	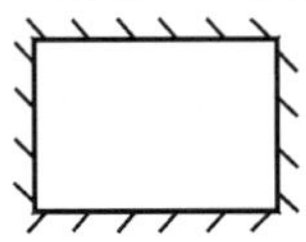	1.08 1.06
2	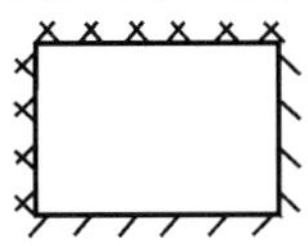	1.06 1.01
3	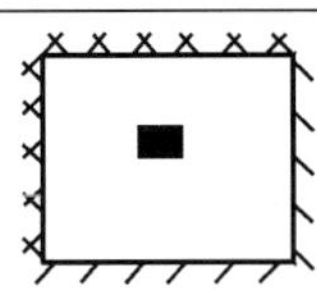	1.14 1.04
4	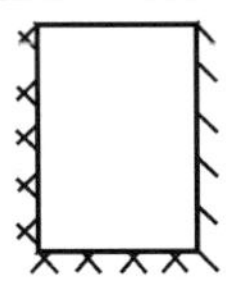	1.02 1.03
5	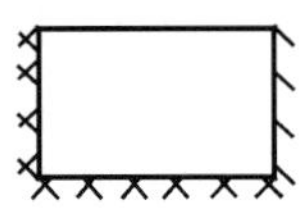	1.03 1.03
6	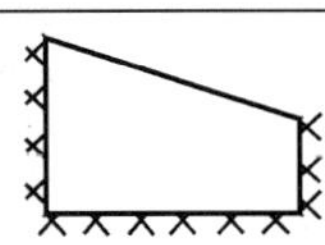	1.13 0.99
7	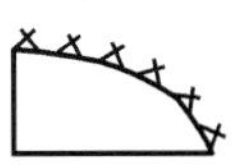	1.00 1.00
8		1.36 1.40
9	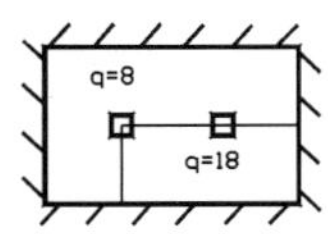 q=8, q=18	1.24 1.29
10	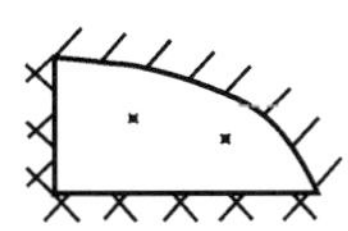	1.10 n/a
11	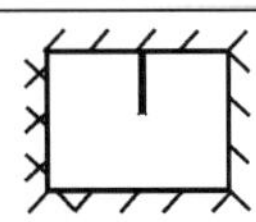	1.13 1.04
12	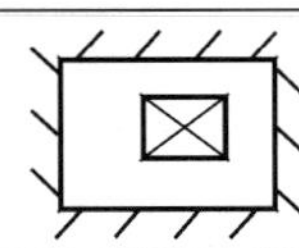	1.22 1.30
13		1.00 0.99
14	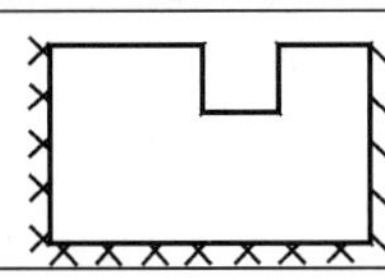	1.06 1.21

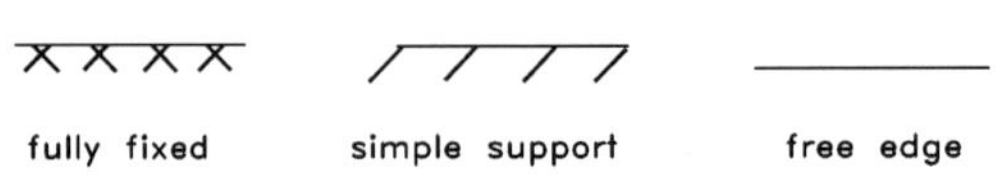

In Table 1 the pairs of load factors (LF) given for each example relate to the initial fine mesh automated yield line solution and the subsequent coarse mesh used to optimise the detected mechanism form. In general, it is expected that the coarse mesh should produce a more critical (lower load factor) than the fine mesh. This may be seen not to be the case in all instances, presumably because in some of the examples the fine mesh allows the generation of a somewhat more complex variation of the mechanism taken as the basis for the coarse analysis. The load factors are based on the reinforcement design generated by the strip method and, taking the yield line results as essentially exact, a value above unity therefore represents a conservative strip design and a below unity represents an unsafe design. On this basis, Table 1 supports Hillerborg's contention that the strip method does not generate unsafe designs for the examples but does indicate significant conservatism for some of the more complex examples, where 20-30% overdesigns are encountered on occasion.

Highly orthotropic slab

The automated lower bound approach routinely makes use of a linear programming algorithm for optimisation purposes. To accommodate the algorithm, a linearization of the yield criterion (Wolfensberger, 1964) is commonly undertaken, so introducing an approximation that tends to become more significant as slabs become more highly orthotropic. To test the degree of approximation involved and also to test an alternative "maximum twist" yield criterion, the highly orthotropic "bridge" type slab shown in Figure 6 has been examined (Johnson, 1999).

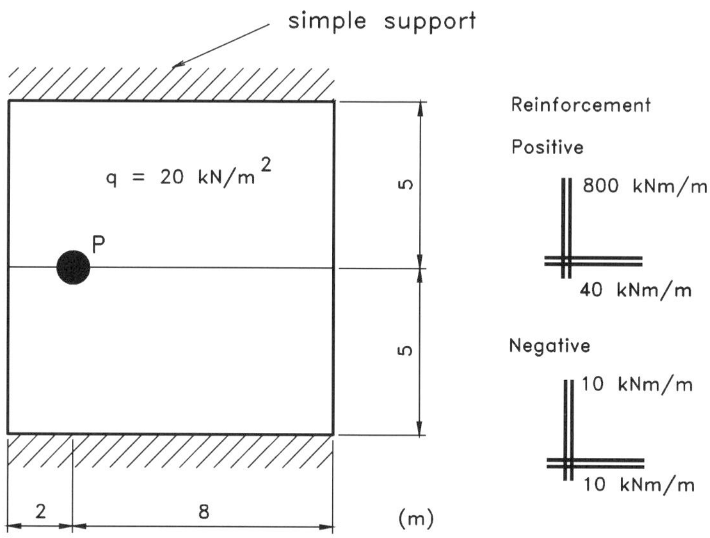

Figure 6: Highly orthotropic slab example

The slab shown in Figure 6 was analysed by the hand yield line approach; the automated yield line method using both a fine and a coarse mesh; and the automated lower bound method using a linearized (Wolfensberger) yield criterion and a revised "maximum twist" approximation to the yield criterion. The results of the various analyses are shown in Table 2, from which it may be seen that the automated yield line method (coarse analysis) and the automated lower bound (maximum twist yield criterion) provide acceptably close bounds on the collapse load. For this case, likely collapse mechanisms are well established and so the hand yield line analysis should also provide an acceptable collapse load estimate. This is indeed the case if the calculated value of 1152 kN is reduced by a conventional 15% to produce a revised value of 979 kN. On the other hand, the automated yield line method (fine analysis) and the automated lower bound (Wolfensberger yield criterion) result in significant over- and under-estimates, respectively, of the collapse load.

Table 2: Comparative analyses for highly orthotropic slab

Analysis	Collapse Load (kN)
Hand yield line analysis	1152
Automated yield line – fine mesh	1743
Automated yield line – coarse mesh	1009
Lower bound analysis (maximum twist yield criteria)	938
Lower bound analysis (Wolfensberger yield criteria)	775

Summary

- The hand approach to yield line (upper bound) analysis can lead to significant unsafe approximations due to the critical mechanism not being examined, as in the Shoemaker and practical design examples. This approach is therefore only reasonably trustworthy in the case of conventional slab layouts and loadings for which the critical yield line patterns have been reliably established.

- The automated yield line analysis will generally be more reliable in identifying critical mechanisms but it involves a somewhat cumbersome two-stage process and can involve uncertainty on the unsafe side. The practical design example, for instance, resulted in a significant discrepancy between automated upper and lower bound solutions.

- The strip (lower bound) hand method will lead to safe designs but can produce conservatism of up to 20-30% if applied to complex configurations and loadings.

- The automated lower bound method has the overriding virtue of guaranteeing a safe solution and is also a single stage analysis, without the need for manual intervention. It is perhaps slightly less intuitive than the automated yield line method, since, by its nature, it does not provide a yield line system. However, a contour plot of the associated collapse mode can be generated. The technique has been shown to produce conservative solutions for highly orthotropically reinforced slabs if the linearized Wolfensberger yield criterion is used, but this can be improved either by using a "maximum twist" linearization of the yield criterion (Johnson, 1999) or by using a formulation that employs the non-linear form of the yield criterion (Krabbenhoft et al, 2002).

Conclusion

Although, in favourable circumstances, the upper and lower bound approaches can be one and the same (just as Heraclitus' roads), in the sense that the same unique collapse load can be found, the chances of going astray and becoming unsafe are substantially greater with the upper bound (yield line) approaches. The lower bound methods are definitely securer and generally quicker (just as the Traditional Song's low road).

References

Hillerborg, A. (1975) *Strip method of design*, Spon, London.
Hillerborg, A. (1991) Letter to *Concrete International*, May, pp 9-10.
Hillerborg, A. (1996) *Strip method design handbook*, Spon, London.

Johansen, K. W. (1964) *Yield line theory*, Cement and Concrete Association, London.

Johnson, D. (1994) Mechanism determination by automated yield-line analysis, *The Structural Engineer*, Vol. 72, No. 19, pp 323-327.

Johnson, D. (1996) Is Yield-Line Analysis Safe? *in Proc. Third Canadian Conference on Computing in Civil and Building Engineering (eds: Moselhi, O., Bedard, C. and Alkass, S).* The Canadian Society for Civil Engineering, Montreal, pp 494-503.

Johnson, D. (1999) Lower Bound Collapse Analysis of Concrete Slabs, *Proc. Concrete Communication Conference (held at the University of Cardiff, July 1999),* BCA, Crowthorne, pp 299-310.

Johnson, D. (2001) On the safety of the strip method for reinforced concrete slab design, *Computers and Structures*, Vol. 79, pp 2425-2430.

Kennedy, C. and Goodchild, C. H. (2003) *Practical yield line design*, The Concrete Centre. Camberley.

Krabbenhoft, K. and Damkilde, L. (2002) Lower bound limit analysis of slabs with non-linear yield criteria, *Computers and Structures*, Vol. 80, pp 2043-2057.

Krenk, S., Damkilde, L. and Høyer, O. (1994) Limit analysis and optimal design of plates with equilibrium elements, *Journal of Engineering Mechanics*, Vol. 120, June, pp 1237-1254.

Munro, J. and Da Fonseca, A. M., (1978) Yield line method by finite elements and linear programming, *The Structural Engineer*, 56B, No. 2, June, pp 37-44.

Shoemaker, W. L. (1989) Computerised yield line analysis of rectangular slabs, *Concrete International*, August, pp 62-65.

Wolfensberger, R. (1964) *Traglast und optimale Bemessung von Platten,* Doctoral Thesis, ETH, Zürich, Switzerland.

Wood, R. H. (1961) *Plastic and elastic design of slabs and plates*, Thames and Hudson, London.

DECISION SUPPORT SYSTEM FOR BRIDGE NETWORK MAINTENANCE PLANNING

Ming Liu and Dan M. Frangopol
Department of Civil, Environmental and Architectural Engineering, University of Colorado, Boulder, Colorado 80309-0428, USA
E-mail: dan.frangopol@colorado.edu

Introduction

Bridge maintenance planning as a part of public policy management is not only a scientific-analytic task (Heineman, 2002), but also involves political, subjective, and sometimes other factors. Decision-makers such as administrators, managers and/or politicians are the key players in the planning process, supported by policy analysts and other technical experts. The ultimate goal of bridge maintenance planning is to find the "best" strategies and/or operational plans that are not only technically feasible, but also are considered optimal by decision makers. This can be achieved by better understanding the real-world situations, identifying all possible objectives and conflicts, evaluating as many alternatives as possible, and finally reaching rational plans. Therefore, Decision Support System (DSS) is necessary for optimal bridge maintenance planning.

This paper presents a DSS for bridge network maintenance planning that involves a group of existing highway bridges with various remaining service lifetimes. The proposed DSS considers five bridge maintenance alternatives, including "do nothing", and the associated cost. Based on the annual bridge maintenance budget and the probabilities that each of the five maintenance alternatives may be conducted at that year on individual bridges in the network, the ultimate goal of this DSS is to find the best combination of the five maintenance alternatives applied to all bridges in the network during certain years. Since the mutual preferential independence requirement can be easily satisfied in this case, the additive form of the multiple attribute utility function can be used to establish the single-objective function for optimization with the weight assignment from the Reliability Importance Factor (RIF) of individual bridges in the network. The RIF reflects the sensitivity of the bridge network reliability in terms of the network connectivity to the change in the individual bridge system reliability due to maintenance actions. The optimization problem in the proposed DSS can be solved by either traditional mathematical programming for combinatorial optimization or the advanced heuristic search methods such as Genetic Algorithms (GAs).

Bridge Maintenance Alternatives and Associated Cost

The bridge maintenance alternatives presented in this paper include both preventive and essential maintenances with actual cost data, as well as "do nothing". The effects of the five different maintenance alternatives, namely, "minor concrete repair", "silane treatment", "cathodic protection", "rebuild" and "do nothing" on individual bridge condition and safety indices over time have been studied as shown in Tables 1 and 2 (Denton, 2002). For example, the "minor concrete repair" results in a decrease of bridge condition index (CI) between 2 and 3 with a triangular probability distribution. The mode of the triangular probability distribution is 2.5, indicating that the most likely decrease of the bridge condition index is 2.5 (see Table 1). Meanwhile, the "minor concrete repair" causes a delay in deterioration of bridge safety index (SI) when the bridge condition index is less than 1.0, in other words, there is no deterioration of the bridge safety index after the "minor concrete repair" maintenance action is applied, and the deterioration of the bridge safety index resumes after the bridge

M. Pandey et al. (eds), Advances in Engineering Structures, Mechanics & Construction, 833–840.

condition index reaches 1.0. The "silane treatment" affects only the deterioration rates during the maintenance effective duration that has a triangular probability distribution between 7.5 and 12.5 years. The bridge condition and safety indices will not change in the first 12.5 years after the "cathodic protection" maintenance action is applied. The "rebuild" is the only essential maintenance actions in this study. If this action is applied, the bridge condition index will be set to zero and the bridge safety index will be assigned to the safety index of the rebuilt bridge. Meanwhile, the deterioration of the bridge condition index will start between 10 and 30 years after "rebuild" with a triangular probability distribution mode of 15 years. The deterioration of the bridge safety index will begin when the bridge condition index reaches 1.0. Table 3 presents the associated cost of the five alternatives considered. Figures 1 to 4 shows the effects of the preventive ("minor concrete repair", "silane treatment" and "cathodic protection") and essential ("rebuild") maintenance alternatives on the mean values of bridge condition and safety indices over time.

Table 1: Effect of Bridge Maintenance Alternatives on Mean Condition Index (after Denton, 2002)

Bridge Maintenance Alternatives	Decrease in Condition Index	Delay in Deterioration (years)	Reduced Deterioration Rate ($year^{-1}$)	Maintenance Effective Duration (years)
Minor Concrete Repair	T(2.0, 2.5, 3.0)	-	-	-
Silane Treatment	-	-	T(0.00, 0.01, 0.03)	T(7.5, 10.0, 12.5)
Cathodic Protection	-	12.5	-	-
Rebuild	set to zero	T(10, 15, 30)	-	-
Do Nothing	-	-	-	-

Note: T (minimum value, mode, maximum value) represents the triangular probability distribution.

Table 2: Effect of Bridge Maintenance Alternatives on Mean Safety Index (after Denton, 2002)

Bridge Maintenance Alternatives	Increase in Safety Index	Delay in Deterioration (years)	Reduced Deterioration Rate ($year^{-1}$)	Maintenance Effective Duration (years)
Minor Concrete Repair	-	CI<1.0	-	-
Silane Treatment	-	-	T(0, 0.007, 0.018)	T(7.5, 10.0, 12.5)
Cathodic Protection	-	12.5	-	-
Rebuild	set to SI	CI<1.0	-	-
Do Nothing	-	-	-	-

Note: T (minimum value, mode, maximum value) represents the triangular probability distribution.

Table 3: Cost for Bridge Maintenance Alternatives (after Denton, 2002)

Bridge Maintenance Alternatives	Cost
Minor Concrete Repair	T(16, 3605, 14437)
Silane Treatment	T(0.3, 39,77)
Cathodic Protection	T(19, 2604, 5189)
Rebuild	T(247, 7410, 28898)
Do Nothing	0.0

Note: T (minimum value, mode, maximum value) represents the triangular probability distribution.

Dynamic Programming for Individual Bridge Maintenance Planning

Bridges without maintenance may not reach a targeted service lifetime due to the aging and deterioration. Therefore, bridge maintenance actions must be applied to extend the remaining service lifetime of individual bridges. Individual bridge maintenance planning needs to answer to questions such as what sequence of maintenance actions and when these maintenance actions should take place in order to minimize the life-cycle maintenance cost throughout the entire targeted lifetime period. The life-cycle maintenance cost can be either construction cost that bridge owners have to pay for or user's cost that includes the time delays and fuel consumption due to detour and/or congestion caused

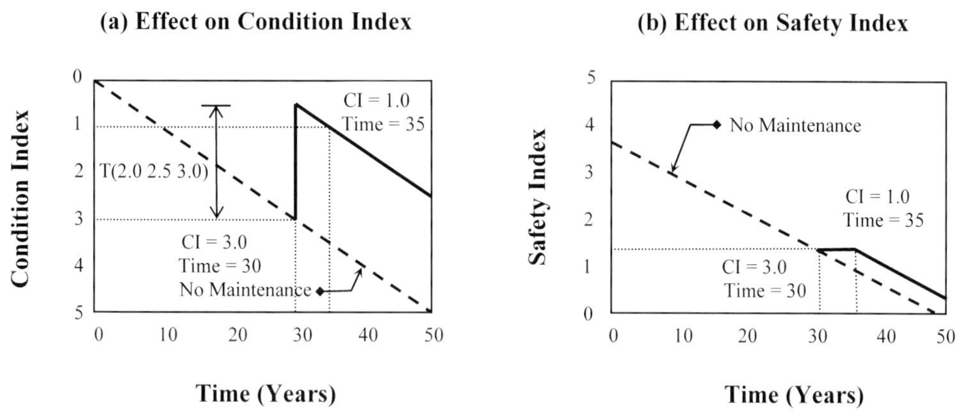

Figure 1: Effects of "Minor Concrete Repair" on Mean Condition and Safety Indices over Time

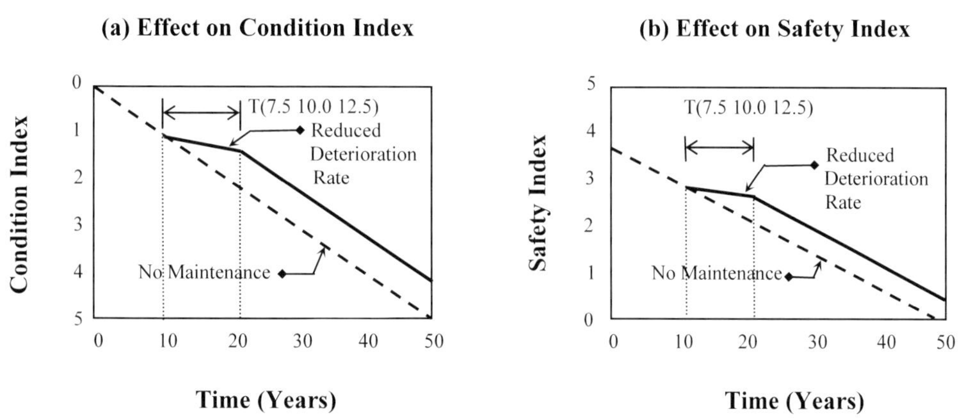

Figure 2: Effects of "Silane Treatment" on Mean Condition and Safety Indices over Time

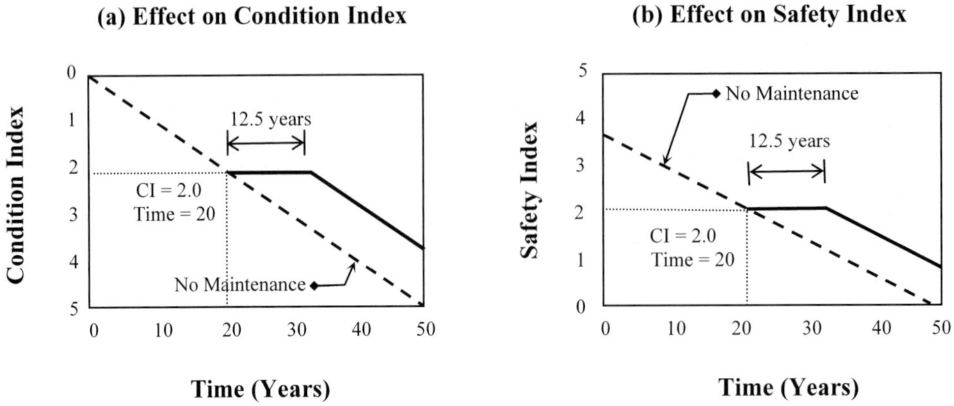

Figure 3: Effects of "Cathodic Protection" on Mean Condition and Safety Indices over Time

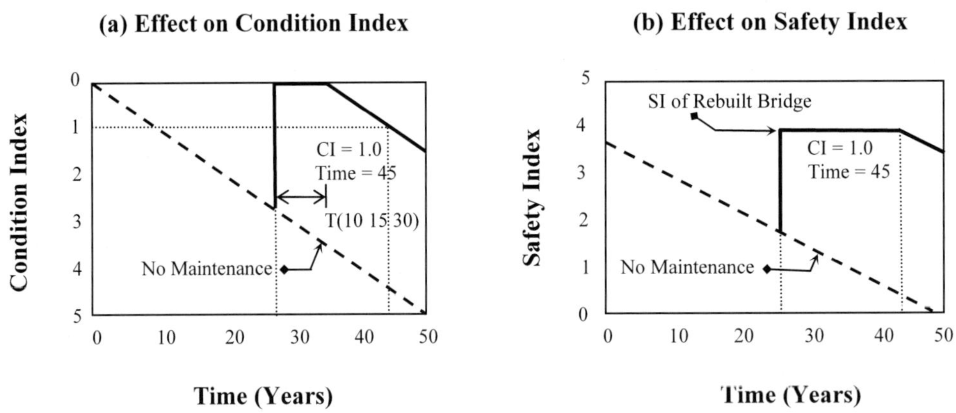

Figure 4: Effects of "Rebuild" on Mean Condition and Safety Indices over Time

by the maintenance actions or combination of both construction and user's cost. In reality, bridge maintenance planning has to consider the maintenance funding limitation as well.

In this study, the prediction of the remaining service lifetime of individual bridges is based on both bridge condition and safety indices. The bridge condition index increases as the bridge deteriorates with time, while the bridge safety index decreases with time. The maximum condition index is set to be 3.0, and the minimum safety index is assigned to be 0.91 (Denton, 2002). In other words, an individual bridge should always have a condition index less than 3.0 and a safety index greater than 0.91 during the entire service lifetime period. The difference between the predicted remaining service lifetime and the targeted service lifetime of a bridge must be covered by applying maintenance alternatives. Any combination of the above five bridge maintenance alternatives that can extend the bridge service lifetime to the targeted level may be regarded as a feasible maintenance plan. These feasible maintenance plans may require performing different combinations of the above five maintenance actions at different application times, resulting in different life-cycle maintenance cost. The life-cycle maintenance cost for each feasible maintenance plan are converted to the net present values (NPV), using the discount rates ranging from 2% to 8%. Thus, an optimal bridge maintenance plan is the feasible plan that has a minimum life-cycle maintenance cost in terms of NPV. A dynamic programming (DP) procedure has been developed to identify the optimal bridge maintenance plans for individual bridges (Liu and Frangopol, 2006). Monte Carlo simulations are integrated within the DP procedure for sensitivity studies, considering the probability distributions of all random variables and parameters. As a result, the probabilities that each of the above five maintenance alternatives (including "Do Nothing") may be conducted at certain time (year) can be obtained for individual bridges. The details of the DP procedure combined with Monte Carlo simulations are presented in Liu and Frangopol (2006).

Multiple Criteria Decision-Making Process

Since almost all real-world decision problems must be addressed on the basis of multi-dimensional approaches, the Multiple Criteria Decision-Making (MCDM) process has been developed long time ago. Pareto first introduced the efficiency concept in 1896 (Doumpos and Zopounidis, 2002). A feasible solution is efficient if and only if there is no other feasible solution which dominates it (Ringuest, 1992). Von Neumann and Morgenstern (1944) developed the utility theory as one of the major methodologies in modern MCDM. The preference structures of a decision maker are represented by multiple attribute utility functions. Charnes and Cooper (1961) extended the traditional mathematical programming theory to the goal programming. In recent decades, more and more user-friendly software has been developed based on advances in information technology and computer science.

Basically, MCDM provides a set of criteria aggregation methodologies that focus on decision maker's preference structures, system values and judgment policy. Since the "optimal" solutions in the traditional mathematical programming usually do not exist in MCDM due to the potential conflicting nature of the multiple objects, MCDM could find an appropriate "compromise" solution that satisfies all of the decision maker's policy. MCDM general procedure consists of (1) identifying decision objectives, all feasible alternatives and participants; (2) developing evaluation criteria that measure the performance of each alternative on decision objectives; (3) modeling criteria aggregation; and (4) providing meaningful recommendations. MCDM approaches include multi-objective mathematical programming (MMP), multiple attribute utility theory (MAUT), outranking relation theory (ORT), interactive methods and preference disaggregation analysis (PDA) (Vincke, 1992; Pardalos et al., 1995).

Optimization for Bridge Network Maintenance Planning

Bridge network maintenance planning has to deal with multiple bridges in a highway network under limited annual maintenance budgets. Thus, MCDM approaches can be used to help bridge owners, authorizers and/or maintenance managers to make rational decisions on maintenance actions applied to each of individual bridges in the highway network. In this study, the multiple attribute utility theory (MAUT) is adopted. As a matter of fact, MAUT focuses on the development of the multiple attribute utility functions to model and represent the decision maker's preferential structures. The multiple attribute utility functions combine all of the marginal utility functions associated with individual attribute of each alternative. The marginal utility functions for each attribute can be built up by either direct interrogation with decision makers or by indirect methods, as well as by using the analytic hierarchy process (Saaty, 1980) that has been mainly used in USA. The decomposition forms of the multiple attribute utility functions may be (1) additive, (2) multiplicative and (3) multi-linear forms (Keeney and Raiffa, 1993). The additive form requires mutual preferential independence, that is, every subset of criteria is preferentially independent from the remaining criteria. A subset of criteria is considered to be preferentially independent form the remaining criteria if and only if the decision maker's preferences on the alternatives differ only with respect to the criteria, and are independent on the remaining other criteria. It must be noted that very complex decomposition forms are of not interest from a practical point of view (Vincke, 1992). MAUT also employs an interactive and iterative procedure involving policy analyst and decision makers to specify the weight and marginal utility function corresponding to each criterion. Finally, the total utility of each alternative can be used as an objective function in traditional mathematical programming in order to make final decisions (Doumpos and Zopounidis, 2002).

In this study, each of individual bridges in a highway network may be treated as a subset of criteria with a marginal utility function associated with the probabilities that each of the above five maintenance alternatives (including "do nothing") may be conducted at certain time (year). Since the mutual preferential independence requirement can be easily satisfied in this case, the additive form of the multiple attribute utility function is used to form a single-objective function for optimization. The objective function of multiple attribute utility may be also weighted by using RIF of individual bridges in the network, where RIF is defined as the sensitivity of the bridge network reliability to the change in the individual bridge system reliability (Liu and Frangopol, 2005). RIF in this paper reflects the sensitivity of the bridge network reliability in terms of the network connectivity to the change in the individual bridge system reliability due to maintenance actions, and must be developed as a function of the bridge system reliability profiles, network reliability, and network topology. RIF may also be expanded to include traffic capacity and impacts of bridge maintenance activities on economy, environment and society, when considering user's satisfaction and critical bridge performance of a bridge network (Liu and Frangopol, 2005). Consequently, the optimization problem in bridge network maintenance planning can be formulated as follows:

$$\text{Maximize} \quad \sum_i D_{ij} \times RIF_i \times P_{ij} \tag{1}$$

$$\text{Subject to:} \quad \sum_i C_i \le C_{budget} \tag{2}$$

where

D_{ij}	is a binary design variable, *i.e.* the value of D_{ij} can be either 0 or 1;
RIF_i	is the reliability importance factor for bridge *i*;
P_{ij}	is the probability of the maintenance alternative *j* applied to bridge *i*;
C_i	is the cost associated with the selected maintenance alternative for bridge *i*;
C_{budget}	is the annual maintenance budget at certain year;

The binary design variable, D_{ij} represents the decision on selecting the maintenance alternative *j* applied to bridge *i*, that is, $D_{ij} = 0$ means the maintenance alternative *j* will not be applied to bridge *i*, and $D_{ij} = 1$ means the maintenance alternative *j* is selected to be applied to bridge *i*. In addition, it should be noted that the values of RIF_i and P_{ij} usually vary during the entire service lifetime of bridge *i*. This is because RIF_i is normalized by considering all bridges in a highway network that experience the aging and deterioration with time (Liu and Frangopol, 2005). On the other hand, P_{ij} is normalized by considering all of the five maintenance alternatives applied to bridge *i* in a certain year, and is dependent on the results from the DP procedure that is combined with Monte Carlo simulations (Liu and Frangopol, 2006). Moreover, C_i is related to the cost in Table 3, but the actual values of C_i should be assigned in order to obtain an optimal bridge network maintenance planning. Finally, this combinatorial optimization problem can be easily solved by either traditional mathematical programming or the advanced heuristic search methods such as Genetic Algorithms (GAs).

Case Study

A numerical example involving five highway bridges is provided to demonstrate the application of the proposed DSS in bridge network maintenance planning. Table 4 presents the values of RIF_i, P_{ij} and C_i for combinatorial optimization that is subject to a budget constraint of C_{budget} = 10,000 The optimization results from a traditional mathematical programming are also summarized in Table 4.

Table 4: Example Values of RIF_i, P_{ij} and C_i

Maintenance Alternatives	Bridge E-17-HE (P_{1j})	Bridge E-17-HR (P_{2j})	Bridge E-17-LE (P_{3j})	Bridge E-16-MU (P_{4j})	Bridge E-16-NM (P_{5j})	Cost (C_i)
Minor Concrete Repair	0.40	0.30	0.15	0.20	0.15	4,500
Silane Treatment	0.20	0.20	0.40	0.10	0.35	48
Cathodic Protection	0.25	0.30	0.20	0.30	0.20	2,600
Rebuild	0.10	0.15	0.05	0.30	0.05	12,000
Do Nothing	0.05	0.05	0.20	0.10	0.25	0
Sum of P_{ij}	*1.00*	*1.00*	*1.00*	*1.00*	*1.00*	-
Reliability Importance Factor (RIF_i)	0.38	0.27	0.19	0.11	0.05	Sum of RIF_i *1.0*
Selected Maintenance Alternative	minor concrete repair	cathodic protection	silane treatment	cathodic protection	silane treatment	C_{budget} 10,000
Cost (C_i)	4,500	2,600	48	2,600	48	Sum of C_i *9,796*

Conclusions

This paper presented a decision support system (DSS) for bridge network maintenance planning using the multiple attribute utility theory (MAUT). The combinatorial optimization problem was developed with a single-objective function of the probabilities that the maintenance alternatives may be applied to each of the individual bridges in a highway network. The probabilities in the single-objective function had to be obtained from a Dynamic Programming (DP) procedure, considering individual bridge condition index, safety index and life-cycle maintenance cost. The single-objective function was also weighted by the Reliability Importance Factors (RIF), which had to be the functions of individual bridge system reliability profiles, bridge network reliability, and network topology. The constraint of the optimization problem was the limited annual maintenance budget. A numerical example was provided to demonstrate the application of the proposed DSS in bridge network maintenance planning.

Acknowledgements

The writers gratefully acknowledge the partial financial support of the U.K. Highways Agency and of the U.S. National Science Foundation through Grant Numbers CMS-0217290 and CMS-0509772.

The opinions and conclusions presented in this paper are those of the writers and do not necessarily reflect the views of the sponsoring organizations.

References

Charnes, A. and Cooper, W.W. (1961) "Management models and industrial applications of linear programming" Wiley Publishers, New York

Denton, S. (2002) "Data estimates for different maintenance options for reinforced concrete cross-heads" (personal communication for Highways Agency, U.K.), Parsons Brinckerhoff Ltd., Bristol, UK

Doumpos, M. and Zopounidis, C. (2002) "Multicriteria decision aid classification methods" Kluwer Academic Publishers, Boston

Heineman, R. A. (2002) "The world of the policy analyst: rationality, values, and politics" Chatham House Publishers, New York

Keeney, R.L. and Raiffa, H. (1993) "Decisions with multiple objectives: preferences and value trade-off" Cambridge University Press, New York

Liu, M. and Frangopol, D. M. (2005) "Time-dependent bridge network reliability: novel approach" Journal of Structural Engineering, ASCE, 131(2), 329-337

Liu, M. and Frangopol, D. M. (2006) "Dynamic programming for optimal bridge maintenance planning" submitted to IABMAS'06, International Conference on Bridge Maintenance and Safety, Porto, Portugal, July

Pardalos, P.M., Siskos, Y. and Zopounidis, C. (1995) "Advances in multicriteria analysis" Kluwer Academic Publishers, Boston

Ringuest, J. L. (1992) "Multiobjective optimization: behavioral and computational considerations" Kluwer Academic Publishers, Boston

Saaty, T. L. (1980) "The analytic hierarchy process: planning, priority setting, resource allocation" McGraw-Hill International Book Co, New York

Vincke, P. (1992) "Multicriteria decision-aid" John Wiley & Sons Publishers, New York

Von Neumann, J. and Morgenstern, O. (1944) "Theory of games and economic behavior" Princeton University Press, Princeton

ANTIMICROBIAL TREATED CONSTRUCTION MATERIALS AND AIR FILTERS REDUCE FACILITY BIOBURDEN AND IMPROVE AIR QUALITY IN A HEALTHCARE ENVIRONMENT

G.M. McNeice[1], O.Z. Tyler[2] and D.W. Blackhurst[3]
[1]Clemson University, [2]OZ Enterprises, [3]Greenville Hospital System[3]

Background

The continued evidence of the presence and spread of pathogenic organisms on environmental surfaces in acute, ambulatory, short-term and long-term healthcare facilities and in the home, suggests that additional cleaning and maintenance protocols are needed. The gold standards of "good hand washing" and daily disinfecting of critical surfaces do not appear to be able to combat the increasing rate of hospital-acquired infections.

Nosocomial Infections

In the twenty-year period between 1975 and 1995, nosocomial or hospital acquired infections increased by 36%. A review of studies reported in two specific conferences on the topic, revealed the magnitude of the problem (APIC 2000; CDC 2000). The papers presented at these conferences provide evidence of increasing infection rates, a shift in infection type and additional locations where these infections are being spread. In addition to being found in most hospitals, nosocomial infections are common among patients admitted to acute rehabilitation units. Rates have been reported as high as 17% and were found to significantly contribute to patient morbidity (Mylotte 2000). Similar concern was voiced in a study involving a VA psychiatric facility (Risa et al. 2000). Nosocomial invasive aspergillosis is associated with high morbidity and mortality in patients with blood related illnesses (Raad et al. 2000). Various fungal infections found in immunocompromised patients were traced to problems during construction renovations of healthcare facilities (Drusin et al. 2000; Donelan et al. 2000; Rebmann et al. 2000). In coastal regions, that are prone to hurricanes, high humidity and moisture levels, patients are subjected to increased risk of nosocomial infections (Ober et al. 2000). Zhan et al. (2003) surveyed 20% of US hospitals and found that hospital acquired infections increase the average patient's stay by almost 10 days, increases the cost per patient by more than $38,000 US and increase their risk of death by more than 4%. This study resulted in the Joint Commission on Accreditation of Hospital Organizations (JCAHO 2004) to revised the standards for 2004. Zoutman et al. (2003) reported a similar study of Canadian hospitals. In 2003 the Province of Ontario was suddenly exposed to a severe epidemic of SARS. Over the six months from the infection's arrival until the last patient was discharged from hospital, 375 cases were recorded (OMH. 2004). At the height of the outbreak thousands of people, including health care workers, mostly in the Toronto area were quarantined for 10-day periods at home and given specific advice on preventing family members from infection. This was one of the extreme examples of a hospital acquired infection.

Resistive Strain Organisms

There is an increasing awareness by infection control personnel that touch surfaces can play a major role in the spread of microorganisms. Of particular concern is the potential spread of antibiotic-resistant organisms such as vancomycin-resistant enterococci (VRE), which is known to be transmitted from surfaces (Temple et al. 2000). This resistant organism is on the increase in hospitals throughout the United States. A study of upholstered chairs found that VRE are capable of prolonged survival on fabric seat cushions. In many hospitals, cloth-styled covering is preferred by management to increase patient comfort while in the hospital or long term care facility. However, these create potential reservoirs for the nosocomial transmission of VRE (Noskin et al. 2000). In another study, Mermel et al. (2000) found that at a number of cultured sites within patients' rooms, the highest rates of VRE were found on patient telephone handsets. Additional VRE organisms were found on bed rails, over-

M. Pandey et al. (eds), Advances in Engineering Structures, Mechanics & Construction, 841–851.

bed tables and sink faucets. This was further supported by studies conducted by Rutala et al. (2000) and by Diekema et al. (2004). Another resistive strain namely, methicillin-resistant staphylococcus aureus (MRSA), was found on computer keyboards and computer mice used in the medicine wards and other acute care areas (Valena et al. 2000). MRSA was also found on hospital bed handsets (Young, J.M. et al. 2005) including call buttons and TV remote control devices. A further study of hospitals in Pennsylvania (Volavka, M. 2005) found that more than 11,600 patients got infections while in hospitals over a one year period. These infections led to 1500 (13%) deaths and over $2 billion in hospital charges.

Additional Factors

In a survey article by Weinstein (1998), nosocomial infections in the US were estimated to have cost $4.5 billion in the year 1995. This article indicates a growing concern based on the increase of antibiotic-resistant microorganisms, such as (MRSA) and (VRE). The shift from inpatient to outpatient care leaves the more vulnerable patients within the hospitals, especially the immunocompromised patients. This increases the opportunities and likelihood of nosocomial infections. These factors, when combined with the renovation of aging hospitals, which increases the risk of airborne fungal and other infections (PHTS 2000), suggest that a better approach is needed to cleaning and maintaining healthcare facilities in the future.

Evidence of Supplementary Protection

With the current industry increase in cost-cutting measures and staffing constraints, some standards are not being practiced as fully as one would hope (Harris et al. 2000). In an attempt to combat this, various approaches have been taken to provide supporting measures to assist the healthcare worker and to provide cleaner surroundings. Indeed, new hand care preparations (Gould 2000) are being developed to assist the healthcare worker. Some of these contain the well-known antimicrobial agent, triclosan (Hoffmann et al. 2000) that has been used in hand soaps for many years. In an attempt to provide continuous antimicrobial action on surfaces, a novel application incorporating triclosan into the surface material used on a hospital over-bed table was reported by Van Enk and Lam (2000). This study showed that such surface treatment was capable of reducing many species of viable or living organisms on the surface within 60 minutes, as compared to more than four hours for the untreated control surface. This was an interesting application of an antimicrobial agent that was integrated into the construction material.

Environmental Surfaces

The realization that environmental surfaces can play a major role in the spread of resistive strains, has led to studies that find inadequacies in current methods of disinfection (Pentella et al. 2000). Many comparative studies have been reported. These studies used standard chemicals as well as new antimicrobial products for surface treatment (Lisay et al. 2000; Tessarin et al. 2000; Rutula et al. 2000). In addition to hospitals, increasing evidence of VRE is being found in long term care facilities (Lai et al. 2000) and in home care settings (Manangan et al. 2000; Greene et al. 2000). Evidence of airborne organisms, inadequate air quality, inadequate hand washing and environmental contamination in Neonatal Intensive Care units (Kassis et al. 2000; Burke et al. 2000) and pediatric care facilities (Goldman et al. 2000) have each been reported.

The Need for Added Protection

With this background and the projection of increased nosocomial infection rates, the question of how to keep the healthcare environment cleaner, and in a more sustainable state of readiness, lead the authors to be challenged beyond simply improving the current cleaning methods. The principal motivation was to provide methods of additional control of the bioburden within a healthcare environment, beyond the regularly scheduled maintenance and cleaning protocols. The idea of providing treatment of the physical environment that would be sustaining and perhaps bridge the periods of use between regular cleaning periods, was worthy of study. The search for various antimicrobial agents that would lend themselves to environmental treatment, led the authors to the use of triclosan, the agent mentioned above. When the authors began the preliminary studies in 1995, a limited number of products containing triclosan were available for application. However, no information existed beyond laboratory findings, that triclosan could control 'real world' bioburden, to the degree needed in a healthcare environment.

Antimicrobial Agent - Triclosan

Triclosan is a diphenyl ether (5-chloro-2-[2,4-dichlorophenoxyl] phenol) and like other chlorine-based chemicals, it acts as a cell wall penetrant. It disrupts the microbial cell wall, disturbing the metabolic process, leading to the death of the organism. Although McMurray et al. (1998) suggest that some organisms may be insensitive to triclosan and could result in the development of resistant strains of organisms, considerable counter evidence and scientific argument dispels this position (FDA 1997, CSMA 1998, Jones 1999). The latter references indicate that currently used antimicrobial agents like triclosan, have very different action mechanisms than antibiotics. Additional studies also indicate that there is no existing clinical evidence to suggest that these agents, as used in 'real world' applications, are either mutagenic or prone to create resistive strains of organisms (Russell 2002, McBain et. al. 2002, Fraise 2002). An extensive review of the literature on bacterial resistance to topical antimicrobial products by Jones (1999), clearly indicates that there is no scientific evidence that triclosan has an influence on the development of resistive strains of organisms. Indeed, Jones et al. (2000) verifies that triclosan was clinically effective in reducing MRSA isolates from surgical wounds as well as reducing the percentage of ciprofloxacin-resistant MRSA strains.

It is worthy to note that triclosoan can be incorporated into the voids of any polymeric structure (Medlin 1997), as well as being suspended in certain liquid compounds. Such potential use of an antimicrobial agent makes it attractive for sustained treatment. This was one of the underlying motivations for using it in the present study. The concept was to have all construction materials, fixtures, medical instruments, and contents within the healthcare delivery areas, either surface treated or constructed with an antimicrobial agent as an integral part of the product. A limited number of polymer products containing triclosan were available for this study. The goal was to test these and where necessary, to develop others. These included surface sprays and coating liquids.

Over a period of three years, the authors used various triclosan-treated products and applied them to different problem areas throughout the main campus. These included concrete floor surfaces in the decontamination and food services areas; carpet underlay in maternity areas where spills created obnoxious odors and in shower areas where dampness led to continuous growth of microorganisms and degradation of carpet backing; in bathrooms on toilet seats, sanitary napkin dispensers, sinks and faucets and waste baskets; on counters, floors and isolettes in the Neonatal Intensive Care Unit; and on floor drain covers and mortar joints on bathroom floors and walls. In every application studied, the use of triclosan-treated items resulted in a reduction of surface bioburden.

With the success of these preliminary 'real world' tests, the authors began planning for a larger application of the treatment. During this planning period, management decided to construct a new outpatient facility, consisting of two floors, each containing 25,000 square feet (2315 m^2) of floor space. This building provided an excellent site for the expanded project. It was scheduled to open in January 1998, at which time this study began.

The Building

The skeletal structure of the building was that of a two storied, standard steel beam system with open web steel joists, erected on reinforced concrete foundation walls and footings. The building envelope consisted of brick veneer with the interior consisting of metal studded partitions with drywall. All concrete floors were concrete slab, either on fill or suspended on metal decking. The building was constructed on a sloped site with the second level serving as the front of the building and providing a ground level main entrance. This level had two identical wings that flanked the main entrance of the building, each with 12500 square feet (1157 m^2) of floor space. In addition to hallways and restrooms, each wing contained patient waiting and exam rooms, faculty and medical resident offices, as well as nursing and administration areas. Since the building had two wings with identical layouts, each being used for identical outpatient care, it provided a unique opportunity to conduct a controlled study. The lower level housed the labs and support services with exit at the rear of the building. The HVAC system consisted of four main areas. The lower level was served by two separate but equally sized AHUs, each rated at 10,000 CFM. The upper level was served by two equally sized but separate units, each rated at 14,000 CFM. All units were balanced, and had separate outside air intake zones. The building was completed in December 1997 and opened on schedule in January 1998. All antimicrobial treatment of the test wing was completed before the building was opened for occupancy.

Treatment

Since the main object of the study was to see what effect the treatment of an entire environment would have on the total bioburden, it was imperative to treat all surfaces including the floors, ceilings, walls, doors, all touch hardware, air filters and all fixtures. The fixtures included sinks, counter tops, examination tables, medical instruments, desks, telephone handsets, computers, filing cabinets, chairs, tables and touch hardware etc. Where possible, existing commercial antimicrobial products were used. For some applications (such as floor wax, protective carpet underlay fabric and paint), special products were developed. No treatment was made to any item within the control wing.

Testing Procedures

Standard microbiological swab and air samples were taken throughout both wings. These tests were scheduled on a rotating basis so that each item and location tested on the treated wing had a corresponding test site on the control wing. Both tests were conducted on the same day and as close as possible in time. Over the 18-month test period, each site was tested approximately every two weeks. There were 45 surface and 10 air sites sampled on each wing, for a total of 110 data point locations. Each week, specific sites were tested. These were rotated on a two-week schedule throughout the 18-month test period.

Surface samples were taken using standard microbiological swabs, covering four square inches (2581 mm^2) and using standard culturing protocols. The results were reported in terms of the total number of colony forming units (cfu) of microorganisms appearing on the agar plates following the incubation period. Air samples were taken using a standard Anderson air sampler. These results were recorded in cfu per cubic foot of air sampled. Although each site on each wing was monitored over the entire test period, only average values for all sites within a wing, for any one period, were used for comparison. For example, at six months the average of the 10 air samples taken in the treated wing at that time, were averaged and compared to the average of the corresponding 10 sites in the control wing. This provided a single value that represented the entire wing, even though the distribution of values within the wing may differ somewhat from month to month. The same comparison of average values was made for the surface swab tests.

Results and Discussion

Surface Samples

A review of biological swab tests for touch and building surfaces in both wings, showed a consistent reduction of colony forming units (cfu) within the treated wing throughout the test period. The range of reductions and the average for each type of surface tested are summarized in Table 1. It is apparent that surfaces that were touched more frequently had greater reduction in the bioburden. This supports earlier tests by others (Medlin, 1997) who reported that triclosan exhibits increased effectiveness when continuously challenged. This was more clearly demonstrated in the present study in a separate set of tests involving items like telephone handsets and computers. These tests were conducted over a six-month period only. Three different telephones and computers located in the control wing were selected for separate study toward the end of the main study. These were treated with a surface spray containing triclosan. The average reductions in bioburden are presented in Table 2 for the 7-day and 6-month test periods. The effect of repeated handling of these devices resulted in an increase in antimicrobial activity. This lead to a greater reduction in surface bioburden as compared to surfaces that were not touched as frequently, such as walls, floors and ceilings as reported in Table 1.

Table 1. Reduction in cfu on Surfaces in Treated Wing

Surface Treated Test Items (number of sites tested)	Range of Reductions in cfu Compared to Control Wing	Average Reduction in cfu Compared to Control Wing
Fixtures and Components (15)	25 to 91 %	55 %
Medical Instruments (10)	0 to 80 %	45 %
Furniture (5)	0 to 85 %	50 %
Wall Paint (5)	10 to 50 %	25 %
Floor tile (5)	30 to 55 %	40 %
Ceiling tile (5)	15 to 35 %	20 %
Total (45)	0 to 91 %	40 %

Table 2. Reduction in cfu on Telephone and Computer Surfaces

Surface Treated Test Items (number of sites tested)	Seven Days Post-Treatment Ave. Reduction in cfu	Six Months Post-Treatment Ave. Reduction in cfu
Telephone Handsets (3)	55 %	99 %
Computer Keyboards (3)	45 %	85 %
Computer Mice (3)	51 %	82 %

Air Samples

Although touch and building surfaces were monitored over the 18-month test period, the main interest was in the potential reduction of airborne bioburden as a result of total environmental treatment. For the treated wing, commercially available reusable air filters containing impregnated triclosan polypropylene filter fibers were selected. The control wing contained regular pleated-type disposable filters, the same type used throughout the healthcare campus in the majority of its buildings. The reusable air filters on the treated wing were removed, washed and reinstalled every four months. The disposable filters on the untreated wing were replaced every two months. A total of ten locations were selected for air testing on the treated wing with a corresponding ten on the control wing. Additional sampling was taken outdoors, as well as in the main entrance vestibule and waiting area. These were used for additional comparison with values obtained in the wings. The results for the two wings are compared in Table 3 in which the average values are reported at three equally spaced time intervals over the 18-month study period. Three reference values are provided at the bottom of Table 3 that can be used for additional comparison. These values were the average values over the entire test period. The laboratory value was normalized to 1.0 cfu/ft^3 and is used in Table 3 as a base comparison. The air quality in the laboratory was considered as one of the best or preferred levels of air quality on the campus.

It is apparent from Table 3 that the treated wing was maintained at a much 'cleaner' level than the control wing throughout the test period. Overall, the results indicate a consistent reduction of total bioburden with time in the treated wing. There was a consistent reduction and stabilization after 12 months. Air quality was remarkably good and remained close to that found in the microbiology lab. In contrast, the bioburden in the untreated wing appeared to increase continuously with time. Although the level of bioburden in the untreated wing at 18 months was only 42% of the value found in the front entrance vestibule, it was 2.4 times higher than the treated wing and continued to increase. Since this was a new building and the level of airborne bioburden could continue to change with age, it is conceivable that the untreated wing could continue to increase in bioburden well above the value reported. In contrast, it is possible that the treated wing could level out, at or about the level found at the end of this study. That level was only 17% of that of the vestibule and only 16% above the laboratory level. The fact that the bioburden in the treated wing continuously decreased with time while that of the untreated wing continuously increased, does indicate that the antimicrobial treatment was very effective in controlling the bioburden.

Table 3. Air Samples (Average cfu/ft^3)

Test Interval	Treated Wing	Untreated Wing	Reduction
6 Months	1.36 (48.3)	1.75 (62.1)	22 %
12 months	1.19 (42.0)	2.26 (79.8)	47 %
18 Months	1.16 (40.9)	2.78 (98.1)	58 %

Reference Values: (Microbiology Lab = 1.00 (35.3)) (Vestibule = 6.69 (236.2)) (Outside = 33.37 (1178.0))

Since the total colony count does not differentiate between species of microorganisms, the air samples taken at the end of the study were examined to determine what viable or living fungi and bacteria were present at that time. For the fungi, when compared to the untreated wing, the treated wing contained 63% less bioburden. This was found to be statistically significant at a P-value = 0.0043 using the Mann-Whitney U test of medians. Also, there was no difference in the distribution of predominant genera of viable fungi between the wings. That is, the same percentage distribution of each species was the same in each of the wings even though the absolute numbers of species was less in the treated wing. For bacteria incubated at 35^0C, the median reduction was only 25% and was found not to be statistically significant. Also, there was a difference in the distribution of the predominant genera on each wing.

Conclusion

To the authors' knowledge, this study is the first of its kind involving a healthcare setting. The purpose was to see if the total bioburden or level of microorganisms present in a healthcare environment could be reduced through antimicrobial treatment of surfaces and the air. Since the building was new and contained two identical wings, each containing 12500 square feet (1157 m^2), it offered the authors a unique opportunity to conduct a controlled 'real world' study.

The treatment consisted of using the well known chlorine based antimicrobial agent triclosan, either through the application of existing products or by developing special purpose products. Physical components of the treated wing included permanent and moveable fixtures, medical instruments, furniture, walls, floors, ceilings and air filters. Included were telephones, computers, filing cabinets, sinks, counters, exam tables, doors and hardware etc. Each item in the treated wing had a corresponding control item on the untreated wing that was tested for comparison. A total of 45 items were selected in each wing for testing surface bioburden. There were 10 locations selected for air sampling on each of the two wings.

The results of the 18-month study showed an average of 40% reduction in colony forming units (cfu) for treated surfaces and a 58% reduction of air borne microorganisms. In the case of air sampling, it was interesting to find that the air quality in the treated wing consistently improved, while that of the untreated wing got progressively worse. The antimicrobial treatment enabled the wing to maintain the level of airborne bioburden to within 16% of that found in the microbiology lab.

Unfortunately, this study cannot be used to determine what effect the triclosan-treated reusable air filters may have had on reducing the airborne bioburden within the treated wing since the triclosan-treated filters were installed at the same time as the other surfaces were treated. Although the authors realized this at the beginning of the study, there was insufficient time and resources to conduct a comparison of the effect of different treatment applications.

During the study period both wings were used for the same type of outpatient care. This was important to the validity of the study, since dissimilar use functions could have affected the test results and negated a direct comparison. Also, since hospital employees conducted all building maintenance and antimicrobial treatment, strict supervision and proper implementation of specific protocols for cleaning and treating was made possible.

The study had to be terminated after the 18-month period, since after one year of occupancy, renovations and functional changes were being scheduled by management. These resulted in the wings being used for very different types of patient care. Also, to reduce operating costs, all maintenance for the building was contracted out and this prevented any further study to be conducted.

What impact this type of environmental treatment could have on the reduction of nosocomial infections in more critical care facilities remains to be seen. The cost of environmental treatment would have to be determined and compared against any reductions in nosocomial infection rates. However, similar treatment approaches could be integrated into infection control activities and studied within a hospital setting. If nothing else, the building and the patient environment could be maintained in a 'cleaner' state of preparedness. Additional comparison of personnel absenteeism rates before and after treatment might also prove interesting, in light of the increasing evidence of the effects of 'sick building syndromes.'

Obviously, as is the case with many studies like the one reported here, more questions can be asked than answers provided. However, the science is here, the products are available, and nosocomial infections are on the increase. Perhaps it is time to consider improving the treatment of the healthcare environment as well as that of the patient.

Dedication

This paper is dedicated to the memory of Robert S. Watterson III, formerly Vice President of Sales, Microban Products Company, Huntersville NC whose dream was to see the construction of cleaner buildings based on the application of 'anitmicrobial treated building products'. Perhaps the results of this study suggest that dreams can come true.

References

APIC. 2000. 27th Annual Educational Conference and International Meeting, Minneapolis, Minnesota, June 18-22.

Burke, R., G. Garvin, C. Korn, K. Perryman, and C. Sulis. 2000. Hospital acquired vancomycin-resistant enterococcus faecium in a neonatal intensive care unit. Abstract of poster presented at the APIC 27th Annual Educational Conference and International Meeting, Minneapolis, Minnesota, June 18-22. American Journal of Infection Control, vol. 28, no. 3, June.

CDC. 2000. 4th Decennial International Conference on Nosocomial Infections and Healthcare-Associated Infections, Atlanta, Georgia, March 3-9. CDC. 4th Decennial International Conference on Nosocomial Infections and Healthcare-Associated Infections, Atlanta, Georgia, March 3-9.

CSMA. 1998. CMSA cites studies countering Tufts Researcher's findings. Chemical Specialties Manufacturers Association. News Release, NR 98-09. Washington, DC, August 10.

Diekema, D.J., Bootsmiller, B.J., Vaughn, T.E., Woolson, R.F., Yankey, J.W. and Ernst, E.J. 2004. Antimicrobial resistance trends and outbreak frequency in United States hospitals. Clinical Infectious Diseases, vol. 38, pp 78-85

Donelan, S., F. Singh, A. Provenzano, M. Kotlas, E. Spitzer, and W.H. Greene. 2000. Active and passive surveillance provide proof of the adequacy of air handling systems during hospital construction. Abstract of poster presented at the 4th Decennial International Conference on Nosocomial and Healthcare Associated infections. Atlanta, Georgia, March 5-9. American Journal of Infection Control, vol. 28, no.1, February

Drusin, L.M., E.A. Bancroft, J. Mintz, A.J. Streifel, and K. Martin. 2000. An unusual cluster of fungal colonization and infections associated with hospital renovation. Abstract of poster presented at the 4th Decennial International Conference on Nosocomial and Healthcare Associated infections. Atlanta, Georgia, March 5-9. American Journal of Infection Control, vol. 28, no.1, February

FDA. 1997. Antibiotic and antiseptic resistance issues related to an industry-proposed healthcare continuum model. Report of the Joint Meeting of the Nonprescription Drugs and Anti-infective Drugs Advisory Committees. Gaithersburg, MD., January 22.

Fraise, A.P. 2002. Susceptibility of antibiotic-resistance cocci to biocides. Journal of Applied Microbiolgy, vol. 92 Supplement, pp. 1585-1625.

Goldman, C., R. Freeman, L. Streitenberger, L. Scott, A. Monteath, C. Sass-Korstak, and A. Matlow. 2000. A cluster of nosocomial varicella in a pediatric institution: opportunities for improvement. Abstract of poster presented at the 4th Decennial International Conference on Nosocomial and Healthcare Associated infections. Atlanta, Georgia, March 5-9. American Journal of Infection Control, vol. 28, no.1, February

Gould, M.A. 2000. The unexpected benefits of a comprehensive hand care program in two critical-care units at a children's hospital. Abstract of poster presented at the APIC 27th Annual Educational Conference and International Meeting, Minneapolis, Minnesota, June 18-22. American Journal of Infection Control, vol. 28, no. 3, June

Greene, L., A. Chodoff, G. Hollick, and J. Schlesinger. 2000. Evidence forcommunity transmission of methicillin-resistant staphylococcus aureus with a propensity for reduced vancomycin susceptibility. Abstract of poster presented at the APIC 27th Annual Educational Conference and International Meeting, Minneapolis, Minnesota, June 18-22. American Journal of Infection Control, vol. 28, no. 3, June

Harris,A.D., M.H. Samore, R. Nafziger, K. DiRosario, M.C. Roghmann, and Y. Carmeli. 2000. Survey on handwashing practices and opinions of healthcare workers in two Boston hospitals. Abstract of poster presented at the 4th Decennial International Conference on Nosocomial and Healthcare Associated infections. Atlanta, Georgia, March 5-9. American Journal of Infection Control, vol. 28, no.1, February

Hoffmann, K., K. Winstead, M. Edmundson, D. Weber and W. Rutula. 2000. Comparison of a novel triclosan hand protectant versus handwashing alone for reducing transient hand bacterial colonization and nosocomial infections. Abstract of poster presented at the APIC 27th Annual Educational Conference and International Meeting, Minneapolis, Minnesota, June 18-22, American Journal of Infection Control, vol. 28, no. 3, June

Jones, R.D. 1999. Bacteria resistance and topical antimicrobial wash products. American Journal of Infection Control, vol. 27, no.4, August, pp. 351-363.

Jones, R.D., H.B. Jampani, J.L. Newman, and A.S. Lee. 2000. Triclosan: A review of effectiveness and safety in health care settings. American Journal of Infection Control, vol. 28, no. 2, April, pp. 184-196.

Kassis, I., I. Makhoul, T. Smolkin, A. Tamir, and P. Sujov. 2000. Neonatal nosocomial candid infections, review of 49 cases in 10 years. Abstract of poster presented at the 4th Decennial International Conference on Nosocomial and Healthcare Associated infections. Atlanta, Georgia, March 5-9, 2000. American Journal of Infection Control, vol. 28, no.1, February

Lai, K.K., S.A. Fontecchio, A.L.Kelley, and Z.S. Melvin. 2000. The changing epidemiology of vancomycin-resistant enterococci. Abstract of poster presented at the 4th Decennial International Conference on Nosocomial and Healthcare Associated infections. Atlanta, Georgia, March 5-9. American Journal of Infection Control, vol. 28, no.1, February

Lisay, C.M., M.J. Brady, D.A. Hale, J.F. Hamberger, S. Garib, G. Manivannan, F. Liu, A. Yurkovetskiy, S. Subramanyam, and S.P. Sawan. 2000. A comparative evaluation of the residual antimicrobial activity of disinfectant products. Abstract of poster presented at the 4th Decennial International Conference on Nosocomial and Healthcare Associated infections. Atlanta, Georgia, March 5-9, American Journal of Infection Control, vol. 28, no.1, February.

Managan, L.P., M. Schantz, M.L. Pearson, J. Taylor, K.E. Greico, K. Stewart, S. Patel, N.A. Mychalak, T.T. Brown, A. Lenar, and W.R. Jarvis.2000. Prevalence of infections among patients in home care. Abstract of poster presented at the 4th Decennial International Conference on Nosocomial and Healthcare Associated infections. Atlanta, Georgia, March 5-9, American Journal of Infection Control, vol. 28, no.1, February

McBain, A.J., A. Rickard, P. Gilbert. 2002. Possible implications of biocide accumulation in the environment on the prevalence of bacterial antibiotic resistance. Journal of Industrial Microbiological Biotechnology, vol.29, no. 6, December, pp.326-330.

McMurray, L.M., M. Oethinger and S.B. Levy. 1998. Triclosan targets lipid synthesis. Nature, August, pp. 531-532.

Medlin, J. 1997. Germ Warfare. Environmental Health Perspectives, vol. 105, no. 3, March, pp. 290-292.

Mermel, L.A., S. Parenteau, J. Dempsey, B. Cifelli, and E. Noteroglu. 2000. Control of nosocomial VRE at a large univ-affiliate teaching hospital. Abstract of poster presented at the 4th Decennial International Conference on Nosocomial and Healthcare Associated infections. Atlanta, Georgia, March 5-9, American Journal of Infection Control, vol. 28, no.1, February

Mylotte, J.M. 2000. Impact of nosocomial infection on length of stay and functional improvement among patients admitted to an acute rehabilitation unit. Abstract of poster presented at the 4th Decennial International Conference on Nosocomial and Healthcare Associated infections. Atlanta, Georgia, March 5-9. American Journal of Infection Control, vol. 28, no.1, February

Noskin, G.A., P. Bednarz, T. Suriano, S. Reiner, and L.R. Peterson. 2000. Persistane contamination of fabric covered furniture by vancomycin-resistant enterococci: implications for upholstery selection in hospitals. Abstract of poster presented at the 4th Decennial International Conference on Nosocomial and Healthcare Associated infections. Atlanta, Georgia, March 5-9. American Journal of Infection Control, vol. 28, no.1, February.

Ober, J.F., G. Hall, M. Hodson, C. Blythe, L. Reynolds, and M. B. Edmond.2000. Hurrican-induced fungal contamination of the hospital environment. Abstract of poster presented at the 4th Decennial

International Conference on Nosocomial and Healthcare Associated infections. Atlanta, Georgia, March 5-9. American Journal of Infection Control, vol. 28, no.1, February

OMH. 2004. Diseases: Severe Acute Respiratory Syndrome (SARS). April Fact Sheet, Ontario Ministry of Health and Long-Term Care.

Pentella, M.A., T. Fisher, S. Chandler, T. Britt-Ohrmund, B.H. Kwa and B.G. Yangco. 2000. Are disinfectants accurately prepared for use in hospital patient care areas. Abstract of poster presented at the 4th Decennial International Conference on Nosocomial and Healthcare Associated infections. Atlanta, Georgia, March 5-9. American Journal of Infection Control, vol. 28, no.1, February

PHTS. 2000. Controlling Biological and Chemical Exposures During Facility Renovation. Educational Program Seminar Report, Palmetto Hospital Trust Services Education Institute, Columbia, SC, November 1.

Raad, I.I., C. Osting, H. Hanna, R. Hachem, J. Umphrey, J. Tarrand and H. Kantarjian. 2000. Masking of nutropenic patients upon transport from patient rooms as associated with decrease in nosocomial aspergillosis during construction. Abstract of poster presented at the 4th Decennial International Conference on Nosocomial and Healthcare Associated infections. Atlanta, Georgia, March 5-9. American Journal of Infection Control, vol. 28, no.1, February

Rebmann, T., J. Mayfield, A. Spizzo, J. DiPersio, S. Johnson, J. Danner, and V. Fraser. 2000. Effective interventions halt outbreak of construction related aspergillosis on a bone marrow transplant unit. Abstract of poster presented at the APIC 27th Annual Educational Conference and International Meeting, Minneapolis, Minnesota, June 18-22, American Journal of Infection Control, vol. 28, no. 3, June.

Risa, K.J., C.Brennen, and R.R. Muder. 2000. Nosocomial infection rates in a VA psychiatric facility. Abstract of poster presented at the APIC 27th Annual Educational Conference and International Meeting, Minneapolis, Minnesota, June 18-22, American Journal of Infection Control, vol. 28, no. 3, June.

Russell, A.D. 2002. Biocides and pharmacologically active drugs as residues and in the environment: Is there a correlation with antibiotic resistance? American Journal of Infection Control, vol. 30, no. 8, December, pp. 495-498

Rutula, W.A., M.F. Gergen, and D.J. Weber. 2000. Antimicrobial. Evaluation of a new surface germicide with antimicrobial persistence. Abstract of poster presented at the APIC 27th Annual Educational Conference and International Meeting, Minneapolis, Minnesota, June 18-22. American Journal of Infection Control, vol. 28, no. 3, June

Rutala, W.A., D.J. Weber and M.F. Gergen. 2000. Vancomycin-resistant enterococcus sp surface disinfection transmissibility via contaminated surfaces. Abstract of poster presented at the 4th Decennial International Conference on Nosocomial and Healthcare Associated infections. Atlanta, Georgia, March 5-9. American Journal of Infection Control, vol. 28, no.1, February

Temple, R.S., W.E. Trick, D. Chen, M.G. Lankford, and D.M. Hacek. 2000. Environmental contamination with vancomycin-resistant enterocci in a rehabilitation hospital. American Journal of Infection Control, vol. 28, no.3, June

Tessarin, M., R. Rigoli, G. Scotton, M. Dametto, R. Ramon, F. Pietrobon, C. Bertic, S. Morciano, and M. Niero. 2000. Abstract of poster presented at the 4th Decennial International Conference on Nosocomial and Healthcare Associated infections. Atlanta, Georgia, March 5-9, American Journal of Infection Control, vol. 28, no.1, February

Valena, F.D., A.V. Simmons, S.M. Smith, O. Badahman, I. Surendran, and R.H.K. Eng. 2000. Are computer terminal keyboards/mice a methicillin-resistant staphylococcus aureus fomite? Abstract of poster presented at the APIC 27th Annual Educational Conference and International Meeting, Minneapolis, Minnesota, June 18-22, American Journal of Infection Control, vol. 28, no. 3, June

Van Enk, R.A. and W.L. Lam. 2000. Effect of a triclosan-containing surface on the survival of nosocomial pathogens on hospital overbed tables. Abstract of poster presented at the APIC 27th Annual Educational Conference and International Meeting, Minneapolis, Minnesota, June 18-22, American Journal of Infection Control, vol. 28, no. 3, June.

Volavka, M. 2005. Report of the Pennsylvania Health Care Cost Containment Council.

Weistein, R.A. 1998. Nosocomial Infection Update. Special Issue, Journal of Emerging Infectious Diseases. Vol. 4, no. 3, July-September. pp. 1-6

Young, J. M., Naqvi, M. and Richards L. 2005. Microbial contamination of hospital bed handsets. American Journal of Infection Control, vol. 33, No. 3

Zhan, C. and Miller, M.R. 2003. Excess length of stay, charges and mortality attributable to medical injuries during hospitalization. JAMA, vol. 290, pp.1868-74

Zoutman, R.E., Ford, B.B., Bryce, E., Gourdeau, M., Hebert, G., Henderson, E. and Paton, S. 2003. The state of infection surveillance and control in Canadian acute care hospitals. American Journal of Infection Control, vol. 31, No. 5

COMPREHENSIVE STRATEGY FOR HSC BEST PERFORMANCE IN EXTENSIVE APPLICATIONS OF LANDMARK WORKS IN ITALY

A. Migliacci[1], P. Ronca[1], P. Crespi[1] and G. Franchi[2]
[1]*Structural Engineering Department, Politecnico of Milan, Milan, Italy*
[2]*AMiS-Structural Engineering Office, Milan, Italy*

Abstract

Centering on worldwidly present urban areas, there have been many high-rise landmark buildings constructed in recent years. It is recognized that reinforced concrete has merit over steel frame construction in high-rise buildings, such as less sway in high winds, better human life protection in case of accidental heavy damage, better noise resistance. The use of high-strength concrete is rising, not only for pillars, in high-rise buildings. The paper points out on the need of classifying the HP-HSC for the different requested characteristic that materials have to exhibit on different structural elements of a complex structure. Among types of concrete, which binds together characteristics of High Strength Concrete (HSC) and High Performance Concrete (HPC), particular reference is made to Limestone Concrete (LSC).

Existing literature provides data on self-levelling, high performance, rapid hardening concrete, able to reach in few days the standard of HPC (Kelham, 1998; Montgomery et al., 1998; Nehdi et al., 1998). In particular the technology here referred for limestone concrete is not the usual one, but it makes reference to a mix design, characterized by an industrially produced limestone aggregates, with total absence of Silica Fume or any other addition of pozzolanic material or accelerating admixture (Cangiano, 2005; Cangiano et al., 2004).

The paper points out the significance of Limestone Concrete, as High Performance Concrete, application, starting from the following key construction requirements: in large public works with characteristic of very high durability, the choice of a technical solution it is not at all dependent on the construction cost only. In fact in this work, life service and safety performances, that slightly increase the construction costs, are of paramount importance. Starting from this key assumption, new materials, and in particular new concretes, may be able to notably cut life service and safety costs, considerably improving the performance/cost ratio of the selected solution, due to the large cut of maintenance costs. The paper wants to briefly explain the state of the art and the today frontier which lead to the material basic choices in structural design of high-rise buildings. In particular the paper refers to a comprehensive campaign of tests, in a starting-up phase, shared among different university and private laboratories in Italy, which aims to draw Guide Lines for different specific uses of Limestone Concrete, as HPC, in different structures typologies and environmental conditions.

Introduction

Italy, like other countries in Europe, has experienced high rise building constructions with significant delay, respect to different countries all over the world. Reasons due to architectural heritage, cultural and educational schools in architecture may be the principal sources of delay. The construction of high rise buildings as new land-mark of our traditional and world-wide known cities is still a debating issue. Nevertheless, or because of that, and due to some episodes unfortunately experienced by important widely known buildings, the today technological background about the best requisites of advanced materials, as well as methods of structural analysis and design, indicates the need of significant and innovative approaches and strategies for achieving the best results for the construction

M. Pandey et al. (eds), Advances in Engineering Structures, Mechanics & Construction, 853–863.

Fig. 1. The sail building phase of the "Dives in Misericordia" Church.

Fig. 2. WTC in San Marino construction site.

of landmark buildings, like the high-rise buildings are (High-Rise Manual, 2003; Simiu et al., 1996; Fairweather, 2004).

The solution of the engineering project of an architecture design has to be the optimal solution among several feasible solutions, often mutually contradictory and conflicting each other. Each different architectural project has own characteristic and demands for the best solution. Recent realizations have been discussed as peculiar examples in Italy, as for example the "Dives in Misericordia" Church in Rome (Figure 1) or the main r.c. structures of the World Trade Center in S. Marino (Figure 2) .

Due to some innovative architectural solutions, the demand due to particular layout of the structural elements or due to geometrical shape and slenderness of the vaulted structure, together with particular demand for the durability of brightness, have stimulate the research toward new technological, engineering and material solutions. The best design of the so called landmark building, as the high-rise buildings mostly are, is today focused, taking advantages from previous experiences, on different main objectives. The objectives to be considered, as far as materials are concerned, are:

(a) mechanical resistance,
(b) stiffness,
(c) ductility and toughness,
(d) durability,
(e) fire resistance.

In addition the objectives to be considered in the construction site are:

(a) simplicity and easiness of the construction processes,
(b) construction immediate costs,
(c) construction time and feasibility to respect the time schedule,
(d) site job environmental impact,
(e) historical and monumental existing constraints.

The new design philosophy for landmark long-lasting architectural buildings has to take account also the objectives for the service life, from which the most significant are:

(a) construction flexibility,
(b) low maintenance costs,
(c) assurance of high safety level (human life safeguards),
(d) environmental sustainability,
(e) urban costs and constraints.

The strategy of choosing the solution has to consider all the costs among the immediate construction, the life service and the safety costs.

As it will be briefly shown in the next section, among different parameters of the optimized solution, the chosen main structural material has a significant role in improving the performances/cost ratio, due to the cut of life service costs and to further advantages, like smaller dimensions of structures, reduction of construction time.

Main Structural Material Performances for New High-Rise Buildings

Some background considerations, from literature and from direct experiences here reported (Namiki, 2005), states that the improved new generation of High Performance Concrete (HPC) may lead the concrete to be highly competitive in respect to different materials, if the material performance is optimized taking account the response of the structure under the following actions:

(a) the structural behavior under wind actions,
(b) the structural behavior under explosion and impact actions,
(c) the behavior under cyclic and/or alternate loadings,
(d) the shrinkage and creep behavior.

Recent investigations comparing the different performances of various building materials can be summarized in Table 1, where the signs +, 0, – indicate respectively good, neutral, bad performance. High Strength Concrete (HSC) shows positive performances respect to the entire spectrum of the chosen criteria (Chew Yit Lin, 2003; High-Rise Manual, 2003; Taranth, 1988).

Table 1. Comparison on different materials frame construction.

Criteria	Reinforced concrete, Normal-strength concrete	Reinforced concrete High-strength concrete	Steel construction	Composite construction method
Construction costs	+	++	0	++
Weight of construction	0	+	++	+
Stiffness	++	++	0	+
Flexibility of plan	0	0	++	+
Behavior in fire	++	++	–	+
Construction time	–	+	++	++
Usable area	–	+	++	+
SCORE	5	9	7	9

Taking moreover a particular look, for example, to the specific criteria identified as "construction cost" and "construction time", Tables 2 and 3 show significative data (High-Rise Manual, 2003). Table 2 indicates sharp favourite performances by using HSC for the structures, not only considering the cost, but even from the point of view of the usable area needed by the structure. Table 3 is self explaining if we consider that HPC, beside improved mechanical properties, have the further relevant feature: formworks can be taken away just 24 hours after the pouring the concrete, while dealing with NSC (Normal Strength Concrete) at least 4–5 days are required. In Table 3 it is easy to observe that, while 13 working days were required to complete the rough work for a standard floor during the construction of the Dresden Bank high-rise in 1974–1979 (obsolete concrete technology + NSC), the same task is currently achieved in mere 4 working days on the Galileo site (modern concrete technology + HPC) (High-Rise Manual, 2003; The Concrete Society, 1997; Fairweather, 2004). The immediate cost construction are primarily related to the time construction. Tough steel structure is traditionally reckoned to be much faster than the concrete structure, but because of significant advantages in concrete technology, this is no more true: in fact, with concrete showing features of self-levelling (SCC), the new pumps equipment assures concrete puring even at heights of 300 m, and also because of modern formwork systems (self-climbing formwork), rapid and safe progress in the rough work is guaranteed.

Research studies carried out, and still in progress, in the field of mix-design of HSC/HPC, and related technology in the construction sites, seem to show favourable results in achieving better performances, i.e. rapid hardening, absence of segregation, better durability (no alcaly-silica reaction), when limestone is used as filler, as it will show in the next section.

It is worth to recall, as mentioned before, due to even the consequences of recent past accidents, that the engineering design of a land-mark buildings can not ignore the possible blast loading effects. As it is well known, the physical action on a wall F, due to a plane shock wave, generates an overpressure Ps and a drag loading P_d, according to the scheme of Figures 3a and 3b.

When an explosion occurs inside a building, then it is the interior surface of the walls and ceiling which are first loaded by the pressure of the shock wave, reflecting therefore and increasing the pressure. The effects on the structure may be devastating, considering in addition that, as consequence of an explosion, even fire may occur. Figure 4 shows the accident occurred in spring 2002 at the Pirelli building, headquarter of the Regional Government in Milan. The explosion of the fuel tanks of the aircraft, among other consequences, caused a permanent deformation of the 26th r.c. floor with a

Table 2. Construction costs and usable area for different structural material.

		Column	Relative Cost
1.	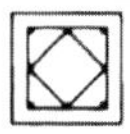	60-MPa (8600-psi) reinforced concrete 840 × 840 6Y24 [33-in² 6–1-in. bars] AS3600 ties (R10–360)	1.0
2.	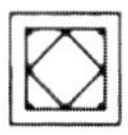	120-MPa (17,300-psi) reinforced concrete 600 × 600 8Y24 [24-in² 8–1-in. bars] crawley (1989) ties (Y32–200)	0.79
3.	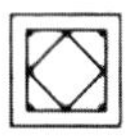	120-MPa (17,300-psi) reinforced concrete 660 × 660 8Y24 [26-in² 8–1-in. bars] AS3600 ties (R10-360)	0.77
4.		60-MPa (8,600-psi) concrete in grade 250 steel tube 740 × 8 CHS [23-in. dia × $\frac{3}{8}$ in. thick]	0.98
5.		120-MPa (17,300-psi) concrete in grade 250 steel tube 570 × 8 CHS (23-in. dia × $\frac{5}{16}$ in. thick)	0.71
6.		steel column grade 350 600 × 40 flanges [24-in. × $1\frac{1}{2}$-in.] 520 × 40 web [21-in. × $1\frac{1}{2}$-in.]	2.21

Usable Area: reduction of column cross section and compressive reinforcement by increasing the grade of concrete

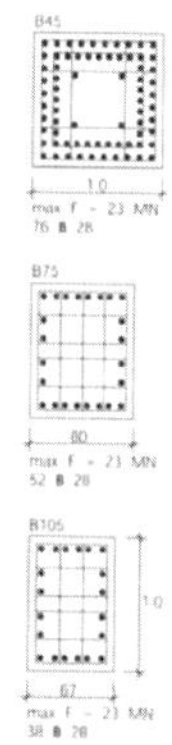

Table 3. Construction time – HPC/HSC versus steel.

Property	Height	Completion	Rough Work per Standard Floor
Business Research Center, Warsaw, Poland	104 m	2000	5 working days
Taunustor Japan Center, Frankfurt, Germany	114 m	1996	4 working days
World Port Center, Rotterdam, The Netherlands	125 m	2001	5 working days
Galileo, Frankfurt, Germany	136 m	2003	4 working days
Dresdner Bank, Frankfurt, Germany	166 m	1979	13 working days
Trianon, Frankfurt, Germany	186 m	1993	5,5 working days
Millennium Tower, Vienna, Austria	202 m	1999	3 working days
Park Tower, Chicago, USA	257 m	2000	3 working days
Trump World Tower, New York, USA	269 m	2001	5 working days
Petronas Tower, Kuala Lumpur, Malaysia	452 m	1998	5 working days

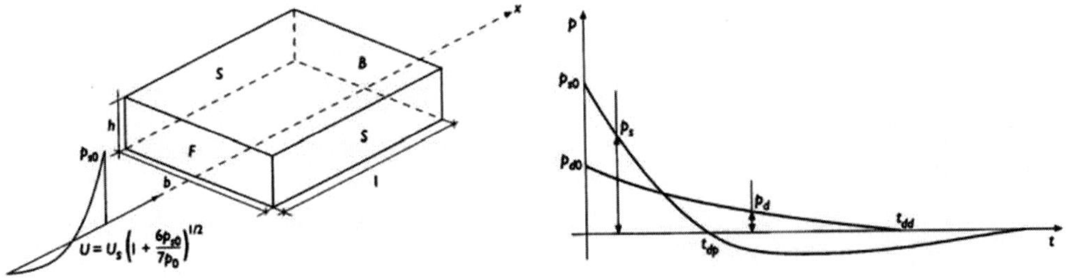

(a) (b)

Fig. 3. Overpressure due to plane shock wave action.

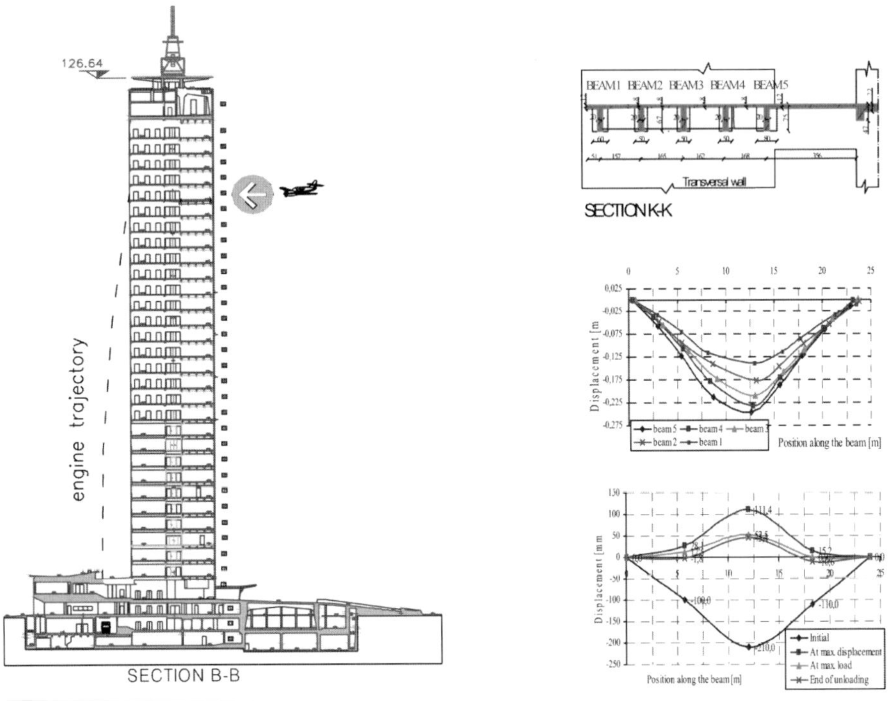

Fig. 4. Draw of the airplane impact and the permanent floor deflection.

deflection of 22 cm, but not the collapse, despite the high temperature due to fire (Migliacci et al., 2005; Kappos, 2002).

The stiffness of the structure is very significant in considering the horizontal loads, i.e. earthquake and wind actions.

- Earthquake effects: stell constructions are highly suitable in areas subjected to earthquake as steel allows the structure to absorb part of the kinetic energy produced by the earthquake in the form of plastic deformation. Technically a similar ductile behaviour can be achieved by using for the structure HSC and reinforcement steel with high strength and ductility.
- Wind effects: reinforced concrete as the merit over steel frame construction in high rise buildings to present a less sway wind. The structural behaviour minimizing the wind effect is at best achieved by concrete displaying a high values of Young modulus.

Fig. 5. Plastic model of "City Life" project in Milan. .

Fig. 6. Plastic model of "Garibaldi-Repubblica" project in Milan.

The points so far briefly recalled seem to highly recommend the choice of using reinforced concrete structure in the engineering design for a high-rise building, with particular attention to the best suitable mix design of HSC/HPC for each different action and structural component of the building.

For these reasons, a comprehensive study on best performing HPC mix design, with particular reference to Limestone Concrete, have been requested, with the purpose to draw Guidelines for use, in the oncoming starting up design phase of two important land mark interventions in the city of Milano. Figures 5 and 6 show the plastic model of the two projects, respectively the so-called "City Life" in the former trade fair area, and the so-called "Garibaldi-Repubblica" area, both in the central part of the town.

Mechanical Characterization of HPC with Limestone Filler

Technical literature on new generation of HS/HPC is available since a decade or more, with several examples of structural applications (Aitcin, 1999; Rols et al., 1999; Person, 1998; Toutanji et al., 1995; Rosati, 1999; Guerrini et al., 1999; Namiki, 2005; Nawy, 2001). The most significant mechanical parameters of the today well known HPC, refer to strength giving less attention to elastic modulus, or other characteristics related more to workability and durability. Because many characteristics of high performance concretes are interrelated, a change in one usually results in changes in one or more of the other characteristics. Consequently, if several characteristics have to be taken into account in producing a concrete for the intended application, each of these characteristics must be clearly specified in the contract documents. That is why the design documents of the two mentioned interventions should refer to a Guideline documents based on experimental investigations

Table 4. some significant figures of Limestone Concrete samples.

Time (Days)	Compressive strength (MPa)	Flexural tensile strength (MPa)	Dynamic elastic modulus (MPa)
1	80	10.8	43,491
2	91	15.2	44,900
7	103	18.9	46,000
28	118	20.6	48,100

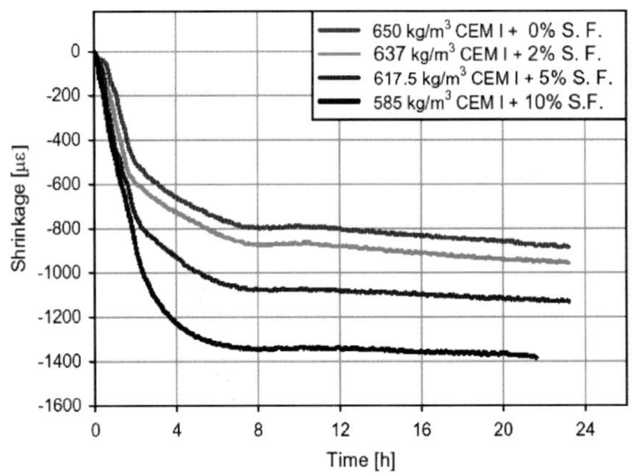

Fig. 7. Technical problems related to silica fume utilisation: plastic shrinkage (constant water/binder ratio) (Cangiano, 2005).

on different mix design best suited for different structural elements of a complex whole building, like are, for examples, underground retaining elements, underground elements in presence of water, floor slab elements, linear horizontal elements, vertical "core" elements. In particular the paper wants to introduce, as mentioned before, the campaign of tests aiming to produce large and well defined spectrum of mix design for particular HS/HPC based on fine aggregate of limestone. LSC's are classified as HS-HP concretes, with the advantages of being characterized by the complete absence of pozzolanic addition, i.e. silica fume, generally present as filling material in practically all type of HPC (De Larrard, 1993; Toutanji, 1995). Table 4 shows some first results of the mean significant mechanical characteristics of Limestone Concrete, and allows us to include Limestone Concrete in the range of HS/HPC, with advantages of being Self-Levelling and Rapid Hardening Concrete (Plizzari et al., 2003).

It is however recognized that the following problems are related to silica fume utilization: (a) the high cost (about 4–5 times the cost of cement), (b) the total shrinkage (plastic and hydraulic) of HPC containing silica fume (10%) may be greater than other HSC based on other mineral admixtures, (c) HPC with silica fume may exhibit a high tendency to desiccation and hence to early micro-cracking, as a consequence the long term durability may suffer, (d) several researchers found a loss of compressive strength between 90 days and 4 years of concrete with silica fume. Figure 7 underline the problems related to shrinkage behaviour when utilizing silica fume and concrete mix designs according to European Code (EN No. 197).

The main task of the experimental research on new generation of HSC using limestone (without any content of silica fume), is to identify a size distribution curve of the system composed by cement and aggregates in order to produce HPC's characterized by: (a) very good rheological properties and

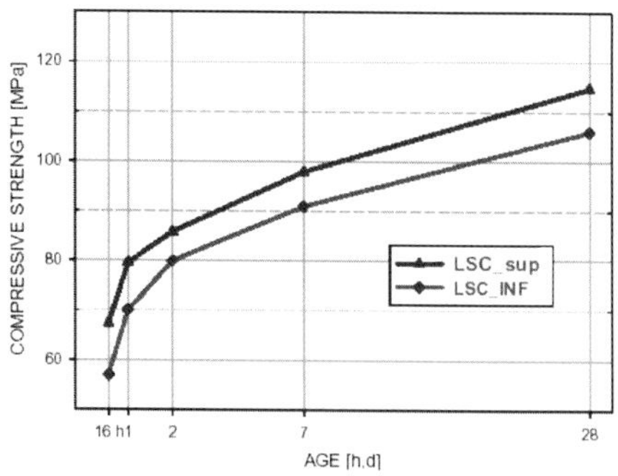

Age	Compressive strength [MPa]	
	LSC_SUP	LSC_INF
16 h	67,3	57,5
1 d	79,6	70,3
2 d	85,8	82,7
7 d	97,9	92,4
28 d	114,7	105,2

Fig. 8. Strength development of two different mix of LSC.

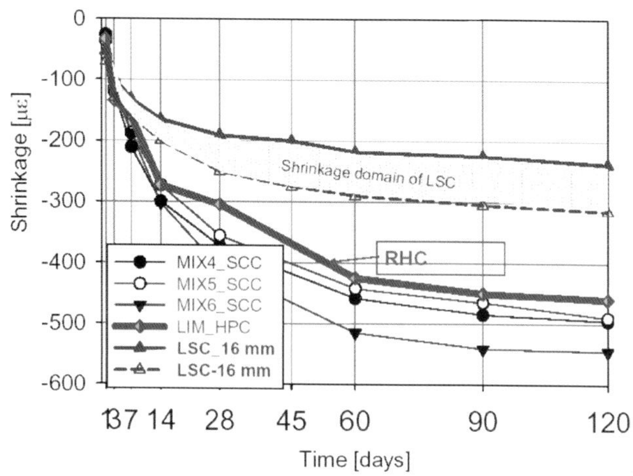

Fig. 9. The favourable performance of LSC respect to hydraulic shrinkage.

(b) rapid development of mechanical performances. In this starting-up phase of the experimental research, just few mix have been proposed and tested with the primar goal to prove the sensibility of the mechanical characteristic to small variations of size aggregates in the early stage.

The graphs and table of Figure 8 show the upper and lower bound of the strength development performance due to slightly different mix, where it is clearly show the rapid hardening characteristic of Limestone Concrete mix.

The favorable performance of LSC respect to shrinkage behavior is summarized in the graphs of Figure 9 where LSC is compaired with different mixes of SCC.

Further development of the research will be focused on obtaining LSC with specific characteristic for the specific used, identified as follows:

- for beams, pillars and floor-slabs → LSC with properties of HSC and RHC;
- for "core" structure → LSC with high resistance to temperature and with high toughness;
- for conteining walls → LSC with properties of durability against salts, very low permeability, low hygrometric shrinkage.

In any case Limestone Concretes, also fiber added, should develop creep characteristics as good as those of SCC.

Conclusions

The paper starts from the assumption that the use of high-strength, high-performance concrete (HS-HPC) is rising up, not only for pillars, in high-rise buildings. The research program undertaken by Politecnico di Milano-Universities of Bergamo and Brescia, which starting-up phase results are here reported, is primarily focused on Limestone Concrete samples tests, to state the mechanical characteristics requested not only by new codes, but also by new building design philosophy in case of high-rise buildings. As a matter of fact, the research on the scaled structural elements (wind test), or/and on numerical models simulating structural elements under impact or blast loading condition will be a necessary completion in the overall knowledge for the optimization of strength and performance capacity of the new HPCŠs and their selected use. The complete range of knowledge may produce a material highly competitive for the bearing structure of the high-rise buildings.

Acknowledgements

The authors acknowledge the researcher team of CTG-Italcementi Group, with particular reference to dott. L. Cassar and eng. S. Cangiano, for their comments regarding this paper.

References

Aitcin, P.C., 1999, *High Performance Concrete*, E&FN SPON.

Cangiano, S., 2005, "The LimeStone Concrete-LSC", CTG Italcementi Group-Scientific Committee, Bergamo, Italy, 14 April.

Cangiano, S. and Plizzari, G., 2004, "The rapid hardening concrete: a new perspective in bridge construction", *Concrete Bridge Conference*, North Carolina, 17–18 May.;

Chew Yit Lin, M., 2003, *Construction Technology for Tall Building*, Singapore University Press.

Concrete Society, 1997, *Proceedings of the Third International Conference "Conquest of Vertical Space in the 21th Century"*, organized by the Concrete Society, London, 7–10 October.

de Larrard, F. and C. Aitcin, P.C., 1993, "Apparent strength retrogression of silica-fume concrete", *ACI Materials Journal*, 90(6).

Eisele, J. and Kloft, E. (eds), 2003, *High-Rise Manual – Typology and Design, Construction and Technology*.

Fairweather, V., 2004, *Expressing Structure – The Technology of Large-Scale Buildings*, Birkhäuser – Publisher for Architecture.

Guerrini, G., Biolzi, L., Cassar, L. and Rosati, G., 1999, "Production and mechanical characterization of very high fibre-reinforced concrete beams", *International Conference "Creating with Concrete"*, Dundee, Scotland, UK, 6–10 September.

Kappos, A.J., 2002, *Dynamic Loading and Design of Structures*, Spon Press.

Kelham, S., 1998, "Portland Limestone Cements", *Concrete (London)*, 32(5), May.

Migliacci, A., Franchi, A., Acito, M. and Crespi, P., 2005, "Analytical and experimental procedures for the realigning of the 26th floor of the 'Pirelli' tall building after the airplane crash on April 2002", presented at the *IASBE Conference*, Lisbon, 14–16 September.

Montgomery, D.G., Van, B.K., Hinczak, I. and Turner, K., 1998, "Limestone modified cement for high-performance concretes", *Fly Ash, Silica Fume, Slag and Natural Pozzolanans in Concrete, Sixth CANMET/ACI/JCI Conference*, Bangkok.

Namiki, S., 2005, "State of the art and future of high-strength concrete technology", *Cement and Concrete*, 69j, January.

Nawy, E.G., 2001, *Fundamentals of High-Performance Concrete*, John Wiley.

Nehdi, M., Mindness S. and Aitcin, P.C., 1998, "Rheology of high-performance concrete: Effect of ultrafine particles", *Cement and Concrete Research*, 28 May 1998.

Person, B., 1998, "Seven-year study on the effect of silica fume in concrete", *Advanced Cement Based Materials*, 7.

Plizzari, G. and Cangiano, S., 2003, "HPC in structural applications and in building codes", *Celebrating Concrete: People and Practice*, Dundee, Scotland, UK, 3–4 September.

Rols, S. et al., 1999, "High performance concrete", in *Proceedings Second CANMET/ACI International Conference*, Gramado, Brazil.

Rosati, G., 1999, "High performance concrete applications in precast and prestressed concrete bridge slabs", *HPFRCC*, Mainz, Germany, 16–19 May.

Simiu, E. and Scanlan, R.H., 1996, *Wind Effects on Structures – Fundamentals and Applications to Design*, John Wiley & Sons.

Taranth, B.S., 1988, *Steel, Concrete and Composite Design of Tall Building*, McGraw-Hill.

Toutanji, H.A. and El Korki, T., 1996, "The influence of silica fume on the compressive strength of cement paste and mortar", *Cement and Concrete Research*, 25.

AUTHORS INDEX

Mechanics

SOLID MECHANICS AND ITS APPLICATIONS

Series Editor: G.M.L. Gladwell

Aims and Scope of the Series

The fundamental questions arising in mechanics are: *Why?*, *How?*, and *How much?* The aim of this series is to provide lucid accounts written by authoritative researchers giving vision and insight in answering these questions on the subject of mechanics as it relates to solids. The scope of the series covers the entire spectrum of solid mechanics. Thus it includes the foundation of mechanics; variational formulations; computational mechanics; statics, kinematics and dynamics of rigid and elastic bodies; vibrations of solids and structures; dynamical systems and chaos; the theories of elasticity, plasticity and viscoelasticity; composite materials; rods, beams, shells and membranes; structural control and stability; soils, rocks and geomechanics; fracture; tribology; experimental mechanics; biomechanics and machine design.

1. R.T. Haftka, Z. Gürdal and M.P. Kamat: *Elements of Structural Optimization*. 2nd rev.ed., 1990 ISBN 0-7923-0608-2
2. J.J. Kalker: *Three-Dimensional Elastic Bodies in Rolling Contact*. 1990 ISBN 0-7923-0712-7
3. P. Karasudhi: *Foundations of Solid Mechanics*. 1991 ISBN 0-7923-0772-0
4. *Not published*
5. *Not published.*
6. J.F. Doyle: *Static and Dynamic Analysis of Structures*. With an Emphasis on Mechanics and Computer Matrix Methods. 1991 ISBN 0-7923-1124-8; Pb 0-7923-1208-2
7. O.O. Ochoa and J.N. Reddy: *Finite Element Analysis of Composite Laminates*. ISBN 0-7923-1125-6
8. M.H. Aliabadi and D.P. Rooke: *Numerical Fracture Mechanics*. ISBN 0-7923-1175-2
9. J. Angeles and C.S. López-Cajún: *Optimization of Cam Mechanisms*. 1991 ISBN 0-7923-1355-0
10. D.E. Grierson, A. Franchi and P. Riva (eds.): *Progress in Structural Engineering*. 1991 ISBN 0-7923-1396-8
11. R.T. Haftka and Z. Gürdal: *Elements of Structural Optimization*. 3rd rev. and exp. ed. 1992 ISBN 0-7923-1504-9; Pb 0-7923-1505-7
12. J.R. Barber: *Elasticity*. 1992 ISBN 0-7923-1609-6; Pb 0-7923-1610-X
13. H.S. Tzou and G.L. Anderson (eds.): *Intelligent Structural Systems*. 1992 ISBN 0-7923-1920-6
14. E.E. Gdoutos: *Fracture Mechanics*. An Introduction. 1993 ISBN 0-7923-1932-X
15. J.P. Ward: *Solid Mechanics*. An Introduction. 1992 ISBN 0-7923-1949-4
16. M. Farshad: *Design and Analysis of Shell Structures*. 1992 ISBN 0-7923-1950-8
17. H.S. Tzou and T. Fukuda (eds.): *Precision Sensors, Actuators and Systems*. 1992 ISBN 0-7923-2015-8
18. J.R. Vinson: *The Behavior of Shells Composed of Isotropic and Composite Materials*. 1993 ISBN 0-7923-2113-8
19. H.S. Tzou: *Piezoelectric Shells*. Distributed Sensing and Control of Continua. 1993 ISBN 0-7923-2186-3
20. W. Schiehlen (ed.): *Advanced Multibody System Dynamics*. Simulation and Software Tools. 1993 ISBN 0-7923-2192-8
21. C.-W. Lee: *Vibration Analysis of Rotors*. 1993 ISBN 0-7923-2300-9
22. D.R. Smith: *An Introduction to Continuum Mechanics*. 1993 ISBN 0-7923-2454-4
23. G.M.L. Gladwell: *Inverse Problems in Scattering*. An Introduction. 1993 ISBN 0-7923-2478-1

Mechanics

SOLID MECHANICS AND ITS APPLICATIONS

Series Editor: G.M.L. Gladwell

24. G. Prathap: *The Finite Element Method in Structural Mechanics*. 1993 ISBN 0-7923-2492-7
25. J. Herskovits (ed.): *Advances in Structural Optimization*. 1995 ISBN 0-7923-2510-9
26. M.A. González-Palacios and J. Angeles: *Cam Synthesis*. 1993 ISBN 0-7923-2536-2
27. W.S. Hall: *The Boundary Element Method*. 1993 ISBN 0-7923-2580-X
28. J. Angeles, G. Hommel and P. Kovács (eds.): *Computational Kinematics*. 1993 ISBN 0-7923-2585-0
29. A. Curnier: *Computational Methods in Solid Mechanics*. 1994 ISBN 0-7923-2761-6
30. D.A. Hills and D. Nowell: *Mechanics of Fretting Fatigue*. 1994 ISBN 0-7923-2866-3
31. B. Tabarrok and F.P.J. Rimrott: *Variational Methods and Complementary Formulations in Dynamics*. 1994 ISBN 0-7923-2923-6
32. E.H. Dowell (ed.), E.F. Crawley, H.C. Curtiss Jr., D.A. Peters, R. H. Scanlan and F. Sisto: *A Modern Course in Aeroelasticity*. Third Revised and Enlarged Edition. 1995 ISBN 0-7923-2788-8; Pb: 0-7923-2789-6
33. A. Preumont: *Random Vibration and Spectral Analysis*. 1994 ISBN 0-7923-3036-6
34. J.N. Reddy (ed.): *Mechanics of Composite Materials*. Selected works of Nicholas J. Pagano. 1994 ISBN 0-7923-3041-2
35. A.P.S. Selvadurai (ed.): *Mechanics of Poroelastic Media*. 1996 ISBN 0-7923-3329-2
36. Z. Mróz, D. Weichert, S. Dorosz (eds.): *Inelastic Behaviour of Structures under Variable Loads*. 1995 ISBN 0-7923-3397-7
37. R. Pyrz (ed.): *IUTAM Symposium on Microstructure-Property Interactions in Composite Materials*. Proceedings of the IUTAM Symposium held in Aalborg, Denmark. 1995 ISBN 0-7923-3427-2
38. M.I. Friswell and J.E. Mottershead: *Finite Element Model Updating in Structural Dynamics*. 1995 ISBN 0-7923-3431-0
39. D.F. Parker and A.H. England (eds.): *IUTAM Symposium on Anisotropy, Inhomogeneity and Nonlinearity in Solid Mechanics*. Proceedings of the IUTAM Symposium held in Nottingham, U.K. 1995 ISBN 0-7923-3594-5
40. J.-P. Merlet and B. Ravani (eds.): *Computational Kinematics '95*. 1995 ISBN 0-7923-3673-9
41. L.P. Lebedev, I.I. Vorovich and G.M.L. Gladwell: *Functional Analysis*. Applications in Mechanics and Inverse Problems. 1996 ISBN 0-7923-3849-9
42. J. Menčik: *Mechanics of Components with Treated or Coated Surfaces*. 1996 ISBN 0-7923-3700-X
43. D. Bestle and W. Schiehlen (eds.): *IUTAM Symposium on Optimization of Mechanical Systems*. Proceedings of the IUTAM Symposium held in Stuttgart, Germany. 1996 ISBN 0-7923-3830-8
44. D.A. Hills, P.A. Kelly, D.N. Dai and A.M. Korsunsky: *Solution of Crack Problems*. The Distributed Dislocation Technique. 1996 ISBN 0-7923-3848-0
45. V.A. Squire, R.J. Hosking, A.D. Kerr and P.J. Langhorne: *Moving Loads on Ice Plates*. 1996 ISBN 0-7923-3953-3
46. A. Pineau and A. Zaoui (eds.): *IUTAM Symposium on Micromechanics of Plasticity and Damage of Multiphase Materials*. Proceedings of the IUTAM Symposium held in Sèvres, Paris, France. 1996 ISBN 0-7923-4188-0
47. A. Naess and S. Krenk (eds.): *IUTAM Symposium on Advances in Nonlinear Stochastic Mechanics*. Proceedings of the IUTAM Symposium held in Trondheim, Norway. 1996 ISBN 0-7923-4193-7
48. D. Ieşan and A. Scalia: *Thermoelastic Deformations*. 1996 ISBN 0-7923-4230-5

Mechanics

SOLID MECHANICS AND ITS APPLICATIONS

Series Editor: G.M.L. Gladwell

49. J.R. Willis (ed.): *IUTAM Symposium on Nonlinear Analysis of Fracture*. Proceedings of the IUTAM Symposium held in Cambridge, U.K. 1997 ISBN 0-7923-4378-6
50. A. Preumont: *Vibration Control of Active Structures*. An Introduction. 1997 ISBN 0-7923-4392-1
51. G.P. Cherepanov: *Methods of Fracture Mechanics: Solid Matter Physics*. 1997 ISBN 0-7923-4408-1
52. D.H. van Campen (ed.): *IUTAM Symposium on Interaction between Dynamics and Control in Advanced Mechanical Systems*. Proceedings of the IUTAM Symposium held in Eindhoven, The Netherlands. 1997 ISBN 0-7923-4429-4
53. N.A. Fleck and A.C.F. Cocks (eds.): *IUTAM Symposium on Mechanics of Granular and Porous Materials*. Proceedings of the IUTAM Symposium held in Cambridge, U.K. 1997 ISBN 0-7923-4553-3
54. J. Roorda and N.K. Srivastava (eds.): *Trends in Structural Mechanics*. Theory, Practice, Education. 1997 ISBN 0-7923-4603-3
55. Yu.A. Mitropolskii and N. Van Dao: *Applied Asymptotic Methods in Nonlinear Oscillations*. 1997 ISBN 0-7923-4605-X
56. C. Guedes Soares (ed.): *Probabilistic Methods for Structural Design*. 1997 ISBN 0-7923-4670-X
57. D. François, A. Pineau and A. Zaoui: *Mechanical Behaviour of Materials*. Volume I: Elasticity and Plasticity. 1998 ISBN 0-7923-4894-X
58. D. François, A. Pineau and A. Zaoui: *Mechanical Behaviour of Materials*. Volume II: Viscoplasticity, Damage, Fracture and Contact Mechanics. 1998 ISBN 0-7923-4895-8
59. L.T. Tenek and J. Argyris: *Finite Element Analysis for Composite Structures*. 1998 ISBN 0-7923-4899-0
60. Y.A. Bahei-El-Din and G.J. Dvorak (eds.): *IUTAM Symposium on Transformation Problems in Composite and Active Materials*. Proceedings of the IUTAM Symposium held in Cairo, Egypt. 1998 ISBN 0-7923-5122-3
61. I.G. Goryacheva: *Contact Mechanics in Tribology*. 1998 ISBN 0-7923-5257-2
62. O.T. Bruhns and E. Stein (eds.): *IUTAM Symposium on Micro- and Macrostructural Aspects of Thermoplasticity*. Proceedings of the IUTAM Symposium held in Bochum, Germany. 1999 ISBN 0-7923-5265-3
63. F.C. Moon: *IUTAM Symposium on New Applications of Nonlinear and Chaotic Dynamics in Mechanics*. Proceedings of the IUTAM Symposium held in Ithaca, NY, USA. 1998 ISBN 0-7923-5276-9
64. R. Wang: *IUTAM Symposium on Rheology of Bodies with Defects*. Proceedings of the IUTAM Symposium held in Beijing, China. 1999 ISBN 0-7923-5297-1
65. Yu.I. Dimitrienko: *Thermomechanics of Composites under High Temperatures*. 1999 ISBN 0-7923-4899-0
66. P. Argoul, M. Frémond and Q.S. Nguyen (eds.): *IUTAM Symposium on Variations of Domains and Free-Boundary Problems in Solid Mechanics*. Proceedings of the IUTAM Symposium held in Paris, France. 1999 ISBN 0-7923-5450-8
67. F.J. Fahy and W.G. Price (eds.): *IUTAM Symposium on Statistical Energy Analysis*. Proceedings of the IUTAM Symposium held in Southampton, U.K. 1999 ISBN 0-7923-5457-5
68. H.A. Mang and F.G. Rammerstorfer (eds.): *IUTAM Symposium on Discretization Methods in Structural Mechanics*. Proceedings of the IUTAM Symposium held in Vienna, Austria. 1999 ISBN 0-7923-5591-1

Mechanics

SOLID MECHANICS AND ITS APPLICATIONS

Series Editor: G.M.L. Gladwell

69. P. Pedersen and M.P. Bendsøe (eds.): *IUTAM Symposium on Synthesis in Bio Solid Mechanics*. Proceedings of the IUTAM Symposium held in Copenhagen, Denmark. 1999 ISBN 0-7923-5615-2
70. S.K. Agrawal and B.C. Fabien: *Optimization of Dynamic Systems*. 1999 ISBN 0-7923-5681-0
71. A. Carpinteri: *Nonlinear Crack Models for Nonmetallic Materials*. 1999 ISBN 0-7923-5750-7
72. F. Pfeifer (ed.): *IUTAM Symposium on Unilateral Multibody Contacts*. Proceedings of the IUTAM Symposium held in Munich, Germany. 1999 ISBN 0-7923-6030-3
73. E. Lavendelis and M. Zakrzhevsky (eds.): *IUTAM/IFToMM Symposium on Synthesis of Nonlinear Dynamical Systems*. Proceedings of the IUTAM/IFToMM Symposium held in Riga, Latvia. 2000 ISBN 0-7923-6106-7
74. J.-P. Merlet: *Parallel Robots*. 2000 ISBN 0-7923-6308-6
75. J.T. Pindera: *Techniques of Tomographic Isodyne Stress Analysis*. 2000 ISBN 0-7923-6388-4
76. G.A. Maugin, R. Drouot and F. Sidoroff (eds.): *Continuum Thermomechanics*. The Art and Science of Modelling Material Behaviour. 2000 ISBN 0-7923-6407-4
77. N. Van Dao and E.J. Kreuzer (eds.): *IUTAM Symposium on Recent Developments in Non-linear Oscillations of Mechanical Systems*. 2000 ISBN 0-7923-6470-8
78. S.D. Akbarov and A.N. Guz: *Mechanics of Curved Composites*. 2000 ISBN 0-7923-6477-5
79. M.B. Rubin: *Cosserat Theories: Shells, Rods and Points*. 2000 ISBN 0-7923-6489-9
80. S. Pellegrino and S.D. Guest (eds.): *IUTAM-IASS Symposium on Deployable Structures: Theory and Applications*. Proceedings of the IUTAM-IASS Symposium held in Cambridge, U.K., 6–9 September 1998. 2000 ISBN 0-7923-6516-X
81. A.D. Rosato and D.L. Blackmore (eds.): *IUTAM Symposium on Segregation in Granular Flows*. Proceedings of the IUTAM Symposium held in Cape May, NJ, U.S.A., June 5–10, 1999. 2000 ISBN 0-7923-6547-X
82. A. Lagarde (ed.): *IUTAM Symposium on Advanced Optical Methods and Applications in Solid Mechanics*. Proceedings of the IUTAM Symposium held in Futuroscope, Poitiers, France, August 31–September 4, 1998. 2000 ISBN 0-7923-6604-2
83. D. Weichert and G. Maier (eds.): *Inelastic Analysis of Structures under Variable Loads*. Theory and Engineering Applications. 2000 ISBN 0-7923-6645-X
84. T.-J. Chuang and J.W. Rudnicki (eds.): *Multiscale Deformation and Fracture in Materials and Structures*. The James R. Rice 60th Anniversary Volume. 2001 ISBN 0-7923-6718-9
85. S. Narayanan and R.N. Iyengar (eds.): *IUTAM Symposium on Nonlinearity and Stochastic Structural Dynamics*. Proceedings of the IUTAM Symposium held in Madras, Chennai, India, 4–8 January 1999 ISBN 0-7923-6733-2
86. S. Murakami and N. Ohno (eds.): *IUTAM Symposium on Creep in Structures*. Proceedings of the IUTAM Symposium held in Nagoya, Japan, 3-7 April 2000. 2001 ISBN 0-7923-6737-5
87. W. Ehlers (ed.): *IUTAM Symposium on Theoretical and Numerical Methods in Continuum Mechanics of Porous Materials*. Proceedings of the IUTAM Symposium held at the University of Stuttgart, Germany, September 5-10, 1999. 2001 ISBN 0-7923-6766-9
88. D. Durban, D. Givoli and J.G. Simmonds (eds.): *Advances in the Mechanis of Plates and Shells The Avinoam Libai Anniversary Volume*. 2001 ISBN 0-7923-6785-5
89. U. Gabbert and H.-S. Tzou (eds.): *IUTAM Symposium on Smart Structures and Structonic Systems*. Proceedings of the IUTAM Symposium held in Magdeburg, Germany, 26–29 September 2000. 2001 ISBN 0-7923-6968-8

Mechanics

SOLID MECHANICS AND ITS APPLICATIONS

Series Editor: G.M.L. Gladwell

90. Y. Ivanov, V. Cheshkov and M. Natova: *Polymer Composite Materials – Interface Phenomena & Processes*. 2001 ISBN 0-7923-7008-2
91. R.C. McPhedran, L.C. Botten and N.A. Nicorovici (eds.): *IUTAM Symposium on Mechanical and Electromagnetic Waves in Structured Media*. Proceedings of the IUTAM Symposium held in Sydney, NSW, Australia, 18-22 Januari 1999. 2001 ISBN 0-7923-7038-4
92. D.A. Sotiropoulos (ed.): *IUTAM Symposium on Mechanical Waves for Composite Structures Characterization*. Proceedings of the IUTAM Symposium held in Chania, Crete, Greece, June 14-17, 2000. 2001 ISBN 0-7923-7164-X
93. V.M. Alexandrov and D.A. Pozharskii: *Three-Dimensional Contact Problems*. 2001 ISBN 0-7923-7165-8
94. J.P. Dempsey and H.H. Shen (eds.): *IUTAM Symposium on Scaling Laws in Ice Mechanics and Ice Dynamics*. Proceedings of the IUTAM Symposium held in Fairbanks, Alaska, U.S.A., 13-16 June 2000. 2001 ISBN 1-4020-0171-1
95. U. Kirsch: *Design-Oriented Analysis of Structures*. A Unified Approach. 2002 ISBN 1-4020-0443-5
96. A. Preumont: *Vibration Control of Active Structures*. An Introduction (2nd Edition). 2002 ISBN 1-4020-0496-6
97. B.L. Karihaloo (ed.): *IUTAM Symposium on Analytical and Computational Fracture Mechanics of Non-Homogeneous Materials*. Proceedings of the IUTAM Symposium held in Cardiff, U.K., 18-22 June 2001. 2002 ISBN 1-4020-0510-5
98. S.M. Han and H. Benaroya: *Nonlinear and Stochastic Dynamics of Compliant Offshore Structures*. 2002 ISBN 1-4020-0573-3
99. A.M. Linkov: *Boundary Integral Equations in Elasticity Theory*. 2002 ISBN 1-4020-0574-1
100. L.P. Lebedev, I.I. Vorovich and G.M.L. Gladwell: *Functional Analysis*. Applications in Mechanics and Inverse Problems (2nd Edition). 2002 ISBN 1-4020-0667-5; Pb: 1-4020-0756-6
101. Q.P. Sun (ed.): *IUTAM Symposium on Mechanics of Martensitic Phase Transformation in Solids*. Proceedings of the IUTAM Symposium held in Hong Kong, China, 11-15 June 2001. 2002 ISBN 1-4020-0741-8
102. M.L. Munjal (ed.): *IUTAM Symposium on Designing for Quietness*. Proceedings of the IUTAM Symposium held in Bangkok, India, 12-14 December 2000. 2002 ISBN 1-4020-0765-5
103. J.A.C. Martins and M.D.P. Monteiro Marques (eds.): *Contact Mechanics*. Proceedings of the 3rd Contact Mechanics International Symposium, Praia da Consolação, Peniche, Portugal, 17-21 June 2001. 2002 ISBN 1-4020-0811-2
104. H.R. Drew and S. Pellegrino (eds.): *New Approaches to Structural Mechanics, Shells and Biological Structures*. 2002 ISBN 1-4020-0862-7
105. J.R. Vinson and R.L. Sierakowski: *The Behavior of Structures Composed of Composite Materials*. Second Edition. 2002 ISBN 1-4020-0904-6
106. Not yet published.
107. J.R. Barber: *Elasticity*. Second Edition. 2002 ISBN Hb 1-4020-0964-X; Pb 1-4020-0966-6
108. C. Miehe (ed.): *IUTAM Symposium on Computational Mechanics of Solid Materials at Large Strains*. Proceedings of the IUTAM Symposium held in Stuttgart, Germany, 20-24 August 2001. 2003 ISBN 1-4020-1170-9

Mechanics

SOLID MECHANICS AND ITS APPLICATIONS

Series Editor: G.M.L. Gladwell

109. P. Ståhle and K.G. Sundin (eds.): *IUTAM Symposium on Field Analyses for Determination of Material Parameters – Experimental and Numerical Aspects.* Proceedings of the IUTAM Symposium held in Abisko National Park, Kiruna, Sweden, July 31 – August 4, 2000. 2003 ISBN 1-4020-1283-7
110. N. Sri Namachchivaya and Y.K. Lin (eds.): *IUTAM Symposium on Nonlinear Stochastic Dynamics.* Proceedings of the IUTAM Symposium held in Monticello, IL, USA, 26 – 30 August, 2000. 2003 ISBN 1-4020-1471-6
111. H. Sobieckzky (ed.): *IUTAM Symposium Transsonicum IV.* Proceedings of the IUTAM Symposium held in Göttingen, Germany, 2–6 September 2002, 2003 ISBN 1-4020-1608-5
112. J.-C. Samin and P. Fisette: *Symbolic Modeling of Multibody Systems.* 2003 ISBN 1-4020-1629-8
113. A.B. Movchan (ed.): *IUTAM Symposium on Asymptotics, Singularities and Homogenisation in Problems of Mechanics.* Proceedings of the IUTAM Symposium held in Liverpool, United Kingdom, 8-11 July 2002. 2003 ISBN 1-4020-1780-4
114. S. Ahzi, M. Cherkaoui, M.A. Khaleel, H.M. Zbib, M.A. Zikry and B. LaMatina (eds.): *IUTAM Symposium on Multiscale Modeling and Characterization of Elastic-Inelastic Behavior of Engineering Materials.* Proceedings of the IUTAM Symposium held in Marrakech, Morocco, 20-25 October 2002. 2004 ISBN 1-4020-1861-4
115. H. Kitagawa and Y. Shibutani (eds.): *IUTAM Symposium on Mesoscopic Dynamics of Fracture Process and Materials Strength.* Proceedings of the IUTAM Symposium held in Osaka, Japan, 6-11 July 2003. Volume in celebration of Professor Kitagawa's retirement. 2004 ISBN 1-4020-2037-6
116. E.H. Dowell, R.L. Clark, D. Cox, H.C. Curtiss, Jr., K.C. Hall, D.A. Peters, R.H. Scanlan, E. Simiu, F. Sisto and D. Tang: *A Modern Course in Aeroelasticity.* 4th Edition, 2004 ISBN 1-4020-2039-2
117. T. Burczyński and A. Osyczka (eds.): *IUTAM Symposium on Evolutionary Methods in Mechanics.* Proceedings of the IUTAM Symposium held in Cracow, Poland, 24-27 September 2002. 2004 ISBN 1-4020-2266-2
118. D. Ieşan: *Thermoelastic Models of Continua.* 2004 ISBN 1-4020-2309-X
119. G.M.L. Gladwell: *Inverse Problems in Vibration.* Second Edition. 2004 ISBN 1-4020-2670-6
120. J.R. Vinson: *Plate and Panel Structures of Isotropic, Composite and Piezoelectric Materials, Including Sandwich Construction.* 2005 ISBN 1-4020-3110-6
121. *Forthcoming*
122. G. Rega and F. Vestroni (eds.): *IUTAM Symposium on Chaotic Dynamics and Control of Systems and Processes in Mechanics.* Proceedings of the IUTAM Symposium held in Rome, Italy, 8–13 June 2003. 2005 ISBN 1-4020-3267-6
123. E.E. Gdoutos: *Fracture Mechanics. An Introduction. 2nd edition.* 2005 ISBN 1-4020-3267-6
124. M.D. Gilchrist (ed.): *IUTAM Symposium on Impact Biomechanics from Fundamental Insights to Applications.* 2005 ISBN 1-4020-3795-3
125. J.M. Huyghe, P.A.C. Raats and S. C. Cowin (eds.): *IUTAM Symposium on Physicochemical and Electromechanical Interactions in Porous Media.* 2005 ISBN 1-4020-3864-X
126. H. Ding, W. Chen and L. Zhang: *Elasticity of Transversely Isotropic Materials.* 2005 ISBN 1-4020-4033-4
127. W. Yang (ed): *IUTAM Symposium on Mechanics and Reliability of Actuating Materials.* Proceedings of the IUTAM Symposium held in Beijing, China, 1–3 September 2004. 2005 ISBN 1-4020-4131-6

Mechanics

SOLID MECHANICS AND ITS APPLICATIONS

Series Editor: G.M.L. Gladwell

128. J.-P. Merlet: *Parallel Robots*. 2006 ISBN 1-4020-4132-2
129. G.E.A. Meier and K.R. Sreenivasan (eds.): *IUTAM Symposium on One Hundred Years of Boundary Layer Research*. Proceedings of the IUTAM Symposium held at DLR-Göttingen, Germany, August 12–14, 2004. 2006 ISBN 1-4020-4149-7
130. H. Ulbrich and W. Günthner (eds.): *IUTAM Symposium on Vibration Control of Nonlinear Mechanisms and Structures*. 2006 ISBN 1-4020-4160-8
131. L. Librescu and O. Song: *Thin-Walled Composite Beams*. Theory and Application. 2006 ISBN 1-4020-3457-1
132. G. Ben-Dor, A. Dubinsky and T. Elperin: *Applied High-Speed Plate Penetration Dynamics*. 2006 ISBN 1-4020-3452-0
133. X. Markenscoff and A. Gupta (eds.): *Collected Works of J. D. Eshelby*. Mechanics and Defects and Heterogeneities. 2006 ISBN 1-4020-4416-X
134. R.W. Snidle and H.P. Evans (eds.): *IUTAM Symposium on Elastohydrodynamics and Microelastohydrodynamics*. Proceedings of the IUTAM Symposium held in Cardiff, UK, 1–3 September, 2004. 2006 ISBN 1-4020-4532-8
135. T. Sadowski (ed.): *IUTAM Symposium on Multiscale Modelling of Damage and Fracture Processes in Composite Materials*. Proceedings of the IUTAM Symposium held in Kazimierz Dolny, Poland, 23–27 May 2005. 2006 ISBN 1-4020-4565-4
136. A. Preumont: *Mechatronics*. Dynamics of Electromechanical and Piezoelectric Systems. 2006 ISBN 1-4020-4695-2
137. M.P. Bendsøe, N. Olhoff and O. Sigmund (eds.): *IUTAM Symposium on Topological Design Optimization of Structures, Machines and Materials*. Status and Perspectives. 2006 ISBN 1-4020-4729-0
138. A. Klarbring: *Models of Mechanics*. 2006 ISBN 1-4020-4834-3
139. H.D. Bui: *Fracture Mechanics*. Inverse Problems and Solutions. 2006 ISBN 1-4020-4836-X
140. M. Pandey, W.-C. Xie and L. Xu (eds.): *Advances in Engineering Structures, Mechanics and Construction*. Proceedings of an International Conference on Advances in Engineering Structures, Mechanics & Construction, held in Waterloo, Ontario, Canada, May 14–17, 2006. 2006 ISBN 1-4020-4890-4
141. G.Q. Zhang, W.D. van Driel and X. J. Fan: *Mechanics of Microelectronics*. 2006 ISBN 1-4020-4934-X
142. Q.P. Sun and P. Tong (eds.): *IUTAM Symposium on Size Effects on Material and Structural Behavior at Micron- and Nano-Scales*. Proceedings of the IUTAM Symposium held in Hong Kong, China, 31 May–4 June, 2004. 2006 ISBN 1-4020-4945-5